Handbook of Behavioral Neurobiology

Volume 12
Circadian Clocks

Edited by

Joseph S. Takahashi
Fred W. Turek
and
Robert Y. Moore

Northwestern University
Evanston, Illinois

KLUWER ACADEMIC / PLENUM PUBLISHERS
New York, Boston, Dordrecht, London, Moscow

The Library of Congress cataloged the first volume of this title as follows:

Sensory integration.

(Handbook of behavioral neurobiology; v. 1)
Includes bibliographies and index.
1. Senses and sensation. 2. Sense-organs. 3. Physiology, Comparative. I. Masterson, R. Bruce. II. Series [DNLM: I. Behavior. 2. Neurophysiology. 3. Psychophysics. WL102.3 H236]
QP431.S4515 596′.01′82 78-17238
ISBN 0-306-35191-9

ISBN 0-306-46504-3

233 Spring Street, New York, N.Y. 10013

http://www.wkap.nl/

10 9 8 7 6 5 4 3 2 1

A C.I.P. record for this book is available from the Library of Congress

Printed in the United States of America

Contributors

PETER ACHERMANN, *Institute of Pharmacology, University of Zürich, Zürich, Switzerland*

T. ÅKERSTEDT, *Laboratory of Clinical Stress Research, Karolinska Institute, Stockholm, Sweden*

JÜRGEN ASCHOFF, *Late of Max Planck Institut für Verhaltensphysologie, Andechs, Germany*

G. ATKINSON, *Research Institute for Sport & Exercise Sciences, Liverpool John Moores University, Liverpool, United Kingdom*

GENE D. BLOCK, *Department of Biology and NSF Center for Biological Timing, University of Virginia, Charlottesville, Virginia*

EDWARD M. BLUMENTHAL, *Department of Biology and NSF Center for Biological Timing, University of Virginia, Charlottesville, Virginia*

ALEXANDER A. BORBELY, *Institute of Pharmacology, University of Zürich, Zürich, Switzerland*

Y. BOUSKILA, *Department of Anatomy and Neurobiology, Colorado State University, Fort Collins, Colorado*

ORFEU M. BUXTON, *Department of Medicine, University of Chicago, Chicago, Illinois*

DANIEL J. BUYSSE, *Sleep and Chronobiology Center, Western Psychiatric Institute and Clinic, University of Pittsburgh School of Medicine, Pittsburgh, Pennsylvania*

JULIE CARRIER, *Sleep and Chronobiology Center, Western Psychiatric Institute and Clinic, University of Pittsburgh School of Medicine, Pittsburgh, Pennsylvania*

CHARLES A. CZEISLER, *Circadian, Neuroendocrine and Sleep Disorders Section, Division of Endocrinology, Department of Medicine, Brigham and Women's Hospital, Harvard Medical School, Boston, Massachusetts*

SERGE DAAN, *Zoological Laboratory, Groningen University, Groningen, The Netherlands*

FREDERICK C. DAVIS, *Department of Biology, Northeastern University, Boston, Massachusetts*

MARY AMANDA DEW, *Sleep and Chronobiology Center, Western Psychiatric Institute and Clinic, University of Pittsburgh School of Medicine, Pittsburgh, Pennsylvania*

DERK-JAN DIJK, *Institute of Pharmacology, University of Zürich, Zürich, Switzerland and Centre for Chronobiology, University of Surrey, Guilford, United Kingdom*

F. E. DUDEK, *Department of Anatomy and Neurobiology, Colorado State University, Fort Collins, Colorado*

JAY C. DUNLAP, *Department of Biochemistry, Dartmouth Medical School, Hanover, New Hampshire*

ARNOLD ESKIN, *Department of Biology and Biochemistry, University of Houston, Houston, Texas*

BRUCE D. GOLDMAN, *Department of Physiology and Neurobiology, University of Connecticut, Storrs, Connecticut*

MICHAEL R. GORMAN, *Department of Psychology, University of California, San Diego, La Jolla, California*

ELLEN FRANK, *Western Psychiatric Institute and Clinic, University of Pittsburgh, Pittsburgh, Pennsylvania*

MARTICA HALL, *Sleep and Chronobiology Center, Western Psychiatric Institute and Clinic, University of Pittsburgh School of Medicine, Pittsburgh, Pennsylvania*

J. WOODLAND HASTINGS, *Department of Molecular and Cellular Biology, Biological Laboratories, Harvard University, Harvard, Massachusetts*

THERESA H. HORTON, *Center for Circadian Biology and Medicine, Department of Neurology and Physiology, Northwestern University, Evanston, Illinois*

CARL HIRSCHIE JOHNSON, *Department of General Biology, Vanderbilt University, Nashville, Tennessee*

TADAO KONDO, *Division of Biolgoical Science, Graduate School of Science, Nagoya University, Chikusa, Nagoya, Japan*

DAVID J. KUPFER, *Sleep and Chronobiology Center, Western Psychiatric Institute and Clinic, University of Pittsburgh School of Medicine, Pittsburgh, Pennsylvania*

SUSAN LABYAK, *Department of Neurobiology and Physiology, Northwestern University, Evanston, Illinois*

REHANA K. LEAK, *Departments of Neurology, Psychiatry and Neuroscience, University of Pittsburgh, Pittsburgh, Pennsylvania*

ELLEN LEIBENLUFT, *Pediatrics and Developmental Neuropsychology Branch, National Institute of Mental Health, Bethesda, Maryland*

JENNIFER J. LOROS, *Department of Biochemistry, Dartmouth Medical School, Hanover, New Hampshire*

J. H. MEIJER, *Department of Physiology, Leiden University Medical Centre, Leiden, The Netherlands*

D. S. MINORS, *School of Biological Sciences, University of Manchester, Manchester, United Kingdom*

TIMOTHY H. MONK, *Sleep and Chronobiology Center, Western Psychiatric Institute and Clinic, University of Pittsburgh School of Medicine, Pittsburgh, Pennsylvania*

ROBERT Y. MOORE, *Departments of Neurology, Psychiatry and Neuroscience, University of Pittsburgh, Pittsburgh, Pennsylvania*

PETER NOWELL, *Sleep and Chronobiology Center, Western Psychiatric Institute and Clinic, University of Pittsburgh School of Medicine, Pittsburgh, Pennsylvania*

TERRY L. PAGE, *Department of Biology, Vanderbilt University, Nashville, Tennessee*

PLAMEN PENEV, *Department of Neurobiology and Physiology, Northwestern University, Evanston, Illinois*

MARTIN R. RALPH, *Department of Psychology, University of Toronto, Toronto, Canada*

T. REILLY, *Research Institute for Sport & Excercise Sciences, Liverpool John Moores University, Liverpool, United Kingdom*

STEVEN M. REPPERT, *Laboratory of Developmental Chronobiology, Children's Service, Massachusetts General Hospital, Harvard Medical School, Boston, Massachusetts*

CHARLES F. REYNOLDS III, *Sleep and Chronobiology Center, Western Psychiatric Institute and Clinic, University of Pittsburgh, Pittsburgh, Pennsylvania*

KATHRYN SCARBROUGH, *Department of Neurobiology and Physiology, Northwestern University, Evanston, Illinois*

G. J. STRECKER, *Department of Anatomy and Neurobiology, Colorado State University, Fort Collins, Colorado*

FRIEDRICH K. STEPHAN, *Neuroscience Program, Department of Psychology, Florida State University, Tallahassee, Florida*

Joseph S. Takahashi, *Center for Biology and Medicine, Department of Neurobiology and Physiology, Northwestern University, Evanston, Illinois*

Irene Tobler, *Institute of Pharmacology, University of Zürich, Zürich, Switzerland*

Fred W. Turek, *Center for Biology and Medicine, Department of Neurobiology and Physiology, Northwestern University, Evanston, Illinois*

Herbert Underwood, *Department of Zoology, North Carolina State University, Raleigh, North Carolina*

Verónica S. Valentinuzzi, *Department of Neurobiology and Physiology, Northwestern University, Evanston, Illinos*

Eve Van Cauter, *Department of Medicine, University of Chicago, Chicago, Illinois*

Olivier Van Reeth, *Department of Neurobiology and Physiology, Northwestern University, Evanston, Illinois*

Martha Hotz Vitaterna, *Department of Neurobiology and Physiology, Northwestern University, Evanston, Illinois*

J. M. Waterhouse, *Research Institute for Sport & Exercise Sciences, Liverpool John Moores University, Liverpool, United Kingdom*

Thomas C. Wehr, *Clinical Psychology Branch, Intramural Research Program, National Institute of Mental Health, Bethesda, Maryland*

Michael W. Young, *National Science Foundation Center for Biological Timing, Laboratory of Genetics, The Rockefeller University, New York, New York*

Phyllis C. Zee, *Department of Neurology, Northwestern University Medical Faculty Foundation, Inc., Chicago, Illinois*

Irving Zucker, *Departments of Psychology and Integrative Biology, University of California, Berkeley, California*

Contents

Part I Introduction

CHAPTER 2

Conceptual Issues in the Ecology and Evolution of Circadian Rhythms

Teresa H. Horton

PART II ORGANIZATION OF CIRCADIAN SYSTEMS

CHAPTER 3

Circadian Rhythms in Unicellular Organisms

Carl Hirschie Johnson and Takao Kondo

CHAPTER 4

Circadian Systems of Invertebrates

Terry L. Page

Part III Regulation of Mammalian Circadian Rhythms

PART IV CELL AND MOLECULAR ANALYSIS OF CIRCADIAN RHYTHMS

CHAPTER 11

Cellular and Molecular Mechanisms of Circadian Regulation in the Unicellular Dinoflagellate *Gonyaulax polyedra*

J. Woodland Hastings

CHAPTER 12

Molecular Genetics of Circadian Rhythms in *Neurospora*, a Prototypic Circadian System

Jay C. Dunlap and Jennifer J. Loros

Part V Interaction of Circadian Clock with Other Rhythmic Functions

Chapter 17

Processes Underlying the Regulation of the Sleep–Wake Cycle

Alexander A. Borbély, Derk-Jan Dijk, Peter Achermann, and Irene Tobler

Chapter 18

Mammalian Photoperiodism

Michael R. Gorman, Bruce D. Borman, and Irving Zucker

CHAPTER 22

Circadian Rhythms, Aging and Dementia

Timothy H. Monk and Phyllis C. Zee

CHAPTER 23

Circadian Rhythms in Affective Disorders

Ellen Leibenluft and Ellen Frank

CHAPTER 24

Sleep Disorders

Daniel J. Buysse, Julie Carrier, Mary Amanda Dew, Martica Hall, Timothy H. Monk, Peter Nowell, Charles F. Reynolds III, and David J. Kupfer

CHAPTER 25

Circadian Modulation of Endocrine Secretion

Eve Van Cauter and Orfeu M. Buxton

CHAPTER 26

Seasonal Photoperiodic Responses of the Human Circadian System

Thomas A. Wehr

PART I

Introduction

Introduction to Circadian Rhythms

FRED W. TUREK AND JOSEPH S. TAKAHASHI

INTRODUCTION

One of the most obvious adaptive features of life on earth is the ability of almost all species to change their behavior on a daily or 24-hour basis (Aschoff, 1981). A remarkable feature of the daily rhythms that are observed in organisms as diverse as algae, fruit flies, and humans is that they are not simply a response to the 24-hour changes in the physical environment imposed by the principles of celestial mechanics, but instead arise from an internal time-keeping system. This time-keeping system, or biological clock(s), allows the organism to predict and prepare in advance for the changes in the physical environment that are associated with night and day. Thus, the organism adapts, both behaviorally and physiologically, to meet the challenges associated with the daily changes in the external environment, and there is temporal synchronization between the organism and the external environment. Of equal, but perhaps less appreciated, importance is the fact that this biological clock, like a conductor of a symphony orchestra, provides internal temporal organization and insures that internal changes take place in coordination with one another. Just as living organisms are organized spatially, they are also organized temporally to insure that there is "internal synchronization" between the myriad biochemical and physiological systems in the body (Pittendrigh, 1993).

FRED W. TUREK AND JOSEPH S. TAKAHASHI Center for Circadian Biology and Medicine, Department of Neurobiology and Physiology, Howard Hughes Medical Institute, Northwestern University, Evanston, Illinois 60208.

Circadian Clocks, Volume 12 of *Handbook of Behavioral Neurobiology*, edited by Joseph S. Takahashi, Fred W. Turek, and R. Y. Moore, Kluwer Academic / Plenum Publishers, New York, 2001.

Virtually all diurnal rhythms that are present under natural conditions continue to be expressed under laboratory conditions devoid of any external time-giving cues from the physical environment. However, under constant environmental conditions, the period of the rhythm rarely remains exactly 24 hours but instead is "about" 24 hours. Because the period of diurnal rhythms is close to but not exactly 24 hours in duration, they are referred to as "circadian rhythms," from the Latin *circa diem* , meaning about a day. When a circadian rhythm is expressed in the absence of any 24-hour signals in the external environment, it is said to be "free-running," (i.e., the rhythm is not synchronized or entrained by any cyclic change in the physical environment). The fact that the endogenous period of the circadian clock is rarely exactly equal to 24 hours demands that changes in the physical environment must somehow synchronize or entrain the internal clock system regulating circadian rhythms. Entrainment serves two purposes: phase control and period adjustment. The light–dark (LD) cycle is clearly the major environmental entraining agent of the circadian clock of most species; indeed the basic features and properties of entrainment by the LD cycle are remarkably similar across species as diverse as single-celled organisms and humans.

Perhaps the greatest strength of the field of biological rhythms is the diversity of organisms that have been utilized to uncover the mechanisms by which circadian rhythms can be generated at various levels of organization. Thus, while the expression of circadian rhythms is clearly regulated by complex neural structures in higher vertebrates, the evidence that single-celled organisms could generate circadian rhythms focused attention on the hypothesis that the fundamental biochemical and cellular mechanism underlying the expression of circadian rhythms would be similar in simple and complex organisms. The ability of living organisms to express circadian rhythmicity occurred early in the history of life on earth since prokaryotic and eukaryotic unicellular organisms show distinct circadian rhythmicity. Much of what we know about the cellular and molecular mechanisms of circadian regulation has come from studies on the marine dinoflagellate *Gonyaulax* and the filamentous fungus *Neurospora.* Indeed, the observation that clocks are a cellular property in single-cell organisms, and the fact that the general properties of circadian rhythms were similar in single-cell and complex multicellular organisms, was a driving force behind the early and now-proven hypothesis that the expression of circadian rhythms in complex organisms, including mammals, would ultimately reside at the cellular level. However, even if the generation of circadian rhythms is at its most fundamental level based within the cell, clearly the generation of rhythms in higher organisms involves the interaction of many cells in the nervous system. A variety of different model systems have been exploited to uncover these interactions in invertebrates and in vertebrates. Such studies have led to the discovery of the circadian pacemakers that drive the expression of circadian rhythms.

While studies of unicellular organisms point to the cellular nature of the circadian rhythm generating system, in higher organisms the circadian pacemaker cells are localized to specific structures including the optic and cerebral lobes of insects, the eye in both invertebrates and vertebrates, and the pineal gland of nonmammalian vertebrates. Clearly a landmark for the field of circadian clocks was the discovery in the early 1970s that lesioning a small region of the anterior hypothalamus in rats, the bilaterally paired suprachiasmatic nucleus (SCN), led to a disruption and abolition of endocrine and behavioral circadian rhythms. The discovery, and the discovery process over the next three decades, of the master circadian pacemaker in mammals represents a truly remarkable story of how literally hundreds of neuroscientists from many different disciplines came together to elucidate the role of this structure for the entrainment, generation, and expression of circadian rhythms at the molecular, cellular and system levels of organization (Klein, Moore, & Reppert, 1991).

In the past few years remarkable progress has been made in elucidating the molecular components of the circadian clock. The origin of this research dates back to the 1970s when mutant organisms were isolated that had either abnormal periods in their circadian clock or were arrhythmic. Pioneering studies in the fruit fly, *Drosophila* and the filamentous fungus, *Neurospora* not only demonstrated that single gene mutations could have dramatic effects on circadian rhythm expression, but also led the way to the eventual cloning of the first circadian clock genes.

More recently a mutagenesis approach was undertaken in mice with the ultimate discovery and cloning of the first mammalian circadian clock gene, called *Clock*, in 1997. In a remarkably short period of time after the discovery of the *Clock* gene in mice, a large number of circadian clock genes were discovered and shown to be members of conserved pathway in mice and flies (Dunlap, 1999; King & Takahashi, 2000). Rapid progress was also made in determining how these genes and their protein products interact with one another to generate 24-hour rhythms. The spectacular success in unraveling the story of the molecular genetic basis of circadian clocks led *Science* magazine to select the "clock story" as one of the top 10 scientific breakthroughs in 1997, and in 1998 the "clock story" was chosen as the major biomedical breakthrough of the year by *Science* (News and Editorial Staff, 1998), an acknowledgement that was a credit to the entire field of biological rhythms.

Importance of Circadian Rhythms for Human Health, Safety, Performance and Productivity

Disturbed circadian rhythmicity, whether due to voluntary (e.g., due to shift work or rapid travel across time zones) or involuntary (e.g., due to illness or advanced age) circumstances, has been associated with numerous mental and physical disorders and can have a negative impact on human safety, performance, and productivity. Much of the adverse effects of disrupted circadian rhythmicity may be linked to disturbances in the sleep–wake cycle. While some rhythmic processes are more affected by the circadian clock than the sleep–wake state, other rhythms are more dependent on the sleep–wake state. For most animals, under natural conditions the timing of sleep and wake is in synchrony with the circadian control of the sleep cycle and all other circadian controlled rhythms. However, humans have the unique capacity to cognitively override their internal biological clock and its rhythmic outputs. When the sleep–wake cycle is out of phase with the circadian clock controlled rhythms, as occurs during rapid travel across time zones or in shift workers, there are adverse effects on human performance which can have severe consequences for health, safety and the ability to work.

The unity of the field of circadian clocks is one of its important and distinguishing characteristics. This unity involves the integration of a variety of experimental approaches in a wide range of species all aimed at uncovering the mechanisms and importance of the circadian clock system in living organisms. As should be apparent from the papers in this volume, a variety of biochemical, molecular, cellular, systems, behavioral, and ecological approaches have been used to unravel how living organisms have evolved to deal with one of the most distinguishing characteristics of the planet earth, the daily cycle of light and darkness. This handbook also emphasizes how our understanding of human circadian rhythms has been and will continue to be built on the "shoulders" of single-celled organisms, slime molds, flies, snails, frogs, birds, rodents, etc. Since the publication of the first handbook (Aschoff, 1981), great advances have been made in our understanding of circadian rhythmicity, particularly at the level of the molecular machinery that allows for the generation of circadian signals.

The field of circadian rhythms is in a constructive positive feedback loop. As the importance of circadian rhythmicity for human health and welfare becomes more recognized, increased interest and support for the study of the basic mechanisms underlying rhythmicity will become available. The ever-accelerating pace of discovery of the molecular-, cellular-, and system-level machinery of the circadian clock system leads to more interest in using information about circadian rhythmicity to improve human rhythms and sleep. As is clear from the papers in this volume, the rates of change in all aspects of this positive feedback loop is increasing, indicating that many new discoveries will be made in the field of circadian rhythms over the next few years. Furthermore, the strong links across the broad community of researchers in the field of biological rhythms will ensure that advances in the field are transitioned into the "real world" at a rapid pace for improving human health, safety, performance and productivity.

References

Aschoff, J. (1981). *Handbook of behavioral neurobiology, Vol. 4: Biological rhythms.* New York: Plenum Press.

Dunlap, J. C. (1999). Molecular bases for circadian clocks. *Cell, 96,* 271–290.

King, D. P., & Takahashi, J. S. (2000). Molecular genetics of circadian rhythms in mammals. *Annual Review of Neuroscience, 23,* 713–742.

Klein, D. C., Moore, R. Y. & Reppert, S. M. (1991), *Suprachiasmatic nucleus—The mind's clock* (pp. 467). New York: Oxford University Press.

News and Editorial Staff. (1998) First runner-up: A remarkable year for Clocks. *Science, 282,* 2157.

Pittendrigh, C. S. (1993) Temporal organization: Reflections of a Darwinian clock-watcher. *Annual Review of Physiology, 55,* 16–54.

1

The Entrainment of Circadian Systems

SERGE DAAN AND JÜRGEN ASCHOFF

INTRODUCTION

The entrainment of circadian systems is essential for their functional significance as well as for our insight into their physiologic organization. Entrainment entails the adjustment of both the frequency and phase of rhythms in the living world to the cycle of the earth's rotation. It is only by virtue of entrainment that programs in behavior and physiology produced by endogenous circadian systems can be properly timed. This is crucial for the advantages in natural selection that in the past gave rise to the evolution and today maintain the genetic basis of these systems. Entrainment requires the sensitivity of endogenous oscillators toward particular environmental cues as well as insensitivity toward others. The sensitivity toward light has been and continues to be a primary guide in probing and unraveling the physiology of circadian systems.

In the past, much of this research made use of the behavior and physiology of whole organisms, as localized circadian pacemakers were unknown prior to 1968. Properties of the driving circadian oscillators were inferred from their behavioral output in response to manipulated sensory input. The present chapter summarizes the general insights derived from such formal ("black box") analysis. It starts with a survey of the phenomena of entrainment, partly based upon Chapter 6 in the first edition of the *Handbook* (Aschoff, 1981b). Then it outlines the mechanism of entrainment for the particular case of brief light pulses, which is understood in rich quantitative detail (Pittendrigh, 1981b). The final section discusses the relevance of Pittendrigh's classical entrainment model for circadian entrainment in nature in the

SERGE DAAN Zoological Laboratory, University of Groningen, Groningen, The Netherlands. JÜRGEN ASCHOFF Late of Max Planck Institut für Verhaltensphysiologie, Andechs, Germany.

Circadian Clocks, Volume 12 of *Handbook of Behavioral Neurobiology*, edited by Joseph S. Takahashi, Fred W. Turek, and R. Y. Moore, Kluwer Academic / Plenum Publishers, New York, 2001.

light of empirical complications that have emerged since the formulation of the model.

This chapter was originally conceived as a revision of Chapter 7 in the first edition of the *Handbook*, to be jointly authored with Colin Pittendrigh. His death on March 18, 1996, precluded finalizing this concept. Pittendrigh's spirit pervades this new chapter, and some of his original text has been retained. It is with deep gratitude for his lifelong stimulating influence and friendship that we dedicate this chapter to the memory of Colin Pittendrigh.

Phenomena of Entrainment

The Experimental Demonstration of Entrainment

A circadian oscillation can be *entrained* by a periodic external signal, which is then called a *zeitgeber* for that oscillation. Entrainment is the adjustment of the period τ of the circadian system such that it equals the period T of the zeitgeber. When entrainment occurs, the system adopts a specific phase relationship with the zeitgeber.

The experimental demonstration of entrainment requires in the first place that the biological rhythm is recorded both in the presence and in the absence of the zeitgeber. In the presence of the zeitgeber, the period should be identical to that of the zeitgeber; in its absence, it should be different. A typical example is presented in Figure 1A. This displays the circadian activity rhythm of a chaffinch maintained in a light–dark (LD) 12:12 cycle of alternating dim (0.5 lux) and bright (200 lux) illumination. On three occasions, the LD cycle is interrupted by episodes of constant dim illumination, and the activity rhythm starts to freerun with a period shorter than 24 hours. After restoring the LD cycle, the rhythm assumes again a 24-hour period through a number of *transient* cycles. These transients can be shorter than T (days 78–80) or longer than T (days 21ff., 45 ff.), depending on the phase of the circadian rhythm where the first LD cycle starts. When the period of the zeitgeber is changed, the biological rhythm should change its period to match that of the new zeitgeber, at least in steady-state entrainment.

The second criterion for entrainment is that the circadian oscillation in the presence of a zeitgeber maintains a unique phase relationship to the latter. This is seen in Figure 1A by the repeated establishment of the same phase position of the chaffinch's activity with respect to the LD cycle, with onset of activity leading lights-on by a few hours. This phase position is attained irrespective of the phase of the system when the first light signal came. Another straightforward demonstration of phase control in entrainment makes use of phase shifts of the zeitgeber. In such a protocol, the circadian system shifts its phase, often gradually via a number of transient cycles, until it reaches steady state with the same phase position relative to the new zeitgeber as before the phase shift. This is seen in another activity record of a chaffinch in a periodically reversed LD 12:12 cycle (Figure 1B).

We emphasize a third criterion for the assessment of entrainment. Environmental variables are known to have direct effects on the output variables of the circadian system which may mimic the entrainment process. Such effects are known as *masking* (Aschoff, 1960; Aschoff & Von Goetz, 1988). For instance, light enhances spontaneous locomotor activity (*positive masking*) in many *diurnal* (i.e., day-active) animals, whereas light suppresses activity (*negative masking*) in most *nocturnal* (night-active) species. Clearly, masking alone might generate an oscillation in the overt rhythm

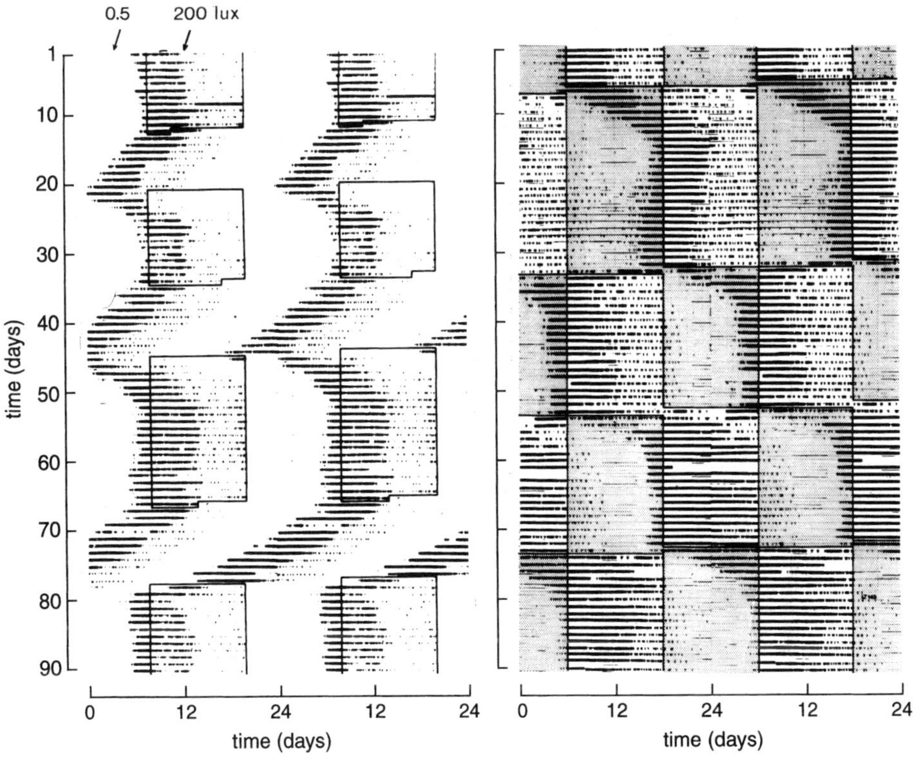

Figure 1. (A) Experimental demonstration of entrainment in a chaffinch, *Fringilla coelebs*, alternatingly kept in a bright–dim light schedule (LD 12:12) and in constant dim illumination. Double-plotted actogram. (From Aschoff, Daan, and Honma, 1982a, Figure 1). (B) Reentrainment after 12-hour phase shifts of the zeitgeber in a chaffinch. After each phase shift, it takes a number of transient cycles for stable entrainment to become reestablished. (From Aschoff, Hoffman, Pohl, and Wever, 1979.)

that precisely matches that of the LD cycle imposed experimentally. Entrainment is distinguished from such induction of a rhythm by several features: First, the phase position of an entrained endogenous rhythm is often, though not necessarily, characterized by anticipation of the zeitgeber: In Figure 1, the chaffinches begin their activity well before the lights are turned on. Second, reentrainment after an episode of freerun, or after a phase shift of the zeitgeber, is usually characterized by a number of transient cycles before steady-state entrainment is reached. Third, the freerun in constant conditions after entrainment would start from a time of day that is uniquely determined by the prior zeitgeber. This last criterion provides the most compelling evidence for true entrainment. In any protocol to study entrainment, it is recommended that the phase of the endogenous circadian oscillation be confirmed by releasing the system into constant conditions after exposure to the zeitgeber tested.

Zeitgebers

The LD cycle is the dominant entraining agent for nearly all circadian systems in animals and plants. This is to be expected on functional grounds. The LD cycle provides the most precise time signal in most environments. In habitats with continuous darkness, such as the deep sea or underground cavities, there is probably also little selection pressure for the maintenance of circadian rhythmicity. In the continuous illumination of the arctic and antarctic summer, other zeitgebers than light may be important, but there is little concrete information on entrainment in those conditions. In addition to the LD cycle, other environmental factors have been shown experimentally to act as zeitgebers. A cycle of low and high temperatures can be a powerful zeitgeber especially for poikilothermic animals. In the lizard *Lacerta sicula,* a cycle with a range of only 0.9°C suffices to entrain the activity rhythms of about one third of the animals tested (Hoffmann, 1969). Zimmerman, Pittendrigh, and Pavlidis (1968) successfully entrained the *Drosophila* pupal eclosion rhythm with cycles of alternatingly 12 hours at 20°C and 12 hours at 28°C. Homeothermic animals are usually less easily entrained by temperature. Eskin (1971) had to use cycles with a range of more than 30°C to obtain entrainment in the house sparrow; for the squirrel monkey, 16°C can be sufficient (Aschoff & Tokura, 1986). In the experiment displayed in Figure 2, two of three monkeys were entrained by such a cycle, with their activity time in the cold phase of the cycle. Note that the free-running rhythm after interruption of the cold–warm cycle starts from the beginning of the cold phase each time, showing that we are dealing with true entrainment, not masking. In contrast, the third monkey (#3) was not entrained by the same cold–warm cycle, although this cycle influenced the rhythm by modulating its frequency (cf. "relative coordination" discussed in the next subsection).

Information on the entraining capacity of other factors is scarce. Convincing demonstrations are available for the zeitgeber action of air pressure (Hayden & Lindberg, 1969), meal timing (see Chapter 12), hoarding opportunity (Rusak, Mistlberger, Losier, & Jones, 1988), wheel-running opportunity (Edgar & Dement, 1991), and species-specific song (Gwinner, 1966; Menaker & Eskin, 1966). Periodic song is one example of a factor potentially involved in social entrainment. Social entrainment has been studied in a number of systems. The best positive evidence is available from natural social systems: bats (*Hipposideros speoris*) roosting colonially in caves in southern India entrain each other, possibly via sound (Marimuthu, Rajan, & Chandrashekaran, 1981). They do not entrain the rhythm of a different species (Marimuthu & Chandrashekaran, 1983). Beavers (*Castor canadensis*) show free-running activity rhythms under the ice of a lake in winter, while the family unit apparently remains in synchrony (Bovet & Oertli, 1974). In laboratory studies on small mammals, it has been notoriously difficult to establish social entrainment by natural interactions [e.g., in the sugar glider (Kleinknecht, 1985), rat (Takahashi *et al.*, 1978), and hamster (Davis, Stice, & Menaker, 1987)]. Periodic exposure of hamsters to a conspecific brought into a cage may phase shift and even entrain the circadian rhythm of the cage resident (Mrosovsky, 1988), but this does not mean that continuous presence of such a conspecific would have the same effect. A special case of social entrainment is the entrainment of circadian rhythms in offspring via the mother, again composing a natural unit. This has been documented for mother–pup interactions, e.g., in rats (Takahashi & Sasaki, 1985) and mice (Viswanathan & Chandrashekaran, 1985). Much of the necessary setting of the pup's rhythm is already achieved *in utero* (Davis & Gorski, 1988; Honma, Honma, & Hiroshige, 1985).

In physiologic analysis of entrainment pathways, numerous substances have been tested and found to phase shift circadian pacemakers (see Chapter 7). These are therefore potential zeitgebers. We restrict our discussion to natural zeitgebers. Some of these chemicals may vary in nature in synchrony with the day–night cycle and require specific mentioning. The endogenous substance melatonin, for instance, has a well-documented zeitgeber function in rats (Armstrong, 1989) and

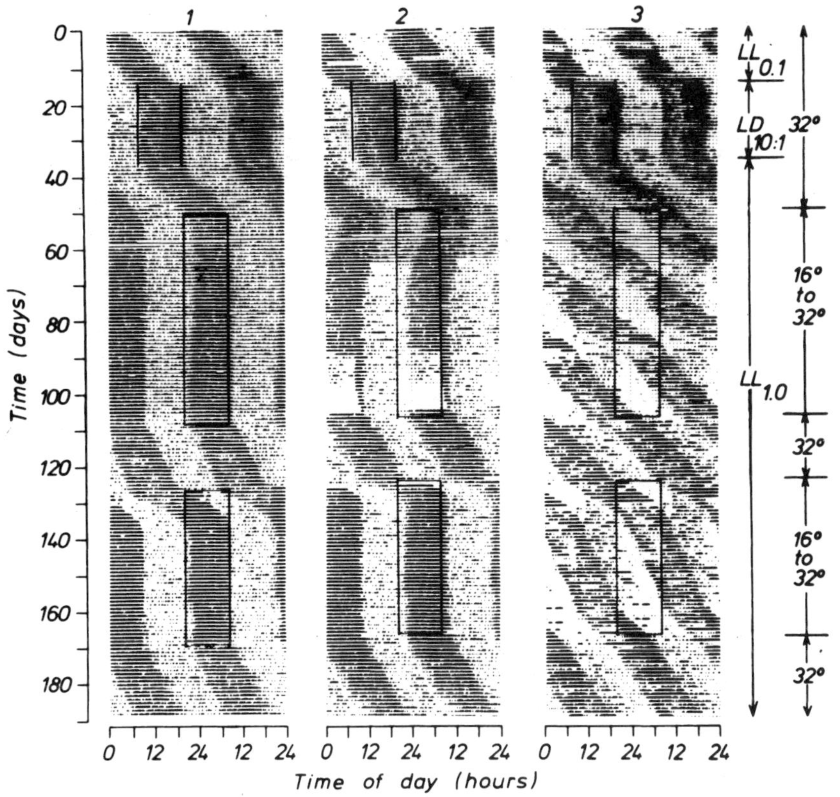

Figure 2. Entrainment by temperature: Circadian activity rhythms of three squirrel monkeys, *Saimiri sciureus*, kept in constant dim light (LL) and temperature (32°C), except for an episode of entrainment to LD, and two episodes of a temperature cycle (16–32°C; dashed line indicates onset of cold phase at 19:00). (Double-plotted actograms from Aschoff and Tokura, 1986.)

humans (Lewy *et al.*, 1994). In songbirds, melatonin can act as a zeitgeber when administered via the drinking water (Heigl & Gwinner, 1994). In hamsters, periodic melatonin can transfer phase from mother to embryo *in utero* (Davis & Mannion, 1988). It is becoming clear that melatonin is widespread in both plants and animals (Hardeland *et al.*, 1995) and may thus well be transferred via food. Hence, melatonin is therefore not only an internal messenger, but potentially also an interindividual zeitgeber both within and between species. Other substances may be important in the aquatic environment. Roenneberg and Rehman (1996) demonstrated the zeitgeber role for nitrate concentrations in the marine alga *Gonyaulax polyedra*. We return to the question of the extent to which nonphotic signals may contribute to natural entrainment in the final section.

Limits of Entrainment

A zeitgeber with period T entrains a circadian rhythm such that its cycle length in steady-state conditions, here denoted by τ^*, is equal to T. This does not mean that entrainment can be achieved for all values of T. Figure 3 summarizes an experiment in which finches were exposed to LD cycles with five different period lengths from

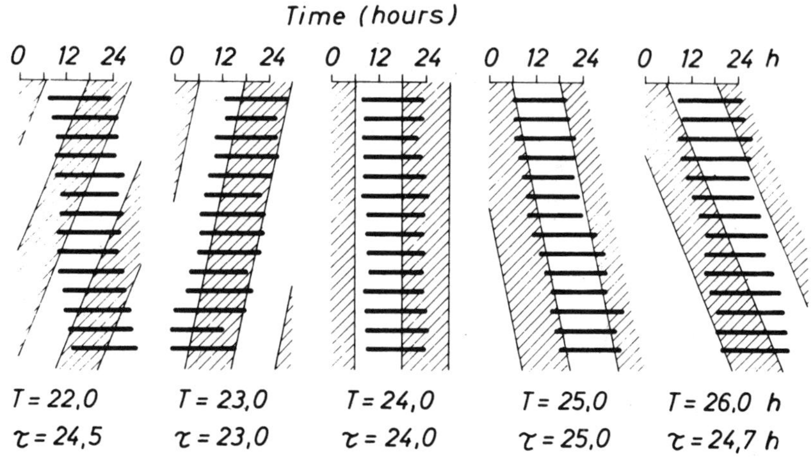

Figure 3. Schamatic activity records of chaffinches (*Fringilla coelebs*) exposed to five different LD cycles with the same 1:1 LD ratio, but with different T. Note the absence of entrainment under $T = 22$ and 26 hours and the different phase position of activity in the LD cycle under entrainment at $T = 23$, 24, and 25 hours. (From Aschoff, 1981a.)

22 to 26 hours. Only in the middle three cases, with T between 23 and 25 hours, was the finches' circadian rhythm entrained; at $T = 22$ and 26 hours, it freeran in spite of the LD cycle. This means that there is a limited *range of entrainment*, i.e., a limited range of T values around the circadian period τ where entrainment is possible. Within this range, there is a systematic variation in the phase position of the rhythm relative to the zeitgeber: With $T = 25$ hours, the activity time of the finches coincided with the light portion of the zeitgeber; with $T = 23$ hours, it overlapped largely with the dark portion, rendering these diurnal birds nocturnal. We return to these phenomena in the next section.

When the period of a zeitgeber is outside the range of entrainment, it may nonetheless exert visible influence on the circadian rhythm studied. The most conspicuous effect is with zeitgebers with T close to $\tau/2$, where full entrainment can occur, but now with $\tau^* = 2T$. An early example found relates to the leaf movements of *Canavalia ensiformis*. Kleinhoonte (1928) exposed these plants, among other treatments, to cycles of LD 6:6, and found that they become entrained with one up- and one downward movement per two zeitgeber cycles (Figure 4). The phenomenon is called *frequency demultiplication*. In principle, similar modes of entrainment may be found with T around 8 and around 6 hours, but with very short cycles, entrainment is no longer possible. For T values close to two or three times the circadian period, again zones of entrainment by *frequency multiplication* have been found ($\tau^* = T/2$, $T/3$), implying that a single external signal once every two, three, or more cycles is sufficient to keep the circadian system entrained. This is not only of theoretical interest, since there may well exist situations in nature where individual animals actually see the zeitgeber only once every few days (DeCoursey, 1986; DeCoursey & Menon, 1990). These phenomena can be summarized in a graph plotting τ as a function of a broad range of T values (see Wever, 1960, Figure 3). In between the primary range of entrainment and the minor secondary ranges for frequency multiplication and demultiplication, there are T ranges where no entrainment ensues.

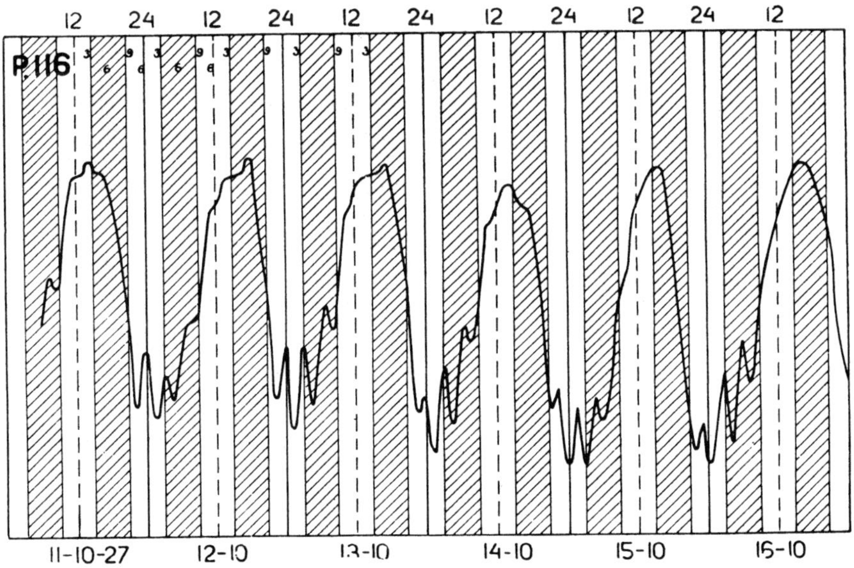

Figure 4. Frequency demultiplication in leaf movements of *Canavalia ensiformis* exposed to an LD 6:6 cycle. Downward movement of the leaf produces an upward movement of the trace. (From Kleinhoonte, 1928, Figure 9.)

Even if there is no entrainment, a zeitgeber can exert its periodic influence. This leads to slow oscillations or beats, where the zeitgeber signal speeds up the biological rhythm when it coincides with a particular phase of the latter and slows it down when coinciding with another phase. This has, for instance, been demonstrated in deermice (*Peromyscus maniculatus*) exposed to sinusoidal cycles of light intensity simulating light conditions in the arctic summer (Swade, 1969). Such zeitgebers may be too weak to entrain the circadian activity rhythm. The oscillatory interference of zeitgeber and rhythm is called *relative coordination*, a term derived from studies in high frequency oscillators in the central nervous system (Von Holst, 1939). An extreme example of relative coordination is the activity pattern of Syrian hamsters (*Mesocricetus auratus*) exposed to another weak zeitgeber, consisting of a 15-minute light pulse once every 25 hours in otherwise continuous darkness (Figure 5). In this situation, the activity rhythm is entrained for some time with the onset of the circadian activity band stably locked to the end of the light pulse. However, the end of activity gradually drifts forward, and the activity time α thus becomes progressively shorter. When α is nearly compressed to zero, entrainment stops, and the system temporarily resumes its free-running state with τ shorter than T, while α decompresses to its original duration of ca. 10 hours. This lasts until the light pulse again starts coinciding with the onset of activity, and the α compression starts anew. Such data suggest that the zeitgeber not only entrains the endogenous rhythm, but also modifies both its state and sensitivity to the zeitgeber (Pittendrigh & Daan, 1976b).

The Phase Angle Difference

The phase relationship between a circadian rhythm and a zeitgeber can be described by selecting arbitrary reference phase angles for both, in standard nota-

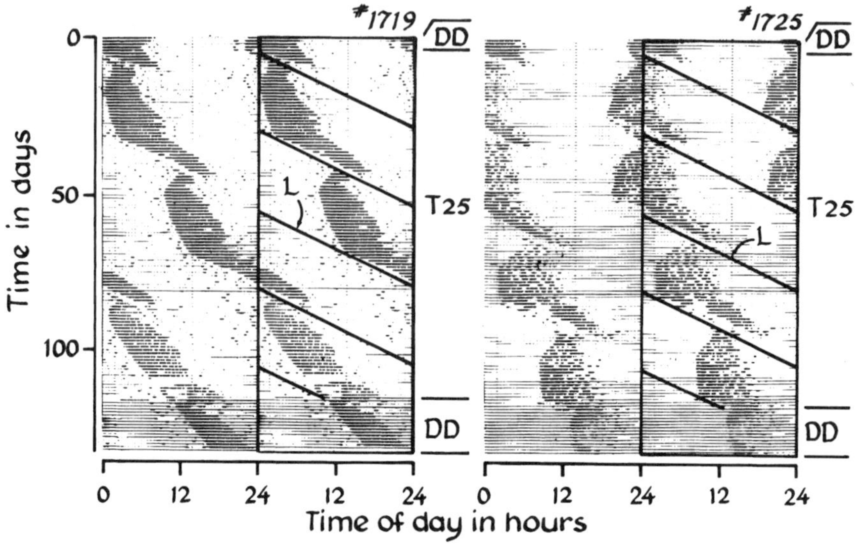

Figure 5. Examples of failing entrainment of Syrian hamster, *Mesocricetus auratus*, by single 15-minute light pulses given with a zeitgeber period $T = 25$ hours. Both animals showed compression and decompression of α associated with the "beats" in relative coordination. (From Pittendrigh and Daan, 1976b, Figure 8.)

tion φ for the rhythm and Φ for the zeitgeber. In an activity rhythm, φ may, for instance, be the time of onset of activity, or the end of activity, or the midpoint between these two. In a rhythm of body temperature, it may be defined as the peak, or as the *acrophase* (the time of the maximum of a sine wave fitted to the rhythm), or the time of maximal rise or decline rate. The phase chosen depends on what is biologically most meaningful and on what yields the most precise, i.e., least variable, estimate. For the zeitgeber, again the best corresponding phase angle is chosen, such as the time of lights-on as the reference for activity onset in a diurnal animal. The phase angle difference $\psi = \Phi - \varphi$ then describes the difference in time between the phase references of zeitgeber and rhythm. In standard notation, ψ is given a positive sign when the φ phase leads Φ, and a negative sign when the φ phase lags Φ.

In steady-state entrainment, the phase angle difference ψ depends on (1) the properties of the zeitgeber and (2) the reponsiveness of the circadian system to the zeitgeber. Together these factors determine the strength of the zeitgeber (Aschoff, 1960). Given a certain strength, the major determinant for ψ is the ratio between the endogenous period τ of the circadian rhythm and the period T of the zeitgeber. The basic qualitative rule describing this relationship, which holds essentially for all oscillating systems, is illustrated in Figure 6. Figure 6A demonstrates the dependence of ψ on τ: when entrained by the same zeitgeber, a relatively fast rhythm phase leads the zeitgeber (ψ positive) and a relatively slow rhythm phase lags it (ψ negative). As shown in Figure 6B, the converse occurs when a rhythm of medium frequency becomes entrained by a zeitgeber of either higher frequency (ψ negative) or of lower frequency (ψ positive). ψ can be expressed either in absolute time or in a dimension relative to the period length of the zeitgeber (which in entrainment equals the period length of the rhythm): e.g., angular degrees ($T/360$) or circadian hours ($T/24$).

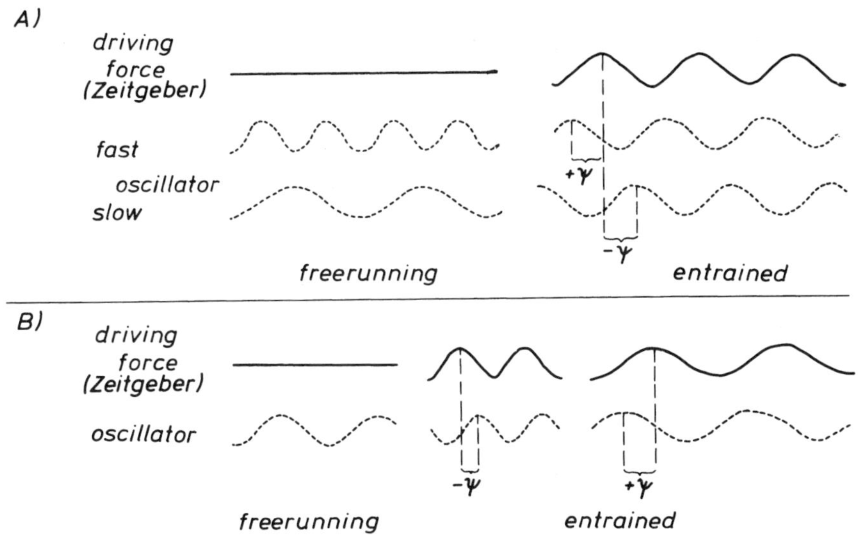

Figure 6. (A) Entrainment of two oscillations (fast and slow) by the same zeitgeber. (B) Entrainment of an oscillation by zeitgebers of high and low frequency. Here ψ is the phase angle difference between peaks in the oscillation and zeitgeber. (From Aschoff, 1979a. Copyright 1974, Lippincott Williams & Wilkins.)

The dependence of ψ on T has been experimentally evaluated for many different organisms simply by recording their circadian rhythms in light–dark cycles of different periods. Representative examples from a host of studies in vertebrates, insects, and plants are shown in Figure 7. In most of these studies, more extreme zeitgeber periods were evaluated than those shown, but these were unable to generate steady-state entrainment. The range of entrainment turns out to be smaller in most vertebrates than in the insects studied, and is often much wider in plants. In every species, ψ is positively associated with T. The slope of this dependence is, however, very different for different circadian systems. As a general rule, the change in ψ per hour of change in T is larger in systems with a narrower range of entrainment. Indeed, the reciprocity between the slope of ψ change and range of T is such that the range of realized phase angle differences approaches $T/2$ or 180° in all these systems (Figure 8). This is also true when studies employing different zeitgeber strengths, different intensities of illumination in the L and D portions of the zeitgeber, are compared. Weak zeitgebers produce a narrow range of entrainment and a steep dependence of ψ on T, strong zeitgebers a broad range and shallow slope. Over this range, a diurnal bird can vary from starting its activity around midnight in long T's to around noon in very short T's (cf. Figure 3). The ψ versus T relationship and range of entrainment reflect the sensitivity of a circadian system to the zeitgeber. Although the studies summarized in Figures 7 and 8 did not employ exactly the same zeitgebers, we may safely conclude that the vertebrate systems are generally less sensitive to LD cycles, or have a more strongly self-sustained pacemaker, than insects and plants.

Just like the dependence of ψ on T, the dependence of ψ on τ has been documented qualitatively in a number of studies (Aschoff & Wever, 1966; Hoffmann, 1969). When entrained by the same zeitgeber, individuals exhibiting a relatively short circadian period in constant conditions tend to phase lead those with a long

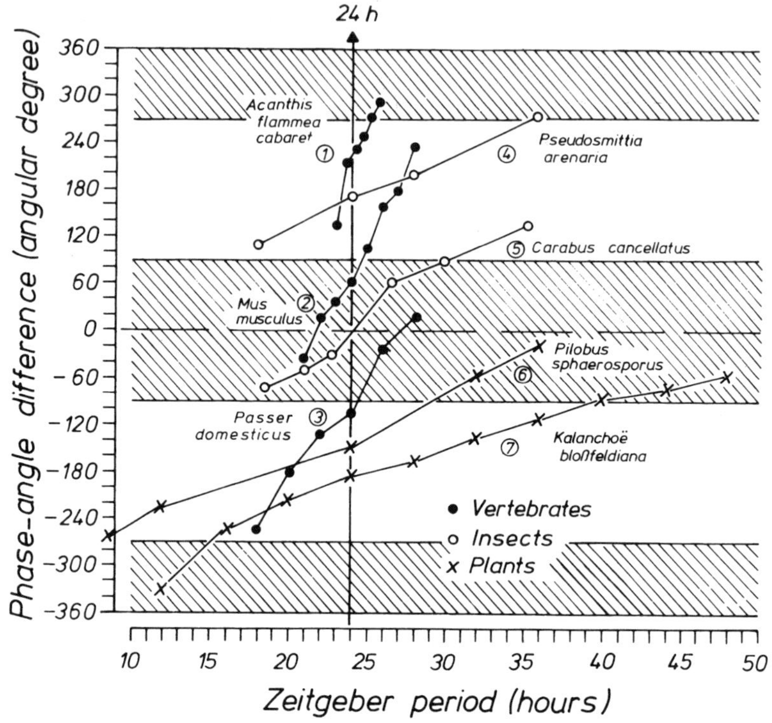

Figure 7. Phase angle differences between entrained circadian systems and zeitgeber with different periods *T* in seven studies. Note the systematic increase in ψ with *T* and the negative association between the steepness of this increase and the range of entrainment. (From Aschoff and Pohl, 1978.)

period. A further important determinant of ψ is the LD ratio of the zeitgeber. However, despite intraspecific consistency in the dependence of ψ on LD ratio, no generalization across species can be made.

Another issue which has been of interest to researchers is the degree of *accuracy* of circadian rhythms under the influence of different zeitgebers. Impressive accuracy is observed especially when there are clear selective benefits of accurate timing, such as in the distinct spawning times of marine algae (Clifton, 1997). Low accuracy is theoretically expected with zeitgebers close to the limits of entrainment, where particular oscillatory patterns in ψ (*relative entrainment*) have been both theoretically predicted (Wever, 1972) and empirically observed in squirrel activity rhythms (DeCoursey, 1972). Accuracy of entrainment, defined as the reciprocal of the day-to-day standard deviation of ψ, has further been analyzed for activity rhythms of songbirds exposed to seasonally varying natural photoperiods (Daan & Aschoff, 1975). Accuracy tends to be highest when activity onset and end coincide, on average, with the maximum rate of change in light intensity, which occurs around civil twilight (Figure 9). This is found usually around the equinoxes, and there is a tendency toward reduced accuracy in midsummer and midwinter, especially at high latitudes. The seasonal and geographic variation may be attributed to increased zeitgeber strength when the LD ratio is close to 12:12. However, no studies have specifically addressed accuracy in relation to zeitgeber strength.

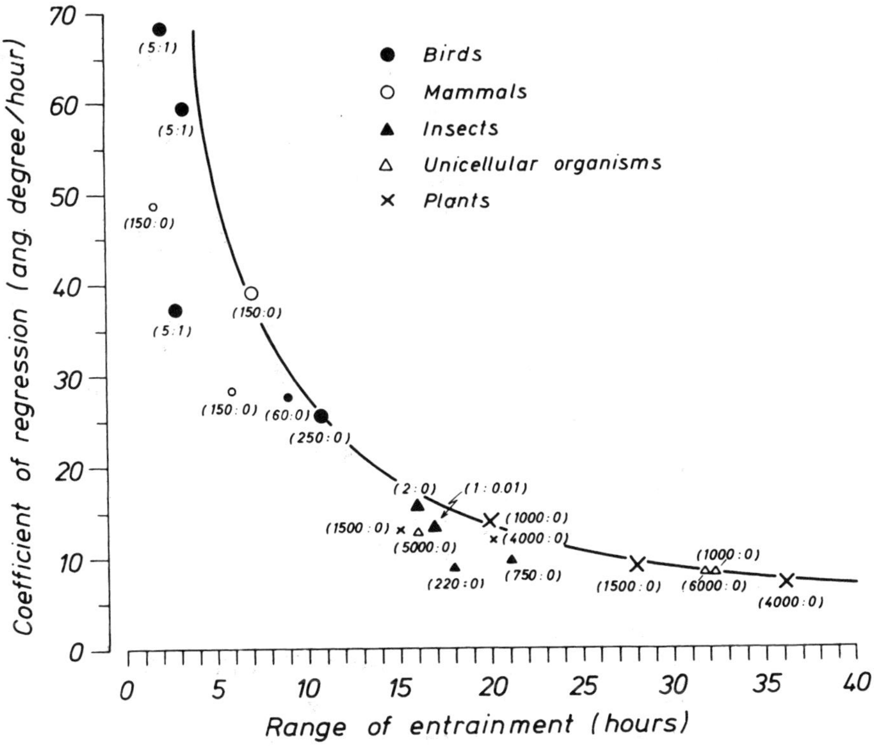

Figure 8. The relationship between the range of entrainment and the slope of the regression of ψ on T. (From Aschoff and Pohl, 1978.)

The Classical Theory of Entrainment: Brief Light Pulses

Phase Response Curves (PRC)

The previous section dealt with some qualitative generalizations on entrainment. A major effort toward a detailed quantitative understanding of entrainment was made by C. S. Pittendrigh (see Pittendrigh, 1958, 1980, 1981a,b; Pittendrigh & Daan, 1976b; Pittendrigh & Minis, 1964). In essence he followed a two-step approach. First, the sensitivity of the circadian system is probed by administering at different phases single brief pulses of light in otherwise continuous darkness (DD) and measuring the ensuing phase shifts. Then a repeated series of identical pulses, together constituting a zeitgeber, is applied, testing the hypothesis that each produces an instantaneous phase shift as dictated by the circadian phase and by the response as assessed in the single-pulse experiment. If this hypothesis is correct, it should be possible fully to predict the pacemaker's time course in any complex experimental design.

When the phase shifts in response to single pulses are plotted as a function of the phase of the circadian system at the time of the pulse, a *phase response curve* (PRC) is obtained. The first PRC was measured for the circadian rhythm of pupal eclosion in the fruitfly *Drosophila pseudoobscura* (Pittendrigh, 1958), and a typical measurement in this system is illustrated in Figure 10. In 25 separate populations of pupae,

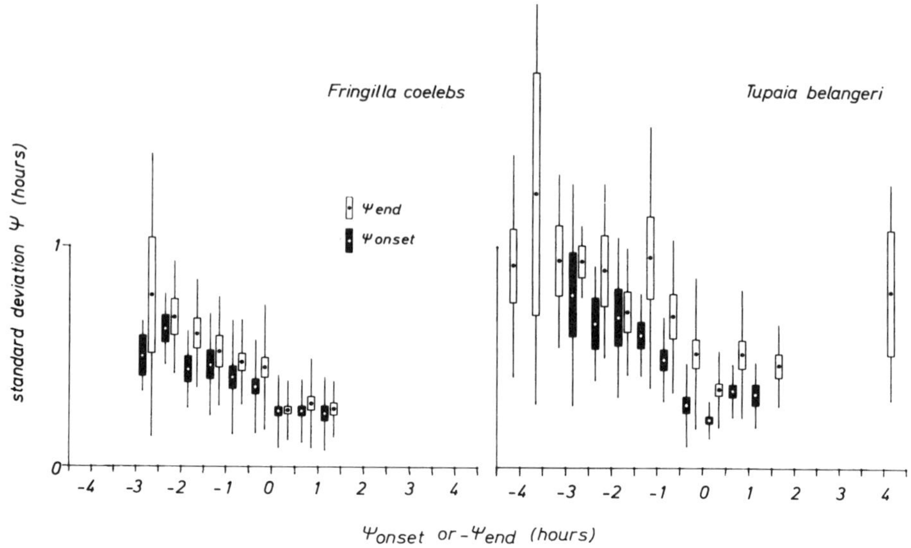

Figure 9. Accuracy of circadian entrainment as a function of ψ in two diurnal species, chaffinch (left) and tree shrew (right), maintained year-round in natural daylight at the arctic circle. Abscissa: Average time of activity onset before (+) or after (−) sunrise, or average time of activity end before (−) or after (+) sunset, in half-hour bins. Ordinate: Corresponding average day-to-day standard deviations, mean ± SEM (boxes), SD (lines). Note that accuracy is maximal when onset and end occur 0–1 hour before sunrise or after sunset. (From Daan and Aschoff, 1975.)

the number of flies emerging from their pupal case was assessed in hourly bins. The eclosion events occur in sharp peaks shortly after lights-on and in ca. 24-hour intervals thereafter when the population is further kept in continuous darkness. One of the 25 populations in Figure 10 served as a control and each of the others received a 15-minute pulse, each at a different hour in the circadian cycle. The phase of the system is measured by the eclosion peak in each successive cycle and the difference with the control population is determined (Δφ; Figure 10A). When the phase shifts are plotted as a function of the circadian phase of the light pulse (with time 0 arbitrarily designated as 24 hours after the last lights-on), a series of PRCs is obtained, one for each successive day. In standard notation, phase advances are expressed with a positive sign, delays with a negative sign. In the *Drosophila* eclosion rhythm, virtually no advance phase shifts are seen on the first day: it takes advance phase shifts up to about five transient circadian cycles to attain steady state. In contrast, phase delays are nearly complete already in the first cycle (Figure 10B). For reasons summarized below, the steady-state phase shifts from cycle 5 onward are assumed to reflect the true phase response of the controlling oscillator. They encompass the whole range from −12 to +12 hours, and a phase advance of 6 hours can be considered also as a phase delay of 18 hours. Therefore, the whole PRC in Figure 10B can also be plotted as a continuous curve as in Figure 10C.

Pittendrigh (1980) showed that the transients, observed especially after phase-advancing pulses, do not reflect the phase-shifting behavior of the pacemaking oscillation that drives the eclosion rhythm. Applying a second light pulse at different times in the first cycle after the first pulse, he assessed the full phase response curve. Two examples are shown in Figure 11. A first pulse at circadian time (ct) 16 (i.e., 4

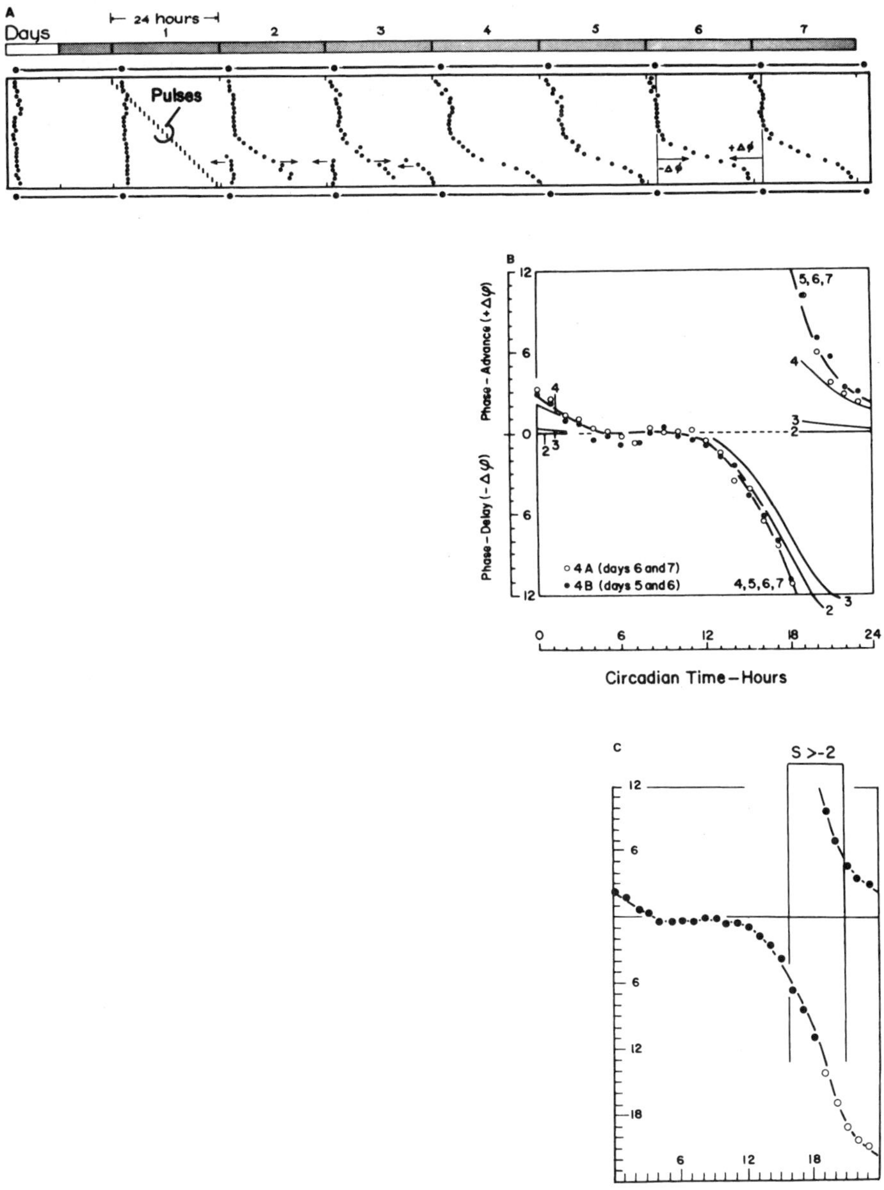

Figure 10. Derivation of a phase response curve (PRC) for the pacemaker of the *Drosophila pseudoobscura* eclosion rhythm. (A) Time course of the rhythm in 24 experimental populations (and 1 control) is given as a series of points marking the midpoint of the eclosion peaks. Each population, otherwise free-running in darkness, receives a single 15-minute light pulse (−100 lux), shown as a vertical black bar; the light pulse is given at a different phase of the rhythm in each population. Following several transient cycles, a steady-state phase shift is realized in each population. (B) The PRC plots the phase shift (Δφ) as a function of the phase (φ) pulsed. Different curves are given for the shifts measured on successive days after the pulse. The advance transients go more slowly than the delays. All phase shifts are in steady state by day 5 after the pulse. (C) The PRC can be given as all delays, yielding a monotonic curve. (From Pittendrigh, 1981a, Figure 1.)

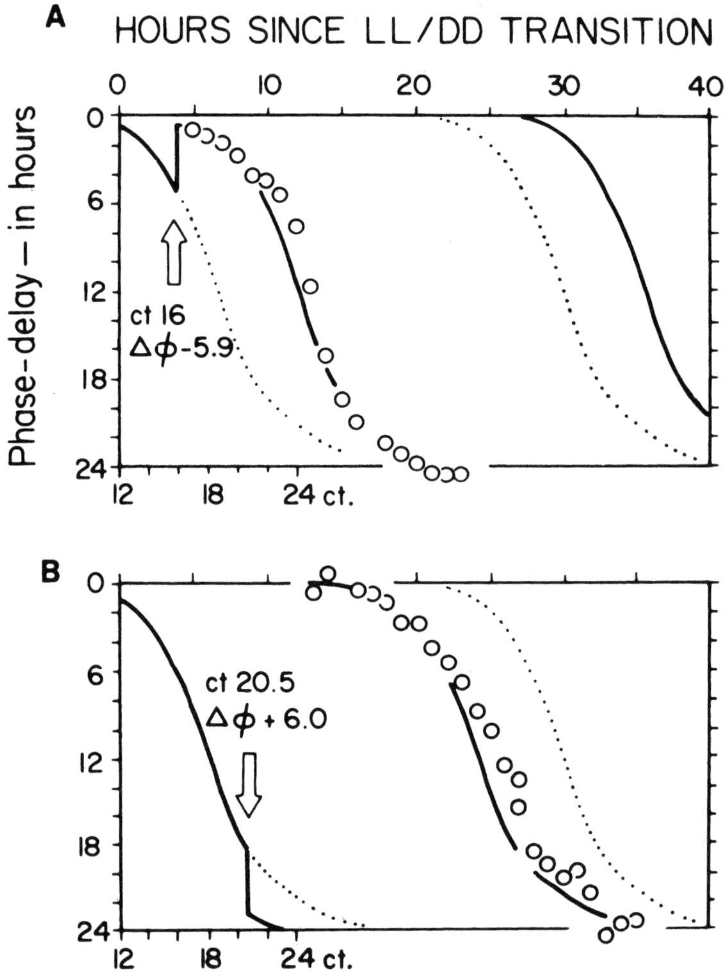

Figure 11. The "instantaneous" nature of the *D. pseudoobscura* pacemaker phase shifts. (A) The first pulse (arrow) is given at ct 16. Subsequent testing pulses are used to track the time course of the pacemaker after the first pulse. Dotted line shows the unshifted PRC, solid line the prediction for the instantaneously shifted PRC, circles the observed phase shifts. (B) The same protocol with a first pulse at ct 20.5 shows that this pulse causes an instantaneous phase advance. (From Pittendrigh, 1981b, Figure 10.)

hours after the last LL/DD transition) produces a steady-state phase delay of −5.9 hours in the eclosion rhythm. If the pacemaking oscillation were to be instantaneously and fully reset by this amount, the whole phase response curve should be phase shifted by −5.9 hours (solid line in Figure 11A). Indeed, the measured shifts due to the second pulse are precisely in accordance with this prediction. When the pacemaker's PRC is advanced by +6 hours as an instantaneous response to a first pulse at ct 20.5, it also accurately predicts the $\Delta\varphi$ responses that test pulses generate at successively later hours after this first pulse. Thus the pacemaker is indeed instantaneously reset, and the slow transients of the observed eclosion rhythm do not reflect this motion of the pacemaker, but the motion of a second subordinate oscillation that only gradually regains a stable steady-state relation to its reset pacemaker (Pittendrigh, 1981a). For this reason, it is important that true steady-state phase shifts, not the initial transients, are used to assess a PRC.

The assay of a PRC for individual circadian rhythms is similar to that used in the *Drosophila* population rhythm. The first system for which this was done was the activity rhythm of the nocturnal flying squirrel, *Glaucomys volans* (DeCoursey, 1960). The procedure and result were basically the same as illustrated in Chapter 7, Figure 1. Again it is emphasized that the PRC describes steady-state phase shifts after transients have subsided. Transients in locomotor activity rhythms may depend on the phase reference chosen. In hamsters, the activity onset phase-delays completely in one cycle and advances via several transient cycles (Daan & Pittendrigh, 1976a), while the reverse is true for the end of activity (Meijer & DeVries, 1995). These differences are related to internal pacemaker structure, possibly composed of two functionally different oscillators, E and M (Pittendrigh & Daan, 1976c). PRCs have now been established for numerous circadian systems (C. H. Johnson, 1992). For light stimuli, they show the basic general pattern illustrated in Figure 12. The circadian cycle is subdivided into 360° of arc, or into 24 circadian hours. Circadian time 0 is arbitrarily defined as that phase in the cycle which under conditions of entrainment by an LD 12:12 cycle coincides with lights-on. In activity rhythms of

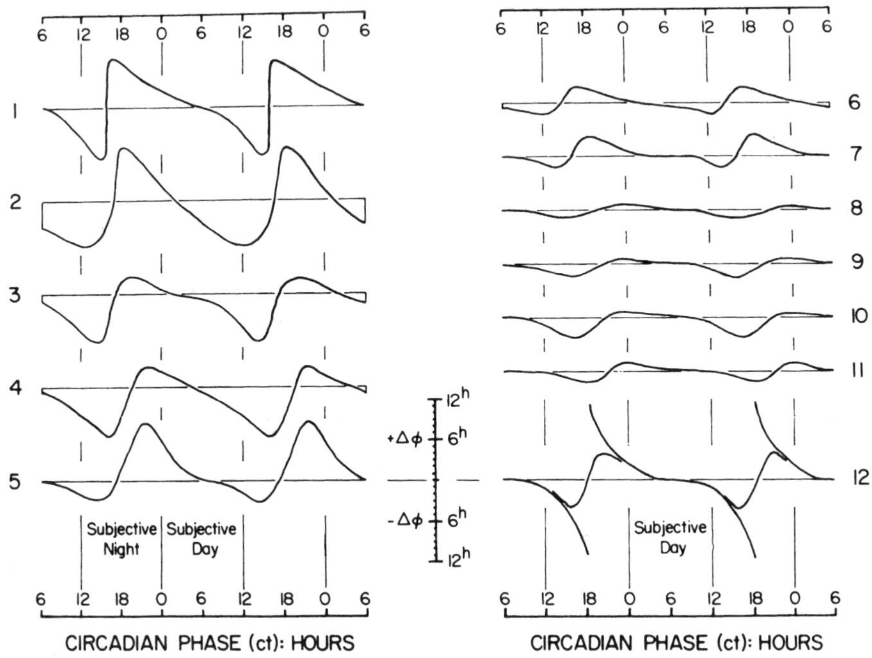

Figure 12. Twelve phase response curves for brief light pulses, plotted on the same circadian time scale, with ct 0–12 as the subjective day and ct 12–24 as the subjective night. Curves derived by Pittendrigh (1981b, and references therein) from (1) fleshfly, *Sarcophaga argyrostoma*, (2) *Coleus*, (3) cockroach, *Leucophaea maderae*, (4) unicellular *Euglena*, (5) marine alga, *Gonyaulax*, (6) mosquito, *Anopheles*, (7) hamster, *Mesocricetus auratus*, (8) white-footed deermouse, *Peromyscus leucopus*, (9) deermouse, *Peromyscus maniculatus*, (10) house mouse, *Mus musculus*, (11) bat, *Taphozous melanopogon*, and (12) fruitfly, *Drosophila pseudoobscura* (pulse 15 minutes: type 0 PRC; pulse 1 msec: type 1 PRC).

nocturnal rodents, where the onset of activity is often the sharpest marker of the circadian system, this is defined as ct 12. The *subjective day* is then defined as the interval ct 0–12, the *subjective night* as the interval ct 12–24. Light pulses presented during the subjective day generally produce small phase shifts; when presented during the subjective night, the phase shifts are larger. In the early part of the subjective night, they are negative (phase delays); in the later subjective night, light pulses induce phase advances (positive). Somewhere in the middle of the subjective night, the PRC crosses over from delays to advances.

Even though the PRC shape for light pulses is qualitatively similar for all circadian systems studied, we emphasize that the quantitative detail of the PRC is defined both by the system studied and by the properties of the probing light pulse: its duration, intensity, and wavelength. When the duration of the light pulse is changed, the PRC may change in an apparently qualitative way, as illustrated by two PRCs for *Drosophila pseudoobscura*, one assessed with a 15-minute pulse, the other with a 1-millisecond light flash (Figure 12). In the middle of the subjective night, the 15-minute pulse induces phase shifts of up to 12 hours, where there is no unequivocal distinction between advances and delays. The 1-millisecond flash causes here a zero phase shift at the crossover from delays to advances. Winfree (1970) pointed out that there is a discrete distinction between these two PRCs. He coined the terms *type 0 PRC* and *type 1 PRC*, based on the slopes of the associated *phase transition curves* (PTC). The PTC plots the new phase φ_n after the pulse as a function of the old phase φ at which the pulse was given (note that $\varphi_n = \varphi - \Delta\varphi$). Figure 13 compares the PRCs and PTCs of *Drosophila pseudoobscura* and *D. melanogaster* pupal eclosion rhythms assessed with the same 15-minute pulses. The *D. melanogaster* PRC is like that of *D. pseudoobscura* responding to weaker (millisecond) pulses. Its PTC has an average slope of 1, and its PRC is therefore referred to as type 1. The *D. pseudoobscura* PTC has an average slope of zero, and its PRC is accordingly referred to as type 0. The PTC is strictly periodic in both directions: φ_n at $\varphi = 0°$ is identical to φ_n at $\varphi = 360°$. Therefore, the slope between these points can only be 0 or 1 (or theoretically 2, 3, etc.), not a fraction in between. We have seen that at least in some systems (*D. pseudoobscura*), weak pulses elicit a type 1 PRC, strong pulses a type 0 PRC. There must then be a pulse strength somewhere in between where a discontinuity occurs. Winfree (1970) predicted theoretically that a pulse of this strength given at a particular phase in the middle of the subjective night would reset the oscillation to its *singularity*. By such a pulse, the amplitude of the pacemaking oscillation is, perhaps temporarily, reset to zero, and the rhythm stops its motion. He subsequently found this singularity indeed experimentally with dim blue light pulses of ca. 50-second duration presented around ct 18.8 in populations of *D. pseudoobscura* pupae (Winfree, 1970). Since then, circadian systems in other organisms have also been put onto their singularity. While this is primarily of theoretical interest and tells us little about function or mechanism of entrainment, it does clarify that type 0 and type 1 PRCs reflect quantitative differences in pulse strength and/or sensitivity to that pulse. The 15-minute pulse clearly is stronger than the 1-millisecond pulse in *D. pseudoobscura*. We further can conclude from the PRC differences that the *D. pseudoobscura* eclosion pacemaker is much more sensitive to light than the *D. melanogaster* pacemaker.

Entrainment by Periodic Pulses

Given an empirically assessed PRC for a specified light pulse and (virtually) instantaneous resetting of the pacemaker, one may ask whether this can accurately

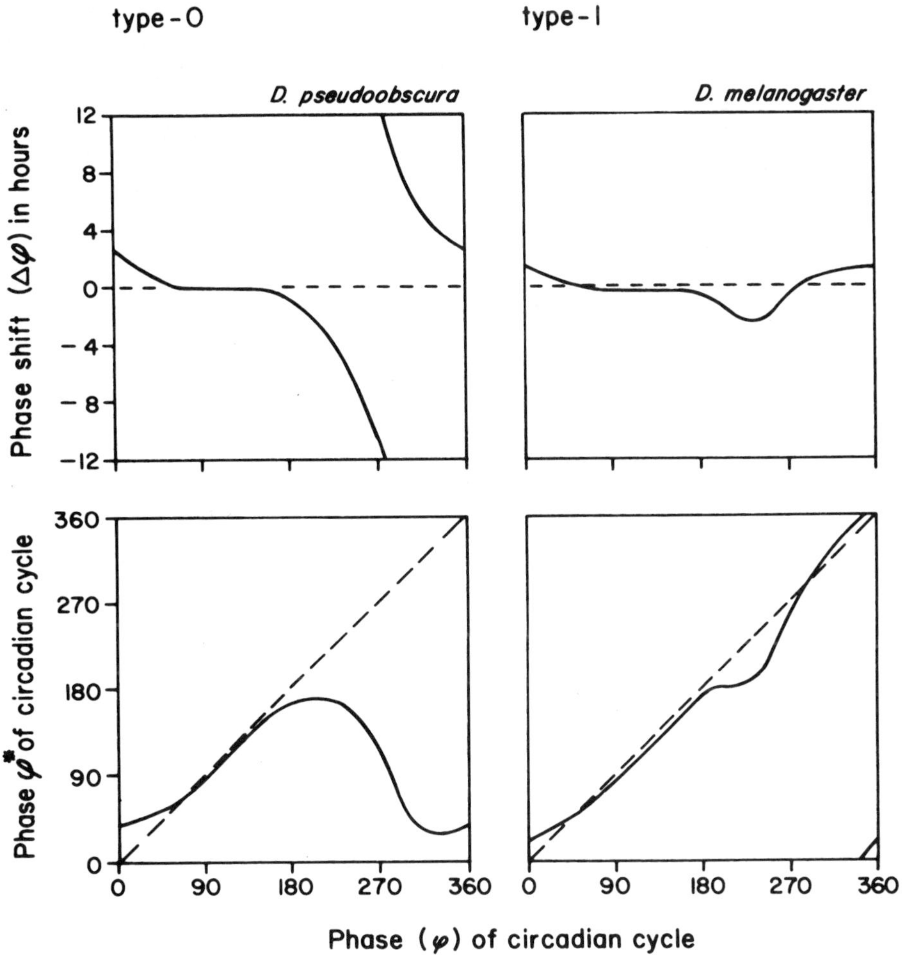

Figure 13. Type 0 and type 1 phase response curves (PRC) and their associated phase transition curves (PTC) observed with the same 15-minute 100-lux light pulse in *Drosophila pseudoobscura* and *D. melanogaster*, respectively. The PRC plots the phase shift for each φ pulsed; the phase transition curve plots the new phase to which the pacemaker is reset when pulsed at a given phase. (From Pittendrigh, 1981b, Figure 7.)

predict the behavior of the circadian system under entrainment by a periodic series of such pulses. In principle, the action of each pulse in steady-state entrainment must be to cause a phase shift given by

$$\Delta\varphi(\varphi) = \tau - T \qquad (1)$$

If T and τ are known along with the PRC describing $\Delta\varphi(\varphi)$, it must be possible to predict at which phase φ the light pulse falls during stable entrainment in each cycle. Thus, when $\tau = 24$ hours and $T = 27$ hours, the pulse recurring every 27 hours causes a 3-hour phase delay, and when $T = 21$ hours, a 3-hour phase advance. According to the PRC for *D. pseudoobscura*, this should be achieved by light pulses falling at ct 14.3 and ct 23.4, respectively. Pittendrigh (1967) tested these predictions by entraining the *Drosophila* eclosion rhythm to light cycles consisting of one 15-minute pulse every

27 hours (or every 21 hours) and by assessing the PRC immediately after the last pulse. In $T = 27$ hours, all pulses, including the last pulse, should hit the system at ct 14.3, and cause a reset to ct 11.3 ($\Delta\varphi = +3$ hours), and the PRC should thus resume its motion from ct 11.3 onward. Similarly, after entrainment by $T = 21$ hours, the system should resume its motion after the last pulse from ct 2.4 onward. Figure 14 compares these predictions with the measurements obtained on separate populations of pupae. The predictions are fully upheld by observations. Note that the assay used here to test the predictions is based on the PRC, which must be considered a property of the controlling oscillator or pacemaker. This does not mean that the overt rhythm can be precisely predicted. In fact, later work by Pittendrigh showed that the eclosion rhythm systematically changes its phase relative to the pacemaker

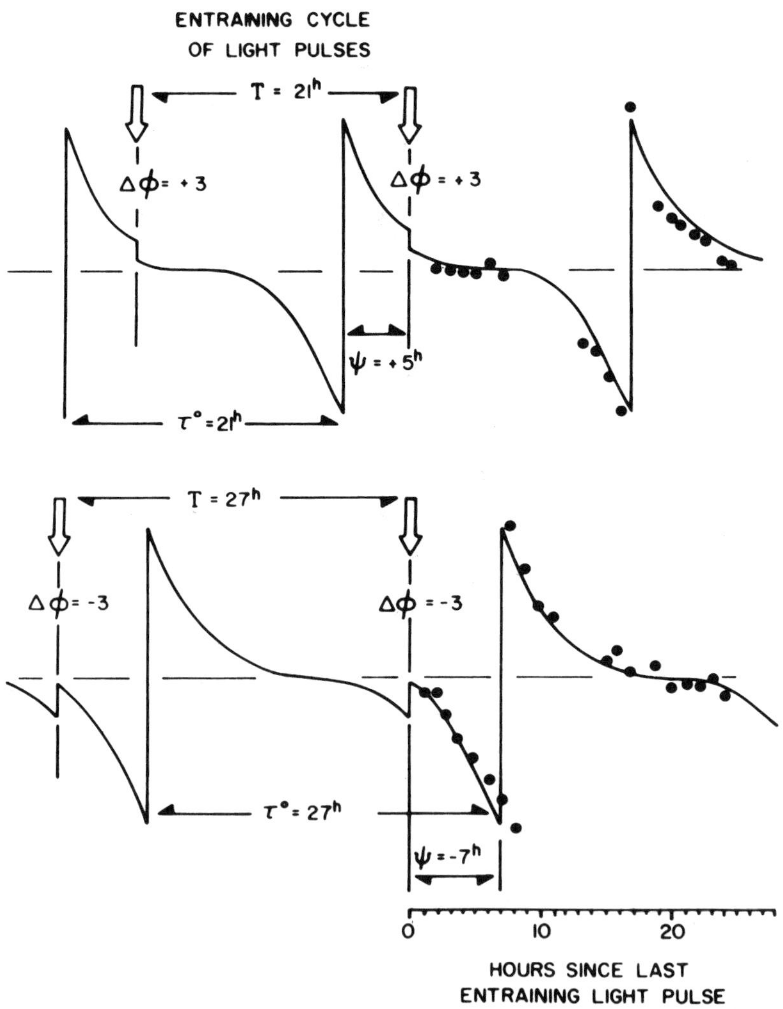

Figure 14. Prediction and observation for *Drosophila pseudoobscura* pacemaker phase in a zeitgeber consisting of one 15-minute light pulse per 21 hours (upper graph) and per 27 hours (lower graph). Solid line is the prediction for PRC time course during entrainment and after the last pulse. Dots are measured phase shifts in the first cycle after the last pulse. (From Pittendrigh, 1981b.)

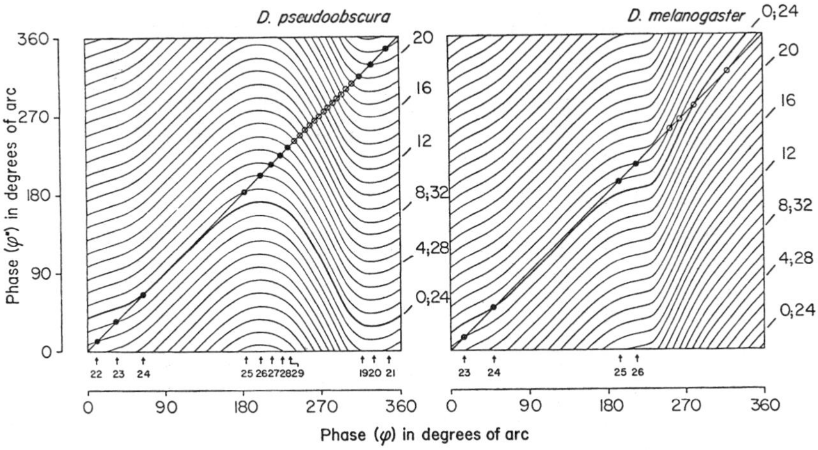

Figure 15. Criteria for stable entrainment from PTCs. See text.

over a wide range of T values, from $T = 18$ to 30 hours (Pittendrigh, 1981a). He attributed this discrepancy, like the transients in Figure 10, to a second subordinate (or "slave") oscillator which actually "gates" the eclosion event and is itself synchronized to the outside world by the pacemaker.

A graphical way to derive predictions for stable entrainment is shown in Figure 15. This figure is derived from Figure 13 by displacing the PTCs of *D. pseudoobscura* and *D. melanogaster* for 15-minute light pulses vertically over δ whole circadian hours ($\delta = -24$ to $+24$). The crossing points of such secondary PTCs with the diagonal are indicated by symbols. These points obey the criteria $\varphi_n = \varphi + \Delta\varphi + \delta$ and also $\varphi_n = \varphi$, or $\Delta\varphi = -\delta$, and thus satisfy entrainment for $T - \tau = \delta$. For each value of T (expressed in circadian hours), one equilibrium solution is found in the case of type 0 PRCs and either no or two solutions in the case of type 1 PRCs. Whether these solutions represent stable or unstable equilibria depends on whether small departures from equilibrium contract or expand. It can be shown that contraction (stable entrainment) requires that the slope of the PTC at the equilibrium point is between -1 and $+1$. This restricts the range of entrainment. There are 11 stable equilibrium points (filled symbols) for *D. pseudoobscura* in Figure 15 and 4 for *D. melanogaster*. Together these represent a range of entrainment from $T = 19$ to $T = 29$ for *D. pseudoobscura* and from $T = 23$ to $T = 26$ for *D. melanogaster* when exposed to single 15-minute light pulses in intervals of T hours. Note that for increasing T, successive equilibrium points where the light pulse hits the circadian cycle in stable entrainment successively shift to the right, i.e., to later phases. This is a topological necessity, true for any form of PRC, and explains the general empirical rule we encountered in Figure 7, that ψ increases monotonically with increase in T.

Note that we have expressed T in this discussion in circadian hours or in degrees of arc. Therefore, a change in T may be brought about either by exposing a system with the same τ to different T in real hours, or by exposing systems with different τ to the same T cycle. A shorter τ has the same effect as a longer T. Thus, when *D. melanogaster* changes its τ from 23 to 25 hours, we predict that the equilibrium phase where the light pulse falls in stable entrainment advances from the beginning of the subjective night (around ct 12) to the beginning of the subjective day (around ct 0).

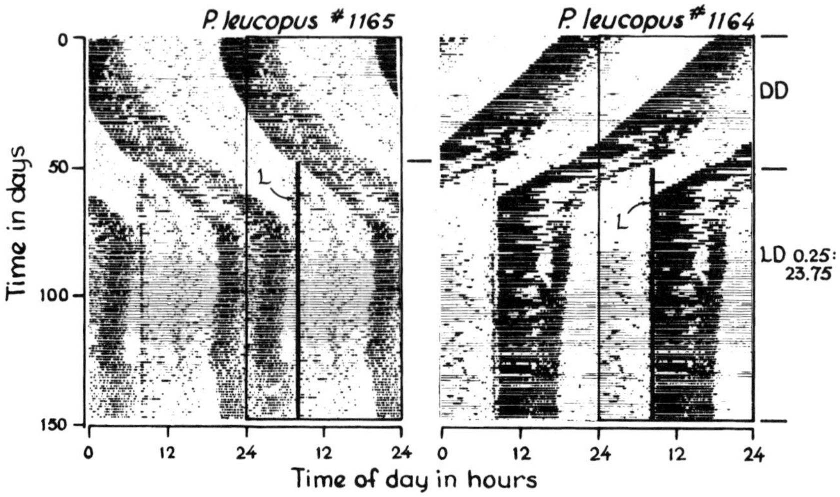

Figure 16. Plot of ψ as a function of τ in *Peromyscus leucopus*. Left: A mouse with a free-running circadian period >24 hours becomes entrained to a 15-minute light pulse (T = 24 hours) with the pulse falling at the end of activity. The pulse is interpreted as dawn. Right: A mouse with a free-running period shorter that T interprets the same signal as dusk. (From Pittendrigh and Daan, 1976b.)

D. melanogaster is not known to do this, but several nocturnal rodent species with a similar PRC for short light pulses have a circadian period varying between individuals from smaller to longer than 24 hours. Figure 16 shows the activity records of two deermice (*Peromyscus leucopus*) kept initially in DD and then exposed to a single 15-minute light pulse recurring every 24 hours. The animal in the left panel had $\tau > 24$ hours and eventually was entrained with the light pulse falling at the end of activity, in the late subjective night; the animal in the right panel had $\tau < 24$ hours and became entrained with the light falling at the onset of activity, in the early subjective night. This variation in behavior between individuals is in full agreement with entrainment theory (Pittendrigh & Daan, 1976b).

Entrainment by "Skeletons"

The theory has been further expanded to generate predictions for artificial LD cycles composed of two brief light pulses per cycle. These pulses may be evenly spaced, as in LD 0.25:11.75:0.25:11.75, henceforth denoted as PP_s 12:12, and then generate entrainment by frequency demultiplication, as encountered above. They may also be unevenly spaced, with one dark interval longer than the other. Such zeitgebers have been called *skeleton photoperiods*. Skeletons are of special interest for several reasons. First, they may give cues to the manner of entrainment by long photoperiods of different duration (Pittendrigh & Minis, 1964). Second, for some animals, such as burrowing nocturnal rodents, they may be realistic representations of what the animal actually sees when it lives in darkness below ground in daytime and above ground at night and only receives brief light exposures around dawn and dusk (Terman, Reme, & Wirz-Justice, 1991). Third, skeletons are employed to study expression of circadian rhythms in conditions of entrainment, but without masking (Strubbe, Spiteri, & Prins, 1986).

Predictions for stable entrainment by skeletons are essentially based on the criterion $\Delta\varphi_1 + \Delta\varphi_2 = \tau - T$, where $\Delta\varphi_1$ and $\Delta\varphi_2$ are the successive phase shifts caused by the two pulses defining the skeleton. There are both graphical and computational ways to find the stable equilibrium solutions. Intuitively, it is clear that for a fully symmetric PP_s, with equal intervals between the pulses, there must be two solutions, with one pulse causing an advance, the other a delay of the system, and the subjective night falls largely within one or the other interval. With an extremely asymmetric PP_s, both pulses must fall shortly after each other, either in the delay section of the PRC or in the advance section, depending on whether $T > \tau$ or $T < \tau$, respectively. For small departures from symmetry, there is a range of bistability where two solutions are possible. For the 15-minute light pulse PRC of *Drosophila pseudoobscura*, this range has been computed to run from PP_s 10.3:13.7 to PP_s 13.7:10.3 (Pittendrigh, 1981b). Which of the two alternative steady-state solutions is realized by skeletons within this range depends on (1) the pacemaker phase hit by the first pulse and (2) the duration of the first dark interval in the zeitgeber cycle. As Figure 17 illustrates for the case of PP_s 11:13 and PP_s 13:11, the predictions from the model are well matched by observation (Pittendrigh, 1981b). Indeed, a wealth of quantitative detail has been collected for the *Drosophila* circadian eclosion rhythm (e.g., Pittendrigh, 1981a, b) which generally supports the notion that entrainment by brief,

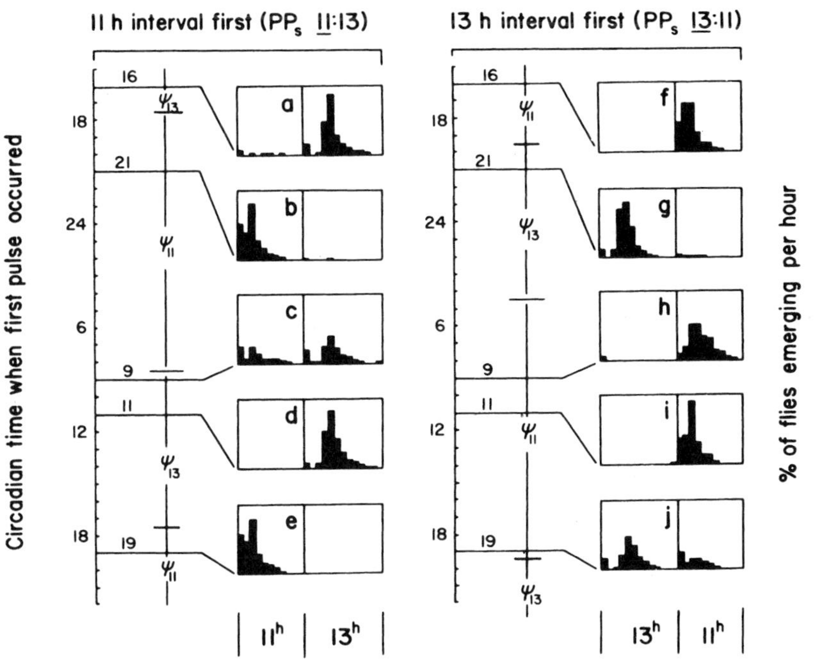

Figure 17. Prediction and observation for the rhythm of pupal eclosion of *Drosophila pseudoobscura* in skeleton photoperiods of two brief light pulses at intervals of alternatingly 11 and 13 hours (PP_s 11:13). Two steady states are possible. Which one is realized depends on the phase of the pacemaker illuminated by the first pulse in the entraining cycle. In one of them (ψ_{11}), the eclosion peak (histogram) falls in the 11-hour interval; in the other (ψ_{13}), it falls in the 13-hour interval. The model of entrainment predicts (line to the left of each panel of data) which steady state will be realized as a function of the phase (ct) of the first pulse. (From Pittendrigh, 1981b.)

bright light pulses can be understood on the basis of instantaneous complete phase resetting.

Skeletons have also been applied to circadian rhythms of locomotor activity in several nocturnal rodent species (Pittendrigh & Daan, 1976b). Also in these systems, a zone of bistability was predicted and found: When activity was in the long dark interval, the onset of activity could be reasonably well predicted. However, with gradual change in PP_s, the zone of bistability turned out to be far wider than predicted. Activity could be compressed in the short interval in skeletons up to PP_s 18:6 before it "jumped" into the long interval. This suggests that with brief intervals between two pulses, the response to the second pulse is not predictable from the standard PRC and instantaneous phase resetting. Also, in *Drosophila*, in the two-pulse experiments discussed above, observations fail to match prediction when the pulses are less than 4 hours apart. Such observations suggest that resetting often is not completely instantaneous and that circadian pacemakers after a single perturbation generally take several hours to get back to steady-state motion on their limit cycle. Such retardation in the kinetics also explains why experiments with fast alternation of light and darkness (e.g., LD 0.25:0.75; Eriksson & Van Veen, 1980) do not stop the pacemaker. Simple PRC theory would predict for such a schedule that the phase where the light produces a 1-hour phase delay once per hour would represent a stable equilibrium. In spite of these interesting complexities, the PRC approach to understanding entrainment has been generally successful as long as the pulses are spaced sufficiently widely apart.

The Complexity of Natural Entrainment

The Variability of τ and PRC

The classical theory of entrainment by brief light pulses as outlined above is based on the empirical determination of τ and PRC, and assumes that both are fixed and exactly known. Several phenomena suggest that exposure to a zeitgeber itself affects τ and PRC. This would restrict the accuracy of the predictions from theory. In the first place, the period of circadian systems is known to be subject to variations, not only spontaneous—perhaps partly genotypic—variations between individuals, but also intraindividual (experimentally induced) variations. The latter are known as *aftereffects*. Pittendrigh and Daan (1976a) exposed mice (*Mus musculus*) to zeitgebers with periods of $T = 20$ and 28 hours and afterward released them in constant darkness. In DD the initial free-running periods of the circadian rhythms were on average 24.35 and 23.15 hours, respectively. The τ values only gradually returned to values closer to 24 hours, but were still significantly different following the two treatments up to 100 days in DD. Conversely, the initial distribution of τ values of deermice released from LD 12:12 into DD was narrowly centered around an average of ca. 24 hours, but gradually widened to much larger variation after 65 days in DD (Pittendrigh & Daan, 1976a, Figure 10).

Apart from aftereffects of T, there are aftereffects of photoperiod. In nocturnal mammals, release into DD after entrainment by a long photoperiod usually leads to much shorter τ values than prior entrainment by a short photoperiod (Pittendrigh & Daan, 1976a). Other species also show consistent influences of prior LD on τ on constant conditions, but there is no general rule concerning the sign of this effect (Aschoff, 1979b). A further serious limitation to the applicability of the classical

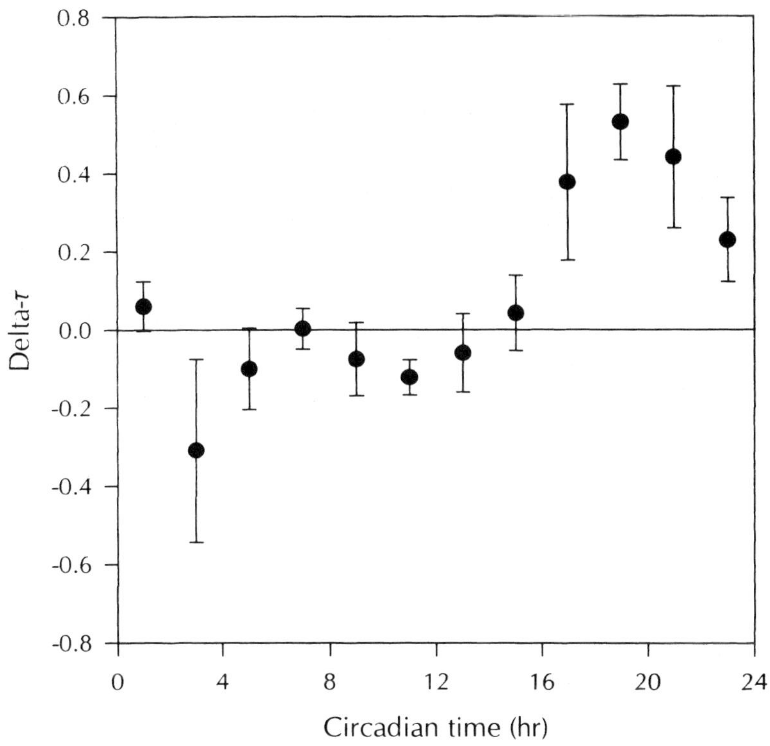

Figure 18. Change in free-running circadian period (τ in DD due to a single 15-minute light pulse in the common vole, *Microtus arvalis*. Plotted symbols indicate mean $\Delta\tau$ (= preceding − subsequent τ) ± 1 SEM. (Modified from Gerkema *et al.*, 1993.)

entrainment theory is provided by aftereffects of single phase shifts. Several authors have observed changes in τ and plotted them as period response curves (Figure 18; see also Aschoff *et al.*, 1982b; Gerkema, Daan, Wilbrink, Hop, & Van der Leest, 1993; Mrosovsky, 1993). Changes in τ usually are in the same direction as the phase shifts: reductions of τ are often associated with phase advances elicited by the same pulse, and increases of τ with phase delays. Figure 18 demonstrates that such changes can be substantial. A reduction of circadian period by 1 hour is not exceptional. This has consequences in the first place for the empirical measurement of the PRC. A steady-state phase shift is only then stable, i.e., measurable in cycle after cycle, when the period of the oscillation is the same before and after the probing pulse. With changes in τ, the determination of the phase shift becomes arbitrary. In the second place, such dramatic changes in τ will have important effects on the prediction of the phase repeatedly hit in stable entrainment according to equation (1). Indeed, in long-term entrainment of deermouse activity rhythms by a single brief light pulse per 24 hours, large spontaneous changes in phase angle differences have been reported (Figure 19). Here the initial entrainment, with activity end locking on to the light pulse, was consistent with the initial free-running τ exceeding 24 hours. Apparently under the influence of repeated phase advances during entrainment, τ spontaneously assumed a shorter period, and activity drifted forward (to the left in the actograms) until the onset locked on to the light pulses. This "bouncing phe-

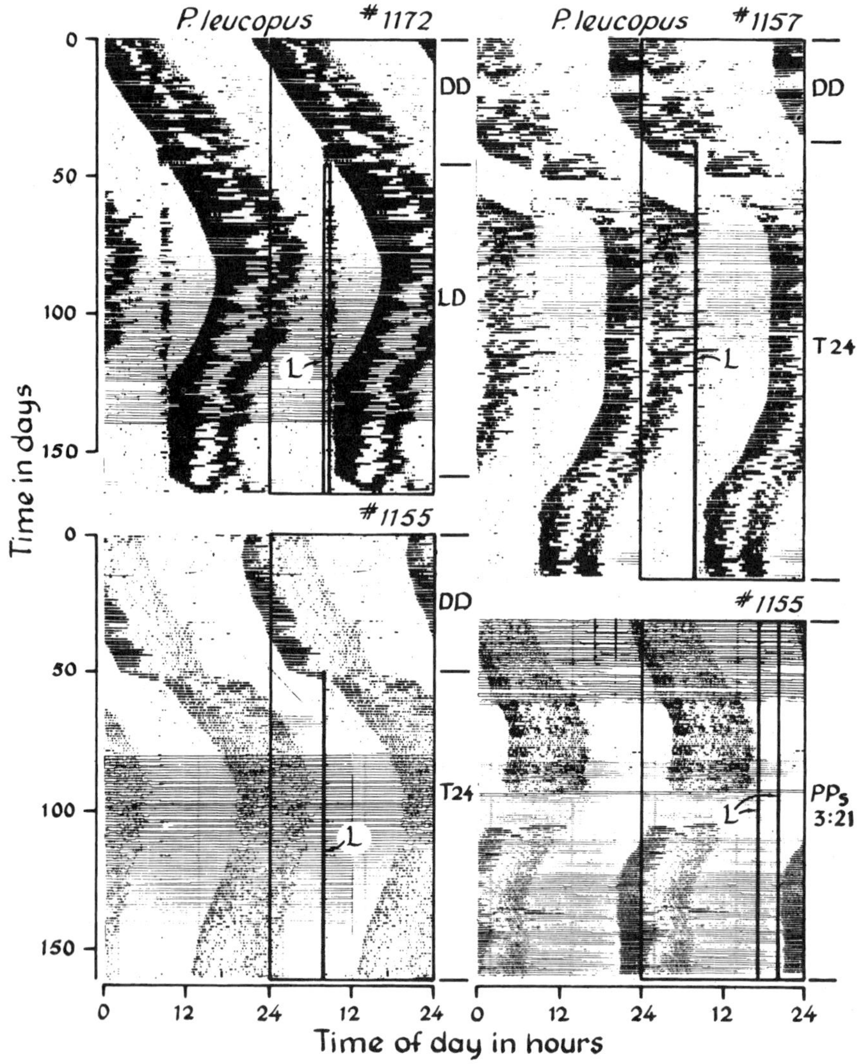

Figure 19. The "bouncing" phenomenon. Four activity records of white-footed mice, showing unstable entrainment by very short photoperiods. After an initial freerun in DD with τ longer than 24 hours, the rhythms are entrained by a single 15-minute light pulse falling at the end of activity time. As τ becomes shorter, the activity "bounces off" and drifts forward every day until entrainment is achieved again, now with the light pulse occurring in the early subjective night. *P. leucopus* #1155 showed two such bounces, the second one in a short skeleton photoperiod (PP$_s$ 3). (From Pittendrigh and Daan, 1976b.)

nomenon" is not inconsistent with the theory of entrainment, but demonstrates that the predictability of phase angle difference breaks down if there are spontaneous changes in period.

A further complication is the lability of the phase response curve itself. J. A. Elliott (personal communication) found in Syrian hamsters that prolonged exposure to continuous darkness leads to long activity time α and a high-amplitude PRC, while exposure to long days in a 24-hour zeitgeber compresses α and suppresses the amplitude of the PRC. These changes are paralleled by modulations of the circadian profile of melatonin secretion and by endocrine changes in the pituitary–gonadal axis. Again, this implies that changes in endocrine condition as occur under the influence of different photoperiods will have repercussions for the pattern of entrainment. Endocrine variations also are known to influence τ of the free-running system. For instance, testosterone shortens τ in laboratory mice (Daan, Damassa, Pittendrigh, & Smith, 1975). Thus, one may expect considerable variation in ψ with seasonal changes in reproductive status. Examples of such changes are the well-known variations in songbirds in the daily onset of activity. During the spring phase of territorial activity and mate acquisition, birds start their morning song and general activity much earlier in the day than at other times of year (Scheer, 1952; Daan & Aschoff, 1975). The midpoint between onset and end of activity also tends to shift forward at this time, but the pattern is highly species- and sex-specific (Figure 20). Such patterns would require precise continuous knowledge of the dynamics of both τ and PRC to be understood quantitatively.

Parametric Contributions to Entrainment

A further problem for the theory of entrainment is the fact that light exerts tonic or parametric influences on the circadian system in addition to its phasic or nonparametric effects which are reflected in the phase responses to brief light pulses. The entrainment model described in the preceding section in essence assumes that the response to pulses can be described by a "phase-only" system, where amplitude and waveform are neglected since the pacemaker is back in steady-state motion on the limit cycle at the time when the next pulse comes. The model has been used with some success in predicting the effect of different photoperiods on pacemaker phase from skeleton photoperiods with the same interval between two pulses (Pittendrigh & Daan, 1976b). In practice, this would mean that the two phasic events which determine entrainment are lights-on and lights-off, and that the responses of the circadian system to these events are instantaneous phase shifts as though these transitions are equivalent to brief light pulses. The level of light intensity in between would thus not appear to matter. The theory takes no parametric effects of light into account.

Parametric effects of light on circadian systems have, however, long been known. They are clearly reflected in the dependence of free-running circadian periods on the intensity of continuous illumination. As first observed by M. S. Johnson (1939), τ lengthens in nocturnal rodents in constant light relative to τ in constant darkness. Indeed, τ monotonically increases with light intensity. In diurnal birds, there is a monotonic decrease with light intensity. While these consistent patterns were originally conceived as a valid functional generalization across all diurnal and nocturnal species (Aschoff, 1960), some taxonomic groups, such as diurnal mammals and arthropods, do not appear to follow this rule (Aschoff, 1979b). Nevertheless, in most species, a parametric effect of light on the velocity of the circadian system has been established (Figure 21). In many, high light intensities eventually lead to total disappearance of rhythmicity. We must thus be aware, as was Pittendrigh (1981b), that the parametric action of light in the entrainment steady state cannot be fully neglected. There are indications that the size of the illumination effect is associated with the phase response curve for brief light pulses. Nocturnal rodent species with a relatively large delay section in the PRC show a strong increase of τ with increasing intensity of illumination and species with a small delay

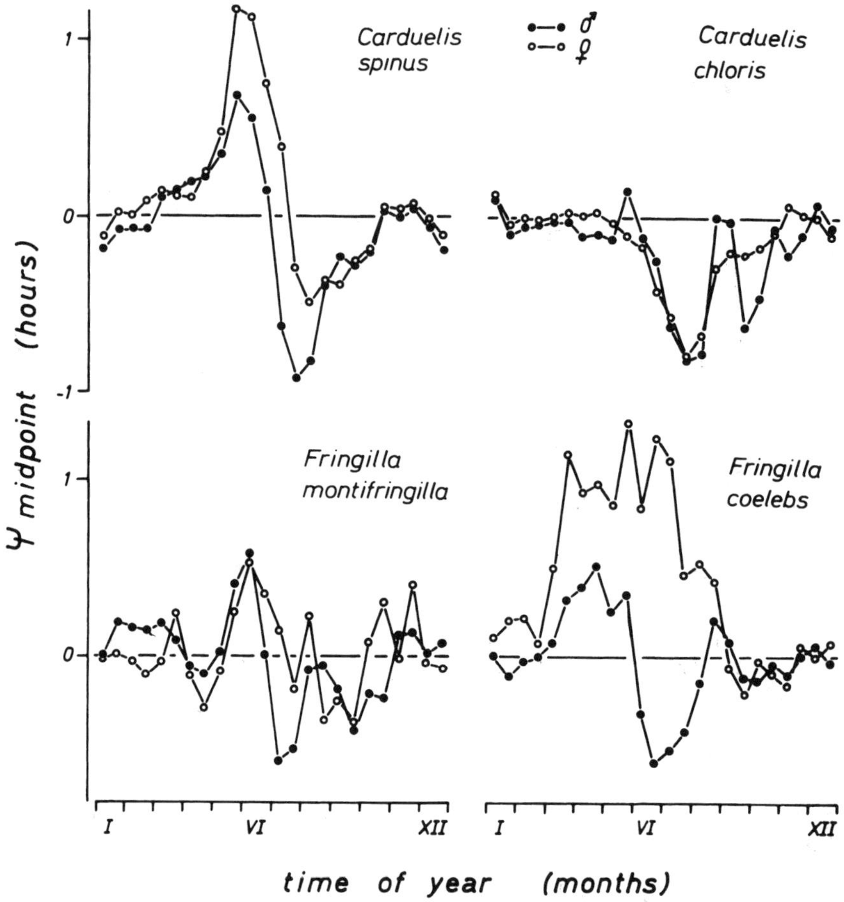

Figure 20. Seasonal change in phase position of daily activity in diurnal songbirds recorded in captivity year-round at the arctic circle (*Carduelis spinus*, siskin; *Carduelis chloris*, greenfinch; *Fringilla montifringilla*, brambling; *Fringilla coelebs*, chaffinch). Phase is indicated by the time interval between midpoint of activity and noon. Note the conspicuous sex- and species-specific difference in the seasonal patterns. (From Daan and Aschoff, 1975.)

section show a small parametric effect. On the basis of such data, it has been suggested that the dependence of τ on light intensity is the net overall effect of phase-dependent parametric accelerations and decelerations in angular velocity of the pacemaker that are functionally associated with the same nonparametric entrainment mechanism reflected in the PRC (Daan & Pittendrigh, 1976b).

The parametric light effects on the circadian system would seem to make a different type of PRC more suitable to predict entrainment by complete photoperiods. This type describes the phase response to single transitions from darkness to light and from light to darkness. Such "step-PRCs" have rarely been measured. A few examples for nocturnal mammals are presented in Figure 22. These suggest that transitions from light to darkness and from darkness to light have nearly mirror-image effects (Subbaraj & Chandrashekaran, 1981). Step-ups into light generate mostly phase delays, while step-downs into darkness generate only advances. The

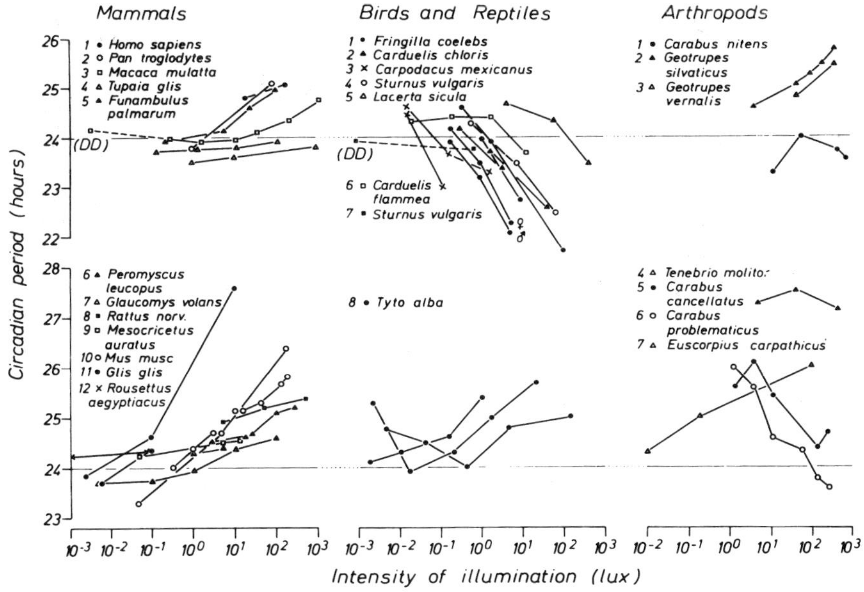

Figure 21. The dependence of free-running circadian period τ of corticosteroid rhythms on the intensity of constant illumination in different vertebrate orders. Upper graphs: nocturnal species. Lower graphs: diurnal species. (From Aschoff, 1979b.)

reverse may be true for diurnal ground squirrels (Kramm, 1974). The maximal responses range from −1.5 to about +6 hours, and represent substantial phase shifts compared to what is known from brief light pulses. A theoretical difficulty with the interpretation of step-PRCs is that the phase reference assessed in LL following a step-up does not need to have the same phase relationship to the pacemaker as the phase reference in DD. If light suppresses or stimulates activity, the onset of activity may well shift phase relative to the pacemaker. The phase shifts measured with step-ups should then be corrected upward in the nocturnal, downward in the diurnal animals, and those after step-downs in the reverse direction. The results would then be consistent with primary light sensitivity in the subjective night. Furthermore, by virtue of the change in τ from DD to LL or vice versa, the phase shift is not unambiguously determined. The phase shift in response to brief light pulses certainly cannot be readily derived from the step-up response followed by a step-down response: the duration of the pulse is usually too brief for the system to revert from steady-state dark motion to steady-state light motion. For longer pulses, this may well be feasible (see final subsection) and separate step-up and step-down PRCs should be more widely applied to predict entrainment of circadian systems under a range of full photoperiods.

For the understanding of entrainment in nature, we should further be aware that the instantaneous on/off switching of light applied in most studies is a poor imitation of the natural signals of dawn and dusk. The transitions are much slower and twilights can last long, particularly in the polar areas of the globe. The consequences of twilights interposed between light and darkness have not received sufficient attention in studies of the entrainment process. However, the significance of twilight duration for the entrainment process has been clearly demonstrated (Asch-

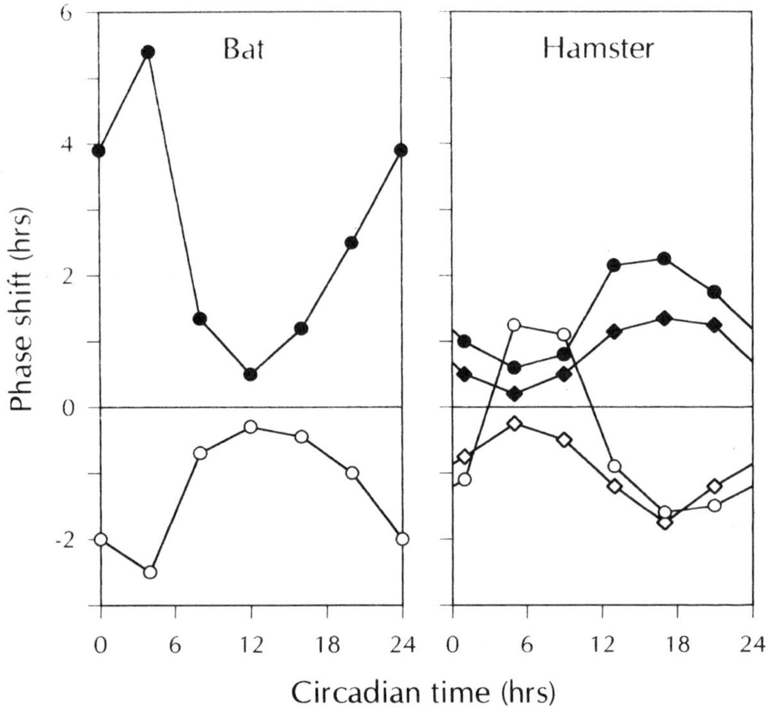

Figure 22. Step phase response curves for bats (*Taphozous melanopogon*; Subbaraj & Chandrashekaran, 1981) and hamsters (*Mesocricetus auratus*; circles: data from Albers, 1986; diamonds: data from Aschoff, 1994). Open symbols represent phase shifts in response to transitions from DD to LL, closed symbols represent phase responses to transitions from LL to DD.

off & Wever, 1965; Wever, 1967; Boulos, Macchi, Houpt, & Terman, 1996). Wever (1967) showed that the phase angle difference of the activity rhythm in finches with the LD cycle increases with longer twilights. This is no masking effect, since activity onset always preceded the beginning of dawn twilight. Wever interpreted these results using a specific mathematical model for both the generation and entrainment of activity rhythms. The phase advance in response to longer twilights may also be the consequence of longer perceived photoperiods, and hence reduction of τ. Be this as it may, the experiments demonstrate that seasonally and latitudinally varying twilights need to be taken into account to derive precise predictions for natural entrainment. The understanding of gradual changes in light intensity and their impact on circadian rhythms becomes particularly important at extreme northern and southern latitudes. Even in the high arctic summer, local birds still maintain entrained circadian rhythms in continuous daylight (e.g., Haarhaus, 1968; Krüll, 1976). The zeitgebers exploited to achieve such entrainment remain largely not understood.

Self Selection of Light

We mentioned that many nocturnal animals see light only at dawn and dusk as they retreat in underground burrows in daytime. Bats are known gradually to congregate in the entrance of their colony before nightfall, awaiting a sufficiently low level of light intensity to take off for nocturnal foraging (Voûte, 1972). This means that the light perceived is controlled by their behavior, which itself is under circadian control. Rats in an experimentally simulated natural environment allowing retreat in darkness in daytime indeed see two brief pulses of light in the beginning and at the end of activity. Terman *et al.* (1991) showed that rats with a short τ tend to see more of the dusk pulse, whereas those with a long τ receive more light at dawn. These facts follow logically from entrainment theory and demonstrate the intricate nature of the behavioral and physiologic responses compensating for deviations of the circadian period from 24 hours.

Some animals are even more strictly nocturnal than laboratory rats and do not leave the burrow as long as there is still some light outside. The flying squirrel, *Glaucomys volans*, retreats into the darkness after a short exposure to the light preceding dusk (Figure 23). This light is sufficient, however, to cause an immediate phase delay of ca. 30 minutes: The animal comes out of the artificial burrow about 1 hour after dusk, and from that moment on, its circadian rhythm freeruns with a short τ over several days (while it does not perceive any light) until again it arrives at the burrow entrance just preceding lights-off. These elegant observations were made by Patricia DeCoursey in flying squirrels in an artificial burrow system (DeCoursey, 1986). Again, the phenomenon of periodic brief light exposures once every 3–7 days is fully compatible with entrainment theory. It is not known whether such periodic exposure is also characteristic of animals in the natural environment with gradual rather than abrupt twilight transitions.

More problematic than the nocturnal burrowers is circadian entrainment in their diurnal counterparts. Most ground squirrel species have activity periods restricted to full daylight. During the night, they remain underground. In the European ground squirrel, *Spermophilus citellus*, emergence from the burrow occurs several hours after sunrise during the summer, and the squirrels retreat underground long before dusk. They thus behaviorally generate their own light–dark cycle, without ever observing the twilight transitions of sunrise and sunset (Hut, Van Oort, & Daan, 1999). There is no light-sampling behavior around the times when the LD transitions present the time signals considered crucial in entrainment theory. This has been confirmed by equipping ground squirrels with photosensors transmitting light intensity to an automatic monitoring system (Figure 24). These results appear incompatible with entrainment theory which, under self selection of the light–dark cycle, would predict a freerun until either onset or end of activity (and light exposure) coincides with the external zeitgeber transitions. Possibly, spectral changes in the natural daylight or also other zeitgeber modalities play a role here. These data attest to the fact that studies on natural entrainment are still in their infancy, and that the demonstration of ubiquitous existence of light sensitivity and phase response curves in animal circadian systems does not yet mean that this solves the problem of how entrainment in nature is achieved.

Nonphotic Contributions to Entrainment

The mechanism of circadian sensitivity to light expressed in the PRC may be attractive by its simplicity, but this should not obscure the fact that it may be only one of many safety provisions helping to keep daily patterns of behavior securely tied to the 24-hour cycle. We have already seen that other zeitgebers such as temperature, air pressure, and sound are in some cases able to entrain circadian rhythmicity. If

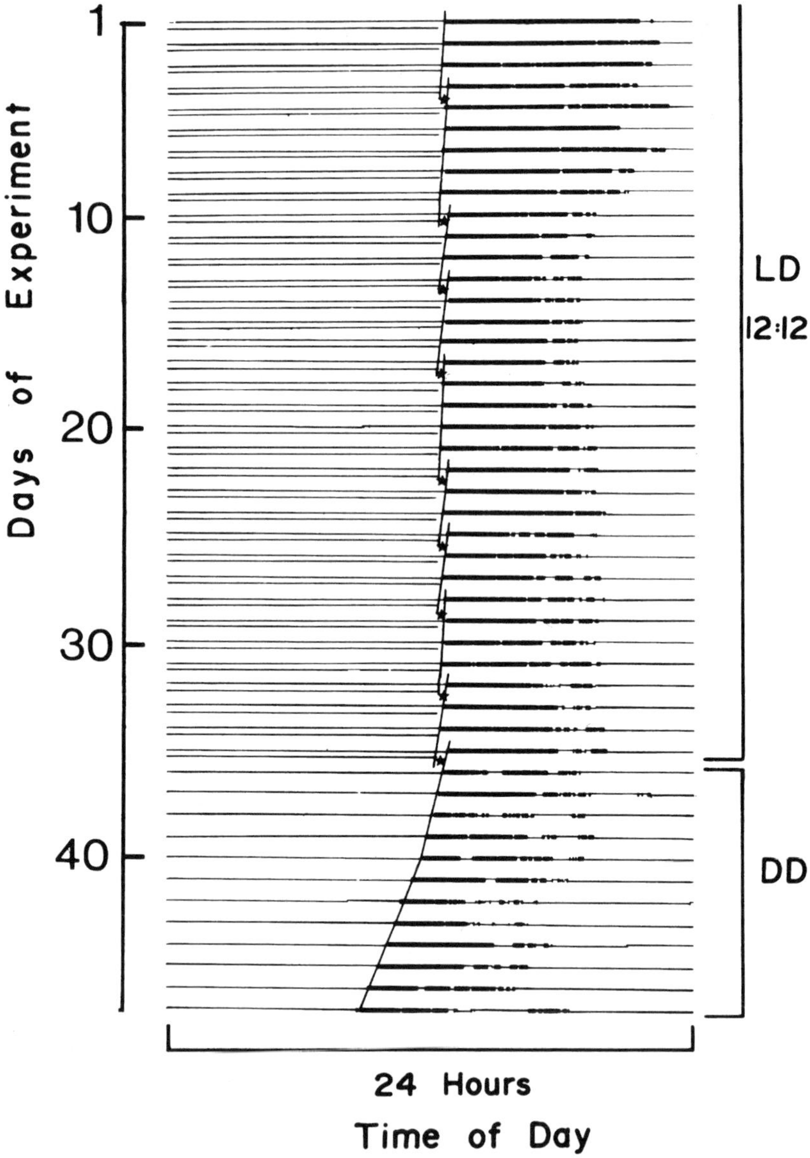

Figure 23. Entrainment of the circadian activity rhythm of a flying squirrel, *Glaucomys volans*, in a simulated den cage. The light of the LD cycle on days 97–131 did not penetrate into the nestbox and was seen by the squirrel only on days marked with an asterisk, when it came to the entrance before lights-off. Each time, the light caused a phase delay shift followed by a 3- to 7-day freerun until the animal again observed the light. (From DeCoursey, 1986, Figure 5.)

they do so, by analytical necessity there must be a phase-dependent response to such stimuli. PRCs for temperature pulses have indeed been measured in half a dozen plant species and usually show both advances and delays (e.g., Engelmann, Eger, Johnsson, & Karlsson, 1974; C. H. Johnson, 1991). In animals, low-temperature pulses are known to produce phase-dependent phase delays in cockroaches (Roberts, 1962;

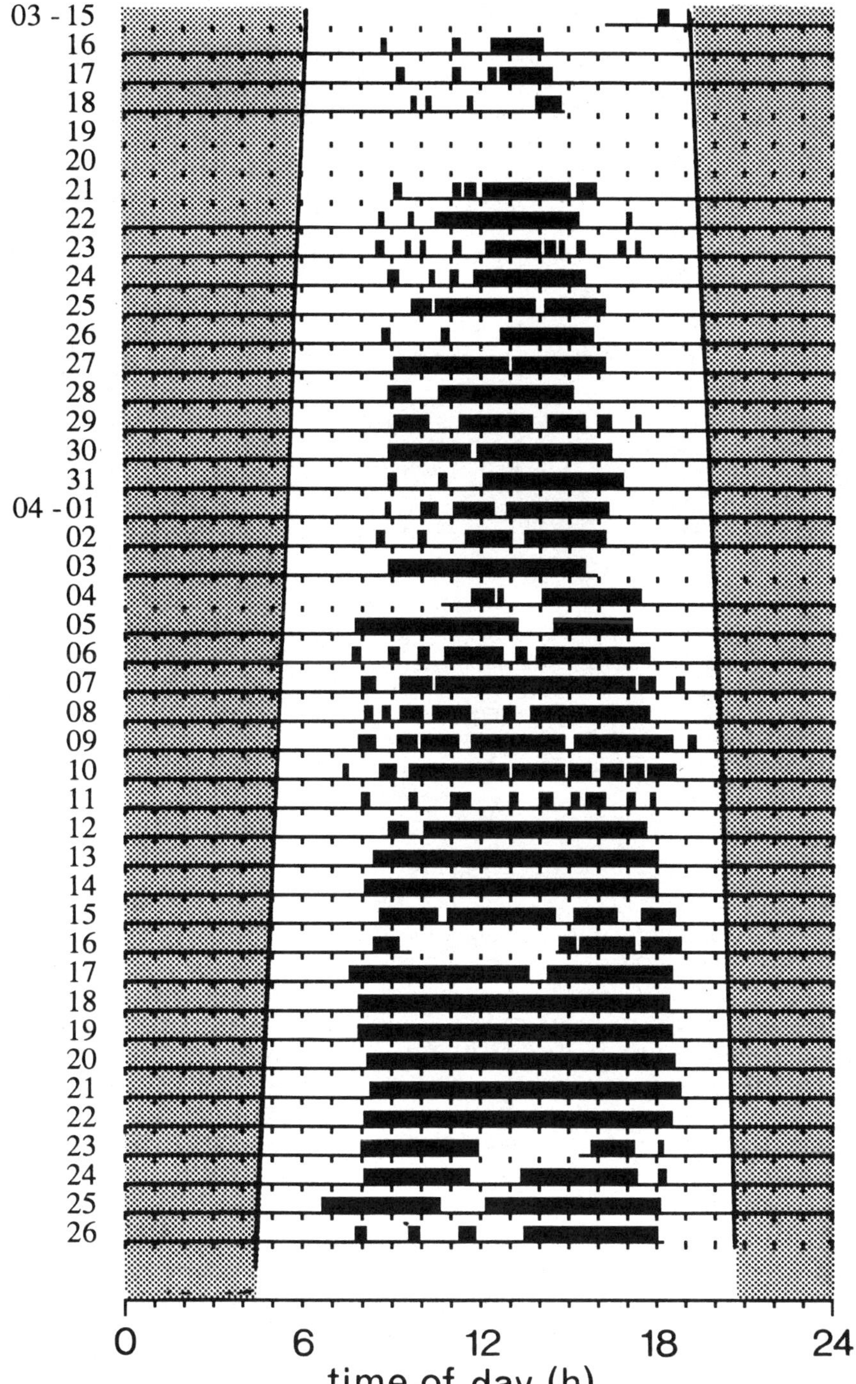

Figure 24. Light exposure of a European ground squirrel, *Spermophilus citellus*, equipped with a light-sensitive radio collar in a large outside pen. Black bars: 2-minute intervals in which the animal was above ground. Oblique lines indicate dawn and dusk civil twilights. Note that the animal never saw the twilight transitions. (From Hut, van Oort, & Daan, 1999.)

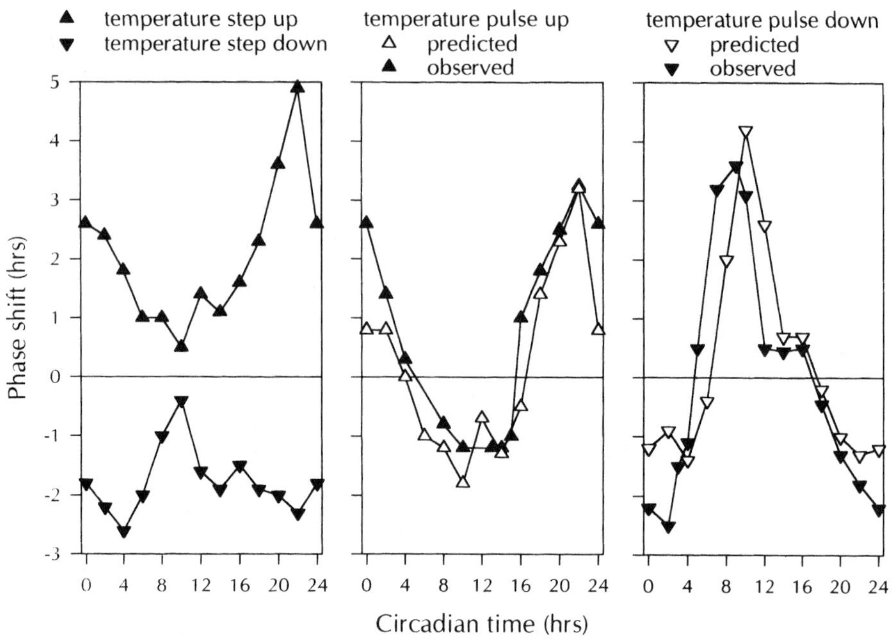

Figure 25. Left: Phase response curves for temperature step-ups (20 → 28°C) and step-downs (28 → 20°C) in *Drosophila pseudoobscura.* Middle: Phase response curves for 12-hour 28°C temperature pulses in 20°C predicted from a step-up followed by a step-down, and observed. Right panel: Phase response curves for 12-hour 20°C temperature pulses in 28°C predicted from a step-down followed by a step-up, and observed. Note that the fit in the right panel is much better than in the original publication by Zimmerman *et al.* (1968, Figure 3), where it was incorrectly plotted.

Wiedenmann, 1977) and in crickets (Gander, 1979). Little is known on phase-shifting effects of temperature in homeotherms. Zimmerman *et al.* (1968) systematically recorded phase response curves for temperature step-ups and step-downs between 20°C and 28°C in the pupal eclosion rhythm of *Drosophila pseudoobscura.* From these two PRCs (Figure 25, left panel), phase shifts in response to pulses of high temperature can be predicted under the assumption that each shift is accomplished instantaneously. The observed PRCs for 12-hour warm (Figure 25, middle panel) and cold pulses (right panel) fit these predictions remarkably well. Zimmerman *et al.* (1968) also derived predictions for temperature pulses of different duration and observed a close correspondence with measured phase angle differences. The success of this rare attempt to understand temperature entrainment suggests that other zeitgebers may act in principle in the same way as light. However, temperature cycles with more naturalistic dynamics than these large, abrupt steps have been hardly explored. In nature, temperature cycles are certainly much less regular and rectangular than the cycle of light and darkness. In view of temperature compensation, it is to be expected that parametric temperature effects are small and that gradual changes in temperature elicit smaller phase-shifting effects. Nevertheless, temperature fluctuations from day to day may well affect circadian entrainment and cause additional variation in phase angle differences under natural circumstances. Similar arguments would apply to other, even less explored, environmental signals.

Nonphotic entrainment often refers to "feedback" effects from activity patterns, themselves generated by the circadian system, on circadian phase, and a separate chapter is devoted to this phenomenon (Chapter 7). We restrict ourselves here to a few remarks on the potential contributions of these effects to natural phase control in the presence of environmental zeitgebers. An association between behavioral activity and circadian period has long been known. Conditions causing a longer τ are generally also those where activity is comparatively suppressed: darkness in diurnal birds, light in nocturnal mammals (Aschoff, 1960), low testosterone levels (Daan & Aschoff, 1975), winter conditions in hibernators (Mrosovsky, Boshes, Hallonquist, & Lang, 1976), etc. A negative correlation between metabolic rate and circadian period associated with the *tau* mutation in hamsters (Oklejewicz, Hut, Daan, Loudon, & Stirland, 1997) points in the same direction. Turek (1989) proposed that much of the association reflects a direct causal effect of amount of activity on the period of the circadian pacemaker. Such modification of τ by activity may functionally contribute to stabilizing entrainment. Both diurnal and nocturnal species typically concentrate maximal activity in the first part of α. They further tend to have τ shorter than 24 hours in constant light (diurnal species) and constant darkness (nocturnal), respectively, and activity often locks on to dawn in diurnals and to dusk in nocturnals (Pittendrigh & Daan, 1976b). In a nocturnal mammal, a reduction of τ by whatever cause will lead to a forward shift of the activity time relative to dusk and cause the activity peak to invade the light. Peak activity is then suppressed and this in turn should slow down the pacemaker. Thus, the direct masking affect of light on activity may contribute indirectly also to the stability of entrainment. The same may be true, albeit with reversed sign, in diurnal animals.

Clearly, this functional argument is dependent on the phase of the circadian rhythm where activity is suppressed or enhanced: Suppression of activity at the phase around activity onset should slow down, or phase delay, the rhythm from a functional point of view. Any factor enhancing activity at this time is expected to speed up, or phase advance, the rhythm. Following the same functional reasoning, we should expect that in the later part of α and the early part of the rest time ρ, suppression of activity should generally advance, and enhancement of activity should delay, the rhythm. This is compatible with results obtained with single activity "pulses" in the nonphotic phase response curves of Mrosovsky, Reebs, Honrado, and Salmon (1989) in hamsters, and with entrainment by periodic access to a running wheel in mice (Edgar & Dement, 1991) (see Chapter 7). We thus infer that nonphotic phase sensitivity of circadian pacemakers adds to the stability of their entrainment in natural circumstances. Nonphotic entrainment reflects the multiplicity of informational pathways which circadian systems use to retain maximal stability with respect to environmental cycles. The light zeitgeber itself may not always be reliably observable day after day, but other information also can be exploited.

Along these same lines, a final behavioral contribution to the maintenance of stable entrainment must be mentioned. Amir and Stewart (1996) showed that in Pavlovian conditioning trials, a conditioned stimulus (gentle air disturbance) was capable of fully replacing light in its phase-shifting effects in rats. Their elegant experiments await confirmation by other laboratories and repetition at other circadian phases than where conditioning took place. If reproducible, this highly important finding would suggest that circadian systems are indeed able to use many sources of information which usually occur in temporal synchrony with light in the natural environment, rather than only light itself. This would lead ultimately to a more kaleidoscopic view of the abundant use pacemakers can make of different zeitgeber modalities, and simultaneously of the tremendous importance which stable entrainment must have for the proper functioning, survival, and reproduction of animals in nature.

References

Albers, H. E. (1986). Response of hamster circadian system to transitions between light and darkness. *American Journal of Physiology, 250,* R708–R711.

Amir, S., & Stewart, J. (1996). Resetting of the circadian clock by a conditioned stimulus. *Nature, 379,* 542–545.

Armstrong, S. M. (1989). Melatonin and circadian control in mammals. *Experientia, 45,* 932.

Aschoff, J. (1960). Exogenous and endogenous components in circadian rhythms. *Cold Spring Harbor Symposia on Quantitative Biology, 25,* 11–28.

Aschoff, J. (1979a). Circadian rhythms: General features and endocrinological aspects. In D. Krieger (Ed.), *Endocrine rhythms* (pp. 1–60). New York: Raven Press.

Aschoff, J. (1979b)). Circadian rhythms: Influences of internal and external factors on the period measured in constant conditions. *Zeitschrift für Tierpsychologie, 49,* 225–249.

Aschoff, J. (1981a). Circadian system properties. In F. Obal & G. Benedek (Eds.), *Advances in physiological sciences—Environmental physiology* (Vol. 18, pp. 1–17). Budapest: Akademiai Kiado.

Aschoff, J. (1981b). Freerunning and entrained circadian rhythms. In J. Aschoff (Ed.), *Handbook of behavioral neurobiology. Vol 4. Biological rhythms* (pp. 81–93). New York: Plenum Press.

Aschoff, J. (1994). On the aging of circadian systems. In T. Hiroshige & K. Honma (Eds.), *Evolution of circadian clock* (pp. 23–44). Sapporo, Japan: Hokkaido University Press.

Aschoff, J., & Pohl, H. (1978). Phase relations between a circadian rhythm and its Zeitgeber within the range of entrainment. *Naturwissenschaften, 65,* 80–84.

Aschoff, J., & Tokura, H. (1986). Circadian activity rhythms in squirrel monkeys: Entrainment by temperature cycles. *Journal of Biological Rhythms, 1,* 91–99.

Aschoff, J., & Von Goetz, C. (1988). Masking of circadian activity rhythms in hamsters by darkness. *Journal of Comparative Physiology A, 162,* 559–562.

Aschoff, J., & Wever, R. (1965). Circadian rhythms of finches in light–dark cycles with interposed twilights. *Comparative Biochemistry and Physiology, 16,* 507–514.

Aschoff, J., & Wever, R. (1966). Circadian period and phase angle difference in chaffinches (*Fringilla coelebs* L.). *Comparative Biochemistry and Physiology, 18,* 397–404.

Aschoff, J., Hoffmann, K., Pohl, H., & Wever, R. (1979). Re-entrainment of circadian rhythms after phase-shifts of the zeitgeber. *Chronobiologia, 2,* 23–78.

Aschoff, J., Daan, S., & Honma, K. (1982a). Zeitgebers, entrainment, and masking: Some unsettled questions. In J. Aschoff, S. Daan, & G. Groos (Eds.), *Vertebrate circadian systems* (pp. 13–24). Berlin: Springer-Verlag.

Aschoff, J., Gerecke, U., Von Goetz, C., Groos, G. A., & Turek, F. W. (1982b). Phase responses and characteristics of free-running activity rhythms in the golden hamster: Independence of the pineal gland. In J. Aschoff, S. Daan, & G. Groos (Eds.), *Vertebrate circadian systems* (pp. 129–140). Berlin: Springer-Verlag.

Boulos, Z., Macchi, M., Houpt, T. A., & Terman, M. (1996). Photic entrainment in hamsters: Effects of simulated twilights and nest box availability. *Journal of Biological Rhythms, 11,* 216–233.

Bovet, J., & Oertli, E. (1974). Free-running circadian activity rhythms in free-living beaver *Castor canadensis*). *Journal of Comparative Physiology, 92,* 1–10.

Clifton, K. E. (1997). Mass spawning by green algae on coral reefs. *Science, 275,* 1116–1118.

Daan, S., & Aschoff, J. (1975). Circadian rhythms of locomotor activity in captive birds and mammals: Their variations with season and latitude. *Oecologia, 18,* 269–316.

Daan, S., & Pittendrigh, C. S. (1976a). A functional analysis of circadian pacemakers in nocturnal rodents II. The variability of phase response curves. *Journal of Comparative Physiology, 106,* 253–266.

Daan, S., & Pittendrigh, C. S. (1976b). A functional analysis of circadian pacemakers in nocturnal rodents III. Heavy water and constant light: Homeostasis of frequency? *Journal of Comparative Physiology, 106,* 267–290.

Daan, S., Damassa, D., Pittendrigh, C., & Smith, E. (1975). An effect of castration and testosterone replacement on a circadian pacemaker in mice (*Mus musculus*). *Proceedings of the National Academy of Sciences of the USA, 72,* 3744–3747.

Davis, F., & Gorski, R. (1988). Development of hamster circadian rhythms: Role of the maternal suprachiasmatic nucleus. *Journal of Comparative Physiology A, 162,* 601–610.

Davis, F., & Mannion, J. (1988). Entrainment of hamster pup circadian rhythms by prenatal melatonin injections to the mother. *American Journal of Physiology, 255,* R439–R448.

Davis, F., Stice, S., & Menaker, M. (1987). Activity and reproductive state in the hamster: Independent control by social stimuli and a circadian pacemaker. *Physiology and Behavior, 40,* 583–590.

DeCoursey, P. (1960). Daily light sensitivity rhythm in a rodent. *Science, 131,* 33–35.

DeCoursey, P. (1972). LD ratios and the entrainment of circadian activity in a nocturnal and a diurnal rodent. *Journal of Comparative Physiology, 78,* 221–235.

DeCoursey, P. J. (1986). Light-sampling behavior in photoentrainment of a rodent circadian rhythm. *Journal of Comparative Physiology A, 159,* 161–169.

DeCoursey, P. J., & Menon, S. A. (1990). Circadian photo-entrainment in a nocturnal rodent: Quantitative measurement of light-sampling activity. *Animal Behaviour, 41,* 781–785.

Edgar, D. M., & Dement, W. C. (1991). Regularly scheduled voluntary exercise synchronizes the mouse circadian clock. *American Journal of Physiology, 261,* R928–R933.

Engelmann, W., Eger, I., Johnsson, A., & Karlsson, H. G. (1974). Effect of temperature pulses on the petal rhythm of Kalanchoe: An experimental and theoretical study. *International Journal of Chronobiology, 2,* 347–358.

Eriksson, L., & Van Veen, T. (1980). Circadian rhythms in the brown bullhead, *Ictalurus nebulosus* (Teleostei). Evidence for an endogenous rhythm in feeding, locomotor, and reaction time behaviour. *Canadian Journal of Zoology, 58,* 1899–1907.

Eskin, A. (1971). Some properties of the system controlling the circadian activity rhythm of sparrows. In M. Menaker (Ed.), *Biochronometry* (pp. 55–80). Washington DC: National Academy of Sciences.

Gander, P. H. (1979). The circadian locomotor activity rhythm of *Hemideina thoracica* (Orthoptera): The effects of temperature perturbations. *International Journal of Chronobiology, 6,* 243–262.

Gerkema, M. P., Daan, S., Wilbrink, M., Hop, M., & Van der Leest, F. (1993). Phase control of ultradian feeding rhythms in the common vole (*Microtus arvalis*): The roles of light and the circadian system. *Journal of Biological Rhythms, 8,* 151–171.

Gwinner, E. (1966). Entrainment of a circadian rhythm in birds by species-specific song cycles (Aves, Fringillidae; *Carduelis spinus, Serinus serinus*). *Experientia, 22,* 1–3.

Haarhaus, H. (1968). Zum Tagesrhythmus des Staren (*Sturnus vulgaris*) und der Schneeammer (*Plectrophenax nivalis*). *Oecologia* (Berlin), *1,* 176–218.

Hardeland, R., Balzer, I., Poeggeler, B., Fuhrberg, B., Uria, H., Behrmann, G., Wolf, R., Meyer, T. J., & Reiter, R. J. (1995). On the primary functions of melatonin in evolution: Mediation of photoperiodic signals in a unicell, photooxidation, and scavenging of free radicals. *Journal of Pineal Research, 18,* 104–111.

Hayden, P., & Lindberg, R. (1969). Circadian rhythm in mammalian body temperature entrained by cyclic pressure changes. *Science, 164,* 1288–1289.

Heigl, S., & Gwinner, E. (1994). Periodic melatonin in the drinking water synchronizes circadian rhythms in sparrows. *Naturwissenschaften, 81,* 83–85.

Hoffmann, K. (1969). Zum Einfluss der Zeitgeberstaerke auf die Phasenlage der synchronisierten Periodik. *Zeitschrift für vergleichende Physiologie, 62,* 93–110.

Honma, S., Honma, K., & Hiroshige, T. (1985). Ontogeny of corticosterone and locomotor rhythms in rats: Effects of maternal rhythms and restricted daily feeding. In T. Hiroshige & K. Honma (Eds.), *Circadian clocks and zeitgebers* (pp. 167–178). Sapporo, Japan: Hokkaido Press.

Hut, R. A., van Oort, B. E. H., & Daan, S. (1999). Natural entrainment without dawn and dusk: The case of the European ground squirrel (*Spermophilus citellus*). *Journal of Biological Rhythms, 14,* 290–299.

Johnson, C. H. (1991). *An atlas of phase response curves for circadian and circatidal rhythms.* Nashville, TN: Vanderbilt University.

Johnson, C. H. (1992). Phase response curves: What can they tell us about circadian clocks? In T. Hiroshige & K. Honma (Eds.), *Circadian clocks: From cell to human* (pp. 209–249). Sapporo, Japan: Hokkaido University Press.

Johnson, M. S. (1939). Effect of continuous light on periodic spontaneous activity of white-footed mice (*Peromyscus*). *Journal of Experimental Zoology, 82,* 315–328.

Kleinhoonte, A. (1928). De door het licht geregelde autonome bewegingen der *Canavalia*-bladeren. Ph.D. dissertation, Utrecht University, Utrecht, The Netherlands.

Kleinknecht, S. (1985). Lack of social entrainment of free-running circadian activity rhythms in the Australian sugar glider (*Petaurus breviceps*: Marsupialia). *Behavioral Ecology and Sociobiology, 16,* 189–193.

Kramm, K. (1974). Phase control of circadian activity rhythms in ground squirrels. *Naturwissenschaften, 61,* 34.

Krüll, F. (1976). Zeitgebers for animals in the continuous daylight of high arctic summer. *Oecologia* (Berlin), *24*, 149–157.

Lewy, A. J., Sack, R. L., Bood, M. L., Bauer, V. K., Cutler, N. S., & Thomas, K. H. (1994). Melatonin marks phase position and resets the endogenous circadian pacemaker in humans. In D. J. Chadwick & K. Ackrill (Eds.), *Circadian clocks and their adjustment* (pp. 303–317). New York: Wiley.

Marimuthu, G., & Chandrashekaran, M. K. (1983). Social cues of a Hipposiderid bat inside a cave fail to entrain the circadian rhythm of an Emballonurid bat. *Naturwissenschaften, 70*, 620.

Marimuthu, G., Rajan, S., & Chandrashekaran, M. 1981). Social entrainment of the circadian rhythm in the flight activity of the microchiropteran bat *Hipposideros speoris. Behavioral Ecology and Sociobiology, 8*, 147–150.

Meijer, J. H., & DeVries, M. J. (1995). Light-induced phase shifts in onset and offset of running-wheel activity in the Syrian hamster. *Journal of Biological Rhythms, 10*, 4–16.

Menaker, M., & Eskin, A. (1966). Entrainment of circadian rhythms by sound in *Passer domesticus. Science, 154*, 1579–1581.

Mrosovsky, N. (1988). Phase response curves for social entrainment. *Journal of Comparative Physiology A, 162*, 35–46.

Mrosovsky, N. (1993). Tau changes after single nonphotic events. *Chronobiology International, 10*, 271–276.

Mrosovsky, N., Boshes, M., Hallonquist, J., & Lang, K. (1976). Circannual cycle of circadian cycles in a golden-mantled ground squirrel. *Naturwissenschaften, 63*, 298–299.

Mrosovsky, N., Reebs, S., Honrado, G., & Salmon, P. (1989). Behavioural entrainment of circadian rhythms. *Experientia, 45*, 696–702.

Oklejewicz, M., Hut, R. A., Daan, S., Loudon, A. S. I., & Stirland, A. J. (1997). Metabolic rate changes proportionally to circadian frequency in tau mutant Syrian hamsters. *Journal of Biological Rhythms, 12*, 413–422.

Pittendrigh, C. (1958). Perspectives in the study of biological clocks. In A. A. Buzatti-Traverso (Ed.), *Perspectives in marine biology* (pp. 239–268). San Francisco: University of California Press.

Pittendrigh, C. (1967). Circadian systems I. The driving oscillation and its assay in *Drosophila pseudoobscura. Proceedings of the National Academy of Sciences of the USA, 58*, 1762–1767.

Pittendrigh, C. S. (1980). Some functional aspects of circadian pacemakers. In M. Suda, O. Hayaishi, & H. Nakagawa (Eds.), *Biological rhythms and their central mechanism* (pp. 3–12). New York: Elsevier Press.

Pittendrigh, C. (1981a). Circadian organization and the photoperiodic phenomena. In B. Follett & D. Follett (Eds.), *Biological clocks in seasonal reproductive cycles* (pp. 1–35). Bristol, England: Scientechnica.

Pittendrigh, C. S., (1981b). Circadian systems: Entrainment. In J. Aschoff (Ed.), *Handbook of behavioral neurobiology. Vol. 4. Biological rhythms* (pp. 95–124). New York: Plenum Press.

Pittendrigh, C. S., & Daan, S. (1976a). A functional analysis of circadian pacemakers in nocturnal rodents I. The stability and lability of spontaneous frequency. *Journal of Comparative Physiology, 106*, 223–252.

Pittendrigh, C. S., & Daan, S. (1976b). A functional analysis of circadian pacemakers in nocturnal rodents IV. Entrainment: Pacemaker as clock. *Journal of Comparative Physiology, 106*, 291–331.

Pittendrigh, C. S., & Daan, S. (1976c). A functional analysis of circadian pacemakers in nocturnal rodents V. Pacemaker structure: A clock for all seasons. *Journal of Comparative Physiology, 106*, 333–355.

Pittendrigh, C., & Minis, D. (1964). The entrainment of circadian oscillations by light and their role as photoperiodic clocks. *American Naturalist, 48*, 261–294.

Roberts, S. (1962). Circadian activity rhythms in cockroaches II. Entrainment and phase shifting. *Journal of Cellular and Comparative Physiology, 59*, 175–186.

Roenneberg, T., & Rehman, J. (1996). Nitrate, a nonphotic signal for the circadian system. *FASEB Journal, 10*, 1443–1447.

Rusak, B., Mistlberger, R. E., Losier, B., & Jones, C. H. (1988). Daily hoarding opportunity entrains the pacemaker for hamster activity rhythms. *Journal of Comparative Physiology, A164*, 165–171.

Scheer, G. (1952). Beobachtungen und Untersuchungen über die Abhängigkeit des Frühgesanges der Vögel von inneren und äusseren Faktoren. *Biologische Abhandlungen, 3/4*, 1–68.

Strubbe, J. H., Spiteri, N. J., & Prins, A. J. A. (1986). Effect of skeleton photoperiod and food availability on the circadian pattern of feeding and drinking in rats. *Physiology and Behavior, 36*, 647–651.

Subbaraj, R., & Chandrashekaran, M. (1981). Mirror imaging phase response curves obtained for the circadian rhythm of a bat with single steps of light and darkness. *Journal of Interdisciplinary Cycle Research, 12*, 305–312.

Swade, R. H. (1969). Circadian rhythms in fluctuating light cycles: Toward a new model of entrainment. *Journal of Theoretical Biology, 24*, 227–239.

Takahashi, K., & Sasaki, Y. (1985). Entraining mechanism of endogenous rhythm of blinded rat pups by the nursing mother. In T. Hiroshige & K. Honma (Eds.), *Circadian clocks and zeitgebers* (pp. 157–166). Sapporo, Japan: Hokkaido University Press.

Takahashi, K., Inoue, K., Kobayashi, K., Hayafuji, C., Nakamura, Y., & Takahashi, Y. (1978). Mutual influence of rats having different circadian rhythm of adrenocortical activity. *American Journal of Physiology, 234,* E515–E520.

Terman, M., Reme, C., & Wirz-Justice, A. (1991). The visual input stage of the mammalian circadian pacemaking system: II. The effect of light and drugs on retinal function. *Journal of Biological Rhythms, 6,* 31–48.

Turek, F. W. (1989). Effects of stimulated physical activity on the circadian pacemaker of vertebrates. *Journal of Biological Rhythms, 4,* 135–148.

Viswanathan, N., & Chandrashekaran, M. (1985). Cycles of presence and absence of mother mouse entrain the circadian clock of pups. *Nature, 317,* 530–531.

Von Holst, E. (1939). Die relative Koordination als Phänomen und als Methode zentralnervöser Funktionsanalyse. *Ergebnisse der Physiologie, 42,* 228–306.

Voûte, A. (1972). Bijdrage tot de oecologie van de Meervleermuis (*Myotis dasycneme* (Boie, 1825)). Ph.D. dissertation, Utrecht University, Utrecht, The Netherlands.

Wever, R. (1960). Possibilities of phase-control, demonstrated by an electronic model. *Cold Spring Harbor Symposia on Quantitative Biology, 25,* 197–206.

Wever, R. (1967). Zum Einfluss der Dämmerung auf die circadiane Periodik. *Zeitschrift für vergleichende Physiologie, 55,* 255–277.

Wever, R. (1972). Virtual synchronisation towards the limits of the range of entrainment. *Journal of Theoretical Biology, 36,* 119–132.

Wiedenmann, G. (1977). Two activity peaks in circadian rhythms of cockroach *Leucophaea maderae. Journal of Interdisciplinary Cycle Research, 8,* 378–383.

Winfree, A. (1970). Integrated view of resetting a circadian clock. *Journal of Theoretical Biology, 28,* 327–374.

Zimmerman, W. F., Pittendrigh, C. S., & Pavlidis, T. (1968). Temperature compensation of the circadian oscillation in *Drosophila pseudoobscura* and its entrainment by temperature cycles. *Journal of Insect Physiology, 14,* 669–684.

2

Conceptual Issues in the Ecology and Evolution of Circadian Rhythms

TERESA H. HORTON

ECOLOGY AND EVOLUTION: DISTINCT, BUT OVERLAPPING VIEWPOINTS

Many species of *Drosophila* eclose from their pupal cases at dawn. Does this eclosion occur in response to an environmental stimulus (i.e., a change in humidity, temperature, or light) or in response to an endogenous timing mechanism? A bird may fly in one direction for a few thousand kilometers, then turn and fly in another direction. Is the bird responding to landmarks or magnetic cues, or does a biological clock determine how long the bird flies in a particular direction during migration? These questions and others like them are important if we are to understand *how* the behavior* of organisms in their natural environment is regulated and *why* it is regulated as it is.

Ecological studies examine the proximate mechanisms underlying the interactions of individuals within their communities or ecosystems. These mechanisms are described at a very different level of biological organization than the molecular and biochemical mechanisms described in other chapters of this book, but the essence of the question is the same: *How* is a system regulated? In this case, the system consists of the organism and its environment rather than the biochemical processes

*For the sake of brevity, "behavior" will be used throughout this chapter to refer both to behavior and internal physiologic processes such as hormonal secretions, metabolism, etc.

TERESA H. HORTON Center for Circadian Biology and Medicine, Department of Neurobiology and Physiology, Northwestern University, Evanston, Illinois 60608.

Circadian Clocks, Volume 12 of *Handbook of Behavioral Neurobiology*, edited by Joseph S. Takahashi, Fred W. Turek, and R. Y. Moore, Kluwer Academic / Plenum Publishers, New York, 2001.

within a single organism or cell. When an ecologist tests hypotheses about which factors regulate the timing of eclosion of fruit flies in the genus *Drosophila* in their natural environment, he or she is studying the mechanism responsible for regulating that behavior in the context of the community to which the flies belong.

Organisms are complex and can adapt to their environments by modifying many physiologic and behavioral mechanisms other than their temporal organization. For example, *Drosophila* are hypothesized to eclose at dawn to avoid desiccation. Most shizophorans, the subfamily of insects that includes the genus *Drosophila*, are day-flying insects with excellent abilities to withstand desiccation (Cloudsley-Thompson, 1960). In contrast, flies in the genus *Drosophila* are hypothesized to have become adapted to arid environments by increasing their visual acuity to low light intensities, which enabled them to confine their activity to darker and more humid times of day (Cloudsley-Thompson, 1960). Evolutionary biologists would ask why this pattern evolved: What were the pressures that resulted in *Drosophila* exhibiting this crepuscular (i.e., active at dawn and dusk) pattern? An evolutionary biologist would also ask whether the crepuscular pattern preceded or followed the invasion of *Drosophila* into an arid habitat to determine whether the crepuscular pattern was an adaptation to life in the arid environment or a preexisting condition that allowed *Drosophila* to invade an open niche. By asking evolutionary questions, we may better understand the fundamental principles which determine the temporal organization of behavior and the principles which determine the mechanisms controlling that behavior. Conversely, studies of circadian rhythms may provide a useful example for the study of complex organisms. Circadian rhythms and the evolution of the temporal organization of behavior can be used to study the fundamental concept of the coevolution of physiology, morphology, and behavior.

Despite distinct intellectual traditions, the fields of ecology and evolutionary biology merge at a broad interface. If an ecologist wishes to test the hypothesis that a particular behavior is an adaptation, she or he needs to understand the phylogeny of the behavior to know whether or not it has been modified from an ancestral condition to meet the demands of the current environment. If an evolutionary biologist wishes to test the hypothesis that the control of a particular behavior by a circadian clock is adaptive, he or she needs to understand the ecology of the organism because it is the interaction of the organism with its environment that will determine the success of the organism and its offspring.

Circadian rhythms have been observed in virtually every living organism, prokaryotic as well as eukaryotic (Crosthwaite, Dunlap, & Loros, 1997; Kay, 1997; Kondo, Golden, Ishiura, & Johnson, 1994). It is beyond the scope of this chapter to provide a comprehensive review of the diverse behaviors and physiologic processes known to be under the control of endogenous circadian pacemakers; in general, circadian controlled behaviors are known to include, but are not limited to, sexual behavior, the sleep–wake cycle, nest building, torpor, feeding, migration (in aquatic and terrestrial environments), luminescence, and several gated developmental rhythms (Daan, 1981; DeCoursey *et al.*, submitted; Enright, 1970).

This chapter summarizes and integrates concepts of ecology and evolutionary biology with observations about circadian rhythms. The goal of this summary and integration is to encourage animal behaviorists, ecologists, evolutionary biologists, and molecular biologists to work hand in hand to elucidate the fundamental biological principles underlying the diversity of both the structure of circadian clocks and the function of circadian rhythms.

The things an organism must do in order to survive and reproduce in nature are find food, find a mate (if it is a sexually reproducing organism), find habitats with optimal conditions for survival (e.g., temperature, humidity, light intensity, pH, oxygen availability, etc.), and avoid predators. The niche occupied by an animal in its community is defined by the interactions of an organism with the biological and physical factors which contribute to its survival and subsequent reproduction. Given the earth's daily rotation, these biological and physical factors are not uniformly distributed throughout the day–night cycle. Thus, the temporal structure of an organism's environment is an equally important component of its niche as the spatial, chemical, energetic, and social components. This temporal structure in the environment contributes to the segregation of the activities of individual animals to specific times of day. The temporal structure of the environment also contributes to the temporal segregation of the activities of entire species which may contribute to the diversity of species that inhabit a community (Schoener, 1970, 1974). Investigations of the mechanisms, including circadian rhythms, by which organisms respond to the temporal structure of their environment should be central to studies of ecology. The question of interest to circadian biologists and ecologists is whether the mechanisms regulating this temporal order (1) rely on a direct stimulus–response relationship between the animal and its environment, (2) rely on an internal clock, or (3) represent the interaction between the external environment and an internal clock.

INFLUENCE ON THE BEHAVIOR OF INDIVIDUALS

One objective of research into the temporal organization of behavior should be to describe correctly the relative contributions of external stimuli and endogenous rhythms when testing hypotheses regarding the organization of behavior under natural conditions (Enright, 1970; Zucker, 1983). A fundamental assumption in ecology and evolution is that individuals which maximize their energy consumption while minimizing their exposure to predation or other stressors are most likely to survive and reproduce in a given environment (Bennett, 1987; Horton & Rowsemitt, 1992). The degree to which external and internal factors regulate the daily temporal program may be determined by the level of correlation (*predictability*) in the occurrence of environmental events from one cycle to the next. Through effects on survival and reproduction, behaviors that coincide with events which are highly predictable from cycle to cycle may be more tightly controlled by an internal program than those associated with environmental conditions having lower levels of predictability (Pittendrigh & Daan, 1976b; Rapp, 1987).

The ability to anticipate a coming change in the environment is provided by a circadian clock. A timing mechanism built on the foundation of an oscillator with a stable cycle length (i.e., *free-running period*, τ) permits behaviors that require the measurement of elapsed time (e.g., time-compensated sun orientation, time-sense [Zeitgedachtnis]; Enright, 1970; Pittendrigh & Daan, 1976b). The presence of an oscillator also provides organisms with the ability to recognize local time. This recognition of local time results in the synchronization of activities with the environmental cycle producing the condition known as *entrainment* (Enright, 1970; Mrosovsky, 1996; Pittendrigh & Daan, 1976b). Entrainment is characterized by a stable relationship (*phase angle* ψ) between specific environmental events (e.g., dawn or dusk) and specific behaviors (e.g., the onset of activity). Desert-dwelling kangaroo rats (genus *Dipodomys*) that emerge from their underground burrows within ½ hour of sunset each day are exhibiting entrainment resulting from exposure to the light–dark cycle and, perhaps, social interactions (Kenagy, 1973; Mrosovsky, 1996).

For behaviors known to be under circadian control, the oscillatory model of circadian rhythms can function as a starting point, or null hypothesis, for predicting patterns of behavior under a variety of environmental conditions. For example, two characteristics of circadian rhythms are that τ is not necessarily equal to 24 hours (i.e., is "*circa*" = about, "*dian*" = a day) and that τ is quite stable. The functional significance of these characteristics becomes apparent when examined from the perspective of the conservation of a stable phase angle (e.g., maintaining a constant onset of activity relative to sunset). A critical analysis of laboratory data from nocturnal rodents and from mathematical models of oscillators reveals that the deviation of τ from 24 hours and its low variability contribute to the conservation of phase angle (Pittendrigh & Daan, 1976b). These characteristics predict that organisms should maintain a stable phase angle under seasonal changes in photoperiod (or day length) (Pittendrigh & Daan, 1976b). The observation of deviation from these predictions in organisms under natural or controlled laboratory conditions would indicate that other factors, either in the external environment or within the organism, are modifying the regulation of the behavior by the circadian system.

There are several examples from nature in which the phase angle is not conserved for activities known to be under circadian control (Daan, 1981; Kenagy, 1976). Kangaroo rats of the genus *Dipodomys* show seasonal changes in ψ (Kenagy, 1976). There is no correlated seasonal change in τ that could account for this change in ψ. The most likely explanation is that the kangaroo rats are responding to the seasonal change in the length of twilight (Kenagy, 1976, 1978). Other examples can be found in the arvicoline rodents (e.g., voles, lemmings, muskrat; Wilson & Reeder, 1993). These rodents frequently shift their activity from a primarily nocturnal pattern in the summer to a crepuscular or diurnal pattern in the winter (Erkinaro, 1969; Hoogenboom, Daan, Dallinga, & Schoenmakers, 1984; Rowsemitt *et al.*, 1982). One proximate mechanism for this shift appears to be a seasonal change in gonadal steroids; high testosterone facilitates nocturnal activity (Rowsemitt, 1986). However, temperature is also a contributing factor; crepuscular voles switch back to nocturnal patterns of activity during the winter if temperatures become unseasonably warm (Claypool, 1984).

These observations indicate that while a circadian clock can regulate temporal patterns of behavior in individuals, animals are not slaves to the physiology of their circadian clocks. As with all models in biology, models of behavior based on circadian principles provide a hypothesis. When behavior is observed to deviate from the hypothesis, new investigations lead us to a better understanding of the actual mechanisms responsible for the regulation of behavior. These new investigations enhance our understanding of both the behavioral ecology of animals and the mechanisms by which circadian rhythms regulate behavior.

Circadian Models and the Organization of Communities

The species within a community may also partition the temporal environment to avoid or reduce competition. Schoener (1970, 1974) observed that different species of lizards of the genus *Anolis* sunned themselves on the same branches, but at different times of day. An additional example can be taken from data collected to test hypotheses derived from the theory of optimal foraging (Kotler, Brown, & Subach, 1993). These data were not collected under conditions necessary to determine the role of an endogenous timing mechanism, but they provide an example for which it may be fruitful to ask such questions. The study examined a community of desert-dwelling, granivorous gerbils (*Gerbillus pyramidum* and *G. allenbyi*) in Israel. The larger species (*G. pyramidum*) prefers to forage in patches with high seed densities. During the course of a night, individuals of the larger species emerge first, forage until the seed densities are reduced to suboptimal levels, then retreat to their burrows. The smaller species then emerges and forages until seed densities are further reduced. If *G. pyramidum* is removed from the community, the timing of activity by *G. allenbyi* shifts to an earlier time. The investigators hypothesized that aggressive encounters between the species maintain the temporal feeding hierarchy (Kotler *et al.*, 1993). A similar pattern of temporal partitioning is seen in a community of desert-dwelling heteromyid rodents (kangaroo rats and pocket mice) in North America (Kenagy, 1973).

From a chronobiologist's point of view, one would test the following alternative hypotheses: (1) that the aggressive encounters regulate a direct stimulus–response system and (2) that the aggressive encounters function as a nonphotic stimulus to entrain the endogenous pacemakers of the various members of the community (Honrado & Mrosovsky, 1991; Mrosovsky, 1996). If an endogenous timing mechanism is involved, it may limit the range of time domains that individuals of a species could occupy if the other species in the community were removed or new ones added (Pittendrigh & Daan, 1976a, c). The implication that the formal properties of an endogenous clock may restrict the possible patterns of temporal organization even when specific environmental stimuli are added or removed is important to ecologists. Since the goal of ecology is to understand the mechanisms organizing the temporal structure of individual behavior or interactions within communities and ecosystems, the appropriate roles must be ascribed to external stimuli and endogenous clocks.

Evolutionary Biology and the Study of Circadian Rhythms

Circadian rhythms are striking in the ubiquity with which they occur, in the similarity of the formal properties of rhythms among diverse groups of organisms, and in the apparent similarities in the molecular structure of pacemaker molecules. These similarities lead to the hypothesis that the cellular pacemakers upon which circadian clocks are built arose early in the history of life on earth and that common elements may be shared among all organisms today (Crosthwaite *et al.*, 1997; Kay, 1997; Kondo *et al.*, 1994). Circadian rhythms are also striking because of the diversity of anatomic structures and biochemical processes that make up complex clock systems (i.e., the integrated system consisting of input mechanisms, pacemaker(s), and output mechanisms) (Menaker & Tosini, 1996; Page, 1989; Roenneberg, 1994). This observation leads to the hypothesis that complex clock systems built on the foundation of a circadian pacemaker have evolved repeatedly (i.e., represent *convergent* evolution) in diverse groups of organisms. Finally, circadian rhythms are striking because of the intricate temporal correlations between the expression of behavior and the occurrence of environmental events in specific habitats (Daan, 1981; Enright, 1970; DeCoursey, 1990). This observation leads to the hypothesis that the expression of rhythmic behavior is easily modified by natural selection as organisms adapt to their specific environments. All three of these evolutionary hypotheses are amenable to testing using comparative methods.

"Evolution" simply means change over time. Starting with the principle of common ancestry, Darwin proposed that there should be a degree of similarity among closely related organisms (Darwin, 1859). However, Darwin also recognized that organisms change over time and that at least some of this change was related to local environmental conditions. Therefore, one must consider both phylogenetic and environmental components if one is to develop a complete understanding of the ubiquity, similarity, and diversity of circadian clocks. While it is not possible to reconstruct the environments in which organisms evolved, we do know what environments organisms inhabit today. The patterns of existing diversity can inform us about the pressures required to maintain circadian pacemakers and the clocks built on their foundation. The information about current conditions can be used to infer what happened in the past (Harvey & Pagel, 1991, p. 40). As with all scientific inquiry, one can never really prove that a hypothesis is correct; however, with careful analysis, it is possible to reject certain alternatives. The persistent testing of hypotheses using comparative methods that consider both phylogenetic and environmental contributions to diversity can be used to build a testable theory of the fundamental organizational principles that underlie the evolution of circadian pacemakers, complex circadian clocks, and the daily temporal expression of behavior.

Comparative Biology and Tests of Hypotheses of Common Origin and Convergence

Two of the hypotheses described above (i.e., regarding the origin of pacemakers and the evolution of complex clock systems) describe questions directed at different levels of analysis of circadian rhythms, but the underlying problem is the same: What is the evolutionary origin of an observed characteristic? The hypotheses that characteristics are derived from a common ancestor or evolved independently by convergent evolution are alternative hypotheses that can be tested at many different levels of analysis.

Evolutionary biologists test alternative hypotheses about the origins of characteristics by looking for patterns of similarity. The theoretical foundation known as *phylogenetic systematics*, which is commonly used by evolutionary biologists to estimate phylogenies, was provided by Hennig (1950, 1966). In principle, the construction of a phylogeny is based on the assumption that there are three types of homologies among traits (Brooks & McLennan, 1991, pp. 24–68; Harvey & Pagel, 1991, pp. 50–77). Shared *general homologies* are common to all members of a taxonomic group (e.g., an internal skeleton is a shared general homology of all vertebrates). Shared *special homologies* are common to specific subgroups within the larger taxonomic group (e.g., fur is a shared special homology of mammals). Finally, *unique homologies* identify individual taxa within groups (e.g., the modification of the forelimbs into wings is a unique character of the Order Chiroptera [bats]).

The same analytical procedures can be used to construct phylogenies of single molecular components of circadian clocks or of entire clock systems. The investigator first needs to define the aspect of circadian clocks under examination (i.e., photoreceptor, input system, pacemaker, output system, or the complete structure of the integrated clock system). He or she then develops a data matrix containing information on the characters (traits) which comprise the system under investigation. The matrix needs to include information about whether or not the characters differ from the presumed ancestral state. The ancestral condition is determined by comparison with an *outgroup*, a group that is very distantly related to the group being examined. The data matrix is then analyzed using computer software specifically designed to generate phylogenetic relationships and establish the statistical significance of the phylogenetic relationships (see Hillis, Moritz, & Mable, 1996; Swofford, Olsen, Wadell, & Hillis, 1996; Wiley, 1981, for details of these procedures). Because the assumptions and mathematical weighting of variables differ among algorithms, evolutionary biologists frequently examine their data using multiple algorithms to look for instabilities in the branching patterns of the phylogenies. Instabilities indicate uncertainties in the predicted phylogenetic relationships.

Tests of hypotheses about the origin of circadian pacemakers must await the accumulation of additional data about the structure and function of the molecules that make up the pacemakers of diverse organisms. However, the current model of circadian pacemakers proposes a cellular feedback loop involving the regulation of gene transcription through the interactions of several different genes and their protein products (Reppert & Weaver, 1997). Examples of these interactions are seen in *Neurospora* (e.g., *frg* and *wc-2*) and in *Drosophila* (e.g., *per* and *tim*) (Crosthwaite *et al.*, 1997; Gekakis *et al.*, 1995; Sehgal *et al.*, 1995). An emerging theme in the molecular biology of circadian pacemakers is the presence of a PAS domain, required for protein dimerization, in at least some pacemaker molecules from *Neurospora*, to *Drosophila*, to humans (Albrecht, Sun, Eichele, & Lee, 1997; Crosthwaite *et al.*, 1997; Huang, Edery, & Rosbash, 1993; Kay, 1997; King *et al.*, 1997; Shearman, Zylka, Weaver, Kolakowski, & Reppert, 1997; Shigeyoshi *et al.*, 1997). Phylogenetic procedures have been employed to examine the phylogenetic relationships of PAS-domain-containing molecules (e.g., those from *a.p.per*, *d.m.per*, *hper1*, *mper1*, *hper2*, *mper2*, *hper3*, and *mClock*; Shearman *et al.*, 1997) and to examine variation in the structure of *per* molecules in different populations of *Drosophila* (e.g., Rosato, Peixoto, Barbujani, Costa, & Kyriacou, 1994). To test the hypothesis that extant pacemakers are derived from a common ancestor will require in-depth assessment of the molecular structure of the different genes and proteins which make up the pacemakers of multiple species each of prokaryotes and eukaryotes.

The function of putative pacemaker molecules should also be subjected to analysis (Sassone-Corsi, 1996). Recent data indicate that structurally related proteins may have different functions in the pacemakers of different species. For example, the cellular patterns of expression of *per* are very different in the silkmoth (*Antheria pernyi*) and in fruitflies (*Drosophila* spp.) (Sassone-Corsi, 1996; Sauman & Reppert, 1996). The different patterns of expression in *A. pernyi* and *Drosophila* provide tantalizing data to suggest an alternative hypothesis for the origin of pacemakers, that certain types of molecules are easily recruited to contribute to a pacemaker mechanism, but that the evolution of the pacemaker itself may have occurred repeatedly by convergent evolution. Differences in expression also are observed among *per*-like proteins in *Drosophila* and mice which suggest significant differences in function (Albrecht *et al.*, 1997; Shearman *et al.*, 1997; Shigeyoshi *et al.*, 1997). The current fountain of molecular data flowing from laboratories should, in time, provide sufficient data on the structure and function of molecules that comprise circadian pacemakers in a variety of organisms to permit testing of hypotheses about the origin of circadian pacemakers.

The same procedures can be employed to analyze the evolution of other components of circadian clocks (i.e., photoreceptive pigments, *N*-acetyltransferase molecules) or the evolution of complete clock systems (Klein *et al.*, 1997; Menaker &

Tosini, 1996; Soni & Foster, 1997). For example, within the vertebrates, mammals appear to be unique in the extent to which pacemaker function is centralized in a single brain region, the suprachiasmatic nucleus. It is hypothesized that this represents the passage of mammals through an evolutionary bottleneck as small nocturnal organisms during the Cretaceous period (Menaker & Tosini, 1996). Several other nocturnal species of vertebrates also show an increased reliance on the hypothalamic pacemaker and a reduction in importance of the retinas or pineal/parietal complex as pacemakers (e.g., owls, geckos, hagfish) (Menaker & Tosini, 1996). Comparison of closely related nocturnal and diurnal species from each of these groups would permit testing of the hypothesis that the reliance on the hypothalamic pacemaker evolved repeatedly by convergence in nocturnal animals.

Comparative Biology and Tests of Adaptation of Circadian Rhythms

The repeated appearance of similar characteristics in diverse groups of organisms is often assumed to indicate that the organisms are adapting to a common environmental pressure. For example, it is hypothesized that differences in sensitivity to light among desert- and forest-dwelling rodents may represent adaptation to the different light environments of these habitats (DeCoursey, 1990). The rodent species compared in this study (forest dwellers: *Glaucomys volans* [flying squirrel] and *Peromyscus leucopus* [white-footed mouse]; desert dwellers: *Dipodomys merriami* [kangaroo rat] and *Mesocricetus auratus* [golden hamster]) are phylogenetically distinct. These ecological data provide the basis for the hypothesis of adaptation. To test the hypothesis that a characteristic is adaptive, it is necessary to demonstrate that (1) variability in the characteristic exists, (2) the observed variation has a genetic basis, (3) variability in the trait leads to nonrandom differences in fitness of the individuals in a population, and (4) the observed trait evolved in response to the hypothesized environmental pressure (Brooks & McLennan, 1991; DeCoursey *et al.*, submitted; Endler, 1986; Gould & Lewontin, 1979; Harvey & Pagel, 1991). These criteria have been met in some studies of circadian rhythms, but data are often insufficient to permit an actual test of the hypothesis.

The fundamental observation that circadian rhythms are an innate, genetically based characteristic of organisms is well established (Aschoff, 1960; Bünning, 1960; Takahashi, 1995). Variability in the length of the free-running period of the circadian pacemaker is frequently observed both within and between individuals. This variability is, in fact, a defining characteristic of the circadian nature of a rhythm (Aschoff, 1960). There is evidence for genetic differences in the responsiveness of the circadian system to light (Pohl, 1983; Pittendrigh, Kyner, & Takamura, 1991; Possidente & Hegmann, 1982; Shimomura *et al.*, 1998). There is evidence of a latitudinal cline in the ability of *Drosophila auria* to phase shift in response to light pulses (Pittendrigh *et al.*, 1991). Variation is also observed in the structure and function of individual components of clock systems of organisms that inhabit different niches. In one of the most thoroughly studied systems, variation in the length of the threonine–glycine repeat of the *per* gene in *Drosophila melanogaster* and *D. simulans* has been shown to occur on a latitudinal gradient. This variation is associated with the stability of the free-running period under different temperatures (i.e., *temperature compensation*) (Costa, Peixoto, Thackeray, Dalgleish, & Kyriacou, 1991; Costa, Peixoto, Barbujani, & Kyriacou, 1992; Rosato *et al.*, 1994; Sawyer *et al.*, 1997).

The selective advantage of circadian rhythms is hypothesized to result from increased *fitness* (i.e., increased reproductive success) that individuals capable of measuring the lapse of time or recognizing local time may accrue (Bünning, 1971; Daan, 1981; Daan & Aschoff, 1982; DeCoursey *et al.*, submitted; Enright, 1970; Pittendrigh & Daan, 1976b). When evaluating the significance of changes in individual clock components, it is important to remember that it is the overt behavior of an organism that interacts with the environment. It is the whole organism that must survive and reproduce, therefore the impact of genetic variation in any single clock component will be determined by its effect on the survival and reproductive fitness of the individual. Therefore, alteration of clock components that result in significant changes in patterns of entrainment are most likely to be subjected to natural selection. Mutations that do not result in a significant change in the pattern of entrainment may accumulate in the genome as neutral mutations.

There are few experimental studies that directly test the hypothesis that the presence of a clock, or a clock exhibiting specific characteristics, actually increases fitness. Early studies suggested that exposing organisms (e.g., tomatoes, fruitflies, and blowflies) to light–dark cycles with periods significantly different from 24 hours was harmful (Hillman, 1956; Pittendrigh & Minis, 1972; Saint Paul & Aschoff, 1978). Recent work using a diurnal mammal, the antelope ground squirrel (*Ammospermophilus leucurus*), demonstrates that animals bearing suprachiasmatic lesions behave normally in the field and laboratory except in one aspect: the animals become arrhythmic (DeCoursey, Krulas, Mele, & Holley, 1997). This arrhythmia results in animals being active at night as well as during the daytime. An example of how this arrhythmia may decrease fitness is the fact that twice as many of the lesioned animals than control animals (60% versus 29%) were killed by the nighttime invasion of a feral cat onto the study site. Field observations of naturally occurring populations of guillemots (*Uria lomvia*) suggest that young birds that fledge from their nests late in the afternoon, in synchrony with the rest of the population, have a greater survival rate than those that do not (Daan & Tinbergen, 1980). While it is an example of a circatidal rhythm rather than a circadian rhythm, marine iguanas that time their movement to the beach to closely match the time of low tide have higher long-term survival than those that do not (Wikelski & Hau, 1995).

Finally, to demonstrate that a character is an adaptation that has arisen in response to a specific environmental pressure, rather than a preexisting permissive condition or a characteristic correlated with some other physiologic change, one must compare ancestral and derived traits (Gould & Lewontin, 1979; Harvey & Pagel, 1991, pp. 13–34). One method currently being used to test the adaptive significance of circadian rhythms is to look for the loss of rhythms under conditions where they appear to present no clear advantage. A number of species have removed themselves from exposure to environmental periodicity by becoming fossorial, inhabiting caves or benthic habitats (Goldman, Goldman, Riccio, & Terkel, 1997; Jegla & Poulson, 1968; Kenagy & Vleck, 1982; Lovegrove & Muir, 1996; Poulson and White, 1969). Kenagy and Vleck (1982) proposed four conditions that could result in relatively uniform day-to-night levels of resting metabolic activity in mammals: (1) continuously fossorial habit, (2) consumption of bulky, low-quality food requiring frequent foraging, (3) small body size, near the lower limit of mammalian body size, and (4) the ability to hibernate.

Patterns of locomotor activity, body temperature, and metabolic activity have been determined for several different species of fossorial mammals from different orders and suborders; the degree to which they express circadian rhythms is highly variable within and between species. Locomotor activity of blind mole rats (*Spalax ehrenbergi*, Order Rodentia, Suborder Sciurognathi, Family Muridae) and the unrelated solitary Cape mole rats (*Georychus capensis*, Order Rodentia, Suborder Hystricognaith, Family Bathyergidae) is primarily expressed at night and can be entrained by the light–dark cycle; however, considerable variation exists among individuals in the strength of their circadian rhythms (Goldman *et al.*, 1997; Lovegrove & Muir, 1996). The Cape mole rat shows little day–night variation in body temperature (Lovegrove & Muir, 1996). In contrast, measurements of metabolic activity demonstrate virtually uniform levels of oxygen consumption in fossorial gophers (*Thomomys talpoides*, Order Rodentia, Suborder Sciurognathi, Family Geomyidae) and true moles (*Scapanus townsendii*, Order Insectivora, Family Talpidae) held on constant conditions (Kenagy & Vleck, 1982; Wilson & Reeder, 1993). It has been hypothesized that this "relaxation" of rhythms represents a relaxation in the selective pressures from the external environment, but that the retention of some rhythmic components reflects the importance of maintaining internal temporal order or perhaps the ability to measure seasonal changes in photoperiod (Goldman *et al.*, 1997; see also Chapter 1 of this volume). A comparative phylogenetic analysis could distinguish between the alternative hypotheses that the expression of relaxed rhythms is an adaptation to life in a fossorial (or otherwise arrhythmic environment) or may have been a permissive character enabling rodents to invade the fossorial habitat. The testing of these and other hypotheses by rigorous analytical procedures will establish the validity of the more general hypotheses, that circadian rhythms are adaptive, and that temporal patterns of behavior represent the adaptation of circadian rhythms to the environment.

Conclusions

The study of circadian rhythms originated with the observation of behavior. The ubiquity of behaviors controlled by circadian rhythms and the seemingly obvious evolutionary advantages that may accrue to organisms that express regular patterns of behavior make it easy to speculate about the origin and ecological roles of rhythms. Such armchair speculation is fun to do, but does not advance scientific understanding. The application of scientific methods of comparative biology can advance our understanding of the ecological consequences and evolutionary history of circadian rhythms.

Acknowledgments

Daniel Bernard, David King, Dan Kolker, Sharon Low-Zeddies, Fred Turek, and Martha Vitaterna provided important comments on previous drafts of the manuscript. John Bates, Patricia DeCoursey, Shannon Hackett, Lawrence Heaney, Paul Heideman, Theresa Lee, Stephan Reebs, Laura Smale, Joseph Takahashi, and Fred Turek provided many useful discussions. G. James Kenagy and Klaus Hoffman introduced me to this area of research as an undergraduate student many years ago. Supported by NSF IBN 9496304.

References

Albrecht, U., Sun, Z. S., Eichele, G., & Lee, C. C. (1997). A differential response of two putative mammalian circadian regulators, *mper1* and *mper2*, to light. *Cell, 91*, 1055–1064.

Aschoff, J. (1960). Exogenous and endogenous components in circadian rhythms. *Cold Spring Harbor Symposia on Quantitative Biology, 25*, 11–28.

Bennett, A. F. (1987). The accomplishments of ecological physiology. In M. E. Feder, A. F. Bennett, W. W. Burggren, & R. B. Huey (Eds.), *New directions in ecological physiology*, (pp. 1–8). Cambridge: Cambridge University Press.

Brooks, D. R., & McLennan, D. A. (1991). *Phylogeny, ecology, and behavior. A research program in comparative biology.* Chicago: University of Chicago Press.

Bünning, E. (1960). Opening address: Biological clocks. *Cold Spring Harbor Symposia on Quantitative Biology, 25*, 1–9.

Bünning, E. (1971). The adaptive value of circadian leaf movements. In M. Menaker (Ed.), *Biochronometry* (pp. 203–211). Washington, DC: National Academy of Sciences.

Claypool, L. E. (1984). The environmental and physiological determinants of activity patterns in *Microtus montanus* the montane vole. Doctoral dissertation, University of Utah, Salt Lake City, UT.

Cloudsley-Thompson, J. L. (1960). Adaptive functions of circadian rhythms. *Cold Spring Harbor Symposia on Quantitative Biology, 25*, 345–355.

Costa, R., Peixoto, A. A., Thackeray, J. R., Dalgleish, R., & Kyriacou, C. P. (1991). Length polymorphism in the threonine-glycine-encoding repeat region of the *period* gene in *Drosophila. Journal of Molecular Evolution, 32*, 238–246.

Costa, R., Peixoto, A. A., Barbujani, G., & Kyriacou, C. P. (1992). A latitudinal cline in a *Drosophila* clock gene. *Proceedings of the Royal Society of London, Series B, 250*, 43–49.

Crosthwaite, S. K., Dunlap, J. C., & Loros, J. J. (1997). Neurospora wc-1 and wc-2: Transcription, photoresponses, and the origins of circadian rhythmicity. *Science, 276*, 763–769.

Daan, S. (1981). Adaptive daily strategies in behavior. In J. Aschoff (Ed.), *Handbook of neurobiology and behavior. Vol. 4. Biological clocks* (pp. 275–298). New York: Plenum Press.

Daan, S., & Aschoff, J. (1982). Circadian contributions to survival. In J. Aschoff, S. Daan, & G. Groos (Eds.), *Vertebrate circadian systems* (pp. 305–321). Berlin: Springer-Verlag.

Daan, S., & Tinbergen, J. M. (1980). Young guillemots (*Uria lomvia*) leaving their Arctic breeding cliffs: A daily rhythm in numbers and risk. *Ardea, 67*, 96–100.

Darwin, C. (1859). *The origin of species by means of natural selection.* London: John Murray.

DeCoursey, P. J. (1990). Circadian photoentrainment in nocturnal mammals: Ecological overtones. *Biology of Behavior, 15*, 213–238.

DeCoursey, P. J., Krulas, J. R., Mele, G., & Holley, D. C. (1997). Circadian performance of suprachiasmatic nuclei (SCN)-lesioned antelope ground squirrels in a desert enclosure. *Physiology and Behavior, 62*, 11099–1108.

DeCoursey, P. J., Heideman, P., Horton, T., Lee, T., Reebs, S., & Smale, L. The behavioral ecology and evolution of biological timing systems. In J. C. Dunlap and J. J. Loros (Eds.), *Biological clocks.* New York: Oxford University Press. (Submitted)

Endler, J. A. (1986). *Natural selection in the wild.* Princeton, NJ: Princeton University Press.

Enright, J. T. (1970). Ecological aspects of endogenous rhythmicity. *Annual Review of Ecology and Systematics, 1*, 221–238.

Erkinaro, E. (1969). Der Phasenwechsel der lokomotorischen Activitat bei *Microtus agrestis* (L.), *M. arvalis* (Pall.) and *M. oeconomus* (Pall.). *Aquilo, Series Zoologica, 8*, 1–31.

Gekakis, N., Saez, L., Delahaye-Brown, A.-M., Myers, M. P., Sehgal, A., Young, M. W., & Weitz, C. J. (1995). Isolation of *timeless* by PER protein interaction: Defective interaction between *timeless* protein and long-period mutant PER^L. *Science, 270*, 811–815.

Goldman, B. D., Goldman, S. L., Riccio, A. P., & Terkel, J. (1997). Circadian patterns of locomotor activity and body temperature in blind mole-rats, *Spalax ehrenbergi. Journal of Biological Rhythms, 12*, 348–361.

Gould, S. J., & Lewontin, R. C. (1979). The spandrels of San Marco and the Panglossian paradigm: A critique of the adaptationist programme. *Proceedings of the Royal Society of London, Series B, 205*, 581–598.

Harvey, P. H., & Pagel, M. D. (1991). *The comparative method in evolutionary biology.* Oxford: Oxford University Press.

Hennig, W. (1950). *Grundzüge einer Theorie der phylogenetischen Systematic.* Berlin: Deutscher Zentralverlag.

Hennig, W. (1966). *Phylogenetic systematics.* Urbana: University of Illinois Press.

Hillis, D. M., Moritz, C., & Mable, B. K. (Eds.) (1996). *Molecular systematics.* Sunderland, MA: Sinauer Associates.

Hillman, W. S. (1956). Injury of tomato plants by continuous light and unfavorable photoperiodic cycles. *American Journal of Botany, 43*, 89–96.

Honrado, G. I., & Mrosovsky, N. (1991). Interaction between periodic socio-sexual cues and light–dark cycles in controlling the phasing of activity rhythms in golden hamsters. *Ethology Ecology and Evolution, 3*, 221–231.

Hoogenboom, I., Daan, S., Dallinga, J. H., & Schoenmakers, M. (1984). Seasonal change in the daily timing of behavior of the common vole, *Microtus arvalis. Oecologia, 61,* 18–31.

Horton, T. H., & Rowsemitt, C. N. (1992). Natural selection and variation in reproductive physiology. In T. E. Tomasi & T. H. Horton (Eds.), *Mammalian energetics. Interdisciplinary views of metabolism and reproduction* (pp. 160–185). Ithaca, NY: Cornell University Press.

Huang, Z. J., Edery, I., & Rosbash, M. (1993). PAS is a dimerization domain common to *Drosophila* period and several transcription factors. *Nature, 364,* 259–262.

Jegla, T. C., & Poulson, T. L. (1968). Evidence of circadian rhythms in a cave crayfish. *Journal of Experimental Zoology, 168,* 273–282.

Kay, S. A. (1997). PAS, present, and future: Clues to the origins of circadian clocks. *Science, 276,* 753–754.

Kenagy, G. J. (1973). Daily and seasonal patterns of activity and energetics in a heteromyid rodent community. *Ecology, 54,* 1201–1219.

Kenagy, G. J. (1976). The periodicity of daily activity and its seasonal changes in free-ranging and captive kangaroo rats. *Oecologia* (Berlin), *24,* 105–140.

Kenagy, G. J. (1978). Seasonality of endogenous circadian rhythms in a diurnal rodent *Ammospermophilus leucurous* and a nocturnal rodent *Dipodomys merriami. Journal of Comparative Physiology, 128,* 21–36.

Kenagy, G. J., & Vleck, D. (1982). Daily temporal organization of metabolism in small mammals: Adaptation and diversity. In J. Aschoff, S. Daan, and G. Groos (Eds.), *Vertebrate circadian systems* (pp. 322–338). Berlin: Springer-Verlag.

King, D. P., Zhao, Y., Sangoram, A. M., Wilsbacher, L. D., Tanaka, M., Antoch, M. P., Steeves, T. D. L., Vitaterna, M. H., Kornhauser, J. M., Lowrey, P. L., Turek, F. W., & Takahashi, J. S. (1997). Positional cloning of the mouse circadian *Clock* gene. *Cell, 89,* 641–653.

Klein, D. C., Coon, S. L., Roseboom, P. H., Weller, J. L., Bernard, M., Gastel, J. A., Zatz, M., Iuvone, P. M., Rodriguez, I. R., Begay, V., Falcon, J., Cahill, G. M., & Cassone, V. M. (1997). The melatonin rhythm-generating enzyme: Molecular regulation of serotonin N-acetyltransferase in the pineal gland. *Recent Progress in Hormone Research, 52,* 307–357 [Discussion, pp. 357–358].

Kondo, T., Golden, S. S., Ishiura, M., & Johnson, C. H. (1994). Circadian rhythms of cyanobacteria expressed from a luciferase reporter gene. In T. Hiroshige & K.-I. Honma (Eds.), *Evolution of circadian clock* (pp. 59–73). Sapporo, Japan: University of Hokkaido Press.

Kotler, B. P., Brown, J. S., & Subach, A. (1993). Mechanisms of species coexistence of optimal foragers: Temporal partitioning by two species of sand dune gerbils. *Oikos, 67,* 548–556.

Lovegrove, B. G., & Muir, A. (1996). Circadian body temperature rhythms of the solitary Cape mole rat *Georychus capensis* (Bathyergidae). *Physiology and Behavior, 60,* 991–998.

Menaker, M., & Tosini, G. (1996). The evolution of vertebrate circadian systems. In K.-I. Honma & S. Honma (Eds.), *Circadian organization and oscillatory coupling* (pp. 39–52). Sapporo, Japan: Hokkaido University Press.

Mrosovsky, N. (1996). Locomotor activity and non-photic influences on circadian clocks. *Biological Review, 71,* 343–372.

Page, T. L. (1989). Physiology and development of a circadian clock. In T. Hiroshige & K.-I. Honma (Eds.), *Circadian clocks and ecology* (pp. 3–17). Sapporo, Japan: Hokkaido University Press.

Pittendrigh, C. S., & Daan, S. (1976a). A functional analysis of circadian pacemakers in nocturnal rodents. I. The stability and lability of spontaneous frequency. *Journal of Comparative Physiology A, 106,* 223–252.

Pittendrigh, C. S., & Daan, S. (1976b). A functional analysis of circadian pacemakers in nocturnal rodents. IV. Entrainment: Pacemaker as clock. *Journal of Comparative Physiology A, 106,* 291–331.

Pittendrigh, C. S., & Daan, S. (1976c). A functional analysis of circadian pacemakers. V. Pacemaker structure a clock for all seasons. *Journal of Comparative Physiology A, 106,* 333–355.

Pittendrigh, C. S., & Minis, D. H. (1972). Circadian systems: Longevity as a function of circadian resonance in *Drosophila melanogaster. Proceedings of the National Academy of Sciences of the USA, 69,* 1537–1539.

Pittendrigh, C. S., Kyner, W. T., & Takamura, T. (1991). The amplitude of circadian oscillations: Temperature dependence, latitudinal clines and photoperiodic time measurement. *Journal of Biological Rhythms, 6,* 299–313.

Pohl, H. (1983). Strain differences in responses of the circadian system to light in the Syrian hamster. *Experientia, 39,* 372–374.

Possidente, B., & Hegmann, J. P. (1982). Gene differences modify Aschoff's rule in mice. *Physiology and Behavior, 28,* 199–200.

Poulson, T. L., & White, W. B. (1969). The cave environment. *Science, 165,* 971–980.

Rapp, P. E. (1987). Why are so many biological systems perioidic? *Progress in Neurobiology, 29,* 261–273.

Reppert, S. M., & Weaver, D. R. (1997). Forward genetic approach strikes gold: Cloning of a mammalian *Clock* gene. *Cell, 89,* 487–490.

Roenneberg, T. (1994). The *Gonyaulax* circadian system: Evidence for two input pathways and two oscillators. In T. Hiroshige & K.-I. Honma (Eds.), *Evolution of circadian clock* (pp. 3–20). Sapporo, Japan: Hokkaido University Press.

Rosato, E., Peixoto, A. A., Barbujani, G., Costa, R., & Kyriacou, C. P. (1994). Molecular polymorphism in the *period* gene of *Drosophila simulans*. *Genetics, 138,* 693–797.

Rowsemitt, C. N. (1986). Seasonal variations in activity rhythms of male voles: Mediation by gonadal hormones. *Physiology and Behavior, 37,* 797–803.

Rowsemitt, C. N., Petterborg, L. J., Claypool, L. E., Hoppenstaedt, F. C., Negus, N. C., & Berger, P. J. (1982). Photoperiodic induction of diurnal locomotor activity in *Microtus montanus*, the montane vole. *Canadian Journal of Zoology, 60,* 2798–2803.

Saint Paul, U. v., & Aschoff, J. (1978). Longevity among blowflies *Phormia terraenovae* R.D. kept in non-24-hour light–dark cycles. *Journal of Comparative Physiology, 127,* 191.

Sassone-Corsi, P. (1996). Same clock, different works. *Nature, 384,* 613–614.

Sauman, I., & Reppert, S. M. (1996). Circadian clock neurons in the silkmoth *Antheraea pernyi*: Novel mechanisms of *period* protein regulation. *Neuron, 17,* 889–900.

Sawyer, L. A., Hennessy, J. M., Peixoto, A. A., Rosato, E., Parkinson, H., Costa, R., & Kyriacou, C. P. (1997). Natural variation in *Drosophila* clock gene and temperature compensation. *Science, 278,* 2117–2120.

Schoener, T. W. (1970). Nonsynchronous spatial overlap of lizards in patchy habitats. *Ecology, 51,* 408–418.

Schoener, T. W. (1974). Resource partitioning in ecological communities. *Science, 185,* 27–58.

Sehgal, A., Rothenfluh-Hilfiker, A., Hunter-Ensor, M., Chen, Y., Myers, M. P., & Young, M. W. (1995). Rhythmic expression of *timeless*: A basis for promoting circadian cycles in *period* gene autoregulation. *Science, 270,* 808–810.

Shearman, L. P., Zylka, M. J., Weaver, D. R., Kolakowski, L. F. J., & Reppert, S. M. (1997). Two *period* homologs: Circadian expression and photic regulation in the suprachiasmatic nuclei. *Neuron, 19,* 1261–1269.

Shigeyoshi, Y., Taguchi, K., Yamamoto, S., Takekida, S., Yan, L., Tei, H., Moriya, T., Shibata, S., Loros, J. J., Dunlap, J. C., & Okamura, H. (1997). Light-induced resetting of a mammalian circadian clock is associated with rapid induction of the *mPer1* transcript. *Cell, 91,* 1043–1053.

Shimomura, K., Vitaterna, M. H., Low-Zeddies, S. S., Whiteley, A. R., & Takahashi, J. S. (1998). Genetic dissection of strain differences in mouse circadian rhythmicity. Presented at the Annual Meeting of the Society for Research on Biological Rhythms, Amelia Island Plantation, June 1998.

Soni, B. G., & Foster, R. G. (1997). A novel and ancient vertebrate opsin. *FEBS Letters, 406,* 279–83.

Swofford, D. L., Olsen, G. J., Waddell, P. J., & Hillis, D. M. (1996). Phylogenetic inference. In D. M. Hillis, C. Moritz, & B. K. Mable (Eds.), *Molecular systematics* (Chapter 12). Sunderland, MA: Sinauer Associates.

Takahashi, J. S. (1995). Molecular neurobiology and genetics of circadian rhythms in mammals. *Annual Review of Neuroscience, 18,* 531–553.

Wikelski, M., & Hau, M. (1995). Is there an endogenous tidal foraging rhythm in marine iguanas. *Journal of Biological Rhythms, 10,* 335–350.

Wiley, E. O. (1981). *Phylogenetics: The theory and practice of phylogenetic systematics.* New York: Wiley.

Wilson, D. E., & Reeder, D. M. (Eds.) (1993). *Mammal species of the world. A taxonomic and geographic reference.* Washington, DC: Smithsonian Institution Press.

Zucker, I. (1983). Motivation, biological clocks, and temporal organization of behavior. In E. Satinoff & P. Teitelbaum (Eds.), *Handbook of behavioral neurobiology.* (Vol. 6, pp. 3–21). New York: Plenum Press.

Part II

Organization of Circadian Systems

3

Circadian Rhythms in Unicellular Organisms

CARL HIRSCHIE JOHNSON AND TAKAO KONDO

CONSERVATION OF CIRCADIAN CLOCK MECHANISMS

This volume of the *Handbook of Behavioral Neurobiology* emphasizes circadian behaviors and mechanisms in higher organisms. The purpose of this chapter is to discuss what studies on simple unicellular systems have told us about clocks that are particularly relevant to clock studies in higher organisms. This chapter does not pretend to review the literature exhaustively; rather, we discuss a few salient points. Other chapters in this volume specifically address clock studies about *Gonyaulax* and *Neurospora* in more detail.

In addition to being fascinating in their own right, the study of clocks in unicellular organisms has enlightened us as to the nature of circadian rhythms in higher organisms. When considering how studies on unicells provide clues to the clock mechanism of higher organisms, we need to examine two underlying assumptions. First, we assume that unicellular systems are indeed more "simple" than multicellular systems. This seems correct, but it is almost certainly the case that some unicellular organisms are more complicated than differentiated cells of a multicellular organism. As Kung and Saimi (1982) noted, "We should be reminded, however, that a protozoan is more complex (having the whole behavioral machinery in one cell) than a specialized receptor cell, neuron, or effector cell in the metazoan." Nevertheless, using unicells should help us to avoid most of the complex intercellular interactions that occur in multicellular organisms. An added benefit is that a population of unicellular organisms is a more homogeneous preparation than is

---

CARL HIRSCHIE JOHNSON Department of Biology, Vanderbilt University, Nashville, Tennessee 37235.
TAKAO KONDO Division of Biological Science, Graduate School of Science, Nagoya University, Chikusa, Nagoya 464-01, Japan.

Circadian Clocks, Volume 12 of *Handbook of Behavioral Neurobiology*, edited by Joseph S. Takahashi, Fred W. Turek, and R. Y. Moore, Kluwer Academic / Plenum Publishers, New York, 2001.

usually possible in multicellular organisms. This is very helpful for biochemical investigations.

A second, more fundamental assumption is that discovering the mechanisms of circadian clocks in unicellular organisms will tell us how a circadian clock works in higher organisms. In particular, a big question is, "Have circadian mechanisms been conserved?" We do not know the answer. Of course, many key metabolic pathways have been conserved throughout evolution, so it is conceivable that the circadian clock mechanism has been conserved. The fundamental properties of circadian oscillators (persistence of an approximately 24-hour rhythm, entrainability, and temperature compensation) must be a part of any functional daily clock, and the fact that these characteristics are shared among all organisms with a circadian clock has encouraged the notion that the biochemical mechanism is also shared. Also, extensive data showing that inhibitors of translation and transcription phase-shift eukaryotic clocks at all phylogenetic levels encourage, but do not demand, the hypothesis that the biochemistry of the central clockwork is similar (Edmunds, 1988; Takahashi, Kornhauser, Koumenis, & Eskin, 1993). Finally and most persuasively, we now know that circadian systems are intrinsic to living organisms from prokaryotes to humans. Given that these clocks must have been adaptive at every stage of evolution, why should their mechanism have been reinvented again and again if they have been in constant use? Ultimately, however, deciding whether circadian mechanisms are kindred or are the result of convergent evolution (Pittendrigh, 1960) will have to wait until we know more about their molecular mechanisms.

What Kinds of Single-Celled Organisms Express Circadian Rhythms?

Eukaryotic Unicells

The first report of a unicellular manifestation of circadian temperature compensation was that of Bruce and Pittendrigh (1956), who extended Pohl's (1948) earlier discovery of a daily rhythm of phototaxis in *Euglena gracilis.* Soon thereafter, other reports appeared which demonstrated temperature-compensated daily rhythms in *Gonyaulax polyedra* (Hastings & Sweeney, 1957; Sweeney & Hastings, 1957) and in *Paramecium bursaria* (Ehret, 1959).

From those roots, our appreciation of the ubiquity of circadian rhythms among unicells has blossomed. A host of eukaryotic unicellular organisms, including many algae, protozoa, and fungi, marches to the daily beat of a circadian drummer (Edmunds, 1988). It is easier to mention which commonly used eukaryotic microorganisms have *not* been reported to exhibit circadian rhythms than the converse. To our knowledge, there have been no reports of circadian rhythms in the cellular slime mold, *Dictyostelium.* In the case of yeast, there is one report of a circadian rhythm of heat tolerance in the fission yeast, *Schizosaccharomyces pombe* (Kippert, 1989). Also, the budding yeast, *Saccharomyces cerevisiae,* has been reported to display circadian rhythms of cell division and amino acid transport (Edmunds, Apter, Rosenthal, Shen, & Woodward, 1979), but the data are not persuasive that this is an authentic circadian rhythm because temperature compensation was not demonstrated. Furthermore, the one temperature that was reported (12°C) is considerably colder than the optimal growth temperature for this organism. Therefore, it is easy to imagine that the cell division cycle has merely been slowed to a period that is close to 24 hours without necessarily being controlled by a circadian clock.

On the topic of temperature and circadian rhythms, several unicells exhibit not only temperature compensation of their free-running circadian rhythm, but temperature *over*compensation, that is, the Q_{10} of their rhythms is less than 1. Therefore, their clocks run *faster* at colder temperatures. This intriguing phenomenon has been observed in the unicells *Gonyaulax* (Sweeney & Hastings, 1960), *Acetabularia* (Berger, Dirk, von Lindern, Wolff, & Mergenhagen, 1992; Karakashian & Schweiger, 1976), *Euglena* (Brinkmann, 1971), and *Chlamydomonas* (Goto & Johnson, 1995).

Prokaryotic Unicells

Can all types of unicellular organisms express circadian rhythms? It used to be thought that circadian rhythms were exclusively a eukaryotic phenomenon (Ehret, 1960; Johnson, Golden, Ishiura, & Kondo, 1996). A few dissenting reports (Halberg & Conner, 1961; Sturtevant, 1973) were not convincing. That prokaryotic cells, either unicellular or multicellular, were too "simple" to express circadian behavior became a dogma (Ehret & Wille, 1970; Kippert, 1986; Schweiger & Schweiger, 1980) despite the fact that there were almost no published reports of rigorous tests of the proposition. Some models of the circadian mechanism relied on this dogma as a necessary factor (Ehret & Trucco, 1967; Kippert, 1986; Schweiger & Schweiger, 1977).

One report of a serious attempt to test whether prokaryotes can exhibit circadian behavior was that of Taylor (1979). Taylor reasoned that photosynthetic cyanobacteria would be the most likely of the prokaryotes to have evolved a circadian system since their niche (i.e., photosynthetic autotrophs) most closely matches that of many eukaryotic unicells that express circadian rhythms, such as the green alga *Chlamydomonas.* Nevertheless, Taylor failed to find persuasive evidence for circadian rhythms of photosynthetic capacity in several species of cyanobacteria. Our later study (of which Taylor was a co-author) shows that at least one of the species that Taylor tested in 1979 (*Anacystis nidulans* R2 = *Synechococcus* sp. PCC 7942) does indeed have a circadian mechanism (Kondo *et al.*, 1993). One possible reason for the discrepancy might be that Taylor's original investigation assayed the endogenous control in DD (constant dark), whereas circadian rhythms of most photosynthetic organisms are more robust in LL (constant light). Also, photosynthetic capacity, which is the parameter that Taylor had chosen to test, might not be under the control of a circadian clock in this species of cyanobacteria.

In 1985–1986, three papers were published that began to unseat the "no circadian clocks in prokaryotes" dogma (Grobbelaar, Huang, Lin, & Chow, 1986; Mitsui, Kumazawa, Takahashi, Ikemoto, & Arai, 1986; Stahl & Krumbein, 1985). These workers discovered that diazotrophic cyanobacteria (*Oscillatoria* sp. strain 23; *Synechococcus* spp. Miami BG 43511 and 43522; *Synechococcus* sp. RF-1) which fix nitrogen display daily rhythms of nitrogen fixation in LD and in LL. One of these studies showed that the cyanobacteria also express a daily rhythm of photosynthesis (Mitsui *et al.*, 1986). The phase relationships of these two rhythms were 180° out of phase; photosynthesis peaks in midday, while nitrogen fixation occurs at night. These phase relationships are particularly interesting from an adaptive perspective. The nitrogen-fixing enzyme nitrogenase is inactive in the presence of even tiny amounts of oxygen, which creates a major design problem for all photosynthetic nitrogen-fixing organisms, since photosynthesis results in the release of oxygen (Fay,

1992). Different organisms have contrived differing solutions to this problem, notably the spatial separation of photosynthesis and nitrogen fixation in filamentous cyanobacteria, which develop specialized cells (heterocysts) for fixing nitrogen. Spatial segregation may not be a practical solution in nonheterocyst cyanobacterial strains. Therefore, these cyanobacteria appear to have separated these two cellular activities *temporally*. Both the photosynthetic and the nitrogen-fixation rhythms free-run in LL and are entrained by the prior LD (Mitsui *et al.*, 1986), so two of the crucial properties of circadian rhythmicity are satisfied. More recent studies using other species of cyanobacteria have confirmed rhythmic modulation of photosynthetic capacity and nitrogen-fixation capabilities (Roenneberg & Carpenter, 1993; Schneegurt, Sherman, Nayar, & Sherman, 1994).

The other crucial circadian characteristic, temperature compensation, has been demonstrated in other strains of *Synechococcus*: the marine *Synechococcus* WH7803 and the freshwater *Synechococcus* RF-1. Sweeney and Borgese (1989) showed that WH7803 displays temperature-compensated rhythms of cell division, and Chen, Chen, Hung, and Huang (1991) found temperature-compensated rhythms of amino acid uptake in the freshwater cyanobacterium isolated from rice fields, *Synechococcus* RF-1. *Synechococcus* RF-1 also expresses a nocturnal circadian rhythm of nitrogen fixation that has been well studied by the Taiwan group, including the demonstration that a rhythm of nitrogenase mRNA abundance probably accounts for the nitrogen-fixation rhythm (Huang & Chow, 1990).

These studies indicated that cyanobacteria have evolved a circadian clock, although Kippert (1991) argued that the evidence for clocks in prokaryotes was not conclusive. His concerns, however, were based on the possible influence of an unidentified exogenous factor that was ruled out by experiments in which we simultaneously measured rhythms from cultures that were 180° out of phase (Kondo *et al.*, 1993). There can be little doubt at this time that at least some cyanobacteria have a bona fide circadian clockwork. Nevertheless, the reports prior to 1993 did not encourage intensive research on the mechanism of the prokaryotic clock, for two reasons. First, molecular and genetic techniques were not established for the strains of cyanobacteria employed for circadian studies before 1993. Second, nitrogen-fixation, amino acid uptake, and cell division assays require laborious manual measurements. Consequently, the prior studies demonstrated the existence of clocks in prokaryotes, but did not identify a suitable organism for an extensive molecular/genetic analysis of the mechanism. If such a prokaryote were found, its advantages for clock analysis would be that (1) knowledge of regulatory mechanisms are generally better understood in prokaryotes than in eukaryotes and (2) the commonly smaller size of prokaryotic genomes would facilitate the goal of saturational mutagenesis for clock-related genes.

To reap the strategic benefits that prokaryotes offer, we, together with our collaborators Susan Golden and Masahiro Ishiura, searched for a cyanobacterium that is amenable to molecular/genetic analyses and which exhibits circadian rhythms of a parameter that can be assayed continuously for many cycles by an automated system (Johnson *et al.*, 1996). We developed such a strain by transforming a genetically tractable cyanobacterium, *Synechococcus* sp. strain PCC 7942, with bacterial luciferase genes that function as a reporter of clock-controlled expression of the endogenous *psbAI* gene (Kondo *et al.*, 1993; Liu, Golden, Kondo, Ishiura, & Johnson, 1995a). The bioluminescence rhythm expressed by this reporter strain is easily assayed by an automated monitoring system, and it conforms to all three salient properties of circadian rhythms: persistence in LL with a period close to 24 hours, entrainability by light/dark signals, and temperature compensation (Kondo *et al.*, 1993). The bioluminescence level rises during the day and falls during the night. This rhythm reflects circadian control over the promoter of the *psbAI* gene (Liu *et al.*, 1995a). What is the adaptive significance of circadian modulation of the *psbAI* gene? This gene encodes the D1 protein, a major component of the reaction center of photosystem II. The D1 protein is rapidly turned over in the light (Clarke, Hurry, Gustafsson, & Oquist, 1993; Mattoo, Hoffman-Falk, Marder, & Edelman, 1984), so it is reasonable to expect that synthesis of D1 must be high in the daytime to replace degraded D1 molecules. Possibly the circadian clock automatically initiates synthesis of D1 to compensate for daytime degradation (Liu *et al.*, 1995a).

A major thrust of our research effort with this cyanobacterium is to isolate mutants of the clock and try to infer the nature of the clockwork by characterizing the genetic lesions. The size of the genome of *Synechococcus* PCC 7942 is comparable to that of *Escherichia coli* (about 2.6×10^6 base pairs), so that saturation mutagenesis is an obtainable goal. Using chemical mutagenesis (ethyl methanesulfonate) and screening for aberrant bioluminescence rhythms with a cooled CCD camera apparatus (Kondo *et al.*, 1994), we isolated over 100 period or apparently arhythmic mutants. The period mutants range from as short as 16 hours to as long as 60 hours, which is the largest range of circadian period phenotypes described from any organism. Four of these mutants are shown in Figure 1: one has a short period (about 22 hours), and three have long periods (about 27, 40, and 60 hours). Despite the mutation, the altered period of each of these mutants remains temperature compensated. Approximately 30 of these mutants have been tested for complementation and are rescued by a single fragment of wild-type *Synechococcus* DNA. This DNA fragment has been sequenced and three open reading frames are found in this

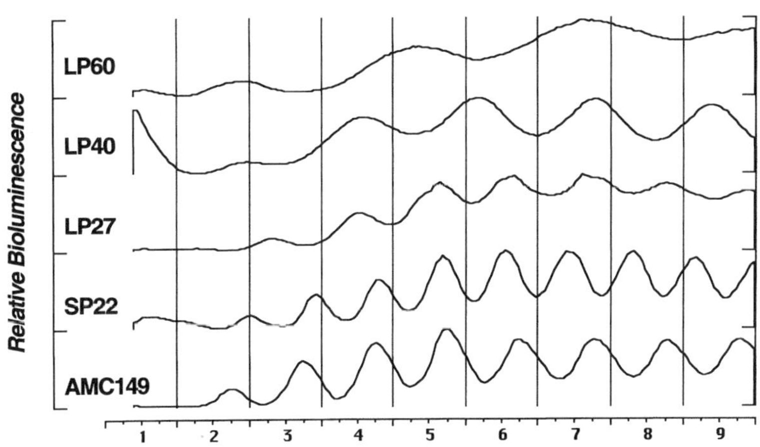

Figure 1. Circadian period mutants of the cyanobacterium *Synechococcus* sp. strain PCC 7942. The wild-type strain that has been transformed with a luciferase reporter construct is strain AMC149, which has a free-running period of approximately 24 hours at 30°C. The other strains were derived by chemical mutagenesis of AMC149. SP22 is a short-period mutant (period of about 22 hours), and the other mutants have free-running periods longer than 24 hours: LP27 (about 27 hours), LP40 (about 40 hours), and LP60 (about 60 hours). The free-running rhythms are expressed in LL of various colors, either white, blue, or red. The experiment illustrated in this figure was performed in red LL (experiment #CY25).

fragment. Over 20 mutations have been mapped to one of those open reading frames, and several mutations map to each of the other two open reading frames. We believe that this constitutes a cluster of three clock genes (M. Ishiura *et al.*, 1998).

We also wondered how extensive circadian control is of gene expression in *Synechococcus* PCC 7942, and devised a strategy to search globally for rhythmic control over promoters (Liu *et al.*, 1995b). Bacterial luciferase genes (*luxAB*) were inserted into the *Synechococcus* genome by conjugal transfer from *E. coli* harboring libraries of promoterless *luxAB* attached to random fragments of *Synechococcus* DNA. After homologous recombination of these libraries into the *Synechococcus* genome, the end result was random insertions of *luxAB* throughout the cyanobacterial chromosome. Luminescence expression patterns from the resulting transformed clones were then monitored using the cooled CCD camera screening apparatus. We screened approximately 30,000 transformed *Synechococcus* colonies and recovered about 800 clones whose luminescence was bright enough to be easily monitored by the apparatus. Unexpectedly, the luminescence expression patterns of almost all of these 800 colonies manifested clear circadian rhythmicity. These rhythms exhibited a range of waveforms and amplitudes and they also showed a variety of phase relationships (Liu *et al.*, 1995b). In *Synechococcus* RF-1, Huang, Chen, Pen, and Chen (1994) conducted a study of protein synthetic patterns as a function of circadian time and found more than 10 polypeptides which exhibit circadian rhythms of translation. As in our study with *Synechococcus* PCC 7942 (Liu *et al.*, 1995b), Huang *et al.* (1994) observed that these synthetic patterns were expressed in a variety of phase relationships. The circadian clock in cyanobacteria does not merely control a specialized set of genes, but dominates the entire metabolism of the cell.

We have found that other genera of cyanobacteria express circadian rhythms, including *Synechocystis* (Aoki, Kondo, & Ishiura, 1995) and *Anabaena* (M. Ishiura & T. Kondo, unpublished observations). Our data and those of others clearly indicate that prokaryotic cyanobacteria are capable of circadian behavior, and thus that circadian mechanisms may have evolved earlier than has been previously assumed. It is difficult to know whether cyanobacteria had circadian clocks when they first emerged in the phylogenetic tree (Woese, Kandler, & Wheelis, 1990) or whether they acquired the clock mechanism later in the course of evolution. Analyzing the distribution of circadian clocks in other prokaryotic organisms may help us to address this question (Johnson *et al.*, 1996).

While the adaptive significance of a daily clock to cyanobacteria may *now* seem self-evident in retrospect, it might still be argued that a 24-hour clock is inconsistent with the rapid-growth lifestyle of many nonphotosynthetic prokaryotes. Is it possible that other prokaryotes have circadian clocks? This is a challenging question, with momentous implications for understanding the early evolution of circadian rhythmicity. The previous suggestion that *E. coli* might have a circadian clock (Halberg & Conner, 1961) was based on a old study (Rogers & Greenbank, 1930) in which *E. coli* cells were growing in rich nutrient medium in an apparatus in which temperature control was almost certainly poor. Therefore, the possible daily trends noted by Halberg and Conner (1961) are likely to be merely a result of a diurnal cycle of temperature, to which the growth rate of *E. coli* would be exquisitely sensitive. Nonetheless, we think that there is a good chance that circadian clocks will be found to be widespread, even among the prokaryotes (Johnson *et al.*, 1996). The keys are to find the proper conditions and to discover an appropriate parameter to measure. For example, *E. coli* cells may not express a circadian rhythm under optimal growth conditions such as those encountered in the rich intestinal environment. But *E. coli* cells live under very different conditions after excretion, in soil or water, barely surviving until they are again ingested. In these suboptimal conditions, a daily clock could again become adaptive.

Can a Single, Isolated Cell Express a Persisting Circadian Rhythm?

Since unicellular organisms express a variety of circadian rhythms, it might seem obvious that isolated single cells therefore can express circadian rhythms. But almost all of the studies on unicellular organisms have been conducted on *populations* of unicellular organisms. Thus, it is conceivable that the circadian rhythms could be an emergent property of intercell interactions within the population. Perhaps cells within the population secrete a factor into the medium which accumulates to a threshold concentration which triggers the secretion of a compensatory second factor, which degrades the first factor, and so on. This is not idle speculation; the intercellular signaling of *Dictyostelium* by secretion of cAMP is famous, and there is an example of a molecule made by an organism which affects the period of its circadian clock: the molecule is "gonyauline" from *Gonyaulax* (Roenneberg *et al.*, 1991). Moreover, there is evidence in *Gonyaulax* that suggests a weak coupling of the circadian rhythms among cells in a population (Broda, Brugge, Homma, & Hastings, 1985).

Although it is probably the case that these population rhythms reflect circadian oscillations occurring independently within each cell, it may be unsettling to realize how few unicellular organisms have been shown to express a persisting circadian rhythm from a single isolated cell for more than one cycle. To our knowledge, only two unicellular organisms meet this rigorous criterion: the alga *Acetabularia* and the protozoan *Paramecium bursaria.* Data from *Gonyaulax* have not yet met the criterion of expression of a persisting circadian rhythm from a single isolated cell for more than one cycle. Sweeney (1960) measured photosynthetic oxygen evolution from single *Gonyaulax* cells at various circadian phases, but no one cell was measured for more than 24 hours. Krasnow *et al.* (1981) monitored the bioluminescence of single *Gonyaulax* cells, but no cell was followed for more than 1 day. The 16-hour trace of spontaneous bioluminescence from a single isolated cell published therein corresponds well with the bioluminescence rhythms of populations, and perhaps we should not doubt that the population rhythm of *Gonyaulax* cells is a direct reflection of a large group of single cells behaving homogeneously. Nevertheless, in the absence of data showing more than one cycle, we should be cautious in concluding that rhythms expressed by populations of unicellular organisms are truly representative of endogenous rhythms which would persist indefinitely in single isolated cells.

That the clocks of at least some unicellular organisms can run without intercellular communication is shown by observations on *Acetabularia* and *Paramecium,* whose cells definitely oscillate in isolation. It was established long ago that single isolated cells of *Acetabularia* express circadian rhythms of photosynthetic oxygen evolution, even in cells without a nucleus (Karakashian & Schweiger, 1976; Sweeney & Haxo, 1961). More recently, other rhythms have been observed to persist for long freeruns in isolated cells of *Acetabularia* (Broda & Schweiger, 1981). Finally, even temperature compensation has been shown using single isolated cells of *Acetabularia*; free-running periods were measured at various temperatures in the same cell (Berger *et al.*, 1992).

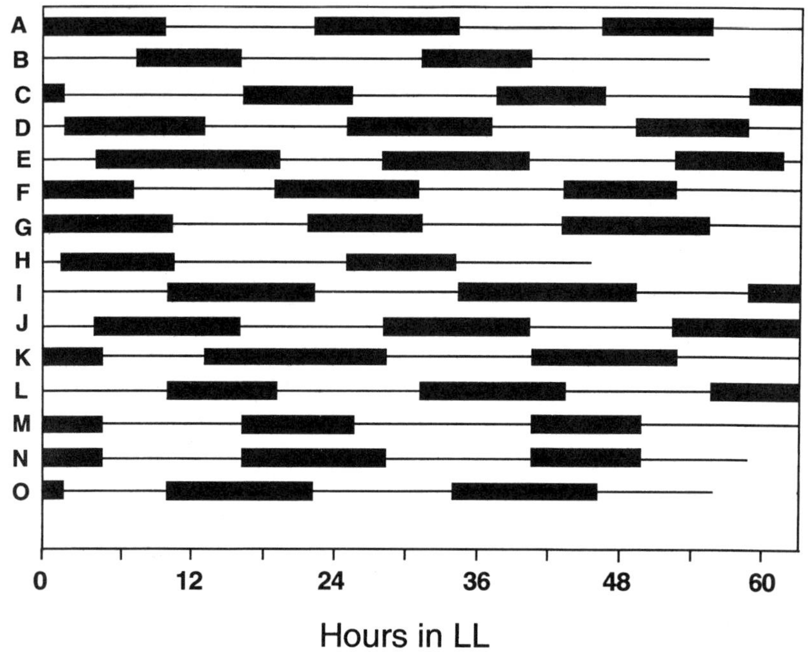

Figure 2. Mating reactivity rhythm of 15 single *Paramecium bursaria* cells isolated randomly from an arhythmic population that had been in LL for 4 weeks. Each single cell was maintained in LL, and its mating reactivity was tested every 3 hours by mixing with about 100 highly reactive cells of the opposite mating type. After a 5-minute testing period, the original cell was reisolated and returned to LL. Each horizontal line represents the data of a single cell; black rectangles mark mating-reactive phases, horizontal lines mark unreactive phases. The letters along the ordinate identify each single cell. Every single cell shows a clear circadian rhythm of mating reactivity, even though the pattern of mating reactivity of the entire population from which these cells were isolated was arhythmic. (Modified from Miwa *et al.*, 1987.)

While it is well known that single isolated cells of *Acetabularia* exhibit sustained circadian rhythmicity, it is not commonly appreciated that the same phenomenon has been demonstrated for *Paramecium bursaria.* By laborious measurements of the mating reactivity rhythms of single isolated *Paramecium* cells, Miwa, Nagatoshi, and Horrie (1987) showed that the rhythm continued for at least three cycles. The data for 15 separate isolated cells are shown in Figure 2. Each horizontal line represents the data of a distinct cell. The cells that were measured in Figure 2 were isolated from a population of cells which had been in LL for a long time, so that the mating reactivity of the population was arhythmic (50% mating reactivity at all phases). Because the population was arhythmic, but the individual cells were rhythmic (Figure 2), Miwa *et al.* (1987) could prove that the arhythmicity of the population was due to the desynchronization of individually rhythmic cells within the population and not due to the clocks "stopping" or otherwise behaving abnormally.

Role of the Nucleus and Other Compartments

The discovery that enucleated *Acetabularia* cells express circadian rhythms of photosynthesis (Karakashian & Schweiger, 1976; Sweeney & Haxo, 1961) indicated that cytoplasmic components in this alga were sufficient for expression of circadian behavior. This is an inappropriate conclusion to extrapolate to other types of cells, however, because *Acetabularia* cells have other bizarre characteristics, such as very long-lived mRNAs (Kloppstech & Schweiger, 1982). An early study of *Acetabularia* rhythms suggested that the role of the nucleus in this alga was to provide phase information to the cytoplasmic oscillator (Schweiger, Wallraff, & Schweiger, 1964). A recent reevaluation of the evidence, however, indicates that the methods used in 1964 to determine the phase were too gross (only two data points per day!) to allow an accurate assessment of the phase of the circadian rhythm of photosynthesis in *Acetabularia* cells, which can show considerable phase variation among individual cells (Woolum, 1991). Doubtless, the nucleus is crucial for circadian expression in most cells, but the conclusions reached in the 1964 study of *Acetabularia* can no longer be taken as conclusive evidence that the role of the nucleus is to provide phase information.

The facts that enucleated *Acetabularia* cells can express a rhythm of photosynthesis in the absence of information emanating from the nucleus, and that the chloroplasts are most directly responsible for the generation of that rhythm, encourage the speculation that the rhythm of enucleated cells of this alga might be an expression of circadian mechanisms residing within the chloroplasts. Support for the notion that chloroplasts have their own circadian clocks also comes from recent studies of *Gonyaulax* rhythms (Roenneberg & Morse, 1993). The speculation that an independent clockwork resides in the chloroplast does not seem so far-fetched now that cyanobacteria, whose ancestors are thought to be the ancestors of chloroplasts (Delwiche, Kuhsel, & Palmer, 1995; Margulis, 1993), have been shown to contain authentic circadian clocks (see above). Perhaps the clock mechanism of a cyanobacterium-like prokaryote was passed by endosymbiosis into a host which had not invented its own clock. If so, the clockwork may have been transferred from organelle to nucleus, just as many plastid genes translocated to the nuclear genome. Could the photosynthetic rhythm observed in enucleated *Acetabularia* cells (Sweeney & Haxo, 1961) be the rhythm of a population of oscillating chloroplasts liberated from nuclear interference, analogous to a rhythm expressed in a population of cyanobacterial cells? This is an intriguing hypothesis. However, direct evidence is needed before we can conclude that organelles have independent circadian mechanisms.

Is There Only One Clockwork in a Single Cell?

Studies prior to 1992 in which two or more overt rhythms were monitored simultaneously from the same population of unicells found that the periods and phase angles of various rhythms appeared to be rigidly constant. The best-documented example was that of the unicell *Gonyaulax*, where the phase relationships among several different rhythms did not waver during long freeruns or after phase resetting (McMurry & Hastings, 1972; Sweeney, 1987). This and similar observations led circadian biologists to believe that each single cell houses only one circadian oscillator mechanism. A few studies provided intriguing, but not convincing, evidence for multioscillators in single cells (Edmunds, 1988; Hoffmans-Hohn, Martin, & Brinkmann, 1984; Johnson & Hastings, 1989). Recent experiments in *Gonyaulax*, however, have demonstrated internal desynchronization of various rhythms under appropriate conditions of illumination and temperature (Roenneberg & Morse, 1993; von der Heyde, Wilkens, & Rensing, 1992). In the decoupled state, the two oscillators exhibit phase responses to light/dark signals that are partially independent (Morse, Hastings, & Roenneberg, 1994). Although alternative explanations may exist, it now seems likely that even single cells can have multiple circadian pacemakers, which are usually coupled, but can decouple if the conditions are right. As mentioned above, unicells are not necessarily simple; for example, *Gonyaulax* has a complicated intracellular organization composed of multiple organelles, such as chloroplasts, vacuoles, mitochondria, etc. Perhaps some of these intracellular compartments have distinct timekeeping mechanisms, as suggested in the preceding section.

What about other evidence for multioscillators in single cells? In addition to internal desynchronization, other observations in multicellular organisms that have been taken to support the hypothesis of multioscillator structure are "transients" and period "aftereffects" after perturbations of the circadian clock (Pittendrigh, 1960). *Transients* lasting more than one cycle have not been often observed in unicellular organisms; in fact, to our knowledge, the only report of reproducible transients of more than one cycle is the study of Dunlap, Taylor, Hastings (1980), which describes transients of the *Gonyaulax* bioluminescence rhythm after pulses of the protein synthesis inhibitor cycloheximide. This might be an exceptional case because the stimulus was a drug that might not have washed out immediately after the pulse was terminated. *Aftereffects* on the free-running period after a perturbation (e.g., a phase-resetting stimulus) have also been evoked in multicellular organisms as evidence for multioscillator organization (Pittendrigh & Daan, 1976). Several studies have reported aftereffects on the free-running period by light pulses that induced phase shifts in the following unicells: *Gonyaulax* (Johnson & Hastings, 1989; Roenneberg & Hastings, 1991), *Paramecium* (Johnson, Miwa, Kondo, & Hastings, 1989), and *Chlamydomonas* (Johnson & Kondo, 1992). Other than the analogy to the phenomenon of aftereffects in multicellular organisms, however, there is no definitive evidence that aftereffects in unicellular organisms are the result of an underlying multioscillator structure; indeed, other types of explanations have been offered (Roenneberg & Hastings, 1991; Johnson & Kondo, 1992).

Circadian versus Cell Division Cycling

Ever since the first report of circadian gating of cell division (Sweeney & Hastings, 1958), the number of examples of circadian control over cell division in unicellular eukaryotes has grown steadily. What is the relationship between cell division and circadian expression in organisms? The "circadian–infradian rule" states that in eukaryotic cells dividing more rapidly than once a day, cellular processes become uncoupled from circadian oscillators; circadian rhythms are expressed only in cells that are dividing once every day or more slowly, i.e., in the "infradian" mode (Ehret & Wille, 1970). While the circadian–infradian rule has successfully predicted the behavior of many eukaryotic unicells and may also apply to some cyanobacteria (Mitsui *et al.*, 1986; Sweeney & Borgese, 1989), it does not hold for the prokaryotic cyanobacterium that we study, *Synechococcus* sp. PCC 7942. We have found that a circadian rhythm of bioluminescence is expressed by cyanobacteria that are dividing in bright LL with doubling times (= DT) as short as 8 hours (Mori, Binder, & Johnson, 1996; Kondo *et al.*, 1997). As shown in Figure 3, this circadian rhythm of bioluminescence continues as cyanobacterial cultures slow from exponential growth (DT = 12.7 hours) to stationary phase (DT = 53 hours). We have

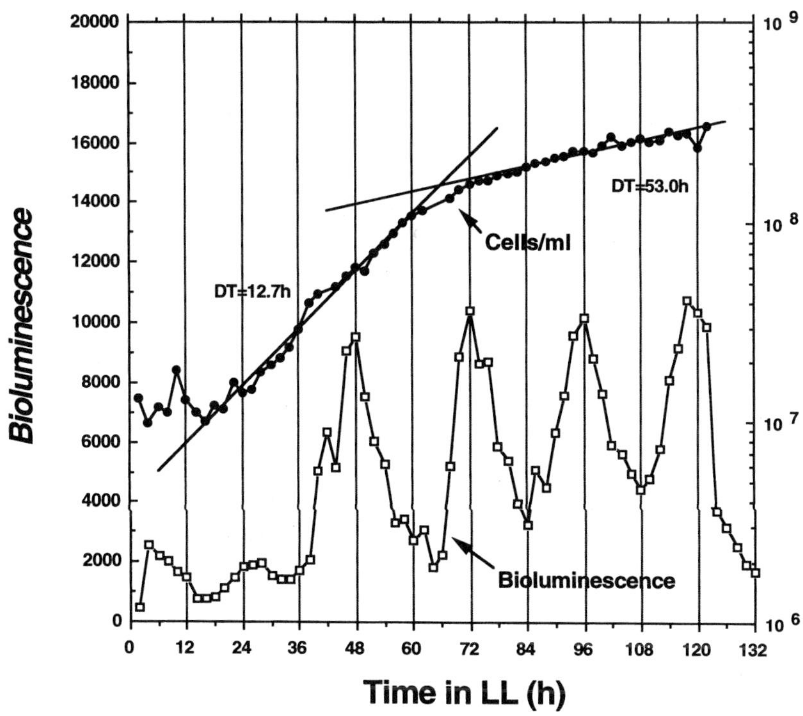

Figure 3. Concurrent measurment of cell division and bioluminescence in the same culture of cyanobacteria that have been genetically engineered to express a circadian rhythm of bioluminescence (= *Synechococcus* PCC 7942, strain AMC149). Cells are growing in LL at a light intensity of 103 microeinsteins/m^2 sec. From hour 16 to hour 60, cells are growing with a doubling time (DT) of about 12.7 hours; from hour 70 to hour 123, the cells are growing with a doubling time of about 53 hours. The cells continue to express a bioluminescence rhythm with a circadian period independent of the dramatic change in division rate that occurs between hour 60 and hour 72. (Data of Tetsuya Mori.)

also found that the rate of cell division in *Synechococcus* has periodic "slowdowns" that recur with a circadian period, even when the average doubling time is less than 24 hours (Mori *et al.*, 1996). Therefore, in the prokaryote *Synechococcus*, a circadian clock regulates cell division and gene expression even when cells are dividing more rapidly than once a day.

Our observations of rapidly dividing *Synechococcus* contradict the usual formulation of the circadian–infradian rule. What does this mean? Apparently the circadian–infradian rule is not as accurate a description of the relationship between the circadian oscillator and the cell division cycle as we once thought. Actually, few have appreciated that there has been evidence for over a decade that the circadian–infradian rule might not always apply to *eukaryotic* unicells either. In particular, marine diatoms have been observed to exhibit daily rhythms when dividing more rapidly than once a day (Brand, 1982), and also to entrain to LD cycles with variable phase relationships that are difficult to reconcile with the simplest version of the circadian–infradian rule (Chisholm, Morel, & Slocum, 1980). These diatoms deserve more intensive study in the laboratory to assess the generality of the circadian–infradian rule. Moreover, in the multiple fission alga *Chlamydomonas*, several cycles of nuclear replication occur during a brief interval (<6 hours) in the subjective night. Therefore, even though the cell division cycle is gated by a circadian clock in these cells (Goto & Johnson, 1995), the data from *Chlamydomonas* show that direct meshing of the circadian cycle and the cycle of one round of nuclear replication is not obligatory.

It is important to realize that the circadian–infradian rule has been based primarily on data from *Euglena*, where the circadian rhythm that was measured was the timing of cell division. Consequently, the assays of circadian cycling versus cell division cycling were not independent in the studies of *Euglena* cell division rhythms. On the other hand, they may be independent in the cyanobacteria of Figure 3, where the circadian rhythm that was measured was of the activity of a promoter for a photosynthetic gene (see also Kondo *et al.*, 1997). Therefore, it is possible that cell division has merely become uncoupled from an ongoing circadian oscillation in rapidly dividing *Euglena* (Edmunds, 1988). An alternative interpretation is based on the recent evidence that multiple circadian oscillators may coexist within a single unicell (Roenneberg & Morse, 1993); it is possible to imagine that under very rapid growth conditions, the circadian clock controlling cell division may be disrupted or inactivated without major effect on other circadian oscillators.

To some, selection for circadian organization has seemed reasonable only if the cell's generation time is as long as or longer than a day (Pittendrigh, 1993). To us, however, it seems equally plausible that the advantages of anticipating, e.g., the onset of solar illumination with its concurrent benefits and deficits would accrue to all light-sensitive organisms, even if they divide more rapidly than the daily cycle. While this alternative proposes the unfamiliar notion that cells have a circadian oscillator precessing with a stately 24-hour cycle even as they frantically divide and subdivide at the behest of a higher frequency cell division timer, the data of Figure 3 persuade us that cyanobacteria accomplish the job (also see Kondo *et al.*, 1997; Mori *et al.*, 1996).

There are at least two ways to think about this enigma. The first is to recognize that unicells that divide asexually are not "organisms" in the same way that sexually reproducing multicellular organisms are. Asexually reproducing unicells can be thought of as a single protoplasm that grows larger and larger and incidentally subdivides. If the fitness of that protoplasm changes as a function of daily alterations in light intensity, temperature, etc., then its ability to adapt optimally to these daily changes in its environment must be useful, even though that protoplasm may have subdivided into different compartments within the day's span.

The second perspective on this puzzle is to analogize (1) the adaptation to the daily cycle of a unicell with a doubling time of 8 hours with (2) the adaptation to the annual cycle of an insect whose generation time is 4 months. For such an insect, only one generation in three is exposed to winter. Nevertheless, if the winter is harsh, it is adaptive for the insect to develop a photoperiodic response to anticipate these annual changes. The insects in summertime do not need the system, but without that photoperiodic response, the insect may not survive in wintertime. This case indicates that an adaptive response does not have to act in every generation in order to be useful. In an analogous fashion, if the fitness of a unicell that divides every 8 hours is a function of the time of day, then it may be adaptive for that unicell to elaborate a circadian program to anticipate the daily transformation of its environment. A specific example is that of photosynthetic cyanobacteria, whose energy metabolism depends entirely upon light. The average growth rate of cyanobacteria may be optimized by the ability to prepare for lights-on and lights-off, even though different generations experience those light/dark transitions.

The major unanswered question about circadian clocks is: What is their molecular mechanism? This question has titillated us ever since the discovery of circadian rhythmicity, yet the answer eludes us. Experimental approaches to this question have ranged from genetic and/or pharmacologic perturbations of the rhythms to tracking input or output pathways back to the mechanism itself (Takahashi *et al.*, 1993). The "input tracking" approach has often involved the identification of photoreceptors and their phototransduction pathways, as reviewed elsewhere (Johnson, 1995). The "output tracking" approach has involved a wide variety of methods (Edmunds, 1988; Takahashi *et al.*, 1993).

Attempts to identify the components of circadian "clockworks" have led to many models, including, among others, (1) "membrane models" (implicating fluxes of ions such as calcium, potassium, protons, and/or changes of membrane potential), (2) cycles of transcription and/or translation, (3) mechanisms implicating cyclic nucleotides and/or protein phosphorylation, (4) the circadian clock as a "counter" of a higher frequency ultradian oscillation, (5) the involvement of stress responses in phase shifting, and (6) an intertwining of the oscillating mechanisms responsible for timing cell division and circadian behaviors. The major liability of all these models is that they have not led to concrete experiments that could conclusively test their validity.

It is superfluous to reiterate the description of these models here, since they are well reviewed by Edmunds (1988) and Rensing and Hardeland (1990). Nevertheless, it may be useful to mention a few relevant investigations that have been reported since Rensing and Hardeland's (1990) review. The following citations are of tests of potential involvement of candidate molecular systems in unicellular organisms; molecular models of the clock derived from studies on multicellular systems (especially *Neurospora*, *Drosophila*, and molluscan eyes) are discussed in other chapters of this volume. A recent test of the involvement of membrane potential in the circadian clockwork of *Paramecium* has been performed by altering extracellular ionic concentrations (Johnson, Nakaoka, & Miwa, 1994). Several reports have tested the role of second messengers (especially cyclic nucleotides and calcium) in circadian expression of *Euglena* (Carre & Edmunds, 1993; Lonergan, 1990; Tamponnet & Edmunds, 1990). Finally, studies with a protein kinase inhibitor that blocks light-induced phase shifting and lengthens the period implicate phosphorylation of proteins as a key event(s) in the clock system of *Gonyaulax* (Comolli, Taylor, & Hastings, 1994). Clearly, unicellular organisms continue to provide fruitful information about circadian mechanisms.

How Did Circadian Clocks Evolve?

No one knows how circadian clocks evolved. Several models have been proposed (Edmunds, 1988). An attractive model is the so-called "escape from light" model, which stresses the fact that many cellular components absorb light and are in some cases adversely affected by that irradiation (Pittendrigh, 1993). Direct or indirect consequences of light absorption might thereby have provided a selective pressure which favored the invention of circadian rhythms. One example would be the ability to diminish and/or repair the damage to DNA caused by ultraviolet light during the daytime. Another lovely example of such pressure is the aforementioned case of some nitrogen-fixing cyanobacteria whose daytime photosynthesis inevitably produces oxygen, which inhibits nitrogenase (Mitsui *et al.*, 1986). In addition, ultraviolet light has a direct deleterious effect on the nitrogenase enzyme (Kumar, Sinha, & Häder, 1996). Therefore, their nitrogen-fixing tasks "moonlight" to escape the indirect and direct consequences of light absorption.

These considerations suggest that the first step toward circadian organization was the initial selection for processes that could directly respond to light/dark/temperature transitions, namely, light (or dark-, or temperature-) regulated genes (Johnson, 1994). The existence of light-regulated genes implies the elaboration of photoreceptors and phototransduction pathways that cascade upon the responsive genes and/or their products. From these beginnings, the elaboration of a circadian oscillator from the transduction pathway allowed (1) anticipation of the environmental transitions and (2) the selection of any phase within the daily cycle for a process to be turned on or off. These advantages of a circadian program must have been adaptive, and therefore favored by selection. A recent study has proven in competition experiments that circadian programs are adaptive in cyanobacteria, such that the adaptiveness is greatest when the period of the biological process "resonates" optimally with the period of the light/dark cycle (Ouyang, Andersson, Kondo, Golden, & Johnson, 1998).

Studies on unicellular organisms have blazed the trail of our understanding of circadian systems. Circadian control permeates all levels of eukaryotic organization, and recent investigations have shown that these timekeepers are found in at least some prokaryotes. Unicellular organisms have been, and will undoubtedly remain, important model systems for revealing the secrets of circadian programs at all levels of biological complexity.

Acknowledgments

We thank Dr. Colin Pittendrigh for his comments on an early version of this paper, and Isoji Miwa and Tetsuya Mori for permission to show their data in Figures 2 and 3. Our research has been supported by grants from the NIMH (U.S.) (MH 43836 and MH01179), the NSF (U.S.) (MCB-9219880 and MCB-9633267), and the Ministry of Education, Science and Culture of Japan (05670079, 07670097, 07554045, and 08404053). Also, the NSF and JSPS (Japan) provided a joint USA–Japan Cooperative Program grant (NSF INT-9218744 and JSPS BSAR382).

References

Aoki, S., Kondo, T., & Ishiura, M. (1995). Circadian expression of the *dnaK* gene in the cyanobacterium *Synechocystis* sp. strain PCC 6803. *Journal of Bacteriology, 177,* 5605–5611.

Berger, S., Dirk, J., von Lindern, L., Wolff, D., & Mergenhagen, D. (1992). Temperature dependency of circadian period in a single cell (*Acetabularia*). *Botanica Acta, 105,* 382–386.

Brand, L. E. (1982). Persistent diel rhythms in the chlorophyll fluorescence of marine phytoplankton species. *Marine Biology, 69,* 253–262.

Brinkmann, K. (1971). Metabolic control of temperature compensation in the circadian rhythm of *Euglena gracilis.* In M. Menaker (Ed.), *Biochronometry* (pp. 567–593). Washington, DC: National Academy of Sciences.

Broda, H., & Schweiger, H. G. (1981). Long-term measurement of endogenous diurnal oscillations of the electrical potential in an individual *Acetabularia* cell. *European Journal of Cell Biology, 26,* 1–4.

Broda, H., Brugge, D., Homma, K., & Hastings, J. W. (1985). Circadian communication between unicells? *Cell Biophysics, 8,* 47–67.

Bruce, V. G., & Pittendrigh, C. S. (1956). Temperature independence in a unicellular "clock." *Proceedings of the National Academy of Sciences of the USA, 42,* 676–682.

Carre, I. A., & Edmunds, L. N. (1993). Oscillator control of cell division in *Euglena*: Cyclic AMP oscillations mediate the phasing of the cell division cycle by the circadian clock. *Journal of Cell Science, 104,* 1163–1173.

Chen, T.-H., Chen, T.-L., Hung, L.-M., & Huang, T.-C. (1991). Circadian rhythm in amino acid uptake by *Synechococcus* RF-1. *Plant Physiology, 97,* 55–59.

Chisholm, S. W., Morel, F. M. M., & Slocum, W. S. (1980). The phasing and distribution of cell division cycles in marine diatoms. In P. Falkowski (Ed.), *Primary productivity in the sea* (pp. 281–300). New York: Plenum Press.

Clarke, A. K., Hurry, V. M., Gustafsson, P., & Oquist, G. (1993). Two functionally distinct forms of the photosystem II reaction-center protein D1 in the cyanobacterium *Synechococcus* sp. PCC 7942. *Proceedings of the National Academy of Sciences of the USA, 90,* 11985–11989.

Comolli, J., Taylor, W., & Hastings, J. W. (1994). An inhibitor of protein phosphorylation stops the circadian oscillator and blocks light phase shifting in *Gonyaulax polyedra*. *Journal of Biological Rhythms, 9,* 13–26.

Delwiche, C. F., Kuhsel, M., & Palmer, J. D. (1995). Phylogenetic analysis of *tufA* sequences indicates a cyanobacterial origin of all plastids. *Molecular Phylogenetics and Evolution, 4,* 110–128.

Dunlap, J. C., Taylor, W., & Hastings, J. W. (1980). The effects of protein synthesis inhibitors on the *Gonyaulax* clock. *Journal of Comparative Physiology, 138,* 1–8.

Edmunds, L. N. (1988). *Cellular and molecular bases of biological clocks.* New York: Springer-Verlag.

Edmunds, L. N., Apter, R. I., Rosenthal, P. J., Shen, W.-K., & Woodward, J. R. (1979). Light effects in yeast: Persisting oscillations in cell division activity and amino acid transport in cultures of *Saccharomyces cerevisiae* entrained by light–dark cycles. *Photochemistry and Photobiology, 30,* 595–601.

Ehret, C. F. (1959). Photobiology and biochemistry of circadian rhythms in non-photosynthesizing cells. *Federation Proceedings, 18,* 1232–1240.

Ehret, C. F. (1960). Action spectra and nucleic acid metabolism in circadian rhythms at the cellular level. *Cold Spring Harbor Symposia on Quantitative Biology, 25,* 149–158.

Ehret, C. F., & Trucco, E. (1967). Molecular models for the circadian clock. I. The chrono concept. *Journal of Theoretical Biology, 15,* 240–262.

Ehret, C. F., & Wille, J. J. (1970). The photobiology of circadian rhythms in protozoa and other eukaryotic microorganisms. In P. Halldal (Ed.), *Photobiology of microorganisms* (pp. 369–416). New York: Wiley.

Fay, P. (1992). Oxygen relations of nitrogen fixation in cyanobacteria. *Microbiological Reviews, 56,* 340–373.

Goto, K., & Johnson, C. H. (1995). Is the cell division cycle gated by a circadian clock? The case of *Chlamydomonas reinhardtii*. *Journal of Cell Biology, 129,* 1061–1069.

Grobbelaar, N., Huang, T.-C., Lin, H. Y., & Chow, T. J. (1986). Dinitrogen-fixing endogenous rhythm *Synechococcus* RF-1. *FEMS Microbiology Letters, 37,* 173–177.

Halberg, F., & Conner, R. L. (1961). Circadian organization and microbiology: Variance spectra and a periodogram on behavior of *Escherichia coli* growing in fluid culture. *Proceedings of the Minnesota Academy of Science, 29,* 227–239.

Hastings, J. W., & Sweeney, B. M. (1957). On the mechanism of temperature independence in a biological clock. *Proceedings of the National Academy of Sciences of the USA, 43,* 804–811.

Hoffmans-Hohn, M., Martin, W., & Brinkmann, K. (1984). Multiple periodicities in the circadian system of unicellular algae. *Zeitschrift für Naturforschung, 39c,* 791–800.

Huang, T.-C., & Chow, T.-J. (1990). Characterization of the rhythmic nitrogen-fixing activity of *Synechococcus* RF-1 at the transcription level. *Current Microbiology, 20,* 23–26.

Huang, T.-C., Chen, H.-M., Pen, S.-Y., & Chen, T.-H. (1994). Biological clock in the prokaryote *Synechococcus* RF-1. *Planta, 193,* 131–136.

Ishiura, M., Kutsuna, S., Aoki, S., Iwasaki, H., Andersson, C. R., Tanabe, A., Golden, S. S., Johnson, C. H., & Kondo, T. (1998). Expression of a gene cluster kaiABC as a circadian feedback process in cyanobacteria. *Science, 281,* 1519–1523.

Johnson, C. H. (1994). Illuminating the clock: Circadian photobiology. *Seminars in Cell Biology, 5,* 355–362.

Johnson, C. H. (1995). Photobiology of circadian rhythms. In W. M. Horspool & P. S. Song (Eds.), *CRC handbook of organic photochemistry and photobiology* (pp. 1602–1610). Boca Raton, FL: CRC Press.

Johnson, C. H., & Hastings, J. W. (1989). Circadian phototransduction: Phase-resetting and frequency of the circadian clock of *Gonyaulax* cells in red light. *Journal of Biological Rhythms, 4,* 417–437.

Johnson, C. H., & Kondo, T. (1992). Light pulses induce "singular" behavior and shorten the period of the circadian phototaxis rhythm in the CW15 strain of *Chlamydomonas*. *Journal of Biological Rhythms, 7,* 313–327.

Johnson, C. H., Miwa, I., Kondo, T., & Hastings, J. W. (1989). Circadian rhythm of photaccumulation in *Paramecium bursaria*. *Journal of Biological Rhythms, 4,* 405–415.

Johnson, C. H., Nakaoka, Y., & Miwa, I. (1994). The effects of altering extracellular potassium ion concentration on the membrane potential and circadian clock of *Paramecium bursaria. Journal of Experimental Biology, 197,* 295–308.

Johnson, C. H., Golden, S. S., Ishiura, M., & Kondo, T. (1996). Circadian clocks in prokaryotes. *Molecular Microbiology, 21,* 5–11.

Karakashian, M. W., & Schweiger, H. G. (1976). Circadian properties of the rhythmic system in individual nucleated and enucleated cells of *Acetabularia mediterranea. Experimental Cell Research, 97,* 366–377.

Kippert, F. (1986). Endocytobiotic coordination, intracellular calcium signalling, and the origin of endogenous rhythms. *Annals of the New York Academy of Sciences, 503,* 476–495.

Kippert, F. (1989). Circadian control of heat tolerance in stationary phase cultures of *Schizosaccharomyces pombe. Archives of Microbiology, 151,* 177–179.

Kippert, F. (1991). Essential clock proteins/circadian rhythms in prokaryotes—What is the evidence? *Botanica Acta, 104,* 2–4.

Kloppstech, K., & Schweiger, H. G. (1982). Stability of poly(A)$^+$ RNA in nucleate and anucleate cells of *Acetabularia. Plant Cell Reports, 1,* 165–167.

Kondo, T., Strayer, C. A., Kulkarni, R. D., Taylor, W., Ishiura, M., Golden, S. S., & Johnson, C. H. (1993). Circadian rhythms in prokaryotes: Luciferase as a reporter of circadian gene expression in cyanobacteria. *Proceedings of the National Academy of Sciences of the USA, 90,* 5672–5676.

Kondo, T., Tsinoremas, N. F., Golden, S. S., Johnson, C. H., Kutsuna, S., & Ishiura, M. (1994). Circadian clock mutants of cyanobacteria. *Science, 266,* 1233–1236.

Kondo, T., Mori, T., Lebedeva, N. V., Aoki, S., Ishiura, M., & Golden, S. S. (1997). Circadian rhythms in rapidly dividing cyanobacteria. *Science, 275,* 224–227.

Krasnow, R., Dunlap, J., Taylor, W., Hastings, J. W., Vetterling, W., & Haas, E. (1981). Measurements of *Gonyaulax* bioluminescence, including that of single cells. In K. H. Nealson (Ed.), *Bioluminescence: Current perspectives* (pp. 52–63). Minneapolis, MN: Burgess.

Kumar, A., Sinha, R. P., & Häder, D.-P. (1996). Effect of UV-B on enzymes of nitrogen metabolism in the cyanobacterium *Nostoc calcicola. Journal of Plant Physiology, 148,* 86–91.

Kung, C., & Saimi, Y. (1982). The physiological basis of taxes in *Paramecium. Annual Review of Physiology, 44,* 519–534.

Liu, Y., Golden, S. S., Kondo, T., Ishiura, M., & Johnson, C. H. (1995a). Bacterial luciferase as a reporter of circadian gene expression in cyanobacteria. *Journal of Bacteriology, 177,* 2080–2086.

Liu, F., Tsinoremas, N. F., Johnson, C. H., Lebedeva, N. V., Golden, S. S., Ishiura, M., & Kondo, T. (1995b). Circadian orchestration of gene expression in cyanobacteria. *Genes and Development, 9,* 1469–1478.

Lonergan, T. A. (1990). Steps linking the photosynthetic light reactions to the biological clock require calcium. *Plant Physiology, 93,* 110–115.

Margulis, L. (1993). *Symbiosis in cell evolution* (2nd ed.). New York: Freeman.

Mattoo, A. K., Hoffman-Falk, H., Marder, J. B., & Edelman, M. (1984). Regulation of protein metabolism: Coupling of photosynthetic electron transport to *in vivo* degradation of the rapidly metabolized 32-kilodalton protein of the chloroplast membranes. *Proceedings of the National Academy of Sciences of the USA, 81,* 1380–1384.

McMurry, L., & Hastings, J. W. (1972). No desynchronization among four different circadian rhythms in the unicellular alga, *Gonyaulax polyedra. Science, 175,* 1137–1139.

Mitsui, A., Kumazawa, S., Takahashi, A., Ikemoto, H., & Arai, T. (1986). Strategy by which nitrogen-fixing unicellular cyanobacteria grow photoautotrophically. *Nature, 323,* 720–722.

Miwa, I., Nagatoshi, H., & Horie, T. (1987). Circadian rhythmicity within single cells of *Paramecium bursaria. Journal of Biological Rhythms, 2,* 57–64.

Mori, T., Binder, B., & Johnson, C. H. (1996). Circadian gating of cell division in cyanobacteria growing with average doubling times of less than 24 hours. *Proceeding of the National Academy of Sciences of the USA, 93,* 10183–10188.

Morse, D., Hastings, J. W., & Roenneberg, T. (1994). Different phase responses of the two circadian oscillators of *Gonyaulax. Journal of Biological Rhythms, 9,* 263–274.

Ouyang, Y., Andersson, C. R., Kondo, T., Golden, S. S., & Johnson, C. H. (1998). Resonating circadian clocks enhance fitness in cyanobacteria. *Proceedings of the National Academy of Sciences, 95,* 8660–8664.

Pohl, R. (1948). Tagesrhythmus im phototaktischen Verhalten der *Euglena gracilis. Zeitschrift für Naturforschung, 3B,* 367–374.

Pittendrigh, C. S. (1960). Circadian rhythms and the circadian organization of living systems. *Cold Spring Harbor Symposia on Quantitative Biology, 25,* 159–184.

Pittendrigh, C. S. (1993). Temporal organization: Reflections of a Darwinian clock-watcher. *Annual Review of Physiology, 55,* 17–54.

Pittendrigh, C. S., & Daan, S. (1976). A functional analysis of circadian pacemakers in nocturnal rodents. *Journal of Comparative Physiology, 106,* 223–252, 333–355.

Rensing, L., & Hardeland, R. (1990). The cellular mechanism of circadian rhythms—A view on evidence, hypotheses and problems. *Chronobiology International, 7,* 353–370.

Roenneberg, T., & Carpenter, E. J. (1993). Daily rhythm of O_2-evolution in the cyanobacterium *Trichodesmium thiebautii* under natural and constant conditions. *Marine Biology, 117,* 693–697.

Roenneberg, T., & Hastings, J. W. (1991). Are the effects of light on phase and period of the *Gonyaulax* clock mediated by different pathways? *Photochemistry and Photobiology, 53,* 525–533.

Roenneberg, T., & Morse, D. (1993). Two circadian oscillators in one cell. *Nature, 362,* 362–364.

Roenneberg, T., Nakamura, H., Cranmer, L. D., III, Ryan, K., Kishi, Y., & Hastings, J. W. (1991). Gonyauline: A novel endogenous substance shortening the period of the circadian clock of a unicellular alga. *Experientia, 47,* 103–106.

Rogers, L. A., & Greenbank, G. R. (1930). The intermittent growth of bacterial cultures. *Journal of Bacteriology, 19,* 181–190.

Schneegurt, M. A., Sherman, D. M., Nayar, S., and Sherman, L. A. (1994). Oscillating behavior of carbohydrate granule formation and dinitrogen fixation in the cyanobacterium *Cyanothece* sp. strain ATCC 51142. *Journal of Bacteriology, 176,* 1586–1597.

Schweiger, H.-G., & Schweiger, M. (1977). Circadian rhythms in unicellular organisms: An endeavor to explain the molecular mechanism. *International Review of Cytology, 51,* 315–342.

Schweiger, H.-G., & Schweiger, M. (1980). Molecular mechanisms of cellular circadian clocks. *European Journal of Cell Biology, 21,* 335–336.

Schweiger, E., Wallraff, H.-G., & Schweiger, H.-G. (1964). Endogenous circadian rhythm in cytoplasm of *Acetabularia*: Influence of the nucleus. *Science, 146,* 658–659.

Stahl, L. J., and Krumbein, W. E. (1985). Nitrogenase activity in the non-heterocystous cyanobacterium *Oscillatoria* sp. grown under alternating light–dark cycles. *Archives of Microbiology, 143,* 67–71.

Sturtevant, R. P. (1973). Circadian variability in *Klebsiella* demonstrated by cosinor analysis. *International Journal of Chronobiology, 1,* 141–146.

Sweeney, B. M. (1960). The photosynthetic rhythm in single cells of *Gonyaulax polyedra. Cold Spring Harbor Symposia on Quantitative Biology, 25,* 145–148.

Sweeney, B. M. (1987). *Rhythmic phenomena in plants* (2nd ed.), San Diego, CA: Academic Press.

Sweeney, B. M., & Borgese, M. B. (1989). A circadian rhythm in cell division in a prokaryote, the cyanobacterium *Synechococcus* WH7803. *Journal of Phycology, 25,* 183–186.

Sweeney, B. M., & Hastings, J. W. (1957). Characteristics of the diurnal rhythm of luminescence in *Gonyaulax polyedra. Journal of Cellular and Comparative Physiology, 49,* 115–128.

Sweeney, B. M., & Hastings, J. W. (1958). Rhythmic cell division in populations of *Gonyaulax polyhedra. Journal of Protozoology, 5,* 217–224.

Sweeney, B. M., & Hastings, J. W. (1960). Effects of temperature on diurnal rhythms. *Cold Spring Harbor Symposia on Quantitative Biology, 25,* 87–104.

Sweeney, B. M., & Haxo, F. T. (1961). Persistence of a photosynthetic rhythm in enucleated *Acetabularia. Science, 134,* 1361–1363.

Takahashi, J. S., Kornhauser, J. M., Koumenis, C., and Eskin, A. (1993). Molecular approaches to understanding circadian oscillators. *Annual Review of Physiology, 55,* 729–753.

Tamponnet, C., & Edmunds, L. N. (1990). Entrainment and phase-shifting of the circadian rhythm of cell division by calcium in synchronous cultures of the wild-type Z strain and of the ZC achlorophyllous mutant of *Euglena gracilis. Plant Physiology, 93,* 425–431.

Taylor, W. R. (1979). Studies on the bioluminescent glow rhythm of *Gonyaulax polyedra.* Ph.D. dissertation, University of Michigan, Ann Arbor, Michigan, Chapter 5, pp. 78–110.

von der Heyde, F., Wilkens, A., & Rensing, L. (1992). The effects of temperature on the circadian rhythms of flashing and glow in *Gonyaulax polyedra*: Are the two rhythms controlled by two oscillators? *Journal of Biological Rhythms, 7,* 115–123.

Woese, C. R., Kandler, O., & Wheelis, M. L. (1990). Towards a natural system of organisms: Proposal for the domains Archaea, Bacteria, and Eucarya. *Proceedings of the National Academy of Sciences of the USA, 87,* 4576–4579.

Woolum, J. C. (1991). A re-examination of the role of the nucleus in generating the circadian rhythm in *Acetabularia. Journal of Biological Rhythms, 6,* 129–136.

4

Circadian Systems of Invertebrates

Terry L. Page

The study of the anatomic and physiologic organization of circadian systems of invertebrates has a long and productive history. Modern research can be traced to the work of Janet Harker in the 1950s, who initiated efforts to localize pacemakers and photoreceptors of the circadian system of the cockroach via lesion and transplantation studies. Ultimately, Harker was not successful, but the questions and approaches she pioneered set the stage for subsequent efforts to identify components of invertebrate circadian systems. These efforts have been directed toward answering several fundamental questions about the organization of the circadian system which are the focus of this review. The first of these concerns the nature of the pacemaking system that generates the timing signal. What are the anatomic loci of component oscillators, what is the significance of multioscillator organization, and how does the pacemaking system emerge in development? Second, what are the pathways and mechanisms by which inputs to the pacemaking system regulate its phase and period? Finally, what are the neural and endocrine signals by which the pacemaking system regulates the various processes under its control?

Localization of Circadian Oscillators

Studies on pacemaker localization in invertebrates (primarily arthropods and mollusks) have focused on the nervous system. In the arthropods, the brain has consistently been found to be important for generation of circadian timing signals, though it is clear, at least in insects, that circadian oscillations arise in structures outside the nervous system as well. In the opisthobranch mollusks, circadian oscilla-

---

Terry L. Page Department of Biology, Vanderbilt University, Nashville, Tennessee 37235.

Circadian Clocks, Volume 12 of *Handbook of Behavioral Neurobiology*, edited by Joseph S. Takahashi, Fred W. Turek, and R. Y. Moore, Kluwer Academic / Plenum Publishers, New York, 2001.

tions are produced by cells of the retina, while limited data indicate that structures outside the eyes are critical in pulmonate gastropods.

Pacemakers in the Molluskan Retina

The eyes of several marine opisthobranch mollusks have been shown to exhibit circadian rhythms of spontaneous nerve impulse activity when removed from the animal and maintained in either artificial-sea seawater or in culture medium (Figure 1). The rhythm in the isolated eye was first discovered by Jacklet (1969) while investigating the visual function in the eye of *Aplysia californica.* Subsequently, similar rhythms were documented in other species of *Aplysia* (Jacklet, 1971, 1974) and in *Navanax inermis* (Eskin & Harcombe, 1977), *Bulla gouldiana* (Block & Wallace, 1982), and *Bursatella leachi plei* (Block & Roberts, 1981). In each of these species, the rhythms are expressed in the frequency of large-amplitude, compound action potentials in the optic nerve that arise from a population of electrically coupled cells in the eye. The impulses are spontaneously generated and occur even when the eye is maintained in complete darkness. In all species, peak impulse activity occurs during the animal's subjective day.

Axons in the optic nerves of these mollusks arise from both classical microvillus photoreceptors and a population of cells originally referred to as "secondary" neurons. Efforts to further localize the pacemaking cells in the eye through localized lesions have been made in *Bulla* (Block & Wallace, 1982) and *Aplysia* (Strumwasser, 1973; Woolum & Strumwasser, 1980) and in both species, the results indicate that the rhythm is generated by cells at the base of the eye. In *Bulla,* these cells, called basal retinal neurons (BRNs), form a distinct cluster (Jacklet & Colquhoun, 1983) that can be surgically isolated from the distal, microvillus photoreceptors that surround the lens (Block & Wallace, 1982). Surgical reduction of the eye did not abolish the circadian rhythm in eye fragments containing as few as six BRN cells, suggesting

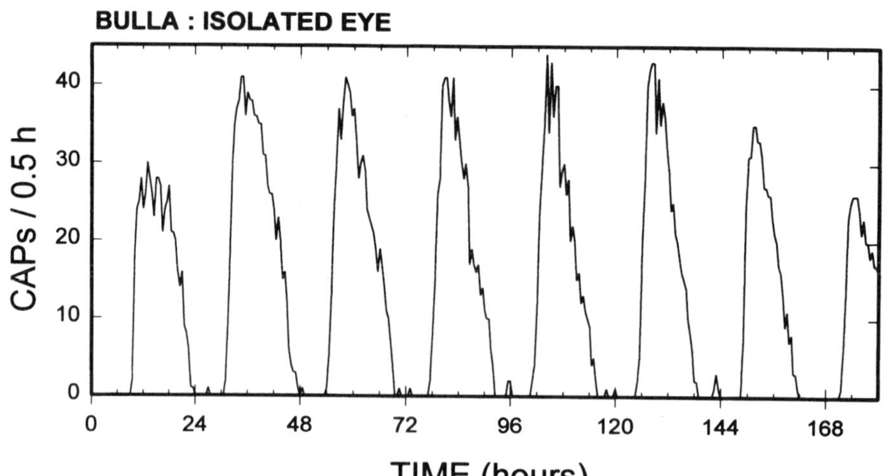

Figure 1. Circadian rhythm in optic nerve impulse activity in the isolated eye of *Bulla gouldiana.* The eye was maintained in constant darkness. The frequency of compound action potentials (CAPs) shows a robust oscillation that persists *in vitro* for many days (T. L. Page, unpublished).

these cells generate the circadian oscillation (Block & Wallace, 1982). This proposition has been recently confirmed in experiments showing periodic conductance changes in single, isolated BRNs (Michel, Geusz, Zaritsky, & Block, 1993).

While the molluskan eye has been primarily exploited as a model system for studies on the cellular and molecular basis of circadian oscillations (see Chapter 14, this volume), there has been an effort to assess the functional role of the ocular pacemaker in controlling behavioral rhythms in the mollusks. In both *Aplysia* (Lickey, Block, Hudson, & Smith, 1976; Lickey, Wozniak, Block, Hudson, & Augter, 1977; Strumwasser, 1973) and *Bulla* (Block & Davenport, 1982), removal of the eyes usually, but not invariably, resulted in disruption of the circadian rhythm in locomotor activity, suggesting these structures may participate as the driving oscillator for the rhythm. However, efforts in *Aplysia* to demonstrate a strong correlation between the phase and amplitude of the rhythm in nerve impulses from the eye and the phase and amplitude of the behavioral rhythm have been equivocal (Jordon, Lickey, & Hiaasen, 1985; Lickey, Hudson, & Hiaasen, 1983). These results, combined with the persistence of locomotor rhythmicity in some eyeless animals, indicate that an additional oscillatory structure outside the eyes may play an important role in the temporal organization of behavior (Jordon *et al.*, 1985).

Effects of eye removal on locomotor activity have been investigated in two pulmonate gastropod species, a terrestrial slug, *Limax maximus* (Beiswanger, Sokolove, & Prior, 1981), and a freshwater snail, *Helisoma trivolvis* (Kavaliers, 1981). Unlike the opisthobranch mollusks, there is no evidence in either of these species that the eye contains a pacemaker, and eye removal appeared to have no significant effect on the ability of animals to generate a circadian rhythm in activity. It seems likely that the pacemaker for the activity rhythms in these mollusks is located in the brain, though that has yet to be demonstrated.

Pacemakers in the Arthropod Brain

Arthropods exhibit circadian rhythmicity in a variety of behaviors and physiologic processes. Three of these—locomotor activity, eclosion, and retinal sensitivity—have been intensively studied. There are now several instances in which circadian pacemakers that regulate these rhythms have been localized to the brain (supraesophageal ganglion). For purposes of discussion, the arthropod brain can be divided into two major areas. First is the optic lobes, which are paired structures on either side of the ganglion. The optic lobes receive and process input via the optic nerves from the compound eyes and transmit information centrally via the optic tracts. Second is the midbrain (or cerebral lobes), which constitutes the central portion of the ganglion. Pacemakers centered in the optic lobes are common in orthopteroid insects (e.g., cockroaches, crickets) and beetles; in lepidopterans (e.g., moths) and dipterans (flies, mosquitoes), the pacemaker appears to reside in the cerebral lobes.

Optic Lobes. Nishiitsutsuji-Uwo and Pittendrigh (1968b) first discovered that surgical ablation of the optic lobes or their neural isolation from the midbrain by section of the optic tracts disrupted the locomotor activity rhythm of the cockroach *Leucophaea maderae*. Similar results have been obtained in other cockroach species (Lukat & Weber, 1979; S. K. Roberts, 1974) and in several species of crickets (Loher, 1972, 1974; Sokolove & Loher, 1975; Tomioka & Chiba, 1984; Waddel, Lewis, & Engelmann, 1990). These studies showed the optic lobes were necessary for the expression of the rhythm. Subsequent studies that showed the optic lobes controlled both the phase and period of the locomotor activity rhythm demonstrated that their functional role was to provide the timing signal.

Although severing the optic tracts invariably disrupted rhythmicity in *L. maderae*, when the optic lobes were left *in situ*, the rhythm consistently reappeared in 3–5 weeks as a consequence of regeneration of the neural connections between the optic lobes and the midbrain (Page, 1983a). The free-running period of the rhythm after regeneration was correlated with the period of the preoperative rhythm. These results prompted a series of experiments in which the optic lobes were transplanted between individuals with quite different free-running periods (Page, 1982b, 1983a). The majority of animals that received transplanted optic lobes recovered rhythmicity, and the preoperative period of the donor and the postoperative period of the host were strongly correlated (Figure 2). Thus, the transplantation of the optic lobes from one animal to an animal whose own optic lobes had been removed both restored the rhythm of locomotor activity and, critically, imposed the period of the donor animal's rhythm on the rhythm of the host.

Utilizing localized low-temperature pulses, it has also been shown that the optic lobes control the phase of the free-running locomotor activity rhythm (Page, 1981a). Cooling the whole animal for several hours can generate large phase shifts of the locomotor activity rhythm (S. K. Roberts, 1962). Assuming that temperature directly affects the pacemaker, it was reasoned that localized cooling of the pacemaker locus should generate large phase shifts, whereas cooling other structures should have no effect (Page, 1981a). After transection of one of the two optic tracts, cooling an intact optic lobe of animals that were free-running in DD (constant dark) consistently caused a phase shift of several hours (Figure 3A), whereas cooling the neurally isolated optic lobe or the midbrain had no effect on the phase of the activity rhythm (Figure 3B). These data indicated that the optic lobes control the phase as well as the period of the activity rhythm (Page, 1981a).

The final and conclusive evidence that the optic lobe contains a competent circadian pacemaker was the demonstration that the tissue could generate an oscillation when isolated from neural or humoral signals from the rest of the animal. In both the cockroach (Colwell & Page, 1990) and the cricket (Tomioka & Chiba, 1986, 1992), when optic lobes were either neurally isolated by section of the optic tract or maintained in organ culture, the spontaneous multiunit impulse activity recorded from the optic tract exhibited a rhythm in constant environmental conditions.

The region within the optic lobe that contains the circadian pacemaker for the locomotor activity rhythm in cockroaches and crickets has been more precisely identified with localized lesions (Page, 1978; S. K. Roberts, 1974; Sokolove, 1975; Tomioka & Chiba, 1984; Waddel *et al.*, 1990). The results suggest that the cells responsible for generating the circadian signal have their somata and/or processes in the proximal half of the optic lobe between the lobula and the medulla. Further work will be necessary to determine whether one, a few, or many cells within this region comprise the oscillator or whether the pacemaking signal can be redundantly generated in several cells.

The localization of circadian pacemakers that regulate locomotor activity raised the question of whether other rhythms are controlled by the same clock. A common feature of arthropod visual systems is that the sensitivity of the eye to light exhibits a circadian variation which is reflected in a daily change in the amplitude of the electroretinogram (ERG) recorded in response to a standard light pulse (Fleissner & Fleissner, 1985; Page, 1981b). Depending on the species, several mechanisms may

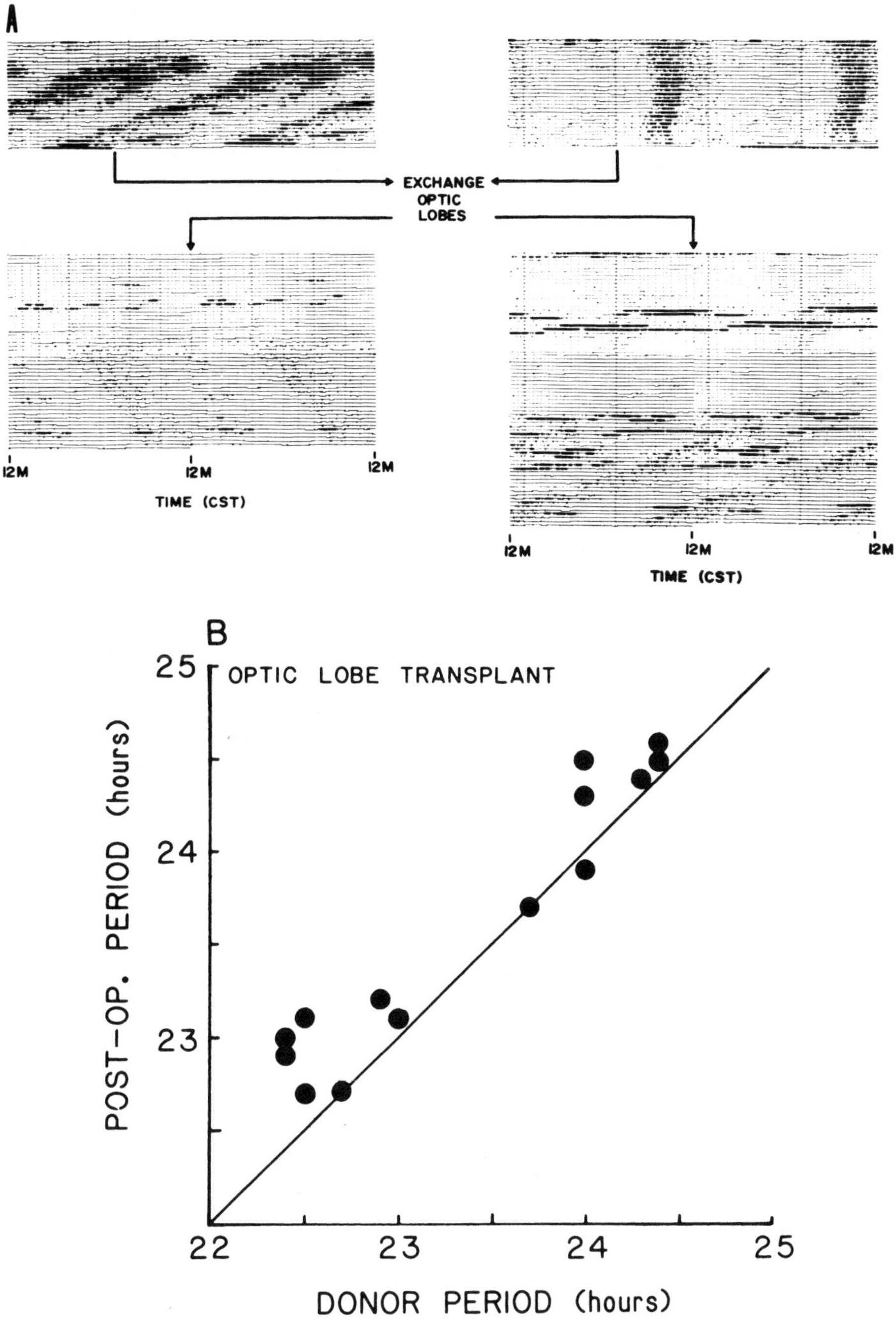

Figure 2. Effects on locomotor activity of optic-lobe transplantation between groups of cockroaches raised in LD 11:11 or LD 13:13. (A) Locomotor activity records showing restoration of rhythmicity after transplantation ("exchange"). The record begins with the animals free-running in constant darkness. After 4 weeks, the optic lobes were exchanged, and after several weeks of low activity, rhythmicity returned in both cases. (B) Plots of free-running period of the activity rhythm after transplantation versus the period of the donor animal before surgery. The diagonal line shows the values expected if the pre- and postoperative period were identical. (Data from Page, 1982b, 1983a.)

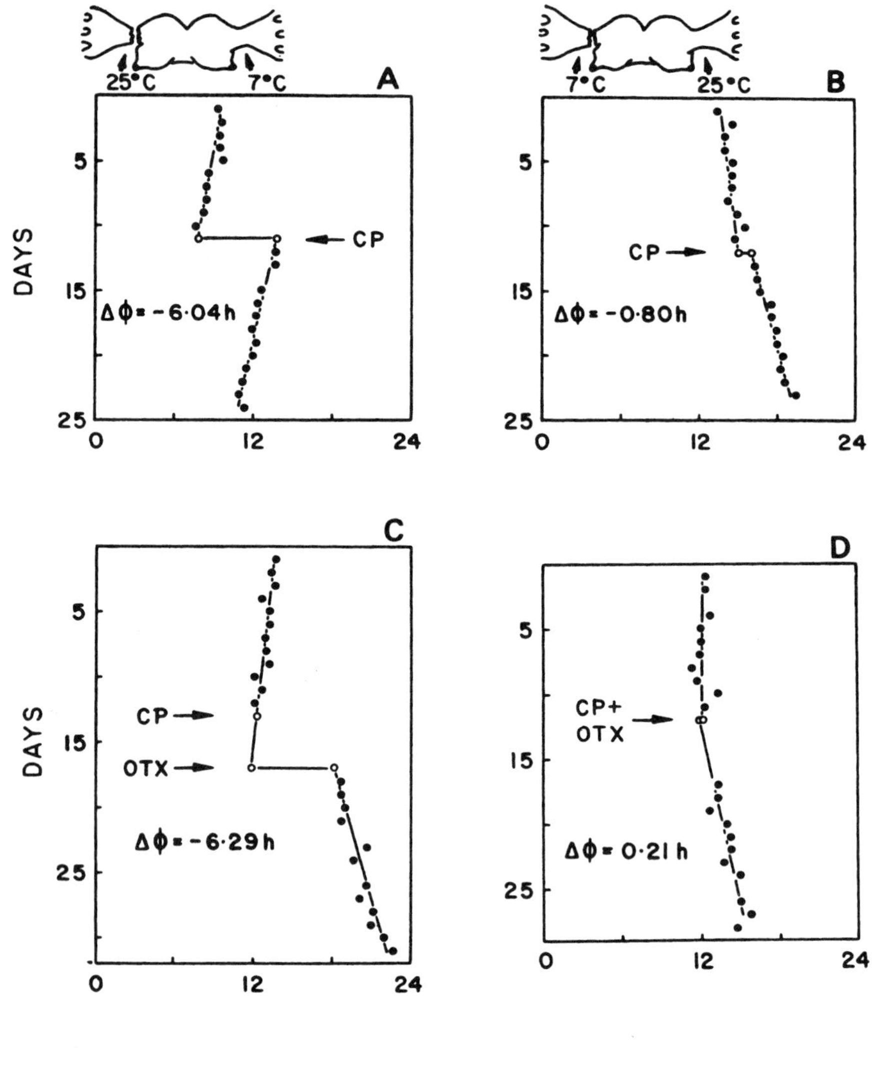

Figure 3. Examples of data from animals treated with localized low-temperature pulses. Solid circles, time of activity onset for each day; open circles, projected phases of the rhythms before and after the pulse; lines are linear regressions; PST, Pacific Standard Time. Pulses were 6 hours in duration and began at activity onset (CT12). (A) The intact optic lobe of an animal with one sectioned optic tract was cooled to 7°C while the isolated lobe was maintained at 25°C. (B) The isolated lobe was cooled. Cooling the intact lobe caused a large phase delay, whereas cooling the isolated lobe had no effect. (C, D) The effects of a low-temperature pulse to one lobe on the rhythm driven by the opposite lobe. (C) The optic tract of the treated lobe was cut 4 days after the pulse, and the subsequent rhythm, driven by the untreated lobe, was phase-delayed by several hours; (D) the optic tract of the treated lobe was section 0.5 hours after the pulse, thus preventing the phase shift of the rhythm. (From Page, 1984.)

contribute to producing the sensitivity change (e.g., Barlow, Kaplan, Renninger, & Saito, 1985); however, it does appear invariably to be associated with structural changes including daily migration of screening pigments which surround the rhabdomeres of the ommatidia (e.g., Ferrell & Reitcheck, 1993). In the cricket *G. bimaculatus* (Tomioka & Chiba, 1982), the beetle *Anthia sexguttata* (Fleissner, 1982), and the cockroach *L. maderae* (Wills, Page, & Colwell, 1985), the pacemaker regulating the ERG amplitude rhythm has been localized to the optic lobe. In each case, rhythms in ERG amplitude persisted after section of the optic tract, neurally isolating the optic lobe/compound eye complex from the rest of the nervous system. On the other hand, the rhythms were abolished by cuts through more distal regions of the optic lobe separating the compound eye from the site of the pacemaker for the activity rhythm. These data indicated that the oscillation that drives the rhythm in ERG amplitude is generated in the optic lobe, and suggest that the same pacemaker controls both the ERG amplitude and activity rhythms.

Identification of Pacemaker Cells. Although the site of the pacemaker in the optic lobe has been localized, the specific cells that generate the oscillation have not been identified. In both cockroaches and crickets, there is a population of cells located in the appropriate region of the optic lobe that are immunostained with antibody against the crustacean peptide hormone pigment-dispersing hormone (PDH) (Homberg, Wurden, Dircksen, & Rao, 1991; Stengl, 1995; Stengl & Homberg, 1994). These cells have processes in the medulla, lamina, and several regions of the midbrain, and to the contralateral optic lobe. On the basis of the anatomy and correlations between the survival or regeneration of these cells and the expression of rhythmicity, it has been suggested that these cells may participate in the generation of circadian rhythms and that PDH may be an important signaling molecule in the circadian system (Stengl, 1995; Stengl & Homberg, 1994). Interestingly, it has been recently shown in *Drosophila* that a group of neurons at the lateral margins of the midbrain that have been implicated in the control of the circadian rhythm of locomotor activity (see below) are also PDH-immunoreactive (Helfrich-Forster, 1995; Helfrich-Forster & Homberg, 1993).

Cerebral Lobes. In contrast to the situation in cockroaches, crickets, and beetles, in a variety of other arthropods, the optic lobes do not appear to be required for rhythmicity and the pacemaker appears instead to reside in the cerebral lobes. Truman and Riddiford (1970; Truman, 1972b) showed that the circadian pacemaker which controls the timing of eclosion in two silkmoth species, *Hyalophora cecropia* and *Antheraea pernyi*, is located in the cerebral lobes of the brain. The time of day that eclosion occurs is different for the two species. When the insects are maintained in LD (light and dark) 17:7, *H. cecropia* emerges shortly after lights-on, whereas *A. pernyi* emerges just before lights-off. Removal of the brain did not prevent eclosion, but did disrupt its timing. However, if the brain was reimplanted in the abdomen, normal rhythmicity was restored in both entrained and free-running conditions. While these results implicated the brain as the site of the pacemaker that controlled the timing of eclosion, the compelling evidence was provided by experiments in which brains removed from one species were transplanted to the abdomen of the other. Moths that had received these "switched brain" transplants exhibited normal species-specific eclosion behavior, but the phase of the rhythm was characteristic of the donor and not the host. *A. pernyi* that had received *H. cecropia* brains emerged just after lights-on, and *H. cecropia* that had received *A. pernyi* brains emerged near lights-off. The demonstration that the transplanted brains not only restored rhythmicity, but also determined the phase of the rhythm leaves little doubt that the circadian pacemaker that regulates the timing of the eclosion rhythm is located in the brains of these moths.

To determine which part of the brain is responsible for timing the eclosion behavior, the brain was subdivided prior to transplantation (Truman, 1972b, 1974a). The results suggested the optic lobes of the brain were unnecessary and that the pacemaker resided in the cerebral lobes. The same conclusion emerged from studies on the activity rhythms of the adult moth (Truman, 1974b). The optic lobes also appear to be dispensable for circadian locomotor rhythms in a variety of other holometabolous insects including a fruitfly, *Drosophila melanogaster* (Helfrich & Engelmann, 1983), the housefly, *Musca domesticus* (Helfrich, Cymborowski, & Engelmann, 1985), the blowfly, *Caliphora vicina* (Cymborowski, Lewis, Hong, & Saunders, 1994), and a mosquito, *Culex pipiens pallens* (Chiba & Tomioka, 1987; Kasai & Chiba, 1987). In each case, the pacemaking oscillation appears to be generated in the cerebral lobes. In *D. melanogaster*, a specific group of neurons near the lateral margin of the cerebral lobes has been implicated (Ewer, Frisch, Hamblen-Coyne, Rosbash, & Hall, 1992).

Efforts at localization of circadian pacemakers have also been made in two chelicerate arthropods, the horseshoe crab, *Limulus polyphemus* (Barlow, 1983; Kass & Barlow, 1992), and a scorpion, *Androctonus australis* (Fleissner, 1983, 1986). In both cases, it appears the pacemaker that drives the circadian rhythm in ERG amplitude is located in the brain and communicates with the eye via efferent neural projections to the retina (discussed in more detail below). In *Limulus*, the circadian rhythm of efferent activity was shown to persist when recorded from the excised protocerebrum *in vitro* (Kass & Barlow, 1992).

A variety of lesion studies have also been carried out on decapod crustacea in an effort to localize pacemakers that regulate both locomotor activity and ERG amplitude rhythms (for a review of early work see Page, 1981b). Much of the work has utilized various species of crayfish, where the results suggest that normal rhythmicity requires both the optic lobes and the cerebral lobes (supraesophageal ganglion) since lesions to either of these structures disrupts, but does not completely abolish, rhythmicity. The evidence indicates that both structures contain circadian oscillators, but more work will be necessary to discover how outputs from the oscillators in the optic lobes and brain are integrated (Barrera-Mera, 1976; Larimer and Smith, 1980; Page & Larimer, 1975a, b; Sanchez & Fuentes-Pardo, 1977).

Circadian Pacemakers Outside the Nervous System

Although the focus of efforts to find circadian oscillators has been on the nervous system, it is interesting and perhaps not widely appreciated that there are well-documented cases, all in insects, of circadian pacemakers that reside outside the brain and retina. In the cockroach, studies on circadian cuticle deposition of cuticle indicated that this rhythm is driven by a pacemaking system that is independent of the brain. In both *Leucophaea maderae* and *Blaberus craniifer*, the cuticle of the newly molted adult is secreted rhythmically in layers. The rhythm persists in constant conditions with a circadian period that is temperature compensated (Wiedenmann, Lukat, & Weber, 1986). Interestingly, this rhythm's phase and period are not influenced by exposure of animals as adults to a variety of light cycles, and the phase appears to be set by the time of the imaginal molt (Wiedenmann *et al.*, 1986). This result indicates the cuticle rhythm is independent of the optic lobe oscillator that drives the locomotor activity and ERG amplitude rhythms which can be entrained by light in both the adult (S. K. Roberts, 1962) and the nymph (Page, 1990b). Further, in *Blaberus*, the rhythm in cuticle deposition persists after complete ablation of the optic lobes (Lukat, 1978) and appears to persist in pieces of leg tissue cultured *in vitro* (Weber, 1985). Thus, this rhythm appears to be driven by an independent circadian pacemaking system, and the results raise the possibility that the epidermal cells that secrete the cuticular material may be autonomously rhythmic.

Recent data have provided a clear demonstration that the complex of testis and seminal ducts in moths also contains a circadian pacemaker. In the flour moth, *Anagasta kueniella*, the release of the sperm from the testis exhibits a daily rhythm that persists in DD in the isolated abdomen (Riemann, Thorson, & Rudd, 1974; Thorson & Riemann, 1977). A similar rhythm of release of mature sperm from the testis into the seminal ducts was found in the gypsy moth, *Lymantria dispar* (Giebultowicz, Bell, & Imberski, 1988). In this species, the rhythm persisted, and could be phase shifted by light *in vitro* in isolated complexes of seminal ducts and testis (Giebultowicz, Riemann, Raina, & Ridgway, 1989). Thus this tissue contains a circadian pacemaker and photoreceptor for entrainment.

In the moth *Samia cynthia*, preparation for pupation and metamorphosis involves a series of morphologic and behavioral changes including an excretion of the contents of the gut which occurs shortly after a cessation in feeding and precedes a "wandering stage." This "gut purge" is regulated by the circadian system (see Ishisaki, Mizoguchi, & Fujishita, 1984, for a review). Timing of the gut purge appears to be controlled by a circadian clock in the prothoracic gland (PG), an endocrine organ which is responsible for the secretion of the insect steroid hormone ecdysone. A small peak of ecdysteroid occurs in the hemolymph just prior to gut purge. The circadian timing of the ecdysteroid peak persists in decapitated larvae, showing that a brain-centered clock is not necessary (Mizoguchi & Ishizaki, 1984a, b). Other studies involving the transplantation of the PG followed by their local illumination indicated that the glands contain a photosensitive circadian clock that can regulate the timed release of ecdysone (Mizoguchi & Ishizaki, 1982). More recent work in the bug *Rhodnius prolixus* (Vafopoulou & Steel, 1991, 1992) and the moth *Galleria mellonella* (Cymborowski, Muszynska-Patel, Porcheron, & Cassier, 1991) also provided evidence that the prothoracic glands contain a photosensitive pacemaker that regulates synthesis of ecdysteroids. In *Rhodnius*, the rhythm has been shown to persist in isolated glands maintained in tissue culture (Vafopoulou & Steel, 1991, 1992).

Summary

Circadian pacemakers have been successfully localized to restricted populations of cell in a phylogenetically diverse group of invertebrates. The body of data calls for generalization. However, the data do not readily yield to general statements about the anatomic location of circadian pacemaking systems. Pacemakers have been found in eyes, brains, endocrine organs, reproductive organs, and epidermis. The one generalization that seems valid (though perhaps not without exception) is that circadian pacemakers are closely associated with photoreceptive structures that mediate entrainment to light. In those cases where the proximity to an organized photoreceptive structure is not obvious, it appears that extraretinal photoreceptors, anatomically near or identical to the pacemaker locus, are involved in entrainment (see section on photoreceptor localization below).

The identification of sites of pacemaker activity also raises the questions of precisely which cells participate in the generation of the oscillation and whether intercellular interactions are a fundamental element of the system. The only case for which there is good information is the *Bulla* eye, where it has been shown that single, isolated BRNs exhibit circadian oscillations (Michel *et al.*, 1993). In view of the evidence that an individual neuron can function as an oscillator in *Bulla* coupled with the fact that a variety of unicellular organisms exhibit circadian rhythms, it seems highly likely that intercellular interactions are not essential, but additional data on this issue are needed.

Finally, it is worth emphasis that tissues outside the central nervous system (CNS) produce robust circadian rhythms. While the number of examples is limited, that may only reflect the focus on the nervous system in searches for pacemaker activity, and at this point it is premature to speculate on the extent to which extra-CNS pacemakers are a common and critical feature of circadian organization. Further, there is a need for information on the relationship between these components of the circadian system and CNS pacemakers. It is unclear whether the extra-CNS pacemakers and CNS pacemakers are completely independent or directly or indirectly coordinate with one another. The next section it part addresses this issue, discussing the multioscillator organization of pacemaking systems by examining the data on the significance of interactions among anatomically distinct oscillators in the determination of the properties of the circadian system.

Multioscillator Organization of the Pacemaking System

Clearly, circadian oscillators can be localized in invertebrates, but in addition, the data show that the circadian *system* in the individual may be composed of several oscillators. That there is more than one oscillator is unambiguous in those cases in which oscillators have been shown to be represented bilaterally (e.g., eyes of mollusks, optic lobes of insects) or localized in distinct tissues (e.g., brain, prothoracic glands, and testis in moths; optic lobes and cuticle in cockroaches). Though generally less compelling, there are also data pointing to the existence of multiple oscillators that arise from observations of "damped" or "residual" rhythmicity following destruction of the primary pacemaker (e.g., Page, 1985b; Rence & Loher, 1975; Stengl, 1995; Tomioka, 1985) or the appearance of multiple components with different periods in activity records (e.g., Christensen & Lewis, 1982; Helfrich, 1986; Wiedenmann & Loher, 1984).

These observations suggest that circadian organization within the individual is derived from a population of oscillators and raise the possibility that the properties of the circadian clock could depend in a significant way on the interactions among component oscillators. A beginning has been made in exploring this hypothesis in studies on coupling relationships between circadian oscillators localized to bilaterally paired structures. There are three general questions: Are bilaterally paired pacemakers in fact coupled? If so, what is the significance of the coupling for the "circadian" properties of the system? Finally, what is the nature of the coupling mechanism?

Variability in Coupling. Early studies on the cockroach showed that destruction of one of the optic lobe oscillators consistently led to a small, but significant increase in free-running period (Page, 1978; Page, Caldarola, & Pittendrigh, 1977). This observation plus the fact that either one of the compound eyes was sufficient to entrain both oscillators (Page, 1978, 1983a; Page *et al.*, 1977) led to the suggestion that bilaterally distributed oscillators in the optic lobes were mutually coupled, and that the free-running period of the coupled pair was shorter than the period of the individual oscillators. Convincing support for this proposition was obtained using localized low-temperature pulses (Page, 1981a). One optic lobe was cooled for several hours, then removed at various times after the pulse to assay the phase of the rhythm driven by the contralateral oscillator. If the treated lobe was removed 4 days after the pulse, the subsequent rhythm (driven by the untreated lobe) was phase delayed by several hours (Figure 3C), but if the treated lobe was removed only 0.5 hour after the pulse, the phase shift in the contralateral lobe was prevented (Figure 3D). The results indicated that the low-temperature pulse caused a phase shift in the treated lobe that was subsequently transmitted to the oscillator in the contralateral optic lobe. Further, cooling one lobe of an intact animal resulted in a phase shift that was nearly 2 hours less than the phase shift obtained when the optic lobe contralateral to the pulse was neurally isolated, indicating that the return of the coupled system to steady state involved a phase advance in the treated oscillator as well as a delay in the oscillator of the untreated optic lobe.

These data showed that phase information was shared between the oscillators and suggested that the coupling was relatively strong. Interestingly, subsequent studies on other insects and on mollusks suggested that the strong coupling in the cockroach may be unusual. In the beetle *Blaps gigas*, the ERG amplitude rhythms from the two eyes of an individual monitored in constant darkness were found to freerun with different periods, and it was shown that the phase and period of the rhythm from one eye could be controlled by local illumination of that eye without any consistent effect on the rhythm from the other eye (Koehler & Fleissner, 1978). The results indicated that each eye was controlled by its own pacemaker, and that the two pacemakers were either completely independent or only very weakly coupled.

Similar results have been obtained in two species of crickets, *Teleogryllus commodus* (Wiedenmann, 1983) and *Gryllus bimaculatus* (Tomioka, Yamada, Yokoyama, & Chiba, 1991). In both instances, following severance of the optic nerves on one side to isolate the optic lobe from light input from its compound eye, exposure to constant light caused the locomotor activity rhythm to split into two components with very different free-running periods. Further, each activity component could be predictably associated with the circadian oscillator in one of the optic lobes (Figure 4). The data indicated that coupling is either absent or weak relative to the light entrainment pathway.

Similar species differences have been observed in the opisthobranch mollusks. In *Aplysia* maintained in constant darkness, the two ocular pacemakers spontaneously desynchronized (Hudson & Lickey, 1980; Lickey *et al.*, 1983), indicating that the coupling between the eyes is weak or absent. In contrast, in *Bulla* maintained in darkness, the eyes remained closely in phase, suggesting that the pacemakers are coupled (M. H. Roberts & Block, 1985). The apparent difference in coupling between the two species was confirmed in studies on the eyes and brain isolated *in vitro*. Following treatments to one eye that induced a phase difference of several hours between the ocular pacemakers, in *Bulla*, the eyes tended to come back into phase at a rate of about 1–1.5 hours/day (Figure 5), whereas in *Aplysia*, phase separations were maintained (M. H. Roberts & Block, 1985; M. H. Roberts, Block, & Lusska, 1987). Subsequent studies on *Bulla* suggested that the coupling between the ocular pacemakers was of moderate strength. In experiments *in vitro* in which the period of one pacemaker was either lengthened (by treatment with lithium) or

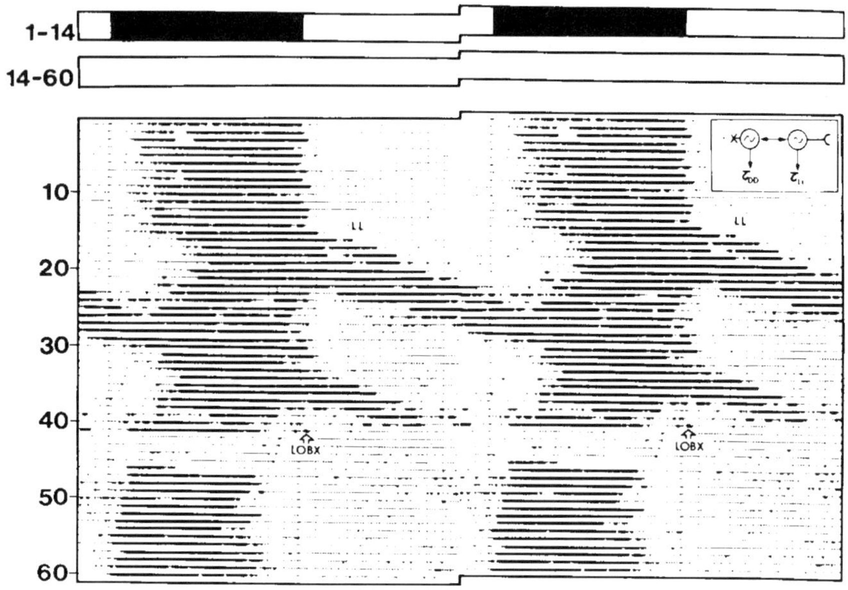

a

Figure 4. Locomotor activity record from the cricket *Teleogryllus commodus*. Prior to the beginning of the experiment, one optic nerve was cut, isolating one optic lobe pacemaker from input from its compound eye (see schematic in inset). After 14 days, the animal was placed in constant light. The treatment lengthens the period of the optic lobe pacemaker that is still attached to its eye, desynchronizing its component of activity. On day 40, that optic lobe is removed (LOBX), eliminating the long-period component. (From Wiedenmann, 1983. Copyright 1983, Springer-Verlag.)

shortened (by reducing the chloride concentration in the seawater), the majority of pairs of eyes were unable to maintain a stable relationship when period difference exceeded about 1 hour (Page & Nalovic, 1992).

Table 1 summarizes studies in which the strength of coupling between bilaterally distributed pacemakers has been at least qualitatively evaluated. The full range of coupling strengths is represented. On one end of the spectrum are *Aplysia* and *Blaps*, where the oscillators spontaneously desynchronize in constant conditions, while at the other end is the cockroach, in which only very transient desynchronization has been achieved. *Bulla* appears to be intermediate.

Fundamental Properties of the Pacemaking System Do Not Depend on Coupling. In the several cases where coupling between bilaterally paired oscillators is weak or nonexistent, coupling must not be critical to normal circadian organization. What about instances where coupling is moderate to strong?

With regard to three measures—the sustainability, amplitude, and period of the oscillation—the data seem clear. Coupling between oscillators does not have a major impact on the properties of the system. In both *Bulla* (e.g., Page & Nalovic, 1992; M. H. Roberts & Block, 1985) and *Leucophaea* (Page *et al.*, 1977), uncoupling by surgical destruction of the coupling pathway or by removal of one of the oscillators has no detectable effect on the sustainability or amplitude of overt rhythms (Figure 5). Further, uncoupling has only minor affects on the free-running period of the oscilla-

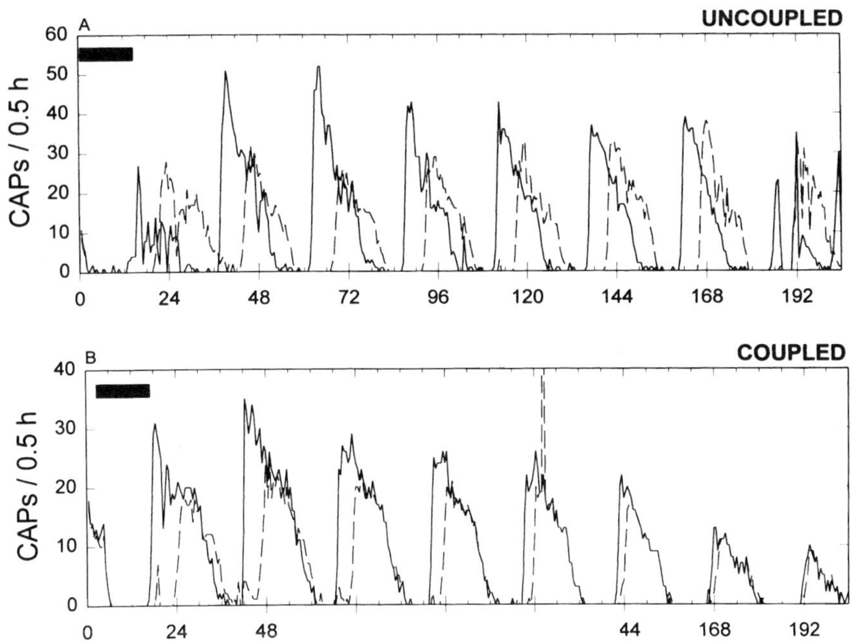

Figure 5. Frequency of spontaneous compound action potentials (CAPs) recorded in constant darkness from the optic nerves of *Bulla gouldiana*. The two eyes and attached central ganglia were isolated *in vitro* in artificial seawater. In the top panel, the ocular pacemaker was uncoupled by severing the cerebral and pedal connective. On the first day of the record, one eye of each animal was treated with a 14-hour pulse of 100 mM lithium, which caused a large phase delay in the treated eye (dashed lines). A constant phase difference is subsequently maintained in the uncoupled eye, while the coupled eye pair regains synchrony over the next several cycles. Note also that uncoupling (top panel) has little effect on the period, amplitude, or persistence of the rhythms in the eyes. (T. L. Page, unpublished.)

tion. In *Leucophaea*, the period of the individual oscillator is about 10–15 minutes shorter than the period of the coupled pair (Page *et al.*, 1977). In *Bulla*, τ of the coupled system is at most only a few minutes longer than τ of the individual eye (Page & Nalovic, 1992; M. H. Roberts & Block, 1985). Furthermore, although it has not been carefully and quantitatively studied, there seems to be little or no effect of

TABLE 1. EXAMPLES OF COUPLING RELATIONSHIPS BETWEEN BILATERALLY DISTRIBUTED CIRCADIAN OSCILLATORS

Species	Pacemaker location	Relative coupling strength
Leucophaea maderae[a]	Optic lobe	Strong
Bulla gouldiana[b]	Retina	Moderate
Teleogryllus commodus[c]	Optic lobe	Moderate
Gryllus bimaculatus[d]	Optic lobe	Moderate
Blaps gigas[e]	Optic lobe(?)	Weak or absent
Aplysia californica[f]	Retina	Weak or absent

[a]Page et al., 1977; Page, 1981a. [b]Roberts & Block, 1985; Page and Nalovic, 1992. [c]Wiedenmann, 1983. [d]Tomioka et al., 1991. [e]Koehler & Fleissner, 1978. [f]Hudson & Lickey, 1980; Roberts et al., 1987.

coupling on phase shifting or entrainment by light. Finally, in *Bulla*, "aftereffects" on τ following entrainment to T cycles appears to be independent of coupling (Page, Wassmer, Fletcher, & Block, 1997). In sum, there is no evidence that the coupling between bilaterally distributed oscillators has functionally important effects on the fundamental properties of the circadian pacemaking system, and it is reasonable to assume that the primary (only?) function of the coupling pathway is to maintain a constant phase relationship between the pacemakers in the absence of environmental cues.

Properties of the Mutual Entrainment Pathway. An additional question of interest concerns the anatomic pathway and physiologic mechanisms by which oscillators are coupled. In the cockroach, section of the optic tract prevents the transmission of phase information from one optic lobe to the other, suggesting that the coupling is neural rather than humoral and that axons in the optic tract mediate the coupling signal (Page, 1983b). It is interesting to note that after surgical section of the optic tracts, the pathway by which the optic lobes control locomotor activity consistently regenerates within a few weeks (Page, 1983a), whereas the coupling pathway between the optic lobe oscillators, equally consistently, does not regenerate (Page, 1983b). The results indicate these two output pathways of the pacemaker are functionally and perhaps structurally distinct. Similarly, in crickets, coupling between optic lobe pacemakers appears to involve neural coupling via fibers that run in the optic tract (Tomioka, Nakamichi, & Yukizane, 1994; Yukizane & Tomioka, 1995).

Detailed information on the nature of the coupling pathway has been obtained in *Bulla*, where it appears that the coupling signal is mediated by direct connections, via the cerebral connectives, between the BRNs that comprise the ocular pacemakers (Block, Roberts, & Lusska, 1986). The view is that impulses originating in each eye phase shift the pacemaker in the other eye via a depolarization of the BRN membrane potential, and that the phase-shifting mechanism beyond the depolarization step is identical to that of other depolarizing treatments such as light pulses. This hypothesis and its predictions were tested in experiments in which the period of one oscillator of the coupled pair was manipulated through changes in the ionic composition of the seawater bathing one eye (Page & Nalovic, 1992). In general, the ocular pacemaker system of *Bulla* conformed to expectations. Following lengthening or shortening of the period of one pacemaker of the coupled pair, several pairs of eyes were able to maintain a stable phase relationship. The average period was intermediate between the periods of the individual oscillators and, as predicted, the eye with the longer period phase lagged the eye with shorter period. The results confirmed that the ocular pacemakers are coupled and supported the inference that pacemakers can interact through both advancing and delaying phase shifts. Interestingly, in a few instances, several pairs of eyes that did not maintain a stable phase relationship and desynchronized by 9.5–10.5 hours resynchronized within a single cycle via a massive phase advance in the rhythm from the phase-lagging eye (Page & Nalovic, 1992). This observation suggests there may be a novel phase-shifting mechanism that is part of the coupling pathway since simple depolarizing treatments are unable to generate a phase shift of this magnitude (McMahon & Block, 1987a, b).

Development of the Pacemaking System

Whereas there has been extensive work in invertebrates on understanding the organization of the circadian pacemaking system in the adult, relatively little attention has been paid to understanding how this organization arises in the course of development. Two questions have been addressed in a limited number of insect species. The first is, at what stage in development is the pacemaker differentiated? Results suggest that a functional pacemaker arises relatively early in development. For example, in the moth *Pectinophora gossypiella*, egg hatching is gated by a circadian oscillation whose phase can be set by temperature steps or pulses approximately midway through embryogenesis (Minis & Pittendrigh, 1968). In the cockroach *Leucophaea maderae*, a circadian rhythm in locomotor activity can be expressed immediately upon hatching from the egg (Page, 1990b). Rhythms were expressed in roaches that had never been exposed to a periodic environmental stimulus, but the phase of the rhythm could be set by light cycles at least 1 week prior to hatching, indicating that the pacemaker becomes functional during embryogenesis (Page, 1990b). Similar results have been obtained in *Drosophila melanogaster*, where it was shown that light pulses become effective in synchronizing the adult locomotor activity rhythm at the end of embryogenesis (Sehgal, Price, & Young, 1992).

The fact that the pacemakers in insects are present and functional early in development prompts the additional question of whether the environment during the subsequent postembryonic development might have an impact on the properties of the circadian system. In both cockroaches (Page, 1990b) and flies (Sehgal *et al.*, 1992), it is clear that exposure to a periodic environment during development is not necessary for the expression of rhythmicity by the adult. However, in the cockroach *L. maderae*, in which the period of postembryonic development is quite extended (about 6 months), the developmental environment can have a major impact on the properties of the circadian system of the adult. Studies in which animals were raised in *T* cycles (LD 11:11, LD 12:12, LD 13:13) or in DD or LL showed that exposure to constant conditions or to non-24-hour light cycles during postembryonic development permanently altered both the free-running period (Figure 6A) and the response to light pulses (Figure 6B) or to constant light (Barrett & Page, 1989; Page, 1991; Page and Barrett, 1989). These results represented the first demonstration that the environment during development may be significant in the determination of fundamental properties of circadian pacemaking systems and suggested that significant developmental events in the ontogeny of circadian organization occur postembryonically. The precise mechanism by which the photic environment during development alters the properties of the circadian system is uncertain, though the question is of considerable interest relative to understanding the cellular organization of the pacemaking system.

Localization of Photoreceptors and Mechanisms of Entrainment

In the invertebrates, efforts to localize the photoreceptors responsible for entrainment of circadian rhythms have produced a diversity of results (Page, 1982a). In a few insects, photoreception for entrainment occurs exclusively in the compound eyes, while in other invertebrates, both extraretinal and retinal photoreception may be important.

Photoreception for Neural Pacemakers

In *Aplysia* (Eskin, 1971) and *Bulla* (Block & Wallace, 1982), eyes isolated *in vitro* can be phase shifted by light, demonstrating that ocular photoreceptors are suffi-

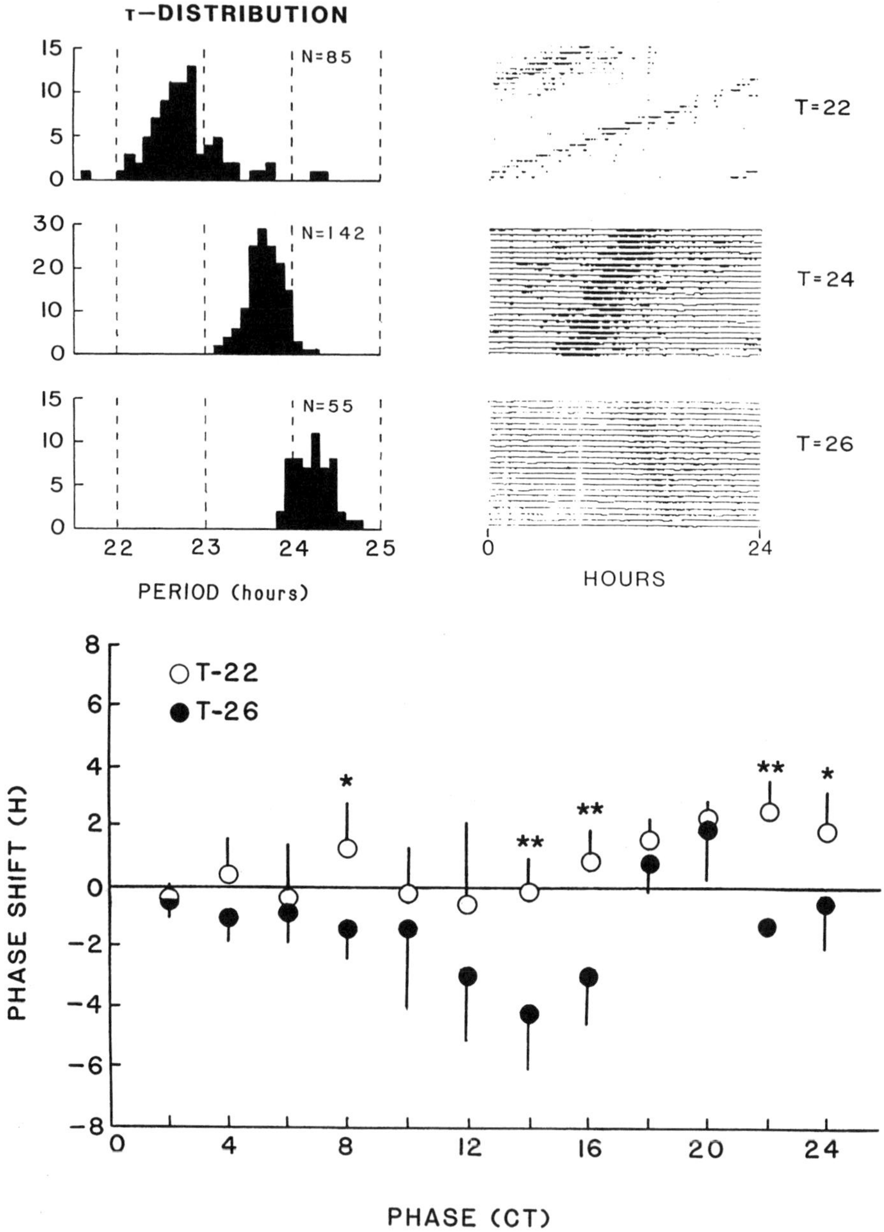

Figure 6. (Top) Examples of activity records of animals raised in either LD 11:11 ($T = 22$), LD 12:12 ($T = 24$), or LD 13:13 ($T = 26$) light cycles. In each case, the period of the individual is near the mean for the group. The distribution of the free-running period values is for animals raised in the same light cycles. (Bottom) Phase response curves to 6-hour light pulses for animals raised in LD 11:11 ($T = 22$) or LD 13:13 ($T = 26$). Each point represents the mean (±SD) of three to nine phase shifts. For every phase, the phase shift in the $T = 22$ animals is larger. Phases where the difference is significant are denoted by asterisks (*$p < .05$, **$p < .01$). (Modified from Page and Barrett, 1989, and Barrett and Page, 1989.)

cient for entrainment of the circadian pacemaker in the eye. On the other hand, although eye removal disrupts the ability of these mollusks to express a free-running rhythm of locomotor activity in constant conditions, eyeless animals were rhythmic in a light cycle (Block & Davenport, 1982; Lickey & Wozniak, 1979). Similar results have been obtained in *Bursatella* (Block & Roberts, 1981). Thus, the mollusks do have extraretinal photoreceptors and these photoreceptors can participate in regulation of the temporal distribution of activity. However, it is uncertain whether the extraocular photoreceptors participate in *entrainment* of either the pacemakers in the eyes or of putative extraocular oscillators, and they may exert control via "masking" (Page, 1989).

Efforts to identify the specific cells within the molluskan eye that are responsible for entrainment of the ocular pacemaker suggested that the pacemaker cells themselves are the site for phototransduction. In *Aplysia*, treatments that block chemical synaptic transmission did not affect the phase shift of the ocular pacemaker by pulses of light, indicating that either the pacemaker cells are photoreceptive or they are electrically coupled to the photoreceptors (Eskin, 1977). In *Bulla*, because the putative pacemaker cells (the BRNs) are anatomically separate from the distal, microvillus photoreceptors, it was possible surgically to remove the distal photoreceptive layer that surrounds the lens and show that BRNs still exhibit an acute response to light and that the pacemaker still can be phase shifted by light pulses (Block & Wallace, 1982). This suggested that the BRNs themselves might be photoreceptive. In support of this interpretation, opsin-like immunoreactivity has been reported in BRN somata and processes (Geusz, Foster, Lawrence, & Block, 1991). Further, the spectral sensitivity of the acute response of BRNs that were isolated by surgical removal of the distal photoreceptors or by blocking chemical synaptic input by treatment with low-calcium/high-magnesium seawater was similar to the spectral sensitivity for phase shifting (Geusz & Page, 1991). In *Aplysia*, Jacklet and Barnes (1993) found that the putative pacemaker cells responded to light in cell culture after dissociation from other retinal cells. The results from both *Bulla* and *Aplysia* therefore indicate that the pacemaker neurons themselves are responsible for the phototransduction step in entrainment.

In the garden slug *Limax maximus*, the circadian rhythm of locomotor activity could still be synchronized by light cycles following removal of the optic tentacles, and on subsequent release into constant darkness, the animals expressed a free-running rhythm that was phased by the prior LD cycle (Beiswanger *et al.*, 1981). However, eye removal did lead to both an increase in the steady-state phase angle between the entrained rhythm and the light cycle and reduced the sensitivity to entrainment by red light (>600 nm; Beiswanger *et al.*, 1981). The results implicate both ocular and extraocular photoreception in the entrainment pathway.

Extraretinal photoreceptors also appear to be sufficient for entrainment of the circadian rhythms of locomotor activity (Page &and Larimer, 1972) and heart rate (Pollard & Larimer, 1977) in the crayfish *Procambarus clarkii* since surgical removal of the eye and most distal neuropile of the optic lobe had no effect on the synchronization of these rhythms to light cycles. Experiments utilizing fiber optics to provide local illumination indicate that the rhythm in ERG amplitude in this species is also extraretinally entrained (Page & Larimer, 1976). The identity of the photoreceptive cells is not certain. Surgical removal of the caudal photoreceptor, a well-known light-sensitive neuron in the sixth abdominal ganglion, had no detectable effect on entrainment of the locomotor activity rhythm, indicating that this photoreceptor is not necessary (Page & Larimer, 1972, 1976), though there is some evidence that it might participate in entrainment (Fuentes-Pardo & Inclan-Rubio, 1987). In general, the data suggest the primary photoreceptor for entrainment is located in the crayfish brain (Page & Larimer, 1976). Interestingly, two clusters of cells with immunohistochemical and electrophysiologic properties of photoreceptors have been identified in the anterior margin of the cerebral lobes of the crayfish brain and are potential candidates for the site of the photoreception for entrainment (Sandeman, Sandeman, & de Covet, 1990).

In insects, extraretinal photoreceptors are typically involved in entrainment. The first demonstration was by Truman (1972b) who showed that the photoreceptor for entrainment of the eclosion rhythm of the silkmoth resides in the brain. Brains were removed from a population of pupae and were either replaced in the head region or were transplanted to the abdomen. The pupae were then placed in holes in a partition which separated two chambers in which the light–dark cycles were out of phase. Whether the pupae entrained the light cycle to which the anterior end of the pupae was exposed or entrained to the light cycle at the posterior end corresponded to the location of the brain.

Additional evidence for extraretinally mediated entrainment to light has been obtained in a variety of other insects including other lepidopterans (Mizoguchi & Ishizaki, 1984a, b; Sasaki, Yamazaki, & Chiba, 1987; Shimizu & Matsui, 1983), dipterans (Cymborowski *et al.*, 1994; Dushay, Rosbash, & Hall, 1989; Engelmann & Honegger, 1966; oHelfrich *et al.*, 1985; Kasai & Chiba, 1987; Zimmerman & Ives, 1971), and orthopterans (Dumortier, 1972; Loher & Chandrashekaran, 1970; Waddel *et al.*, 1990). In those cases where there is evidence on the location of the photoreceptor, the brain appears to be the most likely site. However, more precise identification of the cells involved in the phototransduction has not been accomplished.

Although "organized" photoreceptive structures (eyes, ocelli) are frequently bypassed in entrainment in insects, there are at least two cases, the cockroach and the cricket, in which the compound eyes are the exclusive photoreceptors for entrainment. In the cockroaches *P. americana* and *L. maderae*, section of the optic nerves between the eyes and the optic lobe or painting over the compound eyes eliminated entrainment of the locomotor activity rhythm to light cycles (Nishiitsutsuji-Uwo & Pittendrigh, 1968a; S. K. Roberts, 1965). Similarly in the crickets *T. commodus* and *G. bimaculatus*, severing the optic nerve abolished entrainment by light (Loher, 1972; Tomioka & Chiba, 1984). In both the cricket *G. bimaculatus* (Tomioka, Okada, & Chiba, 1990) and the cockroach *L. maderae* (Page, 1990a), there have been efforts further to localize the specific photoreceptors in the eye that are involved by surgical reduction of the eyes. In both cases, photoreceptors sufficient for entrainment were found to be distributed over the entire surface of the eye. In the cockroach, there are two classes of receptors based on spectral response, the green receptors and the UV receptors (Mote & Goldsmith, 1970). Peak sensitivity for entrainment was found to be near 495 nm, indicating that the green photoreceptors are predominately involved (Mote & Black, 1981).

Of the holometabolous insects, only in coleopterans is there evidence that the compound eyes participate in entrainment. In the beetle *Blaps gigas*, local illumination of the eyes is able to entrain the circadian rhythm in ERG amplitude (Koehler & Fleissner, 1978). However, there is morphologic evidence for extraretinal photoreceptors in the optic lobes of some beetle species and it is likely that with local retinal illumination, light penetrated to these structures (Fleissner, Fleissner, & Frisch, 1993; Schulz, Schluter, & Seifert, 1984). Thus it is possible that in beetles, these extraretinal, optic lobe photoreceptors mediate entrainment.

In chelicerate arthropods, both retinal and extraretinal photoreceptors may be involved in the entrainment of ERG amplitude rhythms recorded from the compound eyes. In the scorpion *Androctonus australis,* localized illumination of either the medial or lateral eyes was sufficient for entrainment (Fleissner, 1977a, b). Scorpions are also known from electrophysiologic experiments to possess extraretinal photoreceptors in the metastoma (tail). Whether these photoreceptors might contribute to entrainment is uncertain (Fleissner, 1977a, b; Hanna, Horne, & Renninger, 1988; Zwicky, 1970). In contrast, in the horseshoe crab, *Limulus polyphemus,* the data are quite clear in that the ERG amplitude rhythm of the lateral eyes can be phase shifted via either the ventral, median, or lateral eyes (Horne & Renninger, 1988) or extraretinal photoreceptors in the telson (Hanna *et al.*, 1988).

Photoreception for "Nonneural" Pacemakers

As detailed above, there are at least three instances where there is convincing evidence for circadian pacemakers outside the nervous system. In the case of the moth testis, since the rhythm measured *in vitro* responds to light, some cells in the testis–seminal duct complex must be photosensitive (Giebultowicz *et al.*, 1989). Similarly, in the saturnid moth *Samia cynthia* (Mizoguchi & Ishizaki, 1982) and in the bug *Rhodnius prolixus* (Vafopoulou and Steel, 1992), the photoreceptor for entrainment of the pacemaker in the prothoracic glands appears to be in the gland itself. The cuticle deposition rhythm of the cockroach presents a particularly unusual case. This rhythm, which appears to be driven by a pacemaker located within the epidermis, was unresponsive to light cycles, and its phase appears to be set by the time of the imaginal molt (Wiedenmann *et al.*, 1986). This is the only circadian rhythm known that cannot be entrained by light.

In summary, efforts to localize photoreceptors for entrainment in invertebrates have produced a diversity of results and no clear generalizations have emerged. While extraretinal photoreception is prevalent, in some species of both arthropods and mollusks, retinal photoreceptors may serve as alternate or sole sources of photic information for the circadian system. The functional significance and consequences of the diversity in photoreception for entrainment in invertebrates is uncertain, as it is for the vertebrates. One difficulty is that the specific cells involved in the phototransduction, which often localize to a restricted region, have not been unequivocally identified. Thus, it has not been possible to tackle many of the detailed questions about the physiologic properties of the photoreceptors that could provide clues to functional importance. One question of interest concerns the general reliance on extraretinal photoreception for entrainment. It seems likely to be significant that extraretinal pathways are quite prevalent even in species that have well-developed visual systems. Further, in *Bulla,* where the photoreceptors for entrainment that reside in the retina have been more precisely localized, the data suggest that photoreception for entrainment of the circadian oscillation generated at the base of the retina does not appear to involve the distal, microvillus photoreceptors that surround the lens of the eye. Instead, the pacemaker cells (BRNs) themselves appear to be the site of phototransduction (Block & Wallace, 1982; Block, McMahon, Wallace, & Friesen, 1984; Geusz & Page, 1991). These observations raise the possibility that the photoreceptors involved in entrainment may possess unique properties relative to photoreceptors that participate in more "classical" visual processes. This suggestion is consistent with the observations in mammals that photoreceptors for entrainment, though located in the retina, may have properties that are distinct from the photoreceptors involved in vision (Foster *et al.*, 1991; Nelson & Takahashi, 1991).

In the invertebrates, specifically what properties may be unique is uncertain. In *Bulla*, the action spectrum for the distal photoreceptors appears to be identical to the action spectrum for the BRNs. Both populations of cells show peak sensitivity near 490 nm and the action spectra suggest an opsin-based photopigment is involved in phototransduction in both cell types (Geusz & Page, 1991). In contrast, in *Drosophila pseudoobscura*, studies on the action spectrum for generating phase shifts and on the effects of carotenoid depletion suggest that photopigments of extraretinal photoreceptors involved in phase shifting may be different from those mediating visual responses of the compound eyes and are likely to be based on a flavin (Frank & Zimmerman, 1969; Klemm & Ninnemann, 1976; Zimmerman & Goldsmith, 1971; Zimmerman & Ives, 1971).

A second question is, why is there a frequent involvement of multiple photoreceptive inputs (e.g., *Limulus* or *Limax*)? It is possible that different photoreceptors provide different types of information about the photic environment. For example, in *Limax*, the spectral sensitivity of the retinal and extraretinal photoreceptors may be different. Another possibility, given the evidence that the structure of the pacemaking system involves several oscillators, is that the different photoreceptors actually subserve different oscillators in the system.

Finally, potential explanations of the diversity in the distribution and organization of the photoreceptive input to the pacemaker may emerge from a more complete understanding of differences among species in both the developmental and evolutionary history that ultimately led to pacemaker structure. For example, in holometabolous insects, extraretinal pathways may be necessary to provide photic information during metamorphosis, a time when organized photoreceptive structures may be absent or not yet functional (Pittendrigh, 1976; Truman, 1976). Alternatively, the widespread distribution of extraretinal pathways could be a reflection of the participation of a photopigment in the molecular machinery of biological clocks (Pittendrigh, 1976). Clearly, more work will be necessary before truly informed judgments can be made about the functional significance of the diversity in photoreception for entrainment. A first step will be the precise identification of photoreceptors and investigation of the mechanisms of transduction, processing, and transmission of temporal information provided by the daily light cycle.

Mechanisms of Regulation of Pacemaker Phase

Cellular Mechanisms of Entrainment by Light. There has been substantial progress in understanding the cellular and molecular mechanisms by which light regulates the phase of the circadian oscillator in the mollusk eye. A detailed account of these studies is beyond the scope of this review, but there are several recent reviews of the topic (Block, Khalsa, Michel, Geusz, & McMahon, 1993; Colwell, Khalsa, & Block, 1992a; Koumenis & Eskin, 1992). Briefly, in *Bulla*, the data suggest that light is absorbed by an opsin-based photopigment (Geusz & Page, 1991). By an unknown mechanism, the absorption of photons leads to a depolarization of the putative pacemaker neurons, the BRNs. Experiments in *Bulla* with direct current injection have demonstrated that the depolarization is both necessary and sufficient for phase shifting the oscillator (McMahon & Block, 1987a). Reduction in extracellular calcium concentration (Khalsa & Block, 1988; McMahon & Block, 1987a) or application of the calcium channel blocker nickel chloride (Khalsa and Block, 1990) blocked the phase shift caused by depolarization, indicating that a voltage-dependent transmembrane calcium flux is a step in the phase-shifting pathway. The role of the increase in intracellular calcium is uncertain.

In *Aplysia*, an opsin-based pigment is also a likely candidate for photon absorption (Jacklen, 1980). There is evidence in *Aplysia*, but not in *Bulla*, that cyclic GMP is involved in the signal transduction mechanism that leads to depolarization (Eskin, Takahashi, Zatz, & Block, 1984). Membrane depolarization (for example, by bathing in artificial seawater with elevated levels of potassium) can also mimic the phase-shifting effects of light, indicating that potential change is a step in the entrainment pathway. As in *Bulla*, a calcium flux may also be an important element, though the exact role of calcium is uncertain (Eskin, 1972, 1977). There is also evidence that protein synthesis is involved in phase shifting by light in *Aplysia* since synthesis inhibition by anisomycin or cycloheximide blocks light-induced advance phase shifts (Raju, Yeung, & Eskin, 1990). Efforts are currently being directed at identifying those proteins whose synthesis is regulated by light (Raju, Nunez-Regueiro, Cook, & Eskin, 1993).

EFFERENT REGULATION OF THE PACEMAKER. There is evidence in invertebrates that efferent signals can modulate pacemaker phase, period, or responsiveness to light. In *Bulla*, efferent activity arising in the brain lengthens the free-running period of the pacemaker in the eye by about 0.5 hour (Page & Nalovic, 1992), and in *Aplysia*, isolation of the eye by cutting the optic nerve altered both the waveform (Eskin, 1971) and the phase angle of entrainment of the rhythm (Block & Page, 1978). The significance of and the precise mechanism by which efferent activity modulates pacemaker activity in these two systems are unclear.

Serotonin serves as a transmitter for efferent fibers from the brain to the eye (Corrent & Eskin, 1982; Takahashi, Nelson, & Eskin, 1989). The functional role of this serotonergic input is not known, but seems to be involved in entrainment. Bath-applied serotonin modulated the effectiveness of light in inducing phase shifts (Colwell, 1990; Nadakavukaran, Lickey, & Jordon, 1986) and pulses of serotonin caused both advance and delay phase shifts in the circadian rhythm from the eye, though the PRC is shifted about 180° relative to the PRC for light (Corrent, McAdoo, & Eskin, 1978). In the absence of information on the timing of endogenous serotonin input, however, it is difficult to interpret the significance of these effects.

FMRFamide has been tentatively identified as an efferent transmitter whose effect on the BRNs is to hyperpolarize membrane potential, likely via an increase in a potassium conductance (Jacklet, Klose, & Goldberg, 1987; M. H. Roberts & Moore, 1987). These results are consistent with the finding that FMRFamide itself can phase shift the ocular pacemaker in *Bulla* and can block phase shifts to light (Colwell *et al.*, 1992b). There are approximately 10–12 neurons that show FMRFamide immunoreactivity projecting from the central ganglia to the eye (Jacklet *et al.*, 1987). Electrical stimulation of an identified FMRFamide neuron suppresses ongoing impulse activity in the BRNs, consistent with the notion that FMRFamide acts via a hyperpolarization. As with serotonin in *Aplysia*, the functional importance of the efferent FMRFamide projection is unknown.

IDENTIFICATION OF OUTPUT PATHWAYS

In addition to discovering the organization of the pacemaking system, another important issue is how circadian oscillators impose periodicity on the various physiologic and behavioral processes they control. The problem is to discover the physiologic parameters that are involved in the representation and transmission of circadian information. *A priori*, several alternative mechanisms are plausible. Phase information within the individual could be represented by the level of a circulating hormone, impulse frequency in specific neural circuits, changes in general levels or neural excitability, or, as the weight of the available evidence suggests, by some combination of these mechanisms (Page, 1988).

The Molluskan Eye

In the opisthobranch mollusks, the output of the ocular pacemaker to the central nervous system is via the optic nerve, whose integrity is necessary for the normal regulation of behavioral rhythms expressed by the mollusks (see p. 4). The most prominent rhythmic signal in the nerve is the rhythm in compound action potentials. However, in *Bulla*, there is an additional rhythm in the frequency of smaller, asynchronous impulses in the optic nerve that is about 180° out of phase with the rhythm in compound action potentials (Geusz & Page, 1990). Recent results indicate that the rhythmic "small-spike" activity is produced by a small subpopulation of distal photoreceptor cells (the "H-cells") and appears to be driven by the BRNs via an inhibitory synaptic connection (Geusz & Block, 1992). How the information arising from the two out-of-phase cell populations is utilized in the CNS is completely unknown.

Rhythms in Insect Development

A large number of studies suggest that secretion of a variety of insect hormones, including ecdysone (e.g., Steel & Ampleford, 1984), prothoracicotropic hormone (Fujushita & Ishizaki, 1981; Truman, 1972a), and eclosion hormone (Truman, 1971a, b), are under the control of the circadian system during development. These hormones are involved in the regulation of various developmental events including ecdysis. Most of the work has been carried out on various species of moths. The experiments involving the transplantation of the silkmoth brain described above provide the clearest demonstration of a hormonal link in the control of behavior by the circadian system (Truman, 1972b; Truman & Riddiford, 1970). The signal for the eclosion behavior is the eclosion hormone which is produced in neurosecretory cells located in a region near the midline of the brain, the pars intercerebralis, and released via the neurohemal organs, the corporal cardiaca (Truman, 1973). The release of the hormone triggers a preprogrammed motor score in the ventral nerve cord that initiates a stereotyped sequence of behavior that ultimately results in the emergence of the adult moth from the pupal case (Truman, 1971a, b, 1974a; Truman & Sokolove, 1972). Studies on the moth *Manduca sexta* suggest that the timing of the release of eclosion hormone depends on two oscillators, one located in the brain that provides an excitatory drive, and the other located outside the brain that is inhibitory (Truman, 1984b). The second oscillator, which regulates the steroid hormone ecdysone, may be located in the prothoracic glands (see above and Mizoguchi & Ishizaki, 1982). During the final phases of development, the titer of ecdysteroid declines in a series of discrete, daily steps. This rhythmic decline appears to control behavioral preparations for eclosion which involve movements that loosen attachments between the old and new cuticle (Truman, 1984b). Once ecdysteroid titers reaches a sufficiently low level, inhibition of the release of eclosion hormone ends, and the excitatory drive from the brain-centered circadian oscillator triggers release (Truman, 1984b). An interesting proposition is that the two oscillators, one located in the prothoracic gland and regulating ecdysone and the other in the brain and regulating eclosion hormone, are the anatomic and physiologic correlates of the A and B oscillators Pittendrigh (1981) proposed as the basis for the *Drosophila* eclosion rhythm (Truman, 1984a).

Rhythms in Adult Behavior

The role of humoral factors in the regulation of adult behaviors in insects (e.g., locomotor activity) is less clear (reviewed in Page, 1985a). In cockroaches, the timing signal that originates in the optic lobe is transmitted to the brain via the optic tracts (Nishiitsutsuji-Uwo & Pittendrigh, 1968b; Page, 1983a, b) and transmission from the brain to the activity centers in the thorax requires the connectives of the ventral nerve cord be intact (S. K. Roberts, Skopik, & Driskill, 1971; also T. L. Page, unpublished). At both levels, nerve impulse activity is rhythmic (Colwell & Page, 1990). Similarly, in crickets, a variety of lesion studies indicate the axons of the optic tracts mediate the pacemaking signal from the optic lobe to the brain (Loher, 1972; Tomioka & Chiba, 1984; Waddel *et al.*, 1990; Wiedenmann, 1983). The target site of the neural signal from the optic lobe is uncertain. Lesion studies have implicated the region of the pars intercerebralis in cockroaches and crickets (Cymborowski, 1970, 1973; Nishiitsutsuji-Uwo, Petropulos, & Pittendrigh, 1967; S. K. Roberts, 1966; Sokolove & Loher, 1975). This region contains cell bodies of both neurosecretory cells and neurons which project to other structures in the brain such as the central body (which has been shown to be involved in regulation of activity; Huber, 1965). The data suggest that the function of the signal from the optic lobe is to modulate the general level of excitability of the central nervous system, and the modulation appears to involve, at least in part, inhibition of neural activity (Page, 1983b, 1988). A reasonable proposition is that the optic lobe pacemaker modulates excitability through direct action on the central body. However, the evidence does not preclude a humoral link at some point in the pathway between the pacemaker and effectors for behavior, and results from transplantation experiments with the cricket *Acheta domesticus* (Cymborowski, 1981) and the fruitfly *Drosophila melanogaster* (Handler & Konopka, 1979), do indicate that the brains of these insects can release a diffusable factor that can drive a rhythm of locomotor activity for at least a few cycles. Whether this reflects the participation of a hormone in the normal control of rhythmicity remains to be demonstrated (Page, 1985a).

A similar situation appears to be found in decapod crustaceans. In crayfish, section of the circumesophageal connectives, isolating the brain from the rest of the nervous system, abolishes the circadian rhythm in locomotor activity, indicating a neural link between the pacemaker and locomotor centers in the thorax (Page & Larimer, 1975a). It has also been shown that there are multiple pathways in the connective capable of mediating the rhythm (Gordon, Larimer, & Page, 1977), suggesting a general modulation of descending neural activity. Circadian variations in hormones that affect neural activity may contribute to this modulation or be the primary mechanism of control (Arechiga, Cortes, Garcia, & Rodriguez-Sosa, 1985).

In summary, the mechanism by which circadian phase information is transmitted to behavioral effectors in the invertebrates is generally not well understood. The emerging picture is that temporal regulation of behavior involves a daily modulation of excitability in the central nervous system (Brady, 1975; Page, 1988). Axonal connections between the brain and lower components of the central nervous system are clearly required for the maintenance of behavioral rhythms in arthropods and mollusks, but the extent to which humoral factors contribute to, or are responsible for, generating the rhythms is uncertain in most instances.

Rhythms in Retinal Sensitivity to Light

A particularly nice demonstration of a neural link in circadian regulation comes from studies on the efferent regulation of retinal sensitivity in the horseshoe crab, *Limulus polyphemus* (Figure 7). In *Limulus*, a circadian pacemaker in the brain drives synchronous activity in efferent fibers to the lateral, ventral, and median eyes, all of which show a circadian rhythm in sensitivity to light (Barlow, 1983; Barlow, Balanowski, & Brachman, 1977; Barlow *et al.*, 1985). The impulse frequency of the efferent fibers exhibits a circadian rhythm *in situ* (Barlow, 1983; Barlow *et al.*, 1983). Octopamine is likely involved in mediating the response, though one or more other transmitters may also participate (Barlow *et al.*, 1985; Kass & Barlow, 1984). A similar role for efferent neural signals has been shown in scorpions, where neurosecretory fibers, controlled by a circadian pacemaker in the brain, appear to regulate retinal sensitivity (Fleissner, 1983; Fleissner & Fleissner, 1985). In cockroaches (Wills, Page, & Colwell, 1985) and beetles (Fleissner, 1982), lesion studies also indicate that neural efferents between the brain and the retina are necessary to mediate rhythms in retinal sensitivity. In contrast, in crayfish, evidence suggests that a circulating hormone is important for control of the movement of retinal pigment which underlies the rhythm in retinal sensitivity (Arechiga *et al.*, 1985).

In summary, the efforts to identify mechanisms by which circadian clocks regulate the effectors under their control indicate that the pacemakers are able to exploit the full range of signaling mechanisms from direct neural transmission to neuromodulatory signals to circulating hormones. There are no clear-cut general-

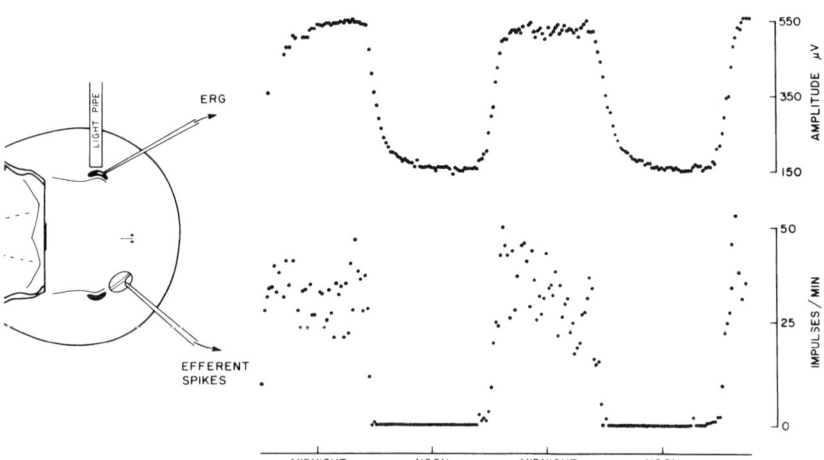

Figure 7. Rhythyms in electroretinogram (ERG) amplitude and efferent impulse activity in the optic nerve of the horseshoe crab, *Limulus polyphemus.* On the left is a schematic of the experimental arrangement. The ERG in response to a brief light pulse delivered via the light pipe was recorded from the left, while an electrode place on the right optic nerve recorded efferent impulse activity. The ERG amplitude (top) and efferent impulse frequency (bottom) show strongly correlated circadian rhythms. (From Barlow, 1983.)

izations suggested by the data other than that it appears that the circadian system is opportunistic in its approach to controlling the timing of the various physiologic and behavioral processes it regulates.

A current focus of research is on the identification of additional molecular signals (neurotransmitters, hormones) used by the circadian system to transmit temporal information. In this context, it is interesting to note recent studies that report circadian rhythms in invertebrates of the hormone melatonin, a molecule well known to participate in daily and seasonal rhythms in vertebrates. Melatonin rhythms have been described in the nervous system of the mollusk *Aplysia californica* (Abran, Anctil, & Ali, 1994), and in the nervous system and hemolymph of the adult cabbage looper moth, *Trichoplusia ni* (Linn, Poole, Wen-Q, & Roelofs, 1995). While the functional significance of melatonin rhythms in invertebrates is not yet clear, the possibility that there may be specific molecular signals for transmitting temporal information that are evolutionarily conserved is quite intriguing.

Acknowledgments

Supported by National Science Foundation Grant No. DCB-9117963 and NIH Grant No. NS 15264.

References

Abran, D., Anctil, M., & Ali, M. A. (1994). Melatonin activity rhythms in the eyes and cerebral ganglia of *Aplysia californica. General and Comparative Endocrinology, 96,* 215–222.

Arechiga, H., Cortes, J. L., Garcia, U., & Rodriguez-Sosa, L. (1985). Neuroendocrine correlates of circadian rhythmicity in crustaceans. *American Zoologist, 25,* 265–274.

Barlow, R. B., Jr. (1983). Circadian rhythms in the *Limulus* visual system. *Journal of Neuroscience, 3,* 856–870.

Barlow, R. B., Jr., Balanowski, S. J., Jr., & Brachman, M. L. (1977). Efferent optic nerve fibers mediate circadian rhythms in the *Limulus* eye. *Science, 197,* 86–89.

Barlow, R. B., Jr., Kaplan, E., Renninger, G. H., & Saito, T. (1985). Efferent control of circadian rhythms in *Limulus* lateral eye. *Neuroscience Research Supplement, 2,* S65–S78.

Barrera-Mera, B. (1976). The effect of cerebroid ganglion lesions on ERG circadian rhythm in the crayfish. *Physiology and Behavior, 17,* 59–64.

Barrett, R. K., & Page, T. L. (1989). Effects of light on circadian pacemaker development. I. The freerunning period. *Journal of Comparative Physiology, 165,* 41–49.

Beiswanger, C. M., Sokolove, P. G., & Prior, D. J. (1981). Extraocular photoentrainment of the circadian locomotor rhythm of the garden slug *Limax. Journal of Comparative Physiology, 216,* 13–23.

Block, G. D., & Davenport, P. A. (1982). Switch from nocturnal to diurnal behavior in the cloudy bubble snail *Bulla gouldiana. Journal of Experimental Zoology, 244,* 57–63.

Block, G. D., & Page, T. L. (1978). Effects of efferent activity on entrainment of the *Aplysia* eye. *Comparative Biochemistry and Physiology A, 62,* 635–638.

Block, G. D., & Roberts, M. H. (1981). Circadian pacemaker in the *Bursatella* eye: Properties of the rhythm and its effect on locomotor behavior. *Journal of Comparative Physiology, 142,* 403–410.

Block, G. D., & Wallace, S. (1982). Localization of a circadian pacemaker in the eye of a mollusk, *Bulla. Science, 217,* 155–157.

Block, G. D., McMahon, D. G., Wallace, S., & Friesen, W. (1984). Cellular analysis of the ocular circadian pacemaker system: A model for retinal organization. *Journal of Comparative Physiology, 155,* 365–378.

Block, G. D., Roberts, M. H., & Lusska, A. E. (1986). Cellular analysis of circadian pacemaker coupling in *Bulla. Journal of Biological Rhythms, 1,* 199–217.

Block, G. D., Khalsa, S., Michel, S., Geusz, M., & McMahon, D. (1993). Cellular basis of biological timekeeping. *International Review of Cytology, 146,* 83–144.

Brady, J. (1975). Circadian changes in central excitability the origin of behavioral rhythms in tsetse flies and other animals? *Journal of Entomology, 50,* 79–95.

Chiba, Y., & Tomioka, K. (1987). Insect circadian activity with special reference to the localization of the pacemaker. *Zoological Science, 4,* 945–954.

Christensen, N. D., & Lewis, R. D. (1982). The circadian locomotor rhythm of *Hemideina thoracica* (Orthoptera; Stenopelmatidae): The circadian clock as a population of interacting oscillators. *Physiological Entomology, 7,* 1–13.

Colwell, C. S. (1990). Light and serotonin interact in affecting the circadian system of *Aplysia. Journal of Comparative Physiology A, 167,* 841–845.

Colwell, C. S., & Page, T. L. (1990). A circadian rhythm in neural activity can be recorded from the central nervous system of the cockroach. *Journal of Comparative Physiology, 166,* 643–649.

Colwell, C. S., Khalsa, S. B. S., & Block, G. D. (1992a). Cellular mechanisms of entrainment. *Chronobiology International, 3,* 163–179.

Colwell, C. S., Khalsa, S. B. S., & Block, G. D. (1992b). FMRFamide modulates the action of phase shifting agents on the ocular circadian pacemakers of *Aplysia* and *Bulla. Journal of Comparative Physiology A, 170,* 211–215.

Corrent, G., & Eskin, A. (1982). Transmitter-like action of serotonin in phase shifting a rhythm from the *Aplysia* eye. *American Journal of Physiology, 242,* R333–R338.

Corrent, G., McAdoo, D. J., & Eskin, A. (1978). Serotonin phase shifts the circadian rhythm from the *Aplysia* eye. *Science, 202,* 977–979.

Cymborowski, B. (1970). Investigations on the neurohormonal factors controlling circadian rhythm of locomotor activity in the house cricket (*Acheta domesticus* L.). I. The role of the brain and the subesophageal ganglion. *Zoologica Poloniae, 20,* 103–126.

Cymborowski, B. (1973). Control of the circadian rhythm of locomotor activity in the house cricket. *Journal of Insect Physiology, 19,* 1423–1440.

Cymborowski, B. (1981). Transplantation of circadian pacemaker in the house cricket, *Acheta domesticus* L. *Journal of Interdisciplinary Cycle Research, 12,* 133–140.

Cymborowski, B., Muszynska-Pytel, M., Porcheron, P., & Cassier, P. (1991). Haemolymph ecdysteroid titres controlled by a circadian clock mechanism in larvae of the wax moth, *Galleria mellonella. Journal of Insect Physiology, 37,* 35–40.

Cymborowski, B., Lewis, R. D., Hong, S. F., & Saunders, D. S. (1994). Circadian locomotor activity rhythms and their entrainment to light–dark cycles continue in flies (*Calliphora vicina*) surgically deprived of their optic lobes. *Journal of Insect Physiology, 40,* 501–510.

Dumortier, B. (1972). Photoreception in the circadian rhythm of stridulatory activity in *Ephippiger* (Ins., Orthoptera): Likely existence of two photoreceptive systems. *Journal of Comparative Physiology, 77,* 80–112.

Dushay, M. S., Rosbash, M., & Hall, J. (1989). The *disconnected* visual system mutations in *Drosophila melanogaster* drastically disrupt circadian rhythms. *Journal of Biological Rhythms, 4,* 1–28.

Engelmann, W., & Honegger, H. W. (1966). Tagesperiodische Schlüpfryhthmik liner augenlosen *Drosophila melanogaster*-Mutante. *Naturwissenschaften, 53,* 588.

Eskin, A. (1971). Properties of the *Aplysia* visual system: *In vitro* entrainment of the circadian rhythm and centrifugal regulation of the eye. *Zeitschrift für Vergleichende Physiologie, 74,* 353–371.

Eskin, A. (1972). Phase shifting a circadian rhythm in the eye of *Aplysia* by high potassium pulses. *Journal of Comparative Physiology, 80,* 353–376.

Eskin, A. (1977). Neurophysiological mechanisms involved in photo-entrainment of the circadian rhythm from the *Aplysia* eye. *Journal of Neurobiology, 8,* 273–299.

Eskin, A., & Harcombe, E. (1977). Eye of *Navanax*: Optic activity, circadian rhythm and morphology. *Comparative Biochemistry and Physiology, 57A,* 443–449.

Eskin, A., Takahashi, J., Zatz, M., & Block, G. D. (1984). Cyclic GMP mimics the effects of light on a circadian pacemaker in the eye of *Aplysia. Journal of Neuroscience, 4,* 2466–2471.

Ewer, J., Frisch, B., Hamblen-Coyle, M. J., Rosbash, M., & Hall, J. (1992). Expression of the *period* clock gene within different cell types in the brain of *Drosophila* adults and mosaic analysis of these cells' influence on circadian behavioral rhythms. *Journal of Neuroscience, 12,* 3321–3349.

Ferrell, B. R., & Reitcheck, B. G. (1993). Circadian changes in cockroach ommatidial structure. *Journal of Comparative Physiology, 173,* 549–556.

Fleissner, G. (1977a). Entrainment of the scorpion's circadian rhythm via the median eyes. *Journal of Comparative Physiology, 118,* 93–99.

Fleissner, G. (1977b). Scorpion lateral eyes: Extremely sensitive receptors of zeitgeber stimuli. *Journal of Comparative Physiology, 118,* 101–108.

Fleissner, G. (1982). Isolation of an insect circadian clock. *Journal of Comparative Physiology, 149,* 311–316.

Fleissner, G. (1983). Efferent neurosecretory fibres as pathways for circadian clock signals in the scorpion. *Naturwissenschaften, 70,* S366.

Fleissner, G. (1986). Die innere Uhr und der Lichtsinn von Skorpionen und Kafern. *Naturwissenschaften, 73,* 78–88.

Fleissner, G., & Fleissner, G. (1985). Neurobiology of a circadian clock in the visual system of scorpions. In F. G. Barth (Ed.), *Neurobiology of arachnids* (pp. 251–375). Berlin: Springer-Verlag.

Fleissner, G., Fleissner, G., & Frisch, B. (1993). A new type of putative non-visual photoreceptor in the optic lobe of beetles. *Cell Tissue Research, 273,* 435–445.

Foster, R. G., Provencio, I., Hudson, D., Fiske, S., De Grip, W., & Menaker, M. (1991). Circadian photoreception in the retinally degenerate mouse (rd/rd). *Journal of Comparative Physiology, 169,* 39–50.

Frank, K. D., & Zimmerman, W. F. (1969). Action spectra for phase shifts of a circadian rhythm in *Drosophila. Science, 163,* 688–689.

Fuentes-Pardo, B., & Inclan-Rubio, V. (1987). Caudal photoreceptors synchronize the circadian rhythms in crayfish—I. Synchronization or the ERG and locomotor circadian rhythms. *Comparative Biochemistry and Physiology, 86A,* 523–527.

Fujishita, M., & Ishizaki, H. (1981). Circadian clock and prothoracicotropic hormone secretion in relation to the larval-larval ecdysis rhythm of the saturnid *Samia cynthia ricini. Journal of Insect Physiology, 27,* 122–128.

Geusz, M. E., & Block, G. D. (1992). The retinal cells generating the circadian small spikes in the *Bulla* optic nerve. *Journal of Biological Rhythms, 7,* 255–268.

Geusz, M. E., & Page, T. L. (1990). The circadian rhythm and photosensitivity of small impulses of the *Bulla* eye. *Journal of Comparative Physiology, 166,* 795–801.

Geusz, M. E., & Page, T. L. (1991). An opsin-based photopigment mediates phase shifts of the *Bulla* circadian pacemaker. *Journal of Comparative Physiology A, 168,* 565–570.

Geusz, M. E., Foster, R. G., Lawrence, M. D., & Block, G. D. (1991). Opsin-like immunoreactivity in the putative pacemaker neurons in the *Bulla* eye. *Society for Neuroscience Abstracts, 17,* 1240.

Giebultowicz, J. M., Bell, R. A., & Imberski, R. B. (1988). Circadian rhythm of sperm movement in the male reproductive tract of the gypsy moth *Lymantria dispar. Journal of Insect Physiology, 34,* 527–532.

Giebultowicz, J. M., Riemann, J. G., Raina, A. K., & Ridgway, R. L. (1989). Circadian system controlling release of sperm in the insect testes. *Science, 245,* 1098–1100.

Gordon, W. H., Larimer, J. L., & Page, T. L. (1977). Circumesophageal interneurons required for reflexive and circadian locomotor behaviors in the crayfish. *Journal of Comparative Physiology A, 116,* 227–238.

Handler, A. M., & Konopka, R. J. (1979). Transplantation of a circadian pacemaker in *Drosophila. Nature, 279,* 236–238.

Hanna, W. J. B., Horne, J. A., & Renninger, G. H. (1988). Circadian photoreceptor organs in *Limulus. Journal of Comparative Physiology A, 162,* 133–140.

Helfrich, C. (1986). Role of the optic lobes in the regulation of the locomotor activity rhythm of *Drosophila melanogaster:* Behavioral analysis of neural mutants. *Journal of Neurogenetics, 3,* 321–343.

Helfrich, C., & Engelmann, W. (1983). Circadian locomotor activity in *Drosophila melanogaster* mutants "sine oculis" and "small optic lobes." *Physiological Entomology, 8,* 257–272.

Helfrich, C., Cymborowski, B., & Engelmann, W. (1985). Circadian activity rhythm of the house fly continues after optic tract severance and lobectomy. *Chronobiology International, 2,* 19–32.

Helfrich-Forster, C. (1995). The *period* clock gene is expressed in central nervous system neurons which also produce a neuropeptide that reveals the projections of circadian pacemaker cells with the brain of *Drosophila melanogaster. Proceedings of the National Academy of Sciences of the USA, 92,* 612–616.

Helfrich-Forster, C., & Homberg, U. (1993). Pigment-dispersing hormone-immunoreactive neurons in the nervous system of wild-type *Drosophila melanogaster* and of several mutants with altered circadian rhythmicity. *Journal of Comparative Neurology, 337,* 177–190.

Homberg, U., Wurden, S., Dircksen, H., & Rao, K. R. (1991). Comparative anatomy of pigment-dispersing hormone-immunoreactive neurons in the brain of orthopteroid insects. *Cell Tissue Research, 266,* 343–357.

Horne, J. A., & Renninger, G. H. (1988). Circadian photoreceptor organs in *Limulus. Journal of Comparative Physiology A, 162,* 127–132.

Huber, F. (1965). Neural integration. In M. Rockstein (Ed.), *The physiology of Insecta* (Vol. 2, pp. 333–406). New York: Academic Press.

Hudson, D., & Lickey, M. (1980). Internal desynchronization between two identified circadian oscillators in *Aplysia. Brain Research, 183,* 481–485.

Ishizaki, H., Mizoguchi, A., & Fujishita, M. (1984). Circadian clock control of hormone secretion in *Samia cynthia ricini.* In R. Porter & G. Collins (Eds.), *Photoperiodic regulation of insect and molluscan hormones* (pp. 136–145). London: Pitman.

Jacklet, J. W. (1969). Circadian rhythm of optic nerve impulses recorded in darkness from isolated eye of *Aplysia. Science, 164,* 562–563.

Jacklet, J. W. (1971). A circadian rhythm in the optic nerve impulses from an isolated eye in darkness. In M. Menaker (Ed.), *Biochronometry* (pp. 351–362). Washington DC: National Academy of Sciences.

Jacklet, J. W. (1974). The effects of constant light and light pulses on the circadian rhythm in the eye of *Aplysia. Journal of Comparative Physiology, 90,* 33–45.

Jacklet, J. W. (1980). Light sensitivity of the rhinophores and eyes of *Aplysia. Journal of Comparative Physiology A, 136,* 257–262.

Jacklet, J. W., & Barnes, S. (1993). Photoresponsive pacemaker neurons from the dissociated retina of *Aplysia. Neuroreport, 5,* 209–212.

Jacklet, J. W., & Colquhoun, W. (1983). Ultrastructure of photoreceptors and circadian pacemaker neurons in the eye of a gastropod, *Bulla. Journal of Neurocytology, 12,* 373–396.

Jacklet, J. W., Klose, M., & Goldberg, M. (1987). FMRF-amide-like immunoreactive efferent fibers and FMRF-amide suppression of pacemaker neurons in eyes of *Bulla. Journal of Neurobiology, 18,* 433–449.

Jordon, W. P., Lickey, M., & Hiaasen, S. (1985). Circadian organization in *Aplysia*: Internal desynchronization and amplitude of locomotor rhythm. *Journal of Comparative Physiology, 156,* 293–303.

Kasai, M., & Chiba, Y. (1987). Effects of optic lobe ablation on circadian activity in the mosquito, *Culex pipiens pallens. Physiological Entomology, 12,* 59–65.

Kass, L., & Barlow, R. B., Jr. (1984). Efferent neurotransmission of circadian rhythms in *Limulus* lateral eye. *Journal of Neuroscience, 4,* 908–917.

Kass, L., & Barlow, R. B., Jr. (1992). A circadian clock in the *Limulus* brain transmits synchronous efferent signals to all eyes. *Visual Neuroscience, 9,* 493–504.

Kavaliers, M. (1981). Circadian and ultradian activity rhythms of a freshwater gastropod, *Helisoma trivolvis*: The effects of social factors and eye removal. *Behavioral and Neural Biology, 32,* 350–363.

Khalsa, S. B. S., & Block, G. D. (1988). Calcium channels mediate phase shifts of the *Bulla* circadian pacemaker. *Journal of Comparative Physiology, 164,* 195–206.

Khalsa, S. B. S., & Block, G. D. (1990). Calcium in phase control of the *Bulla* circadian pacemaker. *Brain Research, 506,* 40–45.

Klemm, E., & Ninnemann, H. (1976). Detailed action spectrum for the delay shift in pupae emergence of *Drosophila pseudoobscura. Photochemistry and Photobiology, 24,* 369–371.

Koehler, W. K., & Fleissner, G. (1978). Internal desynchronization of bilaterally organized circadian oscillators in the visual system of insects. *Nature, 274,* 708–710.

Koumenis, C., & Eskin, A. (1992). The hunt for mechanisms of circadian timing in the eye of *Aplysia. Chronobiology International, 9,* 201–221.

Larimer, J., & Smith, J. T. F. (1980). Circadian rhythm of retinal sensitivity in crayfish: Modulation by the cerebral and optic ganglia. *Journal of Comparative Physiology, 136,* 313–326.

Lickey, M., & Wozniak, J. (1979). Circadian organization in *Aplysia* explored with red light, eye removal, and behavioral recording. *Journal of Comparative Physiology, 131,* 169–177.

Lickey, M. E., Block, G. D., Hudson, D. J., & Smith, J. T. (1976). Circadian oscillators and photoreceptors in the gastropod, *Aplysia. Photochemistry and Photobiology, 23,* 253–273.

Lickey, M., Wozniak, J., Block, G., Hudson, D., & Augter, G. (1977). The consequences of eye removal for the circadian rhythm of behavioral activity in *Aplysia. Journal of Comparative Physiology, 118,* 121–143.

Lickey, M., Hudson, D., & Hiaasen, S. (1983). Circadian organization in *Aplysia*: Relations between locomotor rhythm and eye rhythms after cutting both, one, or neither optic nerves. *Journal of Comparative Physiology, 153,* 133–143.

Linn, C. E., Poole, K. R., Wen-Q, W., & Roelofs, W. L. (1995). Circadian changes in melatonin in the nervous system and hemolymph of the cabbage looper moth, *Tricoplusia ni. Journal of Comparative Physiology, 176,* 761–771.

Loher, W. (1972). Circadian control of stridulation in the cricket, *Teleogryllus commodus* Walker. *Journal of Comparative Physiology, 79,* 173–190.

Loher, W. (1974). Circadian control of spermatophore formation in the cricket *Teleogryllus commodus* Walker. *Journal of Insect Physiology, 20,* 1155–1172.

Loher, W., & Chandrashekaran, M. K. (1970). Circadian rhythmicity in the oviposition of the grasshopper *Chorthippus curtipennis. Journal of Insect Physiology, 16,* 1677–1688.

Lukat, R. (1978). Circadian growth layers in the cuticle of behaviorally arrhythmic cockroaches (*Blaberus fuscus,* Ins., Blattoidea). *Experientia, 34,* 477.

Lukat, R., & Weber, F. (1979). The structure of locomotor activity in bilobectomized cockroaches (*Blaberus fuscus*). *Experientia, 35,* 38–39.

McMahon, D. G., & Block, G. D. (1987a). The *Bulla* ocular circadian pacemaker I: Pacemaker neuron membrane potential controls phase through a calcium dependent mechanism. *Journal of Comparative Physiology, 161,* 35–346.

McMahon, D. G., & Block, G. D. (1987b). The *Bulla* ocular circadian pacemaker II: Pacemaker membrane potential regulates the freerunning period of the oscillator. *Journal of Comparative Physiology, 161,* 347–354.

Michel, S., Geusz, M. E., Zaritsky, J. J., & Block, G. D. (1993). Circadian rhythm in membrane conductance expressed in isolated neurons. *Science, 259,* 239–241.

Minis, D. H., & Pittendrigh, C. S. (1968). Circadian oscillation egg hatching: Its ontogeny during embryogenesis of a moth. *Science, 159,* 534–536.

Mizoguchi, A., & Ishizaki, H. (1982). Prothoracic glands of the saturniid moth *Samia cynthia ricini* possess a circadian clock controlling gut purge timing. *Proceedings of the National Academy Sciences of the USA, 79,* 2726–2730.

Mizoguchi, A., & Ishizaki, H. (1984a). Circadian clock controlling gut-purge rhythm of the saturniid *Samia cynthia ricini*: Its characterization and entrainment mechanism. *Journal of Comparative Physiology A, 155,* 639–647.

Mizoguchi, A., & Ishizaki, H. (1984b). Further evidence for the presence of a circadian clock in the prothoracic glands of the saturnid moth *Samia cynthia ricini*: Decapitated larvae can respond to light–changes. *Development, Growth and Differentiation, 26,* 607–611.

Mote, M. I., & Black, K. R. (1981). Action spectrum and threshold sensitivity of entrainment of circadian running activity in the cockroach *Periplaneta americana. Photochemistry and Photobiology, 34,* 257–265.

Mote, M. I., & Goldsmith, T. H. (1970). Spectral sensitivities of color receptors in the compound eye of *Periplaneta americana. Journal of Experimental Zoology, 173,* 137–146.

Nadakavukaran, J., Lickey, M., & Jordon, W. (1986). Regulation of the circadian clock in the *Aplysia* eye. *Journal of Neuroscience, 6,* 14–21.

Nelson, D. E., & Takahashi, J. S. (1991). Sensitivity and integration in a visual pathway for circadian entrainment in the hamster (*Mesocricetus auratus*). *Journal of Physiology, 439,* 115–145.

Nishiitsutsuji-Uwo, J., & Pittendrigh, C. S. (1968a). Central nervous system control of circadian rhythmicity in the cockroach. II. The pathway of light signals that entrain the rhythms. *Zeitschrift gür Vergleichende Physiologie, 58,* 1–13.

Nishiitsutsuji-Uwo, J., & Pittendrigh, C. S. (1968b). Central nervous system control of circadian rhythmicity in the cockroach. III. The optic lobes, locus of the driving oscillation? *Zeitschrift für Vergleichende Physiologie, 58,* 14–46.

Nishiitsutsuji-Uwo, J., Petropulos, S. F., & Pittendrigh, C. S. (1967). Central nervous system control of circadian rhythmicity in the cockroach. I. Role of the pars intercerebralis. *Biological Bulletin, 133,* 679–696.

Page, T. L. (1978). Interactions between bilaterally paired components of the cockroach circadian system. *Journal of Comparative Physiology, 124,* 225–236.

Page, T. L. (1981a). Effects of localized low-temperature pulses on the cockroach circadian pacemaker. *American Journal of Physiology, 240,* R144–R150.

Page, T. L. (1981b). Neural and endocrine control of circadian rhythmicity in invertebrates. In J. Aschoff (Ed.), *Handbook of behavioral neurobiology. Vol. 4. Biological rhythms* (pp. 145–172). New York: Plenum Press.

Page, T. L. (1982a). Extraretinal photoreception in entrainment and photoperiodism in invertebrates. *Experientia, 38,* 1007–1013.

Page, T. L. (1982b). Transplantation of the cockroach circadian pacemaker. *Science, 216,* 73–75.

Page, T. L. (1983a). Regeneration of the optic tracts and circadian pacemaker activity in the cockroach *Leucophaea maderae. Journal of Comparative Physiology, 152,* 231–240.

Page, T. L. (1983b). Effects of optic-tract regeneration on internal coupling in the circadian system of the cockroach. *Journal of Comparative Physiology, 153,* 231–240.

Page, T. L. (1984). Neuronal organization of a circadian clock in the cockroach *Leucophaea maderae.* In R. Porter & G. Collins (Eds.), *Photoperiodic regulation of insect and molluscan hormones* (pp. 115–135). London: Pitman.

Page, T. L. (1985a). Clocks and circadian rhythms in insects. In G. Kerkut & L. Gilbert (Eds.), *Comprehensive insect biochemistry, physiology, and pharmacology VI. Sensory physiology* (pp. 577–652). Oxford: Pergamon Press.

Page, T. L. (1985b). Circadian organization in the cockroach: Effects of temperature cycles on locomotor activity. *Journal of Insect Physiology, 31,* 235–242.

Page, T. L. (1988). Circadian organization and the representation of circadian information in the nervous systems of invertebrates. In J. M. Hekkens, G. A. Kerkhof, & W. J. Rietveld (Eds.), *Trends in chronobiology* (pp. 67–79). Oxford: Pergamon Press.

Page, T. L. (1989). Masking in invertebrates. *Chronobiology International, 6,* 3–11.

Page, T. L. (1990a). Circadian organization in the cockroach. In I. Huber (Ed.), *Cockroaches as models for neurobiology: Applications in biomedical research* (pp. 225–246). Boca Raton, FL: CRC Press.

Page, T. L. (1990b). Circadian rhythms of locomotor activity in cockroach nymphs: Freerunning and entrainment. *Journal of Biological Rhythms, 5,* 273–290.

Page, T. L. (1991). Developmental manipulation of the circadian pacemaker in the cockroach: Relationship between pacemaker period and response to light. *Physiological Entomology, 16,* 243–248.

Page, T. L., & Barrett, R. K. (1989). Effects of light on circadian pacemaker development. II. The response to light. *Journal of Comparative Physiology, 165,* 51–59.

Page, T. L., & Larimer, J. L. (1972). Entrainment of the circadian locomotor activity rhythm in crayfish. *Journal of Comparative Physiology, 78,* 107–120.

Page, T. L., & Larimer, J. L. (1975a). Neural control of circadian rhythmicity in the crayfish I. The locomotor activity rhythm. *Journal of Comparative Physiology, 97,* 59–80.

Page, T. L., & Larimer, J. L. (1975b). Neural control of circadian rhythmicity in the crayfish II. The ERG amplitude rhythm. *Journal of Comparative Physiology, 97,* 81–96.

Page, T. L., & Larimer, J. L. (1976). Extraretinal photoreception in entrainment of crustacean rhythms. *Photochemistry and Photobiology, 23,* 245–251.

Page, T. L., & Nalovic, K. G. (1992). Properties of mutual coupling between the two circadian pacemakers in the eyes of the mollusk *Bulla gouldiana. Journal of Biological Rhythms, 7,* 213–226.

Page, T. L., Caldarola, P. C., & Pittendrigh, C. S. (1977). Mutual entrainment of bilaterally distributed circadian pacemakers. *Proceedings of the National Academy of Sciences of the USA, 74,* 1277–1281.

Page, T. L., Wassmer, G., Fletcher, J., & Block, G. (1997). Aftereffects of entrainment on the period of the pacemaker in the eye of the mollusk *Bulla gouldiana. Journal of Biological Rhythms, 12,* 218–225.

Pittendrigh, C. S. (1976). Circadian clocks: What are they? In J. W. Hastings & H. Schweiger (Eds.), *Molecular basis of circadian rhythms* (pp. 11–48). Berlin: Dahlem Konferenzen.

Pittendrigh, C. S. (1981). Circadian organization and the photoperiodic phenomena. In B. K. Follett & D. E. Follett (Eds.), *Biological clocks in seasonal reproductive cycles* (pp. 1–35). Bristol, England: Wright.

Pollard, T. G., & Larimer, J. L. (1977). Circadian rhythmicity of heart rate in the crayfish, *Procambarus clarkii. Journal of Comparative Physiology, 57,* 221–226.

Raju, U., Yeung, S., & Eskin, A. (1990). Involvement of proteins in light resetting ocular circadian oscillators of *Aplysia. American Journal of Physiology, 258,* R256–R262.

Raju, U., Nunez-Regueiro, M., Cook, R., & Eskin, A. (1993). Identification of an annexin-like protein and its possible role in the *Aplysia* eye circadian system. *Journal of Neurochemistry, 61,* 1236–1245.

Rence, B., & Loher, W. (1975). Arrhythmically singing crickets: Thermoperiodic reentrainment after bilobectomy. *Science, 190,* 385–387.

Riemann, J. G., Thorson, B. J., & Rudd, R. L. (1974). Daily cycle of release of sperm from the testes of the Mediterranean flour moth. *Journal of Insect Physiology, 20,* 195–207.

Roberts, M. H., & Block, G. D. (1987). Analysis of mutual circadian pacemaker coupling between the two eyes of *Bulla. Journal of Biological Rhythms, 15,* 55–75.

Roberts, M. H., & Moore, R. (1987). Localization of neuropeptides in efferent terminals in the eye in the marine snail, *Bulla. Cell Tissue Research, 248,* 67–73.

Roberts, M. H., Block, G. D., & Lusska, A. E. (1987). Comparative studies of circadian pacemaker coupling in opisthobranch mollusks. *Brain Research, 423,* 286–292.

Roberts, S. K. (1962). Circadian activity in cockroaches. II. Entrainment and phase-shifting. *Journal of Cellular and Comparative Physiology, 59,* 175–186.

Roberts, S. K. (1965). Photoreception and entrainment of cockroach activity rhythms. *Science, 148,* 958–959.

Roberts, S. K. (1966). Circadian activity rhythms in cockroaches. III. The role of endocrine and neural factors. *Journal of Cellular and Comparative Physiology, 67,* 473–486.

Roberts, S. K. (1974). Circadian rhythms in cockroaches: Effects of optic lobe lesions. *Journal of Comparative Physiology, 88,* 21–30.

Roberts, S. K., Skopik, S. D., & Driskill, R. J. (1971). Circadian rhythms in cockroaches: Does brain hormone mediate the locomotor cycle? In M. Menaker (Ed.), *Biochronometry* (pp. 505–515). Washington, DC: National Academy of Sciences.

Sanchez, J. A., & Fuentes-Pardo, B. (1977). Circadian rhythm of the amplitude of the electroretinogram in the isolated eyestalk of the crayfish. *Comparative Biochemistry and Physiology, 56,* 601–605.

Sandeman, D. C., Sandeman, R. E., & de Couet, H. G. (1990). Extraretinal photoreceptors in the brain of the crayfish *Cherax destructor. Journal of Neurobiology, 21,* 619–629.

Sasaki, M., Yamazaki, S., & Chiba, K. (1987). Brain photoreception in the calling rhythm of a noctuid moth, *Anadevidia peponis:* Method of making eyeless moth and micro-irradiation of brain with fiber optics. *Bulletin of the Faculty of Agriculture, 27,* 81–90.

Schulz, W., Schluter, U., & Seifert, G. (1984). Extraocular photoreceptors in the brain of *Epilachna varivestis* (Coleoptera, Coccinellidae). *Cell Tissue Research, 236,* 317–320.

Sehgal, A., Price, J., & Young, M. W. (1992). Ontogeny of a biological clock in *Drosophila melanogaster. Proceedings of the National Academy of Sciences of the USA, 89,* 1423–1427.

Shimizu, I., & Matsui, K. (1983). Photoreceptions in the eclosion of the silkworm *Bombyx mori. Photochemistry and Photobiology, 37,* 409–413.
Sokolove, P. G. (1975). Localization of the cockroach optic lobe circadian pacemaker with microlesions. *Brain Research, 87,* 13–21.
Sokolove, P. G., & Loher, W. (1975). Role of eyes, optic lobes, and pars intercerebralis in locomotory and stridulatory circadian rhythms of *Teleogryllus commodus. Journal of Insect Physiology, 21,* 785–799.
Steel, G. H., & Ampleford, E. J. (1984). Circadian control of haemolymph ecdysteroid titres and the ecdysis rhythm in *Rhodnius prolixus.* In R. Porter & G. Collins (Eds.), *Photoperiodic regulation of insect and molluscan hormones* (pp. 150–163). London: Pitman.
Stengl, M. (1995). Pigment-dispersing hormone-immunoreactive fibers persist in crickets which remain rhythmic after bilateral transection of the optic stalks. *Journal of Comparative Physiology, 176,* 217–228.
Stengl, M., & Homberg, U. (1994). Pigment-dispersing hormone-immunoreactive neurons in the cockroach *Leucophaea maderae* share properties with circadian pacemaker neurons. *Journal of Comparative Physiology, 175,* 203–213.
Strumwasser, F. (1973). Neural and humoral factors in the temporal organization of behavior. *Physiologist, 16,* 9–42.
Takahashi, J. S., Nelson, D., & Eskin, A. (1989). Immunocytochemical localization of serotonergic fibers innervating the ocular circadian system of *Aplysia. Neuroscience, 28,* 139–147.
Thorson, B. J., & Riemann, J. G. (1977). Abdominally entrained periodicities of testis and vas deferens activity in the Mediterranean flour moth. *Journal of Insect Physiology, 23,* 1189–1197.
Tomioka, K. (1985). Residual circadian rhythmicity after bilateral lamina–medulla removal or optic stalk transection in the cricket *Gryllus bimaculatus. Journal of Insect Physiology, 31,* 653–657.
Tomioka, K., & Chiba, Y. (1982). Persistence of circadian ERG rhythm in the cricket with optic tract severed. *Naturwissenschaften, 69,* 395–396.
Tomioka, K., & Chiba, Y. (1984). Effects of nymphal stage optic nerve severance or optic lobe removal on the circadian locomotor rhythm of the cricket, *Gryllus bimaculatus. Zoological Science, 1,* 375–382.
Tomioka, K., & Chiba, Y. (1986). Circadian rhythm in the neurally isolated lamina–medulla complex of the cricket *Gryllus bimaculatus. Journal of Insect Physiology, 32,* 747–755.
Tomioka, K., & Chiba, Y. (1992). Characterization of an optic lobe circadian pacemaker by *in situ* and *in vitro* recording of neural activity in the cricket, *Gryllus bimaculatus. Journal of Comparative Physiology, 171,* 1–8.
Tomioka, K., Okada, Y., & Chiba, Y. (1990). Distribution of circadian photoreceptors in the compound eye of the cricket *Gryllus bimaculatus. Journal of Biological Rhythms, 5,* 303–314.
Tomioka, K., Yamada, K., Yokoyama, S., & Chiba, Y. (1991). Mutual interactions between optic lobe circadian pacemakers in the cricket *Gryllus bimaculatus. Journal of Comparative Physiology, 169,* 291–298.
Tomioka, K., Nakamichi, Y., & Yukizane, M. (1994). Optic lobe circadian pacemaker sends its information to the contralateral optic lobe in the cricket *Gryllus bimaculatus. Journal of Comparative Physiology, 175,* 381–388.
Truman, J. W. (1971a). Circadian rhythms and physiology with special reference to neuroendocrine processes in insects. In *Proceedings of the International Symposium on Circadian Rhythmicity* (pp. 111–135). Wageningen, Netherlands: Pudoc Press.
Truman, J. W. (1971b). Physiology of insect ecdysis. I. The eclosion behavior of silkmoths and its hormonal control. *Journal of Experimental Biology, 54,* 805–814.
Truman, J. W. (1972a). Physiology of insect rhythms. I. Circadian organization of the endocrine events underlying the moulting cycle of larval tobacco hornworms. *Journal of Experimental Biology, 57,* 805–820.
Truman, J. W. (1972b). Physiology of insect rhythms. II. The silk moth brain as the location of the biological clock controlling eclosion. *Journal of Comparative Physiology, 81,* 99–114.
Truman, J. W. (1973). Physiology of insect ecdysis. II. The assay and occurrence of the eclosion hormone in the Chinese oak silkmoth, *Antheraea pernyi. Biological Bulletin, 114,* 200–211.
Truman, J. W. (1974a). Circadian release of a prepatterned neural program in silkmoths. In F. O. Schmitt & F. G. Worden (Eds.), *The neurosciences: Third study program* (pp. 525–529). Cambridge, MA: MIT Press.
Truman, J. W. (1974b). Physiology of insect rhythms. IV. Role of the brain in the regulation of the flight rhythm of the giant silkmoths. *Journal of Comparative Physiology, 95,* 281–296.
Truman, J. W. (1976). Extraretinal photoreception in insects. *Photochemistry and Photobiology, 23,* 215–225.
Truman, J. W. (1984a). Physiological aspects of the two oscillators that regulate timing of eclosion in moths. In R. Porter & G. Collins (Eds.), *Photoperiodic regulation of insect and molluscan hormones* (pp. 221–232). London: Pitman.
Truman, J. W. (1984b). The preparatory behavior rhythms of the moth *Manduca sexta*: An ecdysteroid-

triggered circadian rhythm that is independent of the brain. *Journal of Comparative Physiology, 155,* 521–528.

Truman, J. W., & Riddiford, L. M. (1970). Neuroendocrine control of ecdysis in silkmoths. *Science, 167,* 1624–1626.

Truman, J. W., & Sokolove, P. J. (1972). Silkmoth eclosion: Hormonal triggering of a centrally programmed pattern of behavior. *Science, 175,* 1491–1493.

Vafopoulou, X., & Steel, C. G. H. (1991). Circadian regulation of synthesis of ecdysteroids by prothoracic glands of the insect *Rhodnius prolixus*: Evidence of a dual oscillator system. *General and Comparative Endocrinology, 83,* 27–34.

Vafopoulou, X., & Steel, C. G. H. (1992). *In vitro* photosensitivity of ecdysteroid synthesis by prothoracic glands of *Rhodnius prolixus. General and Comparative Endocrinology, 86,* 1–9.

Waddel, B., Lewis, R. D., & Engelmann, W. (1990). Localization of the circadian pacemakers of *Hemideina thoracica* (Orthoptera; Stenopelmatidae). *Journal of Biological Rhythms, 5,* 131–140.

Weber, F. (1985). Postmolt cuticle growth in a cockroach: *In vitro* deposition of multilamellate and circadian-like layered endocuticle. *Experientia, 41,* 398–400.

Wiedenmann, G. (1983). Splitting in a circadian activity rhythm: The expression of bilaterally paired oscillators. *Journal of Comparative Physiology, 150,* 51–60.

Wiedenmann, G., & Loher, W. (1984). Circadian control of singing in crickets: Two different pacemakers for early-evening and before-dawn activity. *Journal of Insect Physiology, 30,* 145–151.

Wiedenmann, G., Lukat, R., & Weber, F. (1986). Cyclic layer deposition in the cockroach endocuticle: A circadian rhythm? *Journal of Insect Physiology, 32,* 1019–1027.

Wills, S. A., Page, T. L., & Colwell, C. (1985). Circadian rhythms in the electroretinogram of the cockroach. *Journal of Biological Rhythms, 1,* 25–37.

Woolum, J., & Strumwasser, F. (1980). The differential effects of ionizing radiation on the circadian oscillator and other functions in the eye of *Aplysia. Proceedings of the National Academy of Sciences of the USA, 77,* 5542–5546.

Yukizane, M., & Tomioka, K. (1995). Neural pathways involved in mutual interactions between optic lobe circadian pacemakers in the cricket *Gryllus bimaculatus. Journal of Comparative Physiology, 176,* 601–610.

Zimmerman, W. F., & Goldsmith, T. H. (1971). Photosensitivity of the circadian rhythm and of visual receptors in carotenoid depleted *Drosophila. Science, 171,* 1167–1168.

Zimmerman, W. F., & Ives, D. (1971). Some photophysical aspects of circadian rhythmicity in *Drosophila.* In M. Menaker (Ed.), *Biochronometry* (pp. 381–391). Washington, DC: National Academy of Sciences.

Zwicky, K. T. (1970). Behavioral aspects of the extraocular light sense of *Urodacus,* a scorpion. *Experientia* (Basel), *26,* 747–748.

5

Circadian Organization in Nonmammalian Vertebrates

HERBERT UNDERWOOD

INTRODUCTION

Remarkable progress has been made in the last quarter century in our knowledge of the concrete nature of biological clocks both at the levels of tissue and organs and at molecular levels. Studies in vertebrates have focused on the anatomic locations of sites involved in generating circadian rhythms, the ways these sites communicate with one another, and the ways that external stimuli, most notably light, affect the system. These studies have revealed a surprising complexity in the organization of circadian systems in nonmammalian vertebrates. Nonmammalian vertebrates are multioscillator in nature, that is, more than one circadian clock is involved in generating rhythmicity, and there are multiple photic input pathways as well. At least three sites have been shown to exhibit autonomous circadian rhythmicity: the pineal organ, the suprachiasmatic area of the hypothalamus, and the eyes. Circadian "organization" is accomplished by coupling together these rhythmic sites via hormonal and neural pathways so that the multiple clocks can act in a coherent fashion to drive the myriad overt rhythms that the organism possesses. In some cases, persistent rhythmicity in one site may require periodic inputs from another.

The present review summarizes knowledge about the circadian system of nonmammalian vertebrates and includes a description of photic input pathways, the three sites that play a role within the circadian system, and the ways these sites are coupled one to another. Some hypotheses are also presented to explain the complexity of the circadian system based on the evolutionary history of these organisms.

HERBERT UNDERWOOD Department of Zoology, North Carolina State University, Raleigh, North Carolina 27695-7617.

Circadian Clocks, Volume 12 of *Handbook of Behavioral Neurobiology*, edited by Joseph S. Takahashi, Fred W. Turek, and R. Y. Moore, Kluwer Academic / Plenum Publishers, New York, 2001.

In contrast to the exclusively retinal input pathway in mammals, the circadian systems of all nonmammalian vertebrates are characterized by multiple photic inputs; these include the eyes, extrapineal photoreceptors located within the brain (termed "extraretinal photoreceptors"), and pineal photoreceptors.

RETINAL AND EXTRARETINAL PHOTORECEPTORS

A number of laboratories have shown that entrainment to 24-hour light–dark (LD) cycles persists after eye removal in at least 24 species of nonmammalian vertebrates, including representatives of fish, amphibians, reptiles, and birds (e.g., Tabata, 1992; Underwood & Groos, 1982; Underwood & Menaker, 1976) (Figure 1). Furthermore, entrainment of blinded animals can persist after removal of the pineal system (Tabata, 1992; Underwood & Groos, 1982). Studies have shown that significant amounts of light can penetrate to the brain of vertebrates and that longer wavelengths of light penetrate tissue more readily than shorter wavelengths (Hartwig & van Veen, 1979). The extraretinal photoreceptors responsible for entrainment in blinded animals are remarkably sensitive to light; entrainment occurs on LD cycles in which the intensity of light is equivalent to full-moon light, or even less. It is likely that all nonmammalian vertebrates possess functional extraretinal photoreceptors capable of mediating entrainment because no example has been found in which entrainment is lost after blinding, that is, if the intensity of the LD cycle is above a minimal threshold. Information about the exact location of these extraretinal photoreceptors has proven elusive. It is clear that these receptors are located in the brain because blocking light to the head abolishes entrainment in blinded house sparrows (McMillan, Keatts, & Menaker, 1975) (Figure 1), frogs (Taylor & Ferguson, 1970), and lizards (Underwood, 1973). Also, localized illumination of the brain via fiber optics can entrain the activity rhythm of the lizard *Sceloporus olivaceus* (Underwood & Menaker, 1976).

Ignorance of the exact location and number of these brain receptors causes significant difficulties with attempts to identify the photopigments involved. Interestingly, a fairly extensive literature has shown that extraretinal photoreceptors are also involved in the photoperiodic response of nonmammalian vertebrates. In a well-designed study, Foster and Follett (1985)showed that the extraretinal photoreceptors mediating the reproductive response to photoperiod in the Japanese quail likely employ a rhodopsin photopigment. A number of studies in birds have localized the extraretinal photoreceptors mediating the photoperiodic response to the medial basal hypothalamus, although other brain sites have occasionally been implicated as well. It is not known, however, if the extraretinal photoreceptors mediating entrainment are identical to the extraretinal photoreceptors mediating the photoperiodic response.

→

Figure 1. Localization of the extraretinal receptor to the brain of the house sparrow. The bird was exposed to a dim LD cycle ($L = 0.02$ lux) throughout the experiment. (1) Feathers plucked from the bird's back; (2) feathers plucked from the head; (3) feathers (which had regrown) plucked from the head; (4) India ink injected under the skin of the head. The bird entrains after light penetration to the brain is increased by removing head feathers, and freeruns if light is blocked by ink injections. (Modified from Menaker, 1971.)

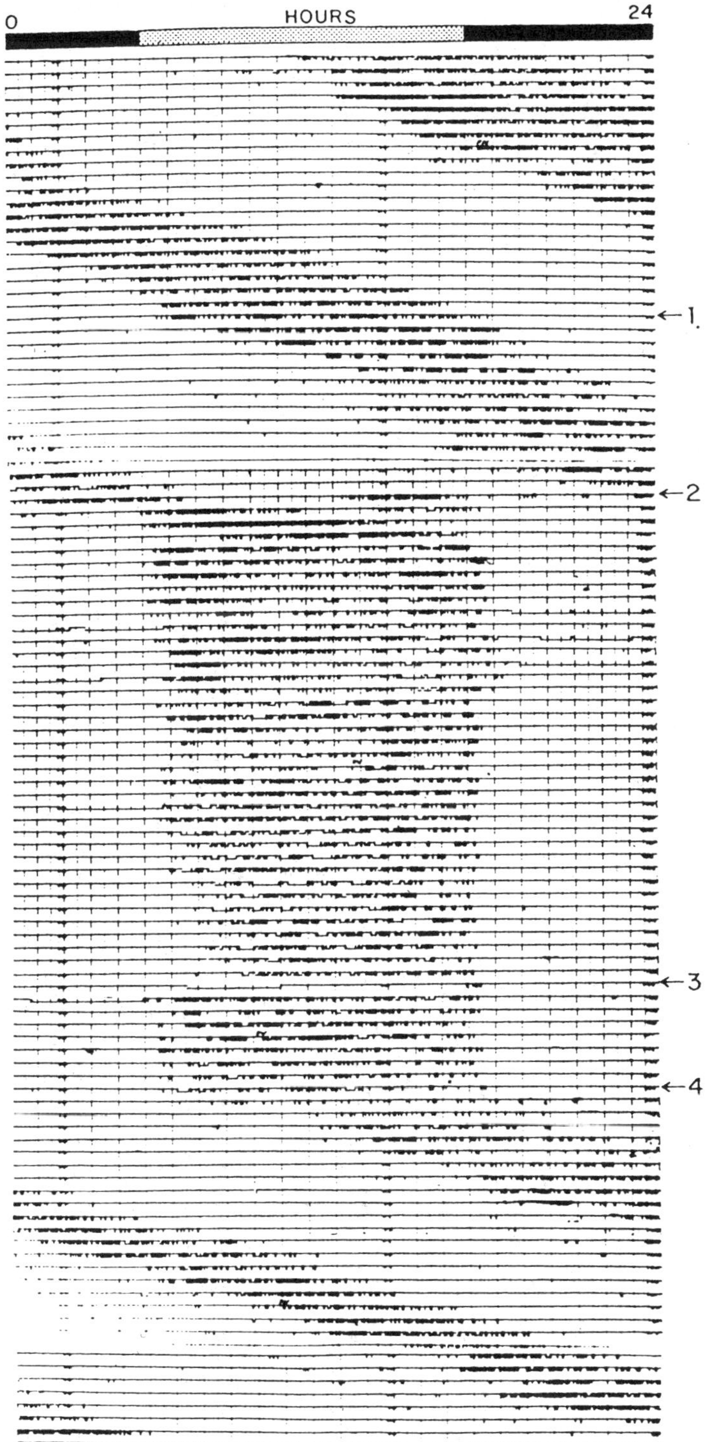
0
HOURS
24
←1
←2
←3
←4

Recent immunohistochemical studies in fish, reptiles, and birds have reignited interest in extraretinal photoreception and may lead to a resolution of the location of the deep brain receptors mediating entrainment and/or photoperiodic responsiveness. In the ring dove, quail, and duck, cerebrospinal fluid (CSF)-contacting neurons within the septal and tuberal areas of the brain can be labeled with an anti-opsin antibody (Silver *et al.*, 1988). In the iguanid lizard *Anolis carolinensis*, three anti-cone opsin antibodies produced an intense immunostaining of CSF-contacting neurons in the lateral ventricle ependyma (adjacent to nucleus paraolfactorius/nucleus ventromedialis and neostriatum/paleostriatum) (Foster, Garcia-Fernandez, Provencio, & DeGrip, 1993). Significantly, on Western blots, the cone opsin antibodies recognized a single 40-kD protein in eye, retinal, anterior brain, and pineal extracts and, furthermore, the vitamin A_2-derived photopigment chromophores 11-*cis*- and all-*trans*-3,4-didehydroretinaldehyde were identified in anterior brain extracts (Foster *et al.*, 1993). Opsin-immunoreactive cells have also been identified in the ependyma of the basal region of the lateral ventricles in the iguanid lizard *Iguana iguana* (Grace, Alones, Menaker, & Foster, 1996). In larval lampreys, anti-rod opsin antisera labeled a population of CSF-contacting neurons within the basal brain (postoptic commissural nucleus and ventral hypothalamic nucleus) (Garcia-Fernandez & Foster, 1994). In addition, opsin-like staining is found within the neurosecretory cells of the nucleus magnocellularis preopticus (NMPO) of fish and amphibians (Foster, Grace, Provencio, DeGrip, & Garcia-Fernandez, 1994). Taken together, these studies suggest multiple types of brain photopigments and at least two types of photoreceptors. Foster and colleagues suggest that CSF-contacting neurons are the strongest candidates for deep-brain photoreceptors in lampreys, reptiles, and birds, whereas classical neurosecretory neurons (NMPO cells) are photosensory in fish and amphibians (Foster *et al.*, 1994). The identification of these sites should provide a focus for studies designed to determine if they are involved in mediating entraining or photoperiodic responses.

Although the data are not extensive, a retinal contribution to entrainment has been demonstrated in some fish, lizards, and birds. For example, in the trout, entrainment to natural LD conditions is weaker in blind fish at certain times of year (Eriksson, 1972). Blinded lake chubs (*Couesius plumbeus*) have a more restricted spectral sensitivity and a higher energy requirement for entrainment than sighted fish (Kavaliers, 1980a). Blinding lizards (*S. olivaceus*) and house sparrows (*Passer domesticus*) exposed to very dim LD cycles can cause them to freerun (McMillan *et al.*, 1975; Menaker, 1971; Underwood, 1973). A significant retinal involvement in entrainment in nonmammalian vertebrates is also supported by the fact that photic input to the circadian system of mammals is exclusively retinal. It seems likely from an evolutionary perspective that light inputs to the circadian system of nonmammalian vertebrates would also involve retinal photoreception.

The relative contribution of retinal and extraretinal photoreceptors to entrainment has not been established. Because the extraretinal photoreceptors have not been definitively localized, it has not been possible to determine if entrainment persists in animals with intact eyes, but without extraretinal (and pineal) photoreceptors. Therefore, while it is clear that the extraretinal photoreceptors can support entrainment in the absence of the eyes, it is not yet clear in most cases if the eyes alone can support entrainment in the absence of extraretinal photoreceptors. Another approach to assessing if the eyes alone are capable of mediating entrainment involves determining if light directed exclusively to the eyes can entrain. Interpretation of this kind of experiment must be made with caution, however, because of the difficulty in insuring that light impinges only on the eyes and does not "leak" to deep-brain photoreceptors.

This approach has been used in both the Japanese quail and the pigeon. In the quail, entrainment of the pineal melatonin rhythm occurred if the eyes were "patched" for 12 of every 24 hours in birds otherwise held in LL (constant light) (Barrett & Underwood, 1991). It is unlikely that entrainment occurred because the patching protocol generated an LD cycle to extraretinal photoreceptors since optic nerve section abolished entrainment in response to the patching regimen. Interestingly, a similar experiment performed in pigeons failed to elicit entrainment of the pineal melatonin rhythm (Hasegawa, Adachi, Yoshimura, & Ebihara, 1994). This may be related to the finding that the eyes of pigeons may be communicating with the central circadian system only via the rhythmic release of melatonin, whereas the eye of the Japanese quail appears to communicate with the rest of the system by both neural and hormonal outputs (see below, section on Eyes as Clocks).

The simplest interpretation of the role of the eyes within the circadian system would assume that the eyes are merely perceiving lighting information and sending this information to circadian clocks located elsewhere. It is likely, however, that in many if not all cases, the role of the eyes is more complex. It is becoming increasingly evident that the eyes themselves are the loci of biological clocks. These intraocular clocks drive numerous circadian rhythms within the eyes, including rhythms in light sensitivity and, in some cases, these intraocular clocks are coupled to, and affect the behavior of, circadian pacemakers located elsewhere. Therefore, intraocular clocks not only modify their own photosensory inputs, but may also interact with other circadian clocks to control the organism's response to light. These roles of the eyes are discussed in greater detail below in a separate section.

Foster and his colleagues have also proposed a theoretical explanation for why both retinal and extraretinal photoreceptors may coexist in nonmammalian vertebrates (Foster *et al.*, 1994). The optical nature of the eye allows a focused representation of the environment: large numbers of photons must be sampled quickly to build a spatial image of the world. The eye measures brightness in a particular point in space (radiance), but not from the whole field of view (irradiance). On the other hand, entrainment of circadian pacemakers (or perception of daylength) requires irradiance detection, but not a spatial imaging capability. Extraretinal photoreceptors are well sited for irradiance detection because overlying tissues scatter light. Because mammals lack extraretinal photoreceptors, Foster and colleagues hypothesize that the random projections of retinal ganglion cells to the location of the mammalian circadian pacemaker (the suprachiasmatic nucleus) could result in a form of irradiance detection in mammals (Foster *et al.*, 1994).

Pineal Organ as a Photoreceptor

The pineal organ is not only a photoreceptor within the circadian system, but it can also act as the locus of a circadian pacemaker. Because of the importance of the pineal organ in circadian systems, a brief description of the anatomy and neurobiology of this organ is provided as background for a discussion of its role as both a photoreceptor and as a biological clock.

Virtually all vertebrates possess a pineal organ. This organ originates as an evagination of the roof of the diencephalon and, in the adult, it remains attached to the brain via a stalk. In most vertebrates, the organ is located on the dorsal surface of the brain, making it readily accessible for investigation. In lower vertebrates (fish, amphibians, and reptiles), there is often a second component to the pineal "system" which is generally termed a parapineal organ (Hamasaki & Eder, 1977). The parapineal organ also originates from the roof of the diencephalon, but usually assumes a more superficial location, moving to a position directly beneath the skull and often underlying an actual opening in the skull. The parapineal component is absent in many species and in some groups, such as the snakes, toads, and turtles. Although some functional connections may occur between the pineal and parapineal organs, a role for the parapineal component within the circadian system has not been shown.

The pineal organs of lower vertebrates contain several types of cells including neurons and photosensory cells (Hamasaki & Eder, 1977; Wiechmann, 1986). In many fish, amphibians, and reptiles, the pineal photoreceptors resemble the photoreceptors of the lateral eyes, possessing a well-defined outer segment and an inner segment with a synaptic pedicle which establishes contact with intrapineal neurons. In some groups, however, such as turtles and snakes, the photosensory cells tend to be rudimentary or "degenerate" in appearance and lack contact with secondary neurons. Birds also can possess pineal cells which exhibit some characteristics of a photosensory cell, but the outer segments are extremely reduced (Collin & Oksche, 1981). The mammalian pineal lacks cells with any photosensory characteristics and appears to be secretory in nature (Karasek, 1983). Immunohistochemical studies, however, have localized some photosensory proteins, such as opsins and S-antigen, in the mammalian pineal organ, although there is no evidence that mammalian pineal organ is actually capable of phototransduction (Foster, Timmers, Schalken, & De Grip, 1989). The pineal photosensory cells have been hypothesized to be the precursors of the secretory cells (termed pinealocytes or parenchymal cells) which are characteristic of the pineal organs of mammals (Collin & Oksche, 1981; Collin *et al.*, 1989).

The abundant ultrastructural and immunocytochemical evidence for photoreception in the parapineal and pineal organs of lower vertebrates is supported by electrophysiologic studies (Hamasaki & Eder, 1977). In general, parapineal organs are capable of wavelength discrimination (chromatic response), whereas pineal organs tend to shown inhibition of on-going electrical activity in response to all wavelengths of visible light (achromatic response), although chromatic responses have been observed in some pineal organs (Morita, Tamotsu, & Uchida, 1989). In fish, amphibians, and reptiles, nerve cells found within the pineal and parapineal organs project axons carrying photic information to pretectal and tegmental centers (Hamasaki & Eder, 1977). Although the avian pineal organ possesses only rudimentary photoreceptors and attempts to record electrophysiologic responses to illumination by the avian pineal organ have proven unsuccessful, functional photoreceptors are indeed present because light can control indoleamine metabolism in the isolated pineal organs of all nonmammalian vertebrates examined so far, including birds (Bolliet, Ali, Lapointe, & Falcón, 1996; Menaker & Wisner, 1983; Takahashi, Murakami, Nikaido, Pratt, & Robertson, 1989).

The decrease in the photosensory aspects of the pineal organs of birds and mammals is paralleled by a decline in intrapineal neurons and their pinealofugal fibers in these groups. In the birds and mammals, the pineal organ is richly innervated by sympathetic nerves originating in the superior cervical ganglia (Cassone & Menaker, 1983; Korf & Møller, 1984). The pineal organs of some lower vertebrates appear to contain sympathetic (efferent) nerves, but the origin of these nerves is uncertain.

In addition to its obvious photosensory capability, the pineal organs of all vertebrates synthesize a variety of biologically active compounds. In particular, a class of compounds called indoleamines has been extensively studied, with most of the interest being focused on the indoleamine melatonin (N-acetyl-5-methoxytryptamine). Immunohistochemical and autoradiographic studies show that the pineal photosensory cells are the sites of indoleamine synthesis (Collin *et al.*, 1989).

The terminal enzyme in the melatonin-synthesizing pathway, hydroxyindole-O-methyltransferase (HIOMT), was originally thought to be found only in pineal tissue, suggesting that melatonin and related methoxyindoles were unique pineal products. More recently, it has been shown that, although the pineal gland is the predominant source of melatonin biosynthesis, other neural and neuroendocrine tissues are capable of melatonin production, including the eyes, the Harderian gland, and enterochromaffin cells located throughout the alimentary tract (Ralph, 1981). In many species, the pineal organ is the principal source of blood-borne melatonin. However, in some species, such as the pigeon and the Japanese quail, the eyes can also contribute significant amounts of melatonin to the blood (Foà & Menaker, 1988; Oshima, Yamada, Goto, Sato, & Ebihara, 1989; Underwood, Binkley, Siopes, & Mosher, 1984). Melatonin is not stored within the pineal: rather, this lipid-soluble molecule is rapidly secreted into the blood.

One of the most notable features of pineal organ metabolism is the presence of daily rhythms in enzyme activities and substrate concentrations, including those involved in indoleamine synthesis. For example, the activity of the enzyme N-acetyltransferase (NAT), which is believed to be the rate-limiting enzyme in the synthesis of melatonin, is higher at night and this is reflected in high nighttime levels of melatonin in the pineal and blood. Interestingly, melatonin levels are always highest at night, regardless of the habits (nocturnal versus diurnal) of the animal. The pineal photosensory cells are not only the sites of melatonin synthesis, but these cells are also the loci of the circadian clocks driving the rhythm of melatonin secretion (Pickard & Tang, 1993, 1994).

Whereas the principal entraining stimulus for the rhythm of melatonin synthesis (and other biochemicals as well) in homeotherms is the daily light cycle, both light and temperature are important in entraining rhythms of indoleamine synthesis in poikilotherms. In poikilotherms, the profile of the melatonin rhythm reflects the nature of the light and temperature cycles to which the organism is exposed. For example, the duration of the light cycle or the amplitude of the temperature cycles can affect the amplitude, phase, or duration of the nocturnal melatonin pulse (Delgado & Vivien-Roels, 1989; Falcón *et al.*, 1994a; Firth & Kennaway, 1989; Underwood, 1990; Vivien-Roels, Arendt, & Bradtke, 1979). Accordingly, the pineal organ can be viewed as a "photothermal transducer" which transduces ambient light and temperature cycles into an internal signal, the melatonin rhythm.

Pineal Organ as a Biological Clock

Several kinds of evidence including (1) pineal ablation and transplant studies, (2) *in vitro* studies of pineal indoleamine metabolism, and (3) studies on the effects of melatonin administration show that the pineal organ in some nonmammalian vertebrates is the locus of a circadian clock that can communicate with the rest of the circadian system via melatonin.

The classical approach of removing the pineal organ and observing the effects on an overt rhythm, usually the rhythm of locomotor activity, has been utilized in fish, reptiles, and birds to assess the pineal organ's role within the circadian system. Table 1 lists the effects of pinealectomy on the free-running activity rhythms of eight species of birds, six species of lizards, one species of amphibian, and five species of fish. Representative activity patterns of the effects of pinealectomy in nine individuals from three classes of vertebrates are shown in Figure 2. Pinealectomy has little effect in one species of bird (Japanese quail) and one species of lizard (desert iguana), but in other species, it causes period changes, splitting, or arrhythmicity. Some species can exhibit more than one kind of response to pinealectomy. For example, some Texas spiny lizards are rendered arrhythmic by pinealectomy, whereas others exhibit period changes or splitting. The range of effects observed is remarkably similar between the reptiles and birds. Splitting behavior, in which an individual can exhibit two different circadian periodicities, is relatively rare and has been observed in one species of lizard, the Texas spiny lizard (*S. olivaceus*), and one species of fish, the white sucker (*Catostomus commersoni*). These data show that the pineal organ must play a significant role within the circadian system and are compatible with the hypothesis that the pineal organ is the site of a circadian pacemaker which controls the period and phase of circadian clocks located elsewhere. Clearly,

Table 1. Effect of Pinealectomy in Nonmammalian Vertebrates

	Effect of pinealectomy	Reference
Fish		
Catfish	Arrhythmicity	Garg & Sundararaj, 1986
Lake chub	Period change	Kavaliers, 1979
Burbot	Period change	Kavaliers, 1980b
White sucker	Period change, splitting	Kavaliers, 1981
Lamprey	Arrhythmicity	Morita, Tabata, Uchida, & Samejima, 1992
Amphibian		
Japanese newt	Arrhythmicity	Chiba *et al.*, 1993
Lizard		
Texas spiny	Period change, splitting, arrhythicity	Underwood, 1977
Green anole	Arrhythmicity	Underwood, 1983
Western fence	Period change, arrhythmicity	Underwood, 1981
Ruin lizard	Period change	Foá, 1991
Desert iguana	No effect	Janik & Menaker, 1990
Gallotia galloti	Arrhythmicity	Molina-Borja, 1996
Bird		
Starling	Period change, arrhythmicity	Gwinner, 1978
House sparrow	Arrhythmicity	Gaston & Menaker, 1968
Java sparrow	Arrhythmicity	Ebihara & Kawamura, 1981
House finch	Arrhythmicity	Fuchs, 1983
White-crowned sparrow	Arrhythmicity	Gaston, 1971
White-throated sparrow	Arrhythmicity	McMillan, 1972
Pigeon	Small period change	Ebihara *et al.*, 1984
Japanese quail	Little or no effect	Simpson & Follett, 1981; Underwood, 1994

the fact that rhythmicity remains after pinealectomy in many cases shows that the pineal organ cannot be the only site generating rhythmicity. The different effects of pinealectomy can be explained by assuming that the strength of coupling differs between the extrapineal oscillators. In the absence of the pineal organ, if coupling is relatively weak, splitting or arrhythmicity will occur, but if coupling is relatively robust, a change in the period of the system will likely occur.

The hypothesis that the pineal organ can act as a circadian *pacemaker* is supported by transplantation studies and *in vitro* studies of pineal metabolism. Transplantation studies have been conducted in one species of bird, the house sparrow, *Passer domesticus* (Zimmerman & Menaker, 1979). Pinealectomy of house sparrows free-running in DD renders them arrhythmic, but transplantation of a pineal organ into the anterior chamber of the eye of these birds will reintroduce rhythmicity within a few days. Significantly, the phase of the donor pineal organ is transplanted along with the pineal organ, supporting a pacemaker function for this organ.

In Vitro Studies

Insight into the role of the pineal organ in the circadian system can be gained by observing the ability of isolated pineal organs to maintain circadian rhythmicity. Studies of organ-cultured pineals of fish, reptiles, and birds have shown that rhythms of NAT activity or melatonin secretion can be entrained by 24-hour LD cycles and, in most cases, these rhythms will persist under constant conditions for at least several cycles (Table 2). Representative examples of rhythms in melatonin secretion from the pineal organs of various species are shown in Figure 3. In five species, a lamprey, *Petromyzon marinus*, an iguanid lizard, *Dipsosaurus dorsalis*, a nocturnal gecko, *Christinus marmoratus*, the trout, *Salmo gairdneri*, and the Japanese quail, *Coturnix coturnix japonica*, a rhythm is observed under LD conditions, but a rhythm is not expressed in constant dark (DD) (Bolliet, Ali, Anctil, & Zachmann, 1993; Gern & Greenhouse, 1988; Janik & Menaker, 1990; Moyer, Firth, & Kennaway, 1995; Murakami, Nakamura, Nishi, Marumoto, & Nasu, 1994). Most of the species tested, however, show rhythmicity in DD, so a biological clock must be driving the rhythm of melatonin secretion. However, rhythmicity in DD tends to dampen out fairly rapidly. In the intact animal, it is possible that pineal organ is only "semiautonomous" and requires a periodic photic or neural input in order to continue to oscillate, or the loss of rhythmicity may be due to uncoupling among multiple oscillators within the pineal organ. A study by Barrett and Takahashi (1997), however, suggests that the damping of rhythmicity observed *in vitro* may be a function of temperature. The rhythm in melatonin production by chick pineal organs persists longer and with a higher amplitude in constant conditions at 40°C than at 37°C.

Although the role of neural inputs in the control of indoleamine rhythmicity in lower vertebrates has been unexplored, sympathetic inputs to the pineal organs of birds (Cassone & Menaker, 1983), and possibly other neural inputs as well (Barrett & Underwood, 1992), have been shown to affect indoleamine metabolism.

Rhythms in NAT activity and melatonin synthesis can be observed in dispersed cell cultures of chick pineal organs and in small fragments of chick pineal organs and pike pineal organs (Barrett & Takahashi, 1997; Bolliet *et al.*, 1994; Takahashi & Menaker, 1984; Takahashi *et al.*, 1989). Furthermore, a study utilizing a reverse hemolytic plaque assay for the detection of melatonin secretion from individual cells demonstrated that individual pineal cells of the iguanid lizard *A. carolinensis* show daily rhythms in melatonin secretion under both LD and DD conditions (Pickard &

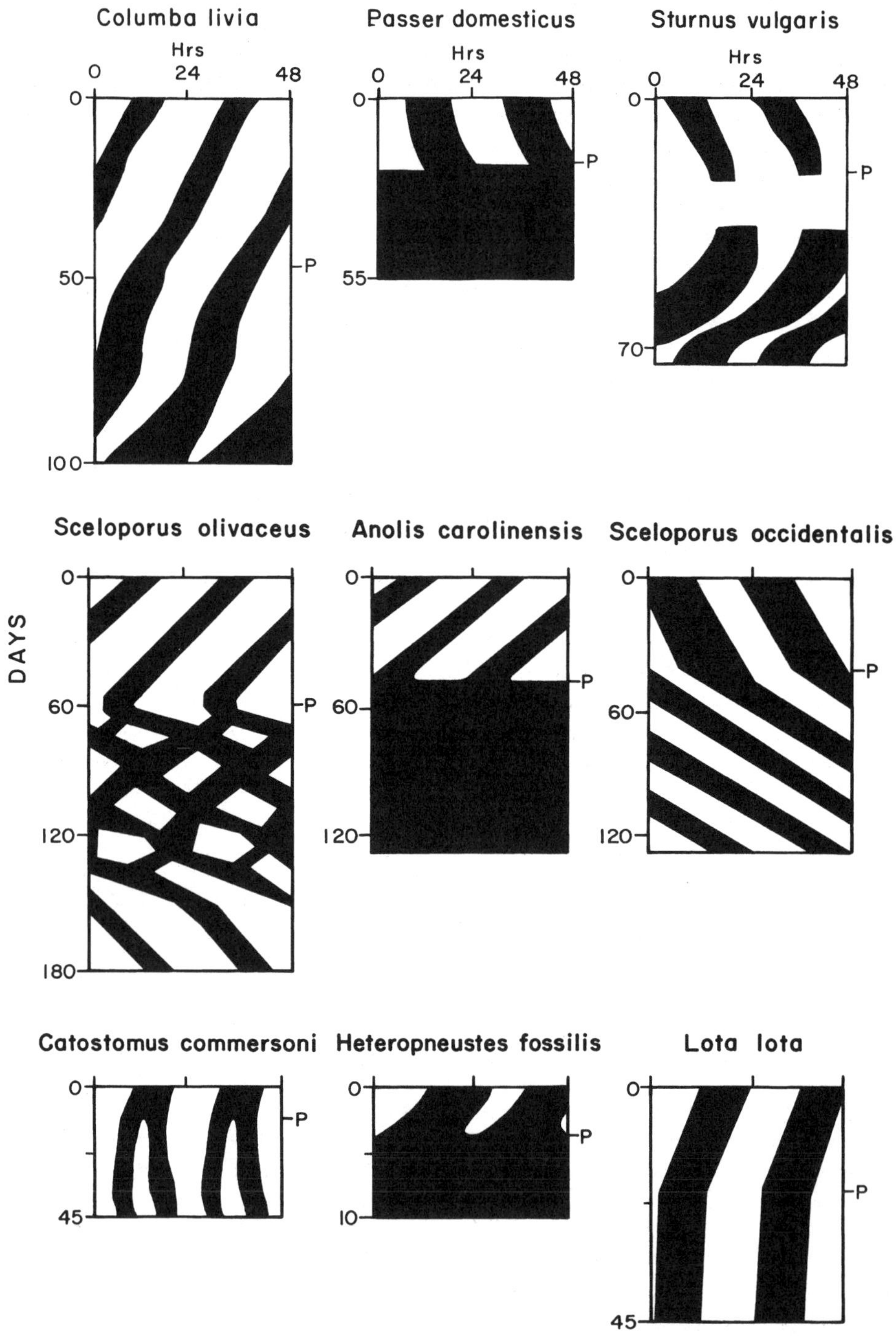

Figure 2. Schema of activity patterns drawn from animals showing effects of pinealectomy in three classes of vertebrates. Top: *Columba livia* (pigeon), *Passer domesticus* (house sparrow), *Sturnus vulgaris* (starling). Middle: *Sceleporus olivaceus* (Texas spiny lizard), *Anolis carolinensis* (green anole), *Scleroporus occidentalis* (Western fence lizard). Lower: *Catostomus commersoni* (white sucker), *Heteropneustes fossilis* (catfish), *Lota lota* (burbot). P, pinealestomized.

TABLE 2. PINEAL MELATONIN SYNTHESIS *IN VITRO*

	LD	DD	Reference
Fish			
White sucker	Rhythmic	Rhythmic	Zachman, Falcón, Knijff, Bolliet, & Ali, 1992
Goldfish	Rhythmic	Rhythmic	Kezuka, Aida, & Hanyu, 1989
Pike	Rhythmic	Rhythmic	Falcón, Marmillon, Claustrat, & Collin, 1989; Falcón *et al.*, 1994a
Trout	Rhythmic	Arrhythmic	Gern & Greenhouse, 1988
Lamprey	Rhythmic	Arrhythmic	Bolliet *et al.*, 1993
Zebrafish	Rhythmic	Rhythmic	Cahill, 1996
Sea bream	Rhythmic	Rhythmic?	Molina-Borja, Falcón, Urquiola, & Ravault, 1996
Catfish	Rhythmic	Rhythmic	Bolliet *et al.*, 1996
Sunfish	Rhythmic	Rhythmic	Bolliet *et al.*, 1996
Bass	Rhythmic	Rhythmic	Bolliet *et al.*, 1996
Shiner	Rhythmic	Rhythmic	Bolliet *et al.*, 1996
Perch	Rhythmic	Rhythmic	Bolliet *et al.*, 1996
Carp	Rhythmic	Rhythmic	Bolliet *et al.*, 1996
Eel	Rhythmic	Rhythmic	Bolliet *et al.*, 1996
Creek chub	Rhythmic	Rhythmic	Bolliet *et al.*, 1996
Alewife	Rhythmic	Rhythmic	Bolliet *et al.*, 1996
Cod	Rhythmic	Rhythmic	Bolliet *et al.*, 1996
Sea raven	Rhythmic	Rhythmic	Bolliet *et al.*, 1996
Flounder	Rhythmic	Rhythmic	Bolliet *et al.*, 1996
Mackerel	Rhythmic	Rhythmic	Bolliet *et al.*, 1996
White hake	Rhythmic	Rhythmic	Bolliet *et al.*, 1996
Lizard			
Desert iguana	Rhythmic	Arrhythmic	Janik & Menaker, 1990
Green anole	Rhythmic	Rhythmic	Menaker & Wisner, 1983
Fence lizard	—	Rhythmic	Menaker, 1985
Gecko	Rhythmic	Arrhythmic	Moyer *et al.*, 1995
Bird			
Chicken	Rhythmic	Rhythmic	Takahashi *et al.*, 1989
House sparrow	Rhythmic	Rhythmic	Murakami *et al.*, 1994
Pigeon	Rhythmic	Rhythmic	Murakami *et al.*, 1994
Japanese quail	Rhythmic	Arrhythmic	Murakami *et al.*, 1994

Tang, 1993). Therefore, the pineal organ itself must be comprised of a number of circadian oscillators which are either normally coupled together, or respond similarly to neural or photic inputs, to generate a single coherent output (i.e., melatonin secretion). The evidence suggests that the pineal photosensory cells are the sites of melatonin synthesis and they may be the sites of the circadian oscillators as well (Collin *et al.*, 1989; Falcón *et al.*, 1994b; Pickard & Tang, 1994; Takahashi *et al.*, 1989).

ROLE OF MELATONIN

A number of studies have implicated a role for melatonin in communicating information between the pineal organ and the rest of the circadian system. Although melatonin can be synthesized in extrapineal sites, in many cases the pineal organ appears to be the main source for blood-borne melatonin. In addition to the observation that removal of the main melatonin-producing organ (the pineal organ) can have dramatic effects on the circadian system of nonmammalian vertebrates, support for a role for melatonin includes observations that exogenous

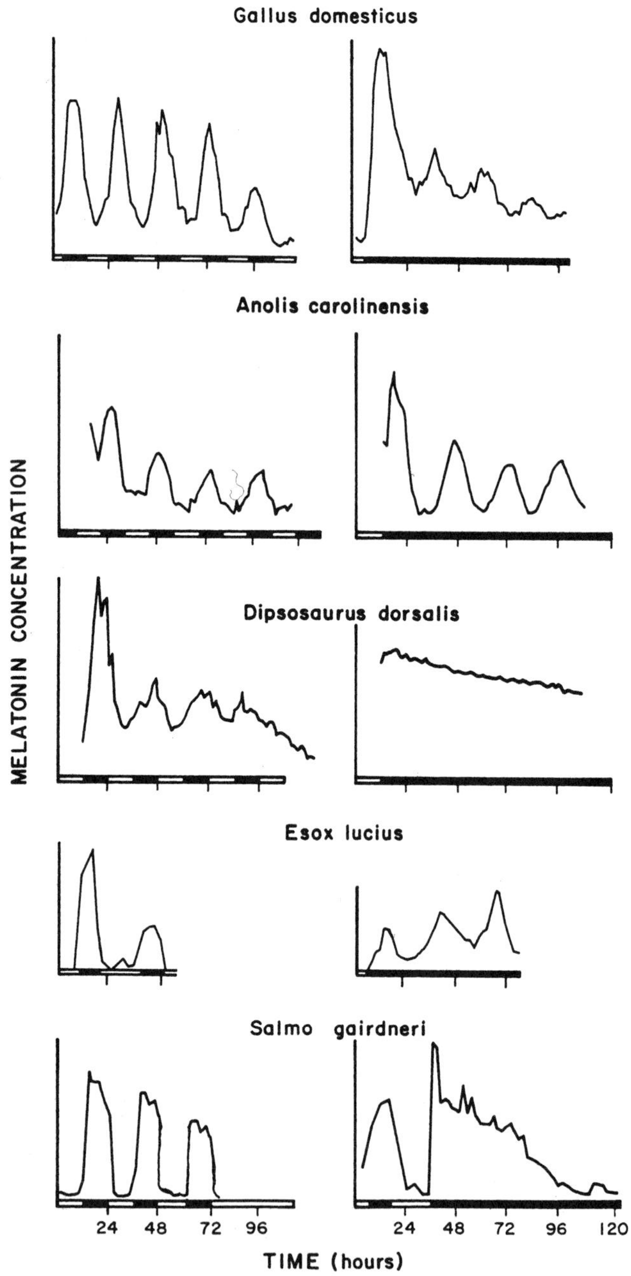

Figure 3. *In vitro* patterns of melatonin secretion from individual pineal glands of the chicken, *G. domesticus*, the green anole, *A. carolinensis*, the desert iguana, *D. dorsalis*, the pike, *E. lucius*, and the trout, *S. gairdneri*. Left panels show melatonin patterns expressed under 24-hour LD cycles and the right panels show patterns expressed in DD.

administration of melatonin can significantly affect the circadian system. Melatonin has been administered to animals continuously via subcutaneous silastic implants, daily via injection or infusion, or daily by ingestion of water containing melatonin. Continuous melatonin administration causes changes in the period (usually a lengthening) of free-running rhythms of birds (house sparrows, Japanese quail) and lizards (western fence lizard, Texas spiny lizard, desert iguana, ruin lizard) and also produces arrhythmicity in birds (house sparrows, Java sparrows, pigeons, Japanese quail) and lizards (Texas spiny lizard) (Chabot & Menaker, 1992a; Foà, Minutini, & Innocenti, 1992; Janik & Menaker, 1990; Turek, McMillan, & Menaker, 1976; Underwood, 1979; Underwood & Edmonds, 1995). A dose-dependent effect of continuous melatonin administration can be seen: increasing doses causes larger period changes or arrhythmicity. Daily injections, daily infusions, or daily presentation of melatonin in the drinking water can entrain the activity rhythms of Japanese newts, western fence lizards, starlings, and house sparrows, the activity and feeding rhythms of pigeons, and the activity and body temperature rhythms of Japanese quail (Chabot & Menaker, 1992b; Chiba, Kikuchi, & Aoki, 1995; Gwinner & Benzinger, 1978; Heigl & Gwinner, 1994, 1995; Hyde & Underwood, 1995; Lu & Cassone, 1993a; Underwood & Edmonds, 1995). Some of these studies used pinealectomized animals, whereas others used intact (nonpinealectomized) animals. Because the dosages of melatonin used in most of these studies were supraphysiologic, it is likely any endogenous melatonin levels were completely swamped by the exogenously administered melatonin. However, in two studies, physiologic dosages of melatonin were administered daily by infusion to pinealectomized animals (Chabot & Menaker, 1992b; Hyde & Underwood, 1995). Figure 4 shows that daily infusions of physiologic

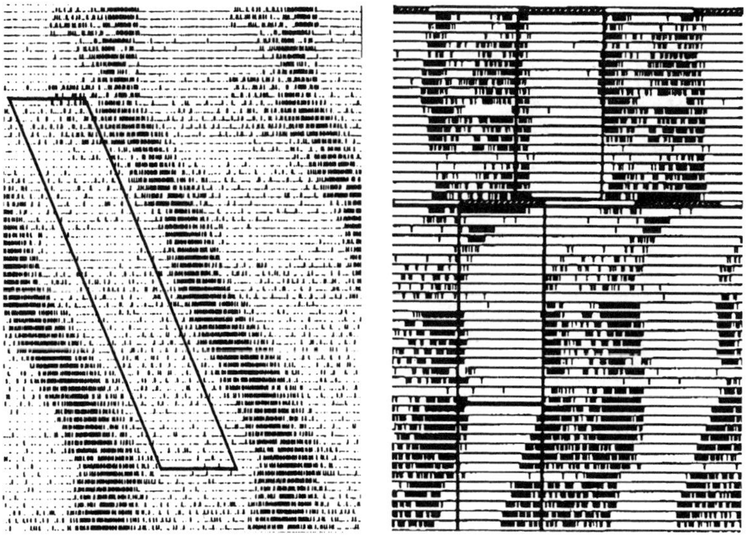

Figure 4. Entrainment of circadian rhythms to melatonin infusions. Left: A melatonin infusion cycle with a period of 24.4 hours entrains the feeding rhythm of a pinealectomized pigeon. The bird was held in DD and melatonin was infused for 10 hours/cycle (during the period enclosed by the polygon). Similar results were seen in pinealectomized and blinded pigeons. (Modified from Chabot and Menaker, 1992b.) Right: A melatonin infusion cycle (12 hours of melatonin infusion/day indicated by the closed rectangles) entrains the activity rhythm of a pinealectomized western fence lizard. Vehicle infusion (bottom enclosed rectangle) fails to entrain. (From Hyde and Underwood, 1995.)

levels of melatonin can entrain the feeding rhythms of pinealectomized pigeons and the activity rhythms of pinealectomized Western fence lizards. A phase response curve (PRC) to melatonin has also been generated for one nonmammalian vertebrate, the Western fence lizard (Underwood, 1986).

Eyes as Clocks

Numerous studies conducted over the last quarter century have shown that the eye is a temporally organized structure. A number of events within the vertebrate eye undergo significant daily fluctuations and many of these rhythms have been shown to persist under constant conditions. These rhythms, therefore, must be controlled by a circadian clock. Ocular rhythms have been described in members of every vertebrate class, and include rhythms in the shedding of disks from photoreceptor outer segments, gene transcription, photomechanical movements of rods and cones, pigment migration, visual sensitivity, synaptic morphology, enzyme activities, and melatonin content (Cahill & Besharse, 1995).

In three classes of vertebrates, the location of the clock driving rhythms in indoleamine metabolism has been definitively localized to the eye itself. Rhythms in NAT activity, tryptophan hydroxylase activity, and melatonin synthesis (Figure 5) can be observed in eye cup preparations of the African clawed frog, *Xenopus laevis*: these rhythms are light-entrainable and will persist in constant darkness (Besharse & Iuvone, 1983; Cahill & Besharse, 1991; Green, Cahill, & Besharse, 1995). The rhythm of tryptophan hydroxylase (TPH) activity is driven by a clock-controlled rhythm in TPH mRNA (Green *et al.*, 1995). An ocular clock has also been shown in the Japanese quail, *Coturnix coturnix japonica* (Underwood, Barrett, & Siopes, 1990) and chicken (Pierce *et al.*, 1993). The quail's eye shows a rhythm in melatonin content *in vivo* which can be entrained by light directed exclusively to the eyes; furthermore, this rhythm is entrainable by retinally perceived light even after transection of the optic nerves. Because optic nerve section would undoubtedly deprive any putative central clocks of entraining information, the fact that entrainment persists after optic nerve section proves that the clock driving the rhythm of melatonin content is located within the eye itself. Studies in the chicken have shown that the mRNA for the cone pigment, iodopsin, fluctuates with a circadian rhythm in primary cultures of dispersed embryonic chick retina and the circadian regulation of iodopsin transcript abundance occurs at the level of gene transcription (Pierce *et al.*, 1993). A self-sustaining oscillator is present in the mammalian eye as well. Circadian rhythms in visual sensitivity and disk shedding persist in the eye of rats after the central pacemaker, the suprachiasmatic nucleus (SCN) has been lesioned (Remé, Wirz-Justice & Terman, 1991; Tosini & Menaker, 1996) and cultured neural retinas of the golden hamster can exhibit light-entrainable circadian rhythms in melatonin synthesis (Tosini and Menaker, 1996). No data are available as to whether or not the ocular clock in mammals drives rhythms outside of the eye itself.

The presence of rhythms in visual sensitivity in the vertebrate eye has significant implications with respect to the way light influences circadian oscillators. Because circadian clocks drive rhythms in light sensitivity, yet light is the primary stimulus causing resetting or entrainment of the clock, the clock is capable of modulating its own primary sensory input.

Several studies in nonmammalian vertebrates suggest that the clock in the eye may play a significant role within the animal's total circadian system; that is, that the

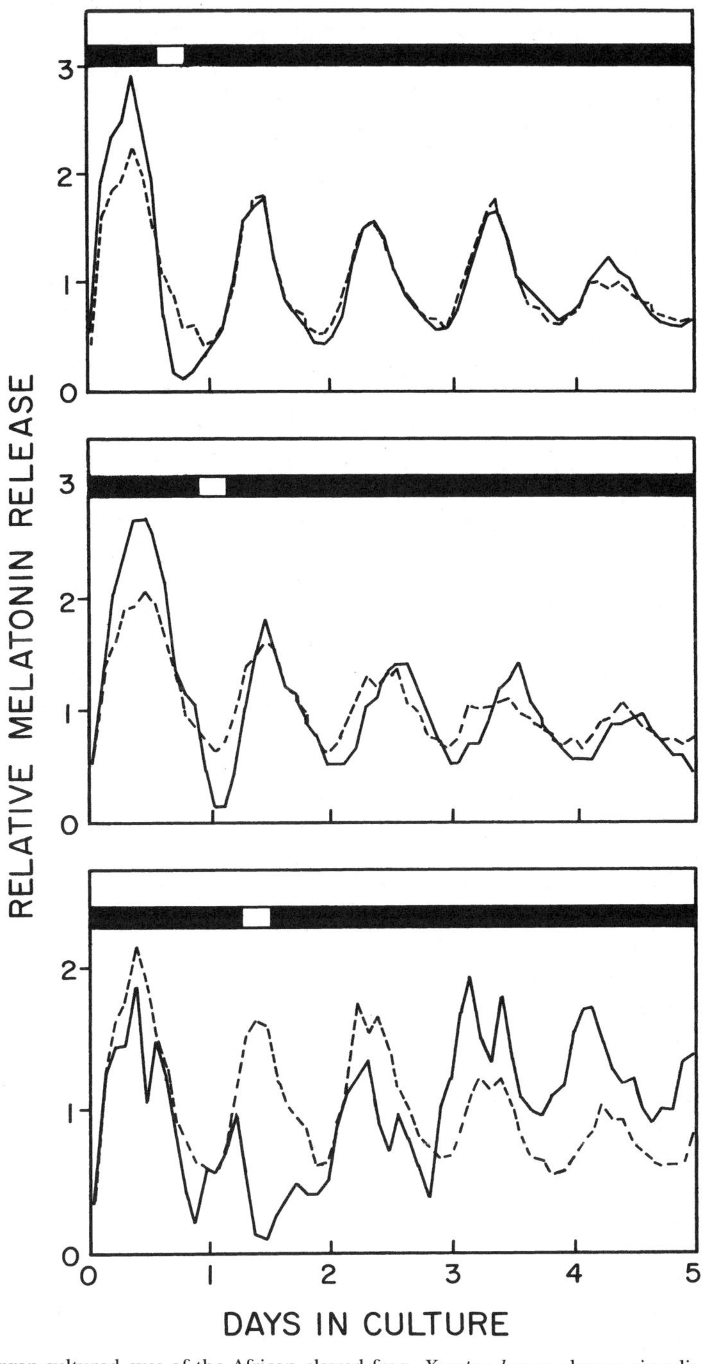

Figure 5. Organ-cultured eyes of the African clawed frog, *Xenopus laevus*, show a circadian rhythm of melatonin synthesis and release that can be reset by light. Control eyecups (dashed lines) were held in DD and the experimental eyecups (solid lines) were exposed to a 6-hour pulse of white light. Pulses presented during the subjective day had no effect, pulses during the early part of the subjective night caused a phase delay, and pulses presented during the late subjective night caused phase advances. (Modified from Cahill and Besharse, 1991.)

eye may act as a circadian pacemaker. Blinding (by eye removal) can cause changes in the free-running period of the lacertid lizard *Podarcis sicula* (Foà, 1991) and the Japanese newt (Chiba, Kikuchi, & Aoki, 1993). In the Japanese quail, blinding by eye removal abolishes the circadian rhythms of locomotor activity and body temperature (Konishi, Ohta, & Homma, 1985; Underwood & Siopes, 1984; Underwood, 1994) (Figure 6). Blinding by eye removal also disrupts the circadian rhythm of activity and body temperature in the pigeon, but only if blinding is combined with pinealectomy. Because blinding affects circadian properties outside of the eye itself, the results in newts, lizards, and birds are compatible with the hypothesis that the eyes of at least some species play a role as pacemakers within a multioscillator system.

In two avian species, the pigeon and the Japanese quail, ocular secretion of melatonin appears to be a mechanism whereby ocular clocks can communicate with the rest of the circadian system. In pigeons, both the pineal and eyes can contribute melatonin to the blood (Foà & Menaker, 1988; Oshima *et al.*, 1989). Furthermore, both the eyes and the pineal of pigeons are involved in circadian organization: blinding alone or pinealectomy alone has little effect, yet blinded pigeons which are also pinealectomized tend to become arrhythmic in DD (Ebihara, Uchiyama, & Oshima, 1984; Oshima *et al.*, 1989). Significantly, daily melatonin infusions which mimic natural levels of melatonin can entrain the activity rhythm of pigeons (Chabot & Menaker, 1992b).

Several observations suggest that the eye of the Japanese quail is coupled to the rest of the circadian system by *both* hormonal (melatonin?) and neural (retinohypothalamic [RH] tract?) pathways. Optic nerve section does not abolish the circadian body temperature rhythm in most Japanese quail, although it does decrease the "robustness" of the rhythm and it causes about 25% of the birds to become arrhythmic in DD (Underwood, 1994). However, complete eye removal abolishes rhythmicity in DD in all quail, suggesting that a hormonal output is involved as well (Underwood, 1994). Melatonin may be the hormonal link involved because (1) the eye contributes significant amounts of melatonin to the blood (Underwood *et al.*, 1984), (2) continuous melatonin administration via subcutaneous silastic capsules either lengthens the free-running period of the quail's body temperature rhythm or produces arrhythmicity, and (3) daily administration of melatonin via the drinking water can entrain the body temperature rhythm of quail (Underwood & Edmonds, 1995).

A neural coupling between putative ocular pacemakers and other components of the circadian system has also been proposed for the lizard *Podarcis sicula*: a significant shortening of the free-running rhythm in DD is caused either by electrolytic lesions that destroy most of both optic nerves at the level of the optic chiasm or by complete removal of both retinas (Foà, 1991; Minutini, Innocenti, Bertolucci, & Foà, 1994).

Although the data from pigeons and Japanese quail suggest that the eyes may contribute to circadian organization via their circadian secretion of melatonin, a role for retinally produced melatonin in the circadian system of several other avian species has not been demonstrated. Pinealectomy in house sparrows, starlings, and chickens appears to reduce plasma melatonin levels to low or nondetectable levels, suggesting that the eyes do not normally contribute melatonin to the blood of these species (Cogburn, Wilson-Placentra, & Letcher, 1987; Janik, Dittami, & Gwinner, 1992). Whether or not the eyes of these species may contain circadian pacemakers that communicate with the rest of the system via neural pathways has not been tested.

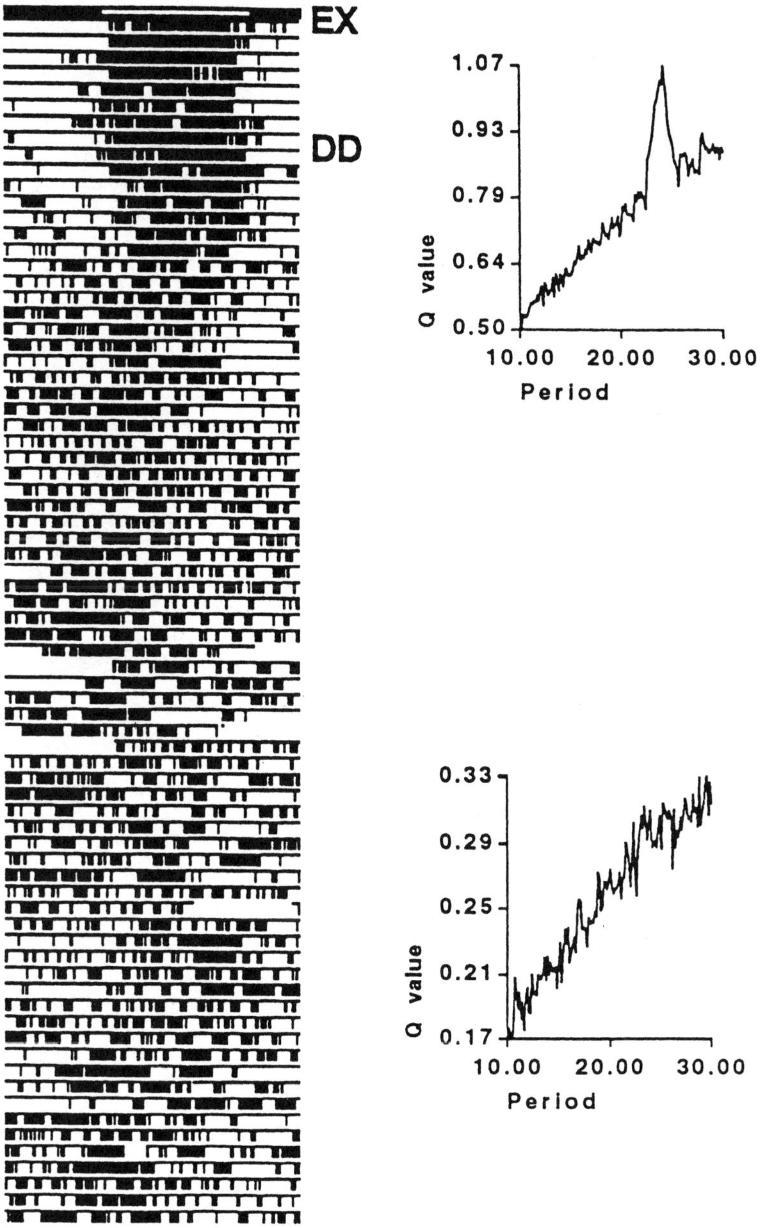

Figure 6. Blinding by complete eye removal (EX) abolishes the circadian rhythm of body temperature in Japanese quail. The blinded bird was rhythmic in LD, but the rhythm decays to arrhythmicity in DD. The F-periodogram analyses show a 24-hour peak during LD, but no rhythmicity during DD.

Melatonin released from the eye, therefore, may play a role within the circadian systems of at least some nonmammalian vertebrates. Melatonin within the eye appears to act as a parahormone controlling some ocular rhythms such as disk shedding, retinomotor movements, and pigment migration (Besharse & Dunis, 1983; Cahill & Besharse, 1995; Pierce & Besharse, 1986; Wiechmann, 1986). Immunohistochemical studies suggest that melatonin is synthesized within retinal photoreceptor cells (e.g., Falcón & Collin, 1991; Wiechmann, 1986, 1996). Further evidence that the photoreceptors are the likely sites of melatonin production include studies in which the retina is chemically or surgically reduced. Destruction of the inner retina of the chick with kainate has little effect on the circadian pattern of N-acetyltransferase in LD and DD (Zawilska and Iuvone, 1992), and removal of the inner retina from isolated eyecup preparations of *Xenopus* with 1% Tris X-100 does not diminish the rhythm of melatonin efflux (Cahill and Besharse, 1992). Cultured photoreceptor layers from *Xenopus* retina which contain mainly (94%) rods or cones can still produce melatonin rhythmically in DD (Cahill & Besharse, 1993).

An interrelationship also exists between melatonin and dopamine, the predominant catecholamine in the retina of most vertebrates (Cahill, Grace, & Besharse, 1991; Zawilska & Nowak, 1992). Within the vertebrate retina, dopamine appears to transduce the light signal, whereas melatonin transduces the dark signal. Light exposure increases the synthesis and release of dopamine in the retina and the subsequent activation of D_2 dopamine receptors on the melatonin-producing cells (the photoreceptors), causing a suppression of NAT activity. Conversely, melatonin released during the dark activates melatonin receptors on the dopamine-producing cells, inhibiting the release of dopamine. Dopamine not only appears to act via D_2 receptors as a mediator for light-induced suppression of melatonin synthesis, but it also acts to mediate light entrainment of the ocular circadian oscillator, at least in *Xenopus*, although a dopamine-independent pathway for light also exists in this system (Cahill & Besharse, 1991; Cahill *et al.*, 1991).

Although plasma melatonin is metabolized by the liver and kidneys, an additional mechanism for melatonin clearance is present in the eyes of at least some vertebrates. In the *Xenopus* eye, melatonin is deacetylated, generating 5-methoxytryptamine, which is then deaminated by monoamine oxidase, producing 5-methoxyindoleacetic acid and 5-methoxytryptophol (Grace & Besharse, 1993). Deacetylation of melatonin has also been shown to occur in the eye of the lizard *Anolis carolinensis*, the fish *Carassius auratus*, and the bird *Gallus domesticus* (Grace, Cahill, & Besharse, 1991). This melatonin-degradation pathway may allow close control over the amount of melatonin present in the eye and determine how much, if any, melatonin is secreted into the blood. The production of methoxyindoles other than melatonin by the retina is also of interest, but nothing is known about the potential roles of these compounds within the circadian system.

In addition to the pineal organ and the eyes, at least one other site, the suprachiasmatic nucleus, has been shown to be an important component of the circadian "system" of nonmammalian vertebrates.

Suprachiasmatic Nucleus

Clearly, the pineal organ and/or eyes cannot be the only loci of circadian pacemakers in nonmammalian vertebrates because, in some individuals, rhythmicity can persist after pinealectomy and/or blinding. Because a substantial literature in mammals has conclusively shown that a circadian pacemaker is present in the hypothalamus, specifically in the suprachiasmatic nuclei, studies on the location(s) of central pacemakers in nonmammalian vertebrates have also focused on the suprachiasmatic area of the hypothalamus. In mammals, the SCN is the principal circadian pacemaker in the system and the main photic input pathway is via a direct monosynaptic retinohypothalamic (RH) tract. The mammalian pineal shows robust rhythms in melatonin synthesis and secretion, but the pineal is not autonomously rhythmic; rather, rhythmicity is driven by neural sympathetic inputs from the SCN. Interestingly, melatonin can affect the circadian system of mammals, so the SCN not only drives melatonin rhythmicity, but it is also sensitive to it.

Unfortunately, far fewer studies have been conducted on the role of the SCN in nonmammalian vertebrates and most of the studies that have been conducted have been done on birds. There is controversy concerning which hypothalamic nucleus in birds is the homologue of the mammalian SCN. Classical anatomic studies suggest that the avian SCN resides adjacent to the preoptic recess of the third ventricle and the optic chiasm in a nucleus termed the "medial hypothalamic nucleus" (MHN) (also termed the periventricular preoptic nucleus), whereas studies of the termination of the RH tract suggest that the SCN resides more caudally, between the supraoptic decussations and the lateroventral geniculate nucleus (vLGN), in a nucleus termed the "lateral hypothalamic retinorecipient nucleus" (LHRN; also termed the visual SCN) (Cassone & Moore, 1987; Norgren & Silver, 1989, 1990). Other terms have also been applied to these nuclei, contributing to the confusion regarding the location of the avian homologue of the mammalian SCN. The LHRN is considered by some to be comparable to the mammalian SCN based on the fact that the LHRN receives significant direct retinal input. However, a number of studies in birds, including cholera toxin mapping of retinal projections in pigeons, show that both the MHN and the LHRN can receive significant retinal input (Norgren & Silver, 1989; Shimizu, Cox, Karten, & Britto, 1994). Attempts to resolve the homologies of the avian hypothalamic nuclei based on immunohistochemical analyses of the various neurochemicals and neurotransmitters found in these nuclei have not been entirely successful. Although a variety of neurochemicals are found in both the MHN and LHRN of birds and in the mammalian SCN, complete homology between the mammalian SCN and either the MHN or the LHRN has not been observed (Norgren & Silver, 1990; Shimizu *et al.*, 1994).

Several investigators have lesioned the suprachiasmatic areas of lizards and birds to determine if these lesions would lead to loss of rhythmicity in the locomotor activity of animals held in DD. Two such studies have been conducted in lizards. Lesions directed at the medial hypothalamic area of the desert iguana, *Dipsosaurus dorsalis*, can cause arrhythmicity of the locomotor activity rhythm in DD (Janik, Pickard, & Menaker, 1990). Janik and coworkers termed this area the SCN and showed that it receives retinohypothalamic inputs, although another hypothalamic nucleus, the supraoptic nucleus, also receives significant retinal input (Janik *et al.*, 1990; Janik, Cassone, Pickard, & Menaker, 1994). Arrhythmicity was observed if 80–90% or more of the SCN was damaged. Although some lesions also damaged the supraoptic nucleus, five of the arrhythmic animals did not show damage to the supraoptic nucleus. Lesions which damaged 80% or more of the SCN also abolished rhythmicity in the ruin lizard, *Podarcis sicula* (Minutini *et al.*, 1995). In birds, most investigators have targeted the MHN for lesion studies, but the total number of species (or individuals) examined is small and, in some cases, the lesions may also have damaged the LHRN as well. Arrhythmicity is caused by MHN lesions in Java sparrows (Ebihara & Kawamura, 1981), Japanese quail (Simpson & Follett, 1981), and house sparrows (Takahashi & Menaker, 1982). Lu and Cassone (1993b) postulate that the central pacemakers in house sparrows reside within the LHRN, and not the MHN, because the uptake of 2-deoxy[^{14}C]glucose (2DG), an indicator of metabolic activity, shows a circadian rhythm in the LHRN, but not in other brain sites. Significantly, pinealectomy abolishes the rhythm of 2DG uptake by the LHRN, but rhythmicity can be restored by daily melatonin administration (Lu & Cassone, 1993a, b). In a preliminary study in the pigeon, Ebihara and coworkers attempted to restrict lesions to either the MHN or the LHRN (Ebihara, Oshima, Yamada, Goto, & Sato, 1987). Lesions of either nucleus alone did not abolish rhythmicity, although some disruption of the activity rhythm could be seen. In chicks, lesions restricted to the LHRN abolished the circadian rhythm of norepinephrine turnover in the pineal gland, whereas lesions of the MHN had no effect (Cassone, Forsyth, & Woodlee, 1990).

One laboratory has used ablation studies to try to localize the central circadian pacemaker in fish (Ooka-Souda & Kabasawa, 1988; Ooka-Souda, Kadota, & Kabasawa, 1993). Ooka-Souda and colleagues showed that surgical ablation of the optic tectum, the anterior portions of the telencephalon, and the dorsal portions of the telencephalon–diencephalon of the hagfish, *Eptatretus burgeri*, did not affect the free-running rhythm of locomotor activity, but ablation of the ventromedial part of the telencephalon–diencephalon, which included the hypothalamus, rendered them arrhythmic (Ooka-Souda & Kabasawa, 1988). Further lesion studies implicated the preoptic nucleus (PON) of the hypothalamus as being the locus of the pacemaker because lesions of the PON rendered hagfish arrhythmic (Ooka-Souda *et al.*, 1993).

In summary, lesion studies have convincingly shown that lesions of the suprachiasmatic area of the hypothalamus of fish, lizards, and birds can disrupt circadian organization, but more studies are needed to determine which hypothalamic nucleus is the homologue of the mammalian SCN. Of course, it is possible that the location of the central pacemakers in nonmammalian vertebrates differs among the vertebrate classes (or even with a class). It would be interesting to determine if the preoptic nucleus of hagfish or the MHN or the LHRN of birds could show circadian rhythmicity *in vitro*, as has been observed in the mammalian SCN.

Because melatonin has major effects on the circadian system of nonmammalian vertebrates, melatonin of pineal (or ocular) origin must be able to affect circadian pacemakers within the suprachiasmatic area either directly or indirectly. The melatonin agonist 2-[^{125}I]iodomelatonin (IMEL) has been extensively used to characterize and map the distribution of melatonin receptors in the brain and these studies have shown that the SCN is a target site for melatonin. Although IMEL binding in the mammalian brain is often restricted to the SCN and the median eminence/pars tuberalis, melatonin-binding sites in the nonmammalian brain are more widespread (Cassone, Brooks, & Kelm, 1995; Rivkees, Cassone, Weaver, & Reppert, 1989; Wiechmann & Wirsig-Wiechmann, 1992). In nonmammalian vertebrates, binding sites are found in the SCN as well as in a number of areas associated with the processing of sensory, particularly lighting, information. Binding is found not only in sites dealing with first-order processing, but also in sites dealing with later order processing of visual information including structures in the tectofugal, thalamofugal, and accessory optic systems. Because melatonin levels in the blood exhibit robust circadian rhythmicities, the widespread distribution of melatonin-binding sties suggests that melatonin exerts temporal control over a broad range of physiologic and behavioral events.

High-affinity melatonin receptors of the guanine nucleotide-binding protein (G protein)-coupled receptor superfamily have been cloned (Ebisawa, Karne, Lerner, & Reppert, 1994). The sites expressing these receptors coincide with the high-affinity melatonin-binding sites found by the use of IMEL. Molecular studies have identified three melatonin receptor subtypes (Mel_{1a}, Mel_{1b}, Mel_{1c}), although these three subtypes are functionally indistinguishable with currently available drugs. Genes encoding all three subtypes are present in some nonmammalian species (e.g., *Xenopus*, chicken) and, in the chicken, both Mel_{1a} and Mel_{1c} receptor subtypes appear to contribute to the pattern of IMEL binding seen (Ebisawa *et al.*, 1994; Reppert, Weaver, Cassone, Godson, & Kolakowski, 1995). Each of the subtypes may mediate physiologic responses to melatonin, but it is not known which of the signal transduction pathways influenced by melatonin actually participate in the entraining/phase shifting effects of melatonin.

The Circadian "System"

With respect to the avian circadian system, a case has yet to be documented in which rhythmicity of central circadian pacemakers persists following both pinealectomy and blinding. In some birds (e.g., house sparrow, Java sparrow, white-crowned sparrow), arrhythmicity is observed in DD following pinealectomy, whereas in others, arrhythmicity follows blinding (Japanese quail) or blinding plus pinealectomy (pigeon). Although rhythmicity can persist following pinealectomy in starlings (Gwinner, 1978), it has yet to be determined if rhythmicity in the starling would persist if the eyes were also removed.

Can the other components of the avian circadian "system" show autonomous rhythmicity? In most cases, organ-cultured pineals (or dispersed cell cultures) show rhythmicity in melatonin synthesis and secretion for only a few cycles in constant conditions. However, the finding by Barrett and Takahashi (1997) that the persistence of rhythmicity in cultured chick pineal cells depends upon temperature raises the possibility that the avian pineal organ may be more autonomous than previously thought. To determine if the avian pineal organ is autonomously rhythmic *in vivo*, it is necessary to examine the behavior of the pineal organ in constant conditions after eliminating inputs from other components of the system. In one study, sympathetic denervation of the chicken pineal organ abolished the plasma melatonin rhythm, whereas daily norepinephrine infusions reestablished the plasma melatonin rhythm, suggesting that daily sympathetic input is necessary to maintain pineal organ rhythmicity (Cassone & Menaker, 1983). Plasma melatonin rhythmicity in DD was also abolished in Japanese quail following blinding (H. Underwood, unpublished). Because eye removal disrupts the central circadian system in quail, it is hypothesized that the central pacemakers were no longer able to provide a coherent daily neural input to the pineal organ, causing it to damp out. This hypothesis is supported by the finding that the quail pineal organ does not show rhythmicity *in vitro* (Murakami *et al.*, 1994). On the other hand, studies in the house sparrow support the hypothesis that the pineal organ can maintain rhythmicity *in vivo* after denervation because circadian locomotor activity persisted following interruption of neural inputs to the pineal organ (Zimmerman & Menaker, 1975, 1979).

Although few in number, the *in vivo* studies in birds suggest that the pineal organ rhythmicity may (chicken, Japanese quail) or may not (house sparrow) depend on neural inputs.

With respect to the third component of the avian circadian "system," the eyes, little data are available to determine if the ocular pacemakers are fully autonomous in nonmammalian vertebrates. One study showed that mRNA levels for the cone pigment iodopsin shows a circadian rhythm for two cycles in primary cultures of dispersed embryonic chick retina (Pierce *et al.*, 1993).

Among the lizards, pinealectomy (or, in some cases, blinding) can also cause arrhythmicity, splitting, or period changes, but, in at least three cases (ruin lizard, western fence lizard, desert iguana) rhythmicity can persist following *both* pinealectomy and blinding (Foà, 1991; Janik & Menaker, 1990; Underwood, 1981), showing that the central circadian pacemakers are capable of autonomous rhythmicity in the absence of inputs from other components of the system. Persistence of rhythmicity is noted after pinealectomy in most fish (Table 1), but it is not known if rhythmicity would persist in fish that were both pinealectomized and blinded.

In summary, in most avian species examined so far and in some lizards and fish, pinealectomy and/or complete eye removal produces arrhythmicity, showing either (1) the rest of the circadian system (SCN?) is not self-sustaining or (2) the remaining oscillators are still self-sustaining, but they have become out of phase with one another. A pineal organ–SCN interaction or an eye–SCN interaction therefore appears necessary to maintain persistent rhythmicity in most birds and some fish and reptiles. This kind of circadian organization can explain the variety of effects noted following pinealectomy or blinding. If the SCN is a relatively robust oscillator (or set of oscillators), pinealectomy or blinding is likely to have either no effect or to affect only the period of the system. If the SCN is a weak oscillator, pinealectomy or blinding will lead to arrhythmicity or splitting. Interestingly, the range of effects of pinealectomy seen in different lizard species is strikingly similar to the range of effects seen in different bird species, implying that the lizard and the bird share very similar systems.

It is also interesting to note that pinealectomy seems to have little effect on the circadian rhythm of activity of the desert iguana or on the activity and body temperature rhythms of Japanese quail (Janik & Menaker, 1990; Underwood, 1994), and the pineal organ of these species does not show rhythmicity *in vitro* (Janik & Menaker, 1990; Murakami *et al.*, 1994). However, caution must be observed in dismissing a role for the pineal only because pinealectomy fails to have a significant effect on rhythmicity. It is possible that the pineal organ is playing a role, but other elements are sufficient to maintain relatively normal behavior in the absence of the pineal organ. Even though the pineal organ may not exhibit autonomous rhythmicity, the pineal organ is rhythmic in the intact animal because it is driven by photic and/or neural inputs. The pineal organ's role in these species may not be that of a pacemaker, but rather that of a photo or thermoendocrine transducer that permits external lighting or temperature conditions to influence the circadian system.

The circadian system of nonmammalian vertebrates is complex: not only are there multiple pacemakers within the system, but multiple photic inputs are present as well (Figure 7). At the level of the organism, therefore, the circadian system is multioscillator in nature, but it must also be recognized that the system is also multioscillator at the level of organs or tissues. For example, individual pineal cells can exhibit circadian rhythmicity, showing that this organ is itself multioscillator in nature.

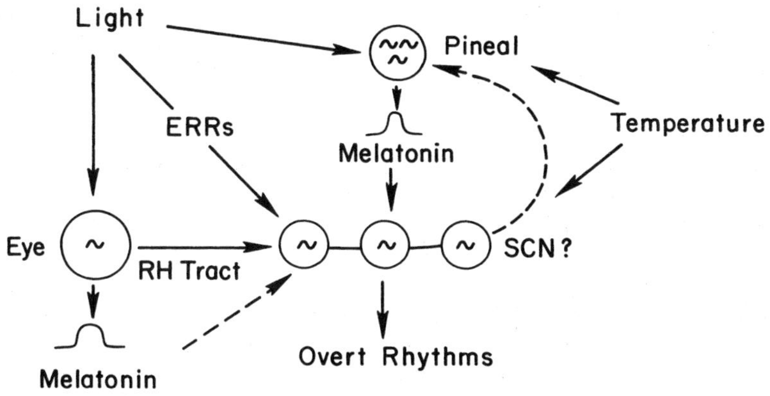

Figure 7. The circadian system of nonmammalian vertebrates. Circadian oscillators reside centrally (probably in the suprachiasmatic area) and more peripherally in the pineal organ and eyes. The pineal organ is linked to the rest of the system via melatonin, whereas the eyes may be linked by melatonin or neurally. Light can entrain the pineal organ and eyes directly, it can entrain the central system via extraretinal receptors (ERRs) in the brain, and it can entrain the central nervous system via the pineal organ and eyes. The circadian oscillators are themselves multioscillator in nature. Rhythmicity in some elements of the system (i.e., SCN, pineal organ) may damp in the absence of inputs from the rest of the system. This behavior may be due to a damping of the oscillators comprising the pacemakers or to an uncoupling of the multiple oscillators comprising the pacemakers.

Furthermore, the eyes and the SCN are also likely comprised of multiple clocks. At this level, virtually nothing is known about how these cellular clocks interact with one another. Somewhat more information is available about how interactions occur at higher levels of organization. One of the main ways in which elements of the circadian system can communicate is via the rhythmic secretion of the hormone melatonin. Although a large number of species have not been examined, melatonin of pineal origin appears to be the principal mechanism by which the pineal organ can control oscillators located in the central nervous system (i.e., the SCN). In addition, melatonin secreted by the eyes has been proposed to link ocular clocks to the central system in two species of birds. Too few species have been examined, however, to generalize about the role of melatonin of retinal origin in controlling extraocular clocks. In addition, ocular clocks could potentially be coupled to the rest of the circadian system by neural routes, as has been suggested for the Japanese quail. Because the RH tract provides a direct neural input to the SCN, this tract presents an attractive possibility for neural coupling between ocular and extraocular clocks.

Light (and probably temperature as well) has multiple inputs to the circadian system of nonmammalian vertebrates. Light can directly affect the clocks in the pineal organ and eyes and it can affect the SCN via the eyes and pineal via extraretinal photoreceptors located in the brain. Furthermore, light can affect indoleamine synthesis in the pineal organ or eyes independently of its effects on the clock by acutely suppressing melatonin production, although a role for the acute effect of light in the natural world is unclear. Interestingly, in animals that have been rendered arrhythmic in DD by pinealectomy or blinding, the presentation of a 24-hour LD cycle will elicit a rhythm (Figure 6). Therefore, light alone is capable of either recoupling the oscillators within the central circadian system that were uncoupled by pinealectomy or blinding, or of engendering rhythmicity in damped oscillators.

It is striking that the circadian "system" of nonmammalian vertebrates appears to be redundant in that it is comprised of multiple (complex) pacemakers, and light can impact the system at several different levels. The reason for such a system is not clear, but the evolutionary history of the animals must have played a significant role. For example, primitive organisms probably utilized extraretinal photoreception before the evolution of the lateral eyes and, for some reason, these extraretinal receptors have been retained. Perhaps retinal and extraretinal receptors are extracting different kinds of information, such as information about the parametric versus the nonparametric content of entraining LD cycles, or, as Foster and his colleagues have suggested (Foster *et al.*, 1994), retinal photoreceptors may be too specialized to provide irradiance detection and this function must be performed by extraretinal receptors. The role of melatonin in transducing photoperiodic information also appears to be an ancient feature of evolution; for example, melatonin appears to play such a role in a variety of present-day invertebrates (Vivien-Roels and Pévet, 1993). In earlier organisms, photoreceptors in the eyes or pineal organ may have utilized melatonin as a transducer of photoperiodic information only within the eye or pineal organ. Subsequently, however, melatonin released into the blood for degradation by the kidneys or liver may have been coopted for a more systemic role within the animal's total circadian system. A "redundant" system with multiple clocks and multiple photic inputs may also have survival value and allow organisms the flexibility to meet the challenges of living in different temporal and physical niches.

The circadian system of mammals presents a number of significant differences from that on nonmammalian vertebrates, although some components of the mammalian system are clearly derived from the nonmammalian vertebrates, such as an important role for the SCN. First, the mammalian pineal organ is not autonomous and pineal organ melatonin synthesis is driven by extrapineal clocks located in the SCN. Consequently, pinealectomy has little or no effect on the mammalian circadian system. Second, extraocular photoreception is absent in mammals: light input to the circadian system occurs exclusively via the eyes. Third, the principal pacemaker in mammals resides in the SCN, although some evidence exists that other oscillators may exist. For example, some species possess a food-entrainable oscillator which appears to be relatively independent of the SCN. Fourth, in many mammalian species, the distribution of melatonin-binding sites in the brain is more restricted than is typical for nonmammalian vertebrates.

Although melatonin production by the mammalian pineal is driven by the SCN, melatonin does have some effects on the circadian system of mammals. For example, daily melatonin administration can entrain the activity rhythm of rats, and pulses of melatonin can phase shift circadian rhythms in humans. It is likely that these effects of melatonin occur at the levels of the SCN, which is richly endowed with melatonin receptors. Because melatonin rhythmicity is driven by the SCN, it appears that pineal melatonin has a feedback effect on the SCN, perhaps to "fine tune" the circadian system to match more precisely prevailing photic conditions.

Undoubtedly, the major components of the mammalian system have been derived from the more complex system possessed by their reptilian ancestors. It is unclear, however, why the birds retained this complex system, whereas the mammals preserved only certain elements of the system. Part of the explanation may reside in the different directions taken for the production of offspring. For example, extraretinal photoreception may have value in mediating photic control of the physiology or biochemistry of embryonic organisms even before the eyes are functional. However, only animals which develop *in ovo* would be able to perceive light extraretinally because light would not be able to reach embryos that develop *in utero.* Furthermore, the thicker skulls and fur of mammals may have precluded the use of extraretinal photoreceptors (brain or pineal organ), with a consequent deemphasis of the role of the pineal organ. Alternatively, the loss of extraretinal photoreception may have occurred during the nocturnal "bottleneck" that occurred during mammalian evolution.

References

Barrett, R. K., & Takahashi, J. S. (1997). Lability of circadian pacemaker amplitude in chick pineal cells: A temperature-dependent process. *Journal of Biological Rhythms, 12,* 309–318.

Barrett, R. K., & Underwood, H. (1991). Retinally perceived light can entrain the pineal melatonin rhythm in Japanese quail. *Brain Research, 563,* 87–93.

Barrett, R. K., & Underwood, H. (1992). The superior cervical ganglia are not necessary for entrainment or persistence of the pineal melatonin rhythm in Japanese quail. *Brain Research, 569,* 249–254.

Besharse, J. C., & Dunis, D. A. (1983). Methoxyindoles and photoreceptor metabolism: activation of rod shedding. *Science, 219,* 1341–1343.

Besharse, J. C., & Iuvone, P. M. (1983). Circadian clock in *Xenopus* eye controlling retinal serotonin N-acetyltransferase. *Nature, 305,* 133–135.

Bolliet, V., Ali, M. A., Anctil, M., & Zachmann, A. (1993). Melatonin secretion *in vitro* from the pineal complex of the lamprey *Petromyzon marinus. General and Comparative Endocrinology, 89,* 101–106.

Bolliet, V., Bégay, V., Rervault, J.-P., Ali, M. A., Collin, J.-P., & Falcón, J. (1994). Multiple circadian oscillators in the photosensitive pike pineal gland: A study using organ and cell culture. *Journal of Pineal Research, 16,* 77–84.

Bolliet, V., Ali, M. A., Lapointe, F.-J., & Falcón, J. (1996). Rhythmic melatonin secretion in different teleost species: An *in vitro* study. *Journal of Comparative Physiology B, 165,* 677–683.

Cahill, G. M. (1996). Circadian regulation of melatonin production in cultured zebrafish pineal and retina. *Brain Research, 708,* 177–181.

Cahill, G. M., & Besharse, J. C. (1991). Resetting the circadian clock in *Xenopus* eyecups: Regulation of retinal melatonin rhythms by light and D_2 dopamine receptors. *Journal of Neuroscience, 11,* 2959–2971.

Cahill, G. M., & Besharse, J. C. (1992). Light-sensitive melatonin synthesis by *Xenopus* photoreceptors after destruction of the inner retina. *Visual Neuroscience, 8,* 487–490.

Cahill, G. M., & Besharse, J. C. (1993). Circadian clock functions localized in *Xenopus* retinal photoreceptors. *Neuron, 10,* 573–577.

Cahill, G. M., & Besharse, J. C. (1995). Circadian rhythmicity in vertebrate retinas: Regulation by a photoreceptor oscillator. *Progress in Retinal and Eye Research, 14,* 267–291.

Cahill, G. M., Grace, M. S., & Besharse, J. C. (1991). Rhythmic regulation of retinal melatonin: Metabolic pathways, neurochemical mechanisms, and the ocular circadian clock. *Cellular and Molecular Neurobiology, 11,* 529–560.

Cassone, V. M., & Menaker, M. (1983). Sympathetic regulation of chicken pineal rhythms. *Brain Research, 272,* 311–317.

Cassone, V. M., & Moore, R. Y. (1987). Retinohypothalamic projection and suprachiasmatic nucleus of the house sparrow, *Passer domesticus. Journal of Comparative Neurology, 266,* 171–182.

Cassone, V. M., Forsyth, A. M., & Woodlee, G. L. (1990). Hypothalamic regulation of circadian noradrenergic input to the chick pineal gland. *Journal of Comparative Physiology A, 167,* 187–192.

Cassone, V. M., Brooks, D. S., & Kelm, T. A. (1995). Comparative distribution of 2[^{125}I]indomelatonin binding in the brains of diurnal birds: Outgroup analysis with turtles. *Brain, Behavior and Evolution, 45,* 241–256.

Chabot, C. C., & Menaker, M. (1992a). Circadian feeding and locomotor rhythms in pigeons and house sparrows. *Journal of Biological Rhythms, 7,* 287–299.

Chabot, C. C., & Menaker, M. (1992b). Effects of physiological cycles of infused melatonin on circadian rhythmicity in pigeons. *Journal of Comparative Physiology A, 170,* 615–622.

Chiba, A., Kikuchi, M., & Aoki, K. (1993). The effects of pinealectomy and blinding on the circadian locomotor activity rhythm in the Japanese newt, *Cynops pyrrhogaster. Journal of Comparative Physiology A, 172,* 683–691.

Chiba, A., Kikuchi, M., & Aoki, K. (1995). Entrainment of the circadian locomotor activity rhythm in the Japanese newt by melatonin injections. *Journal of Comparative Physiology A, 176,* 473–477.

Cogburn, L. A., Wilson-Placentra, S., & Letcher, L. R. (1987). Influence of pinealectomy on plasma and extrapineal melatonin rhythms in young chickens (*Gallus domesticus*). *General and Comparative Endocrinology, 68,* 343–356.

Collin, J.-P., & Oksche, A. (1981). Structural and functional relationships in the nonmammalian pineal organ. In R. J. Reiter (Ed.), *The pineal gland. Vol. 1. Anatomy and biochemistry* (pp. 27–67). Boca Raton, FL: CRC Press.

Collin, J.-P., Voisin, P., Falcón, J., Faure, J.-P., Brisson, P., & Defaye, J.-R. (1989). Pineal transducers in the course of evolution: Molecular organization, rhythmic metabolic activity and role. *Archives of Histology and Cytology, 52,* 441–449.

Delgado, M. J., & Vivien-Roels, B. (1989). Effect of environmental temperature and photoperiod on the melatonin levels in the pineal, lateral eye, and plasma of the frog, *Rana perezi*: Importance of ocular melatonin. *General and Comparative Endocrinology, 75,* 46–53.

Ebihara, S., & Kawamura, H. (1981). The role of the pineal organ and the suprachiasmatic nucleus in the control of circadian locomotor rhythms in the Java sparrow, *Padda oryzivora. Journal of Comparative Physiology, 141,* 207–214.

Ebihara, S., Uchiyama, K., & Oshima, I. (1984). Circadian organization in the pigeon, *Columba livia*: The role of the pineal organ and the eye. *Journal of Comparative Physiology A, 154,* 59–69.

Ebihara, S., Oshima, I., Yamada, H., Goto, M., & Sato, K. (1987). Circadian organization in the pigeon. In T. Hiroshige & K. Honma (Eds.), *Comparative aspects of circadian clocks,* (pp. 84–94). Sapporo, Japan: Hokkaido University Press.

Ebisawa, T., Karne, S., Lerner, M. R., & Reppert, S. M. (1994). Expression cloning of a high-affinity melatonin receptor from *Xenopus* dermal melanophores. *Proceedings of the National Academy of Sciences of the USA, 91,* 6133–6137.

Eriksson, L. O. (1972). Die Jahresperiodik augen- und pinealorganloser Bachsaiblinge *Salvelinus fontinalis* Mitchell. *Aquilo Serie Zoologica, 13,* 8–12.

Falcón, J., & Collin, J.-P. (1991). Pineal–retinal relationships: Rhythmic biosynthesis and immunocytochemical localization of melatonin in the retina of the pike (*Esox lucius*). *Cell and Tissue Research, 265,* 601–609.

Falcón, J., Marmillon, J. B., Claustrat, B., & Collin, J.-P. (1989). Regulation of melatonin secretion in a photoreceptive pineal organ: An *in vitro* study in the pike. *Journal of Neuroscience, 9,* 1943–1950.

Falcón, J., Bolliet, V., Ravault, J. P., Chesneau, D., Ali, M. A., & Collin, J.-P. (1994a). Rhythmic secretion of melatonin by the superfused pike pineal organ: Thermo- and photoperiod interaction. *Neuroendocrinology, 60,* 535–543.

Falcón, J., Bégay, V., Goujon, J. M., Voisin, P., Guerlotté, J., & Collin, J.-P. (1994b). Immunocytochemical localisation of hydroxyindole-O-methyltransferase in pineal photoreceptor cells of several fish species. *Journal of Comparative Neurology, 341,* 559–566.

Firth, B. T., & Kennaway, D. J. (1989). Thermoperiod and photoperiod interact to affect the phase of the plasma melatonin rhythm in the lizard, *Tiliqua rugosa. Neuroscience Letters, 106,* 125–130.

Foà, A. (1991). The role of the pineal and the retinae in the expression of circadian locomotor rhythmicity in the ruin lizard *Podarcis sicula. Journal of Comparative Physiology A, 169,* 201–207.

Foà, A., & Menaker, M. (1988). Contribution of pineal and retinae to the circadian rhythms of circulating melatonin in pigeons. *Journal of Comparative Physiology A, 164,* 25–30.

Foà, A., Minutini, L., & Innocenti, A. (1992). Melatonin: A coupling device between oscillators in the circadian system of the ruin lizard *Podarcis sicula. Comparative Biochemistry and Physiology, 103A,* 719–723.

Foster, R. G., & Follett, B. K. (1985). The involvement of a rhodopsin-like photopigment in the photoperiodic response of the Japanese quail. *Journal of Comparative Physiology A, 157,* 519–528.

Foster, R. G., Timmers, A. M., Schalken, J. J., & De Grip, W. J. (1989). A comparison of some photoreceptor characteristics in the pineal and retina II. The Djungarian hamster (*Phodopus sungorus*). *Journal of Comparative Physiology A, 165,* 565–572.

Foster, R. G., Garcia-Fernandez, J. M., Provencio, I., & DeGrip, W. J. (1993). Opsin localization and chromophore retinoids identified within the basal brain of the lizard *Anolis carolinensis. Journal of Comparative Physiology A, 172,* 33–45.

Foster, R. G., Grace, M. S., Provencio, I., DeGrip, W. J., & Garcia-Fernandez, J. M. (1994). Identification of vertebrate deep brain photoreceptors. *Neuroscience and Biobehavioral Reviews, 18,* 541–546.

Fuchs, J. L. (1983). Effects of pinealectomy and subsequent melatonin implants on activity rhythms in the house finch (*Carpodacus mexicanus*). *Journal of Comparative Physiology, 153,* 413–419.

Garcia-Fernandez, J. M., & Foster, R. G. (1994). Immunocytochemical identification of photoreceptor proteins in hypothalamic cerebrospinal fluid-contacting neurons of the larval lamprey (*Petromyzon marinus*). *Cell and Tissue Research, 275,* 319–326.

Garg, S. K., & Sundararaj, B. I. (1986). Role of pineal in the regulation of some aspects of circadian rhythmicity in the catfish, *Heteropneustes fossilis (Bloch). Chronobiologia, 13,* 1–11.

Gaston, S. (1971). The influence of the pineal organ on the circadian activity rhythm in birds. In M. Menaker (Ed.), *Biochronometry* (pp. 541–548). Washington, DC: National Academy of Sciences.

Gaston, S., & Menaker, M. (1968). Pineal function: The biological clock in the sparrow? *Science, 160,* 1125–1127.

Gern, W. A., & Greenhouse, S. S. (1988). Examination of *in vitro* melatonin secretion from superfused trout (*Salmo gairdneri*) pineal organs maintained under diel illumination or continuous darkness. *General and Comparative Endocrinology, 71,* 163–174.

Grace, M. S., & Besharse, J. C. (1993). Solubilization and biochemical characterization of the melatonin deacetylase from *Xenopus laevis* retina. *Journal of Neurochemistry, 60,* 990–999.

Grace, M. S., Cahill, G. M., & Besharse, J. C. (1991). Melatonin deacetylation: Retinal vertebrate class distribution and *Xenopus laevis* tissue distribution. *Brain Research, 559,* 56–63.

Grace, M. S., Alones, V., Menaker, M., & Foster, R. G. (1996). Light perception in the vertebrate brain: An ultrastructural analysis of opsin- and vasoactive intestinal polypeptide-immunoreactive neurons in iguanid lizards. *Journal of Comparative Neurology, 367,* 575–594.

Green, C. B., Cahill, G. M., & Besharse, J. C. (1995). Regulation of tryptophan hydroxylase expression by a retinal circadian oscillator *in vitro*. *Brain Research, 677,* 283–290.

Gwinner, E. (1978). Effects of pinealectomy on circadian locomotor activity rhythms in European starlings, *Sturnus vulgaris. Journal of Comparative Physiology, 126,* 123–129.

Gwinner, E., & Benzinger, I. (1978). Synchronization of a circadian rhythm in pinealectomized European starlings by daily injections of melatonin. *Journal of Comparative Physiology, 127,* 209–213.

Hamasaki, D. I., & Eder, D. J. (1977). Adaptive radiation of the pineal system. In F. Crescitelli (Ed.), *Handbook of sensory physiology* (pp. 497–548). New York: Springer-Verlag.

Hartwig, H. G., & van Veen, T. (1979). Spectral characteristics of visible radiation penetrating into the brain and stimulating extraretinal photoreceptors. *Journal of Comparative Physiology, 130,* 277–282.

Hasegawa, M., Adachi, A., Yoshimura, T., & Ebihara, S. (1994). Retinally perceived light is not essential for photic regulation of pineal melatonin rhythms in the pigeon: Studies with microdialysis. *Journal of Comparative Physiology A, 175,* 581–586.

Heigl, S., & Gwinner, E. (1994). Periodic melatonin in the drinking water synchronizes circadian rhythms in sparrows. *Naturwissenschaften, 81,* 83–85.

Heigl, S., & Gwinner, E. (1995). Synchronization of circadian rhythms of house sparrows by oral melatonin: Effects of changing period. *Journal of Biological Rhythms, 10,* 225–233.

Hyde, L. L., & Underwood, H. (1995). Daily melatonin infusions entrain the locomotor activity of pinealectomized lizards. *Physiology and Behavior, 58,* 943–951.

Janik, D. S., & Menaker, M. (1990). Circadian locomotor rhythms in the desert iguana I. The role of the eyes and the pineal. *Journal of Comparative Physiology A, 166,* 803–810.

Janik, D. S., Pickard, G. E., & Menaker, M. (1990). Circadian locomotor rhythms in the desert iguana II. Effects of electrolytic lesions to the hypothalamus. *Journal of Comparative Physiology A, 166,* 811–816.

Janik, D. S., Dittami, J., & Gwinner, E. (1992). The effect of pinealectomy on circadian plasma melatonin levels in house sparrows and European starlings. *Journal of Biological Rhythms, 7,* 277–286.

Janik, D., Cassone, V. M., Pickard, G. E., & Menaker, M. (1994). Retinohypothalamic projections and immunocytochemical analysis of the suprachiasmatic region of the desert iguana *Dipsosaurus dorsalis. Cell and Tissue Research, 275,* 399–406.

Karasek, M. (1983). Ultrastructure of the mammalian pineal gland: Its comparative and functional aspects. In R. J. Reiter (Ed.), *Pineal research reviews,* (Vol. 1, pp. 1–48). New York: Liss.

Kavaliers, M. (1979). Pineal involvement in the control of circadian rhythmicity in the lake chub, *Couesius plumbeus. Journal of Experimental Zoology, 209,* 33–40.

Kavaliers, M. (1980a). Retinal and extraretinal entrainment action spectra for the activity rhythms of the lake chub, *Couesius plumbeus. Behavioral and Neural Biology, 30,* 56–67.

Kavaliers, M. (1980b). Circadian locomotor activity rhythms of the burbot, *Lota lota*: Seasonal differences in period length and the effect of pinealectomy. *Journal of Comparative Physiology, 136,* 215–218.

Kavaliers, M. (1981). Circadian organization in white suckers *Catostomus commersoni*: The role of the pineal organ. *Comparative Biochemistry and Physiology, 68A*, 127–129.

Kezuka, H., Aida, K., & Hanyu, I. (1989). Melatonin secretion from goldfish pineal gland in organ culture. *General and Comparative Endocrinology, 75*, 217–221.

Konishi, H., Ohta, M., & Homma, K. (1985). Important role of the eye controlling the locomotor rhythm in quail. *Journal of Interdisciplinary Cycle Research, 16*, 217–226.

Korf, H.-W., & Møller, M. (1984). The innervation of the mammalian pineal gland with special reference to central pinealopetal projections. In R. J. Reiter (Ed.), *Pineal research reviews* (Vol. 2, pp. 41–86). New York: Liss.

Lu, J., & Cassone, V. M. (1993a). Daily melatonin administration synchronizes circadian patterns of brain metabolism and behavior in pinealectomized house sparrows, *Passer domesticus. Journal of Comparative Physiology A, 173*, 775–782.

Lu, J., & Cassone, V. M. (1993b). Pineal regulation of circadian rhythms of 2-deoxy[^{14}C]glucose uptake and 2[^{125}I]iodomelatonin binding in the visual system of the house sparrow, *Passer domesticus. Journal of Comparative Physiology A, 173*, 765–774.

McMillan, J. P. (1972). Pinealectomy abolishes the circadian rhythm of migratory restlessness. *Journal of Comparative Physiology A, 79*, 105–112.

McMillan, J. P., Keatts, H. C., & Menaker, M. (1975). On the role of eyes and brain photoreceptors in the sparrow: Entrainment to light cycles. *Journal of Comparative Physiology, 102*, 251–256.

Menaker, M. (1971). Rhythms, reproduction, and photoreception.*Biology of Reproduction, 4*, 295–308.

Menaker, M. (1985). Eyes—The second (and third) pineal glands? In D. Evered & S. Clark (Eds.), *Photoperiodism, melatonin and the pineal* (pp. 78–87). London: Pitman.

Menaker, M., & Wisner, S. (1983). Temperature-compensated circadian clock in the pineal of *Anolis. Proceedings of the National Academy of Sciences of the USA, 80*, 6119–6121.

Minutini, L., Innocenti, A., Bertolucci, C., & Foà, A. (1994). Electrolytic lesions to the optic chiasm affect circadian locomotor rhythms in lizards. *NeuroReport, 5*, 525–527.

Minutini, L., Innocenti, A., Bertolucci, C., & Foà, A. (1995). Circadian organization in the ruin lizard *Podarcis sicula*: The role of the suprachiasmatic nuclei of the hypothalamus. *Journal of Comparative Physiology A, 176*, 281–288.

Molina-Borja, M. (1996). Pineal gland and circadian locomotor activity rhythm in the lacertid *Gallotia galloti eisentrauti*: Pinealectomy induces arrhythmicity. *Biological Rhythm Research, 27*, 1–11.

Molina-Borja, M., Falcón, J., Urquiola, E., & Ravault, J. P. (1996). Production of melatonin by the gilthead sea bream pineal: An *in vivo* and *in vitro* study. *Fish Physiology and Biochemistry, 15*, 413–419.

Morita, Y., Tamotsu, S., & Uchida, K. (1989). Multiplicity of electrophysiological and immunocytochemical properties in the pineal photosensory system. In R. J. Reiter & S. F. Pang (Eds.), *Advances in pineal research* (Vol. 3, pp. 43–48). London: Libbey.

Morita, Y., Tabata, M., Uchida, K., & Samejima, M. (1992). Pineal-dependent locomotor activity of lamprey, *Lampetra japonica*, measured in relation to LD cycle and circadian rhythmicity. *Journal of Comparative Physiology A, 171*, 555–562.

Moyer, R. W., Firth, B. T., & Kennaway, D. J. (1995). Effect of constant temperatures, darkness and light on the secretion of melatonin by pineal explants and retinas in the gecko *Christinus marmoratus. Brain Research, 675*, 345–348.

Murakami, N., Nakamura, H., Nishi, R., Marumoto, N., & Nasu, T. (1994). Comparison of circadian oscillation of melatonin release in pineal cells of house sparrow, pigeon and Japanese quail, using cell perfusion systems. *Brain Research, 651*, 209–214.

Norgren, R. B., & Silver, R. (1989). Retinohypothalamic projections and the suprachiasmatic nucleus in birds. *Brain, Behavior and Evolution, 34*, 73–83.

Norgren, R. B., & Silver, R. (1990). Distribution of vasoactive intestinal peptide-like and neurophysin-like immunoreactive neurons and acetylcholinesterase staining in the ring dove hypothalamus with emphasis on the question of an avian suprachiasmatic nucleus. *Cell and Tissue Research, 259*, 331–339.

Ooka-Souda, S., & Kabasawa, H. (1988). Circadian rhythms in locomotor activity of the hagfish, *Eptatretus burgeri* III. Hypothalamus: A locus of the circadian pacemaker? *Zoological Science, 5*, 437–442.

Ooka-Souda, S., Kadota, T., & Kabasawa, H. (1993). The preoptic nucleus: The probable location of the circadian pacemaker of the hagfish, *Eptatretus burgeri. Neuroscience Letters, 164*, 33–36.

Oshima, I., Yamada, H., Goto, M., Sato, K., & Ebihara, S. (1989). Pineal and retinal melatonin is involved in the control of circadian locomotor activity and body temperature rhythms in the pigeon. *Journal of Comparative Physiology A, 166*, 217–226.

Pickard, G. E., & Tang, W.-X. (1993). Individual pineal cells exhibit a circadian rhythm in melatonin secretion. *Brain Research, 627*, 141–146.

Pickard, G. E., & Tang, W.-X. (1994). Pineal photoreceptors rhythmically secrete melatonin. *Neuroscience Letters, 171*, 109–112.

Pierce, M. E., & Besharse, J. C. (1986). Melatonin and dopamine interactions in the regulation of rhythmic photoreceptor metabolism. In P. J. O'Brien & D. C. Klein (Eds.), *Pineal and retinal relationships* (pp. 219–237). New York: Academic Press.

Pierce, M. E., Sheshberadaran, H., Zhang, Z., Fox, L. E., Applebury, M. L., & Takahashi, J. S. (1993). Circadian regulation of iodopsin gene expression in embryonic photoreceptors in retinal cell culture. *Neuron, 10,* 579–584.

Ralph, C. L. (1981). Melatonin production by extra-pineal tissues. In N. Birau & W. Chloot (Eds.), *Melatonin—Current status and perspectives* (pp. 35–46). New York: Pergamon Press.

Remé, C. E., Wirz-Justice, A., & Terman, M. (1991). The visual input stage of the mammalian circadian pacemaking system: I. Is there a clock in the mammalian eye? *Journal of Biological Rhythms, 6,* 5–29.

Reppert, S. M., Weaver, D. R., Cassone, V. M., Godson, C., & Kolakowski, L. F. (1995). Melatonin receptors are for the birds: Molecular analysis of two receptor subtypes differentially expressed in chick brain. *Neuron, 15,* 1003–1015.

Rivkees, S. A., Cassone, V. M., Weaver, D. R., & Reppert, S. M. (1989). Melatonin receptors in chick brain: Characterization and localization. *Endocrinology, 125,* 363–368.

Shimizu, T., Cox, K., Karten, H. J., & Britto, L. R. G. (1994). Cholera toxin mapping of retinal projections in pigeons (*Columba livia*), with emphasis on retinohypothalamic connections. *Visual Neuroscience, 11,* 441–446.

Silver, R., Witkovsky, P., Horvath, P., Alones, V., Barnstable, C. J., & Lehman, M. N. (1988). Coexpression of opsin- and VIP-like immunoreactivity in CSF-contacting neurons of the avian brain. *Cell and Tissue Research, 253,* 189–198.

Simpson, S. M., & Follett, B. K. (1981). Pineal and hypothalamic pacemakers: Their role in regulating circadian rhythmicity in Japanese quail. *Journal of Comparative Physiology, 144,* 381–389.

Tabata, M. (1992). Photoreceptor organs and circadian locomotor activity in fishes. In M. A. Ali (Ed.), *Rhythms in fishes* (pp. 223–234). New York: Plenum Press.

Takahashi, J. S., & Menaker, M. (1982). Role of the suprachiasmatic nuclei in the circadian system of the house sparrow, *Passer domesticus. Journal of Neuroscience, 2,* 815–828.

Takahashi, J. S., & Menaker, M. (1984). Multiple redundant circadian oscillators within the isolated avian pineal gland. *Journal of Comparative Physiology A, 154,* 435–440.

Takahashi, J. S., Murakami, N., Nikaido, S. S., Pratt, B. L., & Robertson, L. M. (1989). The avian pineal, a vertebrate model system of the circadian oscillator: Cellular regulation of circadian rhythms by light, second messengers, and macromolecular synthesis. *Recent Progress in Hormone Research, 45,* 279–352.

Taylor, D. H., & Ferguson, D. E. (1970). Extraoptic celestial orientation in the southern cricket frog, *Acris gryllus. Science, 168,* 390–392.

Tosini, G., & Menaker, M. (1996). Circadian rhythms in cultured mammalian retina. *Science, 272,* 419–421.

Turek, F. W., McMillan, J. P., & Menaker, M. (1976). Melatonin: Effects on the circadian locomotor rhythm of sparrows. *Science, 194,* 1441–1443.

Underwood, H. (1973). Retinal and extraretinal photoreceptors mediate entrainment of the circadian locomotor rhythm in lizards. *Journal of Comparative Physiology, 83,* 187–222.

Underwood, H. (1977). Circadian organization in lizards: The role of the pineal organ. *Science, 195,* 587–589.

Underwood, H. (1979). Melatonin affects circadian rhythmicity in lizards. *Journal of Comparative Physiology, 130,* 317–323.

Underwood, H. (1981). Circadian organization in the lizard *Sceloporus occidentalis*: The effect of pinealectomy, blinding, and melatonin. *Journal of Comparative Physiology, 141,* 537–547.

Underwood, H. (1983). Circadian organization in the lizard *Anolis carolinensis*: A multioscillator system. *Journal of Comparative Physiology, 152,* 265–274.

Underwood, H. (1986). Circadian rhythms in lizards: Phase response curve for melatonin. *Journal of Pineal Research, 3,* 187–196.

Underwood, H. (1990). The pineal and melatonin: Regulators of circadian function in lower vertebrates. *Experientia, 46,* 120–128.

Underwood, H. (1994). The circadian rhythm of the thermoregulation in Japanese quail I. Role of the eyes and pineal. *Journal of Comparative Physiology A, 175,* 639–653.

Underwood, H., & Edmonds, K. (1995). The circadian rhythm of thermoregulation in Japanese quail: III. Effects of melatonin administration. *Journal of Biological Rhythms, 10,* 284–298.

Underwood, H., & Groos, G. A. (1982). Vertebrate circadian rhythms: Retinal and extraretinal photoreception. *Experientia, 38,* 1113–1121.

Underwood, H., & Menaker, M. (1976). Extraretinal photoreception in lizards. *Photochemistry and Photobiology, 23,* 227–243.

Underwood, H., & Siopes, T. (1984). Circadian organization in Japanese quail. *Journal of Experimental Zoology, 232,* 557–566.

Underwood, H., Binkley, S., Siopes, T., & Mosher, K. (1984). Melatonin rhythms in the eyes, pineal bodies, and blood of Japanese quail (*Coturnix coturnix japonica*). *General and Comparative Endocrinology, 56,* 70–81.

Underwood, H., Barrett, R. K., & Siopes, T. (1990). The quail's eye: A biological clock. *Journal of Biological Rhythms, 5,* 257–265.

Vivien-Roels, B., & Pévet, P. (1993). Melatonin: Presence and formation in invertebrates. *Experientia, 49,* 642–647.

Vivien-Roels, B., Arendt, J., & Bradtke, J. (1979). Circadian and circannual fluctuations of pineal indoleamines (serotonin and melatonin) in *Testudo hermanni* Gmelin (Reptilia, chelonia). 1. Under natural conditions of photoperiod and temperature. *General and Comparative Endocrinology, 37,* 197–210.

Wiechmann, A. F. (1986). Melatonin: Parallels in pineal gland and retina. *Experimental Eye Research, 42,* 507–527.

Wiechmann, A. F. (1996). Hydroxyindole-O-methyltransferase mRNA expression in a subpopulation of photoreceptors in the chicken retina. *Journal of Pineal Research, 20,* 217–225.

Wiechmann, A. F., & Wirsig-Wiechmann, C. R. (1992). Asymmetric distribution of melatonin receptors in the brain of the lizard *Anolis carolinensis. Brain Research, 593,* 281–286.

Zachmann, A., Falcón, J., Knijff, S. C. M., Bolliet, V., & Ali, M. A. (1992). Effects of photoperiod and temperature on rhythmic melatonin secretion from the pineal organ of the white sucker (*Catostomus commersoni*) *in vitro. General and Comparative Endocrinology, 86,* 26–33.

Zawilska, J. B., & Iuvone, P. M. (1992). Melatonin synthesis in chicken retina: Effect of kainic acid-induced lesions on the diurnal rhythm and D_2-dopamine receptor-mediated regulation of serotonin N-acetyltransferase activity. *Neuroscience Letters, 135,* 71–74.

Zawilska, J. B., & Nowak, J. Z. (1992). Regulatory mechanisms in melatonin biosynthesis in retina. *Neurochemistry International, 20,* 23–36.

Zimmerman, N. H., & Menaker, M. (1975). Neural connections of sparrow pineal: Role in circadian control of activity. *Science, 190,* 477–479.

Zimmerman, N. H., & Menaker, M. (1979). The pineal gland: A pacemaker within the circadian system of the house sparrow. *Proceedings of the National Academy of Sciences of the USA, 76,* 999–1003.

6

Suprachiasmatic Nucleus

ROBERT Y. MOORE AND REHANA K. LEAK

INTRODUCTION

IN THE BEGINNING

The suprachiasmatic nucleus (SCN) is a prominent feature of the anterior hypothalamus. In virtually all mammals, the SCN is a compact group of small neurons dorsal to the optic chiasm along most of its length and just lateral to the periventricular nucleus and third ventricle (Figure 1). The SCN was recognized in early cytoarchitectonic studies (Brockhaus, 1942; Gurdjian, 1927), but its function remained obscure until the early 1970s. Perhaps the first hint was a report by Pate (1937), who observed what must be considered transneuronal atrophy of the SCN after eye removal in the cat. The issue of retinohypothalamic projections in mammals was contentious in that era and the great neuroanatomist W. J. H. Nauta concluded in the late 1960s that there were no compelling data to support such a projection. This conclusion, however, was contrary to the situation in lower vertebrates, where a projection from the retina to the suprachiasmatic area was well established (Ebbeson, 1970). As often happens, new technical advances provided the basis for new understanding. The discovery of anterograde transport by Weiss and colleagues (Weiss, 1972) was dependent upon the fact that tritiated amino acids are incorporated into proteins in neuronal perikarya and then transported through axons to sites of terminal arbors. This phenomenon formed the basis for an axoplasmic tracing method to analyze connections in brain (Cowan, Gottlieb, Hendrickson, Price, & Woolsey, 1972) and this was applied to the study of retinal projections, providing new evidence for a retinohypothalamic tract in mammals (Hendrickson, Wagoner, & Cowan, 1972; Moore, 1973; Moore & Lenn, 1972). In these studies, silver grains were found distributed over the ventral SCN, strongly suggesting that terminals of retinal axons

ROBERT Y. MOORE AND REHANA K. LEAK Departments of Neurology, Psychiatry, and Neuroscience, University of Pittsburgh, Pittsburgh, Pennsylvania 15213.

Circadian Clocks, Volume 12 of *Handbook of Behavioral Neurobiology*, edited by Joseph S. Takahashi, Fred W. Turek, and R. Y. Moore, Kluwer Academic / Plenum Publishers, New York, 2001.

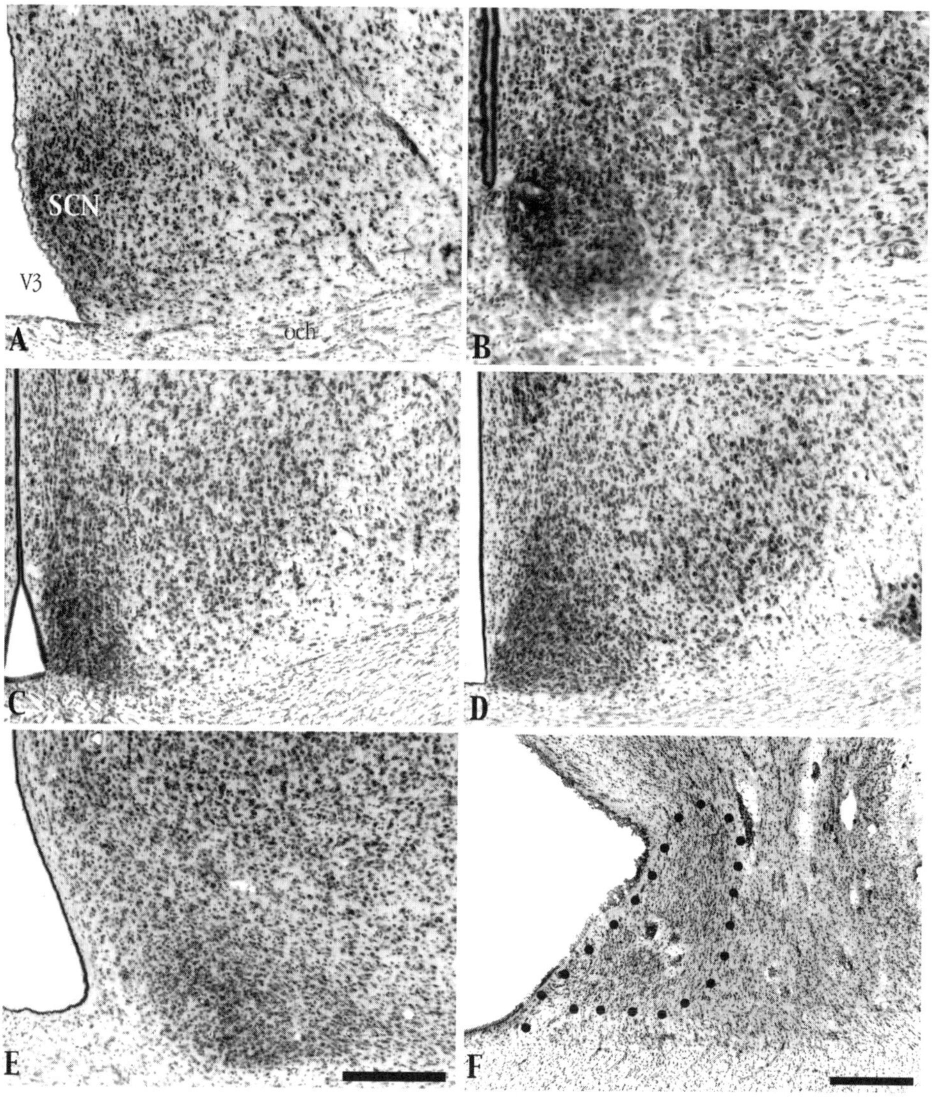

Figure 1. Nissl-stained coronal sections of the suprachiasmatic nucleus (SCN) of six different species. (A) Opossum, (B) albino rat, (C) tree shrew, (D), galago, (E) cebus monkey, and (F) human. Abbreviations: och, optic chiasm; V3, third ventricle. Scale bar in A–E is 250 μm. Scale bar in F is 500 μm.

were present in that location. This was confirmed by the electron microscopic finding of degenerating axon terminals synapsing on SCN neuron dendrites after eye removal (Moore & Lenn, 1972). These data, then, provided definitive evidence for a retinohypothalamic tract terminating in the SCN.

At about the same time, Curt Richter performed experiments indicating that the circadian pacemaker supporting the rest–activity rhythm is located in the anterior hypothalamus. With the evidence that an entrainment pathway, the retinohypothalamic tract, terminates in the SCN, and the suggestion that a circadian pacemaker was localized in the same region, it was logical to hypothesize that the SCN is a circadian pacemaker. This was initially tested by experiments in which the effects of ablation of the SCN were examined. In these, Moore and Eichler (1972) demonstrated that SCN ablation abolished the rhythm in adrenal corticosterone content, and Stephan and Zucker (1972) similarly showed a loss of the rest–activity rhythm after SCN lesions. These observations, the demonstration of a retinohypothalamic tract terminating in the SCN, and a loss of circadian function after SCN lesions, introduced a new era of investigation of the neurobiology of circadian function. This has centered on the SCN, but has extended to the definition of a new neural system, the circadian timing system. The SCN is of particular interest because of its role in circadian timing, but it also represents a component of a system which can be analyzed effectively from the molecular to the behavioral level. The first 25 years of experience with the SCN were reviewed in a symposium held in 1997 (Weaver, 1998).

The SCN Is a Circadian Pacemaker

There is now virtually universal agreement that the SCN is a circadian pacemaker. The data supporting this conclusion evolved slowly over a number of years and a brief summary serves as an appropriate introduction to the remainder of the chapter.

The SCN is the principal site of termination of the retinohypothalamic tract. Since the initial description of the retinohypothalamic tract, it is evident that this pathway is a uniform feature of the mammalian visual system which mediates photic entrainment though its termination in the SCN.

Destruction of the SCN results in a loss of behavioral and physiologic circadian rhythms. Most importantly, SCN ablation produces a loss of sleep–wake rhythmicity and a series of attendant physiologic and endocrine rhythms which form a fundamental basis for the elaboration of adaptive behavior. This is not to suggest that the SCN is the only pacemaker in mammals, only that its role in behavioral adaptation is dominant.

SCN rhythmicity is preserved when the nucleus is isolated from other brain areas *in vivo* and *in vitro*. The initial, landmark demonstration was by Inouye and Kawamura (1979), who recorded rhythmic multiunit neuronal activity persisting in an SCN "island" *in vivo* while the brain outside the SCN lost neuronal rhythmicity. Behavioral rhythmicity was also lost in these animals. Subsequent *in vitro* studies confirmed that SCN neuronal rhythmicity is preserved (Green & Gillette, 1982; Shibata & Moore, 1988; Shibata, Oomura, Kita, & Hattori, 1982). These observations confirmed the view that SCN neurons formed an endogenous oscillator with pacemaker capacity.

SCN transplants restore circadian rhythmicity in hosts that are arrhythmic after SCN lesions. Following the initial studies by Sawaki, Nihonmatsu, and Kawamura (1984) and Drucker-Colin, Auguilar-Roblero, Garcia-Hernandez, Fernandez-Cancino, and Bermudez Rattoni (1984), many groups corroborated the initial observation that the fetal SCN, transplanted into the third ventricle, will restore lesion-abolished rhythmicity (Sollars & Pickard, 1998). The mechanism by which transplants restore rhythmicity remains unknown (LeSauter & Silver, 1998; Sollars & Pickard, 1998), but it is clear that SCN cells are required and that the transplant assumes pacemaker function rather than inducing a new host rhythmicity (Ralph, Foster, Davis, & Menaker, 1990).

A description of SCN development requires that it be placed in the context of development of all components of the circadian timing system and that this, in turn, be considered in the context of development of the structures comprising the system and the ontogeny of circadian regulation. Clearly, structural development is a necessary precedent to the onset of function and both must be timed to meet the demands of the environment for the species under analysis, particularly those controlling the necessary maternal–fetal/newborn interactions. In this section, we shall consider structural and functional development of the circadian system separately for the sake of clarity.

STRUCTURAL DEVELOPMENT OF THE CIRCADIAN TIMING SYSTEM

The SCN develops from the ventral diencephalic germinal epithelium. All neurons in the central nervous system arise from the germinal epithelium located in the subventricular zone adjacent to the embryonic ventricular cavities. Precursor cells divide adjacent to the ventricle, and newly formed cells destined to be neurons migrate out of the subventricular zone to form the mantle zone of the developing neural tube. Early work, particularly by Kuhlenbeck (1931, 1954), noted that the diencephalic vesicle contained four longitudinal cellular zones comprising the walls of the neural tube. The ventral cell zone is the hypothalamus and all hypothalamic neurons are formed from the ventral diencephalic germinal epithelium in three waves that correspond to the major subdivisions of the hypothalamus in the adult brain (Altman and Bayer, 1978, 1986). Most of the developmental work has been done on the rat and this will form the basis for the descriptions below. Work on other species will be noted where it extends our understanding. The first neurons generated in the subventricular zone migrate laterally and become the neurons of the lateral hypothalamic area. The second neurons generated form the nuclei and areas of the medial hypothalamus and those from the final wave become the neurons of the periventricular zone, which includes the SCN and the arcuate nucleus. Within the developing SCN, there is a similar, but slightly more complex pattern of development. The brief description that follows is taken from the work of Altman and Bayer (1978, 1986) and more recent studies from our laboratory (J. C. Speh and R. Y. Moore, unpublished observations).

Studies using the tritiated thymidine method indicate that all neurons in the rat SCN are formed between embryonic day 14 (E14) and E15, with E0 equal to the day of vaginal sperm positivity. The SCN is formed in the third wave of hypothalamic neuron generation, in the periventricular zone. The first group of SCN neurons, generated largely from E14 to E15, lie ventral to a shallow ventromedian sulcus that divides the ventral diencephalic germinal epithelium immediately caudal to the optic recess into dorsomedial and ventrolateral suprachiasmatic primordia (Altman and Bayer, 1986). The neurons formed on E14–E15 migrate ventrally into the SCN core and this is followed on E15–E16 by neurons migrating dorsally. Finally, at E16–E17, a smaller group of neurons is formed to produce the final component of the SCN shell. This pattern is shown diagrammatically in Figure 2. The SCN is evident as a distinct nucleus lying above the optic chiasm by E16, and by E17, cell division in the germinal epithelium is largely complete and the organization of SCN neurons is similar to that in the adult.

The subsequent development of the SCN includes each of the processes in-

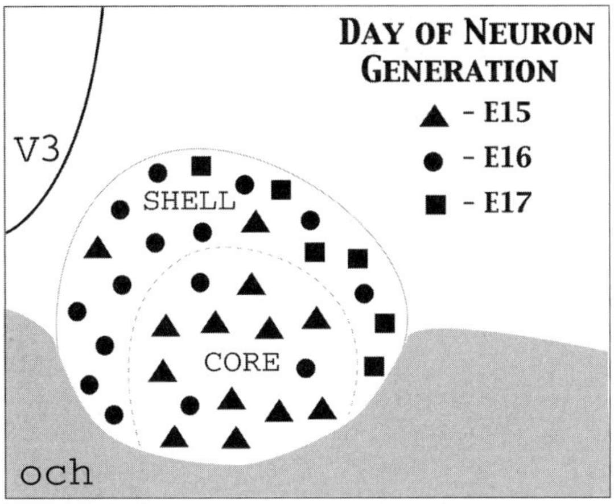

Figure 2. Neuron formation in the core and shell subdivisions of the rat SCN. Abbreviations: och, optic chiasm; V3, third ventricle.

The subsequent development of the SCN includes each of the processes involved in formation of a mature central nervous system area, neuronal and glial maturation, cell death, axonal growth and synapse formation, and development of connections, both afferent and efferent. This last process, development of connections, is reflected in the maturation of SCN neuropil. Part of these processes is reflected in the morphometric growth of the SCN as a nucleus (Table 1). From the end of neuron formation, E18, to the adult, the length of the nucleus increases nearly threefold and its volume increases ninefold. Cell death is prominent in ultrastructural analysis of SCN development form E21 through P6 (Moore & Bernstein, 1989). From comparison of the number of SCN neurons observed in apoptic cell death with studies in other areas, we estimate that 30–50% of SCN neurons are

TABLE 1. DEVELOPMENT OF THE SCN IN THE RAT. MORPHOLOGICAL PARAMETERS

Developmental age	Length (μm + SE)	Volume ($\times10^{-3}$ mm^3 + SE)	Degenerating perikarya[a]
E18	315 ± 25	404 ± 2.4	0
E20	390 ± 82	9.5 ± 2.1	+
E21	—	—	+++
P1	440 ± 33	12.4 ± 2.5	+
P2	500 ± 22	13.0 ± 6.6	+++
P4	485 ± 53	16.1 ± 3.6	+
P6	540 ± 66	16.5 ± 3.7	++
P10	635 ± 70	23.0 ± 2.1	0
Adult	805 ± 73	36.8 ± 7.2	

[a]Degenerating perikarya are identified as dark, shrunken profiles. The numbers are variable among developmental ages and from brain to brain at any age. Therefore, only a rough quantification is given. Electron micrographs from six SCN were examined for each point. 0, No degenerating perikarya noted; +, present in less than half of SCN examined and less tha 3/1000 μm^2 in each SCN; ++, present in more than half of SCN examined and from 1 to 4/1000 μm^2 in each SCN; +++; present in more than half of the SCN examined and as many as 30/1000 μm^2 in an SCN.

subsequently lost. Much of this occurs through the period of retinohypothalamic tract development and arrival of other inputs, and probably reflects the pattern of formation of connections by individual neurons. Shortly after formation in the germinal epithelium, SCN neurons are immature with a very scant cytoplasm and relatively few organelles (Moore & Bernstein, 1989; Reppert & Weaver, 1991). As SCN neuropil develops (see below), implying the development of connections, there is a progressive maturation of the cytoplasm of SCN neurons until the adult appearance by P10. Dendritic development also reflects this time course.

The neuropil of the SCN is composed predominantly of immature dendrites, axons, and growth cones at E18. There is a gradual production of synapses in the prenatal period, from E18 to E21, but the rate of synapse production is low and remains so until about P3, when synapse formation rapidly picks up pace. This is concomitant with development of the retinohypothalamic tract, which is more pronounced in the SCN core than the shell in this period. The rate of synapse formation increases rapidly from P3 to P10 and then slowly continues to the adult state (Figure 3). The development of another important mammalian SCN, that of the Syrian hamster, was described by Muller and Torrealba (1998).

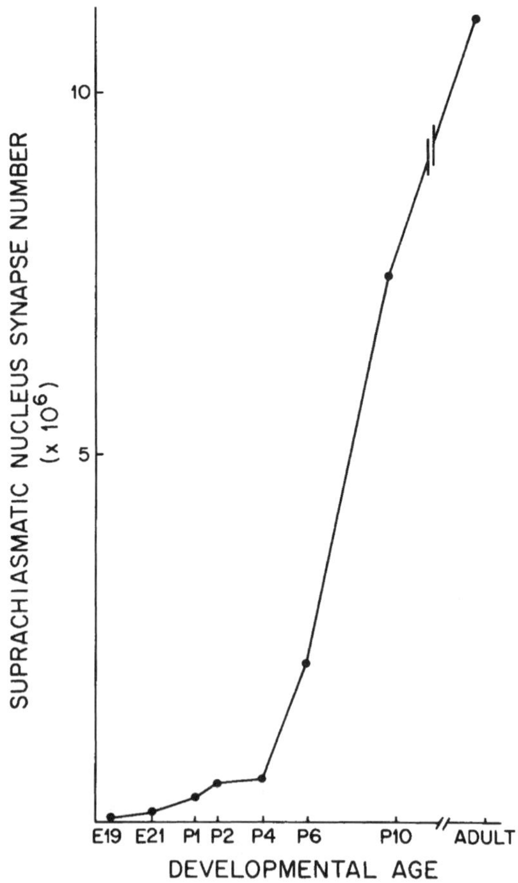

Figure 3. Graph illustrating synaptic development in the rat SCN. The data were obtained by counting the numbers of axon terminals in electron micrographs at each developmental age noted.

There are marked differences in the development of circadian function among mammals. In some species, newborn animals are able to function quite independently, whereas in others, the newborn has a long period of maternal nurturing before it achieves independence. Within this context, we will describe the development of SCN function and SCN control of effector function in a typical mammal that is born quite immature and has a long period of maternal care, the rat (see Reppert & Weaver, 1991, for review). Following the final generation of SCN neurons, there is an abrupt onset of a rhythm in glucose utilization in the fetal SCN. This rhythm develops in the pups of arrhythmic, SCN-lesioned animals and, hence, is independent of maternal influence, an observation in accord with the view that SCN neurons are born as individual circadian oscillators. The rhythm in metabolism is entrained to the maternal rhythm in intact animals. This would appear to be an important feature of adaptation, as it will place the circadian function in the newborn in synchrony with that of the mother. The mechanism by which maternal entrainment is established is unknown, but may involve melatonin in the intact organism (Davis & Mannion, 1988). However, melatonin is not an essential signal, as ablation of the pineal and the pituitary does not alter maternal entrainment of fetal circadian function.

There is another interesting example of communication in the SCN which does not depend upon typical neuronal communication. The onset of rhythmicity in the fetal rat SCN is on E18, a time when there are no synapses in the nucleus. The magnitude of the glucose utilization rhythm argues that individual SCN neurons are synchronized, but the mechanism for the interneuronal communication in the fetal SCN is unknown. It is not a function of maternal synchronization, as the rhythm develops in the absence of maternal rhythmicity. A similar phenomenon occurs in the adult SCN when tetrodotoxin, blocking sodium-dependent action potentials, eliminates the expression of circadian function, while circadian rhythmicity is preserved (Schwartz, Gross, & Morton, 1987).

After birth, the development of the rat SCN progressives rapidly. The firing rate rhythm is present by late gestation and reaches its adult state by P10 (Figure 4). The earliest extrinsic rhythm to develop is that in pineal N-acetyltransferase activity. Rhythms in adrenal corticosterone secretion and locomotor activity develop at much later time points (Figure 5), and it is clear that the expression of circadian function requires both the development of the SCN and its projections to effector sites, and the maturation of the neural mechanisms of effector functions.

ANATOMIC ORGANIZATION

Our knowledge of the anatomy of the SCN is based mainly on the albino rat. Most features of the rat SCN generalize to other species, and the following sections deal mainly with rat studies. This is accompanied by short descriptions of the hamster and human SCN, in which minor differences are pointed out.

CYTOARCHITECTURE, ULTRASTRUCTURE, AND GOLGI STAINS

The SCN is striking in Nissl-stained material because its neurons are small and tightly packed. The neurons, on average about 12 μm in diameter, number between

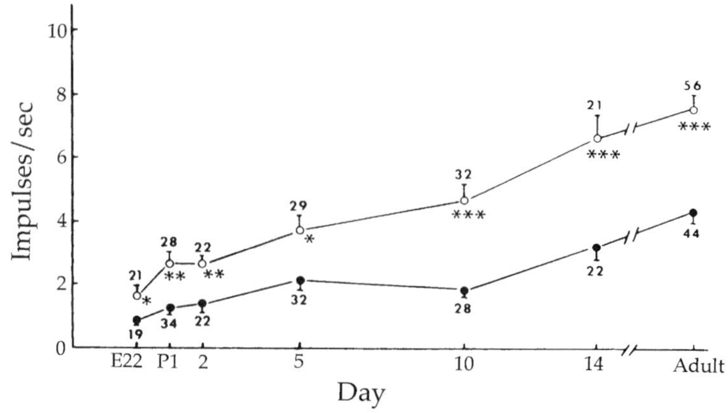

Figure 4. Development of neuronal activity in the rat SCN studies by single-unit recording of neurons in hypothalamic slices *in vitro.* (○) Subjective day recordings; (●) subjective night recordings. For each point, the values are mean firing rate ± SE of neurons exhibiting both type I and type II firing. The numbers next to the data points designate the number of neurons recorded. Asterisks indicate statistically significant differences between subjective day and night points at that developmental age.

10,000 and 12,000 (Moore & Speh, 1993; Van den Pol, 1980). The ventral and medial borders of the nucleus are easy to distinguish from the myelinated fibers of the optic chiasm and the diffuse cells of the periventricular zone, respectively. The dorsal and lateral borders of the SCN merge with the anterior hypothalamus and are difficult to define in the absence of chemoarchitectural criteria (see below). The

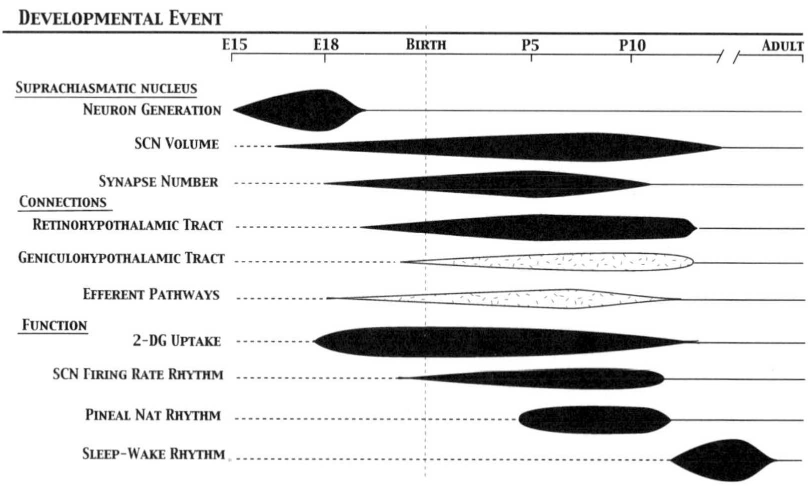

Figure 5. Summary of the major features of the morphologic development of the suprachiasmatic nucleus, entraining pathways, and efferents along with a representation of the development of three functional rhythms. The onset of development for each is indicated by the beginning of a solid or stippled area. Increasing size of the area indicates ongoing maturation of the structure or function. Straight lines indicate no further change into adulthood. The solid areas indicate that development has been studied and the developmental pattern is clearly characterized. The stippled areas indicate that some information is available, but that full details await further analysis.

rostral and caudal limits of the nucleus are recognizable in Nissl-stained coronal sections. As one views the hypothalamus from rostral to caudal, the SCN first appears as a horizontally oriented oval cluster of neurons lying in a slight depression of the optic chiasm. Further caudally, the nucleus expands in size and takes on a more circular shape. At its posterior limits, the SCN consists of a vertically oriented oval cell cluster. The appearance of the SCN in Nissl-stained material is very similar across a wide range of mammalian groups, from metatherian through eutherian species (Figure 1).

The SCN is divisible into a shell and a core based on cytoarchitectonic differences. The shell contains smaller, more highly compacted neurons than the core. As will be seen, the core–shell division is also evident in peptide phenotype, afferent innervation, and pattern of efferent projections. Rostrally, the shell lies dorsal to the core. As one views the nucleus from rostral to caudal, the crescent-shaped shell gives the appearance of rotating toward the midline, from a horizontal position dorsal to the core to a more vertical position medial to the core in the caudal SCN. At mid-levels of the SCN, the thickest portion of the shell lies in the dorsomedial quadrant of the nucleus. Originally, the shell was termed dorsomedial and the core ventrolateral, but this terminology is somewhat misleading because of the shift in the relative positions of the core and shell. The terms core and shell are also more anatomically accurate descriptions of the subdivisions in other species (Cassone, Speh, Card, & Moore, 1988).

At the ultrastructural level, one can see further differences between core and shell neurons. Core neurons have more organelles and rough endoplasmic reticulum, their nuclei have more invaginations, and their nucleoli are less marginated (Van den Pol, 1980). Somal appositions frequently link "chains" of shell neurons, and core neurons occasionally form dendrodendritic contacts (Guldner & Wolff, 1974). There is more neuropil in the core, which is to be expected given that its neurons are not as highly compacted. Gap junctions have not been found between neurons in the SCN, but are present between SCN glia. It is also noteworthy that SCN glia are few in number relative to other brain nuclei.

Golgi studies of SCN neurons reveal that shell neurons have smaller dendritic arbors, which lie in the dorsoventral plane, whereas core neurons have larger dendrites, which extend both dorsoventrally and in the horizontal axis (J. Riley, J. P. Card, & R. Y. Moore, unpublished observations). The larger dendritic arbors of core neurons act to support their denser afferent innervation, whereas the smaller dendrites in the shell accounts for its tighter packing density. SCN dendrites are, for the most part, confined to within subdivision boundaries, but to some degree, the core and shell as well as the SCN and the adjacent anterior hypothalamic area share overlapping dendritic fields.

Chemoarchitecture

All SCN neurons appear to contain the small molecule neurotransmitter gamma-amino butyric acid (GABA): (Buijs, Wortel, & Hou, 1995; Card & Moore, 1984; Moore & Speh, 1993; Okamura *et al.*, 1989; van den Pol, 1986). However, the heterogeneous peptide phenotypes of SCN neurons suggest a highly complex nucleus. Immunohistochemical labeling for virtually all SCN peptide phenotypes can be used to delineate core–shell boundaries. A large number of shell neurons express vasopressin, which is perhaps the most useful peptide for defining core–shell boundaries (Figure 6) (Card & Moore, 1984; Sofroniew & Weindl, 1980; van den Pol &

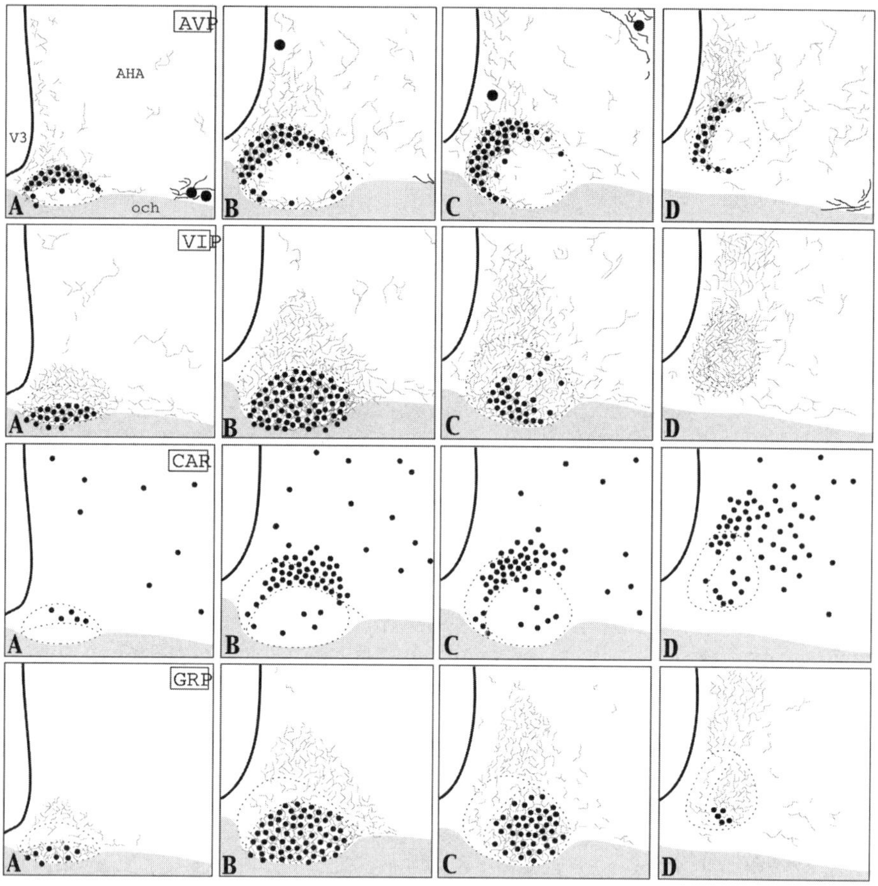

Figure 6. Schematic representations of peptide immunoreactivity in coronal sections of the rat SCN. Sections are shown from rostral to caudal. Abbreviations: AHA, anterior hypothalamic area; AVP, vasopressin; CAL, calretinin; GRP, gastrin-releasing peptide; och, optic chiasm; C3, third ventricle; VIP, vasoactive intestinal polypeptide.

Tsujimoto, 1985; Vandesande, Dierickx, & DeMey, 1975; Watts & Swanson, 1987). Vasopressin is expressed most heavily in medial regions of the shell, but scattered cells and fibers extend out laterally over the core. Lateral regions of the shell also contain enkephalin- and calretinin-immunoreactive neurons (Jacobowitz & Winsky, 1991; Watts & Swanson, 1987; J. C. Speh & R. Y. Moore, unpublished observations). The shell also contains angiotensin II-expressing cells, which are distributed in a pattern similar to vasopressin (Watts & Swanson, 1987; J. C. Speh & R. Y. Moore, unpublished observations).

Cells in the core of the SCN express vasoactive intestinal polypeptide (VIP), peptide histidine isoleucine, gastrin-releasing peptide (GRP), and neurotensin (Card, Brecha, Karten, & Moore, 1981; Card, Fitzpatrick-McElligott, Gozes, & Baldino, 1988; Mikkelsen, Larsen, O'Hare, & Wiegand, 1991). VIP is found in core neurons in the anterior two thirds of the SCN. These neurons colocalize peptide histidine isoleucine, as it arises from the same precursor as VIP (Okamura *et al.*,

1986). VIP neurons occupy the entire core region at rostral SCN levels, with many cells lying between the fibers of the optic chiasm. At mid to caudal levels, the neurons occupy a more ventromedial location within the core (Figure 6). Like VIP, GRP-containing cells are found throughout the core at rostral levels, but they are located more laterally than VIP at mid to caudal levels of the nucleus. In this lateral location within the core, they form a dense, ovoid cluster of cells. This oval region also contains neurotensin cells. A few calretinin neurons are also present in this location, but they are far fewer in number than those in the lateral shell.

The border zone between the core and shell of the rat SCN contains somatostatin and substance P-immunoreactive cells, each organized along a mediolateral axis (Card *et al.*, 1988; Dierickx & Vandesande, 1979; Mikkelsen & Larsen, 1993; Tanaka *et al.*, 1996; van den Pol & Tsujimoto, 1985). A few calretinin neurons are also present in this border zone.

GABA also seems to be the principal transmitter in the hamster SCN (Card & Moore, 1984). Vasopressin neurons also delineate the shell in this species (Figure 7), but extend further laterally, with a few neurons present in the ventrolateral shell. Unlike the rat, the shell of the hamster contains cholecystokinin-immunoreactive neurons. In the anterior lateral shell, numerous cells express the calcium-binding protein calbindin. The ventromedial core of the hamster SCN contains VIP-immunoreactive cells, and GRP cells are densest further dorsolaterally. This region also contains a dense cluster of calbindin-expressing neurons and a few substance P-immunoreactive cells (Silver, LeSauter, Tresco, & Lehman, 1996). The calbindin

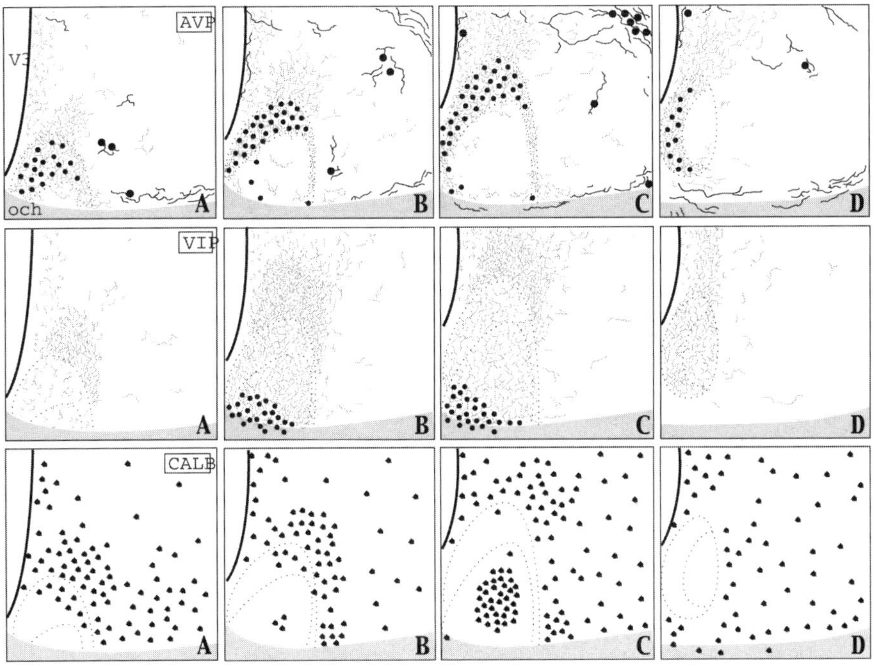

Figure 7. Schematic representations of peptide immunoreactivity in coronal sections of the hamster SCN, from rostral to caudal. Abbreviations: AVP, vasopressin; CALB, calbindin; och, optic chiasm; V3, third ventricle; VIP, vasoactive intestinal polypeptide.

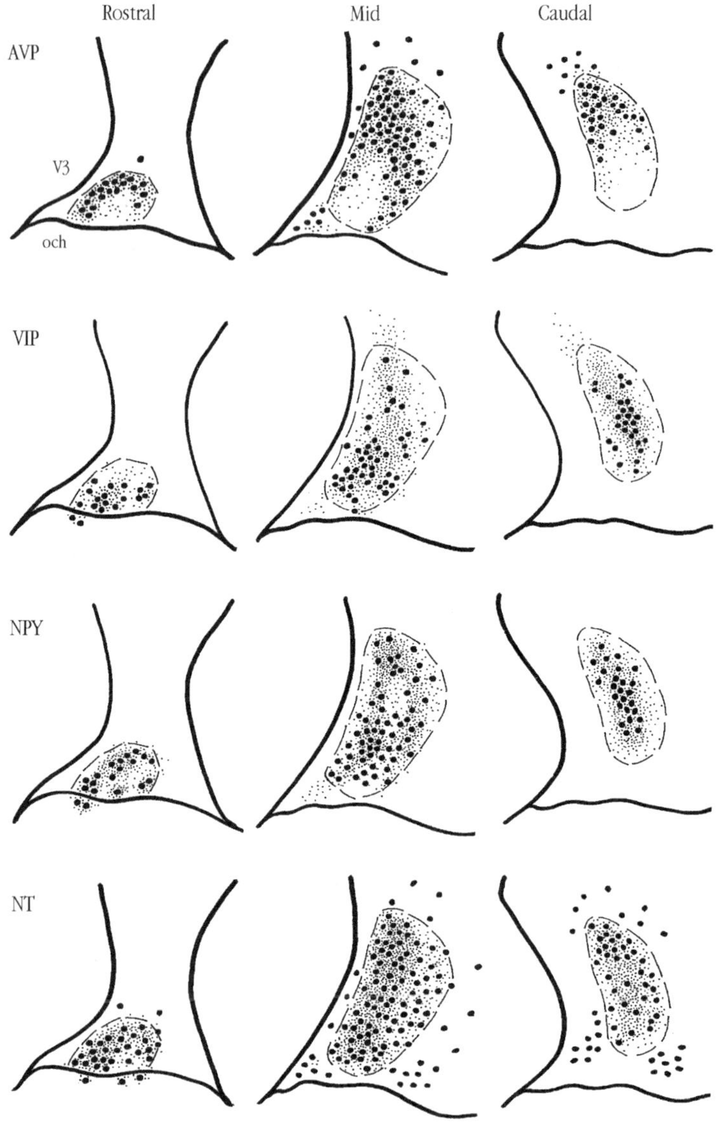

Figure 8. Schematic representations of peptide immunoreactivity in coronal sections of the human SCN, from rostral to caudal. Abbreviations: AVP, vasopressin; och, optic chiasm; NPY, neuropeptide Y, NT, neurotensin; V3, third ventricle; VIP, vasoactive intestinal polypeptide.

cluster of neurons is probably homologous to the oval cluster of neurotensin- and GRP-expressing cells in the rat.

In the human, the organization of peptide-containing neurons is similar to the rodent model (Figure 8). Vasopressin and VIP are expressed in the shell and core, respectively. The human SCN is similar to that of hamster and other species such as cat, monkey, and pig, in that the shell almost completely surrounds the core (Cassone *et al.*, 1988). There are two notable differences between the human SCN and that of other species. Unlike in other mammals, neuropeptide Y is present in a subpopulation of core neurons, and neurotensin is found in cells scattered throughout the human SCN. The functional significance of this difference is not clear. As in other species, all human SCN neurons appear to produce GABA (Gao & Moore, 1996).

Intrinsic Connections

The organization of synaptic connections in the adult SCN has been studied with the Golgi technique (van den Pol, 1980), immunohistochemical analysis of its chemoarchitecture (Card *et al.*, 1981; Castel, Feinstein, Cohen, & Harari, 1990; Daikoku, Hisano, & Kagotani, 1992; Ibata, Tanaka, Ichitani, Takahashi, & Okamura, 1993; van den Pol & Gorcs, 1986; van den Pol & Tsujimoto, 1985), iontophoretic tracer injections confined to the core or shell (Buijs, Hou, Shinn, & Renaud, 1994; Leak, Card, & Moore, 1999), and viral transsynaptic tract tracers (Leak *et al.*, 1999). Golgi stains of the rat SCN reveal a dense local circuitry. This is confirmed in immunohistochemical studies. Vasopressin, angiotensin II, and enkephalin fibers are dense in the shell, suggesting the presence of an extensive local circuitry intrinsic to that subdivision. In contrast, VIP and GRP fibers extend throughout the shell and core, indicating dense circuits intrinsic to the core, as well as projections from the core to the shell (Figure 9). Finally, somatostatin and substance P fibers in the core suggest that projections from the border zone to the core exist. The relative absence of vasopressin, enkephalin, and angiotensin II fibers in the core implies that a significant projection from the shell to the core may not exist. Axons immunoreactive for most peptides expressed by SCN neurons have also been found in fibers crossing the midline. These probably reflect commissural connections between the two SCNs (Figure 9).

Ultrastructural analysis has shown that VIP-immunoreactive terminals are in synaptic contact with vasopressin- and somatostatin-immunoreactive cells. The presence of VIP-, GRP-, somatostatin-, and vasopressin-containing synapses on profiles immunoreactive for the same peptide supports the contention that there are local circuits within each subdivision. The drawback of these studies is that the source of the immunoreactive terminal is difficult to confirm. It could arise in the ipsi- or contralateral SCN or from a source outside the nucleus. Nevertheless, the data suggest that there are projections from the core to the shell, and, like the light microscopic data, provide little evidence for a projection from shell to core. This conclusion is supported by tract-tracing studies with the retrograde transsynaptic viral tracer pseudorabies Bartha and with small, iontophoretic injections of horseradish peroxidase or *Phaseolus* in the core or shell. In the transsynaptic tracing studies, two SCN projection sites, one innervated preferentially by the core (lateral subparaventricular zone) and one by the shell (dorsomedial hypothalamic nucleus), were injected with virus and the survival period was varied so as to follow the transsynaptic passage of virus through local circuitry in the SCN (Leak *et al.*, 1999). Replication within the shell was followed by transneuronal, retrograde passage to the ipsilateral core and contralateral shell, and by an increase in ipsilateral shell labeling. Replication within the core was followed by an increase in core labeling and contralateral core labeling. The increase in the ipsilateral core following shell infections was much greater than the converse, suggesting that a core-to-shell projection is perhaps not reciprocated to a significant degree. The results with viral tracers also point to topographically organized shell-to-shell and core-to-core commissural connections (Figure 9), which is consistent with the restricted pattern of labeling on the contralateral side following small injections of horseradish peroxidase (Leak *et al.*, 1999) or

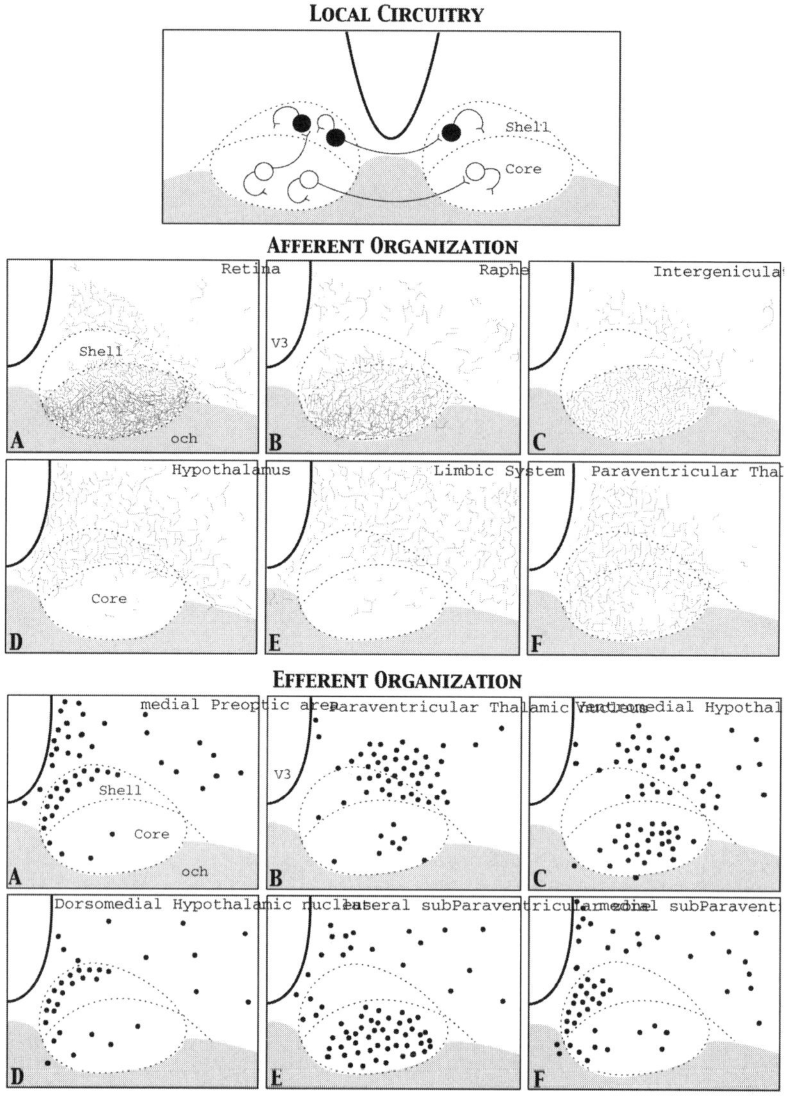

Figure 9. Schematic representations of the anatomic organization of intrinsic circuitry, inputs, and output neurons of the SCN. Coronal sections of midlevels of the rat SCN are shown. See text for details.

Phaseolus (Buijs *et al.*, 1994) in the core or shell. Thus, it appears that information arriving in the core is relayed to the shell but not vice versa, and that communication between the two SCNs is topographically organized.

Afferent Connections

As we noted in the discussion of development, the SCN develops clock function prior to the formation of intrinsic connections and inputs from other brain areas.

The function of afferents, thus, is to establish the precise timing of the SCN clock as an important component of its function as a pacemaker in regulating the temporal organization of function in areas under circadian control. There are approximately 14 sources of afferent neural input into the SCN. These can be divided into four groups on the basis of both the source of the innervation and the SCN subdivisions into which it is distributed.

Visual afferents are the dominant input to the SCN.

The Retinohypothalamic Tract

Early studies demonstrated a direct retinohypothalamic tract terminating in the ventral SCN, the area we are now designating core (Figure 9) (Hendrickson *et al.*, 1972; Moore, 1973; Moore & Lenn, 1972). The retinohypothalamic tract originates from a distinct set of retinal ganglion cells, a subset of type III cells distributed over the entire retina (Moore, Speh, & Card, 1995). These ganglion cells are small with extensive arbors of slim dendrites with relatively few branches. Recent data indicate that the dendrites of retinohypothalamic ganglion cells may receive input from a specific photoreceptor which may be unique to the circadian system (Freedman *et al.*, 1999).

The axons of the retinohypothalamic tract from each eye project bilaterally in the hypothalamus (Johnson, Morin, & Moore, 1988; Levine, Weiss, Rosenwasser, & Miselis, 1991). In most species, there is a slightly greater contralateral than ipsilateral projection. The axons of the retinohypothalamic tract terminate in a dense axonal plexus in the central SCN, in the region occupied by VIP and GRP neurons. In ultrastructural material, the terminals of retinal origin are relatively large, with a large population of small, lucent vesicles in a clear cytoplasm characterized by distinct mitochondria (Guldner, 1978a, b). These terminals synapse predominantly on distal dendrites of SCN neurons, often in glomeruli as are seen in the dorsal lateral geniculate.

The retinohypothalamic tract provides a dense innervation to the SCN, but it also innervates other hypothalamic areas. Retinohypothalamic axons extend rostrally into the preoptic area in front of the SCN, dorsally into the perisuprachiasmatic region and subparaventricular zone, and caudally into the retrochiasmatic area. In each of these areas, the retinohypothalamic fibers overlap output from the SCN. There is also a dense projection to the lateral hypothalamus from a set of retinal ganglion cells which appear to be distinct from those projecting to the SCN (Leak & Moore, 1997).

The Geniculohypothalamic Tract

Although early studies showed a projection from the ventral lateral geniculate area to the SCN (Swanson, Cowan, & Jones, 1974), they preceded the recognition that the intergeniculate leaflet is a distinct morphologic and functional division of the lateral geniculate complex. This was first demonstrated by Hickey and Spear (1976), who showed that the intergeniculate leaflet receives a dense bilateral projection from the retina, in contrast to the dorsal and ventral subdivisions. The retinal afferents projecting to the intergeniculate leaflet arise from the same retinal ganglion cells whose axons form the retinohypothalamic tract (Pickard, 1985), and the terminal plexus on intergeniculate leaflet neurons is virtually identical to that in the

SCN (Moore & Card, 1994). The intergeniculate leaflet lies between the dorsal and ventral lateral geniculate rostrally in the geniculate complex, but swings medially and ventrally further caudally (Card & Moore, 1989; Moore & Card, 1994). All intergeniculate leaflet neurons are GABA-producing (Moore & Card, 1994). Early work showed that a subpopulation of these neurons produce neuropeptide Y and project to the SCN (Card & Moore, 1982; Harrington, Nance, & Rusak, 1985; Moore, Gustafson, & Card, 1984). Subsequent studies demonstrated a separate population of enkephalin-producing neurons, which, in the rat, project exclusively to the contralateral intergeniculate leaflet (Card & Moore, 1989).

The intergeniculate leaflet projects through the geniculohypothalamic tract to the SCN in a pattern that exactly overlaps the retinohypothalamic tract terminal plexus (Card & Moore, 1982, 1989; Moore & Card, 1994). The geniculohypothalamic tract, like the retinohypothalamic input, projects into the perisuprachiasmatic region, the subparaventricular zone, and retrochiasmatic area, as well as the SCN (Horvath, 1998). The afferent and efferent projections of the intergeniculate leaflet are shown diagrammatically in Figure 10. There are two major inputs to the intergeniculate leaflet, from the retina and the contralateral leaflet. Additional secondary visual input comes from the superior colliculus and pretectal area. The intergeniculate leaflet also receives input from the SCN and retrochiasmatic area and from regions involved in behavioral state regulation, the dorsal raphe, locus coeruleus, and zona incerta. Thus, the intergeniculate leaflet is situated to provide a dense secondary visual input to the SCN which is modulated by a series of other areas.

The Pretectohypothalamic Tract

Mikkelsen and Vrang (1994) described a secondary visual projection to the SCN arising in the olivary pretectal and posterior pretectal nuclei which passes along the border of the optic tract to terminate in the SCN core. The function of this pathway has not been studied, but one would assume that, like the input from the raphe to the SCN (see below), it is a modulatory visual pathway affecting entrainment by the retinohypothalamic tract. Like the geniculohypothalamic tract, this projection is probably GABA-containing.

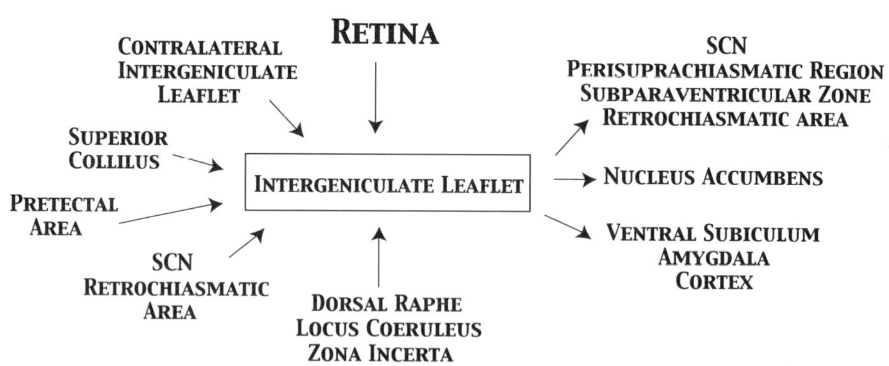

Figure 10. Diagrammatic representation of the inputs and outputs of the intergeniculate leaflet of the lateral geniculate complex.

The SCN receives extensive projections from areas extending from cerebral cortex through basal forebrain to thalamus, hypothalamus, circumventricular organs, and brainstem. The origin and region of termination of these afferents have been reviewed recently (Leak *et al.*, 1999; Moga & Moore, 1997) and are presented in Table 2.

In the past 10 years, there has been extensive work on nonphotic entrainment. Inputs from the raphe and the intergeniculate leaflet, when activated, phase shift the SCN such that phase advances occur during the day and small delays at night. The resultant shifts can be plotted as a "dark" phase response curve, designated as such because dark pulses and other activity-inducing stimuli cause similar responses in rodents. The integrity of the intergeniculate leaflet and the raphe is essential for a response to activity-inducing stimuli. Thus, these regions are thought to feed information on behavioral state back to the pacemaker. The subject of nonphotic entrainment has been reviewed extensively (Hastings *et al.*, 1997; Hastings, Duffield, Smith, Maywood, & Ebling, 1998; Mrosovsky, 1995).

EFFERENT CONNECTIONS

Early studies of SCN projection patterns utilized tritiated amino acid injections (Bark & Finkelstein, 1981; Stephan, Berkley, & Moss, 1981; Swanson & Cowan, 1975)

TABLE 2. DENSITY OF AFFERENT INPUTS TO THE CORE AND SHELL SUBDIVISIONS OF THE SCN[a]

Afferent Source	Core	Shell
Retina	++++	−
Intergeniculate leaflet	++++	−
Olivary pretectal nucleus	+++	−
Median raphe nucleus	++++	−
Paraventricular thalamic Nucleus	++++	++++
Infralimbic cortex	−	+++
Subiculum	−	++
Lateral septal nucleus	−	++++
Substantia inominata	−	++
Ventromedial hypothalamic nucleus	−	+++
Subfornical organ	+	+
Arcuate nucleus	−	+++
Posterior hypothalamic area	−	+++
Medial pretectal nucleus (superior colliculus)	−	++
Parabrachial nucleus	−	++
Contralateral core	++++	−
Ipsilateral core	++++	++++
Contralateral shell	−	++++
Ipsilateral shell	−	+++

[a]The plus and minus sign refect the density of projection. Data from Moga and Moore (1997) and Leak *et al.* (1999).

and anterograde degeneration techniques (Stephan *et al.*, 1981). Most of the information on SCN efferents, however, stems from the *Phaseolus vulgaris* leucoagglutinin anterograde studies of Watts and colleagues (Watts & Swanson, 1987; Watts, Swanson, & Sanchez-Watts, 1987) and from immunohistochemical studies following the route of axons stained for various SCN peptides (Hoorneman & Buijs, 1982; Kalsbeek, Teclemariam-Mesbah, & Pevet, 1993; Sofroniew & Weindl, 1978; van der Beek, Wiegant, van der Donk, van den Hurk, & Buijs, 1993). It was shown that a diffusible signal can reinstate locomotor rhythmicity in SCN-lesioned animals bearing fetal SCN grafts (Silver *et al.*, 1996). However, neural efferents seem essential for all other circadian functions so far investigated, such estrous cyclicity and rhythmic corticosterone and melatonin production (Hakim, DeBernardo, & Silver, 1991; Meyer-Bernstein *et al.*, 1999). The present section deals with neural output pathways of the SCN. The densest projection of the SCN is to the perisuprachiasmatic region and the subparaventricular zone (Swanson, Canteras, & Watts, 1993; Watts, 1991; Watts & Swanson, 1987; Watts *et al.*, 1987). The former lies immediately around the SCN, whereas the subparaventricular zone extends from the caudal SCN dorsally to lie ventral to the posterior magnocellular division of the paraventricular hypothalamic nucleus. SCN efferents continue caudally past the subparaventricular zone and terminate in the dorsomedial hypothalamic nucleus, in regions medial to the ventromedial hypothalamic nucleus, the posterior hypothalamic area, and the periaqueductal gray. A small number of fibers continues dorsally and laterally past the subparaventricular zone into the zona incerta. There are five other major fiber pathways arising in the SCN. One dense fiber pathway travels dorsally to the midline thalamus, where it innervates the paraventricular and paratenial thalamic nuclei. This projection is densest in rostral portions of the thalamus, and consists of both vasopressin and VIP fibers. A second pathway extends anteriorly to ventral and periventricular regions of the medial preoptic area and nuclei within it, such as the anteroventral periventricular nucleus, the anterodorsal and anteroventral preoptic nuclei, and the medial preoptic nucleus. These pathways contain both vasopressin and VIP (Hoorneman & Buijs, 1982; Kalsbeek *et al.*, 1993; van der Beek *et al.*, 1993). Rostrally directed fibers also head to the ventral lateral septum and bed nucleus of stria terminalis. Other fibers continue caudally into the retrochiasmatic area and the region between the ventromedial hypothalamic nucleus and the arcuate. Finally, there is a small projection which passes through the lateral hypothalamus to terminate in the intergeniculate leaflet and ventral lateral geniculate. A few fibers branch off this pathway toward the supraoptic nucleus. Similar SCN projections have been identified in the hamster (Kalsbeek *et al.*, 1993; Morin, Goodless-Sanchez, Smale, & Moore, 1994) and human (Dai, Swaab, & Buijs, 1997; Dai, Swaab, Van der Vliet, & Buijs, 1998a; Dai, Van der Vliet, Swaab, & Buijs, 1998b).

Recently, it has become clear that SCN efferent neurons are topographically organized (Figure 9) (Leak & Moore, 1997). Although there is some overlap, the core and shell innervate the lateral and medial subparaventricular zone, respectively. This is evident in the partial segregation of vasopressin and VIP fibers in this region (Watts, 1991) and in the pattern of retrograde labeling following injections in the lateral subparaventricular zone and injections centered in the medial subparaventricular zone and paraventricular hypothalamic nucleus (Leak & Moore, 1997; Vrang, Larsen, Moller, & Mikkelsen, 1995b). Following retrograde tracer injections in various nuclei in the medial preoptic area and the dorsomedial hypothalamic nucleus, the shell is more densely labeled than the core, whereas injections in the region of the ventromedial hypothalamic nucleus label the core more densely (R. K. Leak & R. Y. Moore, unpublished observations). Preoptic injections which extend dorsally into the lateral septum or ventrally into the preoptic suprachiasmatic nucleus label the core and shell in equal density. Injections in the paraventricular thalamic nucleus label neurons in the lateral shell and perisuprachiasmatic region, and midline thalamic injections spreading into the paratenial nucleus also label the core. Recent anterograde work reveals this topographic organization as well. For example, *phaseolus* injections in the dorsal SCN label more terminals in the dorsomedial hypothalamic nucleus and preoptic area than injections in the core (Buijs, Markman, Nunes-Cardoso, Hou, & Shinn, 1993; van der Beek, Horvath, Wiegant, Van den Hurk, & Buijs, 1997).

In their anterograde studies, Watts and his colleagues found that the subparaventricular zone innervates the same targets as the SCN, but in a denser fashion, and hypothesized that the subparaventricular zone amplifies the output of the pacemaker. However, recent data in our lab (R. K. Leak & R. Y. Moore, unpublished observations) suggest a more complicated situation than a unimodal amplification of SCN output. We find that the lateral subparaventricular zone, innervated by the core, projects more densely to the midline thalamus and ventromedial hypothalamic nucleus, whereas the medial subparaventricular zone, innervated by the shell, projects more densely to the medial preoptic area. This parallel processing of information within the circadian timing system is a new observation. When considered together with the differential innervation of the SCN core and shell, what emerges is a more precise control of the pacemaker than previously thought.

Functional Organization

Neuron versus Network

In the section on development, we suggested that aspects of the pattern of development of the SCN were consistent with the view that SCN neurons are formed as individual circadian oscillators which are coupled by neural connections to form a pacemaker. That is, the individual neuronal oscillators would form intrinsic connections which would couple those neurons, and the population of coupled oscillators would then regulate the temporal pattern of function in other functional systems through efferent connections. This is presented diagrammatically in Figure 11. The alternative interpretation is that individual neurons are not formed as circadian oscillators, but that circadian function arises as an emergent property of the network formed of the individual neuronal elements. The latter was proposed many years ago on purely theoretical grounds (Pavlidis, 1969), and the statistical foundation and limitations of these models have been laid out by Enright (1980).

All of the data favor the individual neuronal circadian oscillator model. First, this is the mechanism used in invertebrates (Michel, Geuss, Zaritsky, & Block, 1993) and recent molecular data argue a conservation of mechanism from invertebrates to mammals. Second, the developmental data indicate that circadian function is initiated in the fetal SCN shortly after the neurons are born and before there is any indication of intrinsic or extrinsic connections in the nucleus. Third, individual SCN

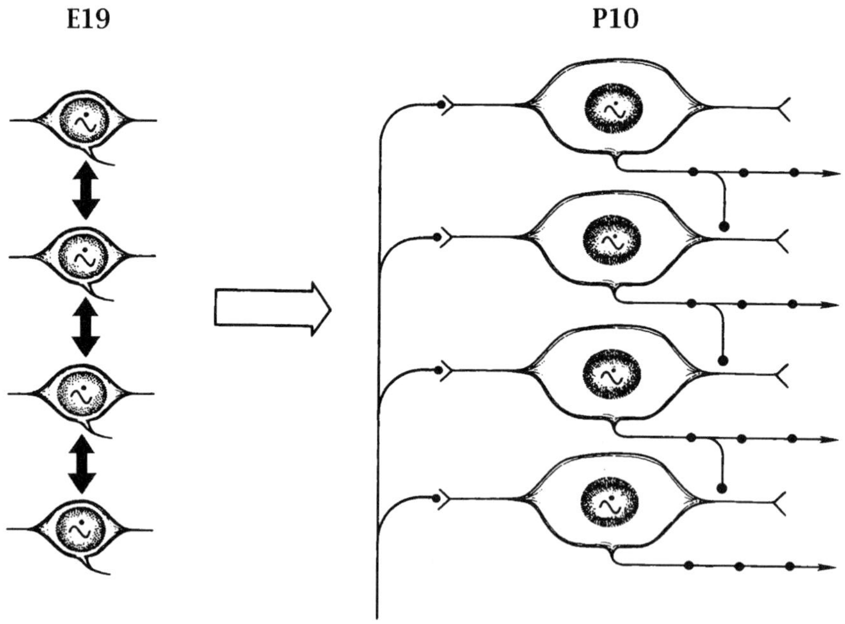

Figure 11. Diagrammatic representation of the development of the SCN from E19 to P10. At E19, the neurons are small with few processes and essentially no connections. Interaction among neurons must occur by a process that is not known, represented by solid arrows. At P10, the SCN is a well-developed neural network with mature neurons having afferent, efferent, and intrinsic connections. The sine wave in the nucleus of each neuron conveys the hypothesis that SCN neurons are born as circadian oscillators.

neurons maintained in long-term culture exhibit individual circadian rhythms in firing rate which differ in phase and period from those of other neurons in the same culture (Welsh *et al.*, 1995).

Firing Rate Rhythm—Is It Informative?

The work of Inouye and colleagues (Inouye & Kawamura, 1979; Kubota, Inouye, & Kawamura, 1981) was the first demonstration of an endogenous rhythm of firing rate in the SCN *in vivo.* With multiunit recording, they showed a high-amplitude rhythm which exhibited a peak at about circadian time 6 and a trough at about circadian time 18, with circadian time 12 designated as onset of activity. The endogenous nature of the rhythm was shown by its persistence after total deafferentation of the SCN from the rest of the brain. Subsequent studies carried out by many investigators on SCN slice preparations *in vitro* provided clear evidence that the rhythm was restricted to SCN neurons and would persist for many days (Green & Gillette, 1982; Groos & Hendriks, 1982; Shibata & Moore, 1988; Shibata *et al.*, 1982). The *in vitro* data also showed a simple rhythm, approximately a sine wave, with mean peak single-unit firing rates at circadian time 6 and trough rates at circadian time 18 in slices obtained from animals in a standard light–dark environment. In the *in vitro* situation, however, the amplitude of the rhythm, measured either by firing rate or glucose utilization (Shibata & Moore, 1988), is small, an approximately twofold difference between trough and peak values. The interpretation of the firing rate rhythm presumably is that this is the signal which is being transmitted to the rest of the brain. If this is the case, how do we account for the differences in phase of various expressed rhythms and for the differences in properties and amplitude of those rhythms? Stated differently, how do we understand the apparent simplicity of the firing rate rhythm in the face of the complexity of SCN organization and projections and of expressed rhythms in the many functions under circadian control?

One answer to these questions is certainly that the firing rate rhythm is probably not as simple as it has been portrayed. Both single- and multiunit recording studies provide information on an average of many units recorded over varying periods of time and over varying parts of the SCN. Data from Gillette (1991) show clearly that individual SCN neurons have very different patterns of activity which are obscured by the averaging methodology. This must be placed in the context of the core and shell organization of the SCN, particularly with respect to efferents (Figure 12). This information provides us with a quite different view, and we are early in developing a full picture of the situation. The view at present is as follows. Photic information is conveyed almost exclusively to the core in a direct retinal pathway and in other pathways which provide secondary visual input. This information is complemented by a dense projection from the midbrain raphe nuclei which appears to function to modify the effects of the visual inputs (Rea, 1998). There is also an input from the paraventricular thalamic nucleus (Moga, Weis, & Moore, 1995) and from selected hypothalamic inputs (Moga & Moore, 1997). The core projects to the shell (Leak *et al.*, 1999), and the shell receives inputs from a series of areas that do not project to the core (Moga & Moore, 1997). Thus, the information received by the core differs from that received by the shell. The core is the only area that can be directly entrained by photic input. Shell output is modulated by core input. With these

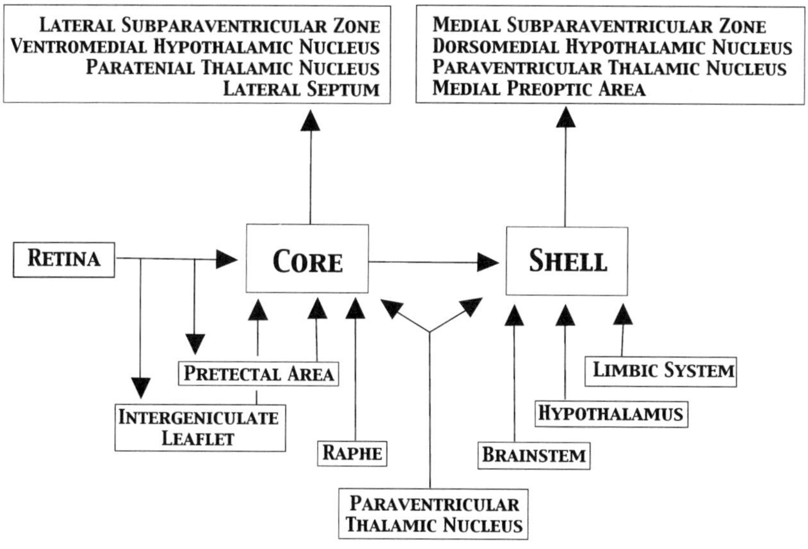

Figure 12. Diagrammatic representation of the inputs and outputs of the core and shell subdivisions of the SCN. Although there is some overlap, the core and shell receive inputs from different sources (with the exception of the paraventricular thalamic nucleus, which innervates both subdivisions), and have largely separate projection sites within the diencephalon.

considerations, and since core and shell share some projection sites, but provide separate inputs to others, this provides for a much more complex pattern of circadian rhythm control than might be inferred from the simplicity of the firing rate rhythm.

Effector Coupling

The "Troublesome" Sleep–Wake Cycle. The fundamental output of the circadian system is the regulation of sleep–wake cycles to compartmentalize sleep in a period which is disadvantageous for waking behavior. In essence, the fundamental function of circadian timing in mammals is to ensure optional adaptive behavior. The outputs of the SCN used for this are unclear, particularly since the work of Silver and associates (LeSauter & Silver, 1998) demonstrates that the locomotor activity rhythm can be restored in arrhythmic, SCN-lesioned hamsters by a diffusible signal arising from SCN cells encapsulated in the third ventricle. The interpretation of these data, however, must be undertaken with great care. They do tell us the following. First, with SCN lesions, the effector sites for the rhythm must be distal to the immediate region of the SCN. Second, the effector sites must be in the near vicinity of the third ventricular lining. That is, the diffusible signal, whether peptide or smaller molecule, would not be expected to diffuse more than a short distance, perhaps a few hundred micrometers, beyond the ependymal lining before it is enzymatically inactivated. Finally, it is important to bear in mind that the rhythmic function assessed, locomotor activity, is most likely not an expression of motor function, but of behavioral state regulation, sleep, and waking. Locomotor activity is an adaptive behavior of the waking state.

How are we to interpret these encapsulated transplant data in the context of the behavioral state regulation? Alternatively, what centers are likely to be involved in behavioral state regulation that are sufficiently close to the third ventricle to be affected by a diffusible factor? The obvious sites are in the periventricular, broadly defined, thalamus, and hypothalamus. Possible structures are outlined in Figure 13. Stated simplistically, waking behavior is a function of the thalamic relay nuclei and the neocortex activated by arousal systems. The brainstem reticular-activating system stimulates arousal and wakefulness through hypothalamic and midline thalamic structures. The posterior hypothalamus contains a major activating system, which, with the lateral hypothalamus, projects diffusely and widely to the limbic forebrain and neocortex. The anterior hypothalamus, particularly the ventrolateral preoptic area, is involved in the generation of slow-wave sleep. The SCN directly innervates this region and midline thalamus, and to a much lesser extent, the lateral hypothalamus and posterior hypothalamic area. Although the exact mechanisms remain to be elucidated, it is reasonable to propose at this time that these effector systems are the ones responsive to SCN input and responsible for the sleep–wake cycle and that they also are the ones affected by the diffusible signal in the encapsulated transplant studies.

Pineal Gland. The pineal gland is a derivative of the caudal portion of the dorsal diencephalic anlage, the epithalamus. It arises as a vesicular structure, but, in mammals, assumes an ovoid, solid glandular appearance. In most mammals, it lies caudal to the epithalamus, connected by a short pineal stalk. In a few mammals, the rat being an example, the pineal gland is situated at a distance from the diencephalon, lying just beneath the confluence of the sinuses, and is attached to the caudal

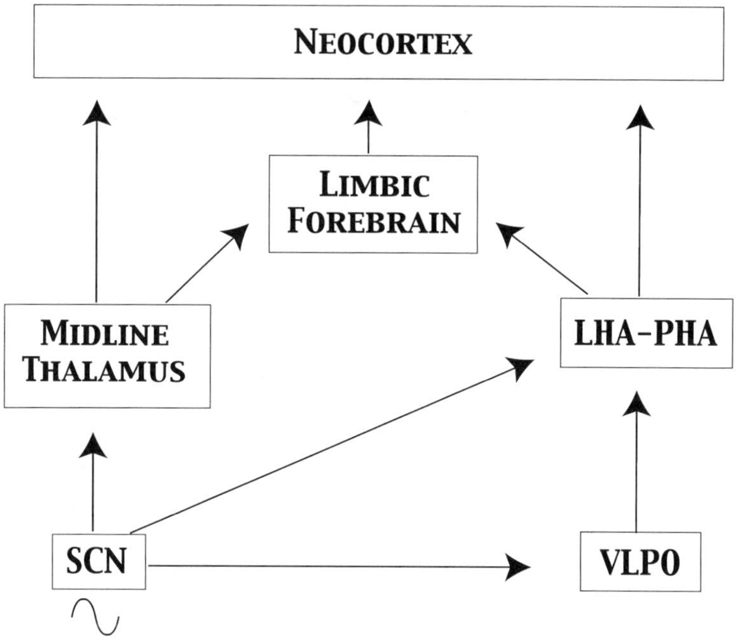

Figure 13. Diencephalic control of sleep–wake behavior. See text for description. Abbreviations: LHA, lateral hypothalamic area; PHA, posterior hypothalamic area; VLPO, ventrolateral preoptic area.

epithalamus by a long stalk. The innervation of the pineal gland is variable among mammals, but all species have a sympathetic innervation arising in the superior cervical ganglion.

The principal functional output of the pineal gland is a phylogenetically ancient derivative of the amino acid tryptophan, melatonin. Since the discovery of melatonin in the late 1950s, it has been shown that there is a circadian rhythm in the synthesis and release of the hormone in nearly all mammals. Unlike many other circadian rhythms, the melatonin rhythm is the same in nocturnal and diurnal species, with higher levels at night and low levels during the day. This rhythm is driven by the SCN (Klein & Moore, 1979; Moore, 1996). The pathways involved in the regulation of the melatonin rhythm are shown in Figure 14.

As noted above, there is a circadian rhythm in firing rate in the SCN which inhibits tonic activity in the autonomic division of the paraventricular nucleus innervating the intermediolateral cell column at the upper thoracic (T1–T2) levels. During the day, the high level of firing in the SCN inhibits activity in the paraventricular nucleus and this, in turn, results in low firing rates in the rest of the pathway and a very low level of melatonin synthesis. As the SCN firing rate goes down in late subjective day, the firing rate of the paraventricular nucleus increases and drives the functional activity of the pre- and postganglionic sympathetic neurons. The release of noradrenaline from postganglionic superior cervical ganglion neuron axons activates alpha- and beta-adrenergic receptors, which then stimulate production of melatonin by control of levels of the enzyme serotonin N-acetyltransferase. As SCN activity increases in early subjective day, the remainder of the pathway is again inhibited and melatonin production drops to very low levels.

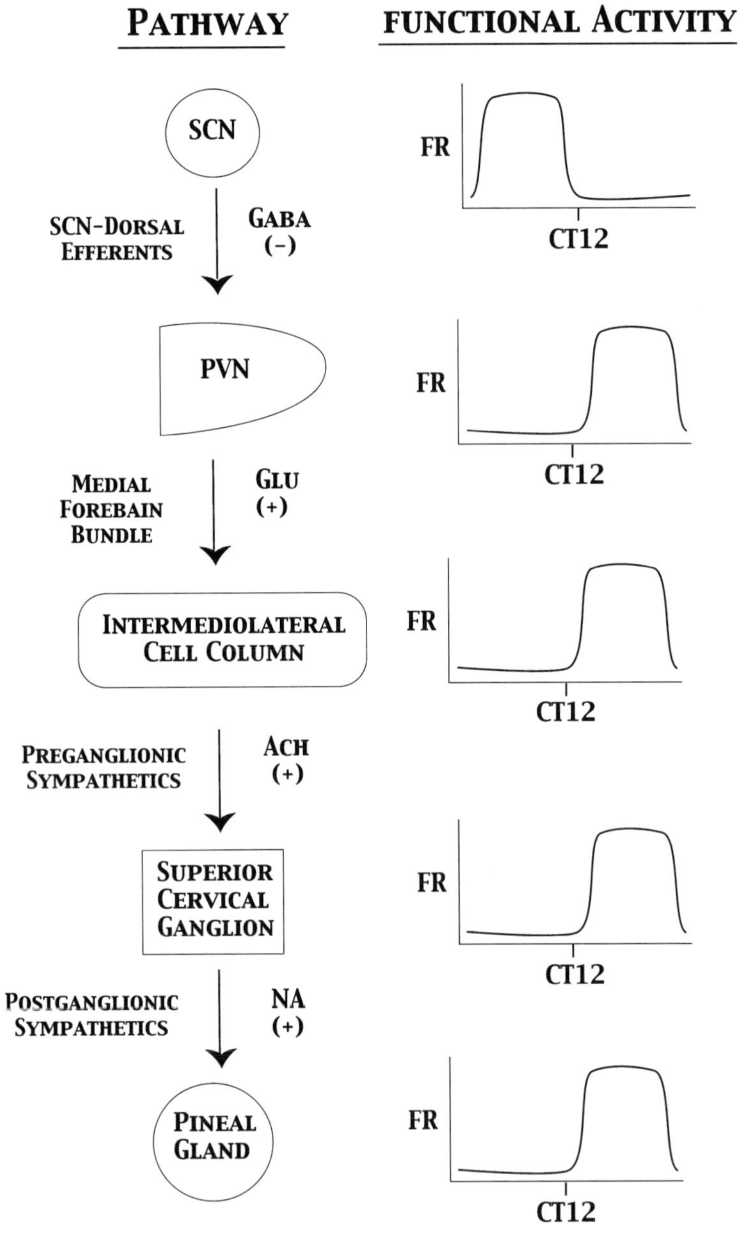

Figure 14. Neural control of the pineal melatonin rhythm. The pathway is shown on the left and the functional activity of each component over time is shown on the right. Abbreviations: GLU, glutamate; ACH, acetylcholine; NA, noradrenaline; FR, firing rate. GABA is an inhibitory transmitter (−), whereas GLU, ACH, and NA are excitatory (+) in this pathway.

Similarly, if light exposure occurs during the night, the SCN is activated, inhibition of the pathway occurs, and melatonin levels drop precipitously.

Hypothalamic–Adrenal–Pituitary (HPA) Axis. The control of corticosterone secretion has two components (Dallman *et al.*, 1987), a systemic response to stress and a daily rhythm, with corticosterone levels highest in the early evening in nocturnal animals and in the early morning in diurnal species. This section deals with the neural control of the latter. The secretion of corticosterone from the zona fasciculata of the adrenal cortex is stimulated by adenocorticotrophic hormone (ACTH), which is released into the blood from the anterior pituitary corticotrophs in response to corticotrophin-releasing hormone (CRH) released at median eminence terminals into the blood of the portal plexus. The source of these terminals is CRH neurons in the medial parvicellular division of the paraventricular hypothalamic nucleus (Swanson, 1987).

The circadian pacemaker in the SCN is responsible for the rhythmicity in corticosterone secretion. Lesions of the SCN abolish rhythmicity in the HPA axis, although the response to stress remains intact (Abe & Rusak, 1994; Krieger, Hauser, & Krey, 1977; Moore & Eichler, 1972; Raisman & Brown-Grant, 1977; Szafarczyk, Ixart, Malaval, Nouguier-Sole, & Assenmacher, 1979). Although it was initially argued that the paraventricular hypothalamic nucleus is not significantly innervated by the SCN (Swanson *et al.*, 1993; Watts, 1991), it is clear from recent studies that SCN fibers do enter the medial, dorsal, and periventricular divisions of the nucleus (Buijs *et al.*, 1993; Vrang, Larsen, & Mikkelsen, 1995a; Vrang, Mikkelsen, & Larsen, 1997). These projections also can be demonstrated electrophysiologically. Parvicellular neurons of the paraventricular hypothalamic nucleus, identified by antidromic activation from the median eminence, respond to SCN stimulation with an increase in firing rate in most cells (Hermes & Renaud, 1993; Hermes, Buijs, & Renaud, 1996). It is likely that these contain CRH, as others have found CRH-immunoreactive cells in close apposition with anterogradely labeled SCN fibers (Vrang *et al.*, 1995a). Furthermore, anterogradely labeled SCN fibers make synaptic contacts with neurons in and around the paraventricular hypothalamic nucleus which express the immediate early c-*fos* gene in response to stressful stimuli (Buijs *et al.*, 1993). Aside from a putative monosynaptic connection with the neural regulators of ACTH secretion from the anterior pituitary, the SCN may indirectly control CRH neurons through a dense projection to the dorsomedial hypothalamic nucleus. The dorsomedial hypothalamic nucleus projects to the paraventricular hypothalamic nucleus (Levin, Sawchenko, Howe, Bloom, & Polak, 1987). Infusions of a vasopressinergic antagonist into the dorsomedial hypothalamic nucleus increase corticosterone levels (Kalsbeek & Buijs, 1992; Kalsbeek, Buijs, van Heerikhuize, Arts, and van der Woude, 1992; Kalsbeek, van der Vliet, & Buijs, 1996a). As this effect is abolished in animals with SCN lesions, it is surmised that the vasopressinergic projections of the SCN to the dorsomedial hypothalamic nucleus normally inhibit the HPA axis. Furthermore, lesions of the vasopressinergic system of the SCN with cytotoxic monoclonal antibodies increase plasma ACTH and CRH mRNA levels (Gomez *et al.*, 1997). VIP may also play an inhibitory role in the circadian control of corticosterone section, as antisense infusion studies have suggested (Scarbrough *et al.*, 1996). An inhibitory role of the SCN in the regulation of the HPA axis is consistent with the observations that all SCN neurons apparently contain GABA (Moore & Speh, 1993) and that infusions of GABA-receptor antagonists into the dorsomedial hypothalamic nucleus increase plasma ACTH and corticosterone (Keim & Shekhar, 1996). The circadian inhibition of the HPA axis is also supported by the observation that both SCN firing rate levels and the release of vasopressin from the SCN drop at the time of the evening rise in ACTH and CRH (Green & Gillette, 1982; Kalsbeek & Buijs, 1992; Kalsbeek *et al.*, 1992; Kalsbeek, Buijs, Engelmann, Wotjak, & Landgraf, 1995; Kalsbeek *et al.*, 1996a; Kalsbeek, van Heerikhuize, Wortel, & Buijs, 1996b ; Shibata *et al.*, 1982). However, CRH and ACTH levels subsequently drop during the night, before the rise in SCN firing rate. Thus, the situation is somewhat more complex than a simple daily inhibition (Kalsbeek *et al.*, 1996b). It may be that the drop in CRH and ACTH at night occurs because circulating corticosterone enters the brain to feed back negatively on CRH and ACTH production (Keller-Wood & Dalman, 1984). Consequently, CRH, ACTH, and corticosterone levels would fall and, because of the rise in SCN firing rate in the early morning, remain low until the CRH neurons were released from inhibition by the SCN in the late afternoon. The neural circuits of this hypothetical scenario are illustrated in Figure 15. Not indicated on the figure is a circadian rhythm in the sensitivity of the adrenal cortex to ACTH which accounts for much of the high amplitude of the corticosterone rhythm and the circadian rhythm in corticosterone feedback efficacy (Kaneko, Hiroshige, Shinsako, & Dallman, 1980; Keller-Wood & Dallman, 1984). The former effect may be mediated by the splanchnic nerve (Jasper & Engeland, 1994). It has been suggested that the splanchnic nerve is controlled through SCN projections to the parvocellular autonomic division of the paraventricular hypothalamic nucleus, which, in turn, sends descending projections to the intermediolateral cell column (Buijs, Wortel, van Heerikhuize, & Kalsbeek, 1997; Buijs *et al.*, 1999). This SCN–paraventricular–spinal cord circuit has been illustrated in transsynaptic viral tract-tracing experiments (Buijs *et al.*, 1999) as well as by analyzing VIPergic SCN efferent fibers on spinal projecting cells of the hypothalamus (Teclemariam-Mesbah, Kalsbeek, Pevet, & Buijs, 1997).

Estrous Cycles. The estrous cycle has a period of 4–5 days in the rat, but it is actually an expression of circadian function. With each estrous cycle, a massive rise in plasma levels of luteinizing hormone (LH) on the afternoon of proestrus causes the rupture of a follicle and consequent ovulation (Freeman, 1994; Kalra, 1993). The release of LH from the anterior pituitary is initiated by two variables: increasing ovarian steroid levels and the release of gonadotropin-releasing hormone (GnRH) into the portal plexus (Levine & Ramirez, 1980; Sarkar, Chiappa, Fink, & Sherwood, 1976). Circulating estrogen enters the brain and feeds back positively on GnRH and LH release, thereby enhancing its own production (Freeman, 1994; Kalra, 1993; Lopez, Merchenthaler, Moretto, & Negro-Vilar, 1998). Estrogen levels increase gradually until right before ovulation, when they peak sharply. Progesterone levels rise sharply at the time of ovulation, and appear to work synergistically with estrogen in exerting a trophic effect on GnRH and LH production (Freeman, 1994; Kalra, 1993; Lopez et al., 1998). Following ovulation, the steroid feedback loop switches to a negative one, and GnRH and LH levels fall sharply. Estrogen receptors are abundant in the bed nucleus of stria terminalis and the ventromedial hypothalamus, areas which project densely into the preoptic area where most GnRH neurons are located (Herbison & Theodosis, 1992; Krieger, Morrell, & Pfaff, 1976; Romano, Knist, & Pfaff, 1989; Shivers, Harlan, Hejtmancik, Conn, & Pfaff, 1986; Witkin, Paden, & Silverman, 1982). Steroid receptors are also expressed by neurons in the preoptic area, particularly in the anteroventral periventricular nucleus, although not on GnRH neurons themselves.

We know that the estrous cycle is an expression of circadian function from the

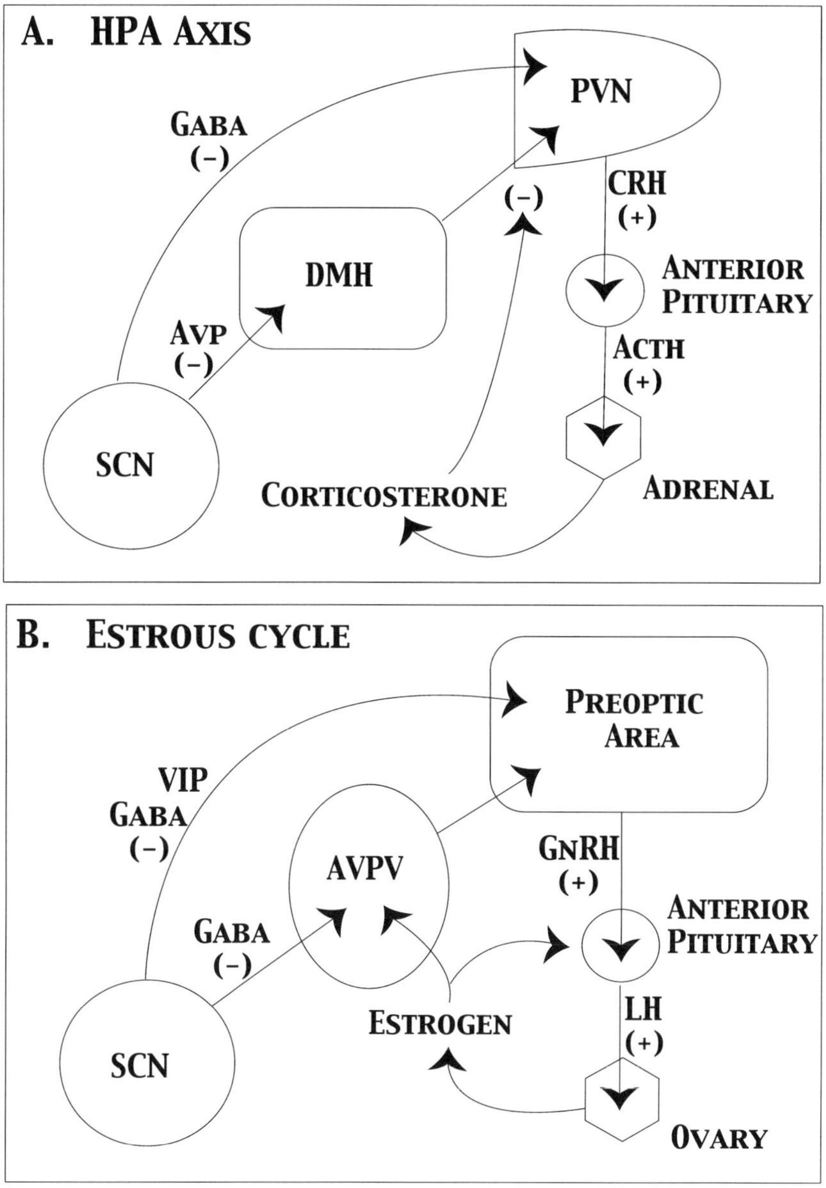

Figure 15. Neural control of (A) the hypothalamic–pituitary–adrenal axis and (B) the hypothalamic–pituitary–gonadal axis. The putative neurotransmitters are indicated. See text for further details. Abbreviations: AVPV, anteroventral periventricular preoptic nucleus; ACTH, adrenocorticotrophic hormone; AVP, vasopressin; CRH, corticotrophin-releasing hormone; DMH, dorsomedial hypothalamic nucleus; GnRH, gonadotrophin-releasing hormone; LH, luteinizing hormone; PVN, paraventricular hypothalamic nucleus; VIP, vasoactive intestinal polypeptide.

elegant experiments of Everett, Sawyer, and colleagues (Everett, 1989; Everett & Sawyer, 1950; Everett, Sawyer, & Markee, 1949), who showed that ovulation can be delayed precisely 24 hours by injecting the rats with sodium pentobarbital during a critical time window before ovulation. Then Legan and colleagues demonstrated that ovariectomized rats treated with estrogen exhibit a daily LH surge (Legan & Karsch, 1975; Legan, Coon, & Karsch, 1975). Thus, a daily permissive signal allows the release of LH to occur only under the appropriate steroidal milieu. The daily LH peak occurs in the late afternoon, about 1–4 hours before activity onset, and phase shifting the pacemaker with a change in the sighting schedule causes the same degree of shift in the timing of the LH surge. The control of the LH surge by the circadian pacemaker is also illustrated in hamsters with split activity rhythms (Swann & Turek, 1985). The locomotor activity rhythms of hamsters placed in constant lighting conditions split apart into two bouts of activity, which eventually lie 12 hours apart. The split hamsters exhibit two daily LH surges, each occurring several hours before activity onset of one of the two bouts. Finally, SCN-lesioned animals do not have estrous cycles, fail to surge even when ovariectomized and primed with estrogen, and their LH levels fall to low levels (Brown-Grant & Raisman, 1977; Coen & MacKinnon, 1980; Stetson & Watson-Whitmyre, 1976). However, when injected with progesterone, they respond with an LH surge whose timing is now determined by the time of injection rather than being keyed to the time of day (Terasawa, Wiegand, & Bridson, 1980; Wiegand & Terasawa, 1982; Wiegand, Teresawa, & Bridson, 1978; Wiegand, Terasawa, Bridson, & Goy, 1980). In contrast, in SCN-intact animals, progesterone injections are able to advance the LH surge by 24 hours, but can never move the surge beyond a critical time window in the early afternoon (Caligaris, Astrada, & Taleisnik, 1971). SCN-lesioned animals also still exhibit basal pulsatile LH release and the negative feedback effects of estrogen remain intact (Arendash & Gallo, 1979; Gray, Soderstein, Tallentire, & Davidson, 1978). Thus, the SCN only seems to be responsible for the temporal organization of the estrous cycle.

The only area which, when lesioned, results in a lack of response to both estrogen and progesterone is the anteroventral periventricular nucleus in the preoptic area (Simerly, 1998; Terasawa *et al.*, 1980; Wiegand & Terasawa, 1982; Wiegand *et al.*, 1978, 1980). This occurs although GnRH pathways are not damaged and GnRH production ability is not compromised (Ma, Kelly, & Ronnekleiv, 1990; Terasawa & Davis, 1983). The anteroventral periventricular nucleus is thus a nodal point which seems necessary for the positive feedback of progesterone and estrogen on the GnRH neuron (Simerly, 1998). Although the anteroventral periventricular nucleus does not contain GnRH wells, it has abundant estrogen receptor neurons which are synaptically contacted by the SCN (de la Iglesia, Blaustein, & Bittman, 1995; Simerly, Chang, Muramatsu, & Swanson, 1990; Simerly, Carr, Zee, & Lorang, 1990; Watson, Langub, Engle, & Maley, 1995; Witkin *et al.*, 1982). The anteroventral periventricular nucleus, in turn, contacts GnRH neurons in the region of the organum vasculosum of lamina terminalis (Gu & Simerly, 1997). VIP-containing fibers from the SCN also make direct synaptic contacts with GnRH neurons around organum vasculosum (Van der Beek *et al.*, 1993, 1994, 1997; Vandesande *et al.*, 1975). One might speculate then that the SCN controls the temporal organization of the LH surge by direct and indirect projections to the GnRH neurons of the preoptic area (Figure 15). There is evidence indicating that VIP plays an inhibitory role in the control of LH secretion (Stobie & Weick, 1989, 1990; van der Beek, Swarts, & Wiegant, 1999; Weick & Stobie, 1992, 1995). The LH surge occurs at a time when SCN firing rate is falling; thus, the GnRH neuron and the anteroventral periventricular nucleus may be released from inhibition by GABA- and VIP-containing SCN projections around this time. This net stimulation of the GnRH neuron would then result in GnRH release into the portal plexus and subsequent ovulation. Following the LH surge, the negative feedback action of steroids on the LH apparatus would keep the system in check until the following day. This hypothetical scenario is similar to the one proposed for the HPA axis in the preceding section. Again, the real story is probably not quite as simple because deafferentation studies show that either rostral or caudal projections of the SCN are sufficient for the maintenance of estrous cyclicity (Nomura, Takahashi, & Kawashima, 1988), whereas horizontal knife cuts of the projections of the SCN to the subparamentricular zone reduce the size of the LH surge (Watts, Sheward, Whale, & Fink, 1989). Thus, the SCN control of estrous cyclicity involves more than simple projections to the GnRH neuron and the anteroventral periventricular nucleus. Nevertheless, it is useful to think of the timing of neuroendocrine rhythms being controlled by the SCN in this two-stage process.

A second component of the estrous cycle which is controlled in a circadian fashion is the secretion of prolactin from the anterior pituitary. Prolactin is a luteotrophic hormone which seems to be involved in female sexual behavior (Freeman, 1994). Prolactin levels peak at the time of the LH surge in intact rats and a daily prolactin surge occurs in ovariectomized, steroid-primed females (Butcher, Collins, & Fugo, 1974; Neill, 1972). SCN-lesioned rats fail to exhibit a prolactin surge (Kawakami, Arita, & Yoshioka, 1980; Pan & Gala, 1985; Ronnekleiv & Kelly, 1988). The secretion of prolactin is inhibited by tuberoinfundibular dopaminergic neurons (Freeman, 1994). Interestingly, a small number of tyrosine-hydroxylase-immunoreactive neurons in the region of the arcuate and the adjacent periventricular nucleus are synapsed upon by SCN efferent terminals (Horvath, 1997). The projection of the SCN to the anteroventral periventricular nucleus may also be critical in the circadian control of prolactin secretion because its neurons also contact dopaminergic cells of the arcuate (Gu & Simerly, 1997).

TEMPERATURE. Core body temperature is controlled by two variables, homeostatic and circadian regulation. The circadian regulation of body temperature is apparent under "unmasking" conditions, where daily variations in locomotor activity levels and subsequent temperature fluctuations are controlled. Thus, in humans subjected to strict bed rest and fed at hourly intervals, a circadian rhythm in basal body temperature is unmasked. In rodents with partial SCN lesions, locomotor activity levels often become arrhythmic, while the more robust temperature cycle remains intact (Eastman, Mistlberger, & Rechtschaffen, 1984; Refinetti, Kaufman, & Menaker, 1994; Satinoff & Prosser, 1988). Thus, the partial lesion unmasks the activity-independent component of the temperature rhythm. However, in animals with complete SCN lesions, both variables become arrhythmic (Eastman *et al.*, 1984; Osborne & Refinetti, 1995; Refinetti *et al.*, 1994; Ruby, Ikuba, Barnes, & Zucker, 1989). SCN lesions, however, do not eliminate homeostatic thermoregulatory responses (Wachulec, Li, Tanaka, Pelosa, & Satinoff, 1997). The daily peak in temperature independent of changes in muscle tone presumably occurs through blood pressure, blood vessel dilation, and cellular metabolic rate changes throughout the body, including in the brain. These variables are under the control of the sympathetic and parasympathetic nervous system, which are ultimately regulated by the hypothalamus. It seems reasonable to assume that the projection of the SCN to the medial preoptic nucleus is the mediator of some of these effects. Animals bearing medial preoptic lesions lose appropriate thermoregulatory reflexes, while their daily temperature rhythms remain intact (Osborne & Refinetti, 1995; Satinoff, Liran, & Chapman, 1982; Szymusiak & Satinoff, 1982). However, the amplitude of the temperature rhythm is enhanced under these conditions, which is consistent with the removal of an inhibitory feedback by preoptic inputs to the SCN. Another route whereby the SCN may impart rhythmicity to core body temperature is through its control of pineal melatonin output. Melatonin is associated with reduced core temperature, increased heat loss, and decreased cardiovascular output (Arendt, 1998; Cagnacci, Krauchi, Wirz-Justice, & Volpe, 1997; Dawson, Rogers, van den Heuvel, Kennaway, & Lushington, 1998). Indeed, the projection of the SCN to spinal-projecting neurons in the parvicellular paraventricular hypothalamic nucleus may be one means of controlling general sympathetic outflow, rather than solely driving melatonin rhythmicity. Combined retro- and anterograde tracing studies reveal closely apposed SCN fibers on spinal-projecting neurons in the dorsal and ventromedial parvicellular divisions of the paraventricular hypothalamic nucleus (Vrang *et al.*, 1997). Pharmacologic depletion of peripheral catecholamines reduces circadian rhythmicity in heart-rate, while SCN lesions completely abolish circadian rhythmicity in heart rate and blood pressure (Janssen, Tyssen, Duindam, & Rietveld, 1994; Sano *et al.*, 1995; Warren, Champney, & Cassone, 1994). A third pathway whereby the SCN may regulate autonomic outflow is by its dense projection to the retrochiasmatic area, where several spinally projecting neurons are located, but the existence of this specific synaptic connection needs to be established.

HUMAN ADAPTATION—PSYCHOMOTOR PERFORMANCE. The primary adaptive behavior of humans is measured in aspects of psychomotor performance. Memory, language, problem solving, and cognition are components of psychomotor performance that determine the success of human adaptation. It has been known for many years that there are circadian rhythms in psychomotor performance (Folkard, Wever, & Wildgruber, 1983; Monk, Fookson, Moline, & Pollak, 1985). This is separate from the control of behavioral state, but reflects, in part, variation in alertness and arousal. The mechanisms and pathways by which the circadian system regulates higher brain functions are unknown. Because of the complexity of these functions, this is a difficult problem to approach and almost certainly requires a complex regulation. Two possible mechanisms are evident. First, the SCN projections to the paraventricular thalamus, which, in turn, projects to the basal forebrain, extended amygdala, and hippocampus (Moga & Moore, 1995), suggest effects on memory and reward. Second, the extensive projections through hypothalamus to the cerebral cortex provide a basis for modulation of cortical function in complex tasks.

CONCLUSIONS

The SCN is a small, paired nucleus arising from the periventricular component of derivatives of the ventral diencephalic germinal epithelium. Its neurons are generated prenatally as individual oscillators and function is initiated prenatally, under maternal control. The control of sleep–wake cycles and associated functions occurs postnatally as SCN connections, intrinsically to form a pacemaker network, and extrinsically with effector systems. Similarly, inputs to the SCN which entrain and modulate its pacemaker function form postnatally.

The SCN is the principal circadian pacemaker in mammals. The primary function of the SCN is to provide a temporal regulation of sleep and waking behavior such that waking behavior, the major feature of mammalian adaptation, is elaborated during the day in animals that use vision as their primary source of information about the environment. In animals which use olfaction and audition to inform them about the environment, waking behavior is typically nocturnal. This distinction is accomplished even though the output of the SCN is the same in both diurnal and nocturnal mammals, a high firing rate during the day and a low firing rate at night. Concomitant with the control of sleep–wake cycles, the SCN provides a temporal organization for a host of neuroendrocrine and physiologic variables and, during waking, provides the underlying mechanisms that promote maximally effective adaptive behaviors. In seasonally reproducing mammals, the SCN provides the control of pineal melatonin production that underlies the photoperiodic regulation of reproduction. Thus, this "master" pacemaker is at the center of the adaptation of mammals to their environment.

The SCN has two subdivisions, (1) a core which receives direct and indirect photic input and a small set of other brain afferents, and (2) a shell which receives input from multiple brain areas and the core. All SCN neurons appear to produce GABA, so that the output of the nucleus is inhibitory. Within the core, large subsets of neurons also contain the peptides VIP or GRP, whereas the shell contains angiotensin II, vasopressin, enkephalin, or calretinin.

The projections of the SCN are organized by core or shell. Core projects to shell and contralateral core and to restricted areas of the hypothalamus, thalamus, and basal forebrain. Shell projects to contralateral shell and to restricted areas of hypothalamus, thalamus, and basal forebrain. The projections of shell partially overlap those of core and are more extensive. These data indicate that the SCN control of effector functions is complex and more highly organized than previously believed.

References

Abe, H., & Rusak, B. (1994). Physiological mechanisms regulating photic induction of Fos-like protein in hamster suprachiasmatic nucleus. *Neuroscience and Biobehavioral Reviews, 18,* 531–536.

Altman, J., & Bayer, S. A. (1978). Development of the diencephalon in the rat. I. Autoradiographic study of the time of origin and settling patterns of neurons of the hypothalamus. *Journal of Comparative Neurology, 182,* 945–971.

Altman, J., & Bayer, S. A. (1986). The development of the rat hypothalamus. *Advances in Anatomy, Embryology, and Cell Biology, 100,* 1–178.

Arendash, G. W., & Gallo, R. V. (1979). Effect of lesions in the suprachiasmatic nucleus–retrochiasmatic area on the inhibition of pulsatile LH release induced by electrical stimulation of the midbrain dorsal raphe nucleus. *Neuroendocrinology, 28,* 349–357.

Arendt, J. (1998). Complex effects of melatonin. *Therapie, 53,* 479–488.

Berk, M. L., & Finkelstein, J. A. (1981). An autoradiographic determination of the efferent projections of the suprachiasmatic nucleus of the hypothalamus. *Brain Research, 226*(1–2), 1–13.

Brockhaus, H. (1942). Beitrag zur normalen anatomie des hypothalamus and der zona incerta beim menschen. *Journal für Psychologie und Neurologie, 51,* 96–196.

Brown-Grant, K., & Raisman, G. (1977). Abnormalities in reproductive function associated with the destruction of the suprachiasmatic nuclei in female rats. *Proceedings of the Royal Society of London Series B. Biological Sciences, 198,* 279–296.

Buijs, R. M., Markman, M., Nunes-Cardoso, B., Hou, Y. X., & Shinn, S. (1993). Projections of the suprachiasmatic nucleus to stress-related areas in the rat hypothalamus: A light and electron microscopic study. *Journal of Comparative Neurology, 335,* 42–54.

Buijs, R. M., Hou, Y. X., Shinn, S., & Renaud, L. P. (1994). Ultrastructural evidence for intra- and extranuclear projections of GABAergic neurons of the suprachiasmatic nucleus. *Journal of Comparative Neurology, 340,* 381–391.

Buijs, R. M., Wortel, J., & Hou, Y. X. (1995). Colocalization of gamma-aminobutyric acid with vasopressin, vasoactive intestinal peptide, and somatostatin in the rat suprachiasmatic nucleus. *Journal of Comparative Neurology, 358,* 343–352.

Buijs, R. M., Wortel, J., Van Heerikhuize, J. J., & Kalsbeek, A. (1997). Novel environment induced inhibition of corticosterone secretion: Physiological evidence for a suprachiasmatic nucleus mediated neuronal hypothalamo-adrenal cortex pathway. *Brain Research, 758,* 229–236.

Buijs, R. M., Wortel, J., Van Heerikhuize, J. J., Feenstra, M. G., Ter Horst, G. J., Romijn, H. J., & Kalsbeek, A. (1999). Anatomical and functional demonstration of a multisynaptic suprachiasmatic nucleus adrenal (cortex) pathway. *European Journal of Neuroscience, 11,* 1535–1544.

Butcher, R. L., Collins, W. E., & Fugo, N. W. (1974). Plasma concentration of LH, FSH, prolactin, progesterone and estradiol-17beta throughout the 4-day estrous cycle of the rat. *Endocrinology, 94,* 1704–1708.

Cagnacci, A. (1997). Influences of melatonin on human circadian rhythms. *Chronobiology International, 14*(2), 205–220.

Cagnacci, A., Krauchi, K., Wirz-Justice, A., & Volpe, A. (1997). Homeostatic versus circadian effects of melatonin on core body temperature in humans. *Journal of Biological Rhythms, 12,* 509–517.

Caligaris, L., Astrada, J. J., & Taleisnik, S. (1971). Biphasic effect of progesterone on the release of gonadotropin in rats. *Endocrinology, 89,* 331–337.

Card, J. P., & Moore, R. Y. (1982). Ventral lateral geniculate nucleus efferents to the rat suprachiasmatic nucleus exhibit avian pancreatic polypeptide-like immunoreactivity. *Journal of Comparative Neurology, 206,* 390–396.

Card, J. P., & Moore, R. Y. (1984). The suprachiasmatic nucleus of the golden hamster: Immunohistochemical analysis of cell and fiber distribution. *Neuroscience, 13,* 415–431.

Card, J. P., & Moore, R. Y. (1989). Organization of lateral geniculate–hypothalamic connections in the rat. *Journal of Comparative Neurology, 284,* 135–147.

Card, J. P., Brecha, N., Karten, H. J., & Moore, R. Y. (1981). Immunocytochemical localization of vasoactive intestinal polypeptide-containing cells and processes in the suprachiasmatic nucleus of the rat: Light and electron microscopic analysis. *Journal of Neuroscience, 1,* 1289–1303.

Card, J. P., Fitzpatrick-McElligott, S., Gozes, I., & Baldino, F., Jr. (1988). Localization of vasopressin-, vasoactive intestinal polypeptide-, peptide histidine isoleucine- somatostatin-mRNA in rat suprachiasmatic nucleus. *Cell Tissue Research, 252,* 307–315.

Cassone, V. M., Speh, J. C., Card, J. P., & Moore, R. Y. (1988). Comparative anatomy of the mammalian hypothalamic suprachiasmatic nucleus. *Journal of Biological Rhythms, 3,* 71–91.

Castel, M., Feinstein, N., Cohen, S., & Harari, N. (1990). Vasopressinergic innervation of the mouse suprachiasmatic nucleus: An immuno-electron microscopic analysis. *Journal of Comparative Neurology, 298,* 172–187.

Coen, C. W., & MacKinnon, P. C. (1980). Lesions of the suprachiasmatic nuclei and the serotonin-dependent phasic release of luteinizing hormone in the rat: Effects on drinking rhythmicity and on the consequences of preoptic area stimulation. *Journal of Endocrinology, 84,* 231–236.

Cowan, W. M., Gottlieb, D. I., Hendrickson, A. E., Price, J. L., & Woolsey, T. A. (1972). The autoradiographic demonstration of axonal connections in the central nervous system. *Brain Research, 37*(1), 21–51.

Dai, J., Swaab, D. F., & Buijs, R. M. (1997). Distribution of vasopressin and vasoactive intestinal polypeptide (VIP) fibers in the human hypothalamus with special emphasis on suprachiasmatic nucleus efferent projections. *Journal of Comparative Neurology, 383,* 397–414.

Dai, J., Swaab, D. F., Van der Vliet, J., & Buijs, R. M. (1998a). Postmortem tracing reveals the organization of hypothalamic projections of the suprachiasmatic nucleus in the human brain. *Journal of Comparative Neurology, 400,* 87–102.

Dai, J., Van der Vliet, J., Swaab, D. F., & Buijs, R. M. (1998b). Postmortem anterograde tracing of intrahypothalamic projections of the human dorsomedial nucleus of the hypothalamus. *Journal of Comparative Neurology, 401,* 16–33.

Daikoku, S., Hisano, S., & Kagotani, Y. (1992). Neuronal associations in the rat suprachiasmatic nucleus demonstrated by immunoelectron microscopy. *Journal of Comparative Neurology, 325,* 559–571.

Dallman, M. F., Akana, S. F., Cascio, C. S., Darlington, D. N., Jacobson, L., & Levin, N. (1987). Regulation of ACTH secretion: Variations on a theme of B. *Recent Progress in Hormone Research, 43,* 113–173.

Davis, F. C., & Mannion, J. (1988). Entrainment of hamster pup circadian rhythms by prenatal melatonin injections to the mother. *American Journal of Physiology, 255,* R439–R448.

Dawson, D., Rogers, N. L., van den Heuvel, C. J., Kennaway, D. J., & Lushington, K. (1998). Effect of sustained nocturnal transbuccal melatonin administration on sleep and temperature in elderly insomniacs. *Journal of Biological Rhythms, 13,* 532–538.

de la Iglesia, H. O., Blaustein, J. D., & Bittman, E. L. (1995). The suprachiasmatic area in the female hamster projects to neurons containing estrogen receptors and GnRH. *Neuroreport, 6*), 1715–1722.
Dierckx, K., & Vandesande, F. (1979). Immunocytochemical localization of somatostatin-containing neurons in the rat hypothalamus. *Cell Tissue Research, 201*, 349–359.
Drucker-Colin, R., Aguilar-Roblero, R., Garcia-Hernandez, F., Fernandez-Cancino, F., & Bermudez Rattoni, F. (1984). Fetal suprachiasmatic nucleus transplants: Diurnal rhythm recovery of lesioned rats. *Brain Research, 311*, 353–357.
Eastman, C. I., Mistlberger, R. E., & Rechtschaffen, A. (1984). Suprachiasmatic nuclei lesions eliminate circadian temperature and sleep rhythms in the rat. *Physiology and Behavior, 32*, 357–368.
Ebbeson, S. V. E. (1970). On the organization of the central visual pathways in vertebrates. *Brain, Behavior and Evolution, 3*, 178–194.
Enright, J. T. (1980). Temporal precision in circadian systems: A reliable neuronal clock from unreliable components? *Science, 209*, 1542–1545.
Everett, J. W. (1989). Neurobiology of reproduction in the female rat. A fifty-year perspective. *Monographs on Endocrinology, 32*, 1–133.
Everett, J. W., & Sawyer, C. H. (1950). A 24 hour periodicity in the LH release apparatus of female rats, disclosed by barbiturate sedation. *Endocrinology, 47*, 198–218.
Everett, J. W., Sawyer, C. H., & Markee, J. E. (1949). A neurogenic timing factor in control of the ovulatory discharge of luteinizing hormone in the cyclic rat. *Endocrinology, 44*, 218–233.
Folkard, S., Wever, R. A., & Wildgruber, C. M. (1983). Multi-oscillatory control of circadian rhythms in human performance. *Nature, 305*, 223–226.
Freedman, M. S., Lucas, R. J., Soni, B., von Schantz, M., Munoz, M., David-Gray, Z., & Foster, R. (1999). Regulation of mammalian circadian behavior by non-rod, non-cone, ocular photoreceptors. *Science, 284*, 502–504.
Freeman, M. E. (1994). The neuroendocrine control of the ovarian cycle of the rat. In E. Knobil & J. D. Neill (Eds.), *The physiology of reproduction* (2nd ed.). New York: Raven Press.
Gao, B., & Moore, R. Y. (1996). Glutamic acid decarboxylase message isoforms in human suprachiasmatic nucleus. *Journal of Biological Rhythms, 11*, 172–179.
Gillette, M. U. (1991). *SCN electrophysiology in vitro: Rhythmic activity and endogenous clock properties.* New York: Oxford University Press.
Gomez, F., Chapleur, M., Fernette, B., Burlet, C., Nicolas, J. P., & Burlet, A. (1997). Arginine vasopressin (AVP) depletion in neurons of the suprachiasmatic nuclei affects the AVP content of the paraventricular neurons and stimulates adrenocorticotrophic hormone release. *Journal of Neuroscience Research, 50*, 565–574.
Gray, G. D., Soderstein, P., Tallentire, D., & Davidson, J. M. (1978). Effects of lesions in various structures of the suprachiasmatic–preoptic region on LH regulation and sexual behavior in female rats. *Neuroendocrinology, 25*, 174–191.
Green, D. J., & Gillette, R. (1982). Circadian rhythm of firing rate recorded from single cells in the rat suprachiasmatic brain slice. *Brain Research, 245*, 198–200.
Groos, G., & Hendriks, J. (1982). Circadian rhythms in electrical discharge of rat suprachiasmatic neurones recorded *in vitro. Neuroscience Letters, 34*, 283–288.
Gu, G. B., & Simerly, R. B. (1997). Projections of the sexually dimorphic anteroventral periventricular nucleus in the female rat. *Journal of Comparative Neurology, 384*, 142–164.
Guldner, F. H. (1978a). Synapses of optic nerve afferents in the rat suprachiasmatic nucleus. I. Identification, qualitative description, development and distribution. *Cell Tissue Research, 194*, 17–35.
Guldner, F. H. (1978b). Synapses of optic nerve afferents in the rat suprachiasmatic nucleus. II. Structural variability as revealed by morphometric examination. *Cell Tissue Research, 194*, 37–54.
Guldner, F. H., & Wolff, J. R. (1974). Dendro-dendritic synapses in the suprachiasmatic nucleus of the rat hypothalamus. *Journal of Neurocytology, 3*, 245–250.
Gurdjian, F. S. (1927). The diencephalon of the albino rat. *Journal of Comparative Neurology, 43*, 1–114.
Hakim, H., DeBernardo, A. P., & Silver, R. (1991). Circadian locomotor rhythms, but not photoperiodic responses, survive surgical isolation of the SCN in hamsters. *Journal of Biological Rhythms, 6*, 97–113.
Harrington, M. E., Nance, D. M., & Rusak, B. (1985). Neuropeptide Y immunoreactivity in the hamster geniculo-suprachiasmatic tract. *Brain Research Bulletin, 15*, 465–472.
Hastings, M. H., Duffield, G. E., Ebling, F. J., Kidd, A., Maywood, E. S., & Schurov, I. (1997). Nonphotic signalling in the suprachiasmatic nucleus. *Biology of the Cell, 89*, 495–503.
Hastings, M. H., Duffield, G. E., Smith, E. J., Maywood, E. S., & Ebling, F. J. (1998). Entrainment of the circadian system of mammals by nonphotic cues. *Chronobiology International, 15*, 425–445.
Hendrickson, A. E., Wagoner, N., & Cowan, W. M. (1972). An autoradiographic and electron microscopic

study of retino-hypothalamic connections. *Zeitschrift für Zellforschung und Mikroskopische Anatomie, 135*(1), 1–26.

Herbison, A. E., & Theodosis, D. T. (1992). Localization of oestrogen receptors in preoptic neurons containing neurotensin but not tyrosine hydroxylase, cholecystokinin or luteinizing hormone-releasing hormone in the male and female rat. *Neuroscience, 50,* 283–298.

Hermes, M. L., & Renaud, L. P. (1993). Differential responses of identified rat hypothalamic paraventricular neurons to suprachiasmatic nucleus stimulation. *Neuroscience, 56,* 823–832.

Hermes, M. L., Buijs, R. M., & Renaud, L. P. (1996). Electrophysiology of suprachiasmatic nucleus projections to hypothalamic paraventricular nucleus neurons. *Progress in Brain Research, 111,* 241–252.

Hickey, T. L., & Spear, P. D. (1976). Retinogeniculate projections in hooded and albino rats: An autoradiographic study. *Experimental Brain Research, 24,* 523–529.

Hoorneman, E. M., & Buijs, R. M. (1982). Vasopressin fiber pathways in the rat brain following suprachiasmatic nucleus lesioning. *Brain Research, 243,* 235–241.

Horvath, T. L. (1997). Suprachiasmatic efferents avoid phenestrated capillaries but innervate neuroendocrine cells, including those producing dopamine. *Endocrinology, 138,* 1312–1320.

Horvath, T. L. (1998). An alternate pathway for visual signal integration into the hypothalamo-pituitary axis: Retinorecipient intergeniculate neurons project to various regions of the hypothalamus and innervate neuroendocrine cells including those producing dopamine. *Journal of Neuroscience, 18,* 1546–1558.

Ibata, Y., Tanaka, M., Ichitani, Y., Takahashi, Y., & Okamura, H. (1993). Neuronal interaction between VIP and vasopressin neurones in the rat suprachiasmatic nucleus. *Neuroreport, 4,* 128–130.

Inouye, S. T., & Kawamura, H. (1979). Persistence of circadian rhythmicity in a mammalian hypothalamic "island" containing the suprachiasmatic nucleus. *Proceedings of the National Academy of Sciences of the USA, 76,* 5962–5966.

Jacobowitz, D. M., & Winsky, L. (1991). Immunocytochemical localization of calretinin in the forebrain of the rat. *Journal of Comparative Neurology, 304,* 198–218.

Janssen, B. J., Tyssen, C. M., Duindam, H., & Rietveld, W. J. (1994). Suprachiasmatic lesions eliminate 24-h blood pressure variability in rats. *Physiology and Behavior, 55,* 307–311.

Jasper, M. S., & Engeland, W. C. (1994). Splanchnic neural activity modulates ultradian and circadian rhythms in adrenocortical secretion in awake rats. *Neuroendocrinology, 59,* 97–109.

Johnson, R. F., Morin, L. P., & Moore, R. Y. (1988). Retinohypothalamic projections in the hamster and rat demonstrated using cholera toxin. *Brain Research, 462,* 301–312.

Kalra, S. P. (1993). Mandatory neuropeptide-steroid signaling for the preovulatory luteinizing hormone-releasing hormone discharge. *Endocrine Reviews, 14,* 507–538.

Kalsbeek, A., & Buijs, R. M. (1992). Peptidergic transmitters of the suprachiasmatic nuclei and the control of circadian rhythmicity. *Progress in Brain Research, 92,* 321–333.

Kalsbeek, A., Buijs, R. M., van Heerikhuize, J. J., Arts, M., & van der Woude, T. P. (1992). Vasopressin-containing neurons of the suprachiasmatic nuclei inhibit corticosterone release. *Brain Research, 580,* 62–67.

Kalsbeek, A., Teclemariam-Mesbah, R., & Pevet, P. (1993). Efferent projections of the suprachiasmatic nucleus in the golden hamster (*Mesocricetus auratus*). *Journal of Comparative Neurology, 332,* 293–314.

Kalsbeek, A., Buijs, R. M., Engelmann, M., Wotjak, C. T., & Landgraf, R. (1995). *In vivo* measurement of a diurnal variation in vasopressin release in the rat suprachiasmatic nucleus. *Brain Research, 682,* 75–82.

Kalsbeek, A., van der Vliet, J., & Buijs, R. M. (1996a). Decrease of endogenous vasopressin release necessary for expression of the circadian rise in plasma corticosterone: A reverse microdialysis study. *Journal of Neuroendocrinology, 8,* 299–307.

Kalsbeek, A., van Heerikhuize, J. J., Wortel, J., & Buijs, R. M. (1996b). A diurnal rhythm of stimulatory input to the hypothalamo-pituitary-adrenal system as revealed by timed intrahypothalamic administration of the vasopressin V1 antagonist. *Journal of Neuroscience, 16,* 5555–5565.

Kaneko, M., Hiroshige, T., Shinsako, J., & Dallman, M. F. (1980). Diurnal changes in amplification of hormone rhythms in the adrenocortical system. *American Journal of Physiology, 239,* R309–R316.

Kawakami, M., Arita, J., & Yoshioka, E. (1980). Loss of estrogen-induced daily surges of prolactin and gonadotropins by suprachiasmatic nucleus lesions in ovariectomized rats. *Endocrinology, 106,* 1087–1092.

Keim, S. R., & Shekhar, A. (1996). The effects of GABAA receptor blockade in the dorsomedial hypothalamic nucleus on corticotrophin (ACTH) and corticosterone secretion in male rats. *Brain Research, 739,* 46–51.

Keller-Wood, M. E., & Dallman, M. F. (1984). Corticosteroid inhibition of ACTH secretion. *Endocrine Reviews, 5,* 1–24.

Klein, D. C., & Moore, R. Y. (1979). Pineal N-acetyltransferase and hydroxyindole-O-methyltransferase: Control by the retinohypothalamic tract and the suprachiasmatic nucleus. *Brain Research, 174,* 245–262.

Krieger, M. S., Morrell, J. I., & Pfaff, D. W. (1976). Autoradiographic localization of estradiol-concentrating cells in the female hamster brain. *Neuroendocrinology, 22,* 193–205.

Krieger, D. T., Hauser, H., & Krey, L. C. (1977). Suprachiasmatic nuclear lesions do not abolish food-shifted circadian adrenal and temperature rhythmicity. *Science, 197,* 398–399.

Kubota, A., Inouye, S. T., & Kawamura, H. (1981). Reversal of multiunit activity within and outside the suprachiasmatic nucleus in the rat. *Neuroscience Letters, 27,* 303–308.

Kuhlenbeck, H. (1981). Uber die Grundbestandteile des Zwischenhirnbauplans bei Reptilien. *Morphologisches Jahrbuch, 66,* 244–317.

Kuhlenbeck, H. (1954). The human diencephalon: A summary of development, structure, function and pathology. *Confin. Neurol. (Suppl.), 14,* 1–230.

Leak, R. K., & Moore, R. Y. (1997). Identification of retinal ganglion cells projecting to the lateral hypothalamic area of the rat. *Brain Research, 770,* 105–114.

Leak, R. K., Card, J. P., & Moore, R. Y. (1999). Suprachiasmatic pacemaker organization analyzed by viral transynaptic transport. *Brain Research, 819,* 23–32.

Legan, S. J., & Karsch, F. J. (1975). A daily signal for the LH surge in the rat. *Endocrinology, 96,* 57–62.

Legan, S. J., Coon, G. A., and Karsch, F. J. (1975). Role of estrogen as initiator of daily LH surges in the ovariectomized rat. *Endocrinology, 96,* 50–56.

LeSauter, J., and Silver, R. (1998). Output signals of the SCN. *Chronobiology International, 15,* 535–550.

Levin, M. C., Sawchenko, P. E., Howe, P. R., Bloom, S. R., & Polak, J. M. (1987). Organization of galanin-immunoreactive inputs to the paraventricular nucleus with special reference to their relationship to catecholaminergic afferents. *Journal of Comparative Neurology, 261,* 562–582.

Levine, J. E., & Ramirez, V. D. (1980). *In vivo* release of luteinizing hormone-releasing hormone estimated with push-pull cannulae from the mediobasal hypothalami of ovariectomized, steroid-primed rats. *Endocrinology, 107,* 1782–1790.

Levine, J. D., Weiss, M. L., Rosenwasser, A. M., & Miselis, R. R. (1991). Retinohypothalamic tract in the female albino rat: A study using horseradish peroxidase conjugated to cholera toxin. *Journal of Comparative Neurology, 306,* 344–360.

Lopez, F. J., Merchenthaler, I. J., Moretto, M., & Negro-Vilar, A. (1998). Modulating mechanisms of neuroendocrine cell activity: The LHRH pulse generator. *Cellular and Molecular Neurobiology, 18,* 125–146.

Ma, Y. J., Kelly, M. J., & Ronnekleiv, O. K. (1990). Pro-gonadotropin-releasing hormone (ProGnRH) and GnRH content in the preoptic area and the basal hypothalamus of anterior medial preoptic nucleus/suprachiasmatic nucleus-lesioned persistent estrous rats. *Endocrinology, 127,* 2654–2664.

Meyer-Bernstein, E. L., Jetton, A. E., Matsumoto, S. I., Markuns, J. F., Lehman, M. N., & Bittman, E. L. (1999). Effects of suprachiasmatic transplants on circadian rhythms of neuroendocrine function in golden hamsters. *Endocrinology, 140,* 207–218.

Michel, S., Geusz, M. E., Zaritsky, J. K., & Block, G. D. (1993). Circadian rhythm in membrane conductance expressed in isolated neurons. *Science, 259,* 239–241.

Mikkelsen, J. D., & Larsen, P. J. (1993). Substance P in the suprachiasmatic nucleus of the rat: An immunohistochemical and *in situ* hybridization study. *Histochemistry, 100*(1), 3–16.

Mikkelsen, J. D., & Vrang, N. (1994). A direct pretectosuprachiasmatic projection in the rat. *Neuroscience, 62,* 497–505.

Mikkelsen, J. D., Larsen, P. J., O'Hare, M. M., & Wiegand, S. J. (1991). Gastrin releasing peptide in the rat suprachiasmatic nucleus: An immunohistochemical, chromatographic and radioimmunological study. *Neuroscience, 40,* 55–66.

Moga, M. M., & Moore, R. Y. (1997). Organization of neural inputs to the suprachiasmatic nucleus in the rat. *Journal of Comparative Neurology, 389,* 508–534.

Moga, M. M., Weis, R. P., & Moore, R. Y. (1995). Efferent projections of the paraventricular thalamic nucleus in the rat. *Journal of Comparative Neurology, 359,* 221–238.

Monk, T. H., Fookson, J. E., Moline, M. L., & Pollak, C. P. (1985). Diurnal variation in mood and performance in a time-isolated environment. *Chronobiology International, 2,* 185–193.

Moore, R. Y. (1973). Retinohypothalamic projection in mammals: A comparative study. *Brain Research, 49,* 403–409.

Moore, R. Y. (1996). Neural control of the pineal gland. *Behavioral Brain Research, 73,* 125–130.

Moore, R. Y., & Bernstein, M. E. (1989). Synaptogenesis in the rat suprachiasmatic nucleus demonstrated by electron microscopy and synapsin I immunoreactivity. *Journal of Neuroscience, 9,* 2151–2162.

Moore, R. Y., & Card, J. P. (1994). Intergeniculate leaflet: An anatomically and functionally distinct subdivision of the lateral geniculate complex. *Journal of Comparative Neurology, 344,* 403–430.

Moore, R. Y., & Eichler, V. B. (1972). Loss of a circadian adrenal corticosterone rhythm following suprachiasmatic lesions in the rat. *Brain Research, 42,* 201–206.

Moore, R. Y., & Lenn, N. J. (1972). A retinohypothalamic projection in the rat. *Journal of Comparative Neurology, 146,* 1–14.

Moore, R. Y., & Speh, J. C. (1993). GABA is the principal neurotransmitter of the circadian system. *Neuroscience Letters, 150,* 112–116.

Moore, R. Y., Gustafson, E. L., & Card, J. P. (1984). Identical immunoreactivity of afferents to the rat suprachiasmatic nucleus with antisera against avian pancreatic polypeptide, molluscan cardioexcitatory peptide and neuropeptide Y. *Cell Tissue Research, 236,* 41–46.

Moore, R. Y., Speh, J. C., & Card, J. P. (1995). The retinohypothalamic tract originates from a distinct subset of retinal ganglion cells. *Journal of Comparative Neurology, 352,* 351–366.

Morin, L. P., Goodless-Sanchez, N., Smale, L., & Moore, R. Y. (1994). Projections of the suprachiasmatic nuclei, subparaventricular zone and retrochiasmatic area in the golden hamster. *Neuroscience, 61,* 391–410.

Mrosovsky, N. (1995). A non-photic gateway to the circadian clock of hamsters. *Ciba Foundation Symposium, 183,* 154–167.

Muller, C., & Torrealba, F. (1998). Postnatal development of neuron number and connections in the suprachiasmatic nucleus of the hamster. *Brain Research. Developmental Brain Research, 110,* 203–213.

Neill, J. D. (1972). Sexual differences in the hypothalamic regulation of prolactin secretion. *Endocrinology, 90,* 1154–1159.

Nomura, K., Takahashi, S., & Kawashima, S. (1988). Maintenance of estrous cycle in female rats with anterior or posterior deafferentation of the suprachiasmatic nucleus. *Neuroendocrinology, 47,* 444–452.

Okamura, H., Murakami, S., Fukui, K., Uda, K., Kawamoto, K., Kawashima, S., Yanaihara, N., & Ibata, Y. (1986). Vasoactive intestinal peptide- and peptide histidine isoleucine amide-like immunoreactivity colocalize with vasopressin-like immunoreactivity in the canine hypothalamo-neurohypophysial neuronal system. *Neuroscience Letters, 69,* 227–232.

Okamura, H., Berod, A., Julien, J. F., Geffard, M., Kitahama, K., Mallet, J., & Bobillier, P. (1989). Demonstration of GABAergic cell bodies in the suprachiasmatic nucleus: *In situ* hybridization of glutamic acid decarboxylase (GAD) mRNA and immunocytochemistry of GAD and GABA. *Neuroscience Letters, 102,* 131–136.

Osborne, A. R., & Refinetti, R. (1995). Effects of hypothalamic lesions on the body temperature rhythm of the golden hamster. *Neuroreport, 6,* 2187–2192.

Pan, J. T., & Gala, R. R. (1985). Central nervous system regions involved in the estrogen-induced afternoon prolactin surge. I. Lesion studies. *Endocrinology, 117,* 382–387.

Pate, J. R. (1937). Transneuronal atrophy of the nucleus ovoidus following eye removal in cats. *Anatomical Record, 67,* 39.

Pavlidis, T. (1969). Populations of interacting oscillators and circadian rhythms. *Journal of Theoretical Biology, 22,* 418–436.

Pickard, G. E. (1985). Bifurcating axons of retinal ganglion cells terminate in the hypothalamic suprachiasmatic nucleus and the intergeniculate leaflet of the thalamus. *Neuroscience Letters, 55,* 211–217.

Raisman, G., & Brown-Grant, K. (1977). The 'suprachiasmatic syndrome': Endocrine and behavioural abnormalities following lesions of the suprachiasmatic nuclei in the female rat. *Proceedings of the Royal Society of London Series B, Biological Sciences, 198,* 297–314.

Ralph, M. R., Foster, R. G., Davis, F. C., & Menaker, M. (1990). Transplanted suprachiasmatic nucleus determines circadian period. *Science, 247,* 975–978.

Rea, M. A. (1998). Photic entrainment of circadian rhythms in rodents. *Chronobiology International, 15,* 395–423.

Refinetti, R., Kaufman, C. M., & Menaker, M. (1994). Complete suprachiasmatic lesions eliminate circadian rhythmicity of body temperature and locomotor activity in golden hamsters. *Journal of Comparative Physiology [A], 175,* 223–232.

Reppert, S. M., & Weaver, D. R. (1991). A biological clock is oscillating in the fetal suprachiasmatic nuclei. In R. Moore, S. Reppert, & D. F. Klein (Eds.), *Suprachiasmatic nucleus: The mind's clock.* New York: Oxford University Press.

Romano, G. J., Krust, A., & Pfaff, D. W. (1989). Expression and estrogen regulation of progesterone receptor mRNA in neurons of the mediobasal hypothalamus: An *in situ* hybridization study. *Molecular Endocrinology, 3,* 1295–1300; Erratum, *3,* 1860.

Ronnekleiv, O. K., & Kelly, M. J. (1988). Plasma prolactin and luteinizing hormone profiles during the estrous cycle of the female rat: Effects of surgically induced persistent estrus. *Neuroendocrinology, 47,* 133–146.

Ruby, N. F., Ibuka, N., Barnes, B. M., & Zucker, I. (1989). Suprachiasmatic nuclei influence torpor and circadian temperature rhythms in hamsters. *American Journal of Physiology, 257,* R210–R215.

Sano, H., Hayashi, H., Makino, M., Takezawa, H., Hirai, M., Saito, H., & Ebihara, S. (1995). Effects of suprachiasmatic lesions on circadian rhythms of blood pressure, heart rate and locomotor activity in the rat. *Japanese Circulation Journal, 59,* 565–573.

Sarkar, D. K., Chiappa, S. A., Fink, G., & Sherwood, N. M. (1976). Gonadotropin-releasing hormone surge in pro-oestrous rats. *Nature, 264,* 461–463.

Satinoff, E., & Prosser, R. A. (1988). Suprachiasmatic nuclear lesions eliminate circadian rhythms of drinking and activity, but not of body temperature, in male rats. *Journal of Biological Rhythms, 3,* 1–22.

Satinoff, E., Liran, J., & Clapman, R. (1982). Aberrations of circadian body temperature rhythms in rats with medial preoptic lesions. *American Journal of Physiology, 242,* R352–R357.

Sawaki, Y., Nihonmatsu, I., & Kawamura, H. (1984). Transplantation of the neonatal suprachiasmatic nuclei into rats with complete bilateral suprachiasmatic lesions. *Neuroscience Research, 1*(1), 67–72.

Scarbrough, K., Harney, J. P., Rosewell, K. L., & Wise, P. M. (1996). Acute effects of antisense antagonism of a single peptide neurotransmitter in the circadian clock. *American Journal of Physiology, 270,* R283–R288.

Schwartz, W. J., Gross, R. A., & Morton, M. T. (1987). The suprachiasmatic nuclei contain a tetrodotoxin-resistant circadian pacemaker. *Proceedings of the National Academy of Sciences of the USA, 84,* 1694–1698.

Shibata, S., & Moore, R. Y. (1988). Electrical and metabolic activity of suprachiasmatic nucleus neurons in hamster hypothalamic slices. *Brain Research, 438,* 374–378.

Shibata, S., Oomura, Y., Kita, H., & Hattori, K. (1982). Circadian rhythmic changes of neuronal activity in the suprachiasmatic nucleus of the rat hypothalamic slice. *Brain Research, 247,* 154–158.

Shivers, B. D., Harlan, R. E., Hejtmancik, J. F., Conn, P. M., & Pfaff, D. W. (1986). Localization of cells containing LHRH-like mRNA in rat forebrain using *in situ* hybridization. *Endocrinology, 118,* 883–885.

Silver, R., LeSauter, J., Tresco, P. A., & Lehman, M. N. (1996). A diffusible coupling signal from the transplanted suprachiasmatic nucleus controlling circadian locomotor rhythms. *Nature, 382,* 810–813.

Simerly, R. B. (1998). Organization and regulation of sexually dimorphic neuroendocrine pathways. *Behavioral Brain Research, 92,* 195–203.

Simerly, R. B., Chang, C., Muramatsu, M., & Swanson, L. W. (1990). Distribution of androgen and estrogen receptor mRNA-containing cells in the rat brain: An *in situ* hybridization study. *Journal of Comparative Neurology, 294,* 76–95.

Simerly, R. B., Carr, A. M., Zee, M. C., & Lorang, D. (1996). Ovarian steroid regulation of estrogen and progesterone receptor messenger ribonucleic acid in the anteroventral periventricular nucleus of the rat. *Journal of Neuroendocrinology, 8,* 45–56.

Sofroniew, M. V., & Weindl, A. (1978). Projections from the parvocellular vasopressin- and neurophysin-containing neurons of the suprachiasmatic nucleus. *American Journal of Anatomy, 153,* 391–429.

Sofroniew, M. V., & Weindl, A. (1980). Identification of parvocellular vasopressin and neurophysin neurons in the suprachiasmatic nucleus of a variety of mammals including primates. *Journal of Comparative Neurology, 193,* 659–675.

Sollars, P. J., & Pickard, G. E. (1998). Restoration of circadian behavior by anterior hypothalamic grafts containing the suprachiasmatic nucleus: Graft/host interconnections. *Chronobiology International, 15,* 513–533.

Stephan, F. K., & Zucker, I. (1972). Circadian rhythms in drinking behavior and locomotor activity of rats are eliminated by hypothalamic lesions. *Proceedings of the National Academy of Sciences of the USA, 69,* 1583–1586.

Stephan, F. K., Berkley, K. J., & Moss, R. L. (1981). Efferent connections of the rat suprachiasmatic nucleus. *Neuroscience, 6,* 2625–2646.

Stetson, M. H., & Watson-Whitmyre, M. (1976). Nucleus suprachiasmaticus: The biological clock in the hamster? *Science, 191,* 197–199.

Stobie, K. M., & Weick, R. F. (1989). Vasoactive intestinal peptide inhibits luteinizing hormone secretion: The inhibition is not mediated by dopamine. *Neuroendocrinology, 49,* 597–603.

Stobie, K. M., & Weick, R. F. (1990). Effects of lesions of the suprachiasmatic and paraventricular nuclei on the inhibition of pulsatile luteinizing hormone release by exogenous vasoactive intestinal peptide in the ovariectomized rat. *Neuroendocrinology, 51,* 649–657.

Swann, J. M., & Turek, F. W. (1985). Multiple circadian oscillators regulate the timing of behavioral and endocrine rhythms in female golden hamsters. *Science, 228,* 898–900.

Swanson, L. W. (1987). The hypothalamus. In *Handbook of chemical neuroanatomy. Vol. 5. Integrated systems of the CNS. Part I* (pp. 1–24). Amsterdam: Elsevier Science.

Swanson, L. W., & Cowan, W. M. (1975). The efferent connections of the suprachiasmatic nucleus of the hypothalamus. *Journal of Comparative Neurology, 160,* 1–12.

Swanson, L. W., Cowan, W. M., & Jones, E. G. (1974). An autoradiographic study of the efferent connections of the ventral lateral geniculate nucleus in the albino rat and the cat. *Journal of Comparative Neurology, 156,* 143–163.

Swanson, L. W., Canteras, N. W., & Watts, A. G. (1993). Organization of suprachiasmatic nucleus inputs to forebrain circuitry. In H. Nakagawa, Y. Oomura, & K. Nagai (Eds.), *New functional aspects of the suprachiasmatic nucleus of the hypothalamus* (pp. 35–44). London: Libbey and Co.

Szafarczyk, A., Ixart, G., Malaval, F., Nouguier-Soule, J., & Assenmacher, I. (1979). Effects of lesions of the suprachiasmatic nuclei and of p-chlorophenylalanine on the circadian rhythms of adrenocorticotrophic hormone and corticosterone in the plasma, and on locomotor activity of rats. *Journal of Endocrinology, 83,* 1–16.

Szymusiak, R., & Satinoff, E. (1982). Acute thermoregulatory effects of unilateral electrolytic lesions of the medial and lateral preoptic area in rats. *Physiology and Behavior, 28,* 161–170.

Tanaka, M., Okamura, H., Matsuda, T., Shigeyoshi, Y., Hisa, Y., Chihara, K., & Ibata, Y. (1996). Somatostatin neurons form a distinct peptidergic neuronal group in the rat suprachiasmatic nucleus: A double labeling *in situ* hybridization study. *Neuroscience Letters, 215,* 119–122.

Teclemariam-Mesbah, R., Kalsbeek, A., Pevet, P., & Buijs, R. M. (1997). Direct vasoactive intestinal polypeptide-containing projection from the suprachiasmatic nucleus to spinal projecting hypothalamic paraventricular neurons. *Brain Research, 748,* 71–76.

Terasawa, E., & Davis, G. A. (1983). The LHRH neuronal system in female rats: Relation to the medial preoptic nucleus. *Endocrinologia Japonica, 30,* 405–417.

Terasawa, E., Wiegand, S. J., & Bridson, W. E. (1980). A role for medial preoptic nucleus on afternoon of proestrus in female rats. *American Journal of Physiology, 238,* E533–E539.

van den Pol, A. N. (1980). The hypothalamic suprachiasmatic nucleus of rat: Intrinsic anatomy. *Journal of Comparative Neurology, 191,* 661–702.

van den Pol, A. N. (1986). Gamma-aminobutyrate, gastrin releasing peptide, serotonin, somatostatin, and vasopressin: Ultrastructural immunocytochemical localization in presynaptic axons in the suprachiasmatic nucleus. *Neuroscience, 17,* 643–659.

van den Pol, A. N., & Gorcs, T. (1986). Synaptic relationships between neurons containing vasopressin, gastrin-releasing peptide, vasoactive intestinal polypeptide, and glutamate decarboxylase immunoreactivity in the suprachiasmatic nucleus: Dual ultrastructural immunocytochemistry with gold-substituted silver peroxidase. *Journal of Comparative Neurology, 252,* 507–526.

van den Pol, A. N., & Tsujimoto, K. L. (1985). Neurotransmitters of the hypothalamic suprachiasmatic nucleus: Immunocytochemical analysis of 25 neuronal antigens. *Neuroscience, 15,* 1049–1086.

van der Beek, E. M., Wiegant, V. M., van der Donk, H. A., van den Hurk, C., & Buijs, R. M. (1993). Lesions of the suprachiasmatic nucleus indicate the presence of a direct vasoactive intestinal polypeptide-containing projection to gonadotrophin-releasing hormone neurons in the female rat. *Journal of Neuroendocrinology, 5,* 137–144.

van der Beek, E. M., van Oudheusden, H. J., Buijs, R. M., van der Donk, H. A., van den Hurk, R., & Wiegant, V. M. (1994). Preferential induction of c-fos immunoreactivity in vasoactive intestinal polypeptide-innervated gonadotropin-releasing hormone neurons during a steroid-induced luteinizing hormone surge in the female rat. *Endocrinology, 134,* 2636–2644.

van der Beek, E. M., Horvath, T. L., Wiegant, V. M., van den Hurk, R., & Buijs, R. M. (1997). Evidence for a direct neuronal pathway from the suprachiasmatic nucleus to the gonadotropin-releasing hormone system: Combined tracing and light and electron microscopic immunocytochemical studies. *Journal of Comparative Neurology, 384,* 569–579.

van der Beek, E. M., Swarts, H. J., & Wiegant, V. M. (1999). Central administration of antiserum to vasoactive intestinal peptide delays and reduces luteinizing hormone and prolactin surges in ovariectomized, estrogen-treated rats. *Neuroendocrinology, 69,* 227–237.

Vandesande, F., Dierickx, K., & DeMey, J. (1975). Identification of the vasopressin-neurophysin producing neurons of the rat suprachiasmatic nuclei. *Cell Tissue Research, 156,* 377–380.

Vrang, N., Larsen, P. J., & Mikkelsen, J. D. (1995a). Direct projection from the suprachiasmatic nucleus to hypophysiotrophic corticotropin-releasing factor immunoreactive cells in the paraventricular nucleus of the hypothalamus demonstrated by means of *Phaseolus vulgaris*-leucoagglutinin tract tracing. *Brain Research, 684,* 61–69.

Vrang, N., Larsen, P. J., Moller, M., & Mikkelsen, J. D. (1995b). Topographical organization of the rat suprachiasmatic-paraventricular projection. *Journal of Comparative Neurology, 353,* 585–603.

Vrang, N., Mikkelsen, J. D., & Larsen, P. J. (1997). Direct link from the suprachiasmatic nucleus to hypothalamic neurons projecting to the spinal cord: A combined tracing study using cholera toxin subunit B and *Phaseolus vulgaris*-leucoagglutinin. *Brain Research Bulletin, 44,* 671–680.

Wachulec, M., Li, H., Tanaka, H., Peloso, E., & Satinoff, E. (1997). Suprachiasmatic nuclei lesions do not eliminate homeostatic thermoregulatory responses in rats. *Journal of Biological Rhythms, 12,* 226–234.

Warren, W. S., Champney, T. H., & Cassone, V. M. (1994). The suprachiasmatic nucleus controls the circadian rhythm of heart rate via the sympathetic nervous system. *Physiology and Behavior, 55,* 1091–1099.

Watson, R. E., Jr., Langub, M. C., Jr., Engle, M. G., & Maley, B. E. (1995). Estrogen-receptive neurons in the anteroventral periventricular nucleus are synaptic targets of the suprachiasmatic nucleus and peri-suprachiasmatic region. *Brain Research, 689,* 254–264.

Watts, A. G. (1991). The efferent projections of the suprachiasmatic nucleus. Anatomical insights into the control of circadian rhythms. In D. C. Klein, R. Y. Moore, & S. M. Reppert (Eds.), *Suprachiasmatic nucleus: The mind's clock* (pp. 77–106). New York: Oxford University Press.

Watts, A. G., & Swanson, L. W. (1987). Efferent projections of the suprachiasmatic nucleus: II. Studies using retrograde transport of fluorescent dyes and simultaneous peptide immunohistochemistry in the rat. *Journal of Comparative Neurology, 258,* 230–252.

Watts, A. G., Swanson, L. W., & Sanchez-Watts, G. (1987). Efferent projections of the suprachiasmatic nucleus: I. Studies using anterograde transport of *Phaseolus vulgaris* leucoagglutinin in the rat. *Journal of Comparative Neurology, 258,* 204–229.

Watts, A. G., Sheward, W. J., Whale, D., & Fink, G. (1989). The effects of knife cuts in the sub-paraventricular zone of the female rat hypothalamus on oestrogen-induced diurnal surges of plasma prolactin and LH, and circadian wheel-running activity. *Journal of Endocrinology, 122,* 593–604.

Weaver, D. R. (1998). The suprachiasmatic nucleus: A 25-year retrospective. *Journal of Biological Rhythms, 13,* 100–112.

Weick, R. F., & Stobie, K. M. (1992). Vasoactive intestinal peptide inhibits the steroid-induced LH surge in the ovariectomized rat. *Journal of Endocrinology, 133,* 433–437.

Weick, R. F., & Stobie, K. M. (1995). Role of VIP in the regulation of LH secretion in the female rat. *Neuroscience and Biobehavioral Reviews, 19,* 251–259.

Weiss, P. A. (1972). Neuronal dynamics and axonal flow: Axonal peristalsis. *Proceedings of the National Academy of Sciences of the USA, 69,* 1309–1312.

Welsh, D. K., Logothetis, D. E., Meister, M., & Reppert, S. M. (1995). Individual neurons dissociated from rat suprachiasmatic nucleus express independently phased circadian firing rhythms. *Neuron, 14,* 697–706.

Wiegand, S. J., & Terasawa, E. (1982). Discrete lesions reveal functional heterogeneity of suprachiasmatic structures in regulation of gonadotropin secretion in the female rat. *Neuroendocrinology, 34,* 395–404.

Wiegand, S. J., Terasawa, E., & Bridson, W. E. (1978). Persistent estrus and blockade of progesterone-induced LH release follows lesions which do not damage the suprachiasmatic nucleus. *Endocrinology, 102,* 1645–1648.

Wiegand, S. J., Terasawa, E., Bridson, W. E., & Goy, R. W. (1980). Effects of discrete lesions of preoptic and suprachiasmatic structures in the female rat. Alterations in the feedback regulation of gonadotropin secretion. *Neuroendocrinology, 31,* 147–157.

Witkin, J. W., Paden, C. M., & Silverman, A. J. (1982). The luteinizing hormone-releasing hormone (LHRH) systems in the rat brain. *Neuroendocrinology, 35,* 429–438.

PART III

Regulation of Mammalian Circadian Rhythms

7

Photic Entrainment in Mammals

J. H. MEIJER

INTRODUCTION

The rotating earth imposes a rhythm of more or less 24 hours upon environmental conditions. The resulting changes in physical environment are drastic, but predictable. Organisms time their behavior in synchrony with the environmental changes and show clear rhythmic patterns in activity, temperature, food intake, etc. For a long time it was assumed that the daily timing of animal behavior was dictated by the 24-hour light–dark cycle. However, when Johnson (1939) recorded mice that were kept in constant darkness, he noticed that their activity pattern was still rhythmic, but the period of the cycle was now slightly different from 24 hours. He therefore concluded that the origin of circadian rhythms should be sought inside animals, and postulated a "self-winding" physiologic clock.

The existence of an endogenous circadian pacemaker is now widely accepted and has been shown in many prokaryotic and eukaryotic species. In mammals, the circadian pacemaker is located at the base of the hypothalamus in the suprachiasmatic nuclei (SCN), just above the optic chiasm. When animals are housed under constant experimental conditions, their rhythms deviate from 24 hours and reflect the period of the endogenous pacemaker. When exposed, however, to the daily environmental light–dark cycle, animals' circadian rhythms adopt this cycle. This process is called photic entrainment. Nocturnal animals entrain to the environmental cycle in such a way that their activity is restricted mainly to the night. Diurnal animals have their active period during the day when they are entrained.

Photic entrainment is a crucial aspect of the circadian timekeeping system because it ensures that animals time their behavior properly to accord with the 24-hour changes in the environment. The study of photic entrainment is interesting also from a more strategic point of view since it was realized that the entrainment

J. H. MEIJER Department of Physiology, Leiden University Medical Centre, 2300 RC Leiden, The Netherlands.

Circadian Clocks, Volume 12 of *Handbook of Behavioral Neurobiology*, edited by Joseph S. Takahashi, Fred W. Turek, and R. Y. Moore, Kluwer Academic / Plenum Publishers, New York, 2001.

pathways leading to the circadian pacemaker will eventually help to identify the circadian pacemaker itself. In this chapter, the effects of light on circadian behavioral rhythms will first be described. This will be followed by a description of the neuroanatomic pathways that mediate photic entrainment, including the photoreceptors and neurotransmitters involved. After that, the electrophysiologic responses of SCN neurons to light will be discussed and the extent to which electrophysiologic attributes of SCN neurons correlate with the characteristics of photic entrainment will be analyzed.

Effect of Light on Behavioral Circadian Rhythms

Effect of Light on Circadian Period

In constant conditions, the period or revolution time, deviating as it does from 24 hours, is determined by the light intensity in which an animal is kept (Daan & Pittendrigh, 1976). In the golden hamster (*Mesocricetus auratus*), house mouse (*Mus musculus*), white-footed mouse (*Peromyscus leucopus*), flying squirrel (*Glaucomys volans*), rat (*Rattus norvegicus*), and several more nocturnal species of mammals, the circadian period is longer in constant light than in constant darkness (Daan & Pittendrigh, 1976), whereas in several diurnal mammals such as the Siberian chipmunk (*Eutamias sibiricus*), antelope ground squirrel (*Amnospermophilus leucorus*), red squirrel (*Tamiasciurus hudsonicus*), and tree shrew (*Tupaia belangeri*), the free-running period gets shorter or remains the same in constant light as compared to constant darkness (Daan & Pittendrigh, 1976; Meijer, Daan, Overkamp & Hermann, 1990; Pohl, 1982). A general rule ("Aschoff's rule") is that in nocturnal animals when kept in constant conditions, the period increases with light intensity, whereas in diurnal animals, the period decreases (Pittendrigh, 1960). This rule appears to hold especially in nocturnal species, but there are many exceptions in diurnal species.

The free-running period not only depends on the light-intensity to which animals are exposed, but is also history dependent. This can be observed when a particular light-intensity at which an animal is kept is changed from one level to another. The long lasting, slowly decaying changes in pacemaker period are called after effects (Pittendrigh & Daan, 1976a). They can be observed not only when light intensity is changed from one level to another, but also when animals are released in constant darkness following exposure to particular photoperiods.

Enduring exposure to continuous light or darkness may alter behavioral circadian patterns dramatically. Nocturnal animals show "splitting" of their circadian activity, food, or drinking pattern when they are kept in constant bright light (Hoffmann, 1971; Meijer, van der Zee, & Dietz, 1988a; Meijer *et al.*, 1990; Pittendrigh & Daan, 1976b, Pohl, 1972; Shibuya, Melnyk, & Mrosovsky, 1980; Turek, Earnest, & Swan, 1982). This means that they become active twice a day instead of once (Figure 1). The activity components are often roughly 12 hours apart and the free-running period of the split activity rhythm is usually shorter than the previous "intact" rhythm under the same light intensity. Diurnal animals show splitting in several behavioral functions when they are kept in constant dim light or darkness and the resulting period is often longer than the previous period in the same lighting condition (Hoffmann, 1969; Meijer *et al.*, 1990; Pittendrigh, 1960; Swade & Pittendrigh, 1967).

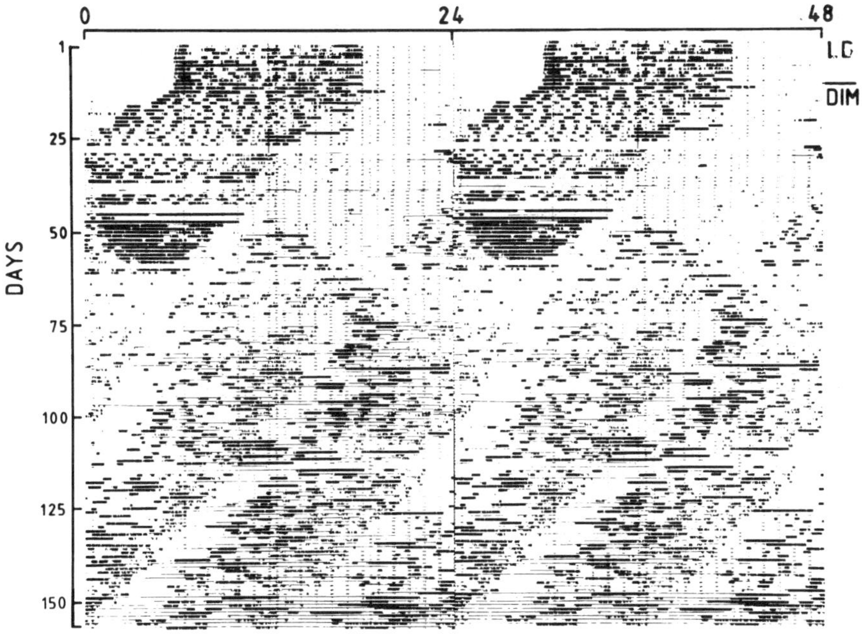

Figure 1. Double-plotted actogram for a tree shrew (*Tupaia belangeri*). The animal is first entrained to a light–dark cycle (LD) and then released in continuous dim light. In dim light, the free-running period is typically shorter than 24 hours. At day 40, an increase in activity is seen in the middle of the active period. When the period of increased activity becomes shorter, two activity components become visible that reach a stable anti-phase relationship that is characteristic for splitting at day 75. (Reprinted from Meijer *et al.*, 1990, with permission. Copyright 1990, Sage Publications.)

Pittendrigh and Daan (1976b) suggested that the two components reflect the action of two functionally separate oscillators in the mammalian pacemaker with different light sensitivities. Several attempts have been made to distinguish the two components in their responsiveness to various stimuli. However, the responsiveness to light pulses was indistinguishable and their intensity dependence appeared the same (Earnest & Turek, 1985; Meijer *et al.*, 1990). The same behavior of the two components was moreover observed after dark pulses (Boulos & Rusak, 1982; Lees, Hallonquist, & Mrosovsky, 1983) and carbachol injections (Meijer, van der Zee, & Dietz, 1988b). The onset of lordosis can be associated with either of the two components or with both (Swann & Turek, 1982). It remains open whether the two components represent distinct units with a separate role in photic entrainment.

Effect of Light on Circadian Phase

When animals are kept in constant darkness, their circadian system appears very sensitive to short pulses of light. It is less sensitive to short pulses of darkness when animals are housed in constant light (Boulos & Rusak, 1982; Meijer *et al.*, 1990).[1] The major effect of a light pulse is a phase advance or phase delay of the circadian activity rhythm. A phase shift is a single displacement of an oscillation along the time axis. This means that an animal starts its activity at an earlier or later time, respectively, than was expected from its initial free-running rhythm (Figure 2).

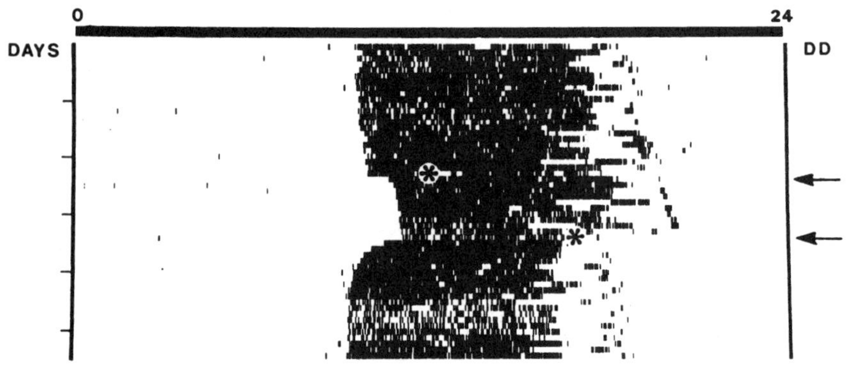

Figure 2. Wheel running activity rhythm of a hamster kept in constant darkness. Days are plotted underneath each other, and the time of the day is indicated above the record. Per minute, it is indicated whether the animal is active (black) or not (white). At the day indicated by arrows, light pulses of 60 minutes are presented. The timing of the light pulses is indicated by an asterisk in the actogram. The first light pulse during early subjective night induces a phase delay, and the second light pulse during late subjective night induces a phase advance. (Reprinted from Meijer *et al.*, 1996, with permission. Copyright 1996, Elsevier Science.)

For example, in hamsters, the onset of running-wheel activity is a very precise marker of the circadian phase. A usual protocol is to keep the hamster in constant darkness for 7 days and mark its activity onsets. On day 7, a light pulse is presented at a selected phase of the activity rhythm, and the onsets of activity are again marked on the following 2 weeks in constant darkness. Straight lines are fitted through the activity onsets prior to the day of the light pulse and through the activity onsets following the light pulse once a stable free-running rhythm is reestablished. By measuring the difference between the fitted lines before and after the light pulse, one determines the magnitude of the phase shift.

It has been observed that phase advances are usually not maximal at the first day after the light pulse, but that a number of transient cycles will pass before the phase shift reaches its steady-state value. A circadian transient cycle has been defined as a cycle of the pacemaker system with a period longer or shorter than that of the steady-state period; it occurs during the transition from one steady state to another (Daan & Pittendrigh, 1976). Advancing phase shifts are accompanied by one or more transient cycles, each of which has a duration shorter than the circadian period. Delaying phase shifts are effected by transient cycles which last longer than the circadian period (Daan & Pittendrigh, 1976). Once a phase shift has reached its steady-state value, the free-running activity rhythm remains shifted.

Phase shifts are normally determined at the basis of daily onsets in activity, which show a great precision from day to day. The offset of circadian activity is often not so precise and therefore not valuable as a marker of circadian phase of the underlying oscillator. A few studies exist in which light-induced phase shifts in activity onset and offset were studied in parallel (Elliot & Tamarkin, 1994; Honma, Watanabe, & Heroshige, 1985; Meijer & De Vries, 1995). In one study, hamsters with lesions of the geniculate nucleus were used in addition to intact animals (Meijer & De Vries, 1995) because such lesions have been shown to decrease the variability in running-wheel activity offset in hamsters (De Vries & De Vries, 1995). In contrast to the expectation, the initial shifts in onset and offset of circadian activity appeared to differ both in intact and in geniculate-lesioned animals (Elliot & Tamarkin, 1994;

Honma *et al.*, 1985; Meijer & De Vries, 1995). When light elicited a phase advance, the onset shifted gradually whereas the offset shift appeared immediate (see also Figure 2). When light-induced a phase delay on the other hand, the initial transient shift in activity onset appeared to exceed the shift in the offset (Meijer & De Vries, 1995). As a result, the duration of the animals' activity shortened for almost 10 days both following phase-advancing and phase-delaying shifts. Differences in free-running periods of activity onsets versus activity offsets during this period led to a restoration of the initial activity duration.

The differential effect of light pulses on onset and offset of activity is a complicating factor in the analysis of light induced phase shifts. *A priori*, there is no reason to attribute more value to activity onset as a phase marker than to activity offset.

The direction and magnitude of a light-induced phase shift in activity onset is dependent on several variables.

Intensity of the Light Pulse. In the golden hamster, the effect of light intensity, determined at a fixed circadian phase, has been studied in detail (Meijer, Rusak, & Gänshirt, 1992; Nelson & Takahashi, 1991; Takahashi, Decoursey, Bauman, & Menaker, 1984). The relation between light intensity and phase shift is sigmoidal. The threshold for a phase shift is 0.1 lux (Meijer *et al.*, 1992), 10^{11} photons cm^{-2} $second^{-1}$ or 0.1 cd m^{-2} (Nelson & Takahashi, 1991; Takahashi, *et al.*, 1984). These values are high compared with the threshold for vision (Remé, Wirz-Justice, & Terman, 1991). Above the threshold, increasing light intensities elicit increasing phase shifts up to a saturation intensity of about 1000 lux, above which there is no further increase.

In diurnal animals, including humans, much higher light intensities have often been used to obtain a phase shift based on the assumption that high light intensities were required (Czeisler *et al.*, 1989; Drennan, Kripke, & Gillin, 1989; Lewy, Wehr, Goodwin, Newsome, & Markey, 1980; Meijer, *et al.*, 1990; Minors, Waterhouse, & Wirz-Justice, 1991; Pohl, 1982). However, it was found that intensities of about 200 lux are also sufficient for phase shifting in diurnal *Tupaia* (Meijer *et al.*, 1990) and humans (Boivin, Duffy, Kronauer, & Czeisler, 1996; Terman & Schlager, 1990).

Duration of a Light Pulse. Longer light pulses produce larger phase shifts. In nocturnal animals, light pulses of several minutes or more result in a phase shift of ca. 1 hour (Daan & Pittendrigh, 1976; DeCoursey, 1960; Nelson & Takahashi, 1991). In diurnal mammals, longer light pulses have often been used to induce phase shifts in circadian activity rhythms (Meijer *et al.*, 1990; Pohl, 1982). In the hamster, the sensitivity to pulse duration has been determined by keeping the total number of photons in a stimulus constant and varying the pulse duration. The circadian system appears maximally sensitive to 300-second stimuli (Nelson & Takahashi, 1991).

Wavelength of the Light. The circadian system of the hamster is most sensitive to green light and less so to blue or red light (Boulos, 1995; Takahashi *et al.*, 1984). No indication for differential spectral sensitivity for phase advances and delays has been found. The spectral sensitivity curve resembles that of rhodopsin, the rod photopigment. However, it is unlikely that rods mediate photic entrainment (see below, subsection on Photoreceptors of the Circadian System).

Circadian Phase of the Light Pulse. In nocturnal animals, the onset of activity is the onset of subjective night and is defined as circadian time (CT) 12. Light pulses that are presented at the beginning of the animals' subjective night produce a phase delay of the free-running rhythm, whereas they produce an advance when presented towards the end of the subjective night. Light pulses during the animals' inactive period (or subjective day) do not result in a phase shift. The phase response curve summarizes the relation between the circadian time at which a light pulse is applied relative to activity onset and the direction and magnitude of a phase shift, and is useful for predicting the effect of light pulses (Figure 3). Phase advances are plotted in the positive direction and delays in the negative direction (Daan & Pittendrigh, 1976).

In diurnal animals, the onset of activity is defined as CT0 (Pohl, 1982; Meijer *et al.*, 1990). Light pulses at the beginning of the inactive period produce phase delays, and at the end of the animals' inactive period, they produce phase advances. The phase response curve of diurnal animals, including humans, often has no clear dead zone (Jewett *et al.*, 1997; Pohl, 1982).

The phase-dependent light responsiveness will cause the pacemaker to entrain to the external light–dark cycle: when nocturnal animals leave their burrow too early in the evening, they will be exposed to environmental light at the beginning of their active period. This induces a phase delay which causes them to leave their burrow later the next day. However, if they return to their burrow too late, morning light will induce a phase advance, leading to an earlier return on the next day. In a behavioral study in which the natural habitat was mimicked, DeCoursey (1986) showed that entrainment of the flying squirrel (with a period shorter than 24 hours) occurs by such repeated (but not necessarily daily) exposures to the evening light (Figure 4).

The example of the flying squirrel illustrates entrainment on the basis of nonparametric entrainment. This means that entrainment is accomplished by discrete actions of light which result predominantly in instantaneous phase shifts. In

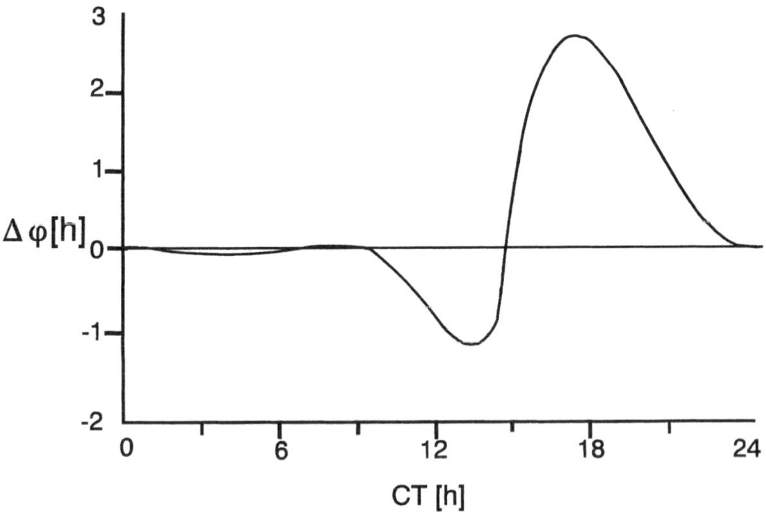

Figure 3. Phase response curve (modified from Takahashi *et al.*, 1984) for the effects of light pulses (1 hour, 350 lux) on the phase of the running-wheel activity rhythm of the hamster. Timing of the light pulse relative to activity onset (CT12) is indicated on the x axis; phase advances and delays are plotted in the positive and negative direction respectively. (Modified from Takahashi *et al.*, 1984.)

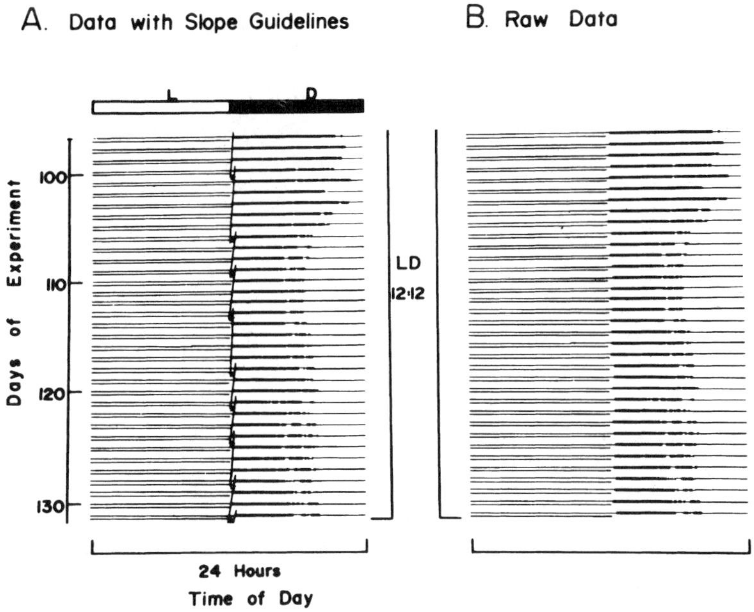

Figure 4. Photoentrainment of the nocturnal flying squirrel (*Glaucomys volans*) to an LD (12:12) cycle in simulated den cages. Left: Data with visually fitted lines through activity onsets. Right: Raw data. The animal is exposed to a few minutes of light only once every 3–5 days. In between, a free-running activity pattern is visible. Only a few minutes of light every 3–5 days was sufficient for the animal to become entrained. (Part of Figure 5 in DeCoursey, 1986, reprinted with permision. Copyright 1986, Springer-Verlag.)

addition, it has been observed that light induces small adjustments in the free-running period (Beersma *et al.*, 1999; DeCoursey, 1989; Hut *et al.*, 1999). These light induced changes in circadian period are dependent on the phase of the circadian cycle (Beersma *et al.*, 1999). Adjustments in period may allow animals to tune their intrinsic period to the environmental cycle without having to perturb the system each day (Daan, 2000).

Afferent Connections

The Retina with Its Photoreceptors

In several nonmammalian species of vertebrates, light can be perceived by (para-) pineal and encephalic photoreceptors (Menaker, 1982; Rusak & Zucker, 1979; Underwood & Groos, 1982). It has recently been demonstrated that skin illumination at the back of the knee can phase shift circadian temperature rhythms in humans (Campbell & Murphy, 1998). In adult mammals, however, extraretinal photoreception has not been shown to play a role, and entrainment of the circadian pacemaker to the light–dark cycle seems to be mediated via the retina only (Underwood & Groos, 1982; Meijer *et al.*, 1999; Yamazak, Goto, & Menaker, 1999).

A maximum sensitivity of the circadian system has been described for wavelengths of about 500 nm, which is similar to the peak sensitivity of rhodopsin

(Boulos, 1995; Takahashi *et al.*, 1984). However, rats entrain to red light pulses of a wavelength beyond the sensitivity of rods (McCormack & Sontag, 1980) and degeneration of the rod cell layer does not prevent rats from entraining to a light–dark cycle (Dunn, Dyer, & Bennett, 1972). The same applies to mutant mice (*Mus domesticus*) with almost complete loss of rod photoreceptors (Foster *et al.*, 1991; Freedman *et al.*, 1999). Suprisingly, they do not show any decrease in light sensitivity.

Furthermore, a considerable difference in light thresholds exists between rods and the circadian system. The light threshold of the circadian system is high as compared to the threshold of rhodopsin. This difference may not be too important as it has been shown that postsynaptic density of optic synapses in the SCN and light sensitivity of the circadian system vary as a consequence of lighting history (Güldner & Phillips, 1985; Reiter *et al.*, 1983). Light sensitivity may therefore be regulated postsynaptically.

An alternative candidate to mediate photic entrainment is a middle wavelength-sensitive (M) cone pigment which has, similar to rhodopsin, a peak sensitivity at 500 nm (Jacobs, 1993; Jacobs, Neitz, & Deegan, 1991; von Schantz, Argamaso-Hernan, Szél, & Foster, 1997). The presence of this cone opsin has been demonstrated in the retina of the Syrian hamster both by immunocytochemistry and reverse-transcription polymerase chain reaction. Recent studies however have indicated that mice with genetically degenerated cones show normal phase shifting responses to 500 nm light (Freedman *et al.*, 1999). The studies of Freedman indicate therefore that opsin based pigments are not required for entrainment.

Spectral sensitivity of the circadian system has not only been studied by measuring phase shift as the dependent variable, but also by investigating melatonin suppression (Reiter, 1985). Melatonin production is driven by the SCN and is low in daytime and high at night (Moore & Klein, 1974). The nighttime elevation can be suppressed by exposing animals to light, an effect which is also mediated by the SCN (Klein & Weller, 1970).

In the rat, green light (510–550 nm), and in the Syrian hamster, blue light (435–500 nm) appear most effective for suppressing melatonin (Brainard, Richardson, King, & Reifer, 1984; Brainard *et al.*, 1986; Cardalini, Larin, & Wurtman, 1972). Data on pineal electrical activity in rats, guinea pigs, and hamsters indicated highest sensitivity in dark-adapted animals for 500 nm. In light-adapted animals, additional sensitivity peaks were observed in the blue (450 nm, guinea pigs) and red (560 nm, rats and hamsters; Thiele & Meissl, 1987). The proposed role of additional red or blue cone-like pigments is consistent with the observation that the pineal gland and circadian system of animals with light-damaged or rod-degenerated retinas still respond to light (Dunn *et al.*, 1972; Foster *et al.*, 1991; Reiter, 1985; Webb, Champney, Lewinski, & Reiter, 1985).

Several species of rodents respond to a wavelength of 370 nm, which would indicate the existence of an ultraviolet light-sensitive pigment. The involvement of this pigment in phototransduction to the SCN is indicated by the observation that UV-A radiation can influence melatonin production in the Syrian hamster and rat (Brainard *et al.*, 1984, 1986; Podolin, Rollag, & Brainard, 1987) and phase shifting in the hamster (von Schantz *et al.*, 1997). Therefore, it has been questioned whether a short-wavelength-sensitive (S) cone pigment that has been described for the rodent retina (Jacobs, 1993) is involved in entrainment. However, no S/ UV cone opsin mRNA or translocated protein has been found.

Cryptochromes are photosensitive pigments that could play an essential role in the generation of circadian rhythmicity as it has been demonstrated that cryptochrome deficient mice show a-rhythmic activity patterns (Miyamoto & Sancar, 1998; Thresher *et al.*, 1998; van der Horst *et al.*, 1999). The role of cryptochromes for photic entrainment remained unclear for some time because double knock out mice are still responsive to photic stimulation (van der Horst, 1999). However, triple knock out mice, lacking both classical photoreceptors and cryptochromes, exhibit nearly a-rhythmic activity patterns under a light-dark cycle (Selby *et al.*, 2000), indicating a crucial role for cryptochromes in photic entrainment.

Electrophysiological recordings have revealed that different light responsive cells prefer different wavelengths. In the rat, action spectra of SCN neurons showed a peak at 510 nm and a secondary peak in the near-ultraviolet region of the spectrum (Aggelopoulos and Meissl, 2000). In the cat, SCN projecting ganglion cells show maximum responsiveness to 500 nm wavelengths whereas other ganglion cells responded to 560 and 575 nm (Pu, 2000). These data indicate therefore that different light receptors are involved in transmitting photic information to the SCN. In view of the finding that triple knock out mice are nearly a-rhythmic, it is unlikely that these different pigments play a major role in photic entrainment. Instead they may subserve other photic functions of the circadian system.

Circadian Rhythms in the Retina

There are now clear indications that the mammalian retina has endogenous oscillator properties (Tosini & Menaker, 1996; see Remé, Wirz-Justice, & Terman, 1991, for review). In constant conditions, circadian rhythms have been measured in disk shedding and autophagy, dopamine and melatonin synthesis, and opsin mRNA synthesis (Remé *et al.*, 1991). Moreover, a circadian rhythm has been observed in light-induced c-Fos production in retinal ganglion cells (Chambille, Doyle, & Serviere, 1993) and a visual sensitivity rhythm has been measured psychophysically (Remé *et al.*, 1991). In SCN-lesioned animals, the circadian rhythms remain in visual sensitivity and in rod outer segment disk shedding (Remé *et al.*, 1991). Isolated retinas of the golden hamster show circadian rhythms in melatonin production, which demonstrates convincingly that the retinal rhythm is not driven by other brain structures (Tosini & Menaker, 1996). An endogenous circadian rhythm in retinal processes may have consequences for the light sensitivity of the eye and may thereby introduce a circadian rhythm in the processing of light information to the SCN and other visual brain areas.

Retinal Ganglion Cells

The retinal photoreceptors absorb light and convert the photic signal with large amplification to a change in membrane potential. The potential change is transmitted from the photoreceptors to the ganglion cells, which respond to this with a change in discharge rate, enabling the signal to be transmitted to the brain.

Three classes of ganglion cells exist, which have been ranked according to their conduction velocities as W-, X-, and Y-cells (Rowe & Stone, 1977).[1] None of these has properties that correspond completely to those of light-responsive SCN cells. The W-, X-, and Y-cells differ in cell body size, arborization, response latency, and their intensity–response relation. The conspicious response characteristics of SCN cells (discussed in a later section of this chapter) closely resembles those of a subclass of W-cells which respond to light with a transient increase in discharge before a steady-state discharge level is obtained (Barlow & Levick, 1969; Berson, 1988; Stone &

Fukuda, 1974). The other W-cells respond to light with a tonic decrease in discharge. The close correspondence between the properties of the tonic W-cells and light-responsive SCN cells suggests that these project to the SCN (Groos & Mason, 1980; Groos & Meijer, 1985; Meijer, 1991). Moreover, SCN cells have long response latencies, which also correspond to the slow W-cell pathway. Anatomic studies confirmed the small neuron size, typical for type III cells, and indicated that they are distributed across the whole retina, with a slightly higher density in the temporal quadrants (Cooper, Herbin, & Nevo, 1993; Moore, Speh, & Card, 1995; Pu, 1999). A discrepancy between SCN and W-cells is that receptive fields of W-cells are smaller than those of the SCN. It is possible that W-cells converge on SCN cells, thereby increasing the receptive fields of the secondary neurons. This possibility is realized, for instance, in the colliculus superior, where convergence of W-cells produces receptive fields very similar to those of the SCN (Berson, 1988).

Photic Afferents of the Suprachiasmatic Nucleus

From the retinal ganglion cells, several pathways diverge before they enter the SCN. The shortest, most direct route by which light information is transduced to the SCN is the retinohypothalamic tract (RHT). The RHT consists of unmyelinated fibers which contact the SCN mainly along its lateral and ventral parts. Its presence has been shown in many species, including humans (Cassone, Speh, Card, & Moore, 1988; Moore, 1973; Moore & Lenn, 1972; Pickard, 1982; Sadun, Schaechter, & Smith, 1984). The RHT also projects to adjacent hypothalamic areas, but here the density of terminals is lower than in the SCN (Johnson, Morin, & Moore, 1988a; Leevine, Weiss, Rosenwasser, & Miselis, 191; Seph & Moore, 1993; Youngström, Weiss, & Nunez, 1991).

A second pathway is formed in part by collaterals of the RHT (Pickard, 1985) and projects to the intergeniculate leaflet (IGL) and ventral lateral geniculate nucleus (vLGN) (Hickey & Spear, 1976; Swanson, Cowan, & Jones, 1975). Here neurons are contacted that form the geniculohypothalamic tract (GHT) (Card & Moore, 1984; Harrington, Nance, & Rusak, 1985, 1987; Morin, Blanchard, & Moore, 1992; Pickard, 1985; Ribak & Peters, 1975).

A third photic input pathway may be formed by the median and dorsal raphe nuclei. Mosko and Jacobs (1974) found evidence for light responsiveness of the raphe of the rat. The retinal input was later confirmed by anatomic studies in the rat and cat (Foote, Taber-Pierce, & Edwards, 1978; Shen & Semba, 1994). The raphe projects to the SCN (Azmitia & Segal, 1978; Card & Moore, 1984; Moore, Halaris, & Jones, 1978; for review, see Meijer & Rietveld, 1989; Morin, 1994), but it remains to be determined whether retinorecipient raphe cells also make up this projection. Only then would the raphe form a photic pathway to the SCN. Finally, the pretectum, an area involved in pupillary reflexes, projects directly to the SCN (Mikkelsen & Vrang, 1994; Moga & Moore, 1996, 1997).

Neurotransmitters of Photic Afferents and Their Role in Photic Entrainment

Retinohypothalamic Tract. The function of the RHT and GHT in photic entrainment has been investigated by lesion or transsection studies and by studies in which various transmitter agents were applied. Lesions of the RHT by semicircular SCN cuts in the hamster disrupt entrainment to a light–dark cycle (Johnson, Moore,

& Morin, 1988b). This is suggestive evidence that the RHT is necessary for entrainment, but not conclusive since the possibility that the GHT may have been damaged by these cuts as well cannot be ruled out (Johnson *et al.*, 1988b). The use of specific agonists and antagonists could be an alternative strategy to determine whether the RHT is required for photic entrainment.

Excitatory amino acids (EAAs) are the most likely transmitter candidate for the RHT. *In vitro* SCN measurements indicate an increase in the release of glutamate and aspartate following optic nerve stimulation (Liou, Shibata, & Ueki, 1986b). Retinal terminals inside the SCN show increased glutamate immunoreactivity (Castell, Belenky, Cohen, Ottersen, & Storm-Mathisen, 1993; DeVries, Nunes Cardozo, van der Want, de Wolf, & Meijer, 1993) and several N-methyl-D-aspartate (NMDA), non-NMDA, and metabotropic glutamate receptor subunits have been demonstrated (Gannon & Rea, 1993; Ishida, Matsui, Mitsui, & Mishina, 1994; Meeker, Greenwood, & Hayward, 1994; Mick, Yoshimura, Ohno, Kiyama, & Tohyama, 1995; Mikkelsen, Larsen, & Ebling, 1993; van den Pol, Hermans-Borgmeyer, Hofer, Ghosh, & Heinemann, 1994a; van den Pohl, Kogelman, Ghosh, Liljelund, & Blackstone, 1994b; for review, see Ebling, 1996). An EAA-related dipeptide which could have a transmitter function in the RHT is N-acetylaspartylglutamate (NAAG), which acts on a NMDAR1 receptor subtype (Sekiguchi, Wada, & Wenthold, 1992). The RHT and SCN show immunoreactivity to NAAG, which decreases following optic nerve transection (Moffett, Williamson, Palkovits, & Namboordiri, 1990).

Three classes of ionotropic postsynaptic EAA receptors have been well described. EAA receptors have been distinguished based on the action of selective receptor agonists, NMDA, AMPA, and kainate. Electrophysiologic measurements revealed several differences in the properties of these receptor types (Nicoll, Malenka, & Kauer, 1990). AMPA and kainate receptors are often grouped together and referred to as non-NMDA receptors. Glutamate, aspartate, and NAAG result in activation of all of these receptor types.

EAA receptors differ in their activation and deactivation time courses, desensitization kinetics, and ion permeabilities. The associated channel of the NMDA receptor is highly permeable to monovalent, positively charged ions and to Ca^{2+}, displays slow kinetics and is blocked in a voltage-dependent way by Mg^{2+} (Mayer, Westbrook, & Guthrie, 1984). At resting membrane potentials, the channel is blocked by Mg^{2+} and has to sense membrane depolarization before it can be activated. The AMPA receptor is associated with a fast channel with high permeability to monovalent (but not divalent) ions. The kainate receptor has similar properties, but gates faster desensitizing currents (Huettner, 1990). The three classes of EAA receptors often coexist in the same cell (Nicoll *et al.*, 1990; Stern, Edwards, & Sakmann, 1992).

Reverting to circadian rhythms, various EAA agents have been used and their role in light-induced phase shifts and SCN discharge rate has been determined. The first aim of these studies was often not to determine whether the RHT is a sufficient entrainment pathway, but rather to determine which of the EAA receptor types is involved in light transduction to the SCN. This is of ultimate importance when entrainment pathways are being followed in order to approach the pacemaker mechanism.

Most SCN cells are excited by glutamate, aspartate, NAAG, quisqualate, kainate, and NMDA, and some are inhibited (Bos & Mirmiran, 1993; Ito, Yamatodani, Nagai, & Wada, 1991; Kim & Dudek, 1991; Mason & Rusak, 1991; Meijer, Albus, Weidema, & Ravesloot, 1993; Nishino & Koizumi, 1977; Shibata, Liou, & Ueki, 1983). In those studies where transsynaptic effects on the recorded cell can be excluded, glutamate appeared only excitatory. Aspartate and NAAG are somewhat less effective than glutamate (Bos & Mirmiran, 1993; Mason, Piggins, & Rusak, 1993; Meijer *et al.*, 1993).

NMDA and glutamate produce a sustained activation of SCN cells and no desensitization has been observed either in SCN slices (Mason & Rusak, 1991; Meijer *et al.*, 1993) or in cultured hypothalamic cells (Meijer *et al.*, 1993). Patch-clamp recordings in cultured SCN neurons have shown that glutamate is always depolarizing (Meijer *et al.*, 1993). Intracellular recordings of SCN cells *in vitro* reveal that at resting membrane potentials, non-NMDA receptors mediate excitatory input of the RHT to the SCN, whereas at less negative potentials NMDA receptors also play a role (Kim & Dudek, 1991).

A puzzling fact is that glutamate injections (1 mM) *in vivo* in the SCN induce phase shifts unlike those induced by light pulses (Meijer *et al.*, 1988b), whereas glutamate (10 mM) application to brain slices which contain the SCN mimic the effects of light (Ding *et al.*, 1994; Gillette *et al.*, 1993; Shirakawa & Moore, 1994a). One explanation is that glutamate, in the concentration of 1 mM, causes a depolarization block (Meijer *et al*, 1988b). Another explanation could be that *in vivo* gross application may stimulate efferent pathways which in turn project to the SCN and cause phase shifts during the day (Meijer *et al.*, 1988b). Recent NMDA injection in small volumes *in vivo* has indeed mimicked the effects of light on the circadian system (Mintz and Albers, 1997).

Nonselective antagonists of EAA receptors block the SCN response to optic nerve stimulation *in vitro* (Cahill & Menaker, 1987, 1989). Intraperitoneal administration of MK-801, a noncompetitive antagonist of the NMDA receptor, blocks the phase-shifting effect of light pulses (Colwell, Ralph, & Menaker, 1990; Colwell, Foster, & Menaker, 1991) and of electrical stimulation of the RHT *in vivo* (De Vries, Treep, de Pauw, & Meijer, 1994). The latter study ruled out the possibility of MK-801 blocking retinal processes instead of synaptic transmission in the SCN by bypassing the retina with stimulation of the optic nerve (Figure 5). The competitive NMDA receptor antagonist CPP and the non-NMDA receptor antagonists DNQX and CNQX, which act nonspecifically on non-NMDA receptor types, also block phase shifts (Colwell & Menaker, 1992; Rea, Buckley, & Lutton, 1993a; Vindlacheruvu, Ebling, Maywood, & Hasting, 1992).

Measurements of immediate-early gene expression have also been used as a tool to investigate photic activation of the SCN (Ebling *et al.*, 1994; Kornhauser, Nelson, Mayo, & Takahasi, 1990; Kornhauser, Mayo, & Takahashi, 1993; Rea, Michel, & Lutton, 1993b; Rusak, Robertson, Wisden, & Hunt, 1990). Intracerebroventricular administration of NMDA increases c-Fos immunoreactivity in SCN cells (Ebling *et al.*, 1991; Ebling, Hui, Mirakhur, Maywoord, & Hastings, 1993), whereas glutamate antagonists reduce c-Fos immunoreactivity (Abe & Rusak, 1994; Abe, Rusak, & Robertson, 1991, 1992; Colwell, Foster, & Menaker, 1993; Ebling *et al.*, 1991; Rea *et al.*, 1993a; Vindlacheruvu *et al.*, 1992). However, reduction of c-Fos expression was absent in the dorsolateral SCN. It has been suggested that a subclass of retinal ganglion cells project selectively to the dorsolateral SCN, which may be neurochemically different from the other innervating fibers (Treep, Abe, Rusak, & Goguen, 1995).

A second transmitter candidate of the RHT is substance P. In the rat, substance P-containing ganglion cells project to the SCN and after enucleation, the density of substance P-containing fibers in the SCN drops (Mikkelsen & Larsen, 1993; Takatsuji, Miguel-Hidalgo, & Tohyama, 1991). The effect of substance P on SCN discharge is mainly excitatory, which is similar to glutamate (Shirakawa & Moore, 1994b) and phase shifts induced *in vitro* by substance P resemble the light-induced

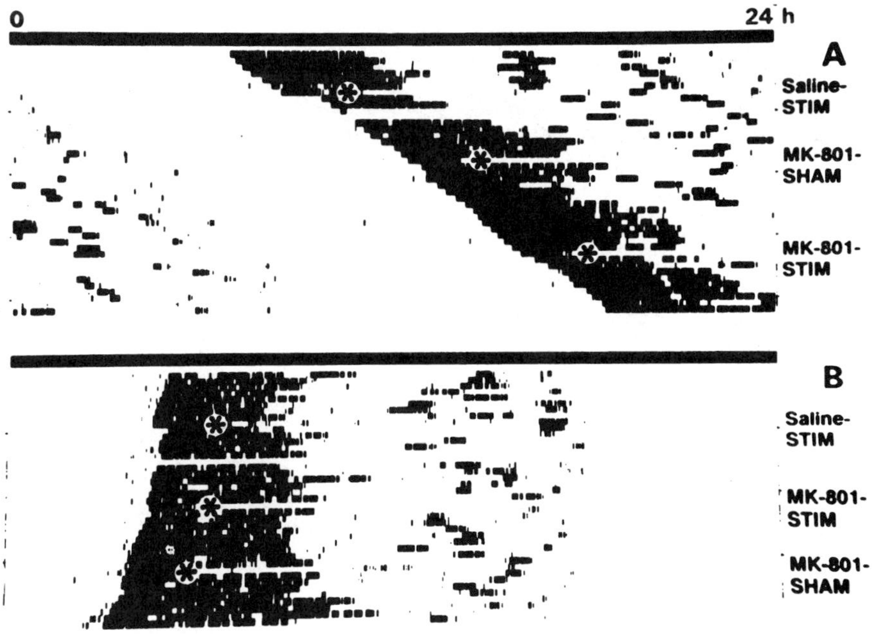

Figure 5. Running-wheel activity records of two hamsters with implanted electrodes in the optic nerve and anterior chiasm. Time of day is indicated above the record, days are plotted beneath each other. Saline injection (i.p.) followed by electrical stimulation at CT14 induces phase delays. I.p. injection of the noncompetitive antagonist of the NMDA receptor MK-801 at 30 minutes prior to optic nerve stimulation attenuates stimulation-induced phase shifts, whereas MK-801 injection in combination with sham stimulation was not effective. Results indicate that the NMDA receptor is important for phase shifting by light. (Reprinted from DeVries *et al.*, 1994, with permission. Copyright 1994, Elsevier Science.)

phase shifts (Shibata, Tsuneyoski, Hameda, Tominaga, & Watanabe, 1992a) but *in vivo* substance P has almost no effect (Piggins & Rusak, 1997). In conclusion, both RHT transmitter candidates have an excitatory effect on SCN discharge and *in vitro* they produce phase shifts that are similar to those of light.

A third candidate is pituitary adenylate cyclase-activating peptide (PACAP), which stimulates adenylate cyclase activity (Arimura, 1992). PACAP-immunoreactive fibers project both to the SCN and to the intergeniculate leaflet (Hannibal *et al.*, 1997; Piggins, Stamp, Burns, Rusak, & Semba, 1996). PACAP application *in vitro* resets the phase of the neuronal activity rhythm of the SCN when it is administered during the day, but not during the night (Hannibal *et al.*, 1997). It is unknown what conditions trigger release of PACAP and what its role is in entrainment.

GENICULOHYPOTHALAMIC TRACT. Lesions of the lateral geniculate nucleus slow the rate of reentrainment to a shifted light–dark cycle and alter the magnitude of phase shifts induced by short pulses of light (Dark & Asdourian, 1975; Harrington & Rusak, 1986; Pickard, Ralph, & Menaker, 1987; Rusak, 1977; Zucker, Rusak, & King, 1976). Ongoing entrainment to a light–dark cycle is not affected by lesions of the lateral geniculate nucleus and IGL (Dark & Asdourian, 1975; Zucker *et al.*, 1976), indicating that the GHT may alter the fine tuning of photic input, but is not required for entrainment to occur.

The GHT contains neuropeptide Y (NPY) and gamma amino butyric acid

(GABA), and in the rat, also corticosteroid-releasing factor and met-enkephalines (Harrington *et al.*, 1985, 1987; for review, see Harrington, 1997; Moore & Speh, 1993; Morin *et al.*, 1992; Takatsuji & Tohyama, 1989). Electrical stimulation of the GHT produces phase shifts in the circadian activity rhythm of hamsters. Phase advances are evoked in the subjective day and phase delays in the late subjective night (Rusak, Meijer, & Harrington, 1989). Of the GHT transmitters, NPY and GABA have been most extensively studied.

NPY application to SCN slices mostly inhibits neuronal discharge (Liou & Albers, 1991; Gribkoff *et al.*, 1998; Schmahl & Bohme, 1997; Shibata & Moore, 1988). NPY injections in the SCN produces phase advances in the subjective day and phase delays during the whole subjective night (Albers & Ferris, 1984). NPY bath application *in vitro* produces advances during the subjective day, which is similar to above, but differs in its effects during the subjective night (Golombec, Biello, Rendon, & Harrington, 1996; Medanic & Gillette, 1993; Shibata & Moore, 1993). Phase shifts are mediated by the Y_2 receptor and may not involve membrane hyperpolarization (Golombec *et al.*, 1996; Hall *et al.*, 1999).

NPY seems to counteract effects of light on the circadian pacemaker because NPY antisera increase the phase-advancing effects of light (Biello, 1995). NPY reduces the phase-advancing effects of glutamate *in vitro* (Biello, Golombek, & Harrington, 1996) and the phase-shifting effects of light *in vivo* (Biello, 1995; Weber & Rea, 1997). Moreover, it has been demonstrated that neuropeptide Y can block glutamate-induced shifts during the night and that glutamate can block neuropeptide Y-induced shifts during the day (Biello, Golombek, & Harrington, 1997). Finally, NPY can block phase shifts to PACAP via the Y_5 and Y_1 receptor but not via the Y_2 receptor that mediates phase shifting to NPY (Harrington & Hoque, 1997).

GABA colocalizes with NPY in all GHT terminals (Moore & Speh, 1993). In addition, all SCN neurons are probably GABA-ergic (Albers, Liou, Ferris, Stopa, & Zoellner, 1991; Moore & Speh, 1993), and all SCN neurons are innervated by GABA-ergic terminals (Liou, Shibata, Albers, & Ueki, 1990; Mason, Biello, & Harrington, 1991; van den Pol & Tsujimoto, 1985). The investigated effects of several GABA-ergic agents are therefore not specifically related to activation of the GHT.

The response of SCN cells to GABA is inhibitory (Bos & Mirmiran, 1993; Mason *et al.*, 1991). GABA receptors are divided into $GABA_A$ and $GABA_B$ receptors. The $GABA_A$-receptor agonist muscimiol induces phase advances during the subjective day (Tominaga, Shibata, Hamada, & Watanabe, 1994a). The benzodiazepine agonist diazepam, which modulates the $GABA_A$ receptor, blocks light-induced phase advances, but not delays (Ralph & Menaker, 1986). The $GABA_A$-receptor antagonist bicuculline blocks delays (Ralph & Menaker, 1985) and advances (Smith, Turek, & Slater, 1990). Baclofen ($GABA_B$ receptor agonist) changes the SCN field potentials which are induced by optic nerve stimulation (Gannon, Cato, Hart, Armstrong, & Rea, 1995; Jiang & Allen, 1995; Shibata, Liou, & Ueki, 1986) and decreases release of glutamate and aspartate within the SCN (Liou *et al.*, 1986b). Both light-induced phase advances and delays are blocked by baclofen (Gillespie, Mintz, Marvel, Huhman, & Albers, 1997; Ralph & Menaker, 1989). Phase-advancing shifts that are induced by a coctail containing vasoactive intestinal peptide (VIP), peptide histidine isoleucine (PHI), and gastrin-releasing peptide (GRP) can be blocked by $GABA_A$ and $GABA_B$ agonists (Gillepsie *et al.*, 1997).

A role for enkephalines in phase shifting has recently been indicated. Morphine injections in mice produce phase shifts during subjective day. However these shifts depended on increased activity levels of the animals (Marchant & Mistlberger, 1995). More clear is the case for fentanyl, a mu receptor agonist, which also affects the delta and kappa receptor. Fentanyl injections during the day induce phase advances in hamster running wheel activity rhythms. These shifts do not depend on increased levels of running wheel activity (Meijer *et al.*, 2000). Moreover, local injection of a delta agonist in the SCN produces similar phase shifts during the day and delta opioid receptor have been demonstrated inside the SCN (Byku & Gannon, 2000; Byku *et al.*, 2000). The role of endogenous opioids for phase adjustments of the circadian pacemaker remains to be determined.

INNERVATION BY THE RAPHE. Most serotonin afferents end in the ventrolateral SCN, and thus overlap with the terminal field of the RHT and GHT (Card & Moore, 1984). Serotonin and NPY project to the same neurons in the SCN and in some cases they appose presynaptically (Guy, Bosler Dusticier, Pelletier, & Galas, 1987). Therefore, a neural substrate exists for possible interactions between the RHT and input coming from the raphe and IGL. About 50% of the SCN cells appear responsive to serotonin (Meijer & Groos, 1988). The effect of serotonin on SCN discharge is mainly inhibitory (Bloom, Hoffer, Siggins, Barjer, & Nicoll, 1972; Mason, 1986; Meijer & Groos, 1988; Nishino & Koizumi, 1977; Ying & Rusak, 1994) as is the response to raphe stimulation (Bloom *et al.*, 1972; Groos, Mason, & Meijer, 1983). The postsynaptic sensitivity to serotonin displays a diurnal variation with a two- to threefold increase in serotonin sensitivity during the subjective night (Mason, 1986).

Raphe lesions reduce the amplitude of the activity rhythm (Block & Zucker, 1976), an effect that is even more pronounced when animals are kept in constant light (Levine, Rosenwasser, Yanovski, & Adler, 1986). A reduction of amplitude is also observed following depletion of brain serotonin by parachlorophenylalanine (Honma *et al.*, 1979). This has led to a conjecture that the raphe is important for obtaining a coherent circadian rhythm.

PHASE SHIFTING EFFECTS OF SEROTONIN. Lesions of the raphe nucleus do not prevent entrainment to a light-dark cycle (Kam & Moberg, 1977; Levine *et al.*, 1986). Hence, this input pathway is not required for entrainment to occur. On the other hand, it has now been shown that serotonin induces phase shifts and modulates photic input to the SCN in the hamster (Duncan, Johnson, & Wehr, 1992; Morin & Blanchard, 1991) but not in the rat (Kohler *et al.*, 2000). The serotonin (5-HT) agonist quipazine induces advances of the circadian discharge pattern in brain slices when administered during the subjective day and induces delays during the subjective night (Prosser, Miller, & Heller, 1990). Serotonin and the 5-HT_{1A}-receptor agonist (8-OH-DPAT) do not induce delays, but induce advances when applied during the subjective day (Edgar, Miller, Prossner, Dean, & Dement, 1993; Medianic & Gillette, 1992; Prosser, Dean, Edgar, Heller, & Miller, 1993; Shibata, Tsuneyoshi, Hamada, Tominaga, & Watanabe, 1992b; Tominaga, Shibata, Ueki, & Watanabe, 1992). This suggests that the 5-HT_{1A} receptor type mediates advances. On the other hand, there is accumulating evidence that the 5HT_7 receptor is involved in mediating phase shifts (Prosser, 2000). For example, ritanserin, which antagonizes the 5HT_7 receptor blocks 8-OH-DPAT induced shifts (Lovenberg *et al.*, 1993). The receptors may be located on the clock cells because similar phase advances are found when neuronal interactions between cells are blocked by tetrodoxin or high Mg^{2+} (Prosser, Hiller, & Miller, 1992).

Light-Blocking Effects of Serotonin. Many experiments now indicate that serotonin inhibits the photic responsiveness of the SCN (Rea & Pickard, 2000). SCN field potentials, which can be recorded following optic nerve stimulation, decrease when serotonin is added to the bath (Liou, Shibata, & Ueki, 1986a; Rea, Glass, & Colwell, 1994) and light-responsive SCN cells are inhibited by 5HT (Nishino & Koizumi, 1977; Ying & Rusak, 1994). When the raphe nuclei are lesioned, the threshold to optic nerve stimulation is reduced (Liou *et al.*, 1986c). Moreover, quipazine blocks light-induced c-Fos during the late subjective night (Rea *et al.*, 1994; Selim, Glass, Hauser, & Rea, 1993).

Behavioral experiments have shown that 5-HT depletion (with 5,7-dihydroxy-tryptamine) increases light-induced phase delays (Morin & Blanchard, 1991), whereas chlorgyline, which elevates brain serotonin (as well as norepinephrine and dopamine) slows down the rate of reentrainment (Duncan *et al.*, 1992). Finally, the 5-$HT_{1A,7}$ agonist 8-OH-DPAT blocks light-induced phase shifts of the activity rhythm (Rea *et al.*, 1994) and light-induced c-Fos expression (Glass, Selim, & Rea, 1994; Rea *et al.*, 1994), whereas the 5-HT antagonist NAN-190 increases the phase-shifting effects of light. As $GABA_B$ agonists increase the release of serotonin in the SCN (Francois-Bellan, Hery, Barrit, Faudon, & Fery, 1987; Francois-Bellan, Hery, Faudon, & Fery, 1988; Rea, Barrera, Glass, & Gannon, 1995), GABA may potentiate the serotonergic action and suppress light-induced phase shifts. It has been suggested by Morin (1994) that the large phase shifts that can be observed *in vitro* and not so much *in vivo* may reflect an absence of inhibitory input from, among others, the raphe.

Together, the data indicate that stimulation of the GHT and raphe afferents not only phase shift the circadian pacemaker during the day, but also suppress the light response of the circadian pacemaker during the night. Glutamate and NPY counteract each other; GABA potentiates serotonin; serotonin as well as NPY block light-induced phase shifts.

Postsynaptic Response of SCN Neurons to Light

Electrophysiologic Responses and Light Transduction Pathways

Incoming photic information affects membrane potential and discharge rate in a subpopulation of SCN neurons, which will be discussed in the remainder of this chapter. The light responsiveness of visual SCN cells is not only expressed by electrophysiologic changes; also by a complex cascade of intracellular events. A few elements of the intracellular signal transduction pathway have now been identified that probably lead to transcriptional activation (Gillette, 1997). Light-induced binding of glutamate to different glutamate receptors leads to an increase in intracellular calcium (Tominaga, Guesz, Michel, & Inouye, 1994b; van den Pol, Finkbeiner, & Cornell-Bell, 1992) and to a secondary activation of nitric oxide (NO) (Amir, 1992; Decker & Reus, 1994; Ding *et al.*, 1994; Watanabe, Hamada, Shibata, & Watanabe, 1994; Weber, Gannon, Michel, Gillette, & Rea, 1995b). The cAMP response element-binding protein (CREB) has recently been recognized as an important phase-dependent protein that is activated (phosphorylated to its active form) only at night, when phase shifts occur (Ding, Faiman, Hurst, Kuriashkina, & Gillette, 1997; Ginty *et al.*, 1993).

Newly discovered mouse genes, *mPer1*, *mPer2*, and *mPer3*, have recently been recognized as important components of the circadian pacemaker mechanism.

These genes are homologues of the Drosophila circadian clock gene period (*Per*) (Sun *et al.*, 1997; Tei *et al.*, 1997; Albrecht *et al.*, 1997; Shearman *et al.*, 1997; Takumi *et al.*, 1998a; Zylka *et al.*, 1998; Takumi *et al.*, 1998b; Zeng *et al.*, 1999). In continuous darkness, the three genes are expressed rhythmically and show peak values during subjective day (Albrecht *et al.*, 1997; Takumi *et al.*, 1998b). The peak in *mPer1* expression precedes the peak in *mPer2* by 4 hours. It has been suggested that *mPer1* corresponds with the morning oscillator and *mPer2* with the evening oscillator (Takumi *et al.*, 1998b). *mPer3* shows a broader expression profile, and peaks at mid-subjective day (Takumi *et al.*, 1998b).

There are strong indications that *mPer1* and *mPer2*, but not *mPer3*, are critically involved in phase shifting by light. Light exposure leads to an increment of both *mPer1* and *mPer2* expression, 60 and 90 minutes after light exposure respectively (Albrecht *et al.*, 1997; Shearman *et al.*, 1997; Takumi *et al.*, 1998b). *mPer3* expression is not responsive to light stimulation (Takumi *et al.*, 1998b). Pretreatment of mice with *mPer1* antisense phosphotioate oligonucleotide (ODN) inhibits light-induced phase delays of behavioural activity rhythms at CT16 (Akiyama *et al.*, 1999). Antisense ODN moreover attenuates glutamate induced phase delays in neuronal discharge in SCN slices, indicating a clear role for *mPer* in light-induced phase delays (Akiyama *et al.*, 1999).

The electrophysiologic membrane properties of SCN neurons have been studied by intracellular recordings with sharp microelectrodes or by whole-cell recordings with patch pipettes. These experiments have been performed in brain slices that contain the SCN and in cultured or acutely dissociated SCN neurons. Due to the difficulty of obtaining intracellular recordings from the very small SCN neurons, relatively few intracellular studies with sharp microelectrodes have been published. The action potential and membrane characteristics of SCN neurons were studied for the first time by Wheal and Thomson (1984). This group described resting potentials of −60mV, high input resistances, short time constants, and short duration action potentials.

Patch clamp recordings in acutely prepared slices of the rat hypothalamus have been assessed to determine membrane properties as a function of circadian time (Schaap *et al.*, 1999). Significant differences were observed in spontaneous firing rate, input resistance and frequency adaptation (Schaap *et al.*, 1999). Frequency adaptation is stronger during the day. As a result, reponses to excitatory stimuli are weaker during this time. Input resistance is high during the day and low during the night (de Jeu *et al.*, 1998; Schaap *et al.*, 1999). This is consistent with data from the invertebrate Bulla Gouldiana (Michel *et al.*, 1993; Block *et al.*, 1993). Together with the rhythm in membrane potential, which peaks during the day (de Jeu *et al.*, 1998), it can be predicted that hyperpolarizing channels drive the circadain rhythm in membrane potential.

Evidence was obtained for the existence of different clusters of SCN neurons on the basis of the electrophysiological properties (Schaap *et al.*, 1999; Pennartz *et al.*, 1998). Cluster I neurons are characterized by irregular firing patterns and by a monophasic spike afterpolarization. Cluster II neurons have regular firing patterns and a slow biphasic spike afterpolarization. Cluster III neurons shows biphasic spike afterpolarization and a large low threshold calcium potential (Pennartz *et al.*, 1998). Schaap *et al.* (1999) observed rhythmicity in firing rate and input resistance in Cluster I but not in Cluster II neurones. This raises the possibility that not all neurons of the SCN are pacemaker neurons.

Attempts have been made to get insight into the essential currents which bring the cell to firing threshold and which determine tonic firing (Kim & Dudek, 1993; Thomson & West, 1990). With the patch-clamp technique, the time- and voltage-dependent currents which determine the spontaneous firing rate have been characterized (Walsh, van den Berg, & Rietveld, 1995) and its circadian rhythmicity have been characterized (De Jeu & Pennartz, 1997; Jiang, Yang, Liu, & Allen, 1997; Walsh *et al.*, 1990). Circadian rhythms have been observed both in input conductance and in holding currents in SCN neurons (Jiang *et al.*, 1997). Special attention has been paid to the I_H current, which contributes to spontaneous firing in SCN neurons (Akasu, Shoji, & Hasuo, 1993). This is a hyperpolarization-activated Na^+/K^+ current that is probably coupled to a second messenger. However, the I_H current does not contribute to circadian rhythmicity in spontaneous firing (De Jeu & Pennartz, 1997; Jiang *et al.*, 1997).

In brain slices, one can try to mimic the light stimulus by application of transmitter, or better, by stimulation of the optic nerve. Although this approach introduces a few problems of interpretation, it is functionally reliable in that the light-induced phase response curve can be mimicked by optic nerve stimulations *in vitro* as well as *in vivo* (De Vries *et al.*, 1994; Shibata & Moore, 1993). Intracellular recordings of SCN neurons during optic nerve stimulation have indicated that the majority of responsive neurons depolarize (Jiang *et al.*, 1997; Kim & Dudek, 1991, 1993). These depolarizing responses are also seen when glutamate is applied to cultured SCN cells with patch-clamp techniques (Ito *et al.*, 1991; Meijer *et al.*, 1993). SCN neurons sum excitatory inputs from both optic nerves and receive innervation from at least 4.8 RHT axons per neuron (Jiang *et al.*, 1997). Both NMDA and non-NMDA receptors mediate inward currents under voltage clamp conditions (Jiang *et al.*, 1997; Kim & Dudek, 1991). GABA antagonists block a small subpopulation of cells that responded with outward currents to optic nerve stimulation (Jiang *et al.*, 1997). Inward recification may be related to enhanced excitability and firing rate (Kim & Dudek, 1993).

Extracellular recordings of SCN neurons in brain slices during optic nerve stimulation have shown that most SCN cells respond with an increase, but some with a decrease, in discharge rate (Nishino, Koizumi, & Brooks, 1976; Sawaki, 1979; Shibata & Moore, 1993). To study photic responsiveness of the SCN to a true or natural light stimulus, research on intact animals is required. In and near the SCN, cells have been recorded that appeared responsive to retinal illumination (Aggelopolous & Meissl, 2000; Cui & Dyball, 1996; Groos & Mason, 1980; Groos & Rusak, 1982; Groos & Meijer, 1985; Inouye, 1984; Lincoln, Church & Mason, 1975; Meijer, 1991; Meijer, Gross, & Rusak, 1986; Meijer, Rusak, & Harrington, 1989; Miller, Murakami, & Fuller, 1987; Nishino *et al.*, 1976; Sawaki, 1977). SCN cells that respond to a photic stimulus by a change in discharge rate have been called light-responsive or visual SCN cells.

Light-Responsive SCN Neurons in Nocturnal Species

SCN cells of the rat, hamster, and cat show changes in discharge when light intensity is increased for several minutes or more (Figure 6). In the cat, the proportion of light-responsive cells has not been assessed reliably (Groos & Mason, 1980), but in the rat, 32%, and in the hamster, 38% of SCN neurons respond to light (Meijer *et al.*, 1986). Most light-responsive cells increase discharge in response to light. A smaller number of cells respond to light with a decrease in discharge rate. These cell types have been classified as light-activated and light-suppressed cells, respectively.

Long (and variable) latencies have been observed following retinal stimulation (Groos & Mason, 1980; Meijer *et al.*, 1986). In the rat, latencies of 100 milliseconds to 20 seconds (Groos & Mason, 1980; Meijer *et al.*, 1986) are observed, and in the hamster, latencies of 400 milliseconds to 10 seconds (Meijer *et al.*, 1986). The RHT consists mainly of unmyelinated fibers which are known to have low conduction velocities (Mason, Sparrow, & Lincoln, 1977), and which may contribute in part to the long latencies observed after optic nerve stimulation. Response latencies following retinal illumination far exceed those induced by electrical stimulation of the optic nerve (Sawaki, 1977), suggesting that the long latencies mainly originate in the retina.

Many light-activated cells respond to a stepwise increase in light intensity with a transient "overshoot" in discharge before a somewhat lower sustained discharge level is finally reached (Figure 6). In the hamster and cat, the transient response at light onset is more common than in the rat. In light-suppressed cells, this discharge pattern is not observed (Meijer *et al.*, 1986).

The sustained discharge level is a function of the light intensity presented and has been studied in the rat and hamster. Below 0.1–1 lux, no responses to light are observed. Above these values, light-responsive cells code for light intensity monotonically; increasing light intensities lead to increasing discharge rates in light-activated cells and to decreasing discharge rates in light-suppressed cells. Above a certain light intensity (1000 lux), saturation occurs and a further increase in intensity does not lead to a further change in SCN discharge (Meijer *et al.*, 1986).

When electrophysiologic responses to slowly increasing light intensities are compared with responses to slowly decreasing light intensities, there appears to be a small difference. This phenomenon, which is an example of hysteresis, demonstrates that light responses of the SCN are to some extent history dependent (Meijer *et al.*, 1986).

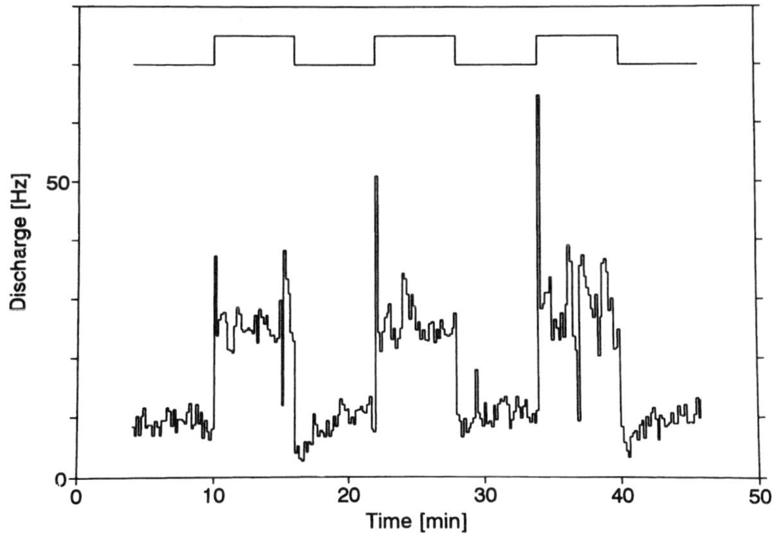

Figure 6. Response of a small number of SCN neurons in a freely moving unanesthetized rat to repeated 6-minute light presentations of 0.15 lux. The timing of the light pulse is indicated above the record. Time is given on the *x* axis and discharge on the *y* axis.

The majority of visually responsive SCN neurons shows maximum sensitivity at around 505 nm (Aggelopoulos & Meissl, 2000). When studied under light adaptation, the action spectrum of SCN cells resembles that of green cone opsin, with one peak at 510 nm, and a secondary peak in the near ultraviolet wavelengths (Aggelopoulos & Meissl, 2000). Long wavelength sensitivity that was observed in some SCN projeting ganglion cells (Pu, 2000) has not yet been found in the SCN.

The receptive field properties of SCN cells have been investigated only in the cat (Groos & Mason, 1980), where most light-responsive cells appear to have large receptive fields (>20° diameter). No antagonistic center–surround organization has been observed, but throughout the receptive field, illumination causes either a tonic activation or tonic suppression of discharge, depending on the cell recorded.

Single-Unit Recordings in a Diurnal Species

The response of SCN cells to retinal illumination has also been recorded in the diurnal 13-lined ground squirrel (*Spermophilus tridecemlineatus* (Meijer *et al.*, 1989). As in the nocturnal species, the responses in the squirrel SCN are mainly of the sustained type. However, the total proportion of light-responsive cells in the squirrel is only 8%, compared with 32% in the rat and 38% in the hamster (Meijer *et al.*, 1986). Although the proportions found in nocturnal animals differ strongly among several studies (20–50%), they have never been as low as 8%. Possible effects of anesthetics used in squirrels cannot be ruled out (Colwell *et al.*, 1993). Of the 8% light-responsive cells, about half are activated by light and the others are light-suppressed. This contrasts with the situation in nocturnal species, where light-activated cells are much more common than light-suppressed cells (Meijer *et al.*, 1986). A second difference from the nocturnal species is that light intensities of more than 1000 lux are required to stimulate light-responsive SCN cells (Meijer *et al.*, 1986).

More recently, light-responsive SCN neurons of the diurnal *Octodon degus* were investigated (Jiao *et al.*, 1999). In contrast to the thirteen lined ground squirrel, the degu tolerates the anaestetic urethane, which is commonly used in studies in nocturnal species. Moreover, the degu was raised in laboratory conditions, unlike the squirrels that were captured in the wild. As a result, the background and recording conditions in the degue experiments are more similar to those of nocturnal animals and differences in response characteristics are interpretable as differences relating to nocturnality versus diurnality.

The study in the degu confirmed a number of light-response characteristics that were observed in the thirteen lined ground squirrel. The majority of light-responsive neurons was suppressed by light (73%) and high light-intensities (of about 1000 lux) were generally required to stimulate these cells. The latter finding is also consistent with the high light intensities that are required for c-Fos expression in the SCN of degus (Krajnak *et al.*, 1997). Inconsistent with squirrel experiments is the proportion of light responsive neurons, being 25% in the degu. This is considerably higher than the proportion of responsive cells in squirrels.

Light-Responsive Cells in the Raphe, IGL, and Pretectum

The location of light-responsive cells corresponds with the terminal field of the RHT in the rat, hamster, and squirrel. Visual cells have also been observed outside the boundaries of the SCN, although here the percentage of responsive units is lower (Meijer *et al.*, 1986, 1989). This is consistent with the observation that the RHT ends not only in the SCN, but also, somewhat less densely, outside the SCN (Johnson, *et al.*, 1988a; Meijer *et al.*, 1989). Despite the close correspondence between the locations of light-responsive SCN cells and RHT terminals, the recorded units may not have been driven by the RHT. Three other input pathways from the raphe, from the vLGN/IGL, and from the pretectum have terminal fields which overlap strongly with the projection area of the RHT (Moga & Moore, 1996, 1997).

Raphe cells of the rat, like SCN cells, are unresponsive to light flashes (Mosko & Jacobs, 1974). Long light presentations (5 minutes), on the other hand, are effective in changing raphe discharge. Of the light-responsive cells, about two thirds respond to whole retinal illumination with an increase in discharge and one third respond to light with a decrease in discharge (Heym, Trulson, & Jacobs, 1982).

In the vLGN of the rat, hamster and cat, cells have variable responses to light (Groos & Rusak, 1982; Hale & Sefton, 1978; Harrington & Rusak, 1989; Spear, Smith, & Williams, 1977; Zhang & Rusak, 1989; for review, see Harrington, 1997). A subpopulation of vLGN cells that respond in a sustained way to maintained illumination levels also have large receptive fields (Spear *et al.*, 1977). It has been demonstrated that these sustained responders are mainly located in the IGL, whereas vLGN units respond mostly phasically to light (Harrington & Rusak, 1989). By a collision test, it was proven that mostly sustained responders project to the SCN (Groos & Rusak, 1982; Zhang & Rusak, 1989). The ratio of sustained on:off responses in the hamster IGL is approximately 5:1 when the contralateral eye is stimulated, which is similar to the ratio found in the hamster SCN (Harrington & Rusak, 1989). Stimulation of the ipsilateral eye produced mainly suppression of discharge. IGL neurons code for luminance with an intensity–response relation resembling that of SCN neurons. The half saturating light intensity is approximately 35 $\mu W/cm^2$.

Pretectum recordings in rats have indicated that about 50% of the neurons respond in a sustained way to light, whereas others respond predominantly phasically to light (Trejo & Cicèrone, 1984). Two thirds of the sustained responders are light-activated and one third are light-suppressed. Pretectum cells are furthermore characterized by large receptive fields. It is unknown which of the pretectum cells project to the SCN.

It is concluded that the response properties of the raphe, IGL, and pretectum correspond in part with those of SCN neurons, and for the IGL it was demonstrated specifically that these corresponding visual neurons also project to the SCN. The question arises of the extent to which these light-responsive brain areas determine the overall light responsiveness of the SCN and the ability to phase shift. When the IGL is deafferented from the retina (Sawaki, 1979) or when the GHT is ablated (Groos & Rusak, 1982), both light-activated and light-suppressed cells can be recorded. In deafferented SCN slices, both activated and suppressed responses can be observed following optic nerve stimulation (Kim & Dudek, 1991; Nishino *et al.*, 1976; Sawaki, 1979). Both phase delays and phase advances can be induced by optic nerve stimulation *in vitro* (Shibata & Moore, 1993). Light-induced phase shifts are almost completely blocked by antagonists of glutamate (e.g., Colwell *et al.*, 1990, Colwell & Menaker, 1992; DeVries *et al.*, 1994). These findings all indicate that the RHT is sufficient for photic entrainment. It is expected that the other light input pathways support this function, in view of their neural wiring and visual properties.

SCN CELLS MONITOR GLOBAL LIGHT INTENSITY

The sustained response pattern is characteristic for SCN cells and is almost absent in areas which are involved in spatial discrimination. Hence, light-responsive SCN cells appear particularly suitable to monitor enduring light intensity in the environment, the more so since these cells have large receptive fields. The resulting ability of light-responsive SCN cells is consistent with their function in entraining the circadian pacemaker. The SCN also controls the circadian pattern of pineal melatonin synthesis by an efferent neuronal connection to the sympathetic nervous system (Reiter, 1980). The temporal profile of melatonin provides the reproductive system with a humoral signal that reflects the environmental photoperiod. The sustained response pattern of SCN neurons allows them to record daylength and annual changes in the length of day. In this respect, the SCN with its characteristic light-responsive neurons is an integral component of the photoperiodic time measurement sysem.

LIGHT-ACTIVATED AND LIGHT-SUPPRESSED CELLS

Cells either increase or decrease their mean discharge rate in response to light. The two distinct response types do not originate in the retina only, because electrical stimulation of the RHT induces both light-activated and light-suppressed responses. Input pathways from the raphe or IGL may contribute to the existence of opposite responses in intact animals, but do not wholly explain them. Even in brain slices in the absence of raphe and IGL, both response types can be encountered following electrical stimulation of the RHT (Nishino *et al.*, 1976; Sawaki, 1979; Shibata, Oomura, Hattori, & Kita, 1984).

The question arises of where these opposing responses originate. One possibility lies in the RHT itself: the locations of light-activated and light-suppressed cells correspond to the terminal field of the RHT. Receptor studies by Güldner (1978) indicated inhibitory and excitatory synapses at the optic terminals in the SCN. However, no inhibitory transmitter has been identified for the RHT. It is also possible that inhibitory interneurons in the SCN mediate light-suppressed responses (Jiang *et al.*, 1997).

LUMINANCE CODING PROPERTIES, MELATONIN SUPPRESSION, AND NOCTURNALITY VERSUS DIURNALITY

The intensity–response relation of light-responsive SCN cells has been described by a sigmoid function with a half saturation value of 70 lux and a working range of 2–3 log units (Meijer *et al.*, 1986). This is a very narrow range as compared to the environmental illuminance range (Remé *et al.*, 1991). The light intensities between threshold and saturation value correspond more or less to light intensities occurring during dawn and dusk. The intensity–response relation of SCN cells, together with their sustained response properties, make light-responsive SCN cells particularly equipped to inform the pacemaker about whether it is day or night.

A qualitatively similar curve has been used to plot the magnitude of a phase shift as a function of light intensity, and is discussed in the next subsection: the phase response curve. The same curve was also used to plot the suppression of melatonin as a function of light intensity. Melatonin production in the pineal gland increases during the night, whereas it is low during the day in both nocturnal and diurnal species of mammals (Reiter, 1985). The nighttime elevation of melatonin can be suppressed by exposing animals to light. The SCN mediates the effects of light on the pineal gland (Moore & Klein, 1974).

The intensity response curve for melatonin suppression is sigmoidal with a small working range and a threshold that is high compared with that for vision (Brainard, Richardson, King, Matthews, & Reiter, 1983; Minneman, Lynuh, & Wurtman, 1974; Nelson & Takahashi, 1991). It is thus very similar to the intensity–response relation of SCN cells. The properties of light-responsive SCN cells could therefore directly explain the responsiveness of the pineal gland to light.

This idea is further substantiated by studies in diurnal animals. In the diurnal 13-lined ground squirrel and degu much higher light intensities appeared necessary to stimulate the SCN (Meijer *et al.*, 1989; Jiao *et al.*, 1999). Melatonin suppression in this and other species of squirrels also occurs only at high light intensities (Reiter, 1985). Although it is tempting to speculate that diurnal animals exhibit an intensity–response relation which is shifted toward higher light intensities, the difference between the intensity–response curves in nocturnal and diurnal species should be carefully interpreted. When Reiter, Steinlecher, Richardson, and King (1983) bred 13-lined ground squirrels in an environment of much lower light intensity, the threshold for melatonin suppression in these animals appeared much lower than in the same species when they were captured in the wild. Hence the intensity–response relation of these squirrels appears to be determined by their lighting history and may be independent of their diurnality.

Electrophysiologic recordings in the rat and hamster SCN showed thresholds to light corresponding to the intensity–response relation for melatonin suppression in rats. Recordings in diurnal squirrels that were captured in the wild showed much higher light thresholds in the SCN, corresponding to the very high threshold for melatonin suppression in squirrels. These results suggest that the variability in light sensitivity does not originate in the pineal gland, but in the SCN or in the photic input pathway to the SCN. It can then be predicted that diurnal mammals (with a normal bright lighting history) exhibit a high threshold for phase shifting as well. Some evidence for this comes from work on tree shrews with split activity rhythms (Meijer *et al.*, 1990). In these animals, light pulses of at least 200 lux were required to phase shift activity components. In humans, light intensities of 180 lux are required to shift the temperature rhythm (Boivin *et al.*, 1996). Although these values are not as high as was initially expected for humans, it is still two to three orders of magnitude higher than in nocturnal species.

The proportion of light-responsive cells in the rat and hamster is about equal, and in both species a majority are light-activated and a minority are light-suppressed. In the diurnal squirrel, the total proportion of light-responsive cells is only 8%, which is primarily due to a strong reduction in light-activated cells. In the diurnal degus, 25% of the neurons are light-responsive and most of these are suppressed by light (Jiao *et al.*, 1999). This finding is correlated with behavioral differences between the squirrel and degu on the one hand and the hamster and rat on the other. In the nocturnal species, free-running period lengthens with increasing light intensities and may result in splitting. In diurnal species, an increase in light intensity does not have a predictable effect on the free-running period, unless, in degus, light intensity is raised to sufficiently high levels (Lee & Labyak, 1997). Moreover, in nocturnal animals, light pulses of several minutes are sufficient to phase shift their rhythm, whereas in diurnal animals, longer light pulses are required (Meijer *et al.*, 1990).

THE INTENSITY OF A LIGHT PULSE AND THE MAGNITUDE OF A PHASE SHIFT. The intensity–response relation of light-responsive SCN cells is characterized by a threshold at 0.1–10 lux and a saturation intensity at 1000 lux. In between these values, SCN cells can code for luminance. The intensity–response curve of visual SCN cells of the hamster has been described by a sigmoid curve (Meijer *et al.*, 1986). It can be fitted by a Michaelis function, $y = x^a/(x^a + b^a)$, with $a = 0.95$ and $b = 70$ lux. This function was obtained by pooling all light-activated SCN cells of the hamster together after their response to light was normalized.

In a behavioral study, the magnitude of phase advances of the circadian activity rhythm of hamsters has been expressed as a function of light intensity (Meijer *et al.*, 1992). This intensity–response curve could be fitted by a Michaelis equation with $a = 0.8$ and $b = 1$ lux. The steepness a of this function is very similar to the one for light-responsive SCN cells. However, the half saturating light intensity b is much lower (Figure 7).

Several factors may account for the difference in light sensitivity between the electrophysiologic and behavioral experiments: 1) In the electrophysiologic experiments, animals were exposed to multiple light pulses in search for light-responsive cells, whereas in the phase-shifting experiment, only one light pulse was presented after 1 week of constant darkness. (2) In the behavioral experiments, animals were anesthetized. (3) The electrophysiologic experiments were performed throughout the circadian cycle, while phase-shifting experiments were performed during subjective night at CT18. During subjective night, sensitivity to light is higher than during the day (see below, p. 45). For these reasons, changes in sensitivity might be expected. The similarity in curvature, on the other hand, is striking (Meijer *et al.*, 1992).

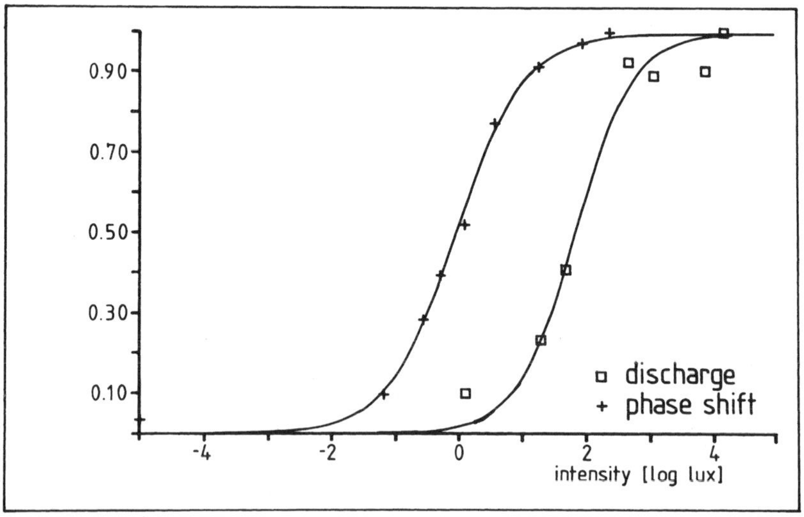

Figure 7. Fitted intensity–response curves for light induced phase shifting (mean values are indicated by squares) and for light induced discharge rate (plus signs). The data have been fitted by $y = x^a/(x^a + b^a)$. For phase shifting, a = 0.8 and $b = 1$ lux. For light-induced discharge, $a = 0.95$ and $b = 70$ lux. (Reprinted with permission from Meijer *et al.*, 1992. Copyright 1992, Elsevier Science.)

THE DURATION OF A LIGHT PULSE AND THE MAGNITUDE OF A PHASE SHIFT. Typical for light-responsive SCN cells is that their response is sustained. This is qualitatively consistent with the effect of pulse duration on the magnitude of a phase shift. Long light pulses trigger more electrical activity in the SCN, which correlates with a larger phase shift. It has been investigated whether the change in SCN discharge is a direct predictor for the magnitude of a phase shift (Meijer *et al.*, 1992). In electrophysiologic experiments, the effect of a series of 1-minute light pulses on SCN discharge was investigated (range: one to eight light pulses). It appeared that succeeding light pulses contributed a similar amount of action potentials. Thus, the effectiveness of light pulses on SCN discharge did not diminish with the number of pulses. In a parallel series of behavioral experiments, the effects of series consisting of either one, two, four, or eight light pulses on the magnitude of a phase shift were determined. It appeared that additional light pulses had a continually decreasing effect on the total phase shift, in other words, each additional light pulse contributed less to the total magnitude of a phase shift than the prior light pulse in the same series (Figure 8). The electrophysiologic experiments described a linear relation between the number of light pulses and the magnitude of a phase shift. The behavioral (phase-shifting) experiments showed, in contrast, that doubling the amount of light pulses did not double the magnitude of a phase shift. Hence, light-induced discharge in the SCN does not predict the magnitude of a phase shift linearly (Meijer *et al.*, 1992).

PHASE-DEPENDENT RESPONSIVENESS OF SCN NEURONS. One of the most distinctive properties of the circadian system is that its responsiveness to light changes as a function of the circadian time. The phase-dependent responsiveness of the pacemaker may lead one to suspect that light-responsive SCN cells respond to light in a phase-dependent way. To study light responsiveness of SCN cells as a function of circadian time, recordings should be performed in freely moving, unanesthetized

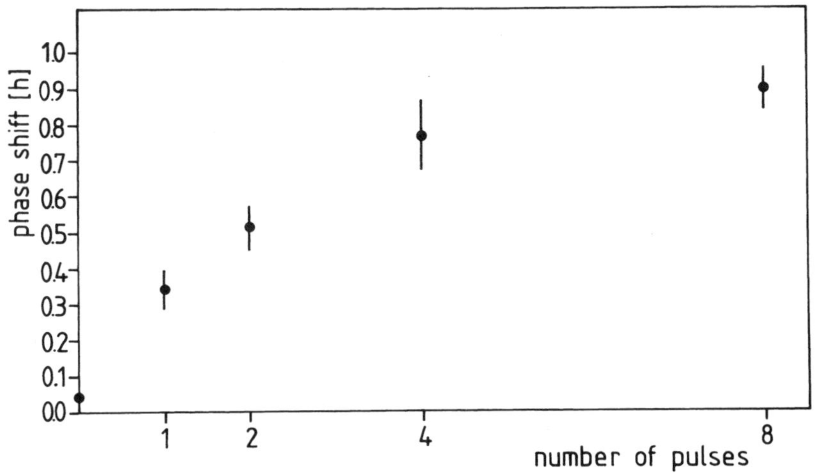

Figure 8. Phase-advancing shifts (± SEM) of the running-wheel activity rhythm as a function of the number of 1-minute light pulses. Mean light intensity of the pulses was 2.7 lux, which is about the half saturating light intensity of the circadian system in the hamster. (Reprinted with permission from Meijer *et al.*, 1992. Copyright 1992, Elsevier Science.)

animals because anesthetics affect the pacemaker's functioning (Colwell *et al.*, 1993). Such a setup allows recordings of firing rates of groups of neurons.

Long-term recordings in freely moving rats and hamsters have indicated that in the night, baseline discharge is low, whereas in the day, it is high (Inouye, 1984; Inouye and Kawamura, 1979; 1982; Meijer, Watanabe, Détári, & Schaap, 1996; Meijer, Schaap, Watanabe, & Albus, 1997; Meijer, Watanabe, Schaap, Albue, & Détari, 1998; Yamazaki *et al.*, 1999). To study light responsiveness as a function of circadian time, light pulses of one fixed intensity have been given hourly throughout the circadian cycle, while the animal's activity rhythm was recorded to confirm that this protocol did not interfere with ongoing rhythmicity (Meijer *et al.*, 1996, 1998). Throughout the circadian cycle, light-activated responses could be recorded. Light-induced changes in SCN discharge appeared high during the night, whereas during the day, only small responses were observed (Figure 9). The difference between light response during the night versus the day was largest when intermediate light intensities were presented, but the difference became smaller when light intensity was increased. This indicates that the difference in light response is a difference in sensitivity to the photic stimulus between day and night (Figure 10). Thus, maximum sensitivity coincides with the time of the cycle when light induces a phase shift. Although light of sufficient intensity induces changes in neuronal discharge also during the day, phase shifts remain absent.

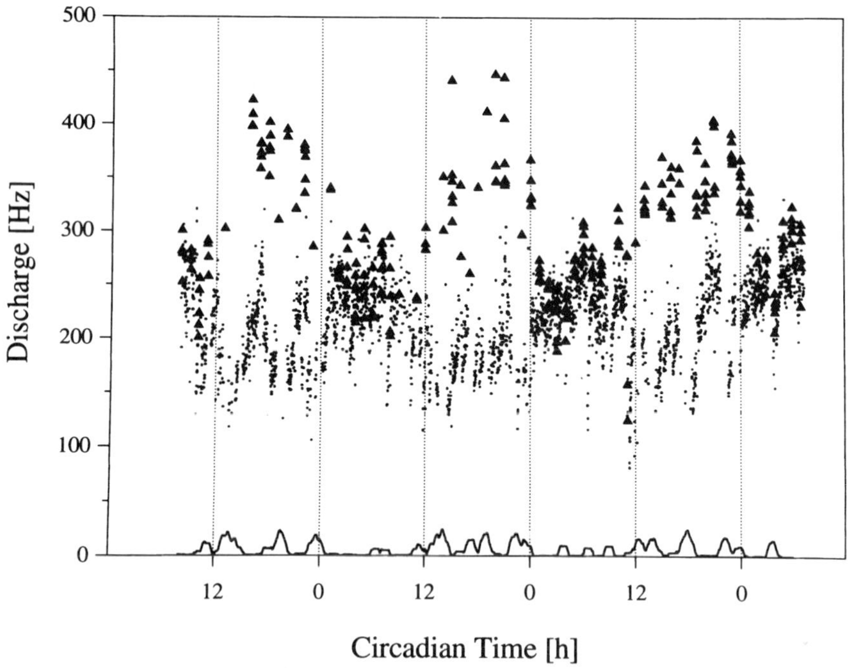

Figure 9. Light response of a population of SCN neurons as a function of circadian time, recorded from a freely moving rat for 72 hours (three cycles). Every hour, a 6-minute light pulse of 0.15 lux was presented. Discharge was counted every 10 seconds and results are given as mean per minute during darkness (circles) and during the light (triangles). Those 10-second bins in which movement artifacts were detected were denied. Movements are depicted at the bottom of the figure. (Reprinted with permission from Meijer *et al.*, 1996. Copyright 1996, Elsevier, Science.)

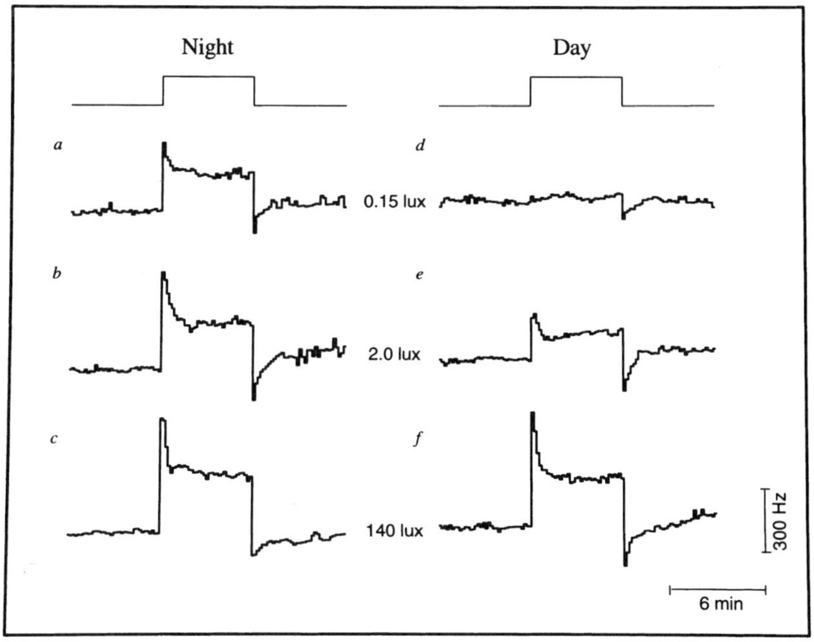

Figure 10. Examples of light-activated responses of a population of SCN neurons during the night and during the day to various 6-minute light pulses of 0.15, 2.0, and 140 lux. (Reprinted with permission from Meijer *et al.*, 1996 Copyright 1996, Elsevier, Science.)

It is not clear whether electrophysiologic responses to light are required for the induction of phase shifts. Changes in membrane potential and discharge rate are accompanied by ionic currents and as a result by an increase in, among others, intracellular calcium. Electrophysiologic membrane proceses therefore could represent the first step in the intracellular transduction cascade involved in photic entrainment.

Conclusion

Clearly, we have some way to go before we fully understand the process of photic entrainment. The progress that has been made in the last two decades nevertheless seems very promising and many parts of the light input pathway have been elucidated. Light is first perceived by the retina. The signal is conducted via one or more photic pigments most likely to tonic W-cells. From here, the signal is transmitted to the SCN via two or three neuronal pathways. Important neurotransmitters of these pathways have been described and their effect on the pacemaker are under investigation. The excitatory transmitter glutamate produces phase shifts during the night, whereas the inhibitory tranmitters of the GHT and raphe produce shifts during the day. SCN neurons respond to the incoming light information in a sustained way and can code for luminance. Their characteristics correlate with the suppression of melatonin, which is mediated by the SCN, and with the effects of light intensity and light duration on the magnitude of a phase shift at a given circadian time.

The sensitivity of SCN neurons to light changes in the course of the circadian cycle, which becomes apparent especially when low light intensities are applied. During the night, sensitivity to light is larger than during the day. As a result, the system is optimized to detect light at those phases where phase shifts should occur in order to entrain to the light–dark cycle. However, when light intensity is raised, light responses can be obtained also during daytime, while phase shifts remain absent. The phase-dependent responsiveness of the pacemaker to light remains a most important and unsolved part of the process of photic entrainment. The search for intracellular components has only recently begun, and calcium, NO, CREB, and *mPer* have been identified as putative mediators of the photic entrainment pathway. At the intracellular level, the process of photic entrainment will probably intermingle with the pacemaker mechanism. A complete understanding of photic entrainment will therefore go hand in hand with understanding the pacemaker mechanism itself.

Acknowledgments

I thank Dr. T. de Boer, Dr. G. A. Kerkhof, Dr. C. E. Remé, Dr. M. H. Rowe, Dr. J. Schaap, and Dr. J. M. Waterhouse for answering some specific questions, and Dr. H. Albus, Dr. M. E. Harrington, Dr. M. J. de Vries, and Dr. M. E. Wise for their valuable comments on the manuscript.

Note

1. The classification into Y-, X-, and W-cells corresponds closely with the morphologic classification into α, β, and γ-cells or into I, II, and III cells, respectively (Boycott and Wässle, 1974; Perry, 1979).

References

Abe, H., & Rusak, B. (1994). Physiological mechanisms regulating photic induction of Fos-like protein in hamster suprachiasmatic nucleus. *Neuroscience and Biobehavioral Reviews, 18,* 531–536.

Abe, H., Rusak, B., & Robertson, H. A. (1991). Photic induction of Fos protein in the suprachiasmatic nucleus is inhibited by the NMDA receptor antagonist MK801. *Neuroscience Letters, 127,* 9–12.

Abe, B., Rusak, B., & Robertson, H. A. (1992). NMDA and non-NMDA receptor antagonists inhibit photic induction of Fos protein in the hamster suprachismatic nucleus. *Brain Research, 28,* 831–835.

Aggelopoulos, N. C. & Meissl, H. (2000). Responses of neurons of the rat suprachiasmatic nucleus to retinal illumination under photopic and scotopic conditions. *Journal of Physiology, 523* Pt 1, 211–222.

Akasu, T., Shoji, S., & Hasuo, H. (1993). Inward rectifier and low-threshold calcium currents contribute to the spontaneous firing mechanism in neurons of the rat suprachiasmatic nucleus. *Pflügers Archiv, 425,* 109–116.

Albers, H. E., & Ferris, C. F. (1984). Neuropeptide Y: Role in light–dark cycle entrainment of hamster circadian rhythms. *Neuroscience Letters, 50,* 163–168.

Albers, H. E., Liou, S. Y., Stopa, E. G., & Zoeller, R. T. (1991). Interaction of colocalized neuropeptides: Functional significance in the circadian timing system. *Journal of Neuroscience, 11*(3), 846–851.

Albrecht, U., Sun, Z. S., Eichele, G., & Lee, C. C. (1997). A differential response of two putative mammalian circadian regulators, mperl and mper2, to light. *Cell, 91,* 1055–1064.

Amir, S. (1992). Blocking NMDA receptors or nitric oxide production disrupts light transmission to the suprachiasmatic nucleus. *Brain Research., 586,* 336–339.

Arimura, A., (1992). Pituitary adenylate cyclase activating polypeptide (PACAP): Discovery and current status of research. *Regulatory Peptides, 37,* 287–303.

Azmitia, E. C., & Segal, M. (1978). An autoradiographic analysis of the differential ascending projections of the dorsal and median raphe nuclei of the rat. *Journal of Comparative Neurology, 179,* 641–668.

Barlow, H. B., & Levick, W. R. (1969). Changes in the maintained discharge with adaptation level in the cat retina. *Journal of Physiology, 202,* 699–718.

Berson, D. M. (1988). Converence of retinal W-cell and corticotectal input of the cat superior colliculus. *Journal of Neurophysiology, 60,* 1861–1873.

Biello, S. M. (1995). Enhanced photic phase shifting after treatment with antiserum to neuropeptide Y. *Brain Research, 673,* 25–29.

Biello, S. M., Golombek, D., & Harrington, M. E. (1996). Interactions between neuropeptide Y and glutamate in the hypothalamic slice preparation. In *Fifth meeting of the Society for Research on Biological Rhythms,* 79.

Biello, S. M., Golombek, D. A., & Harrington, M. E. (1997). Neuropeptide Y and glutamate block each other's phase shifts in the suprachiasmatic nucleus *in vitro. Neuroscience, 77,* 1049–1057.

Block, G. D., Khalsa, S. B. S., McMahon, D. G., Michel, S., & Guesz, M. Biological clocks in the retina: Cellular mechanisms of biological timekeeping. *International Review of Cytology, 146,* 83–143.

Block, M., & Zucker, I. (1976). Circadian rhythms of rat locomotor activity after lesions of the midbrain raphe nuclei. *Journal of Comparative Physiology, 109,* 235–247.

Bloom, F. E., Hoffer, B. J., Siggins, G. R., Barjer, J. L., & Nicoll, R. A. (1972). Effects of serotonin on central neurons: Microiontophoretic administration. *Federation Proceedings, 31,* 97–106.

Boivin, D. B., Duffy, J. F., Kronauer, R. E., & Czeisler Ch. A. (1996). Dose–response relationships for resetting of human circadian clock by light. *Nature, 379,* 540–542.

Bos, N. P. A., & Mirmiran, M. (1993). Effects of excitatory and inhibitory amino acids on neuronal discharges in the cultured suprachismatic nucleus. *Brain Research Bulletin, 31,* 67–72.

Boulos, Z. (1995). Wavelenth dependence of light-induced phase shifts and period changes in hamsters. *Physiology and Behavior 57,* 1025–1033.

Boulos, Z., & Rusak, B. (1982). Circadian phase response curves for dark pulses in the hamster. *Journal of Comparative Physiology, 146,* 411–417.

Boycott, B. B., & Wässle, H. (1974). The morphological types of ganglion cells of the domestic cat's retina. *Journal of Physiology, 240,* 397–419.

Brainard, G. C., Richardson, B. A., King, T. S., Matthews, S. A., & Reiter, R. J. (1983). The suppression of pineal melatonin content and N-acetyltransferase activity by different light irradiances in the Syrian hamster: A dose–response relationship. *Endocrinology, 113,* 293–296.

Brainard, G. C., Richardson, B. A., King, T. S., & Reiter, R. J. (1984). The influence of different light spectra on the suppression of pineal melatonin content in the syrian hamster. *Brain Research, 294,* 333–339.

Brainard, G. C., Podolin, P. L., Leivy, S. W., Rollag, M. D., Cole, C., & Barker, F. M. (1986). Near-ultraviolet radiation suppresses pineal melatonin content. *Endocrinology, 119,* 2201–2205.

Byku, M., & Gannon, R. L. (2000). SNC 80, a delta opioid agonist, elicits phase advances in hamsters circadian activity rhythms. *NeuroReport, 11,* 1449–1452.

Byku, M., Legutko, R., & Gannon, R. L. (2000). Distribution of delta opioid receptor immunoreactivity in the hamster suprachiasmatic nucleus and intergeniculate leaflet. *Brain Research, 857,* 1–7.

Cahill, G. M., & Menaker, M. (1987). Kynurenic acid blocks suprachismatic nucleus responses to optic nerve stimulation. *Brain Research, 410,* 125–129.

Cahill, G. H., & Menaker, M. (1989). Effects of excitatory amino acid receptor antagonists and agonists on suprachiasmatic nucleus to retinohypothalamic tract volleys. *Brain Research, 479,* 76–82.

Campbell, S. S., & Murphy, P. J. (1998). Extraocular circadian phototransduction in humans. *Science, 279,* 396–399.

Card, J. P., & Moore, R. Y. (1984). The suprachiasmatic nucleus of the golden hamster: Immunohistochemical analysis of cell and fiber distribution. *Neuroscience, 13,* 425–431.

Cardalini, D. P., Larin, F., & Wurtman, R. J. (1972). Control of the rat pineal gland by light spectra. *Proceedings of the National Academy of Sciences of the USA, 69,* 2003.

Cassone, V. M., Speh, J. C., Card, J. P., & Moore, R. Y. (1988). Comparative anatomy of the mammalian hypothalamic suprachiasmatic nucleus. *Journal of Biological Rhythms, 3,* 71–91.

Castell, M., Belenky, M., Cohen, S., Ottersen, O. P., & Storm-Mathisen, I. (1993). Glutamate-like immunoreactivity in retinal terminals of the mouse suprachiasmatic nucleus. *European Journal Neuroscience, 5,* 368–381.

Chambille, I., Doyle, S., & Serviere, J. (1993) Photic induction and circadian expression of Fos-like protein. Immunohistochemical study in the retina and suprachiasmatic nuclei of hamster. *Brain Research, 612,* 138–150.

Colwell, C. S., & Menaker, M. (1992). NMDA as well as non-NMDA receptor antagonists can prevent the phase-shifting effects of light on the circadian system of the golden hamster. *Journal of Biological Rhythms, 7,* 125–136.

Colwell, C. S., Ralph, M. R. & Menaker, M. (1990). Do NMDA receptors mediate the effects of light on circadian behavior? *Brain Research, 523,* 117–120.

Colwell, C. S., Foster, R. G., & Menaker, M. (1991). NMDA receptor antagonists block the effects of light on circadian behavior in the mouse. *Brain Research, 554,* 105–110.

Colwell, C. S., Foster, R. G., & Menaker, M. (1993). Light-induced phase shifts and Fos expression in the hamster circadian system: The effects of anesthetics. *Journal of Biological Rhythms, 8*(3), 179–188.

Cooper, H. M., Herbin, M., & Nevo, E. (1993). Occular regression conceals adaptive progression of the visual system in a blind subterranean mammal. *Nature, 361,* 156–159.

Cui, L.-N., & Dyball, R. E. J. (1996). Synaptic input from the retina to the suprachiasmatic nucleus changes with the light–dark cycle in the Syrian hamster. *Journal of Physiology, 492,* 483–493.

Czeisler, C. A., Kronauer, R. E., Allan, S., Duffy, J. F., Jewett, M. E., Brown, E. N., & Ronda, J. M. (1989). Bright light induction of strong (type 0) resetting of the human circadian pacemaker. *Science, 244,* 1328–1333.

Daan, S., & Pittendrigh, C. S. (1976). A functional analysis of circadian pacemaker in nocturnal rodents, II The variability of phase response curves. *Journal of Comparative Physiology, 106,* 253–266.

Daan, S., Pittendrigh, C. S., & Aschoff, J. (2000). The natural entrainment of circadian systems. *Journal of Biological Rhythms, 15,* 195–207.

Dark, J. G., & Asdourian, D. (1975). Entrainment of the rat's activity by cyclic light following lateral geniculate nucleus lesions. *Physiology and Behavior, 15,* 295–301.

Decker, K., & Reuss, S. (1994). Nitric oxide-synthesizing neurons in the hamster suprachiasmatic nucleus: a combined NOS- and NADPH-staining and retinohypothalamic tract tracing study. *Brain Research, 666,* 284–288.

DeCoursey, P. J. (1960). Daily light sensitivity rhythm in a rodent. *Science, 131,* 33–35.

DeCoursey, P. J. (1986). Light-sampling behavior in photoentrainment of a rodent circadian rhythm. *Journal of Comparative Physiology A, 159,* 161–169.

DeCoursey, P. J. (1989). Photoentrainment of circadian rhythms: An ecologist's viewpoint. In T. Hiroshige and K. Honma, (Eds), *Circadian clocks and ecology* (pp. 187–206). Sapporo: Hokkaido University Press.

De Jeu, M. T. G., & Pennartz, C. M. A. (1997) Functional characterization of the H-current in SCN neurons in subjective day and night: A whole-cell patch-clamp study in acutely prepared brain slices. *Brain Research, 767,* 72–80

De Vries, M. J., & De Vries, A. A. (1995). Change in the pattern in circadian running-wheel activity after lesions of the intergeniculate leaflet and ventral lateral geniculate nucleus in the syrian hamster. *Biological Rhythm Research, 26,* 331–341.

De Vries, M. J., Nunes Cardozo, B., van der Want, J., de Wolf, A., & Meijer, J. H. (1993). Glutamate immunoreactivity in terminals of the retinohypothalamic tract of the brown Norwegian rat. *Brain Research, 612,* 231–237.

De Vries, M. J., Treep, J. A., de Pauw, E. S. D., & Meijer, J. H. (1994). The effects of electrical stimulation of the optic nerves and anterior optic chiasm on the circadian activity rhythm of the Syrian hamster: Involvement of excitatory amino acids. *Brain Research, 642,* 206–212.

Ding, J. M., Chen, D., Weber, E. T., Faiman, L. E., Rea, M. A., & Gillette M. U. (1994). Resetting the biological clock: Mediation of nocturnal circadian shifts by glutamate and NO. *Science, 266,* 1713–1717.

Ding, J. M., Faiman, L. E., Hurst, W. J., Kuriashkina, L. R., & Gillette M. U. (1997). Resetting the biological clock: Mediation of nocturnal CREB phosphorylation via light, glutamate, and nitric oxide. *Journal of Neuroscience, 17,* 667–675.

Drennan, M., Kripke, D. F., & Gillin, J. C. (1989). Bright light can delay human temperature rhythm independent of sleep. *American Journal of Physiology. Regulatory, Integrative and Comparative, 11,* R136–R141.

Duncan, W. C., Johnson, K. A., & Wehr, T. A. (1992). Altered sensitivity of the photic entrainment pathway by selected antidepressant drugs. In *Third Meeting of the Society for Research on Biological Rhythms, 3,* p. 62.

Dunn, J., Dyer, R., & Bennett, M. (1972). Diurnal variation in plasma corticosterone following long term exposure to continuous illumination. *Endocrinology, 90,* 1660–1663.

Earnest, W. C., & Turek, F. W. (1985). Neurochemical basis for the photic control of circadian rhythms and seasonal reproductive cycles. *Proceedings of the National Academy of Sciences of the USA, 82,* 4277–4281.

Ebling, F. J. P. (1996). The role of glutamate in the photic regulation of the suprachiasmatic nucleus. *Progress in Neurobiology, 50,* 109–132.

Ebling, F. J. P., Staley, K., Maywood, E. S., Humby, T., Hancock, D. C., Waters, C. M., Evan, G. I., & Hastings, M. H. (1991). The role of NMDA-type glutamatergic neurotransmission in the photic induction of immediate-early gene expression in the suprachiasmatic nuclei of the Syrian hamster. *Journal of Neuroendocrinology, 3*, 641–652.

Ebling, F. J. P., Hui, Y., Mirakhur, A., Maywoord, E. S., & Hastings, M. H. (1993). Photoperiod regulates the LH response to central glutamatergic stimulation in the male Syrian hamster. *Journal of Neuroendocrinology, 5*, 609–618.

Ebling, F. J. P., Mikkelsen, J. D., Grosse, J., Dickerson, J. M., Maywood, E. S., Bittman, E. L., & Hastings, M. H. (1994). Immediate-early gene expression as a marker of neuronal activity in the suprachiasmatic nucleus. In M. Moller & P. Pévet (Eds.), *Advances in Pineal Research*, (Vol. 8, pp. 77–87). London: Libbey.

Edgar, D. M., Miller, J. D., Prossner, R. A., Dean, R. R., & Dement, W. C. (1993). Serotonin and the mammalian circadian system: II. Pase-shifting rat behavioral rhythms with serotonergic agonists. *Journal of Biological Rhythms, 8*, 17–31.

Elliott, J. A., & Tamarkin, L. (1994). Complex circadian regulation of pineal melatonin and wheel-running in Syrian hamsters. *Journal of Comparative Physiology A, 174*, 469–484.

Foote, W. E., Taber-Pierce, E., & Edwards, L. (1978). Evidence for a retina projection to the midbrain of the cat. *Brain Research, 156*, 121–124.

Foster, R. G., Provencio, I., Hudson, D., Fiske, S., DeGrip, W., & Menaker, M. (1991). Circadian photoreception in the retinally degenerate mouse (rd/rd). *Journal of Comparative Physiology, 169*, 39–50.

Francois-Bellan, A. M., Hery, M., Barrit, M. C., Faudon, M., & Fery, F. (1987). The stimulation of $GABA_B$ receptors increases serotonin release in the rat suprachiasmatic area. *Neurochemistry International, 11*, 55–62.

Francois-Bellan, A. M., Hery, M., Faudon, M., & Fery, F. (1988). Evidence for GABA control of serotonin metabolism in the rat suprachiasmatic area. *Neurochemistry International,13*, 455–462.

Freedman, M. S., Lucas, R. J., Soni, B., von Schantz, M., Munoz, M., David Gray, Z., & Foster, R. (1999). Regulation of mammalian circadian behavior by non-rod, non-cone, ocular photoreceptors. *Science, 284*, 502–504.

Gannon, R. L., & Rea, M. A. (1993). Glutamate receptor immunoreactivity in the rat suprachiasmatic nucleus. *Brain Research, 622*, 337–342.

Gannon, R. L., Cato, M. J., Hart, K., Armstrong, D. L., & Rea, M. A.(1995). GABAergic modulation of optic nerve-evoked field potentials in the rat suprachiasmatic nucleus. *Brain Research, 694*, 264–270.

Gillespie, C. F., Mintz, E. M., Marvel, C. L., Huhman, K. L., & Albers, H. E. (1997). $GABA_A$ and $GABA_B$ agonists and antagonists alter the phase-shifting effects of light when microinjected into the suprachiasmatic region. *Brain Research, 759*, 181–189.

Gillette, M. U. (1997). Cellular and biochemical mechanisms underlying circadian rhythms in vertebrates. *Current Opinion in Neurobiology., 7*, 797–804.

Gillette, M. U., Demarco, S. J., Ding, J. M., Gallman, E. A., Faiman, L. E., Liu, C., McArthur, A. J., Medianic, M., Richard, D., Tcheng, T. K., & Weber, E. T. (1993). The organization of the suprachiasmatic circadian pacemaker of the rat and its regulation by neurotransmitters and modulators. *Journal of Biological Rhythms, 8*, 53–58.

Ginty, D. D., Kornhauser, J. M., Thompson, M. A., Bading, H., Mayo, K. E., Takahashi, J. S., & Greenberg, M. E. (1993). Regulation of CREB phosphorylation in the suprachiasmatic nucleus by light and a circadian clock. *Science, 260*, 238–241.

Glass, J. D., Selim, M., & Rea, M. A. (1994). Modulation of light-induced c-Fos expression in the suprachiasmatic nuclei by 5-HT-1A receptor agonists. *Brain Research, 638*, 235–242.

Golombek, D. A., Biello, S. M., Rendon, R. A., & Harrington, M. E. (1996). Neuropeptide Y phase shifts the circadian clock *in vitro* via a Y2 receptor. *NeuroReport, 7*, 1315–1319.

Gribkoff, V. K., Pieschl, R. L., Wisialowski, T. A., van den Pol, A. N., & Yocca, F. D. (1998). Phase shifting of circadian rhythms and depression of neuronal activity in the rat suprachiasmatic nucleus by neuropeptide Y: Mediation by different receptor subtypes. *Journal of Neuroscience, 18*, 3014–3022.

Groos, G. A., & Mason, R. (1980). The visual properties of rat and cat suprachiasmatic neurons. *Journal of Comparative Physiology, 135*, 349–356.

Groos, G. A., & Meijer, J. H. (1985). Effects of illumination on suprachiasmatic nucleus electrical discharge. *Annals of the New York Academy of Sciences, 453*, 134–146.

Groos, G. A., & Rusak, B. (1982). Neurophysiological studies of the ventral lateral geniculate–suprachiasmatic nucleus projection in the rat. *Society for Neuroscience Abstracts, 8*, 543.

Groos, G. A., Mason, R., & Meijer, J. H. (1983). Electrical and pharmacological properties of the suprachiasmatic nuclei. *Federation Proceedings, 42*, 2790-2795.

Güldner, F. H. (1978). Synapses of optic nerve afferents in the rat suprachiasmatic nucleus. II. Structural variability as revealed by morphometric examination. *Cell Tissue Research 194*, 37–54.

Güldner, F. H., & Phillips, J. C. (1985). Structural plasticity of developing optic synapses under different lighting conditions. *Neuroscience Letters, 55,* 225–228.

Guy, J., Bosler, O., Dusticier, G., Pelletier, G., & Galas, A. (1987). Morphological correlates of serotonin–neuropeptide Y interactions in the rat suprachiasmatic nucleus: Combined radioautographic and immunocytochemical data. *Cell Tissue Research, 250,* 657–662.

Hale, P. T., & Sefton, A. J. (1978). A comparison of the visual and electrical response properties of cells in the dorsal and ventral lateral geniculate nuclei. *Brain Research 153,* 591–595.

Hall, A. C., Earle-Cruikshanks, C., & Harrington, M. E. (1999). Role of membrane conductances and protein synthesis in subjective day phase advances of the hamster circadian clock by neuropeptide. Y. *European Journal of Neuroscience, 11,* 3424–3432.

Hannibal, J., Ding, J. M., Chen, D., Fahrenkrug, J., Larsen, P. J., Gillette, M. U., Mikkelsen, J. D. (1997). Pituitary adenylate cyclase-activating peptide (PACAP) in the retinohypothalamic tract: A potential daytime regulator of the biological clock. *Journal of Neuroscience, 17,* 2637–2644.

Harrington, M. E. (1997) The ventral lateral geniculate nucleus and the intergeniculate leaflet: Interrelated structures in the visual and circadian systems. *Neuroscience and Biobehavioral Reviews, 21,* 705–727.

Harrington, M. E., & Hoque, S. (1997). NPY opposes PACAP phase shifts via receptors different from those involved in NPY phase shifts. *NeuroReport, 8,* 2677–2680.

Harrington, M. E., & Rusak, B. (1986). Lesions of the geniculo-hypothalamic tract alters circadian activity rhythms. *Journal of Biological Rhythms, 1,* 309-325.

Harrington, M. E., & Rusak, B. (1989). Photic responses of geniculo-hypothalamic tract neurons in the Syrian hamster. *Visual Neuroscience, 2,* 367–375.

Harrington, M. E., Nance, D. M., & Rusak, B. (1985). Neuropeptide Y immunoreactivity in the hamster circadian rhythms. *Brain Research Bulletin, 15,* 465–472.

Harrington, M. E., Nance, D. M., & Rusak, B. (1987). Double-labeling of neuropeptide Y-immunoreactive neurons which project from the geniculate to the suprachiasmatic nuclei. *Brain Research, 410,* 275–282.

Heym, J., Trulson, M. E., & Jacobs, B. L. (1982). Raphe unit activity in freely moving cats: Effects of phasic auditory and visual stimuli. *Brain Research, 232,* 29–39.

Hickey, T. L., & Spear, P. D. (1976). Retinogeniculate projections in hooded and albino rats: An autoradiographic study. *Experimental Brain Research, 24,* 523–529.

Hoffmann, K. (1969). Circadiane Periodik bei Tupaias (*Tupaia glis*) in konstanten Bedingungen. *Zoologischer Anzeiger, 33* (Supplement), 171–177.

Hoffmann, K. (1971). Splitting the circadian rhythm as a function of light intensity. In M. Memaker (ed.), *Biochronometry* (pp. 143–150). Washington DC: National Academy of Sciences.

Honma, K. I., Watanabe, K., & Heroshige, T. (1979). Effects of parachlorophenylalanine and 5,6 dihydroxytryptamine on the freerunning rhythms of locomotor activity and plasma corticosterone. *Brain Research, 169,* 531–544.

Honma, K. I., Honma, S., & Hiroshige, T. (1985). Response curve, free-running period, and activity time in circadian locomotor rhythm of rats. *Japanese Journal of Physiology, 35,* 643–658.

Horst van der, G. T., Muijtjens, M., Kobayashi, K., Takano, R., Kanno, S., Takao, M., de Wit, J., Verkerk, A., Eker, A. P., van Leenen, D., Buijs, R., Bootsma, D., Hoeijmakers, J. H., & Yasui, A. (1999). Mammalian Cry1 and Cry2 are essential for maintenance of circadian rhythms. *Nature, 398,* 627–630.

Huettner, J. E. (1990). Glutamate receptor channels in rat DRG neurons: Activation by kainate and quisqualate and blockade of desensitization by Con A. *Neuron, 5,* 255–266.

Hut, R. A., Van Oort B. E. H., & Daan, S. (1999). Natural entrainment without dawn and dusk: The case of the European ground squirrel. *Journal of Biological Rhythms, 14,* 290–299.

Inouye, S. T. (1984). Light responsiveness of the suprachiasmatic nucleus with the island with the retinohypothalamic tract spared. *Brain Research, 294,* 263–268.

Inouye, S. I., & Kawamura, H. (1979). Persistence of circadian rhythmicity in a mammalian hypothalamic "island" containing the suprachiasmatic nucleus. *Proceedings of the National Academy of Sciences of the USA, 76,* 5962–5966.

Inouye, S. I., & Kawamura, H. (1982). Characteristics of a circadian pacemaker in the suprachiasmatic nucleus. *Journal of Comparative Physiology, 146,* 153–160.

Ishida, N., Matsui, M., Mitsui, Y., & Mishina, M. (1994). Circadian expression of NMDA receptor mRNAs, E3 and zeta 1 in the suprachiasmatic nucleus of rat brain. *Neuroscience Letters, 166,* 211–215.

Ito, N., Yamatodani, A., Nagai, K., & Wada, H. (1991). Dual effect of glycine on isolated rat suprachiasmatic neurons. *American Journal of Physiology 260* , 213–218.

Jacobs, G. H. (1993). The distribution and nature of color vision among the mammals. *Biological Review, 68,* 413–471.

Jacobs, G. H., Neitz, J., & Deegan, J. F. (1991). Retinal receptors in rodents maximally sensitive to ultraviolet light. *Nature, 353*, 655–656.

Jeu, M. de, Hermes, M., & Pennartz, C. (1998). Circadian modulation of membrane properties in slices of rat suprachiasmatic nucleus. *NeuroReport, 9*, 3725–3729.

Jewett, M. E., Rimmer, D. W., Duffy, J. F., Klerman, E. B., Kronauer, R. E., & Czeisler, A. (1997). Human circadian pacemaker is sensitive to light throughout subjective day without evidence of transients. *American Journal of Physiology, 273*, R1800–R1809.

Jiang, Z. G., & Allen R. A. (1995). Presynaptic inhibition by baclofen of retinohypothalamic excitatory synpatic transmission in rat suprachiasmatic nucleus. *Neuroscience, 64*, 813–819.

Jiang, Z. G., Yang, Y., Liu, Z., & Allen, C. N. (1997). Membrane properties and synaptic inputs of suprachiasmatic nucleus neurons in rat brain slices. *Journal of Physiology, 499*, 141–159.

Jiao, Y., Lee, T. M., & Rusak, B. (1999). Photic responses of suprachiasmatic area neurons in diurnal degus (*Octodon degus*) and nocturnal rats (*Rattus norvegicus*). *Brain Research, 817*, 93–103.

Johnson, M. S. (1939). Effect of continuous light on periodic spontaneous activity of white-footed mice (*Peromyscus*). *Journal of Experimental Zoology, 82*, 315–328.

Johnson, F. J., Morin, L. P., & Moore, R. Y. (1988a). Retinohypothalamic projections in the hamster and rat using a cholera bungarotoxin. *Brain Research, 462*, 301–312.

Johnson, R. F., Moore, R. Y., & Morin, L. P. (1988b). Loss of entrainment and anatomical plasticity after lesions of the hamster retinohypothalamic tract. *Brain Research, 460*, 297–313.

Kam, L. M., & Moberg, G. P. (1977). Effects of raphe lesions on the circadian pattern of wheelrunning in the rat. *Physiology and Behavior, 18*, 213–217.

Kim, Y. I., & Dudek, F. E. (1991). Intracellular electrophysiological study of suprachiasmatic nucleus neurons in rodents: Excitatory synaptic mechanisms. *Journal of Physiology, 444*, 269–287.

Kim, Y. I., & Dudek, F. E. (1993). Membrane properties of rat suprachiasmatic nucleus neurons receiving optic nerve input. *Journal of Physiology, 464*, 229–243.

Klein, D. C., & Weller, J. L. (1970). Indole metabolism in the pineal gland: A circadian rhythm in N-acetyltransferase. *Science, 169*, 1093–1095.

Kohler, M., Kalkowski, A., & Wollnik, F. (2000). Species differences between hamsters and rats with regard to the putative role of serotonin in the circadian system. *Biological Rhythm Research, 31*(3), 340–354.

Kornhauser, J. M., Nelson, D. E., Mayo, K. E., & Takahashi, J. S. (1990). Photic and circadian regulation of c-*fos* gene expression in the hamster suprachiasmatic nucleus. *Neuron, 5*, 127–134.

Kornhauser, J. M., Mayo, K. E., & Takahashi, J. S. (1993). Immediate-early gene expression in a mammalian circadin pacemaker, the suprachiasmatic nucleus. In M. W. Young (Ed.), *Molecular genetics of biological rhythms, cellular clock series* (Vol. 4, pp. 271–307). New York: Dekker.

Lee, T. M., & Labyak, S. E. (1997). Free-running rhythms and light- and dark-pulse phase response curves for diurnal *Octodon degus* (rodentia). *American Journal of Physiology, 273*, R278–R286.

Lees, J. G., Hallonquist, J. D., & Mrosovsky, N. (1983). Differential effects of dark pulses on the two components of split circadian activity rhythms in golden hamsters. *Journal of Comparative Physiology, 153*, 123–132.

Levine, J. D., Rosenwasser, A. M., Yanovski, J. A., & Adler, N. T. (1986). Circadian activity rhythms in rats with midbrain raphe lesions. *Brain Research, 384*, 240–249.

Levine, J. D., Weiss, M. L., Rosenwasser, A. M., & Miselis, R. R. (1991). Retinohypothalamic tract in the female albino rat: A study using horseradish peroxydase conjugated to cholera toxin. *Journal of Comparative Neurology, 306*, 344–360.

Lewy, A. J., Wehr, T. A., Goodwin, F. K., Newsome, D. A., & Markey, S. P. (1980). Light suppresses melatonin secretion in humans. *Science, 210*, 1267–1269.

Lincoln, D. W., Church, J., & Mason, C. A. (1975). Electrophysiological activation of suprachiasmatic neurones by changes in retinal illumination. *Acta Endocrinologica Supplementum, 199*, 184.

Liou, S. Y., & Albers, H. E. (1991). Single unit response of neurons within the hamster suprachiasmatic nucleus to neuropeptide Y in the hypothalamic slice preparation. *Brain Research Bulletin., 27*, 825–828.

Liou, S. Y., Shibata, S., & Ueki, S. (1986a). Effects of monoamines on field potentials in the suprachiasmatic nucleus of slices of hypothalamus of the rat evoked by stimulation of the optic nerve. *Neuropharmacology, 15*, 1009–1014.

Liou, S. Y., Shibata, S., & Ueki, S. (1986b). Optic nerve stimulation-induced increase of release of ^{3}H-glutamate and ^{3}H-asparate but not ^{3}H-GABA from the suprachiasmatic nucleus in slices of rat hypothalamus. *Brain Research Bulletin, 16*, 527–531.

Liou, S. Y., Minamoto, Y., Shibata, S., Shiratsuchi, A., Miyamoto, K., & Ueki, S. (1986c). The effects of raphe lesions on the neuronal activity of the suprachiasmatic nucleus in rat hypothalamic slice preparation. *Japanese Journal of Pharmacology, 40*, 188P.

Liou, S. Y., Shibata, S., Albers, H. E., & Ueki, S. (1990). Effects of GABA and anxiolytics on the single unit discharge of suprachiasmatic neurons in the rat hypothalamic slices. *Brain Research Bulletin, 25,* 103–107.

Lovenberg, T. W., Baron, B. M., De Lecea, L., Miller, J. D., Prosser, R. A., Rea, M. A., Foye, P. E., Racke, M., Slone, A. L., Siegel, B. W., Danielson, P. E., Sutcliffe, J. G., & Erlander, M. G. (1993). A novel adenylate cyclase-activating serotonin receptor (5-HT_7) implicated in the regulation of mammalian circadian rhythms. *Neuron, 11,* 449–458.

Marchant, E. G., & Mistlberger, R. E. (1995). Morphine phase-shifts circadian rhythms in mice: Role of behavioural activation. *NeuroReport, 7,* 209–212.

Mason, C. A., Sparrow, N., & Lincoln, D. W. (1977). Structural features of the retinohypothalamic projection in the rat during normal development. *Brain Research, 132* 141–148.

Mason, R. (1986). Circadian variation in sensitivity of suprachiasmatic and lateral geniculate neurons to 5-hydroxytriptamine in the rat. *Journal of Physiology, 377,* 1–13.

Mason, R., & Rusak, B. (1991). NMDA-evoked responses in the Syrian hamster suprachiasmatic nucleus *in vitro. Journal of Physiology, 435,* 39.

Mason, R., Biello, S. M., & Harrington, M. E. (1991). The effects of GABA and benzodiazepines on neurons in the suprachiasmatic nucleus (SCN) of Syrian hamsters. *Brain Research, 525,* 53–57.

Mason, R., Piggins, H., & Rusak, B. (1993). The effect of the putative retinohypothalamic neurotransmitter NAAG (N-acetylaspartylglutamate) on rat and hamster suprachiasmatic neurons *in vitro. Journal of Physiology, 459,* 482.

Mayer, M. L., Westbrook, G. L., & Guthrie, P. B. (1984). Voltage-dependent block by Mg^{2+} of NMDA responses in spinal cord neurones. *Nature, 309,* 261–263.

McCormack, C. E., & Sontag, C. R. (1980). Entrainment by red light of running activity and ovulation rhythms of rats. *American Journal of Physiology, 239,* R450–R454.

Medanic, M., & Gillette, M. U. (1992). Serotonin regulates the phase of the rat suprachiasmatic circadian pacemaker *in vitro* only during the subjective day. *Journal of Physiology, 450,* 629–642.

Medanic, M., & Gillette, M. U. (1993). Suprachiasmatic circadian pacemaker of rat shows two windows of sensitivity to neuropeptide Y *in vitro. Brain Research, 620,* 281–286.

Meeker, R. B., Greenwood, R. S., & Hayward, J. N. (1994). Glutamate receptors in the rat hypothalamus and pituitary. *Endocrinology, 134,* 621–629.

Meijer, J. H. (1991). Integration of visual information by the suprachiasmatic nucleus. In D. C. Klein, R. Y. Moore & S. M. Reppert (Eds.), *The suprachiasmatic nucleus: The mind's clock* (pp. 107–119). New York: Oxford University Press.

Meijer, J. H., & Groos, G. A. (1988). Responsiveness of suprachiasmatic and ventral lateral geniculate neurons to serotonin and imipramine: A micro-iontophoretic study in normal and imipramine-treated rats. *Brain Research Bulletin, 20,* 89–96.

Meijer, J. H., & Rietveld, W. J. (1989). Neurophysiology of the suprachiasmatic circadian pacemaker in rodents. *Physiological Review, 69,* 671–707.

Meijer, J. H., & De Vries, M. J. (1995). Light induced phase shifts in onset and offset of running-wheel activity in the Syrian hamster. *Journal of Biological Rhythms, 10,* 4–16.

Meijer, J. H., Groos, G. A., & Rusak, B. (1986). Luminance coding in a circadian pacemaker: The suprachiasmatic nucleus of the rat and the hamster. *Brain Research, 382,* 109–118.

Meijer, J. H. , van der Zee, E.A., & Dietz, M. (1988a). The effects of intraventricular carbachol injections on the free running activity rhythm of the hamster. *Journal of Biological Rhythms, 3,* 333–348.

Meijer, J. H., van der Zee, E. A., & Dietz, M. (1988b). Glutamate phase shifts circadian activity rhythms in hamsters. *Neuroscience Letters, 86,* 177–183.

Meijer, J. H., Rusak, B., & Harrington, M. E. (1989). Photically responsive neurons in the hypothalamus of a diurnal ground squirrel. *Brain Research, 501,* 315–323.

Meijer, J. H., Daan, S., Overkamp, G. J. F., & Hermann, P. M. (1990). The two-oscillator circadian system of tree shrews (*Tupaia belangeri*) and its response to light and dark pulses. *Journal of Biological Rhythms, 5,* 1–16.

Meijer, J. H., Ruis, A., Albus, H., & Dahan, A. (1999). Functional absence of extraocular photoreception in rodents. *Brain Research, 831,* 337–339.

Meijer, J. H. Ruijs, A. C. J., Albus, H., van der Geest, B., Duindam, H., Zwinderman, A. H., & Dahan, A. (2000). Fentanyl, a mu-opioid receptor agonist, phase shifts the hamster circadian pacemaker. *Brain Research, 868,* 135–140.

Meijer, J. H., Rusak, B., & Gänshirt, G. (1992). The relation between light-induced discharge in the suprachiasmatic nucleus and phase shifts of hamster circadian rhythms. *Brain Research, 598,* 257–288.

Meijer, J. H., Albus, H., Weidema, F., & Ravesloot, J. H. (1993). The effects of glutamate on membrane potential and discharge rate of suprachiasmatic neurons. *Brain Research, 603,* 284–288.

Meijer, J. H., Watanabe, K., Détári, L., & Schaap, J. (1996). Circadian rhythm in light response in suprachiasmatic nucleus neurons of freely moving rats. *Brain Research, 741,* 352–355.

Meijer, J. H., Schaap, J., Watanabe, K., & Albus, H. (1997). Multiunit activity recordings in the suprachiasmatic nuclei: *In vivo* versus *in vitro* models. *Brain Research, 753,* 322–327.

Meijer, J. H., Watanabe, K., Schaap, J., Albus, H., & Détári, L. (1998). Light responsiveness of the suprachiasmatic nucleus: Long-term multiunit and single-unit recordings in freely moving rats. *Journal of Neuroscience, 18,* 9078–9087.

Melo, L., Golombek, D. A., Ralph, M. R. (1997). Regulation of circadian photic responses by nitric oxide. *Journal of Biological Rhythms, 12,* 319–326.

Menaker, M. (1982). The search for principles of physiological organization in verbrate circadian systems. In J. Aschoff, S. Daan, & G. A. Groos (Eds.), *Vertebrate circadian systems* (pp. 1–12). Berlin: Springer-Verlag.

Michel, S., Geusz, M. E., Zaritsky, J. J., & block. G. D. (1993). Circadian rhythm in membrane conductance expressed in isolated neurons. *Science, 259,* 239–241.

Mick, G., Yoshimura, R., Ohno, K., Kiyama, H., & Tohyama, M. (1995). The messenger RNAs encoding metabotropic glutamate receptor subtypes are expressed in different neuronal subpopulations of the rat suprachiasmatic nucleus. *Neuroscience, 66,* 161–173.

Mikkelsen, J. D., & Larsen, P. J. (1993). Sustance P in the suprachiasmatic nucleus of the rat: An immunihistochemical and *in situ* hybridization study. *Histochemistry, 100,* 3–16.

Mikkelsen, J. D., & Vrang, N. (1994). A direct pretectosuprachiasmatic projection in the rat. *Neuroscience, 62,* 497–505.

Mikkelsen, J. D., Larsen, P. J., & Ebling, F. J. P. (1993). Distribution of N-methyl D-aspartate (NMDA) receptor mRNAs in the rat suprachiasmatic nucleus. *Brain Research, 632,* 329–333.

Millar, A. J. (2000). Clock proteins: Turned over after hours? *Current Biology, 10,* R529–R531.

Miller, J. D., Murakami, D. M., & Fuller, C. A. J. (1987). The response of suprachiasmatic neurons of the rat hypothalamus to photic and nicotinic stimuli. *Neuroscience, 7,* 978–986.

Minneman, K. P., Lynuh, H. J., & Wurtman, R. J. (1974). Relationship between environmental light intensity and retina-mediated suppression of rat pineal serotonin-N-acetyl-transferase. *Life Science, 15,* 1791–1796.

Minors, D. S., Waterhouse, J. M., & Wirz-Justice, A. (1991). A human phase-response curve to light. *Neuroscience Letters, 133,* 36–40.

Mintz, E. M., & Albers, H. E. (1997). Microinjection of NMDA into the SCN region mimics the phase shifting effect of light hamsters. *Brain Research, 758,* 245–249.

Miyamoto, Y., & Sancar, A. (1998). Vitamin B2-based blue light photoreceptors in the retinohypothalamic tract as the photoactive pigments for setting the circadian clock in mammals. *Proceedings of the National Academy of Sciences of the USA, 95,* 6097–6102.

Moffett, J. R., Williamson, L., Palkovits, M., & Namboodiri, M. A. A. (1990). N-Acetylaspartylglutamate: A transmitter candidate for the retinohypothalamic tract. *Proceedings of the National Academy of Sciences of the USA, 87,* 8065–8069.

Moga, M. M., & Moore, R. Y. (1996). Putative excitatory amino acid projections to the rat suprachiasmatic nucleus in the rat. *Brain Research 743,* 171–177

Moga, M. M., & Moore, R. Y. (1997). Organization of neural inputs to the suprachiasmatic nucleus in the rat. *Journal of Comparative Neurology, 389,* 508–534.

Moore, R. Y. (1973). Retinohypothalamic projection in mammals: A comparative study. *Brain Research, 49,* 403–409.

Moore, R. Y., & Klein, D. C. (1974). Visual pathways and the central neural control of a circadian rhythm in pineal serotonin N-acetyl transferase activity. *Brain Research, 71,* 17–33.

Moore, R. Y., & Lenn, N. J. (1972). A retinohypothlamic tract in the rat. *Journal of Comparative Neurology, 146,* 1–14.

Moore, R. Y., & Speh, J. (1993). GABA is the principal neurotransmitter of the circadian system. *Neuroscience Letters, 150,* 112–116.

Moore, R. Y., Halaris, A. E., & Jones, B. A. (1978). Serotonin neurons of the midbrain raphe: Ascending projections. *Journal of Comparative Neurology, 180,* 417–438.

Moore, R. Y., Speh, J. C., & Card, J. P. (1995). The retinohypothalamic tract originates from a distinct subset of retinal ganglion cells. *Journal of Comparative Neurology, 352,* 351–366.

Morin, L. P. (1994). The circadian visual system. *Brain Research Reviews, 67,* 102–127.

Morin, L. P., & Blanchard, J. (1991). Depletion of brain serotonin by 5,7-DHT modifies hamster circadian rhythm response to light. *Brian Research, 566,* 173–185.

Morin, L. P., Blanchard, J., & Moore, R. Y. (1992). Intergeniculate leaflet and suprachiasmatic nucleus organisation and connections in the golden hamster. *Visual Neuroscience, 8,* 219–230.

Mosko, S. S., & Jacobs, B. L. (1974). Midbrain raphe neurons: Spontaneous activity and response to light. *Physiology Behavior, 13,* 589–593.

Nelson, D. E., & Takahashi, J. S. (1991). Sensitivity and integration in a visual pathway for circadian entrainment in the hamster (*Mesocricetus auratus*). *Journal of Physiology, 439,* 115–145.

Nicoll, R. A., Malenka, R. C., & Kauer, J. A. (1990). Functional comparison of neurotransmitter receptor subtypes in mammalian central nervous system. *Physiological Review, 70,* 513.

Nishino, H., & Koizumi, K. (1977). Responses of neurons in the suprachiasmatic nuclei of the hypothalamus to putative transmitters. *Brain Research, 120,* 167–172.

Nishino, H., Koizumi, K., & Brooks, C. M. (1976). The role of the suprachiasmatic nuclei of the hypothalamus in the production of circadian rhythm. *Brain Research, 112,* 45–59.

Pennartz, C. M. A., De Jeu, M. T. G., Geurtsen, A. M. S., Sluiter, A. A., & Hermes, M. L. H. J. (1998). Electrophysiological and morphological heterogeneity of neurons in slices of rat suprachiasmatic nucleus. *Journal of Physiology, 506,* 775–793.

Pennartz, C. M. A., Bierlaagh, M. A., & Geurtsen, A. M. S. (1997). Cellular mechanisms underlying spontaneous firing in rat suprachiasmatic nucleus: Involvement of a slowly inactivating component of sodium current. *Journal of Neurophysiology, 78,* 1811–1825.

Perry, V. H. (1979). The ganglion cell layer of the retina of the rat: A Golgi study. *Proceedings of the Royal Society B, 204,* 363–375.

Pickard, G. A. (1982). The afferent connections of the suprachiasmatic nucleus of the golden hamster with emphasis on the retinohypothalamic tract. *Journal of Comparative Neurology, 1982,* 65–83.

Pickard, G. E. (1985). Bifurcating axons of retinal ganglion cells terminate in the hypothalamic suprachiasmatic nucleus and the intergeniculate leaflet of the thalamus. *Neuroscience Letters, 55,* 211–217.

Pickard, G. E., Ralph, M. R., & Menaker, M. (1987). The intergeniculate leaflet partially mediates effects of light on circadian rhythms. *Journal of Biological Rhythms, 2,* 35–56.

Piggins, H. D., & Rusak, B. (1997). Effects of microinjections of substance P into the suprachiasmatic nucleus region on hamster wheel-running rhythms. *Brain Research Bulletin, 42,* 451–455.

Piggins, H. D., Stamp, J. A., Burns, J., Rusak, B., & Semba, K. (1996). Distribution of pituitary adenylate cyclase activating polypeptide (PACAP) immunoreactivity in the hypothalamus and extended amygdala of the rat. *Journal of Comparative Neurology, 376,* 278–294.

Pittendrigh, C. S. (1960). Circadian rhythms and the circadian organization of living systems. *Cold Spring Harbor Symposia on Quantitative Biology, 25,* 159–184.

Pittendrigh, C. S., & Daan, S. (1976a). A functional analysis of circadian pacemakers in nocturnal rodents. IV. Entrainment: Pacemaker as clock. *Journal of Comparative Physiology, 106,* 291–331.

Pittendrigh, C. S., & Daan, S. (1976b). A functional analysis of circadian pacemakers in nocturnal rodents V. Pacemaker structure: A clock for all seasons. *Journal of Comparative Physiology, 106,* 333–355.

Podolin, P. L., Rollag, M. D., & Brainard, G. C. (1987). The suppression of nocturnal pineal melatonin in the syrian hamster: Dose–response curves at 500 and 360 nm. *Endocrinology, 121,* 266–270.

Pohl, H. (1972). Die Aktivitätsperiodik von zwei tagakiven nagern—*Funambulus palmarum* und *Eutamias sibiricus* unter dauerlicht Bedingungen. *Journal of Comparative Physiology, 78,* 60–74.

Pohl, H. (1982). Characteristics and variability in entrainment of circadian rhythms to light in diurnal rodents. In A. Aschoff, S. Daan, & G. A. Groos (Eds.), *Vertebate circadian systems* (pp. 339–346). Berlin: Springer-Verlag.

Prosser, R. A. (2000). Serotonergic actions and interactions on the SCN circadian pacemaker: In vitro investigations. *Biological Rhythm Research, 31*(3), 315–339.

Prosser, R. A., Miller, J. D., & Heller, H. C. (1990). A serotonergic agonist phase-shifts the circadian clock in the suprachiasmatic nuclei *in vitro. Brain Research, 534,* 336–339.

Prosser, R. A., Heller, H. C., & Miller, J. D. (1992). Serotonergic phase shifts of the mammalian circadian clock: Effects of tetrodotoxin and high Mg+. *Brain Research, 573,* 336–340.

Prosser, R. A., Dean, R. R., Edgar, D. M., Heller, H. C., & Miller, J. D. (1993). Serotonin and the mammalian circadian system: I. *In vitro* phase shifts by serotonergic agonists and antagonists. *Journal of Biological Rhythms, 8,* 1–16.

Pu, M. (1999). Dendritic morphology of cat retinal ganglion cells projecting to suprachiasmatic nucleus. *Journal of Comparative Neuroology, 414,* 267–274.

Pu, M. (2000). Physiological response properties of cat retinal ganglion cells projecting to suprachiasmatic nucleus. *Journal of Biological Rhythms, 15,* 31–36.

Ralph, M. R., & Menaker, M. (1985). Bicuculline blocks circadian phase delays but not advances. *Brain Research, 325,* 362–365.

Ralph, M. R., & Menaker, M. (1986). Effects of diazepam on circadian phase advances and delays. *Brain Research, 372,* 405–408.

Ralph, M. R., & Menaker, M. (1988). A mutation of the circadian system in golden hamsters. *Science, 247,* 1225–1227.

Ralph, M., & Menaker, M. (1989) GABA regulation of circadian responses to light I. Involvement of $GABA_A$-benzodiazepine and $GABA_B$ receptors. *Journal of Neuroscience, 9,* 2858–2865.

Rea, M. A., Buckley, B., & Lutton, L. M. (1993a). Local administration of EAA antogonists blocks light-induced phase shifts and c-*fos* expression in hamster SCN. *American Journal of Physiology, 265,* R1191–R1198.

Rea, M. A., Michel, A. M. & Lutton, L. M. (1993b). Is *fos* expression necessary and sufficient to mediate light-induced phase advances of the suprachiasmatic circadian oscillator? *Journal of Biological Rhythms, 8,* S59–S64.

Rea, M. A., Glass, J. D., & Colwell, C. S. (1994). Serotonin modulates photic responses in the hamster suprachiasmatic nuclei. *Journal of Neuroscience, 14,* 3635–3642.

Rea, M. A., Barrera, J., Glass, J. D., & Gannon, R. L. (1995). Serotonergic potentiation of photic phase shifts of the circadian activity rhythm. *NeuroReport, 6,* 1289–1292.

Rea, M. A., & Pickard, G. E. (2000). Serotonergic modulation of photic entrainment in the Syrian hamster. *Biological Rhythm Research, 31*(3), 284–314.

Reiter, R. J. (1980). The pineal and its hormones in the control of reproduction in mammals. *Endocrine Reviews, 2,* 109–131.

Reiter, R. J. (1985) Action spectra, dose response relationships and temporal aspects of light's effects on the pineal gland. *Annals of the New York Academy of Sciences, 453,* 215–230.

Reiter, R. J., Steinlecher, B. A., Richardson, B. A., & King, T. S. (1983). Differential response to pineal melatonin levels to light at night in laboratory raised and wild-captured 13-lined ground squirrels *Spermophilus tridecemlineatus. Life Science, 32,* 2625–2629.

Remé, C. E., Wirz-Justice, A., & Terman, M. (1991). The visual input stage of the mammalian circadian pacemaking system: I. Is there a clock in the mammalian eye? *Journal of Biological Rhythms, 6,* 1–29.

Ribak, C. E., & Peters, A. (1975). An autoradiographic study of the projections from the lateral geniculate body of the rat. *Brain Research, 92,* 341–368.

Rowe, M. H., & Stone, J. (1977). Naming of neurons. Classification and naming of cat retinal ganglion cells. *Brain, Behavior and Evolution, 14,* 185–216.

Rusak, B. (1977). The role of the suprachiasmatic nuclei in the generation of the circadian rhythms in the golden hamster *Mesocricetus auratus. Journal of Comparative Physiology, 118,* 145–164.

Rusak, B., & Zucker, I. (1979). Neural regulation of circadian rhythms. *Physiological Review, 59,* 449–526.

Rusak, B., Meijer, J. H., & Harrington, M. E. (1989). Hamster circadian rhythms are phaseshifted by electrical stimulation of the geniculo-hypothalamic tract. *Brain Research, 493,* 283–291.

Rusak, B., Robertson, H. A., Wisden, W., & Hunt, S. P. (1990). Light pulses that shift rhythms induce gene expression in the suprachiasmatic nucleus. *Science, 248,* 1237–1240.

Sadun, A., Schaechter, J. D., & Smith, L. E. H. (1984). A retinohypothalamic pathway in man: Light mediation of circadian rhythms. *Brain Research, 302,* 371–377.

Sawaki, Y. (1977). Retinohypothalamic projection: Electrophysiological evidence for the existence in female rats. *Brain Research, 120,* 336–341.

Sawaki, Y. (1979). Suprachiasmatic nucleus neurones, excitation and inhibition mediated by the direct retino hypothalamic projection in female rats. *Experimental Brain Research, 37,* 127–138.

Schaap, J., Bos, N. P. A., de Jeu, M. T. G., Geurtsen, A. M. S., Meijer, J. H. & Pennartz, C. M. A., (1998). Neurons of the rat suprachiasmatic nucleus show a cercadian rhythm in membrane properties that is lost during prologed whole-cell recording. *Brain Research, 19018,* 154–166.

Schmahl, C., & Bohmer, G. (1997). Effects of excitatory amino acids and neuropeptide Y on the discharge activity of suprachiasmatic neurons in rat brain slices. *Brain Research, 746,* 151–163.

Sekiguchi, M., Wada, K., & Wenthold, R. J. (1992). N-acetylaspartylglutamate acts as an agonist upon homomeric NMDA receptor (NMDARI) expressed in *Xenopus* oocytes. *FEBS Letters, 311,* 285–289.

Selim, M., Glass, J. D., Hauser, U. E., & Rea, M. A. (1993). Serotonergic inhibition of light-induced *fos* protein expression and extracellular glutamate in the suprachiasmatic nucleus. *Brain Research, 621,* 181–188.

Shearman, L. P., Zylka, M. J., Weaver, D. R., Kolakowski, L. F. Jr., & Reppert, S. M. (1997). Two period homologs: Circadian expression and photic regulation in the supraschiasmatic nuclei. *Neuron, 19,* 1261–1269.

Shen, H., & Semba, K. (1994). A direct retinal projection to the dorsal raphe nucleus in the rat. *Brain Research, 635,* 159–168.

Shibata, S., & Moore, R. Y. (1988). Neuropeptide Y and vasopressin effects on rat suprachiasmatic nucleus neurons *in vitro. Brain Research, 3,* 265–276.

Shibata, S., & Moore, R. Y. (1993). Neuropeptide Y and optic chiasm stimulation affect suprachiasmatic nucleus circadian function *in vitro. Brain Research, 615*, 95–100.

Shibata, S., Liou, S. Y., & Ueki, S. (1983). Different effects of amino acids, acetylcholine and monoamines on neuronal activity of suprachiasmatic nucleus in rat pups and adults. *Neuroscience Letters, 39*, 187–192.

Shibata, S., Oomura, Y., Hattori, K., & Kita, H. (1984). Responses of suprachiasmatic nucleus to optic nerve stimulation in rat hypothalamic slice preparation. *Brain Research, 302*, 83–89.

Shibata, S., Liou, S. Y., & Ueki, S. (1986). Influence of excitatory amino acid receptor antagonists and of baclofen on synaptic transmission in the optic nerve to the suprachiasmatic nucleus in slices of rat hypothalamus. *Neuropharmacology, 25*, 403–409.

Shibata, S., Tsuneyoshi, A., Hamada, T., Tominaga, K., & Watanabe, S. (1992a). Effect of substance P on circadian rhythms of firing activity and the 2-deoxyglucose uptake in the rat suprachiasmatic nucleus. *Brain Research, 597*, 257–263.

Shibata, S., Tsuneyoshi, A., Hamada, T., Tominaga, K., & Watanabe, S. (1992b). Phase resetting effect of 8-OH-DPAT, a serotonin la receptor antagonist, on the circadian rhythm of firing rate in the rat suprachiasmatic nucleus *in vitro. Brain Research, 582*, 353–356.

Shibuya, C. A., Melnyk, R. B., & Mrosovsky, N. (1980). Simultaneous splitting of drinking and locomotor activity in a golden hamster. *Naturwissenschaften, 67*, 45–47.

Shirakawa, T., & Moore, R. Y. (1994a). Glutamate shifts the phase of the circadian neuronal firing rhythm in the rat suprachiasmatic nucleus *in vitro. Neuroscience Letters, 178*, 47–50.

Shirakawa, T., & Moore, R. Y. (1994b). Responses of rat suprachiasmatic nucleus neurons to substance P and glutamate *in vitro. Brain Research, 642*, 213–220.

Smith, R. D., Turek, F. W., & Slater, N. T. (1990). Bicuculline and picrotoxin block phase advances induced by GABA agonists in the circadian rhythm of locomotor activity in the golden hamster by a phaclofen-intensive mechanism. *Brain Research, 530*, 275–282.

Spear, P. D., Smith, D. C., & Williams, L. L. (1977) Visual receptive-field properties of single neuron in cat's ventral lateral geniculate nucleus. *Journal of Neurophysiology, 40*, 390–409.

Speh, I. C., & Moore, R. Y. (1993). Retinohypothalamic tract development in hamsters and rat. *Developmental Brain Research, 76*, 171–181.

Stern, P., Edwards, F. A., & Sakmann, B. (1992). *Journal of Physiology, 449*, 247–278.

Stone, J., & Fukuna, Y. (1974). Properties of cat retinal ganglion cells: A comparison of W-cells with X- and Y-cells. *Journal of Neurophysiology, 37*, 722–748.

Sun, Z. S., *et al.* (1997). RIGUI, a putative mammalian ortholog of the Drosophila period gene. *Cell, 90*, 1003–1011.

Swade, R. H., & Pittendrigh, C. S. (1967). Circadian locomotor rhythms of rodents in the Arctic. *American Naturalist, 101*, 431–466.

Swann, J., & Turek, F. W. (1982). Cycle of lordosis behavior in female hamsters whose circadian activity rhythm has split into two components. *American Journal of Physiology, 423*, R112–R118.

Swanson, L. W., Cowan, W. M., & Jones, E. G. (1975). An autoradiographic study of the efferent connections of the ventral geniculate nucleus in the albino rat and the cat. *Journal of Comparative Neurology, 156*, 143–164.

Takahashi, J. S., Decoursey, P. J., Bauman, L., & Menaker, M. (1984). Spectral sensitivity of a novel photoreceptive system mediating entrainment of mammalian circadian rhythms. *Nature, 308*, 186 188.

Takatsuji, K., & Tohyama, M. (1989). The organization of the rat lateral geniculate body by immunohistochemical analysis of neuroactive substances. *Brain Research, 480*, 198–209.

Takatsuji, K., Miguel-Hidalgo, J. J., & Tohyama, M. (1991). Substance P-immunoreactive innervation from the retina of the suprachiasmatic nucleus in the rat. *Brain Research, 568*, 223–229.

Takumi, T., *et al.* (1998). A new mammaliam period gene predominantly expressed in the suprachiasmatic nucleus. *Genes Cells, 3*, 167–176.

Takumi, T., *et al.* (1998). A light-independent oscillatory gene mPer3 in mouse SCN and OVLT. *The EMBO Journal, 17*, 4753–4759.

Tei, H., *et al.* (1997). Circadian oscillation of a mammalian homologue of the Drosophila period gene. *Nature, 389*, 512–516.

Terman, M., & Schlager, D. S. (1990). Twilight therapeutics, winter depression, melatonin, and sleep. In J. Montplaisir & R. Godbout (Eds.), *Sleep and biological rhythms* (pp. 113–128). New York: Oxford University Press.

Thielen, G., & Meissl, H. (1987). Action spectra of the lateral eyes recorded from mammalian pineal glands. *Brain Research, 424*, 10–16.

Thresher, R. J., Hotz Vitaterna, Y., Miyamoto, A., Kazantsev, D. S., Hsu, C., Petit, C. P., Selby, L., Dawut, O.,

Smithies and Takahashi, J. S., & Sancar, A. (1998). Role of mouse cryptochrome blue-light photoreceptor in circadian photoresponses. *Science, 282,* 1490–1494.

Thomson, A. M., & West, D. C. (1990) Factors affecting slow regular firing in the suprachiasmatic nucleus *in vitro. Journal of Biological Rhythms, 5,* 59–75.

Tominaga, K., Shibata, S., Ueki, S., & Watanabe, S. (1992). Effects of 5-HT1A receptor agonists on the circadian rhythm of wheel running activity in hamsters. *European Journal of Pharmacology, 214,* 79–84.

Tominaga, K., Shibata, S., Hamada, T., & Watanabe, S. (1994a). GABAA receptor agonist muscimol can reset the phase of neural activity rhythm in the rat suprachiasmatic nucleus *in vitro. Neuroscience Letters, 166,* 81–4.

Tominaga, K., Geusz, M. E., Michel, S., & Inouye, S. T. (1994b). Calcium imaging in organotypic cultures of the rat suprachiasmatic nucleus. *NeuroReport, 5,* 1901–1905.

Tosini, G., & Menaker, M. (1996). Circadian rhythms in cultured mammalian retina. *Science, 272,* 419–421.

Treep, J. A., Abe, H., Rusak, B., & Goguen, D. M. (1995). Two distinct retinal projections to the hamster suprachiasmatic nucleus. *Journal of Biological Rhythms, 10,* 299–307.

Trejo, L. J., & Cicèrone, C. M. (1984). Cells in the pretectal olivary nucleus are in the pathway for the direct light reflex of the pupil in the rat. *Brain Research, 300,* 49–62.

Turek, F. W., Earnest, D. J. & Swann, I. (1982). Splitting of the circadian rhythm of activity in hamsters. In I. Aschoff, S. Daan, & G. A. Groos (Eds.), *Vertebrate circadian systems. Structure and physiology* (pp. 203–214). Berlin: Springer-Verlag.

Underwood, H., & Groos, G. A. (1982). Vertebrate circadian rhythms: Retinal and extraretinal photoreceptors. *Experientia, 38,* 1013–1021.

van den Pol, A. N., & Tsujimoto, K. L. (1985). Neurotransmitters of the hypothalamic suprachiasmatic nucleus: Immunohistochemical analysis of 25 neuronal antigens. *Neuroscience, 1985,* 1049–1086.

van den Pol, A. N., Finkbeiner, S. M., & Cornell-Bell, A. H.(1992). Calcium excitability and oscillations in suprachiasmatic nucleus neurons and glia *in vitro. Journal of Neuroscience, 12,* 2648–2664.

van den Pol, A. N., Hermans-Borgmeyer, I., Hofer, M., Ghosh, P., & Heinemann, S. (1994a). Ionotropic glutamate-receptor gene expression in hypothalamus: Localization of AMPA, kainate and NMDA receptor RNA with *in situ* hybridization. *Journal of Comparative Neurology, 343,* 428–444.

van den Pol, A. N., Kogelman, L., Ghosh, P., Liljelund, P., & Blackstone, C. (1994b). Developmental regulation of the hypothalamic metabotropic glutamate receptor mGluRI. *Journal of Neuroscience, 14,* 3816–3834.

Vindlacheruvu, R. R., Ebling, F. J. P., Maywood, E. S., & Hasting, M. H. (1992). Blockade of glutamatergic neurotransmission in the suprachiasmatic nucleus prevents cellular and behavioural responses of the circadian system to light. *European Journal of Neuroscience, 4,* 673–679.

von Schantz, M., Argamaso-Hernan, S. M., Szél, Á., & Foster, R. G. (1997). Photopigments and photoentrainment in the Syrian golden hamster. *Brain Research,* 770, 131–138.

Walsh, I. B., van den Berg, R. J., & Rietveld, W. J. (1995). Ionic currents in cultured rat suprachiasmatic neurons. *Neuroscience, 89* 915–929.

Watanabe, A., Hamada, T., Shibata, S., & Watanabe, S. (1994). Effects of nitric oxide synthase inhibitors on N-methyl-D-aspartate-induced phase delay of circadian rhythm of neuronal activity in the rat suprachiasmatic nucleus *in vitro. Brain Research, 646,* 161–164.

Webb, S. M., Champney, T. H., Lewinski, A. K., & Reiter, R. J. (1985). Photoreceptor damage and eye pigmentation. Influence on the sensitivity of rat pineal-N-acetyltransferase activity and melatonin levels to light at night. *Neuroendocrinology, 40,* 205–209.

Weber, E. T., & Rea, M. A. (1997). Neuropeptide Y blocks light-induced phase advances but not delays of the circadian activity rhythm in hamsters. *Neuroscience Letters, 231,* 159–162.

Weber, E. T., Gannon, R. L., Michel, A. M., Gillette, M. U., & Rea, M. A. (1995b). Nitric oxide synthase inhibitor blocks light-induced phase shifts of circadian activity rhythm, but not c-*fos* expression in the suprachiasmatic nucleus of the Syrian hamster. *Brain Research, 692;* 137–142.

Wheal, H. V., & Thomson, A. M. (1984). The electrical properties of neurones of the rat suprachiasmatic nucleus recorded *in vitro. Neuroscience, 13,* 97–104.

Yamazaki, S., Kerbeshian, M. C., Hocker, C. G., Block, G. D. & Menaker, M. (1998). Rhythmic properties of the hamster suprachiasmatic nucleus in vivo. *Journal of Neuroscience, 18,* 10709–10723.

Yamazaki, S., Goto, M. & Menaker, M. (1999). No evidence for extraocular photoreceptors in the circadian system of the Syrian hamster. *Journal of Biological of Rhythms, 14,* 197–201.

Ying, S. W., & Rusak, B. (1994). Effects of serotonergic agonists on firing rates of photically responsive cells in the hamster suprachiasmatic nucleus. *Brain Research, 651,* 37–46.

Youngstrom, T. G., Weiss, M. L., & Nunez, A. A. (1991). Retinofungal projections to the hypothalamus, anterior thalamus and basal forebrain in hamsters. *Brain Research Bulletin, 26,* 403–411.

Zheng, B., Larkin, D. W., Albrecht, U., Sun, Z. S., Sage, M., Eichele, G., Lee, C. C., & Bradley, A. (1999).

The mPer2 gene encodes a functional component of the mammalian circadian clock. *Nature, 400,* 169–173.

Zhang, Y., & Rusak, B. (1989). Photic sensitivity of geniculate neurons that project to the suprachiasmatic nuclei or the contralateral geniculate. *Brain Research, 504,* 161–164.

Zucker, I., Rusak, B., & King, R. G. (1976). Neural bases for circadian rhythms in rodent behavior. In A. H. Riesen & R. F. Thompson (Eds.), *Advances in psychobiology* (Vol. III, pp. 35–74). New York: Wiley.

Zylka, M. J., Shearman, L. P., Weaver, D. R., & Reppert, S. M. (1998). Three period homologs in mammals: Differential light responses in the suprachiasmatic circadian clock and oscillating transcripts outside of brain. *Neuron, 20,* 1103–1110.

8

Food-Entrainable Oscillators in Mammals

FRIEDRICH K. STEPHAN

INTRODUCTION

FEEDING IN THE TIME DOMAIN

In the battle against entropy, living organisms require sufficient energy intake to survive long enough to reproduce. Consequently, strong selective pressures must have shaped efficient feeding strategies. For foraging omnivores and carnivores, some food sources may be available only during temporal windows within the day and these may change during seasons. For herbivores, food sources are more constant on a daily basis, but it may be advantageous to feed only at certain times of day to avoid predators. Thus, the time of food availability and the time of feeding can be important factors in survival. In addition to relying on external geophysical cues, animals can use endogenous circadian clocks to generate optimal temporal patterns of behavior, including feeding. The purpose of this chapter is to present evidence that many mammals have a separate circadian clock system that responds to food, rather than to light, as a zeitgeber.

SEARCH FOR A FEEDING-ENTRAINABLE OSCILLATOR (FEO)

During the first half of this century, observations and experiments established that bees had a memory of feeding time (*Zeitgedächtniss*) which was based on an endogenous circadian clock (Bünning, 1973). For mammals, Richter's (1922) innovative investigations revealed that rats fed daily meals increased their locomotor activity several hours prior to meal time, i.e., rats "anticipated" the time of food

---

FRIEDRICH K. STEPHAN Neuroscience Program, Department of Psychology, Florida State University, Tallahassee, Florida 32306-1270.

Circadian Clocks, Volume 12 of *Handbook of Behavioral Neurobiology*, edited by Joseph S. Takahashi, Fred W. Turek, and R. Y. Moore, Kluwer Academic / Plenum Publishers, New York, 2001.

availability. The first hint that this anticipatory activity (AA) might be timed by a *separate* circadian clock appeared in the work of Meyer-Lohmann (1955). Mice kept in constant darkness (DD) and fed once per day became active prior to food access, but simultaneously showed free-running activity rhythms with a period shorter than 24 hours. While this implied that two oscillators were involved, several decades passed before this interpretation became accepted. Subsequent studies provided evidence that AA of rats was a circadian phenomenon; in particular, the demonstration that rats could not anticipate meals presented at 19- or 29-hours intervals (Bolles & Stokes, 1965) suggested that there were limits of entrainment in the circadian range for this behavior.

Investigations of the effects of restricted daily feeding on physiologic processes revealed that core temperature and serum corticosterone levels rose *prior* to food access (Krieger, 1974) and that food appeared to compete with light as a zeitgeber for these functions (Krieger & Hauser, 1978; Nelson, Nichols, Halberg, & Kottke, 1973). Conversely, other processes (e.g., the pattern of melatonin secretion; Holloway, Tsui, Grota, & Brown, 1979) were less affected by restricted feeding, although a recent study reports that the reentrainment of the melatonin rhythm in response to a phase advance of the light–dark cyle is retarded in rats maintained on eat-fast (EF) cycles (Kalsbeek, Barassin, van Heerikhuize, van der Vliet, & Buijs, 2000). However, such studies do not resolve the question as to whether food and light act on the same or on a separate circadian oscillator. That the latter might be the case was suggested by Edmonds and Adler (1977a) after observing that one component of wheel running of rats became entrained to a 24-hour feeding schedule while another component freeran with a period greater than 24 hours. A similar conclusion was drawn by Sulzman, Fuller, Hiles, and Moore-Ede (1978) after finding that drinking and urinary rhythms of squirrel monkeys were entrained by meals presented at 24-hour intervals, while core temperature followed a 23-hour (LD) light–dark cycle. On the other hand, the failure of daily meals to entrain free-running rhythms was interpreted by others to mean that anticipatory wheel running was not a circadian phenomenon(e.g., Gibbs, 1979)

A major impetus to resolve this question was provided by a study which indicated that lesions of the suprachiasmatic nuclei (SCN) did not abolish the anticipatory rise of core temperature and corticosterone in rats kept on EF cycles (Krieger, Hauser, & Krey, 1977). It was suggested that entrainment to meals was therefore mediated by an oscillator outside the SCN. However, since food was presented at 24-hour intervals, alternative explanations could not be ruled out. For example, rats with SCN lesions might have responded to external laboratory cucs or to a feedback signal from the gastrointestinal system which provided information about elapsed time.

An obvious approach to test these alternatives was to determine whether or not circadian properties of anticipatory behavior wcrc retained in rats with SCN lesions. Proof of an endogenous oscillator required that such rats could anticipate meals presented at non-24-hour intervals (to rule out exogenous cues) as well as failure to anticipate when the period *T* of meals was outside the circadian range. This was accomplished in rapid succession (Boulos, Rosenwasser, & Terman, 1980; Stephan, 1981; Stephan, Swann & Sisk, 1979a, b). Since SCN lesions had been shown to abolish most circadian rhythms that are entrained by LD cycles, and the innervation of the SCN by the retinohypothalamic tract suggested that both the generation and entrainment of circadian rhythms by light are the function of these nuclei (Klein, Moore, & Reppert, 1991), it now seemed reasonable to distinguish between feeding-

and light-entrainable rhythms (FER and LER) when considering the driven functions (e.g., activity, core temperature, etc.) and between feeding- and light-entrainable oscillators (FEO and LEO) when considering the underlying mechanisms.

Functional Properties of the FEO

Because of various interactions between the two circadian systems which will be considered later, the properties of the FEO in isolation from the LEO (i.e., in animals with SCN lesions) are discussed first. The vast majority of these studies have been conducted using rats, although hamsters (Abe & Rusak, 1992) and mice (Marchant & Mistlberger, 1997) with SCN lesions retain the ability to entrain to EF cycles as well.

Limits of Entrainment

In our initial studies (Stephan *et al.*, 1979a, b), rats with SCN lesions (SCNX) were unable to anticipate meals at 18-hour intervals, but could anticipate a 23-hour schedule. More extensive investigations, using unreinforced bar pressing (Boulos *et al.*, 1980) or wheel running (Stephan, 1981) as measures of anticipation, found clear entrainment to EF cycles with periods T between 23 and about 31 hours, but not to shorter or longer T cycles (Figure 1). The upper limit of entrainment was longer when T was increased in gradual steps. When T was changed abruptly from 24 to 31 hours none of the rats was able to anticipate meal time. The period of AA during food deprivation or outside the limit of entrainment also was influenced by entrainment to T-cycles (Stephan, 1981). These results imply that the FEO shows history dependence, i.e., entrainment probably has lasting effects on the period τ and the phase response curve (PRC) of the oscillator (Pittendrigh & Daan, 1976a, b). A recent study did not detect a difference in τ after rats with SCN lesions had been entrained to 24-hour versus 25-hour T-cycles (Yoshihara, Honma, Mitome, & Honma, 1997). However, AA did not appear to follow the 25-hour T-cycle, making it difficult to interpret the result.

Although food and light cannot be compared directly in terms of zeitgeber strength, intact rats exposed to 2-hour light pulses show limits of entrainment at 23 and 26 hours (Stephan, 1983b). Thus, the lower limits of entrainment appear to be similar for the LEO and FEO, but the latter has a considerably broader upper limit.

In intact rats, Aschoff, von Goetz, and Honma (1983) noticed that the onset of AA relative to meal time increased systematically with T. This is an expected result based on oscillator theory, which indicates that the onset of AA represents the phase angle of entrainment Ψ. This relationship appears in a number of studies that used intact or SCNX animals exposed to non-24-hour feeding cycles (Aschoff, 1987). An interesting aspect of the FEO is that Ψ appears to be positive throughout its range of entrainment. With light pulses, Ψ reverses as T becomes longer or shorter than τ (Pittendrigh & Daan, 1976b). This special constraint on the FEO makes sense from an evolutionary point of view in that, assuming that AA is a type of food-seeking or foraging behavior, becoming active after the food source has disappeared would be futile. In part, failure to observe a small, negative Ψ (i.e., activity after meal onset time) may be due to the fact that once food becomes available, eating competes with other behaviors (e.g., wheel running), or if bar pressing for food or approaches to a food cup are the behavioral measures, eating can no longer be distinguished from

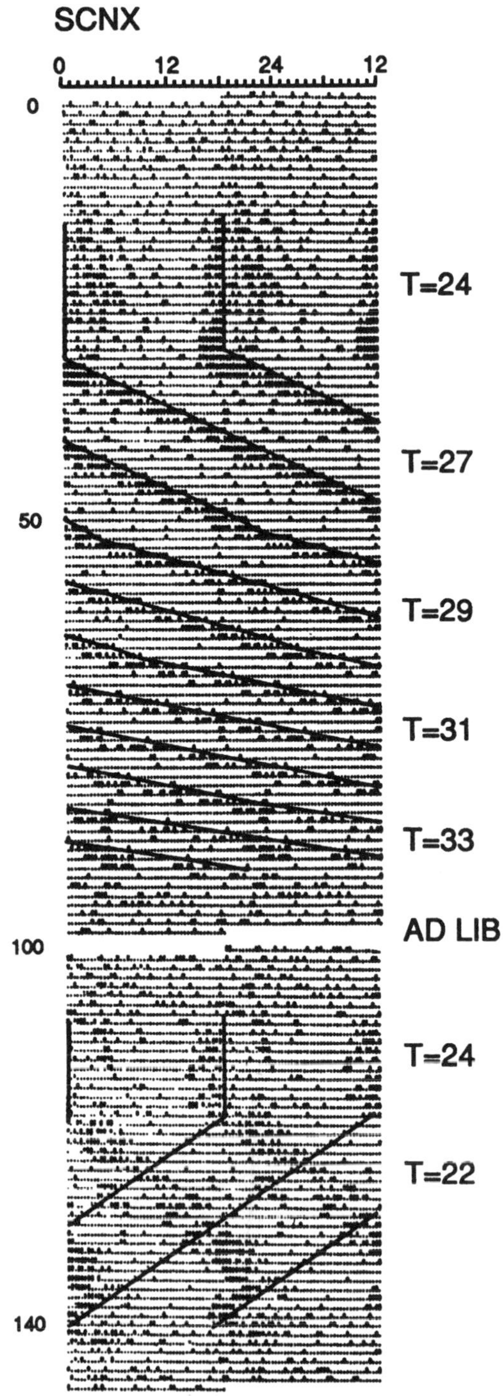

Figure 1. Event record of wheel running for a rat with a lesion of the SCN. The onset of restricted food access is marked by lines and the period is indicated in the right margin. Entrainment fails at $T = 33$ and 22 hours. Note the free-running rhythm at $T = 22$ (Modified from Stephan, 1981.)

anticipation. Nevertheless, in a few cases with $T < 24$ hours, we have observed rats that do not increase activity prior to meal time, but alternate between wheel running and eating during the food access time (Stephan, 1989a).

Some rats increase activity both before and after food access, the latter being called "succeeding activity" by Aschoff *et al.* (1983). The onset of AA and the end of succeeding activity are correlated and the total span of activity (6–8 hours) could represent the active phase α of the FEO's circadian cycle (Aschoff, 1991). A similar span of activity is observed when restricted feeding is followed by food deprivation. Rats continue to run past the previous access time and several hours beyond, for a total of about 6–8 hours (e.g., Figure 2, FD [food deprivation]). In intact rats, the anticipatory rise in corticosterone shows almost identical limits of entrainment and changes in Ψ on non-24-hour EF cycles as behavioral measures (Honma, Honma, & Hiroshige, 1984).

Phase Shifts of Food Access

After wheel running of SCNX rats has been entrained to a 24-hour EF cycle, phase shifts of meal time induce transients in the onset of AA until reentrainment occurs (Figure 2). While transients are expected in oscillator -mediated processes, those in AA have some unusual features. The most predictable transients are observed when food access is delayed by 4–10 hours. The onset of AA delays rapidly (1–4 hours per cycle) and reentrainment occurs typically within 4–5 days (Stephan, 1984, 1992a, b). These transients are much more rapid than those observed after phase shifts of the LD cycle, but the large delays are consistent with the broad upper limits of entrainment, e.g., on a 29-hour *T*-cycle, a daily delay of 5 hours is required. In response to phase advances, some rats show advancing transients, but only after AA precedes the new meal time (Figure 2, +8HR). In those cases, wheel running may be suppressed during and after the meal time, so that transients are masked for the first few days after the phase shift. When phase advances are 8–10 hours, most rats show delaying, rather then advancing transients. This indicates that the FEO is more readily decelerated then accelerated, a characteristic that applies to LEOs as well (Pittendrigh & Daan, 1976b). Some rats display advancing and delaying transients simultaneously (Stephan, 1984, 1992a) (Figure 2), suggesting that the FEO is composed of at least two suboscillators that are driven in opposite directions when the zeitgeber is shifted to a critical phase.

An attempt to establish a phase response curve for the FEO uncovered another unusual feature of this system (Stephan, 1992b). A phase-shifted meal was presented for only 1 or 2 days. This was followed by ad lib feeding and 3 days of food deprivation. Since AA is expressed during food deprivation (see below), it should be possible to infer the phase of the FEO by the time at which AA was expressed. Surprisingly, in most cases, AA reappeared at the pre-shift phase, i.e., despite initial delaying transients, the FEO apparently had not been reset. When the phase-shifted meal was presented for three or four consecutive days, AA appeared at or near the new phase position during food deprivation probes.

An additional result was that after phase shifts of meal time, some rats showed reentrainment of AA, but also expressed a bout of activity near the pre-shift phase. In view of the split transients described above, one speculative interpretation is that the FEO can retain a "memory" of phase displacement, so that one oscillator can drive activity at a previous phase position, while another entrains to the new meal time. This would be of adaptive significance when a food source that was available at a particular time may have disappeared only temporarily.

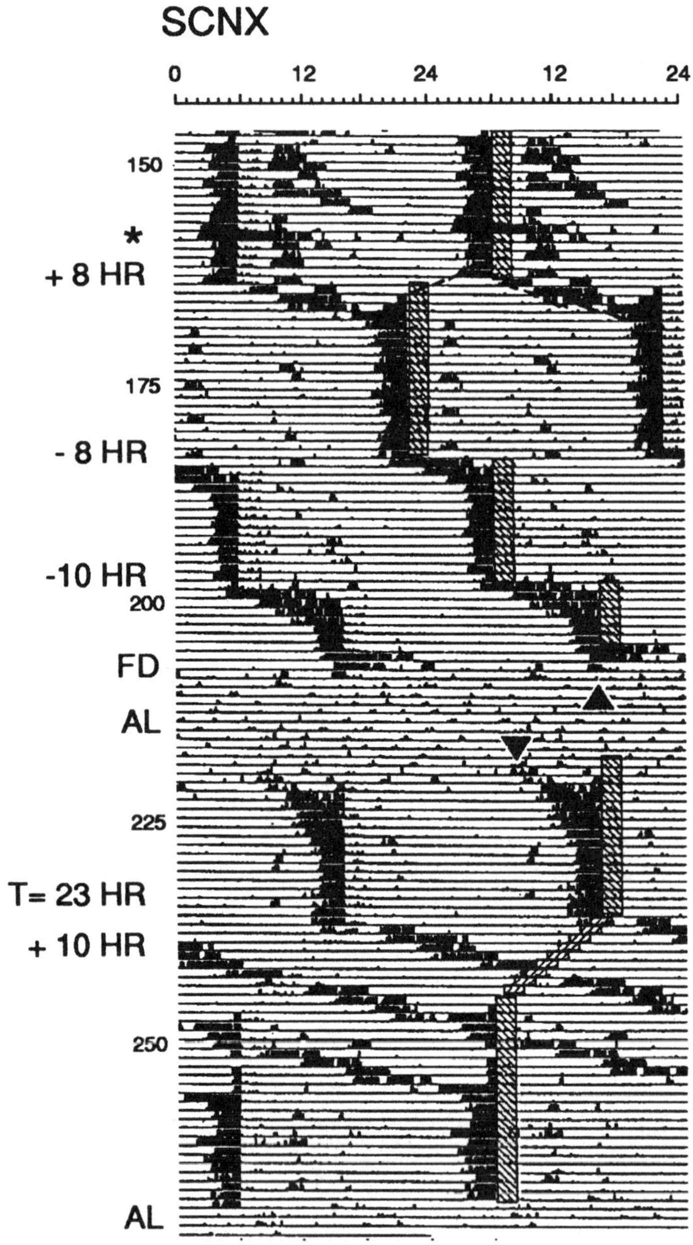

Figure 2. Event record of wheel running for a rat with a lesion of the SCN exposed to phase shifts of food access. Phase shifts are indicated in the left margin. The 8-hour phase advance induces both advancing and delaying transients. Both 8- and 10-hour delays induce delaying transients and activity freeruns during a 10-hour advance in 1-hour steps (23-hour *T*-cycle). For 3 days, the rat simultaneously anticipates the last 24-hour eat-fast cycle and displays delaying transients. Anticipatory activity persists during 3 days of food deprivation (FD), but not in ad lib (AL) conditions. (Modified from Stephan, 1992a.)

If EF cycles are followed by ad lib feeding, AA usually disappears within 1 or a few days. In intact rats maintained on a skeleton photoperiod with food limited to the subjective day, feeding related activity persisted into ad lib feeding conditions for 6–10 days (Brinkhof, Daan, & Strubbe, 1998). However, if EF cycles are followed by food deprivation, AA persists at or near the previous phase for up to 5 days (Boulos *et al.*, 1980). Unfortunately, the self-limiting aspects of food deprivation make it impossible to observe prolonged freeruns. Initially, this was interpreted to mean that either AA was not circadian or was mediated by a damped oscillator. However, a clever procedure used in two different labs (Clarke & Coleman, 1986; Coleman, Harper, Clarke, & Armstrong, 1982; Rosenwasser, Pelchat, & Adler 1984) established an explanation for the disappearance of AA during ad lib feeding. Rats were entrained to EF cycles, followed by ad lib feeding for days or weeks, and then deprived of food for several days. During food deprivation, AA was expressed at or near the phase of the previously entrained AA. The most plausible interpretation of this result is that activity becomes decoupled from the FEO in ad lib conditions, but that the FEO continues to freerun with a period of about 24 hours. Food deprivation restores the coupling to activity. When intact rats are placed on EF cycles for a prolonged time, similar decoupling of the light-entrainable activity rhythm is observed (see Figure 3). Upon return to ad lib feeding, the LER appears at a phase that can be roughly extrapolated back to the pre-EF phase. This indicates that the outputs of the LEOs and FEOs compete for the expression of activity (as well as other functions, see Figure 5).

Additional compelling evidence that the FEO is a self-sustained oscillator comes from prolonged freeruns observed in some rats when T is outside the limits of entrainment (see Figure 2 during 23-hour T-cycle).

At least for the rat, the circadian limits of entrainment, the transients after phase shifts, and the freeruns during food deprivation or outside the limits of entrainment clearly establish the FEO as a self-sustained circadian oscillator that is anatomically independent of the SCN. A critique of alternative models can be found in a review paper by Mistlberger (1994).

ONE OR SEVERAL OSCILLATORS?

Although rats are unable to anticipate meals at 18- or 21-hour intervals, when two meals per day are presented at intervals of 6–12 hours to intact or SCNX rats, anticipation is observed (Bolles & Moot, 1973; Boulos & Logothetis, 1990; Stephan, 1989b; Stephan *et al.*, 1979a). Since two meals occur within one circadian cycle, one possible explanation is that the FEO is a "continuously consultable" clock which enables the animal to estimate time of day. An alternative explanation is that multiple circadian oscillators that can be entrained at different phases are involved.

In rats, these alternatives have been tested by presenting two meals per day, each with a different period, to intact (Edmonds & Adler, 1977b) and to SCNX rats. At least some SCNX rats entrain AA to 24- and 24.5-hour meals (Stephan, 1983a), to 23.75- and 24-hour meals (Figure 3C), as well as to 25- and 26-hour meals (Stephan, 1989a). Since one oscillator cannot generate two periods simultaneously (unless some unusual assumptions are made), it seems likely that two or more oscillators are involved. Furthermore, this interpretation is supported by the split transients described above. However, when three meals per day are presented, SCNX rats antici-

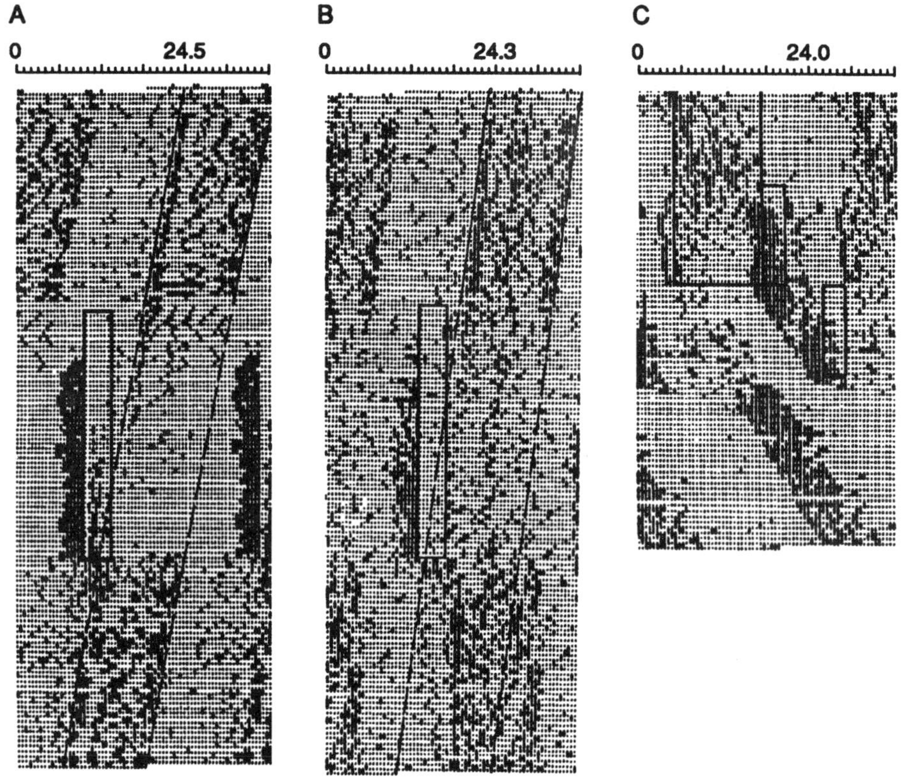

Figure 3. (A, B) Wheel running for enucleated rats. Restricted feeding times are enclosed by rectangles. The dashed lines indicate the projected free-running rhythm. For rat A, the period of the free-running rhythm was not affected by restricted feeding. A rare case of synchronization of the free-running rhythm is shown for rat B. The change in period persisted into ad lib condition. Note the suppression of activity associated with the free-running rhythm during restricted feeding. The records are plotted modulo the period of feeding, 24.5 hours in panel A and 24.3 hours in panel B (Modified from Stephan, 1986a.) (C) A rat maintained on an LD 12:12 cycle and then exposed to restricted feeding (narrow rectangle) at lights-on. This was followed by DD and an 8-hour delay of food access. Anticipatory activity does not delay rapidly (cf. Figure 2), but delays in phase with the free-running light-entrainable rhythm. (Modified from Stephan, 1986c.)

pate only two of the meals, although they clearly eat food at the meal time that is not preceded by activity (Stephan, 1989b). Of course, in *T*-cycle experiments, animals also eat when *T* is outside the limits of entrainment. These observations indicate that while AA may be accompanied by a hunger signal, the output of the FEO is not simply a hunger response since the two can become dissociated. The evidence suggests that the FEO consists of at least two coupled oscillators, perhaps analogous to the E and M oscillators of the LEO (Pittendrigh and Daan, 1976c).

A related question is whether or not rats given two meals per day at different locations are able to associate meal time with meal location. One study used two food levers on opposite sides of a cage (Boulos & Logothetis, 1990) and another used a T-maze (Mistlberger, de Groot, Bossert, & Marchant, 1996). While both studies report positive results, the time–place discriminations were far from perfect. This may be because the animals are in a state of high arousal and the cost of pressing the "wrong" lever is minimal.

INTERACTIONS BETWEEN THE LIGHT-ENTRAINABLE OSCILLATOR AND THE FEEDING-ENTRAINABLE OSCILLATOR

In intact rats, the two circadian systems appear to be only weakly coupled. For example, when rats are maintained in DD and then placed on EF cycles, AA entrains to meal time while simultaneously a free-running rhythm with $\tau > 24$ hours continues. This lack of entrainment of the free-running rhythm gave rise to considerable confusion and misinterpretation of AA. However, in studies which concluded that the LER was totally unaffected by EF cycles (e.g.,Gibbs, 1979), closer inspection shows small changes in of the free-running LER. More recent experiments indicate that the probability of synchronization or near synchronization is increased when is close to T and at certain phase relations between the LER and the FER. However, in most cases, EF cycles do not entrain the free-running rhythm of rats (Stephan, 1986a, b). Figure 4 shows one rat in which the LER is totally unaffected (Figure 4A) and another rat in which the LER became synchronized to the EF cycle (Figure 4B). A higher proportion of rats that entrained wheel-running and body temperature rhythms to EF cycles was obtained when rats were kept in DD and fed 50% of ad lib intake early in the subjective day (Challet, Malan, & Pevet, 1996a).

When food is restricted to the early light phase, the onset of pineal and plasma melatonin secretion as well as the acrophase of body temperature are phase advanced (Challet, Pévet, Vivien-Roels & Malan, 1997c). The phase advance is attenuated by 5,7-dihydroxytryptamine (DHT) lesions of serotonergic terminals in the SCN (Challet, Pévet, & Malan, 1997a),by lesions of the intergeniculate leaflet (Challet, Pévet, & Malan, 1996b), or by lesions of the ventromedial hypothalamic nuclei (Challet, Pévet, Lakhdar-Ghazal & Malan, 1997b). These results suggest that these structures might be involved in coupling some aspect of the FEO to the SCN.

In the context of two endogenous oscillators which drive activity, the results are best interpreted as a consequence of weak coupling between the oscillators, i.e., rather than assuming that food or some aspect of metabolism affects the SCN

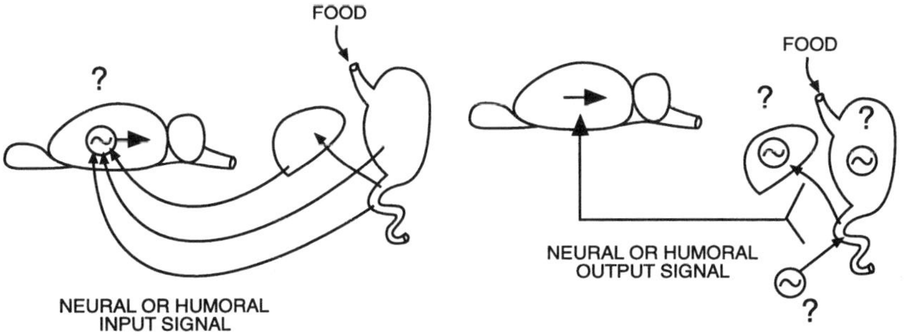

Figure 4. Two plausible alternative models for the locus of the FEO. If the oscillator is in the CNS (left), then its phase is most likely set by neural or humoral signals from the liver, stomach, or small intestines. If the oscillator is located in peripheral organs (right), an output signal to the CNS is required to trigger activity and other processes which precede food access.

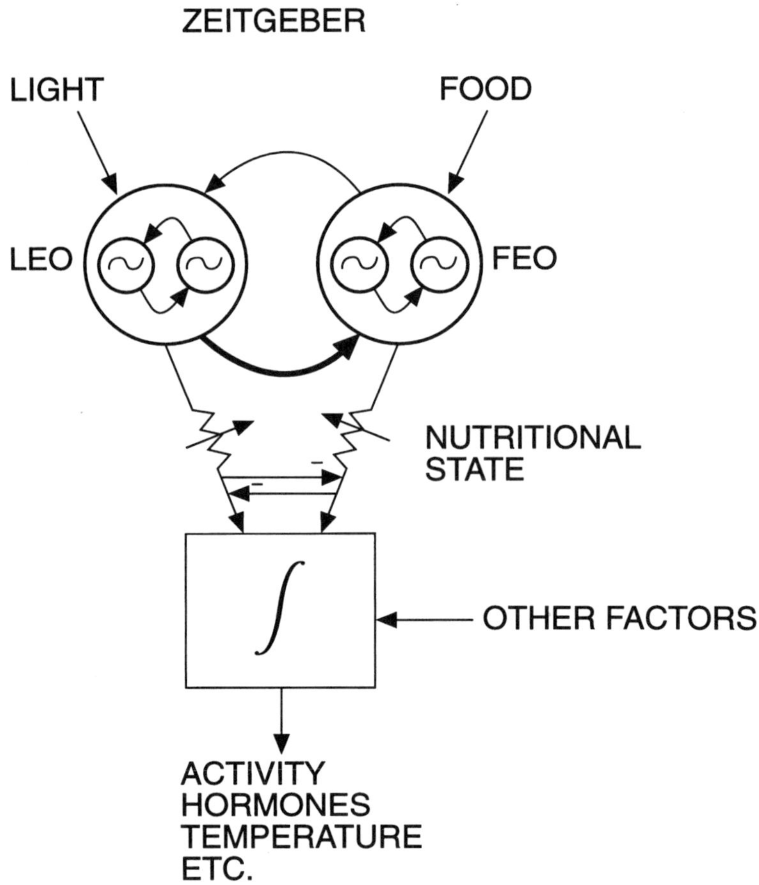

Figure 5. A minimal conceptual model required to account for light- and feeding-entrainable circadian rhythms. Both systems have at least two coupled suboscillators and the two systems are reciprocally coupled (curved arrows). The outputs of the oscillators converge on various functions with variable and competing strengths. The integration of oscillator outputs and other factors produces the empirically observed temporal patterns.

directly, the FEO may have a weak and indirect functional influence on the SCN (see Figure 5). Relative coordination between the LER and FER has also been described for other species, but in some species, EF cycles entrain the whole circadian system. These apparent species differences will be discussed later.

The coupling of the FEO to the LEO appears to be somewhat stronger. For example, if EF cycles are followed by ad lib feeding and a phase shift of the LD cycle, AA observed during subsequent food deprivation probes maintains a near-constant phase relationship to the LER, rather than being expressed near the previous meal time (Ottenweller, Tapp, & Natelson, 1990; Rosenwasser *et al.*, 1984). Additional evidence for asymmetrical coupling was obtained by phase shifting food access and placing the animals in DD. The rapidly delaying transients observed in SCNX rats were not observed; rather, AA delayed in near synchrony with the free-running LER (Stephan, 1986c; see Figure 3C).

The relatively weak coupling between the FEO and the LEO may be of adaptive significance since food sources may shift suddenly, whereas seasonal changes in sunrise and sunset are gradual. This would enable animals to reset the phase of the FEO without phase shifting the circadian system *in toto.*

Nature of the Zeitgeber

Caloric Foods and Macronutrients

Nutritive foods are clearly sufficient, but perhaps not necessary, to entrain the FEO. While most rats given ad lib chow and a palatable nutritive meal for 2 hours/ day anticipate the nutritive meals, when the meal size is reduced to 4 g, only a few will do so, and none anticipate palatable, but nonnutritive meals, suggesting that both meal size and caloric value influence the expression of AA (Mistlberger & Rusak, 1987). When a diet lacking two macronutrients was presented ad lib and the missing macronutrients were available for 2 hours/day, no AA was observed. However, presentation of the missing macronutrients at two different times per day (first meal, protein; second meal, carbohydrate or fat), rats anticipated both meals regardless of macronutrient type (Mistlberger, Houpt, & Moore-Ede, 1990b).

Palatable nutritive meals are also effective in eliciting anticipatory wheel running in hamsters, and those with SCN lesions show more robust AA then intacts (Abe & Rusak, 1992; Mistlberger, 1992a). One important aspect of these studies is that the animals were not, or only mildly, food deprived, i.e., entrainment of AA does not require conditions of severe food deprivation. A similar conclusion can be drawn from a study which varied food access duration. Anticipatory wheel running was maintained nearly unaltered with duration ranging from 4 to 10 hours (Stephan & Becker, 1989).

Some evidence indicates that AA can be induced by less nutritive or even nonnutritive substances. For example, some rats display anticipatory licking to sucrose or glucose solutions, but only when the solutions are the only source of calories (Bays & Stephan, 1992). By asking whether phase-delayed meals of different size or caloric composition could induce transients, we recently found that about 7 g of chow was required to trigger delaying transients, whereas up to 20 g of cellulose was ineffective (Stephan, 1997). Also, 10 g of glucose in 25 ml of saccharin phase shifted the FEO, whereas 25 ml of saccharine alone was ineffective. Surprisingly, thus far, we have been unable reliably to phase shift the FEO with 5.8 g of corn oil given on two consecutive days after an 8-hour phase delay of meal time (Stephan & Davidson, 1998). In this particular paradigm, the FEO appears to be less responsive to calories from fat than from glucose, but it is possible that fat meals given for more days might eventually phase shift the FEO.

Non-nutritive Substances

Restricted access to water appears to be a relatively ineffective way to induce AA and is confounded by the problem that rats eat little in the absence of water (Mistlberger & Rechtschaffen, 1985). Hamsters with SCN lesions show anticipatory wheel running when water is restricted to 2 hours/day, but consume up to 55% of their daily food intake during the water-access time (Mistlberger, 1993). By using a different paradigm in which food was restricted to one time of day and water to a different time, some rats showed anticipatory approaches to the water delivery bin, and this activity persisted for several days when rats were totally deprived of food and water (Mistlberger, 1992b).

A different approach was to investigate whether induced states of biological need can elicit AA. Rats were adrenalectomized or treated with deoxycorticosterone acetate (DOCA) to induce salt appetite and then placed on daily restricted saline access. DOCA-treated rats did not show anticipatory wheel running (Rosenwasser, Schulkin, & Adler, 1985), but adrenalectomized rats showed anticipatory lever pressing for NaCl (Rosenwasser, Schulkin, & Adler, 1988). The anticipatory responses diminished when rats were placed in constant light (LL) and did not reappear during salt deprivation. As noted by the authors, it is difficult to equate the motivational states induced by EF cycles and by adrenalectomy. Nevertheless, it seems that stimuli other than caloric nutrients are capable of inducing anticipatory behavior, and these studies again point out the importance of choosing an appropriate behavioral measure of anticipation.

Rats rendered anosmic by intranasal infusions of zinc sulfate show anticipatory wheel running comparable to controls (Coleman & Hay, 1990) and we found no deficit in bulbectomized rats (F. K. Stephan, in preparation). Thus, neither taste (e.g., saccharin) nor olfaction is essential for entraining the FEO.

Outputs of the FEO

Anticipatory Behaviors

Like the LEO, the FEO controls rhythmicity in a variety of behaviors and physiologic processes and most of these appear to be under the control of both endogenous oscillators. The studies reviewed above indicate that a variety of behaviors can be mediated by the FEO. Some of these are rather specific food-directed behaviors (e.g., bar pressing, approaches to a food cup), whereas others are more general (e.g., wheel running, tilt cage activity). In squirrel monkeys, anticipation can be manifested by a pause in tree-climbing activity or by increased attempts to open the feeder (Aschoff & von Goetz, 1986).

For rats, a case can be made that wheel running is related to foraging activity (Mather, 1981). For foraging animals, food sources may be located at considerable distances from their burrows so that anticipatory behavior would provide sufficient lead time to get to the food. About one third of rats with SCN lesions show anticipatory licking on a drinking spout, although water is available ad lib. When free-running rhythms in AA are observed, drinking is elevated around the same time as wheel running (e.g., Stephan, 1992a). This suggests that the output of the FEO induces a state of arousal that can be expressed in several different behaviors, albeit to different degrees. It has already been mentioned that the behavioral output of the FEO usually becomes uncoupled in ad lib conditions. Its output also appears to be attenuated in obese rats (Persons, Stephan, & Bays, 1993), in aged rats (Mistlberger, Houpt, & Moore-Ede, 1990c; Shibata, Minamoto, Ono, & Watanabe, 1994) and in constant light (Mistlberger *et al.*, 1990a). Surprisingly, compared to lean controls, feeder approaches are elevated in genetically obese Zucker rats, which have a mutation in the leptin receptor gene, and this finding suggests that the reduced AA in obese rats may be mediated by leptin (Mistlberger & Marchant, 1999).

Physiology

Core temperature and serum corticosterone levels increase prior to meals with a time course roughly similar to that of anticipatory behavior. These two variables have been studied in rats with SCN lesions on 24-hour (Krieger *et al.*, 1977) as well as non-24-hour EF cycles (Honma *et al.*, 1984). Thus, one can be confident that these functions are controlled by an SCN-independent clock, i.e., the FEO. The circadian rhythm of cell proliferation in the gastric mucosa can be influenced by EF cycles and appears to persist in rats with SCN lesions (Scheving *et al.*, 1983a). Rats with SCN lesions that are sacrificed 1 hour before meal time (during AA) have lower serum glucagon levels than rats sacrificed about 6 hours before meal time (prior to AA). No changes were found in serum insulin, motilin, or glucose levels between these time points. The drop in glucagon was not observed in rats that were food deprived and sacrificed at the same times after their last meal (Davidson & Stephan, 1999a). Whether or not this represents a disinhibitory signal to the CNS remains to be established. In intact rats entrained to EF cycles, serum levels of free fatty acids and ketone bodies increase prior to meal time, whereas triacylglycerides and liver glycogen decline (Escobar, Diaz-Munoz, Encinas, & Aguilar-Roblero, 1998). Again, it is not known whether these changes are signals to the central nervous system (CNS) that could trigger AA.

Less direct evidence for the involvement of the FEO is obtained in studies which demonstrate that a function becomes displaced from its normal phase position by daily meals, precedes meal time, and persists at that phase during food deprivation. The increase of duodenal disaccharidases before meal time meets these criteria, as well as entrainment to non-24-hour meal schedules (Saito, Murakami, Nishida, Fujisawa & Suda, 1976; Saito, Kato & Suda, 1980). There is some indication that circadian enzyme rhythms can be maintained by periodic intravenous infusions of nutrients (Stevenson, Sitren, & Furuya, 1980). However, in view of the long infusion times (14 hour), the presence of an LD cycle, and low data resolution, it is not clear whether or not these rhythms represent entrainment to nutrients. Overall, these results indicate that the gastrointestinal (GI) system displays anticipatory digestive responses which are undoubtedly of adaptive significance.

Other experimental paradigms that permit some inference about the control of the FEO versus the LEO are to present meals at a different period than the LD cycle or to phase shift the LD cycle without changing the meal time (e.g., Krieger & Hauser, 1978). Some functions are more likely to be entrained by food and others by light. However, negative results are often difficult to interpret because the entrainment and expression of anticipatory responses depend to some extent on the periods and phase relationship between the two zeitgebers (see above).

As might be expected, many functions of the GI system display circadian rhythms that are influenced by EF cycles (for reviews, see Boulos & Terman, 1980; Philippens, von Mayersbach, & Scheving, 1977; Scheving *et al.*, 1983a), but in the absence of circadian experimental manipulations, it is difficult to assess whether or not these are controlled by the FEO, the LEO, both, or noncircadian mechanisms. Nevertheless, it appears that the output of the FEO has a profound influence on digestive physiology. The ease with which many GI functions can be displaced from their typical phase within the circadian cycle by EF cycles again suggests relatively weak coupling between the LEO and FEO systems.

Physiologic Substrate of the FEO System

Pituitary and Adrenal Glands

While the ablation–behavior approach led to the identification of the SCN as a circadian oscillator (presumably the LEO), the locus (or loci) of the FEO is unknown and only a few studies have identified critical components of this system. Since serum corticosterone levels rise prior to feeding (Krieger, 1974) and there is some evidence that hamster adrenal glands have a circadian rhythm *in vitro* (Andrews, 1968), the earliest attempts to localize the FEO were to assess the effects of adrenalectomy on AA in SCNX rats (Boulos *et al.*, 1980; Stephan *et al.*, 1979a). However, AA was basically unaffected by adrenalectomy. Another endocrine gland that might be relevant to the FEO system because of metabolically important hormones is the adenohypophysis. However, despite low body weights and chronic hypothermia, hypophysectomized rats expressed AA as well as controls, but failed to increase core temperature before meal time, thus dissociating the increase in activity from that in core temperature (Davidson & Stephan, 1999b).

Central Nervous System Structures

A number of studies have investigated the role of hypothalamic nuclei known to be involved in the regulation of food intake. Initially, it appeared that lesions of the ventromedial nuclei (VMN) abolished anticipatory wheel running (Inouye, 1982b) as well as the anticipatory rise in corticosterone and core temperature (Krieger, 1980). However, a subsequent study showed that AA recovered about 18 weeks after the lesions (Mistlberger & Rechtschaffen, 1984). The reason for the long latency to recovery is not clear. However, a selective and striking reduction of anticipatory wheel running is observed in rats made obese by a high-fat diet (Persons *et al.*, 1993). Thus, one possibility is that VMNX rats do not show AA because they are obese. A reinvestigation of the effects of VMN lesions on corticosterone rhythms indicated that entrainment recovered 8–10 weeks after the lesions when rats were given 4 hours of food access per day and even earlier when the access time was reduced to 1 hour (Honma, Honma, Nagasaka, & Hiroshige, 1987). In rats with colchicine-induced lesions of the VMH, restricted feeding caused a change in c-*fos* expression in the subparaventricular zone and other targets of SCN efferents (Choi, Wong, Yamat, & Dallman, 1998), but the significance of this finding is not clear.

Lesions of the paraventricular nuclei (PVN) with or without SCN lesions attenuated or abolished AA measured in tilt cages in most rats, but all rats displayed clear anticipatory approaches to the food bin (Mistlberger & Rusak, 1988). This result indicates that the output signal of the FEO does not have an equal effect on all behaviors and that loss of a particular behavior (e.g., wheel running) does not necessarily indicate the destruction of the FEO. Ibotenic acid lesions of the lateral hypothalamic area fail to abolish AA in tilt cages, and large lesions of the hippocampal formation, amygdala, or nucleus accumbens or administration of the dopamine receptor blocker haloperidol do not abolish anticipatory wheel running (Mistlberger & Mumby, 1992).

Chronic multiunit activity recordings in the SCN show little or no change associated with EF cycles (Inouye, 1982a). In the VMN, multiunit activity increased during feeding, but did not show an anticipatory increase (Inouye, 1983). On the other hand, neural activity in the lateral hypothalamus was elevated at the time that motor activity increased prior to food access and this increase persisted for several days during total food deprivation (Kurumiya & Kawamura, 1991). Neurotoxic lesions of the ventral dopamine bundle abolished this response. This result suggests that neural activity in the LH may play some role in the anticipation of food access. However, the interpretation of this finding remains uncertain. One possibility is that the LH and its fiber systems is part of the output pathway of the FEO. Perhaps the most severe disruption of AA yet reported results from ibotenic acid or electrolytic lesions of the parabrachial nucleus region. Some rats displayed no AA, and core temperature did not rise before meal time (Davidson, Cappendijk, & Stephan, 2000). There was no indication of recovery of function and body weights were similar to controls. It is not known whether this region contains an FEO or is a critical element in the FEO system. The latter appears more likely.

Peripheral Structures

Another potential site of the FEO is a peripheral structure. Since all known LEOs are in or near photoreceptive structures, it is conceivable that a clock that responds to food is associated with the digestive system. Some hypothetical configurations of the FEO and its inputs and outputs are shown in Figure 4.

Regardless of a CNS or peripheral locus, the vagus nerve (consisting primarily of afferents) is a potential path of communication between the GI system and the brain. However, total subdiaphragmatic vagotomy has no effect on anticipatory wheel running in SCN-intact or SCNX rats (Comperatore & Stephan, 1990). The entrainment of serum corticosterone by EF cycles is also not impaired by vagotomy (Moreira & Krieger, 1982). To rule out nonvagal visceral afferents, rats with capsaicin induced deafferentation were maintained on EF cycles. They anticipated as well as control rats (Davidson & Stephan, 1998).

Another attempt to assess the role of the GI system in entrainment to EF cycles was to record the electrical activity of the stomach and duodenum (Comperatore & Stephan, 1987). At 4–5 hours prior to food access, a marked increase in the motility of the duodenum, presumably the migrating myoelectric complex (MMC), was recorded. When food access was phase shifted, transients in the motility pattern were observed, suggesting an underlying circadian control. Of particular interest was the finding that the increase in motility always preceded behavioral anticipation (unreinforced bar pressing). In individual rats, bar pressing followed duodenal activation by several minutes and was never observed during the quiescent phase of the MMC.

The temporal relationship between duodenal motility and bar pressing suggests the possibility of a humoral signal released by the gut which has an activating effect on the CNS. A duodenal locus is an attractive hypothesis because the complexity of the enteric nervous system as well as the presence of endocrine and paracrine cells which could be the substrate for an FEO. A number of mitotic rhythms in the alimentary canal are strongly affected by EF cycles (Scheving, Tsai, & Scheving, 1983b) and the amplitude of the cell proliferation rhythm in the stomach appears to increase after SCN lesions, implying independence from the LEO (Scheving *et al.*, 1983b).

Considering the critical role of the liver in processing metabolic signals, this organ is another viable candidate. Circadian rhythms in liver function are strongly affected by EF cycles (Boulos *et al.*, 1980; Phillipens, 1980; Phillipens *et al.*, 1977). Circadian rhythms in the liver enzyme tyrosine amino transferase appear to persist in *in vitro* cultures (Hardeland, 1973a, b). The cloning of mammalian clock genes has provided a powerful new tool to investigate circadian oscillators. Surprisingly, these genes are expressed not only in the SCN and retinas, but in other CNS structures and also in peripheral organs, including the liver and pancreas (e.g., Shearman, Zylka, Weaver, Kolakowski, & Reppert, 1997). Although these gene products could have nonclock functions, it has been shown that gene expression in the liver displays a circadian rhythm both *in vivo* and *in vitro* (Balsalobre, Damiola, & Schibler, 1998). This indicates that the liver has a circadian clock, but whether its function is strictly intrinsic or its output is communicated to the CNS remains to be established.

A puzzling finding is that arrhythmic SCNX rats given methamphetamine in their drinking water display free-running rhythms in locomotor activity which become entrained to EF cycles, suggesting activation of a circadian clock outside the SCN (Honma, Kanematsu, & Honma, 1992). However, the activity coincides with or starts after food access and continues for 10–12 hours and therefore differs considerably from anticipatory activity. If, as suggested by the authors, the methamphetamine-induced rhythm is not generated by the FEO, then it appears that one effect of methamphetamine is to suppress the anticipatory activity normally expected in SCNX rats.

Species Differences

Anticipation of periodic meals has been observed across a wide range of species and may therefore have a long phylogenetic history. FEOs could have evolved in response to circadian rhythms in other organisms, making them periodic food sources, to exploit a particular temporal niche in feeding behavior, or to reduce predation.

Rodents

In addition to rats, anticipatory behavior has been reported in intact mice as well as in mice with SCN lesions (Marchant & Mistlberger, 1997). In the early study of Meyer-Lohman (1955), EF cycles did not entrain the free-running activity rhythm in DD. In a recent study, the same result was reported for the C57BL/6J strain; however, in mice of the CS strain, the LER became synchronized to the EF cycle (Abe, Kida, Tsuji, & Mamo, 1989). Such differences do not necessarily imply a fundamental difference in circadian organization, as several factors could influence the effect of EF cycles on the LER. One is a difference in the coupling strength between the FEO and the LEO. Furthermore, coupling strength is itself affected by the difference between T and τ. Other potential factors are whether animals are maintained in LL or DD, the conditions of food restriction (e.g., diet and duration), and the selection of dependent variables.

Hamsters (*Mesocricetus auratus*) present a different problem in that they fail to increase food intake sufficiently to survive when maintained on restricted feeding (Silverman & Zucker, 1976). However, when given a fixed time of day to hoard food, the LER of some hamsters became entrained (Rusak, Mistlberger, Losier, & Jones, 1988). Interestingly, anticipatory hoarding was observed after, but not before SCN lesion. In another study (Abe and Rusak, 1992), hamsters were given chow ad lib and a highly palatable diet for 2 hours/day. Hamsters failed to display anticipatory wheel running, but after SCN lesions, two thirds animals showed elevated activity prior to the palatable diet. When chow was limited to 70% of ad lib intake, some intact hamsters showed anticipatory activity and changes in τ of the LER, but the phase of the latter was not set by meal time. The combination of reduced chow and a restricted palatable diet proved the most effective in eliciting anticipatory wheel running in intact and in SCNX hamsters. The LER became synchronous with EF in some but not all hamsters.

Thus, the circadian organization of hamsters with regard to the LEO and FEO is not unlike that of rats in particular experimental conditions. This is of considerable interest because hamsters are primarily grain and seed eaters (i.e., nonperiodic food sources) and hoard food, whereas rats are opportunistic omnivores.

In a squirrel (*Funambulus palmarum*), the free-running activity rhythm in dim, constant light became entrained to a 24-hour EF cycle (Chandrashekaran, 1982). Unfortunately, diurnal rodents have been much neglected in this area of circadian biology.

Lagomorphs

A number of studies have investigated the effects of EF cycles on circadian rhythms of the herbivorous rabbit. In constant light, rabbits showed clear anticipatory locomotor activity before food access (Jilge, 1991). The free-running activity rhythm became entrained to the EF cycle and the development of AA depended on the phase angle difference between the free-running rhythm and time of food availability. When ad lib feeding was reinstated, the phase of the free-running rhythm was determined by the previous feeding time. While this result did not provide evidence for a separate FEO, a more recent study indicates that this is the case. When EF was presented with $T = 24.17$ hours and an LD cycle with $T = 23.84$ hours, one component of activity entrained to food and another to light, indicating that two oscillators are involved (Jilge & Stahle, 1993). Furthermore, in the only developmental study of the FEO to date, it was shown that rabbit pups anticipate the once-daily nursing visit of the doe as early as 5 days after birth (Jilge, 1995).

Carnivores

A priori, one might expect that carnivores which predate on small rodents would have a strongly developed FEO. Unfortunately, the effects of EF cycles on carnivore behavior have not been extensively studied and it is difficult to simulate natural carnivore behavior in the laboratory (e.g., predation). One study restricted mink (*Mustela vison*) and two species of weasels (*M. nivalis* and *M. freneta*) to 85% of ad lib intake. Both in DD and in LL, these animals showed anticipatory wheel running or increased activity in a tilt cage several hours before feeding time. The data suggest that in some animals the free-running rhythm was entrained by EF cycles, whereas it continued to freerun in others (Zielinski, 1986).

Marsupials

In a nocturnal predatory marsupial, *Sminthopsis macroura froggatti*, limiting food access to 3 hours/day induced anticipatory wheel running in most animals while they were on an LD cycle and food was presented in the light phase (Coleman, O'Reilly, & Armstrong, 1989). However, unlike rats, these animals did not show activity around feeding time during food deprivation probes. Instead, food deprivation caused phase shifts in the free-running rhythm.

In the kowari (*Dasyuroides byrnei*), anticipatory activity was observed in LD cycles as well as in DD. As is the case for rats, free-running activity rhythms were not entrained to EF cycles, but showed considerable relative coordination (O'Reilly, Armstrong, & Coleman, 1986). In a subsequent study, several kowaris showed entrainment of the free-running rhythm to meal time, and this may have been the result of presenting a more natural and preferred food source (Kennedy, Coleman, & Armstrong, 1991). Similar results were reported for other marsupials, the quoll (*Dasyuras viverrinus*) (Kennedy, Coleman, & Armstrong, 1990) and the bandicoot (*Isoodon macrourus*) (Kennedy, Coleman, & Armstrong, 1995). However, the stripe-faced dunnart (*Sminthopsis macroura*) failed to display AA, although the EF cycle clearly affected the free-running activity rhythm and in some cases entrained it (Kennedy, Coleman, & Armstrong, 1996). In view of the many factors that influence the expression of AA and the coupling of the FEO to the LEO, caution must be exercised when fundamental species differences are postulated.

Primates

An early study using squirrel monkeys (*Saimiri sciureus*) kept in LL found that 3 hours of food access per day entrained free-running rhythms of activity and body temperature (Sulzman, Fuller, & Moore-Ede, 1977). However, a later report indicates that light-entrainable rhythms freeran when food access was controlled automatically (Boulos, Frim, Dewey, & Moore-Ede, 1989). Intragastric feeding induced some relative coordination and in one case apparent entrainment of the activity and temperature rhythms (Aplegren, Frimm, Harling-Berg, Gander, & Moore-Ede, 1985). A long-term study of squirrel monkeys kept in LL and exposed to *T*-cycles of food access observed clear relative coordination of free-running activity rhythm, and when *T* was close to τ, the free-running rhythm became entrained (Aschoff & von Goetz, 1986). No increase in tree climbing was observed prior to meal time; however, attempts at pulling the food cup increased prior to meal time when the animals were on an LD 12:12 cycle. Interestingly, in LL, this behavior freeran in phase with tree-climbing activity rather than staying entrained to meal time. While this suggests that anticipatory behavior was not independently entrained to the EF cycle, it should be noted that anticipatory wheel running in rats can be driven out of entrainment when the free-running activity rhythm crosses food access (Stephan & Becker, 1989).

Whether or not humans have a separate FEO has not been ascertained. However, it is a common experience that hunger sensations occur around customary mealtime. If a meal is omitted, hunger subsides until the next mealtime. This would not be expected if hunger were strictly a deficit signal. Although the FEO clearly cannot account for hunger under all conditions (see above), the FEO could contribute a timing signal to hunger sensations in the same way that the LEO may generate a feeling of sleepiness in the absence of sleep deprivation. An interesting observation is that the spacing of meals in time-free environments appears to be strictly proportional to the time of wakefulness, even in states of internal desynchronization with waking times of some 40 hours (Aschoff, von Goetz, Wildgruber, & Wever, 1986). While a number of factors (e.g., a change in metabolic rate) could be responsible, it is not inconceivable that FEOs freerun with the circadian oscillator that mediates the sleep–wake cycle and maintain their relative phase angle differences.

Nonmammals

The ability of honeybees (*Apis mellifera*) to anticipate feeding times (Frisch & Aschoff, 1987) has already been pointed out. Several recent studies have investigated the effects of EF cycles on birds. In the house sparrow, EF cycles cause relative coordination of free-running activity rhythms and entrainment when *T* is close to

(Hau & Gwinner, 1992). Garden warblers have the ability to visit four feeding stations at different times of day and the pattern of visits persists in constant light (Biebach, Falk, & Krebs, 1991), but whether or not an FEO is involved is not clear. In a similar experiment with starlings, the feeding pattern established in LD persisted and freeran in LL although food was ad lib (Wenger, Biebach, & Krebs, 1991). Anticipatory key pecking has been demonstrated in pigeons and the key-pecking rhythm in ad lib feeding freeran in LL from the phase of the EF cycle, indicating entrainment of the circadian system (Abe & Sugimoto, 1987). On the other hand, body temperature of pigeons rises prior to food access while at the same time a free-running temperature rhythm is observed (Phillips, Rautenberg, Rashotte, & Stephan, 1992), suggesting separate control by the FEO and LEO. Furthermore, the temperature increases before light onset and before feeding are easily dissociated by phase shifting one or the other zeitgeber or by presenting them with different periods (Rashotte & Stephan, 1996).

Based on internal desynchronization of the perch-hopping and feeding rhythms of starlings in LL, Ebihara and Gwinner (1992) recently concluded that these behaviors are controlled by separate circadian oscillators.

Several species of fish also appear to anticipate EF cycles by increased activity (Spieler, 1992). As is the case for mammals, some behaviors (e.g., fighting) became entrained to feeding, whereas others (courtship and egg laying) remained entrained to the LD cycle (Weber & Spieler, 1987). Goldfish anticipate scheduled meals, and the presence of two periods after release into DD without food suggests that two oscillators might be involved (Sanchez-Vasquez, Madrid, Zamora, & Tabata, 1997).

Summary

The ability to anticipate periodic meals appears to be widespread among animals, suggesting a long phylogenetic history. Despite some species differences concerning the effects of EF cycles on light-entrainable circadian rhythms, these may not reflect fundamental differences in circadian organization, i.e., under particular experimental conditions, entrainment to food and to light appear to be mediated by separate, but coupled oscillators. Differences in coupling strength, in the phase and period of the two rhythms, and the choice of behavioral or physiologic processes selected for study may influence the particular results obtained. A schematic representation of the LEO and the FEO in mammals is shown in Figure 5. Undoubtedly, future research will reveal additional complexities and lead to refinement of this minimal model. Nevertheless, it is clear that EF cycles must be regarded as a powerful zeitgeber to the FEO and the behaviors and physiologic processes it controls. When meal time and the LD cycle conflict, it is not surprising that food often wins.

References

Abe, H., & Sugimoto, S. (1987). Food-anticipatory response to restricted food access based on the pigeon's biological clock. *Animal Learning Behavior, 15*, 353–359.

Abe, H., & Rusak, B. (1992). Anticipatory activity and entrainment of circadian rhythms in Syrian hamsters exposed to restricted palatable diets. *American Journal of Physiology, 263*, R116–R124.

Abe, H., Kida, M., Tsuji, K., & Mano, T. (1989). Feeding cycles entrain circadian rhythms of locomotor activity in CS mice but not in C57BL/6J mice. *Physiology and Behavior, 45*, 397–401.

Andrews, R. V. (1968). Temporal secretory responses of cultured hamster adrenal glands. *Comparative Biochemistry and Physiology, 26*, 179–193.

Apelgren, K. N., Frim, D. M., Harling-Berg, C. J., Gander, P. H., & Moore-Ede, M. C. (1985). Effectiveness of cyclic intragastric feeding as a circadian zeitgeber in the squirrel monkey. *Physiology and Behavior, 34*, 335–340.

Aschoff, J. (1987). Effects of periodic availability of food on circadian rhythms. In T. Hiroshige & K. Honma (Eds.), *Comparative aspects of circadian clocks* (pp. 19–40). Sapporo, Japan: Hokkaido University Press.

Aschoff, J. (1991). Activity in anticipation and in succession of a daily meal. *Bolletino Societa Italiana di Biologia Sperimentale, 67*, 213–228.

Aschoff, J., & von Goetz, C. (1986). Effects of feeding cycles on circadian rhythms in squirrel monkeys. *Journal of Biological Rhythms, 1*, 267–276.

Aschoff, J., von Goetz, C., & Honma, K. (1983). Restricted feeding in rats: Effects of varying feeding cycles. *Zeitschrift für Tierpsychologie, 63*, 91–111.

Aschoff, J., von Goetz, C., Wildgruber, C. & Wever, R. A. (1986). Meal timing in humans during isolation without time cues. *Journal of Biological Rhythms, 1*, 151–162.

Balsalobre, A., Damiola, F., & Schibler, U. (1998). A serum shock induces circadian gene expression in mammalian tissue culture cells. *Cell, 93*, 929–937.

Bays, M. E., & Stephan, F. K. (1992). Entrainment of anticipatory licking to nutritious solutions. *Society for Research on Biological Rhythms Abstracts, 83*, 82.

Biebach, H., Falk, H., & Krebs, J. R. (1991). The effect of constant light and phase shifts on a learned time–place association in garden warblers (*Sylvia borin*): Hourglass or circadian clock? *Journal of Biological Rhythms, 6*, 353–365.

Bolles, R. C., & Moot, S. A. (1973). The rat's anticipation of two meals a day. *Journal of Comparative Physiology and Psychology, 83*, 510–514.

Bolles, R. C., & Stokes, L. W. (1965). Rat's anticipation of diurnal and a-diurnal feeding. *Journal of Comparative Physiology and Psychology, 60*, 290–294.

Boulos, Z., & Logothetis, D. E. (1990). Rats anticipate and discriminate between two daily feeding times. *Physiology and Behavior, 48*, 523–529.

Boulos, Z., & Terman, M. (1980). Food availability and daily biological rhythms. *Neuroscience and Biobehavioral Reviews, 4*, 119–131.

Boulos, Z., Rosenwasser, A. M., & Terman, M. (1980). Feeding schedules and the circadian organization of behavior in the rat. *Behavioral Brain Research, 1*, 39–65.

Boulos, Z., Frim, D. M., Dewey, L. K., & Moore-Ede, M. C. (1989). Effects of restricted feeding schedules on circadian organization in squirrel monkeys. *Physiology and Behavior, 45*, 507–516.

Brinkhof, M. W. G., Daan, S., & Strubbe, J.H. (1998). Forced dissociation of food- and light-entrainable circadian rhythms of rats in a skeleton photoperiod. *Physiology and Behavior, 65*, 225–231.

Bünning, E. (1973). *The physiological clock.* New York: Springer-Verlag.

Challet, E., Malan, A., & Pevet, P. (1996a). Daily hypocaloric feeding entrains circadian rhythms of wheel-running and body temperature in rats kept in constant darkness. *Neuroscience Letters, 211*, 1–4.

Challet, E., Pévet, P., & Malan, A. (1996b). Intergeniculate leaflet lesion and daily rhythms in food-restricted rats fed during daytime. *Neuroscience Letters, 216*, 214–218.

Challet, E., Pévet, P., & Malan, A. (1997a). Lesions of the serotonergic terminals in the suprachiasmatic nuclei limits the phase advance of body temperature rhythm in food-restricted rats fed during daytime. *Journal of Biological Rhythms, 12*, 235–244.

Challet, E., Pévet, P., Lakhdar-Ghazal, N., & Malan, A. (1997b). Ventromedial nuclei of the hypothalamus are invoved in the phase advance of temperature and activity rhythms in food-restricted rats fed during daytime. *Brain Research Bulletin, 43*, 209–218.

Challet, E., Pévet, P., Vivien-Roels, B., & Malan, A. (1997c). Phase-advanced daily rhythms of melatonin, body temperature, and locomotor activity in food-restricted rats fed during daytime. *Journal of Biological Ryhthms, 12*, 65–79.

Chandrashekaran, M. K. (1982). Social cues and circadian rhythms. *Current Science, 51*, 158–167.

Choi, S., Wong, L. S., Yamat, C., & Dallman, M. F. (1998). Hypothalamic ventromedial nuclei amplify circadian rhythms: Do they contain a food-entrained endogenous oscillator? *Journal of Neuroscience, 18*, 3843–3852.

Clarke, J. D., & Coleman, G. J. (1986). Persistent meal-associated rhythms in SCN-lesioned rats. *Physiology and Behavior, 36*, 105–113.

Coleman, G. J., & Hay, M. (1990). Anticipatory wheel running in behaviorally anosmic rats. *Physiology and Behavior, 47,* 1145–1151.

Coleman, G. J., Harper, S., Clarke, J. D., & Armstrong, S. (1982). Evidence for a separate meal-associated oscillator in the rat. *Physiology and Behavior, 29,* 107–115.

Coleman, G. J., O'Reilly, H. M., & Armstrong, S. M. (1989). Food-deprivation-induced phase shifts in *Sminthopsis macroura frogatti. Journal of Biological Rhythms, 4,* 49–60.

Comperatore, C. A., & Stephan, F. K. (1987). Entrainment of duodenal activity to periodic feeding. *Journal of Biological Rhythms, 2,* 227–242.

Comperatore, C. A., & Stephan, F. K. (1990). Effects of vagotomy on entrainment of activity rhythms to food access. *Physiology and Behavior, 47,* 671–678.

Davidson, A. J. & Stephan, F. K. (1998). Circadian food anticipation persists in capsaicin deafferented rats. *Journal of Biological Rhythms, 13,* 422–429.

Davidson, A. J., & Stephan, F. K. (1999a). Plasma glucagon, glucose, insulin, and motilin in rats anticipating daily meals. *Physiology and Behavior, 66,* 309–315.

Davidson, A. J., & Stephan, F. K. (1999b). Feeding-entrained circadian rhythms in hypophysectomized rats with suprachiasmatic nucleus lesions. *American Journal of Physiology, 277,* R1376–R1384.

Davidson, A. J., Cappendijk, S. L. T., & Stephan, F. K. (2000). Feeding-entrained circadian rhythms are attenuated by lesions of the parabrachial region in rats. *American Journal of Physiology, 278,* R1296–R1304.

Ebihara, S., & Gwinner, E. (1992). Different circadian pacemakers control feeding and locomotor activity rhythms in European starlings. *Journal of Comparative Physiology A, 171,* 63–67.

Edmonds, S. C., & Adler, N. T. (1977a). Food and light as entrainers of circadian running activity in the rat. *Physiology and Behavior, 18,* 915–919.

Edmonds, S. C., & Adler, N. T. (1977b). The multiplicity of biological oscillators in the control of circadian running activity in the rat. *Physiology and Behavior, 18,* 921–930.

Escobar, C., Diaz-Munoz, M., Encinas, F., & Aguilar-Roblero, R. (1998). Persistence of metabolic rhythmicity during fasting and its entrainment by restricted feeding schedules in rats. *American Journal of Physiology, 274,* R1309–R1316.

Frisch, B., & Aschoff, J. (1987). Circadian rhythms in honeybees: Entrainment by feeding cycles. *Physiological Entomology, 12,* 41–49.

Gibbs, F. P. (1979). Fixed interval feeding does not entrain the circadian pacemaker in blind rats. *American Journal of Physiology, 236,* R249–R253.

Hardeland, R. (1973a). Circadian rhythmicity in cultured liver cells. I. Rhythms in tyrosine aminotransferase activity and inducibility and in [^{3}H] leucine incorporation. *International Journal of Biochemistry, 4,* 589–590.

Hardeland, R. (1973b). Circadian rhythmicity in cultured liver cells. II. Re-induction of rhythmicity in tyrosine transaminase activity. *International Journal of Biochemistry, 4,* 591–595.

Hau, M., & Gwinner, E. (1992). Circadian entrainment by feeding cycles in house sparrows, *Passer domesticus. Journal of Comparative Physiology, 170,* 403–409.

Holloway, W. R., Tsui, H. W., Grota, L. J., & Brown, G. M. (1979). Melatonin and corticosterone regulation: Feeding time or the light:dark cycle? *Life Science, 25,* 1837–1842.

Honma, K., Honma, S., & Hiroshige, T. (1984). Feeding-associated corticosterone peak in rats under various feeding cycles. *American Journal of Physiology, 246,* R721–R726.

Honma, S., Honma, K., Nagasaka, T., & Hiroshige, T. (1987). The ventromedial hypothalamic nucleus is not essential for the prefeeding corticosterone peak in rats under restricted daily feeding. *Physiology and Behavior, 39,* 211–215.

Honma, S., Kanematsu, N., & Honma, K. (1992). Entrainment of methamphetamine-induced locomotor activity rhythm to feeding cycles in SCN-lesioned rats. *Physiology and Behavior, 52,* 843–850.

Inouye, S. T. (1982a). Restricted daily feeding does not entrain circadian rhythms of the suprachiasmatic nucleus in the rat. *Brain Research, 232,* 194–199.

Inouye, S. T. (1982b). Ventromedial hypothalamic lesions eliminate anticipatory activities of restricted daily feeding schedules in the rat. *Brain Research, 250,* 183–187.

Inouye, S. T. (1983). Does the ventromedial hypothalamic nucleus contain a self-sustained circadian oscillator associated with periodic feedings? *Brain Research, 279,* 53–63.

Jilge, B. (1991). Restricted feeding: A nonphotic zeitgeber in the rabbit. *Physiology and Behavior, 51,* 157–166.

Jilge, B. (1995). Ontogeny of the rabbit's circadian rhythms without an external zeitgeber. *Physiology and Behavior, 58,* 131–140.

Jilge, B., & Stahle, H. (1993). Restricted food access and light–dark: Impact of conflicting zeitgebers on circadian rhythms of the rabbit. *American Journal of Physiology, 264,* R708–R715.

Kalsbeek, A., Barassin, S, van Heerikhuize, J. J., van der Vliet, & Buijs, R. M. (2000) Restricted daytime feeding attenuates reentrainment of the circadian melatonin rhythm after an 8-h phase advance of the light–dark cycle. *Journal of Biology Rhythms, 15*, 57–66.

Kennedy, G. A., Coleman, G. J., & Armstrong, S. M. (1990). The effect of restricted feeding on the wheel-running activity rhythms of the predatory marsupial *Dasyurus viverrinus. Journal of Comparative Physiology, 166*, 607–618.

Kennedy, G. A., Coleman, G. J., & Armstrong, S. M. (1991). Restricted feeding entrains circadian wheel-running activity rhythms of the kowari. *American Journal of Physiology, 261*, R819–R827.

Kennedy, G. A., Coleman, G. J. & Armstrong, S. M. (1995). Entrainment of circadian wheel-running rhythms of the northern brown bandicoot, *Isoodon macrourus*, by daily restricted feeding schedules. *Chronobiology International, 12*, 176–187.

Kennedy, G. A., Coleman, G. J., & Armstrong, S. M. (1996). Daily restricted feeding effects on the circadian activity rhythms of the stripe-faced dunnart, *Sminthopsis macrura. Journal of Biological Rhythms, 11*, 188–195.

Klein, D. C., Moore, R. Y., & Reppert, S. M. (1991). *Suprachiasmatic nucleus: The mind's clock*. New York: Oxford University Press.

Krieger, D. T. (1974). Food and water restriction shifts corticosterone, temperature, activity and brain amine periodicity. *Endocrinology, 95*, 1195–1201.

Krieger, D. T. (1980). Ventromedial hypothalamic lesions abolish food-shifted circadian adrenal and temperature rhythmicity. *Endocrinology, 106*, 649–654.

Krieger, D. T., & Hauser, H. (1978). Comparison of synchronization of circadian corticosteroid rhythms by photoperiod and food. *Proceedings of the National Academy of Sciences of the USA, 75*, 1577–1581.

Krieger, D. T., Hauser, H., & Krey, L. C. (1977). Suprachiasmatic nuclear lesions do not abolish food-shifted circadian adrenal and temperature rhythmicity. *Science, 197*, 398–399.

Kurumiya, S., & Kawamura, H. (1991). Damped oscillation of the lateral hypothalamic multineuronal activity synchronized to daily feeding schedules in rats with suprachiasmatic nucleus lesions. *Journal of Biological Rhythms, 6*, 115–127.

Marchant, E. G., & Mistlberger, R. E. (1997). Anticipation and entrainment to feeding time in intact and SCN-ablated C57BL/6j mice. *Brain Research, 765*, 273–282.

Mather, J. E. (1981). Wheel-running activity: a new interpretation. *Mammal Review, 11*, 41–51.

Meyer-Lohman, J. (1955). Über den einfluss taglicher futtergaben auf die 24-stunden periodik der lokomotorischen activität weisser mäuse. *Pflügers Archiv, 260*, 292–305.

Mistlberger, R. E. (1992a). Non-photic entrainment of circadian activity rhythms in suprachiasmatic nuclei-ablated hamsters. *Behavioral Neuroscience, 106*, 192–202.

Mistlberger, R. E. (1992b). Anticipatory activity rhythms under daily schedules of water access in the rat. *Journal of Biological Rhythms, 7*, 149–160.

Mistlberger, R. E. (1993). Circadian properties of anticipatory activity to restricted water access in suprachiasmatic nuclei-ablated hamsters. *American Journal of Physiology, 264*, R22–R29.

Mistlberger, R. E. (1994). Circadian food-anticipatory activity: Formal models and physiological mechanisms. *Neuroscience Biobehavioral Reviews, 18*, 171–195.

Mistlberger, R. E., & Marchant, E. G. (1999). Enhanced food-anticipatory circadian rhythms in the genetically obese Zucker rat. *Physiology and Behavior, 66*, 329–335.

Mistlberger, R. E., & Mumby, D. G. (1992). The limbic system and food-anticipatory circadian rhythms in the rat: Ablation and dopamine blocking studies. *Behavioral Brain Research, 47*, 159 168.

Mistlberger, R. E., & Rechtschaffen, A. (1984). Recovery of anticipatory activity to restricted feeding in rats with ventromedial hypothalamic lesions. *Physiology and Behavior, 33*, 227–235.

Mistlberger, R. E., & Rechtschaffen, A. (1985). Periodic water availability is not a potent zeitgeber for entrainment of circadian locomotor rhythms in rats. *Physiology and Behavior, 34*, 17–22.

Mistlberger, R. E., & Rusak, B. (1987). Palatable daily meals entrain anticipatory activity rhythms in free-feeding rats: Dependence on meal size and nutrient content. *Physiology and Behavior, 41*, 219–226.

Mistlberger, R. E., & Rusak, B. (1988). Food-anticipatory circadian rhythms in rats with paraventricular and lateral hypothalamic ablations. *Journal of Biological Rhythms, 3*, 277–291.

Mistlberger, R. E., Houpt, T. A., & Moore-Ede, M. C. (1990a). Characteristics of food-entrained circadian rhythms in rats during long-term exposure to constant light. *Chronobiology International, 7*, 383–391.

Mistlberger, R. E., Houpt, T. A., & Moore-Ede, M. C. (1990b). Food-anticipatory rhythms under 24-hour schedules of limited access to single macronutrients. *Journal of Biological Rhythms, 5*, 35–46.

Mistlberger, R. E., Houpt, T. A., & Moore-Ede, M. C. (1990c). Effects of aging on food-entrained circadian rhythms in the rat. *Neurobiology of Aging, 11*, 619–624.

Mistlberger, R. E., de Groot, M. H. M., Bossert, J. M., & Marchant, E. G. (1996). Discrimination of circadian phase in the rat: Role of light- and food-entrainable pacemakers. *Brain Research, 739*, 12–18.

Moreira, A. C., & Krieger, D. T. (1982). The effects of subdiaphragmatic vagotomy on circadian corticosterone rhythmicity in rats with continuous or restricted food access. *Physiology and Behavior, 28,* 789–790.

Nelson, W., Nichols, G., Halberg, F., & Kottke, G. (1973). Interacting effects of lighting [LD (12:12)] and restricted feeding (4h–24h) on circadian temperature rhythms of mice. *International Journal of Chronobiology, 1,* 347.

O'Reilly, H., Armstrong, S. M., & Coleman, G. J. (1986). Restricted feeding and circadian activity rhythms of a predatory marsupial, *Dasyuriodes byrnei. Physiology and Behavior, 38,* 471–476.

Ottenweller, J. E., Tapp, W. N., & Natelson, B. N. (1990). Phase shifting the light–dark cycle resets the food-entrainable circadian pacemaker. *American Journal of Physiology, 258,* R994–R1000.

Persons, J. E., Stephan, F. K., & Bays, M. E. (1993). Diet-induced obesity attenuates anticipation of food access in rats. *Physiology and Behavior, 54,* 55–64.

Philippens, K. M. H. (1980). Synchronization of rhythms to meal timing. In L. E. Scheving & F. Halberg (Eds.), *Chronobiology: Principles and applications to shifts in schedules* (pp. 403–416). Rockville, MD: Sijthoff & Noordhoff.

Philippens, K. M. H., von Mayersbach, H., & Scheving, L. E. (1977). Effects of scheduling of meal-feeding at different phases of the circadian system in rats. *Journal of Nutrition., 107,* 176–193.

Phillips, D. L., Rautenberg, W., Rashotte, M. E., & Stephan, F. K. (1993). Evidence for a separate food-entrainable circadian oscillator in the pigeon. *Physiology Behavior, 53,* 1105–1113.

Pittendrigh, C. S., & Daan, S. (1976a). A functional analysis of circadian pacemakers in nocturnal rodents I. The stability and lability of spontaneous frequency. *Journal of Comparative Physiology, 106,* 223–252.

Pittendrigh, C. S., & Daan, S. (1976b). A functional analysis of circadian pacemakers in nocturnal rodents IV. Entrainment: pacemaker as a clock. *Journal of Comparative Physiology, 106,* 291–331.

Pittendrigh, C. S., & Daan, S. (1976c). A functional analysis of circadian pacemakers in nocturnal rodents V. Pacemaker structure: A clock for all seasons. *Journal of Comparative Physiology, 106,* 333–355.

Rashotte, M. E., & Stephan, F. K. (1996). Coupling between light- and food-entrainable oscillators in pigeons. *Physiology and Behavior, 59,* 1005–1010.

Richter, C. P. (1922). A behavioristic study of the activity of the rat. *Comparative Psychology Monographs, 1,* 1–54.

Rosenwasser, A. M., Pelchat, R. J., & Adler, N. T. (1984). Memory for feeding time: Possible dependence on coupled circadian oscillators. *Physiology and Behavior, 32,* 25–30.

Rosenwasser, A. M., Schulkin, J., & Adler, N. T. (1985). Circadian wheel-running activity of rats under schedules of limited daily access to salt. *Chronobiology International, 2,* 115–119.

Rosenwasser, A. M., Schulkin, J., & Adler, N. T. (1988). Anticipatory appetitive behavior of adrenalectomized rats under circadian salt-access schedules. *Animal Learning Behavior, 16,* 324–329.

Rusak, B., Mistlberger, R. E., Losier, B., & Jones, C. H. (1988). Daily hoarding opportunity entrains the pacemaker for hamster activity rhythms. *Journal of Comparative Physiology, 164,* 165–171.

Saito, M., Murakami, E., Nishida, T., Fujisawa, Y., & Suda, M. (1976). Circadian rhythms of digestive enzymes in the small intestine of the rat. II. Effects of fasting and refeeding. *Journal of Biochemistry, 80,* 563–568.

Saito, M., Kato, H., & Suda, M. (1980). Circadian rhythm of intestinal disaccharidases of rats fed with adiurnal periodicity. *American Journal of Physiology, 238,* G97–G101.

Sanchez-Vazquez, F. J., Madrid, J. A., Zamora, S., & Tabata, M. (1997). Feeding entrainment of locomotor activity rhythms in the goldfish is mediated by a feeding-entrainable circadian oscillator. *Journal of Comparative Physiology A, 181,* 121–131.

Scheving, L. E., Tsai, T. H., Powell, E. W., Pasley, J. N., Halberg, F., & Dunn, J. (1983a). Bilateral lesions of suprachiasmatic nuclei affect circadian rhythms in [^{3}H]-thymidine incorporation into deoxyribonucleic acid in mouse intestinal tract, mitotic index of corneal epithelium, and serum corticosterone. *Anatomical Record, 205,* 239–249.

Scheving, L. E., Tsai, T. H., & Scheving, L. A. (1983b). Chronobiology of the intestinal tract of the mouse. *American Journal of Anatomy, 168,* 433–465.

Shearman, L. P., Zylka, M. J., Weaver, D. R., Kolakowski, Jr., L. F., & Reppert, S. M. (1997). Two period homologs: Circadian expression and photic regulation in the suprachiasmatic nuclei. *Neuron, 19,* 1261–1269.

Shibata, S., Minamoto, Y., Ono, M., & Watanabe, S. (1994). Age-related impairment of food anticipatory locomotor activity in rats. *Physiology and Behavior, 55,* 875–878.

Silverman, H. J., & Zucker, I. (1976). Absence of post-fast food compensation in the golden hamster (*Mesocricetus auratus*). *Physiology and Behavior, 17,* 271–285.

Spieler, R. E. (1992). Feeding-entrained circadian rhythms in fishes. In M. A. Ali (Ed.), *Rhythms in fishes* (pp. 137–147). New York: Plenum Press.

Stephan, F. K. (1981). Limits of entrainment to periodic feeding in rats with suprachiasmatic lesions. *Journal of Comparative Physiology, 143,* 401–410.

Stephan, F. K. (1983a). Circadian rhythm dissociation induced by periodic feeding in rats with suprachiasmatic lesions. *Behavioral Brain Research, 7,* 81–98.

Stephan, F. K. (1983b). Circadian rhythms in the rat: Constant darkness, entrainment to T cycles and to skeleton photoperiods. *Physiology and Behavior, 30,* 451–462.

Stephan, F. K. (1984). Phase shifts of circadian rhythms of activity entrained to food access. *Physiology and Behavior, 32,* 663–671.

Stephan, F. K. (1986a). The role of period and phase in interactions between feeding- and light-entrainable circadian rhythms. *Physiology and Behavior, 36,* 151–158.

Stephan, F. K. (1986b). Interaction between light- and feeding-entrainable circadian rhythms in the rat. *Physiology and Behavior. 38,* 127–133.

Stephan, F. K. (1986c). Coupling between feeding- and light-entrainable circadian pacemakers in the rat. *Physiology and Behavior, 38,* 537–546.

Stephan, F. K. (1989a). Forced dissociation of activity entrained to T cycles of food access in rats with suprachiasmatic lesions. *Journal of Biological Rhythms, 4,* 467–479.

Stephan, F. K. (1989b). Entrainment of activity to multiple feeding times in rats with suprachiasmatic lesions. *Physiology and Behavior, 46,* 489–497.

Stephan, F. K. (1992a). Resetting of a feeding-entrainable circadian clock in the rat. *Physiology and Behavior, 52,* 985–995.

Stephan, F. K. (1992b). Resetting of a circadian clock by food pulses. *Physiology and Behavior, 52,* 997–1008.

Stephan, F. K. (1997). Calories affect zeitgeber properties of the feeding entrained circadian oscillator. *Physiology and Behavior, 62,* 995–1002.

Stephan, F. K., & Becker, G. (1989). Entrainment of anticipatory activity to various durations of food access. *Physiology Behavior, 46,* 731–741.

Stephan, F. K., & Davidson, A. J. (1998). Glucose, but not fat, phase shifts the feeding-entrained circadian clock. *Physiology and Behavior, 65,* 277–288.

Stephan, F. K., Swann, J. M., & Sisk, C. L. (1979a). Anticipation of 24 hr feeding schedules in rats with lesions of the suprachiasmatic nucleus. *Behavior Neural Biology, 25,* 346–363.

Stephan, F. K., Swann, J. M. & Sisk, C. L. (1979b). Entrainment of circadian rhythms by feeding schedules in rats with suprachiasmatic lesions. *Behavior and Neural Biology, 25,* 545–554.

Stevenson, N. R., Sitren, H. S., & Furuya, S. (1980). Circadian rhythmicity in several small intestinal functions is independent of use of the intestine. *American Journal of Physiology, 238,* G203–G207.

Sulzman, F. M., Fuller, C. A., & Moore-Ede, M. C. (1977). Feeding time synchronizes primate circadian rhythms. *Physiology and Behavior, 18,* 775–779.

Sulzman, F. M., Fuller, C. A., Hiles, L. G., & Moore-Ede, M. C. (1978). Circadian rhythm dissociation in an environment with conflicting temporal information. *American Journal of Physiology, 235,* R175–R180.

Weber, D. N., & Spieler, R. E. (1987). Effects of the light-dark cycle and scheduled feeding on behavioral and reproductive rhythms of the cyprinodontfish, medaka, *Oryzias latipes. Experientia, 43,* 621–624.

Wenger, D., Biebach, H., & Krebs, J. R. (1991). Free-running circadian rhythm of a learned feeding pattern in starlings. *Naturwissenschaften, 78,* 87–89.

Yoshihara, T., Honma, S., Mitome, M., & Honma, K. (1997). Independence of feeding-associated circadian rhythm from light conditions and meal intervals in SCN lesioned rats. *Neuroscience Letters, 222,* 95–98.

Zielinski, W. J. (1986). Circadian rhythms of small carnivores and the effect of restricted feeding on daily activity. *Physiology and Behavior, 38,* 613–620.

9

Development of Mammalian Circadian Rhythms

FRED C. DAVIS AND STEVEN M. REPPERT

INTRODUCTION

The goal of developmental biology is to understand the origins of biological organization as it unfolds within each generation. Investigations of development often begin with describing how normal development proceeds, then ask how the observed processes or events are regulated. At the level of the whole organism, the most fundamental of such questions (even if oversimplified) is to ask whether the behavior, structure, or physiologic process of interest is inborn or is in some way shaped by the external environment under which the organism develops. This was one of the earliest questions to be asked about circadian rhythms. Charles Darwin contested the view he attributed to Wilhelm Pfeffer that the persistent rhythms of leaf movements in plants under constant conditions should be attributed to " 'Nachwirkung' or the aftereffects of light and darkness." Darwin concluded instead that "the periodicity of their movements is to a certain extent inherited" (Darwin, 1896). Seventy-five years later, as the modern field of circadian biology was being established, experiments with several different organisms investigated whether organisms needed to be exposed to 24-hour cycles in light and dark during development in order to express circadian rhythms when mature (Aschoff, 1960; Pittendrigh, 1954). The conclusion from these studies was that the expression of circadian rhythms is "independent of any ontogenetic learning process" (Pittendrigh, 1954). For the whole organism, this conclusion is still appropriate today, but as the regulation of circadian rhythmicity continues to be elucidated, the general question persists.

FRED C. DAVIS Department of Biology, Northeastern University, Boston, Massachusetts 02115. STEVEN M. REPPERT Laboratory of Developmental Chronobiology, Children's Service, Massachusetts General Hospital, Harvard Medical School, Boston, Massachusetts 02114.

Circadian Clocks, Volume 12 of *Handbook of Behavioral Neurobiology,* edited by Joseph S. Takahashi, Fred W. Turek, and R. Y. Moore, Kluwer Academic / Plenum Publishers, New York, 2001.

Understanding the relative contributions of intrinsic programs and environmental effects in guiding the differentiation of specific features of circadian organization remains a goal of developmental studies.

There are several reasons for trying to understand the development of any particular aspect of biological organization, including circadian rhythms: (1) To identify general mechanisms of development, such as the role of the environment; (2) to better understand the particular organization of interest as it exists and functions in the mature organism (using development as a "natural" experimental perturbation); (3) to learn whether the organization of interest has adaptive functions specific to stages of development; and (4) specifically to understand the etiology of normal and abnormal variations of the organization in humans. Each of these goals requires, as a starting point, descriptive information about development. In this chapter, we review the development of mammalian circadian rhythms, both to summarize the descriptive information and, where possible, discuss the relevance of the information to the above goals.

As with the field of circadian biology as a whole, the pace of research on the development of circadian rhythms has increased in the last 25 years, fueled in part by the discovery of specific tissues that function as circadian pacemakers. It is now established that the suprachiasmatic nucleus (SCN) of the hypothalamus is a circadian pacemaker in mammals. The term "circadian pacemaker" when used in reference to mammals is almost synonymous with the SCN. However, this linkage should be made with caution. "Circadian pacemaker" has a formal definition referring to properties which may or may not be attributable to the activity of the SCN alone. Defined formally, a circadian pacemaker is the mechanism which regulates the timing properties, in particular the phase and period, of a circadian rhythm. Developmental studies may address the development of a circadian pacemaker as it is formally defined, whereas others may address the development of the SCN, the anatomic locus of a circadian pacemaker. It is likely that the results of one type of study are directly relevant to the other, but this cannot always be assumed.

Development of Overt Rhythms

Although conceptually straightforward, the determination of when during development organisms first express circadian rhythms is not straightforward in its execution. The developing animal may not be amenable to methods of measurement used in adults, and, especially in the case of mammals, it may not be possible to isolate the animal from the influence of external rhythmicity. In the presence of either maternal or other environmental rhythmicity, the initial appearance of a rhythm may reflect the response of the animal to that rhythmicity rather than the expression of an endogenously generated rhythm. When the measurement of a rhythm depends on sampling from a population or when values from several individuals are combined for other reasons, the appearance of rhythmicity requires synchrony among the individuals. Such synchrony may not be possible until well after rhythmicity is expressed by individuals. Even when a rhythm is clearly expressed by an individual, the age of first expression may reflect the development of output mechanisms rather than initiation of the underlying pacemaker.

With few exceptions, the expression of circadian rhythms does not occur until after birth in mammals. This is true for the most commonly measured rhythms of plasma or pineal melatonin, body temperature, adrenal steroid secretion, and activity and rest (Davis, 1981). Although some of these rhythms have been measured in fetuses, it is uncertain whether the rhythms are regulated by an active fetal circadian pacemaker or are responses to maternal rhythms. These are not mutually exclusive alternatives and are further discussed below with respect to rhythms in humans. In some animals, rhythms have been measured within the fetal SCN itself, and these rhythms are therefore likely to reflect endogenously generated oscillations (see below).

Development of the Suprachiasmatic Nucleus (SCN)

In all mammals, the SCN is a bilaterally represented cluster of several thousand neurons and glia located near the midline within the anterior hypothalamus. Although a great deal is known about the anatomy of the SCN in a variety of mammals, including humans, the relationship between various anatomic features and the the function of the SCN is poorly understood. At least eight neuropeptides and neurotransmitters have been localized to cell soma within the SCN (Mai, Kedziora, Teckhaus, & Sofroniew, 1991; J. D. Miller, Morin, Schwartz, & Moore, 1996; Van den Pol & Tsujimoto, 1985), and although some of these may be colocalized within cells, the number of different neurons within the SCN is at least as great. A given type of neuron, defined by specific immunoreactivity, is generally not uniformly distributed within the SCN, and the SCN have been described as comprising distinct subdivisions, either ventrolateral and dorsal medial divisions or a core and shell (Card & Moore, 1984; Cassone, Speh, Card, & Moore, 1988; Decavel & Van den Pol, 1990; Mai *et al.*, 1991; J. D. Miller *et al.*, 1996; Van den Pol & Tsujimoto, 1985). Additional neuropeptides and neurotransmitters are present in fibers afferent to the SCN, including projections from the retina (retinohypothalamic tract, RHT; glutamate), the intergeniculate leaflet (IGL) of the lateral geniculate (geniculohypothalamic tract, GHT; neuropeptide Y), midbrain raphe nuclei (serotonin), basal forebrain (cholinergic), and mammillary hypothalamus (histamine) (J. D. Miller *et al.*, 1996). These projections also show regional distributions.

Formation of the SCN

The SCN arises from proliferating cells of the neuroepithelium in the anterior, ventral diencephalon. It arises at the boundary between the preoptic area, the most rostral extent of the hypothalamus, and the anterior hypothalamic area. The neuroepithelium which gives rise to the SCN appears to be located just at the caudal edge of the preoptic recess and optic fissures (Altman & Bayer, 1978a, b, 1986, 1995) (Figure 1). Altman and Bayer (1986) concluded on the basis of sequential thymidine autoradiograms that the SCN arises from rostral anterior hypothalamic neuroepithelium located within a ventromedian evagination of the ventral third ventricle. Laterally and rostrally, this neuroepithelium is continuous with optic stalk neuroepithelium (Altman & Bayer, 1986, 1995). Circadian pacemaker cells may become localized to the suprachiasmatic region of the hypothalamus because the neuroepithelium which gives rise to the SCN shares a common history with that which gives rise to the retina. The mammalian retina also contains cells capable of generating circadian oscillations (Tosini & Menaker, 1996), and the SCN and retina may arise from a common, or at least contiguous, neuroepithelium with a restricted fate that includes the ability to produce circadian pacemaker cells.

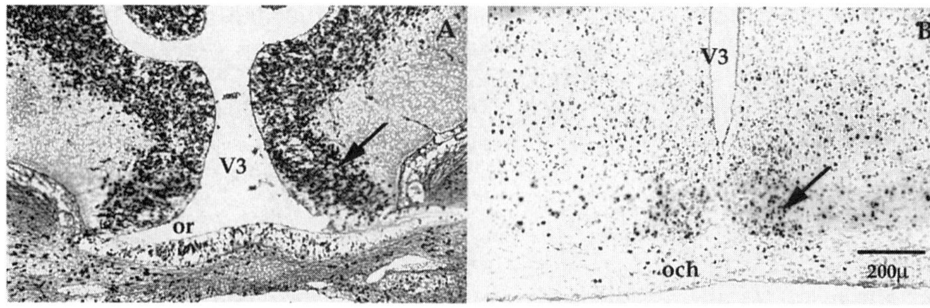

Figure 1. Neurogenesis of the SCN. Photomicrographs of coronal sections through the brains of (A) fetal and (B) 21-day-old mouse (C57BL/6) brains. Both mice received single injections of the thymidine analog bromodeoxyuridine (BrdU) on embryonic day 13 (E13). The mouse on the left was sacrificed 2 hours later, and that on the right was sacrificed 3 weeks after birth. The brains were processed for immunoreactivity to BrdU, and cell nuclei staining for BrdU appear dark. Stained cells were undergoing DNA synthesis at the time of the injections, and in the older brain are cells that became postmitotic (were "born") soon after the injection. The large number of stained cells in the SCN (arrow) indicates that SCN neurogenesis was occurring on E12. The section from a fetal brain shows third ventricle and optic stalk neuroepithelium at the time of SCN neurogenesis. Areas of dark staining around the ventricle (arrow) indicate high proliferative activity. The SCN are likely to arise from third ventricle neuroepithelium just caudal to the section shown (V3, third ventricle; och, optic chiasm; or, optic recess). (Courtesy of Colette Kabrita.)

The time in development when SCN neurons become postmitotic has been determined using thymidine autoradiography in rats, mice, hamsters, and the gray short-tailed opossum. As shown in Table 1, neurogenesis of the SCN occurs before birth in rats, mice, and hamsters and after birth in the opossum. In rats, the SCN has been generalized by Altman and Bayer (1986) to arise as part of a third wave of hypothalamic development. The first wave of cell production gives rise to the "reticular hypothalamus," which includes the lateral hypothalamus; the second gives rise to the "core," which includes structures such as the medial preoptic area, anterior hypothalamic area, ventromedial nucleus, and the mammillary complex; and third gives rise to the "midline," which includes the SCN and arcuate nucleus.

TABLE 1. NEUROGENESIS OF THE RODENT SUPRACHIASMATIC NUCLEUS USING THYMIDINE AUTORADIOGRAPHY[a]

Rodent	Range	Peak	Gestation (days)	Reference
Norway rat (*Rattus norvegicus*) Purdue-Wistar	E14–E17	E15	22	Altman & Bayer, 1978a
Syrian hamster (*Mesocricetus auratus*)	E10.5–E13.0	E11.5	15.5	Davis *et al.*, 1990
House mouse (*Mus musculus*) ICR-JCL	E10–E14	E12	19	Shimada & Nakamura, 1973
Gray short-tailed opossum (*Monodelphis domestica*)	P4–P8	P4	14–15	Rivkees *et al.*, 1988

[a]The first 24 hours after fertilization is embryonic day 1 (E1) and the day of birth is postnatal day 1 (P1).

In mice and rats, these developmentally defined regions correspond roughly to functional/anatomic divisions of the developed hypothalamus, the lateral, medial, and periventricular zones (Niimi, Harada, Kusaka, & Kishi, 1962; Okamura *et al.*, 1983; Shimada & Nakamura, 1973; Swanson, 1987).

Within the SCN itself, gradients in neurogenesis have been described indicating that different regions of the SCN and possibly, therefore, different cell types have different birthdates. In particular, in rats and hamsters, ventrolateral cells tend to become postmitotic before dorsomedial cells, and caudal cells are produced somewhat earlier than rostral cells (Altman & Bayer, 1986; Davis, Boada, & LeDeaux, 1990). In mice, the production of neurons expressing vasopressin is greatest near the end of neurogenesis (Okamura *et al.*, 1983; Shimada & Nakamura, 1973). In both rats and hamsters, there are exceptions to the predominant gradients. In the rat, late-produced cells settle ventrally in a region described by Altman and Bayer as the "basal subnucleus" of the SCN (Altman & Bayer, 1986). In the hamster, late-produced cells can be found on the lateral margin of the nucleus as well as in the dorsomedial region (Davis *et al.*, 1990). The settling patterns of SCN neurons may reflect both passive, lateral displacement of early cells by late-produced cells as well as active migration of late-produced cells around the margins of the nucleus.

The sequential production of different cell types is a common feature of nervous system development. For example, in the rat, retinal ganglion cells are produced before birth and rod photoreceptors are produced primarily after birth. The type of cells produced by progenitors appears to be progressively restricted under the influence of the microenvironment provided by cells already produced (Reh, 1992). How the fate of particular cell types within the SCN is determined is not known. Altman and Bayer (1986) suggested that the ventrolateral and dorsomedial SCN arise from different regions within a ventromedian evagination of the ventral third ventricle. Specifically, they suggested that earlier mitotic activity in the ventral portion of the evagination gives rise to the ventrolateral SCN followed by increased mitotic activity dorsally giving rise to the dorsomedial SCN. The more dorsal mitotic activity was suggested to also produce cells of the basal subnucleus. It is not known, however, whether progenitor cells near the end of neurogenesis can only produce cells characteristic of the dorsomedial SCN, such as those immunoreactive for vasopressin, or whether the fate of the cells produced at that time is unrestricted and determined by the environment into which they migrate.

The divergence of neuronal and glial fates has also been extensively examined in other regions of the nervous system. Glial cells of the SCN may be produced at the same time that neurons are produced, but may also continue to be produced after neurogenesis is complete. Altman and Bayer (1986) suggested that cells labeled by thymidine and found to be away from the neuroepithelium immediately after exposure to the label are glia produced by cells capable of dividing away from the ventricular epithelium. Such labeled cells were found during SCN neurogenesis, but increased in the hypothalamus as a whole after neurogenesis, suggesting that most glia are produced after neurons. In Syrian hamsters, astrocytes within the SCN begin to express glial fibrillary acidic protein (GFAP) on day 15 of gestation and the expression continues to increase for 3 weeks (Botchkina & Morin, 1995a). Thus, GFAP expression occurs after SCN neurogenesis and after SCN neurons begin to express phenotypic characteristics (see below). On the other hand, radial glia (identified by their morphology and immunoreactivity to vimentin), which may later become astrocytes, are already present in the SCN at the time of SCN neurogenesis (Botchkina & Morin, 1995a).

A common feature of neuronal development in many parts of the nervous system is the overproduction of neurons followed by a period of cell death. Although some cell death may occur in the SCN during development, its magnitude and significance are unclear. In the rat, neurogenesis of the SCN is complete by E17 (day 17 of embryonic development) and only occasional degenerating perikarya can be seen between E21 and P6 (postnatal day 6) (Moore & Bernstein, 1989). In three strains of mice showing abnormal development, SCN smaller than those of control mice were reported (Noguchi, Sugisaki, Kudo, & Satoh, 1986; Silver, 1977). There was no evidence, however, that these effects resulted from increased cell death. The greatest such effect was in anophthalmic mice (strain ZRDCT-AN), in which 30% of the strain showed at least one SCN which was significantly reduced in size (as low as 15% of normal cell number). The reduced cell number was suggested to result from disruptions of the epithelium caused by the resorption of the eye rudiments on E13, possibly reducing the population of proliferating cells or the rate of proliferation. Cell death did not appear to be a factor in the normal or abnormal development of the SCN (Scheuch & Silver, 1982; Silver, 1977). Three of four mutants with SCN that were 50% or less the size of normal showed noncircadian patterns of activity, whereas only 1 of 16 mutants with normal-sized SCN showed a noncircadian pattern. The loss of circadian function appeared to related to the overall size of the SCN, but it was also possible that the mutation affected the function of pacemaker cells even when the gross morphology of the SCN was not severely affected (Laemle & Rusa, 1992; Silver, 1977).

Conditions during development which might be expected to promote cell death have not been found to do so in the SCN. Deafferentation of the SCN by perinatal enucleation does not have gross morphologic effects on the SCN or on SCN cell number (Lenn, Beebe, & Moore, 1977), and no effect on SCN morphology was observed in a strain of anophthalmic rats (Ibuka, 1987). Thus, it is unlikely that the overproduction of cells followed by cell death has a significant role in SCN development. It is more likely that cell number is regulated primarily by the rate and duration of proliferation and by the coalescence of cells into nuclei.

Differentiation of the SCN

SCN neurons begin to express phenotypic characteristics soon after becoming postmitotic. Within 2 or 3 days of SCN neurogenesis, mRNAs (vasopressin, VP; vasoactive intestinal peptide, VIP) proteins (VP; peptide histidine isoleucine PHI; VIP), and specific ligand binding (VIP, melatonin), characteristic of SCN cells, can be detected (Ban, Shigeyoshi, & Okamura, 1997; Botchkina & Morin, 1995a; Duncan & Davis, 1993; Kuhlman, Watts, Sanchez-Watts, & Davis, 1995; Reppert & Uhl, 1987; Robinson & Fuchs, 1993; Romero & silver, 1990; 1991; Terman, Remé, & Terman, 1993; Whitnall, Key, Ben-Barak, Ozato, & gainer, 1985; Williams, Martinoli, Titchener, & Pelletier, 1991). VIP peptide expression is observed before VP expression, consistent with the possibility that some cell types are produced before others. For a given peptide, the initial expression of the peptide and its mRNA occur at approximately the same time, indicating that appropriate translation machinery is in place at the time of initial gene expression (Reppert & Uhl, 1987). Also, when specific mRNAs or peptides are first expressed, their distributions are roughly similar to those observed in adults. There is some evidence, however, that VIP expression and the number and distribution of VIP cells are influenced by retinal afferents; expression of VIP mRNA was found to be elevated after neonatal blinding in rats (Holtzman, Malach, &

Gozes, 1989), there is a large increase in mRNA after postnatal day 10 (Ban *et al.*, 1997), and the number and distribution of VIP immunoreactive cells are greater in anophthalmic mice than sighted mice (Laemle & Rusa, 1992).

No SCN-specific protein or mRNA has been observed to be expressed before or during SCN neurogenesis. The hypothalamic expression of *Oct-2*, a gene coding for a transcriptional activating protein of the POU family, is restricted to the SCN in the adult (He *et al*, 1989), and it is expressed both before and during SCN neurogenesis. However, its distribution at those times is throughout the ventral hypothalamus (Rivkees, Weaver, & Reppert, 1992), and it is not essential for SCN development (Ninkina, 1995).

A gene (*clock*) has been identified in mice which, when mutated, affects free-running period and, in homozygotes, the expression of rhythmicity (Vitaterna *et al.*, 1994). This gene has been recently cloned and it and its protein products are likely to be involved in the generation of circadian oscillations (Antoch *et al.*, 1997; King *et al.*, 1997). The developmental expression of this gene has not been examined, but when accomplished, may provide insight into the initiation of pacemaker function.

Although there is rapid differentiation of SCN neurons during the first few days after the cells are produced, the differentiation of the SCN as a whole continues over at least 2 weeks. This is based on functional studies (see below) as well as on cytologic studies. Using both ultrastructure and immunoreactivity to the synapse-associated protein synapsin, Moore and Bernstein (1989) found that synaptogenesis within the rat SCN occurs primarily between embryonic day 21 and postnatal day 10. This general time frame for synaptogenesis is supported by other studies (Güldner, 1978; Koritsanszky, 1981; Laemle, Repke, Hawkes, & Rice, 1991; Lenn *et al.*, 1977). Synaptogenesis in the Syrian hamster appears to begin and end somewhat earlier (E15–P4) (Speh & Moore, 1993).

The SCN of mammals receives three principal afferent projections: from the retina (RHT), from the intergeniculate leaflet of the lateral geniculate nucleus (GHT), and a serotonergic projection from the midbrain raphe nuclei (Morin, 1994). The development of synapses from these projections contributes to SCN synaptogenesis; the development of the RHT in rats and hamsters based on anterograde transport of cholera toxin conjugated to horseradish peroxidase occurs most intensively between postnatal days 1 and 10 and days 4 and 15, respectively. Based on immunocytochemistry for serotonin, the serotonergic innervation of the SCN develops between E22 and P21 in rats (Ugrumov *et al.*, 1994a) and between P3 and P21 in Syrian hamsters (Botchkina & Morin, 1993). Like the effects of retinal deafferentation during development, the prenatal depletion of serotonin using *p*-chlorphenylalanine in rats causes an elevated expression of VIP in the SCN (Ugrumov, Trembleau, & Calas, 1994c). In hamsters, the GHT, based on neuropeptide Y (NPY) immunoreactivity, develops between P4 and P11 (Botchkina & Morin, 1995b). The development of this projection in rats is probably similar (Kagotani *et al.*, 1989). The IGL itself is formed in the hamster around embryonic day 14 (Botchkina & Morin, 1995b). Interestingly, the IGL appears to have an embryonic origin which is different from that of the dorsal and ventral lateral geniculate nuclei. The latter arise from the lateral ventricle, whereas at least some IGL neurons which express NPY during their migration are produced in the dorsal third ventricle. These cells migrate along radial glia that appear subsequently to become astrocytes withing the IGL (Botchkina & Morin, 1995b).

Retinal projections of the visual system are topographically organized and the mechanisms by which the correspondence between retinal and brain maps is achieved has been greatly studied. The RHT, however, does not appear to be topographically organized. The rat and hamster SCN receives input from ganglion cells throughout the retina (Moore, Speh, & Card, 1995; Pickard, 1980), and in sheep, any particular region of the RHT projection field within the SCN may receive input from any of the widely dispersed ganglion cells that project to the SCN (Cooper *et al.*, 1993). Even if there is a lack of topographic organization within the RHT, there must be other types of specification. For example, in some species (including rats) the RHT projection from one retina is not bilaterally symmetric (Magnin, Cooper, & Mick, 1989; Moore, 1973), i.e, there is preference for either ipsilateral or contralateral projections. Furthermore, the RHT often projects more strongly to the ventrolateral SCN (Güldner, 1978; Johnson, Morin, & Moore, 1988; Magnin *et al.*, 1989), and there is evidence that within the projection field, ganglion cells are segregated, not topographically, but by whether they are collaterals of axons projecting to the geniculate or are axons projecting only to the SCN (Treep, Abe, Rusak, & Goguen, 1995).

The mechanisms by which RHT axons identify their targets is not known. In rats (but not in hamsters) the RHT initially projects to a larger field which includes areas outside of the SCN. These ectopic projections disappear between P4 and P10 (Speh & Moore, 1993), a time when ganglion cell death is occurring in the retina (Sengelaub & Finlay, 1982). Thus it is possible that, as elsewhere in the nervous system, the survival of afferents depends on competition for a limited resource provided by functionally appropriate targets. In rats, the RHT from each eye is asymmetric, projecting more heavily to the contralateral SCN (Johnson *et al.*, 1988; Moore & Lenn, 1972). If one eye is removed early in development (P2), projections of the remaining eye to the ipsilateral SCN increase (Stanfield & Cowan, 1976), presumably in response to the availability of additional synaptic sites and possibly to an increased availability of trophic substances. The projections of the remaining eye continue to be asymmetric (Stanfield & Cowan, 1976), suggesting that other SCN afferents, such as the GHT, also occupied some of the newly available synaptic sites. In hamsters, plasticity of the RHT may occur in adults; anomalous hypothalamic projections have been observed following knife cuts of the RHT (Johnson, Moore, & Morin, 1988).

While it is likely that interactions with target cells influence the development of the RHT, there is little evidence for the converse, i.e., that SCN afferents influence the development of the SCN. In particular, there is no evidence that the RHT specifies which hypothalamic cells will become pacemaker cells or that the RHT is required for pacemaker cell differentiation. For example, ablations of the SCN before RHT development do not result in the respecification of another tissue as pacemaker tissue; circadian rhythms in rats are permanently abolished by lesions early in development just as they are when lesions are performed in adults, and the early lesions prevent the development of the RHT (Mosko & Moore, 1978, 1979). In addition, grafts of fetal tissue containing the SCN restore rhythmicity in SCN-lesioned host even though the tissue is transplanted before the SCN would have been innervated by the RHT (Drucker-Colin, Aguilar-Roblero, Garcia-Hernandez, Fernandez-Cancino, & Rattoni, 1984; Lehman *et al.*, 1987). Furthermore, no difference was observed in the average free-running periods of neonatally enucleated and intact hamsters (Davis, 1982). As previously noted, however, VIP expression may be elevated by neonatal deafferentation (Holtzman *et al.*, 1989), and in anophthalmic mice, dendritic process were found to be abnormal for most SCN neurons (Silver, 1977).

Little is known about the development of SCN efferents. When an SCN-controlled rhythm appears during development, this means that not only is the pacemaker functional at that time, but the output pathway to that function has also developed. The best-characterized efferent pathway of the SCN is that for the control of the pineal melatonin rhythm. In rats, a rhythm in activity of *n*-acetyltransferase (NAT, the rate-limiting enzyme for the synthesis of melatonin) begins on postnatal day 2–4, so that by this age, functional SCN efferents must have developed (Ellison, Weller, & Klein, 1972). Other evidence for the development of SCN output comes from transplantation studies in which fetal tissue is transplanted. It is possible for rhythmicity to be restored in the hosts within 1 week, suggesting that in an intact hamster during normal development, SCN output is sufficient to drive a rhythm in activity and rest at least by the first postnatal week. The restoration of rhythmicity by grafts transplanted within capsules suggests that, at least for locomotor activity, the output of the SCN is a diffusible signal that may not require the development of synapses (Silver, LeSauter, Tresco, & Lehman, 1996).

Specification of Fate

Within complex metozoans such as vertebrates, it is likely that the ability to generate circadian oscillations is a property restricted to specialized cells. In mammals, for example, SCN neurons generate circadian oscillations, whereas most other hypothalamic and other central nervous system (CNS) neurons probably do not (Inouye & Kawamura, 1979; LeSauter, Lehman, & Silver, 1996; Welsh, Logothetis, Meister, & Reppert, 1995). While it is certain that some cells other than those of the SCN (the retina, for example) generate circadian oscillations, it is nevertheless appropriate to view this ability as a specialized property of certain cells that arises during cellular differentiation. This differentiation likely involves specific gene expression. How many such genes might be involved and how cellular fate, as defined by these genes, is regulated is not known. For example, the recently cloned *clock* gene of mammals is expressed in areas of the brain other than the SCN (Antoch *et al.*, 1997). Either cells in other areas generate circadian oscillations or *clock* is expressed as part of other cellular phenotypes.

The mechanism of fate determination is a fundamental question in developmental biology and there are usually two general choices for the answer. Either the cell and its lineage are programmed from some early time in development eventually to achieve a certain fate or, alternatively, even up to the time of and after final differentiation, the cell's fate can be specified by its environment. Although this question has yet to be adequately addressed for circadian pacemaker cells, the differentiation of the mammalian SCN as a whole appears to be little affected by the environment in which it develops. In rats, fetal anterior hypothalamic tissue has been transplanted to the anterior chamber of the eye in adult hosts before SCN neurogenesis (E13). Clusters of cells immunoreactive for VIP and vasopressin, characteristic of the SCN, were observed in the tissues despite this development in an abnormal environment. Thus, before SCN cells have become postmitotic and before SCN afferents develop, the fate of anterior hypothalamic tissue to produce SCN-like neurons has been determined. It is possible, however, that the fate of SCN cells themselves was not already determined at the time of transplantation, but instead was subsequently influenced by signals from the other hypothalamic tissue within the grafts. Development within the anterior chamber was not entirely normal; grafts transplanted at E13 did not express peptides for 6 weeks and those transplanted at

E16 never expressed them. In contrast, grafts from E13–E17 fetuses transplanted to the third or lateral ventricles express characteristic peptides within 3–6 weeks (Griffioen, Duindam, Van der Woude, Rietveld, & Boer, 1993; Wiegand & Gash, 1988). Even in these grafts, however, peptide expression is delayed relative to the *in situ* SCN by 1–2 weeks (Griffioen *et al.*, 1993; Romero & Silver, 1990). It appears, therefore, that the environment, although crudely manipulated in these studies, can to some extent affect the normal timing of phenotypic expression.

Cultured SCN cells or tissue are also capable of expressing phenotypes characteristic of the *in situ* SCN. SCN cells dispersed from 2- to 4-day old rat pups express the same variety of peptides as the *in situ* SCN, with different cell types expressed in approximately the same proportions (Welsh *et al.*, 1995). The dispersed neurons also generate circadian oscillations with a range of periods from 21.5 to 26.5 hours. Whether the range of periods observed in culture reflects the variation that exists among pacemaker cells *in vivo* is not known. In some transplantation studies, fetal tissue was minced or cells dispersed prior to transplantation (Silver, Lehman, Gibson, Gladstone, & Bittman, 1990; Wiegand & Gash, 1988), yet SCN phenotypes developed, indicating that specific spatial relationships among SCN cells during differentiation are not required for phenotypic expression. Rat brain slices containing the SCN have also been maintained in culture for several weeks, during which time the slices thin out ("organotypic"). Although most of the cells die during the maintenance of these slices the phenotypic organization of the SCN is remarkably well preserved, and some peptides (VP, VIP) are secreted with a circadian rhythm (Tominaga, Inouye, & Okamura, 1994; Wray, Castel, & Gainer, 1993). These results indicate that by at least postnatal day 4 in rats, a "program" for SCN development appears to be underway and expressed in culture; synaptic interactions develop and subpopulations of cells identified by peptide and mRNA expression become regionally organized.

Functional Development of the Pacemaker

The functional development of the SCN refers to the developmental onset of circadian oscillations generated by the SCN, the development, by functional criteria, of input to and output from the SCN, and developmental changes in properties such as free-running period. Most of what is known about the functional development of the SCN is from work on rats and Syrian hamsters and is briefly summarized below and in Figure 2, followed by more detailed discussion of specific topics.

Circadian oscillations are probably first generated by the SCN within 1 or 2 days after the neurons become postmitotic, but possibly earlier. Part of the evidence for this is that the oscillations become entrainable before birth, not by light, but by maternal rhythmicity or various experimental treatments. In addition, rhythmicity expressed by the fetal SCN has been directly measured both *in vivo* and *in vitro*. After becoming postmitotic, individual SCN neurons, possibly all SCN neurons (Welsh *et al.*, 1995), begin to generate oscillations. Individually these cells would probably express different phases and periods, but they normally become coupled by mechanisms that are not yet understood (Welsh *et al.*, 1995). Early in development, however, coupling is probably mediated, at least in part, by signals from the mother.

As already described, synaptogenesis within the SCN, probably involving both intrinsic and afferent connections, is most intense in the first few days after birth. During this time, in rats, the sensitivity of the SCN to maternal signals decreases and

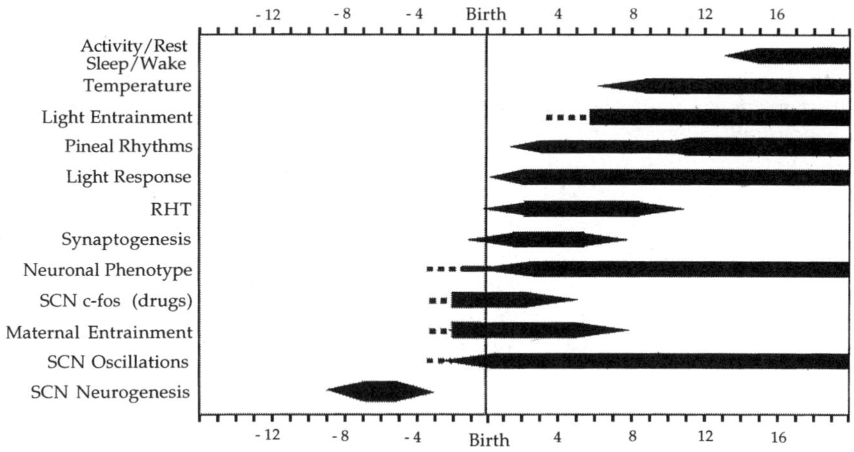

Figure 2. Development of the rodent circadian system, especially the suprachiasmatic nucleus (SCN). The time scale shows days relative to birth. The black bars indicate the times in development after which structures or functions exist or during which developmental processes are occurring. Dashed lines indicate uncertainty about the beginning of a feature, and variation in line thickness roughly indicates the maturation of the feature. Maternal entrainment includes entrainment by exogenous melatonin. Drug effects on *c-fos* refers to the effects of dopamine agonists and nicotine. Light responses refers to acute changes in *c-fos* expression or 2DG uptake in the SCN.

the sensitivity to photic stimuli increases, so that by the end of the first postnatal week, the SCN is entrainable by light (Duncan, Banister, & Reppert, 1986). In mice, sensitivity to maternal signals may continue for several weeks after birth (Viswanathan, 1989). SCN synaptogenesis in the first postnatal week might also be related to an increase in the amplitude of SCN rhythmicity (Shibata & Moore, 1987). In contrast to these changes in sensitivity to stimuli and in amplitude, there is no evidence for a systematic change in the free-running period of circadian oscillations as the SCN develops. It is also during the first postnatal week that the SCN begins to regulate overt rhythmicity, in particular the rhythm in pineal NAT activity (Ellison *et al.*, 1972). The development of other overt rhythms, such as those in sleep and wakefulness and hormone secretion, also occur after the initiation of pacemaker function, probably reflecting the ongoing maturation of output pathways and systems.

Initiation of Oscillations

Deguchi (1975) reported that the postnatal phase of the rhythm in rat pineal NAT activity was determined in part by the phase of maternal rhythmicity experienced before birth; the NAT rhythm in a population of 10-day old pups was out of phase with foster mothers if the prenatal phase to which the pups were exposed was different from that of the foster mothers. This result was substantiated and extended by Reppert, Coleman, Heath, & Swedlow (1984), who demonstrated that not only was phase initially established before birth, but also that maternal rhythmicity was responsible since it was the phase of maternal rhythms rather than the external light/dark cycle that was critical. Furthermore, they demonstrated that the ability of a foster mother to reset phase postnatally diminished after the first postnatal week. Thus, while phase is initially established before birth, the dam is also able to cause entrainment after birth (Deguchi, 1975; Takahashi & Deguchi, 1983; Takahashi, Hayafuji, & Murakami, 1982). In the mouse *Mus budooga*, entrainment by artificial cycles in the presence and absence of the dam continues for up to 5 weeks after birth (Viswanathan & Chandrashekaran, 1985).

Evidence for prenatal entrainment similar to that initially obtained using the pineal NAT rhythm in rats was also obtained for the plasma corticosterone (Hiroshige, Honma, & Watanabe, 1982b) and temperature (Nuesslein & Schmidt, 1990) rhythms in rats and for the wheel-running activity rhythm in Syrian hamsters (Davis & Gorski, 1986) and the spiny mouse (Weaver & Reppert, 1987). Because these rhythms are regulated by the SCN in adults, the prenatal establishment of phase in them is likely to be accomplished by the prenatal entrainment of oscillations generated by the SCN, thus providing evidence that the SCN is generating oscillations before birth.

The direct measurement of rhythmicity in the fetal SCN was first accomplished in rats by Reppert and Schwartz (1983, 1984). They measured 2-deoxy-D-glucose (2DG) uptake in the fetal SCN at two times of day. Pregnant dams were entrained to different light/dark cycles so that fetuses in different groups were of the same development age during either the dam's subjective day or subjective night. A clear difference in 2DG uptake was observed between day and night in both the fetuses and dams, with uptake being higher during the subjective day. They also showed that 2DG uptake in the fetal SCN was not directly affected by light, indicating that the rhythm was either directly caused by maternal rhythms or was regulated by endogenous oscillations that had been entrained by maternal rhythms. The latter would be consistent with the evidence for prenatal entrainment of the pacemaker described above. Shibata and Moore (1988) measured 2DG uptake in slices from E22 rats at two times of day, 1 hour after taking the fetus from the dam. A day/night difference was observed even though the slices were no longer exposed to maternal rhythms.

The day/night difference in 2DG uptake *in vivo* was first seen at about E19.5 and at all subsequent fetal ages (E20.5 and E21.5) (Reppert & Schwartz, 1984). E19.5 appeared to be transitional in that detectable uptake during the subjective day was observed in about one half of the fetuses, but was not observed in any fetuses at E18.5. It is possible that oscillations started before E19.5, but were not yet expressed in 2DG utilization.

In addition to 2DG utilization, rhythms in electrophysiologic activity and vasopressin (VP) mRNA levels have also been measured in the fetal rat SCN. Shibata and Moore (1987) sampled single-unit activity of SCN neurons in slices of fetal, early postnatal, and adult rat brains at two times of day beginning ½ hour after taking the fetuses from the mothers. The earliest age examined, E22 (1 day before birth), showed a low amplitude, but significant day/night difference. The magnitude of the difference increased steadily during the first 2 postnatal weeks and continued to increase to adult values. Reppert and Uhl (1987) first detected VP mRNA in some sections of fetal rat SCN on E20.5 and observed a significant day/night difference on E21.5. Message levels were higher during subjective day and the magnitude of the difference increased through at least postnatal day 11.

Rhythms measured in the fetal rat SCN are likely be the expressions of oscillations generated by the fetal SCN rather than responses to maternal rhythmicity. Not only is this consistent with prenatal entrainment and with the observation of day/night differences measured *in vitro*, but 2DG uptake has also been measured in fetuses whose dams received SCN lesions early in gestation (Reppert & Schwartz, 1986; Shibata & Moore, 1988). While the fetuses from such mothers, as a group, did

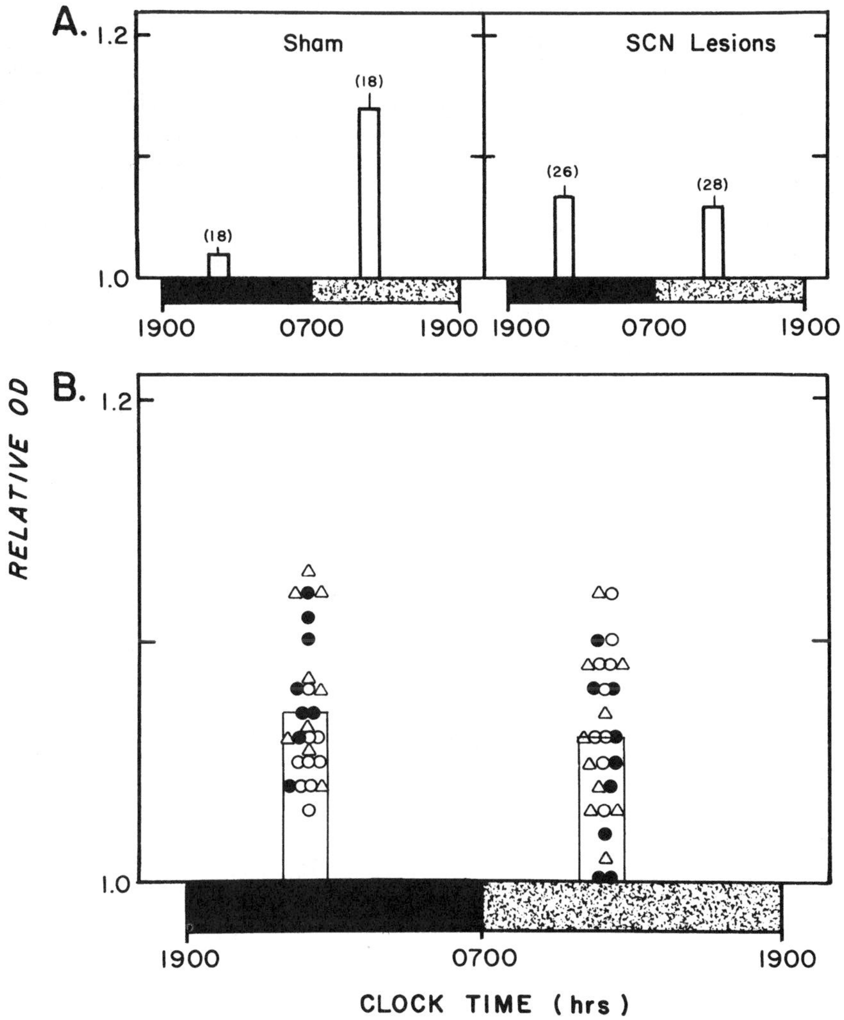

Figure 3. Variation in 2DG uptake in the fetal rat SCN (E20–E21) in the absence of maternal rhythmicity (SCN-lesioned). (A) The average values of fetuses from lesioned and sham-treated dams; (B) the values for the individual fetuses from the lesioned dams. At a given time of day, similar symbols indicate fetuses from the same litter. The within-litter range of values in the fetuses born to SCN-lesioned dams suggests that individual fetuses continued to express oscillations, but that these oscillations were no longer synchronized among the fetuses, thereby producing intermediate average values. (From Reppert & Schwartz, 1986, with permission.)

not express a rhythm, they showed a range of values, suggesting that individual fetuses were expressing rhythms which are out of phase with each other (Figure 3). Maternal SCN lesions also disrupted a population rhythm in pineal NAT activity in pups as well as the synchrony normally observed among the drinking rhythms of pups. Similarly, when pregnant hamsters receive SCN lesions early in gestation (day 7), the pups expressed normal wheel-running activity rhythms at weaning, but the synchrony among the pups within a litter was disrupted (Davis & Gorski, 1988). Thus, in the absence of maternal rhythmicity, the fetal SCN begins to generate oscillations, but the SCN of different fetuses are not entrained to a common signal. The 2DG results (Reppert & Schwartz, 1986) in which all cells appear to be in approximately the same state at any particular time, suggest, however, that even from early in development and in the absence of a common signal, the cells within an SCN are coupled.

In some studies in which maternal rhythms have been disrupted by SCN lesions or constant light during gestation, some nonrandomness among the phases of pups has been observed (Honma, Honma, Shirakawa, & Hiroshige, 1984; Nuesslein-Hildesheim & Schmidt, 1996; Reppert & Schwartz, 1986). Although this could result from direct interactions among pups prior to isolation, the evidence does not support this (Hiroshige, Honma, & Watanabe, 1982a; Reppert *et al.*, 1984). Alternatively, such "residual" synchrony could result from developmental synchrony. This possibility would require that the initiation of oscillations occurs at a specific time in development and at a specific phase of the oscillations.

Development of Entrainment

It appears that as soon as circadian oscillations begin to be generated, they are entrained, not by the light/dark cycle, but instead by some aspect of maternal rhythmicity. When rhythms are first measurable within the fetal SCN, they are in phase with rhythmicity of the dam, and the entrainment of postnatal rhythms by maternal rhythmicity or by exogenous treatments is possible before birth. For example, the within-litter synchrony among hamster pups at weaning is less disrupted by SCN lesions of the dam on day 14 of gestation than by lesions on day 7 of gestation, suggesting that entrainment occurred before day 14 (Davis & Gorski, 1988). If circadian oscillations do not begin until some time after SCN neurogenesis, then entrainment must occur between E17 and E19.5 (2DG rhythm) in the rat and between E12 and 14 in the hamster. Entrainment must therefore be accomplished rapidly, suggesting that oscillations when they are first generated can be easily reset. Alternatively, or perhaps in addition, the initiation of oscillations might in some way be tied to maternal rhythmicity so that as oscillations are initiated, they are also entrained.

Maternal Entrainment. The principal evidence for maternal entrainment is the synchrony observed between rhythmicity of the dam and rhythmicity of the fetal SCN or the rhythmicity of pups when there is no other periodicity (other than that of the pups themselves) available (Figure 4). Since there is no evidence that pups entrain each other (Reppert *et al.*, 1984; Hiroshige *et al.*, 1982a), the synchrony between the dam and her offspring is likely to be imposed by the dam. Furthermore, when the dam's SCN is ablated early in gestation, the normal synchrony among pups is disrupted, indicating that maternal rhythmicity was required. The maternal rhythm or rhythms which mediate entrainment have not been identified, but the case for some candidate signals is strong.

The first attempt to identify specific maternal rhythms was by disrupting them. Reppert and Schwartz (1986) removed, in separate experimental groups, the adrenal, pituitary, or pineal glands or the ovaries and saw no effect on entrainment of the fetal SCN 2DG rhythm or the postnatal pineal NAT rhythm, which was equivalent to the effect of ablating the dam's SCN. The only possible effect of these ablations was an effect of pinealectomy on the 2DG rhythm, where some abnormally low values were observed.

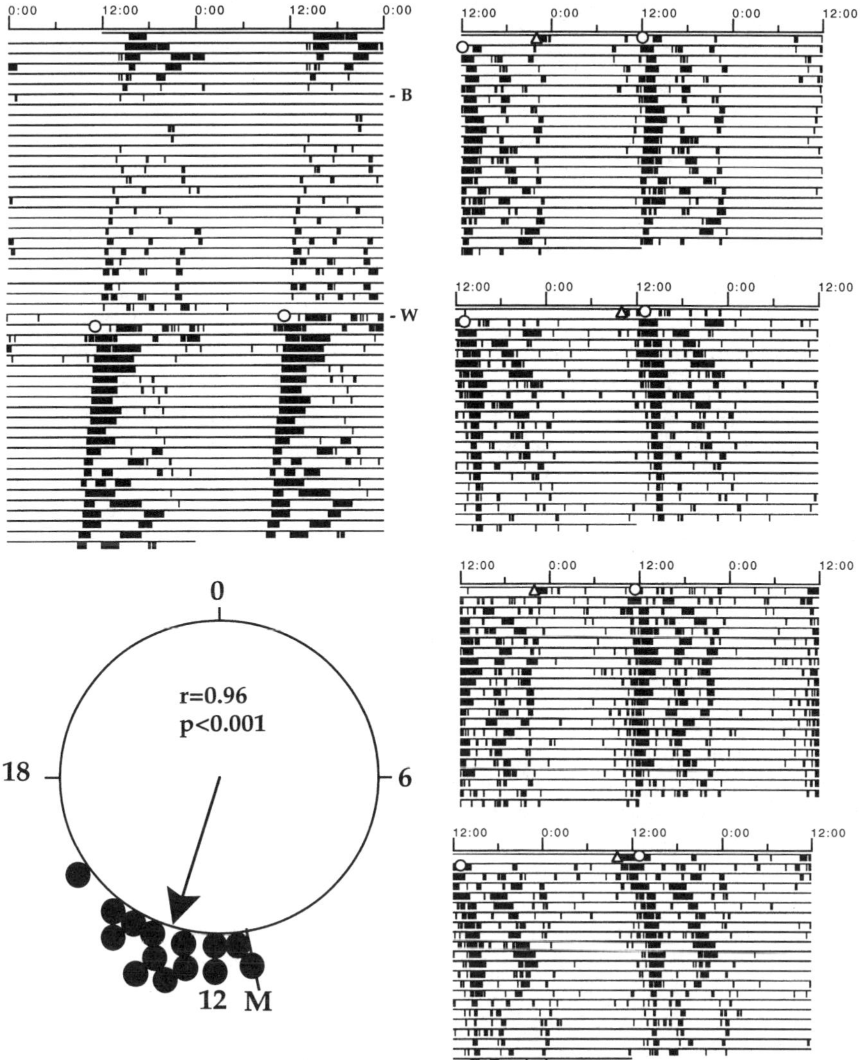

Figure 4. Maternal entrainment of Syrian hamster behavioral rhythms. Top left: The double-plotted wheel-running activity record of a pregnant hamster in dim, constant light. Her litter was born (B) and weaned (W) on the days indicated. The records of four of her pups, also kept in constant light, are shown on the right. The pups' records begin on the day they were separated from the dam, and on each of the pups' records, the time of weaning is indicated with a triangle. The estimated phase of activity onset on that day is indicated with a circle. The activity-onset phases of the four pups and those from the other pups in the litter ($n = 14$) are also plotted on a circle representing the 24 hours of the day of weaning (filled circles). The pups' phases were clustered among themselves and around the dam's activity onset (M). The arrow within the circle indicates the pups' average phase, and its length, also given by r, indicates the degree of synchrony among them. The p value is based on the Rayleigh test and indicates that the distribution of pups' phases is significantly different from uniform.

Another approach to testing candidate signals has been experimentally to administer such signals to pregnant rats or hamsters. Three such treatments have been shown to entrain the rhythms of the pups measured postnatally: in rats, restricted feeding schedules imposed on the pregnant dam (Weaver & Reppert, 1989a), and in Syrian hamsters, melatonin injections or injections of a D1-dopamine receptor agonist (SKF 38393) to the pregnant dams (Davis & Mannion, 1988; Viswanathan, Weaver, Reppert, & Davis, 1994). In no case has it been established that natural rhythms mimicked by these treatments are physiologic entraining signals, but at the same time, it is possible that all three are in some way involved.

Restricted Feeding. It was demonstrated by Reppert and Schwartz (1984) that SCN lesions of pregnant rats disrupted the synchrony among pups that is normally observed postnatally. Weaver and Reppert (1989a) then imposed restricted feeding schedules (4 hours with, 20 hours without) on SCN-lesioned dams between days 8 and 19 of gestation. Two groups of dams received schedules that were 180° out of phase. The average phases of the pups from the two groups were clustered approximately 180° apart, indicating that the phases had been set by the schedules. The restricted feeding schedule could have had its effect in either of two general ways, by producing a rhythm that directly affected the fetuses, such as in nutrient availability, or by organizing and entraining circadian oscillations in the dam that then provided an entraining signal. There are probably several rhythms that could be influenced by restricted feeding and could also affect the fetus (Krieger & Hauser, 1977). It remains unclear what specific signal was responsible for the entrainment by restricted feeding.

Melatonin. There is strong precedence for the pineal gland hormone melatonin to function as a circadian signal. Throughout vertebrates, including mammals, a rhythm in plasma melatonin is likely to function as an internal signal for nighttime (elevated melatonin levels) and for daytime (low melatonin levels) (S. M. Armstrong, 1989). Maternal melatonin passes the placenta, producing within the fetus a rhythm in melatonin (Klein, 1972; McMillen & Nowak, 1989; Reppert, Shea, Anderson, & Klein, 1979; Yellon & Longo, 1988). Exogenous melatonin can entrain the circadian rhythms of adult rats (Redman, Armstrong, & Ng, 1983) and possibly those of humans as well (Deacon & Arendt, 1995; Lewy, 1992). Furthermore, in Siberian hamsters, the maternal pineal melatonin rhythm conveys photoperiodic information to the fetus (Elliott & Goldman, 1989; Weaver Reppert, 1986). To test the hypothesis that melatonin is a maternal entraining signal, Davis and Mannion (1988) injected SCN-lesioned, pregnant hamsters with melatonin on days 7–15 of gestation. All of the hamsters actually received two injections a day, one with melatonin and one with the vehicle only. Two groups received the melatonin at different times of day, 12 hours apart, and the phases of the pups' wheel-running activity rhythms at weaning were measured. The melatonin injections restored synchrony among the pups within individual litters and set the average phases of the two melatonin groups 180° apart (Figure 5). Because the dams were handled and given injections twice a day, the cause of the different average phases must have been the timing of melatonin exposure. These initial studies used large doses of melatonin (100, 50, or 10 μ), but subsequent studies have found that injections of only 1 μ are adequate to cause entrainment (N. Viswanathan and F. C. Davis, unpublished observations). This dose is the threshold for the reproductive effects of melatonin in Syrian hamsters when given as injections (Tamarkin, Westrom, Hamill, & Goldman, 1976),

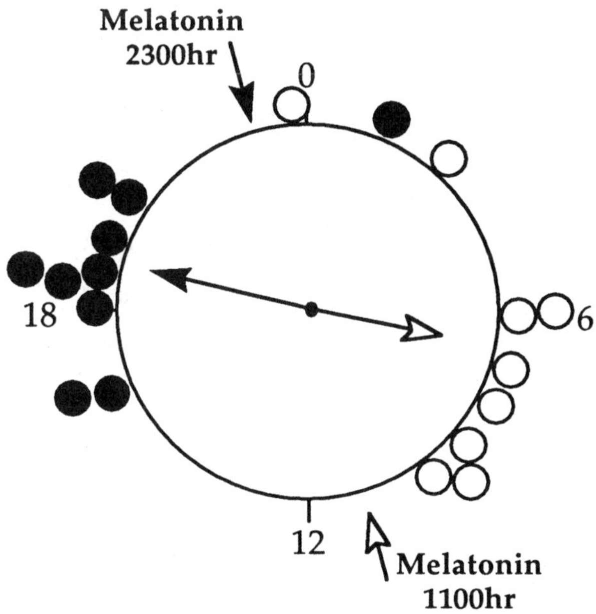

Figure 5. Entrainment of Syrian hamster pups by prenatal melatonin injections to SCN-lesioned dams during the last week of gestation. The large circle represents the 24 hours of the day of weaning and the small symbols are the average phases of entire litters. The dams of the litters represented by closed symbols received melatonin at night and vehicle alone in the morning; those of the litters represented by open symbols received melatonin in the morning and vehicle alone at night.The arrows indicate the average phases of the two groups, which differed by 180°. (Redrawn from Davis and Mannion, 1988.)

suggesting that, like the reproductive effects of melatonin, entrainment by melatonin is a physiologic effect. The fetal SCN is a likely target of melatonin since melatonin receptors have been localized to the fetal SCN in several species including Syrian hamsters, rats, and humans (Carlson, Weaver, & Reppert, 1991; Duncan & Davis, 1993; Reppert, Weaver, Rivkees, & Stopa, 1988; Rivkees & Reppert, 1991; Roca, Godson, Weaver, & Reppert, 1996; Weaver, Rivkees, & Reppert, 1989). Since the melatonin injections in the entrainment studies were given to the pregnant dams, it is possible, despite the presence of melatonin receptors in the fetal SCN, that melatonin caused entrainment indirectly by affecting the dam in some way. However, although melatonin has not been given directly to fetal hamsters, melatonin injections given directly to pups on postnatal days 1–5 causes entrainment, suggesting that fetuses are also directly affected by melatonin (Grosse, Velickovic, & Davis, 1996).

The case for melatonin as a signal mediating the initial entrainment of the circadian pacemaker in mammals is strong, yet, as previously mentioned, pinealectomy in rats does not disrupt synchrony among pups. Pinealectomy in Syrian hamsters also does not disrupt synchrony (N. Viswanathan and F. C. Davis, unpublished observations). There may, however, be a subtle effect of pinealectomy even if there remains a redundant signal capable of maintaining maternal/fetal synchrony. For example, pinealectomy could alter the phase relationship between the mother and fetus, an effect which would be consistent with the observations in rats by Reppert and Schwartz (1986) noted earlier.

D1-Dopamine Receptor Agonist. Using *in situ* hybridization to detect specific mRNAs, Weaver, Rivkees, and Reppert (1992) discovered that the fetal rat SCN expresses a D1-dopamine receptor. It was demonstrated that the D1 agonist SKF 38393 or cocaine injected into pregnant rats or mice induced expression of the immediate-early genes c-*fos* and *jun-B* in the fetal SCN. In rats, the induction of c-*fos* by cocaine could be partially blocked by the D1 antagonist SCH (Weaver *et al.*, 1992; Weaver, Roca, & Reppert, 1995). Prenatal injections of SKF38393 were subsequently shown to cause entrainment of the offspring in hamsters as well as to cause an induction of c-*fos* expression in the fetal hamster SCN. The fetal hamster SCN was also found to contain mRNA for the D1 receptor (Viswanathan *et al.*, 1994). These results raise the possibility that a maternal signal causes entrainment by dopaminergic activation within the fetal SCN. An unidentified maternal signal could stimulate dopamine release within the fetal hypothalamus since immunoreactivity for tyrosine hydroxylase with the fetal rat hypothalamus is consistent with the synthesis and secretion of dopamine near the SCN (Ugrumov, Popov, Vladimirov, Kasmambetova, & Thibault, 1994b).

An interesting aspect of prenatal entrainment by SKF is that the average phase of the pups' rhythms is approximately 180° different from that produced by melatonin (Viswanathan & Davis, 1997; Viswanathan *et al.*, 1994) (Figure 6). Melatonin sets the oscillations in such a way that the subjective night phases of the pups' rhythms become coincident with the time of the injection, while SKF sets the oscillations in such a way that the time of the subjective day phases of the pups' rhythms become coincident with the time of the injection. Also, unlike prenatal SKF and light in adults, there is no evidence that prenatal melatonin induces c-*fos* (Viswanathan & Davis, 1997). These observations suggest the hypothesis that elevated maternal melatonin is a signal representing nighttime, whereas dopaminergic activation is a signal representing daytime. In the retina, melatonin and dopamine have antagonistic effects on retinal physiology, and at the cellular level, have antagonistic effects on cAMP levels (Iuvone & Gan, 1995). Similar antagonistic effects at the cellular level may occur within the SCN resulting in the establishment of opposite phases.

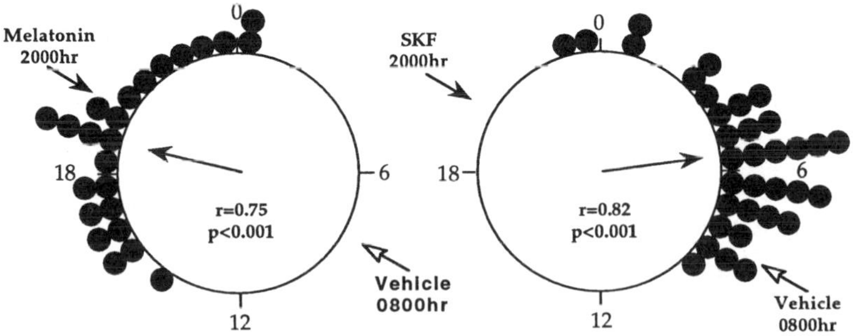

Figure 6. Prenatal entrainment of Syrian hamster circadian rhythms by single prenatal injections of melatonin (Mel) or the D1-dopamine-receptor agonist SKF 38393 (SKF). Injections were given to SCN-lesioned dams on E15, and the filled symbols represent the activity-onset phases of individual pups. The single injections were able to establish synchrony within the groups, and although the two substances were given at the same time of day, they established average phases differing by approximately 180°. (Redrawn from Viswanathan & Davis, 1997.)

The entraining effects of melatonin and dopamine were initially demonstrated by giving injections over several days of gestation. In a subsequent study, it was shown in Syrian hamsters that single injections on day 15 of gestation can also cause entrainment (Viswanathan & Davis, 1997) (Figure 6). Since these injections were given to SCN-lesioned dams, it is likely that the phases of the fetal pacemakers were widely distributed prior to the injections. The production of synchrony from such distributions by single injections would require phase shifts much larger than those produced by light in adult hamsters or by melatonin in rats, suggesting that the fetal SCN is highly sensitive to phase-setting stimuli. In addition, in contrast to c-*fos* induction by light in adults, it appears that the fetal SCN is responsive to SKF at any phase (Viswanathan *et al.*, 1994). Interestingly, the sensitivity of the fetal SCN to these drugs is lost soon after birth. Both melatonin and SKF injections given directly to hamster pups on postnatal days 1–5 cause entrainment, but injections on days 6–10 do not (Grosse *et al.*, 1996; J. Grosse and F. C. Davis, unpublished observations). The loss in sensitivity to melatonin could be the result of reduced expression of melatonin receptors, since iodomelatonin binding in the SCN declines after the first postnatal week in Syrian hamsters (Duncan & Davis, 1993). The rhythmicity which is restored by fetal grafts can also be entrained by melatonin injections to the host if the injections are given immediately after transplantation, but injections several weeks after transplantation do not affect the restored rhythms (Grosse and Davis, 1997). The induction of c-*fos* expression in the fetal rat SCN by SKF also ends between postnatal days 3 and 5 (Weaver & Reppert, 1995). Interestingly, at the same time, the induction of c-*fos* by light begins (see below). The loss of c-*fos* induction by SKF is not caused by the development of retinal afferents since the loss occurs in rats enucleated on the day of birth (Weaver, 1995).

Although there are a number of other maternal rhythms which could play a role in entrainment of the fetus, such as those in adrenal steroids, body temperature and uterine contractions, there is no evidence for the involvement of any specific physiologic rhythm other than those in melatonin and food intake. However, whether related to the physiologic mechanisms of entrainment or not, the fetal SCN may be susceptible to influence from a range of substances, including drugs of abuse. Cocaine was already noted in connection with dopaminergic activation, but nicotine has also been shown to induce c-*fos* expression in the fetal rat SCN (Clegg, O'Hara, Heller, & Kilduff, 1995). It is not known, however, whether either cocaine or nicotine can cause or alter entrainment.

Postnatal Maternal Entrainment. Maternal entrainment begins before birth, and continues to varying degrees after birth. The signals mediating postnatal entrainment could be the same or different from those mediating prenatal entrainment. For example melatonin has been found in human milk (Illnerová, Buresová, & Presl, 1993) and has been shown to pass from the mother to her pups via this route in rats (Reppert & Klein, 1978). Thus melatonin could mediate both a pre- and postnatal maternal entrainment. On the other hand, postnatal maternal entrainment in the mouse *Mus budooga* can be mediated by artificial cycles in the presence and absence of the dam. The presence of the dam is always interpreted as day irrespective of the phase in mother's circadian cycle when she is with the pups (Viswanathan & Chandrashekaran, 1985). Of the animals studied (mice, rats, and hamsters) postnatal maternal entrainment appears to be strongest in mice, and persists for up to 5 weeks, even when the pups are no longer dependent on the dam for nourishment. Pups have also been shown to entrain to non-24-hour cycles in the presence and absence of the dam. The minimum duration of maternal presence that is required for entrainment is 8 hours, but the specific signal causing entrainment has yet to be identified (Viswanathan, 1990).

Postnatal maternal entrainment in rats has been studied primarily using cross-fostering studies in which the phase of maternal rhythms is different between the prenatal natural mother and a postnatal foster mother. If pups are crossed to a foster mother within 1 day after birth, the foster mother can reentrain them to her cycle. If the exchange is made later, the foster mother does not entrain the pups' rhythms and the pups express phases which can be related to the prenatal phase of their natural mothers. When this ability ends has been examined, but the answer differs somewhat among studies. The ability to reentrain may end as early as 5 days after birth or extend to the end of the second postnatal week (Hiroshige *et al.*, 1982a; Reppert *et al.*, 1984; Takahashi, Ohi, Shimoda, Tamada, & Hayashi, 1989). Cross-fostering between litters of different ages indicates that the decline in postnatal maternal entrainment is due to development of the pups rather than to changes occurring in the dam, and the strength of postnatal maternal entrainment might be affected by the number of pups in the litter. The signals mediating postnatal entrainment have not been identified, but, as in mice, an artificial rhythm in the presence and absence of the dam can cause entrainment. Imposing a restricted feeding schedule on the dam can also cause entrainment and the significant effect might be the absence of the mother during the time of food availability (Takahashi, Westrom, Hamill, & Goldman, 1989). In a marsupial, the gray short-tailed opossum (*Monodelphis domestica*), the SCN are already innervated by the RHT when the SCN first expresses a rhythm (postnatal day 20), and there is no evidence for maternal entrainment (Rivkees & Reppert, 1990).

In rabbits, the doe visits and feeds her young only once a day and for only a few minutes each time. During these visits, the pups receive all of their nourishment. These visits occur at predictable times of days and the pups anticipate the arrival of the doe by uncovering themselves before she arrives. They show this anticipation even when the doe has been prevented from visiting for 1 day (Figure 7). Thus, a behavioral rhythm expressed by the pups appears to be endogenously generated and is probably entrained by the doe since no other cue would normally be available (Hudson & Distel, 1989; Jilge, 1993).

Entrainment by Light. In rats, the ability of the circadian pacemaker within the SCN to be entrained by light develops during the first postnatal week, coincident with the declining strength of maternal entrainment (Duncan *et al.*, 1986; Reppert *et al.*, 1984) and the ability of dopaminergic or nicotinergic activation to induce c-*fos* in the SCN (Clegg *et al.*, 1995; Weaver & Reppert, 1995). The first postnatal week is also the time in Syrian hamster development when entrainment by exogenous treatment with melatonin and the dopamine agonist SKF 38393 is lost.

The earliest age for entrainment by light in rats is 6 days after birth (Duncan *et al.*, 1986). When rat pups were exposed to a shifted light/dark cycle on postnatal days 1–9, their pineal NAT rhythms (measured on day 11) were also shifted. If exposed on days 1–7, they were partially shifted. Exposure to the shifted cycle on days 7–9 only was not adequate to accomplish the shift, providing additional evidence that light before day 7 has an effect. Although exposure to the shifted cycle on days 1–5 was ineffective, entrainment by light might still be possible during this time, but not strong enough to override the influence of maternal rhythmicity, which would still have been in phase with the unshifted cycle. The RHT begins to inner-

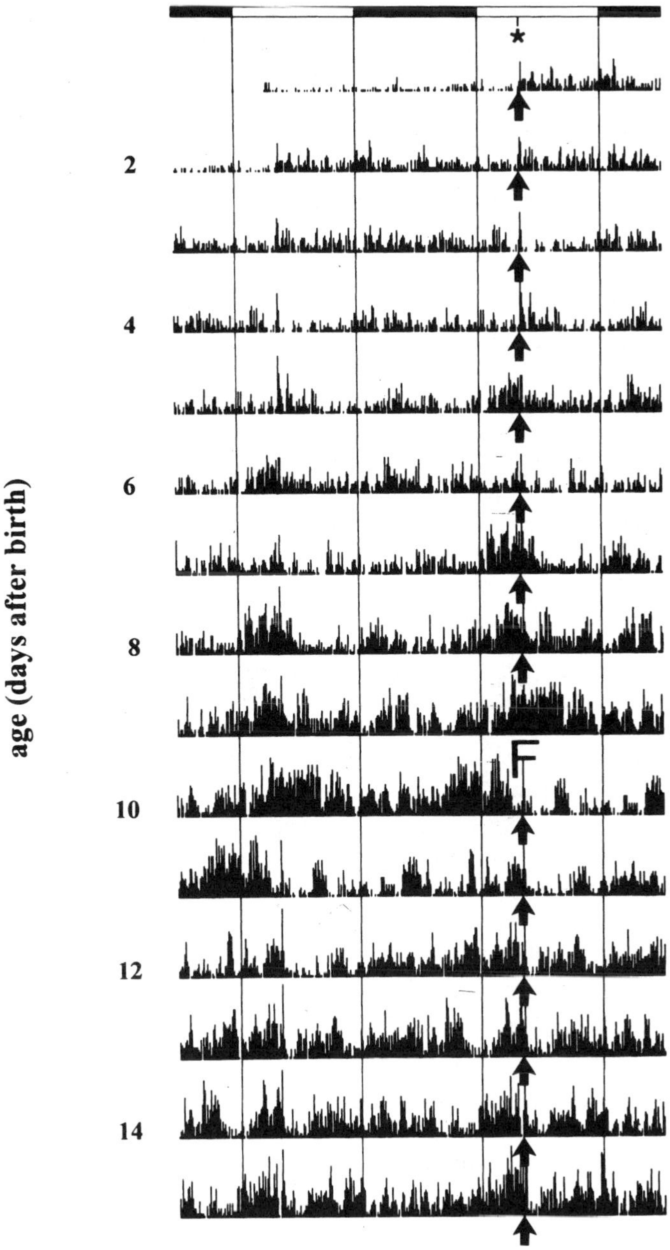

Figure 7. Rhythms of litter activity in rabbits when feeding by the doe was experimentally scheduled to occur at 1000 hours (arrow). On the day marked with an F, the doe was not allowed to nurse. The pups showed activity which anticipated the feeding times even on the day after the day when feeding did not occur. The results suggest an endogenous rhythm in anticipatory activity which is normally entrained by the doe. (From Jilge, 1993, with permission. Copyright 1993, Sage Publications.)

vate the SCN on postnatal day 1 and the SCN is well innervated by day 7 when entrainment is first observed.

The possibility that entrainment by light can occur earlier than postnatal day 6 is suggested by the responsiveness of the SCN to light on day 1. This was initially indicated by Fuchs and Moore (1980), who measured a light-induced increase in 2DG uptake on day 1. This has been substantiated by studies of c-*fos* induction. Leard, Macdonald, Heller, and Kilduff (1994) observed c-*fos* mRNA induction by light in the fetal rat SCN at circadian time 22 (CT22) on postnatal day 2, but not at CT14 on day 2 or at any time on day 1. Weaver and Reppert (1995) showed induction on day 1 and both studies observed a progressive increase in induced expression over the next 3 or 4 days (Figure 8). The development of light responsiveness has also been examined by measuring the induction of Fos protein. Kaufman and Menaker (1994) observed significant induction of Fos-immunoreactive cells on postnatal day 4, but not on day 3 in Syrian hamsters. The number of immunoreactive cells increased steadily up to day 15. The distribution of cells also changed, with most cells initially located midway on the dorsoventral axis along the lateral margin of the nucleus. Between days 13 and 15, cells also appear in the ventral region, resulting in a pattern of expression typical of adults. Duffield *et al.* (1995) observed significant induction of Fos immunoreactive cells on postnatal day 3, but not on day 2 in Siberian hamsters. The number of immunoreactive cells also increased on subsequent days, and, although always located in the general ventrolateral region (as in the adult), the density of cells appeared to increase selectively in the most ventral region between 12 and 24 days postnatal. These changes in the distribution of immunoreactive cells could be related to ongoing changes in the innervation pattern of the RHT and/or to eye opening, which occurs around day 15 in Syrian hamster and around day 11 in Siberian hamsters. The slightly later development of a response to light in Syrian hamsters relative to rats could be related to differences in the development of the RHT. RHT development is delayed by 3–5 days in hamsters relative to rats (Speh & Moore, 1993). As already noted, a response to light in the gray short-tailed opossum does not occur until 20 days after birth, but this is soon after formation of the SCN and is the age when SCN oscillations are first observed (Rivkees & Reppert, 1990).

Light-induced c-*fos* expression in the adult SCN is "gated" so that expression is only induced during subjective night. Light-induced c-*fos* expression does not appear to be gated on postnatal day 2 in rats (Weaver & Reppert, 1995). Similarly, the induction of c-*fos* by dopaminergic activation is not gated in the fetus or neonate (Leard *et al.*, 1994; Viswanathan *et al.*, 1994; Weaver *et al.*, 1992). Gating of the light-induced response, and possibly of the SKF response as well (just before the response disappears), appears to begin on postnatal day 3 (Weaver & Reppert, 1995). A qualification to the conclusion that gating is not present early in development is that in any given study, only two times of day were examined. Although it is also possible that a response in a subset of cells is gated earlier, but masked by an ungated

Figure 8. Development of responsiveness to light in the rat SCN. Hybridization of a radiolabeled riboprobe to c-*fos* mRNA is shown in emulsion autoradiographs form newborn rat SCN photographed under darkfield illumination. Hybridization signal appears white, and dotted lines indicate the borders of the SCN. Pups were sacrificed after 40 minutes of bright light (LIGHT) or remained in dim red light (CON). Some response to light is seen on the first night following birth (PD 0) and a gated response in the ventrolateral SCN is seen by the third night (PD 2). (From Weaver & Reppert, 1995, with permission. Copyright 1995, Elsevier Science.)

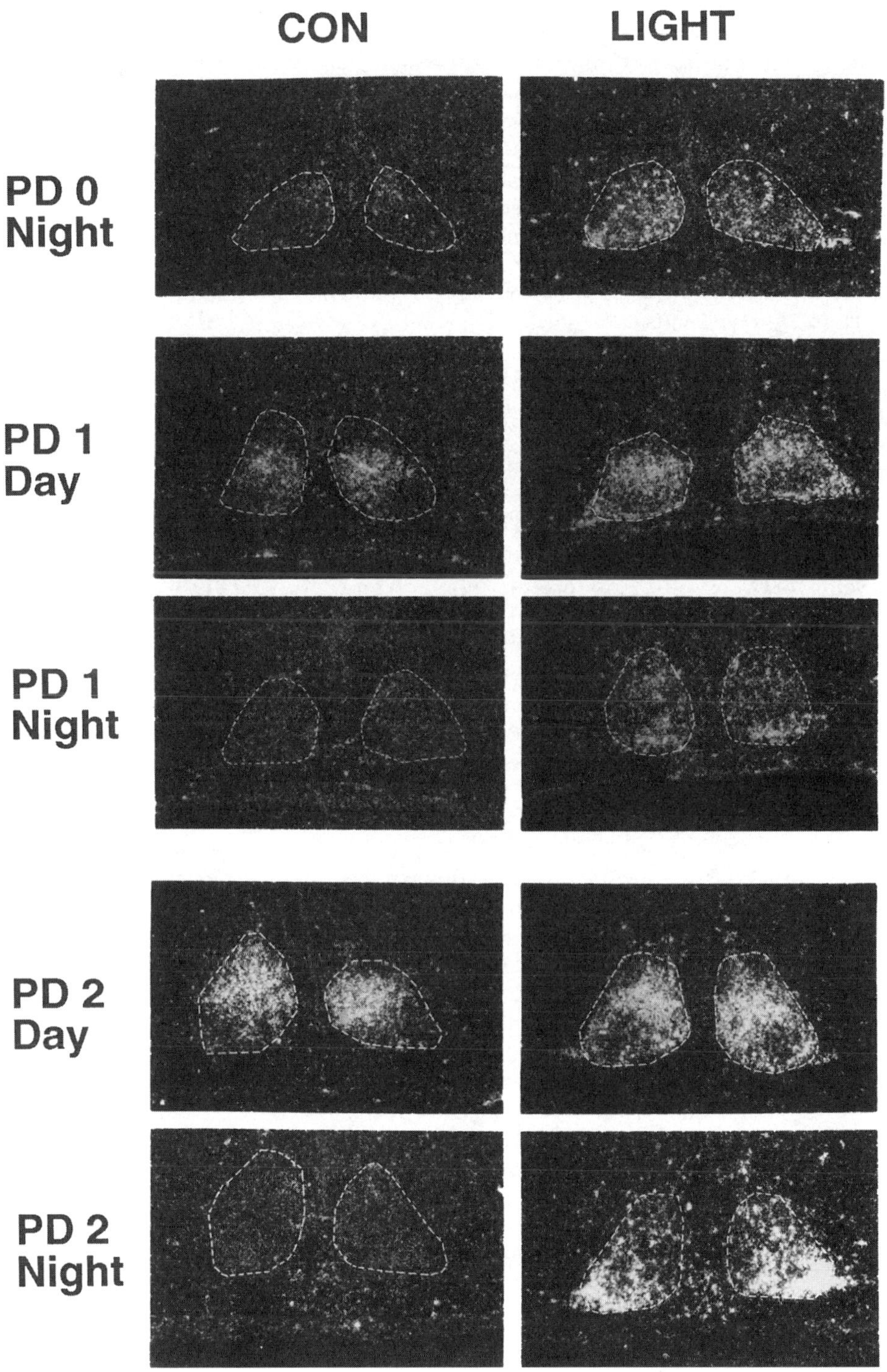
CON
LIGHT
PD 0
Night
PD 1
Day
PD 1
Night
PD 2
Day
PD 2
Night

response in other cells, gating clearly becomes more pronounced during development. If the gating of light-induced *c-fos* expression is an output of SCN-generated oscillations, it appears to develop after other outputs such as 2DG uptake and VP gene expression. It is also possible that the gating is controlled by a rhythm in the retina. Circadian rhythms are endogenously generated in the mammalian retina, but it is not known when this rhythm develops.

It is clear from the early development of responsiveness to light that the opening of the eyes in these altricial rodents is not required for the perception of light. The correlation of responsiveness with the development of the RHT suggest, however, that these afferents and the retina are required. Indeed, enucleation has been shown to prevent both the entrainment and the response to light that occur in the first postnatal weeks in rats, hamsters, and a marsupial (Kaufman & Menaker, 1994; Reppert *et al.*, 1984; Rivkees & Reppert, 1990; Weaver, 1985). Not only can light pass through the closed eyelids, but light can pass through to the intrauterine environment as well. Weaver and Reppert (1989b) demonstrated that the the SCN is innervated by the RHT on the day of birth in the precocious rodent the spiny mouse, and that light exposure at night increased 2DG uptake in the fetal SCN while still *in utero*. It is unclear whether this response also means that light could cause entrainment at this time. In attempts to test this, maternal entrainment appeared to be stronger than any *in utero* effect of light (Weaver & Reppert, 1989b).

At the time the SCN is first responsive to light in rats, the retina is still poorly developed. For example, photoreceptor outer segments do not begin to appear until about postnatal day 5 (Reh, 1992). However, in mice with degenerated retinas containing only a reduced population of cones that lack outer segments, normal circadian responses to light still occur (Provencio, Wong, Lederman, Argamaso, & Foster, 1994). Because in rodents, cones are produced before birth, it is possible that by the time of birth, immature cone photoreceptors lacking outer segments are sufficient to mediate responses to light.

Specification of Pacemaker Properties

As noted in the Introduction, it is unnecessary for an organism to experience periodicity in the environment in order to generate and express circadian oscillations when mature. Hamsters or rats born to SCN-lesioned dams in constant conditions develop normal circadian rhythms (Davis & Gorski, 1988; Nuesslein-Hildesheim & Schmidt, 1996; Reppert & Schwartz, 1986) and SCN grafts transplanted to SCN-lesioned hosts as early as embryonic day 11 in hamsters still express rhythmicity (Kaufman & Menaker, 1993). In addition, the free-running period of mice was uninfluenced by raising the mice on non-24-hour maternal and light/dark cycles (Davis & Menaker, 1981). Thus the absence of environmental rhythmicity or the presence of abnormal environmental rhythmicity appears to be without effect on the development of pacemaker properties in mammals. Similarly, deafferentation appears to have little or no effect on circadian rhythms; neonatal enucleation had no effect on free-running period in hamsters (Davis, 1982) and the periods of the rhythms restored by grafts (which are abnormal in many ways including the absence of retinal afferents) are similar to the periods expressed by intact animals (Davis & Viswanathan, 1996; Viswanathan & Davis, 1995).

An aspect of SCN development which could influence pacemaker properties is the proliferation of its cells. Variation in the rate or duration of proliferation could affect the relative or total numbers of SCN cells and thereby affect properties of the population. For example, lesion and transplantation studies suggest that cell number can affect the free-running period expressed by the organism (Davis & Gorski, 1984; Davis & Viswanathan, 1996; Silver *et al.*, 1990). Hormones might also affect pacemaker development and its properties since sex differences have been reported in circadian rhythms and in SCN anatomy both in humans other animals (Davis, Darrow, & Menaker, 1983; De Vries, Buijs, & Swaab, 1981; Swaab, Zhou, Ehlhart, & Hofman, 1994; Wever, 1984; Zucker, Fitzgerald, & Morin, 1980).

Functional Aspects of Circadian Rhythmicity during Development

The mammalian circadian pacemaker begins to function and is entrained at a time when coordination with 24-hour periodicity in the environment would seem to be irrelevant. Even *in utero*, however, the embryo and fetus are exposed to 24-hour periodicity in features of its environment which may be critical for normal growth and development, including rhythms in nutrients, hormones, temperature (Serón-Ferré, Ducsay, & Valenzuela, 1993), and possibly toxic substances (Ganzhorn & Wright, 1994; Sauerbier, 1986, 1987). It may be adaptive for aspects of fetal physiology or development to occur only at certain times within the mother's circadian cycle. Daily rhythms in maternal weight gain have been seen in rats, mice, and Syrian hamsters, and in rats, fetal weight gain during the last 3 days of gestation is greater during the night than during the day (Barr, 1973). In mice, fetal weight gain, measured earlier in development, is also rhythmic, but with 12-hour periodicity (Davis, 1989). Since the fetal rhythms were in dry fetal weight, they indicate daily variation in the growth and/or proliferation of cells. A rhythm in proliferation was reported in the telencephalic ventricular zone of fetal rats, with the proliferation index higher during the day than night, regardless of the time of day when the dam ate most of her food (M. W. Miller, 1992).

The possibility that fetal growth and development is adapted to the rhythmic intrauterine environment does not offer a satisfactory explanation for why the fetal pacemaker should function *in utero* and be entrained by the mother. In rodents, where the evidence for a fetal pacemaker is strongest, there is no evidence of a rhythm, outside of the SCN itself, which is regulated by the pacemaker. Thus, there is no known output through which the pacemaker might play a role in coordinating fetal physiology with that of the mother. This may be because it is difficult to identify and measure such rhythms in rodents; in other animals, including humans (see below), there is substantial evidence for rhythms in fetal physiology. To what extent these rhythms are controlled by an endogenous fetal pacemaker is, however, uncertain; rhythms observed in the fetus are often absent in the newborn or disrupted in the absence of maternal rhythmicity. Therefore, fetal rhythms may result primarily from direct effects of maternal rhythms rather than from entrainment of an endogenous fetal pacemaker.

A likely function for prenatal entrainment of the pacemaker is that it is important for an animal's overt rhythms to be in phase with the environment when the rhythms appear some time after birth. The environment in this case would continue to include maternal rhythmicity, but also other rhythmicity such as the light/dark cycle. Maternal rhythmicity appears to provide the fetus with important information about the environment, in particular, information about time of day as well as about time of year. In several seasonally breeding mammals, reproductive development occurs at a different rate depending on the time of year when the offspring are born (Horton, 1983; Stetson, Elliott, & Goldman, 1986). The photoperiod experienced by the dam provides the offspring with information about changing photoperiod by influencing how the offspring will respond to the photoperiod experienced after birth. In Siberian hamsters where this has been most thoroughly studied, the prenatal effect of photoperiod is mediated by the maternal melatonin rhythm (Carlson *et al.*, 1991; Weaver & Reppert, 1986; Weaver, Keohan, & Reppert, 1987). Although the mechanism by which maternal melatonin affects the fetal neuroendocrine system is not understood, it may include entrainment of the fetal pacemaker (Shaw & Goldman, 1995). Thus one function for the early development of the pacemaker and its entrainment may be to mediate the transfer of photoperiodic information via the dam.

The feeding behavior of rabbits, already described, represents a strong case for functionally significant coordination between the rhythms of the mother and her offspring (Hudson & Distel, 1989; Jilge, 1993). A pup that does not anticipate the doe's arrival may miss feedings, with consequences for its growth and survival. The anticipatory rhythm is probably entrained by postnatal maternal rhythms (Jilge, 1993; 1995) (Figure 7), but entrainment of an underlying pacemaker may begin before birth as well. In hares, the doe and leverts come together once a day for feeding at times that are different for each litter (Broekhuizen & Maaskamp, 1980). Since the hares are exposed to a light/dark cycle during this time, it is likely that the light/dark cycle mediates entrainment. Nevertheless, a maternal effect might influence the precise time for the meeting.

Another suggestion for a functional fetal circadian pacemaker is the involvement of the fetus in the timing of birth. In many mammals, including humans, birth occurs at restricted times of day (Honnebier, Swaab, & Mirmiran, 1989). Because the fetus is known to play a role in the initiation of parturition, and through maternal entrainment has time-of-day information, the fetus could also influence the time of day when birth occurs. There is compelling evidence both for and against this. For example, in rats, removal of the fetal brain disrupts the circadian timing of birth (Reppert, Henshaw, Schwartz, & Weaver, 1987). In Syrian hamsters, however, exogenous prenatal melatonin can entrain the circadian pacemaker, but at the same time, these injections have no effect on the timing of birth (Viswanathan & Davis, 1993). It is possible that fetal rhythms contribute to the timing of birth, but that the strength of this contribution varies among species or under different circumstances.

Prenatal entrainment could be important for the normal development of the pacemaker itself, but there is little evidence for this. For example, without entrainment during development, the pacemaker may not function normally or there may be a delay in the development of the pacemaker's regulation of normally rhythmic functions. Various conditions during development, such as constant light, have been noted to delay the expression of rhythms in animals, but it is not known if these effects are on the pacemaker and, if so, whether they are long-term effects (Cambras & Diez-Noguera, 1991; Krieger, 1972). Nevertheless, additional research is needed, especially in humans, where disruption of rhythmicity early in development may become increasingly common. Furthermore, small differences in the development of rhythms such as the sleep/wake cycle in infants could have a large impact on the general health and well-being of both infants and parents.

Humans begin to express rhythmic physiology and behavior *in utero* (Table 2). It is uncertain, however, whether these rhythms are endogenously generated by a fetal circadian pacemaker and could freerun in the absence of maternal rhythmicity or are passive responses to maternal rhythms. Because it is not possible to measure fetal rhythms in isolation from those of the mother, studies aimed at addressing the endogenous nature of fetal rhythms have focused on preterm infants observed in neonatal intensive care units under relatively constant conditions, especially with respect to temperature, lighting, and feeding. The most compelling examples of rhythms expressed under these conditions were reported by Mirmiran and Kok (1991). In 12 infants born at 26–32 weeks, they observed body temperature rhythms in 7, heart rate rhythms in 5, and a motility rhythm in 1 (Figure 9). In four cases of body temperature, periodogram analysis gave periods different from 24 hours. Given that the majority of periods were 24 hours and that the data contained considerable variability, it is difficult to conclude that rhythms were free-running. Tenreiro *et al.* (1991) obtained long-term records of heart rate and skin temperature from preterm infants born at 24–29 weeks. A wide range of periodicities were observed in the data, with those in the circadian range only transiently expressed even when a light/dark and feeding schedule was imposed. Glotzbach, Edgar, and

TABLE 2. PRESENCE OF TWENTY-FOUR-HOUR RHYTHMS DURING HUMAN DEVELOPMENT[a]

Function	Fetus	Preterm	Postnatal appearance[b]	Reference
Activity or sleep/wake	+	±	3–8 weeks	Coons & Guilleminault, 1984; Martin du Pan, 1974; Ehrstrom, 1984; Glotzbach *et al.*, 1994, 1995; Hoffman, 1959; McMillen *et al.*, 1991; Mirmiran & Kok, 1991; Parmelee *et al.*, 1964; Patrick *et al.*, 1982a; Patrick *et al.*, 1982b
Breathing movement	+	±	—	Patrick *et al.*, 1982a; Updike *et al.*, 1985
Heart Rate	+	±	2 days–20 weeks	Glotzbach *et al.*, 1994, 1995; Helbrugge *et al.*, 1964; Mirmiran & Kok, 1991; Patrick *et al.*, 1981; Sitka *et al.*, 1994; Tenreiro *et al.*, 1991; Visser *et al.*, 1982
Bladder volume	+			Chamberlain *et al.*, 1984
Temperature		±	2 days–4 weeks	Glotzbach *et al.*, 1994, 1995; Helbrugge *et al.*, 1964; Mirmiran & Kok, 1991; Sitka *et al.*, 1994; Tenreiro *et al.*, 1991; Updike *et al.*, 1985
Melatonin[c]		—	12 weeks	Kennaway *et al.* 1992
Mitotic Rate (epidermis)			6–11 days	Cooper, 1932
Urine, Na^+, and K^+ excretion			4–20 weeks	Helbrugge *et al.*, 1964
Electrical Skin resistance			1 week	Helbrugge *et al.*, 1964

[a]+, Consistently observed; ±, inconsistently observed.
[b]Postnatal age of term infants.
[c]Urinary sulfatoxymelatonin.

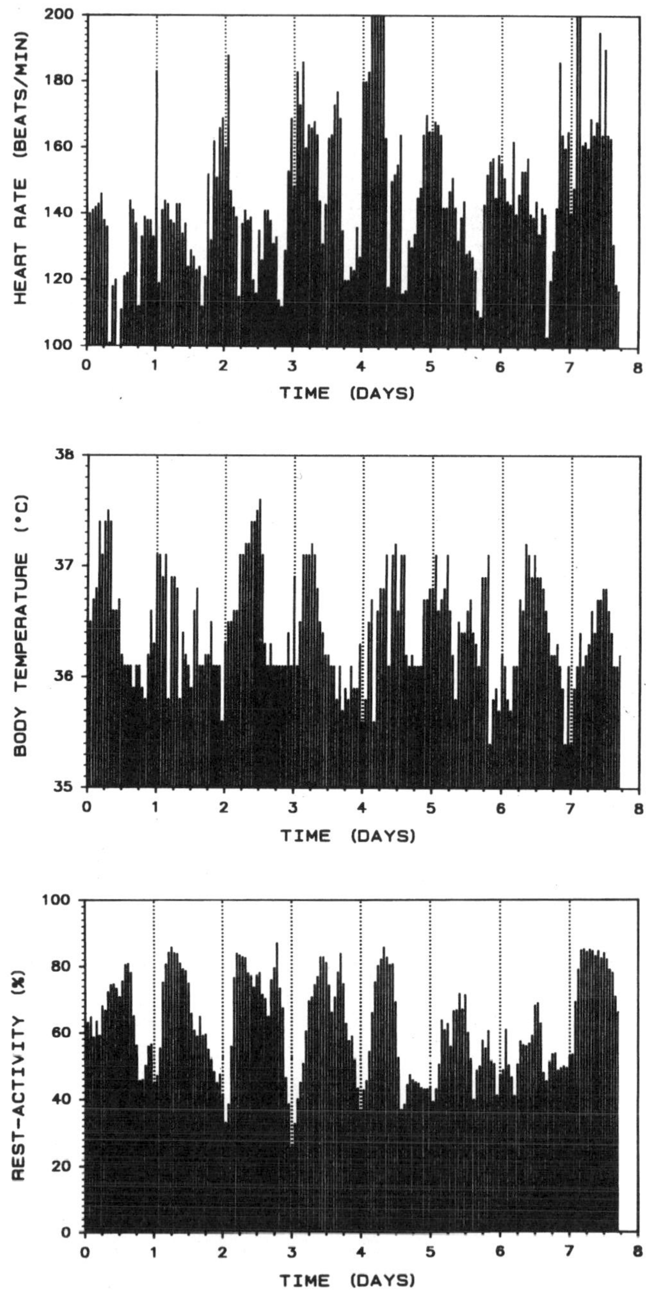

Figure 9. Eight days of continuous recording of heart rate, skin temperature, and rest/activity cycles of a female infant born at 26 weeks gestation and recorded at 29 weeks postconceptional age. The infant was kept in an incubator in a neonatal intensive care unit under constant light and temperature and with intragastric feeding every 2 hours. (From Mirmiran and Kok, 1991, with permission. Copyright Elsevier Science, 1991.)

Ariagno (1995) observed significant rectal and skin temperature rhythms in preterm infants. Both the conceptional age and time since birth, regardless of conceptional age, appeared to influence the amplitude of the rhythm. The data were aligned to obtain group averages according to the rectal temperature variation, thus maximizing rhythmicity in this variable. Rhythms in heart rate and temperature were not seen. Updike, Accurso, and Jones (1985) also observed a weak rhythm in skin temperature in six preterm infants (34–37 weeks), but these were averaged across clock time, suggesting an environmental influence.

Although a rhythm in heart rate has been reported in some infants 2 days after birth and, as noted, an activity rhythm has been reported in one preterm infant (Mirmiran & Kok, 1991; Sitka, Weinert, Berle, Rumler, & Schuh, 1994), these rhythms are in general more clearly expressed by the *in utero* fetus than by preterm and newborn infants (Ehrstrom, 1984; Hellbrugge, Lange, Rutenfranz, & Stehr, 1964; Patrick, Campbell, Carmichael, Natale, & Richardson, 1982a; Patrick, Campbell, Carmichael, & Probert, 1982b; Visser, Goodman, Levine, & Dawes, 1982). It is therefore likely that these rhythms result from passive responses to maternal rhythms. Consistent with this are reports in which rhythms in fetal heart rate and movement were absent in the fetus of a woman lacking adrenal glands and, consequently, rhythms in adrenal steroids (Arduini *et al.*, 1987), and rhythms in heart rate failed to develop in continuously fed infants (Salzarulo, Fagioli, & Ricour, 1985). Fetal rhythms in heart rate and activity appear to begin between 13 and 22 weeks of gestation (de Vries, Visser, Mulder, & Prechtl, 1987), with fetal activity highest in the evening and heart rate lowest in the early morning.

A few variables have been reported to show rhythmicity within the first postnatal month in term infants, but it is around the third postnatal month that most rhythms are clearly established, including those in behavioral state, rectal temperature, and the urinary metabolite of melatonin, 6-sulfatoxymelatonin (Table 2) (Figures 10–12).

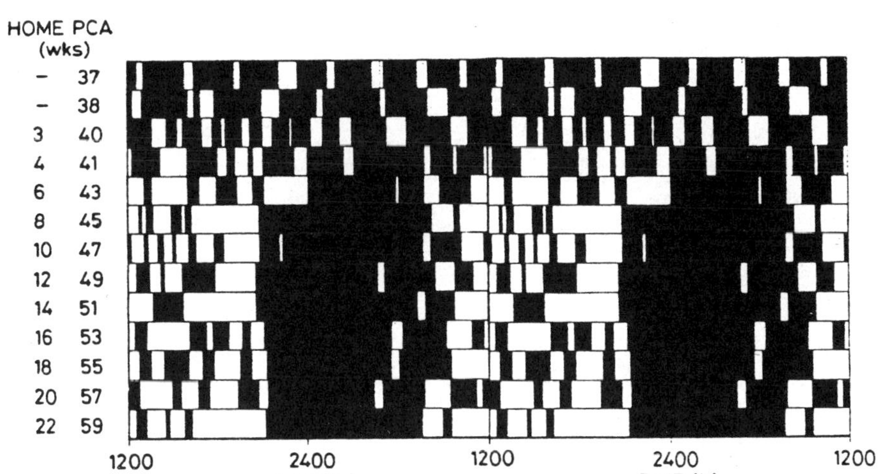

Figure 10. The double-plotted sleep/wake activity record of a female infant born at gestational age 34 weeks and discharged home at postconceptional age of 38 weeks. Behavioral state was determined by observation for 24 hours every 1–2 weeks. The dark bars indicate times asleep. The sleep/wake activity rhythm of this infant was considered to have been entrained after 8 weeks in the home environment. (From McMillen *et al.*, 1991, with permission.)

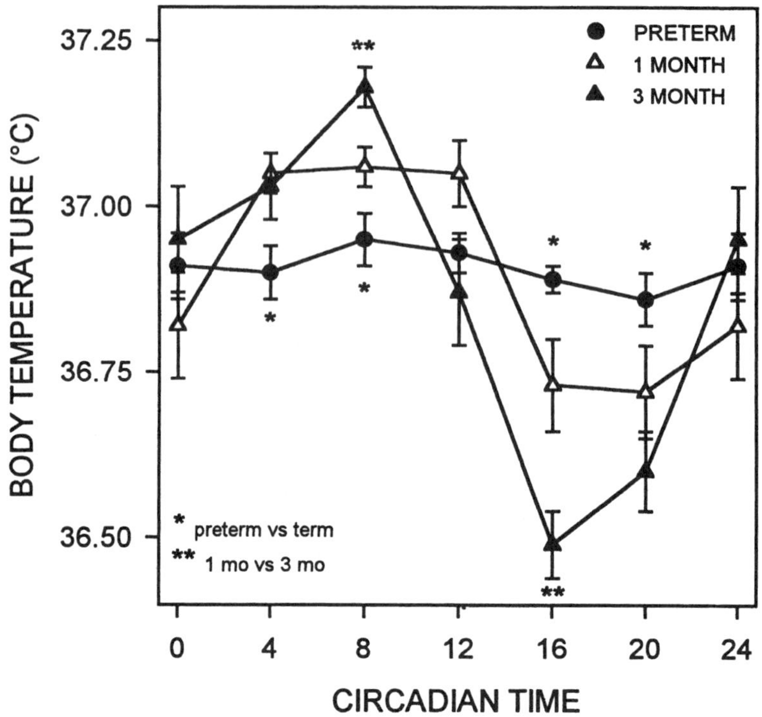

Figure 11. Rectal temperature rhythms of infants recorded at three different ages. Preterm infants were approximately 31 weeks gestational age at birth and 35 weeks postconceptional age at recording, and were recorded in a neonatal intensive care unit under constant light and temperature. An arbitrary phase reference (CT0) was determined for each infant from 72-hour recordings. To obtain group averages, the records were then aligned according to the phase reference. Recordings from older infants were obtained at home over 3 days at each of the two ages. Averages were obtained by aligning individual records according to clock time (CT0 = 0800). (From Glotzbach *et al.*, 1995, with permission. Copyright 1995, American Acadmey of Pediatrics.)

The absence of clear circadian rhythms earlier than this does not mean, however, that newborn or premature infants lack a functional circadian pacemaker. The pacemaker may be functioning but the output mechanisms which mediate its influence over a particular rhythm might not have matured. The presence of melatonin-binding sites with the human fetal SCN (18–19 weeks of gestation; Reppert *et al.*, 1988) (Figure 13) and the observation of a 2DG rhythm in the fetal SCN of squirrel monkeys (Reppert & Schwartz, 1984) (Figure 14), indicate that, like rodents, the human fetus has a functional and entrainable pacemaker within the SCN. There is also suggestive evidence for a fetal pacemaker in sheep. A rhythm in the number of cells showing Fos-like immunoreactivity has been described in fetal sheep (Constandil, Parraguez, Torrealba, Valenzuela, & Serón-Ferré, 1995) and a rhythm in cerebrospinal fluid vasopression is likely to originate from the SCN (Stark & Daniel, 1989). The latter rhythm was disrupted, however, when ewes were kept in constant light. In a study of fetal baboons, fetal heart rate and fetal breathing movements were both rhythmic. The fetal heart rate rhythm showed a consistent phase relationship to the mother, whereas the breathing movement did not, suggesting that the

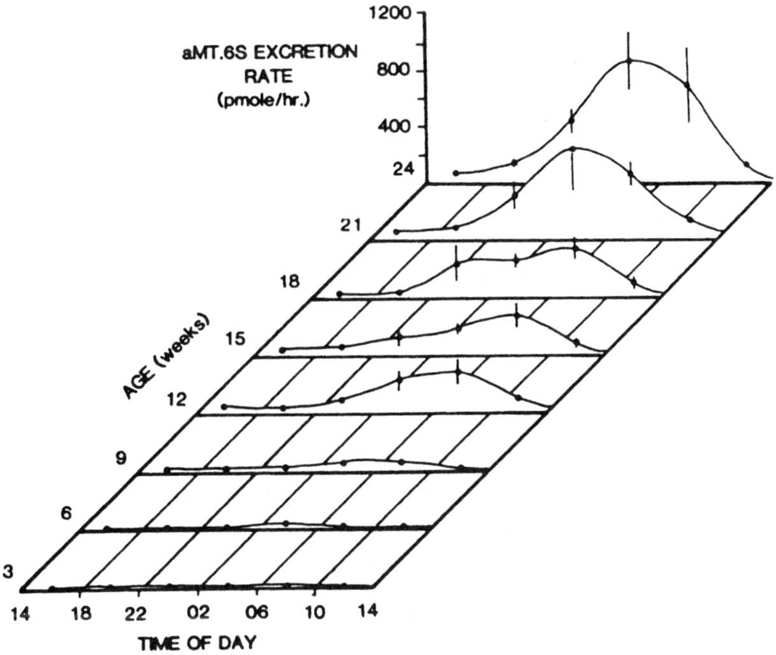

Figure 12. Development of the daily rhythm in 6-sulfatoxymelatonin excretion in full-term human infants. (From Kennaway *et al.*, 1992. Copyright 1992, The Endocrine Society.)

latter was endogenously generated by the fetus and not entrained or caused by the mother (Fletcher, Leung, Myers, & Stark, 1996). However, the periods of the rhythms must have been close to 24 hours since several days of recording were summed for each fetus with a folding period of 24 hours.

If there is a functional fetal pacemaker in humans, it is likely to be entrained by maternal rhythms. Based on evidence in rodents and the presence of melatonin-binding sites in the human fetal SCN, it is also possible that melatonin is a signal for this entrainment. As in rodents, however, it is also possible that a number of other substances could affect the human fetal SCN, including dopamine agonists, cocaine, and nicotine. Regardless of what the normal maternal signal for entrainment is, exposing the fetus to such substances may phase shift the fetal pacemaker with long-term consequences for the eventual normal entrainment of the pacemaker and the development of circadian rhythms.

After birth, the human pacemaker, as in rodents, is likely to make a transition from entrainment by maternal rhythms to entrainment by other environmental cycles, in particular the light/dark cycle. Although it is not known when the human circadian pacemaker becomes entrainable, it is likely that it is entrainable by light soon after birth; the retinal hypothalamic projection appears to be present before birth (Glotzbach, Sollars, Ariagno, & Pickard, 1992), plasma melatonin, although not yet rhythmic, may be influenced by light within the first 2 days after birth (Jaldo-Alba, Munoz-Hoyos, Molina-Carballo, Molina-Font, & Acuna-Castroviejo, 1993), and in newborn baboons, light entrains circadian rhythms and induces c-*fos* expression in the SCN (Rivkees, Hofman, & Fortman, 1997). Furthermore, in several studies

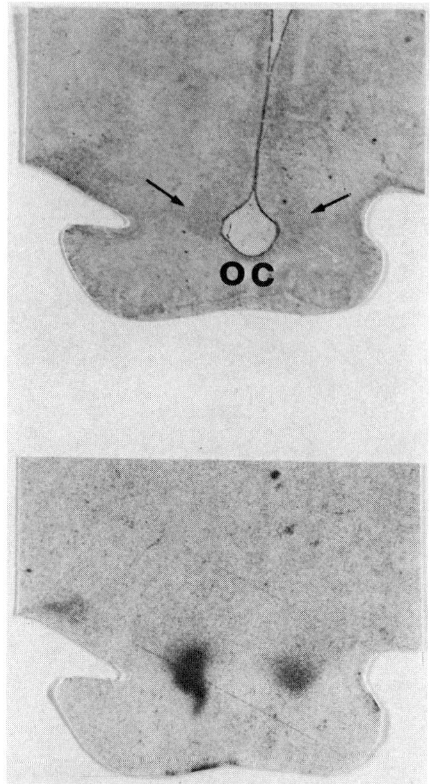

Figure 13. Specific ^{125}I-labeled melatonin binding in the SCN of the human fetus (18–19 weeks of gestation). A section through the SCN (top) and its autoradiographic image are shown. Specific binding is indicated in the autoradiographic image as dark densities over the SCN. (From Reppert *et al.*, 1988, with permission. Copyright 1988, American Association for the Advancement of Science.)

describing the development of human circadian rhythms, the rhythms are in phase with the environment when they appear, suggesting prior entrainment of the underlying pacemaker or the influence of an uncontrolled environmental factor (Coons & Guilleminault, 1984; Hellbrugge *et al.*, 1964; Kleitman & Engelmann, 1953; McMillen, Kok, Adamson, Deayton, & Nowak, 1991; Parmelee, Wenner, & Schulz, 1964; Spangler, 1991; Updike *et al.*, 1994; Weinert, Sitka, Minors, & Waterhouse, 1994). In one study, the phases of body temperature rhythms, although not random on day 2, became more synchronized by week 4 (Sitka *et al.*, 1994).

The development of sleep and wakefulness is of particular interest because of the impact their timing has on caregivers and because sudden infant death syndrome (SIDS) occurs predominantly at night during sleep and during the time in development when the circadian rhythm in sleep and wakefulness is becoming established (Gibson, 1992). In humans, the pattern of sleep and wakefulness changes from one of multiple, relatively short sleep-and-wake bouts per day with no distinction between day and night, to one of consolidated sleep at night and continuous waking during the day (Figure 10). An increase in the average duration of sleep episodes is more pronounced than an increase in the average duration of waking, and the average duration of sleep episodes may be the first aspect of sleep/wake

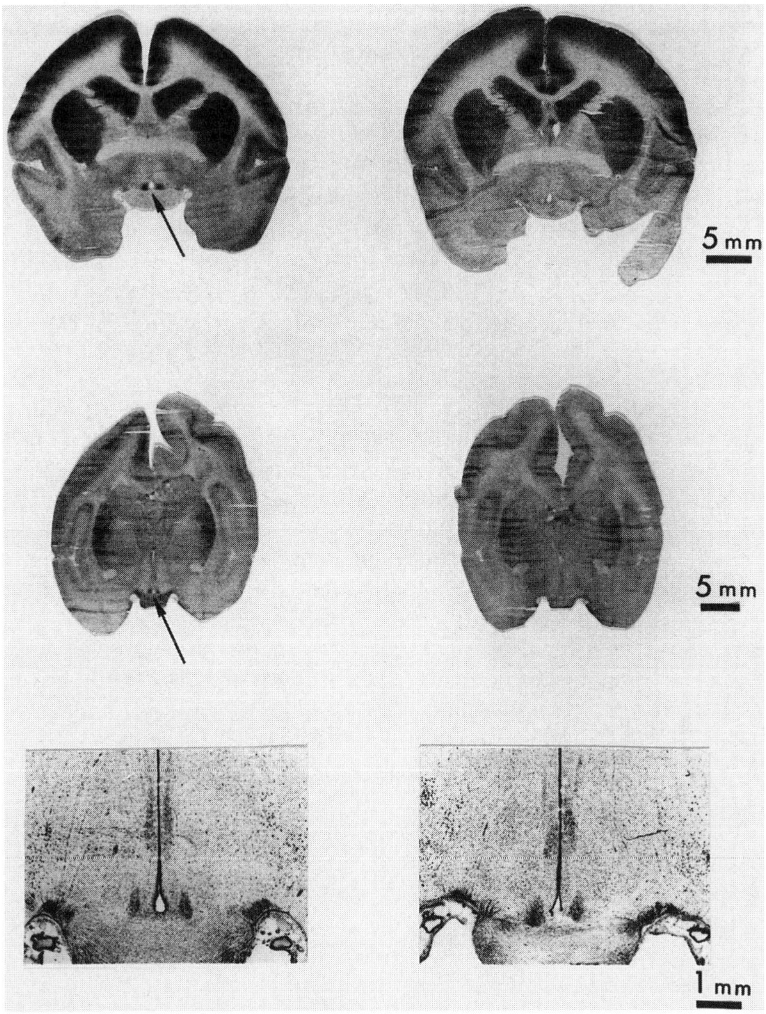

Figure 14. A rhythm in deoxyglucose (DG) uptake in the SCN of a fetal primate and its mother (squirrel monkey, *Saimiri sciureus*). Autoradiographic images of maternal and fetal sections are shown above histologic sections from the fetus (approx 90% of term). Images on the left are from animals given 2DG during the subjective day and those on the right from animals given DG during subjective night. The metabolically active SCN are indicated on the subjective-day autoradiographs by arrows. (From Reppert & Schwartz, 1984. Copyright 1984, Elsevier Science.)

patterns to be influenced by circadian timing (Coons & Guilleminault, 1984; Kleitman & Engelmann, 1953; Parmelee *et al.*, 1964; Pollak, 1994). Although the consolidation of sleep and waking may continue for several months or years after birth, clear differences between day and night are established by at least the third postnatal month (Coons & Guilleminault, 1984; Glotzbach, Edgar, Boeddiker, & Ariagno, 1994; Kleitman & Engelmann, 1953; McMillen *et al.*, 1991; Parmelee *et al.*, 1964). Small, but significant differences in the activity of infants or in average sleep episode duration have also been seen within the first month after birth (Glotzbach *et al.*, 1994; Kleitman & Engelmann, 1953; Parmelee *et al.*, 1964; Sitka *et al.*, 1994), and, as noted, an activity rhythm was reported in a preterm infant recorded at 33 weeks postconception (Mirmiran & Kok, 1991).

The consolidation of sleep with the simultaneous occurrence of the consolidated sleep at night has been referred to as the development of entrainment of the rhythm (Coons & Guilleminault, 1984; McMillen *et al.*, 1991; Pollak, 1994). In most instances, however, as a 24-hour rhythm emerges, the longest period of uninterrupted sleep is already coincident with night (Coons & Guilleminault, 1984; Kleitman & Engelmann, 1953; McMillen *et al.*, 1991) and the criterion for whether there is a rhythm may even be that there is a significant difference between day and night. Thus, the emergence of a sleep/wake rhythm may indicate little about entrainment of an underlying circadian pacemaker, but may instead reflect the increasing influence over sleep and arousal mechanisms by a pacemaker that is already entrained.

Although Kleitman and Engleman (1953) reported that most infants are in phase with the environment when the sleep/wake rhythm develops, they highlighted one exceptional case. The sleep/wake record of an infant from 2 to 26 weeks of age showed a clear circadian rhythm with a period of greater than 24 hours beginning at about 8 weeks. It was not until 24 weeks that the rhythm showed a stable 24-hour period with the normal phase relationship to the light/dark cycle (Figure 15). The light and dark exposure of this infant was not described, but it was maintained on a self-demand feeding schedule. In this case, therefore, it would appear that a rhythm was present before it was entrained, suggesting that the occurrence of sleep at night was limited at least in part by entrainment of the pacemaker. It is uncertain, however, whether the observed free-running rhythm was an expression of the circadian pacemaker. Because it is possible for sleep/wake rhythms in adults to express a circadian period which is clearly distinct from that of the circadian pacemaker which continues to regulate other rhythms, such as the body temperature rhythm (Aschoff, Gerecke, & Wever, 1967), it is possible that the rhythm initially expressed in this infant was not regulated by the SCN. Instead, it may have been generated by the homeostatic aspects of sleep/wake regulation which gradually came under control of the pacemaker. Sleep was most consolidated when it was coincident with what would have been both environmental night as well as subjective night of the pacemaker.

McMillen *et al.* (1991) found that the appearance of the sleep/wake rhythm is determined by the length of time infants have been home and exposed to stronger rhythms in light and darkness than were present in the hospital. This environmental effect could be due in part to entrainment of the underlying pacemaker, but could also result from environmental influences that promote sleep at the appropriate phase of the already entrained pacemaker. Whatever the underlying mechanism, exposure to environmental cycles early in development may have long-term effects. Mann *et al.* (1986) studied two groups of preterm infants that were kept in nurseries differing in the extent of environmental periodicity, especially in the light/dark cycle. Infants from the more rhythmic environment spent more time asleep and less time feeding and gained more weight after discharge than those from the relatively arrhythmic environment. In addition, Kennaway, Goble, and Stamp (1996) found that the delayed development of a melatonin rhythm in some preterm infants could be accelerated by exposing the infants to dark at night. They did not , however, see any effect of this treatment on nocturnal sleep or on growth.

At birth, infants are removed from rhythmicity of the intrauterine environment. Term infants are subsequently exposed to other environmental rhythmicity such as

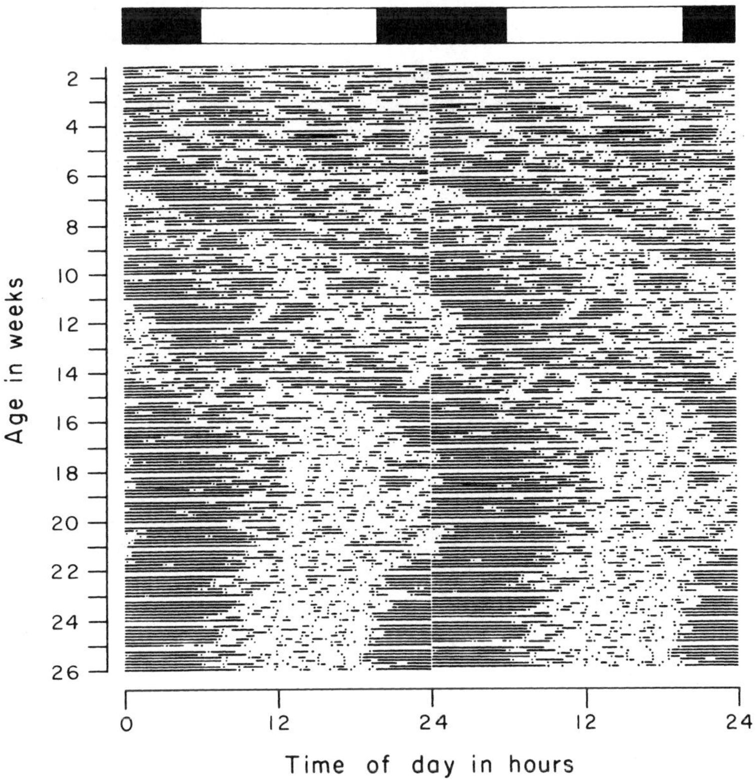

Figure 15. The double-plotted sleep/wakefulness record of a female infant recorded by observation from 2 to 26 weeks of age. The lines indicate time asleep, and dots are feedings. The dark portion of the bar at the top indicates nighttime as defined by the infant's time of uninterrupted sleep at the end of the record. (From Davis, 1981; redrawn from Kleitman & Engelmann, 1953.)

the light/dark cycle. Preterm infants, however, may be deprived of environmental periodicity if maintained in a neonatal intensive care unit. As already noted, this may have detrimental effects. Consistent with this, McMillen *et al.* (1991) noted that an infant exposed to bright light during night feedings at home over the course of their study never developed a sleep/wake rhythm. On the other hand, Kenneway *et al.* (1996) found that infants that were preterm because of premature labor developed a melatonin rhythm at the same time as term infants when corrected for postconception age, indicating that their were no detrimental effects of the arrhythmic environment while in the hospital. Whether or not environmental rhythmicity has effects on subsequent development is likely to depend on the maturational state of the infant (Kennaway *et al.*, 1996).

The impact of circadian rhythm development on the general health and well-being of a family and the possible linkage of circadian rhythms to rarer but more devastating outcomes such as SIDS point to the need for understanding the consequences of abnormal rhythmicity during development. In addition to the arrhythmic conditions of hospital nurseries, other aspects of modern life might also contribute to the disruption of both maternal and fetal rhythmicity. Drugs such as cocaine or nicotine may directly affect the fetal pacemaker and shift work performed by pregnant women may contribute to low birth weight (B. G. Armstrong, Nolin, & McDonald, 1989; Axelsson, Rylander, & Molin, 1989; Nurminen, 1989). Other potential, but poorly documented disruptions of rhythmicity during the neonatal period may include excessive use of bright light during the night at home and the feeding of breast milk collected at one time of day but given at another. Human milk is known to contain melatonin and may differ in other ways between day and night as well (Illnerová *et al.*, 1993). A lesson learned form the last three decades of research on circadian rhythms is that human physiology functions within a 24-hour framework and that this cannot be ignored with respect to either health maintenance or the treatment of disease. Because the human fetus develops within a rhythmic environment and is likely to have a functional circadian pacemaker which can be reset by a variety of stimuli, this lesson is likely to be as applicable to human development as it is to the rest of life.

Acknowledgments

The preparation of this chapter was supported in part by National Institute of Child Health and Human Development Grant HD18686 to F. C. Davis. Portions of the text appeared in Davis, Frank, and Heller, (1999). Material provided by Colette Kabrita was produced with support from National Institute of Child Health and Human Development Grant HD18686 to F. C. Davis.

References

Altman, J., & Bayer, S. A. (1978a). Development of the diencephalon in the rat I. Autoradiographic study of the time of origin and settling patterns of neurons of the hypothalamus. *Journal of Comparative Neurology, 182,* 945–972.

Altman, J., & Bayer, S. A. (1978b). Development of the diencephalon in the rat II. Correlation of the embryonic development of the hypothalamus with the time of origin of its neurons. *Journal of Comparative Neurology, 182,* 973–994.

Altman, J., & Bayer, S. A. (1986). The development of the rat hypothalamus. *Advances in Anatomy Embryology and Cell Biology, 100,* 1–178.

Altman, J., & Bayer, S. A. (1995). *Atlas of prenatal rat brain development.* Boca Raton, FL: CRC Press.

Antoch, M. P., Song, E. J., Chang, A. M., Vitaterna, M. H., Zhao, Y. L., Wilsbacher, L. D., Sangoram, A.M., King, D. P., Pinto, L. H., & Takahashi, J. S. (1997). Functional identification of the mouse circadian *clock* gene by transgenic BAC rescue. *Cell, 89,* 655–667.

Arduini, D., Rizzo, G., Parlati, E., Dell'Acqua, S., Romanini, C., & Mancuso, S. (1987). Loss of circadian rhythms of fetal behaviour in a totally adrenalectomized pregnant woman. *Gynecological and Obstetrical Investigations, 23,* 226–229.

Armstrong, B. G., Nolin, A. D., & McDonald, A. D. (1989). Work in pregnancy and birth weight for gestational age. *British Journal of Industrial Medicine, 46,* 196–199.

Armstrong, S. M. (1989). Melatonin and circadian control in mammals. *Experientia, 45,* 933–938.

Aschoff, J. (1960). Exogenous and endogenous components in circadian rhythms. *Cold Spring Harbor Symposium on Quantitative Biology, 25,* 11–28.

Aschoff, J., Gerecke, U., & Wever, R. (1967). Desynchronization of human circadian rhythms. *Japanese Journal of Physiology, 17,* 450–457.

Axelsson, G., Rylander, R., & Molin, I. (1989). Outcome of pregnancy in relation to irregular and inconvenient work schedules. *British Journal of Industrial Medicine, 46,* 393–398.

Ban, Y., Shigeyoshi, Y., & Okamura, H. (1997). Development of vasoactive intestinal peptide mRNA rhythm in the rat suprchiasmatic nucleus. *Journal of Neuroscience, 17,* 3920–3931.

Barr, M. J. (1973). Prenatal growth of wistar rats: Circadian periodicity of fetal growth late in gestation. *Teratology, 7,* 283–287.

Botchkina, G. I. & Morin, L. P. (1993). Development of the hamster serotoninergic system: Cell groups and diencephalic projections. *Journal of Comparative Neurology, 338,* 405–431.

Botchkina, G. I., & Morin, L. P. (1995a). Ontogeny of radial glia, astrocytes and vasoactive intestinal peptide immunoreactive neurons in hamster suprachiasmatic nucleus. *Developmental Brain Research, 86*, 48–56.

Botchkina, G. I., & Morin, L. P. (1995b). Organization of permanent and transient neuropeptide Y-immunoreactive neuron groups and fiber systems in the developing hamster diencephalon. *Journal of Comparative Neurology, 357*, 573–602.

Broekhuizen, S., & Maaskamp, F. (1980). Behaviour of does and leverets of the European hare (*Lepus europaeus*) whilst nursing. *Journal of Zoology* (London), *191*, 487–501.

Cambras, T., & Diez-Noguera, A. (1991). Evolution of rat motor activity circadian rhythm under three different light patterns. *Physiology and Behavior, 49*, 63–68.

Card, J. P., & Moore, R. Y. (1984). The suprachiasmatic nucleus of the golden hamster: Immunohistochemical analysis of cell and fiber distribution. *Neuroscience, 13*, 415–431.

Carlson, L. L., Weaver, D. R., & Reppert, S. M. (1991). Melatonin receptors and signal transduction during development in Siberian hamsters (*Phodopus sungorus*). *Developmental Brain Research, 59*, 83–88.

Cassone, V. M., Speh, J. C., Card, J. P., & Moore, R. Y. (1988). Comparative anatomy of the mammalian hypothalmic suprachiasmatic nucleus. *Journal of Biological Rhythms, 3*, 71–91.

Chamberlain, P. F., Manning, F. A., Morrison, I., & Lange, I. R. (1984). Circadian rhythm in bladder volumes in the term human fetus. *Obstetrics and Gynecology, 64*, 657–660.

Clegg, D. A., O'Hara, B. F., Heller, H. C., & Kilduff, T. S. (1995). Nicotine administration differentially affects gene expression in the maternal and fetal circadian clock. *Developmental Brain Research, 84*, 46–54.

Constandil, L., Parraguez, V. H., Torrealba, F., Valenzuela, G., & Serón-Ferré, M. (1995). Day–night changes in c-*fos* expression in the fetal sheep suprachiasmatic nucleus at late gestation. *Reproduction Fertility and Development, 7*, 411–413.

Coons, S., & Guilleminault, C. (1984). Development of consolidated sleep and wakeful periods in relation to the day/night cycle in infancy. *Developmental Medicine and Child Neurology, 26*, 169–176.

Cooper, Z. K. (1939). Mitotic rhythm in human epidermis. *Journal of Investigative Dermatology, 2*, 289–300.

Cooper, H. M., Tessonneaud, A., Caldani, A., Locatelli, A., Richard, S., & Viguier-Martinez, M.-C. (1993). Morphology and distribution of retinal ganglion cells (RGC) projecting to the suprachiasmatic nucleus in the sheep. *Society for Neuroscience Abstracts, 11*, 1704.

Darwin, C. (1896). *The power of movement in plants.* New York: Appleton.

Davis, F, C. (1981). Ontogeny of circadian rhythms. In J. Aschoff (Ed.), *Handbook of behavioral neurobiology*, Vol. 4, *Biological rhythms* (pp. 257–274). New York: Plenum Press.

Davis, F. C. (1982). Development of the suprachiasmatic nuclei and other circadian pacemakers. In D. C. Klein (Ed.), *Melatonin rhythm generating system: Developmental aspects* (pp. 1–19). Basel: Karger

Davis, F. C. (1989). Daily variation in maternal and fetal weight gain in mice and hamsters. *Journal of Experimental Zoology, 250*, 273–282.

Davis, F. C., & Gorski, R. A. (1984). Unilateral lesions of the hamster suprachiasmatic nuclei: Evidence for redundant control of circadian rhythms. *Journal of Comparative Physiology A, 154*, 221–232.

Davis, F. C., & Gorski, R. A. (1986). Development of hamster circadian rhythms: Prenatal entrainment of the pacemaker. *Journal of Biological Rhythms, 1*, 77–89.

Davis, F. C., & Gorski, R. A. (1988). Development of hamster circadian rhythms: Role of the maternal suprachiasmatic nucleus. *Journal of Comparative Physiology A, 162*, 601– 610.

Davis, F. C., & Mannion, J. (1988). Entrainment of hamster pup circadian rhythms by prenatal melatonin injections to the mother. *American Journal of Physiology, 255*, R439–R448.

Davis, F. C., & Menaker, M. (1981). Development of the mouse circadian pacemaker: Independence from environmental cycles. *Journal of Comparative Physiology A, 143*, 527–539.

Davis, F. C., & Viswanathan, N. (1996). The effect of transplanting one or two suprachiasmatic nuclei on the period of the restored rhythm. *Journal of Biological Rhythms, 11*, 291–301.

Davis, F. C., Darrow, J. M., & Menaker, M. (1983). Sex differencs in the circadian control of hamster wheel-running activity. *American Journal of Physiology, 244*, R93–R105.

Davis, F. C., Boada, R., & LeDeaux, J. (1990). Neurogenesis of the hamster suprachiasmatic nucleus. *Brain Research, 519*, 192–199.

Davis, F. C., Frank, M. G., & Heller, H. C. (1999). Ontogeny of sleep and circadian rhythms. In F. W. Turek & P. C. Zee (Eds.), *Regulation of sleep and circadian rhythms*, Vol. 133, *Lung biology in health and disease.* New York: Marcel Dekker.

Deacon, S., & Arendt, J. (1995). Melatonin-induced temperature suppression and its acute phase-shifting effects correlate in a dose-dependent manner in humans. *Brain Research, 688*, 77–85.

Decavel, C., & Van den Pol, A. N. (1990). GABA: A dominant neurotransmitter in the hypothalamus. *Journal of Comparative Neurology, 302*, 1019–1037.

Deguchi, T. (1975). Ontogenesis of a biological clock for serotonin: Acetyl coenzyme A N-acetyltransferase in pineal gland of rat. *Proceedings of the National Academy of Sciences of the USA, 72,* 2814–2818.

De Vries, G. J., Buijs, R. M., & Swaab, D. F. (1981). Ontogeny of the vasopressinergic neurons of the suprachiasmatic nucleus and their extrahypothalamic projections in the rat brain—Presence of a sex difference in the lateral septum. *Brain Research, 218,* 67–78.

de Vries, J. I. P., Visser, G. H. A., Mulder, E. J. H., & Prechtl, H. F. R. (1987). Diurnal and other variations in fetal movement and heart rate patterns at 20–22 weeks. *Early Human Development, 15,* 333–348.

Drucker-Colin, R., Aguilar-Roblero, R., Garcia-Hernandez, F., Fernandez-Cancino, F., & Rattoni, F. B. (1984). Fetal suprachiasmatic nucleus transplants: Diurnal rhythm recovery of lesioned rats. *Brain Research, 311,* 353–357.

Duffield, G. E., Dickerson, J. M. Alexander, I. H. M., & Ebling, F. J. P. (1995). Ontogeny of a photic response in the suprachiasmatic nucleus in the Siberian hamster (phodopus sungorus). *European Journal of Neuroscience, 7,* 1089–1096.

Duncan, M. J., & Davis, F. C. (1993). Developmental appearance and age related changes in specific 2-[125I]iodomelatonin binding sites in the suprachiasmatic nuclei of female Syrian hamsters. *Developmental Brain Research, 73,* 205–212.

Duncan, M. J., Banister, M. J., & Reppert, S. M. (1986). Developmental appearance of light–dark entrainment in the rat. *Brain Research, 369,* 326–330.

Ehrstrom, C. (1984). Circadian rhythm of fetal movements. *Acta Obstetrica et Gynecologica Scandinavica, 63,* 539–541.

Elliott, J. A., & Goldman, B. D. (1989). Reception of photoperiodic information by fetal Siberian hamsters: Role of the mother's pineal gland. *Journal of Experimental Zoology, 252,* 237–244.

Ellison, N., Weller, J. L., & Klein, D. C. (1972). Development of a circadian rhythm in the activity of pineal seritonin N-acetyltransferase. *Journal of Neurochemistry, 19,* 1335–1341.

Fletcher, K. L., Leung, K., Myers, M. M., & Stark, R. I. (1996). Diurnal rhythms in cardiorespiratory function of the fetal baboon. *Early Human Development, 46,* 27–42.

Fuchs, J. L., & Moore, R. Y. (1980). Development of circadian rhythmicity and light responsiveness in the rat suprachiasmatic nucleus: A study using the 2-deoxy[1-^{14}C] glucose method. *Proceedings of the National Academy of Sciences, 77,* 1204–1208.

Ganzhorn, J. U., & Wright, P. C. (1994). Temporal patterns in primate leaf eating: The possible role of leaf chemistry. *Folia Primatologica, 63,* 203–208.

Gibson, A. A. M. (1992). Current epidemiology of SIDS. *Journal of Clinical Pathology, 45* (Supplement), 7–10.

Glotzbach, S. T., Sollars, P., Ariagno, R. L., & Pickard, G. E. (1992). Development of the human retinohypothalamic tract. *Society for Neuroscience Abstracts, 18,* 875.

Glotzbach, S. F., Edgar, D. M., Boeddiker, M., & Ariagno, R. L. (1994). Biological rhythmicity in normal infants during the first 3 months of life. *Pediatrics, 94,* 482–488.

Glotzbach, S. F., Edgar, D. M., & Ariagno, R. L. (1995). Biological rhythmicity in preterm infants prior to discharge from neonatal intensive care. *Pediatrics, 95,* 4231–237.

Griffioen, H. A., Duindam, H., Van der Woude, T. P., Rietveld, W. J., & Boer, G. J. (1993). Functional development of fetal suprachiasmatic nucleus grafts in suprachiasmatic nucleus-lesioned rats. *Brain Research Bulletin, 31,* 145–160.

Grosse, J., & Davis, F. C. (1998). Melatonin entrains restored circadian activity rhythms of Syrian hamsters bearing fetal SCN grafts. *Journal of Neuroscience, 18,* 8032–8037.

Grosse, J., Velickovic, A., & Davis, F. C. (1996). Entrainment of Syrian hamster circadian activity rhythms by neonatal melatonin injections. *American Journal of Physiology, 270,* R533–R540.

Güldner, F.-H. (1978). Synapses of optic nerve afferents in the rat suprachiasmatic nucleus. I. Identification, qualitative description, development and distribution. *Cell and Tissue Research, 194,* 17–35.

He, X., Treacy, M. N., Simmons, D. M., Ingraham, H. A., Swanson, L. W., & Rosenfeld, M. G. (1989). Expression of a large family of POU-domain regulatory genes in mammalian brain development. *Nature, 340,* 35–42.

Hellbrugge, T., Lange, J. E., Rutenfranz, J., & Stehr, K. (1964). Circadian periodicity of physiological functions in different stages of infancy and childhood. *Annals of the New York Academy of Sciences, 117,* 361–373.

Hiroshige, T., Honma, K., & Watanabe, K. (1982a). Possible zeitgebers for external entrainment of the circadain rhythm of plasma corticosterone in blind infantile rats. *Journal of Physiology, 325,* 507–519.

Hiroshige, T., Honma, K., & Watanabe, K. (1982b). Prenatal onset and maternal modifications of the circadian rhythm of plasma corticosterone in blind infantile rats. *Journal of Physiology, 325,* 521–532.

Hoffman, K. (1959). Die aktivitatsperiodik von im 18- und 36-stunden-tag erfruteten eidechsen. *Zeitschrift fur vergleichende Physiologie, 42,* 422–432.

Holtzman, R. L., Malach, R., & Gozes, I. (1989). Disruption of the optic pathway during development affects vasoactive intestinal peptide mRNA expression. *New Biologist, 1,* 215–221.

Honma, S., Honma, K., Shirakawa, T., & Hiroshige, T. (1984). Effects of elimination of maternal ceircadian rhythms during pregnancy on the postnatal development of circadian corticosterone rhythm in blinded infantile rats. *Endocrinology, 114,* 44–50.

Honnebier, M. B. O. M., Swaab, D. F & Mirmiran, M. (1989). Diurnal rhythmicity during early human development. In S. M. Reppert (Ed.), *Development of circadian rhythmicity and photoperiodism in mammals* (pp. 221–244). Ithaca, NY: Perinatology Press

Horton, T. H. (1983). Growth and maturation in *Microtus montanus*: Effects of photoperiods before and after weaning. *Canadian Journal of Zoology, 62,* 1741–1746.

Hudson, R., & Distel, H. (1989). Temporal pattern of suckling in rabbit pups: A model of circadian synchrony between mother and young. In S. M. Reppert (Ed.), *Development of rhythmicity and photoperiodism in mammals* (pp. 83–102). Ithaca, NY: Perinatology Press

Ibuka, N. (1987). Circadian rhythms in sleep–wakefullness and wheel-running activity in a congenitally anophthalmic rat mutant. *Physiology and Behavior, 39,* 321–326.

Illnerová, H., Buresová, M., & Presl, J. (1993). Melatonin rhythm in human milk. *Journal of Clinical Endocrinology and Metabolism, 77,* 838–841.

Inouye, S.-I. T., & Kawamura, H. (1979). Persistence of circadian rhythmicity in a mammalian hypothalamic "island" containing the suprachiasmatic nucleus. *Proceedings of the National Academy of Sciences of the USA, 76,* 5962–5966.

Iuvone, P. M. & Gan, J. (1995). Functional interaction of melatonin receptors and D1 dopamine receptors in cultured chick retinal neurons. *Journal of Neuroscience, 15,* 2179–2185.

Jaldo-Alba, F., Munoz-Hoyos, A., Molina-Carballo, A., Molina-Font, J. A., & Acuna-Castroviejo, D. (1993). Light deprivation increases plasma levels of melatonin during the first 72 h of life in human infants. *Acta Endocrinologica, 129,* 442–445.

Jilge, B. (1993). The ontogeny of circadian rhythms in the rabbit. *Journal of Biological Rhythms, 8,* 247–260.

Jilge, B. (1995). Ontogeny of the rabbit's circadian rhythms without an external zeitgeber. *Physiology and Behavior, 58,* 131–140.

Johnson, R. F., Moore, R. Y., & Morin, L. P. (1988). Loss of entrainment and anatomical plasticity after lesions of the hamster retinohypothalamic tract. *Brain Research, 460,* 297–313.

Johnson, R. F., Morin, L. P., & Moore, R. Y. (1988). Retinohypothalamic projections in the hamster and rat demonstrated using cholera toxin. *Brain Research, 462,* 301–312.

Kagotani, Y., Hashimoto, T., Tsuruo, Y., Kawano, H., Daikoku, S., & Chihara, K. (1989). Development of the neuronal system containing neuropeptide Y in the rat hypothalamus. International *Journal of Developmental Neuroscience, 7,* 359–374.

Kaufman, C. M., & Menaker, M. (1993). Effect of transplanting suprachiasmatic nuclei from donors of different ages into completely SCN lesioned hamsters. *Journal of Neural Transplantation and Plasticity, 4,* 257–265.

Kaufman, C. M., & Menaker, M. (1994). Ontogeny of light-induced Fos-like immunoreactivity in the hamster suprachiasmatic nucleus. *Brain Research, 633,* 162–166.

Kennaway, D. J., Stamp, G. E., & Goble, F. C. (1992). Development of melatonin production in infants and the impact of prematurity. *Journal of Clinical Endocrinology and Metabolism, 75,* 367–369.

Kennaway, D. J., Goble, F. C., & Stamp, G. E. (1996). Factors influencing the development of melatonin rhythmicity in humans. *Journal of Clinical Endocrinology and Metabolism, 81,* 1525–1532.

King, D. P., Zhao, Y. L., Sangoram, A. M., Wilsbacher, L. D., Tanaka, M., Antoch, M. P., Steeves, T. D. L., Vitaterna, M. H., Kornhauser, J. M., Lowrey, P. L., Turek, F. W., & Takahashi, J. S. (1997). Positional cloning of the mouse circadian *clock* gene. *Cell, 89,* 641–653.

Klein, D. C. (1972). Evidence for placental transfer of 3H-acetyl-melatonin. Nature, 237, 117–119.

Kleitman, N. & Engelmann, T. G. (1953). Sleep characteristics of infants. *Journal of Applied Physiology, 6,* 269–282.

Koritsanszky, S. (1981). Fetal and early postnatal cyto- and synaptogenesis in the suprachiasmatic nucleus of the rat hypothalamus. *Acta Morphologica Academiae Scientiarum Hungaricae, 29,* 227–239.

Krieger, D. T. (1972). Circadian corticosteroid periodicity: Critical period for abolition by neonatal injection of corticosteroid. *Science, 178,* 1205–1207.

Krieger, D. T., & Hauser, H. (1977). Suprachiasmatic nuclear lesions do not abolish food-shifted circadian adrenal and temperature rhythmicity. *Science, 197,* 398–399.

Kuhlman, S., Watts, A. G., Sanchez-Watts, G., & Davis, F. C. (1995). Developmental expression of preprovasoactive intestinal polypeptide (VIP) mRNA in the Syrian hamster suprachiasmatic nucleus. *Society for Neuroscience Abstracts, 21,* 452.

Laemle, L. K., & Rusa, R. (1992). VIP-like immunoreactivity in the suprachiasmatic nuclei of a mutant anophthalmic mouse. *Brain Research, 589*, 124–128.

Laemle, L. K., Repke, K. B., Hawkes, R., & Rice, F. L. (1991). Synaptogenesis in the rat suprachiasmatic nucleus: A light microscopic immunocytochemical survey. *Brain Research, 544*, 108–117.

Leard, L. E., Macdonald, E. S., Heller, H. C., & Kilduff, T. S. (1994). Ontogeny of photic-induced *c-fos* mRNA expression in rat suprachiasmatic nuclei. *Neuroreport, 5*, 2683–2687.

Lehman, M. N., Silver, R., Gladstone, W. R., Kahn, R. M., Gibson, M., & Bittman, E. L. (1987). Circadian rhythmicity restored by neural transplant. Immunocytochemical characterization of the graft and its integration with the host brain. *Journal of Neuroscience, 7*, 1626–1638.

Lenn, N. J., Beebe, B., & Moore, R. Y. (1977). Postnatal development of the suprachiasmatic hypothalamic nucleus of the rat. *Cell and Tissue Research, 178*, 463–475.

LeSauter, J., Lehman, M. N., & Silver, R. (1996). Restoration of circadian rhythmicity by transplants of SCN "micropunches." *Journal of Biological Rhythms, 11*, 163–171.

Lewy, A. J. (1992). Melatonin shifts human circadian rhythms according to a phase-response curve. *Chronobiology International, 9*, 380–392.

Magnin, M., Cooper, H. M., & Mick, G. (1989). Retinohypothalamic pathway: A breach in the law of Newton–Muller–Gudden? *Brain Research, 488*, 390–397.

Mai, J. K., Kedziora, O., Teckhaus, L., & Sofroniew, M. V. (1991). Evidence for subdivisions in the human suprachiasmatic nucleus. *Journal of Comparative Neurology, 305*, 508–525.

Mann, N. P., Haddow, R., Stokes, L., Goodley, S. & Rutter, N. (1986). Effect of night and day on preterm infants in a newborn nursery: Randomised trial. *British Journal of Medicine, 293* 1265–1267.

Martin du Pan, R. (1974). Some clinical applications of our knowledge of the evolution of the circadian rhythm in infants. In L. E. Scheving, F. Halberg, & J. E. Pauly (Eds.), *Chronobiology* (pp. 342–347). Tokyo: Iguku Shoin.

McMillen, I. C., & Nowak, R. (1989). Maternal pinealectomy abolishes the diurnal rhythm in plasma melatonin concentrations in the fetal sheep and pregnant ewe during late gestation. *Journal of Endocrinology, 120*, 459–464.

McMillen, I. C., Kok, J. S. M., Adamson, T.M., Deayton, J. M., & Nowak, R. (1991). Development of circadian sleep–wake rhythms in preterm and full-term infants. *Pediatric Research, 29*, 381–384.

Miller, J. D., Morin, L. P., Schwartz, F. J., & Moore, R. Y. (1996). New insights into the mammalian circadian clock. *Sleep, 19*, 641–667.

Miller, M. W. (1992). Circadian rhythm of cell proliferation in the telencephalic ventricular zone: Effect of *in utero* exposure to ethanol. *Brain Research, 595*, 17–24.

Mirmiran, M., & Kok, J. H. (1991). Circadian rhythms in early human development. *Early Human Development, 26*, 121–128.

Moore, R. Y. (1973). Retinohypothalamic projection in mammals: A comparative study. *Brain Research, 49*, 403–409.

Moore, R. Y., & Bernstein, M. E. (1989). Synaptogenesis in the rat suprachiasmatic nucleus demonstrated by electron microscopy and synapsin I immunoreactivity. *Journal of Neuroscience, 9*, 2151–2162.

Moore, R. Y., & Lenn, N. J. (1972). A retinohypothalamic projection in the rat. *Journal of Comparative Neurology, 146*, 1–14.

Moore, R. Y., Speh, J. C., & Card, J. P. (1995). The retinohypothalamic tract originates from a distinct subset of retinal ganglion cells. *Journal of Comparative Neurology, 352*, 351–366.

Morin, L. P. (1994). The circadian visual system. *Brain Research Reviews, 19*, 102–127.

Mosko, S., & Moore, R. Y. (1978). Neonatal suprachiasmatic nucleus ablation: Absence of functional and morphological plasticity. *Proceedings of the National Academy of Sciences of the USA, 75*, 6243–6246.

Mosko, S., & Moore, R. Y. (1979). Retinohypothalamic tract development: Alteration by suprachiasmatic lesions in the neonatal rat. *Brain Research, 164*, 1–15.

Niimi, K., Harada, I., Kusaka, Y., & Kishi, S. (1962). The ontogenetic development of the diencephalon of the mouse. *Tokushima Journal of Experimental Medicine, 8*, 203–238.

Ninkina, N. N. (1995). Nerve growth factor-regulated properties of sensory neurones in *Oct-2* null mutant mice. *Molecular Brain Research, 33*, 233–244.

Noguchi, T., Sugisaki, T., Kudo, M., & Satoh, I. (1986). Retarded growth of the suprachiasmatic nucleus and pineal body in dw and lit dwarf mice. *Developmental Brain Research, 26*, 161–172.

Nuesslein, B., & Schmidt, I. (1990). Development of circadian cycle of core temperature in juvenile rats. *American Journal of Physiology, 259*, R270–R276.

Nuesslein-Hildesheim, B., & Schmidt, I. (1996). Manipulation of potential perinatal zeitgebers for the juvenile circadian temperature rhythm in rats. *American Journal of Physiology, 271*, R1388–R1395.

Nurminen, T. (1989). Shift work, fetal development and course of pregnancy. *Scandinavian Journal of Work and Environmental Health, 15*, 395–403.

Okamura, H., Fukui, K., Koyama, E., Tsutou, H. L. O., Tsutou, T., Terubayashi, H., Fujisawa, H., & Ibata, Y. (1983). Time of vasopressin neuron origin in the mouse hypothalamus: Examination by combined technique of immunocytochemistry and [3H]thymidine autoradiography. *Developmental Brain Research, 9,* 223–226.

Parmelee, A. H., Wenner, W. H., & Schulz, H. R. (1964). Infant sleep patterns: From birth to 16 weeks of age. *Journal of Pediatrics, 65,* 576–582.

Patrick, J., Campbell, K., Carmichael, L., Natale, R., & Richardson, B. (1981). Daily relationships between fetal and maternal heart rates at 38 to 40 weeks of pregnancy. *CMA Journal, 124,* 1177–1178.

Patrick, J., Campbell, K., Carmichael, L., Natale, R., & Richardson, B. (1982a). Patterns of gross fetal body movements over 24-hour observation intervals during the last 10 weeks of pregnancy. *American Journal of Obstetrics and Gynecology, 142,* 363–371.

Patrick, J., Campbell, K., Carmichael, L., & Probert, C. (1982b). Influence of maternal heart rate and gross fetal body movements on the daily pattern of fetal heart rate near term. *American Journal of Obstetrics and Gynecology, 144,* 533–538.

Pickard, G. E. (1980). Morphological characteristics of retinal ganglion cells projecting to the suprachiasmatic nucleus: A horseradish peroxidase study. *Brain Research, 183,* 458–465.

Pittendrigh, C. S. (1954). On temperature independence in the clock system controlling emergence time in *Drosophila. Proceedings of the National Academy of Sciences of the USA, 40,* 1018–1029.

Pollak, C. P. (1994). Regulation of sleep rate and circadian consolidation of sleep and wakefulness in an infant. *Sleep, 17,* 567–575.

Provencio, I., Wong, S., Lederman, A. B., Argamaso, S. M., & Foster, R. G. (1994). Visual and circadian responses to light in aged retinally degenerate mice. *Vision Research, 34,* 1799–1806.

Redman, J., Armstrong, S., & Ng, K. T. (1983). Free-running activity rhythms in the rat: Entrainment by melatonin. *Science, 219,* 1089–1091.

Reh, T. A. (1992). Generation of neuronal diversity in the vertebrate retina. In *Determinants of neural identity* (pp. 433–467). New York: Academic Press.

Reppert, S. M., & Klein, D. C. (1978). Transport of maternal [3H]melatonin to suckling rats and the fate of [3H]melatonin in the neonatal rat. *Endocrinology, 102,* 582–588.

Reppert, S. M., & Schwartz, W. J. (1983). Maternal coordination of the fetal biological clock *in utero. Science, 220,* 969–971.

Reppert, S. M., & Schwartz, W. J. (1984). Functional activity of the suprachiasmatic nuclei in the fetal primate. *Neuroscience Letters, 46,* 145–149.

Reppert, S. M., & Schwartz, W. J. (1986). Maternal suprachiasmatic nuclei are necessary for maternal coordination of the developing circadian system. *Journal of Neuroscience, 6,* 2724–2729.

Reppert, S. M. & Uhl, G. R. (1987). Vasopressin messenger ribonucleic acid in supraoptic and suprachiasmatic nuclei: Appearance and circadian regulation during development. *Endocrinology, 120,* 2483–2487.

Reppert, S. M., Shea, R. A., Anderson, A., & Klein, D. C. (1979). Maternal–fetal transfer of melatonin in a non-human primate. *Pediatric Research, 13,* 788–791.

Reppert, S. M., Coleman, R. J., Heath, H. W., & Swedlow, J. R. (1984). Pineal N-acetyltransferase activity in 10-day-old rats: A paradigm for studying the developing circadian system. *Endocrinology, 115,* 918–925.

Reppert, S. M., Henshaw, D., Schwartz, W. J., & Weaver, D. R. (1987). The circadian-gated timing of birth in rats: Disruption by maternal SCN lesions or by removal of the fetal brain. *Brain Research, 403,* 398–402.

Reppert, S. M., Weaver, D. R., Rivkees, S. A., & Stopa, E. G. (1988). Putative melatonin receptors in a human biological clock. *Science, 242,* 78–84.

Rivkees, S. A., & Reppert, S. M. (1990). Entrainment of circadian phase in developing gray short-tailed opossums: Mother vs. environment. *American Journal of Physiology, 259,* E384–E388.

Rivkees, S. A. & Reppert, S. M. (1991). Appearance of melatonin receptors during embryonic life in Siberian hamsters (*Phodopus sungorous*). *Brain Research, 568,* 345–349.

Rivkees, S. A., Weaver, D. R., & Reppert, S. M. (1992). Circadian and developmental regulation of Oct-2 gene expression in the suprachiasmatic nuclei. *Brain Research, 598,* 332–336.

Rivkees, S. A., Hofman, P. L., & Fortman, J. (1997). Newborn primate infants are entrained by low intensity lighting. *Proceedings of National Academy of Sciences of the USA, 94,* 292–297.

Rivkees, S. A., Fox, C. A., Jacobsen, C. D., & Reppert, S. M. (1988). Anatomic and functional development of the suprachiasmatic nuclei in the gray short-tailed opposum. *Journal of Neuroscience, 8,* 4269–4276.

Robinson, M. L. & Fuchs, J. L. (1993). [125I]Vasoactive intestinal peptide binding in rodent suprachiasmatic nucleus: Developmental and circadian studies. *Brain Research, 605,* 271–279.

Roca, A. L., Godson, C., Weaver, D. R., & Reppert, S. M. (1996). Structure, characterization, and expression of the gene encoding the mouse Mella melatonin receptor. *Endocrinology, 137,* 3469–3477.

Romero, M.-T., & Silver, R. (1990). Time course of peptidergic expression in fetal suprachiasmatic nucleus transplanted into adult hamster. *Developmental Brain Research, 57,* 1–6.

Salzarulo, P., Fagioli, I., & Ricour, C. (1985). Long term continuously fed infants do not develop heart rate circadian rhythm. *Early Human Development, 12,* 285–289.

Sauerbier, I. (1986). Circadian variation in teratogenic response to dexamethasone in mice. *Drug and Chemical Toxicology, 9,* 25–31.

Sauerbier, I. (1987). Circadian modification of ethanol damage in utero to mice. *American Journal of Anatomy, 178,* 170–174.

Scheuch, G. C., & Silver, J. (1982). Ontogeny of the suprachiasmatic nuclei in genetically anopthalmic mice: Anatomical and behavioral studies. In D. C. Klein (Ed.), *Melatonin rhythm generating system: Developmental aspects* (pp. 20–41). Basel: Karger.

Sengelaub, D. R., & Finlay, B. L. (1982). Cell death in the mammalian visual system during normal development: I. Retinal ganglion cells. *Journal of Comparative Neurology, 204,* 311–317.

Serón-Ferré, M., Ducsay, C. A., & Valenzuela, G. J. (1993). Circadian rhythms during pregnancy. *Endocrine Reviews, 14,* 594–609.

Shaw, D., & Goldman, B. D. (1995). Gender differences in influence of prenatal photoperiods on postnatal pineal melatonin rhythms and serum prolactin and follicle-stimulating hormone in the Siberian hamster (*Phodopus sungorus*). *Endocrinology, 136,* 4237–4246.

Shibata, S., & Moore, R. Y. (1987). Development of neuronal activity in the rat suprachiasmatic nucleus. *Developmental Brain Research, 34,* 311–315.

Shibata, S., & Moore, R. Y. (1988). Development of a fetal circadian rhythm after disruption of the maternal circadian system. *Developmental Brain Research, 41,* 313–317.

Shimada, M., & Nakamura, T. (1973). Time of neuron origin in mouse hypothalamic nuclei. *Experimental Neurology, 41,* 163–173.

Silver, J. (1977). Abnormal development of the suprachiasmatic nuclei of the hypothalamus in a strain of genetically anophthalmic mice. *Journal of Comparative Neurology, 176,* 589–606.

Silver, R., Lehman, M. N., Gibson, M., Gladstone, W. R., & Bittman, E. L. (1990). Dispersed cell suspensions of fetal SCN restore circadian rhythmicity in SCN-lesioned adult hamsters. *Brain Research, 525,* 45–58.

Silver, R., LeSauter, J., Tresco, P. A., & Lehman, M. N. (1996). A diffusible coupling signal from the transplanted suprachiasmatic nucleus controlling circadian locomotor rhythms. *Nature, 382,* 810–813.

Sitka, U., Weinert, D., Berle, K., Rumler, W., & Schuh, J. (1994). Investigations of the rhythmic function of heart rate, blood pressure and temperature in neonates. *European Journal of Pediatrics, 153,* 117–122.

Spangler, G. (1991). The emergence of adrenocortical circadian function in newborns and infants and its relationship to sleep, feeding and material adrenocortical activity. *Early Human Development, 25,* 197–208.

Speh, J. C., & Moore, R. Y. (1993). Retinohypothalamic tract development in the hamster and rat. *Developmental Brain Research, 76,* 171–181.

Stanfield, B., & Cowan, W. M. (1976). Evidence for a change in the retino-hypothalamic projection in the rat following early removal of one eye. *Brain Research, 104,* 129–136.

Stark, R. I., & Daniel, S. S. (1989). Circadian rhythm of vasopressin levels in cerebrospinal fluid of the fetus: Effect of continuous light. *Endocrinology, 124,* 3095–3101.

Stetson, M. H., Elliott, J. A., & Goldman, B. D. (1986). Maternal transfer of photoperiodic information influences the photoperiodic response of prepubertal djungarian hamsters. *Biology of Reproduction, 34,* 664–669.

Swaab, D. F., Zhou, J. N., Ehlhart, T., & Hofman, M. A. (1994). Development of vasoactive intestinal polypeptide neurons in the human suprachiasmatic nucleus in relation to birth and sex. *Developmental Brain Research, 79,* 249–259.

Swanson, L. W. (1987). The hypothalamus. In A. Björkland, T. Hökfelt, & L. W. Swanson (Eds.), *Handbook of chemical neuroanatomy, Vol. 5: Integrated systems of the CNS,* (Part I, pp. 1–124). Amsterdam: Elsevier.

Takahashi, K., & Deguchi, T. (1983). Entrainment of the circadian rhythms of blinded infant rats by nursing mothers. *Physiology and Behavior, 31,* 373–378.

Takahashi, K., Hayafuji, C., & Murakami, N. (1982). Foster mother rat entrains circadian adrenocortical rhythm in blinded pups. *American Journal of Physiology, 243,* 443–449.

Takahashi, K., Ohi, K., Shimoda, K., Tamada, N., & Hayashi, S. (1989). Postnatal maternal entrainment of circadian rhythms. In S. M. Reppert (Ed.), *Development of circadian rhythmicity and photoperiodism in mammals* (pp. 67–82). Ithaca, NY: Perinatology Press.

Tamarkin, L., Westrom, W. K., Hamill, A. I., & Goldman, B. D. (1976). Effect of melatonin on the reproductive systems of male and female Syrian hamsters: A diurnal rhythm in sensitivity to melatonin. *Endocrinology, 99,* 1534–1541.

Tenreiro, S., Dowse, H. B., D'Souza, S., Minors, D., Chiswick, M., Simms, D., & Waterhouse, J. (1991). The development of ultradian and circadain rhythms in premature babies maintained in constant conditions. *Early Human Development, 27,* 33–52.

Terman, J. S., Remé, C. E., & Terman, M. (1993). Rod outer segment disk shedding in rats with lesions of the suprachiasmatic nucleus. *Brain Research, 605,* 256–264.

Tominaga, K., Inouye, S.-I. T., & Okamura, H. (1994). Organotypic slice culture of the rat suprachiasmatic nucleus: Sustenance of cellular architecture and circadian rhythm. *Neuroscience, 59,* 1025–1042.

Tosini, G. & Menaker, M. (1996). Circadian rhythms in cultured mammalian retina. *Science, 272,* 419–421.

Treep, J. A., Abe, H., Rusak, B., & Goguen, D. M. (1995). Two distinct retinal projections to the hamster suprachiasmatic nucleus. *Journal of Biological Rhythms, 10,* 299–307.

Ugrumov, M. V., Popov, A. P., Vladimirov, S. V., Kasmambetova, S., Novodjilova, A. P., & Tramu, G. (1994a). Development of the suprachiasmatic nucleus in rats during ontogenesis: Serotonin-immunopositive fibers. *Neuroscience, 58,* 161–165.

Ugrumov, M. V., Popov, A. P., Vladimirov, S. V., Kasmambetova, S., & Thibault, J. (1994b). Development of the suprachiasmatic nucleus in rats during ontogenesis: Tyrosine hydroxylase immunopositive cell bodies and fibers. *Neuroscience, 58,* 151–160.

Ugrumov, M. V., Trembleau, A., & Calas, A. (1994c). Altered vasoactive intestinal polypeptide gene expression in the fetal rat suprachiasmatic nucleus following prenatal serotonin deficiency. *International Journal of Developmental Neuroscience, 12,* 143–149.

Updike, P. A., Accurso, F. J., & Jones, R. H. (1985). Physiologic ciradian rhythmicity in preterm infants. *Nursing Research, 34,* 160–163.

Van den Pol, A. N. & Tsujimoto, K. L. (1985). Neurotransmitters of the hypothalamic suprachiasmatic nucleus: immunocytochemical analysis of 25 neuronal antigens. *Neuroscience, 15,* 1049–1086.

Visser, G. H. A., Goodman, J. D. S., Levine, D. H., & Dawes, G. S. (1982). Diurnal and other cyclic variations in human fetal heart rate near term. *American Journal of Obstetrics and Gynecology, 142,* 535–544.

Viswanathan, N. (1989). Presence–absence cycles of the mother and not light–darkness are the zeitgeber for the circadian rhythm of newborn mice. *Experientia, 45,* 383–385.

Viswanathan, N. (1990). Role of relative durations of presence/absence of mother mouse (*Mus booduga*) in circadian rhythm of pups. *Current Science, 59,* 409–411.

Viswanathan, N., & Chandrashekaran, M. K. (1985). Cycles of presence and absence of mother mouse entrain the circadian clock of pups. *Nature, 317,* 530–531.

Viswanathan, N., & Davis, F. C. (1993). The fetal circadian pacemaker is not involved in the timing of birth in hamsters. *Biology of Reproduction, 48,* 530–537.

Viswanathan, N., & Davis, F. C. (1995). Suprachiasmatic nucleus grafts restore circadian function in aged hamsters. *Brain Research, 686,* 10–16.

Viswanathan, N., & Davis, F. C. (1997). Single prenatal injections of melatonin or the D1-dopamine receptor agonist SKF 38393 to pregnant hamsters sets the offsprings' circadian rhythms to phases 180° apart. *Journal of Comparative Physiology A, 180,* 339–346.

Viswanathan, N., Weaver, D. R., Reppert, S. M., & Davis, F. C. (1994). Entrainment of the fetal hamster circadian pacemaker by prenatal injections of the dopamine agonist, SKF 38393. *Journal of Neuroscience, 14,* 5393–5398.

Vitaterna, M. H., King, D. P., Chang, A.-M., Kornhauser, J. M., Lowrey, P. L., McDonald, J. D., Dove, W. F., Pinto, L. H., Turek, F. W., & Takahashi, J. S. (1994). Mutagenesis and mapping of a mouse gene, *clock,* essential for circadian behavior. *Science, 264,* 719–725.

Weaver, D. R., & Reppert, S. M. (1995). Definition of the developmental transition from dopaminergic to photic regulation of c-*fos* gene expression in the rat suprachiasmatic nucleus. *Molecular Brain Research, 33,* 136–148.

Weaver, D. R., & Reppert, S. M. (1986). Maternal melatonin communicates daylength to the fetus in djungarian hamsters. *Endocrinology, 119,* 2861–2863.

Weaver, D. R., & Reppert, S. M. (1987). Maternal–fetal communication of circadian phase in a precocious rodent, the spiny mouse. *American Journal of Physiology, 253,* E401–E409.

Weaver, D. R. & Reppert, S. M. (1989a). Periodic feeding of SCN-lesioned pregnant rats entrains the fetal biological clock. *Developmental Brain Research, 46,* 291–296.

Weaver, D. R. & Reppert, S. M. (1989b). Direct *in utero* perception of light by the mammalian fetus. *Developmental Brain Research, 47,* 151–155.

Weaver, D. R., Keohan, J. T., & Reppert, S. M. (1987). Definition of a prenatal sensitive period for maternal–fetal communication of day length. *American Journal of Physiology, 253,* E701–E704.

Weaver, D. R., Rivkees, S. A., & Reppert, S. M. (1989). Localization and characterization of melatonin receptors in rodent brain by *in vitro* autoradiography. *Journal of Neuroscience, 9,* 2581–2590.

Weaver, D. R., Rivkees, S. A., & Reppert, S. M. (1992). Dl-dopamine receptors activate *c-fos* expression in the fetal suprachiasmatic nuclei. *Proceedings of National Academy of Sciences of the USA, 89,* 9201–9204.

Weaver, D. R., Roca, A. L., & Reppert, S. M. (1995). *c-fos* and *jun-B* mRNAs are transiently expressed in fetal rodent suprachiasmatic nucleus following dopaminergic stimulation. *Developmental Brain Research, 85,* 293–297.

Weinert, D., Sitka, U., Minors, D. S., & Waterhouse, J. M. (1994). The development of circadian rhythmicity in neonates. *Early Human Development, 36,* 117–126.

Welsh, D. K., Logothetis, D. E., Meister, M., & Reppert, S. M. (1995). Individual neurons dissociated from rat suprachiasmatic nucleus express independently phased circadian firing rhythms. *Neuron, 14,* 697–706.

Wever, R. A. (1984). Sex differences in human circadian rhythms: Intrinsic periods and sleep fractions. *Experientia, 40,* 1226–1234.

Whitnall, M. H., Key, S., Ben-Barak, Y., Ozato, K., & Gainer, H. (1985). Neurophysin in the hypothalamo-neurohypophysial system. *Journal of Neuroscience, 5,* 98–109.

Wiegand, S. J., & Gash, D. M. (1988). Organization and efferent connections of transplanted suprachiasmatic nuclei. *Journal of Comparative Neurology, 267,* 562–579.

Williams, L. M., Martinoli, M. G., Titchener, L. T., & Pelletier, G. (1991). The ontogeny of central melatonin binding sites in the rat. *Endocrinology, 128,* 2083–2090.

Wray, S., Castel, M., & Gainer, H. (1993). Characterization of the suprachiasmatic nucleus in organotypic slice explant cultures. *Microscopy Research and Technique, 25,* 46–60.

Yellon, S. M., & Longo, L. D. (1988). Effect of maternal pinealectomy and reverse photoperiod on the circadian melatonin rhythm in the sheep and fetus during the last trimester of pregnancy. *Biology of Reproduction, 39,* 1093–1099.

Zucker, I., Fitzgerald, K. M., & Morin, L. P. (1980). Sex differentiation of the circadian system in the golden hamster. *American Journal of Physiology, 238,* R97–R101.

10

Aging of the Mammalian Circadian System

Fred W. Turek, Kathryn Scarbrough, Plamen Penev, Susan Labyak, Verónica S. Valentinuzzi, and Olivier Van Reeth

Numerous studies have demonstrated that as animals reach advanced age for the species, there are pronounced changes in the expression of diverse circadian rhythms. Many of the observed changes may reflect an overall deterioration of the health of the organism. On the other hand, changes in the circadian system may not simply reflect such deterioration, but may represent one of the underlying causes for the negative health effects associated with advanced age. Most studies of the effects of advanced age on the circadian system have been carried out on rodents and humans. Since Chapter 22 of this volume reviews the literature on the effects of aging on the human circadian system, this chapter will focus primarily on studies of rodents, particularly the best-studied animals in this regard: laboratory rats, mice, and golden (Syrian) hamsters. After first describing the multitude of changes that have been observed in the aging circadian system of rodents, this review will (1) examine the underlying physiologic mechanisms that lead to these changes, (2) provide an overview of attempts to attenuate or reverse age-related effects on circadian rhythmicity, and (3) speculate on the functional significance of such changes.

Fred W. Turek, Kathryn Scarbrough, Plamen Penev, Susan Labyak, Verónica S. Valentinuzzi, and Olivier Van Reeth Department of Neurobiology and Physiology, Northwestern University, Evanston, Illinois 60208-3520.

Circadian Clocks, Volume 12 of *Handbook of Behavioral Neurobiology*, edited by Joseph S. Takahashi, Fred W. Turek, and R. Y. Moore, Kluwer Academic / Plenum Publishers, New York, 2001.

Characteristics of Age-Related Changes in the Circadian Clock System

The aging of the mammalian circadian system is associated with a number of changes in overt rhythms, including a reduction in amplitude, increased day-to-day variability and fragmentation under both entrained and free-running conditions, alterations in the phase angle of entrainment and in the free-running period, and a decreased sensitivity to the phase-shifting effects of both photic and nonphotic stimuli. Most studies of aging and circadian rhythmicity in rodents have focused on the effects of aging on the rhythm of locomotor activity because of the ease with which this rhythm can be measured for long periods of time without disturbing the animal. Studies involving other behavioral, as well as endocrine and metabolic rhythms, indicate that similar changes are occurring in diverse physiologic systems. Only a few studies have examined the effects of aging on sleep in rodents, and more studies are clearly needed to determine how age-related changes in the timing of the activity–rest cycle and on sleep itself interact with one another.

Amplitude

A prominent change that is observed in many overt 24-hour rhythms in both rodents and humans is a reduction in amplitude (for review, see Myers & Badia, 1995). For example, circadian rhythms of temperature (Halberg, Halberg, Regal, & Halberg, 1981; Sacher & Duffy, 1978), corticosterone (Nicolau & Milcu, 1977), growth hormone (Laartz, Losee-Olsen, Ge, & Turek, 1994), testosterone (Miller & Riegle, 1982; Tenover, Matsumoto, Clifton, & Bremner, 1988), and melatonin (Reiter *et al.*, 1981; Tang, Hadjiconstantinov, & Pang, 1985) are all dampened in old rats and/or hamsters. An age-related reduction in the amplitude of the rhythm of locomotor activity is a universal change observed in rodents (J. R. Martin, Fuchs, Bender, & Harting, 1985; Mosko, Erickson, & Moore, 1980; Peng, Jiang, & Hsu, 1980; Sacher & Duffy, 1978). Old rats also show a reduction in the amplitude of circadian fluctuations of slow-wave sleep and wakefulness (Rosenberg, Zepelin, & Rechtschaffen, 1979; Van Gool & Mirmiran, 1986), and in some old rats, the rhythms of drinking and locomotor activity have been reported to become arrhythmic, i.e., have no amplitude (Peng & Kang, 1984). Compared to young mice, old mice spend more time asleep during the normal active period and more time awake during the normal sleep period (Welsh, Richardson, & Dement, 1986). It should be noted that changes in the amplitude of rhythms occur well before "old" age. For example, in hamsters, which normally live about 2 years, the decrease in the amplitude of the growth hormone and locomotor activity rhythms is already apparent by 11–12 months of age (Laartz *et al.*, 1994; Labyak, Zee, Wallen, & Turek, 1996; Scarbrough, Losee-Olson, Wallen, & Turek, 1997) (Figure 1). The impact of the change in the amplitude of specific endocrine, metabolic, and behavioral rhythms in "middle" age on the cascade of events which is part of the aging process remains to be determined.

Of course, a reduction in the amplitude of overt circadian rhythms does not necessarily indicate that there has been a reduction in the output signal from the central circadian clock. Changes in amplitude (as well as other changes in overt rhythmicity) could be due to changes that are either "upstream" or "downstream" from the circadian clock. (See the next section for further discussion of the location of age-related changes in the circadian clock system and for data indicating an

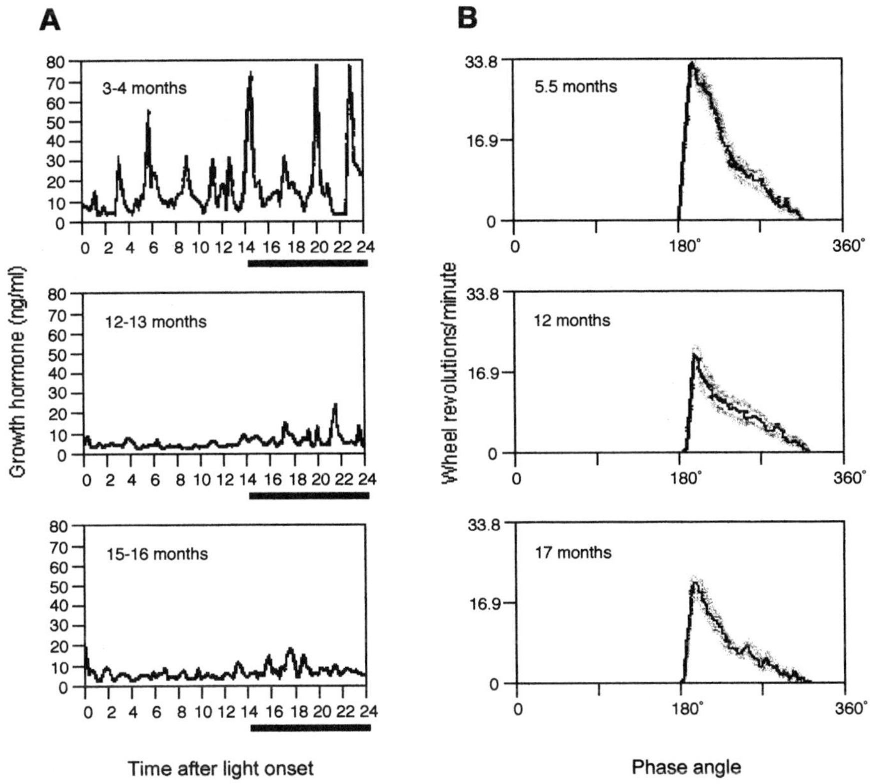

Figure 1. Age-related changes in the 24-hour profiles of (A) growth hormone (GH) secretion and (B) wheel-running behavior. Blood was sampled every 15 minutes for 24 hours and analyzed for GH by radioimmunoassay to generate the profiles of representative animals at 3–4 months, 12–13 months, or 15–16 months of age. Dark bar under the *X* axis represents the period of darkness when blood was collected under dim, red light. Wheel-running activity per day over a 10-day period was analyzed for groups of hamsters at age 5.5, 12, and 17 months (n = 20, 20, and 18, respectively). Phase angle of 180° represents the time of light offset, and the dotted curves represent the SE of the mean number of revolutions performed in the wheel per minute by each group. (Left: from Laartz *et al.*, 1994; with permission. Copyright 1994, Sage Publications. Right: from Labyak, Turek, Wallen, & Zee, 1998, with permission. Copyright 1998, American Physiological Society.)

alteration in the amplitude of the circadian pacemaker.) If changes in amplitude are due to changes in the input, output, or end-organ response, increasing the strength of the circadian signal might restore the normal amplitude of rhythms in older animals.

Variability and Fragmentation

Another universal change in rhythmicity that accompanies aging is that the precision of rhythms becomes more variable and the integrity of the rhythm becomes more fragmented. That is, the timing of a phase marker of a particular rhythm (e.g., onset of wake or locomotor activity) shows much more day-to-day variation under both entrained and free-running conditions, and normally consolidated rhythms show a loss of consolidation (for reviews, see Brock, 1991; Richardson,

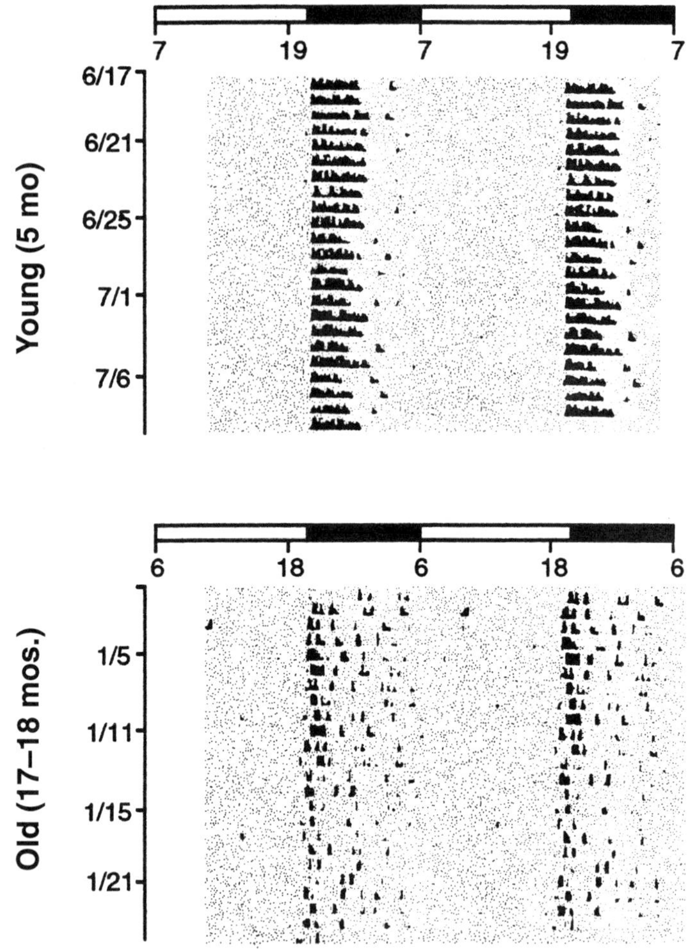

Figure 2. Examples of wheel-running activity records from one young (3 months) and one old (17–18 months) golden hamster maintained in a 14:10 light–dark cycle as indicated by the bar above the records. Data are double-plotted (each line represents 48 hours) for better visual demonstration of the fragmentation of the locomotor activity rhythm which occurs with age.

1990). Day-to-day variability in a given rhythm has been correlated with amplitude such that a decrease in rhythm amplitude is associated with more variability (Reynolds *et al.*, 1991). While such loss of precision and fragmentation are often easily apparent by the visual examination of rhythmic profiles (Figure 2), we have recently taken various approaches to quantify age-related changes in these variables (Penev, Zee, & Turek, 1997b). The wheel-running activity of rodents occurs in a series of temporal clusters separated by various intervals of inactivity. A well-established technique for the identification of discrete bouts of behavior in a continuous sequence of recorded events known as "log-survivorship analysis" (P. Slater & Lester, 1982) was used to analyze the wheel-running activity patterns in a reference group of 20 young male golden hamsters over a 7-day period under a 14:10 light–dark cycle. The graphical analysis of the resulting log-survivorship function confirmed the suitability of this approach for the analysis of hamster wheel-running behavior and was used to determine the appropriate bout criterion interval (J. R. Martin *et al.*, 1985). All of the identified activity bouts were further characterized by a set of parameters including times of bout onset and offset, bout duration, average and maximal wheel-running intensity, and total wheel revolutions (wheel-running activity) per bout. Once developed, this method of analysis was used to compare the 24-hour activity patterns of young (10–14 weeks) versus old (17–18 months) male golden hamsters. The bout characteristics of each individual animal were used to calculate a set of 18 parameters reflecting the average pattern of daily wheel-running activity. Analysis of variance detected significant differences in 15 of these variables ($p < .05$), indicating a substantial degree of fragmentation in the 24-hour pattern of wheel-running activity in the old hamsters (Figure 2). Furthermore, the use of discriminant analysis, based on the same set of 18 variables, provided a powerful means of distinguishing between the activity patterns of young versus old animals. Indeed, the resulting discriminant function correctly classified 95% of the animals in each group. In addition to providing such quantitative results in golden hamsters, this bout analysis of 24-hour activity patterns has been successfully used with other species, including Djungarian hamsters and several strains of laboratory mice. Taken together, the results indicate that the method can provide useful information for quantifying age-related changes in circadian rhythmicity in a variety of experimental paradigms using genetic, pharmacologic, and behavioral approaches in aging rodents.

Pattern of Entrainment and Free-running Period

In general, a phase advance in rhythms entrained to a light–dark (LD) cycle is observed in older animals, including humans (Myers & Badia, 1995; Zee, Rosenberg, & Turek, 1992; also see this volume, Chapter 22). Such a phase advance is consistent with the observation that the endogenous period of the free-running rhythms of locomotor activity during exposure to constant light or constant darkness is reduced in older rats and hamsters (Davis & Menaker, 1980; Morin, 1988; Pittendrigh & Daan, 1974; Witting, Mirmiran, Bos, & Swaab, 1994), as is the period of the sleep–wake cycle (Van Gool, Wittig, & Mirmiran, 1987). It should be noted that the full extent of the effects of aging on advancing the phase of the entrained rhythm of locomotor activity may be even greater than has so far been observed since any advance in the rhythm of activity in a nocturnal rodent is subject to the negative masking effect light may have on the expression of locomotor activity. That is, since activity onset occurs within a few minutes after lights-off in young nocturnal rodents entrained to normal laboratory lighting conditions (e.g., LD 12:12 or LD 14:10), any phase advance in the activity rhythm of an older animal would require the animal to begin activity in the light phase. To examine the effects of aging on the phase angle of entrainment independent of possible masking effects, we recently determined the effects of age on hamsters entrained to a short-day, LD 6:18, cycle. During exposure to such a short day, activity onset occurs about 5 hours after lights-off in young animals in this species (Figure 3). Older animals (16 months of age) showed nearly a 2-hour phase advance in the onset of their activity rhythm under LD 6:18 conditions (Scarbrough *et al.*, 1997). This nearly 2-hour difference between young and old animals raises a number of interesting questions about the functional significance of such changes, given the adaptive significance that has been associated with the way animals entrain to the LD cycle (see final section). Furthermore, these results indicate that the

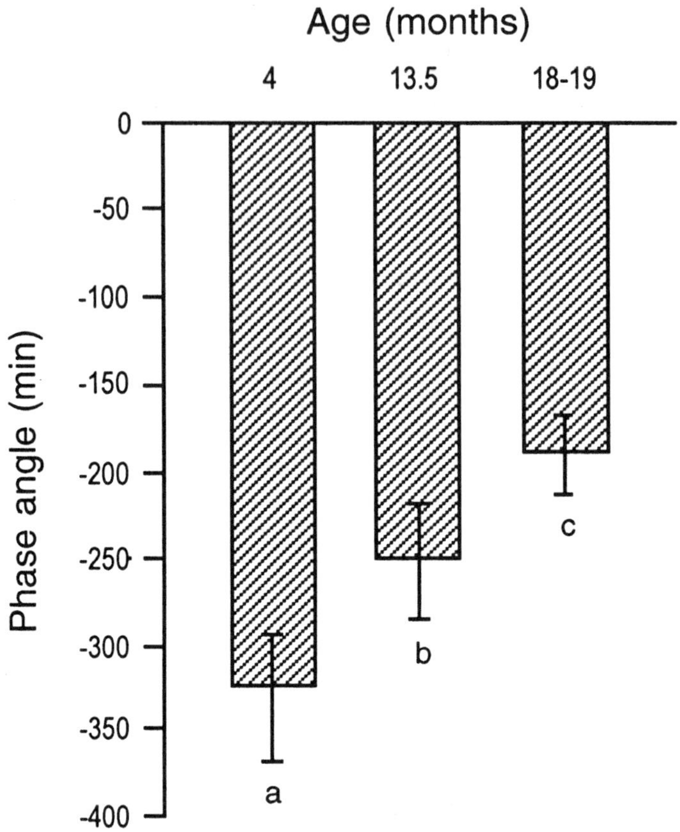

Figure 3. Age-related change in the phase angle of entrainment among groups of hamsters entrained to a 6:18 light–dark cycle. The time of activity onset in minutes following the onset of darkness (averaged over a period of 10 days) was analyzed by ANOVA followed by Newman–Keuls multiple comparison test. Both middle-aged and old animals showed a significant phase advance in activity onset compared to young animals. (From Scarbrough *et al.*, 1997, with permission. Copyright 1997, American Physiological Society.)

effects of aging on the circadian clock system may vary as a function of the seasonally changing environment. Such effects may have implications for humans because the length of the day alters the pattern of expression of human endocrine rhythms (Wehr, 1991; Wehr *et al.*, 1993).

While changes in the phase angle of entrainment and a reduction in the free-running period of circadian rhythms are observed in hamsters, rats, and humans, too few species have been examined to conclude that this is a universal feature associated with aging of the circadian clock system. Indeed, studies in mice indicate that, at least for some strains, the period of the rhythm of locomotor activity lengthens with age (Possidente, McEldowney, & Pabon, 1995; Valentinuzzi, Scarbrough, Takahashi, & Turek, 1997). Associated with an increase in the free-running period is a delay of about 30–40 min in the onset of locomotor activity in older mice during entrainment to LD 12:12 cycle conditions. Although previous lighting conditions and/or selection of specific phenotypes, due to the attrition that occurs before the testing of older animals, may account for the observed differences between species, the direction of age-related changes in the endogenous period of the circadian clock may indeed vary among species/strains of animals.

The phase angle of entrainment to the light–dark cycle depends not only on the period of the endogenous clock, but also on the responsiveness of the clock to the phase-shifting effects of light. Characterization of the phase-shifting effects of light depends on the phase response curve (PRC) to light pulses presented at various circadian times. Few attempts have been made to determine if the shape and/or amplitude of the PRC to light changes with age. In one study in hamsters, the magnitude of the phase shifts in response to a 1-hour pulse of bright light that induces maximum phase delays and advances at two circadian times (CT) tested (i.e., CT14 and CT19, respectively) was similar between young and old animals. However, at one circadian time (CT16), extremely large phase shifts were observed in older animals, i.e., up to 12-hour phase shifts (Rosenberg, Zee, & Turek, 1991). Such large phase shifts are never observed in young, wild-type hamsters, but have been observed in young hamsters carrying the *tau* mutation, which shortens the free-running period of the circadian clock (Menaker & Refinetti, 1992; Scarbrough & Turek, 1996; Shimomura & Menaker, 1994). Interestingly, circadian amplitude is a predictor of phase-shift magnitude (Czeisler *et al.*, 1989; Czeisler, Kronauer, & Allan, 1992), with a decrease in amplitude being associated with the induction of larger phase shifts. Thus, the extremely large phase shifts, that are observed when light is presented at the transition phase between delays and advances of the phase response curve in old animals, indicates that the amplitude of the circadian pacemaker itself may be reduced in old hamsters.

Rate of Reentrainment

Only a few studies have examined the effects of aging on the pattern of resynchronization of circadian rhythms following a phase shift in the LD cycle. Early studies in rats yielded inconsistent results, with some authors reporting no change in the rate of reentrainment between young and old animals and others observing that young animals reentrained faster (Peng & Kang, 1984; Peng *et al.*, 1980; Quay, 1972; Rosenberg, Winter, & Rechtschaffen, 1980). Interpretation of these studies is difficult because they involved the complete reversal of the LD 12:12 light cycle, which can lead to either a sequence of phase advances or phase delays until steady-state entrainment is reachieved. More recently, in both hamsters and mice, it has been found that the effects of age on the rate of reentrainment of the locomotor activity rhythm depends on the direction of the shift in the LD cycle (Valentinuzzi *et al.*, 1997; Zee *et al.*, 1992) and was correlated with age-related changes in the free-running period. Comparing the results between studies on the effects of aging on reentrainment are difficult since the time to reentrain to a new light–dark cycle will depend on a combination of external factors (e.g., direction and magnitude of phase shifts, intensity of light) and internal factors (e.g., direction of change in period, altered responsiveness to light, change in PRC to light).

Sensitivity to Light

Recent studies in hamsters have demonstrated that old animals (18–22 months) are 20 times less sensitive to the phase-shifting effects of subsaturating light pulses on the free-running rhythm of locomotor activity than are young animals (3–4 months) (Zhang *et al.*, 1996). The threshold of light needed to induce a detectable

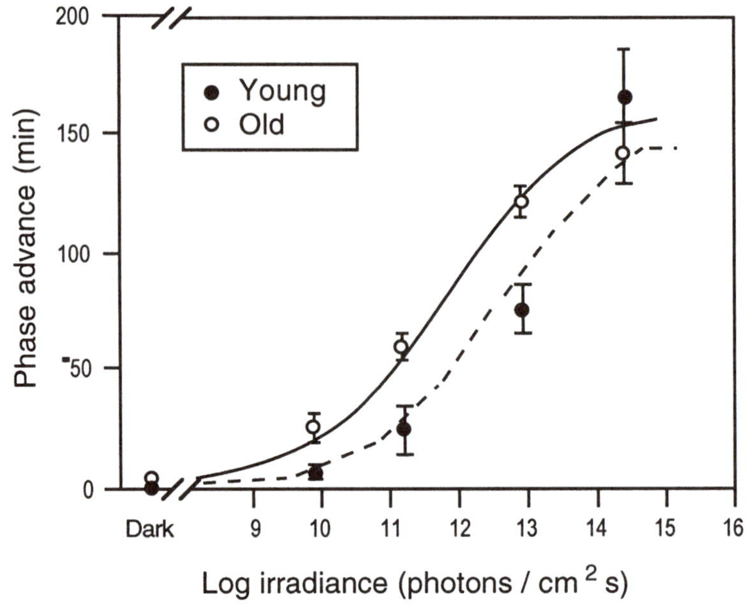

Figure 4. Mean phase advances of the locomotor activity rhythm (±SEM) in groups of young and old hamsters exposed to a 5-minute pulse of monochromatic light (503 nm) at one of four irradiances at CT19. Control animals (points labeled "dark") were handled, but not exposed to any light at the same CT. Each point represents the response of eight young and six old hamsters. The two curves differ significantly in the half-saturation point, but do not differ in slope or the minimum or maximum phase shifts induced by light. The phase shifts in the activity rhythm were significantly greater in young than old animals at the three lower light intensities ($p < .001$). No significant differences ($p > .05$) were observed between young and old animals that were not exposed to light or that were exposed to the highest irradiance level. (From Zhang *et al.*, 1996, with permission. Copyright 1996, Elsevier Science.)

phase shift was lower and the magnitude of the phase shift induced by light pulses of intermediate intensity was greater in young than in old animals (Figure 4) (Zhang *et al.*, 1996). Consistent with previous results, no age-related differences were observed in response to high-intensity light pulses that induce maximum phase shifts at the circadian time tested (Rosenberg *et al.*, 1991).

Studies in rodents indicate that the coupling of light information to the circadian clock within the SCN involves the protooncogene c-*fos* and its protein product Fos (Kornhauser, Nelson, Mayo, & Takahashi, 1990). The photic threshold and the amount of light-induced c-*fos* expression is correlated with the threshold and the magnitude of light-induced phase shifts of the circadian rhythm of locomotor activity (Kornhauser *et al.*, 1990). Consistent with the decreased behavioral response to the phase shifting effects of light in old animals is the finding that the induction of c-*fos* by pulses of light of varying intensity is also reduced in old hamsters and rats (Sutin, Dement, Heller, & Kilduff, 1993; Zhang *et al.*, 1996). Aging is also associated with a decline in light-stimulated phosphorylation of CREB, the cyclic AMP response element-binding protein (Zhang *et al.*, 1996). The phosphorylation of CREB has been shown to mediate the transcriptional activation of c-*fos* in other systems (Sheng, McFadden, & Greenberg, 1990). Thus, the effects of age on the molecular response of immediate-early genes in the suprachiasmatic nucleus (SCN) to light are correlated with an age-related decrease in the response to the phase-shifting effects of light on the circadian clock. A decreased response to the phase-shifting (and, by implication, entraining) effects of light on the circadian clock could underlie at least some of the age-related changes that are observed in entrained circadian rhythms.

Sensitivity to Nonphotic Stimuli

A number of nonphotic stimuli, both pharmacologic and nonpharmacologic, have been identified which can induce phase shifts in the mammalian circadian clock (Mrosovsky, Reebs, Honrado, & Salmon, 1989; Rusak & Bina, 1990; Turek, 1987; Turek *et al.*, 1995; Van Reeth *et al.*, 1994). Many of the nonphotic stimuli that induce phase shifts in the circadian clock also induce an acute increase in locomotor activity and it appears that at least some of the phase-shifting effects of these agents are due to the induction of activity and/or arousal in the animal (Mrosovsky, 1996). The phase-shifting response to two activity-inducing stimuli, dark pulses on a background of constant light and the injection of the short-acting benzodiazepine triazolam, are both greatly attenuated in older hamsters compared to the response of young animals (Van Reeth, Zhang, Zee, & Turek, 1992). Furthermore, aging also attenuates the ability of old animals to entrain to daily injections of triazolam (Van Reeth *et al.*, 1993). Taken together, these results indicate that the circadian system of old hamsters becomes less responsive to the synchronizing signals mediated by the activity–rest cycle, and suggest that aging is associated with a weakened coupling between the activity–rest cycle and the circadian clock.

Physiologic Bases for Age-Related Changes in Circadian Clock System

As noted earlier, alterations in the circadian clock system that are observed in older animals could be due to the effects of aging at one or more levels, including input pathways to the clock, changes within the circadian clock itself, as well as to changes in the strength of the output signals or the end-organ response to these signals. Indeed, given the multitude of changes that have been observed in both the expression of 24-hour rhythms and in the response to environmental stimuli, it is expected that age-related changes will be occurring at many different levels within the circadian clock system. In addition, changes in the expression of one rhythm may in turn alter the expression of other overt circadian rhythms. This latter possibility will be discussed in the next section, while this section will focus on potential age-related changes between the input signals to the clock and the end organs that express circadian rhythms in response to signals from the clock. This section also includes a discussion of pharmacologic and genetic approaches for elucidating the underlying physiologic bases for age-related changes in the circadian clock system.

Input Pathways

Photic information reaches the circadian clock in the SCN via both a direct retinohypothalamic tract (RHT) as well as an indirect retinal–geniculohypothalamic tract (Miller, Morin, Schwartz, & Moore, 1996) that involves the lateral geniculate nucleus (LGN). Information about the overall rest–activity state of the animal may feed back upon the circadian clock via serotonin (5-HT) projections from the midbrain raphe nuclei to the SCN itself or through projections to the LGN. Recent studies indicate that the rhythmic release of the pineal hormone, melatonin, itself under rhythmic control by the SDN, may also have feedback effects on the circadian clock within the SCN (Morgan, Howell, & Helliwell, 1994). Thus, aging could be altering the input signals to the circadian clock at multiple loci.

Evidence that changes in photic sensitivity at the level of the eye itself may mediate the age-related decreased response to the phase-shifting effects of light come from the finding that there is a 10–20% decrease in the amount of light transmitted through the lens of 18-month-old hamsters compared to young controls (Zhang *et al.*, 1996). However, it is unlikely that this decrease in lens transmittance could account for the near 20-bold decrease of the sensitivity to the phase-shifting effects of light in old hamsters. Alternatively, an age-related decline in the number of retinal photoreceptors or ganglion cells projecting to the SCN could account for an altered response of the circadian clock to light. As one approach to study this possibility, we used an anterograde tracer and found no difference in the volume of SCN tissue that is innervated by the RHT in young and old animals (Zhang, 1995). The two most likely causes for an age-related decline in sensitivity of the circadian clock to light are a decline in neurotransmission in the RHT projecting to the SCN and/or in the responsiveness of SCN neurons to RHT signals. No data exist to distinguish between these two hypotheses.

Recently, a dramatic attenuation of the phase shifting effects of a 5-HT agonist in old hamsters was observed (Penev, Zee, Wallen, & Turek, 1995), providing support for the hypothesis that aging is associated with impaired serotonergic modulation of circadian rhythmicity. In addition, treatment with a 5-HT agonist during the late subjective night attenuates the phase-advancing effects of a light pulse in young hamsters free-running in constant dark (DD), but this treatment is considerably less effective in older animals (Penev, Turek, Wallen & Zee, 1997a) (Figure 5). Taken together, these findings raise the possibility that age-related changes in the serotonergic inputs to the circadian clock may interfere with the optimal response of the organism to photic and nonphotic stimuli. Interestingly, no age-related decline in responsiveness to 5-HT agonists has been observed in *in vitro* SCN studies. Indeed, phase advances in the neuronal rhythm of single-unit activity in the SCN from old animals in response to a 5-HT agonist were the same as or greater than those observed in the SCN from young rats (Cao, Edgar, Heller, Dement, & Miller, 1995) and hamsters (P. D. Penev, unpublished results) (Figure 6). Thus, the age-related decline in the phase-shifting effects of activity-inducing stimuli and 5-HT agonists may lie on the pathway by which 5-HT information reaches the circadian system. Whether these changes involve the 5-HT projection to the SCN or the indirect projection to the LGN remains to be determined.

In addition to receiving information about the external and internal environment via neuronal signals, the SCN may also receive feedback signals humorally via the pineal hormone, melatonin. While it has been known for many years that the SCN controls the circadian rhythm of melatonin synthesis and release into the circulation, the findings that (1) daily injection of melatonin can entrain rats (Redman, Armstrong, & Ng, 1983) and blind humans (Sack, Levy, Blood, & Stevenson, 1991), (2) the SCN contains a high concentration of melatonin receptors (Dubocovich, Benloucif, & Masana, 1996; Morgan *et al.*, 1994), and (3) melatonin can induce phase shifts in the SCN neuronal rhythm *in vitro* (McArthur, Gillette, & Prosser, 1991) have focused attention on the possible role of melatonin in the regulation of

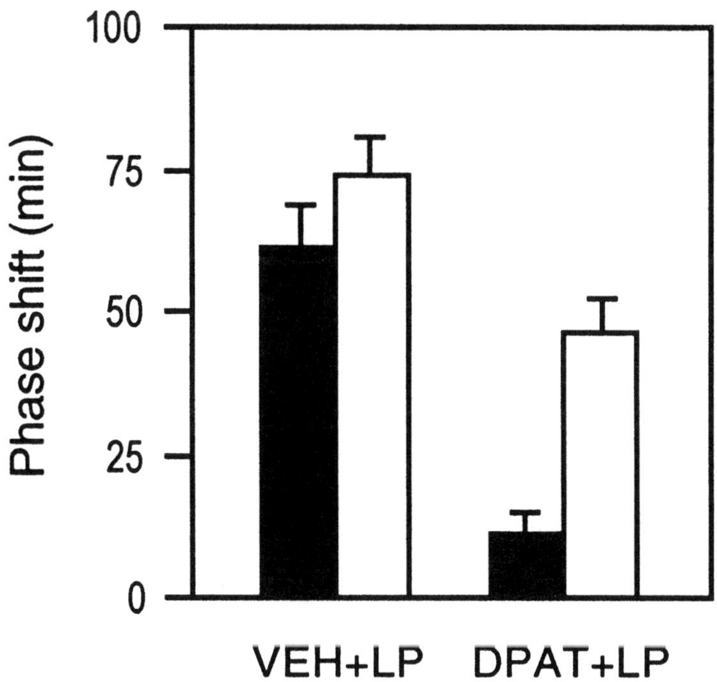

Figure 5. Mean (±SEM) phase shifts in the circadian activity rhythm of 13 young (solid bars) and 11 old (open bars) hamsters in response to light pulses at CT19. Photic stimulus was given to the same animals on two separate occasions 30 minutes after they were treated i.p. with 5 mg/kg 8-OH 8-hydroxy-2(di-*n*-propylamino)tetralin (DPAT) or vehicle (VEH; saline). (From Penev *et al.*, 1997, with permission. Copyright 1997, American Physiological Society.)

circadian rhythms by the SCN. The nighttime release of melatonin has the potential of providing the organism with a reliable chemical equivalent of the temporal position and duration of darkness. The fact that nighttime melatonin levels are greatly reduced in old animals, including humans (Humbert & Pevet, 1994; Reiter, 1992; Schmid, 1993), raises the possibility that the decreased strength of the melatonin feedback signal to the SCN may play a role in the observed age-related changes in circadian rhythmicity (Armstrong & Redman, 1991; Reiter, 1995). Indeed, recent studies indicate that decreased binding of melatonin to the SCN of old rats is correlated with a disruption of the expression of overt circadian rhythms (Whealin, Burwell, & Gallagher, 1993). On the other hand, other studies have failed to find any age-related changes in melatonin-binding sites in the SCN (Duncan & Purvis, 1994).

Recent studies in animals indicating that melatonin can have effects on the immune system, suppress tumor formation, and act as a powerful free-radical scavenger, coupled with the well-known decline in circulating melatonin levels in old animals, have led to a number of hypotheses about melatonin being an antiaging hormone (Arendt, 1994; Armstrong & Redman, 1991; Grad & Rozencwaig, 1993; Reiter, 1992; Reiter *et al.*, 1994). The sleep-inducing effects of melatonin (Dawson & Encel, 1993; Dollins, Zhdanova, Wurtman, Lynch, & Dens, 1994; Lavie, 1997; Zhdanova & Wurtman, 1997), while somewhat controversial, have nevertheless raised the possibility that the age-related decline in melatonin may also underlie

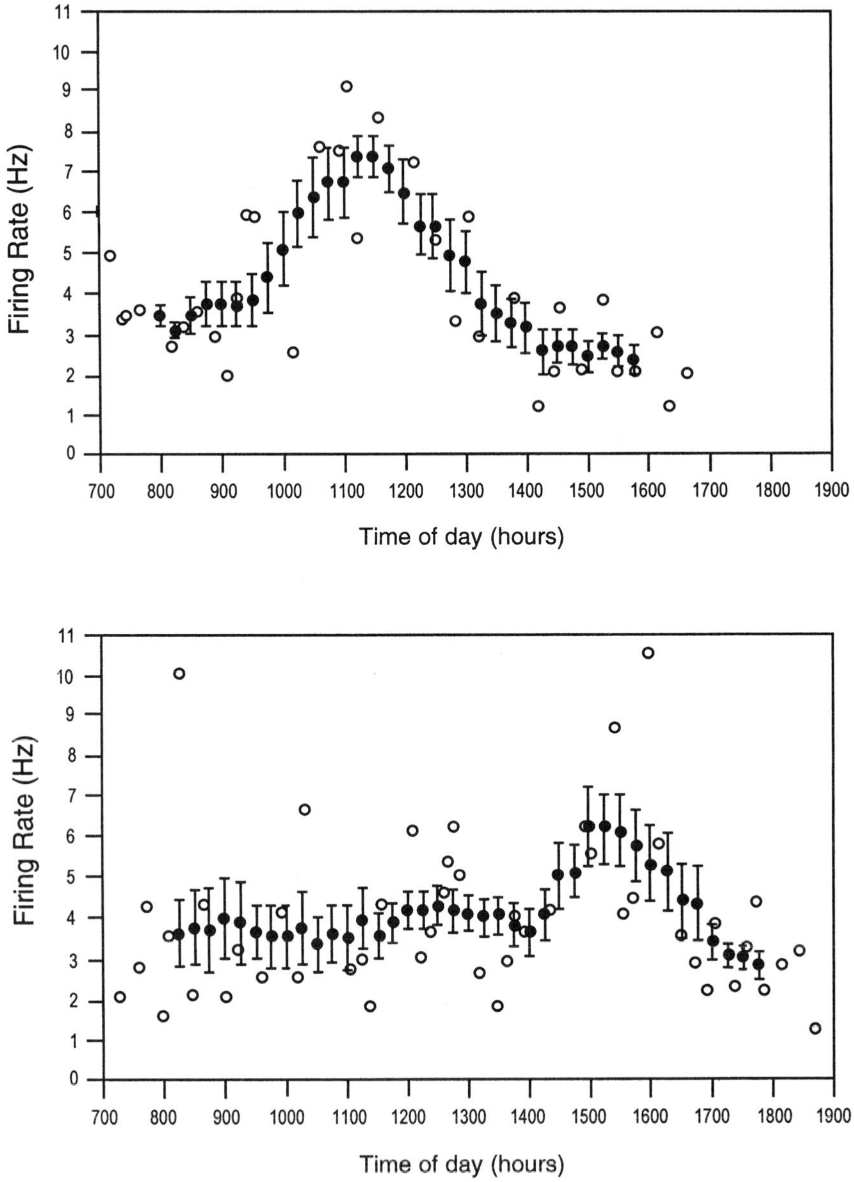

Figure 6. 8-OH DPAT is still capable of phase shifting the circadian rhythm of single-unit activity in an SCN slice derived from an old hamster that did not respond to 5-HT agonist treatment *in vivo*. Animals were entrained to a 14:10 light–dark cycle with light offset at 21:00 hours prior to sacrifice. Filled circles represent the 2-hour running average (±SEM) of the single neurons recorded *in vitro*. Open circles are the average firing rate of each neuron sampled. Top: Treatment with 8-OH DPAT *in vitro* at CT8 the day before recording caused a phase shift in peak neuronal firing rate to CT2.5 (11:30 hours). Bottom: A vehicle-treated slice from a different animal shows the expected peak in firing rate at CT6 (15:00 hours).

some of the decrements in sleep that have been associated with aging (see Chapter 22). The recent report that the administration of melatonin on a circadian basis significantly prolongs the life span of mice raises intriguing questions about the properties of this indolamine during senescence (Pierpaoli & Regelson, 1994). However, the finding that pineal grafts have similar effects on longevity (Pierpaoli & Regelson, 1994) makes the role of melatonin unclear, since the inbred strains of laboratory mice used in these studies lack pineal melatonin (Ebihara, Marks, Hudson, & Menaker, 1986). Interestingly, increased longevity in rats resulting from lifelong food restriction is associated with improved pineal function and increased melatonin levels (Reiter, 1992). These intriguing, but somewhat controversial findings (Reppert & Weaver, 1995; Turek, 1996) indicate a need for more rigorous research approaches for determining the antiaging effects of melatonin and the importance of the decline in melatonin with age for alterations in the circadian clock and sleep–wake systems.

Suprachiasmatic Nucleus

There is now substantial evidence for age-related changes in the SCN itself. A significant reduction in the mean firing rate of SCN neurons *in vitro* was observed in old rats (Satinoff *et al.*, 1993). In this study, the SCN firing patterns were also more erratic in the SCN from old rats, and the authors concluded that aging could either disrupt the coupling between SCN pacemaker cells or their outputs, or cause deterioration of the pacemaking properties of SCN cells. However, in another study involving the recording of neuronal rhythms from the SCN of young and old rats *in vitro*, no major differences in basal firing rates were observed (Cao *et al.*, 1995), indicating that age-related degeneration of rhythmicity may be due to deficits in the SCN afferents, efferents, or downstream clock effector mechanisms. Other studies in rats have found that aging alters the circadian rhythms of glucose utilization (Wise *et al.*, 1992) and α1-adrenergic receptor levels in the SCN, and that these changes are correlated with changes in the circadian rhythm of luteinizing hormone (LH) release (i.e., the preovulatory "LH surge") that are observed with aging in female rats (Weiland & Wise, 1990; Wise, Walovitch, Cohen, Weiland, & London, 1987; Wise, Cohen, Weiland, & London, 1988). In addition, changes in neuropeptide levels in the human SCN have been associated with senility (Swaab, Fisser, Kamphorst, & Troust, 1988), and while no vasopressin cell loss was associated with aging of the human hypothalamus (Van der Woude *et al.*, 1995), an alteration in the seasonal detection of vasopressin-producing cells has been reported in the human SCN (Hofman & Swaab, 1994). Changes in either the number or the size of VIP and vasopressin cells have also been associated with the aging of the rat SCN (Chee, Roozendaal, Swaab, Goudsmit, & Mirmiran, 1988; Roozendaal, van Gool, Swaab, Hoogendyk, & Mirmirian, 1987).

As noted in the previous section, light-induced gene expression in the SCN is markedly reduced in old rats and hamsters (Sutin *et al.*, 1993; Zhang *et al.*, 1996). While no significant decrease in the total number of CREB-immunoreactive cells in the SCN of aged hamsters was observed, there was a marked decrease in the number of cells in which CREB was activated (i.e., phosphorylated) by light, suggesting a specific age-associated deficit in cellular signaling pathways upstream of c-*fos* gene regulation. Since CREB is rapidly phosphorylated in response to the influx of Ca^{2+} triggered by membrane depolarization (Bahr, Godshall, Hall, & Lynch, 1992), as well as through a pathway involving cAMP, it has been suggested that the reduction of Fos and phospho-CREB expression in the SCN of aged hamsters might be attributed to age-related deficits in Ca^{2+} and/or cyclic-AMP regulation (Zhang *et al.*, 1996). Indeed, there is evidence suggesting that aging is associated with deficits in the regulation of neuronal Ca^{2+} levels, specifically in reduced uptake and clearance of Ca^{2+}, indicating that cytosolic Ca^{2+} levels might be reduced (Smith, 1988). Therefore, age-related neurophysiologic changes, such as abnormal Ca^{2+} metabolism in the SCN, could provide a potential mechanism for age-related changes in Fos and phospho-CREB induction by light.

Output Pathways and End-Organ Response to Circadian Signals

Any discussion of possible age-related effects on either circadian outputs from the SCN, or on the end-organ response to these signals, is only speculative since no studies have addressed this question. This is not surprising since how the SCN communicates with the rest of the organism remains one of the major unknowns in the field (Turek, 1991). Even though a number of SCN efferent pathways have been mapped to various areas of the brain (Miller 1996; Moore, 1996), the only known functional efferent pathway for carrying circadian information is the multisynaptic pathway from the SCN to the pineal gland (Illnerova, 1991; Moore, 1996). However, whether the neural signal from the SCN or the pineal's response to the signal is reduced in old age, and thus could contribute to the age-related decline in melatonin production, is not known.

Although many investigators have speculated for a number of years that the SCN may communicate at least some circadian information to the rest of the organism through neurohumoral factors, only recently has clear evidence been provided to support this hypothesis. Silver and her colleagues have been able to restore circadian rhythmicity in SCN-lesioned animals via fetal SCN transplants that have been "packaged" inside a membrane that allows for the release of diffusable substances from the SCN, but not any neural outgrowth (Silver, LeSauter, Tresco, & Lehman, 1996). Identification of possible neurohumoral circadian signals from the SCN will enable investigators to determine if the strength of these signals is altered in advanced age.

In addition to studies that focus directly on possible age-related changes in the known components of the circadian clock system, including end organ responsiveness to circadian signals, an alternative strategy for uncovering the physiologic basis for such changes involves the use of pharmacologic and/or genetic approaches to mimic the effects of aging.

Pharmacologic Approaches

In view of the fact that catecholamine activity declines in the brain (including the hypothalamus) of old animals (Meites, 1991), we have initiated a series of studies to determine if depleting monoamine levels in the brain of young animals can mimic any of the age-related changes in circadian clock function of hamsters. In initial studies, we used reserpine, a potent, but nonselective depletor of monoamines in the brain (Penev, Turek, & Zee, 1993; Penev, Zee, & Turek, 1994). Because of our uncertainty as to which monoamine, or combination of monoamines, might be involved in age-related changes in the circadian system, our initial strategy was to use this nonselective approach. In response to treatment with reserpine (2.5–4.0 mg/kg), young hamsters showed (1) a decrease in locomotor activity, (2) a phase advance in the onset of activity during entrainment to an LD 14:10 cycle, (3) a loss in responsiveness to the phase shifting effects of triazolam, (4) an enhanced response to the phase-shifting effects of light pulses presented around CT16, and (5) a reduction in the levels of serotonin, norepinephrine, and dopamine in the hypothalamus, striatum, and pons/medulla. The changes in the circadian rhythm of activity and its response to phase-shifting stimuli are all similar to those that occur spontaneously in old animals (see above). These initial findings support the hypothesis that a decrease in monoaminergic activity in the brain may underlie some of the age-related changes in the circadian clock system of rodents, and suggest that monoamines are important determinants of the shape and amplitude of the PRC to light.

Other studies have demonstrated that the specific depletion of serotonergic activity in the brain also induces changes in the circadian clock system that are similar to those that occur spontaneously in old animals (Morin & Blanchard, 1991; Penev *et al.*, 1994; Smale, Michels, Moore, & Morin, 1990; Sumova, Maywood, Selvage, Ebling, & Hastings, 1996). These findings, coupled with an altered response of old animals to the phase-shifting effects of a 5-HT agonist (see above), once again suggest a role for an alteration in serotonergic activity in the aging circadian clock system. Moreover, changes in the 5-HT fiber system associated with serotonergic neurotoxicity are similar to those occurring spontaneously with advanced age (Davidoff & Lolova, 1991; Van Luijtelaar, Tonnaer, & Steinbusch, 1989).

Genetic Approaches

While no attempts have been made to select for specific genes that may play a role in age-related changes in the circadian clock system, the two single-gene clock mutations that have been identified in mammals (the *tau* gene in hamsters and the *clock* gene in mice) have some effects on rhythmicity in young animals that are similar to age-related changes in wild type animals. The *clock* mutation in mice results in about a 1-hour increase in the free-running period in animals heterozygous for the mutation. While animals homozygous for the *clock* mutation initially show a 3–4-hour increase in the free-running period, after a few weeks in constant darkness, rhythmicity disappears (Vitaterna *et al.*, 1994). In addition to changes in the free-running period, two other changes in circadian phenotype are observed in mice heterozygous for the *clock* mutation: the stability of period is decreased and the amplitude of the resetting response to light pulses is increased (Vitaterna *et al.*, 1996). Both of these changes are consistent with the idea that the circadian clock of *clock* heterozygous animals has a lower amplitude than in young, wild-type animals; a decrease in the amplitude of the circadian clock with age has been hypothesized for rodents (see above). The fact that young heterozygous *clock* mutant mice display some of the changes in circadian properties seen with aging suggests the possibility that the *clock* mutation and the aging process act upon the circadian system in a similar manner (e.g., by lowering circadian amplitude). The isolation of the *clock* mutant and the recent cloning of the gene responsible for the aberrant phenotype (King *et al.*, 1997) open up a new set of questions that can be addressed with respect to the interaction between the circadian system and aging.

The *tau* mutation in the hamster reduces the free-running period of the circadian clock to about 22 and 20 hours in the heterozygous and homozygous conditions, respectively (Ralph & Menaker, 1988). Young homozygous *tau*-mutant hamsters show a number of phenotypic differences compared to young wild-type hamsters that are similar to changes which occur spontaneously in older wild-type animals. For example, *tau*-mutant animals show phase shifts in response to a single 1-hour pulse of light presented at CT15 that average about 10 hours, whereas in response to the same treatment, wild-type animals show phase shifts of only about 1–2 hours (Menaker & Refinetti, 1992; Scarbrough & Turek, 1996; Shimomura & Menaker, 1994). In addition, total wheel revolutions per circadian cycle are reduced, the precision of activity onset is less, and the overall activity pattern is more fragmented in some young *tau* mutants compared to young wild-type animals, all changes in phenotype that are characteristic of aging in wild-type animals (Scarbrough & Turek, 1996).

Attempts to Attenuate or Reverse the Effects of Aging on Circadian Clock System

Despite the important medical, social, and economic consequences of disrupted circadian rhythms in older adults (see next section), surprisingly few strategies designed to restore normal 24-hour rhythms in animal model systems have been developed. There are two primary reasons why it is of interest to find countermeasures for the effects of aging on the circadian clock system. First, finding effective countermeasures may yield new insights not only into the underlying causes for age-related changes in circadian function, but also into basic clock mechanisms. The second reason, of course, is that the discovery of countermeasures for the effects of aging on the circadian system in animal models could lead to the development of similar measures in humans.

The few attempts that have been made to reverse the effects of aging on circadian rhythmicity in rodents have followed a standard protocol: take animals of a specific age that are showing an age-related change in rhythmicity and determine if a given intervention can restore rhythmicity to a more youthful state. Few attempts have been made to begin an intervention early in life, i.e., before age-related changes in rhythmicity are readily apparent, to determine if treatment prevents, retards, or delays the onset of such changes. Our finding (described below) that the use of bright light to improve rhythmicity in middle-age and older animals was more successful in the middle-age animals raises the possibility that countermeasures will be more effective when started earlier in life, perhaps even before age-related changes in rhythmicity are observed.

In the literature, three different approaches have been taken with "moderate" success to attenuate the effects of aging on circadian rhythms in rodents: bright light, SCN transplants, and limited pharmacologic interventions.

Bright Light

Since the light–dark cycle is the major synchronizing agent of the circadian clock, an obvious strategy for attempting to improve rhythmicity in old animals is to increase the strength of the entraining light–dark cycle and examine whether this intervention reverses or attenuates some of the effects of aging on the circadian rhythmicity. Indeed, strategies involving the use of bright light exposure for this end have been employed in both rodents and humans (see Chapter 22). In rats, the amplitude of the locomotor activity rhythm was increased by exposing old animals to bright light during the daytime (Witting, Mirmiran, Bos, & Swaab, 1993). Similarly, in hamsters normally maintained on an LD 14:10 light cycle with the intensity of light being about 300 lux, an increase in the daytime light intensity to 1500 lux resulted in a significant change in the activity rhythm which was more pronounced in the middle-aged (≈12–13 months of age) than in the older (19 months) animals

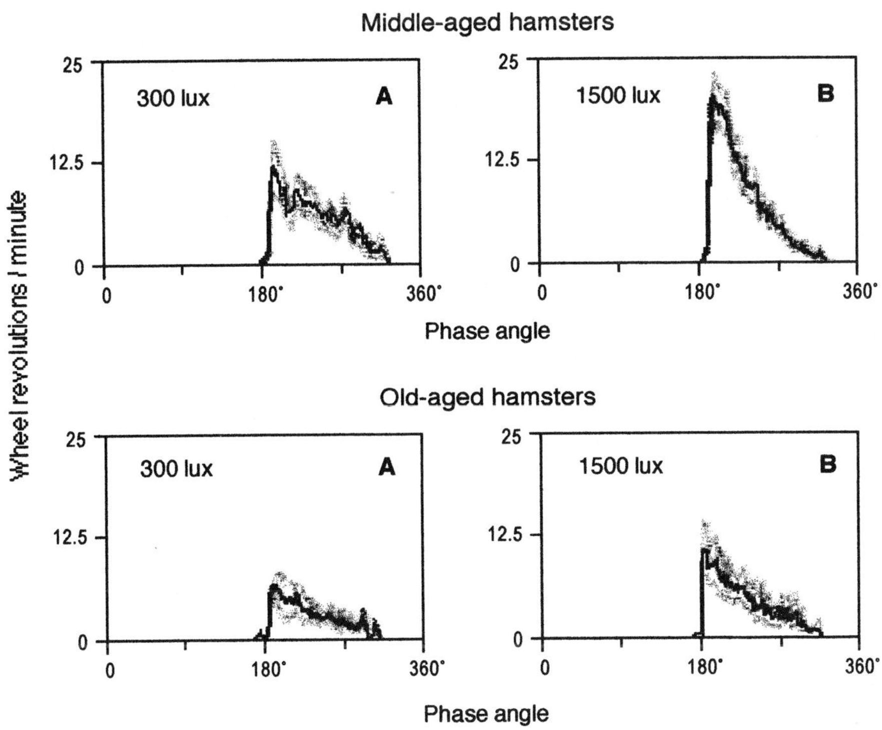

Figure 7. Effects of increased light intensity on the daily activity profile of middle-aged and old hamsters. Mean number of wheel revolutions performed per minute (±SEM) over a 24-hour period for middle-aged and old hamsters maintained under different lighting conditions for 6 weeks are (A) recorded under conditions where the light phase of the 14:10 light–dark cycles was measured as 300 lux or (B) when the light phase measured 1500 lux. Phase angle of 180° denotes the time when the dark phase began. (From Labyak *et al.*, 1998, with permission. Copyright 1998, American Physiological Society.)

(Figure 7). Thus, exposure to bright light led to a more "youthful" profile of the overall 24-hour activity pattern where there is a significantly higher amplitude of activity during the early hours of the dark period.

Our finding that bright light was more effective in increasing the amplitude of the activity rhythm of middle-age versus older hamsters focuses attention on the use of countermeasures at an earlier age. Indeed, the aging of the circadian clock system, as with other physiologic systems, is a continual process. As in humans, changes in rhythmicity in rodents can be expected to occur gradually throughout the adult life and not all at once at some particular "old age." For example, serum growth hormone levels in both humans and hamsters are dramatically reduced at an age that is well below the half-maximum of the life span for the species (Laartz *et al.*, 1994; Van Cauter & Plat, 1996). Although difficult and expensive to carry out, it will be of great interest to determine if specific countermeasures can be utilized beginning at an early age to delay the effects of aging on the circadian clock system.

SCN Transplants

At least four groups have attempted to restore youthful characteristics of circadian rhythmicity by transplanting fetal SCN tissue into SCN-intact older animals.

One rationale behind these studies is the fact that such transplants can restore rhythmicity in adult animals that had previously been made arrhythmic due to lesioning of the SCN (Lehman *et al.*, 1987). Positive effects of fetal SCN transplants include (1) restoration of the response to the phase-shifting effects of a single injection of triazolam in old hamsters (Van Reeth *et al.*, 1994), (2) the blockade of the normal alteration in the free-running period that is observed in old hamsters (Viswanathan & Davis, 1995), and (3) an increase in the amplitude of the locomotor activity, temperature, and/or drinking rhythm in old rats (Satinoff & Li, 1996). Interestingly, fetal SCN transplanted into middle-aged hosts restore light-induced Fos expression in the host (Cai, Lehman, Lloyd, & Wise, 1997). The transplants also reduced the nocturnal expression of Fos in middle-aged rats such that the diurnal pattern of baseline Fos expression became similar to that observed in young rats. The authors concluded that the SCN transplants are able to provide signals that enable the cellular system of the host to regain functional capabilities.

These positive effects of fetal SCN grafts on circadian rhythmicity in old rats and hamsters are encouraging because they (1) demonstrate that age-related changes in circadian rhythmicity are not fixed and can be reversed or delayed and (2) raise the possibility that neuropharmacologic interventions that can simulate the effects of fetal SCN grafts might be utilized to treat age-related disorders in circadian function. Nevertheless, it should be noted that there are limitations in the use of fetal SCN implants for restoring rhythmicity in old animals, including the high mortality in old animals following surgery and the continued aging of the animal following surgery. Our own strategy now is to find pharmacologic interventions, rather than use SCN transplants, to restore rhythmicity in older animals.

The mechanisms by which fetal SCN tissue can impact on the circadian organization of old animals with an intact SCN are not known. Grafts could be functioning as new "independent SCN" or the graft may be inducing changes in the activity of the older host SCN. A "rejuvenation" of the aged host SCN might be due to the release of trophic factors by fetal grafts and/or an increase in trophic activity in the surrounding tissue. Alternatively, the fetal and host SCN could perhaps establish connections with one another that lead to a chimeric regulation of circadian function. Recent studies by Hurd, Zimmer, Lehman, and Ralph (1995) indicate that transplanting fetal SCN of one genotype into old hamsters with a different SCN genotype can result in both SCN functioning relatively independently of one another in controlling the period of the free-running circadian rhythm of locomotor activities. While such transplants can contribute to the expressed period in old host animals, they are not effective in young SCN-intact animals. Interestingly, SCN transplants into young animals with partial lesions of the SCN, lesions which lead to changes in the activity rhythm characteristic of those that occur spontaneously in advanced age, result in the expression of both host and donor phenotypes. These results support the hypothesis that the aging SCN is not functioning to control circadian rhythmicity as well as in young animals.

Melatonin

The use of melatonin as an antiaging agent has received a great deal of attention recently (Turek, 1996), particularly in the public media. Two hypothesis for the antiaging effects of melatonin that relate to its role in the overall circadian organization involve melatonin's effects as hypnotic in humans and/or as an agent which can induce phase shifts in the circadian clock of both rodents and humans. Thus, at least on theoretical grounds, treatment with melatonin might improve sleep in the elderly and/or normalize circadian organization in older individuals. Few studies have actually been performed in older humans, and even in young adults, the evidence that melatonin can have effects on sleep or on phase-shifting human rhythms that are therapeutically useful is limited (Turek & Czeisler, 1998).

For all the interest that has been generated around the possible beneficial effects of melatonin in aging, surprisingly few studies have been carried out to determine if treatment with melatonin can normalize circadian rhythmicity in old rodents. In one study, treatment with the melatonin agonist S-20242 partially reversed the age-related decrease in the amplitude of the circadian temperature rhythm in old rats (Koster-van Hoffen *et al.*, 1993). Similarly, we have found that chronic treatment with the melatonin agonist S-20098, via administration in the food, induced a significant increase in the amplitude of the activity rhythm of old (26–27 months) hamsters and in the cohesion of the activity rhythms (O. Van Reeth *et al.*, , unpublished). Melatonin-treated hamsters had a more "youthful"-looking overall pattern of activity, where there is a large peak in activity during the early part of the dark period. Clearly, more studies need to be carried out before any conclusions about the beneficial effects of melatonin (or agonist) treatment on the circadian organization of animals can be made.

While the "fountain of youth" for the aging circadian clock system has clearly not been found, the results from the preliminary results described above raise the possibility that future studies will lead to either the prevention, delay, or reversal of age-related changes in the circadian clock system. Just as age-related changes in the circadian clock system may occur at various levels in the circadian organization, intervention to reverse these changes could also be directed at the input pathways, at the level of the clock itself, at the output pathways, or at end organs ultimately responsible for expressing a particular rhythm. An intervention that affects many diverse rhythms would probably be acting at the level of the SCN itself or on an input pathway to the clock. As discussed above, the finding of a high density of melatonin receptors in the SCN makes the SCN itself a possible target site for any beneficial effects of melatonin on the aged circadian system.

Circadian information from the SCN to an end organ expressing an overt rhythm undoubtedly follows many pathways, pathways that may involve a number of complex steps. Indeed, the ultimate expression of a rhythm may involve a cascade of neural, neuroendocrine, and endocrine events. Some of the events (e.g., possibly melatonin) may in turn have feedback effects on the circadian clock in the SCN which itself initiated the cascade. Effective countermeasures for age-related changes in the expression of circadian rhythms could be targeted at specific output rhythms which themselves affect many other rhythmic processes. Again, in view of melatonin's potential role as a mediator of circadian information from the SCN to other physiologic systems, one could speculate that any improvement in rhythmicity in old animals by melatonin could be due to its effects on processes that are downstream from the circadian clock.

Functional Significance of Age-Related Changes in the Circadian System

A basic tenet of the field of circadian rhythms is that internal temporal organization is central to the health and well-being of the organism. The observation that circadian temporal organization is disrupted in advanced age, and that older animals become less sensitive to the entraining signals in the environment, leads to the central question: Does the breakdown in circadian organization lead or contribute to the aging process, or is circadian disorganization a byproduct of the normal aging process? In view of the central role played by the circadian clock in the regulation of diverse biological systems, one could expect that age-related changes in biochemical, metabolic, hormonal, and/or behavioral rhythms will in fact have an impact on the health and well-being of the organism. Numerous investigators have speculated that the progressive deterioration of both physiologic function and adaptability that occur with advanced age may be the result of the loss of coordination among various circadian rhythms and that a decay in circadian organization may be involved in the physiologic deterioration that limits the life span of the species. It has been hypothesized that deficits in the internal temporal organization and its synchronization to the environment might be important limiting factors for the longevity of the organism (Pittendrigh, 1974). It is of obvious importance to determine if the normalization of circadian rhythmicity in older animals will have a positive effect on the health and well-being of the organism.

The recent findings that the effects of age on the way the circadian clock system is entrained to the LD cycle depends upon the length of the day (Scarbrough *et al.*, 1997) raises some intriguing questions about the functional significance for age-related changes. The phase angle of entrainment to the environmental light–dark cycle is of important adaptive significance for the survival of the organism (Pittendrigh & Daan, 1976). As the length of the day varies on a seasonal basis, animals adjust the timing of their active period (and presumably also the timing of other behavioral and metabolic rhythms) to meet the demands of the changing environment. The finding of nearly a 2-hour difference between the entrainment pattern of young and old animals to a short-day photoperiod of LD 6:18 indicates that age has a major effect on the timing of daily activities, and the extent of this effect depends on the season of the year. Whether such changes reflect a selectively adaptive change for animals of different ages, or instead is a nonadaptive change that is a reflection of an alteration of the circadian clock system with age, is not known. Similarly, the length of the day has a major influence on the effects of age on the overall pattern of locomotor activity, indicating that during different seasons of the year, the influence of age on the temporal organization will be different. There is now significant evidence to indicate that human rhythms are influenced by changes in day length (Wehr, 1991; Wehr *et al.*, 1993) and recently it has been reported that seasonal changes in vasopressin content in the human SCN is influenced by age (Hofman & Swaab, 1995). In any case, it is clear that at least in a photoperiodic rodent species, there are interactions between day length and age that have a major effect on the overall temporal organization of the animal relative to the entraining light–dark cycle.

If disturbances of normal circadian temporal organization are important for the health and well-being of the organism, then one would predict that disrupting rhythmicity in early life would have a negative effect on the health and perhaps longevity of the organism. Surprisingly few attempts have been made to determine if disrupting normal circadian rhythmicity is deleterious and/or could lead to an advancement of the aging process in animals. Pittendrigh and Minis (1972) demonstrated that fruit flies reared on LD 12:12 live longer than flies exposed to light–dark cycles significantly shorter (e.g., 21 hours) or longer (e.g., 27 hours) than the normal 24 hours. Likewise, in the blowfly, repeated exposure to 5 hour phase shifts in the

LD 12:12 cycle reduced longevity (Aschoff *et al.*, 1971; Saint Paul, 1978). These deleterious effects on life span have been attributed to the disruption of the organism's normal internal temporal organization (Pittendrigh, 1974). Other studies in invertebrates have failed to confirm the results of these early studies and, as noted by Brock (1991), there is probably a complex interplay of light intensity, frequency, and direction of phase shifting which must be taken into account in interpreting such longevity studies.

While there is some evidence in mammals to indicate that a similar disruption of normal circadian organization decreases longevity, other studies have failed to find any such effects (Brock, 1991). Most of the longevity studies in mammals have been limited to studies involving repeated phase shifts in the light–dark cycle (Nelson & Halberg, 1986), and little is known about the effects of changes in either the endogenous period of the clock or changes in the period of the entraining agent on longevity in mammals. There are no data on whether or not SCN-lesioned animals have an increased rate of morbidity or a decreased life span. Such studies would be costly and difficult to carry out since in order truly to test for the necessity of the circadian clock system for normal health during aging, one should carry out such studies in animals maintained in their "normal" environment, i.e., an environment that is more challenging to the survival of the organism than that found in a sterile laboratory.

It should be noted that the robust quantitative differences between circadian rhythms in young and old animals observed in the laboratory probably represent the *minimal* nature of the changes that actually occur with aging. That is, when one compares the characteristics of circadian rhythmicity in, for example, 18-month-old *healthy* hamsters with young (e.g., 3–4-month-old) animals, one is using animals that have lived a longer life than many other animals, and this may be due in part to a circadian clock system that is functioning at better than normal levels. This point was made by Myers and Badia (1995) in their excellent review of the effects of age on sleep and circadian rhythms, where they concluded, "The older subjects that are available for inclusion may have characteristics which are not representative of their original group. For example, those subjects remaining are likely to be more healthy [than their deceased cohort]. Therefore, many of the differences [or lack thereof] described may be artificial." This is an important point since in studies on circadian rhythmicity and aging in both rodents and humans, we are selecting those individuals who have aged well. By definition, older individuals are survivors, and these survivors may have an unusually healthy circadian system, which is one of the reasons they have survived to be included in our studies. Indeed, well-documented studies of the rhythm of body temperature under constant routine conditions have indicated that completely healthy, self-sufficient older subjects with active lifestyles (i.e., successful agers) may have little evidence of disrupted circadian function (Monk, Buysse, Reynolds, & Kupfer, 1993). Thus, as with other physiologic systems, there is undoubtedly a wide interindividual variation in the magnitude of disturbance of markers of circadian rhythmicity. Disturbance may be severe in people who are not aging well, moderate in "usual" aging, and minimal in "successful" aging. It is obviously a great challenge to determine the importance of the circadian clock system for successful aging.

The most widely acknowledged consequence of age-related changes in circadian rhythms in humans is the disruption of sleep in the elderly (Czeisler, Chiasera, & Duffy, 1991). Unfortunately, few studies have probed how age-related changes in circadian rhythmicity affect sleep, and vice versa, in animal models. However, animal models have been used to characterize age-related changes in neurotransmitter activity in the area of the brain that is central to the control of circadian rhythms and sleep, the hypothalamus (Meites, 1991). Dysfunction of hypothalamic activity with advanced age has been associated with a number of age-related changes in physiology and metabolism (Meites, 1991). Aging of the circadian clock system may well be a cause as well as an effect of these changes in hypothalamic function, which in turn alters other areas of the brain as well as the periphery. Aging may represent a positive feedback loop, with an aging clock system causing other systems to be disturbed, while these disturbances, particularly as they impact on the quality of sleep, may cause further disruption of normal circadian clock functions. Perhaps the faster the events in the positive feedback loop take place, the faster the organism ages. It will be of great interest to determine if environmental and pharmacologic interventions that interfere with the development of such a positive feedback loop are able to increase the quality and/or duration of life.

References

Arendt, J. (1994). The pineal gland: Basic physiology and clinical implications. In L. J. DeGroot (Ed.), *Endocrinology* (pp. 432–444). Philadelphia: Saunders.

Armstrong, S. M., & Redman, J. R. (1991). Melatonin: A chronobiotic with anti-aging properties? *Medical Hypotheses, 34,* 300–309.

Aschoff, J., Fatranska, M., Giedke, H., Doerr, P., Stamm, D., & Wisser, H. (1971). Human circadian rhythms in continuous darkness: Entrainment by social cues. *Science, 171,* 213–215.

Bahr, B. A., Godshall, A. C., Hall, R. A., & Lynch, G. (1992). Mouse telencephalon exhibits an age-related decrease in glutamate (AMPA) receptors but no change in nerve terminal markers. *Brain Research, 589,* 320–326.

Brock, M. A. (1991). Chronobiology and aging. *Journal of the American Geriatrics Society, 39,* 74–91.

Cai, A., Lehman, M. N., Lloyd, J. M., & Wise, P. M. (1997). Transplantation of fetal suprachiasmatic nuclei into middle-aged rats restores diurnal Fos expression in host. *American Journal of Physiology, 272,* R422–R428.

Cao, V. H., Edgar, D. M., Heller, H. C., Dement, W. C., & Miller, J. D. (1995). Basal and phase-shifted neuronal rhythms in the aged SCN *in vitro. Society for Neuroscience, Abstracts, 21,* 1235.

Chee, C. A., Roozendaal, B., Swaab, D. F., Goudsmit, E., & Mirmiran, M. (1988). Vasoactive intestinal polypeptide neuron changes in the senile rat suprachiasmatic nucleus. *Neurobiology of Aging, 9,* 307–312.

Czeisler, C. A., Kronauer, R. E., Allan, J. S., Duffy, J. F., Jewett, M. E., Brown, E. N., & Ronda, J. M. (1989). Bright light induction of strong (type 0) resetting of the human circadian pacemaker. *Science, 244,* 1328–1333.

Czeisler, C. A., Chiasera, A. J., & Duffy, J. F. (1991). Research on sleep, circadian rhythms and aging: Applications to manned spaceflight. *Experimental Gerontology, 26,* 217–232.

Czeisler, C. A., Kronauer, R. E., & Allan, J. S. (1992). *Assessment of modification of a subject's endogenous circadian cycle.* U.S. Patent No. 5,163,426.

Davidoff, M. S., & Lolova, I. S. (1991). Age-related changes in the serotonin immunoreactivity in the telencephalon and diencephalon of rats. *Journal fur Hirnforschung, 32,* 745–753.

Davis, F. C., & Menaker, M. (1980). Hamsters through time's window: Temporal structure of hamster locomotor rhythmicity. *American Journal of Physiology, 239,* R149–R155.

Dawson, D., & Encel, N. (1993). Melatonin and sleep in humans. *Journal of Pineal Research, 15,* 1–12.

Dollins, A. B., Zhdanova, I. V., Wurtman, R. J., Lynch, H. J., & Deng, M. H. (1994). Effect of inducing nocturnal serum melatonin concentrations in daytime on sleep, mood, body temperature, and performance. *Proceedings of the National Academy of Sciences of the USA, 91,* 1824–1828.

Dubocovich, M. L., Benloucif, S., & Masana, M. I. (1996). Melatonin receptors in the mammalian suprachiasmatic nucleus. *Behavioural Brain Research, 73,* 141–147.

Duncan, M. J., & Purvis, C. C. (1994). Effects of aging on photoperiodic responsiveness and specific 2-[125I]-iodomelatonin binding sites in the pars tuberalis and suprachiasmatic nuclei of Siberian hamsters. *Journal of Pineal Research, 16,* 184–187.

Ebihara, S., Marks, T., Hudson, D. J., & Menaker, M. (1986). Genetic control of melatonin synthesis in the pineal gland of the mouse. *Science, 231,* 491–493.

Grad, B. R., & Rozencwaig, R. (1993). The role of melatonin and serotonin in aging: Update. *Psychoneuroendocrinology, 18,* 283–295.

Halberg, J., Halberg, E., Regal, P., & Halberg, F. (1981). Changes with age characterize circadian rhythm in telemetered core temperature of stroke prone rats. *Journal of Gerontology, 36,* 28–30.

Hofman, M. A., & Swaab, D. F. (1994). Alterations in circadian rhythmicity of the vasopressin-producing neurons of the human suprachiasmatic nucleus (SCN) with aging. *Brain Research, 651,* 134–142.

Hofman, M., & Swaab, D. (1995). Influence of aging on the seasonal rhythm of the vasopressin-expressing neurons in the human suprachiasmatic nucleus. *Neurobiology of Aging, 16,* 965–971.

Humbert, W., & Pevet, P. (1994). The decrease of pineal melatonin production with age. *Annals of the New York Academy of Sciences, 719,* 43–63.

Hurd, M. W., Zimmer, K. A., Lehman, M. N., & Ralph, M. R. (1995). Circadian locomotor rhythms in aged hamsters following suprachiasmatic transplant. *American Journal of Physiology, 269,* R958–R968.

Illnerova, H. (1991). The suprachiasmatic nucleus and rhythmic pineal melatonin production. In D. C. Klein, R. Y. Moore, & S. M. Reppert (Eds.), *Suprachiasmatic nucleus—The mind's clock* (pp. 197–216). New York: Oxford University Press.

King, D. P., Zhao, Y., Sangoram, A. M., Wilsbacher, L. D., Tanaka, M., Antoch, M. P., Steeves, T. D. L., Vitaterna, M. H., Kornhauser, J. M., Lowery, P. L., Turek, F. W., & Takahashi, J. S. (1997). Positional cloning of the mouse circadian *Clock* gene. *Cell, 89,* 641–653.

Kornhauser, J. M., Nelson, D. E., Mayo, K. E., & Takahashi, J. S. (1990). Photic and circadian regulation of *c-fos* gene expression in the hamster suprachiasmatic nucleus. *Neuron, 5,* 127–134.

Koster-van Hoffen, G. C., Mirmiran, M., Bos, N. P., Witting, W., Delagrange, P., & Guardiola-Lemaitre, B. (1993). Effects of a novel melatonin analog on circadian rhythms of body temperature and activity in young, middle-aged, and old rats. *Neurobiology of Aging, 14,* 565–569.

Laartz, B., Losee-Olson, S., Ge, Y.-R., & Turek, F. W. (1994). Diurnal, photoperiodic and age-related changes in plasma growth hormone levels in the golden hamster. *Journal of Biological Rhythms, 9,* 111–123.

Labyak, S. E., Zee, P. C., Wallen, E. P., & Turek, F. W. (1996). An evaluation of circadian locomotor activity in young, middle aged, and old golden hamsters. *Society for Research on Biological Rhythms, Abstracts, 173,* 113.

Labyak, S. E., Turek, F. W., Wallen, E. P., & Zee, P. C. (1998). The effects of bright light on age-related changes in the locomotor activity of Syrian hamsters. *American Journal of Physiology, 274,* R830–R839.

Lavie, P. (1997). Melatonin: Role in gating nocturnal rise in sleep propensity. *Journal of Biological Rhythms, 12,* 657–665.

Lehman, M. N., Silver, R., Gladstone, W. R., Kahn, R. M., Gibson, M., & Bittman, E. L. (1987). Circadian rhythmicity restored by neural transplant: Immunocytochemical characterization of the graft and its integration with the host brain. *Journal of Neuroscience, 7,* 1626–1638.

Martin, J. R., Fuchs, A., Bender, R., & Harting, J. (1985). Altered light–dark activity difference with aging in two rat strains. *Journal of Gerontology, 44,* 2–7.

Martin, P., & Bateson, P. (1997). *Measuring behaviour: An introductory guide.* Cambridge: Cambridge University Press.

McArthur, A. J., Gillette, M. U., & Prosser, R. A. (1991). Melatonin directly resets the rat suprachiasmatic circadian clock *in vitro. Brain Research, 565,* 158–161.

Meites, J. (1991). Role of hypothalamic catecholamines in aging processes. *Acta Endocrinologica, 125,* 98–103.

Menaker, M., & Refinetti, R. (1992). The *tau* mutation in golden hamsters. In M. Young (Ed.), *Molecular genetics of biological rhythms* (pp. 255–269). New York: Marcel Dekker.

Miller, A. E., & Riegle, G. D. (1982). Temporal patterns of serum luteinizing hormone and testosterone and endocrine response to luteinizing hormone in aging male rats. *Journal of Gerontology, 37,* 522–528.

Miller, J. D., Morin, L. P., Schwartz, W. J., & Moore, R. Y. (1996). New insights into the mammalian circadian clock. *Sleep, 19,* 641–667.

Monk, T. H., Buysse, D. J., Reynolds, C. F. I., & Kupfer, D. J. (1993). Inducing jet lag in older people: Adjusting to a 6-hour phase advance in routine. *Experimental Gerontology, 28,* 119–133.

Moore, R. Y. (1996). Neural control of the pineal gland. *Behavioural Brain Research, 73,* 125–130.

Morgan, P. J., Howell, H. E., & Helliwell, R. (1994). Melatonin receptors: Localization, molecular pharmacology and physiological significance. *Neurochemistry International, 24,* 101–146.

Morin, L. P. (1988). Age-related changes in hamster circadian period, entrainment and rhythm splitting. *Journal of Biological Rhythms, 3,* 237–248.

Morin, L. P., & Blanchard, J. (1991). Depletion of brain serotonin by 5,7-DHT modifies hamster circadian rhythm response to light. *Brain Research, 566,* 173–185.

Mosko, S. S., Erickson, G. F., & Moore, R. Y. (1980). Dampened circadian rhythms in reproductively senescent female rats. *Behavioral Neural Biology, 28,* 1–14.

Mrosovsky, N. (1996). Locomotor activity and non-photic influences on circadian clocks. *Biological Reviews of the Cambridge Philosophical Society, 71,* 343–372.

Mrosovsky, N., Reebs, S. G., Honrado, G. I., & Salmon, P. A. (1989). Behavioural entrainment of circadian rhythms. *Experientia, 45,* 696–702.

Myers, B. L., & Badia, P. (1995). Changes in circadian rhythms and sleep quality with aging: Mechanisms and interventions. *Neuroscience and Biobehavioral Reviews, 19,* 553–571.

Nelson, W., & Halberg, F. (1986). Schedule-shifts, circadian rhythms and lifespan of freely-feeding and meal-fed mice. *Physiology and Behavior, 38,* 781.

Nicolau, G. Y., & Milcu, S. (1977). Circadian rhythm of corticosterone and nucleic acids in the rat adrenals in relation to age. *Chronobiologia, 4,* 136.

Penev, P. D., Turek, F. W., & Zee, P. C. (1993). Monoamine depletion alters the entrainment and the response to light of the circadian activity rhythm in hamsters. *Brain Research, 612,* 156–164.

Penev, P. D., Zee, P. C., & Turek, F. W. (1994). Reserpine abolishes the phase-shifting effect of triazolam on locomotor activity rhythms of golden hamsters. *Brain Research, 637,* 255–261.

Penev, P. D., Zee, P. C., Wallen, E. P., & Turek, F. W. (1995). Aging alters the phase-resetting properties of a serotonin agonist on hamster circadian rhythmicity. *American Journal of Physiology, 268,* R293–R298.

Penev, P. D., Turek, F. W., Wallen, E. P., & Zee, P. C. (1997a). Aging alters the serotonergic modulation of light-induced phase advances in golden hamsters. *American Journal of Physiology, 272,* R509–R513.

Penev, P. D., Zee, P. C., & Turek, F. W. (1997b). Quantitative analysis of the age-related fragmentation of hamster 24-h activity rhythms. *American Journal of Physiology, 273,* R2132–R2137.

Peng, M. T., & Kang, M. (1984). Circadian rhythms and patterns of running-wheel activity, feeding and drinking behaviors of old male rats. *Physiology and Behavior, 33,* 615–620.

Peng, M. T., Jiang, M. J., & Hsu, H. K. (1980). Changes in running-wheel activity, eating and drinking and their day/night distribution throughout the life span of the rat. *Journal of Gerontology, 35,* 339–347.

Pierpaoli, W., & Regelson, W. (1994). Pineal control of aging: Effect of melatonin and pineal grafting on aging mice. *Proceedings of the National Academy of Sciences of the USA, 94,* 787–791.

Pittendrigh, C. S. (1974). Circadian oscillations in cells and the circadian organization of multicellular systems. In F. C. Schmitt & F. G. Worden (Eds.), *The neurosciences, third study program* (pp. 437–458). Cambridge, MA: MIT Press.

Pittendrigh, C. S., & Daan, S. (1974). Circadian oscillations in rodents: A systematic increase of their frequency with age. *Science, 186,* 548–550.

Pittendrigh, C. S, & Daan, S. (1976). A functional analysis of circadian pacemakers in nocturnal rodents. II. The variability of phase response curves. *Journal of Comparative Physiology, 106,* 253–266.

Pittendrigh, C. S., & Minis, D. H. (1972). Circadian systems: Longevity as a function of circadian resonance in *Drosophila melanogaster. Proceedings of the National Academy of Sciences of the USA, 69,* 1537.

Possidente, B., McEldowney, S., & Pabon, A. (1995). Aging lengthens circadian period of wheel-running activity in C57BL mice. *Physiology and Behavior, 57,* 575–579.

Quay, W. B. (1972). Pineal homeostatic regulation of shifts in the circadian activity rhythm during maturation and aging. *Transactions of the New York Academy of Sciences, 34,* 239–254.

Ralph, M. R., & Menaker, M. (1988). A mutation of the circadian system in golden hamster. *Science, 241,* 1225–1227.

Redman, J., Armstrong, S., & Ng, K. T. (1983). Free-running activity rhythms in the rat: Entrainment by melatonin. *Science, 219,* 1089–1091.

Reiter, R. J. (1992). The aging pineal gland and its physiological consequences. *Bioessays, 14,* 169–175.

Reiter, R. J. (1995). The pineal gland and melatonin in relation to aging: A summary of the theories and the data. *Experimental Gerontology, 30,* 199–212.

Reiter, R. J., Craft, C. M., Johnson, J. E., Jr., King, T. S., Richardson, B. A., Vaughan, G. M., & Vaughan, M. K. (1981). Age associated reduction in nocturnal pineal melatonin levels in female rats. *Endocrinology, 109,* 1295–1297.

Reiter, R. J., Tan, D. X., Poeggeler, B., Menendez-Pelaez, A., Chen, L. D., & Saarela, S. (1994). Melatonin as a free radical scavenger: Implications for aging and age-related diseases. *Annals of the New York Academy of Sciences, 719,* 1–12.

Reppert, S. M., & Weaver, D. R. (1995). Melatonin madness. *Cell, 83,* 1059–1062.

Reynolds, C. F., III, Jennings, J. R., Hoch, C. C., Monk, T. H., Berman, S. R., Hall, F. T., Matzzie, J. V., Buysse, D. J., & Kupfer, D. J. (1991). Daytime sleepiness in the healthy "old old": A comparison with young adults. *Journal of the American Geriatrics Society, 39,* 957–962.

Richardson, G. S. (1990). Circadian rhythms and aging. In E. L. Schneider & J. W. Rowe (Eds.), *Handbook of the biology of aging* (pp. 275–305). San Diego, CA: Academic Press.

Roozendaal, B., Van Gool, W. A., Swaab, D. F., Hoogendyk, J. E., & Mirmirian, M. (1987). Changes in vasopressin cells of the rat suprachiasmatic nucleus with aging. *Brain Research, 409,* 259–264.

Rosenberg, R. S., Zepelin, H., & Rechtschaffen, A. (1979). Sleep in young and old rats. *Journal of Gerontology, 34,* 525–532.

Rosenberg, R. S., Winter, J., & Rechtschaffen, A. (1980). Effects of light cycle phase reversal on sleep rhythms of young and old rats. *Sleep Research, 9,* 105.

Rosenberg, R. S., Zee, P. C., & Turek, F. W. (1991). Phase response curves to light in young and old hamsters. *American Journal of Physiology, 261,* R491–R495.

Rusak, B., & Bina, K. G. (1990). Neurotransmitters in the mammalian circadian system. *Annual Review of Neuroscience, 13,* 387–401.

Sacher, G. A, & Duffy, P. H. (1978). Age changes in rhythms of energy metabolism, activity and body core temperature in *Mus musculus* and *Peromyscus.* In H. V. Samis & S. Copobianco (Eds.), *Aging and biological rhythms* (pp. 105–124). New York: Plenum Press.

Sack, R. L., Lewy, A. J., Blood, M. L., & Stevenson, J. (1991). Melatonin administration to blind people: Phase advances and entrainment. *Journal of Biological Rhythms, 6,* 249–261.

Saint Paul, U. V. (1978). Longevity among blowflies *Phormia terraenovae* R.D. kept in non-24 hour light–dark cycles. *Journal of Comparative Physiology, 127,* 191.

Satinoff, E., & Li, H. (1996). Fetal tissue containing the suprachiasmatic nucleus restores circadian rhythms in old rats. *Society for Research on Biological Rhythms, Abstracts, 110,* 82.

Satinoff, E., Li, H., Tcheng, T. K., Liu, C., McArthur, A. J., Medanic, M., & Gillette, M. U. (1993). Do the suprachiasmitc nuclei oscillate in old rats as they do in young ones? *American Journal of Physiology, 265,* R1216–R1222.

Scarbrough, K., & Turek, F. W. (1996). Quantitative differences in the circadian rhythm of locomotor activity and vasopressin and vasoactive intestinal peptide gene expression in the suprachiasmatic nucleus of *tau* mutant compared to wildtype hamsters. *Brain Research, 736,* 251–259.

Scarbrough, K., Losee-Olson, S., Wallen, E. P., & Turek, F. W. (1997). Aging and photoperiod effect on entrainment and quantitative aspects of locomotor behavior in Syrian hamsters. *American Journal of Physiology, 272,* R1219–R1225.

Schmid, H. A. (1993). Decreased melatonin biosynthesis, calcium flux, pineal gland calcification and aging: A hypothetical framework. *Gerontology, 39,* 189–199.

Sheng, M., McFadden, G., & Greenberg, M. E. (1990). Membrane depolarization and calcium induce c-*fos* transcription via phosphorylation of transcription factor CREB. *Neuron, 4,* 571–582.

Shimomura, K., & Menaker, M. (1994). Light-induced phase shifts in *tau* mutant hamsters. *Journal of Biological Rhythms, 9,* 97–110.

Silver, R., LeSauter, J., Tresco, P. A., & Lehman, M. N. (1996). A diffusible coupling signal from the transplanted suprachiasmatic nucleus controlling circadian locomotor rhythms. *Nature, 382,* 810–813.

Slater, P. J. B., & Lester, N. P. (1982). Minimising errors in splitting behaviour into bouts. *Behaviour, 79,* 153–162.

Smale, L., Michels, K. M., Moore, R. Y., & Morin, L. P. (1990). Destruction of the hamster serotoninergic system by 5,7-DHT: Effects on circadian rhythm phase, entrainment and response to triazolam. *Brain Research, 515,* 9–19.

Smith, D. O. (1988). Cellular and molecular correlates of aging in the nervous system. *Experimental Gerontology, 23,* 399–412.

Sumova, A., Maywood, E. S., Selvage, D., Ebling, F. J., & Hastings, M. (1996). Serotonergic antagonists impair arousal-induced phase shifts of the circadian system of the Syrian hamster. *Brain Research, 709,* 88–96.

Sutin, E. L., Dement, W. C., Heller, H. C., & Kilduff, T. S. (1993). Light-induced gene expression in the suprachiasmatic nucleus of young and aging rats. *Neurobiology of Aging, 14,* 441–446.

Swaab, D. F., Fisser, B., Kamphorst, W., & Troust, D. (1988). The human suprachiasmatic nucleus: Neuropeptide changes in senium and Alzheimer's disease. *Basic and Applied Histochemistry, 32,* 43–54.

Tang, F., Hadjiconstantinov, M., & Pang, S. F. (1985). Aging and diurnal rhythms of pineal serotonin, 5-hydroxy-indoleacetic acid, norepinephrine, dopamine and serum melatonin in the rat. *Neuroendocrinology, 40,* 160–164.

Tenover, J. S., Matsumoto, A. M., Clifton, D. K., & Bremner, W. J. (1988). Age-related alterations in the circadian rhythm of pulsatile luteinizing hormone and testosterone secretion in healthy men. *Journal of Gerontology, 43,* 163–169.

Turek, F. W. (1987). Pharmacological probes of the mammalian circadian clock: Use of the phase response curve approach. *Trends in Pharmacological Science, 8,* 212–217.

Turek, F. W. (1991). Introduction: Chapter IV, circadian SCN outputs. In D. C. Klein, R. Y. Moore, & S. M. Reppert (Eds.), *Suprachiasmatic nucleus: The mind's clock* (pp. 191). New York: Oxford University Press.

Turek, F. W. (1996). Melatonin hype hard to swallow. *Nature, 379,* 295–296.

Turek, F. W., & Czeisler, C. A. (1998). Role of melatonin in the regulation of sleep. In F. W. Turek & P. C. Zee (Eds.), *Neurobiology of sleep and circadian rhythms* (pp. 181–195). New York: Marcel Dekker.

Turek, F. W., Pinto, L., Vitaterna, M., Penev, P., Zee, P. C., & Takahashi, J. S. (1995). Pharmacological and genetic approaches for the study of circadian rhythms in mammals. *Frontiers in Neuroendocrinology, 16,* 191–223.

Valentinuzzi, V. S., Scarbrough, K., Takahashi, J. S., & Turek, F. W. (1997). Effects of aging on the circadian rhythm of wheel-running activity in C57BL/6 mice. *American Journal of Physiology, 273,* R1957–R1964.

Van Cauter, E., & Plat, L. (1996). Physiology of growth hormone secretion during sleep. *Journal of Pediatrics, 128,* S32–S37.

Van der Woude, P. F., Goudsmith, E., Wierda, M., Purba, J. S., Hofman, M. A., Bogte, H., & Swaab, D. F. (1995). No vasopressin cell loss in the human hypothalamus in aging and Alzheimer's disease. *Neurobiology of Aging, 16,* 11–18.

Van Gool, W. A., & Mirmiran, M. (1986). Effects of aging and housing in an enriched environment upon sleep–wake patterns in rats. *Sleep, 9,* 335–347.

Van Gool, W. A., Witting, W., & Mirmiran, M. (1987). Age-related changes in circadian sleep–wakefulness rhythms in male rats isolated from time cues. *Brain Research, 413,* 384–387.

Van Luijtelaar, M. G. P. A., Tonnaer, J. A. D. M., & Steinbusch, H. W. M. (1989). Similarities between aberrant serotonergic fibers in the aged and 5,7-DHT denervated young adult rat brain. *Experimental Brain Research, 78,* 81–89.

Van Reeth, O., Zhang, Y., Zee, P. C., & Turek, F. W. (1992). Aging alters feedback effects of the activity–rest cycle on the circadian clock. *American Journal of Physiology, 263,* R981–R986.

Van Reeth, O., Zhang, Y., Reddy, A., Zee, P. C., & Turek, F. W. (1993). Aging alters the entraining effects of an activity-inducing stimulus on the circadian clock. *Brain Research, 607,* 286–292.

Van Reeth, O., Sturis, J., Bryne, M. M., Blackman, J. D., L'Hermite-Balériaux, M., Leproult, R., Oliner, C., Refetoff, S., Turek, F. W., & Van Cauter, E. (1994). Nocturnal exercise phase-delays the circadian rhythms of melatonin and thyrotropin secretion in normal men. *American Journal of Physiology, 266,* E964–E974.

Viswanathan, N., & Davis, F. C. (1995). Suprachiasmatic nucleus grafts restore circadian function in aged hamsters. *Brain Research, 686,* 10–16.

Vitaterna, M. H., King, D. P., Chang, A.-M., Kornhauser, J. M., Lowrey, P. L., McDonald, J. D., Dove, W. F., Pinto, L. H., Turek, F. W., & Takahashi, J. S. (1994). Mutagenesis and mapping of a mouse gene, *clock,* essential for circadian behavior. *Science, 264,* 719–725.

Vitaterna, M. H., Chang, A.-M., King, D. P., Pinto, L. H., Turek, F. W., & Takahashi, J. S. (1996). Heterozygosity at the *clock* locus alters phase response curves to light in mice. *Society for Research on Biological Rhythms, Abstracts, 200,* 127.

Wehr, T. A. (1991). The durations of human melatonin secretion and sleep respond to changes in day length (photoperiod). *Journal of Clinical Endocrinology and Metabolism, 73,* 1276–1280.

Wehr, T. A., Moul, D. E., Barbato, G., Giesen, H. A., Seidel, J. A., Barker, C., & Bender, C. (1993). Conservation of photoperiod-responsive mechanisms in humans. *American Journal of Physiology, 265,* R846–R857.

Weiland, N. G., & Wise, P. M. (1990). Aging progressively decreases the densities and alters the diurnal rhythms of alpha-1 adrenergic receptors in selected hypothalamic regions. *Endocrinology, 126,* 2392–2397.

Welsh, D. K., Richardson, G. S., & Dement, W. C. (1986). Effect of age on the circadian pattern of sleep and wakefulness in the mouse. *Journal of Gerontology, 41,* 579–586.

Whealin, J. M., Burwell, R. D., & Gallagher, M. (1993). The effects of aging on diurnal water intake and melatonin binding in the suprachiasmatic nucleus. *Neuroscience Letters, 154,* 149–152.

Wise, P. M., Walovitch, R. C., Cohen, I. R., Weiland, N. G., & London, D. E. (1987). Diurnal rhythmicity and hypothalamic deficits in glucose utilization in aged ovariectomized rats. *Journal of Neuroscience, 7,* 3469–3473.

Wise, P. M., Cohen, I. R., Weiland, N. G., & London, D. E. (1988). Aging alters the circadian rhythm of glucose utilization in the suprachiasmatic nucleus. *Proceedings of the National Academy of Sciences of the USA, 85,* 5305–5309.

Wise, P. M., Scarbrough, K., Larson, G. H., Lloyd, J. M., Weiland, N. G., & Chiu, S. (1991). Neuroendocrine influences on aging of the female reproductive system. *Frontiers in Neuroendocrinology, 12,* 323–356.

Witting, W., Mirmiran, M., Bos, N. P. A., & Swaab, D. F. (1993). Effect of light intensity on diurnal sleep–wake distribution in young and old rats. *Brain Research Bulletin, 30,* 157–162.

Witting, W., Mirmiran, M., Bos, N. P. A., & Swaab, D. F. (1994). The effect of old age on the free-running period of circadian rhythms in rat. *Chronobiology International, 11,* 103–112.

Zee, P. C., Rosenberg, R. S., & Turek, F. W. (1992). Effects of aging on entrainment and rate of resynchronization of the circadian locomotor activity. *American Journal of Physiology, 263,* 1099–1103.

Zhang, Y. (1995). *Photic and nonphotic manipulation of the circadian clock of young and aged golden hamsters.* Ph.D. Dissertation, Northwestern University, Evanston, Illinois.

Zhang, Y., Kornhauser, J. M., Zee, P. C., Mayo, K. E., Takahashi, J. S., & Turek, F. W. (1996). Effects of aging on light-induced phase-shifting of circadian behavioral rhythms, *fos* expression, and CREB phosphorylation in the hamster suprachiasmatic nucleus. *Neuroscience, 70,* 951–961.

Zhdanova, I. V., & Wurtman, R. J. (1997). Efficacy of melatonin as a sleep-promoting agent. *Journal of Biological Rhythms, 12,* 644–650.

Part IV

Cell and Molecular Analysis of Circadian Rhythms

11

Cellular and Molecular Mechanisms of Circadian Regulation in the Unicellular Dinoflagellate *Gonyaulax polyedra*

J. WOODLAND HASTINGS

GONYAULAX AS A MODEL: MEASUREMENTS OF SEVERAL CLOCK-CONTROLLED RHYTHMS

Although neural mechanisms are centrally involved in the generation and functioning of circadian rhythms in animals, notably the suprachiasmatic nuclei (SCN) in the mammalian brain, robust circadian rhythms also occur in unicellular organisms, including prokaryotes (Golden, Ishiura, Johnson, & Kondo, 1997; Hastings, Boulos, & Rusak, 1991; Johnson, Golden, Ishiura, & Kondo, 1996; Kondo *et al.*, 1994; Sweeney, 1987). Such rhythms, for example in motility, represent good analogs of rhythms in higher organisms. Indeed, individual cells in the SCN also exhibit autonomous circadian oscillations (Liu, Weaver, Strogatz, & Reppert, 1997; Welsh, Logothetis, Meister, & Reppert, 1995).

Single cells are thus models of minimal complexity for understanding the basic clock mechanism and its molecular components (Hastings, 1992). The marine dinoflagellate *Gonyaulax polyedra* has a number of accessible circadian-regulated cellular processes whose peaks occur at different phases of the circadian cycle, but with similar free-running periods (Figures 1A, 1B). Several rhythms have some defined biochemical components, such as bioluminescence, photosynthesis, and

---

J. WOODLAND HASTINGS Department of Molecular and Cellular Biology, Biological Laboratories, Harvard University, Cambridge, Massachusetts 02138.

Circadian Clocks, Volume 12 of *Handbook of Behavioral Neurobiology*, edited by Joseph S. Takahashi, Fred W. Turek, and R. Y. Moore, Kluwer Academic / Plenum Publishers, New York, 2001.

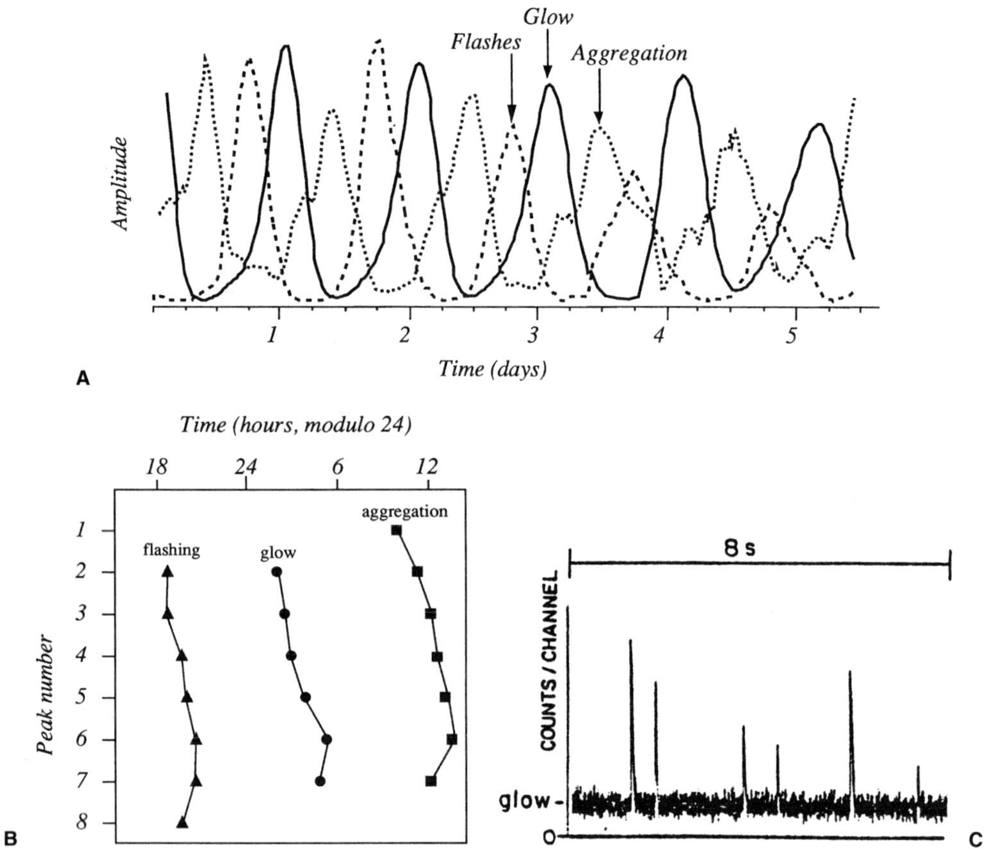

Figure 1. (A) Concurrent recordings of three different circadian rhythms from a single *Gonyaulax* culture vial in white light at 19° C. (Courtesy of Dr. Till Roenneberg.) The glow measures the luminescence intensity (excluding the flashes); flashing is the number per unit time. Aggregation, which peaks in the day phase, is calculated as the number of pixels above an arbitrary intensity threshold (Roenneberg, Colfax, & Hastings, 1989). Not illustrated are rhythms of photosynthesis and cell division. (B) A phase plot shows that free-running periods of the three are about the same and slightly greater than 24 hours under the conditions of the measurement. (C) Recording showing the time-resolved 100-msec flashes superimposed on the steady glow emission.

cell division, whereas for others (motility, aggregation), biochemical correlates are not easily specified. Melatonin, a classic marker of night phase in animals (Touitou, Arendt, & Pévet, 1993; Underwood, 1989), also exhibits a rhythm peaking at night in *Gonyaulax* (Figure 2) (Balzer & Hardeland, 1991; Pöggler, Balzer, & Hardeland, 1991).

Bioluminescence has two components, flashing and glow (Figure 1C), which can be monitored separately. Flashes (~100 msec) are triggered by an action potential conducted by the vacuolar membrane (tonoplast) enveloping the light-emitting organelles (scintillons), which project like balloons into the acidic vacuole. The action potential initiates a proton influx and the pH-dependent biochemical steps leading to flashing, which occurs with a maximum spontaneous frequency centered in the night phase. The glow is a dim emission that rises and falls gradually over

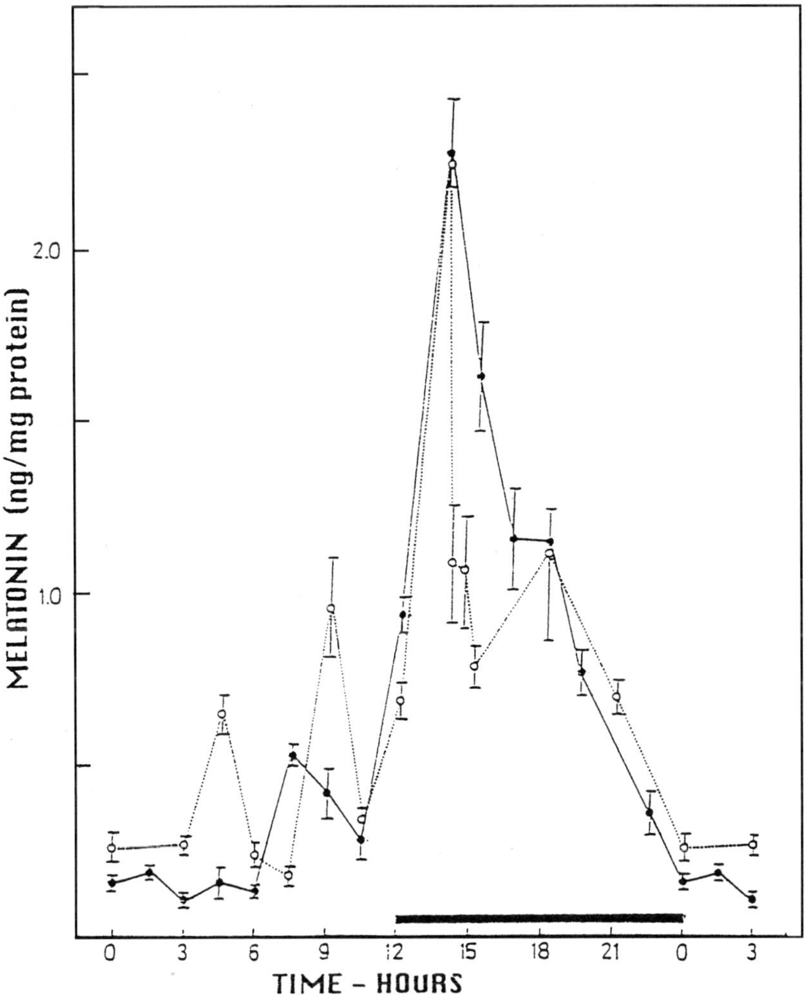

Figure 2. Melatonin concentration (ng/mg total protein) in *Gonyaulax* extracts made at different times of the 12:12 light:dark cycle (dark, black bar), determination by high-performance liquid chromatography (solid circles), or by radioimmunoassay (open circles; Pöggler *et al.*, 1991).

several hours at the end of the night phase and is postulated to result from the breakdown of the scintillons, which are formed and discarded each cycle (Fritz, Morse, & Hastings, 1990). Unlike the flashes, which may serve to deter or frighten predators (Abrams and Townsend, 1993), the glow is not believed to have ecological significance.

Effects of Environmental Conditions on Period and Phase

The behavior of the *Gonyaulax* system suggests that the temperature insensitivity of circadian oscillators is due to a compensation mechanism: the free-running period is somewhat longer at higher temperatures ($Q_{10} \approx 0.85$), indicative of overcompensation (Hastings and Sweeney, 1957). Circadian rhythmicity may also be conditional upon temperature: below 12°C, *Gonyaulax* luminescence exhibits no rhythmicity, but it resumes spontaneously upon return to a higher temperature, phased by the time of the temperature shift (Njus, McMurry, & Hastings, 1977). Higher organisms also exhibit this phenomenon (Pittendrigh, 1974, Sweeney & Hastings, 1960).

Light is typically the dominant circadian zeitgeber; indeed, circadian pacemakers in animals are related to the visual system. As in many organisms, the free-running period τ in constant light is dependent on intensity. But in *Gonyaulax*, red and blue light have opposite effects: with higher intensities, τ increases in red light, but decreases in blue light (Figure 3), implicating two different photoreceptors and transduction pathways (Roenneberg and Hastings, 1988). Blue light is 10 times more effective than red in this regard (Roenneberg and Hastings, 1991).

The phase of the clock in most organisms can also be shifted by a brief exposure to light, the responses differing depending on the time in the cycle when the light pulse is given, which defines the phase response curve (PRC) (Johnson, 1992). In *Gonyaulax*, the action spectrum for advance phase shifts by 3-hour light pulses has two peaks, at 475 and 650 nm (Figure 4); however, in this case, blue light is only about three times more effective than red (Hastings and Sweeney, 1960). But the *Gonyaulax* PRC is very dependent on background light and chemical environment.

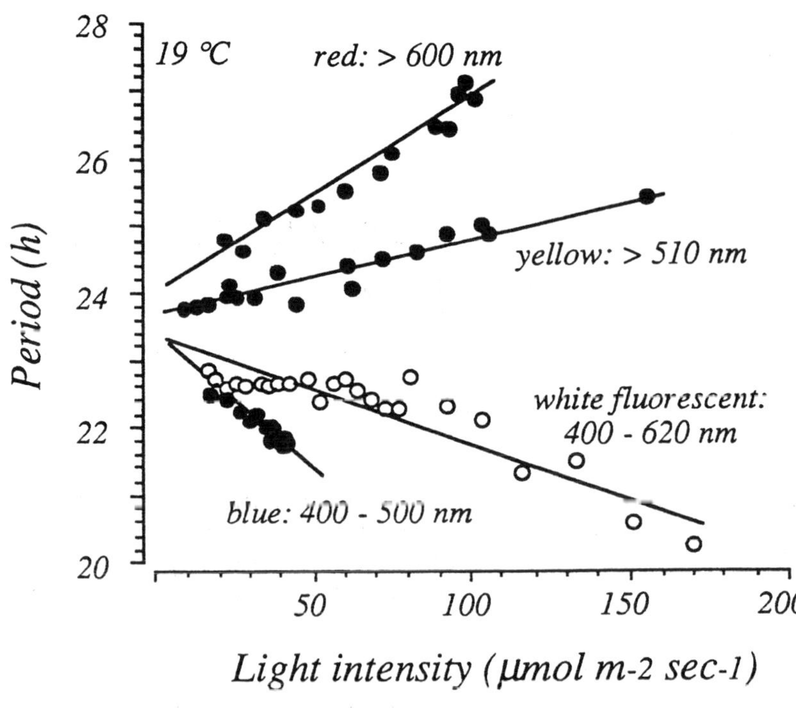

Figure 3. Effects of the intensity (abscissa) of the continuous light to which *Gonyaulax* are exposed on τ (ordinate). τ is lengthened with increasing intensities of red light, but shortened under blue light. (Roenneberg and Hastings, 1988.)

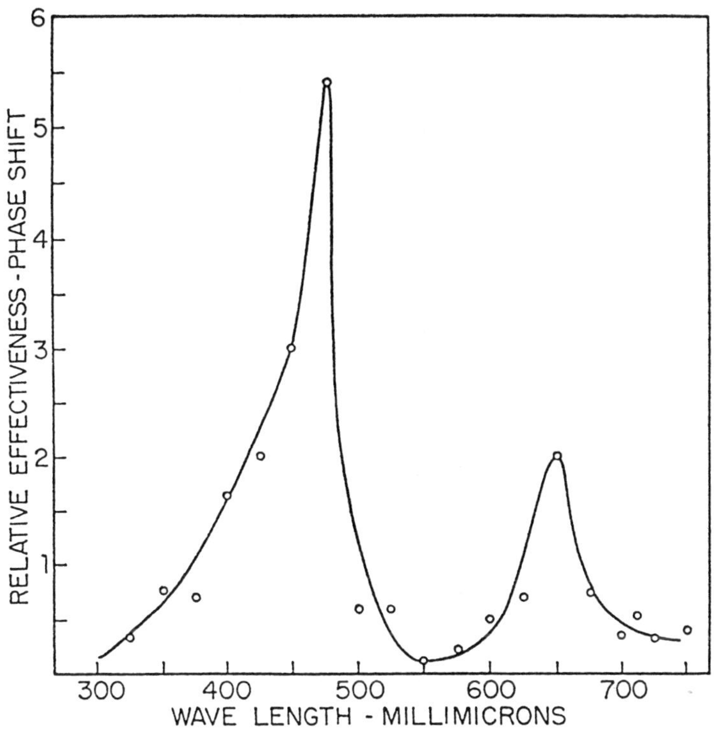

Figure 4. Action spectrum for shifting the phase of the luminescence rhythm in *Gonyaulax* by single 3-hour exposures to light of different wavelengths. (Hastings and Sweeney, 1960.)

Pulses of bright light (with cells kept otherwise in darkness) give a PRC with large (~12 hours) advances and modest (up to 5 hours) delays (Hastings and Sweeney, 1958). With a dim white light background, however, advances remain large, whereas delays are very small (< 1 hour; Johnson and Hastings, 1989). However, light pulses cause delays as great as 12 hours if the animal phosphagen creatine is present in the medium (Figure 5) (Roenneberg and Taylor, 1994).

The activity of creatine on the *Gonyaulax* circadian system was discovered in an attempt to find substances in the mammalian SCN that might affect the *Gonyaulax* clock. Added to the medium, extracts from all regions of the bovine brain resulted in large (4–5 hours) decreases in τ (Roenneberg, Nakamura, & Hastings, 1988). The substance was identified as creatine; its effect on τ (Figure 6) occurs in cells kept in constant white or blue light, but the effect is smaller in red light. Creatine is not endogenous to *Gonyaulax*, but a novel compound (gonyauline) having a similar effect has been isolated (Roenneberg *et al.*, 1991).

The protein kinase inhibitor 6-dimethylaminopurine (6-DMAP) completely blocks phase shifting by light and lengthens τ (Figures 5 and 6) (Comolli, Taylor, & Hastings, 1994), suggesting a role for protein phosphorylation in the clock. At a high concentration, the rhythm is lost, but restarts upon removal of the drug, phased to the time of drug removal. Many of the more specific kinase inhibitors are without effect, but staurosporine is effective; it appears to target only one or a few kinases

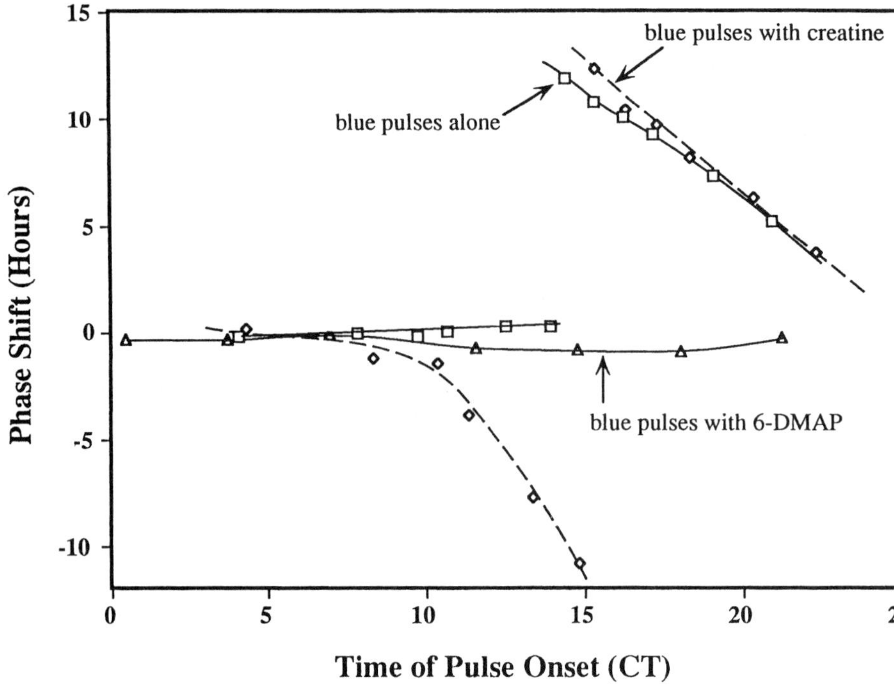

Figure 5. PRCs for 3-hour exposures to bright blue light of cells kept otherwise in dim white light at 19°C with no additions (circles), 15 μM creatine (squares), and 5 mM 6-DMAP (triangles). (Roenneberg and Taylor, 1994.)

and also increases the amplitude of the light PRC (Comolli and Hastings, 1999). Okadaic acid and other inhibitors of protein phosphatases also alter progression of the circadian oscillator, and also alter the light PRC (Comolli, Taylor, Rehman, & Hastings, 1996).

Inhibitors of 80S ribosomal protein synthesis also affect circadian rhythms in *Gonyaulax* (Taylor, Dunlap, & Hastings, 1982), as in many other species. The effects are primarily phasic in nature: pulses of anisomycin result in permanent shifts of up to 12 hours, both advances and delays (Taylor and Hastings, 1982), giving a strong (type 0) PRC at 5 μM and weak (type 1) at 2 μM (Figure 7). Protein synthesis by the cells may take hours to recover after drug treatment, but the clock is shifted promptly, as judged by the effects of second pulses of the drug (Hobohm, Cornelius, Taylor, & Rensing, 1984). Other protein synthesis inhibitors (cycloheximide, puromycin, streptimidone) have similar phase-shifting effects.

Two (or More?) Oscillators in a Unicell

In earlier studies, the phase relationships of four rhythms remained the same for about 3 weeks under constant conditions (McMurry and Hastings, 1972), as is also illustrated in Figure 1 for a week-long measurement. Later studies reported cases of different τ values for flashing and glow rhythms (Von der Heyde, Wilkens, & Rensing, 1992) and for aggregation and (Figure 8) bioluminescence rhythms (Roen-

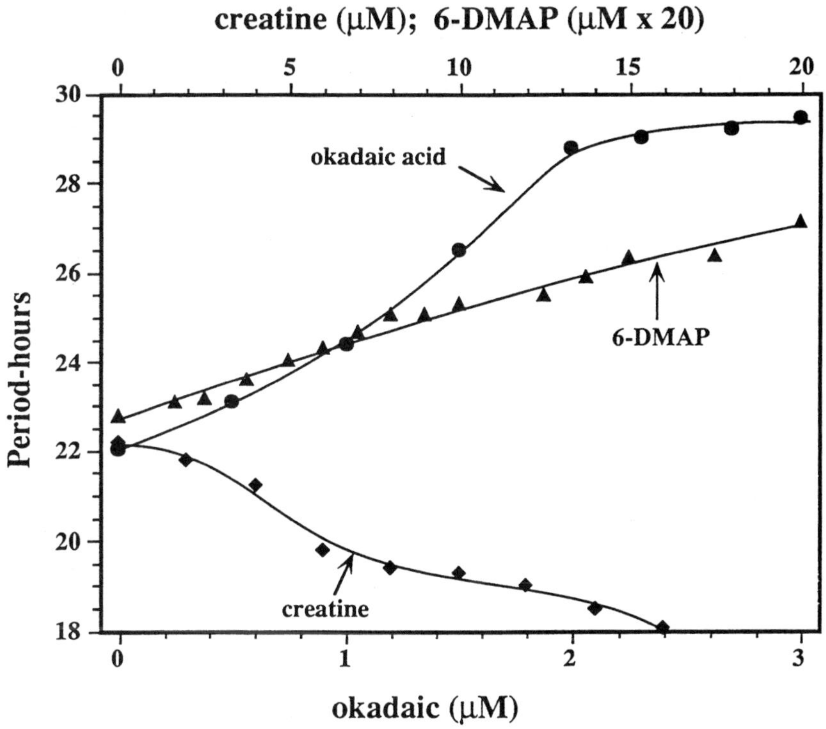

Figure 6. Effects of the concentration (abscissa) of creatine, okadaic acid, and 6-DMAP on τ (ordinate) at 19°C. (Comolli *et al.*, 1994; 1996.)

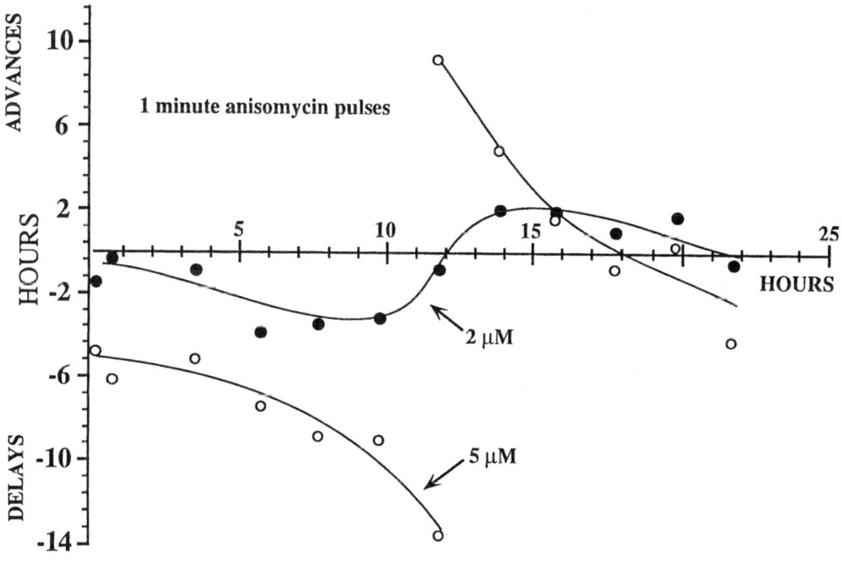

Figure 7. Drug PRCs for pulses of anisomycin on the *Gonyaulax* luminescence rhythm in constant light. A "weak" (type 1) response occurred at 2 μM, but a "strong" (type 0) response at 5 μM. (Taylor and Hastings, 1982.)

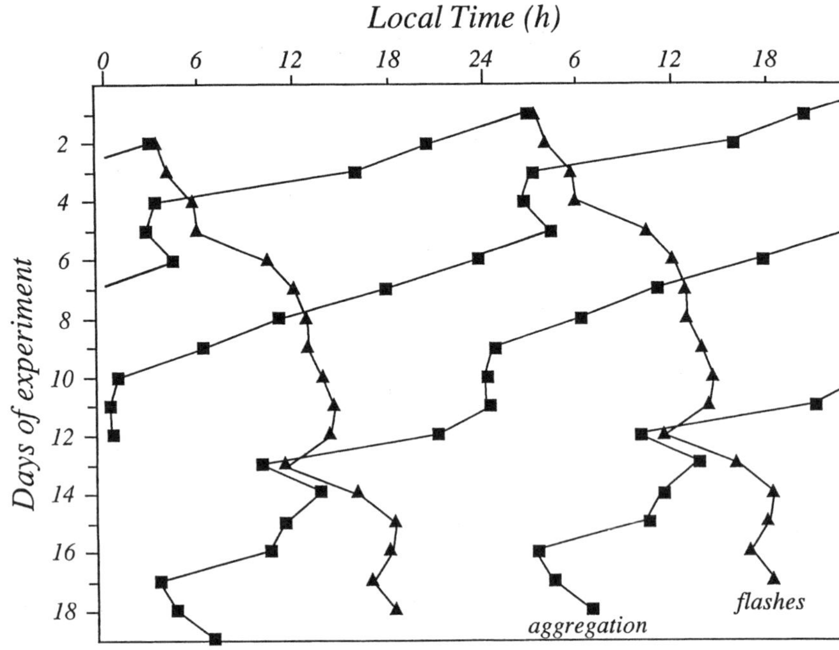

Figure 8. Dissociation of the bioluminescence and aggregation rhythms, shown by a phase plot of peak times from an experiment similar to Figure 1 except that the culture was kept in dim red light. Aggregation occurred earlier on successive days ($\tau \approx 21$ hours), whereas bioluminescence occurred slightly later each day ($\tau \approx 24.6$ hours). (Courtesy of Dr. Till Roenneberg.)

neberg and Morse, 1993). However, conditions for causing rhythm dissociation were not established. In the measurement method, the light is not truly continuous; it is interrupted every 30 minutes for about 3 minutes during measurements. These dark pulses have a significant effect on the period in white light (Broda, Gooch, Taylor, Aiuto, & Hastings, 1986). On the assumption that there are two oscillators, they might differ in sensitivity to the dark pulses and thus exhibit different free-running periods (Roenneberg and Morse, 1993).

As suggested by temporary compromises in the periods of both rhythms (Figure 8), the two putative oscillators do interact (transient coupling). Rhythms of sleep/wake and of body temperature in humans are also known to similarly dissociate and interact (relative coordination; Wever, 1979). Although the interpretation of the data is not without controversy, different oscillators (X and Y), controlling the two rhythms, have been postulated (Czeisler & Jewett, 1990; Kronauer, Czeisler, Pilato, Moore-Ede, & Weitzman, 1982).

Circadian Regulation of Bioluminescence in *Gonyaulax* Involves Translational Control

The nature of circadian oscillations may be elucidated by knowledge of how an oscillator controls an "output" system (Dunlap, 1993; Friesen, Block, & Hocker, 1993; Takahashi & Kornhauser, 1993). For many rhythms, it is difficult to specify the biochemical identity of clock-controlled component(s), but bioluminescence proteins in *Gonyaulax* are known and readily assayed. The *in vitro* activities of luciferin-binding protein (LBP) and luciferase (LCF) are correlated with the *in vivo* rhythm, peaking at night. Both proteins cycle in amount (Johnson *et al.*, 1984; Morse, Milos, Roux, & Hastings, 1989), indicative of *de novo* synthesis and destruction (Figure 9).

Not only are these proteins degraded; the entire organelle is destroyed and reformed during each cycle. There are about 400 scintillons per cell at night, but 10-fold fewer by day (Fritz *et al.*, 1990). This represents an important aspect of circadian control, namely ultrastructural changes. Cellular architecture can be strikingly altered, as, for example the chloroplast morphology in *Gonyaulax* (Rensing, Taylor, Dunlap, & Hastings, 1980).

Regulation of protein synthesis often occurs at the transcriptional level, mediated by the timed production and degradation of specific mRNAs. However, Northern blots have shown that the amounts of *lbp* and *lcf* mRNA remain constant during the circadian cycle (Figure 9) (Mittag, Eckerskorn, Strupat, & Hastings, 1997; Morse, Milos, Roux, & Hastings, 1989). It is thus concluded that the circadian control of these proteins is exerted at the level of translation.

Even though the amount of message remains the same, structural changes over time could alter its ability to be translated. However, *in vitro* synthesis rates in a heterologous (rabbit reticulocyte) system are the same with *lbp* mRNAs isolated at

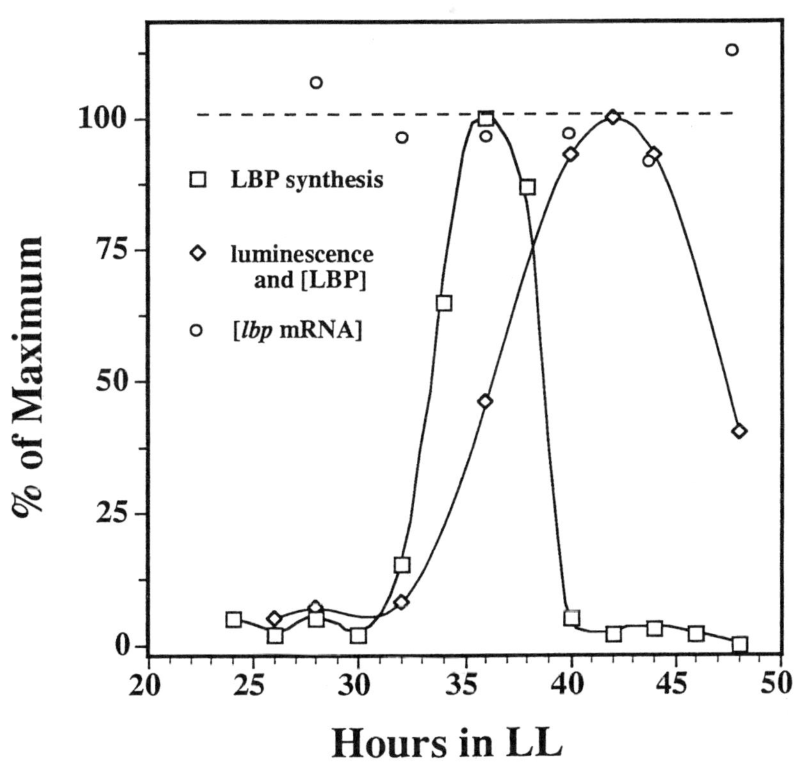

Figure 9. The circadian rhythm in the amount of *Gonyaulax* LBP parallels the *in vivo* bioluminescence. A pulse of LBP synthesis occurs during the late subjective day and early night, but *lbp* mRNA levels remain constant. (Morse *et al.*, 1989.)

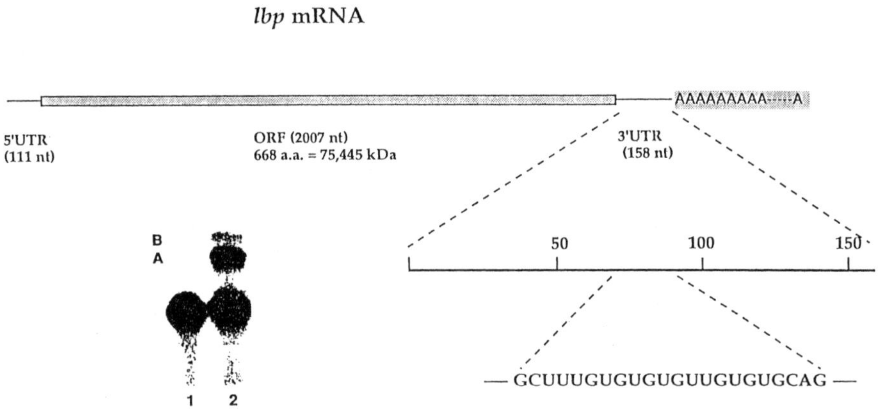

Figure 10. The *lbp* mRNA structure showing the location and nucleotide sequence of the region in the 3′ UTR to which a 45-kD protein binds. Inset: Gel retardation assay showing the electrophoretic migration of a ^{32}P-labeled RNA transcript of the *lbp* 3′ UTR without (lane 1) and with (lane 2) crude *Gonyaulax* extract. The binding of protein (A) is specific. (M. Mittag and J. W. Hastings, unpublished.)

different time points. This indicates that a cytoplasmic activator or repressor may be involved in the translation of *lbp* mRNA.

To investigate this possibility, the complete *lbp* cDNA, including the 5′ and 3′ untranslated regions (UTRs) (Figure 10), was cloned and sequenced (Lee, Mittag, Sczekan, Morse, & Hastings, 1993). Mobility shift assays were used to look for protein factor(s) binding to the UTRs of the mRNA (Mittag, Lee, & Hastings, 1994). With a transcript covering the entire 5′ UTR, no specific binding proteins were found. However, two proteins were detected which bind to the *in vitro* transcript of the 3′ UTR (Figure 10, inset). The more prominent of these (A) binds specifically, has a molecular weight of ~45 kD, and may exist as a dimer. Its RNA-binding site was localized to within a 22-nucleotide (nt)-long region of the *lbp* 3′ UTR, distinctive by virtue of an unusual sequence containing seven U(U)G repeats (not found in any other *Gonyaulax* clock-controlled proteins sequenced to date). The binding activity of protein A cycles on a daily basis and is greatest during the day phase, suggesting that it functions as a clock-controlled repressor, preventing the translation of *lbp* mRNA during the day. This might be expected to act at the level of initiation, and thus involve the 5′ region of the message. Indeed, differential translational initiation of *lbp* mRNA appears to occur due to an open reading frame in the 5′ UTR (Mittag *et al.*, 1997).

Pulse labeling of *Gonyaulax* proteins at times 12 hours apart, followed by two-dimensional gel analysis, showed that many are preferentially synthesized at one phase or the other, and regulated translationally (Milos *et al.*, 1990). Determined at more frequent time points (Markovic, Roenneberg, & Morse, 1996), 10 of these were shown to fall into three time frames, with about 10-fold differences from peak to trough (Figure 11). But while their synthesis rates are circadian controlled, the cellular concentrations of some of the proteins do not exhibit pronounced rhythms, indicating that they have long turnover times.

Although the expression and/or activities of many proteins in different organisms are circadian controlled, they are surely not all regulated in the same way. In addition to translational control, there may be posttranslational (Edery, Zwiebel,

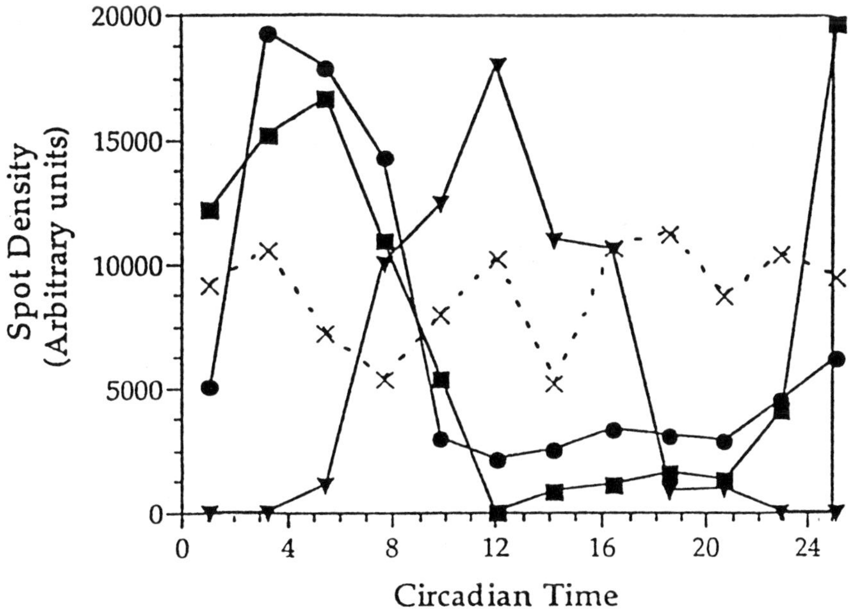

Figure 11. Pulse labeling of four proteins resolved by two-dimensional polyacrylamide gel electrophoresis, determined by radiolabeling for 20 minutes. One is a nonregulated protein (X); the triangles (p75) are LBP; the squares are ribulose 1,5-biphosphate carboxylase (p55); and the circles are peridinin-chlorphyll-*a*-binding protein (p32). (Marcovic *et al.*, 1996.)

Dembinska, & Rosbash, 1994) or transcriptional regulation (Aronson, Johnson, Loros, & Dunlap, 1994; Hardin, Hall, & Rosbash, 1990; Sehgal, Price, Man, & Young, 1994); translocation and compartmentalization may also be involved. Indeed, in a different dinoflagellate, *Pyrocystis lunula*, whose luminescence is also circadian controlled (Colepicolo, Roenneberg, Morse, Taylor, & Hastings, 1993), it has recently been shown that the luciferase is not synthesized and is destroyed each day (Knaust, Urbig, Li, Taylor, & Hastings, 1998). Instead, it is translocated to a different cellular compartment, where it is not coupled to the flash generation mechanism (Widder & Case, 1982). The circadian control must therefore be mediated via different protein(s). An understanding of the circadian clock mechanism itself will be greatly enhanced as the circadian regulatory mechanisms for different specific proteins, whatever they may be, become known.

Acknowledgments

I am indebted to the members of my research group, past and present, for their many contributions to the work described. The studies have been supported in part by grants from the NIH and NSF.

References

Abrams, M. W., Townsend, L. D. (1993). Bioluminescence in dinoflagellates: A test of the burglar alarm hypothesis. *Ecology, 74*, 258–260.

Aronson, B. D., Johnson, K. A., Loros, J. J., & Dunlap, J. C. (1994). Negative feedback defining a circadian clock: Autoregulation of the clock gene *frequency*. *Science, 263*, 1578–1584.

Balzer, I., & Hardeland, R. (1991). Photoperiodism and effects of indoleamines in a unicellular alga, *Gonyaulax polyedra*. *Science, 253*, 795–797.

Broda, H., Gooch, V. D., Taylor, W., Aiuto, N., & Hastings, J. W. (1986). Acquisition of circadian bioluminescence data in *Gonyaulax* and an effect of the measurement procedure on the period of the rhythm. *Journal of Biological Rhythms, 1*, 251–263.

Colepicolo, P., Roenneberg, T, Morse, D., Taylor, W., & Hastings, J. W. (1993). Circadian regulation of bioluminescence in the dinoflagellate *Pyrocystis lunula*. *Journal of Phycology, 29*, 173–179.

Comolli, J., & Hastings, J. W. (1999). Novel effects on the *Gonyaulax* circadian system produced by the protein kinase inhibitor staurosporine. *Journal of Biological Rhythms, 14*, 11–19.

Comolli, J., Taylor, W., & Hastings, J. W. (1994). Inhibitor of protein phosphorylation lengthens τ and blocks light-induced phase shifting in *Gonyaulax polyedra*. *Journal of Biological Rhythms, 9*, 13–26.

Comolli, J., Taylor, W., Rehman, J., & Hastings, J. W. (1996). Inhibitors of serine/threonine phosphoprotein phosphatases alter circadian properties in *Gonyaulax polyedra*. *Plant Physiology, 111*, 285–291.

Czeisler, C. A., & Jewett, M. (1990). Human circadian physiology: Interaction of the behaviorial rest–activity cycle with the output of the endogenous circadian pacemaker. In M. Thorpy (Ed.), *Handbook of sleep disorders* (pp. 117–137). New York: Marcel Dekker.

Dunlap, J. C. (1993). Genetic analysis of circadian clocks. *Annual Review of Physiology, 55*, 683–728.

Edery, I., Zwiebel, L. J., Dembinska, M. E., & Rosbash, M. (1994). Temporal phosphorylation of the *Drosophila* period protein. *Proceedings of the National Academy of Sciences of the USA, 91*, 2260–2264.

Friesen, W. O., Block, G. D., & Hocker, C. G. (1993). Formal approaches to understanding biological oscillators. *Annual Review of Physiology,55*, 661–681.

Fritz, L., Morse, D., & Hastings, J. W. (1990). The circadian bioluminescence rhythm of *Gonyaulax* is related to daily variations in the number of light emitting organelles. *Journal of Cell Science, 95*, 321–328.

Golden, S. S., Ishiura, M., Johnson, C. H., & Kondo, T. (1997) Cyanobacterial circadian rhythms. *Annual Review of Plant Physiology and Plant Molecular Biology, 48*, 327–354.

Hardin, P. E., Hall, M., & Rosbash, M. (1990). Feedback of the *Drosophila* period gene product on circadian cycling of its messenger RNA levels. *Nature, 343*, 536–540.

Hastings, J. W. (1992). The cellular clock in a dinoflagellate: Translational control of a circadian regulated protein. *Discussions in Neurosciences, 8*, 98–102.

Hastings, J. W., & Sweeney, B. M. (1957). On the mechanism of temperature independence in a biological clock. *Proceedings of the National Academy of Sciences of the USA, 43*, 804–811.

Hastings, J. W., & Sweeney, B. M. (1958). A persistent diurnal rhythm of luminescence in *Gonyaulax polyedra*. *Biological Bulletin, 115*, 440–458.

Hastings, J. W., & Sweeney, B. M. (1960). The action spectrum for shifting the phase of the rhythm of luminescence in *Gonyaulax polyedra*. *Journal of General Physiology, 43*, 697–706.

Hastings, J. W., Boulos, Z., & Rusak, B. (1991). Circadian rhythms. In C. L. Prosser (Ed.), *Neural and integrative animal physiology* (pp. 435–546). New York: Wiley Interscience.

Hobohm, U., Cornelius, G., Taylor, W., & Rensing, L. (1984). Is the circadian clock of *Gonyaulax* held stationary after a strong pulse of anisomycin? *Comparative Biochemistry and Physiology, 79A*, 371–378.

Johnson, C. H. (1992). Phase response curves: What can they tell us about circadian clocks? In T. Hiroshige and K. Honma (Eds.), *Circadian clocks from cell to human* (pp. 209–249). Sapporo, Japan: Hokkaido University Press.

Johnson, C. H., & Hastings, J. W. (1989). Circadian phototransduction: Phase resetting and frequency of the circadian clock of *Gonyaulax* cells in red light. *Journal of Biological Rhythms, 4*, 417–437.

Johnson, C. H., Roeber, J., & Hastings, J. W. (1984). Circadian changes in enzyme concentration account for rhythm of enzyme activity in *Gonyaulax*. *Science, 223*, 1428–1430.

Johnson, C. H., Golden, S. S., Ishiura, M., & Kondo, T. (1996). Circadian clocks in prokaryotes. *Molecular Microbiology, 21*, 5–11.

Knaust, R., Urbig, T., Li, L., Taylor, W., & Hastings, J. W. (1998). The circadian rhythm of bioluminescence in *Pyrocystis* is not due to differences in the amount of luciferase: A comparative study of the three bioluminescent marine dinoflagellates. *Journal of Phycology, 34*, 167–172.

Kondo, T., Tsinoremas, Golden, S. S., Johnson, C. H., Kutsuna, S., & Ishiura, M. (1994). Circadian clock mutants of cyanobacteria. *Science, 266*, 1233–1236.

Kronauer, R. E., Czeisler, C. A., Pilato, S. F., Moore-Ede, M. C., & Weitzman, E. D. (1982). Mathematical model of the human circadian system with two interacting oscillators. *American Journal of Physiology, 242*, R3–R17.

Lee, D., Mittag, M., Sczekan, S., Morse, D., & Hastings, J. W. (1993). Molecular cloning and genomic organization of a gene for luciferin-binding protein from the dinoflagellate *Gonyaulax polyedra. Journal of Biological Chemistry, 268,* 8842–8850.

Liu, C., Weaver, D. R., Strogatz, S. H., & Reppert, S. M. (1997). Cellular construction of a circadian clock: Period determination in the suprachiasmatic nuclei. *Cell, 91,* 855–860.

Markovic, P., Roenneberg, T., & Morse, D. (1996). Phased protein synthesis at several circadian times does not change protein levels in Gonyaulax. *Journal of Biological Rhythms, 11,* 57–67.

McMurry, L., & Hastings, J. W. (1972). No desynchronization among four circadian rhythms in the unicellular alga, *Gonyaulax polyedra. Science, 175,* 1137–1139.

Milos, P., Morse, D., & Hastings, J. W. (1990). Circadian control over synthesis of many *Gonyaulax* proteins is at the translational level. *Naturwissenschaften, 77,* 87–89.

Mittag, M., Lee, D.-H., & Hastings, J. W. (1994). Circadian expression of the luciferin-binding protein correlates with the binding of a protein to its 3′ untranslated region. *Proceedings of the National Academy of Sciences of the USA, 91,* 5257–5261.

Mittag, M., Eckerskorn, C., Strupat, K., & Hastings, J. W. (1997). Differential translational initiation of lbp mRNA is caused by a 5′ open reading frame. *FEBS Letters, 411,* 245–250.

Morse, D., Milos, P. M., Roux, E., & Hastings, J. W. (1989). Circadian regulation of the synthesis of substrate binding protein in the *Gonyaulax* bioluminescent system involves translational control. *Proceedings of the National Academy of Sciences of the USA, 86,* 172–176.

Njus, D., McMurry, L., & Hastings, J. W. (1977). Conditionality of circadian rhythmicity: Synergistic effect of light and temperature. *Journal of Comparative Physiology, 117,* 335–344.

Pittendrigh, C. S. (1974). Circadian oscillations in cells and the circadian organization of multicellular systems. In F. O. Schmitt and F. G. Warden (Eds.), *The neurosciences, Third study program* (pp. 437–458). Cambridge, MA: MIT Press.

Pöggler, B., Balzer, I., & Hardeland, R. (1991). Pineal hormone melatonin oscillates also in the dinoflagellate *Gonyaulax polyedra. Naturwissenschaften, 78,* 268–269.

Rensing, L., Taylor, W., Dunlap J. C., & Hastings, J. W. (1980). The effects of protein synthesis inhibitors on the *Gonyaulax* clock. II: The effect of cyclohexamide on ultrastructural parameters. *Journal of Comparative Physiology, 138,* 9–18.

Roenneberg, T., & Hastings, J. W. (1988). Two photoreceptors control the circadian clock of a unicellular alga. *Naturwissenschaften, 75,* 206–207.

Roenneberg, T., & Hastings, J. W. (1991). Are the effects of light on the phase and period of the *Gonyaulax* clock mediated by different pathways? *Photochemistry and Photobiology, 53,* 525–533.

Roenneberg, T., & Morse, D. (1993). Two circadian oscillators in one cell. *Nature, 362,* 362–364.

Roenneberg, T., & Taylor, W. (1994). Light induced phase responses in *Gonyaulax* are drastically altered by creatine. *Journal of Biological Rhythms, 9,* 1–12.

Roenneberg, T., Nakamura, H., & Hastings, J. W. (1988). Creatine accelerates the circadian clock in the unicellular alga *Gonyaulax polyedra. Nature, 334,* 432–434.

Roenneberg, T., Colfax, G. N., & Hastings, J. W. (1989) A circadian rhythm of population behavior in *Gonyaulax polyedra. Journal of Biological Rhythms, 4,* 201–216.

Roenneberg, T., Nakamura, H., Cranmer, L. D., Ryan, K., Kishi, Y., & Hastings, J. W. (1991). Gonyauline: A novel endogenous substance shortening the period of the circadian clock of a unicellular alga. *Experientia, 47,* 103–106.

Sehgal, A., Price, J. L., Man, B., & Young, M. W. (1994). Loss of circadian behavioral rhythms and *per* RNA oscillations in the *Drosophila* mutant *timeless. Science, 263,* 1603–1606.

Sweeney, B. M. (1987) *Rhythmic phenomena in plants.* New York: Academic Press.

Sweeney, B. M., & Hastings, J. W. (1960). Effects of temperature upon diurnal rhythms. *Cold Spring Harbor Symposia on Quantitative Biology, 25,* 87–104.

Takahashi, J. S., & Kornhauser, J. M. (1993). Molecular approaches to understanding circadian oscillations. *Annual Review of Physiology, 55,* 729–753.

Taylor, W. R., & Hastings, J. W. (1982). Minute-long pulses of anisomycin phase-shift the biological clock in *Gonyaulax* by hours. *Naturwissenschaften, 69,* 94–96.

Taylor, W. R., Dunlap, J. C., & Hastings, J. W. (1982). Inhibitors of protein synthesis on 80s ribosomes phase shift the *Gonyaulax* clock. *Journal of Experimental Biology, 97,* 121–136.

Touitou, Y., Arendt, J., & Pévet, P. (Eds.). (1993). *Melatonin and the pineal gland. From basic science to clinical application.* New York: Elsevier.

Underwood, H. (1989). The pineal and melatonin: Regulators of circadian function in lower vertebrates. *Experientia, 45,* 914–922.

Von der Heyde, F., Wilkens, A., & Rensing, L. (1992). The effects of temperature on the circadian rhythms of flashing and glow in *Gonyaulax polyedra*: Are the two rhythms controlled by two oscillators? *Journal of Biological Rhythms, 7,* 115–123.

Welsh, D. K., Logothetis, D. E., Meister, M., & Reppert, S. M. (1995). Individual neurons dissociated from rat suprachiasmatic nucleus express independenlty phased circadian firing rhythms. *Neuron, 144,* 697–706.

Wever, R. A. (1979). *The circadian system of man.* Heidelberg: Springer-Verlag.

Widder, E. A., & Case, J. F. (1982). Distribution of subcellular bioluminescent sources in a dinoflagellate, *Pyrocystis fusiformis. Biological Bulletin, 162,* 423–448.

12

Molecular Genetics of Circadian Rhythms in *Neurospora*, a Prototypic Circadian System

JAY C. DUNLAP AND JENNIFER J. LOROS

Earlier sections of this book have established that circadian systems are complicated. With this section, we try to cut through the complexity to the basic molecular issues defining and governing circadian timing at the biochemical level. We recall the three great questions in circadian biology: (1) *Input*—how do environmental signals reset the clock? (2) *The oscillator*—what are the gears and cogs in the feedback loop generating time and how is the loop assembled? (3) *Output*—how is time information that is generated by the clock then transduced to effect changes in the behavior of the organism? Answers to all of these questions are attainable at the level of proteins, RNAs, genes, and small molecules. It is widely accepted as a tenet of circadian biology that intercellular communication plays no obligate role in the generation of circadian rhythmicity, although it is surely involved in the entrainment and expression of rhythms in multicellular organisms. Clearly the cell is the basic unit timer. Current work on prototypic cellular circadian systems promises to establish the paradigms that will govern the way we approach the molecular dissection of more complex systems in the same way that Pittendrigh's early work on the resetting responses of *Drosophila pseudoobscura* governed the way an entire generation thought about how to determine the formal characteristics of circadian pacemakers. Thus, model systems are of interest not only because that tell us concretely how more complex systems will work, but also because they teach us how to think about problems in more complex systems.

Over the past few years there has continued to be rapid progress in the molecu-

JAY C. DUNLAP AND JENNIFER J. LOROS Department of Biochemistry, Dartmouth Medical School, Hanover, New Hampshire 03755-3844.

Circadian Clocks, Volume 12 of *Handbook of Behavioral Neurobiology*, edited by Joseph S. Takahashi, Fred W. Turek, and R. Y. Moore, Kluwer Academic / Plenum Publishers, New York, 2001.

lar dissection of circadian clocks, and our expectation that the insights derived from model systems would be informative has been amply borne out. Although parts of this review would be written differently now, many of the perspectives here are still pertinent even though the literature survey for this review, with the exception of this paragraph, was completed with the text in 1997. Among the advances since this time, a mechanism for temperature resetting has been proposed based on post-transcriptional regulation of FRQ (Liu *et al.*, 1998), the central importance of FRQ phosphorylation to its degradation and the determination of the period length of the clock has been established (Liu *et al.*, 2000), and the direct protein interactions among FRQ, WC-1, and WC-2 predicted in the models now formally demonstrated (Denault *et al.*, 2001). More broadly, clear parallels have been shown between the light resetting mechanism first elucidated for Neurospora and that now understood for mammals (Shigeyoshi *et al.*, 1997), and a second interconnected feedback loop, based on the positive regulation of WC-1 synthesis by FRQ, has been revealed (Lee *et al.*, 2000). Reviews of this progress can be found in (Dunlap, 1999; Iwasaki and Dunlap, 2000; Loros and Dunlap, 2001).

A Prototypic System for Understanding the Clock at the Level of the Cell

Given the cellular basis of the clock, the ubiquitous nature of these rhythms, and their shared characteristics, any student of modern molecular biology would be comfortable in asserting that common elements and mechanisms will be found working in all circadian clocks. For this reason, microbial systems, and systems amenable to genetic analysis, have long been appreciated for the insights they can provide into the mechanism of the clock and the means by which clocks act to regulate the metabolism and behavior of cells. The heterothallic filamentous fungus *Neurospora crassa* is one of a handful of valuable model organisms that have been extensively investigated, especially in genetic analysis, yielding numerous insights into the workings of the cell, including Beadle and Tatum's (1945) original genetic observation that a single genetic locus corresponded to a single enzyme, the "one gene, one polypeptide" hypothesis that underlies our understanding of biochemistry and molecular regulation. A tantalizing 1953 report on a zonation in the growth habits of *Neurospora* caught the attention of Colin Pittendrigh and his colleagues, who in 1959 reported that this "zonation" in growth was present in constant conditions of temperature and darkness, but not in constant light (Pittendrigh, Bruce, Rosenzweig, & Rubin, 1959). Moreover the periodicity of about 22 hours was maintained at both 24°C and 31°C. In this paper they state "The *Neurospora* system appears potentially valuable for an attack on the basic problem of the nature of oscillations that constitute the circadian clock. We do not interpret the presence or absence of zonation as a reflection of the presence or absence of the clock, but rather as a coupling or uncoupling from the clock of the particular processes leading to zonation." These were prophetic words, as *Neurospora* has indeed been developed into a basic genetic and molecular model for the biological clock whose study has resulted in a substantial understanding of how oscillators are built, how they interact with environmental signals like light and temperature that reset the oscillator, and, finally, how an oscillator feeds time information out into the cell to control cell behavior through control of gene regulation.

The growth zonation is now understood as a time-dependent change in development whereby undifferentiated fungal mycelia growing at a relatively constant rate on the surface of a growth medium will differentiate, yielding a second cell type, aerial hyphae, directed perpendicular to the surface and to the plane of mycelial growth. These can then elaborate a third cell type, fungal spores (conidiospores or conidia). The traditional assay for the clock in *Neurospora* is by growth in long, medium-containing glass "race tubes" (Figure 1). When grown on these race tubes, during the circadian cycle in the late subjective night, a recurrent developmental switch for conidiation is activated that eventually results in the production of asexual conidial spores. Once initiated by the clock, this developmental process is largely inflexible and will proceed with its own kinetics resulting in the production of conidiospores some hours later. Meanwhile circadian time progresses, and later, at a phase roughly corresponding to the middle of the subjective day, this developmental switch is turned off, surface mycelia are no longer triggered to differentiate, and a region of undifferentiated mycelia is laid down on the solid medium of the race tube. Each day, this cycle of growth and differentiation recurs in the absence of environmental input; the period length of the wild-type rhythm in *Neurospora* is about 22 hours at 25°C. This clock control of asexual development displays all the formal characteristics of clocks found in other systems including temperature compensation and entrainment by light. Although other daily rhythms in *Neurospora* have been identified including spore shooting, CO_2 evolution, and various metabolic rhythms (reviewed in Dunlap, 1993; Feldman & Dunlap, 1983; Lakin-Thomas, Coté, & Brody, 1990), the race tube assay remains the principal tool of choice for routine monitoring of rhythmicity due to its ease of execution. This developmental program, however, is exhibited only when cultures are grown on solid media, and these are conditions that are not suitable for the preparation of tissue for biochemical or molecular biological analysis. For these sorts of experiments, rhythmic liquid cultures (Aronson, Johnson, Loros, & Dunlap, 1994a; Loros & Dunlap, 1991; Loros, Denome, & Dunlap, 1989; Nakashima, 1981) are generated first by growing a my-

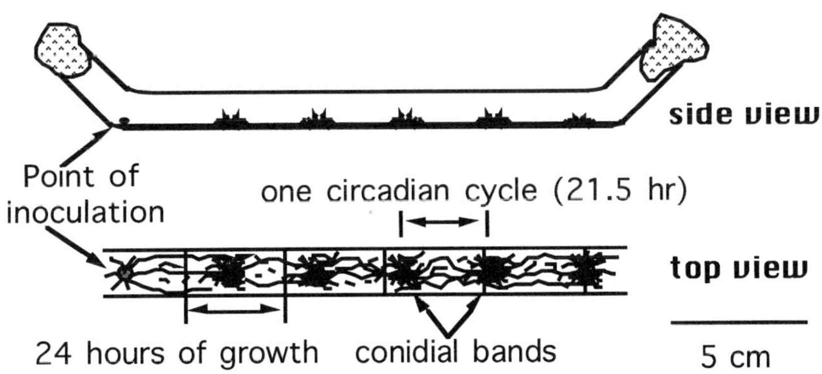

Figure 1. A schematic view of a race tube. Asexual spores (conidia) are inoculated onto an agar medium at one end of the 40-cm-long glass tube. Following a period of growth in constant light, transfer of the tube to darkness sets the clock to subjective dusk. Development and the production of asexual spores occurs periodically. Since growth rate is relatively constant, daily marks at 24-hour intervals allow distance along the tube to be correlated with time, thus allowing the calculation of period length and phase.

celial mat in liquid and then cutting disks from the mat and culturing these in nutrient-poor medium. Each disk contains an identical clock that can be reset by light or temperature, placed on a race tube to initiate an in-phase race tube culture, or fractionated for the examination of RNAs and proteins as a function of time.

The cells and tissues executing these temporal programs contain the circadian clocks under study. Figure 2 is used to keep sight of the larger picture of molecular timing in the rhythmic cell, and can be used to describe each of the separate research approaches to understanding molecular rhythmicity. One avenue has been to identify and characterize clock components through the identification, cloning, and characterization of genes that, when mutated, affect the expression of rhythmicity. To this end, seven genes have been identified in genetic screens for clock-affecting genes, two other genes required for rhythmicity have been found, and the involvement of three of these (*frequency*, *white collar-1*, and *white collar-2*) have been characterized in some detail (Dunlap, 1993, 1996; Loros, 1995) A second approach has been aimed at understanding how environmental cues such as light and temperature act to synchronize internals cycles with external time cues. In this work, existing strains defective in blue-light photoresponses in *Neurospora* have yielded insights into how light acts to reset the pacemaker mechanism and also how the mechanism itself is assembled. Third, the pathways through which clocks act to control cellular metabolism and behavior are being dissected. Initially, genes whose transcript levels are controlled by the clock were identified, and two such genes, designated *clock controlled gene-1* (*ccg-1*) and *clock-controlled gene-2* (*ccg-2*), arose from a

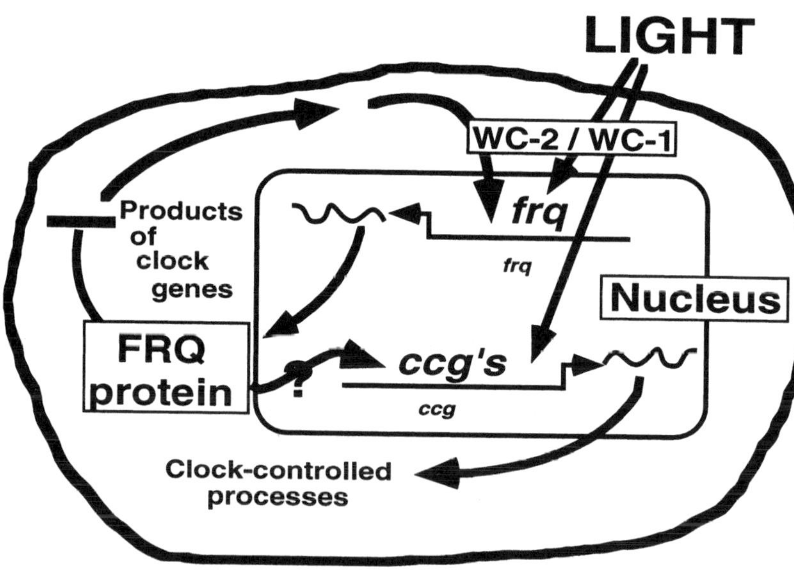

Figure 2. A schematic view of the rhythmic cell. The clock mechanism is a negative feedback oscillator in which the FRQ proteins may have several functions including downregulation of *frq* mRNA. FRQ acts either directly or indirectly to transcriptionally regulate rhythmic clock-controlled gene expression. Identification of components in the light entrainment pathway such as the *wc* genes has provided clues to how clocks perceive light and how light acts to reset the clock. The *wc* genes are also required for sustained rhythmicity and act as positive elements in the feedback loop, promoting *frq* expression in both light and dark. (Figure adapted from Aronson *et al.*, 1994a.)

subtractive hybridization procedure (Loros *et al.*, 1989). Six additional genes have now been identified through the use of differential screening of time-specific libraries (Bell-Pedersen, Shinohara, Loros, & Dunlap, 1996b).

The Oscillator Includes a Negative Autoregulatory Feedback Loop Involving the Products of the *Frequency* Gene

A variety of genes have been identified in *Neurospora* having the characteristic that when these genes are mutated, the period length and sometimes the temperature compensation properties of the clock are altered. Many were identified by Feldman and colleagues in screens for genes affecting the operation of the clock (Feldman, 1982; Feldman, Gardner, & Dennison, 1979; Morgan & Feldman, 1997), although nearly all workers in the *Neurospora* rhythms field have contributed genes or alleles to this set (reviewed in Dunlap, 1993, 1996). One locus, called *frequency* (*frq*), is identified by seven independently isolated alleles, and the other six genes, *chr*, and *prd 1, 2, 3, 4,* and *6,* are identified by single alleles. A number of additional genes isolated in other contexts might be classified as clock affecting (see, for instance, Dunlap, 1993), although mutations in these genes result in generally small, albeit real, effects on the clock.

How the Oscillator Works

Most of what is known in *Neurospora* about how the oscillator actually works is based on genetic and molecular studies that have illuminated the actions and regulation of *frq,* a gene encoding central components of a feedback loop that is essential for normal operation of this circadian clock (Aronson *et al.*, 1994a; Aronson, Johnson, & Dunlap, 1994b; Dunlap, 1996). The circadian oscillator in *Neurospora* includes negative autoregulatory feedback cycle (Aronson *et al.*, 1994a) in which rhythmic transcription of the *frq* gene yields transcripts that encode two forms of the protein FRQ, a longer form of 989 amino acids and a shorter form of 890 amino acids resulting from alternative initiation of translation at an internal ATG codon (Garceau, Liu, Loros, & Dunlap, 1997). Both forms of FRQ contain a functional nuclear localization signal (Luo, Loros, & Dunlap, 1998), serine-rich and hyperacidic regions (Aronson *et al.*, 1994b), and a region predicted to have structural similarity to a helix–turn–helix DNA–binding domain (Lewis & Feldman, 1993) ; all these domains are phylogenetically conserved (Lewis & Feldman, 1997; Lewis, Morgan, & Feldman, 1997; Merrow & Dunlap, 1994). No biochemical functions have been assigned to either form of FRQ and no functional distinction can be made between the two forms, so FRQ is used to refer to both forms unless otherwise specified. Both the *frq* RNAs and FRQ cycle in amount (Aronson*et al.*, 1994a; Garceau *et al.*, 1997), and FRQ acts at least indirectly to depress the level of the *frq* mRNA (Aronson *et al.*, 1994a; Merrow, Garceau, & Dunlap, 1997). Importantly, rhythmic change in the amount of *frq* transcript is required for the overt circadian rhythm on race tubes (no amount of constant *frq* expression supports the overt rhythm) and step changes in the expression of *frq* reset the clock (Aronson *et al.*, 1994a). Since rhythmic autoregulated expression, not merely unregulated *frq* expression, is essential for the circadian clock, the levels of *frq* and FRQ can be said to be not just components but state variables of the circadian oscillator, elements whose oscillation defines the clock.

In a cell kept in constant darkness and constant temperature where only internal cellular processes will affect its time course, the *Neurospora* clock cycle can be imagined beginning at subjective midnight. *frq* and FRQ levels are low, but *frq* transcript is rising, a process that takes about 10–12 hours to reach a peak and that requires the WC-2 (and maybe WC-1) proteins (Crosthwaite, Dunlap, & Loros, 1997) (see also below). Following a lag of about 4 hours, FRQ protein begins to appear (Garceau *et al.*, 1997). Since FRQ expressed from a heterologous promoter (Merrow *et al.*, 1997), or following light induction (Garceau, 1996), appears with little delay, the 4-hour lag seen in the normal cycle is probably a regulated part of the circadian cycle (Merrow *et al.*, 1997). Almost immediately after its synthesis, FRQ enters the nucleus (Luo *et al.*, 1998) where it acts either directly or indirectly to depress the expression of its own (Aronson *et al.*, 1994a; Merrow *et al.*, 1997) and possibly other genes (including perhaps some of the eight known clock-controlled genes (*ccg*'s), Bell-Pedersen *et al.*, 1996b; Loros, 1995). Nuclear localization is a required step for FRQ to function (Luo *et al.*, 1998). *frq* transcript levels peak in the midmorning (Aronson *et al.*, 1994a; Crosthwaite, Loros, & Dunlap, 1995) about 4 hours before the peak of FRQ (both forms simultaneously), which occurs in the early afternoon (Garceau, *et al.*, 1997). As soon as either form of FRQ can be seen, it is partially phosphorylated, and forms become more phosphorylated during the day. At midday, *frq* RNA levels peak, they then fall during the afternoon, and the amount of FRQ, which is now becoming extensively phosphorylated, declines through the early night. This is consistent with the phosphorylation triggering FRQ turnover. By midnight, the levels of both RNA and protein are low and the cycle begins again.

ENVIRONMENTAL INPUT ENTRAINS THE CLOCK BY AFFECTING THE LEVELS OF *FRQ* MRNA AND FRQ PROTEIN

LIGHT

Light is perceived in *Neurospora* through the action of an as-yet-unidentified blue light photoreceptor that apparently utilizes flavin as a chromophore (Paietta & Sargent, 1981). Light resets the clock by acting within minutes to increase the level of *frq* transcript level; the dose response for clock resetting matches the dose response for this light induction and shows reciprocity between intensity and duration (Crosthwaite *et al.*, 1995). For successful entrainment of an internal clock with the environment, the same light signal must elicit different responses at different times of day: light in the early subjective morning should be interpreted as an early dawn and advance the clock into the next day, whereas light in the early subjective evening should be interpreted as a long day and delay the clock by moving it back into the previous daytime (Crosthwaite *et al.*, 1995; Pittendrigh, 1976, 1993). In broad outline, the response of *frq* to light explains the organismal resetting response—light in the late night/early morning acts rapidly to bring *frq* transcript and FRQ levels up to their peak, a process that, in the absence of light, takes several hours (Figure 3). This precocious rise causes an advance in the rhythm. Conversely, late-day and early-evening delays in the rhythm result from the light induction of *frq* that returns *frq* levels to the peak amounts from which they have declined over the past hours. Newly synthesized *frq* mRNA gives rise to new FRQ protein, which begins the slow process of phosphorylation and eventual turnover.

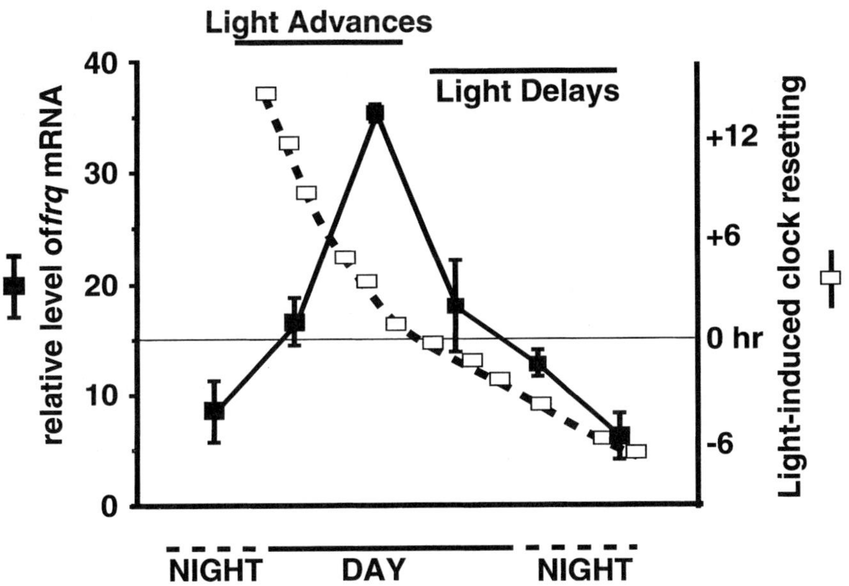

Figure 3. The *Neurospora* phase response curve can be predicted from the response of *frq* to light. The response of the circadian oscillator to light (the PRC; thick solid line from Dharmananda, 1980) is correlated in time with the molecular cycling in the level of expression of the *frq* locus (thin solid line, *frq* mRNA as measured, plotted as mean ± SEM; $n = 3$). The times when light causes an advance or a delay can be predicted simply from the known response of the *frq* gene to light, and the knowledge that the clock is reset by changes in the level of *frq* expression. See text for details. (Figure adapted from Crosthwaite *et al.*, 1995.)

An Intimate Connection between the Entrainment Pathway and the Oscillator Provides Clues to the Origins of Circadian Rhythmicity and the Identity of Clock Components

Efforts to dissect the signal transduction pathway through which light triggers this activation of *frq* has led to the identification of two additional proteins required for the execution of the feedback loop (Crosthwaite *et al.*, 1997), and has allowed us, at least conceptually, to "close the loop" in the *Neurospora* oscillator by identifying positively acting components in the clock in addition to the (at least formally) negatively acting FRQ proteins. The novel components are encoded by the global regulatory genes *white collar-1* (*wc-1*) and *white collar-2* (*wc-2*), which govern nearly all known photoresponses in *Neurospora*, including the light induction of *frq* RNA (Ballario *et al.*, 1996; Crosthwaite *et al.*, 1997; Linden & Macino, 1997). However, no functions for these proteins in the dark were anticipated. The mRNA levels of *frq* are very low in the dark in the *white collar* strains under all conditions, the mutant strains display no overt rhythms, and (importantly) rhythmicity cannot be initiated by nonphotic zeitgebers such as temperature treatments that act independently of the light response pathway (Crosthwaite *et al.*, 1997). In short, *wc-1* and *wc-2* appear to encode novel clock components and/or clock-associated functions, and for this reason can be included both in the feedback loop and within the light input pathway in Figure 2. Both *wc-1* and *wc-2* are Zn-finger type DNA-binding transcription factors of the GATA factor family (Ballario *et al.*, 1996; Linden & Macino, 1997), and each contains a PAS domain associated with protein–protein interactions. Although the two molecularly characterized circadian clocks in *Neurospora* and in *Drosophila* are both believed to be transcription/translation-based negative feedback loops, these WC proteins, known to be transcription factors, are the first clock proteins having a known biochemical function.

The identification of PAS domains in both WC proteins was exciting and unanticipated. These are functional domains previously identified in other classes of proteins, including photoresponse proteins like the bacterial protein photoactive yellow protein (PYP) and the plant photoreceptor phytochrome (Lagarias, Shu-Hsing, & Lagarias, 1995; Quail *et al.*,1995) and insect and mammalian transcription factors. Interestingly it had also been found in the *Drosophila* clock protein Period (PER; Huang, Edery, & Rosbash, 1993). The identification of PAS domains in the WC proteins associated both with photoresponsivity and with clock function thus suggests an evolutionary origin for the biochemistry that became the circadian oscillator—that clock proteins may have evolved from ancient proteins involved with photoresponses (Crosthwaite *et al.*, 1997), a conclusion supported by a variety of physiologic and anatomic data from many organisms (Crosthwaite *et al.*, 1997). Further, the association of PAS with two classes of clock molecules suggested that this would be a theme and that other clock-associated proteins might be found having this characteristic, a prediction soon confirmed with the identification of the mouse clock-associated protein *Clock* (CLOCK) (Antoch *et al.*, 1997; King *et al.*, 1997). An attractive conjecture is that the mouse CLOCK protein is, in addition to being a sequence homolog, an actual functional homolog of the WC proteins, having PAS domains, DNA-binding regions, and a glutamine (Q)-rich putative transcriptional activation domain (Figure 4). The similarity in the PAS region among PER, CLOCK, and WC proteins is the first example of sequence similarity among clock-associated proteins and, given the distant evolutionary connections between flies, fungi, and mice, suggests that other putative clock components might be identified having this similarity (Crosthwaite *et al.*, 1997).

These data have provided strong support for the notion that the molecular mechanisms of many, if not all, circadian oscillators will be similar and that *Neurospora* is an excellent model system for describing the underpinnings of vertebrate

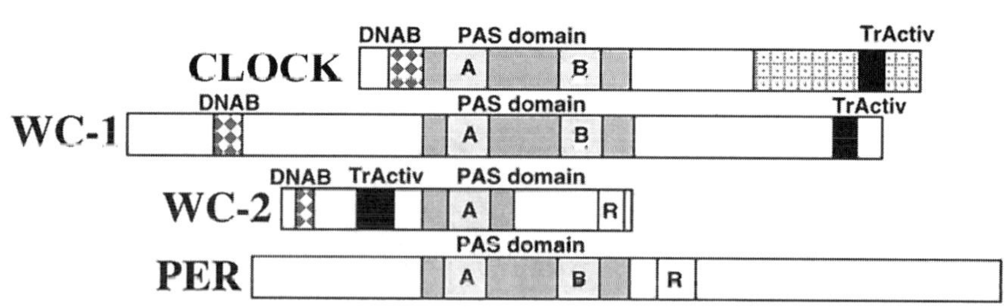

Figure 4. Common structural domains among clock components and clock-associated genes. PAS domains are found in the clock components WC-2 (as a single PAS) and PER (as a repeat of A and B) and WC-1 and CLOCK (again as a repeat). WC-1 and WC-2 are known to bind to DNA via their DNA-binding domains (DNAB); a putative DNAB is also found in CLOCK. Similarly, the transcription factors WC-1 and WC-2 contain putative transcriptional activation domains (TrActiv), as does CLOCK. R stands for repeat region, a direct repeat of Thr/Gly in PER and a mixed repeat of Thr/Gly, Ser/Gly, and Met/Gly in WC-2; a similar region is found in FRQ. The proteins and regions are roughly proportional to size, and the four proteins are lined up so that their PAS domains coincide.

and mammalian clocks as well as microbial clocks. A working model consistent with the data finds *frq* normally activated at night by one or both WC proteins acting perhaps in concert (Crosthwaite *et al.*, 1997), although this has yet to be shown. *frq* can always be light induced, so WC-1 is probably always there, but not necessarily always active.

Temperature Influences the Clock

Temperature is, along with light, the other universal zeitgeber. There are three aspects of circadian temperature responses—physiologic limits to rhythmicity, resetting induced by temperature steps, and temperature compensation—and the response of FRQ synthesis to ambient temperature speaks to all three of these. As noted above, *frq* gives rise to two forms of FRQ through alternatives in frame initiation of translation. Either form alone is sufficient to complete a functional clock at some temperatures, but both forms of FRQ are necessary for robust overt rhythmicity (Liu, Garceau, Loros, & Dunlap, 1997). This is due to the facts that, as the temperature rises, more of both forms of FRQ are made, and also that temperature regulates the ratio of the two FRQ forms by favoring different initiation codons at different temperatures. Importantly, when either initiation codon is eliminated, the temperature range permissive for rhythmicity is demonstrably reduced. This temperature-influenced choice of translation initiation site represents a novel adaptive mechanism that extends the physiologic temperature range over which the clock can function. Data further suggest that there is a threshold level of FRQ required to establish the feedback loop comprising the oscillator, and that this threshold level increases as the temperature increases (Liu *et al.*, 1997). Thus, the temperature limits permissive for rhythmicity, one of the basic characteristics of circadian clocks (Liu *et al.*, 1997; Pittendrigh, 1993), are established by a temperature response that determines how much FRQ is made at what temperature. Temperature resetting is achieved through abrupt changes in the level of FRQ with temperature. Finally, it is clear that temperature compensation is not achieved by keeping the level of FRQ constant over the physiologic range, but rather the temperature compensation mechanism must embrace a large normal variation in the level of this central clock component.

What We Know and Don't Know about the Oscillator and Input

Figure 5 describes just the oscillator-specific aspects of the circadian cycle. The general path and characteristics of the oscillator cycle are now known with some confidence, and some specifics are known: The negative feedback loop comprising the oscillator involves transcription and translation, and with the appearance of the WC proteins, we now know transcription factors acting as clock components. In addition to the WC proteins, FRQ, PER, TIM (from insects), and CLOCK (from mammals) are also expected to have roles in transcription, the first three as negative elements and CLOCK, like the WC proteins, acting positively. The mechanism of light-induced resetting is largely understood, and major aspects of temperature responses have been revealed. Still, there are a large number of issues under study: we do not know what FRQ, PER, or TIM actually do or how their activity or stability is regulated, and a role for CLOCK in the oscillator is yet to be confirmed (Reppert & Weaver, 1997). Although the identification of the WC genes provides the expected positive elements in the clock and closes the clock loop conceptually, we do not know

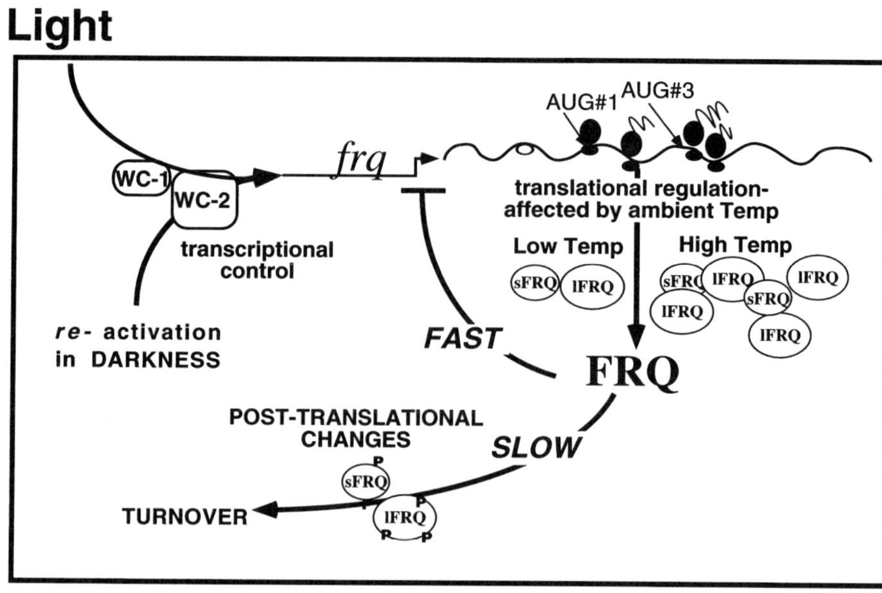

Figure 5. Transcriptional and posttranscriptional regulation in the oscillator—the effect of temperature on FRQ production. At dawn, FRQ is low and *frq* mRNA (horizontal line with arrow) levels are rising, so there is a natural delay between the rise in RNA and protein. FRQ acts very rapidly, within just a few hours, to depress *frq* transcript levels (Merrow *et al.*, 1997). Following its synthesis, posttranslational processing of FRQ through phosphorylation is slow, occupying most of the day, and is thought to govern FRQ turnover; all phosphorylated forms of FRQ appear to be active in maintaining depression of *frq* transcript levels. Light at any time restarts the cycle, but progression into a new cycle (involving WC-1 and WC-2) can also occur in complete darkness (Crosthwaite *et al.*, 1997). *frq* mRNA (wavy line) contains an open reading frame (ORF) having both upstream and downstream AUGs in a good context for translation initiation. Ribosomes (ovals, with the small subunit by itself being open) are thought to scan the *frq* transcript, beginning upstream of the 5′-most (upstream) AUG, continuing, and eventually reaching the downstream one. Ambient temperature influences (a) the relative efficiency of the ribosomes' two possible initiation events and (b) the absolute amount of total FRQ protein. As the temperature rises, ribosomes successfully initiate at the first AUG more often, so more large FRQ is made. The findings and the summary figures of Garceau *et al.* (1997) and Liu *et al.* (1997) led to this figure, which was adapted from Hall (1997).

for sure what the immediate regulators of the WC proteins are nor their immediate targets, and the loop has yet actually to be closed physically (that is, by identifying the protein–protein or protein–nucleic acid interactions at each step).

Describing Clock Output at the Level of Molecules

Identifying Clock-Controlled Genes

The first global screen for clock-regulated genes was executed in *Neurospora* in the mid-1980s using subtractive hybridization (Loros *et al.*, 1989), and resulted in the identification of two "*clock-controlled genes*" or *ccg*'s (Loros *et al.*, 1989). The term "*ccg*" coined at that time has become a useful generic term in the field of rhythms to refer to components or targets of the output pathway as distinct from elements within the

Gene	Peak	Ampl.	Location	Light	Devel.	Identity
ccg-1	CT3	8X	LGVR	+	+	stress ?
ccg-2	CT22	10X	LGIIR	+	+	hydrophobin
ccg-4	CT5	22X	LGIL	+	+	ER trafficking ?
ccg-6	CT19	18X	LGVR	+	+	regulator ?
ccg-7	CT21	9X	LGIIR	-	-	GAPDH
ccg-8	CT20	7X	LGVC	-	-	?
ccg-9	CT19	9X	LGVIIL	+	+	regulator ?
ccg-12	CT18	4X	LGVIL	-	-	CuMT

Figure 6. A summary of the *Neurospora* clock-controlled genes. Peak denotes the time of day at which the amount of transcript from the gene reaches its maximum, and amplitude refers here to the ratio of peak-to-trough levels. Location is genetic location of each *ccg*, that is, on which linkage group (LG I–VII) and which chromosome arm (L or R) it is found; the *ccg*'s are dispersed all over the genome and do not appear in clusters. Light and development refer to the types of regulation displayed by the different *ccg*'s, and "Identity" to the possible cellular role (with a question mark) or known identity of the gene products. Other *ccg*'s have also been identified in other contexts; see text for details.

oscillator. More recently, differential hybridization has been used to identify six additional *ccg*'s (Bell-Pedersen *et al.*, 1996b) (Figure 6). Although daily regulation of development by the clock as seen on race tubes (Figure 1) is the major overt rhythm in *Neurospora*, not all of the *ccg* genes are required for or even associated with the execution of this daily developmental program. For instance, CCG-7, is the single *Neurospora* glyceraldehyde-3-phosphate dehydrogenase (GAPDH) (Shinohara, Loros, & Dunlap, 1998), a pivotal control enzyme in the glycolytic pathway of all living things. This provides the first evidence for circadian regulation of a gene and protein found in organisms across all phylogenetic groups. In *Neurospora*, GAPDH (CCG-7) is not induced by development, stress, or light, suggesting that circadian regulation may affect the core of metabolism and cannot be thought of a simply an epiphenomenon that lies at the periphery of the life of the cell. Additional genes identified in different contexts now known to be clock regulated, including several developmental stage-specific *con* genes (for instance, Lauter & Yanofsky, 1993) and *al-3* (Arpaia, Carattoli, & Macino, 1995a) involved in cell pigmentation, so that nearly a dozen different circadianly regulated output genes are known in *Neurospora.* They display a diverse set of functions involving development, stress, and intermediary metabolism and, as the largest collection of clock-regulated genes known in any organism, provide a window into the role of the clock in the life of the organism. They provide an excellent context in which to study the mechanisms through which clock regulation is achieved; it is not hard to imagine an ion channel being among the *ccg*'s in an oscillator cell in the suprachiasmatic nucleus (D'Souza & Dryer, 1996), thus providing the basis for circadian regulation in a mammal.

Regulation of the *ccg*'s

It would be expected that genes that are important to the life of an organism and that interface with the environment will be acutely regulated by a variety of factors in addition to the clock; some of these may be independent of the clock or codependent with the clock, and all aspects of regulation must be integrated at the level of the target controlled gene. The initial clock-controlled genes identified, *ccg-1* and *ccg-2* (Loros *et al.*, 1989), are controlled by the clock at the level of transcription (Loros & Dunlap, 1991), and it is expected that this mode of regulation will predominate in most, although not necessarily all organisms (see, for instance, Hastings, 1995). For this reason, major efforts in recent years have been devoted to developing an understanding of this complex regulatory network, the first steps of which have involved the dissection of clock-regulated promoters with the goal of identifying each of the constituent parts of the promoter (Bell-Pedersen, Dunlap, & Loros, 1996a). Most promoter resection work on *ccg*'s has been done in plant and fungal systems. Figure 7 shows the results of this work for *eas* (*ccg-2*), which along with the *Arabidopsis* CAB2 gene (Anderson & Kay, 1995), is the best-described circadianly regulated promoter. Moving in a 5′ to 3′ direction from the more distal regions of the promoter towards the start site of transcription, there are the following: (1) a general transcriptional enhancer that is disrupted in the canonical *eas* allele (Bell-Pedersen *et al.*, 1992) that lies within (2) a mycelial repressor which keeps *ccg-2* repressed prior to terminal differentiation; (3) a positive-acting light element required to mediate the acute activation of the gene by light, followed by (4) a light-responsive element that acts to modify the kinetics of the light-driven gene expression but that is not required for the light response; and finally (5) the ACE (activating circadian element; positive clock element) lies closest to the transcription start site, localized within a small region between −57 and −102 (Bell-Pedersen *et al.*, 1996a). The ACE is both necessary and sufficient for conferring circadian regulation: removal of the ACE eliminates circadian regulation (reducing it to the level characteristic of the low points in the cycle) but leaves other aspects of regulation. Further, insertion of the ACE into the 5′ (upstream) region of a basal non-clock-regulated promoter confers circadian clock regulation on a non-clock-regulated

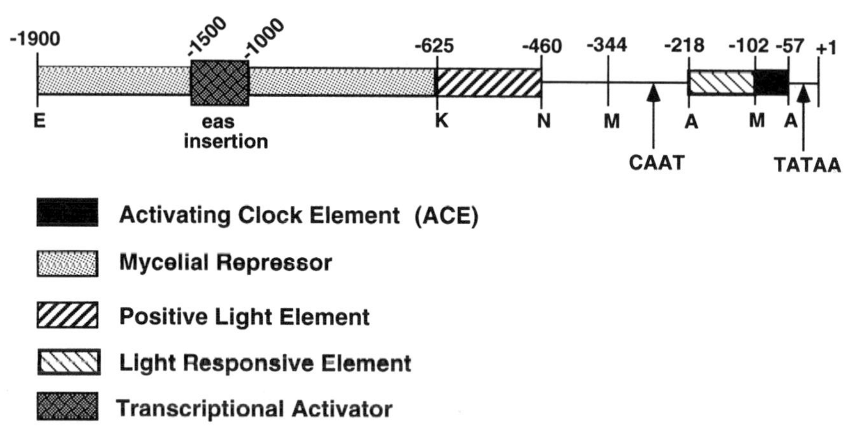

Figure 7. Regulatory sequences governing the expression of the clock-controlled gene *eas* (*ccg-2*). Pieces of promoter DNA were selectively removed and the effects of the deletions on the regulation of the gene subsequently determined. When the activating circadian element (ACE) was localized in this way through promoter resection analysis, its identity was further confirmed by placing it in front of a constitutively expressed (non-clock-regulated) promoter, and showing that just 60 base pairs of the ACE region are necessary and sufficient to confer morning specific clock regulation (Bell-Pedersen *et al.*, 1996a). K (*Kpn*I), N (*Nde*I), M (*Msc*I), and A (*Ava*II) refer to the positions of restriction endonuclease cleavage sites within the promoter of the *ccg-2* gene. (Figure adapted from Bell-Pedersen *et al.*, 1996a.)

reporter gene (β-galactosidase) driven by the naive promoter (Bell-Pedersen *et al.*, 1996a). Sequences within this region are similar, but not obviously identical to sequences found in other clock-regulated and light-regulated genes. A protein factor from nuclear extracts that binds to the ACE region has now been identified (D. Bell-Pedersen, J., Loros, & J., Dunlap, unpublished); this is a candidate for the clock-regulated *trans*-acting factor that is the active element in conferring circadian regulation on *ccg-2* and potentially on other clock-regulated genes in *Neurospora.*

Separate collaborative work has focused on the light regulation of the *ccg* genes, many of which are rapidly induced following brief exposure to blue light (Arpaia, Loros, Dunlap, Morelli, & Macino, 1993; Arpaia, Loros, Dunlap, Morelli, & Macino, 1995b). The light induction signal is not obligately transduced through the clock since these genes remain photoinducible in an *frq*-null (clock-disrupted) strain. Light regulation is blocked in strains bearing single gene mutations in the *wc-1* and *wc-2* genes which are known to result in photoblindness just as discussed above for *frq*. These *ccg* data show that clock-regulated morning-specific genes are acutely regulated by light; in this way, external (e.g., environmental) and internal (e.g., circadian) signals reinforce each other in driving the expression of downstream target genes at appropriate times of day.

Conclusions from a Prototypic Model System

The greatest advances in understanding the molecular bases of circadian rhythmicity have come from genetic and molecular genetic analysis. The *Neurospora* system has already provided an excellent model system in which to pursue this, and the recent identification of a clock-associated gene from the mouse that is a sequence and potentially a functional homolog of the *wc* genes is added, convincing proof of the utility of this model and approach for understanding circadian oscillations in more complex organisms. Genetics originally identified the *frequency* locus as encoding a possible clock component, and molecular analysis has shown the products of this locus to be clock components and provided along the way an example of how to establish that a protein does act within the oscillator. The *Neurospora* circadian oscillator is understood to include an autoregulatory feedback loop wherein the products of the *frq* locus act to regulate their own behavior and expression and thereby help to regulate the behavior of the cell as a function of time. The products of the *wc-1* and *wc-2* genes participate in this feedback loop, apparently by helping to promote the expression of *frq* at an appropriate time. Light resets the clock by acting rapidly to promote the transcription of *frq*; here again, the *wc* gene products play an essential role. Temperature acts principally at the level of translation to influence both the total amount of FRQ made and also the relative amount of the two forms of FRQ. In this way, temperature steps can act to reset the clock, and the physiologic temperature limits for rhythmicity are set. A dozen genes involved strictly in output are known in *Neurospora,* and they are involved in a variety of cellular functions including development, stress responses, and the core of intermediary metabolism. The *cis*-acting elements mediating circadian regulation of target genes are now being elucidated.

From work in *Neurospora,* the identity of several components of circadian oscillators have been revealed within the past few years along with the overall scheme governing the assembly of these components to create this and perhaps all circadian oscillators. This knowledge in the context of existing theory paved the way for discovering the means through which these oscillators respond to light and temperature. Since the behavioral characteristics of all clocks are so similar, it is expected that this knowledge of clock mechanism and resetting will pave the way for understanding microbial as well as animal and plant clocks.

Acknowledgments

This work was supported by grants from the National Institutes of Health (GM 34985 and MH01186 to J.C.D and MH44651 to J.C.D and J.J.L.), the National Science Foundation (MCB-0084509 to J.J.L.), and the Norris Cotton Cancer Center core grant at Dartmouth Medical School.

References

Anderson, S. L., & Kay, S. A. (1995). Functional dissection of circadian clock and phytochrome-regulated transcription of the *Arabidopsis CAB2* gene. *Proceedings of the National Academy of Sciences of the USA, 92,* 1500–1504.

Antoch, M., Soog, E., Chang, A., Vitaterna, M., Zhao, Y., Wilsbacher, L., Sangoram, A., King, D., Pinto, L., & Takahashi, J. (1997). Functional identification of the mouse circadian *clock* gene by transgenic BAC rescue. *Cell, 89,* 655–667.

Aronson, B., Johnson, K., Loros, J. J., & Dunlap, J. C. (1994a). Negative feedback defining a circadian clock: Autoregulation in the clock gene *frequency. Science, 263,* 1578–1584.

Aronson, B. D., Johnson, K. A., & Dunlap, J. C. (1994b). The circadian clock locus *frequency*: A single ORF defines period length and temperature compensation. *Proceedings of the National Academy of Sciences of the USA, 91,* 7683–7687.

Arpaia, G., Loros, J. J., Dunlap, J. C., Morelli, G., & Macino, G. (1993). The interplay of light and the circadian clock: Independent dual regulation of clock-controlled gene *ccg-2* (*eas*). *Plant Physiology, 102,* 1299–1305.

Arpaia, G., Carattoli, A., & Macino, G. (1995a). Light and development regulate the expression of the *albino-3* gene in *Neurospora crassa. Developmental Biology, 170,* 626–635.

Arpaia, G., Loros, J. J., Dunlap, J. C., Morelli, G., & Macino, G. (1995b). The circadian clock-controlled gene *ccg-1* is induced by light. *Molecular and General Genetics, 247,* 157–163.

Ballario, P., Vittorioso, P., Magrelli, A., Talora, C., Cabibbo, A., & Macino, G. (1996). *White collar-1,* a central regulator of blue-light responses in *Neurospora crassa,* is a zinc-finger protein. *EMBO Journal, 15,* 1650–1657.

Beadle, G. W., & Tatum, E. L. (1945). *Neurospora* II. Methods of producing and detecting mutations concerned with nutritional requirements. *American Journal of Botany, 32,* 678–686.

Bell-Pedersen, D., Dunlap, J. C., & Loros, J. J. (1996a). Distinct *cis*-acting elements mediate clock, light, and developmental regulation of the *Neurospora crassa eas* (*ccg-2*) gene. *Molecular and Cellular Biology, 16,* 513–521.

Bell-Pedersen, D., Shinohara, M., Loros, J., & Dunlap, J. C. (1996b). Circadian clock-controlled genes isolated from *Neurospora crassa* are late night to early morning specific. *Proceedings of the National Academy of Sciences of the USA, 93,* 13096–13101.

Bell-Pedersen, D., Dunlap, J. C., & Loros, J. J. (1992). The *Neurospora* circadian clock-controlled gene, *ccg 2,* is allelic to *eas* and encodes a fungal hydrophobin required for formation of the conidial rodlet layer, *Genes & Dev. 6,* 2382–2394.

Crosthwaite, S. C., Loros, J. J., & Dunlap, J. C. (1995). Light-induced resetting of a circadian clock is mediated by a rapid increase in *frequency* transcript. *Cell, 81,* 1003–1012.

Crosthwaite, S. C., Dunlap, J. C., & Loros, J. J. (1997). *Neurospora wc-1* and *wc-2*: Transcription, photoresponses, and the origins of circadian rhythmicity. *Science, 276,* 763–769.

Denault, D. L., Loros, J. J., and Dunlap, J. C. WC-2 mediates WC-1-FRQ interaction within the PAS protein-linked circadian feedback loop of *Neurospora crassa, EMBO Journal.* In press.

Dharmananda, S. (1980). *Studies of the circadian clock of* Neurospora crassa*: Light-induced phase shifting.* Ph.D. dissertation, University of California, Santa Cruz, California.

D'Souza, T., & Dryer, S. (1996). A cationic channel regulated by a vertebrate intrinsic circadian oscillator. *Nature, 382,* 165–167.

Dunlap, J. C. (1993). Genetic analysis of circadian clocks. *Annual Review of Physiology, 55,* 683–728.

Dunlap, J. C. (1996). Genetic and molecular analysis of circadian rhythms. *Annual Review of Genetics, 30,* 579–601.

Dunlap, J. C. (1999). Molecular bases for circadian clocks, *Cell 96,* 271–290.

Feldman, J. F. (1982). Genetic approaches to circadian clocks. *Annual Review of Plant Physiology, 33,* 583–608.

Feldman, J. F., & Dunlap, J. C. (1983). *Neurospora crassa*: A unique system for studying circadian rhythms. *Photochemistry and Photobiology Review, 7,* 319–368.

Feldman, J. F., Gardner, G. F., & Dennison, R. A. (1979). Genetic analysis of the circadian clock of *Neurospora.* In M. Suda (Ed.), *Biological rhythms and their central mechanism* (pp. 57–66). Amsterdam: Elsevier.

Garceau, N. (1996). *Molecular and genetic studies on the* frq *and* ccg-1 *loci of* Neurospora. Ph.D. dissertation, Dartmouth College, Hanover, New Hampshire.

Garceau, N., Liu, Y., Loros, J. J., & Dunlap, J. C. (1997). Alternative initiation of translation and time-specific phosphorylation yield multiple forms of the essential clock protein Frequency. *Cell, 89,* 469–476.

Hall, J. (1997). Circadian pacemakers blowing hot & cold—But they're clocks, not thermometers. *Cell, 90,* 9–12.

Hastings, M. (1995). Circadian rhythms: Peering into the molecular clockwork. *Journal of Neuroendocrinology, 7,* 331–340.

Huang, Z. J., Edery, I., & Rosbash, M. (1993). PAS is a dimerization domain common to *Drosophila* Period and several transcription factors. *Nature, 364,* 259–262.

Iwasaki, H., and Dunlap, J. C. (2000). Microbial circadian oscillatory systems in *Neurospora* and *Synechococcus*: models for cellular clocks, *Currrent Opinions in Microbiology, 3,* 189–196.

King, D., Zhao, Y., Sangoram, A., Wilsbacher, L., Tanaka, M., Antoch, M., Steeves, T., Vitaterna, M., Kornhauser, J., Lowrey, P., Turek, F., & Takahashi, J. (1997). Positional cloning of the mouse circadian *clock* gene. *Cell, 89,* 641–653.

Lagarias, D. M., Shu-Hsing, W., & Lagarias, J. C. (1995). Atypical phytochrome gene structure in the green alga *Mesotaenium caldariorum. Plant Molecular Biology, 29,* 1127–1142.

Lakin-Thomas, P., Coté, G., & Brody, S. (1990). Circadian rhythms in *Neurospora. CRC Critical Reviews in Microbiology, 17,* 365–416.

Lauter, F.-R., & Yanofsky, C. (1993). Day/night and circadian rhythm control of *con* gene expression in *Neurospora. Proceedings of the National Academy of Sciences of the USA, 90,* 8249–8253.

Lee, K., Loros, J. J., and Dunlap, J. C. (2000). Interconnected feedback loops in the Neurospora circadian system, *Science, 289,* 107–110.

Lewis, M. T., & Feldman, J. F. (1993). The putative *frq* clock protein of *Neurospora crassa* contains sequence elements that suggest a nuclear transcriptional regulatory role. *Protein Sequences and Data Analysis, 5,* 315–323.

Lewis, M., & Feldman, J. F. (1997). Evolution of the *frequency* clock locus in ascomycete fungi. *Molecular Biology and Evolution, 13,* 1233–1241.

Lewis, M., Morgan, L., & Feldman, J. F. (1997). Cloning of (*frq*) clock gene homologs from the *Neurospora sitophila* and *Neurospora tetrasperma, Chromocrea spinulosa* and *Leptosphaeria australiensis. Molecular and General Genetics, 253,* 401–414.

Linden, H., & Macino, G. (1997). White collar-2, a partner in blue-light signal transduction, controlling expression of light-regulated genes in *Neurospora crassa. EMBO Journal, 16,* 98–109.

Liu, Y., Garceau, N., Loros, J. J., & Dunlap, J. C. (1997). Thermally regulated translational control mediates an aspect of temperature compensation in the *Neurospora* circadian clock. *Cell, 89,* 477–486.

Liu, Y., Merrow, M., Loros, J. J., and Dunlap, J. C. (1998). How temperature changes reset a circadian oscillator, *Science, 281,* 825–829.

Loros, J. (1995). The molecular basis of the *Neurospora* clock. *Seminars in Neuroscience, 7,* 3–13.

Loros, J., & Dunlap, J. C. (1991). *Neurospora crassa* clock-controlled genes are regulated at the level of transcription. *Molecular and Cellular Biology, 11,* 558–563.

Loros, J. J., and Dunlap, J. C. (2001). Genetic and molecular analysis of circadian rhythms in *Neurospora, Annual Reviews of Physiology, 63,* 757–794.

Loros, J. J., Denome, S. A., & Dunlap, J. C. (1989). Molecular cloning of genes under the control of the circadian clock in *Neurospora. Science, 243,* 385–388.

Luo, C., Loros, J. J., & Dunlap, J. C. (1998). Nuclear localization is required for function of the essential clock protein Frequency. *EMBO Journal, 17,* 1228–1235.

Merrow, M., & Dunlap, J. C. (1994). Intergeneric complementation of a circadian rhythmicity defect: Phylogenetic conservation of the 989 amino acid open reading frame in the clock gene *frequency*. *EMBO Journal, 13*, 2257–2266.

Merrow, M., Garceau, N., & Dunlap, J. C. (1997). Dissection of a circadian oscillation into discrete domains. *Proceedings of the National Academy of Sciences of the USA, 94*, 3877–3882.

Morgan, L., & Feldman, J. (1997). Isolation and characterization of a temperature-sensitive circadian clock mutant in *Neurospora crassa. Genetics, 146*, 525–530.

Nakashima, H. (1981). A liquid culture system for the biochemical analysis of the circadian clock of *Neurospora. Plant and Cell Physiology, 22*, 231–238.

Paietta, J., & Sargent, M. (1981). Riboflavin mutants are blind. *Proceedings of the National Academy of Sciences of the USA, 78*, 5573–5577.

Pittendrigh, C. S. (1976). Circadian clocks: What are they? In J. W. Hastings & H.-G. Schweiger (Eds.), *The molecular basis of circadian rhythms* (pp. 11–48). Berlin: Abakon.

Pittendrigh, C. S. (1993). Temporal organization: Reflections of a Darwinian clock-watcher. *Annual Review of Physiology, 55*, 17–54.

Pittendrigh, C. S., Bruce, V. G., Rosenzweig, N. S., & Rubin, M. L. (1959). A biological clock in *Neurospora. Nature, 184*, 169–170.

Quail, P., Boylan, M. T., Parks, B. M., Short, T. W., Xu, Y., & Wagner, D. (1995). Phytochromes: photosensory perception and signal transduction. *Science, 268*, 675–680.

Reppert, S., & Weaver, D. R. (1997). Forward genetic approach strikes gold: Cloning of mammalian *clock* gene. *Cell, 89*, 487–490.

Shigeyoshi, Y., Taguchi, K., Yamamoto, S, Takeida, S., Yan, L., Tei, H., Moriya, S, Shibata, S., Loros, J. J., Dunlap, J. C., and Okamura, H. (1997). Light-induced resetting of a mammalian circadian clock is associated with rapid induction of the *mPer1* transcript, *Cell, 91*, 1043–1053.

Shinohara, M., Loros, J. J., & Dunlap, J. C. (1998). Glyceraldehyds-3-phosphate dehydrogenase is regulated on a daily basis by the circadian clock. *Journal of Biological Chemistry, 273*, 446–452.

13

Circadian Timekeeping in *Drosophila*

MICHAEL W. YOUNG

INTRODUCTORY NOTE

This chapter, written in the spring of 1997, describes the first two clock genes to be characterized in *Drosophila*. A study of the interactions of *period* and *timeless* pointed to a simple mechanism that generates self-sustained molecular oscillations, but that can be reset by exposure to cycles of day and night. Four years later, it is remarkable to see how quickly answers to many of the questions posed in this chapter have emerged. Missing pieces of the clock have been found, and insights from the fruit fly have allowed molecular dissection of human cycles of sleep and wakefulness.

ORIGINS OF THE GENETIC APPROACH

Functional genetic studies of circadian behavioral rhythms were initiated by Konopka and Benzer (1971), who focused on the X chromosome of *Drosophila melanogaster*. The importance of heredity had been recognized earlier in another *Drosophila* species, *D. pseudoobscura*. Genetic differences affected the phase of certain rhythmic behaviors (Pittendrigh, 1967), but no connection to pacemaker organization was established. Heritable variations in a developmental rhythm with a circadian periodicity also had been observed in *Neurospora* by Tatum (A. Chovnick, personal communication), but the role of single genes in the control of these rhythms was not established until Feldman and Hoyle's pioneering studies of the *Neurospora* conidiation rhythm (Feldman and Hoyle, 1973).

MICHAEL W. YOUNG National Science Foundation Center for Biological Timing, Laboratory of Genetics, The Rockefeller University, New York, New York 10021.

Circadian Clocks, Volume 12 of *Handbook of Behavioral Neurobiology*, edited by Joseph S. Takahashi, Fred W. Turek, and R. Y. Moore, Kluwer Academic / Plenum Publishers, New York, 2001.

Populations of wild-type adult flies emerge from pupation according to a rhythm. High levels of eclosion are seen in the morning, with little emergence in the afternoon or at night. Adult flies also produce locomotor activity rhythms with activity high during the day and low at night. Both classes of rhythms persist with a circadian periodicity in constant darkness, pointing to an endogenous oscillator. Konopka and Benzer searched for mutations that affected eclosion and discovered a series of mutations at a single genetic locus that simultaneously affected eclosion and locomotor activity rhythms.

Konopka and Benzer's report had an enormous impact. The work established a prominent role for single genes in the control of these behavioral rhythms and suggested that one such gene, *period* (*per*), might encode a component of a circadian oscillator. The evidence came from the behavioral phenotypes associated with different *per* alleles: Arrhythmic mutations might block a link between pacemaker and overt behavior, but period-altering alleles of the same gene had also been recovered. At least one of the mutants, the short-period allele per^S, also affected entrainment, so that the amplitude of the fly's response to phase-resetting pulses of light was altered (reviewed by Konopka, 1981). These genetic approaches made it possible to organize biochemical studies without prior knowledge of the functional properties of pacemaker molecules. With evidence that mutations can affect pacemaker machinery, it became clear that the object of study should be the product of an affected gene.

Genetic complementation tests suggested that the arrhythmic allele, per^0, was likely null (Konopka and Benzer, 1971). This was confirmed by Young and Judd (1978), who demonstrated that a viable combination of chromosomal deletions removed *per* and produced arrhythmic flies. All *per* alleles are semidominant. per^0/per^+ flies have 24.5- rather than 24-hour rhythms. The long-period mutation (per^L) produces 28-hour rhythms when hemi- or homozygous, and per^L/per^+ heterozygotes have rhythms about 1 hour longer than wild type. The short-period mutation, per^S, gives 19-hour rhythms, with ~22-hour rhythms produced by per^S/per^+ heterozygotes (Konopka and Benzer, 1971). This semidominance is now found to be a dependable feature of clock mutations in *Neurospora*, the hamster, and the mouse (Feldman and Hoyle 1973; Ralph and Menaker, 1988, Vitaterna *et al.*, 1994), and constitutes an important approach for contemporary genetic screening.

These early genetic studies also mapped the anatomic site of action of *per*. Two classes of investigations agreed in placing the focus in the head. Transplantation studies showed that an adult brain from a per^S donor could restore circadian rhythmicity when transferred to the abdomen of a per^0 recipient. Such "two-brained" flies gave locomotor activity rhythms with 19-hour rather than 24-hour periodicities, consistent with the expected donor phenotype (Handler and Konopka, 1979). Genetically mosaic flies, carrying mixtures of wild-type and mutant tissue, also showed that the genotype of the head determines rhythmic behavior. Arrhythmicity was only seen if per^0 tissue was found in the head (Konopka, Wells, & Lee, 1983). These studies showed that a clock dependent on local *per* function resides in the *Drosophila* brain in keeping with localizations of circadian pacemaker tissues in other insects (Truman, 1974) and in vertebrates (reviewed by Takahashi and Zatz, 1982).

Gene Isolation and Physical Characterization of *per*

Molecular cloning of *per* provided another landmark, as this was the first isolation of a gene controlling animal behavior (Bargiello and Young, 1984; Reddy *et al.*, 1984). It was also demonstrated that the behavior could be restored by transferring cloned copies of the wild-type gene to arrhythmic, *per*^{0} flies (Bargiello, Jackson, & Young, 1984; Zehring *et al.*, 1984). Sequencing showed that the gene encoded a large protein, PER (~1200 amino acids), and that long- and short-period mutations both affected PER structure through single amino acid substitutions (Baylies, Bargiello, Jackson & Young, 1987; Citri *et al.*, 1987; Jackson, Bargiello, Yun, & Young, 1986; Yu *et al.*, 1987). Studies of transgenic flies also confirmed earlier genetic evidence that the period of *Drosophila*'s rhythms is sensitive to PER dosage (Smith and Konopka, 1982): decreasing overall levels of per *expression* lengthens period. In fact, the period of the rhythm appears to be a function of the log of RNA titer (Baylies *et al.*, 1987; Cote and Brody, 1986). The sensitivity of period length to the level of functional clock gene product is likely to contribute to the semidominant phenotypes produced by arrhythmic, long-period, and short-period mutants. Proteins with modified activity will compete with the wild type to give a semidominant modification of period, while null mutations will reduce the overall level of the gene product.

PER includes a sequence of ~260 amino acids that is shared by a small family of transcription factors (Burbach, Poland, & Bradfield; Crews, Thomas, & Goodman, 1988; Hoffman *et al.*, 1991; Reisz-Porszasa, Reyes, & Hankinson, 1992). These include two subunits of the dioxin receptor, Ah receptor (AHR) and Ah receptor nuclear transporter (ARNT), and single minded (SIM), a protein essential for early embryonic central nervous system development. This domain, now referred to as PAS, was shown to support protein dimerization *in vitro* (Huang, Edery, & Rosbash, 1993). The biological significance of PAS is clear with respect to AHR and ARNT, which must physically associate to form an active complex that regulates target gene expression. Finding PAS in PER suggested that the protein's activity might be regulated by homodimerization or physical interaction with an unidentified partner protein.

Molecular Cycles of *per* Expression

PER is expressed in a variety of cell types with different subcellular localizations in different tissues (Liu *et al.*, 1992; Saez and Young, 1988; Siwicki, Eastman, Petersen, Rosbash, & Hall, 1988). In the head, the protein is produced in the eyes, ocelli (dorsal external photoreceptors), and brain. In most cells of the head, the protein accumulates in cell nuclei (Liu *et al.*, 1992; Saez and Young, 1988; Siwicki *et al.*, 1988). There is also a circadian rhythm of this nuclear accumulation, so that nuclear PER is seen late at night, but not late in the day (Siwicki *et al.*, 1988). Studies of genetically mosaic and of transgenic *Drosophila* have shown that *per* expression limited to 20–30 neurons of the central brain is sufficient to drive circadian behavioral rhythms (Ewer, Frisch, Hamblen-Coyle, Rosbash, & Hall, 1992; Frisch, Hardin, Hamblen-Coyle, Rosbash, & Hall, 1994; Vosshall and Young, 1995). Rhythms produced by such flies can be entrained by light cycles even in the absence of the eyes and ocelli (Vosshall and Young, 1995).

PER is similarly found in nuclei outside the head in certain cells composing the thoracic ganglia, ring gland complex, gut, gonads, and Malpighian tubules (Liu *et al.*, 1992; Emery, Noveral, Jamison, & Siwicki, 1997; Giebultowicz & Hege, 1997; Hege, Staniewski, & Hall, 1996; Saez and Young, 1988; Siwicki *et al.*, 1988). Cycles of PER nuclear localization have been observed in Malpighian tubules and in the ring gland, and such cycling persists when these tissues are maintained *in vitro* (Emery *et al.*, 1997; Giebultowicz & Hege, 1997). For Malpighian tubules, rhythms of PER accumulation occur with the same phase as rhythms exhibited in the head. Nevertheless, light can reset the phase of the Malpighian tubule rhythm in headless flies (J. Giebultowicz & Hege, 1997). These results suggest that autonomous, entrainable circadian pacemakers exist outside the head.

In contrast to the predominantly nuclear accumulation of PER in the head, cytoplasmic localization is observed in the ovary (Liu *et al.*, 1992; Saez and Young, 1988; Siwicki *et al.*, 1988). Studies of *per* transcription also indicate a difference in regulation in ovaries. Whereas tissues expressing nuclear PER show circadian cycling of *per* RNA levels (Hardin, Hall, & Rosbash, 1990, 1992), constitutive *per* transcription occurs in ovaries (Hardin, 1994). Thus, nuclear accumulation may reflect autonomous pacemaker function. While the function of cytoplasmic PER proteins in the ovary is unknown, per^0 mutants alter ovarian diapause by establishing a critical daylength that is 5 hours shorter than that seen in wild-type flies (Saunders, 1990). Since per^S and per^L do not affect diapause (Saunders, 1990), it is possible that PER has a novel function in ovaries that does not involve action in a circadian pacemaker (reviewed by Hall, 1996).

Because *per* RNA and proteins are produced with a circadian rhythm it has been proposed that PER proteins somehow regulate *per* transcription through an autoregulatory feedback loop (Hardin *et al.*, 1990, 1992). Three early observations supported the suggestion. First, *per* mutations produce corresponding effects on behavioral and molecular rhythms (Hardin *et al.*, 1990; 1992). Second, the timing of PER protein accumulation lags behind the timing of *per* RNA accumulation by 4–6 hours indicating that PER might negatively affect its own transcription (Hardin *et al.*, 1990, 1992). Third, constitutive overexpression of a *per* transgene in the eye blocked cycling of *per* transcription in that tissue, but not in the brain (Zeng, Hardin, & Rosbash, 1994).

A similar hypothesis has emerged from studies of the *Neurospora* clock gene *frequency* (*frq*). *frq* is required for the circadian regulation of a developmental output, conidiation (Feldman and Hoyle, 1973). FRQ protein is required for circadian cycling of *frq* transcription (Aronson, Johnson, Loros, & Dunlap, 1994). Ectopic overexpression of *frq* eliminates circadian cycling of transcription from the endogenous gene, and *frq* null mutations induce constitutive overexpression. Moreover, step reductions in the FRQ protein will completely reset the phase of the conidiation rhythm, no matter what the initial phase (Aronson *et al.*, 1994; see this volume, Chapter 12, for a full description of the *Neurospora* studies).

These studies suggested that fluctuating concentrations of PER in the fly and FRQ in *Neurospora* each compose a part of a circadian oscillator. Evidently the concentration of such molecules at different points within a molecular cycle can be read to give information about time of day. Molecules with such properties have long been postulated, and have been referred to as "state variables" (reviewed by Aronson *et al.*, 1994; see also this volume, Chapter 12).

Still, work on *per* and *frq* did not suggest a self-sustaining mechanism by which oscillations of these components are achieved. In simplest form, feedback regulation will generate stable, constitutive production of both RNA and protein. How is clock-like regulation of these gene products brought about? Another puzzle concerns the mechanism by which time of day would be established from a single state variable. If the phase of an oscillator is somehow derived from the concentration of a key molecule, only two reference points, coinciding with highest and lowest concentrations, can be interpreted unambiguously. Intermediate concentrations can only provide information about time of day if there is an additional mechanism that determines whether the level is rising or falling. If it is proposed that high concentrations of a protein downregulate gene expression to drive a molecular oscillator, what keeps intermediate concentrations of such a molecular regulator from propagating reversals and truncations of the cycle? Renewed genetic screening allowed a resolution of these issues. *per* has a partner. Molecular cycles are derived from the functional interaction of two state variables.

Discovery of *Timeless*, a Regulator of PER Accumulation, Cycling, and Subcellular Localization

Sehgal, Price, Man, and Young (1994) searched for clock mutations in a screen of homozygous second and third chromosomes. A mutant line was recovered that was arrhythmic for both eclosion and locomotor activity rhythms. These behavioral phenotypes were indistinguishable from per^0 mutants. The new mutant, *timeless* (tim^0), mapped by genetic recombination to the second chromosome (Sehgal *et al.*, 1994). Because tim^0 abolished all behavioral rhythms in the fly, effects on molecular rhythms of *per* expression were also assessed. Circadian cycling of *per* RNA accumulation was eliminated in tim^0 flies, suggesting a close functional interaction between *per* and *tim* (Sehgal *et al.*, 1994). Subsequent genetic screening also led to the recovery of three long period mutants of *tim* (A. Rothenfluh-Hilfiker and M. Young, unpublished) and a *tim* mutant that suppresses the long period phenotype of per^L (Rutila *et al.*, 1996).

Vosshall, Price, Sehgal, Saez, and Young (1994) extended the study of the tim^0 mutant by examining expression of PER proteins in the adult head. Whereas wild-type flies show robust accumulation of PER in brain and photoreceptor cell nuclei late at night, tim^0 flies failed to show nuclear PER accumulation at any time of day or night. In fact, immunocytochemical studies and protein blotting both indicate that very little PER accumulates in the eyes and brains of tim^0 flies (Price, Dembinska, Young, & Rosbash, 1995; Vosshall *et al.*, 1994). The defect in PER accumulation was further examined in transgenic flies expressing fusions of PER and β-galactosidase (β-gal) that are stable in tim^0 flies. Such fusion proteins, expressed under the control of the *per* promoter, accumulated in the same tissues in wild-type and tim^0 flies, but failed to enter nuclei in any cells of the head (Figures 1A, 1B). Therefore *tim* function is required for both accumulation and nuclear localization of the PER protein (Price *et al.*, 1995; Vosshall *et al.*, 1994). Substantial, constitutive expression of *per* RNA is observed in tim^0 flies, and no difference has been detected in the rates of synthesis of PER–β-gal fusion proteins in tim^+ versus tim^0 flies (Vosshall *et al.*, 1994). Thus, the low levels of PER found in tim^0 mutants suggest that function of *tim* is required to stabilize PER (Price *et al.*, 1995; Vosshall *et al.*, 1994). Nuclear localization of PER–β-gal fusion proteins also occurs in per^0 flies. This proves that correct subcellular localization of PER depends directly on a product of *tim* rather than being influenced by signals from an active circadian clock (Vosshall *et al.*, 1994) (Figure 1C).

Vosshall *et al.* (1994) found that PER–β-gal reporters missing sequences including the PAS domain no longer required *tim* for nuclear localization. This nuclear accumulation can be conferred by a nuclear localization sequence that maps to the N-terminal 95 amino acids of the PER protein. It was also shown that β-gal reporter proteins that carried heterologous nuclear localization sequences and no sequence related to PER were unaffected by the *tim* mutation; these reporters accumulated in nuclei in both tim^0 and tim^+ flies (Vosshall *et al.*, 1994). The observations indicated that *tim*'s regulation of nuclear localization is specific for PER, and that PER contains sequences that inhibit its nuclear localization in the absence of *tim* function. One reason for the behavioral and molecular arrhythmicity of tim^0 flies should be absence

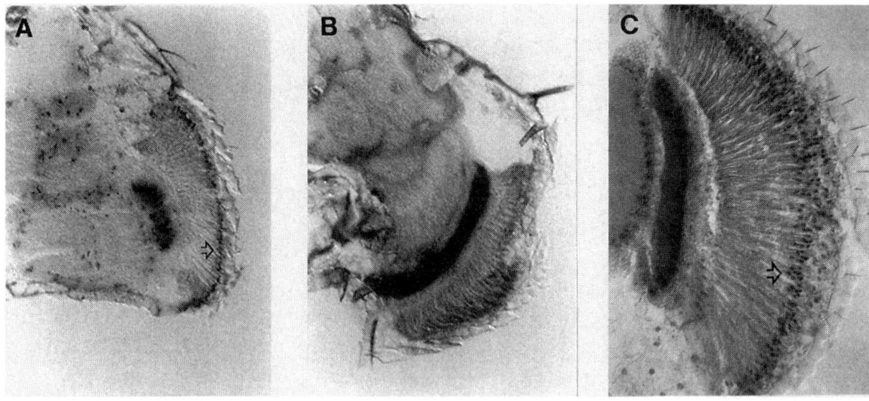

Figure 1. The tim^0 (null) mutation blocks nuclear localization of a PER–β-galactosidase reporter protein. (A) The reporter protein accumulates in photoreceptor nuclei (open arrow) and brain cell nuclei in wild-type flies, but is restricted to the cytoplasm of these cells in tim^0 mutants (B). The reporter protein continues to localize to nuclei in per^0; tim^+ flies (open arrow), indicating nuclear localization depends on *tim* activity, but not on a functional circadian clock (C). (Immunocytochemistry by L. Vosshall; see also Vosshall *et al.*, 1994.)

of PER nuclear localization (Vosshall *et al.*, 1994). Cloning of *tim* verified much of the above, and further suggested that a complex of the PER and TIM proteins is the functional unit of feedback regulation.

Cloning of *tim* Reveals Rhythmic Expression and a Physical Interaction of the PER and TIM Proteins

A novel deletion, Df(2L)tim^{02}, was recovered that mapped *tim* to a short chromosomal interval corresponding to ~65 kb of cloned genomic DNA (Myers, Wager-Smith, Wesley, Young, & Sehgal, 1995). Since tim^0 arose spontaneously in a screen involving hybrid dysgenesis (Sehgal *et al.*, 1994), DNA corresponding to that deleted by Df(2L)tim^{02} was examined for evidence of a polymorphism. Comparison of tim^0 flies and the nonmutant parental strain from which tim^0 was derived showed a deletion in tim^0. Sequencing cDNA and genomic DNA from both strains revealed a 64-bp deletion that introduced a frame shift to give a partial TIM product (~800 instead of ~1400 amino acids) (Myers *et al.*, 1995). No significant homologies to TIM were detected by sequence database comparisons (Myers *et al.*, 1995).

Expression studies confirmed that the cloned gene was *tim*. Oscillations in *tim* transcription were observed, and circadian rhythms of *tim* RNA accumulation had the same phase as *per*. In addition, the *tim* and *per* mutants produced corresponding changes in patterns of *per* and *tim* transcription: tim^0 and per^0 each eliminated cycling of both RNAs. Period-lengthening mutations (per^L, tim^L) lengthened both RNA rhythms and per^S shortened the period of both RNA cycles (Sehgal *et al.*, 1995; A. Rothenfluh-Hilfiker and M. Young, unpublished) (Figure 2). These studies proved that *per* and *tim* were subject to two feedback loops, one providing autoregulation and another reciprocal regulation (Sehgal *et al.*, 1995). The work also indicated that timing of PER's stabilization and nuclear localization must be gated by the temporal pattern of *tim* expression; accumulation and nuclear translocation of PER would be permitted only at times of day when TIM proteins are available (Sehgal *et al.*, 1995).

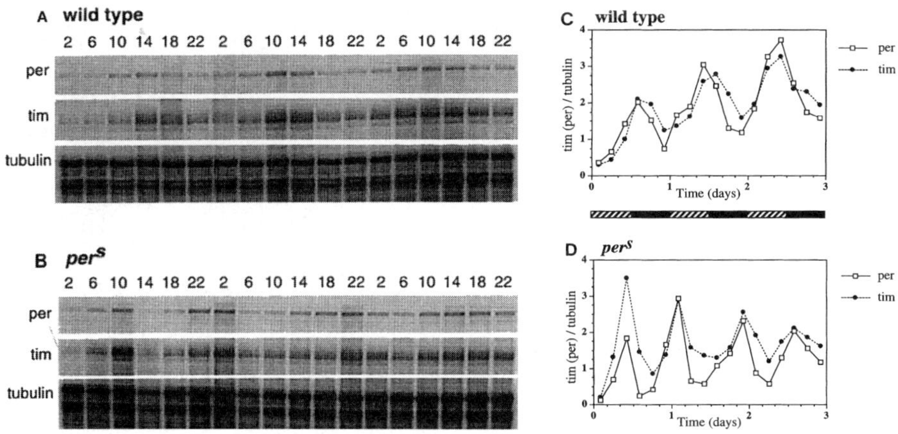

Figure 2. Mutations that alter period of the behavioral rhythm produce corresponding changes in the period of *per* and *tim* RNA cycles. Rhythms of RNA cycling have ~24-hour periods in wild-type flies (A, C), but ~18-hour periods are seen for *per*^{S} mutants (B, D). Cycling of both RNAs is coordinately affected by the mutation at *per*. Data plotted (C, D) are RNAse protections of *per* and *tim* versus tubulin (A, D). Flies were entrained to a light/dark cycle (LD 12:12) for 3 days and tranferred to constant darkness (DD). RNA collections began 14 hours after lights-off. For panels A and B, times of RNA collection (circadian time, CT) are indicated and correspond to timepoints shown in panels C, D. (From Sehgal *et al.*, 1995, with permission. Copyright 1995, American Association for the Advancement of Science.)

Two studies initially showed that TIM and PER can physically associate. TIM proteins translated from the above cloned cDNA were affinity purified by adsorption to PER–GST (Glutathione-S-transferase) fusion proteins *in vitro* (Gekakis *et al.*, 1995; Saez and Young, 1996). Two binding sites have been mapped on each protein by this method, and interaction of PER and TIM through these sequences affects subcellular localization of both proteins (Figure 3) (Saez and Young, 1996; discussed further below). The second assay involved a search for PER-interacting proteins in a yeast two-hybrid screen (Gekakis *et al.*, 1995). A total of 2×10^7 transformants, each

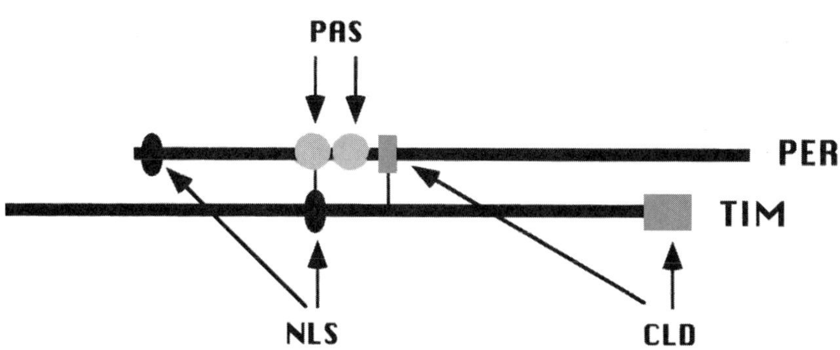

Figure 3. Two sites on PER interact with two sites on TIM. TIM and PER each carry nuclear localization sequences (NLS, red) and cytoplasmic localization domains (CLD, green). CLDs prevent monomeric TIM and PER proteins from entering nuclei. PER's PAS region is composed of two weakly homologous domains (yellow). The N-terminal domain of PAS binds to the NLS of TIM, and a second region on TIM binds to the PER CLD. The physical interaction of PER and TIM appears to inhibit CLD activities on both proteins and allows nuclear localization of the heterodimer.

producing a PER bait and a prey library protein (derived from a *Drosophila* head cDNA library), gave 48 interacting clones. Blots of plasmid DNA from each of the interacting clones were probed with *tim* cDNA. Sixteen of the clones carried *tim* (Gekakis *et al.*, 1995). In addition to providing a second line of evidence supporting a physical association the PER and TIM proteins, the yeast screens indicated that the interaction is quite specific.

Further experiments in yeast suggested that changes in the affinity of PER for TIM can affect period length of the behavioral rhythm. The mutation per^L lengthens the period of molecular and behavioral rhythms to 28 hours (Konopka and Benzer, 1971). Period length is also temperature sensitive in the mutant, so that increasing temperature lengthens period (Konopka, Pittendrigh, & Orr, 1989). Curtin *et al.* (1995) found that in per^L mutants, the timing of PER nuclear localization is delayed. There is a delay of ~4 hours at room temperature, and further delays are seen with increasing temperature. Since *tim* null mutants block nuclear localization of PER, one of PER's binding sites for TIM involves the PAS domain, and the per^L mutation is an amino acid substitution in PER's PAS domain (Baylies *et al.*, 1987), Gekakis *et al.* (1995) tested the effects of the per^L mutation on PER/TIM associations in yeast. PER^L binding was strongly reduced by shifting yeast from 22°C to 37°C. These studies indicate that reduced affinity of TIM for PER^L delays nuclear entry and lengthens period in the mutant. As described above, earlier work showed that reducing gene dosage or levels of expression for *per* increases period length (Baylies *et al.*, 1987; Cote and Brody, 1986; Smith and Konopka, 1982). Presumably, lowering the cellular concentration of PER also delays PER/TIM heterodimer formation to lengthen period (Sehgal *et al.*, 1995). Period length appears to be less sensitive to changes in the dosage of TIM because period length is not detectably altered in tim^0/tim^+ flies (J. Price and M. Young, unpublished). The different responses of period length to altered PER and TIM dosage may reflect the instability of monomeric PER proteins: PER proteins fail to accumulate in tim^0 flies, indicating dependence on PER/TIM heterodimerization for PER stability. TIM proteins do not depend on interaction with PER for stability as high levels of TIM accumulate in per^0 flies (Myers, Wagner-Smith, Rothenfluh-Hilfiker, & Young, 1996).

The PER/TIM interaction originally observed in yeast and *in vitro* was subsequently confirmed *in vivo*. PER proteins are predominantly recovered from the *Drosophila* head in a heterodimeric complex with TIM (Lee, Parikh, Itsukaichi, Bae, & Edery, 1996; Zeng, Qian, Myers, & Rosbash, 1996). That is, one molecule of PER associates with a single molecule of TIM (Zeng *et al.*, 1996). These studies also showed that little or no PER homodimerization occurs *in vivo* (Zeng *et al.*, 1996). PER and TIM show the same temporal program of nuclear accumulation in wild type flies (Hunter-Ensor, Ousley, & Sehgal, 1996; Myers, Price, Sehgal, Saez, & Young, 1996), and just as PER-reporter proteins failed to localize to nuclei in tim^0 flies (Vosshall *et al.*, 1994), TIM proteins accumulate in the cytoplasm rather than in nuclei in per^0 mutants (Myers *et al.*, 1996), indicating a requirement for interaction of the proteins to allow movement of either to nuclei.

Assembly of the PER/TIM Complex is Required for Nuclear Localization

The role of PER/TIM complex formation in nuclear localization, although first recognized *in vivo* (Hunter-Ensor *et al.*, 1996; Myers *et al.*, 1996; Vosshall *et al.*, 1994), has been most extensively characterized in cultured *Drosophila* cells. Regulation of subcellular localization involves suppression of sequences on both PER and TIM that promote cytoplasmic localization (Saez and Young, 1996). As depicted in Figure 3, functional nuclear localization signals (NLS) have been recognized on both TIM and PER. When isolated, these regions of PER and TIM will promote nuclear localization of reporter proteins in the fly and in cultured cells (Saez and Young, 1996; Vosshall *et al.*, 1994). Elsewhere on TIM and PER are found sequences that inhibit NLS function. These were discovered by generating truncated versions of TIM and PER. Monomeric proteins carrying the PER or TIM NLS will only accumulate in nuclei if a certain segment of each protein is removed. Since the latter sequences promote cytoplasmic localization of PER and TIM monomers, they are referred to as cytoplasmic localization domains (CLD; Saez and Young, 1996). Figure 3 shows that two sites on PER bind to two sites on TIM to allow PER/TIM complex formation. The NLS of TIM binds to the PAS domain of PER, and a second TIM region binds directly to the PER CLD. This suggests a mechanism in which CLDs are inactivated by the close physical association of the PER and TIM proteins (Saez and Young, 1996).

To further examine the nature of CLD activity, reporter proteins have been constructed that carry the TIM and PER NLS, but only one CLD (either TIM or PER). In both configurations, such reporter proteins are localized to the cytoplasm of cultured cells. This dominance of CLD activity may indicate that CLDs interact with factors in the cytoplasm or at the nuclear pore to prevent translocation of the proteins to nuclei (L. Saez and M. Young, unpublished).

The dominant activities of the PER and TIM CLDs suggest that assembly of PER/TIM complexes can only take place in the cytoplasm. One result of this mechanism is that rates of cytoplasmic PER and TIM accumulation will influence the timing of nuclear localization and thus period length. This restriction of PER/TIM assembly to the cytoplasm should also be important during entrainment. As discussed below, certain conditions of entrainment cause elimination of TIM from PER/TIM nuclear complexes. By requiring heterodimerization for nuclear translocation of either protein, a mechanism is established whereby nuclear PER monomers, once formed, cannot subsequently renew a partnership with TIM. Any function lost in nuclei due to dissociation of the complex can only be reinstated by import of newly synthesized, cytoplasmically derived heterodimers of the PER and TIM proteins. In this view, an entraining stimulus stops progression of the pacemaker. Further motion is dictated by the sizes of cytoplasmic pools of the PER and TIM proteins.

TIM is a Light Sensor and Couples the Pacemaker to the Environment

The entraining stimulus of most interest is light. Circadian rhythms adapt to the solar day. Period length and the phase of the rhythm are reset. Studies of the *Drosophila* pacemaker have suggested that phase resetting occurs within seconds to minutes of exposure to light (Pittendrigh, 1981). Presumably, the molecular pacemaker emerging from studies of *per* and *tim* is realigned in response to a phase-shifting light pulse, but how quickly does this occur and through what pathway do entrainment functions affect the phase of the pacemaker?

A first indication that TIM might play a central role in entrainment came from the observation that constant light produces a *tim*^{0} phenocopy; constant light suppresses the accumulation and phosphorylation of PER proteins and results in behavioral arrhythmicity, as for tim^0 (Price *et al.*, 1995). Because the low level of hypophosphorylated PER was not further affected by light in tim^0 flies, it was suggested that in wild-type flies, light might be affecting an activity of TIM (Price *et al.*, 1995). Light's effect on TIM was confirmed by following the protein's accumulation following light pulses and in a light/dark cycle (Hunter-Ensor *et al.*, 1996; Lee *et al.*, 1996; Myers *et al.*, 1996; Zeng *et al.*, 1996). Although in per^0 flies, the level of *tim* RNA does not cycle (Sehgal *et al.*, 1995), a rhythm of TIM accumulation was nevertheless observed in these arrhythmic flies, with high levels of the protein found at night and low levels produced during the day (Myers *et al.*, 1996; Zeng *et al.*, 1996). Pulses of light rapidly diminished TIM levels in dark-reared per^0 flies (Myers *et al.*, 1996) or in wild-type flies prematurely exposed to light at night (Hunter-Ensor *et al.*, 1996; Myers *et al.*, 1996; Zeng *et al.*, 1996). In wild-type flies, TIM is lost from cell nuclei and cytoplasm (Figure 4) indicating sensitivity of both PER/TIM complexes and monomeric TIM proteins to light. per^0 flies given a single pulse of light and returned to darkness begin to reaccumulate TIM within 1 hour, with pretreatment levels of TIM accumulating within ~5 hours (Myers *et al.*, 1996) (Figure 5). These results suggested that light rapidly degrades TIM, but in the absence of light, the protein can be resynthesized from constitutively maintained pools of *tim* RNA in per^0 flies. The equally rapid elimination of TIM by light treatment of wild-type or per^0 flies indicated that loss of TIM was independent of the clock (Myers *et al.*, 1996; Zeng *et al.*, 1996): Light's effect on TIM was direct, and not a consequence of first resetting the pacemaker. Evidently TIM is the light sensor that links the pacemaker to the environment.

The phase response curve (PRC) of Figure 6A shows reaction of the *D. melanogaster* locomotor activity rhythm to phase-resetting pulses of light as a function of time of day. The wild-type PRC is of low amplitude (type 1, Winfree, 1970; reviewed by Pittendrigh, 1981). For the conditions shown (Figure 6, legend), maximum delays in the rhythm (~4 hours) are produced by light pulses delivered between circadian time (CT) 14 and CT17. Phase advances (up to ~2 hours) are generated by pulses at CT20–CT23. If light acts uniformly to effect loss of TIM, what is the molecular

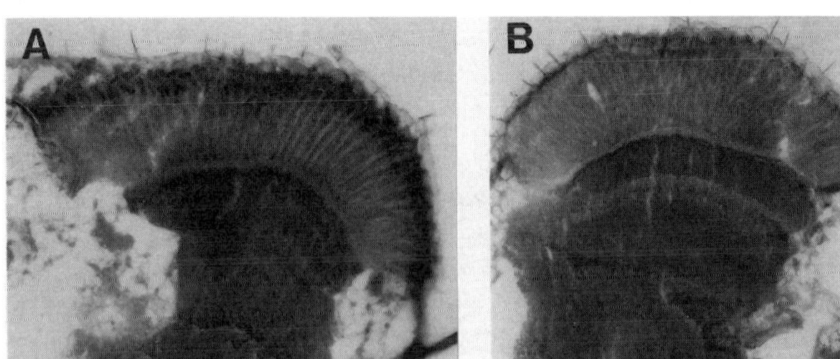

Figure 4. TIM is lost from all cells of the head following a 10 minute pulse of light. TIM accumulation is observed in eyes (photoreceptor nuclei, open arrow) and brain (lateral neurons, arrowhead) in control flies at CT19 (A). TIM is not detected in the eyes or brain at CT18 for flies exposed to light at CT16 (B). Absence of all nuclear staining in light-pulsed flies indicates sensitivity of nuclear TIM/PER complexes. Cytoplasmic TIM also is not detected. Stronger evidence for cytoplasmic TIM degradation comes from the loss of TIM in per^0 flies after a light pulse (see Figure 5) because TIM is constitutively cytoplasmic in per^0 flies (Myers *et al.*, 1996). (Immunocytochemistry by K. Wager-Smith.)

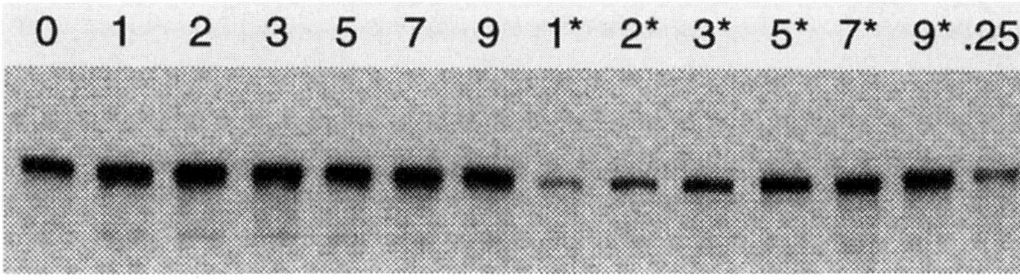

Figure 5. TIM is eliminated by a light treatment in per^0 flies, but reaccumulates when the flies are returned to darkness. per^0 *Drosophila* were maintained in constant darkness for 4 days prior to administration of a 1 hour light pulse (~8000 lux). Numbers above each lane indicate time of protein collection in hours after initiation of the light pulse. Asterisks indicate protein preparations from experimental (light-pulsed) flies. Absence of an asterisk indicates that the sample was taken at the indicated time following the light pulse, but from control (nonpulsed) flies. Rapid loss of TIM was seen at the end of the 1-hour light pulse (1*) or immediately following a shorter light pulse lasting 15 minutes (.25*). When light-pulsed flies are returned to constant darkness, TIM reaccumulation is observed within the first hour of dark (2*). Reaccumulation reaches pretreatment levels after ~4–6 hours (see lanes 5* and 7*). The reaccumulation of TIM in darkness should reflect new synthesis from high levels of constitutively produced *tim* RNA in per^0 flies. TIM's light sensitivity in per^0 flies shows that the response is clock independent. Thus, TIM links the pacemaker to the entraining stimulus. (From Meyers *et al.*, 1996, with permission. Copyright 1996, American Academy for the Advancement of Science.)

mechanism underlying the production of advances versus delays? Myers *et al.* (1996) followed TIM levels in flies exposed to phase-delaying and phase-advancing pulses of light. While comparable rates of TIM degradation were observed for light pulses delivered at CT16 (delay) and CT23 (advance), patterns of TIM reaccumulation differed.

For the delaying pulse, loss of TIM was followed by reaccumulation within the same circadian cycle (Figure 6B). TIM levels begin to rise within 2 hours, and produced a second peak of TIM accumulation ~7 hours after the end of the light pulse. This light-induced secondary accumulation matched a level of TIM seen in nonpulsed, control flies 4–5 hours earlier (compare light-pulsed CT23* sample to nonpulsed samples at CT18 and CT19). Over the remainder of the cycle, TIM levels drop in both pulsed and nonpulsed flies, but with different phases, reflecting the secondary accumulation of TIM in the light-pulsed samples (Figure 6B).

For advancing pulses, TIM reaccumulation is not seen in a comparable time frame. TIM prematurely and monotonically declines with no evidence of reaccumulation until the following day's cycle (Myers *et al.*, 1996) (Figure 6C). The phase of TIM accumulation in light-pulsed flies then precedes the expected phase by 1–2 hours. Advances and delays are thus seen in the molecular cycling of TIM that correspond well with the behavioral responses measured by the PRC.

The different TIM reaccumulation profiles following delaying and advancing light pulses should reflect the very different levels of accumulated *tim* RNA at CT16 versus CT23. The rate of replacement of TIM after either light pulse would be a function of *tim* RNA pools, which are large at CT16 and small at CT23. Evidently, delays occur when TIM can be replaced and advances occur when it cannot (Myers *et al.*, 1996).

The light sensitivity of TIM makes it possible further to evaluate the role of the PER/TIM complex in the regulation of *per* and *tim* transcription. If the complex directly or indirectly suppresses *per* and *tim* transcription, transient elimination of TIM with light should briefly stimulate RNA synthesis. Both delaying and advancing

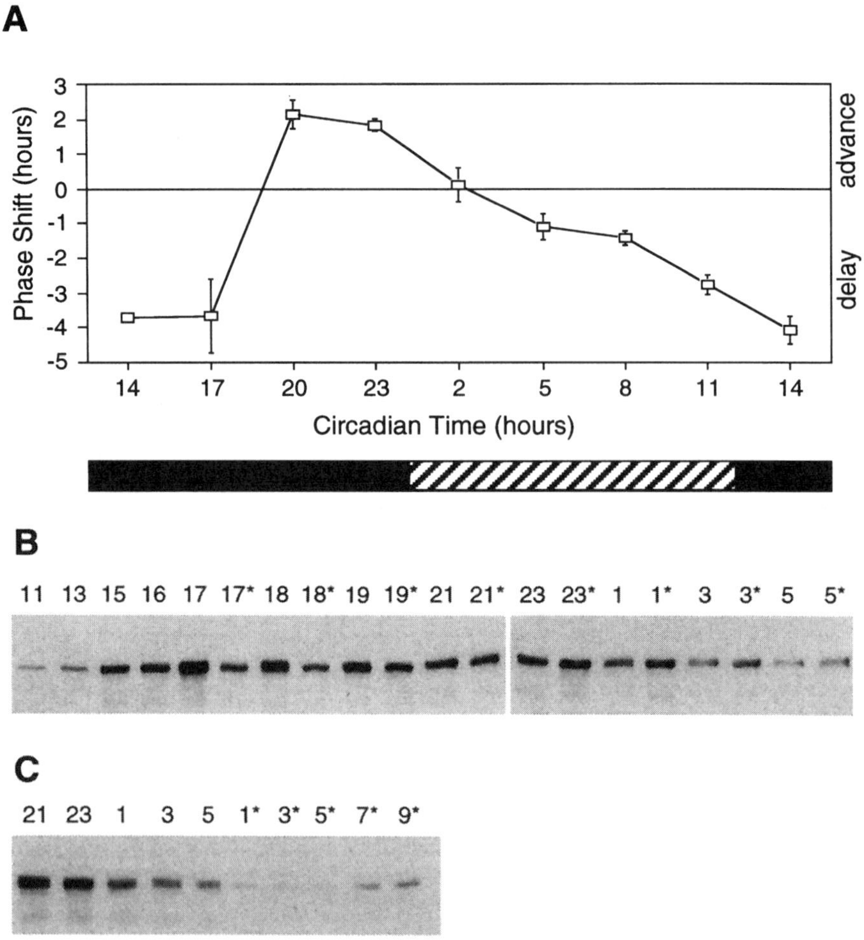

Figure 6. Pulses of light produce corresponding phase shifts of the locomotor activity rhythm and molecular rhythm of TIM accumulation. (A) Phase response curve (PRC) of the locomotor activity rhythm of wild-type *D. melanogaster*. Magnitude of advances and delays is indicated for 10-minute light pulses (~8000 lux) given at the indicated times of day (CT). Under these conditions, transients are not observed in *D. melanogaster* locomotor rhythms. Data reflect a behavioral response that appears to be complete within 24 hours of the light pulse. Maximum phase delays (~4 hours) occur between CT14 and CT17. Maximum advances (~2 hours) occurs between CT20 and CT23. Flies were entrained to a light/dark cycle (LD 12:12) for ≥3 days and transferred to constant darkness (DD) to obtain the PRC. Bar at bottom depicts phase of entrainment (hatching is subjective day). (B) A 1 hour light pulse began at CT16 and TIM protein levels were assessed at later times for light-pulsed flies (asterisks) and nonpulsed controls. For the pulsed flies, TIM is diminished by light (17*), but reaccumulates with a secondary peak at CT23. At CT23*, the absolute level of TIM in CT16-pulsed flies is greater than in nonpulsed controls (compare 23* and 23). The new peak (23*) corresponds to a level of TIM seen in control flies 4–5 hours earlier (compare to time points 18 and 19). Accordingly, the light pulse produced a delay in the molecular rhythm of 4–5 hours. (C) A 1 hour light pulse began at CT23. TIM is reduced by the light pulse and fails to reaccumulate within the same molecular cycle. Reaccumulation occurs on the following day with a slight phase advance (compare 9* in panel C to 11 in panel B). The potential for TIM reaccumulation reflects size of the *tim* RNA pool; the pool is large at CT16 and small at CT23 (see text and Figure 7). (From Myers *et al.*, 1996, with permission. Copyright 1996, American Association for the Advancement of Science.)

light pulses stimulate *per* and *tim* transcription, but only hours after the diminution of TIM proteins (Lee *et al.*, 1996; Young, Wager-Smith, Vosshall, Saez, & Myers, 1996). The lag in the transcriptional response suggests a role for unknown factors in mediating PER/TIM signaling in the nucleus.

PER and TIM Homologues in Other Species

A search outside the Insecta has not yet revealed clear homologues for *per* and *tim*, but striking analogies are emerging in studies of the *Drosophila* and *Neurospora* (*frq*) pacemakers. In both organisms, cycling proteins regulate their own synthesis by negatively regulating transcription (Aronson *et al.*, 1994). Light stimulates transcription of *per*, *tim*, and *frq* to reset the molecular rhythm. Although for *frq* the light-sensitive factor is probably not an oscillating component of the pacemaker (Crosthwaite, Loros, & Dunlap, 1995), these studies make a compelling argument for a universal molecular strategy, especially as the comparisons involve organisms representing different kingdoms.

The *per* gene has now been cloned from insect species representing three orders, Diptera, Lepidoptera, and Orthoptera (Reppert, Tsai, Rosa, & Sauman, 1994). Outside *Diptera*, studies of cellular and molecular patterns of activity have only been reported in the silkmoth *Antheraea pernyi*. Here, patterns of cycling *per* RNA and protein synthesis are seen in the eyes of the adult and in certain cells of the embryonic gut. These patterns are indistinguishable from those previously observed in the eyes and brain of *D. melanogaster*, indicating a highly conserved molecular mechanism (Sauman and Reppert, 1996; Sauman, Tsai, Roca, & Reppert, 1996).

At the same time, intriguing patterns of PER and TIM expression have been reported in the adult brain, and it has been suggested that the proteins may contribute to a second, distinct molecular pacemaker in these cells (Sauman and Reppert, 1996). Here, both proteins accumulate only in the cytoplasm of the studied cells. This is reminiscent of the cytoplasmic PER expression found in the ovaries of *Drosophila*. In the silkmoth brain, rhythms of expression are detected, but unlike the rhythms of the eyes and gut, *per* RNA and protein in these brain cells accumulate with the same phase, possibly indicating a lack of autoregulation. An antisense RNA is detected in the brain cells, and it is expressed with an altered phase. If a cell autonomous feedback loop promotes PER cycling in these cells, control might be manifest through effects on the antisense transcript (Sauman and Reppert, 1996). Thus, the silkmoth may present a view of evolving, alternative strategies for pacemaker organization around common, cycling clock proteins, PER and TIM.

An important caveat is that there is so far no evidence that the brain cells autonomously generate the observed molecular oscillations. Recall that PER is abundantly produced in cells having no clear pacemaker activity in *Drosophila* (Hardin, 1994; Liu *et al.*, 1988; Siwicki *et al.*, 1988; Saez and Young, 1988). A comparison of TIM and PER responses to dawn in a light/dark cycle raises further doubts because TIM shows no distinct light sensitivity in the adult silkmoth brain cells (Sauman and Reppert, 1996). If the cells are autonomous pacemakers, a novel entrainment mechanism is also required. Another concern is that embryonic studies of the brain cells do not show molecular oscillations (and once again TIM is not degraded by light). The rhythms must emerge later in development. Nevertheless, the embryos possess a light-sensitive circadian pacemaker that regulates the time of hatching, and this pacemaker depends on the function of PER (Sauman *et al.*, 1996). The only molecular oscillations detected during embryogenesis are restricted to the gut, where PER nuclear cycling is observed as in *Drosophila* (Sauman *et al.*, 1996).

Modeling a Light-Sensitive Circadian Pacemaker

Sehgal *et al.* (1995) put forward a model for a self-sustaining molecular pacemaker based on the known interactions of *tim* and *per* in *Drosophila.* This model was extended to incorporate light's effects on TIM stability (Hunter-Ensor *et al.*, 1996; Lee *et al.*, 1996; Myers *et al.*, 1996; Zeng *et al.*, 1996). If action of the pacemaker in the presence of a solar day is considered, much of the mechanism is simplified by TIM's light sensitivity (Figure 7). At dawn, TIM is destroyed by light either as a monomeric protein or in PER/TIM complex. Absence of PER/TIM complexes permits transcription of both *tim* and *per.* The daytime rise of RNA is not accompanied by accumulation of either protein, as TIM is continuously destroyed by light and PER is unstable in the cytoplasm when not complexed with TIM. At sunset, TIM proteins begin to accumulate from enlarged *tim* RNA pools. When sufficiently abundant, TIM begins to stabilize nascent PER proteins through PER/TIM heterodimerization. Interaction with PER also suppresses CLD activities on PER and TIM, allowing nuclear localization of the complexes. As nuclear PER/TIM heterodimers accumu-

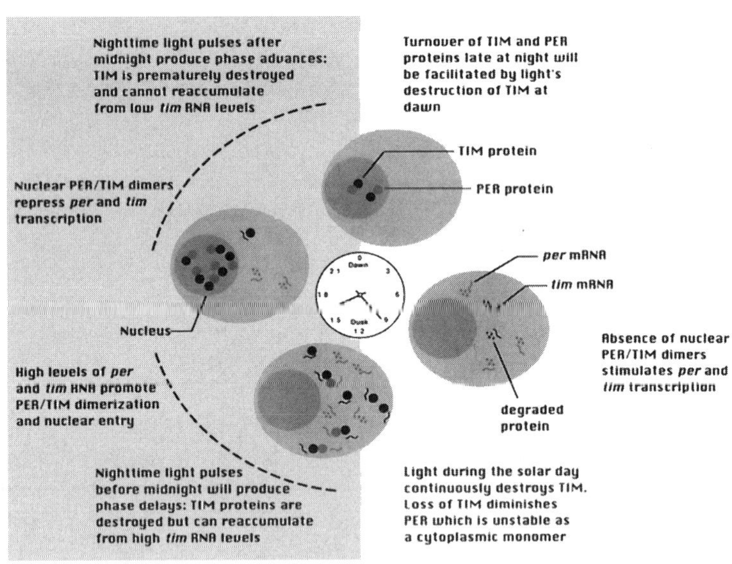

Figure 7. Model for the PER/TIM circadian pacemaker. Cycles of *per* (green) and *tim* (red) RNA and protein acummulation are depicted. As indicated (see also text), TIM is degraded by light, which holds molecular cycles to the phase and period of the solar day. At sunset, light-sensitive TIM proteins begin to accumulate and can bind to and stabilize PER. PER/TIM complexes are translocated to nuclei and this relocalization is associated with suppression of further *per* and *tim* transcription. Pulses of light at night eliminate TIM to reset the phase of the molecular oscillation, giving advances when delivered between CT18 and CT0, and delays when administered between CT12 and CT18. The phase of *per* and *tim* RNA accumulation and the phase of physical association of the PER and TIM proteins differ by ~4–6 hours, which sustains molecular oscillations in constant darkness (see text).

late, *per* and *tim* transcription are coordinately repressed. Declining RNA pools no longer favor heterodimerization of newly translated PER and TIM proteins. This further accelerates the drop in PER protein levels due to instability of cytoplasmic PER monomers. At least two factors contribute to completion of a molecular cycle: (1) the nuclear turnover of PER/TIM complexes as they fail to be replaced by newly synthesized proteins, and (2) the end, due to sunrise, of any residual activity of nuclear PER/TIM complexes.

While a light/dark cycle can insure molecular cycling with a corresponding period length, self-sustaining oscillations are expected in constant darkness because of the separate phases of RNA synthesis and PER/TIM complex formation. The timing of PER/TIM heterodimerization should be determined by rates of accumulation of *per* and *tim* RNAs, affinity of PER for TIM, and the rate of elimination of monomeric PER proteins. Each of these elements of the pacemaker mechanism promotes phase differences between the timing of RNA synthesis and autoregulatory activities associated with the PER/TIM complex. For example, it was shown that modification of only one of these parameters, depressed binding of PER to TIM in the *per*^{L} mutant, significantly lengthened the period of the molecular and behavioral rhythm (Gekakis *et al.*, 1995).

Unsettled Questions

A challenge in future work will be to determine how a circadian rhythm, rather than a higher or lower frequency rhythm, is produced by these interactions in the absence of entrainment. The specific period of the molecular oscillation is likely to be controlled by unidentified factors that specify rates of RNA and protein synthesis and degradation. This is already supported by the differential sensitivities of PER and TIM to light and the different patterns of PER and TIM monomer accumulation. What are the factors responsible for TIM degradation following exposure to light? Does a cytoplasmic protease specifically degrade monomeric PER proteins? What dictates the half-life of the PER/TIM complex in the nucleus? Cycling phosphorylation is found for both PER and TIM (Edery, Zweibel, Dembinska, & Rosbash, 1994; Lee *et al.*, 1996; Zeng *et al.*, 1996). Do postranslational modifications affect stability or activity?

And what of temperature compensation? Circadian rhythms are notoriously insensitive to changes in temperature. Is there a single element of the molecular cycle that is a target for thermoregulation? One possibility is raised by studies of the PER/TIM interaction in yeast. Although Gekakis *et al.* (1995) observed a substantial, temperature-sensitive change in PER/TIM binding when testing the PER^{L} form of the protein, and a smaller temperature effect can be seen even for the wild-type proteins. In a yeast PER/TIM interaction assay that followed levels of enzyme activity for a β-galactosidase reporter, enzyme activity (reflecting binding strength) rose twofold as temperature was decreased from 35°C to 28°C (Gekakis *et al.*, 1995). If the interaction of wild type PER and TIM proteins is similarly temperature dependent in flies, it follows that with increasing temperature, PER/TIM complexes may become more active, but less easily formed.

A new set of questions arises when behaviors regulated by the *Drosophila* clock are considered. The brain transplantation experiments of Truman (1994) in *Manduca* and Handler and Konopka (1979) in *Drosophila* showed that an insect's clock could produce rhythmic behavior through action of a diffusible substance: A donor brain transplanted to the abdomen of a recipient fly generated rhythmic output from the host nervous system (Handler and Konopka, 1979). There is now comparable evidence for such factors in mammals (Silver, LeSauter, Tresco, & Lehman, 1996). The substance(s) has not been identified in *Manduca* or *Drosophila* (or mammals), but possibly involves the product of a clock-controlled gene (CCG). These have been described in *Neurospora* (Loros, Denome, & Dunlap, 1989), *Drosophila* (Van Gelder and Krasnow, 1996; Van Gelder, Bae, Palazzolo, & Krasnow, 1995), cyanobacteria (Liu *et al.*, 1995), and a variety of vertebrates (Borjigin, Wang, & Snyder, 1995; Coon *et al.*, 1995; Foulkes, Duval, & Sassone-Corsi, 1996; Korenbrot and Fernald, 1989; Pierce *et al.*, 1993) and plants (Giuliano, Hoffman, Ko, Scolnik, & Cashmore, 1988; Nagy, Kay, & Chua, 1988). CCGs are targets of the circadian pacemaker, because transcription can be rhythmic and variously phased, but loss of a CCG fails to block pacemaker function. In some cases, the encoded product is already known to influence rhythmic behavior; rhythmic serotonin N-acetyltransferase expression controls timed synthesis of melatonin (Borjigin *et al.*, 1995; Coon *et al.*, 1995). Perhaps the mechanism underlying autoregulation of *per* and *tim* initiates a cascade of gene expression that includes control of intrinsic and extracellular signals. As the list of CCGs grows, a network of timed functions should be revealed. These may begin to explain how a multitude of developmental and behavioral signals can be generated in response to a relatively simple molecular oscillator.

References

Aronson, B. D., Johnson, K. A., Loros, J. J., & Dunlap, J. C. (1994). Negative feedback defining a circadian clock: autoregulation of the clock gene *frequency*. *Science, 263*, 1578–1584.

Bargiello, T. A., & Young, M. W. (1984). Molecular genetics of a biological clock in Drosophila. *Proceedings of the National Academy of Sciences of the USA, 81*, 2142–2146.

Bargiello, T. A., Jackson, F. R., & Young, M. W. (1984). Restoration of circadian behavioral rhythms by gene transfer in *Drosophila. Nature, 312*, 752–754.

Baylies, M. K., Bargiello, T. A., Jackson, F. R., and Young, M. W. (1987). Changes in abundance or structure of the *per* gene product can affect periodicity of the *Drosophila* clock. *Nature, 326*, 390–392.

Borjigin, J., Wang, M. M., and Snyder, S. H. (1995). Diurnal variation in mRNA encoding serotonin *N*-acetyltransferase in pineal gland. *Nature, 378*, 783–785.

Burbach, K. M., Poland, A., & Bradfield, C. A. (1992). Cloning of the Ah-receptor cDNA reveals a distinctive ligand-activated transcription factor. *Proceedings of the National Academy of Sciences of the USA, 89*, 8185–8189.

Citri, Y. V., Colot, H. V., Jacquier, A. C. Yu, Q., Hall, J. C., Baltimore, D., & Rosbash, M. (1987). A family of unusually spliced biologically active transcripts encoded by a *Drosophila* clock gene. *Nature, 326*, 42–47.

Coon, S. L., Roseboom, P. H., Ruben, B., Weller, J. L., Namboodiri, M. A. A., Koonin, E. V., & Klein, D. C. (1995). Pineal serotonin *N*-acetyltransferase: Expression cloning and molecular analysis. *Science, 270*, 1681–1683

Cote, G. C., & Brody, S. (1986). Circadian rhythms in *D. melanogaster*. Analysis of period as a function of gene dosage at the *per* (*period*) locus. *Journal of Theoretical Biology, 121*, 487–503.

Crews, S. T., Thomas, J. B., & Goodman, C. S. (1988). The *Drosophila single-minded* gene encodes a nuclear protein with sequence similarity to the *per* gene product. *Cell, 52*, 143–151.

Crosthwaite, S. K., Loros, J. J., & Dunlap, J. C. (1995). Light-induced resetting of a circadian clock is mediated by a rapid increase in frequency transcript. *Cell, 81*, 1003–1012.

Curtin, K. D., Huang, Z. J., & Rosbash, M. (1995). Temporally regulated nuclear entry of the Drosophila period protein contributes to the circadian clock. *Neuron, 14*, 365–372.

Edery, I., Zweibel, L. J., Dembinska, M. E., & Rosbash, M. (1994). Temporal phosphorylation of the *Drosophila period* protein. *Proceedings of the National Academy of Sciences of the USA, 91*, 2260–2264.

Emery, I. F., Noveral, J. M., Jamison, C. F., and Siwicki, K. K., (1997). Rhythms of Drosophila period gene expression in culture. *Proceedings of the National Academy of Sciences of the USA, 94*, 4092–4096.

Ewer, J., Frisch, B., Hamblen-Coyle, M. J., Rosbash, M., & Hall, J. C. (1992). Expression of the *period* clock gene within different cell types in the brain of *Drosophila* adults and mosaic analysis of the cells' influence on circadian behavioral rhythms. *Journal of Neuroscience, 12,* 3321–3349.

Feldman, J. F., & Hoyle, M. N. (1973). Isolation of circadian clock mutants of *Neurospora crassa. Genetics, 75,* 605–613.

Foulkes, N. S., Duval, G., and Sassone-Corsi, P. (1996). Adaptive inducibility of CREM as transcriptional memory of circadian rhythms. *Nature, 381,* 83–85

Frisch, B., Hardin, P. E., Hamblen-Coyle, M. J., Rosbash, M., & Hall, J. C. (1994). A promoterless period gene mediates behavioral rhythmicity and cyclical per expression in a restricted subset of the Drosophila Nervous system. *Neuron, 12,* 555–570.

Gekakis, N., Saez, L., Delahaye-Brown, A-M., Myers, M. P., Sehgal, A., Young, M. W., & Weitz, C. J. (1995). Isolation of *timeless* by PER protein interaction: Defective interaction between *timeless* protein and long-period Mutant PER^L. *Science, 270,* 811–815.

Giebultowicz, J. M. & Hege, D. M. (1997). Circadian clock in malphigian tubules. *Nature, 386,* 664.

Giuliano, G., Hoffman, N. E., Ko, K., Scolnik, P. A., & Cashmore, A. R. (1988). A light-entrained circadian clock controls transcription of several plant genes. *EMBO Journal. 7,* 3635–3642.

Hall, J. C. (1996). Are cycling gene products as internal zeitgebers no longer the zeitgeist of chronobiology? *Neuron, 17,* 799–802.

Handler, A. M., & Konopka, R. J. (1979). Transplantation of a circadian pacemaker in *Drosophila. Nature, 279,* 236–238.

Hardin, P. E. (1994). Analysis of period mRNA cycling in Drosophila head and body tissues indicates that body oscillators behave differently from head oscillators. *Molecular and Cellular Biology, 14,* 7211–7218.

Hardin, P. E., Hall, J. C., & Rosbash, M. (1990). Feedback of the *Drosophila period* gene product on circadian cycling of its messenger RNA levels. *Nature, 343,* 536–540.

Hardin, P. E., Hall, J. C., & Rosbash, M. (1992). Circadian oscillations in *period* gene mRNA levels are transcriptionally regulated. *Proceedings of the National Academy of Sciences of the USA, 89,* 11711–11715.

Hoffman, E. C., Reyes, J., Chu, F. F., Sander, F., Conley, L. H., Brooks, B. A., & Hankinson, O. (1991). Cloning of a factor required for activity of the Ah (dioxin) receptor. *Science, 252,* 954–958.

Huang, Z. J., Edery, I., & Rosbash, M. (1993). PAS is a dimerization domain common to *Drosophila* period and several transcription factors. *Nature, 364,* 259–262.

Hunter-Ensor, M., Ousley, A., & Sehgal, A. (1996). Regulation of the *Drosophila* protein Timeless suggests a mechanism for resetting the circadian clock by light. *Cell, 84,* 677–685.

Jackson, F. R., Bargiello, T. A., Yun, S.-H., & Young, M. W. (1986). Product of *per* locus of *Drosophila* shares homology with proteoglycans. *Nature 320,* 185–188.

Konopka, R. J., & Benzer, S. (1971). Clock mutants of *Drosophila melanogaster. Proceedings of the National Academy of Sciences of the USA, 68,* 2112–2116.

Konopka, R. J. (1981). Genetics and development of circadian rhythms in invertebrates. In J. Aschoff (Ed.), *Handbook of behavioral neurobiology* (pp. 173–182). New York: Plenum Press.

Konopka, R. J., Wells, S., and Lee, T. (1983). Mosaic analysis of a Drosophila clock mutant. *Molecular and General Genetics, 190,* 284–288.

Konopka, R. J., Pittendrigh, C. S., & Orr, D. (1989). Reciprocal behavior associated with altered homeostasis and photosensitivity of *Drosophila* clock mutants. *Journal of Neurogenetics, 6,* 1–10.

Korenbrot, J. I., & Fernald, R. D. (1989). Circadian rhythm and light regulate opsin mRNA in rod photoreceptors. *Nature, 337,* 454–457.

Lee, C., Parikh, V., Itsukaichi, T., Bae, K., & Edery, I. (1996). Resetting the *Drosophila* clock by photic regulation of PER and PER-TIM complex. *Science, 271,* 1740–1744.

Liu, X., Lorenz, L. J., Yu, Q., Hall, J. C., & Rosbash, M. (1988). Spatial and temporal expression of the *per* gene in *D. melanogaster. Genes and Development, 2,* 228–238.

Liu, X., Zweibel, L. J., Hinton, D., Benzer, S., Hall, J. C., & Rosbash, M. (1992). The *period* gene encodes a predominantly nuclear protein in adult *Drosophila. Journal of Neuroscience, 12,* 2735–2744.

Liu, Y., Tsinoremas, N. F., Johnson, C. H., Lebedeva, N. V., Golden, S. S., Ishiura, M., & Kondo, T. (1995). Circadian orchestration of gene expression in cyanobacteria. *Genes and Development, 9,* 1469–1478.

Loros, J. J., Denome, S. A., & Dunlap, J. C. (1989). Molecular cloning of genes under control of the circadian clock in Neurospora. *Science, 243,* 385–388

Millar, A. J., Carre, I. A., Strayer, C. A., Chua, N.-H., & Kay, S. (1995). Circadian clock mutants in Arabidopsis identified by luciferase imaging. *Science, 267,* 1161–1163

Myers, M. P., Wager-Smith, K., Wesley, C. S., Young, M. W., & Sehgal, A. (1995). Positional cloning and sequence analysis of the *Drosophila* clock gene, *timeless. Science, 270,* 805–808.

Myers, M. P., Wager-Smith, K., Rothenfluh-Hilfiker, A., & Young, M. W. (1996). Light-induced degradation of Timeless and entrainment of the *Drosophila* circadian clock. *Science, 271,* 1736–1740.

Nagy, F., Kay, S. A., & Chua, N.-H. (1988). A circadian clock regulates transcription of the wheat *Cab-1* gene. *Genes and Development, 2,* 376–382.

Pierce, M. E., Sheshberadaran, H., Zhang, Z., Fox, L. E., Applebury, M. L., & Takahashi, J. S. (1993). Circadian regulation of iodopsin gene expression in embryonic photoreceptors in retinal cell culture. *Neuron, 10,* 579–584.

Pittendrigh, C. S. (1967). Circadian systems I. The driving oscillation and its assay in *Drosophila pseudoobscura. Proceedings of the National Academy of Sciences of the USA, 58,* 1762–1767.

Pittendrigh, C. S. (1981). Circadian systems: Entrainment. In J. Aschoff (Ed.), *Handbook of Behavioral Neurobiology* (pp. 95–124). New York: Plenum Press.

Price, J. L., Dembinska, M. E., Young, M. W., & Rosbash, M. (1995). Suppression of PERIOD protein abundance and circadian cycling by the *Drosophila* clock mutation timeless. *EMBO Journal, 14,* 4044–4047.

Ralph, M. R., & Menaker, M. (1988). A mutation of the circadian system in golden hamsters. *Science, 241,* 1225–1227.

Reddy, P., Zehring, W. A., Wheeler, D. A., Pirrota, V., Hadfield, C., Hall, J. C., & Rosbash, M. (1984). Molecular analysis of the *period* locus of *Drosophila melanogaster* and identification of a transcript involved in biological rhythms. *Cell, 38,* 701–710.

Reppert, S. M., Tsai, T., Roca, A. L., & Sauman, I. (1994) . Cloning of a structural and functional homolog of the circadian clock gene *period* from the giant silkmoth *Antheraea pernyi. Neuron, 13,* 1167–1176

Reyes, H., Reisz-Porszasa, S., & Hankinson, O. (1992). Identification of the Ah receptor nuclear translocator protein (Arnt) as a component of the DNA binding form of the Ah receptor. *Science, 256,* 1193–1195.

Rutila, J. E., Zeng, H., Le, M., Curtin, K. D., Hall, J. C., & Rosbash, M. (1996). The tim^{SL} mutant of the Drosophila rhythm gene *timeless* manifests allele-specific interactions with *period* gene mutants. *Neuron, 17,* 921–929.

Saez, L., & Young, M. W. (1988). *In situ* localization of the *per* clock protein during development of *Drosophila melanogaster.* Mol. Cell. Biol.. 8, 5378–5385.

Saez, L., & Young, M. W. (1996). Regulation of nuclear entry of the *Drosophila* clock proteins Period and Timeless. *Neuron 17,* 911–920.

Sauman, I., & Reppert, S. M. (1996). Circadian clock neurons in the silkmoth *Antheraea pernyi*: novel mechanisms of period protein regulation. *Neuron, 17,* 889–900.

Sauman, I., Tsai, T., Roca, A. L., & Reppert, S. M. (1996). Period protein is necessary for circadian control of egg hatching behavior in the silkmoth Antheraea pernyi. *Neuron 17,* 901–909.

Saunders, D. S. (1990). The Circadian Basis of Ovarian Diapause regulation in *Drosophila melanogaster.* Is the *period* gene causally involved in photoperiodic time measurement? *Journal of Biological Rhythms, 5,* 315 331.

Sehgal, A., Price, J. L., Man, B., and Young, M. W. (1994). Loss of circadian behavioral rhythms and *per* RNA oscillations in the *Drosophila* mutant *timeless. Science, 263,* 1603–1606.

Sehgal, A., Rothenfluh-Hilfer, A., Hunter-Ensor, M., Chen, Y., Myers, M. P., & Young, M. W. (1995). Rhythmic expression of *timeless*: A basis for promoting circadian cycles in *period* gene autoreguation. *Science, 270,* 808–810.

Silver, R., LeSauter, J., Tresco, P. A., & Lehman, M. N. (1996). A diffusible coupling signal from the transplanted suprachiasmatic nucleus controlling circadian locomotor rhythms. *Nature, 382,* 810–813.

Siwicki, K. K., Eastman, C., Petersen, G., Rosbash, M., & Hall, J. C. (1988). Antibodies to the period gene product of *Drosophila* reveal diverse tissue distribution and rhythmic changes in the visual system. *Neuron, 1,* 141–150.

Smith, R. F., & Konopka, R. J. (1982). Effects of dosage alterations at the *per* locus on the period of the circadian clock of *Drosophila. Molecular and General Genetics, 185,* 30–36.

Takahashi, J. S., & Zatz, M. (1982). Regulation of circadian rhythmicity. *Science, 217,* 1104–1111.

Truman, J. W. (1974). Physiology of insect rhythms IV. Role of the brain in the regulation of the flight rhythm of the giant silkmoth. *Journal of Comparative Physiology, 95,* 281–296.

Van Gelder, R., & Krasnow, M (1996). A novel circadianly expressed Drosophila melanogaster gene dependent on the period gene for its rhythmic expression. *EMBO Journal, 15,* 1625–1631.

Van Gelder, R., Bae, H., Palazzolo, M., and Krasnow, M. (1995). Extent and character of circadian gene expression in *Drosophila melanogaster.* Identification of 20 oscillating mRNAs in the fly head. *Current Biology, 5,* 1424–1436.

Vitaterna, M. H., King, D. P., Chang, A.-M., Kornhauser, J. M., Lowrey, P. L., McDonald, J. D., Dove, W. F., Pinto, L. H., Turek, F. W., & Takahashi, J. S. (1994). Mutagenesis and mapping of a mouse gene, *Clock,* essential for circadian behavior. *Science, 264,* 719–725.

Vosshall L. B., Price, J. L., Sehgal, A., Saez, L., & Young, M. W. (1994). Block in nuclear localization of *period* protein by a second clock mutation, *timeless. Science, 263,* 1606–1609.

Vosshall, L. B., & Young, M. W. (1995). Circadian rhythms in *Drosophila* can be driven by *period* expression in a restricted group of central brain cells. *Neuron, 15,* 345–360.

Winfree, A. T. (1970). The temporal morphology of a biological clock. In M. Gerstenhaber (Ed.), *Lectures on Mathematics in the Life Sciences* (vol. 2, p. 109). Rhode Island: American Mathematical Society.

Young, M. W., & Judd, B. (1978). Nonessential sequences, genes, and the polytene chromosome bands of *Drosophila melanogaster. Genetics, 88,* 723–742.

Young, M. W., Wager-Smith, K., Vosshall, L. B., Saez, L., & Myers, M. P. (1996). Molecular anatomy of a light-sensitive circadian pacemaker in *Drosophila. Cold Spring Harbor Symposia on Quantitaive Biology, 61,* 279–284.

Yu, Q., Jacquier, A. C., Colot, H. V., Citri, Y., Hamblen, M., Hall, J. C., & Rosbash, M. (1987). Molecular mapping of point mutations in the *period* gene that stop or speed up biological clocks in *Drosophila melanogaster. Proceedings of the National Academy of Sciences of the USA, 84,* 784–788.

Zehring, W. A., Wheeler, D. A., Reddy, P, Konopka, R. J., Kyriacou, C. P, Rosbash, M., & Hall, J. C. (1984). P-element transformation with *period* locus DNA restores rhythmicity to mutant, arrhythmic *Drosophila melanogaster. Cell, 39,* 369–376.

Zeng, H., Hardin , P. E., & Rosbash, M. (1994). Constitutive overexpression of the *Drosophila period* protein inhibits *period* mRNA cycling. *EMBO Journal, 13,* 3590–3598.

Zeng, H., Qian, Z., Myers, M. P., & Rosbash, M. (1996). A light-entrainment mechanism for the Drosophila circadian clock. *Nature, 380,* 129–135.

14

Cellular and Molecular Analysis of Molluscan Circadian Pacemakers

EDWARD M. BLUMENTHAL, GENE D. BLOCK, AND ARNOLD ESKIN

INTRODUCTION

A detailed understanding of how cells and tissues generate 24-hour periodicities remains one of the great challenges to contemporary cell and molecular science. Nonetheless, there has been near spectacular progress in the past 5 years in our knowledge about the molecular and cellular machinery responsible for generating circadian rhythms. This progress has come about in large part through the use of molecular genetic approaches in mature (e.g., *Drosophila, Neurospora*) and new (e.g., *Arabidopsis*, mouse) chronobiological model systems.

Amid the impressive progress achieved in systems allowing for genetic manipulations, the opisthobranch mollusks remain unique as the only preparations where a positive identification of pacemaker neurons has been achieved within nervous tissue (Barnes & Jacklet, 1997; Michel, Geusz, Zaritsky, & Block, 1993). Moreover, the ability to study circadian rhythmicity at many levels within a single model (Block, Geusz, Khalsa, Michel, & Whitmore, 1996; Koumenis & Eskin, 1992; see also Chapter 4) has provided unparalleled opportunities to gain a *comprehensive* understanding of biological timekeeping. Through the use of the opisthobranch model our knowledge has been deepened in many areas, from the control of circadian behaviors by cellular pacemakers, to the cellular dynamics of multioscillator organization, to our

EDWARD M. BLUMENTHAL AND GENE D. BLOCK Department of Biology and NSF Center for Biological Timing, University of Virginia, Charlottesville, Virginia 22903. ARNOLD ESKIN Department of Biology and Biochemistry, University of Houston, Houston, Texas 77204.

Circadian Clocks, Volume 12 of *Handbook of Behavioral Neurobiology*, edited by Joseph S. Takahashi, Fred W. Turek, and R. Y. Moore, Kluwer Academic / Plenum Publishers, New York, 2001.

ability to measure and manipulate circadian parameters within individual neurons. Indeed, much of what we currently understand about the synchronization, generation, and expression of circadian rhythms derives from work on molluscan preparations.

A Circadian Pacemaker in the Eye

The eye of opisthobranch mollusks contains a circadian pacemaker. The behavior of this pacemaker can be most easily monitored by removing the eye from the animal and recording spontaneous electrical activity from the optic nerve. The isolated eye remains viable and active for more than 1 week in seawater; this is obviously important when studying low-frequency oscillations such as circadian rhythms. A typical example of an optic nerve activity record is shown in Figure 1A. Prominent in such recordings are large amplitude impulses called compound action potentials (CAPs), which are so named because their variable amplitude indicates that they arise from a population of synchronously firing neurons (Jacklet, 1969a). Long-term recordings from eyes kept in constant darkness (DD) show a clear circadian rhythm in CAP frequency with a period very close to 24 hours (Figure 1B). Activity increases rapidly around subjective dawn, plateaus during the early day, declines, and eventually ceases around subjective dusk. While the waveform of the rhythm varies from animal to animal, the waveforms of the two eyes of any given animal are remarkably reproducible, allowing precise measurements of the effects of altered ionic conditions or pharmacologic manipulations. Although an exhaustive study of opisthobranch species has not been conducted, a CAP rhythm has been observed in all five members of the class that have been tested: *Aplysia californica* (Jacklet, 1969b), *Bulla gouldiana* (Block & Wallace, 1982), *Bursatella leachi plei* (Block & Roberts, 1981), *Navanax inermis* (Eskin & Harcombe, 1977), and *Haemonea vesicula* (McMahon & Block, 1982); the first two, *Aplysia* and *Bulla,* have been most intensively studied and are the focus of the remainder of this chapter. The isolated eye fulfills all of the criteria for a biological clock: the CAP rhythm persists in constant darkness (Block & Wallace, 1982; Jacklet, 1969b), can be phase shifted and entrained by light (Block & Wallace, 1982; Eskin, 1971; Jacklet, 1974), and is temperature compensated, with the period of the rhythm in *Aplysia* exhibiting a Q_{10} of 1.2 between 10°C and 23°C (Benson & Jacklet, 1977) and in *Bulla,* a Q_{10} of 1.0 between 12°C and 25°C (Bogart, 1992).

In *Aplysia* and *Bulla,* it has been possible to record optic nerve activity in the intact animal and to demonstrate that a circadian rhythm in CAP activity also exists *in vivo* (Block, 1981; Block & Davenport, 1982). In the intact system, the waveform of the rhythm is not as reproducible or precise as in the isolated eye due to modulation of impulse activity by efferent signals from the central nervous system, and, in *Bulla,* by signals from the contralateral eye (Eskin, 1971; Roberts & Block, 1985; see below and also Chapter 4). In other respects, however, such as free-running period and the phase relationship between the light cycle and the CAP rhythm, the rhythms recorded *in vivo* and *in vitro* are quite similar.

The ocular pacemaker is an important component of the circadian system in the intact mollusk (see also Chapter 4). *Aplysia* show a clear circadian rhythm in locomotor behavior; they are diurnally active and typically begin moving several hours before dawn (Block & Lickey, 1973; Kupfermann, 1967; Lickey, Wozniak, Block, Hudson, & Augter, 1977; Strumwasser, 1967). Following eye removal, animals continue to show rhythmic activity on an LD cycle, although the predawn "anticipatory activity" is no longer seen, and most eyeless animals become arrhythmic after a few days in constant darkness (Block & Lickey, 1973; Lickey *et al.*, 1977; Strumwasser, 1973). These observations demonstrate several aspects of the molluscan circadian system. First, continued diurnal activity in eyeless animals implies the existence of extraocular photoreceptors. It has been shown directly that the ganglia and several peripheral tissues are photoreceptive (Arvanitaki & Chalazonitis, 1961; Block & Smith, 1973; Chase, 1979; Eskin, 1971; Jacklet, 1980; Roberts & Block, 1982). Indeed, Block, Hudson, and Lickey (1974) demonstrated in *Aplysia* that a cycle of red light, which does not activate ocular photoreceptors, can entrain the ocular pacemaker (measured upon subsequent isolation of the eyes) via efferent optic nerve activity.

The deterioration of free-running behavior in eyeless *Aplysia* shows that the eyes are necessary for sustained rhythmicity. However, the persistence of circadian

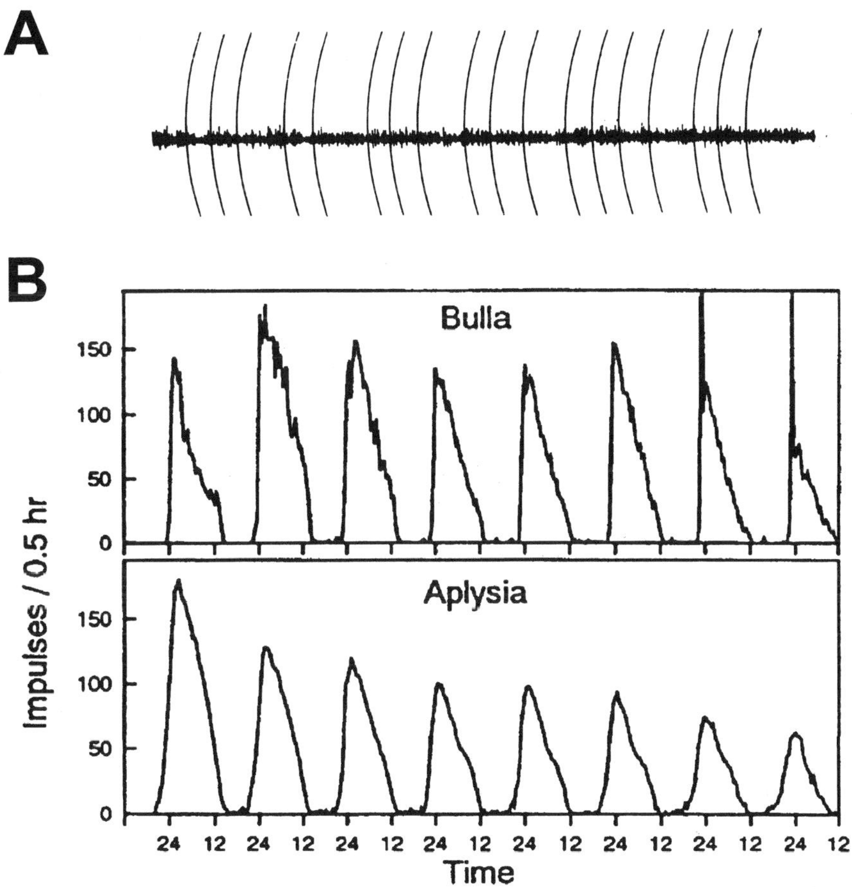

Figure 1. (A) Spontaneous optic nerve electrical activity from a *Bulla* eye recorded in constant darkness. The trace spans approximately 8 minutes, and the CAPs measure approximately 120 μV peak to peak. (From Block *et al.*, 1984; with permission. Copyright 1984, Springer-Verlag.) (B) Ocular circadian rhythms from *Bulla* and *Aplysia*. Spontaneous CAP frequency from eyes maintained in constant darkness at 15°C is plotted against time (in hours). (From Block, Khalsa, McMahon, Michel, & Guesz, 1993; with permission. Copyright 1984, Springer-Verlag.)

rhythms in some animals following eye removal or severance of the optic nerves indicates that at least one additional oscillator must be present elsewhere in the animal (Block & Lickey, 1973; Lickey *et al.*, 1977; Strumwasser, Alvarez, Viele, & Woolum, 1979). In the absence of the eyes, this oscillator is unable to control the locomotor rhythm, either because it is not self-sustaining for more than a few cycles or because the amplitude of its rhythm stabilizes at a level insufficient to influence behavior measurably. Lesion experiments implicate the cerebral ganglia as the location of this extraocular oscillator; cutting the nerve that connects the pedal ganglia, which contain the output neurons controlling locomotion, with the cerebral ganglia completely eliminates locomotor rhythmicity, even in an LD cycle (Roberts & Block, 1982). The precise cellular location of this oscillator, as well as the mechanism by which it is influenced by the eyes remain to be determined.

The organization of the *Bulla* circadian system appears to be quite similar to *Aplysia*. Intact *Bulla* exhibit a circadian rhythm in locomotor activity, although this species is nocturnally active. Interestingly, while eye removal again reduces or eliminates free-running rhythms, eyeless animals become diurnally active in an LD cycle (Block & Davenport, 1982). This demonstrates the existence of extraocular photoreceptors and suggests that the output from the ocular pacemakers governs locomotor rhythmicity by suppressing both the behavior as well as the excitatory signal from the extraocular photoreceptors during the day. The locations of the extraocular photoreceptors and oscillators in *Bulla* are not yet known, although the importance of the extraocular pacemaker in the intact animal has been demonstrated by comparing the phase relationship between the ocular CAP rhythm and the locomotor rhythm in different photoperiods. Because the time between onset of CAP activity (measured following subsequent isolation of the eye) and onset of locomotor activity varies depending upon photoperiod length, an extraocular oscillator must be critical in controlling locomotor behavior (Roberts & Xie, 1996). In addition, there is preliminary evidence that some animals continue to exhibit circadian locomotor rhythmicity following eye removal (Raiford, 1984).

Localization of the Pacemaker within the Eye

The *Bulla* and *Aplysia* eyes contain all of the components of a biological clock: input, oscillator, and output. In attempting to localize these functions within the eye, there is no *a priori* reason to assume that a single cell type would be responsible for phototransduction, circadian timekeeping, and generation of CAPs. Early experimental results were interpreted with the assumption that the three functions were subserved by separate cell types (Audesirk, 1973); in fact, many models predicted that generation of a long period rhythm did not occur within a single cell at all, but rather was an emergent property of a network of higher frequency oscillators (Jacklet & Geronimo, 1971).

The identification of the cells responsible for each aspect of pacemaker function was more straightforward in *Bulla*, which has a simpler retina than *Aplysia* and lends itself more readily to the long-term intracellular recordings that were necessary to assign functions to particular cell types. The *Bulla* eye consists of a lens surrounded by a layer of photoreceptors and a basal region from which the optic nerve runs (Figure 2). The photoreceptor layer contains approximately 1000 cells (Block, McMahon, Wallace, & Friesen, 1984; Jacklet & Colquhoun, 1983); intracellular recording showed that the most prevalent photoreceptor (R type) does not spike

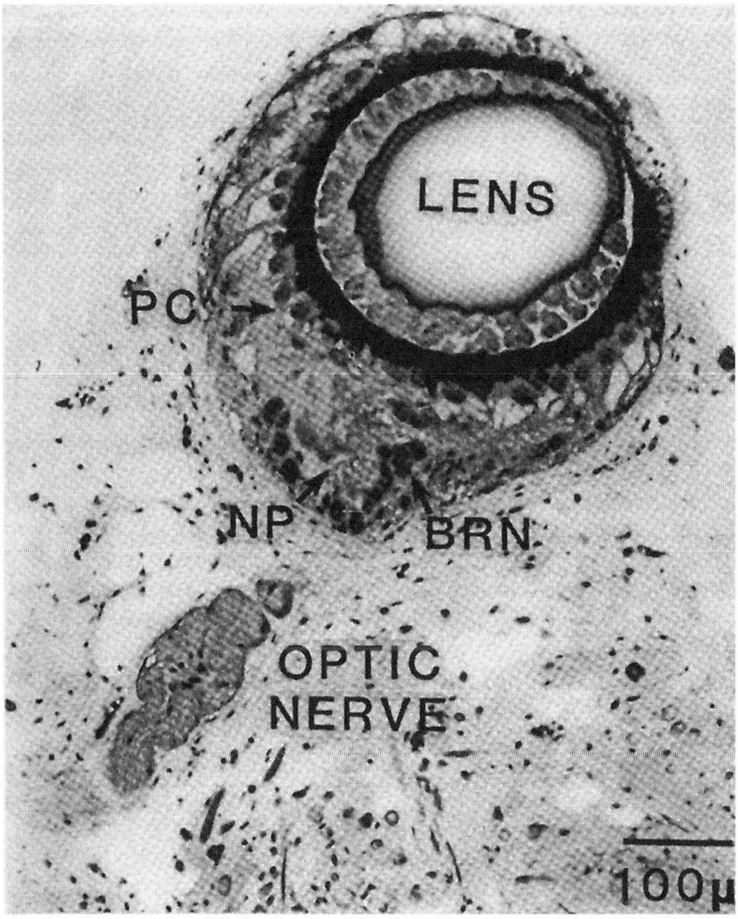

Figure 2. Thick section of a *Bulla* eye, showing discret localization of photoreceptors (PC), neuropil (NP), and basal retinal neurons (BRN). (From Block *et al.*, 1984; with permission. Copyright 1984, Springer-Verlag).

and does not exhibit a circadian rhythm in its resting membrane potential (Block & Wallace, 1982; Block *et al.*, 1984; McMahon, Wallace, & Block, 1984), while the spiking activity of the rarer H-type photoreceptors was not correlated with CAP activity (Block *et al.*, 1984). In contrast, a population of approximately 130 neurons surrounding the neuropil at the base of the retina was found to be electrically coupled and to fire action potentials in 1:1 correspondence with CAPs (Block *et al.*, 1984; Jacklet & Colquhoun, 1983; McMahon *et al.*, 1984). These neurons, called basal retinal neurons (BRNs) (Block *et al.*, 1984), also manifest a circadian rhythm in resting membrane potential (see below) (McMahon *et al.*, 1984). The BRNs are a homogeneous group of neurons, 15–25 μm in diameter, that extends neurites into the neuropil and axons out the optic nerve. The neurites are presumed to be the sites of electrical coupling as no gap junctions have been observed on BRN somata (Jacklet & Colquhoun, 1983).

The BRNs make up the entire circadian pacemaker in the *Bulla* eye. Surgical removal of the photoreceptor layer fails to disrupt the circadian rhythm in CAP frequency, and the resulting rhythm can still be shifted by light pulses, suggesting that the BRNs themselves are photoresponsive (Block & McMahon, 1984; Block & Wallace, 1982). Additional support for this idea comes from the finding that the BRNs express an opsin-like protein (Geusz, Foster, Degrip, & Block, 1997). Further reduction of the *Bulla* retina demonstrated that a fragment containing as few as six intact BRN somata is still capable of generating a circadian rhythm in CAP frequency (Block & McMahon, 1984). That the period of this rhythm appears to be similar to that of the intact eye strongly suggests that the 24-hour periodicity of the eye is not an emergent property of interactions among the BRNs but rather is intrinsic to single BRNs. The ability of an isolated BRN to act as a circadian oscillator has now been shown directly (Michel *et al.*, 1993; see below).

The localization of pacemaker functions within the *Aplysia* eye has been more difficult due to its greater anatomic complexity. The *Aplysia* eye contains at least five morphologically distinct photoreceptor cell types and at least two neuronal subtypes in the basal retina (Herman & Strumwasser, 1984). As in *Bulla*, surgical reduction of the eye suggested that the pacemaker is contained in the basal retina (Jacklet & Geronimo, 1971; Strumwasser, 1973), as did experiments showing that selective irradiation of the basal retina eliminates the CAP rhythm, whereas irradiation of the photoreceptor layer has no effect (Woolum & Strumwasser, 1980).

The neurons responsible for CAP production in *Aplysia* were identified as a population of approximately 1000 secondary neurons (also called neurosecretory cells, and now known as pacemaker neurons), 15–23 μm in diameter, in the basal portion of the eye (Herman & Strumwasser, 1984; Jacklet, Schuster, & Rolerson, 1982; Luborsky-Moore & Jacklet, 1977). By showing that localized illumination of the base of the eye was able to elicit CAP activity whereas illumination of the photoreceptor layer was not, Block and McMahon (1983) demonstrated that the pacemaker neurons are also likely to be directly photosensitive. Subsequent work has shown conclusively that these neurons are photosensitive in dispersed cell culture (Jacklet & Barnes, 1993) and most likely behave as autonomous circadian oscillators similar to *Bulla* BRNs (see below).

Mechanisms of Entrainment

Light

The phase of the circadian pacemaker in the eye can be shifted by two input signals: light and efferent transmitters from the central nervous system. As we will discuss below, evidence suggests that both of these pathways converge on a common second messenger, calcium, which regulates the phase of the pacemaker by a still unknown mechanism.

Exposing a *Bulla* or *Aplysia* eye to light has two effects. Acutely, light excites the retina, increasing CAP activity at all times of the circadian cycle (Block *et al.*, 1984; Jacklet, 1969a), although the degree of excitation differs somewhat depending on phase (Ralph & Block, 1990). Light pulses also can have a long-term effect of producing stable shifts in the CAP rhythm (Eskin, 1971). The magnitude and direction of these phase shifts is dependent upon the phase of the rhythm at which the light pulse is given; a typical phase response curve (PRC) to light is shown in Figure 3. Light pulses during the early subjective night delay the rhythm and light during

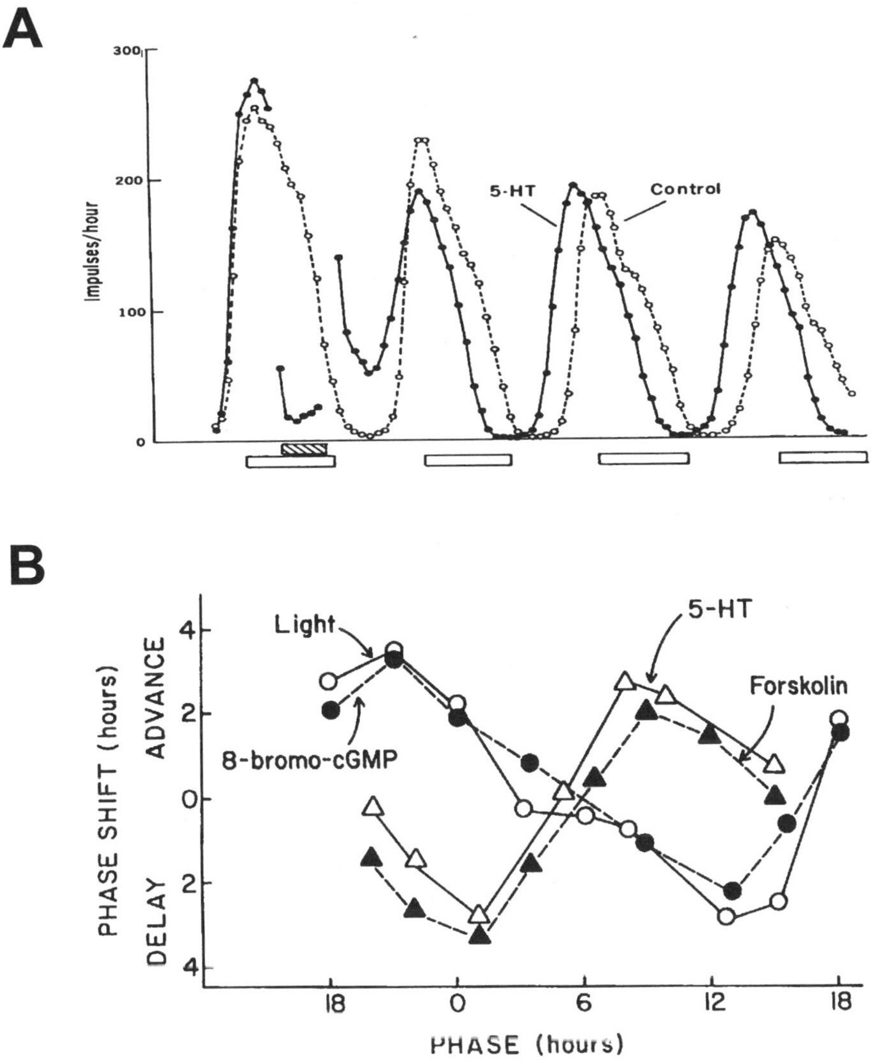

Figure 3. (A) Stable phase shift in the CAP rhythm of an *Aplysia* eye induced by serotonin. The frequency of spontaneous CAP activity from two isolated eyes from the same animal maintained in constant darkness is shown. The hatched bar indicates the 6-hour interval during which one eye was exposed to 100 nM serotonin. The open bars denote the projected light portion of the LD cycle to which the animals had been entrained. (From Corrent *et al.*, 1978; with permission. Copyright 1978, American Association for the Advancement of Science.) (B) PRCs of the *Aplysia* eye to 6-hour treatments with light (1200 lux), 8Br-cGMP (2 mM), serotonin (10 μM), and forskolin (1 μM). (From Koumenis & Eskin, 1992; with permission. Copyright 1992, Marcel Dekker.)

late night causes phase advances (Block & McMahon, 1984; Corrent, Eskin, & Kay, 1982; Jacklet, 1974).

The proximal second messenger for light, at least in *Aplysia*, appears to be cGMP. Pulses of the membrane permeant analog 8-bromo-cGMP (8Br-cGMP) have effects very similar to those of light, producing acute increases in CAP activity and long-term phase shifts (Eskin, Takahashi, Zatz, & Block, 1984a). The PRC for 8Br-cGMP is very similar to that of light, and combinations of light and 8Br-cGMP do not produce larger phase shifts than either agent alone. Finally, light pulses have been shown to increase the amount of cGMP present in the eye (Eskin *et al.*, 1984a). In *Bulla*, the role of cGMP is not clear. Application of 8Br-cGMP neither increases CAP activity nor shifts the phase of the CAP rhythm, but it is possible that the membrane permeability of 8Br-cGMP in the *Bulla* system is too low for this particular analog to have effects at the concentration tested (Whitmore & Block, 1996).

In both *Bulla* and *Aplysia*, light causes depolarization of the pacemaker neurons. This was first demonstrated in *Aplysia* with the observation that pulses of seawater containing a higher than normal potassium concentration, which should depolarize all of the cells in the eye, causes phase shifts with a PRC very similar to light (Eskin, 1972). Further studies showed that the effects of both light and 8Br-cGMP can be blocked by lowering the concentration of sodium in the bathing solution, thereby preventing the depolarization caused by either of these agents (Eskin, 1977; Eskin *et al.*, 1984a). This was shown more directly in *Bulla* BRNs. Sharp electrode recording of neurons in the intact eye showed that light pulses do depolarize the BRNs, stimulating CAP activity (McMahon & Block, 1987a). The phase-shifting effects of light in the *Bulla* eye can be mimicked either with high-potassium seawater or by directly injecting depolarizing current into a single BRN (because the BRNs are electrically coupled, this will depolarize the entire cell population). In addition, injecting hyperpolarizing current into the BRNs attenuates light-induced phase shifts (McMahon & Block, 1987a). Taken as a whole, these experiments demonstrate that depolarization of the pacemaker neurons is both necessary and sufficient for light to shift the phase of the ocular clock.

Membrane depolarization causes phase shifts by allowing calcium entry into the pacemaker neurons. Early experiments suggested that calcium entry was not necessary for light-induced phase shifts because bathing the *Aplysia* eye in a low-calcium/high-magnesium medium does not block light's effects (Eskin, 1972). In this study, however, a low millimolar concentration of calcium was still present; further reduction of the extracellular calcium concentration into the nanomolar range by addition of ethylene-bis(oxyethylenenitrilo) tetraacetic acid (EGTA) blocks light-induced phase shifts in both *Bulla* and *Aplysia* (Colwell, Whitmore, Michel, & Block, 1994; Khalsa & Block, 1988a; McMahon & Block, 1987a). Removal of extracellular calcium also blocks phase shifts induced by high potassium, demonstrating that calcium entry occurs subsequent to membrane depolarization and suggesting that voltage-gated calcium channels might be involved in the entrainment pathway (Khalsa & Block, 1988a). Consistent with this, application of Ni^{2+}, a specific blocker of voltage-gated calcium channels, attenuates light-induced phase shifts in both *Bulla* and *Aplysia* eyes (Colwell *et al.*, 1994; Khalsa & Block, 1988a). Finally, direct measurement of intracellular calcium concentrations in cultured BRNs confirmed that high potassium induces a prolonged increase in intracellular calcium, and this increase is blocked by Ni^{2+} (Geusz, Michel, & Block, 1994).

Efferent Transmitters

The second entrainment pathway appears to be activated by efferent signals from the central nervous system (see also Chapter 4). In *Aplysia*, the evidence suggests that the neurotransmitter for this pathway is serotonin. Fibers immunoreactive to serotonin are found in the optic and accessory optic nerves and terminate in and around the eye (Takahashi, Nelson, & Eskin, 1989). The isolated *Aplysia* eye is able to synthesize serotonin, presumably due to enzymes present in serotonergic nerve terminals (Corrent & Eskin, 1982). The actions of serotonin on the CAP rhythm, described below, can be mimicked by serotonergic agonists and are seen under ionic conditions which should block synaptic transmission, showing that serotonin acts on receptors on the pacemaker neurons themselves (Corrent & Eskin, 1982). Finally, bath application of serotonin is able to reproduce the effects of the cerebral ganglia in modulating the phase shifts of the CAP rhythm caused by LL to DD transitions (Nadakavukaren, Lickey, & Jordan, 1986).

Pulses of serotonin cause stable phase shifts in the CAP rhythm of the isolated *Aplysia* eye. Like light, serotonin can advance or delay the rhythm, depending on the phase of application; however, the PRC of serotonin is quite different from that of light (Figure 3). Serotonin causes phase delays in the early subjective day, advances in the late day, and has a dead zone during the night (Corrent, McAdoo, & Eskin, 1978; Corrent *et al.*, 1982). The dissimilarity between the PRCs of light and serotonin is strong evidence that the two treatments cause different biochemical changes within the pacemaker neurons.

In the *Bulla* eye, the efferent transmitter appears to be the tetrapeptide FMRFamide (phenylalanine-methionine-arginine-phenylalanine-amide), which can be detected immunohistochemically in fibers entering the eye and in cell bodies in the cerebral ganglia (Jacklet, Klose, & Goldberg, 1987; Roberts & Moore, 1987). Although in an early report FMRFamide (1–5 μM) was found not to phase shift the *Bulla* eye (Jacklet *et al.*, 1987), a later study using a relatively high concentration (50 μM) reported that FMRFamide can cause small phase delays during the early day and advances during late day (Colwell, Khalsa, & Block, 1992a). Thus, it appears that FMRFamide in *Bulla* may have the same role as serotonin in *Aplysia*, and, as discussed below, this similarity extends, more or less, to their mechanism of action.

Serotonin acts in *Aplysia* by stimulating the production of the second messenger cAMP. This has been shown directly by demonstrating an increase in cAMP levels in the eye following application of the transmitter (Eskin & Takahashi, 1983; Eskin, Corrent, Lin, & McAdoo, 1982). In addition, serotonin-induced phase shifts can be mimicked by the adenylate cyclase activator forskolin or the membrane-permeant analog 8-benzylthio-cAMP (8BzT-cAMP) and can be potentiated by phosphodiester ase inhibitors (Eskin & Takahashi, 1983; Eskin *et al.*, 1982). A cAMP analog has been reported to cause phase shifts in *Bulla* as well with a PRC similar to FMRFamide (Ralph, Khalsa, & Block, 1992), but there is no direct evidence that cAMP is a second messenger for FMRFamide in *Bulla*.

The efferent pathway phase-shifting agents (serotonin in *Aplysia*, FMRFamide in *Bulla*, and cAMP analogs in both species) inhibit spontaneous CAP activity (Benson, 1980; Eskin *et al.*, 1982). Intracellular recordings from BRNs show that both FMRFamide and cAMP analogs hyperpolarize the cell membrane (Jacklet *et al.*, 1987; McMahon & Block, 1987a); this change in membrane potential eliminates CAP activity by inhibiting the voltage-gated ion channels underlying the action potentials. The same effect can be obtained in a semiintact preparation by stimulating the area of the cerebral ganglia that contains FMRFamidergic cell bodies (Jacklet *et al.*, 1987). The membrane hyperpolarization appears to be due to the opening of ion channels, as it is associated with an increase in conductance across the cell membrane (Jacklet *et al.*, 1987).

The change in membrane potential is both necessary and sufficient for phase shifting by the efferent pathway just as it is in the light entrainment pathway. Direct injection of hyperpolarizing current into BRNs produces phase shifts similar to those caused by efferent stimulation (McMahon & Block, 1987a). Depolarization of the *Aplysia* eye with high potassium blocks serotonin-induced phase shifts (Eskin, 1982a), as does the potassium channel blocker Ba^{2+} (Colwell, Michel, & Block, 1992b). In *Bulla*, cAMP-mediated phase shifts are eliminated by the potassium channel blocker tetraethylammonium (TEA) (Ralph *et al.*, 1992); therefore the opening of potassium channels appears to mediate the hyperpolarization seen in both species. The failure of Ba^{2+} to block the increase in cAMP mediated by serotonin confirms that potassium channel activation and the subsequent hyperpolarization lie downstream of cAMP production (Colwell *et al.*, 1992b). It is not yet known whether cAMP activates potassium channels directly, by binding to the channel proteins, or indirectly, by activating the cAMP-dependent protein kinase; examples of both mechanisms have been reported in invertebrates, although action through kinase activation is much more common (Adams & Levitan, 1982; Delgado, Hidalgo, Diaz, Latorre, & Labarca, 1991; Kaczmarek & Strumwasser, 1984).

Continuing the parallel with the light entrainment pathway, intracellular calcium may also be a key messenger in the efferent pathway. In this case, hyperpolarization appears to act by inhibiting calcium entry; the most likely mechanism for this is the inhibition of action potentials, as calcium probably enters the cell when voltage-dependent calcium channels open during the action potential. In both *Bulla* and *Aplysia*, removing extracellular calcium causes phase shifts similar to those of hyperpolarizing treatments (Colwell *et al.*, 1994; Khalsa & Block, 1990). This result is especially convincing because while hyperpolarization inhibits CAP activity, removing extracellular calcium causes the eye to become hyperexcitable. That these two treatments should have the same effect on the rhythm suggests that it is not the hyperpolarizing effect of efferent transmitters per se that is important, but rather that hyperpolarizing the cell membrane inhibits the entry of calcium. For this model to be completely convincing, however, it still must be shown directly that efferent transmitters lower the intracellular calcium concentration of pacemaker neurons.

A Common Input Pathway

In light of the experimental evidence presented above, it seems clear that the two entrainment pathways in the molluscan eye are not separate pathways at all, but rather opposite sides of the same coin. Membrane potential and control of calcium entry appear to be the points of convergence for all known input signals in this system. This finding is able to explain the different phase dependences of light and efferent signals. During the subjective day, the pacemaker neurons are relatively depolarized and actively spiking (see discussion of output pathways below). At this time, additional depolarization by light does not shift the oscillator, either because there is no further increase in intracellular calcium or because any additional increase in calcium is beyond the dynamic sensitivity of the clock and so has no effect. Conversely, serotonin- or FMRFamide-mediated hyperpolarization during the subjective night, when the pacemaker neurons are not active, would not be expected to affect the clock either. The obvious corollary to this model is that the predicted rise in intracellular calcium that occurs when the pacemaker neurons spontaneously begin firing at subjective dawn would be expected to impact the motion of the clock. This feedback between input and output is critical to the understanding of this and other circadian systems and will be discussed at length below (see Figure 5).

The relationship between calcium and phase shifts is probably not quite as straightforward as the simplest model would predict, as suggested by the effects of pentylenetetrazol. The convulsant drug induces phase delays in the *Bulla* eye when given during early night, but does not cause phase advances at any time (Khalsa & Block, 1992). Because pentylenetetrazol is thought to raise intracellular calcium and is known to stimulate CAP activity, one would expect it to have a light-like PRC. Clearly, delaying and advancing phase shifts must differ mechanistically in some way and are not the result of simple elevations of calcium levels in the early and late night, respectively. Perhaps the particular spatial or temporal dynamics of the pentylenetetrazol-induced calcium increase could be the explanation; direct measurements of these parameters with calcium-sensitive dyes should prove to be enlightening.

An interesting question raised by this model is why multiple second messengers should be necessary for the transduction of a given stimulus; for example, serotonin causes the production of cAMP, which changes membrane potential and affects the concentration of intracellular calcium, the ultimate messenger. Why does cAMP not shift the pacemaker directly? Two possible explanations present themselves. First, it is possible that cAMP does affect the clock in some way that, along with the change in calcium levels, causes a different phase shift than calcium changes alone might. Because changes in calcium levels alone have not been shown to be sufficient for phase shifts, some synergistic interactions among the different second messenger systems is definitely imaginable. A second explanation is that calcium was linked to clock function before the particular entrainment pathways evolved. The evolution of the input pathways could then have been a matter of linking preexisting transmitter/second messenger systems to the clock via calcium (serotonin, for example, causes cAMP production in other *Aplysia* tissues; Kandel & Schwartz, 1982). If the role of calcium in the circadian oscillator is indeed ancient, it would exist in distantly related organisms; indeed, calcium is an entraining signal in mammals (Ding *et al.*, 1994), and intracellular calcium concentrations oscillate with a circadian rhythm in plants (Johnson *et al.*, 1995). We look forward to further progress in this area.

Interaction among Entraining Signals

One would predict from the calcium model that simultaneous application of light and an efferent signal would result in no net phase shift. In *Bulla*, this turns out to be the case; FMRFamide blocks both phase advances and delays induced by light pulses (Colwell *et al.*, 1992a). In *Aplysia*, however, it is not as simple. Serotonin does block light-induced phase delays, but, surprisingly, it potentiates light-induced advances (Colwell, 1990). This could possibly be due to the known ability of serotonin to modulate the sensitivity of the eye to light (Eskin & Maresh, 1982; Jacklet, 1991), or it could be another indication of a mechanistic difference between phase advances and delays.

Interestingly, FMRFamide has effects in the *Aplysia* clock as well (Colwell *et al.*, 1992a). By itself, FMRFamide does not shift the CAP rhythm in the *Aplysia* eye, but it affects light-induced phase shifts in a similar manner to serotonin, blocking delays and potentiating advances. Furthermore, FMRFamide blocks all phase shifts induced by serotonin. This last observation is consistent with data from the *Aplysia* ganglia, where FMRFamide, coupled to arachidonic acid metabolism, antagonizes the effects of serotonin on neuronal excitability (Belardetti, Kandel, & Siegelbaum, 1987; Belardetti, Campbell, Falck, Demontis, & Rosolowsky, 1989; Sweatt *et al.*, 1989); interestingly, an inhibitor of arachidonic acid metabolism has been shown to phase shift the *Aplysia* eye (Raju *et al.*, 1993; see below). An explanation of all of these interactions must await the identification of the effectors of FMRFamide action and will doubtless further illuminate the biochemical pathways involved in the generation of phase shifts.

Actions Downstream of Membrane Potential and Calcium

In attempting to identify the steps in the entrainment pathway downstream of intracellular calcium, we reach the limits of what can readily be learned from the physiologic/pharmacologic approach used thus far. One study has reported that light-induced phase shifts in *Bulla* are not blocked by calmodulin antagonists (Khalsa & Block, 1988b), but other targets of calcium in molluscan cells are not well understood. An alternate approach is to try to identify specific proteins that are affected, in either activity or expression level, by phase-shifting agents.

One possible way that entrainment information could be expressed downstream is through protein phosphorylation. This is particularly likely as calcium and the other second messengers in the entrainment pathway are known to alter the activity of protein kinases or phosphatases. Candidate proteins important for entrainment then would be those whose phosphorylation states are affected by entraining agents. In an initial study of protein phosphorylation in the eye of *Aplysia* using two-dimensional gel electrophoresis, Zwartjes and Eskin (1990) found 14 proteins whose phosphorylation is increased and 2 proteins whose phosphorylation is decreased by the application of serotonin at a phase during which it advances the CAP rhythm. Application of a cAMP analog mimics all but one of these changes in phosphorylation. In contrast, light, which at this phase has little or no effect on the rhythm, only increases the phosphorylation of one of these proteins. In another study on *Aplysia*, 2 other proteins were found whose phosphorylation is altered by both serotonin and a cAMP analog (Lotshaw & Jacklet, 1987). Although none of these proteins has been identified, the feasibility of purifying and identifying a phosphoprotein from a two-dimensional gel has recently been demonstrated in *Aplysia* ganglia (Homayouni, Nunez-Regueiro, Byrne, & Eskin, 1997).

Protein synthesis, at least under some circumstances, is necessary for phase-shifting agents to affect the clock. In *Aplysia*, the phase advances caused by application of serotonin during the late subjective day are blocked by coapplication of translation inhibitors (Eskin, Yeung, & Klass, 1984b). The inhibitors do not affect the acute reduction in CAP frequency caused by serotonin. Furthermore, the phase advances caused by cAMP analogs (Eskin *et al.*, 1984b) or by removal of extracellular calcium (Colwell *et al.*, 1994) are also blocked by protein synthesis inhibitors, showing that translation is necessary for entrainment at a point downstream of these messengers. The dependence on protein synthesis of light-induced phase shifts is less clear; inhibitors of protein synthesis appear to block light-induced phase shifts under some conditions in *Aplysia* (Raju, Yeung, & Eskin, 1990) but not in *Bulla* (Bogart, 1992). Because translational inhibitors by themselves induce phase shifts during the subjective night (see below), their effects in combination with light pulses remain controversial, although light has been shown to increase the production of specific proteins in the *Aplysia* eye during the night (see below).

The proximal output of the ocular pacemaker in *Aplysia* and *Bulla* is electrical activity. This output can be communicated to the animal either through CAPs in the optic nerve or by hormonal secretion from the eye. Isolated *Aplysia* eyes have been shown to release both peptides and melatonin (Abran, Anctil, & Ali, 1994; Harf, Arch, & Eskin, 1976), and in *Bulla,* the eyes are sometimes able to organize locomotor rhythmicity following severance of the optic nerves (Raiford, 1984). In order to produce a rhythm in either CAPs or hormone release, the electrophysiologic characteristics of the pacemaker neurons must be controlled by the clock in such a way as to permit the neurons to fire action potentials spontaneously only during the subjective day. The details of this output pathway, both the identities of the ion channels controlling electrical excitability and the mechanism by which these channels are modulated by the clock, are unclear.

BRNs in the *Bulla* eye exhibit a circadian rhythm in resting membrane potential that underlies, at least in part, the rhythm of CAP production. Long-term intracellular recording from BRNs in the semiintact eye show, that these neurons are depolarized by an average of approximately 13 mV during the subjective day compared to the subjective night (McMahon *et al.*, 1984). During the night, when the cells are silent, their membrane potential is around −70 mV. They begin to depolarize before firing commences at subjective dawn and remain depolarized throughout the day, even as CAP frequency decreases. Finally, the BRNs hyperpolarize again around dusk. This rhythm in membrane potential is associated with changes in resting membrane conductance, which is substantially higher at night than during the day (Ralph & Block, 1990). Because the day–night conductance difference is not seen in the presence of TEA, it is likely to be the result of changes in potassium channel activity (Michel *et al.*, 1993). These data support a model in which a population of clock-controlled potassium channels closes at dawn and opens at dusk, causing the rhythm in membrane potential.

When *Bulla* retinal cells are dispersed and grown in cell culture, BRNs continue to exhibit a circadian rhythm in membrane conductance similar to that seen in the intact eye. Furthermore, this conductance rhythm persists in fully isolated BRNs (no other cells present in the culture dish; Michel *et al.*, 1993) (Figure 4). These results prove unequivocally that each BRN is a fully competent circadian pacemaker, able to generate long-period rhythms solely from processes contained within the cell.

Aplysia pacemaker neurons grown in cell culture are also rhythmic. While resting membrane conductance has not been measured in these neurons, the amplitude of a specific potassium current recorded under voltage clamp shows a circadian rhythm (Barnes & Jacklet, 1997). The delayed rectifier current, which is a noninactivating, voltage-gated potassium current, is larger during the subjective night than during the day. A similar rhythm in delayed rectifier amplitude has also been observed in BRNs (Michel, Manivannan, Zaritsk, & Block, 1999). Because this current activates only upon substantial depolarization of the membrane potential from rest, it is unlikely to underlie the rhythm in either resting potential or conductance. Nonetheless, the rhythm in current amplitude suggests that *Aplysia* pacemaker neurons, like *Bulla* BRNs, are autonomous oscillators, although the pacemaker neurons in this study were dispersed in culture, but not fully isolated.

The potassium conductance governing resting membrane potential is unlikely to be the only ionic conductance regulated by the clock in BRNs. This hypothesis

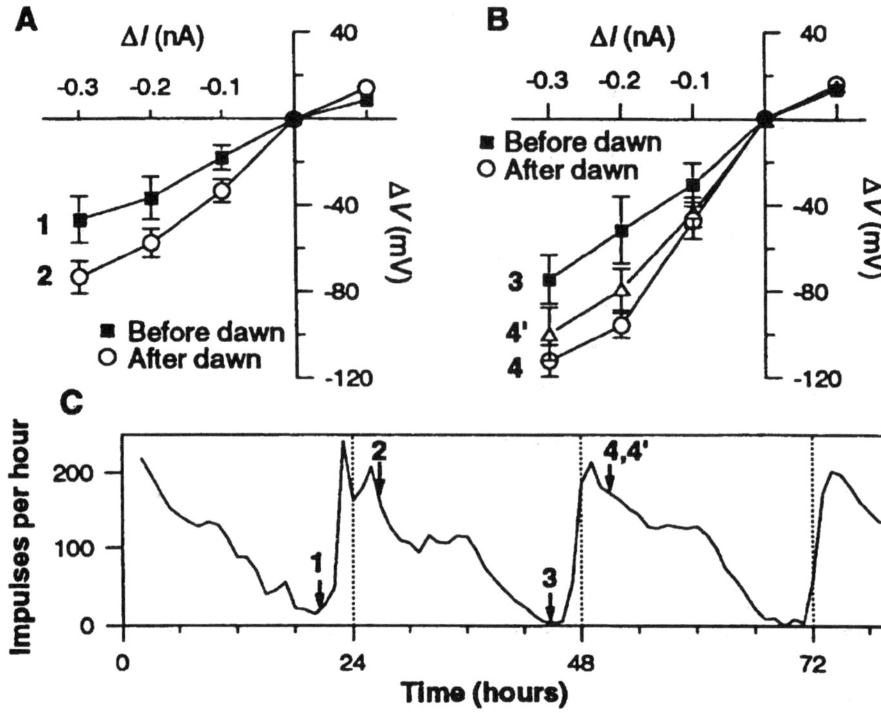

Figure 4. Circadian rhythm in membrane conductance of fully isolated *Bulla* BRNs. Conductance was measured at −65 mV using discontinuous current clamp. (A, B) The decrease in conductance through the first and second projected dawns in culture. In both cases, the decrease was highly significant. (C) Rhythm in CAP frequency from an isolated *Bulla* eye, with arrows indicating the relative recording times from the isolated cells (From Michel *et al.*, 1993; with permission. Copyright 1993, American Association for the Advancement of Science.)

stems from the observation that the relationship between membrane potential and CAP frequency is variable depending on the phase of the circadian cycle (McMahon *et al.*, 1984). During the late night and early day, CAP frequency and membrane potential increase together. Later in the day, however, CAP frequency drops while the membrane potential stays relatively constant at its depolarized value. This suggests that another ion channel, possibly one that initiates each action potential, is also clock regulated, and that its rhythm phase leads the potassium conductance rhythm. In other excitable cells with pacemaker activity, inward currents such as the h-current in cardiac myocytes or a slow sodium current in rat suprachiasmatic nucleus neurons regulate spike frequency (DiFrancesco, 1991; Pennartz, Bierlaagh, & Geurtsen, 1997). Such a current could potentially be important in the generation of the CAP rhythm.

A chloride current may be important for stabilizing the membrane potential of BRNs during the late subjective night. Removing extracellular chloride or adding a specific blocker of chloride channels shortens the free-running period of the *Bulla* eye (Khalsa, Ralph, & Block, 1990). This effect appears to be entirely due to a shortening of the subjective night when CAP activity is absent, and a PRC to low chloride shows large phase advances during the late night. These data have been interpreted as suggesting that a resting chloride conductance acts as a current shunt in the BRN membrane; blocking these channels would consequently make the BRNs hyperexcitable by increasing the membrane response to small depolarizing currents (Michel, Khalsa, & Block, 1992). This might accentuate the action of a clock-regulated inward current and hasten the onset of CAP activity, which would be the equivalent of a depolarizing stimulus and would cause a phase advance. In the continued presence of chloride channel blockers, these phase advances would sum to produce a shortening of the free-running period.

The biochemical pathways by which the clock regulates ion channel activity are not known. One common mechanism for channel modulation is phosphorylation by protein kinases. A large number of drugs that inhibit different serine/threonine kinases have been tested on isolated *Bulla* eyes, and none affect the CAP rhythm (Roberts, Towles, & Leader, 1992; Whitmore, 1997). In contrast, agents that alter tyrosine phosphorylation can perturb CAP activity. Application of genistein, a tyrosine kinase inhibitor, eliminates CAP activity (Roberts *et al.*, 1992). Conversely, vanadate, a tyrosine phosphatase inhibitor, reversibly enhances CAP frequency (Herzog, Silva, Whitmore, & Block, 1997). Because the channels modulated by these treatments have not been identified, it is unclear whether tyrosine phosphorylation is part of the output of the circadian clock or may be regulating a separate population of channels and masking the CAP rhythm.

Mechanism of the Circadian Oscillator

Role of Transcription and Translation

Bulla BRNs and *Aplysia* pacemaker neurons are autonomous circadian pacemakers; therefore, all of the components of the circadian clock must be contained within a single cell. In other species, particularly *Drosophila* and *Neurospora*, the circadian clock is thought to be composed of autoregulatory feedback loops regulating the expression of specific gene products (Aronson, Johnson, Loros, & Dunlap, 1994; Darlington *et al.*, 1998; Hardin, Hall, & Rosbash, 1990). The molecular basis of the molluscan clock is far less well understood, although translation and transcription are known to be necessary, and a number of potential oscillator proteins have been identified.

The molluscan clock is sensitive to inhibitors of translation and transcription. Pulses of the translation inhibitors puromycin, cycloheximide, or anisomycin cause phase delays when applied to *Aplysia* eyes during the subjective night (Jacklet, 1977; Lotshaw & Jacklet, 1986; Rothman & Strumwasser, 1976). Early studies showed that puromycin and anisomycin also advance the clock when applied during the day (Jacklet, 1977), but this effect was later shown to be nonspecific and is only seen at high inhibitor concentrations (Lotshaw & Jacklet, 1986). The sensitive period of the clock to inhibition of translation was explored with 1-hour pulses of cycloheximide, the inhibitor that reverses most quickly, and was found to be restricted to the late night and early morning (Yeung & Eskin, 1988). Temporary inhibition of protein synthesis at other times of the cycle have no effect on the motion of the pacemaker.

Experiments in *Bulla* have demonstrated that translational inhibitors actually stop the clock (Khalsa, Whitmore, & Block, 1992). In a so-called "wedge" experiment, *Bulla* eyes were incubated with cycloheximide or anisomycin for different lengths of time, ranging from 5 to 52 hours. The CAP activity, which is inhibited by the drugs, resumed following washout and the phase of the rhythm was assessed. It was observed that if an inhibitor is present through subjective dawn, the phase of the rhythm after the drug pulse is dependent only on the time at which the drug was removed. This result shows that the circadian clock is stopped around dawn by translational inhibitors; if the drug is applied before dawn, the clock continues to run until dawn and then is stopped for the duration of the treatment. In light of these data, it seems likely that the phase delays seen in *Aplysia* are the result of the clock stopping during the short pulses of inhibitor rather than of the clock being reset. In both species, translation seems to be necessary only around subjective dawn.

Transcriptional inhibitors appear to stop the clock as well, although the sensitive period for transcription is longer than for protein synthesis. The reversible transcriptional inhibitor 5,6-dichlorobenzimidazole riboside (DRB) delays the phase of the *Aplysia* eye rhythm when it is present any time during a 12-hour period ranging from Zeitgeber Time (ZT) 0 to 12, where lights-on is defined as ZT 0 (Koumenis, Tran, & Eskin, 1996; Raju, Koumenis, Nunez-Regueiro, & Eskin, 1991). Wedge experiments in *Bulla* show that DRB does stop the clock, with a sensitive period that also encompasses the end of the night and most of the subjective day (Khalsa, Whitmore, Bogart, & Block, 1996).

One striking feature of the effect of transcription inhibitors is that the sensitive period to them is substantially longer than that for translation inhibitors, suggesting either that transcription itself is part of the circadian timing loop or that the critical transcripts are stored throughout the subjective night to be translated into protein at dawn. To distinguish between these possibilities, a double-pulse experiment was performed on *Bulla* eyes based on experiments done previously in *Drosophila* by Pittendrigh (1967). Groups of eyes were given a 6-hour pulse of DRB from ZT 6 to 12, a phase at which transcription but not translation is necessary for pacemaker function, followed at various times by a light pulse. The resulting PRC was shifted approximately 7 hours compared to a control PRC, demonstrating that DRB produces an immediate effect on the motion of the clock (Khalsa *et al.*, 1996). This, in turn, shows that the rate of transcription itself is part of the circadian timing loop. The behavior of the oscillator in *Aplysia* is likely to be quite similar. Short pulses of DRB in the early day, when both transcription and translation are necessary, cause an immediate shift in the CAP rhythm (Koumenis *et al.*, 1996). This result demonstrates that at this phase, both oscillator motion and output are dependent upon RNA synthesis, although translation could be important as well. Interestingly, short (30 minute) pulses of DRB in the late day, when translational inhibitors have no effect, cause phase delays that do not appear in the CAP rhythm until the following day. This observation could mean that the phase delay itself is not immediate, which would be different from *Bulla*, or that the shift in the phase of the pacemaker is not immediately reflected in the output. Long (6 hour) DRB pulses during the late day in *Bulla* do affect the CAP rhythm right away (Khalsa *et al.*, 1996), but it is not clear if this is a real difference in the role of transcription in the output pathway between the two species or simply a result of differences in the experimental protocols.

The Search for Specific Oscillator Proteins

At some point on an input or entrainment pathway, the entrainment information must perturb the activity or level of a component of the oscillator. This point is both the end of the input pathway and an element of the oscillator. Since circadian oscillators appear to be composed of feedback loops involving transcription and translation, entrainment information will ultimately lead to changes in the synthesis of oscillator components. In addition, transcription and translation could be involved on the input pathway separate from the actual oscillator. The roles of transcription and translation in the oscillator mechanism, together with the blockade of some phase shifts with inhibitors of protein synthesis, predict that entraining agents such as serotonin, and possibly light, should affect the synthesis of some proteins and perhaps mRNAs as well. A series of experiments using two-dimensional gel electrophoresis was performed in *Aplysia* to find proteins whose synthesis is regulated by entraining agents and other treatments that perturb the rhythm. Light, serotonin, cAMP analogs, and DRB were used as treatments in these experiments (Colwell *et al.*, 1994; Eskin *et al.*, 1984b; Koumenis, Nunez-Regueiro, Raju, Cook, & Eskin, 1995; Koumenis *et al.*, 1996; Raju *et al.*, 1990, 1993; Yeung & Eskin, 1987).

The synthesis of a number of proteins is altered by various phase shifting treatments. Five of these proteins have been identified. A 31-kD protein belongs to a family of proteins containing auxin-induced protein, stringent starvation protein, and glutathione S-transferase (Koumenis, Nunez-Regueiro, & Eskin, 1992; Noel *et al.*, 1994). These proteins may be involved in transcription and cellular protective mechanisms (Toung & Tu, 1992). A 36-kD protein was identified as a channel protein belonging to the porin family of voltage-dependent anion channels (Koumenis *et al.*, 1995). Two other proteins are potential molecular chaperones: a 70-kD protein that belongs to the HSP-70 (heat shock protein) family and a 78-kD protein that belongs to a family including human immunoglobulin binding protein (BiP) and glucose regulated protein 78 (GRP) (Koumenis *et al.*, 1995). HSP-70 functions as a molecular chaperone in the cytoplasm and aids in the entry of proteins into the nucleus (Hartl, 1996), whereas BiP resides in the endoplasmic reticulum and also functions as a chaperone (Brodsky, 1996). The fifth protein to be identified is discussed below. Other proteins with interesting properties remain to be identified, most notably a 34-kD protein whose levels are modulated by the efferent pathway in a phase-dependent manner and which is expressed in the half of the *Aplysia* eye containing the pacemaker neurons (Eskin *et al.*, 1984b; Yeung & Eskin, 1987).

The fifth protein to be identified is a 40-kD light- and serotonin-regulated protein that belongs to a family called the annexins (Raju *et al.*, 1993). The annexins are calcium-binding proteins that mediate the effects of intracellular calcium in a number of different ways, including regulation of arachidonic acid (AA) metabolism and of calcium channels (Heizmann, 1992; Pepinsky *et al.*, 1986). Given the significance of calcium in pacemaker entrainment in the mollusk, the annexin-like protein seems particularly interesting. The annexins can affect AA metabolism by inhibiting phospholipase A_2; to test further the role of the AA, an inhibitor of AA-metabolizing enzyme lipoxygenase was added to *Aplysia* eyes. Two-hour pulses of the inhibitor cause significant phase delays when applied during the subjective night and have no effect during the day (Raju *et al.*, 1993). It seems quite possible, therefore, that AA is somehow involved in the entrainment of the *Aplysia* ocular pacemaker.

What are the roles of the other identified proteins in the circadian system, and do they function in the entrainment pathway, oscillator, or even the output pathway? The specific functions of porins, HSP-70, and the AIP-SSP-like proteins are not known in *Aplysia*, but these proteins could have regulatory roles and, as such, be on the input or output pathways. It is particularly interesting that the AIP-like and HSP-70 proteins may regulate transcription. Conceivably, they could play a role in the oscillator mechanism. BiP, because of its location in the endoplasmic reticulum, probably does not function in the input pathway or the oscillator, but could be involved in output regulation. Each protein still needs to be thoroughly characterized in *Aplysia*, and it remains to be determined whether the levels of these proteins or their transcripts exhibit circadian rhythms. It is critical to determine whether these proteins are essential for some function of the ocular circadian rhythm. In addition, it is likely that more proteins whose synthesis is regulated by light and serotonin remain to be found.

Relatively little work has been done in the mollusk to identify transcripts that are affected by entraining agents or are themselves rhythmic. It is known that the levels of porin and BiP transcripts are altered by light and serotonin in opposite directions (Sloan, Koumenis, & Eskin, 1995). One interesting transcript codes for the transcription factor and immediate-early gene C/EBP (Alberini, Ghirardi, Metz, & Kandel, 1994). Light treatment of intact *Aplysia* produces large increases in the level of C/EBP mRNA in the eyes (Hattar & Eskin, 1996). These effects of light are phase dependent; much larger effects on the levels of C/EBP mRNA are seen during the subjective night, when light induces phase shifts, than during the day, when it does not. The C/EBP transcript can also be induced by serotonin pulses in isolated eyes (Whitmore, 1997). In the intact animal, C/EBP mRNA levels in the eye are rhythmic, peaking at dawn on an LD cycle and at the same relative phase in constant darkness (Hattar & Eskin, 1996). In the isolated eye, however, C/EBP mRNA is not rhythmic (Hattar, Levenson, & Eskin, 1997; Whitmore, 1997). As C/EBP is known to be induced by injury, it is possible that a rhythm exists in certain cells in the isolated eye but is masked by a general induction resulting from the surgery. The properties of C/EBP in the intact animal are interesting, but a particular function for C/EBP in the eye circadian system has not been established. C/EBP might function in the entrainment pathway with its activity gated by the clock in a manner similar to c-Fos in the mammalian system (Rusak, Robertson, Wisden, & Hunt, 1990). Alternatively, C/EBP might function on the output side and regulate the circadian synthesis of a number of proteins. The data also do not rule out a function for C/EBP in the oscillator mechanism. Finally, it is important to note that no information is available on the behavior of the C/EBP protein, which might not be regulated in parallel with its message.

Other Treatments That Affect the Motion of the Clock

Several reports have implicated protein kinases as being necessary for pacemaker function. The tyrosine kinase inhibitor genistein can cause large phase delays in *Bulla* (Roberts *et al.*, 1992). A wedge experiment to determine whether genistein actually stops the clock is not feasible due to the irreversible effects of extended treatments with the drug. However, a recent report demonstrates that inhibition of tyrosine phosphorylation does not stop the clock; even though spontaneous CAP activity is inhibited, the ability of light pulses to elicit CAPs remains rhythmic in the presence of genistein (Herzog *et al.*, 1997). Thus, genistein is likely to inhibit a step on the input pathway to the clock. Application of the broad-spectrum kinase inhibitor H-8 or the cyclin-dependent kinase inhibitors roscovitine and olomoucine has been reported to slow or stop the *Bulla* clock (Krucher, Meijer, & Roberts, 1997; Roberts, Bedian, & Chen, 1989), but these drugs are known to inhibit transcription (Dubois, Nguyen, Bellier, & Bensaude, 1994; Sankrithi & Eskin, 1999) and so could be affecting the clock in that manner.

The concentration of certain amino acids in the culture medium appears to modulate the velocity of the ocular pacemaker. Early investigators noticed that the free-running period of the *Aplysia* eye rhythm was more than 2 hours longer when measured in tissue culture medium compared to seawater. This difference was found to be due entirely to the amino acids in the tissue culture medium; many amino acids, most notably tyrosine and tryptophan, lengthen the period of the clock (Eskin, 1982b). Short pulses of amino acids do not cause phase shifts, suggesting that the effect develops only upon prolonged incubation. The mechanism of this period lengthening is uncertain.

Finally, the motion of the clock is sensitive to treatments that affect overall cellular metabolism. In *Aplysia*, exposing the eye to the metabolic poisons NaCN or DNP, to Mn^{2+}, or to the calcium ionophore A23187 in combination with EGTA results in large phase delays at most times of the cycle (Eskin & Corrent, 1977); Mn^{2+} also causes phase delays in *Bulla* (Roberts & Block, 1983). While these four agents can affect multiple cellular processes, they have in common the ability to inhibit mitochondrial oxidative phosphorylation. This result is reminiscent of the finding by Pittendrigh (1974) that anoxic conditions stop the *Drosophila* clock. Another possibly related result is that treatments that lower intracellular pH in the *Bulla* eye stop the motion of the pacemaker around subjective dawn (Khalsa, Ralph, & Block, 1991). While the specific intracellular target of decreased pH has not been identified, it could possibly be a relatively nonspecific metabolic effect.

Input–Output Feedback Loop

Extensive analysis of entrainment has shown that membrane potential and intracellular calcium concentration are key parameters that influence the phase of the molluscan circadian clock. In addition, the clock output is electrical activity, which certainly must influence intracellular calcium concentration because the pacemaker neurons contain voltage-gated calcium channels that open during action potentials. It necessarily follows, therefore, that the output of the clock feeds back and may then affect the oscillator in some way (Figure 5). The precise role that this feedback has on the function of the whole system remains unclear.

Long-term treatments that alter membrane potential have relatively modest effects on the rhythm of the *Bulla* eye. Chronic treatment of the eye with seawater containing either high potassium or low sodium concentrations, which depolarize and hyperpolarize the BRNs, respectively, causes a slight (<1 hr) lengthening of the free-running period (McMahon & Block, 1987b). In contrast, complete removal of extracellular sodium, which should hyperpolarize the BRNs even more and which eliminates CAP activity, does not affect the motion of the clock (Khalsa, Michel, & Block, 1997). This was determined retrospectively in a wedge experiment, since the elimination of CAP activity meant that phase and period could not be assessed during the treatment with sodium-free seawater.

A more direct test of the role of calcium feedback on the clock was performed by effectively removing extracellular calcium for extended periods by incubating the eyes with seawater containing EGTA (Khalsa, Ralph, & Block, 1993). While this treatment usually eliminates rhythmic CAP activity, retrospective phase analysis following restoration of calcium indicated that the clock continues to run during the treatment. Occasionally, eyes show weak 24-hour CAP rhythms in the absence of

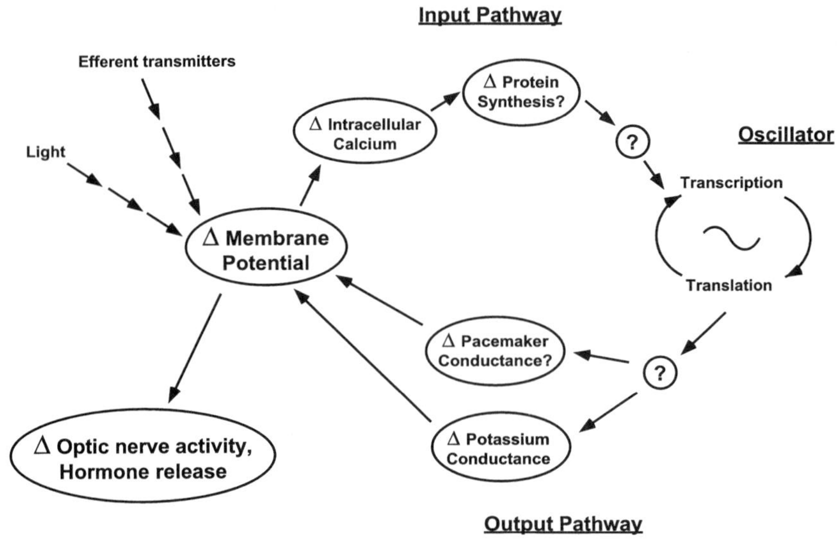

Figure 5. A model of the feedback from output to input in the molluscan ocular pacemaker.

extracellular calcium, proving more directly that clock function continues with a period indistinguishable from control eyes. Extended treatments with EGTA-seawater do cause phase delays if they last through subjective dawn, but the data indicate that this is more likely to be an effect of the restoration of normal calcium at the end of the pulse than of the absence of calcium during the pulse. Thus calcium entry is not required for normal function of the BRN circadian oscillator.

What, then, is the role of the membrane potential and calcium feedback loop? Under an LD cycle, intracellular calcium concentrations presumably are higher in the light, when the pacemaker neurons are firing, then in the dark, although this has not been measured directly. Even when the eye is maintained in constant darkness, the intracellular clock presumably is exposed to a circadian cycle of high calcium/low calcium due to calcium entry during spontaneous activity. This rhythm in the "internal environment" of the cell could potentially help regulate the angular velocity of the molecular clock, though it is clearly not an absolute requirement for clock function. Further insight into the role of the feedback loop must await a more complete understanding both of calcium dynamics within the pacemaker neurons and of the mechanism of calcium action in phase resetting.

The feedback from output back to the entrainment pathway is without a doubt not unique to mollusks. Indeed, the feedback loop described above should function in any neuronal pacemaker in which calcium is an input element, as electrical activity will affect intracellular calcium concentration in any neuron. In the rodent suprachiasmatic nucleus (SCN), calcium is believed to be a messenger in the entrainment response to optic nerve stimulation (Ding *et al.*, 1994). How neuronal activity might also activate this pathway has not been examined, but recent progress in recording circadian rhythms from cultured SCN neurons should lead to an appreciation of the role of output feedback in this system (Welsh, Logothetis, Meister, & Reppert, 1995).

With a detailed cellular model of the molluscan ocular pacemaker in place, it is now possible to begin analyzing the changes underlying long-term plasticity in the function of the system as a whole. Two general phenomena common to most or all circadian systems seem especially amenable to analysis using the molluscan model: aging and aftereffects.

In aging animals, some deterioration of the circadian system is usually observed; external manifestations of this process can include changes in the amplitude and period of an output rhythm or a change in the phase angle between the output rhythm and the environmental cycle (Turek *et al.*, 1995; see also Chapter 10). *Aplysia* is no exception to this rule. Recordings of ocular CAP rhythms both in vivo and in vitro from animals ranging in age from 4 to 14 months show significant chages in all of these parameters (Hattar, Eskin, & Levenson, 1998; Sloan, Levenson, Tran, Kerbeshian, Block, & Eskin, 1999). In progressively older animals, maximum firing rate is decreased, free-running period is increased, and the phase angle relative to the previous light cycle is altered. In addition, a significant number of eyes from the oldest animals become arrhythmic after one or two cycles in constant darkness; this is rarely if ever seen in younger animals. These changes are reflected at the molecular level as well; the amplitude of the C/EBP RNA rhythm is also damped in older animals (Hattar *et al.*, 1998).

The causes of age-dependent loss of circadian function have not been identified in any system. Most interesting is the question of whether these changes are cellular, stemming from decreased robustness of the oscillator within each pacemaker neuron, or are the result of deterioration of the intracellular interactions which synchronize the entire tissue. In the second scenario, individual pacemakers would continue to cycle normally but would drift out of phase with one another, and if the output reflects the average of the population of pacemakers, it would gradually become arrhythmic. It should be possible to study individual *Aplysia* pacemaker neurons, either at the electrophysiologic or molecular level, to determine whether the intracellular clock is affected by age. Furthermore, pharmacologic treatments known to target specific steps on the input and output pathways could be employed to identify the specific deficits in older neurons. As age-dependent deterioration of the circadian system appears to be universal, knowledge gained from study of the *Aplysia* eye could be quite relevant to understanding aging of the circadian system in other organisms.

Aftereffects are changes in the free-running period of a circadian oscillator induced by its previous history of exposure to light. For example, if an organism is entrained to an LD cycle with a period of greater than 24 hours, upon release into constant conditions it will freerun with a longer period than if the entraining cycle had been 24 hours. These aftereffects can persist for many weeks, are seen in all organisms that have been examined, and are among the most mysterious behaviors of the circadian system. Aftereffects can be seen in the *Bulla* eye; if *Bulla* are entrained to 22- or 26-hour LD cycles, the period of the CAP rhythm in subsequently isolated eyes is significantly shorter or longer, respectively, than that measured following entrainment to a 24-hour cycle (Page, Wassmer, Fletcher, & Block, 1997). This experiment demonstrates that all of the machinery required for maintenance (though not necessarily acquisition) of altered periods exists in the isolated eye. As in the aging experiments, it should now be possible to determine whether these aftereffects persist in dispersed BRNs and to look for underlying changes in cellular functions, such as the dynamics of intracellular calcium or the amplitudes of specific ionic currents.

Conclusions

Advantages and Disadvantages of the System

The opisthobranch eye is an excellent system for studying the function of the circadian clock. The primary advantage of the system by far is the existence within the eye of a defined, homogeneous population of autonomous circadian pacemaker neurons. Whether one is studying the clock at the level of single ion channels, changes in gene expression, or protein–protein interactions, interpreting one's experimental results is much more straightforward in this context. The ocular pacemaker can be studied in dispersed cells, in the isolated eye, and in the intact animal in order to determine the relative effects of intracellular, cell-cell, and tissue-level interactions on clock function.

The *Bulla* and *Aplysia* eyes afford the additional advantage that the organismal circadian systems of these mollusks are also fairly well characterized, as described in detail in Chapter 4. As a result, one should be able to translate system properties into cellular terms with relative ease. For example, the opisthobranchs differ significantly in the strength of coupling between the two ocular pacemakers (Hudson & Lickey, 1980; Roberts & Block, 1985; Roberts, Block, & Lusska, 1987). Is this reflected in single-cell function, or is it due solely to the strength of the synaptic connections between the eyes? Interoscillator coupling is a key component in modeling the behavior of a variety of circadian systems (Page, 1981; Pittendrigh, 1960; Pittendrigh & Daan, 1976), and the mollusk appears to be an excellent preparation for explaining this phenomenon in cellular detail. In another example, the existence of extraocular pacemakers in these animals is well known, as is the dominant role of the ocular pacemakers. Is there a cellular correlate to the hierarchical organization of the system, or is it entirely determined by intercellular coupling strength? Why do the extraocular oscillators damp when the eyes are removed? Such questions should be tractable once the locations of these additional oscillators is determined, as their cellular mechanisms are unlikely to be significantly different from those of the ocular pacemaker neurons. In these and other areas, the mollusk offers the opportunity to move from single pacemaker cells to a greater understanding of the system as a whole.

The primary disadvantages of the molluscan system are the lack of genetic tools with which to study the clock and the small amounts of RNA and protein in the eyes. The explosive progress over the last few years in identifying clock genes in arthropods, vertebrates, and fungi has highlighted the advantages of genetically approachable model systems. With the description of a virtually complete molecular feedback loop in *Drosophila* (Allada, White, So, Hall, & Rosbash, 1998; Darlington *et al.*, 1998; Gekakis *et al.*, 1995; Hardin *et al.*, 1990; Myers, Wager-Smith, Wesley, Young, & Sehgal, 1995; Rutila *et al.*, 1998), however, it is clear that a major upcoming challenge is going to be connecting this molecular loop with the input and output pathways that have been best characterized in the mollusk. As a consequence of the expanding knowledge of clock gene structure in multiple organisms, it should not be prohibitively difficult to isolate molluscan homologues of these genes and to begin to make the connections among the input, output, and oscillator pathways.

Future Directions

As has been discussed above, the connections among the different components of the ocular circadian system will likely be a major focus of future work in *Aplysia* and *Bulla*. The identification of the proteins involved in transducing changes along the entrainment pathway into phase shifts will be a major area of work. In addition, since timing is so critical to circadian rhythms, the temporal relationships between the different steps will have to be determined in detail. The dynamics of calcium entry, buffering, and sequestration as well as the kinetics of subsequent events will have to be well understood before all of the experimental data can be fit together. On the output side, the key issues remain the identities of the clock-regulated ion channels and the mechanisms of that regulation. After the channels are identified and characterized, it may be possible, using specific channel blockers or kinase inhibitors, to disrupt selectively specific components of the output pathway. Such experiments could help to define the role of output feedback on the function of the circadian clock.

A number of interesting and important issues also remain at the system level. Where are other circadian oscillators located in *Aplysia* and *Bulla*, how are they entrained, and how do the extraocular and ocular circadian oscillators interact to govern the output rhythms? Another important question is why so few physiologic circadian rhythms have been found so far in *Aplysia* and *Bulla* although multiple rhythmic outputs are usually found in other organisms. Does this indicate a different type of circadian organization in the mollusks? Finally, the presence of a circadian rhythm of CBP in the eyes *in vivo* but not *in vitro* raises the interesting possibility that, as it has been very fruitful to study the ocular oscillators *in vitro* in the past, a great deal may also be learned in the future by additional studies of the ocular oscillator *in vivo*.

It seems likely that a satisfactory explanation of the molecular processes responsible for generating circadian rhythms is within our immediate reach. Recent progress in mouse, *Neurospora*, and *Drosophila* in the identification of additional components of the feedback loop (Allada *et al.*, 1998; Crosthwaite, Dunlap, & Loros, 1997; Darlington *et al.*, 1998; Gekakis *et al.*, 1998; Rutila *et al.*, 1998) brings the potential for a relatively complete explanation within view. However, as has been emphasized (Dunlap, 1998), such disclosure is more of a beginning than an end. We now recognize that the *circadian system* will be a good deal more complicated than any autoregulatory feedback loop that produces a circa 24-hour oscillation. The ability of biological rhythms to be modified by the prior light cycle history, the complex behavioral transients associated with shifts in the phase of the light cycle, and the demonstrated interactions among discrete circadian oscillators when acting as an ensemble to control rhythmic behaviors reveal a complexity that will most likely require analysis beyond the molecular level to the study of cellular and tissue interactions within the circadian systems. As these questions come into focus, it is unclear whether many of the existing genetic models will prove adequate to the task. For example, the difficulty of neurophysiologic study in *Drosophila* will continue to drive efforts to identify molecular rhythms in larger insects such as *Antheraea*. It seems most likely that as additional molecular mechanisms are revealed and the research focus shifts to cells and tissues, the opisthobranch mollusks will continue to play an important role in the discovery process.

Acknowledgments

Supported by NRSA NS10210 to E.M.B., NIH grant NS15264 and support from the NSF Center for Biological Timing to G.D.B., and NIH grant MH41978 to A.E.

References

Abran, D., Anctil, M., & Ali, M. A. (1994). Melatonin activity rhythms in eyes and cerebral ganglia of *Aplysia californica*. *General & Comparative Endocrinology, 96*, 215–222.

Adams, W. B., & Levitan, I. B. (1982). Intracellular injection of protein kinase inhibitor blocks the serotonin-induced increase in K^+ conductance in *Aplysia* neuron R15. *Proceedings of the National Academy of Sciences of the USA, 79*, 3877–3880.

Alberini, C. M., Ghirardi, M., Metz, R., & Kandel, E. R. (1994). C/EBP is an immediate-early gene required for the consolidation of long-term facilitation in *Aplysia*. *Cell, 76*, 1099–1114.

Allada, R., White, N. E., So, W. V., Hall, J. C., & Rosbash, M. (1998). A mutant *Drosophila* homolog of mammalian *Clock* disrupts circadian rhythms and transcription of *period* and *timeless*. *Cell, 93*, 791–804.

Aronson, B. D., Johnson, K. A., Loros, J. J., & Dunlap, J. C. (1994). Negative feedback defining a circadian clock: Autoregulation of the clock gene frequency. *Science, 263*, 1578–1584.

Arvanitaki A., & Chalazonitis N. (1961). Excitatory and inhibitory processes initiated by light and infrared radiations in single identifiable nerve cells (giant ganglion cells in *Aplysia*). In E. Florey (Ed.), *Nervous inhibition* (pp. 194–231). New York: Pergamon Press.

Audesirk, G. (1973). Spontaneous and light-induced compound action potentials in the isolated eye of *Aplysia*: Initiation and synchronization. *Brain Research, 59*, 229–242.

Barnes, S., & Jacklet, J. W. (1997). Ionic currents of isolated retinal pacemaker neurons: Projected daily phase differences and selective enhancement by a phase-shifting neurotransmitter. *Journal of Neurophysiology, 77*, 3075–3084.

Belardetti, F., Kandel, E. R., & Siegelbaum, S. A. (1987). Neuronal inhibition by the peptide FMRFamide involves opening of S K^+ channels. *Nature, 325*, 153–156.

Belardetti, F., Campbell, W. B., Falck, J. R., Demontis, G., & Rosolowsky, M. (1989). Products of heme-catalyzed transformation of the arachidonate derivative 12-HPETE open S-type K^+ channels in *Aplysia*. *Neuron, 3*, 497–505.

Benson, J. A. (1980). A cAMP analogue inhibits compound action potential production and phase shifts the circadian clock in the isolated eye of *Aplysia californica*. *Comparative Biochemistry and Physiology—C: Comparative Pharmacology, 67C*, 195–198.

Benson, J. A., & Jacklet, J. W. (1977). Circadian rhythm of output from neurones in the eye of *Aplysia*: I. Effects of deuterium oxide and temperature. *Journal of Experimental Biology, 70*, 151–166.

Block, G. D. (1981). *In vivo* recording of the ocular circadian rhythm in *Aplysia*. *Brain Research, 222*, 138–143.

Block, G. D., & Davenport, P. A. (1982). Circadian rhythm in *Bulla gouldiana*: Role of the eyes in controlling locomotor behavior. *Journal of Experimental Zoology, 224*, 57–63.

Block, G. D., & Lickey, M. E. (1973). Extraocular photoreceptors and oscillators can control the circadian rhythm of behavioral activity in *Aplysia*. *Journal of Comparative Physiology, 84*, 367–374.

Block, G. D., & McMahon, D. G. (1983). Localized illumination of the *Aplysia* and *Bulla* eye reveals new relationships between retinal layers. *Brain Research, 265*, 134–137.

Block, G. D., & McMahon, D. G. (1984). Cellular analysis of the *Bulla* ocular circadian pacemaker system: III. Localization of the circadian pacemaker. *Journal of Comparative Physiology A—Sensory, Neural and Behavioral Physiology, 155*, 387–395.

Block, G. D., & Roberts, M. H. (1981). Circadian pacemaker in the *Bursatella* eye: Properties of the rhythm and its effect on locomotor behavior. *Journal of Comparative Physiology A—Sensory, Neural and Behavioral Physiology, 142*, 403–410.

Block, G., & Smith, J. T. (1973). Cerebral photoreceptors in *Aplysia*. *Comparative Biochemistry and Physiology A—Comparative Physiology, 46A*, 115–121.

Block, G. D., & Wallace, S. F. (1982). Localization of a circadian pacemaker in the eye of a mollusc, *Bulla*. *Science, 217*, 155–157.

Block, G. D., Hudson, D. J., & Lickey, M. E. (1974). Extraocular photoreceptors can entrain the circadian oscillator in the eye of *Aplysia*. *Journal of Comparative Physiology, 89,* 237–249.

Block, G. D., McMahon, D. G., Wallace, S. F., & Friesen, W. O. (1984). Cellular analysis of the *Bulla* ocular circadian pacemaker system: I. A model for retinal organization. *Journal of Comparative Physiology A—Sensory, Neural and Behavioral Physiology, 155,* 365–378.

Block, G. D., Khalsa, S. B., McMahon, D. G., Michel, S., & Guesz, M. (1993). Biological clocks in the retina: Cellular mechanisms of biological timekeeping. *International Review of Cytology, 146,* 83–144.

Block, G. D., Geusz, M., Khalsa, S. B., Michel, S., & Whitmore, D. (1996). Circadian rhythm generation, expression and entrainment in a molluscan model system. *Progress in Brain Research, 111,* 93–102.

Bogart B. L. (1992). *The effects of light and temperature on the circadian system of the marine mollusk Bulla gouldiana.* Ph.D. dissertation, University of Virginia, Charlottesville, Virginia.

Brodsky, J. L. (1996). Post-translational protein translocation: Not all hsc70s are created equal. *Trends in Biochemical Sciences, 21,* 122–126.

Chase, R. (1979). Photic sensitivity of the rhinophore in *Aplysia*. *Canadian Journal of Zoology, 57,* 698–701.

Colwell, C. S. (1990). Light and serotonin interact in affecting the circadian system of *Aplysia*. *Journal of Comparative Physiology A—Sensory, Neural and Behavioral Physiology, 167,* 841–845.

Colwell, C. S., Khalsa, S. B., & Block, G. D. (1992a). FMRFamide modulates the action of phase shifting agents on the ocular circadian pacemakers of *Aplysia* and *Bulla*. *Journal of Comparative Physiology A—Sensory, Neural and Behavioral Physiology, 170,* 211–215.

Colwell, C. S., Michel, S., & Block, G. D. (1992b). Evidence that potassium channels mediate the effects of serotonin on the ocular circadian pacemaker of *Aplysia*. *Journal of Comparative Physiology A—Sensory, Neural and Behavioral Physiology, 171,* 651–656.

Colwell, C. S., Whitmore, D., Michel, S., & Block, G. D. (1994). Calcium plays a central role in phase shifting the ocular circadian pacemaker of *Aplysia*. *Journal of Comparative Physiology A—Sensory, Neural and Behavioral Physiology, 175,* 415–423.

Corrent, G., & Eskin, A. (1982). Transmitterlike action of serotonin in phase shifting a rhythm from the *Aplysia* eye. *American Journal of Physiology, 242,* R333–R338.

Corrent, G., McAdoo, D. J., & Eskin, A. (1978). Serotonin shifts the phase of the circadian rhythm from the *Aplysia* eye. *Science, 202,* 977–979.

Corrent, G., Eskin, A., & Kay, I. (1982). Entrainment of the circadian rhythm from the eye of *Aplysia*: Role of serotonin. *American Journal of Physiology, 242,* R326–R332.

Crosthwaite, S. K., Dunlap, J. C., & Loros, J. J. (1997). *Neurospora wc-1* and *wc-2*: Transcription, photoresponses, and the origins of circadian rhythmicity. *Science, 276,* 763–769.

Darlington, T. K., Wager-Smith, K., Ceriani, M. F., Staknis, D., Gekakis, N., Steeves, T. D. L., Weitz, C. J., Takahashi, J. S., & Kay, S. A. (1998). Closing the circadian loop: CLOCK-induced transcription of its own inhibitors *per* and *tim*. *Science, 280,* 1599–1603.

Delgado, R., Hidalgo, P., Diaz, F., Latorre, R., & Labarca, P. (1991). A cyclic AMP-activated K^+ channel in *Drosophila* larval muscle is persistently activated in *dunce*. *Proceedings of the National Academy of Sciences of the USA, 88,* 557–560.

DiFrancesco, D. (1991). The contribution of the 'pacemaker' current (i_f) to generation of spontaneous activity in rabbit sino-atrial node myocytes. *Journal of Physiology, 434,* 23–40.

Ding, J. M., Chen, D., Weber, E. T., Faiman, L. E., Rea, M. A., & Gillette, M. U. (1994). Resetting the biological clock: Mediation of nocturnal circadian shifts by glutamate and NO. *Science, 266,* 1713–1717.

Dubois, M. F., Nguyen, V. T., Bellier, S., & Bensaude, O. (1994). Inhibitors of transcription such as 5,6-dichloro-1-beta-D-ribofuranosylbenzimidazole and isoquinoline sulfonamide derivatives (H-8 and H-7) promote dephosphorylation of the carboxyl-terminal domain of RNA polymerase II largest subunit. *Journal of Biological Chemistry, 269,* 13331–13336.

Dunlap, J. (1998). An end in the beginning. *Science, 280,* 1548–1549.

Eskin, A. (1971). Properties of the *Aplysia* visual system: *In vitro* entrainment of the circadian rhythm and centrifugal regulation of the eye. *Zeitschrift fur vergleichende Physiologie, 74,* 353–371.

Eskin, A. (1972). Phase shifting a circadian rhythm in the eye of *Aplysia* by high potassium pulses. *Journal of Comparative Physiology, 80,* 353–376.

Eskin, A. (1977). Neurophysiological mechanisms involved in photo-entrainment of the circadian rhythm from the *Aplysia* eye. *Journal of Neurobiology, 8,* 273–299.

Eskin, A. (1982a). Increasing external K^+ blocks phase shifts in a circadian rhythm produced by serotonin or 8-benzylthio-cAMP. *Journal of Neurobiology, 13,* 241–249.

Eskin, A. (1982b). Differential effects of amino acids on the period of the circadian rhythm from the *Aplysia* eye. *Journal of Neurobiology, 13,* 231–239.

Eskin, A., & Corrent, G. (1977). Effects of divalent cations and metabolic poisons on the circadian rhythm from the *Aplysia* eye. *Journal of Comparative Physiology A—Sensory, Neural and Behavioral Physiology, 117,* 1–21.

Eskin, A., & Harcombe, E. (1977). Eye of *Navanax*: Optic activity, circadian rhythm and morphology. *Comparative Biochemistry & Physiology A-Comparative Physiology, 57A,* 443–449.

Eskin, A., & Maresh, R. D. (1982). Serotonin or electrical optic nerve stimulation increases the photosensitivity of the *Aplysia* eye. *Comparative Biochemistry & Physiology-C: Comparative Pharmacology & Toxicology, 73C,* 27–31.

Eskin, A., & Takahashi, J. S. (1983). Adenylate cyclase activation shifts the phase of a circadian pacemaker. *Science, 220,* 82–84.

Eskin, A., Corrent, G., Lin, C. Y., & McAdoo, D. J. (1982). Mechanism for shifting the phase of a circadian rhythm by serotonin: Involvement of cAMP. *Proceedings of the National Academy of Sciences of the USA, 79,* 660–664.

Eskin, A., Takahashi, J. S., Zatz, M., & Block, G. D. (1984a). Cyclic guanosine 3′:5′-monophosphate mimics the effects of light on a circadian pacemaker in the eye of *Aplysia. Journal of Neuroscience, 4,* 2466–2471.

Eskin, A., Yeung, S. J., & Klass, M. R. (1984b). Requirement for protein synthesis in the regulation of a circadian rhythm by serotonin. *Proceedings of the National Academy of Sciences of the USA, 81,* 7637–7641.

Gekakis, N., Saez, L., Delahaye-Brown, A. M., Myers, M. P., Sehgal, A., Young, M. W., & Weitz, C. J. (1995). Isolation of *timeless* by PER protein interaction: Defective interaction between *timeless* protein and long-period mutant PERL. *Science, 270,* 811–815.

Gekakis, N., Staknis, D., Nguyen, H. B., Davis, F. C., Wilsbacher, L. D., King, D. P., Takahashi, J. S., & Weitz, C. J. (1998). Role of the CLOCK protein in the mammalian circadian mechanism. *Science, 280,* 1564–1569.

Geusz, M. E., Michel, S., & Block, G. D. (1994). Intracellular calcium responses of circadian pacemaker neurons measured with fura-2. *Brain Research, 638,* 109–116.

Geusz, M. E., Foster, R. G., Degrip, W. J., & Block, G. D. (1997). Opsin-like immunoreactivity in the circadian pacemaker neurons and photoreceptors of the eye of the opisthobranch mollusc *Bulla gouldiana. Cell and Tissue Research, 287,* 203–210.

Hardin, P. E., Hall, J. C., & Rosbash, M. (1990). Feedback of the *Drosophila period* gene product on circadian cycling of its messenger RNA levels. *Nature, 343,* 536–540.

Harf, L., Arch, S., & Eskin, A. (1976). Polypeptide secretion from the eye of *Aplysia californica. Brain Research, 111,* 295–299.

Hartl, F. U. (1996). Molecular chaperones in cellular protein folding. *Nature, 381,* 571–579.

Hattar, S., & Eskin, A. (1996). ApC/EBP mRNA has a circadian rhythm and it is affected by light in the eye of *Aplysia. Society of Neuroscience Abstracts, 22,* 551.3.

Hattar, S., Levenson, J., & Eskin, A. (1997). Rhythm of ApC/EBP mRNA in the eye of *Aplysia*: *In vivo* vs *in vitro. Society of Neuroscience Abstracts, 23,* 521.16.

Hattar, S., Eskin, A., & Levenson, J. (1998). Effects of age on the circadian rhythm and eye of *Aplysia californica. Society of Neuroscience Abstracts, 24,* 761.7.

Heizmann, C. W. (1992). Calcium-binding proteins: basic concepts and clinical implications. *General Physiology and Biophysics, 11,* 411–425.

Herman, K. G., & Strumwasser, F. (1984). Regional specializations in the eye of *Aplysia*, a neuronal circadian oscillator. *Journal of Comparative Neurology, 230,* 593–613.

Herzog, E. D., Silva, C. M., Whitmore, D., & Block, G. D. (1997). The *Bulla* retinal circadian clock regulates spontaneous firing through tyrosine kinase and phosphatase activity. *Society of Neuroscience Abstracts, 23,* 521.15.

Homayouni, R., Nunez-Regueiro, M., Byrne, J. H., & Eskin, A. (1997). Identification of two phosphoproteins affected by serotonin in *Aplysia* sensory neurons. *Brain Research, 750,* 87–94.

Hudson, D. J., & Lickey, M. E. (1980). Internal desynchronization between two identified circadian oscillators in *Aplysia. Brain Research, 183,* 481–485.

Jacklet, J. W. (1969a). Electrophysiological organization of the eye of *Aplysia. Journal of General Physiology, 53,* 21–42.

Jacklet, J. W. (1969b). Circadian rhythm of optic nerve impulses recorded in darkness from isolated eye of *Aplysia. Science, 164,* 562–563.

Jacklet, J. W. (1974). The effects of constant light and light pulses on the circadian rhythm in the eye of *Aplysia. Journal of Comparative Physiology, 90,* 33–45.

Jacklet, J. W. (1977). Neuronal circadian rhythm: Phase shifting by a protein synthesis inhibitor. *Science, 198,* 69–71.

Jacklet, J. W. (1980). Light sensitivity of the rhinophores and eyes of *Aplysia. Journal of Comparative Physiology A—Sensory, Neural and Behavioral Physiology, 136,* 257–262.

Jacklet, J. W. (1991). Photoresponsiveness of *Aplysia* eye is modulated by the ocular pacemaker and serotonin. *Biological Bulletin, 180,* 284–294.

Jacklet, J. W., & Barnes, S. (1993). Photoresponsive pacemaker neurons from the dissociated retina of *Aplysia. Neuroreport, 5,* 209–212.

Jacklet, J. W., & Colquhoun, W. (1983). Ultrastructure of photoreceptors and circadian pacemaker neurons in the eye of a gastropod, *Bulla. Journal of Neurocytology, 12,* 673–696.

Jacklet, J. W., & Geronimo, J. (1971). Circadian rhythm: Population of interacting neurons. *Science, 174,* 299–302.

Jacklet, J. W., Schuster, L., & Rolerson, C. (1982). Electrical activity and structure of retinal cells of the *Aplysia* eye: I. Secondary neurones. *Journal of Experimental Biology, 99,* 369–380.

Jacklet, J. W., Klose, M., & Goldberg, M. (1987). FMRF-amide-like immunoreactive efferent fibers and FMRF-amide suppression of pacemaker neurons in eyes of *Bulla. Journal of Neurobiology, 18,* 433–449.

Johnson, C. H., Knight, M. R., Kondo, T., Masson, P., Sedbrook, J., Haley, A., & Trewavas, A. (1995). Circadian oscillations of cytosolic and chloroplastic free calcium in plants. *Science, 269,* 1863–1865.

Kaczmarek, L. K., & Strumwasser, F. (1984). A voltage-clamp analysis of currents underlying cyclic AMP-induced membrane modulation in isolated peptidergic neurons of *Aplysia. Journal of Neurophysiology, 52,* 340–349.

Kandel, E. R., & Schwartz, J. H. (1982). Molecular biology of learning: Modulation of transmitter release. *Science, 218,* 433–443.

Khalsa, S. B., & Block, G. D. (1988a). Calcium channels mediate phase shifts of the *Bulla* circadian pacemaker. *Journal of Comparative Physiology A—Sensory, Neural and Behavioral Physiology, 164,* 195–206.

Khalsa, S. B., & Block, G. D. (1988b). Phase-shifts of the *Bulla* ocular circadian pacemaker in the presence of calmodulin antagonists. *Life Sciences, 43,* 1551–1556.

Khalsa, S. B., & Block, G. D. (1990). Calcium in phase control of the *Bulla* circadian pacemaker. *Brain Research, 506,* 40–45.

Khalsa, S. B., & Block, G. D. (1992). Phase-shifting of a neuronal circadian pacemaker in *Bulla gouldiana* by pentylenetetrazol. *Comparative Biochemistry and Physiology—C: Comparative Pharmacology and Toxicology, 101,* 557–560.

Khalsa, S. B., Ralph, M. R., & Block, G. D. (1990). Chloride conductance contributes to period determination of a neuronal circadian pacemaker. *Brain Research, 520,* 166–169.

Khalsa, S. B., Ralph, M. R., & Block, G. D. (1991). Does low intracellular pH stop the motion of the *Bulla* circadian pacemaker? *Journal of Neuroscience, 11,* 2672–2679.

Khalsa, S. B., Whitmore, D., & Block, G. D. (1992). Stopping the circadian pacemaker with inhibitors of protein synthesis. *Proceedings of the National Academy of Sciences of the USA, 89,* 10862–10866.

Khalsa, S. B., Ralph, M. R., & Block, G. D. (1993). The role of extracellular calcium in generating and in phase-shifting the *Bulla* ocular circadian rhythm. *Journal of Biological Rhythms, 8,* 125–139.

Khalsa, S. B., Whitmore, D., Bogart, B., & Block, G. D. (1996). Evidence for a central role of transcription in the timing mechanism of a circadian clock. *American Journal of Physiology, 271,* 1, C1646–1651.

Khalsa, S. B., Michel, S., & Block, G. D. (1997). The role of extracellular sodium in the mechanism of a neuronal *in vitro* circadian pacemaker. *Chronobiology International, 14,* 1–8.

Koumenis, C., & Eskin, A. (1992). The hunt for mechanisms of circadian timing in the eye of *Aplysia. Chronobiology International, 9,* 201–221.

Koumenis C., Nunez-Regueiro M., & Eskin A. (1992). Identification of three additional putative oscillator proteins (POPs) from the eye of *Aplysia* as stress-related proteins. *Society of Neuroscience Abstracts, 18,* 881.

Koumenis, C., Nunez-Regueiro, M., Raju, U., Cook, R., & Eskin, A. (1995). Identification of three proteins in the eye of *Aplysia,* whose synthesis is altered by serotonin (5-HT). Possible involvement of these proteins in the ocular circadian system. *Journal of Biological Chemistry, 270,* 14619–14627.

Koumenis, C., Tran, Q., & Eskin, A. (1996). The use of a reversible transcription inhibitor, DRB, to investigate the involvement of specific proteins in the ocular circadian system of *Aplysia. Journal of Biological Rhythms, 11,* 45–56.

Krucher, N. A., Meijer, L., & Roberts, M. H. (1997). The cyclin-dependent kinase (cdk) inhibitors, olomoucine and roscovitine, alter the expression of a molluscan circadian pacemaker. *Cellular and Molecular Neurobiology, 17,* 495–507.

Kupfermann, I. (1967). A circadian locomotor rhythm in *Aplysia californica. Physiology and Behavior, 3,* 179–182.

Lickey, M. E., Wozniak, J. A., Block, G. D., Hudson, D. J., & Augter, G. K. (1977). The consequences of eye removal for the circadian rhythm of behavioral activity in *Aplysia. Journal of Comparative Physiology A—Sensory, Neural and Behavioral Physiology, 118,* 121–143.

Lotshaw, D. P., & Jacklet, J. W. (1986). Involvement of protein synthesis in circadian clock of *Aplysia* eye. *American Journal of Physiology, 250,* R5–R17.

Lotshaw, D. P., & Jacklet, J. W. (1987). Serotonin induced protein phosphorylation in the *Aplysia* eye. *Comparative Biochemistry and Physiology—C: Comparative Pharmacology and Toxicology, 86,* 27–32.

Luborsky-Moore, J. L., & Jacklet, J. W. (1977). Ultrastructure of the secondary cells in the *Aplysia* eye. *Journal of Ultrastructure Research, 60,* 235–245.

McMahon D. G., & Block, G. D. (1982). Organized photoreceptor layer is not required for light responses in three opisthobranch eyes. *Society of Neuroscience Abstracts, 8,* 33.

McMahon, D. G., & Block, G. D. (1987a). The *Bulla* ocular circadian pacemaker. I. Pacemaker neuron membrane potential controls phase through a calcium-dependent mechanism. *Journal of Comparative Physiology A—Sensory, Neural and Behavioral Physiology, 161,* 335–346.

McMahon, D. G., & Block, G. D. (1987b). The *Bulla* ocular circadian pacemaker. II. Chronic changes in membrane potential lengthen free running period. *Journal of Comparative Physiology A—Sensory, Neural and Behavioral Physiology, 161,* 347–354.

McMahon, D. G., Wallace, S. F., & Block, G. D. (1984). Cellular analysis of the *Bulla* ocular circadian pacemaker system: II. Neurophysiological basis of circadian rhythmicity. *Journal of Comparative Physiology A—Sensory, Neural and Behavioral Physiology, 155,* 379–385.

Michel, S., Khalsa, S. B., & Block, G. D. (1992). Phase shifting of the circadian rhythm in the eye of *Bulla* by inhibition of chloride conductance. *Neuroscience Letters, 146,* 219–222.

Michel, S., Geusz, M. E., Zaritsky, J. J., & Block, G. D. (1993). Circadian rhythm in membrane conductance expressed in isolated neurons. *Science, 259,* 239–241.

Michel, S., Manivannan, K., Zaritsky, J. J., & Block, G. D. (1999). A delayed rectifier current is modulated by the circadian pacemaker in *Bulla. Journal of Biological Rhythms, 14,* 141–150.

Myers, M. P., Wager-Smith, K., Wesley, C. S., Young, M. W., & Sehgal, A. (1995). Positional cloning and sequence analysis of the *Drosophila* clock gene, *timeless. Science, 270,* 805–808.

Nadakavukaren, J. J., Lickey, M. E., & Jordan, W. P. (1986). Regulation of the circadian clock in the *Aplysia* eye: Mimicry of neural action by serotonin. *Journal of Neuroscience, 6,* 14–21.

Noel, F., Koumenis, C., Nunez-Regueiro, M., Raju, U., Byrne, J. H., & Eskin, A. (1994). Effects on protein synthesis produced by pairing depolarization with serotonin, an analogue of associative learning in *Aplysia. Proceedings of the National Academy of Sciences of the USA, 91,* 4150–4154.

Page, T. L. (1981). Effects of localized low-temperature pulses on the cockroach circadian pacemaker. *American Journal of Physiology, 240,* R144–R150.

Page, T. L., Wassmer, G. T., Fletcher, J., & Block, G. D. (1997). Aftereffects of entrainment on the period of the pacemaker in the eye of the mollusk *Bulla gouldiana. Journal of Biological Rhythms, 12,* 218–225.

Pennartz, C. M., Bierlaagh, M. A., & Geurtsen, A. M. (1997). Cellular mechanisms underlying spontaneous firing in rat suprachiasmatic nucleus: Involvement of a slowly inactivating component of sodium current. *Journal of Neurophysiology, 78,* 1811–1825.

Pepinsky, R. B., Sinclair, L. K., Browning, J. L., Mattaliano, R. J., Smart, J. E., Chow, E. P., Falbel, T., Ribolini, A., Garwin, J. L., & Wallner, B. P. (1986). Purification and partial sequence analysis of a 37-kDa protein that inhibits phospholipase A_2 activity from rat peritoneal exudates. *Journal of Biological Chemistry, 261,* 4239–4246.

Pittendrigh, C. S. (1960). Circadian rhythms and the circadian organization of living systems. *Cold Spring Harbor Symposia on Quantitative Biology, 25,* 159–184.

Pittendrigh, C. S. (1967). Circadian systems I: The driving oscillation and its assay in *Drosophila pseudoobscura. Proceedings of the National Academy of Sciences of the USA, 58,* 1762–1767.

Pittendrigh C. S. (1974). Circadian oscillations in cells and the circadian organization of multicellular systems. In F. O. Schmidt & F. G. Worden (Eds.), *The neurosciences, Third study program* (pp. 437–458). Cambridge, MA: MIT Press.

Pittendrigh, C. S., & Daan, S. (1976). A functional analysis of circadian pacemakers in nocturnal rodents: V. Pacemaker structure: A clock for all seasons. *Journal of Comparative Physiology A—Sensory, Neural and Behavioral Physiology, 106,* 333–355.

Raiford P. R. (1984). Locomotor control by an ocular pacemaker in *Bulla gouldiana.* Masters thesis, University of Virginia, Charlottesville, Virginia.

Raju, U., Yeung, S. J., & Eskin, A. (1990). Involvement of proteins in light resetting ocular circadian oscillators of *Aplysia. American Journal of Physiology, 258,* R256–R262.

Raju, U., Koumenis, C., Nunez-Regueiro, M., & Eskin, A. (1991). Alteration of the phase and period of a circadian oscillator by a reversible transcription inhibitor. *Science, 253,* 673–675.

Raju, U., Nunez-Regueiro, M., Cook, R., Kaetzel, M. A., Yeung, S. C., & Eskin, A. (1993). Identification of an annexin-like protein and its possible role in the *Aplysia* eye circadian system. *Journal of Neurochemistry, 61,* 1236–1245.

Ralph, M. R., & Block, G. D. (1990). Circadian and light-induced conductance changes in putative pacemaker cells of *Bulla gouldiana. Journal of Comparative Physiology A—Sensory, Neural and Behavioral Physiology, 166,* 589–595.

Ralph, M. R., Khalsa, S. B. S., & Block, G. D. (1992). Cyclic nucleotides and circadian rhythm generation in *Bulla gouldiana. Comparative Biochemistry and Physiology A—Comparative Physiology, 101A,* 813–817.

Roberts, M. H., & Block, G. D. (1982). Dissection of circadian organization of *Aplysia* through connective lesions and electrophysiological recording. *Journal of Experimental Zoology, 219,* 39–50.

Roberts, M. H., & Block, G. D. (1983). Mutual coupling between the ocular circadian pacemakers of *Bulla gouldiana. Science, 221,* 87–89.

Roberts, M. H., & Block, G. D. (1985). Analysis of mutual circadian pacemaker coupling between the two eyes of *Bulla. Journal of Biological Rhythms, 1,* 55–75.

Roberts, M. H., & Moore, R. Y. (1987). Localization of neuropeptides in efferent terminals of the eye in the marine snail, *Bulla gouldiana. Cell and Tissue Research, 248,* 67–73.

Roberts, M. H., & Xie, X. (1996). Phase relationship between ocular and behavioral circadian rhythms in *Bulla gouldiana* exposed to different photoperiods. *Physiology and Behavior, 59,* 703–708.

Roberts, M. H., Block, G. D., & Lusska, A. E. (1987). Comparative studies of circadian pacemaker coupling in Opisthobranch molluscs. *Brain Research, 423,* 286–292.

Roberts, M. H., Bedian, V., & Chen, Y. L. (1989). Kinase inhibition lengthens the period of the circadian pacemaker in the eye of *Bulla gouldiana. Brain Research, 504,* 211–215.

Roberts, M. H., Towles, J. A., & Leader, N. K. (1992). Tyrosine kinase regulation of a molluscan circadian clock. *Brain Research, 592,* 170–174.

Rothman, B. S., & Strumwasser, F. (1976). Phase shifting the circadian rhythm of neuronal activity in the isolated *Aplysia* eye with puromycin and cycloheximide. Electrophysiological and biochemical studies. *Journal of General Physiology, 68,* 359–384.

Rusak, B., Robertson, H. A., Wisden, W., & Hunt, S. P. (1990). Light pulses that shift rhythms induce gene expression in the suprachiasmatic nucleus. *Science, 248,* 1237–1240.

Rutila, J. E., Suri, V., Le, M., So, W. V., Rosbash, M., & Hall, J. C. (1998). CYCLE is a second bHLH-PAS clock protein essential for circadian rhythmicity and transcription of *Drosophila period* and *timeless. Cell, 93,* 805–814.

Sankrithi, N., & Eskin, A. (1999). Effects of cyclin-dependent kinase inhibitors on transcription and ocular circadian rhythm of *Aplysia. Journal of Neurochemistry, 72,* 605–613.

Sloan M., Koumenis C., & Eskin A. (1995). Effects of phase-shifting treatments on the mRNA levels of putative oscillator proteins, BiP and porin, in *Aplysia. Society of Neuroscience Abstracts, 21,* 658.1.

Sloan, M. A., Levenson, J., Tran, Q., Kerbeshian, M., Block, G. D., & Eskin, A. (1999). Aging affects the ocular circadian pacemaker of *Aplysia california. Journal of Biological Rhythms, 14,* 151–159.

Strumwasser, F. (1967). Neurophysiological aspects of rhythms. In G. Quarton, T. Melnechuk, & F. O. Schmitt (Eds.), *The neurosciences* (pp. 516–528). New York: Rockefeller University Press.

Strumwasser, F. (1973). Neural and humoral factors in the temporal organization of behavior. *Physiologist, 16,* 9–12.

Strumwasser, F., Alvarez, R. B., Viele, D. V., & Woolum, J. C. (1979). Structure and function of a neuronal circadian oscillator system. In H. Nakagawa (Ed.), *Biological rhythms and their central mechanism* (pp. 41–56). Amsterdam: Elsevier/North-Holland.

Sweatt, J. D., Volterra, A., Edmonds, B., Karl, K. A., Siegelbaum, S. A., & Kandel, E. R. (1989). FMRFamide reverses protein phosphorylation produced by 5-HT and cAMP in *Aplysia* sensory neurons. *Nature, 342,* 275–278.

Takahashi, J. S., Nelson, D. E., & Eskin, A. (1989). Immunocytochemical localization of serotonergic fibers innervating the ocular circadian system of *Aplysia. Neuroscience, 28,* 139–147.

Toung, Y. P., & Tu, C. P. (1992). *Drosophila* glutathione S-transferases have sequence homology to the stringent starvation protein of *Escherichia coli. Biochemical and Biophysical Research Communications, 182,* 355–360.

Turek, F. W., Penev, P., Zhang, Y., Van Reeth, O., Takahashi, J. S., & Zee, P. (1995). Alterations in the circadian system in advanced age. *Ciba Foundation Symposium, 183,* 212–226.

Welsh, D. K., Logothetis, D. E., Meister, M., & Reppert, S. M. (1995). Individual neurons dissociated from rat suprachiasmatic nucleus express independently phased circadian firing rhythms. *Neuron, 14,* 697–706.

Whitmore, D. (1997). *The role of transcription and translation in the circadian pacemaker of the marine mollusks Bulla gouldiana and Aplysia californica.* Ph.D. dissertation, University of Virginia, Charlottesville, Virginia.

Whitmore, D., & Block, G. D. (1996). Cellular aspects of molluskan biochronometry. *Seminars in Cell and Developmental Biology, 7,* 781–789.

Woolum, J. C., & Strumwasser, F. (1980). The differential effects of ionizing radiation on the circadian oscillator and other functions in the eye of *Aplysia*. *Proceedings of the National Academy of Sciences of the USA, 77,* 5542–5546.

Yeung, S. J., & Eskin, A. (1987). Involvement of a specific protein in the regulation of a circadian rhythm in *Aplysia* eye. *Proceedings of the National Academy of Sciences of the USA, 84,* 279–283.

Yeung, S. J., & Eskin, A. (1988). Responses of the circadian system in the *Aplysia* eye to inhibitors of protein synthesis. *Journal of Biological Rhythms, 3,* 225–236.

Zwartjes, R. E., & Eskin, A. (1990). Changes in protein phosphorylation in the eye of *Aplysia* associated with circadian rhythm regulation by serotonin. *Journal of Neurobiology, 21,* 376–383.

15

Cellular Mechanisms of Circadian Function in the Suprachiasmatic Nucleus

Y. Bouskila, G. J. Strecker, and F. E. Dudek

Introduction

This chapter deals with a fundamental component of the biological clock in mammals: the circadian rhythm of electrical activity in neurons of the suprachiasmatic nucleus (SCN). Even though the rhythm of electrical activity is probably not responsible for the generation of the intrinsic time-keeping mechanism (see below and other chapters), an understanding of the electrophysiologic changes associated with the circadian rhythm should eventually lead us to the cellular mechanism of the circadian clock. The circadian rhythm of spontaneous neuronal discharge in the SCN is the primary cellular expression of the clock and a well-established phenomenon. Electrical activity (i.e., action potentials and synaptic potentials) serves as a critical link between light input and the clock in the SCN, and mediates expressions of the clock such as behavioral and physiological circadian rhythms. Understanding the oscillations in neuronal excitability that underlie the circadian rhythm of firing rate in the SCN should provide a solid basis for studies of other cellular mechanisms (e.g., biochemical or genetic) that may drive the electrical output of the biological clock. Among the different circadian rhythms, the rhythm of neuronal activity in the mammalian SCN can be compared to the better understood ocular pacemakers of the marine mollusks *Bulla* and *Aplysia* to gain additional insight into its mechanism. In the following sections, we summarize information regarding the cellular mechanisms of generation and communication of the electrical rhythm in the SCN, as well

---

Y. Bouskila, G. J. Strecker, and F. E. Dudek Department of Anatomy and Neurobiology, Colorado State University, Fort Collins, Colorado 80523.

Circadian Clocks, Volume 12 of *Handbook of Behavioral Neurobiology*, edited by Joseph S. Takahashi, Fred W. Turek, and R. Y. Moore, Kluwer Academic / Plenum Publishers, New York, 2001.

as the mechanisms of resetting the clock and the transmission of information from the SCN. Possible future directions will also be discussed.

The Endogenous Rhythm of Neuronal Discharge

The neuronal population of the SCN exhibits oscillations in firing rate with a period close to 24 hours. Peak activity is registered during the day and the trough occurs at night. This rhythm of neuronal activity was first described in the "SCN island" preparation (i.e., an SCN isolated by knife cuts) in freely behaving animals (Inouye & Kawamura, 1979). The circadian rhythm of firing could persist in the SCN for about 35 days after isolation (Inouye & Kawamura, 1982) (Figure 1). Later reports indicated that the rhythm of firing rate could persist for one to three cycles in hypothalamic slices containing the SCN (Green & Gillette, 1982; Groos & Hendriks, 1982; Prosser & Gillette, 1989; Shibata, Oomura, Kita, & Hattori, 1982) (Figure 2). More recently, a similar circadian rhythm was shown to persist in cultured SCN slices for up to 66 hours (Bos & Mirmiran, 1990) and in dissociated SCN neuronal cultures for over 30 days (Welsh, Logothetis, Meister, & Reppert, 1995). The persistence of the neuronal rhythm in different types of isolated SCN preparations indicates the endogenous nature of the clock and shows that input to the nucleus is not required for the generation of the circadian rhythm.

The slice preparation has been used extensively for the study of circadian rhythms in the SCN. When the firing frequency of individual SCN neurons in

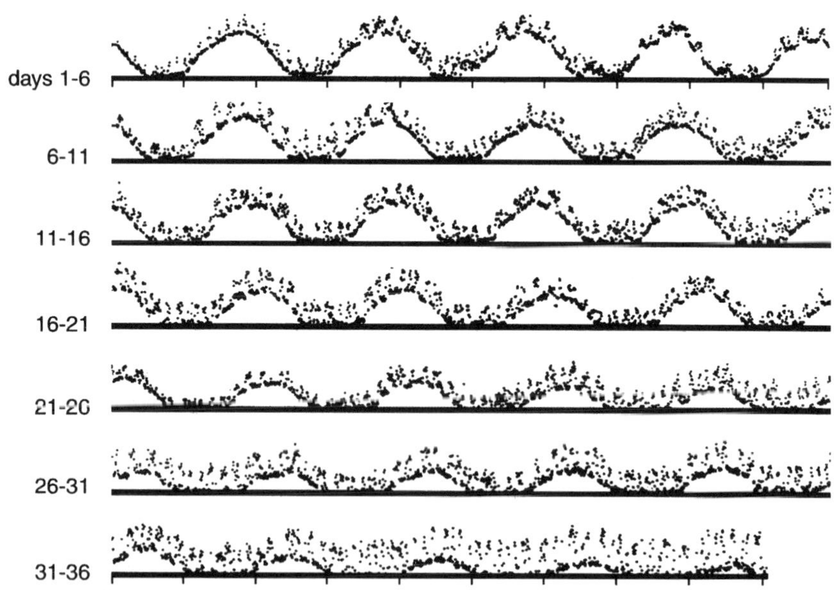

Figure 1. Circadian rhythm of neuronal activity *in vivo*. Rhythm of SCN multiple-unit activity inside a hypothalamic "island" from rats kept in constant darkness. Each dot represents the count of multiple-unit activity in 5 minutes, measured over 35 days. Each mark on horizontal axis equals 12 hours. The period of this rhythm was calculated to be 24 hours and 19 minutes. (Adapted from Inouye & Kawamura, 1982.)

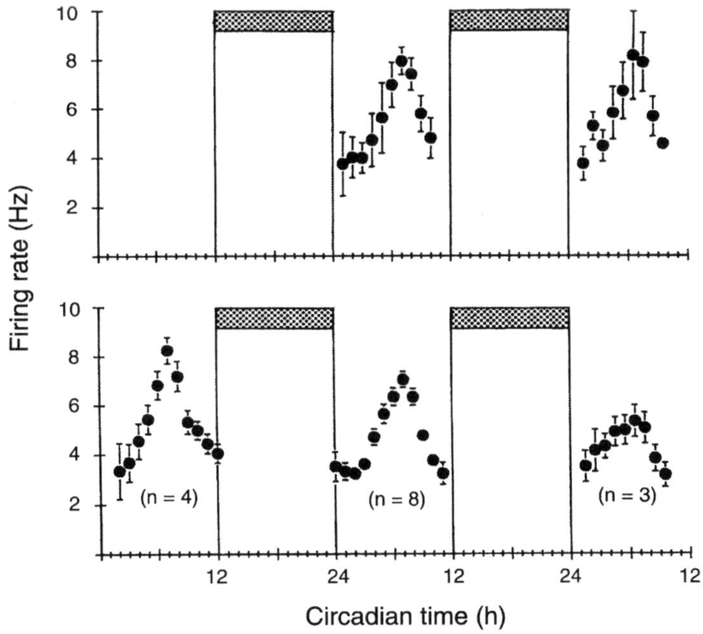

Figure 2. Circadian rhythm of neuronal activity *in vitro.* Top: Recording from a single SCN on day 2 (peak = CT7.0) and day 3 (peak = CT6.5). Lower: Mean of firing rates on successive days from several animals (*n* equals the number of animals). The means (±SEM) of all single units recorded from 2-hour intervals are plotted with 1-hour lags. Horizontal bars represent subjective night. (Adapted from Prosser & Gillette, 1989.)

hypothalamic slices is recorded for a few minutes (typically, 2 minutes) and averaged in 2-hours bins, a circadian rhythm of neuronal activity with a sinusoidal waveform emerges. During the subjective day, the range of firing frequency is high (i.e., from below 1 Hz to 14 Hz), while during the subjective night the range is narrower (0.5–3.0 Hz). The average activity at the peak reaches 7–8 Hz and falls about threefold at the trough. The peak activity occurs around circadian time (CT) 6–7 (Gillette, 1991; Prosser, McArthur, & Gillette, 1989) CT0 is defined as when light is turned on and experimental animals are typically entrained to a schedule with 12 hours of light and 12 of hours dark). Comparison between multiunit activity *in vivo* versus *in vitro* reveals that the variability of the discharge is larger and the rhythm amplitude is smaller *in vivo,* which is partly related to behavioral activity (Meijer, Schapp, Watanabe, & Albus, 1997). The following sections describe the electrophysiological properties of SCN neurons and the communication between them that may underlie the expression of the biological clock.

The Neuronal Rhythm—Expression or Generation of the Clock?

Neuronal activity is clearly an expression of the clock but does it also generate the time-keeping mechanism? Several studies have addressed this question by perturbing various rhythms with tetrodotoxin (TTX), a potent blocker of voltage-dependent Na^+ current that was reported to block spontaneous and induced action potentials in the SCN (Walsh, van den Berg, Marani, & Rietveld, 1992). When TTX was applied to the SCN, the circadian rhythm of drinking was blocked (Schwartz, Gross & Morton, 1987). In SCN explants, TTX blocked the rhythm of vasopressin secretion (Earnest, Digiorgio, & Sladek, 1991), electrical activity, and glucose utilization (Shibata & Moore, 1993). A similar result has been obtained with single-cell recordings of dissociated cell cultures of SCN (Welsh *et al.*, 1995). Upon TTX removal, however, these rhythms resumed at precisely the phase that would be predicted if no TTX had been administered, thus indicating that blockade of electrical activity did not disturb the time-keeping mechanism. Therefore, electrical activity does not seem to be critical for the generation of the time-keeping mechanism of the circadian clock.

A related question is whether the rhythm of electrical activity drives other circadian rhythms. Some of the experiments mentioned above (Earnest *et al.*, 1991; Schwartz *et al.*, 1987) also demonstrated that TTX application blocks several circadian rhythms, supporting the notion that Na^+-dependent action potentials are necessary for the expression of at least some behavioral and hormonal rhythms. Additional evidence comes from the demonstration of a link between electrical activity and behavior. Some mutant hamsters exhibit no circadian rhythm in wheel-running behavior, and brain slices prepared from these animals show no circadian variation in SCN electrical activity, in contrast to animals that show behavioral rhythm (Margraf, Puchalski, & Lynch, 1992). Furthermore, hamsters with a split rhythm (i.e., two activity periods per day instead of one) of wheel-running behavior induced by constant light have been shown to exhibit bimodal peaks of SCN firing rate, where the two troughs coincide with the projected time of wheel-running behavior (Zlomanczuk, Margraf, & Lynch, 1991). In view of the endogenous nature of the neuronal rhythm, these results support the hypothesis that electrical activity in the SCN mediates the expression of behavioral and hormonal rhythms, but is not part of the clock generation mechanism.

Characterization of SCN Neurons

The ability of SCN neurons to express a circadian rhythm of firing rate presumably depends on the modulation of their excitability. In order to gain insight into the mechanisms of modulation, an understanding of the membrane properties, ionic currents, and firing patterns of SCN neurons is needed.

Passive Membrane Properties. The values of input resistance reported in SCN neurons vary widely depending on the preparation and recording technique used. In slice preparations studied with sharp electrodes, average values of about 125 MΩ (Sugimori, Shibata, & Oomura, 1986), 147 MΩ (Wheal & Thomson, 1984), 183 MΩ (Akasu, Shoji, & Hasuo, 1993) and 301 MΩ (Kim & Dudek, 1993) were reported. Using the whole-cell patch-clamp technique, the reported values were higher; mean input resistance in the slice preparation was 657 MΩ (Bouskila & Dudek, 1993) and 741 MΩ (1.35 nsec; Jiang, Yang, Liu, & Allen, 1997b); in dissociated, cultured SCN neurons, the mean was 1600 MΩ (Walsh *et al.*, 1992); in acutely dissociated cells, the mean was 5400 MΩ (R. C.Huang, 1993). The lower values found with sharp electrodes may be underestimations due to the leak around the electrode whereas the extremely high values in dissociated cells are probably an overestimation due to the extensive loss of dendritic and axonal arbors during the dissociation process, and possibly to changes in conductance due to chemical treatments. However, it was suggested that an additional specific ionic conductance may be present in sharp electrode recordings and not seen in whole-cell recordings, probably due to washout (Staley, Otis, & Mody, 1992). The higher input resistance typical of small cells, such as SCN neurons, means that small currents can produce large voltage changes, thus making these cells more sensitive to synaptic input.

The average membrane time constant using sharp electrodes was 11 msec (Kim & Dudek, 1993), whereas data from R. C. Huang (1993) obtained with patch pipettes suggest a time constant of 37 msec. The higher value reflects the higher input resistance. The resting membrane potential of SCN neurons was reported to fall within the range of −50 to −70 mV, with an average around −60 mV in all preparations and recording techniques used (Akasu *et al.*, 1993; R. C. Huang, 1993; Kim & Dudek, 1993; Sugimori *et al.*, 1986; Walsh *et al.*, 1992; Wheal & Thomson, 1984). Passive membrane properties of SCN neurons, therefore, appear to have values similar to those of other brain areas.

IONIC CURRENTS. Information on the distribution and characterization of the ionic currents present in SCN neurons is critical for the understanding of neuronal excitability and its role in the expression of circadian rhythms. Recent studies have concentrated on the description of the ionic currents in the SCN using voltage-clamp techniques. Among the ionic currents that influence neuronal excitability are K^+, Ca^{2+} and Na^+, and mixed ionic currents.

Potassium. K^+ currents in the SCN deserve special attention because of their critical roles in determining firing pattern, firing frequency, and action potential waveform and their wide range of modulation possibilities (for review, see Storm, 1990). Also notable is the circadian oscillation of a tetraethylammonium (TEA)-sensitive K^+ conductance in the *Bulla* circadian pacemaker cells (Michel, Geusz, Zaritsky, & Block, 1993). As in many other neurons, K^+ currents in SCN neurons contribute to repolarization of the membrane after an action potential, as indicated by the reduction of the hyperpolarizing afterpotential (HAP) after the nonspecific K^+ channel blocker TEA or 4-aminopyridine was applied (Sugimori *et al.*, 1986). Four K^+ currents have been described in the SCN: a transient A-current (Bouskila & Dudek, 1995b; R. C. Huang, Peng, & Yau, 1993; Walsh, van den Berg, & Rietveld, 1995), a novel type of delayed rectifier (Bouskila & Dudek, 1995b), a calcium-dependent current (Walsh *et al.*, 1995) and a nonselective cationic inward rectifier (Akasu *et al.*, 1993); all the tested SCN neurons in the respective studies exhibited the first two current types, but two thirds of the neurons exhibited the last current.

The A-current in slices (Bouskila & Dudek, 1995b) was activated above −60 mV with half-activation at −24 mV and half-inactivation at −65 mV. Similar results were obtained from cultured neurons (Walsh *et al.*, 1995), but the latter value was more negative (−80 mV) in acutely dissociated neurons (R. C. Huang *et al.*, 1993). Activation was rapid (time-to-peak of a few milliseconds) and voltage-dependent, while inactivation was slower and composed of voltage-dependent and voltage-independent components (time constants of about 10 and 50 msec, respectively). Onset and recovery from inactivation had similar time constants of 40 msec. This current was blocked by 5 mM 4-aminopyridine (R. C. Huang *et al.*, 1993) and was resistant to 30 mM tetraethylammonium. In contrast, 4-aminopyridine had only a small blocking effect in cultured neurons (Walsh *et al.*, 1995)

Around the reported resting membrane potential (about −60 mV), A-current is mostly inactivated, but this inactivation may be removed in at least two ways. Hyperpolarization, such as during the HAP (Bouskila & Dudek, 1995b), or a low dose of

Zn^{2+}, which might be released by synaptic terminals in the SCN (R. C. Huang *et al.*, 1993), would result in a larger effect of A-current on the membrane potential. A-currents are generally known to delay the onset of firing, to permit slow repetitive firing, and to repolarize the action potential (Storm, 1990). Analysis of the A-current kinetics (Y. Bouskila & Dudek, 1995b), as well as application of 4-aminopyridine during evoked action potentials (Bouskila, unpublished) in the SCN, suggest that this current affects mainly the firing rate, but also the action potential waveform.

The rapidly activating outward rectifier ($IK_{(FR)}$) belongs to the delayed rectifier family. It was activated above −40 mV with half-activation at 14 mV. The rates of activation and deactivation were rapid (time constants of a few milliseconds) and voltage-dependent, while the main component of the inactivation was very slow (time constant of a few seconds) and voltage-independent. A higher sensitivity to TEA was reported in cultured neurons (Walsh *et al.*, 1995). This current was reduced by TEA (IC_{50} = 20 mM), but was resistant to 10 mM 4-aminopyridine. Its rapid kinetics and its high threshold of activation suggest a major role in shaping the action potential waveform and little role in determining the firing rate (Bouskila & Dudek, 1995b).

Evidence for calcium-dependent K^+ current was found in most SCN neurons. This current was voltage-dependent and could be activated above 0 mV in 2 mM extracellular Ca^{2+}. It showed no inactivation up to 500 msec (Walsh *et al.*, 1995). Calcium-dependent K^+ channels are activated by a rise in intracellular Ca^{2+} either from an intra- or extracellular source. Their function may be to regulate spike frequency and accommodation. These channels are frequent targets of modulation by neurotransmitters and second messengers (Rudy, 1988).

Inward rectification of the voltage–current relation in SCN neurons in response to hyperpolarizing currents has sometimes been observed (Kim & Dudek, 1993; Sugimori *et al.*, 1986), which suggests the involvement of an inward rectifier current in the generation of firing rate. Voltage-clamp recordings showed that this inward current in SCN neurons is activated slowly during hyperpolarizing steps more negative than −60 mV with a half-activation voltage of around −87 mV, and shows no inactivation. Consistent with an H-current (I_H), it is blocked by 1 mM extracellular Cs^+, but not Ba^{2+}, and is carried primarily by K^+ and Na^+ (Akasu & Shoji, 1994; de Jeu & Pennartz, 1997). When recorded with sharp electrodes, bath application of Cs^+ depressed the slope of the depolarizing potential before the action potential, resulting in a lowered firing rate of tonically active SCN neurons (Figure 3A). In contrast, when recorded in the whole-cell configuration, extracellular Cs^+ had no effect on spike waveform or frequency despite the presence of I_H in the majority of SCN neurons (de Jeu & Pennartz, 1997; Pennartz, Bierlaagh & Geurtsen, 1997). Dialysis of the cell content may be responsible for these negative results, especially since I_H is modulated by cAMP in SCN neurons (Akasu & Shoji, 1994). In view of these results, I_H deserves special attention as a possible modulator of firing rate in the SCN.

Calcium. Ca^{2+} currents play a role in the generation of action potentials, transmitter release, and modulation of firing patterns (for review, see Bertolino & Llinàs, 1992). In the SCN, low-threshold and high-threshold Ca^{2+} current types have been described. Sugimori *et al.* (1986) and Akasu *et al.* (1993) demonstrated rebound depolarizing potentials upon return to resting membrane potential from a hyperpolarizing current pulse, suggesting the presence of a low-threshold Ca^{2+} current in SCN neurons. These potentials were resistant to TTX, enhanced by replacing Ca^{2+}

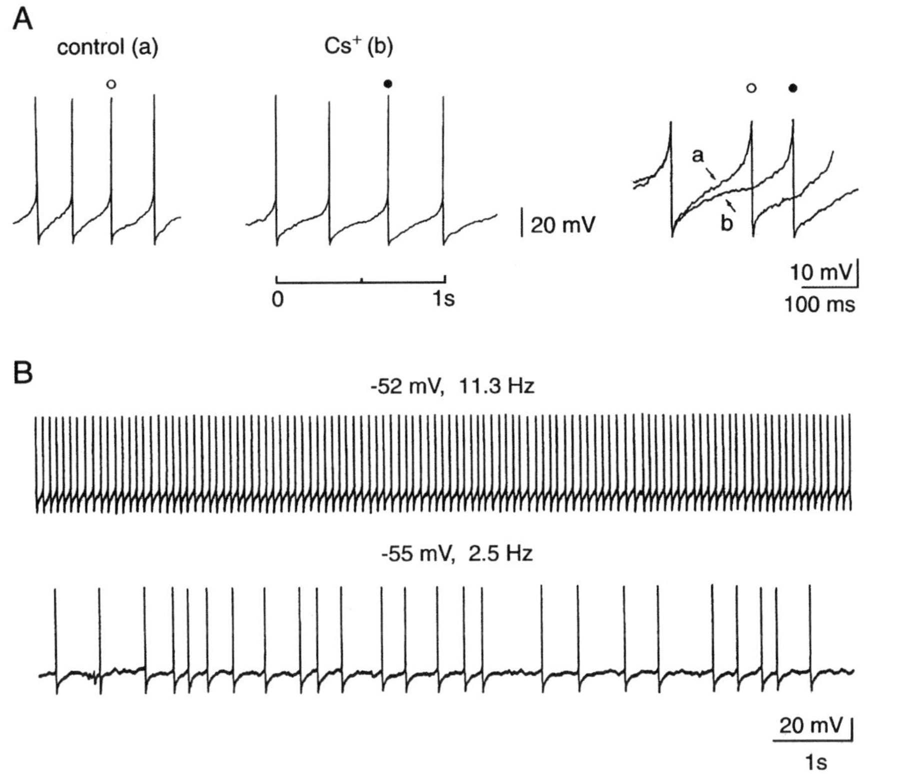

Figure 3. Modulation of firing rate in SCN neurons. (A) Effect of Cs^+ (1 mM) on a spontaneously firing SCN neuron. Action potentials are shown before and after application of Cs^+. Superimposed traces illustrate the change in waveform and duration of the interspike interval. The action potentials marked with open and closed circles are compared directly in the right panel. Action potentials were partially truncated. (Adapted from Akasu *et al.*, 1993.) (B) Effect of membrane potential on firing pattern. A regular firing pattern observed in a neuron at rest (upper trace) and an irregular pattern recorded from the same neuron at a more negative potential (lower trace) imposed by passage of steady current. (Adapted from Kim & Dudek, 1993.)

with Ba^{2+}, and blocked by the Ca^{2+} channel blockers Co^{2+} and Cd^{2+}. A more direct demonstration of the low-threshold Ca^{2+} current was provided by voltage-clamp experiments using dissociated SCN cells (R. C. Huang, 1993). This inward current was activated by depolarizations above -70 mV and inactivated with a time constant of about 25 msec. It was blocked more effectively by Ni^{2+} than Cd^{2+}, and in many respects, this current is similar to the T-type current described in other cells (Bertolino & Llinàs, 1992). Since the inactivation is complete at about -70 mV in the steady state, this current could be activated only if its inactivation is removed by hyperpolarization and subsequent depolarization. This condition can occur after an HAP, and therefore the low-threshold Ca^{2+} current could contribute to repetitive firing. This calcium current is present in other central neurons and appears to play a role in pacemaking activity (Llinàs, 1988). In the slice preparation, the majority of SCN neurons exhibited low-threshold spikes after a depolarizing pulse from a hyperpolarized potential (Kim & Dudek, 1993).

A high-threshold, slowly inactivating Ca^{2+} current was also observed in dissociated SCN neurons. It was activated by depolarizations positive to −30 mV and blocked more effectively by Cd^{2+} than by Ni^{2+}, consistent with a high-threshold Ca^{2+} current. The complete blockade of evoked trains of action potentials by low Cd^{2+} but not by Ni^{2+} suggests that the high-threshold Ca^{2+} current underlies these spikes (R. C. Huang, 1993). This Ca^{2+} current inactivates slowly and therefore has been suggested to contribute to repetitive firing during prolonged depolarizations when Na^+ current could no longer evoke action potentials due to its rapid inactivation.

Sodium. The voltage-dependent Na^+ current appears to be the main current underlying the depolarizing phase of action potentials in the SCN (but see next section). In dissociated SCN neurons this current was activated above −40 mV, and peaked at −10 mV (R. C. Huang, 1993). It was rapidly inactivated with a time constant of about 0.5 msec at 0 mV and in the steady state it was half inactivated at −60 mV. Recovery of the current from inactivation showed two components with the main one having a time constant of 30 msec at −60 mV. The characteristics of this Na^+ current suggest that it can support a fast firing rate in SCN neurons if a sufficiently long and large-amplitude interspike hyperpolarization occurs to remove inactivation. Evidence suggests that a slowly inactivating component of the Na^+ current contributes to the depolarizing ramp preceding spontaneous and evoked action potentials. The slow component is activated at a potential 13 mV more hyperpolarized than the fast component. Its voltage-dependence and kinetics suggest a role in determining the spontaneous firing rate (Pennartz *et al.*, 1997). The same study found that a persistent Na^+ current was absent or small in amplitude. Future studies should test whether this slowly inactivating component exhibits circadian oscillations.

Action Potentials and Firing Patterns. Action potentials in the SCN appear to be similar to those found in other areas of the brain. The average spike amplitude was 60–80 mV from threshold, spike duration at half amplitude was 0.8 msec, spike threshold was around −48 mV, and HAP amplitude was about 24 mV more negative than threshold. Evoked spike trains always showed spike inactivation, spike broadening, and frequency accommodation to some degree (Kim & Dudek, 1993; Wheal & Thomson, 1984). The slow after hyperpolarization that followed spike trains was present in almost all SCN neurons. It lasted up to 1 second and had a reversal potential close to the K^+ equilibrium potential, suggesting that K^+ was the dominant underlying current (Kim & Dudek, 1993), as in many other neuronal types.

Some evidence suggests that Ca^{2+} currents can also generate spikes. In dissociated SCN cells, depolarizing pulses generated a few small-amplitude and long-duration spikes in the presence of TTX (R. C. Huang, 1993). However, it is not clear whether these action potentials could occur spontaneously or be evoked without TTX. Activation of the high-threshold Ca^{2+} current required depolarization above −30 mV, whereas the low-threshold current required hyperpolarization to below −80 mV to remove its inactivation. The voltage-dependence of these Ca^{2+} currents excludes them from supporting repetitive firing alone, without Na^+ spikes. Indeed, in cultured SCN cells, application of TTX blocked spontaneous or evoked action potentials from resting membrane potential (Walsh *et al.*, 1992). Evoked synaptic potentials, however, could generate a low-threshold Ca^{2+} spike that triggered Na^+ spikes (Kim & Dudek, 1993). Taken together, the evidence suggests that Ca^{2+} currents may shape firing patterns, but cannot sustain repetitive firing on their own.

SCN neurons show a variety of firing patterns, which can be categorized into regular, irregular, and bursting types, based on extracellular recordings in slice preparations. However, reports of the relative occurrence of the different types vary. For example, an *in vitro* study by Shibata, Shiratsuchi, Liou, and Ueki (1984b) reported 77%, 20% and 3%, respectively, whereas Groos and Hendriks (1979) found that only 16% of the cells fired regularly and 84% were either irregular or bursting. The definition of "regularity" may have been different in these two studies but, in contrast to the *in vitro* recordings, *in vivo* recordings showed that only 2 of 397 neurons fired regularly using the latter definition (Groos & Hendriks, 1979). Intracellular recordings, however, revealed that depolarization of the membrane by current injection not only increased the firing rate (Kim & Dudek, 1993; Wheal & Thomson, 1984), but also changed the firing pattern from irregular (below 4 Hz) to regular (above 6 Hz, Kim & Dudek, 1993) (Figure 3B). These results suggest that the spontaneous firing pattern of SCN neurons is a function of membrane potential rather than cell type and raise the possibility that a circadian change in the resting membrane potential may underlie the circadian rhythm in firing rate.

The firing pattern of SCN neurons is also influenced by extracellular Ca^{2+}. Replacing Ca^{2+} with nonpermeable divalent cations such as Mg^{2+}, Co^{2+}, or Mn^{2+} resulted in a disruption of the regular firing pattern. Glutamate application caused SCN neurons to fire more regularly in the presence of extracellular Ca^{2+}, but not in the absence of Ca^{2+} (Thomson, 1984). Lowering of extracellular $[Ca^{2+}]$ caused most regularly firing cells to fire irregularly (Shibata *et al.*, 1984b). In contrast, incubation of hypothalamic slices for several hours in Ca^{2+}-free solutions with 1,2-bis(2-aminophenoxy)ethane-*N*,*N*,*N*′,*N*′-tetraacetic acid (BAPTA), a strong Ca^{2+} chelator, induced bursting activity in most SCN neurons (Bouskila & Dudek, 1993). Thus, it appears that extracellular Ca^{2+} is necessary to maintain continuous regular firing in SCN neurons but the mechanism is unclear. Possible mechanisms include membrane repolarization after an action potential [as suggested by the Ca^{2+}-dependent early component of the HAP (Thomson & West, 1990) and the Ca-dependent K^+ current (Walsh *et al.*, 1995)], involvement of Ca^{2+} current in depolarizing the membrane, maintenance of intracellular Ca^{2+}, or screening of membrane surface charges (Frankenhaeuser & Hodgkin, 1957).

The spontaneous firing rate of SCN neurons recorded with sharp electrodes in slice preparations varied widely in frequency, ranging from 0.5 to 10 Hz (Akasu *et al.*, 1993) or <4 to 34 Hz (mean = 9 Hz, Kim & Dudek, 1993). Extracellular recordings gave similar results (see above). In contrast, studies using whole-cell patch-clamp techniques in cultured or dissociated cells reported very low spontaneous firing rates, typically below 1 Hz (R. C. Huang, 1993; Walsh *et al.*, 1992). Though the whole-cell experiments were conducted at lower temperatures (by about 10°C), thermosensitivity of firing rate alone (Derambure & Boulant, 1992; Miller, Cau & Heller, 1994) is unlikely to account for these large differences since the majority of SCN cells exhibit insufficient thermosensitivity. An intriguing possibility is that a diffusible substance is necessary to maintain high and more physiological spontaneous firing rate. It would be interesting to test this hypothesis by comparing the spontaneous firing rate in slices using patch pipettes in the whole-cell versus cell-attached configuration.

Modulation of Neuronal Excitability

Some modulation of excitability must underlie the circadian rhythm of firing rate in SCN neurons. The possible mechanisms for modulation are numerous and may include alterations in passive membrane properties, neurotransmission, ionic conductances or concentrations, or second messenger systems.

NEUROTRANSMITTERS AND NEUROMODULATORS. The method most widely used to study the effects of substances such as neurotransmitters, neuromodulators, and hormones on SCN neurons has been extracellular recordings of single neurons during application of agonists and antagonists. This method is relatively simple technically and has been used to test over 20 different substances (for review, see Meijer & Rietveld, 1989; Rusak & Bina, 1990). However, despite the advantage of technical ease, the interpretation of data gathered in this way can be problematic. One problem is determining the site of action, which may be presynaptic, postsynaptic, or even outside the SCN. Another problem is determining the nature of the action, which may be direct (e.g., activation of a receptor), indirect (e.g., modulation of another substance), or even nonspecific. With extracellular recordings, the determination of the effect of a substance is usually limited to noting whether it produces an increase or decrease in firing rate. In contrast, intracellular recordings, although technically more difficult, can reduce some of these problems.

Glutamate appears to be the main neurotransmitter that mediates excitatory synaptic events in SCN neurons. Optic nerve stimulation, as well as stimulation of sites lateral or dorsocaudal to the SCN, evoked fast excitatory postsynaptic potentials (EPSPs), which were mediated by the α-amno-3-hydroxy-5-methylisoxazole proprionic acid (AMPA)/kainate class of glutamate receptors at resting and hyperpolarized membrane potentials (Figure 4A). NMDA (*N*-methyl-D-aspartate) receptors contributed to the synaptic response at more depolarized potentials (Jiang *et al.*, 1997b; Kim & Dudek, 1991), consistent with the removal of Mg^{2+} block during depolarization. NMDA and AMPA/kainate antagonists also blocked the fast EPSPs evoked by stimulation of a site dorsal or lateral to the SCN (Akasu *et al.*, 1993; Bouskila & Dudek, 1993). Spontaneous synaptic events may arise from fibers originating inside or outside the SCN, and therefore, they may represent more fully the different transmitters operating on SCN neurons. Spontaneous excitatory synaptic activity, however, was also blocked by the broad-spectrum excitatory amino acid (EAA) antagonist γ-D-glutamylglycine (Thomson & West, 1990) or by a mixture of

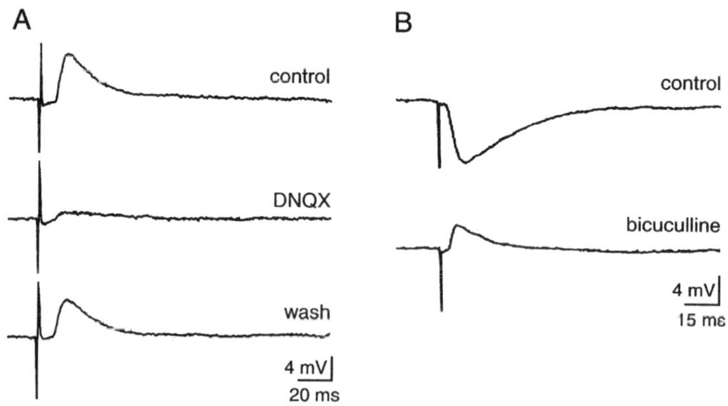

Figure 4. Excitatory and inhibitory postsynaptic potentials (EPSPs and IPSPs) recorded intracellularly from SCN neurons. (A) EPSPs evoked by stimulation of the optic nerve (control) were blocked by bath application of the AMPA/kainate antagonist DNQX (10 mM). This block was reversible (wash). (Adapted from Kim & Dudek, 1991.) (B) IPSPs evoked by stimulation of the medial hypothalamus (dorsocaudal to the SCN) were blocked by the $GABA_A$ antagonistbicuculline methiodide (20 mM), which occasionally revealed EPSPs. (Adapted from Kim & Dudek, 1992.)

NMDA, AMPA/kainate, and γ-aminobutyric acid type A ($GABA_A$) antagonists (Bouskila & Dudek, 1993). Northern analysis revealed five NMDA receptor subunits in the mouse SCN, with NR1 the most highly expressed. Among AMPA/kainate receptors, expression of GluR5 was highest, followed by GluR6 (O'Hara, Andretic, Heller, Carter, & Kilduff, 1995).

Fast inhibitory postsynaptic potentials (IPSPs) are also found frequently in SCN neurons. In slice experiments, electrical stimulation dorsocaudal to the SCN evoked fast IPSPs in nearly all SCN neurons tested. Spontaneous and evoked IPSPs were hyperpolarizing at resting potential, rose quickly to peak in about 9 msec, decayed with a time constant of 25 msec, and had a reversal potential of about -75 mV. With KCl electrodes, the IPSPs were positive-going, consistent with a shift in Cl^- equilibrium potential. The IPSPs were blocked by bicuculline, a $GABA_A$-receptor antagonist (Figure 4B; Kim & Dudek, 1992). Thomson and West (1990) also reported that evoked IPSPs in slices were sensitive to bicuculline. The responses of cultured SCN neurons to applied GABA were blocked by bicuculline also, and potently by zinc, but were not potentiated by the benzodiazepine diazepam (Kawahara, Saito & Katsuki, 1993), thus suggesting the absence of γ_2 subunits in the $GABA_A$ receptors. However, dissociated SCN neurons from immature rats did reveal diazepam sensitivity (Shimura, Harata, Tamai, & Araike, 1996), so there is little consensus on these aspects of $GABA_A$-receptor pharmacology in the SCN. Nevertheless, SCN neurons in acute hypothalamic slices from mature rats showed moderate zinc sensitivity and benzodiazepine sensitivity (Strecker, Park, & Dudek, 1999) similar to phenotypes found with specific subunit combinations that include both δ and γ_2 (Saxena and Macdonald, 1994) and receptors found in rat dentate granule cells (Gibbs, Shumate, & Coulter,, 1996; Kapur and Macdonald, 1996) and CA1 pyramidal cells (Gibbs *et al.*, 1996).

Using Northern analysis, several $GABA_A$-receptor subunit mRNAs were detected in the mouse SCN including subtypes of α, β, and γ (O'Hara *et al.*, 1995). An immunocytochemical study (Gao, Fritschy, & Moore, 1995), however, failed to detect either of the β subtypes examined ($\beta_{2,3,}$). Nevertheless, both groups found strong evidence for γ_2 subunits in the SCN, which is important for sensitivity to benzodiazepine (Pritchett, Lüddens, & Seeburg, 1989). The diversity of subunits might provide for a diversity in neuronal responses to GABA in the SCN since different assemblies are possible. Together these results indicate that SCN neurons receive extensive GABAergic input, that the IPSPs represent an increase in Cl^- conductance mediated by $GABA_A$ receptors, and that these receptors are likely substrates for various modulatory influences.

Evidence suggests GABA actions in the SCN switch from inhibition during subjective night to excitation during subjective day (Wagner, Castel, Gainer, & Yarom, 1997), presumably through a circadian change in intracellular chloride concentration. This finding carries interesting implications regarding the function of GABAergic circuitry in the SCN (see below). This shift, however, was not evident in some previous studies (Liou & Albers, 1990; Mason, 1986) nor in a preliminary report (Gribkoff, Pieschl, Wisialowski, & Dudek, 1997). Nevertheless, the importance of this result, in light of plausible GABAergic plasticity and the preponderance of GABAergic receptors in the SCN, warrants further investigation into the mechanisms responsible for these different effects of GABA. $GABA_A$-mediated currents can demonstrate substantial lability when intensely activated in some preparations, switching from inhibitory to excitatory, due to changes in a relatively fragile chloride ion gradient in the face of a more robust bicarbonate ion gradient (Staley, Soldo, & Proctor 1995). SCN neurons in culture appear to utilize GABA primarily as an excitatory neurotransmitter early in development, before switching to an inhibitory role (Obrietan and van den Pol, 1995). Any changes in GABA function in mature SCN could profoundly affect the activity of this nucleus.

Evidence for functional $GABA_B$ receptors in the SCN also exists. Baclofen, a $GABA_B$ agonist, induced an outward current and a conductance increase in one third of SCN cells. In addition, baclofen reduced the amplitude of EPSCs evoked by optic nerve or focal stimulation. It is plausible that $GABA_B$ receptor activation has pre- as well as postsynaptic inhibitory effects on SCN neurons (Jiang, Allen, & North, 1995a). However, there is no evidence for $GABA_B$-mediated postsynaptic currents or potentials in the SCN.

Although all evoked and spontaneous inhibitory synaptic responses could be blocked by $GABA_A$ antagonists, functional glycine receptor channels also appear to be present on SCN neurons. In acutely dissociated SCN cells, glycine-induced currents were primarily Cl^- sensitive and competitively blocked by strychnine. In addition, responses to glutamate and aspartate were enhanced by adding glycine (Ito, Wakamori, & Akaike, 1991). Thus, glycine may act in the SCN as an inhibitory neurotransmitter and an excitatory neuromodulator of NMDA receptors.

The discrepancy between the vast number of putative neurotransmitters and neuromodulators that affect firing rate and the limited number of receptors that mediate the detected synaptic events using intracellular recordings may be explained in several ways. Some of the agents tested may have modulatory actions on glutamate- and GABA-mediated neurotransmission. Others may have direct effects, which suggests that functional receptors are present on SCN neurons, and that the terminals secreting the endogenous ligands were isolated from their cell bodies during the slice preparation. An accumulating body of evidence suggests that a variety of substances intrinsic to the SCN can act as modulators on glutamate- and GABA-mediated synaptic responses. For example, serotonin and vasoactive intestinal peptide (VIP) appeared to potentiate the effects of exogenously applied glutamate (S. H. Huang & Pan, 1993). In contrast, preliminary data suggest that serotonin inhibits glutamatergic synaptic currents (Jiang, Yang, & Allen, 1995b), possibly via a presynaptic mechanism, while both serotonin (Kawahawa *et al.*, 1994) and NPY (Chen & van den Pol, 1996) attenuate responses to GABA in cultured SCN cells.

Modulation of Ionic Conductances. Since SCN cells can generate a circadian rhythm (Bos & Mirmiran, 1990), and a TEA-sensitive membrane conductance exhibits daily fluctuations in the *Bulla* circadian pacemaker (Michel *et al.*, 1993), it is possible that a circadian modulation of ionic currents underlies the rhythm of firing in the SCN. Very little is known about modulation of ionic currents in SCN neurons. The two specific ionic conductances that have been tested for oscillations in the SCN, namely the delayed rectifier (Bouskila & Dudek, 1995b) and the H-current (de Jeu & Pennartz, 1997; Jiang *et al.*, 1997b), showed no variations of amplitude with circadian time. Nevertheless, these studies used whole-cell recordings, a configuration that most likely dialyzed possible modulatory factors. Since this recording configuration produced an unusually low spontaneous firing rate (see section on Action potential and firing rate), which also failed to demonstrate a significant day–night difference (de Jeu & Pennartz, 1997; Jiang *et al.*, 1997b), it may not be suitable for detecting circadian oscillations. However, a few modulators of ionic currents in the SCN have been identified and are briefly described below.

Extracellular Zn^{2+} at low concentrations shifts the steady-state inactivation of the A-current to more positive voltages. The steady-state half-inactivation voltage showed a shift from −80 mV to −60 mV in 30 μM Zn^{2+} and to −35 mV in 200 μM Zn^{2+}. Since the A-current can be expected to modulate firing rate (Bouskila & Dudek, 1995b), perhaps Zn^{2+} plays a regulatory role. Although histologic evidence suggests that Zn^{2+} is present in the SCN (R. C. Huang *et al.*, 1993), it is unknown whether its concentration fluctuates.

cAMP and adenylate cyclase activator enhance the amplitude of I_H without altering the kinetics or voltage-dependence of the current (Akasu & Shoji, 1994). Since I_H can modulate the firing rate (Akasu *et al.*, 1993) and contribute to the resting membrane potential (Akasu & Shoji, 1994) in SCN neurons, and cAMP level oscillates in the nucleus (see below), it is possible that I_H contributes to the oscillations in neuronal excitability.

Adenosine was found to inhibit reversibly the high-threshold current in most SCN neurons tested. The currents were reduced one third without any change in kinetics or current–voltage relations (Chen & van den Pol, 1997).

Second messenger systems are known to modulate many types of ionic currents (for review of K^+ current modulation, see Rudy, 1988). Among the second messenger systems that modulate K^+ current are cAMP (cyclic adenosine 3′,5′-monophosphate), Ca-calmodulin kinases, and kinase C (Rudy, 1988). Changes in cAMP concentration have been reported in the SCN with a peak at late subjective day (Murakami & Takahashi, 1983; Prosser & Gillette, 1991), whereas another study on freshly dissected hypothalami reported a peak in early day (Valases, Wright, & Catravas, 1980). The finding that the rhythm of phosphodiesterase activity is opposite in phase to the level of cAMP and the lack of significant changes in adenylate cyclase activity in the SCN suggest that the circadian rhythm of cAMP level originates from a rhythm in the rate of cAMP breakdown rather than in the rate of its synthesis (Prosser & Gillette, 1991). An increase in cGMP (cyclic guanosine 3′,5′-monophosphate) level has been reported to occur around the middle of the night (Valases *et al.*, 1980). However, cAMP and cGMP analogs have been shown to decrease and increase, respectively, the firing rate of some SCN cells *in vitro* (Liou, Shibata, Shiratsuchi & Ueki, 1986). These results may suggest a role for cAMP and cGMP in the modulation of neuronal excitability in the SCN, but more data are needed to support this hypothesis. It is not clear whether the oscillations in cAMP and cGMP occur in glia or neurons, or whether the rhythm in neuronal activity induces these oscillations. Furthermore, no direct modulation of currents by these second messengers has been demonstrated in the SCN.

Another second messenger system that has been reported to operate in the SCN induces inositol trisphosphate (IP_3) formation (Nadakavukaren, Welsh & Reppert, 1990). Lithium, which is known to deplete the intracellular pool of inositol and lead to a decrease in intracellular second messengers IP_3 and diacylglycerol, suppresses neuronal firing in a dose-dependent manner. This effect is reversed by *myo*-inositol but not *epi*-inositol, the latter of which is not incorporated into phospholipids (Mason & Biello, 1992). Currently, no evidence is available for circadian changes in this system in the SCN.

The Circadian Clock—Network or Single Cell Phenomenon?

The generation of circadian signals may require the interaction of many neurons within the SCN. Alternatively, circadian rhythmicity may be a property of individual cells within the nucleus, and the interaction between them may serve only to achieve overall synchronization (Turek, 1985). The capacity to generate circadian rhythms is not restricted to higher plants and animals or even to multicellular organisms. A number of eukaryotic microorganisms have been documented to display a wide spectrum of behavioral, physiologic, and biochemical circadian rhythms (for review, see Edmunds, 1988). It is quite possible that the capacity of single cells to generate a circadian rhythm is preserved in the biological clock system of multicellular organisms. The best approach to test the hypothesis of a single-cell rhythm would be long-term recording of rhythms from completely isolated SCN neurons. Two studies using recordings from single SCN neurons in culture have provided interesting information. Several days of continuous electrical recordings from single SCN cells grown in cultured slices, where the cells were clearly not isolated, revealed a wide range of periods of their firing rate rhythm, with periods ranging from 16 to 32 hours and average period close to 24 hours (Bos & Mirmiran, 1990). A recent study where SCN neurons were cultured on recording electrodes showed a circadian rhythm (i.e., period near 24 hours) of firing rate in different phases which could persist over 30 days (Welsh *et al.*, 1995). A periodogram analysis of firing rhythm amplitude for various period lengths from nearly 200 units in acute slices revealed that in addition to the large peak at 24 h, smaller peaks at 6-, 10, and 18-hours periods also existed (Green & Gillette, 1982). Multiple-unit recordings from the SCN *in vivo,* which represent the overall electrical activity of a large population of neurons, displayed a slightly longer period than 24 hours (Inouye & Kawamura, 1982). Thus, it appears that single neurons in cultured SCN can display rhythms with a wide range of periods and/or different phases, whereas the combined neuronal population in more intact preparations displays a period of around 24 hours. It is not clear, however, whether an active synchronization between the cells takes place in the intact SCN, or whether the combined output from a population of independent oscillators can produce a period of nearly 24 hours. Reanalysis of data from Bos and Mirmiran has suggested that the first option is more likely (Bouskila & Dudek, 1995a). The ability of completely isolated SCN neurons to generate a rhythm has never been tested. In all the published studies, cells were grown in proximity, which allows for numerous intercellular interactions. It would be interesting to test for the presence of the rhythm in a low-density and purely neuronal culture. A suitable preparation may be dissociated SCN neurons from adult rats, which do not grow processes, survive for more than 2 weeks without glia, and exhibit normal currents (Y. Bouskila, unpublished). Destruction of at least 50% of the SCN was sometimes associated with the appearance of a significant 8-hour ultradian behavioral rhythm (van den Pol & Powley, 1979). It is interesting to note that mathematical models have suggested that a precise aggregate rhythm can be produced by coupled oscillators with faster or slower individual rhythms (Winfree, 1967). Although SCN neurons in culture can express rhythms with periods deviating from 24 hours, an interaction between the cells in the intact SCN probably leads to a more precise rhythm where the neurons are kept in phase, as will be discussed in the following section.

Cellular Communication within the SCN

The output of the SCN consists of a singular and precise circadian period under normal conditions. Since other circadian pacemaker systems involve electrically coupled cells (e.g., for *Bulla,* see Block, McMahon, Wallace & Friesen, 1984), it is possible that neurons in the SCN are synchronized to some degree. Histologic evidence suggests the existence of extensive GABAergic synaptic connectivity among

SCN neurons (Belenky *et al.*, 1996; van den Pol & Gorcs, 1986). Furthermore, functional evidence indicates that selective stimulation of SCN neurons with various forms of glutamate microapplication can produce barrages of $GABA_A$-mediated IPSCs in other SCN neurons (Figure 5). In contrast no glutamatergic EPSCs could be evoked in SCN neurons by local stimulation using this technique (Strecker *et al.*, 1997). The existence of GABAergic local circuitry implies a degree of feedback regulation among SCN neurons. The result of Wagner *et al.* (1997) suggesting that

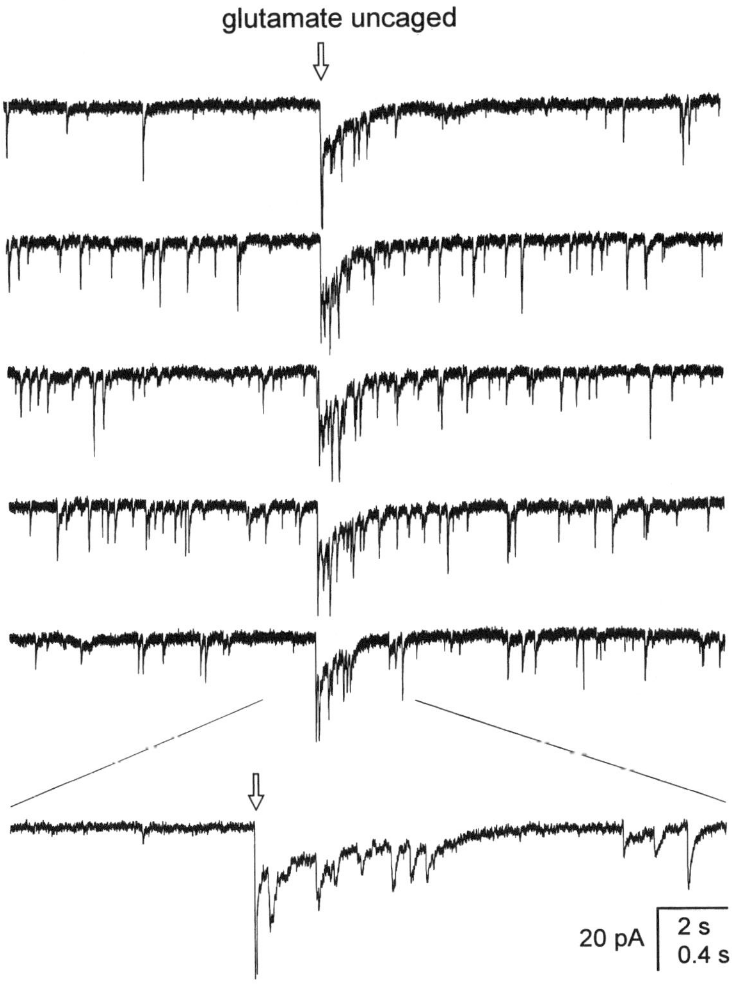

Figure 5. Photolysis of caged glutamate within the SCN evoked barrages of IPSCs. Glutamate was uncaged within the SCN with a brief flash from a xenon lamp, which evoked action potentials in a neuron(s) in the SCN that produced IPSCs in another recorded SCN neuron. Five consecutive glutamate uncaging events are shown in 20-second sequential traces. The last trace is shown in expanded time. This SCN neuron displayed a series of nine consecutive trials with significant ($p < .05$) increases in IPSC frequency ($V_h = -60$ mV, KCl intracellular solution, hence inward GABAergic IPSCs). The large inward current of approximately 2-second duration underlying the barrage of evoked IPSCs reflects the additional effect of uncaged glutamate on the recorded SCN neuron. IPSCs could also be evoked in other SCN neurons without a direct effect. (Adapted from Strecker *et al.*, 1997.)

GABA switches from excitation to inhibition with circadian regularity (see above) raises the possibility that the intrinsic synaptic circuitry acts to increase the amplitude of the well-known circadian rhythm in firing rate. In general, any change in GABAergic function may have profound consequences on SCN activity given such local synaptic connectivity. The apparent synchronization in electrical activity cannot, however, be fully explained by conventional chemical synaptic transmission mediated by action potentials (for review, see van den Pol & Dudek, 1993). The lack of effect of TTX application on the time-keeping mechanism suggests that Na^+ spikes and chemical synaptic transmission are not necessary for synchronization. However, since Ca^{2+} spikes could be evoked in the SCN (J. C. Huang, 1993; Sugimori *et al.*, 1986) and may be conducted along the dendrites (e.g., see Llinàs, 1988), it is possible that Ca^{2+} spikes mediate synaptic release in the presence of TTX. This seems unlikely, as discussed earlier, since Ca^{2+} spikes do not appear capable of supporting regular firing in the SCN, due to the voltage changes required for their activation.

There is evidence that SCN neurons can synchronize their activity by a mechanism that does not involve chemical synaptic transmission (Bouskila & Dudek, 1993). Dual multiple-unit recordings in Ca^{2+}-free solution containing a mixture of glutamate and $GABA_A$-receptor antagonists showed that two groups of neurons in different parts of the same SCN showed synchronized bursts of activity (Figure 6A, 6B). The synchronization was of a "loose" type (Figure 6C) and not one-for-one, which differs from the *Bulla* circadian pacemaker (McMahon, Wallace & Block, 1984). The elimination of spontaneous and evoked synaptic events detected with whole-cell patch-clamp recordings indicated that the experimental conditions were sufficient to block synaptic transmission (Bouskila & Dudek, 1993).

Several alternative mechanisms may be involved to produce coordination of the cellular elements in the SCN. Electrical field effects, where current flowing along an active cell excites an adjacent cell, have been demonstrated in the hippocampus (for review of original literature, see Dudek, Snow & Taylor, 1986), but in the SCN, these appear unlikely. A combination of high extracellular resistance and laminar organization of neurons is critical for enhancing electrical field effects. In the SCN, although the extracellular space appears to be small, the dendritic arbors are randomly oriented (van den Pol, 1980), and therefore their electrical fields would not be expected to summate effectively.

Electrotonic coupling through gap junctions between cells plays a role in other circadian pacemakers, for example, in *Bulla* (Block *et al.*, 1984) and *Aplysia* (Jacklet, 1969). Gap junctions allow not only electrotonic coupling between cells, but also diffusion of chemical signals with a molecular weight smaller than 1000 Daltons; therefore, gap junctions could synchronize the activity of the coupled cells (Pitts & Finbow, 1986; Spray & Bennett, 1985). Diffusion of chemical signals, such as cAMP (Sheridan & Atkinson, 1985), through gap junctions may explain the insensitivity of the clock to TTX, while electrical coupling may explain the synchronization under blockade of chemical synaptic transmission. Multiple neurobiotin-stained neurons in close proximity to each other and a small number of neurons exhibiting evidence of electrotonic coupling suggest the existence of gap junctions between SCN neurons (Jiang, Yang, & Allen, 1997a). A previous thin-section ultrastructural study found that although gap junctions were common between astroglia, they were not detected between neurons in the SCN (van den Pol, 1980).

Ionic interactions are another means by which neurons may synchronize their activity independently of synaptic transmission. Increases in extracellular [K^+] or

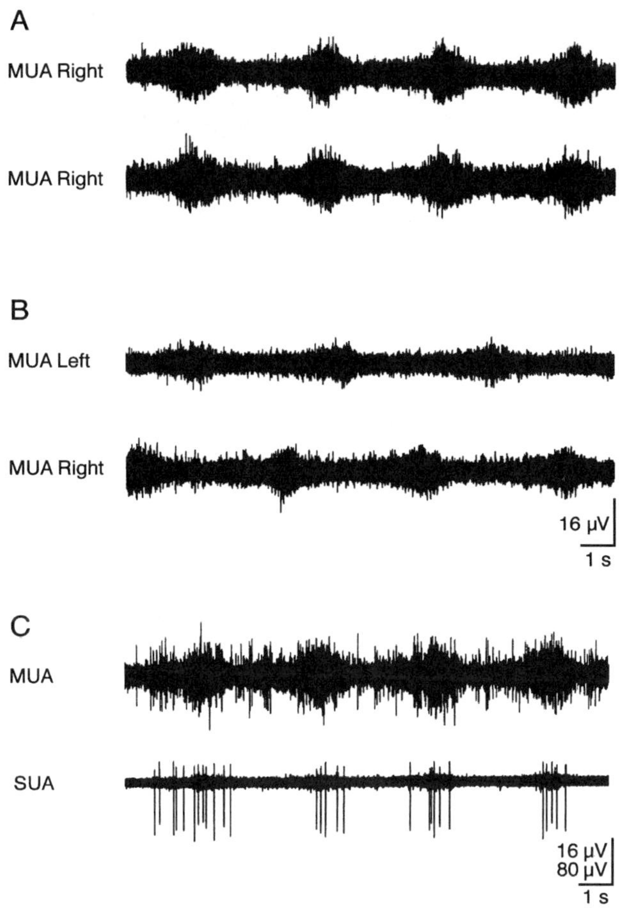

Figure 6. Synchronization in the SCN. Simultaneous recordings of extracellular spike bursts in Ca^{2+}-free medium containing BAPTA. (A) Simultaneous recordings of multiple-unit activity (MUA) from two sites separated by 150 μm in one SCN (right side) demonstrated synchronous bursts of action potentials. (B) Similar recordings from left and right SCNs showed that although the bursts were present on both sides, they were not synchronous and burst intervals were different. (C) Single-unit activity (SUA) recording during a burst of MUA. The bursts of single-unit activity coincided with the bursts of multiple-unit activity. (Adapted from Bouskila & Dudek, 1993.)

decreases in extracellular $[Ca^{2+}]$ can affect neuronal excitability (see, for example, Dudek *et al.*, 1986; Schweitzer, Patrylo & Dudek, 1992). The large regions of membrane apposition between SCN neurons and the small volume of extracellular space (van den Pol, 1980) would maximize the changes in ionic concentrations introduced by ionic currents. Changes in extracellular ionic concentrations may explain the neuronal synchronization that is independent of synaptic transmission, but in order to explain the insensitivity of the clock to TTX, the ionic changes must be independent of electrical activity. Such a mechanism could originate in glial cells.

Synchronized oscillations of intracellular $[Ca^{2+}]$ in glial cells have been demonstrated in the SCN (van den Pol, Finkbeiner & Cornell-Bell, 1992). This synchronized activity in glia could result in changes of neuronal excitability via their

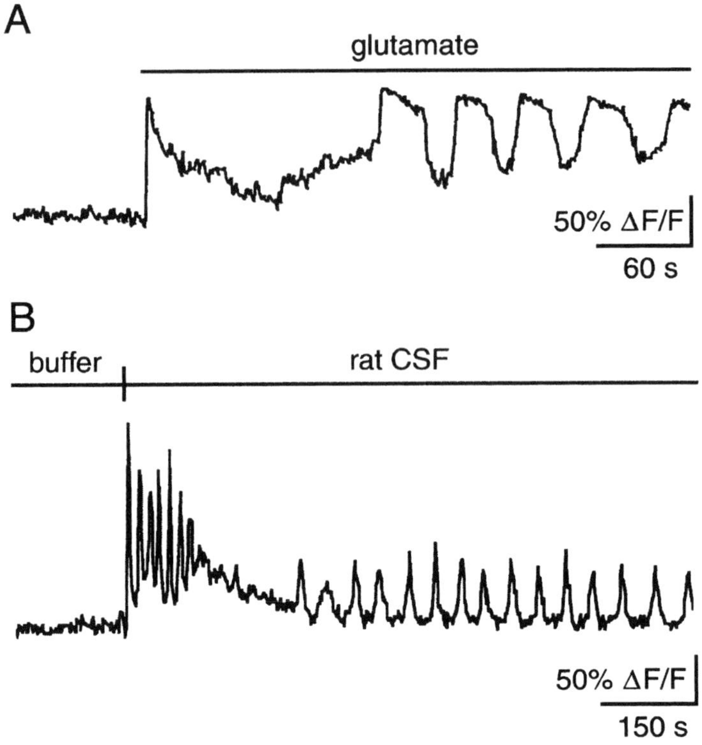

Figure 7. Oscillations of intracellular Ca^{2+} concentration in astrocytes of the SCN. A. Glutamate induced Ca^{2+} waves that propagated among astrocytes. Ca^{2+} increases occurred with a period of about 55 seconds. (B) In rat cerebrospinal fluid (CSF), an astrocyte exhibited a regular oscillation of intracellular Ca^{2+} that changed from a fast to a slow oscillation after about 200 seconds. (Adapted from van den Pol *et al.*, 1992.)

regulation of extracellular ions or other chemical signals. Regular intracellular Ca^{2+} oscillations were seen in astrocytes using the Ca^{2+} indicator fluo-3; these Ca^{2+} fluctuations could be induced by glutamate, serotonin and ATP, and had a period of 7–20 sec in cultured SCN cells (Figure 7A). Long-distance communication between glial cells in cultures has been seen in the form of waves of elevated Ca^{2+} moving from cell to cell, possibly through gap junctions, in response to stimulation with neurotransmitter. Spontaneously oscillating astrocytes were common in the presence of rat cerebrospinal fluid (Figure 7B). Similar results were found in astrocytes in organotypic cultures of hippocampal slices (Dani, Chernjavsky & Smith, 1992). Recent studies have shown that stimulation of glia in mixed glial–neuronal cultures triggered a cytosolic Ca^{2+} increase in overlying neurons (Hassinger *et al.*, 1995; Nedergaard, 1994; Parpura, 1994). The mechanism of glia-to-neuron signalling remains unclear; the former study implicated a role for gap junctions between glia and neurons, but the latter two studies provided evidence for a glutamate-mediated mechanism.

These data show that glial cells in the SCN can communicate, and they raise the possibility that neuronal activity may also be synchronized via the glial cells. Taken together, the evidence suggests that SCN neurons can synchronize their activity via a mechanism independent of chemical synaptic transmission. The alternative mechanisms may include electrotonic coupling and/or ionic interaction, or possibly a mechanism driven by the synchronized activity of glia.

Conclusions

The circadian rhythm of neuronal firing rate in the SCN is generated endogenously. This rhythm does not drive the primary biological clock in the SCN; instead, it is an expression of the clock, as indicated by the lack of perturbation of rhythm after a temporary block of Na^+-dependent spikes. Although individual SCN neurons are capable of generating rhythms of activity with different periods, it appears that some interaction between these oscillators is essential for the production of a stable and singular circadian rhythm that is in phase throughout the population of neurons. The nature of this interaction is unclear, but it appears that if an active synchronization takes place, it is of a weak or "loose" type, since at any phase of the rhythm, a range of firing frequencies is present. Synchronization between SCN neurons can occur without active chemical synaptic transmission, but the mechanism is unknown. This mechanism may coordinate the oscillators during TTX application and in the intact SCN. The plausible mechanisms of nonsynaptic synchronization that have been proposed are electrical coupling between neurons, ionic interactions, or glia-to-neuron interactions. Synaptic transmission plays important roles in the SCN, among them the transmission of photic information from the retina to regulate the rhythm phase. In the SCN as in other parts of the hypothalamus, synaptic transmission is mediated primarily via GABA and glutamate, with the majority of local synaptic activity being apparently GABAergic. While many other agents tested have not been shown to cause direct synaptic actions on SCN neurons, they may have a modulatory role.

The mechanism of the rhythm of neuronal excitability remains unknown. Data on the excitability of SCN neurons have become available only recently. Several ionic currents have been described recently in the SCN, but their precise function in firing rate is not well defined. Even less is known about their modulation. It seems that the passive membrane properties and the ionic currents that have been found are not unique to SCN neurons (but see Wagner *et al.*, 1997). Therefore, the circadian rhythm may originate from an interaction between certain currents or from modulation of specific currents. Ca^{2+}, Na^+, and K^+ currents that are present in SCN neurons play a role in regulating the firing rate and its pattern, and may be modulated in a circadian manner. On the other hand, it is possible that the circadian rhythm is generated in glial cells and expressed in the firing rate of neurons via glia-to-neuron interactions. Glial cells may regulate neuronal firing rate in several ways; for example, changes in extracellular $[K^+]$ or $[Ca^{2+}]$ can influence firing rate and pattern.

Modulation of one or several of the ionic conductances that regulate the firing threshold and/or membrane potential can be expected to underlie the circadian rhythm of firing rate. Several approaches can be taken in future research. One avenue involves further investigations concerning the excitability of SCN neurons; that is, describing each current in terms of its kinetics, voltage-dependence, and role in firing. Another approach could be a search for circadian changes in specific membrane conductances, in the levels of modulators of these conductances, or in ionic concentrations. The role of an oscillating ionic conductance (as found in the pacemaker of *Bulla*) in the firing rhythm could be tested by application of a specific channel blocker for that current. Modification of the firing rate at certain periods of the rhythm but not at others would suggest an involvement of that current in the generation of the firing rhythm.

Resetting of the Clock

Circadian clocks permit organisms to anticipate periodic events in the environment and to initiate slow physiologic processes before they are required. To achieve the necessary synchronization with the environment, circadian rhythms are regulated by exogenous cycles even though they are generated endogenously. The dominant environmental variable that entrains the circadian rhythm is light, although temperature and social cues may also be important (Takahashi & Zatz, 1982). In understanding the ability of light to reset the clock, two questions should be addressed. How is photic information transmitted to the SCN, and what are the cellular mechanisms that enable this information to reset the clock?

Immediate Effects of Light

In the rat SCN, 32% of the neurons responded to stimulation of the retina by light (Meijer *et al.*, 1986). Most of these photic-responsive neurons (73%) were light activated, whereas a minority of the cells (27%) were light suppressed. Similarly, single-unit recordings in the slice revealed that following optic nerve stimulation, most responsive neurons were excited, a smaller proportion were transiently excited and then inhibited, and a few others were only inhibited (Shibata, Oomura, Hattori, & Kita, 1984). A monotonic increase in discharge rate as a function of luminance (2–3 log units of dynamic range) characterizes the intensity–response curve of light-activated cells (Meijer *et al.*, 1986). Earlier studies found no relation between the light-responsiveness of SCN neurons and the environmental light–dark cycle, suggesting that phase-dependent shifts of the SCN rhythm after a light exposure are not induced by a rhythm of light-responsiveness in the SCN (Inouye, 1984; Meijer *et al.*, 1986). However, a more recent study found evidence for a daily variation of synaptic function in SCN neurons receiving optic input *in vivo*. The proportion of neurons responding to optic nerve stimulation during the dark period was more than twice the proportion responding during the light period (Cui & Dyball, 1996). Although the mechanism of these day–night variations is not known, it could be driven by reported oscillations in GABA function (Wagner *et al.*, 1997) in the local synaptic circuits. Extracellular recordings in the slice preparation revealed that electrical stimulation of the optic nerve produced in the SCN two brief positive waves, followed by a slow negative field potential. The negative wave disappeared in low-[Ca^{2+}] and high- [Mg^{2+}] medium (which would disrupt synaptic transmission), and all responses were blocked in TTX (Shibata, Oomura, Hattori, & Kita, 1984). In the mouse SCN, a similar Ca^{2+}-sensitive component of the field potential was found (Cahill & Menaker, 1989a). These results suggest that photic signals from the retina to the SCN are mainly excitatory, are mediated by Na^+-dependent action potentials, and involve chemical synaptic transmission.

Early results obtained with extracellular recording suggested that glutamate receptors of the NMDA type mediate retinohypothalamic transmission (Shibata, Liou & Ueki, 1986), whereas Cahill and Menaker (1989b) provided evidence that AMPA/kainate receptors are involved. The nature of retinohypothalamic synaptic transmission has been determined with intracellular recordings (Kim & Dudek, 1991). Optic nerve stimulation evoked fast EPSPs and never a pure IPSP. The EPSPs rose quickly to peak (in about 6 msec) and decayed gradually over 50–250 msec. At membrane potentials more negative than −60 mV, an antagonist for AMPA/kainate-type EAA receptors, 6,7-dinitroquinoxaline-2,3-dione (DNQX), depressed in a concentration-dependent and reversible manner the EPSPs evoked by optic nerve stimulation On the other hand, an NMDA antagonist, DL-2-amino-5-phosphono-pentanoic acid (AP5), blocked or depressed the slow depolarizing component of the EPSPs evoked at membrane potentials more positive than −55 mV. These results indicate that at resting or more negative membrane potentials, AMPA/kainate receptors mediate excitatory synaptic transmission for retinal input to SCN neurons. At less negative potentials, however, NMDA receptors may also play a role. A study using whole-cell recordings has confirmed the above findings (Jiang *et al.*, 1997b). In view of the abundance of synaptic terminals that are immunoreactive for GABA (Decavel & van den Pol, 1990), the lack of purely inhibitory synaptic responses to optic nerve stimulation (Kim & Dudek, 1992), and the local GABAergic synaptic circuits (Strecker, Wuarin, & Dudek, 1997), it is likely that the inhibitory responses exhibited by some SCN neurons following light or optic nerve stimulation arose from activation of inhibitory interneurons, whereas the direct input from the retina to the SCN is excitatory and mediated by glutamate.

Phase Shifts as a Model for Resetting the Clock

Circadian rhythms can be advanced or delayed by some stimuli; the type and magnitude of a shift in the phase of a circadian rhythm depend on the time of stimulus presentation. A phase delay in motor activity, for example, means that an animal commences its period of motor activity later in the day, whereas initiation of activity earlier in the day represents phase advance. In nocturnal rodents, phase delays occurred after light pulses presented during the early portion of the subjective night, and phase advances occurred after light pulses presented during the late subjective night; little or no shift in phase occurred when light pulses were given during most of the subjective day (DeCoursey, 1964). In contrast, dark pulses in otherwise constant illumination produced opposite phase shifts of motor activity rhythm in hamsters (i.e., phase was advanced instead of delayed, and vice versa; Boulos & Rusak, 1982). Two pathways delivering photic information from the retina to the SCN are the retinohypothalamic tract (RHT), a direct projection from the retinal ganglion cells to SCN neurons, and the geniculohypothalamic tract (GHT), an indirect photic afferent arising from the intergeniculate leaflet of the thalamus. The RHT projection is sufficient for the entrainment of circadian rhythms, whereas the GHT may also be involved (Card & Moore, 1991).

In the slice preparation, phase shifts in the rhythm of neuronal activity in the SCN have been used in order to gain insight into the mechanism of resetting the clock. Changes in the time of peak activity are used as an indication for phase shifts following various treatments (Figure 8A). The time of peak (as determined from single-unit recordings) was relatively stable for up to three cycles of the rhythm in SCN slice preparations, however, the standard deviation of the average peak time was up to 40 minutes (Prosser & Gillette, 1989); therefore, small phase shifts may not be detected. Various treatments have been shown to produce significant phase shifts in the neuronal activity rhythm in the slice, among them cAMP, cGMP, serotonin, melatonin, neuropeptide Y (NPY), acetylcholine (Ach), high extracellular [K^+], and even the cutting of the slices themselves. Like the effect of light pulses, the magnitude of the phase shift induced by a chemical agent depends on the circadian time of treatment. In the following two sections, we discuss various phase-shifting

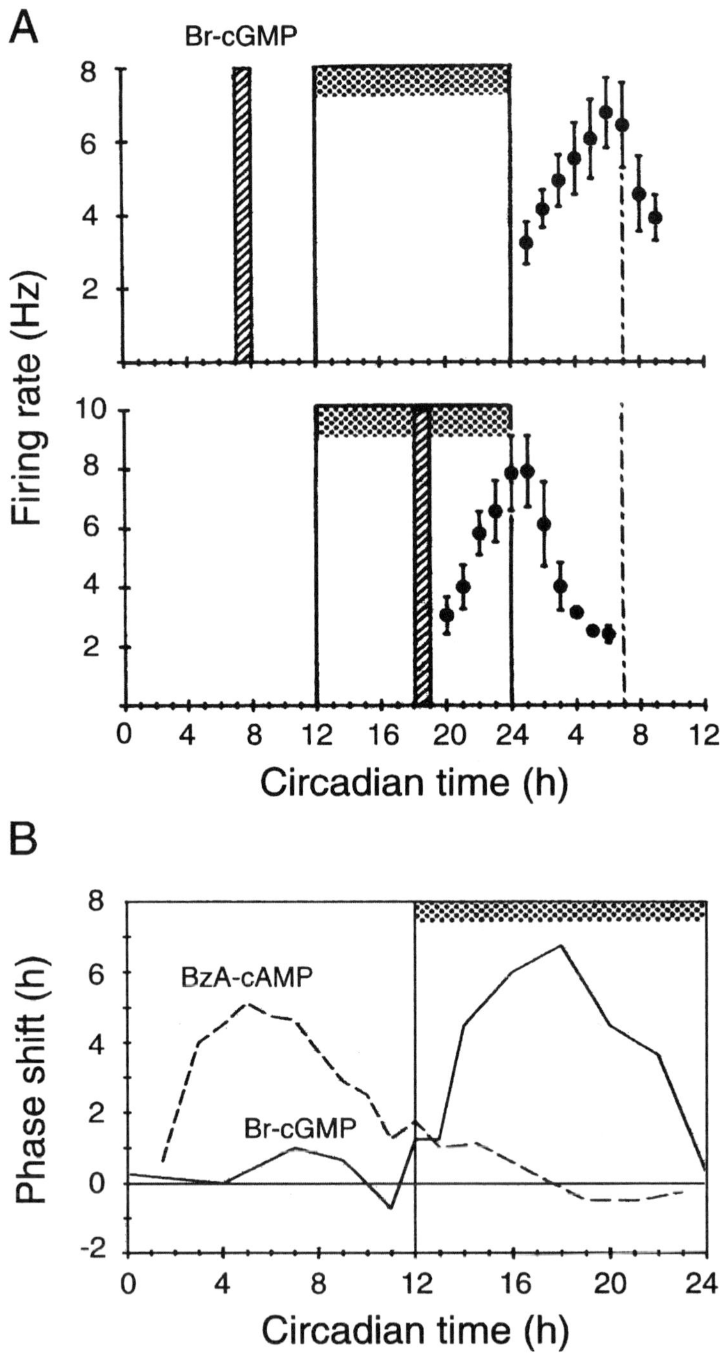

Figure 8. Phase shifts of the circadian rhythm of firing rate in SCN. (A) Phase-dependent effect of a 1-hour pulse of 8-bromo-cGMP (Br-cGMP, vertical bar). Treatment during subjective day (CT7–8) had little effect on the time-of-peak (top panel). Treatment during subjective night (CT18–19) induced a 6.5-hour phase advance (lower panel). See Figure 2 for explanation of the symbols. B. Superimposed phase-response curves for 1-hour treatments with 8-benzylamino-cAMP (BzA-cAMP, dashed line) and Br-cGMP (solid line). Changes in the time-of-peak in the rhythm of firing rate are plotted against circadian time of treatment. (Adapted from Prosser *et al.*, 1989.)

agents categorized according to the most effective time of application that induces a phase shift (i.e., day or night).

Night-Sensitive Phase Shifts

Despite the extensive knowledge regarding phase shifts of behavioral rhythms induced by light, no study has verified that the phase shifting effects of light on neuronal rhythm are preserved *in vitro.* Gillette (1986) found that preparing the slice had similar phase shifting effects on peak neuronal activity as light pulses had on wheel-running behavior (DeCoursey, 1964) and on SCN electrical activity *in vivo* (Inouye & Kawamura, 1982). Similar phase shifts of motor activity rhythm could be induced by electrical stimulation of the SCN in hamsters (Rusak & Groos, 1982), although similar data from rats are lacking. The shifts *in vitro,* however, were twice as large and reached up to 5 hours. The SCN is exposed to several potential phase-shifting stimuli during slice preparation, including stimulation of the optic nerve (both by light and by cutting) and elevated activity caused by the slicing procedure, and it is not clear which is responsible for the shifts (Gillette, 1986). The similarity of the phase shifts induced by light and by preparation of the slices is mainly in waveform, but the magnitude of these shifts differs. Treatment of rat SCN *in vitro* with a brief application of glutamate produced dose-dependent light-like phase shifts. Antagonists of NMDA or nitric oxide synthase blocked the glutamate effects *in vitro.* These results suggest that glutamate release and activation of the nitric oxide pathway are involved in light-induced phase shifts (Ding *et al.*, 1994).

Another stimulus that induced phase shifts similar to those by light was the second messenger cGMP. Bath application of cGMP analogs (0.5 mM) that are membrane soluble and resist degradation by phosphodiesterase induced large phase advances (up to 7 hours in magnitude) when applied at mid-subjective night and had no effect during the subjective day (Figure 8A) in contrast to a cAMP analog (Figure 8B). These phase advances appeared to be complete within 6 hours of treatment and represented stable shifts of the underlying pacemaker, as demonstrated by the lack of further shift on the second day after treatment (Prosser *et al.*, 1989). In these respects, the phase shifts induced by cGMP analogs were similar to those induced by a series of three 10-minute pulses of high $[K^+]$ (50 mM) in the slice, which also phase advanced the rhythm by 6–7 hours when applied at mid-subjective night (Gillette, 1987). In freely behaving animals, the phase of the circadian rhythm of drinking was shifted by the infusion of high $[K^+]$ (20 mM) or by the Na^+ channel activator veratridine (Schwartz, 1991). The phase-shifting effect of veratridine was blocked by simultaneous infusion of TTX. However, application of TTX alone did not induce phase shifts *in vitro* (Shibata & Moore, 1993). These results suggest that an increase in firing rate or membrane depolarization is necessary for the induction of night-sensitive phase shifts.

Another effective phase-shifting agent is the neurotransmitter ACh. A microdrop of Ach or the cholinergic agonist carbachol, applied directly on the SCN in brain slices during the subjective night, induced a robust phase advance of 6 hours in the circadian rhythm of firing rate. Coapplication of cholinergic agonists and specific antagonists revealed that carbachol reset the phase of the firing rhythm via an M1-like muscarinic ACh receptor (Liu & Gillette, 1996).

Day-Sensitive Phase

Among the stimuli that phase shift the neuronal rhythm, serotonin has been the most studied. Depending on the circadian time of treatment, microdrop application of serotonin *in vitro* resulted in phase advances of electrical rhythm in the SCN. Application at mid-subjective day resulted in robust phase advances of up to 7 hours, while treatment during the subjective night had no effect (Medanic & Gillette, 1992). The nonspecific serotonin agonist quipazine induced somewhat different phase shifts than serotonin. Treatments of 1 hour during subjective day induced phase advances, whereas treatments at night induced phase delays. Treatments at subjective dawn and dusk had little or no effect on the phase. The maximal phase shifts were 4 hours in each direction, and the magnitude of phase advances exhibited dose dependenc with an EC_{50} of 0.5 μM (Prosser, Miller, & Heller, 1990). These results by themselves do not indicate the site of serotonin action, but another study showed that quipazine reset the clock even when it was coapplied with TTX or high [Mg^{2+}], indicating that serotonin acts directly on SCN cells, since Na^+-mediated action potentials and synaptic transmission were presumably blocked by this treatment (Prosser, Heller, & Miller, 1992).

In an attempt to determine the receptor subtype(s) by which serotonin exerts its effect, specific agonists and antagonists of serotonin receptor subtypes were tested. The results of these studies suggest that serotonin (5-HT) phase shifts are mediated via 5-HT_1 or 5-HT_7 receptor subtypes (Lovenberg *et al.*, 1993; Medanic & Gillette, 1992). Serotonin was thought to attenuate $GABA_A$-induced currents in cultured neurons (Kawahara *et al.*, 1994), and to inhibit glutamatergic synapses (Jiang *et al.*, 1995b) via a $5HT_7$ receptor. In view of the complexity of serotonin pharmacology and molecular biology, a conclusive statement regarding the receptor identity and its mechanism of action is not possible at this time. Serotonin, and its analogs 5-carboxyamidotryptamine maleate (5-CT) and 8-hydroxydipropylaminotetralin hydroxybromide (8-OH-DPAT), all increased cAMP accumulation in REC20-transfected cells in a dose-dependent manner (REC20 is the $5HT_7$-receptor clone isolated from a brain cDNA library, and has some homology to known G-protein-coupled 5-HT receptors). This is particularly interesting since elevated cAMP (exogenously or endogenously) produced similar phase response effects as serotonin (Prosser & Gillette, 1989). Although it is tempting to link the serotonin effect to an increase in cAMP, no direct evidence exists to support this hypothesis. The physiologic significance of serotonin-induced phase shifts may come from behavioral studies. Hamsters underwent phase shifting by behavioral arousal, forced activity, or social interaction at midday (Mrosovky, 1988), a time when serotonin could induce phase shifts *in vitro* (Medanic & Gillette, 1992). Mrosovky (1988) hypothesized that the state of arousal is communicated to the SCN via serotonergic projections from the raphe nuclei.

Another phase-shifting agent that has received increasing attention is NPY. When applied at CT6 on the SCN in hypothalamic slices, NPY introduced a phase advance of about 3 hours in the rhythm of firing rate on the following day. These shifts were unaffected by pretreatment with Na^+ and Ca^{2+} channel blockers or antagonists of glutamate, $GABA_A$, or $GABA_B$ receptors, but were blocked by protein kinase C (PKC) inhibitors. These results suggest a direct postsynaptic action of NPY on pacemaker neurons which involves PCK activity (Biello, Golombek, Schak, & Harrington, 1997). Interestingly, NPY induces long-term depression in both electrical activity and intracellular calcium in SCN neurons, which may involve presynaptic sites on glutamatergic axons (van den Pol, Obrietan, Chen, & Belousov, 1996).

Conclusions

Although the cellular mechanism of resetting the biological clock is unclear, a partial picture does emerge from the available information. SCN neurons respond to photic stimulation of the retina primarily with an increased firing rate as a function of light intensity. This information is transmitted via excitatory synaptic transmission mediated by AMPA/kainate and by NMDA receptors, depending on the membrane potential. Activation of these receptors leads to a cascade of subcellular events, which may involve, for example, Ca^{2+} entry and activation of the nitric oxide pathway, that eventually results in a phase shift. In addition to light, behavioral and pharmacologic treatments also produce significant phase shifts and can be broadly categorized according to the circadian time of application at which they induce these shifts. Accordingly, night application of glutamate, cGMP, or high extracellular K^+ and electrical stimulation of the SCN have similar phase-shifting effects to those of light pulses *in vivo*, while day application of cAMP and serotonin have similar effects to dark pulses *in vivo*. Apparently, all the stimuli that phase shift during daytime also decrease spontaneous firing rate, whereas the nighttime phase shifters increase firing rate in the SCN (see above, and Mason, 1986). The similar physiologic effects of each group of phase-shifting treatments suggest common cellular mechanisms, which remain unknown. Although both cAMP and serotonin are known to inhibit firing in the SCN, the ability of serotonin to phase shift in the presence of TTX suggests that the resetting of the clock is not induced by a simple transient inhibition of firing rate.

Future work may dissect the subcellular changes that occur after induction of phase shifts. More data are needed on the role of second messengers in phase shifting. Coapplication of different treatments and testing for consequent augmentation may be a useful way to dissect the cellular pathways of action. In addition, more information is needed on the immediate subcellular changes of light or optic nerve stimulation at different circadian phases. Another approach would be to concentrate on rhythmic cellular changes (e.g., specific membrane conductances) and their modulation that are responsible for the rhythm of neuronal excitability. This last approach, although more focused, will rely on knowledge of the specific circadian oscillations in membrane excitability, which is unavailable for the SCN at this time.

Output from the SCN

The Nature of the Output

An important question for understanding the function of the biological clock is the nature of its output. In other words, how is the information from the SCN transmitted to other brain areas to control the overt circadian rhythms? At least two possible communication pathways have been postulated, an electrical signal and a humoral signal. As described earlier, TTX blocked several circadian rhythms at the time of application. The rhythm of vasopressin secretion showed a peak at mid-subjective day, the same time that the rhythm of electrical activity reached its peak. Elevated K^+ increased vasopressin secretion in the presence of TTX (Earnest *et al.*, 1991). These observations strongly suggest that the outputs from the SCN rely on Na^+-dependent spikes, but cannot exclude a humoral secretion hypothesis. The reason is that the common mechanism for stimulus–secretion coupling in neurons involves voltage-dependent Ca^{2+} influx that relies on Na^+-dependent spikes (Douglas, 1978). An important target for SCN efferents is believed to be in or near the hypothalamic paraventricular nucleus, since this nucleus regulates neuroendocrine and autonomic functions which display circadian rhythmicity. Evidence has been presented that the efferents employ both GABAergic and glutamatergic neurotransmission (Hermes, Coderre, Buijers, & Renaud, 1996). The supraoptic nucleus is also thought to receive both GABAergic and glutamatergic input from SCN efferents (Cui *et al.*, 1997). These results suggest that an appreciable number of glutamatergic neurons exist in the SCN, even though most neurons and local circuitry there appear to be GABAergic (see above). It is possible that local circuits use GABA, whereas efferent projections have more glutamatergic fibers, or that stimulation methods using electrical stimulation may involve fibers both exogenous and endogenous to the SCN (although see Hermes *et al.* 1996).

Isolation of the SCN by knife cuts abolished the electrical rhythm outside the SCN but not inside (Inouye & Kawamura, 1979). Knife cuts anterior, ventral or lateral to the SCN had little effect on circadian rhythms of eating and drinking, whereas damage directly posterior to the SCN abolished the circadian rhythms (van den Pol & Powley, 1979). Since knife cuts are expected to remove efferent projections from the SCN rather than prevent local secretion from the nucleus, these results suggest that the output from the SCN is mediated by electrical signals propagated along efferents to the nucleus. In view of this conclusion, it is surprising that the SCN provides efferents to rather restricted regions of the hypothalamus, none of which have any direct motor function. Nevertheless, these SCN efferents, which appear to be mostly GABAergic, may affect circadian rhythms via a wide variety of neuroendocrine functions (Watts, 1991).

Conclusions

It appears that Na^+-dependent spikes are essential for the expression of several circadian rhythms, such as drinking, electrical activity, vasopressin secretion, and glucose utilization; however, it is not clear if they are sufficient. The temporal relation between the rhythm of electrical activity and the rhythm of vasopressin secretion, as well as the blockade of both by TTX, suggest that the former drives the latter within the SCN. It is possible that the expression of some overt circadian rhythms depends indirectly on electrical signals transmitted inside or outside the SCN with a consequent secretion of humoral signal. Future experiments should further test the hypothesis that humoral signals play a role in the expression of circadian rhythms, whereas electrical activity drives the rhythm of secretion. Since the SCN is known to secrete vasopressin in a circadian manner, one could test this hypothesis by applying antibodies against vasopressin or its receptor antagonists. Results showing block of some overt circadian rhythms while leaving an intact rhythm of neuronal activity would suggest a verification of this hypothesis.

Overview

SCN neurons exhibit an endogenous circadian rhythm of firing rate, where peak activity occurs at mid-subjective day and the trough at mid-subjective night. The neuronal activity rhythm was first observed *in vivo*, but since the rhythm is preserved *in vitro*, it has been studied extensively in the slice preparation and more recently in cultures. This cellular expression of the biological clock in mammals drives several behavioral and hormonal circadian rhythms, but does not generate the time-keeping mechanism itself, since blockade of electrical activity appears to have no effect on the phase of the rhythm. The neuronal oscillations are driven by rhythmic changes in neuronal excitability that remain unknown. SCN neurons seem to exhibit typical electrophysiological characteristics. All the ionic currents that have been described so far in the SCN share common features with neuronal currents present in other brain areas. Several mechanisms capable of modulating firing rate have been demonstrated in SCN neurons and many more are possible, but so far, none of these has been shown to oscillate in a circadian manner except possibly the $GABA_A$ current polarity, which is still controversial. Although the second messengers cAMP and cGMP have been found to exhibit circadian rhythms of concentration in the SCN, no direct modulatory effect on neuronal excitability has been demonstrated. Furthermore, it remains unclear whether the circadian rhythm originates in neurons or in glial cells, which may in turn communicate with neurons.

Information to the SCN in the form of synaptic events appears to be mediated primarily by glutamate and GABA. Light input from the retina, which can reset the clock, is transmitted to the SCN via EPSPs. These fast synaptic events are mediated by NMDA and AMPA/kainate receptors and cause a transient increase of firing rate in most light-responsive cells. In contrast, synchronization of activity within the SCN can occur even without chemical synaptic transmission. Since communication between glial cells seems to occur in the SCN, it is possible that glia-to-neuron interactions mediate the synchronization among the neurons via glial buffering of extracellular ion concentrations, which in turn may regulate firing rate. Since single cells exhibited a wide range of circadian period lengths and/or phases, but the intact SCN shows a stable and singular period, it is possible that communication within the SCN serves to coordinate the different oscillators.

The resetting ability of the clock phase in the SCN is preserved *in vitro*. Two groups of phase-shifting agents or treatments can be broadly distinguished, based on day- or night-activated shifts. The day-sensitive treatments are characterized by an inhibitory effect on firing rate, whereas the night-sensitive treatments share an excitatory effect, but not much is known about the cellular pathways of the resetting mechanism.

Although a great deal has been learned about the circadian rhythm of neuronal activity since the recognition that the SCN contains the mammalian biological clock, many mechanisms involving the generation, synchronization, and resetting of the clock remain unclear.

Acknowledgments

We thank Dr. G. Fain for constructive comments on the manuscript. We are also grateful to A. Bienvenu for assistance with word processing. This work was supported by Air Force Office for Scientific Research and to F.E.D.

References

Akasu, T., & Shoji, S. (1994). cAMP-dependent inward rectifier current in neurons of the rat suprachiasmatic nucleus. *Pflugers Archives European Journal of Physiology, 429*, 117–125.

Akasu, T., Shoji, S., & Hasuo, H. (1993). Inward rectifier and low-threshold calcium currents contribute to the spontaneous firing mechanism in neurons of the rat suprachiasmatic nucleus. *Pflugers Archives European Journal of Physiology, 425*, 109–116.

Belenky, M., Wagner, S., Yarom, Y., Matzner, H., Cohen, S., and Castel, M. (1996). The suprachiasmatic nucleus in stationary organotypic culture. *Neuroscience, 70*, 127–143.

Biello, S. M., Golombek, D. A., Schak, K. M., & Harrington, M. E. (1997). Circadian phase shifts to neuropeptide Y *in vitro*: Cellular communication and signal transduction. *Journal of Neuroscience, 17*, 8468–8475.

Bertolino, M., & Llinás, R. (1992). The central role of voltage-activated and receptor-operated calcium channels in neuronal cells. *Annual Review of Pharmacological Toxicology, 32,* 399–421.

Block, G. D., McMahon, D. G., Wallace, S. F., & Friesen, W. O. (1984). Cellular analysis of the *Bulla* ocular circadian pacemaker system I. A model for retinal organization. *Journal of Comparative Physiology A, 155,* 365–378.

Bos, N. P. A., & Mirmiran, M. (1990). Circadian rhythms in spontaneous neuronal discharge of the cultured suprachiasmatic nucleus. *Brain Research, 511,* 158–162.

Boulos, Z., & Rusak, B. (1982). Circadian phase response curves for dark pulse in the hamster. *Journal of Comparative Physiology, 146,* 411–417.

Bouskila, Y., & Dudek, F. E. (1993). Neuronal synchronization without calcium-dependent synaptic transmission in the hypothalamus. *Proceedings of the National Academy of Sciences of the USA, 90,* 3207–3210.

Bouskila, Y., & Dudek, F. E. (1995a). Can a population of suprachiasmatic nucleus neurons with different period lengths produce a stable circadian rhythm? *Brain Research, 650,* 333–336.

Bouskila, Y., & Dudek, F. E. (1995b). A rapidly activating type of outward rectifier K^+ current and A-current in rat suprachiasmatic nucleus neurones. *Journal of Physiology, 488,* 339–350.

Cahill, G. M., & Menaker, M. (1989a). Responses of the suprachiasmatic nucleus to retinohypothalamic tract volleys in a slice preparation of the mouse hypothalamus. *Brain Research, 479,* 65–75.

Cahill, G. M., & Menaker, M. (1989b). Effects of excitatory amino acid receptor antagonists and agonists on suprachiasmatic nucleus responses to retinohypothalamic tract volleys. *Brain Research, 479,* 76–82.

Card, J. P., & Moore, R. Y. (1991). The organization of visual circuits influencing the circadian activity of the suprachiasmatic nucleus. In D. C. Klein, R. Y. Moore, & S. M. Reppert (Eds.), *Suprachiasmatic nucleus, the mind's clock* (pp. 51–76). New York: Oxford University Press.

Chen, G., & van den Pol, A. N. (1996). Multiple NPY receptors coexist in pre- and postsynaptic sites: Inhibition of GABA release in isolated self-innervating SCN neurons. Journal of Neuroscience, 16, 7711–7724.

Chen, G., & van den Pol, A. N. (1997). Adenosine modulation of calcium currents and presynaptic inhibition of GABA release in suprachiasmatic and arcuate nucleus neurons. *Journal of Neurophysiology, 77,* 3035–3047.

Cui, L.-N., & Dyball, R. E. J. (1996). Synaptic input from the retina to the suprachiasmatic nucleus changes with the light–dark cycle in the Syrian hamster. *Journal of Physiology, 497,* 483–493.

Cui, L. N., Saeb-Parsy, K. & Dyball, R. E. (1997). Neurones in the supraoptic nucleus of the rat regulated by a projection from the suprachiasmatic nucleus. *Journal of Physiology, 502,* 149–159.

Dani, J. W., Chernjavsky, A., & Smith, S. J. (1992). Neuronal activity triggers calcium waves in hippocampal astrocyte networks. *Neuron, 8,* 249–440.

Decavel, C., & van den Pol, A. N. (1990). GABA: A dominant neurotransmitter in the hypothalamus. *Journal of Comparative Neurology, 302,* 1019–1037.

DeCoursey, P. J. (1964). Function of a light response rhythm in hamsters. *Journal of Comparative Physiology, 63,* 189–196.

de Jeu, M. T. G., & Pennartz, C. M. A. (1997). Functional characterization of the H-current in SCN neurons in subjective day and night: A whole-cell patch-clamp study in acutely prepared brain slices. *Brain Research, 767,* 72–80.

Derambure, P. S., & Boulant, J. A. (1992). Circadian rhythm of neuronal temperature sensitivity in rat suprachiasmatic nucleus. *Society for Neuroscience Abstracts, 18,* 877.

Ding, J. M., Chen, D., Weber, E. T., Faiman, L. E., Rea, M. A., & Gillette, M. U. (1994). Resetting the biological clock: Mediation of nocturnal circadian shifts by glutamate and NO. *Science, 266,* 1713–1717.

Douglas, W. W. (1978). Stimulus–secretion coupling: Variations on the theme of calcium-activated exocytosis involving cellular and extracellular sources of calcium. *CIBA Foundation Symposium, 61,* 87.

Dudek, F. E., Snow, R. W., & Taylor, C. P. (1986). Role of electrical interactions in synchronization of epileptiform bursts. In A. V. Delgado-Escueta, A. A. Ward, Jr., D. M. Woodbury, & R. J. Porter (Eds.), *Basic mechanisms of the epilepsies: molecular and cellular approaches* (pp. 593–617). New York: Raven Press.

Earnest, D. J., Digiorgio, S. M., & Sladek, C. D. (1991). Effects of tetrodotoxin on the circadian pacemaker mechanism in suprachiasmatic explants *in vitro. Brain Research Bulletin, 26,* 677–682.

Edmunds, L. N. (1988). Circadian organization in eukaryotic microorganisms. In *Cellular and molecular bases of biological clocks* (pp. 12–74). New York: Springer-Verlag.

Frankenhaeuser, B., & Hodgkin, A. L. (1957). The action of calcium on the electrical properties of squid axons. *Journal of Physiology, 137,* 218–244.

Gao, B., Fritschy, J.-M., & Moore, R. Y. (1995). $GABA_A$-receptor subunit composition in the circadian timing system. *Brain Research, 700,* 142–156.

Gibbs, J. W., III, Shumate, M. D., & Coulter, D. A. (1996). Differential epilepsy-associated alterations in postsynaptic $GABA_A$ receptor function in dentate granule and CA1 neurons. *Journal of Neurophysiology, 77,* 1924–1938.
Gillette, M. U. (1986). The suprachiasmatic nuclei: Circadian phase-shifts induced at the time of hypothalamic slice preparation are preserved *in vitro. Brain Research, 379,* 176–181.
Gillette, M. U. (1987). Effects of ionic manipulation on the circadian rhythm of neuronal firing rate in the suprachiasmatic brain slice. *Society for Neuroscience Abstracts, 13,* 51.
Gillette, M. U. (1991). SCN electrophysiology *in vitro*: Rhythmic activity and endogenous clock properties. In D. C. Klein, R. Y. Moore, & S. M. Reppert (Eds.), *Suprachiasmatic nucleus, the mind's clock* (pp. 77–106). New York: Oxford University Press.
Green, D. J., & Gillette, R. (1982). Circadian rhythm of firing rate recorded from single cells in the rat suprachiasmatic brain slice. *Brain Research, 245,* 198–200.
Gribkoff, V. K., Pieschl, R. L., Wisialowski, T. A., and Dudek, F. E. (1997). Effects of GABA and GABA-receptor antagonists on neuronal discharge and circadian rhythmicity in the suprachiasmatic nucleus *in vitro. Society for Neuroscience Abstracts, 23,* 241.
Groos, G. A., & Hendriks, J. (1979). Regularly firing neurones in the rat suprachiasmatic nucleus. *Experientia, 35,* 1597–1598.
Groos, G. A., & Hendriks, J. (1982). Circadian rhythms in electrical discharge of rat suprachiasmatic neurones recorded *in vitro. Neuroscience Letters, 34,* 283–288.
Hassinger, T. D., Atkinson, P. B., Strecker, G. J., Whalen, L. R., Dudek, F. E., Kossel, A. H., & Kater, S. B. (1995). Evidence for glutamate-mediated activation of hippocampal neurons by glial calcium waves. *Journal of Neurobiology, 28,* 159–170.
Hermes, M. L. H. J., Coderre, E. M., Buijes, R. M., & Renaud, L. P. (1996). GABA and glutamate mediate rapid neurotransmission from suprachiasmatic nucleus to hypothalamic paraventricular nucleus in rat. *Journal of Physiology, 496,* 749–757.
Huang, R. C. (1993). Sodium and calcium currents in acutely dissociated neurons from the rat suprachiasmatic nucleus. *Journal of Neurophysiology, 70,* 1692–1703.
Huang, R. C., Peng, Y. W., & Yau, K. W. (1993). Zinc modulation of a transient potassium current and histochemical localization of the metal in neurons of the suprachiasmatic nucleus. *Proceedings of the National Academy of Sciences of the USA, 90,* 11806–11810.
Huang, S. H., & Pan, J. T. (1993). Potentiating effects of serotonin and vasoactive intestinal peptide on the action of glutamate on suprachiasmatic neurons in brain slices. *Neuroscience Letters, 159,* 1–4.
Inouye, S. T. (1984). Light responsiveness of the suprachiasmatic nucleus within the island with the retino-hypothalamic tract spared. *Brain Research, 294,* 263–268.
Inouye, S. T., & Kawamura, H. (1979). Persistence of circadian rhythmicity in mammalian hypothalamic island containing the suprachiasmatic nucleus. *Proceedings of the National Academy of Sciences of the USA, 76,* 5962–5966.
Inouye, S. T., & Kawamura, H. (1982). Characteristics of a circadian pacemaker in the suprachiasmatic nucleus. *Journal of Comparative Physiology, 146,* 153–160.
Ito, C., Wakamori, M., & Akaike, N. (1991). Dual effect of glycine on isolated rat suprachiasmatic neurons. *American Journal of Physiology, 260,* C213–C218.
Jacklet, J. W. (1969). Circadian rhythm of optic nerve impulses recorded in darkness from isolated eye of *Aplysia. Science, 164,* 562–563.
Jiang, Z.-G., Allen, C. N., & North, R. A. (1995a). Presynaptic inhibition by baclofen of retinohypothalamic excitatory synaptic transmission in rat suprachiasmatic nucleus. *Neuroscience, 64,* 813–819.
Jiang, Z. G., Yang, Y. O., and Allen, C. N. (1995b). Pre- and postsynaptic inhibition by serotonin in rat suprachiasmatic nucleus (SCN) *in vitro. Society for Neuroscience Abstracts, 21,* 1370.
Jiang, Z.-G., Yang, Y.-Q., & Allen, C. N. (1997a). Tracer and electrical coupling of rat suprachiasmatic nucleus neurons. *Neuroscience, 77,* 1059–1066.
Jiang, Z.-G., Yang, Y.-Q., Liu, Z.-P., & Allen, C. N. (1997b). Membrane properties and synaptic inputs of suprachiasmatic nucleus neurons in rat brain slices. *Journal of Physiology, 499,* 141–159.
Kapur, J., & Macdonald, R. L. (1996). Pharmacological properties of γ-aminobutyric $acid_A$ receptors from acutely dissociated rat dentate granule cells. *Molecular Pharmacology, 50,* 458–466
Kawahara, F., Saito, H., & Katsuki, H. (1993). Pharmacological characteristics of $GABA_A$ responses in postnatal suprachiasmatic neurons in culture. *Neuroscience Letters, 160,* 45–48.
Kawahara, F., Saito, H., and Katsui, H. (1994). Inhibition by 5-HT_7 receptor stimulation of $GABA_A$ receptor-activated current in cultured rat suprachiasmatic neurones. *Journal of Physiology, 478,* 67–73.
Kim, Y. I., & Dudek, F. E. (1991). Intracellular electrophysiological study of suprachiasmatic nucleus neurons in rodents: Excitatory synaptic mechanisms. *Journal of Physiology, 444,* 270–287.

Kim, Y. I., & Dudek, F. E. (1992). Intracellular electrophysiological study of suprachiasmatic nucleus neurons in rodents: Inhibitory synaptic mechanisms. *Journal of Physiology, 458*, 247–260.

Kim, Y. I., & Dudek, F. E. (1993). Membrane properties of rat suprachiasmatic nucleus neurons receiving optic nerve input. *Journal of Physiology, 464*, 229–243.

Liou, S. Y., and Albers, H. E. (1990). Single unit response of neurons within the hamster suprachiasmatic nucleus to GABA and low chloride perfusate during the day and night. *Brain Research Bulletin, 25*, 93–98.

Liou, S. Y., Shibata, S., Shiratsuchi, A., & Ueki, S. (1986). Effects of dibutyryl cyclic adenosine monophosphate and dibutyryl cyclic guanosine monophophate on neuron activity of suprachiasmatic nucleus in rat hypothalamic slice preparation. *Neuroscience Letters, 67*, 339–343.

Liu, C., & Gillette, M. U. (1996) Cholinergic regulation of the suprachiasmatic nucleus circadian rhythm via a muscarinic mechanism at night. *Journal of Neuroscience, 15*, 744–751.

Llinàs, R. (1988). The intrinsic electrophysiological properties of mammalian neurons: Insights into central nervous system function. *Science, 242*, 1654–1664.

Lovenberg, T. W., Baron, B. M., de Lecea L., Miller, J. D., Prosser, R. A., Rea, M. A., Foye, P. E., Racke, M., Slone, A. L., Seigel, B. W., Danielson, P. E., Sutcliffe, J. G., & Erlander, M. G. (1993). A novel adenyl cyclase-activating serotonin receptor (5-HT_7) implicated in the regulation of mammalian circadian rhythms. *Neuron, 11*, 449–458.

Margraf, R. R., Puchalski, W., & Lynch, G. R. (1992). Absence of a daily neuronal rhythm in the suprachiasmatic nuclei of circadian Djungarian hamsters. *Neuroscience Letters, 142*, 175–178.

Mason, R. (1986). Circadian variation in sensitivity of suprachiasmatic and lateral geniculate neurones to 5-hydroxytryptamine in the rat. *Journal of Physiology, 377*, 1–13.

Mason, R., & Biello, S. M. (1992). A neurophysiological study of a lithium-sensitive photoinositide system in the hamster suprachiasmatic (SCN) biological clock *in vitro. Neuroscience Letters, 144*, 135–138.

McMahon, D. G., Wallace, S. F., & Block, G. D. (1984). Cellular analysis of the *Bulla* ocular circadian pacemaker system II. Neurophysiological basis of circadian rhythmicity. *Journal of Comparative Physiology A, 155*, 379–385.

Medanic, M., & Gillette, M. U. (1992). Serotonin regulates the phase of the rat suprachiasmatic circadian pacemaker *in vitro* only during subjective day. *Journal of Physiology, 450*, 629–642.

Meijer, J. H., & Rietveld, W. J. (1989). Neurophysiology of the suprachiasmatic circadian pacemaker in rodents. *Physiological Reviews, 69*, 671–707.

Meijer, J. H., Groos, G. A., & Rusak, B. (1986). Luminance coding in a circadian pacemaker: The suprachiasmtic nucleus of the rat and the hamster. *Brain Research, 382*, 109–118.

Meijer, J. H., Schapp, J., Watanabe, K., & Albus, H. (1997). Multinunit activity recordings in suprachiasmatic nuclei: *In vivo* versus *in vitro* models. *Brain Research, 753*, 322–327.

Michel, S., Geusz, M. E., Zaritsky, J. J., & Block, G. D. (1993). Circadian rhythm in membrane conductance expressed in isolated neurons. *Science, 259*, 239–241.

Miller, J. D., Cao, V. H., & Heller, C. (1994). Thermal effects on neuronal activity in suprachiasmatic nuclei of hibernators and nonhibernators. *American Journal of Physiology, 266*, R1259–R1266.

Mrosovsky, N. (1988). Phase response curves for social entrainment. *Journal of Comparative Physiology A, 162*, 35–46.

Murakami, N., & Takahashi, K. (1983). Circadian rhythm of adenosine-3′,5′-monophosphate content in the suprachiasmatic nucleus (SCN) and ventromedial hypothalamus (VMH) in the rat. *Brain Research, 276*, 297–304.

Nadakavukaren, J. J., Welsh, D. K., & Reppert, S. M. (1990). Aluminum fluoride reveals a phosphoinositide system within the suprachiasmatic region of rat hypothalamus. *Brain Research, 507*, 181–188.

Nedergaard, M. (1994). Direct signaling from astrocytes to neurons in cultures of mammalian brain cells. *Science, 263*, 1768–1771.

Obrietan, K., and van den Pol, A. (1995). GABA neurotransmission in the hypothalamus: Developmental reversal from Ca2+ elevating to depressing. *Journal of Neuroscience, 15*, 5065–5077.

O'Hara, B. R., Andretic, R., Heller, H. C., Carter, D. B., & Kilduff, T. S. (1995) $GABA_A$, $GABA_C$, and NMDA receptor subunit expression in the suprachiasmatic nucleus and other brain regions. *Molecular Brain Research, 28*, 239–250.

Parpura, v., Basarky, T. A., Liu, F., Jeftinija, K., Jeftenija, S. & Haydon, P. G. (1994). Glutamate-medicated astrocyte-neuron signalling, *Nature, 369*, 744–747.

Pennartz, C. M. A., Bierlaagh, M. A., & Geurtsen, A. M. S. (1997). Cellular mechanisms underlying spontaneous firing in rat suprachiasmatic nucleus: Involvement of a slowly inactivating component of sodium current. *Journal of Neurophysiology, 78*, 1811–1825.

Pitts, J. D., & Finbow, M. E. (1986). The gap junction. *Journal of Cell Science, 4*, 239–266.

Pritchett, D.B., Láddens, H., & Seeburg, P.H. (1989). Type I and type II $GABA_A$-benzodiazepine receptors produced in transfected cells. *Science, 245*, 1389–1392.

Prosser, R. A., & Gillette, M. U. (1989). The mammalian circadian clock in the suprachiasmatic nuclei is reset *in vitro* by cAMP. *Journal of Neuroscience, 9*, 1073–1081.

Prosser, R. A., & Gillette, M. U. (1991). Cyclic changes in cAMP concentration and phosphodiesterase activity in mammalian circadian clock studied *in vitro. Brain Research, 568*, 185–192.

Prosser, R. A., McArthur, A. J., & Gillette, M. U. (1989). cGMP induces phase shifts of a mammalian circadian pacemaker at night, in antiphase to cAMP effects. *Proceedings of the National Academy of Sciences of the USA, 86*, 6812–6815.

Prosser, R. A., Miller, J. D., & Heller, H. C. (1990). A serotonin agonist phase-shifts the circadian clock in the suprachiasmatic nuclei in vitro. *Brain Research, 534*, 336–339.

Prosser, R. A., Heller, H. C., & Miller, J. D. (1992). Serotonergic phase shifts of the mammalian circadian clock: Effects of tetrodotoxin and high Mg^{2+}. *Brain Research, 573*, 336–340.

Rudy, B. (1988). Diversity and ubiquity of K channels. *Neuroscience, 25*, 729–749.

Rusak, B., & Bina, K. G. (1990). Neurotransmitters in the mammalian circadian system. *Annual Review of Neuroscience, 13*, 387–401.

Rusak, B., & Groos, G. (1982). Suprachiasmatic stimulation phase shifts rodent circadian rhythms. *Science, 215*, 1407–1409.

Saxena, N. C., & Macdonald, R. L. (1994). Assembly of $GABA_A$ receptor subunits: Role of the δ-subunit. *Journal of Neuroscience, 14*, 7077–7086.

Schwartz, W. J. (1991). Further evaluation of the tetrodotoxin-resistant circadian pacemaker in the suprachiasmatic nuclei. *Journal of Biological Rhythms, 6*, 149–158.

Schwartz, W. J., Gross, R. A., & Morton, M. T. (1987). The suprachiasmatic nuclei contain a tetrodotoxin-resistant circadian pacemaker. *Proceedings of the National Academy of Sciences of the USA, 84*, 1694–1698.

Schweitzer, J. S., Patrylo, P. R., & Dudek, F. E. (1992). Prolonged field bursts in the dentate gyrus: Dependence on low calcium, high potassium, and nonsynaptic mechanisms. *Journal of Neurophysiology, 68*, 2016–2025.

Sheridan, J. D., & Atkinson, M. M. (1985). Physiological roles of permeable junctions: Some possibilities. *Annual Review of Physiology, 47*, 337–353.

Shibata, S., & Moore, R. Y. (1993). Tetrodotoxin does not affect circadian rhythms in neuronal activity and metabolism in rodent suprachiasmatic nucleus *in vitro. Brain Research, 606*, 259–266.

Shibata, S., Oomura, Y., Kita, H., & Hattori, K. D. (1982). Circadian rhythmic changes of neuronal activity in the suprachiasmatic nucleus of the rat hypothalamic slice. *Brain Research, 247*, 154–158.

Shibata, S., Oomura, Y., Hattori, K., & Kita, H. (1984a). Responses of suprachiasmatic nucleus neurons to optic nerve stimulation in rat hypothalamic slice preparation. *Brain Research, 302*, 83–89.

Shibata, S., Shiratsuchi, A., Liou, S. Y., & Ueki, S. (1984b). The role of calcium ions in circadian rhythm of suprachiasmatic nucleus neuron activity in rat hypothalamic slices. *Neuroscience Letters, 52*, 181–184.

Shibata, S., Liou, S. Y. & Ueki, S. (1986). Influence of excitatory amino acid receptor antagonists and of baclofen on synaptic transmission in the optic nerve to the suprachiasmatic nucleus in slices of rat hypothalamus. *Neuropharmacology, 25*, 403–409.

Shimura, M., Harata, N., Tamai, M., & Akaike, N. (1996). Allosteric modulation of $GABA_A$ receptors in acutely dissociated neurons of the suprachiasmatic nucleus. *American Journal of Physiology, 270*, C1726–C1734.

Spray, D. C., & Bennett, M. V. L. (1985). Physiology and pharmacology of gap junctions. *Annual Review of Physiology, 47*, 281–303.

Staley, K. J., Otis, T. S., & Mody, I. (1992). Membrane properties of dentate gyrus granule cells: Comparison of sharp microelectrode and whole-cell recordings. *Journal of Neurophysiology, 67*, 1346–1358.

Staley, K. J., Soldo, B. L., & Proctor, W. R. (1995). Ionic mechanisms of neuronal excitation by inhibitory $GABA_A$ receptors. *Science, 269*, 977–929.

Storm, J. F. (1990). Potassium currents in hippocampal pryamidal cells. *Progress in Brain Research, 83*, 161–187.

Strecker, G. J., Park, W. K., and Dudek, F. E. (1997a). Flunitrazepam potentiates $GABA_A$ receptor-mediated currents in the rat suprachiasmatic nucleus. *Society for Neuroscience Abstracts, 23*, 949.

Strecker, G. J., Wuarin, J.-P., & Dudek, F. E. (1997b). $GABA_A$-mediated local synaptic pathways connect neurons in the rat suprachiasmatic nucleus. *Journal of Neurophysiology, 78*, 2217–2220.

Strecker, G. J., Park, W. K., & Dudek, F. E. (1999). Zinc and flunitrazepam modulation of GABA-mediated currents in rat suprachiasmatic neurons. *Journal of Neurophysiology, 81*, 184–191.

Sugimori, M., Shibata, S., & Oomura, Y. (1986). Electrophysiological bases for rhythmic activity in the suprachiasmatic nucleus of the rat: An *in vitro* study. In Y. Oomura (Ed.), *Emotions: Neuronal and chemical control* (pp. 199–206). Basel: Karger.

Takahashi, J. S., & Zatz, M. (1982). Regulation of circadian rhythmicity. *Science, 217,* 1104–1111.

Thomson, A. M. (1984). Slow, regular discharge in suprachiasmatic neurones is calcium dependent, in slices of rat brain. *Neuroscience, 13,* 761–767.

Thomson, A. M., & West, D. C. (1990). Factors affecting slow regular firing in the suprachiasmatic nucleus *in vitro. Journal of Biological Rhythms, 5,* 59–75.

Turek, F. W. (1985). Circadian rhythms in mammals. *Annual Review of Physiology, 47,* 49–64.

Valases, C., Wright, S. J., & Catravas, G. N. (1980). Diurnal changes in cyclic nucleotide levels in the hypothalamus of the rat. *Experimental Brain Research, 40,* 261–264.

van den Pol, A. N., (1980). The hypothalamic suprachiasmatic nucleus of the rat: Intrinsic anatomy. *Journal of Comparative Neurology, 191,* 661–702.

van den Pol, A. N. & Dudek F. E. (1993). Cellular communication in the circadian clock, the suprachiasmatic nucleus. *Neuroscience, 56,* 793–811.

van den Pol, A. N., and Gorcs, T. (1986). Synaptic relationships between neurons containing vasopressin, gastrin-releasing peptide, vasoactive intestinal polypeptide, and glutamate decarboxylase immunoreactivity in the suprachiasmatic nucleus: Dual ultrastructural immunocytochemistry with gold-substituted silver peroxidase. *Journal of Comparative Neurology, 252,* 507–521.

van den Pol, A. N., & Powley, T. (1979). A fine grained anatomical analysis of the role of the rat suprachiasmatic nucleus in circadian rhythms of feeding and drinking. *Brain Research, 160,* 307–326.

van den Pol, A. N., Finkbeiner, S. M., & Cornell-Bell, A. H. (1992). Calcium excitability and oscillations in suprachiasmatic nucleus neurons and glia *in vitro. Journal of Neuroscience, 12,* 2648–2664.

van den Pol, A. N., Obrietan, K., Chen, G., & Belousov, A. B. (1996). Neuropeptide Y-mediated long-term depression of excitatory activity in suprachiasmatic nucleus neurons. *Journal of Neuroscience, 16* , 5883–5895.

Wagner, S., Castel, M., Gainer, H., & Yarom, Y. (1997). GABA in the mammalian suprachiasmatic nucleus and its role in diurnal rhythmicity. *Nature, 387,* 598–603. Walsh, I. B., van den Berg, R. J., Marani, E., & Rietveld, W. J. (1992). Spontaneous and stimulated firing in cultured rat suprachiasmatic neurons. *Brain Research, 588,* 120–131.

Walsh, I. B., van den Berg, R. J., & Rietveld, W. J. (1995). Ionic currents in cultured rat suprachiasmatic neurons. *Neuroscience, 69,* 915–929.

Watts, A. G. (1991). The efferent projections of the suprachiasmatic nucleus: Anatomical insights into the control of circadian rhythms. In D. C. Klein, R. Y. Moore, & S. M. Reppert (Eds.), *Suprachiasmatic nucleus, the mind's clock* (pp. 77–106). New York: Oxford University Press.

Welsh, D. K., Logothetis, D. E., Meister, M., and Reppert, S. M. (1995). Individual neurons from rat suprachiasmatic nucleus express independently phased circadian firing rhythms. *Neuron, 14,* 697–706.

Wheal, H. V., & Thomson, A. M. (1984). The electrical properties of neurones of the rat suprachiasmatic nucleus recorded intracellularly *in vitro. Neuroscience, 13,* 97–104.

Winfree, A. T. (1967). Biological rhythms and the behavior of populations of coupled oscillators. *Journal of Theoretical Biology, 16,* 15–42.

Zlomanczuk, P., Margraf, R. R., & Lynch, G. R. (1991). *In vitro* electrical activity in the suprachiasmatic nucleus following splitting and masking of wheel-running behavior. *Brain Research, 559,* 94–99.

16

Mammalian Clock Genetics

MARTIN R. RALPH AND MARTHA HOTZ VITATERNA

INTRODUCTION

Very early in the modern history of biological rhythm research, Erwin Bünning demonstrated that the circadian period was under genetic control (Bünning, 1935). The experiments not only verified the biological (endogenous) nature of rhythm generation, but also demonstrated the innate determination of circadian periodicity. At the start of a new millennium, the 65 years of research into clock genetics until recently had brought out only a rudimentary understanding of the direct role of genes in this process. Nonetheless, with the discovery of *timeless* in *Drosophila*, which has provided new insight into the activity and attributes of clock genes (see Chapter 14), the cloning of *Clock* in the mouse (King *et al.*, 1997b; Vitaterna *et al.*, 1994), and the identification of *per* homologues in the mammalian suprachiasmatic nucleus (SCN) (Sun *et al.*, 1997; Tei *et al.*, 1997), the field is poised for a significant assault on the genetic mechanisms of rhythm generation.

While a genetic foundation for circadian characteristics is not in dispute, the number of genes and the nature of their involvement in rhythmic processes are problems of considerable interest. This is particularly true of research into mammalian systems, where emphasis for decades has been on biochemical/physiologic mechanisms and only very recently has shifted to the molecular genetic. Yet mammalian clock genetics (for much the same reason as the biochemistry of entrainment pathways) has been more often than not a pursuit of the mundane. Mammalian clocks, after all, are ensembles of pacemaker cells. These communicate with one another and receive information about the environment via general cellular mechanisms. Not surprisingly, rhythm regulation involves common intracellular pathways, which, when influenced by pharmacologic or genetic manipulations,

---

MARTIN R. RALPH Department of Psychology, University of Toronto, Toronto, M5S 3G3, Canada. MARTHA HOTZ VITATERNA Department of Neurobiology and Physiology, Northwestern University, Evanston, Illinois 60208.

Circadian Clocks, Volume 12 of *Handbook of Behavioral Neurobiology*, edited by Joseph S. Takahashi, Fred W. Turek, and R. Y. Moore, Kluwer Academic / Plenum Publishers, New York, 2001.

produce perturbations of rhythm expression (see below). It would not be surprising, therefore, to find that a considerable number of polymorphic loci contribute to total phenotypic variance within any species.

However, Konopka's discovery of the *period* mutations in *Drosophila* (Konopka & Benzer, 1971) took genetic analysis to another level. Not only could cellular processes and their genetic underpinnings account for phenotypic variance, but these results indicated that specific genes were vested with a primary responsibility for setting circadian period. The finding also set a standard for the field, so that phenotypes generally attract very little interest unless they are well outside the range for the population. In mammals, the *tau* mutation, discovered in 1985, was the first to meet the new criterion, producing a profound alteration of circadian period comparable to the per^s phenotype (Ralph & Menaker, 1988).

In the annals of pure genetic research, the discovery itself was of little importance. Because it was discovered in the golden hamster (*Mesocricetus auratus*), an organism for which almost no genetic information was available (and not likely to be forthcoming), the mutation could easily have been relegated to the credenza of obscure scientific curiosities. However, two circumstances conspired to ensure that this would not happen. First, the hamster, while not a model genetic system, was nonetheless a preeminent organism for biological rhythm research. Hence a new gateway into the structure and function of the clock, if not the gene itself, was presented. Second, the discovery extended to mammals the notion that the mechanism of rhythm generation could be known at the molecular level, and that perhaps a universal clock structure existed with possibly genetic conservation throughout phylogeny.

Since this discovery, the first (putative) mammalian clock gene has been isolated and cloned from the mouse (King *et al.*, 1997b; Vitaterna *et al.*, 1997). The *Clock* phenotypes, like those of the *tau* mutation, resemble effects of mutations in other organisms, extending the notion of a unifying design for circadian clock mechanisms. Together, these discoveries provide a new perspective from which to view the state of mammalian clock genetics. In this chapter, we examine the various influences, both subtle and profound, that genes are known to have on mammalian circadian systems.

Strain Differences and Selected Lines

One of the simplest approaches to studying genetic control is to examine the influence of genetic background. In different species, various strains have been retained for reasons absolutely unrelated to circadian rhythmicity. Comparisons have been made primarily with regard to circadian period and photic responsiveness, but the pattern of rhythmicity—or daily profile—may also vary (Table 1).

Inbred Strain Comparisons

Inbred strains are genetically defined strains of mice or other experimental species which have been systematically inbred (most typically by brother–sister matings) for at least 20 generations so that individuals of a strain are more than 99% homozygous and genetically identical to one another. An important caveat in the interpretation of such data is that inbred strains represent random, not selected genotypes; in fact, inbreeding can be presumed to "fix" genes at random, since on

TABLE 1. GENETIC DETERMINANTS OF CIRCADIAN PROPERTIES

Gene/trait	Species	Circadian properties affected	Reference
Strain differences and selected traits			
—	Rabbit	Locomotor rhythm amplitude; Dutch-Himalayan > English outbred	Kennedy *et al.*, 1994
—	Rat	Period response to lithium: ACI > BH and Long Evans	Kafen & Wollnick, 1994
—	Rat	Rate of reentrainment; Wistar albino > hooded	Kennaway, 1994
—	Rat	Period and circadian pattern; ACI/Ztm vs. LEW/Ztm	Siebert & Wollnick, 1991
—	Rat	Pattern differences in three strains	Bauer, 1990
—	Rat	Differences in amount, duration, and timing of activity in three strains	Wollnick, 1991
—	Mouse	Period; C57BL/6J > BALB/cByJ: polygenic inheritance	Schwartz & Zimmerman, 1990
—	Mouse	Polygenic inheritance, not maternal inheritance	Beau, 1988, 1991
—	Mouse	Rate of reentrainment; C57BL/6 and BALB/c > C3H/He and CBA	Kopp *et al.*, 1998
—	Mouse	Different inheritance patterns for period, mesor, acrophase, amplitude	Peleg *et al.*, 1989, 1995
—	Mouse	Food entrainment in CS, but not C57BL/6J	Abe *et al.*, 1989
—	Mouse	Different circadian responses to GABA-active agents in SK and C57BL/6	Ebihara *et al.*, 1988a, b
—	Golden hamster	Critical photoperiod for four inbred strains	Vitaterna & Turek, 1993
—	Golden hamster	Phase response curve to benzodiazepine, amount and duration of activity	Turek *et al.*, 1995
Nest-building behavior	Mouse	Different phase response to light between high, low, and control selected lines	Bult *et al.*, 1993
Accelerated senescence	Mouse	Rate of reentrainment; senescence-prone > senescence-resistant	Sanchez-Barcelo *et al.*, 1998; Miyamoto, 1997
Diabetes	Rat	Circadian variation; diabetes-prone > diabetes-resistant	Berg *et al.*, 1997
Diabetes	Mouse	Circadian variation	Roesler *et al.*, 1985
Hyperprolactinemia	Rat	Dissociated rhythmic parameters in sleep	Valatz & Jouvet, 1988
Athymic (nude)	Mouse	Normal rhythms; reduced activity	Beau *et al.*, 1990
Staggerer	Mouse	Loss of diurnal feeding rhythm	Guastavino *et al.*, 1991
Sparse fur (*spf*)	Mouse	Loss of locomotor rhythm during hyperammonemia	Minakami, 1994
Asebia	Mouse	Abnormal rhythm of epidermal cell division	Brown *et al.*, 1988
Retinal degeneration (*rds*)	Mouse	Altered phase angle of entrainment	Sanyal & Hawkins, 1989
CBA/J (*rd/rd*)	Mouse	Reduced circadian photosensitivity	Yoshimura *et al.*, 1994

(continued)

TABLE 1. (*CONTINUED*)

Gene/trait	Species	Circadian properties affected	Reference
Spontaneous hypertension	Rat	Shortened period	Peters *et al.*, 1994; Rosenwaser & Plante, 1993
Vasopressin deficiency	Brattleboro rat	Reduced photic response	Murphy *et al.*, 1996
AchR muscarinic deficiency	Rat	Shortened period	Shiromani & Overstreet, 1994
Microphthalmia	Rat	Arrhythmic feeding and ornithine decarboxylase activity	Shim *et al.*, 1997
Anophthalmia	Rat	Normal rhythms; loss of entrainment in LD	Ibuka, 1987
Anophthalmia	Rat	Loss of locomotor rhythms in DD and LD	Rao & Diwan, 1995
Anophthalmia	Mouse	Intact circadian rhythms; loss of entrainment	Faradji-Prevautel *et al.*, 1990; Laemle *et al.*, 1995
Pink-eyed dilution (*p/p*)	Mouse	Normal rhythms; reduced responses to light	Vitaterna *et al.*, 1993
Melatonin 1bR (−/−)	Siberian hamster	Normal reproductive and circadian responses to melatonin	Weaver *et al.*, 1996
Spontaneous rhythm mutations and mutagenesis screening			
tau	Golden hamster	Shortened period; increased light and nonphobic phase shifts	Ralph & Menaker, 1988; Shimomura & Menaker, 1994; Mrosovsky *et al.*, 1992
Clock	Mouse	Lengthened period; arrhythmic in DD	Vitaterna *et al.*, 1994
		Increased phobic responses	Chang *et al.*, 1998
Whl	Mouse	Lengthened period; abnormal rhythms and responses to light	Pickard *et al.*, 1995
Targeted mutations and transgenics (tg)			
DBP	Mouse (−/−)	Shortened period; reduced locomotor activity	Lopez-Molina *et al.*, 1997
Interleukin-1β	Mouse (−/−)	Normal	Kozak *et al.*, 1995
$Ncam^{tmlCwr}$	Mouse (−/−)	Shortened period; arrhythmicity in DD	Shen *et al.*, 1995
Corticotropin-releasing hormone-binding protein	Mouse (−/−)	Altered circadian feeding pattern; increased locomotor activity	Burrows *et al.*, 1998
Prion protein (PrP)	Mouse (−/−)	Lengthened period	Tobler *et al.*, 1996
Wocko	Mouse (tg)	Shortened period	Sollars *et al.*, 1996
Histamine H1R	Mouse (tg)	Increased daytime activity	Inoue *et al.*, 1996
Growth hormone	Mouse (tg)	Shortened period in LL	Ferraro *et al.*, 1994
Hypertension	Rat (tg)	Abornomal circadian profile	Lemmer *et al.*, 1994

Light-regulated and entrainment-related genes in the SCN			
c-fos	Mouse (−/−)	Normal rhythms; reduced photic response	Honrado *et al.*, 1996
Jun-B; c-*jun*	Mouse (−/−)	Heterozygote normal (homozygous lethal)	Ralph, M. R., unpublished
nNos; *eNOS*	Mouse (−/−)	Heterozygote normal (homozygous lethal)	Ralph, M. R., unpublished
NGFI-A; *NGFI-B*	Mouse (−/−)	Normal rhythms and entrainment; double knockout normal	Kilduff *et al.*, 1998
Connexin-43	Mouse (−/−)	Heterozygote normal (homozygous lethal)	Ralph, M. R., unpublished
CFTR	Mouse (−/−)	Normal rhythms and entrainment	Ralph, M. R., unpublished
NGF	Mouse (−/−)	Heterozygote normal (homozygous lethal)	Ralph, M. R., unpublished
p75 (*NGFR*)	Mouse (−/−)	Normal rhythms; reduced responses to light	Golombeck *et al.*, 1996
BDNF	Mouse (−/−)	Normal rhythms; reduced responses to light	Liang *et al.*, 2000
Glu2 (*AMPA*)	Mouse (−/−)	Rescued activity; highly fragmented behavior; normal period	Walters, D., unpublished
Grm5 (*mGluR5*)	Mouse (−/−)	Normal rhythms and entrainment	Walters, D., *et al.*, unpublished
CaM kinase II	Mouse (−/−)	Lengthened period	Zhao *et al.*, 1998
vgf	Golden hamster	Not tested as mutant	Wisor and Takahashi, 1997
Homer	Rat	Not tested as mutant	Park *et al.*, 1997
egr-3	Mouse	Not tested as mutant	Morris *et al.*, 1998
mper1, *mper 2*, *mper3*	Mouse	Not tested as mutant	Albrecht *et al.*, 1997; Shearman *et al.*, 1998; Shigeyoshi *et al.*, 1998; Sun *et al.*, 1997; Takumi *et al.*, 1998; Tei *et al.*, 1997; Zylka *et al.*, 1998
Quantitative analyses			
—	Human	Similarities between monozygotic twins	Reinberg *et al.*, 1985
—	Mouse	Polygenic nature of six parameters	Beau, 1988, 1991
Periodicity	Mouse	Inheritance of period	Diez-Noguera *et al.*, 1989
Periodicity	Mouse	QTL (BALB/cBY × C57BL/6J)	Mayeda *et al.*, 1996
Periodicity	Mouse	QTL (CS × C57Bl/6)	Suzuki *et al.*, 1998
Periodicity	Mouse	QTL (BALB/cJ × C57BL/6J)	Shimomura *et al.*, 1998
Fatal familial insomnia	Human	Loss of circadian locomotor rhythms; hyperactivity	Piazzi *et al.*, 1997
		Loss of circadian melatonin rhythm	Portaluppi *et al.*, 1994
Ethnicity	Human	Variance in rhythm profiles for 11 hormones in three ethnic groups and three ages	Ticher *et al.*, 1995
Gender	Human	Meta-analysis, gender differences in rhythmic chemical and hormone profiles	Ticher *et al.*, 1994

average 50% of the genetic variability will be lost with each generation. Further, the essentially complete homozygosity means a highly unusual (i.e., abnormal) genotype.

A number of alterations in different properties of circadian rhythmicity have become fixed in the genomes of inbred strains of rats, mice, and hamsters including the length of the free-running period (Ebihara, Tsuji, & Kondo, 1978; Possidente & Hegmann, 1982; Schwartz & Zimmerman, 1990; Turek *et al.*, 1995), the phase angle of entrainment in LD conditions (Castellano, Puglisi-Allegra, Renzi, & Oliviero, 1985; Connolly & Lynch, 1981; Lucas & Eleftheriou, 1980; Peleg, Zvulunov, & Ashkenazi, 1989; Possidente & Hegmann, 1980; Turek *et al.*, 1995), the response to various environmental manipulations (Abe *et al.*, 1989; Hotz, Connolly, & Lynch, 1987; Roussel, Turrillot, & Kitahama, 1984; Wax, 1977), and changes in the amplitude and waveform of circadian rhythms (Bauer, 1990; Beau, 1988; Büttner & Wollnik, 1984; Kempf, Mandel, Oliviero, & Puglisi-Allegra, 1982). Only a few attempts at examining the phase response curve to light or other resetting stimuli have been made. Inbred strains of hamsters clearly differ in their phase-shifting responses to triazolam administration (Turek *et al.*, 1995), but not in response to light pulses, whereas the BALB/cByJ and C57BL/6J strains of mice are distinct in the shapes of their phase response curves to light pulses (Schwartz & Zimmerman, 1990). Differences between mouse strains in phase shifting response to pharmacologic agents have also been reported (Ebihara, Goto, & Oshima, 1998a, b). Characteristics of particular inbred strains are reviewed in Table 1.

Relatively few studies have examined multiple strains for multiple circadian characteristics so that evaluation of degree of genetic correlation among traits can be assessed (Hegmann & Possidente, 1981). Because of the random and abnormal genetic makeup of an inbred strain, tests of relationships between characteristics which are based on a single pairwise comparison are highly suspect. Any given pair of inbred strains will differ in a number of traits, some related and some not. Genetic correlations between phase angle of entrainment and endogenous period have been reported for inbred strains of mice (Possidente & Hegmann, 1980). A similar relationship is seen among inbred strains of golden hamsters (Turek *et al.*, 1995). Surprisingly, the differences in phase angle of entrainment did not translate to differences in the so-called "critical daylength" to maintain reproductive function, although differences among hamster strains in reproductive responses to photoperiod clearly were observed (Vitaterna & Turek, 1993).

Crosses between Strains

In order to obtain more detail concerning the nature of genetic control of a circadian trait, such as the number of loci, direction of dominance, sex linkage, maternal effects, etc., one needs to make crosses between strains. Genetic correlations are more powerfully assessed by examining a segregating generation as well. Discouragingly few studies involving crosses between strains have been carried out. Examination of a segregating intercross generation as well as a systematically maintained heterozygous population indicated a potential linkage between endogenous period and pigmentation loci (Possidente & Hegmann, 1982; Possidente, Hegmann, Carlson, & Elder, 1982), and studies of reciprocal hybrids between BALB/c and C57BL/6 strains of mice indicate no maternal effects on circadian period in mice (Possidente & Stephan, 1988).

A genetic method which has begun to be explored with respect to rhythmicity involves the use of recombinant inbred (RI) lines and quantitative trait loci (QTL) analysis. RI lines are created by inbreeding progeny of a segregating intercross generation (such as F2 or F3 generations) (Takahashi, Pinto, & Vitaterna, 1994). Each RI strain has a different, but reproducible, recombined sampling of the genomes of the parental inbred strains. Examination of a panel of RI lines can be likened to examination of an F2 panel, and such mouse panels have been studied with respect to circadian period, clearly demonstrating that there is polygenic control of this trait (Hofstetter, Mayeda, Possidente, & Nurnberger, 1995; Mayeda, Hofstetter, Belknap, & Nurnberger, 1996; Schwartz & Zimmerman, 1990; Shimomura, Vitaterna, Low-Zeddies, Whitely, & Takahashi, 1998; Suzuki *et al.*, 1998). QTL analysis represents a statistical correlative method by which linkage of the phenotypic trait to particular chromosomal regions can be detected. The power of such a method depends on a number of factors including, but not limited to, the number of genotypes and phenotypes determined, the map resolution, and the number of genes contributing to the phenotypic trait under examination. While QTL analysis can be initiated with an RI panel, it is severely limited in its power to identify candidate gene regions. Hence, the power of QTL analysis to identify genes involved in circadian rhythmicity has yet to be realized.

The use of artificial selection (i.e., selective breeding for a particular phenotype) for specific traits provides a somewhat more refined approach to the question in that specific features may be selected that are hypothesized from other findings to have a relationship with circadian timekeeping. The majority of rhythm disturbances that have been reported can be most easily attributed to changes in the effector mechanisms, but remain important for discussions of clock genetics because of the potential for feedback regulation. Some of the specific lines that have been examined for their circadian rhythmicity are listed in Table 1.

Two major limitations of these comparative approaches need to be addressed when examining the genetic basis of circadian rhythmicity. First, the circadian parameters usually measured are interrelated. Factors such as environmental input to the clock (e.g., light intensity) or feedback regulation (e.g., locomotor behavior) can alter fundamental clock properties such as period and phase shift responsiveness (Edgar, Kilduff, Martin, & Dement, 1991; Reebs & Mrosovsky, 1989; Van Reeth & Turek, 1989). Arguably, any allele that significantly alters any of these influences carries the potential of altering basic circadian parameters and presenting the appearance of a "clock" gene. For example, an allele that affects the transmission of photic information anywhere along the entrainment pathway(s) may contribute to measured period, either by affecting angular velocity (Aschoff's rule) or influencing aftereffects of entrainment. Similarly, a change in activity level might affect periodicity through changes in nonphotic responsiveness.

Second, the isolation of a genetically defined strain by definition involves a genetic bottleneck. Assuming that a polygenic foundation exists for clock structure and regulation, the development of any inbred strain retains an inadvertent selection of a distinct allelic complement for all the genes that might influence rhythmicity in any way. Predictably, once in a while, one complement will produce a phenotype that is distinct from another.

Spontaneous Mutations

While the search for genes that affect rhythm generation focuses naturally on those that alter fundamental properties of rhythm generation, a number of spontaneous mutations have been used to some advantage to address questions about the control of circadian rhythms.

The *rd* mutation in the mouse results in retinal degeneration so that virtually all rod photoreceptors are lost by the time the animal is an adult, and it thus becomes blind. This mutation is present in most C3H strains of mice. C3H mice, as well as C57BL/6 congenic lines carrying the mutation, have virtually no deficit in circadian clock responses to light, although visual responses are clearly impaired (Foster *et al.*, 1991). These results have led researchers to question what the photoreceptors involved in circadian entrainment may be, since rods appear unnecessary.

An apparent association of free-running period to pigmentation led to the suggestion that in mice, the albino (*c*) and pink-eyed dilution (*p*) loci might contribute to strain differences in period. Since such hypopigmentation mutations affect innervation of retinorecipient brain areas, such mutations might also affect circadian responses to light (Vitaterna, Wu, Turek, & Pinto, 1993). Although the total volume of the SCN is reduced in pink-eyed dilution (*p/p*) mice, the reduction in retinal innervation of the SCN is disproportionate. The animals exhibit no differences in free-running period in constant darkness, phase angle of entrainment, or phase shifting response to saturating light pulses. However, the phase shifts in response to subsaturating light pulses and the period-lengthening effects of constant light are reduced by the mutation. The mutation does not disrupt clock function per se, but does apparently diminish photic sensitivity.

The hormone melatonin is associated with circadian rhythms and photoreception in all vertebrate classes. In mammals, it carries temporal information from the pineal organ to the rest of the organism, and can function as a feedback signal to the clock. As such, alterations in the production of melatonin might be expected to affect rhythm generation. Most common laboratory strains of mice appear to lack the ability to synthesize melatonin. Ebihara and colleagues demonstrated that virtually all common laboratory strains except C3H lack activity in the two critical synthetic enzymes in the pineal organ, serotonin N-acetyltransferase (NAT) and hydroxyindole-O-methyltransferase (HIOMT) (Ebihara *et al.*, 1986; Ebihara *et al.*, 1987; Goto, Oshima, Tomita, & Ebihara, 1989; Goto, Oshima, Hasegawa, & Ebihara, 1994) apparently due to mutations in the genes encoding each of these enzymes. Even some wild species of mice appear to lack these enzyme functions. Behavioral rhythmicity is clearly unimpaired in these mice, and so melatonin appears unnecessary for clock function in this species. Interestingly, the circadian system of mice lacking the Mel1a receptor is still responsive to melatonin (Liu *et al.*, 1997), as are the circadian and reproductive systems of Siberian hamsters, which lack the Mel1b receptor (Weaver, Liu, & Reppert, 1996).

Altogether, these comparative approaches to the genetics of circadian rhythmicity are predisposed to the discovery of significant but subtle genetic influences on circadian rhythms. In our hands, interindividual free-running period of CD-1 mice, an outbred strain, ranges from 22.0 to 24.2 hours. Yet strain differences on the order of tenths of an hour are usually reported as significant differences between inbred lines. The number of genes and alleles that may contribute to rhythmicity in significant but subtle ways is incalculable. Still, the value of any difference depends upon whether or not the cause of the distinctiveness can be understood or the difference itself can be exploited.

Transgenics, Gene Targeting, and Candidate Genes

The interpretation of differences among strains and selected lines is limited by the extent to which genetic background either can be maintained or controlled.

This problem can be eluded if the loci contributing to phenotypic differences are identified. An alternative approach, therefore, is the use of identified mutations or candidate genes, which has the procedural advantage of allowing each candidate to be associated with a discrete hypothesis. The two common approaches that are used are transgene expression and gene targeting. Both are theoretically elegant and highly specific means of altering gene expression. Unfortunately, both are beset with problems that can defeat meaningful interpretation.

The primary problem with the use of transgenes is that spatial and temporal expression is difficult to control. Since there are no known genes that are expressed exclusively in the SCN, there is no promoter that will reliably direct gene expression to the nucleus, and the chance of expression in the SCN is often a matter of serendipity. A few transgenic animals whose rhythms have been examined are included in Table 1. Apart from providing protein markers, though, the transgene approach has not provided much insight into clock mechanisms. Exceptions to this arise when genetic deficits are produced that may be rescued by transgene expression. An elegant example of this is the rescue of the wild-type circadian phenotype in *Clock* mutant mice with the insertion of a 140-kb bacterial artificial chromosome (BAC) construct containing the wild-type *Clock* locus (Antoch *et al.*, 1997; see description below).

The second general approach is gene targeting. In this method, a DNA construct replaces by homologous recombination part of the gene of interest. The replacement introduces a number of stop codons that effectively truncate or eliminate the production of the targeted protein. For candidate genes, the knockout tests the hypothesis that the gene is required for "normal" rhythm generation and control. Unfortunately, the elegant technique has problems that often render effects impossible to interpret.

For rhythm analysis, gene targeting has been used mainly in an effort to verify the cascade of biochemical events by which environmental cues reach and alter circadian phase. At some point, the entrainment pathway must intersect and affect a step in the biochemical circadian oscillator; from studies in nonmammalian species, this appears to involve proteins and genes whose activities are light and circadian time dependent. The demonstration that immediate-early genes (IEGs) in rodent SCN cells are induced by light pulses, but only during the subjective night, raised the possibility that not only could the entrainment pathways be followed within SCN cells, but perhaps the biochemistry of the rhythm generation mechanism was very close. Such hope was built on the reasonable but simplistic notion that the intracellular leg of the entrainment pathways was a linear cascade of events. If so, then null mutations produced in the genes encoding these components should eliminate entrainment and possibly disrupt the clock. Perhaps the most attractive of these events has been the photic induction of the IEGs c-*fos* and *jun*-B in the SCN. These are induced following phase-shifting light pulses during the subjective night, and light-induced phase shifts are blocked by a cocktail of anti-c-*fos* and anti-*jun*-B oligonucleotides (Wollnick *et al.*, 1995). The cellular response is initiated with the release of glutamate from RHT terminals, and involves the receptor-mediated stimulation of Ca^{2+}-calmodulin dependent pathways (Golombek & Ralph, 1994, 1995; Shibta & Moore, 1994) that are reviewed in Chapter 18. Because similar biochemical pathways are thought to underlie associative learning and the development of hippocampal long-term potentiation (LTP) in rodents, numerous null mutations have been produced that disrupt these pathways, including knockouts at most of the glutamate receptor loci, nitric oxide synthase (NOS), various kinases, cAMP response element-binding (CREB) protein, and other transcription factors (Kornhauser, Mayo, & Takahashi, 1992). Many of these mice have been examined for effects on circadian rhythms, and for the most part, changes in period or entrainment are small or nonexistent (see Table 1).

In the case of *jun*-B, the homozygous mutants do not survive, hence the specific hypothesis for *jun*-B cannot be tested using this method. The knockout of c-*fos*, though, produces a viable homozygote (Johnson, Spiegelman, & Papaioannou, 1992). These animals, with no teeth, severe scoliosis, and other skeletal defects, and eyes bulging from their too-small crania, display *a priori* too many abnormalities to allow the source of any rhythm aberration to be identified with confidence. Fortunately, this problem is not presented: mutants, if they can stagger to the running wheel, display circadian periods within the range for the species, and only a heroic experimental effort has enabled the detection of a marginally decreased responsiveness to light (Honrado *et al.*, 1996).

With very few exceptions, knockouts at other loci thought to be or to encode components of the entrainment pathways have produced a similar lack of interpretable results. Lines that have been tested, and are known to us, are compiled in Table 1. The genotypes examined and their respective phenotypes reveal the apparent inadequacy of the approach. Like the transgenics, the knockouts present the problems of spatial and temporal control. Mutants develop from fertilization in the absence of the gene activity and its protein product. The result can be an organism that is no longer viable, making testing impossible, or one whose own genetic makeup may compensate for the loss of one of its complement.

Still, a few mutations may provide some insight into the mechanism of rhythm generation. First is a loss of the *Clock* locus that is discussed in detail below. Two others are null mutations at neural cell adhesion molecule (NCAM; Shen *et al.*, 1997) and prion protein (PrP; Tobler *et al.*, 1996) loci. Phenotypes associated with the last two are changes in the circadian period. Since the proteins are thought to be involved in cell–cell communication in the adult central nervous system, the phenotypes suggest the importance of pacemaker coupling in setting overt period. Interestingly, in neither case are the periods outside the range of the species as a whole.

Another class of mutation has been produced at loci that appear to regulate the photic sensitivity or responsiveness of the clock. Studies of mouse lines with congenital retinal degeneration indicate that the loss of most, if not all, visual photoreception has little or no effect on circadian photoreception (Yoshimura, Nishio, Goto, & Ebihara, 1994). Changes in light responsiveness have also been reported in mice carrying null mutations at the low-affinity nerve growth factor (NGF) ($p75^{NGFR}$; Golombek, Hurd, Lee, & Ralph, 1996) and brain-derived neurotrophic factor (BDNF) (Liang, Allen, & Earnest, 2000) receptor loci.

Finally, a fascinating problem is introduced when using two strains to produce null mutant lines. The mutation is usually inserted into the genome of stem cells from one strain, which are then introduced into an early embryo from a different strain. Gametes produced by the chimera are from either one or the other strain. The problem, as discussed by Gerlai (1996), is that alleles and phenotypes that are linked with the mutant locus will segregate with the mutant phenotype for a predictable number of generations. We examined a knockout at *Grm5* (metabotropic glutamate receptor 5, mGluR5; Lu *et al.*, 1997) for effects on locomotor rhythms, and found that circadian amplitude was significantly greater and activity onsets more predictable in homozygous mutants compared with the wild-type siblings. Heterozygotes were intermediate. This initial line was derived from a 129-stem-cell donor on a CD-1 background. We then examined the same knockout on a C57BL/6 background, and obtained pattern segregation with the mutant allele, but this time, it was the mutant rhythms that were reduced and less stable. The simplest conclusion is that mGluR5 does not contribute significantly to the expression of locomotor rhythmicity; however, the segregation pattern needs explaining. A reasonable hypothesis for now is that genes that are linked with *Grm5* influence the pattern of rhythmic locomotor behavior, and donor alleles are expressed in the mutant line because they are selected along with the mutant locus (D. Walters, in preparation). The problem is a concern for any research that uses gene-targeted animals derived from two different genetic backgrounds. Linked genes may harbor alleles that influence phenotype, but the appropriate controls, derived from the chimera in parallel with the mutant line, are rarely retained or reported.

In summary, all of these mutations produce relatively small changes, if any, in either free-running period or sensitivity to environmental input; in no case is the alteration outside the species range. One cannot conclude, therefore, that any are direct effects on the basic rhythm-generating mechanisms, but in fact may alter in subtle ways the numerous devices that regulate circadian systems or allow them to adapt to environmental circumstances.

Profound Influences: *Tau* and *Clock*

Arguably, a mutation that pushes a phenotype outside of the species range would, under natural conditions, be subject to intense selection. An extension of this is that while the locus does not have to encode a state variable, it is nonetheless likely to be central to the clock structure. In mammals, mutations at two loci, *tau* and *Clock*, meet this criterion.

Tau

The original *tau* mutant was discovered among a group of golden hamsters purchased in 1985 from the Charles River colony (Cambridge, MA) (Figure 1). The mutant was recognized initially because of its unusually early onset of locomotor behavior in a 24-hour light-dark (LD) cycle. The defining phenotype was determined subsequently to be a 22-hour period in continuous dark (DD), compared with the shortest period (23.5 hours) that had been reported previously for the hamster. Behavioral records of the first F_1 litters verified the genetic root of the aberrant behavior, and indicated that 22 hours was actually the phenotype of the heterozygote animals. Subsequent intercrosses produced a distinct 20-hour phenotype for the homozygotes, and suggested a single, autosomal locus for this period gene (Ralph & Menaker, 1988).

However, as mentioned above, the value of the *tau* mutation does not lie in the genetics per se, but in the nature and usefulness of the mutation. As a model organism for the study of circadian rhythms, the hamster provides an abundance of information about clocks and clock-regulated behavior and physiology.

Among the most fundamental properties of circadian systems are the relationships among free-running period, photic entrainment, and the photic phase response curve (PRC). Early behavioral studies with *tau* indicated that heterozygotes could entrain to 24-hour LD cycles, but with phase angles of +4–5 hours; it was questionable whether homozygotes could entrain at all. Considering that the range

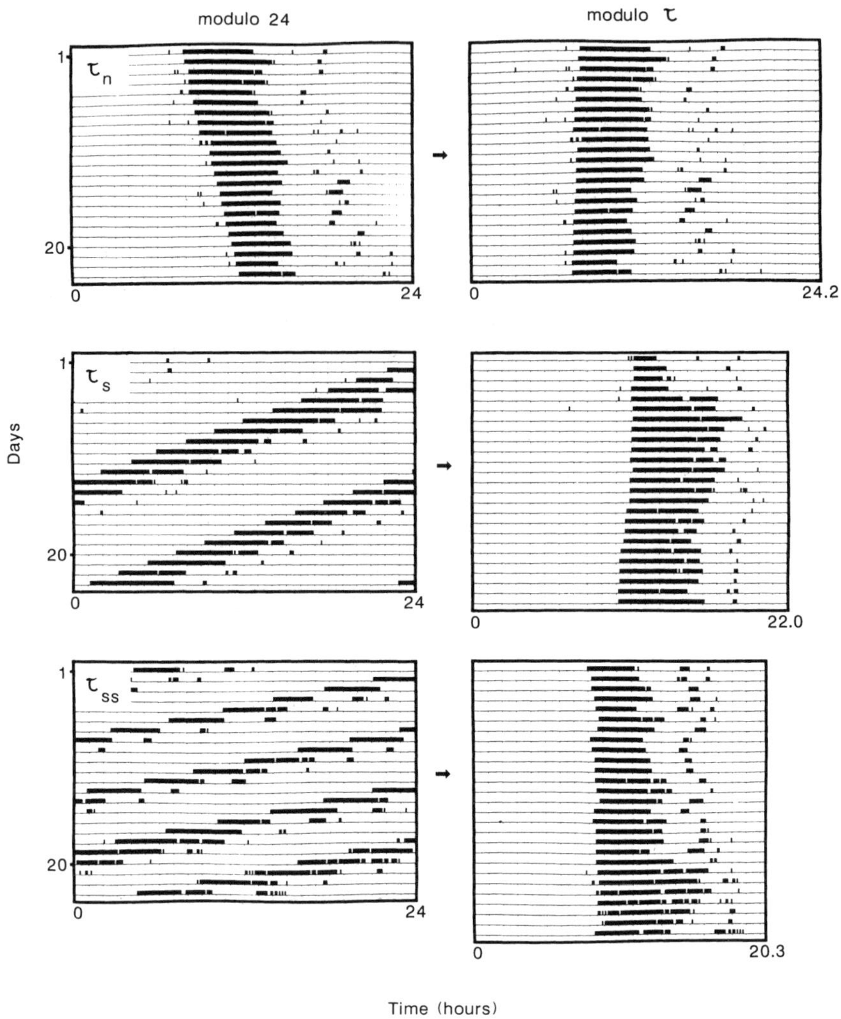

Figure 1. Effects of the *tau* mutation on circadian locomotor behavior in the golden hamster.

of entrainment exhibited by wild-type hamsters would not accommodate the 2 hour difference between T and heterozygous τ, it was not surprising to discover than PRC amplitude was increased in these mutants. A careful study by Shimomura and Menaker (1994) revealed, however, that the amplitude of the mutant PRC increased over time in DD, and was similar to the wild type in the first few days following a period of 24-hour photic entrainment. The relatively low amplitude of this PRC could explain why phase angle is so positive when entrainment to LD 14:10 is achieved.

Interestingly, not only is the phase angle changed, but in LD conditions, the entire behavioral profile is altered (Osiel, Golombek, & Ralph, 1998). In DD, the amplitude of the behavioral rhythm and the temporal organization of behavior are similar among the three *tau* genotypes. However, in LD 14:10, heterozygotes show variable onset times, lower rhythm amplitude, reduced total activity, and a fragmented pattern. The locomotor behavior is mirrored by similar changes in sleep–wake architecture, body temperature, and general activity rhythms, and is associated with reduced reproductive vigor and performance on learning tasks (Antoniadis, Ko, Ralph, & McDonald, 2000).

Perhaps the most important contribution of this mutation has been its use as a tool in the functional dissection of the hamster clock. Tissue transplantation studies have revealed a number of important functions of the hamster SCN that should help explain how this nucleus operated in the control of rhythmic behavior and physiology. The first and most basic of these was a test of the hypothesis that the SCN had primary responsibility in mammals for setting the period of the overt circadian rhythm.

Reciprocal transplantation of SCN tissue into the brains of arrhythmic, SCN-lesioned hosts of different *tau* genotype showed that not only did the SCN restore overt rhythmicity, but the range of period was the same as that predicted for intact adults of the donor genotype (Ralph, Foster, Davis, & Menaker, 1990). Thus, it appears that the SCN has essentially the sole responsibility for period determination.

This basic result has been confirmed from the activity of intact hosts bearing heterologous SCN tissue grafts. The expression of graft-derived periodicity varies considerably with the conditions of the experiment. The SCN produces no overt (behavioral) rhythmicity when grafted into young, intact hosts; however, a second, donor-derived rhythm emerges when the host SCN is damaged. This new rhythm is superimposed over the host's own rhythm without influencing host periodicity (Vogelbaum & Menaker, 1992).

Different effects from this are obtained when the grafts are made into intact hosts whose own rhythms have become disrupted due to aging. In these animals, the graft-derived rhythm is not only superimposed on the host behavior, but the host period may be modulated by or entrained to that of the donor. These interactions may be useful in eventually determining when the two clocks are actively influencing each other (Hurd, Zimmer, Lehman & Ralph, 1995; Ralph & Hurd, 1996; Ralph *et al.*, 1996). Finally, even in the absence of any donor-linked periodicity, the presence of an SCN graft is associated with a restoration of a more robust (youthful?) circadian organization (Cai, Lehman, Lloyd, & Wise, 1997).

The *tau* mutation makes it possible, therefore, to distinguish three potentially distinct influences of the SCN on the behavior of the animal: (1) A diffusible factor may be responsible for the general restoration of rhythmicity in arrhythmic hosts with SCN ablations (Silver, LeSauter, Tresco, & Lehman, 1996); (2) a second factor may be responsible for the SCN–SCN coupling required to produce relative coordination and entrainment in aged hosts (Hurd *et al.*, 1995; Ralph & Hurd, 1996); and (3) a third factor may produce the recovery of the aged hosts' rhythm in the absence of graft-generated rhythmicity (Hurd & Ralph, 1998).

Clock

The first putative clock locus in a mammal to be described at the molecular genetic level was discovered in a mutagenesis screen for circadian period (Figure 2). Mouse #25 had a free-running period 1 hour longer (more than six standard deviations) than the mean for the population (Vitaterna *et al.*, 1994). The mutation proved to be semidominant, and homozygous mutants have periods that are about 3 hours longer than normal, but rapidly become arrhythmic in DD. The homozygous

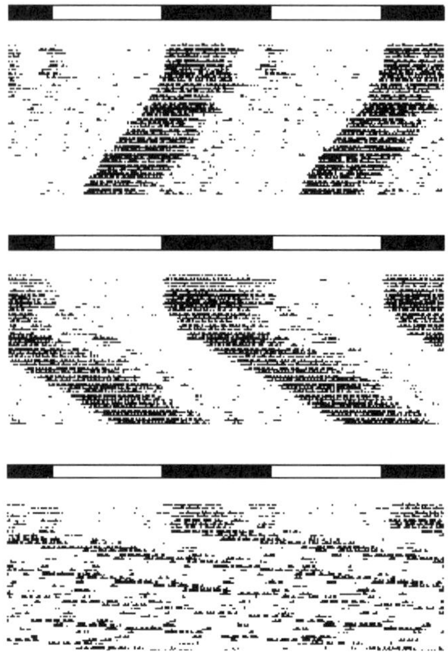

Figure 2. Effects of the *clock* mutation on circadian locomotor behavior in the laboratory mouse.

animals are able to entrain to a 24-hour LD cycle, and exhibit increase phase shift responses to light (Vitaterna *et al.*, 1996).

The locus was genetically mapped to chromosome 5, 0.7 cM distal to the c-*kit* (*W*) locus with the W^{19H} deletion (King *et al.*, 1997a, b; Vitaterna *et al.*, 1994). Using a combination of approaches, transcripts in the minimal genetically defined region were identified (King *et al.*, 1997b). Transgenic BAC rescue provided functional evidence of the critical *Clock* transcript (Antoch *et al.*, 1997). The transcription unit itself is about 100,000 bp in 24 exons, and encodes a member of the bHLH-PAS class of transcription factors. The mutation is an A–T nucleotide transversion that results in the loss of a splice donor site and the consequent deletion of 51 amino acids from the CLOCK protein (King *et al.*, 1997b). Functionally, the protein behaves as an antimorph with the mutation producing a greater effect on period than a deletion that includes *Clock* (W^{19}H), that is, *Clock/Clock* is more severe than Clock/− (King *et al.*, 1997a). The mutant phenotype was rescued by expressing a 140-kb BAC spanning the locus in transgenic mice (Antoch *et al.*, 1997). Transgene overexpression shortened the circadian period beyond the wild-type range.

The *Clock* gene encodes a bHLH-PAS transcription factor that places the protein in a family that includes clock proteins in other species. Since bHLH-PAS-containing proteins act typically as heterodimeric pairs, the dominant-negative characteristic of the mutant CLOCK protein suggests that it acts in concern with another protein of this family. It is significant, therefore, that a number of bHLH-PAS proteins, three of which are homologues of *Drosophila per*, have been identified in mammals, and are expressed in the SCN; *mper3* (Takumi *et al.*, 1998; Zylka, Shearman, Weaver, & Reppert, 1998), *mper2* (Albrecht, Sun, Eichele, & Lee, 1997; Shearman, Zylka, Weaver, Kolakowski, & Reppert, 1997), and *mper1* (Sun *et al.*, 1997;

Tei *et al.*, 1997), along with BMAL1 (Abe *et al.*, 1998; Gekakis *et al.*, 1998; Ikeda *et al.*, 1997). Yeast 2-hybrid and interaction screens have identified BMAL1 as forming a dimer with CLOCK, and being capable of DNA binding and transactivating transcription of mPer1 (Gekakis *et al.*, 1998; Hoganesch *et al.*, 1998). In *Drosophila*, the proteins PER and TIM dimerize, then enter the nucleus, where they negate their own gene transcription, possibly through another bHLH-PAS factor. Recently, the gene responsible for the *tau* mutation in the hamster was identified as casein kinase 1 epsilon, the mammalian homolog of *Drosophila* doubletime (dbt) (Lowrey & Takahashi, 2000). The biochemistry of rhythm generation appears to be widely conserved across phylogeny.

Summary

Over 20 years since the discovery of the first bona fide candidate for a clock gene, it behooves workers in this field to define exactly what is meant by this term. The narrowest definition places the gene product or the gene activity itself as a state variable that defines phase in the chemical mechanism that produces the circadian oscillation. Yet it is clear that other factors may be critical to cycle progression or period determination without being vested with the requisite attributes of a state variable.

In the broadest sense, clock genes could include virtually any locus whose alleles have distinct influences on circadian period or entrainment characteristics. This leaves open the possibility that genes that encode components of the most ubiquitous cellular processes (metabolism, neurotransmission, signal transduction) might contribute their distinct influences on phenotypic variance. After all, pharmacologic manipulations of these processes produce profound, but not surprising, changes in clock responses without providing any direct insight into clock structure or function.

Yet, most mutations produce very little effect on circadian parameters (assuming that the organism survives long enough for these measurements to be made), and trait and strain differences are relatively small compared with the effects of mutations at a small number of what appear to be highly conserved loci. This means that either (1) the number of genes that comprise a core group of "clock" genes is relatively small, (2) important "clock" genes may also be crucial for the animal's survival, hence are rarely seen as significantly altered alleles, or (3) circadian systems are sufficiently well regulated that resilience in the face of single-gene mutations is a systemic property. Currently, the best evidence for an elite collection of clock genes is the paucity of distinct loci that have been produced from mutagenesis screens in many species, mammals in particular. Screens have hardly begun to saturate the mouse genome, and mutations, spontaneous and induced, may prove to be lethal at important loci. So, while the phenotypic changes that are produced by the two major rhythm mutations in mammals are remarkably similar to those in other species, mammalian clock genetics may yet prove to be highly complex.

References

Abe, H., Kida, M., Tsuji, K., & Mano, T. (1989). Feeding cycles entrain circadian rhythms of locomotor activity in CS mice but not in C57BL/6J mice. *Physiology and Behavior, 45*, 397–401.

Abe, H., Honma, S., Ikeda, M., Tanahashi, Y., Namihira, K., & Honma, K. (1998). Expression of a novel bHLH/PAS related gene, BMAL1, in the rat brain. *Society for Research on Biological Rhythms, 10*, 49.

Abe, H., Honma, S., Honma, K., Suzuki, T., & Ebihara, S. (1999). Functional diversities of two activity components of circadian rhythm in genetical splitting mice (CS strain). *Journal of Comparative Physiology, A184,* 243–251.

Albrecht, U., Sun, Z. S., Eichele, G., & Lee, C. C. (1997). A differential response of two putative mammalian circadian regulators, mper1 and mper2, to light. *Cell, 91,* 1055–1064.

Antoch, M. P., Song, E.-J., Chang, A. M., Vitaterna, M. H., Zhao, Y., Wilsbacher, L. D., Sangoram, A. M., King, D. P., Pinto, L. H., & Takahashi, J. S. (1997). Functional identification of the mouse circadian *Clock* gene by transgenic BAC rescue. *Cell, 89,* 655–667.

Antoniadis, E. A., Ko, C. H., Ralph, M. R., & McDonald, R. J. (2000). Circadian rhythms, aging and memory. *Behavior and Brain Research, 111,* 25–37.

Bauer, M. S. (1990). Intensity and precision of circadian wheel running in three outbred rat strains. *Physiology and Behavior, 47,* 397–401.

Beau, J. (1988). Mise en evidence de correlates polygeniques des caracteristiques du rythme de l'activité chez un Mammifere: Etude de deux lignées de souris consanguines C57 BL/6 et BALB/c By. *Comptes Rendus l'Academie de Science III, 307,* 37–40.

Beau, J. (1991). Activity rhythms in inbred mice. I. Genetic analysis with recombinant inbred strains. *Behavioral Genetics, 21,* 117–129.

Beau, J., Levi, F., & Morra, R. (1990). The influence of the athymic mutation *nude* on the components of the circadian rhythm of activity in mice. *Chronobiology International, 7,* 371–376.

Berg, S., Dunger, A., Vogt, L., Kloting, I., & Schmidt, S. (1997). Circadian variations in blood pressure and heart rate in diabetes prone and resistant rat strains compared with spontaneously hypertensive rats. *Experimental and Clinical Endocrinology, 105,* 7–9.

Brown, M. H., & Nunez, A. A. (1989). Vasopressin-deficient rats show a reduced amplitude of the circadian sleep rhythm. *Physiology and Behavior, 46,* 759–762.

Brown, W. R., Furukawa, R. D., & Ramsay, C. A. (1988). Circadian rhythms are suppressed in hyperproliferative mouse epidermis. *Cell Tissue Kinetics, 21,* 159–167.

Bult, A., Hiestand, L., Van der Zee, R., & Lynch, C. (1993). Circadian rhythms differ between selected mouse lines: A model to study the role of vasopressin neurons in the suprachiasmatic nuclei. *Brain Research Bulletin, 32,* 623–627.

Bünning, E. (1935). Zur kenntnis der erblichen tagesperiodiztat bei den primarblattern von *Phaseolus multiflorus. Jahrbucher für wissenschaftliche Botanik, 81,* 411–418.

Burrows, H. L., Nakajima, M., Lesh, J. S., Goosens, K. A., Samuelson, L. C., Inui, A., Camper, S. A., & Seashotz, A. F. (1998). Excess corticotropin releasing hormone-binding protein in the hypothalamic–pituitary–adrenal axis in transgenic mice. *Journal of Clinical Investigation, 101,* 1439–1447.

Buttner, D., & Wollnick, F. (1984). Strain-differentiated circadian and ultradian rhythms in locomotor activity of the laboratory rat. *Behavioral Genetics, 14,* 137–152.

Cai, A., Lehman, M. N., Lloyd, J. M., & Wise, P. M. (1997). Transplantation of fetal suprachiasmatic nuclei into middle-aged rats restores diurnal Fos expression in host. *American Journal of Physiology, 272,* R422–R428.

Castellano, C., Puglisi-Allegra, S., Renzi, P., & Oliviero, A. (1985). Genetic differences in daily rhythms of pain sensitivity in mice. *Pharmacology and Biochemistry of Behavior, 23,* 91–92.

Chang, A.-M., Vitaterna, M. H., Turek, F. W., & Takahashi, J. S. (1998). Altered response to constant light conditions in *clock* mutant mice. *Society for Research on Biological Rhythms Abstracts, 10,* 50.

Connolly, M., & Lynch, C. (1981). Circadian variation of strain differences in body temperature and activity in mice. *Physiology and Behavior, 27,* 1045–1049.

Connolly, M., & Lynch, C. (1983). Classical genetic analysis of circadian body temperature rhythms in mice. *Behavioral Genetics, 13,* 491–500.

Diez-Noguera, A. C. T., Ribot, M., & Torralba, A. (1989). Hereditary nature of the pattern of the motor activity circadian rhythm in mice. *Physiology and Behavior, 45,* 307–311.

Ebihara, S., Tsuji, K., & Kondo, K. (1978). Strain differences of the mouse's free-running circadian rhythm in continuous darkness. *Physiology and Behavior, 20,* 795–799.

Ebihara, S., Marks, T., Hudson, D. J., & Menaker, M. (1986). Genetic control of melatonin synthesis in the pineal gland of the mouse. *Science, 231,* 491–493.

Ebihara, S., Hudson, D. J., Marks, T., & Menaker, M. (1987). Pineal indole metabolism in the mouse. *Brain Research, 416,* 136–140.

Ebihara, S., Goto, M., & Oshima, I. (1988a). The phase-shifting effects of pentobarbital on the circadian rhythm of locomotor activity: Strain differences. *Brain Research, 454,* 404–407.

Ebihara, S., Goto, M., & Oshima, I. (1988b). Different responses of the circadian system to GABA-active drugs in two strains of mice. *Journal of Biological Rhythms, 3,* 357–364.

Edgar, D. M., Kilduff, T. S., Martin, C. E., & Dement, W. C. (1991). Influence of running wheel on free-running sleep–wake and drinking circadian rhythms in mice. *Physiology and Behavior, 50,* 373–378.

Faradji-Prevautel, H., Cespuglio, R., & Jouvet, M. (1990). Circadian rest–activity rhythms in the anophthalmic, monocular and binocular ZRDCT/An mice. Retinal and serotonergic influences. *Brain Research, 526,* 207–216.

Ferraro, J. S., Dorsett, J. A., Wagner, T. E., Yun, J. S., & Bartke, A. (1994). Overexpression of growth hormone genes in transgenic mice shortens free-running periods in constant light. *Biological Rhythms Research, 25,* 315–328.

Foster, R. G., Provencio, I., Hudson, D. J., Fiske, S., De Grip, W., & Menaker, M. (1991). Circadian photoreception in the retinally degenerate mouse (*rd/rd*). *Journal of Comparative Physiology A, 149,* 39–50.

Gekakis, N., Staknis, D., Nguyen, H. B., Davis, F. C., Wilsbacher, L. D., King, D. P., Takahashi, J. S., & Weitz, C. J. (1998). Role of the CLOCK protein in the mammalian circadian mechanism. *Science, 280,* 1564–1569.

Gerlai, R. (1996). Gene-targeting studies of mammalian behavior: Is it the mutation or the background genotype? *Trends in Neuroscience, 19,* 177–181.

Golombek, D. A., & Ralph, M. R. (1994). KN-62, an inhibitor of Ca^{2+}/calmodulin kinase II, attenuates circadian responses to light. *Neuroreport, 13,* 1638–1640.

Golombek, D. A., & Ralph, M. R. (1995). Circadian responses to light: The calmodulin connection. *Neuroscience Letters, 192,* 101–104.

Golombek, D. A., Hurd, M. W., Lee, K.-F., & Ralph, M. R. (1996). Mice lacking the $p75^{NGFR}$ receptor exhibit abnormal circadian responses to light. *Biological Rhythms Research, 27,* 409–418.

Goto, M., Oshima, I., Tomita, T., & Ebihara, S. (1989). Melatonin content of the pineal gland in different mouse strains. *Journal of Pineal Research, 7,* 195–204.

Goto, M., Oshima, I., Hasegawa, M., & Ebihara, S. (1994). The locus controlling pineal serotonin N-acetyltransferase activity (Nat-2) is located on mouse chromosome 11. *Molecular Brain Research, 21,* 349–354.

Guastavino, J. M., Bertin, R., & Portet, R. (1991). Effects of the rearing temperature on the temporal feeding pattern of the staggerer mutant mouse. *Physiology and Behavior, 49,* 405–409.

Hafen, T., & Wollnick, F. (1994). Effect of lithium carbonate on activity level and circadian period in different strains of rats. *Pharmacology and Biochemistry of Behavior, 49,* 975–983.

Hegmann, J., & Possidente, B. (1981). Estimating genetic correlations from inbred strains. *Behavioral Genetics, 11,* 103–114.

Hofstetter, J., Mayeda, A., Possidente, B., & Nurnberger, J. J. (1995). Quantitative trait loci (QTL) for circadian rhythms of locomotor activity in mice. *Behavioral Genetics, 52,* 545–556.

Honrado, G. I., Johnson, R. S., Golombek, D. A., Spiegelman, B. M., Papaioannou, V. E., & Ralph, M. R. (1996). The circadian system of c-*fos* deficient mice. *Journal of Comparative Physiology A, 178,* 563–570.

Hotz, M. M., Connolly, M. S., & Lynch, C. B. (1987). Adaptation to daily meal-timing and its effect on circadian temperature rhythms in two inbred strains of mice. *Behavioral Genetics, 17,* 37–51.

Hurd, M. W., & Ralph, M. R. (1998). The significance of circadian organization for longevity in the golden hamster. *Journal of Biological Rhythms, 13,* 430–436.

Hurd, M. W., Zimmer, K. A, Lehman, M. N., & Ralph, M. R. (1995). Circadian locomotor rhythms in aged hamsters following suprachiasmatic transplant. *American Journal of Physiology, 269,* R958–R968.

Ibuka, N. (1987). Circadian rhythms in sleep–wakefulness and wheel-running activity in a congenitally anophthalmic rat mutant. *Physiology and Behavior, 39,* 321–326.

Ikeda, M., & Nomura, M. (1997). cDNA cloning and tissue-specific expression of a novel basic helix–loop–helix/PAS protein (BMAL1) and identification of alternatively spliced variants with alternative translation initiation site usage. *Biochemical and Biophysical Research Communications, 233,* 258–264.

Inoue, I., Yanai, K., Kitamura, D., Taniuchi, Y., Kobayashi, T., Nimura, K., Watanabe, T., & Watanabe, T. (1996). Impaired locomotor activity and exploratory behavior in mice lacking histamine H1 receptors. *Proceedings of the National Academy of Sciences of the USA, 93,* 13316–13320.

Johnson, R. S., Spiegelman, B. M., & Papaioannou, V. E. (1992). Pleiotropic effects of a null mutation in the c-*fos* proto-oncogene. *Cell,* 71, 577–586.

Kempf, E., Mandel, P., Oliviero, A., & Puglisi-Allegra, S. (1982). Circadian variations in noradrenaline, 5-hydroxytryptamine and dopamine in specific brain areas of C57Bl/6 and BALB/c mice. *Brain Research, 232,* 472–478.

Kennaway, D. J. (1994). Effect of phase advance of the light/dark cycle on pineal function and circadian running activity in individual rats. *Brain Research Bulletin, 33,* 639–644.

Kennedy, G. A., Hudson, R., & Armstrong, S. M. (1994). Circadian wheel running activity rhythms in two strains of domestic rabbit. *Physiology and Behavior, 55,* 385–389.

Kilduff, T. S., Vergrinic, C., Lee, S. L., Milbrandt, J. D., Mikkelson, J. D., O'Hara, B. F., & Heller, H. C. (1998). Characterization of the circadian system of NGFI-A and NGFI-A/NGFI-B deficient mice. *Journal of Biological Rhythms, 13,* 347–357.

King, D. P., Vitaterna, M. H., Chang, A. M., Dove, W. F., Pinto, L. H., Turek, F. W., & Takahashi, J. S. (1997a). The mouse *Clock* mutation behaves as an antimorph and maps within the W^{19H} deletion, distal of *Kit. Genetics, 146,* 1049–1060.

King, D. P., Zhao, Y., Sangoram, A. M., Wilsbacher, L. D., Tanaka, M., Antoch, M. P., Steeves, T. D. L., Vitaterna, M. H., Kornhauser, J. M., Lowery, P. L., Turek, F. W., & Takahashi, J. S. (1997b). Positional cloning of the mouse circadian *Clock* gene. *Cell, 89,* 641–653.

Konopka, R. J., & Benzer, S. (1971). Clock mutants of *Drosophila melanogaster. Proceedings of the National Academy of Sciences of the USA, 68,* 2112–2116.

Kopp, C., Vogel, E., Rettori, M. C., Delagrange, P., Guardiola-Lemaitre, B., & Misslin, R. (1998). Effects of daylight cycle reversal on locomotor activity in several inbred strains of mice. *Physiology and Behavior, 63,* 577–585.

Kornhauser, J. M., Mayo, K. E., & Takahashi, J. S. (1992). Immediate-early gene expression in a mammalian circadian pacemaker: The suprachiasmatic nucleus. In M. Young (Ed.), *Molecular genetics of circadian rhythms* (pp. 271–307). New York: Marcel Dekker.

Kozak, W., Zheng, H., Conn, C. A., Soszynski, D., van der Ploeg, D. H., & Kluger, M. J. (1995). Thermal and behavioral effects of lipopolysaccharide and influenza in interleukin-1 beta-deficient mice. *American Journal of Physiology, 269,* R969–R977.

Laemle, L. K., Ottenweller, J. E., & Fugaro, C. (1995). Diurnal variations in vasoactive intestinal polypeptide-like immunoreactivity in the suprachiasmatic nucleus of congenitally anophthalmic mice. *Brain Research, 688,* 203– 208.

Lemmer, B., Witte, K., Makabe, T., Ganten, D., & Mattes, A. (1994). Effects of enalaprilat on circadian profiles in blood pressure and heart rate of spontaneously and transgenic hypertensive rats. *Journal of Cardiovascular Pharmacology, 23,* 311–314.

Liang, F. Q., Allen, G., & Earnest, D. J. (2000). Role of brain-derived neurotrophic factor in the circadian regulation of the suprachiasmatic pacemaker by light. *Journal of Neuroscience, 20,* 2978–2987.

Liu, C., Weaver, D. R., Jin, X., Shearman, L. P., Pieschl, R. L., Gribkof, V. K., & Reppert, S. M. (1997). Molecular dissection of two distinct actions of melatonin on the suprachiasmatic circadian clock. *Neuron, 19,* 91–102.

Lopez-Molina, L., Conquet, F., Dubois-Dauphin, M., & Schibler, U. (1997). The DBP gene is expressed according to the circadian rhythm in the suprachiasmatic nucleus and influences circadian behavior. *EMBO Journal, 16,* 6762–6771.

Lowrey, P. L., & Takahashi, J. S. (1998). Identification of genetic markers linked to the Syrian hamster *tau* mutation by genetically directed representational difference analysis. *Society for Research on Biological Rhythms Abstracts, 10,* 46.

Lowrey, P. L., & Takahashi, J. S. (2000). Genetics of the mammalian circadian system. *Annual Review of Genetics, 34,* 533–562.

Lu, Y. M., Jia, Z., Janus, C., Henderson, J. T., Gerlai, R., Wojtowicz, J. M., & Roder, J. C. (1997). Mice lacking metabotropic glutamate receptor 5 show impaired learning and reduced CA1 long-term potentiation (LTP) but normal CA3 LTP. *Journal of Neuroscience, 17,* 5196–5205.

Lucas, L. A., & Eleftheriou, B. E. (1980). Circadian variation in concentrations of testosterone in the plasma of male mice: A difference between BALB/cBy and C57BL/6By inbred strains. *Journal of Endocrinology, 87,* 37–46.

Mayeda, A., Hofstetter, J., Belknap, J., & Nurnberger, J. J. (1996). Hypothetical quantitative trait loci for circadian period of locomotor activity in CXB recombinant inbred strains of mice. *Behavioral Genetics, 26,* 505–511.

Minakami, K. (1994). Brain monoamines and behavior in hyperammonemic sparse-fur mice. *Nippon Yakurigaku Zasshi, 103,* 219–229.

Morris, M. E., Viswanathan, N., Kuhlman, S., Davis, F. C., & Weitz, C. J. (1998). A screen for genes induced in the suprachiasmatic nucleus by light. *Science, 279,* 1544–1577.

Mrosovsky, N., Salmon, P. A, Menaker, M., & Ralph, M. R. (1992). Nonphotic phase shifting in hamster clock mutants. *Journal of Biological Rhythms, 7,* 41–49.

Murphy, H. M., Wideman, C. H., & Nazdam, G. R. (1996). The interaction of vasopressin and the photic oscillator in circadian rhythms. *Peptides, 17,* 467–475.

Myamoto, M. (1997). Characteristics of age-related behavioral changes in senescence-accelerated mouse SAMP8 and SAMP10. *Experimental Gerontology, 32,* 139–148.

Osiel, S., Golombek, D. A., & Ralph, M. R. (1998). Conservation of locomotor behavior in the hamster: Effects of light cycle and a circadian period mutation. *Physiology and Behavior, 65,* 123–131.

Park, H. T., Kang, E. K., & Bae, K. W. (1997). Light regulates *Homer* mRNA expression in the rat suprachiasmatic nucleus. *Molecular Brain Research, 52,* 318–322.

Peleg, L., Nesbitt, M., & Ashkenazi, I. (1989). Strain dependent response of circadian rhythms during exposure to continuous illumination. *Life Science, 44,* 893–900.

Peleg, L., Zvulunov, A., & Ashkenazi, I. (1995). Genetic control of biological rhythms: Independent expression of each rhythm parameter. *Life Science, 56,* 1143–1149.

Peters, R. V., Zoeller, R. T., Hennessey, A. C., Stopa, E. G., Anderson, G., & Albers, H. E. (1994). The control of circadian rhythms and the levels of vasoactive intestinal peptide mRNA in the suprachiasmatic nucleus are altered in spontaneously hypertensive rats. *Brain Research, 639,* 217–227.

Pickard, G. E., Sollars, P. J., Rinchik, E. M., Nolan, P. M., & Bucan, M. (1995). Mutagenesis and behavioral screening for altered circadian activity identifies the mouse mutant, *wheels. Brain Research, 705,* 255–266.

Plazzi, G., Schutz, Y., Cortelli, P., Provini, F., Heikkila, E., Solieri, L., Lugaresi, E., & Montagna, P. (1997). Motor overactivity and loss of motor circadian rhythm in fatal familial insomnia: An actigraphic study. *Sleep, 20,* 739–742.

Portaluppi, F., Cortelli, P., Avoni, P., Vergnani, L., Maltoni, P., Pavani, A., Sforza, E., Degli Uberti, E. C., Gambetti, P., & Lugaresi, E. (1994). Progressive disruption of the circadian rhythm of melatonin in fatal familial insomnia. *Journal of Clinical Endocrinology and Metabolism, 78,* 1075–1078.

Possidente, B., & Hegmann, J. P. (1980). Circadian complexes: Circadian rhythms under common gene control. *Journal of Comparative Physiology B, 139,* 121–125.

Possidente, B., & Hegmann, J. (1982). Gene differences modify Aschoff's rule in mice. *Physiology and Behavior, 28,* 199–200.

Possidente, B., & Stephan, F. K. (1988). Circadian period in mice: Analysis of genetic and maternal contributions to inbred strain differences. *Behavioral Genetics, 18,* 109–117.

Possidente, B., Hegmann, J. P., Carlson, L., & Elder, B. (1982). Pigment mutations associated with altered circadian rhythms in mice. *Physiology and Behavior, 28,* 389–392.

Ralph, M. R., & Hurd, M. W. (1996). Pacemaker interactions in the mammalian circadian system. *Brazilian Journal of Medical and Biological Research, 29,* 77–85.

Ralph, M. R., & Menaker, M. (1988). A mutation of the circadian system in golden hamsters. *Science, 241,* 1225–1227.

Ralph, M. R., Foster, R. G., Davis, F. C., & Menaker, M. (1990). Transplanted suprachiasmatic nucleus determines circadian period. *Science, 247,* 975–978.

Ralph, M. R., Hurd, M. W., Takeuchi, J., Melo, L., Mathur, A., & Golombek, D. A. (1996). Regulation and integration in the mammalian circadian system. In R. M. Buijs, A. Kalsbeek, H. J. Romijn, C. M. A. Pennartz, & M. Mirmiran (Eds.), *Progress in brain research* (pp. 189–201). Amsterdam: Elsevier.

Rao, S. H., & Diwan, P. V. (1995). A behavioral profile of bilateral anophthalmic mutant rats. *Experimental Animals, 44,* 67–69.

Reebs, S. G., & Mrosovsky, N. (1989). Effects of wheel running on the circadian activity rhythms of Syrian hamsters: Entrainment and phase response curve. *Journal of Biological Rhythms, 4,* 39–48.

Reinberg, A., Touitou, Y., Restoin, A., Migraine, C., Levi, F., & Montagner, H. (1985). The genetic background of circadian and ultradian patterns of 17-hydroxycorticosteroids: A cross-twin study. *Journal of Endocrinology, 105,* 247–253.

Roesler, W. J., Helgason, C., Gulka, M., & Khandelwal, R. L. (1985). Aberrations in the diurnal rhythms of plasma glucose, plasma insulin, liver glycogen, and hepatic glycogen synthase and phosphorylase activities in genetically diabetic (*db/db*) mice. *Hormones and Metabolism Research, 17,* 572–575.

Rosenwasser, A. M., & Plante, L. (1993). Circadian activity rhythms in SHR and WKY rats: Strain differences and effects of clonidine. *Physiology and Behavior, 53,* 23–29.

Roussel, B., Turrillot, P., & Kitahama, K. (1984). Effect of ambient temperature on the sleep-waking cycle in two strains of mice. *Brain Research, 294,* 67–73.

Sanchez-Barcelo, E. J., Megias, M., Verduga, R., & Crespo, D. (1997). Differences between the circadian system of two strains of senescence-accelerated mice (SAM). *Physiology and Behavior, 62,* 1225–1229.

Sanyal, S., & Hawkins, R. K. (1989). Development and degeneration of retina in *rds* mutant mice: Altered disc shedding pattern in the heterozygotes and its relation to ocular pigmentation. *Current Eye Research, 8,* 1093–1101.

Schwartz, W. J., & Zimmerman, P. (1990). Circadian timekeeping in BALB/c and C57BL/6 inbred mouse strains. *Journal of Neuroscience, 10,* 3685–3694.

Semikhoskii, A. G., Ball, A. J., Stirland, J. A., Mohammad, Y. N., Bruford, M., Powell, J. F., & Loudon, A. S. I. (1998). Mapping the *tau* mutation of the Syrian hamster using amplified fragments length polymorphism (AFLP). *Society for Research on Biological Rhythms Abstracts, 10,* 51.

Shearman, L. P., Zylka, M. J., Weaver, D. R., Kolakowski, L. F., Jr., & Reppert, S. M. (1977). Two period homologs: Circadian expression and photic regulation in the suprachiasmatic nucleus. *Neuron, 19,* 1261–1269.

Shen, H., Watanabe, M., Tomasiewicz, H., Rutishauser, U., Magnuson, T., & Glass, J. D. (1997). Role of neural cell adhesion molecule and polysialic acid in mouse circadian clock function. *Journal of Neuroscience, 17,* 5221–5229.

Shibata, S., & Moore, R. Y. (1994). Calmodulin inhibitors produce phase shifts of circadian rhythms *in vivo* and *in vitro. Journal of Biological Rhythms, 9,* 27–41.

Shigeyoshi, Y., Taguchi, K., Yamamoto, S., Yan, L., Tei, H., Moriya, T., Shibata, S., Loros, J. J., Dunlap, J. C., & Okamura, H. (1997). Light-induced resetting of a mammalian circadian clock is associated with rapid induction of the mPer1 transcript. *Cell, 91,* 1043–1053.

Shim, S., Sugita, S., Sugahara, K., & Tanaka, H. (1997). Feeding rhythm and ornithine decarboxylase activity in hereditary microphthalmic rats. *Physiology and Behavior, 62,* 1365–1369.

Shimomura, K., & Menaker, M. (1994). Light-induced phase shifts in *tau* mutant hamsters. *Journal of Biological Rhythms, 9,* 97–110.

Shimomura, K., Vitaterna, M. H., Low-Zeddies, S. S., Whitely, A. R., & Takahashi, J. S. (1998). Genetic dissection of strain differences in mouse circadian rhythmicity. *Society for Research on Biological Rhythms Abstracts, 10,* 47.

Shiromani, P. J., & Overstreet, D. (1994). Free-running period of circadian rhythms is shorter in rats with a genetically upregulated cholinergic system. *Biological Psychiatry, 36,* 622–626.

Siebert, U., & Wollnick, F. (1991). Wheel-running activity rhythms in two inbred strains of laboratory rats under different photoperiods. *Physiology and Behavior, 50,* 1137–1143.

Silver, R., LeSauter, J., Tresco, P. A., & Lehman, M. N. (1996). A diffusible coupling signal from the transplanted suprachiasmatic nucleus controlling circadian locomotor rhythms. *Nature, 382,* 810–813.

Sollars, P. J., Ryan, A., Ogilvie, M. D., & Pickard, G. E. (1996). Altered circadian rhythmicity in the Wocko mouse, a hyperactive transgenic mutant. *Neuroreport, 7,* 1245–1248.

Sun, Z. S., Albrecht, U., Zhuchenko, O., Bailey, J., Eichele, G., & Lee, C. C. (1997). RIGUI, a putative mammalian ortholog of the *Drosophila period* gene. *Cell, 90,* 1003–1011.

Takahashi, J. S., Pinto, L. P., & Vitaterna, M. H. (1994). Forward and reverse genetic approaches to behavior in the mouse. *Science, 26,* 1724–1733.

Takumi, T., Matsubara, C., Shigeyoshi, Y., Taguchi, K., Yagita, K., Maebayashi, Y., Sakakida, Y., Okumura, K., Takashim, N., & Okamura, H. (1998). A new mammalian *period* gene predominantly expressed in the suprachiasmatic nucleus. *Genes Cells, 3,* 167–176.

Tei, H., Okamura, H., Shigeyoshi, Y., Fukuhara, C., Ozawa, R., Hirose, M., & Sakaki, Y. (1997). Circadian oscillation of a mammalian homologue of the *Drosophila period* gene. *Nature, 389,* 512–516.

Ticher, A., Sackett-Lundeen, L., Ashkenazi, I. E., & Haus, E. (1994). Human circadian time structure in subjects of different gender and age. *Chronobiology International, 11,* 349–355.

Ticher, A., Zoossman-Diskin, A., Haus, E., Sackett-Lundeen, L., & Ashkenazi, I. E. (1995). Prevalence of genetic versus environmental factors in human female temporal organization. *Human Biology, 67,* 769–778.

Tobler, I., Gaus, S. F., Deboer, T., Achermann, P., Fischer, M., Rulicke, T., Moser, M., Oesch, B., McBride, P. A., & Manson, J. C. (1996). Altered circadian activity rhythms and sleep in mice devoid of prion protein. *Nature, 380,* 639–642.

Turek, F. W., Pinto, L., Vitaterna, M., Penev, P., Zee, P. C., & Takahashi, J. S. (1995). Pharmacological and genetic approaches for the study of circadian rhythms in mammals. *Frontiers in Neuroendocrinology, 16,* 191–223.

Valaty, J. L., & Jouvet, M. (1988). Circadian rhythms of slow-wave sleep are in opposite phase in genetically hypoprolactinemic rats. *Comptes Rendues de l'Academie de Science III, 307,* 789–794.

Van Reeth, O., & Turek, F. W. (1989). Stimulated activity mediates phase shifts in the hamster circadian clock induced by light pulses or benzodiazepines. *Nature, 339,* 49–51.

Vitaterna, M. H., & Turek, F. W. (1993). Photoperiodic responses differ among inbred strains of golden hamsters (*Mesocricetus auratus*). *Biology of Reproduction, 49,* 496–501.

Vitaterna, M. H., Wu, J. C., Turek, F. W., & Pinto, L. H. (1993). Reduced light sensitivity of the circadian clock, in a hypopigmented mouse mutant. *Experimental Brain Research, 95,* 436–442.

Vitaterna, M. H., King, D. P., Chang, A.-M., Kornhauser, J. M., Lowrey, P. L., McDonald, J. D., Dove, W. F., Pinto, L. H., Turek, F. W., & Takahashi, J. S. (1994). Mutagenesis and mapping of a mouse gene, *clock,* essential for circadian behavior. *Science, 264,* 719–725.

Vitaterna, M. H., Chang, A.-M., King, D. P., Pinto, L. H., Turek, F. W., & Takahashi, J. S. (1996). Heterozygosity at the *clock* locus alters phase response curves to light. *Society for Research on Biological Rhythms, Abstracts, 200,* 127.

Vogelbaum, M., & Menaker, M. (1992). Temporal chimeras produced by hypothalamic transplants. *Journal of Neuroscience, 12,* 3619–3627.

Wax, T. M. (1977). Effects of age, strain, and illumination intensity on activity and self-selection of light–dark schedules in mice. *Journal of Comparative Physiology and Psychology, 91,* 51–62.

Weaver, D. R., Liu, C., & Reppert, S. M. (1996). Nature's knockout: The Mel 1b receptor is not necessary for reproductive and circadian responses to melatonin in Siberian hamsters. *Molecular Endocrinology, 10,* 1478–1487.

Wisor, J. P., & Takahashi, J. S. (1997). Regulation of the *vgf* gene in the golden hamster suprachiasmatic nucleus by light and by the circadian clock. *Journal of Comparative Neurology, 378,* 229–238.

Wollnick, F. (1991). Strain differences in the pattern and intensity of wheel running activity in laboratory rats. *Experientia, 47,* 593–598.

Wollnick, F., Gärtner, K., & Büttner, D. (1987). Genetic analysis of circadian rhythms and ultradian locomotor activity rhythms in laboratory rats. *Behavioral Genetics, 17,* 167–178.

Wollnick, F., Brysch, W., Uhlmann, E., Gillardon, F., Bravo, R., Zimmerman, M., Schlingensiepen, K. H., & Herdegen, T. (1995). Block of c-*fos* and *jun-b* expression by antisense oligonucleotides inhibits light-induced phase shifts of the mammalian circadian clock. *European Journal of Neuroscience, 7,* 388–393.

Yoshimura, T., Nishio, M., Goto, M., & Ebihara, S. (1994). Differences in circadian photosensitivity between retinally degenerate CBA/J mice and normal CBA/N mice (+/+). *Journal of Biological Rhythms, 9,* 51–60.

Zhao, Y., Vitaterna, M. H., Shimomura, K., Giese, K. P., Mayford, M., Silva, A. J., Kandel, E. R., Tonegawa, S., & Takahashi, J. S. (1998). Role of calcium/calmodulin dependent protein kinase II in the circadian system. *Society for Research on Biological Rhythms Abstracts, 10,* 48.

Zylka, M. J., Shearman, L. P., Weaver, D. R., & Reppert, S. M. (1998). Three *period* homologs in mammals: Differential light responses in the suprachiasmatic circadian clock and oscillating transcripts outside the brain. *Neuron, 20,* 1103–1110.

PART V

Interaction of Circadian Clock with Other Rhythmic Functions

17

Processes Underlying the Regulation of the Sleep–Wake Cycle

ALEXANDER A. BORBÉLY, DERK-JAN DIJK, PETER ACHERMANN, AND IRENE TOBLER

Sleep and waking are the two major functional states which in homeotherms can be unambiguously identified by both behavioral and electrophysiologic criteria. Sleep and waking are closely associated with rest and activity. The circadian rest–activity rhythm is present in organisms in which the usual criteria for sleep and waking cannot be applied (e.g., in invertebrates). In fact, sleep and waking may have evolved from the states of rest and activity whose propensity is modulated by the circadian pacemaker. It has been argued that the emergence of sleep has brought an increased flexibility by loosening the control of the circadian pacemaker over behavior and enabling a need-dependent recovery (Borbély and Neuhaus, 1979). It is in particular the intensity dimension of sleep which can be viewed as allowing sleep to fulfil its putative need-dependent functions without disrupting the circadian sleep–wake rhythm. As will be shown, there is evidence that the circadian and sleep–wake-dependent aspects of sleep regulation are based on separate mechanisms.

Three distinct processes can be identified which underlie the regulation of the sleep-wake cycle in humans and other mammals (Figure 1): (1) A sleep-wake dependent, homeostatic process determined by prior sleep and waking, (2) a clock-like circadian process that is largely independent of sleep and waking, and (3) an

ALEXANDER A. BORBÉLY, PETER ACHERMANN, AND IRENE TOBLER Institute of Pharmacology, University of Zürich, CH-8057 Zürich, Switzerland. DERK-JAN DIJK Institute of Pharmacology, University of Zürich, CH-8057 Zürich, Switzerland and Center for Chronobiology, University of Surrey, Guilford GU27XH United Kingdom.

Circadian Clocks, Volume 12 of *Handbook of Behavioral Neurobiology*, edited by Joseph S. Takahashi, Fred W. Turek, and R. Y. Moore, Kluwer Academic / Plenum Publishers, New York, 2001.

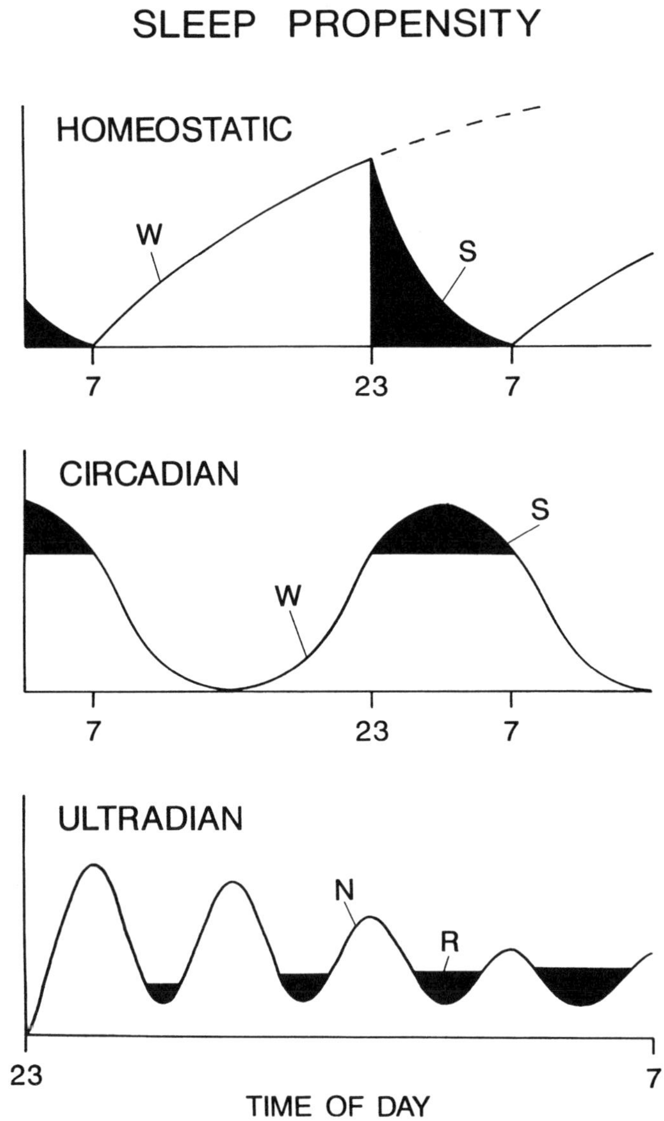

Figure 1. Schematic representation of the three major processes underlying sleep regulation. W, Waking; S, sleep; N, nonREM sleep; R, REM sleep. The progressive decline in nonREM sleep intensity is represented both in the top and bottom panels (decline of ultradian amplitude). Note the progressive increase in the duration of REM sleep episodes.

ultradian process that is characterized by the alternation of the two basic sleep states, non-rapid-eye-movement (nonREM) sleep and REM sleep.

Sleep Homeostasis

Definition, Conceptual Basis and Models

Homeostasis has been defined as "the coordinated physiological processes which maintain most of the steady states in the organism" (Cannon, 1939). The term "sleep homeostasis" (Borbély, 1980) refers to the sleep–wake-dependent aspect of sleep regulation. Thus, homeostatic mechanisms counteract deviations of sleep from an average "reference level" by augmenting sleep propensity when sleep is curtailed or absent, and reducing sleep propensity in response to excess sleep. The two-process model, originally proposed to account for sleep regulation in the rat (Borbély, 1980; 1982a), postulates that the homeostatic Process S rises during waking and declines during sleep. Its time course in humans was derived from electroencephalographic (EEG) slow-wave activity (i.e., spectral power density in the delta band) recorded during sleep. S interacts with a second, circadian process (Process C). Various aspects of human sleep regulation were addressed in the first qualitative version of the two-process model (Borbély, 1982b). A subsequent, quantitative version (Daan & Beersma, 1984; Daan, Beersma, & Borbély, 1984) predicted the timing of human sleep under various experimental conditions. In the following years, Achermann and colleagues (Achermann, 1988; Achermann & Borbély, 1990, 1992; Achermann, Beerma, & Borbély, 1990, Achermann, Dijk, Brunner, & Borbély, 1993) extended the homeostatic part of the model by incorporating the ultradian variations of slow-wave activity. A combination of an ultradian and a homeostatic process was proposed by Massaquoi and McCarley (1992). Finally, Folkard and Åkerstedt (1987, 1989, 1992; Åkerstedt & Folkard, 1990, 1995) and Achermann and Borbély (1994) simulated the changes of daytime alertness on the basis of the combined action of a homeostatic process, a circadian process, and a sleep inertia process.

In the initial qualitative version of the two-process model (Borbély, 1982b), the homeostatic facet of REM sleep regulation was recognized, but not incorporated. Based on the reciprocal interaction of nonREM sleep and REM sleep (McCarley & Hobson, 1975), REM sleep was assumed to be under the inhibitory influence of slow-wave propensity. New efforts were made to include REM sleep regulation in composite models (Achermann & Borbély,1992; Massaquoi & McCarley, 1992).

Although the two-process model had originated from animal data (Borbély, 1982a), it was elaborated for human sleep. A quantitative simulation of nonREM sleep homeostasis in the rat was performed for baseline, sleep deprivation and recovery (Franken, Tobler, & Borbély, 1991b; Tobler, Franken, Trachsel, & Borbély, 1992). In particular, the typical changes of slow-wave activity such as its biphasic time course during baseline, its initial increase after sleep deprivation, and the subsequent prolonged negative rebound could be reproduced.

EEG Slow-Wave Activity as a Marker of NonREM Sleep Homeostasis

The predominance of slow-wave sleep in the early part of the sleep episode was recognized already in early studies (Blake & Gerard, 1937) and was repeatedly confirmed (Dement & Kleitman, 1957; Williams, Hammack, Daly, Dement, & Lubin, 1964; Webb & Agnew, 1971). A quantitative measure of the EEG can be obtained by spectral analysis (e.g., Borbély, Baumann, Brandeis, Strauch, & Lehmann, 1981), period amplitude analysis (e.g. Church, March, Hibi, Cavness, & Feinberg, 1975), or similar methods of signal analysis. Although each method has its inherent limitations (Geering, Achermann, Eggimann, & Borbély, 1993), they all allow the quantitative determination of EEG components in the low-frequency range referred to as "slow-wave activity." This measure, which is derived directly from the EEG signal, yields a more faithful representation of the variations occurring across the nonREM–REM sleep cycle and nonREM sleep episodes than the sleep profile based on the conventional sleep stage scoring. Moreover, some EEG variables, such as slow waves and sleep spindles, can be directly related to their generating processes at the cellular level.

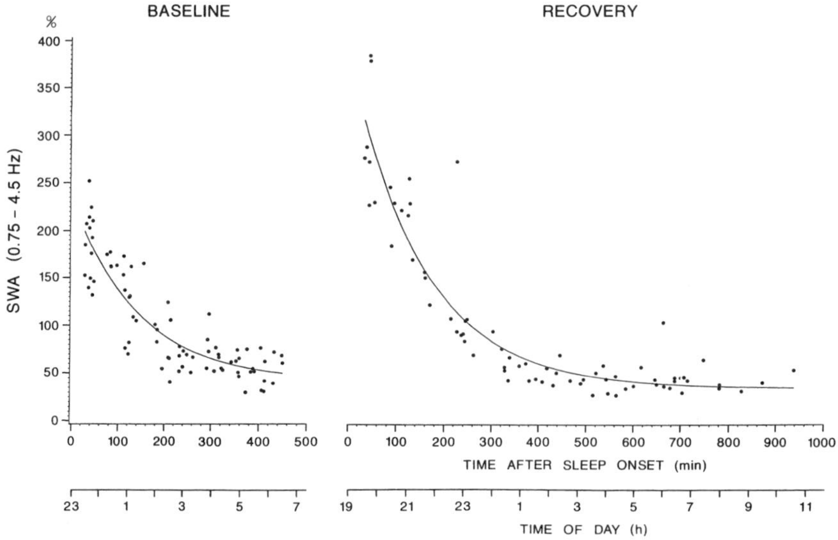

Figure 2. Time course of slow-wave activity (SWA) in nonREM sleep for two baseline nights and a recovery night after 36 hours of waking. Data points represent mean SWA per nonREM sleep episode after standardizing the values with respect to mean SWA in baseline night 2 (= 100%). The curves represent fits of exponential functions to the data. (From Dijk *et al.*, 1990a, with permission.)

Mean EEG slow-wave activity per nonREM sleep episode exhibits typically a monotonic decline during sleep, a time course that can be approximated by an exponential function (Figure 2, left) (Borbély *et al.*, 1981; Dijk, Brunner, & Borbély, 1990a). This progressive decrease of slow-wave activity reflects the gradual reduction of the "depth" or intensity of nonREM sleep as defined by arousal threshold in the course of the sleep episode. The association between slow waves and sleep intensity had been already recognized by Blake and Gerard (1937) and was confirmed in subsequent studies (e.g., Williams *et al.*, 1964). This relationship applies not only to humans, but also to other animals. Thus, in the rat, EEG slow-wave activity was closely correlated with sleep continuity as indexed by the rate of spontaneous brief arousals (Franken, Dijk, Tobler, & Borbély, 1991a; Tobler & Franken, 1993; Tobler, Franken, Gao, Jaggi, & Borbély, 1994), as well as with the arousal threshold to stimuli (Neckelmann & Ursin, 1993).

Manipulation of Sleep Propensity: Effects on Slow-Wave Activity

Sleep propensity can be enhanced by extending the waking episode. This is a powerful tool for exploring the processes underlying sleep regulation. Slow-wave sleep and EEG slow-wave activity in nonREM sleep are typically enhanced when sleep occurs after extended waking. In their early study, Webb and Agnew (1971) obtained compelling evidence that slow-wave sleep increases as a function of prior waking. All-night spectral analysis of the sleep EEG revealed that after a night without sleep (i.e., 40.5 hours of wakefulness), slow-wave activity was enhanced (Figure 2, right), and that the extent of the increase is a function of the duration of prior waking (Borbély *et al.*, 1981; Dijk,Brunner, Bersma, & Borbély, 1990b; for overviews, see Borbély & Achermann, 1992, Borbély, 1994; Dijk, 1995).

Effects of sleep deprivation on slow-wave activity have been investigated also in nonhuman mammals. Thus, an increase of slow-wave activity after sleep deprivation was demonstrated in the rabbit (Opp & Krueger, 1994; Pappenheimer, Koski, Fencl; Karnovsky, & Krueger, 1975; Tobler, Franken, & Scherschlicht, 1990a) and extensively documented in the rat (Borbély & Neuhaus, 1979; Friedman, Bergmann, & Rechtschaffen, 1979). By varying the duration of sleep deprivation it was shown that the increase in slow-wave activity is a function of the duration of prior waking (Tobler & Borbély, 1986, 1990). This relationship was reported not only for the rat but also for various other species, including primates, cat, chipmunk, Syrian hamster, Djungarian hamster, guinea pig, and dolphin (for references, see Deboer, Franken, & Tobler, 1994; Dijk and Daan, 1989; Lancel, van Riezen, & Glatt, 1991; Oleksenko, Mukhametov, Polyakova, Supin, & Kovalzon, 1992; Tobler, 1985; Tobler and Jaggi, 1987; Tobler and Scherschlicht, 1990).

Sleep propensity can be not only experimentally enhanced, but also reduced by having subjects sleep prior to their habitual bedtime. The time course of the gradual buildup of "sleep pressure" during the waking hours can be investigated by analyzing the EEG during naps taken after different intervals of waking. Early studies showing that naps situated early in the day contain less slow-wave sleep (SWS) than late naps (Maron, Rechtscaffen, & Wolpert, 1964) were recently confirmed (Knowles, Maclean, Brunet, & Coulter, 1990). Direct evidence for a monotonic rise of slow-wave activity (SWA) in the course of waking was obtained in a study in which naps were scheduled at 2-hour intervals throughout the day (Beersma, Daan, & Dijk, 1987; Dijk, Beersma, & Daan, 1987). The rise of slow-wave activity could be approximated by a saturating exponential function.

A reduction of SWS and SWA in nocturnal sleep has been reported after naps in the late afternoon or early evening (Feinberg *et al.*, 1985; Williams, Finley, & Hursch, Karacan, 1970), but not after a nap in the morning (Karacan *et al.*, 1970; Knowles, MacLean, Salem, Vetere, & Coutler, 1986; Knowles *et al.*, 1990). The smaller effect of the morning nap was predicted by the two-process model (Daan, Beersma, Dijk, Åkerstedt, & Gillberg, 1988). According to this model, the sum of the cumulated SWA (i.e., slow-wave energy) during the nap and the postnap sleep episode is equal to the cumulated value of the baseline sleep episode. This prediction was confirmed in two studies (Feinberg, Maloney, & March, 1992; Werth, Dijk, Achermann, & Borbély, 1996).

Do Specific Aspects of Waking Promote Slow-Wave Activity?

The close relationship between waking and sleep propensity raises the question of which aspect of the waking state is responsible for the effect on sleep. In humans, experimental sleep deprivation usually entails not only a lack of sleep, but also the maintenance of the upright body posture during the habitual sleep period, an increased level of sensory stimulation, motor and cognitive activity, social interaction as well as eating and drinking. In animal experiments, the sleep deprivation procedure may involve forced locomotion and stress. To what extent are such factors affecting sleep? There is evidence that it is waking per se and not its concomitant elements which cause the typical changes of sleep. One important argument is that sleep propensity (as indexed by slow-wave activity during naps) gradually builds up during the habitual waking episode in the absence of sleep deprivation (Beersma *et al.*, 1987; Dijk, 1995; Dijk *et al.*, 1987). Therefore the extension of the waking episode during sleep deprivation does not differ qualitatively from usual waking in its effects on sleep. Conversely, if sleep deprivation is performed under conditions which resemble those of sleep (i.e., physical inactivity, semirecumbent posture, constant dim light), EEG slow-wave activity is increased to a similar level as after sleep deprivation with normal waking activities (Dijk & Czeisler, 1993).

The effect of locomotion and stress on sleep was investigated in animal studies. Varying the rotation rate of the sleep deprivation apparatus (Borbély & Neuhaus, 1979; Friedman *et al.*, 1979) or allowing the rat to engage in voluntary running activity (Hanagasioglu & Borbély, 1982) had no major effect on the slow-wave fraction of nonREM sleep during recovery. Moreover, in both rat (Franken *et al.*, 1991a) and hamster (Tobler & Jaggi, 1987), sleep deprivation by forced locomotion and by gentle handling had similar effects. However, even the 24-hour forced locomotion procedure did not seem to constitute a major stress for the rat, since the level of plasma corticosterone was not significantly increased after 21.5 hours (Tobler, Murison, Ursin, Ursin, & Borbély, 1983a). The results of these studies do not indicate that motor activity or stress is a major factor in promoting slow-wave activity in nonREM sleep. Nevertheless, an influence of motor activity on sleep cannot be excluded because slow-wave incidence in the rat was reported to be affected by the degree of forced locomotion (Mistlberger, Bergmann, & Rechtschaffen, 1987).

The evidence that the increased brain metabolism during waking is a critical variable for the effect on sleep is inconclusive. In humans, slow wave-sleep was enhanced after raising body temperature in the late afternoon, in the evening, or after two cycles of nocturnal sleep (Bunnell and Horvath, 1985; Horne and Shackell, 1987; Horne and Staff, 1983). Rats kept awake either in a warm environment or at room temperature showed no significant difference in slow-wave activity (Tobler *et al.*, 1994). However, in another study, slow-wave activity was reported to be enhanced after exposing the animals to a considerable heat load (Morairty, Szymusiak, Thomson, & McGinty, 1993). In young rats, a moderate heat load induced a marked increase in slow-wave activity (Obál, Alföldi, & Rubicsek, 1995). In the cat, preoptic/anterior hypothalamic warming enhanced slow-wave activity (McGinty, Szymusiak, & Thomson, 1994).

A provocative hypothesis postulates that sleep is a local, use-dependent phenomenon (Krueger & Obál, 1993). The hypothesis was tested by investigating whether a local activation of a brain region during wakefulness affects the EEG recorded from this site during sleep (Kattler, Dijk, & Borbély, 1994). An intermittent vibratory stimulus was applied to the left or right hand during the 6-hour period prior to sleep to activate the contralateral somatosensory cortex. Stimulation of the dominant hand resulted in a shift of power in the nonREM sleep EEG toward the left hemisphere. This effect was limited to the first hour of sleep and to the central EEG derivation overlying the somatosensory cortex. It was most prominent in the delta band. Although the effect was small and no significant changes were observed following left-hand stimulation, it is consistent with the hypothesis that the activation of specific neuronal populations during wakefulness may have repercussions on their electrical activity pattern during subsequent sleep.

Developmental Aspects

Animal studies indicate that the responsiveness to sleep deprivation is higher during the early stages of development than during adulthood. Thus, in the 23-day old rat, an enforced waking episode as short as 2 hours induced a massive increase in slow-wave activity (Figures 3b, 3d), whereas at the age of 30 day, a much smaller

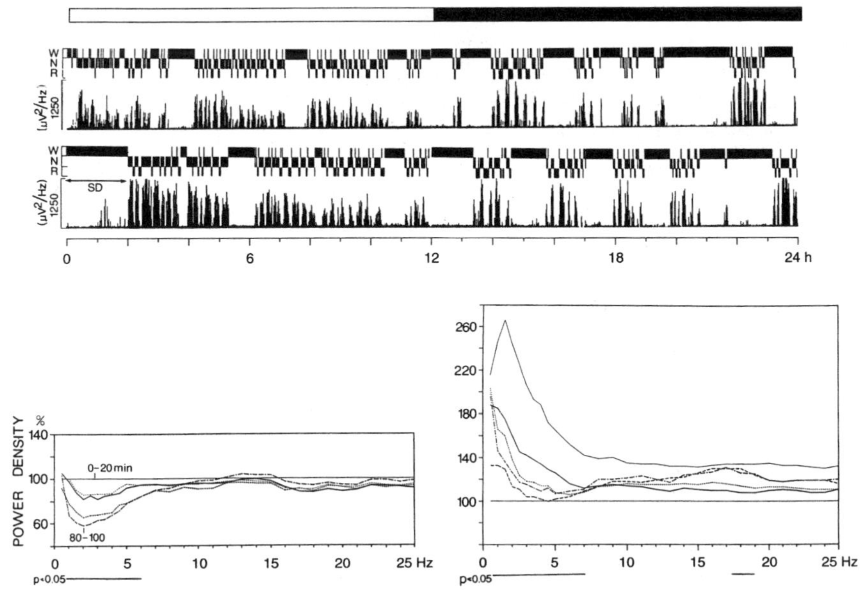

Figure 3. Vigilance states and EEG slow-wave activity (SWA) in the young rat (23–24 days) under baseline conditions and after 2-hour sleep deprivation. *Top*: Continuous 24-hours records of vigilance states (W, waking; N, nonREM sleep; R, REM sleep) and SWA in a rat kept under baseline conditions (upper plots), and continuation of the record during and after 2-hour sleep deprivation (SD) and recovery (lower plots). Note the increase of SWA after the short enforced waking period. Light from 0 to 12 hours as indicated by top bar. *Lower left*: Evolution of EEG power density in nonREM sleep within ultradian nonREM sleep episodes. Each curve connects mean power density values computed for 0.5-Hz or 1-Hz bins. Plots are derived from 13 episodes (seven animals) and represent values for 20-minute intervals. The power density of each bin is expressed as the percentage of the corresponding 0- to 20-minute value (= 100%). The bar below the panel indicates the frequency range with significant changes between the different 20-min values ($p < .05$; one-way ANOVA). Note the decline of SWA during the "minidays." *Lower right*: Evolution of EEG power density in nonREM sleep during recovery from 2-hour sleep deprivation. For each 2-hour interval of recovery, the values are expressed as a percentage of the corresponding baseline interval starting at lights on. Statistics as in lower left panel. Note the massive increase of SWA after only 2 hours of enforced waking. (From Alföldi *et al.*, 1990, with permission).

effect was seen following a 6-hour deprivation (Alföldi, Tobler, & Borbély, 1990). The rapidly increasing "slow-wave pressure" in the young rat became apparent in the second hour of the enforced waking period by the intrusion of slow waves into the waking EEG. This phenomenon can be also seen in the adult rat (Borbély, Tobler, & Hanagasioglu, 1984; Friedman *et al.*, 1979; Mistlberger *et al.*, 1987). However, in the adult animal, a significant increase of slow-wave activity was seen only after 9-12 hours of enforced waking in one study (Franken, Tobler, & Borbély, 1993) and after 12-18 hours of waking in another study (Franken *et al.*, 1991a).

In the 23-day-old rat, ultradian episodes of sleep and waking are predominant (Figure 3, top). During consolidated ultradian sleep episodes in the light period, slow-wave activity declined by 40% within 100 minutes (Figure 3, bottom left). Thus, at this early age, "sleep pressure" not only builds up rapidly during waking, but also dissipates rapidly during sleep. In contrast to later stages of development, there is no global trend of slow-wave activity during the light period. The rapid buildup and decay of slow-wave activity during ultradian sleep/wake cycles is in balance, and, in this respect, the ultradian periods can be regarded as miniature versions of the circadian sleep/wake period ("minidays"). As maturation proceeds, the time constants of the homeostatic processes seem to increase. Consequently, the accumulation of a "sleep debt" during the dark period and its dissipation during the light period are reflected in the rising and declining trends of slow-wave activity (Dijk, & Borbély, Tobler, 1990b). Accordingly, the guinea pig, a species with little or no 24-hour rhythmicity, exhibits no trend of slow-wave activity (Tobler, Franken, & Jaggi, 1993).

Neurophysiologic Basis of the EEG in NonREM Sleep

Recently, spectacular advances were made in elucidating the neurophysiologic mechanisms in the thalamocortical system in relation to sleep and the sleep EEG (for reviews, see McCormick, & Sejnowski, 1993a; Steriade, Contreras, & Amzica, 1994). Slow waves and sleep spindles, the two salient types of oscillation in the sleep EEG, have been shown to be closely related to cellular changes at the level of thalamic and cortical neurons. Although the generating mechanisms of sleep spindles with the reticular thalamic nucleus playing a central role have been known for some time (Steriade, Jones, Llinás, 1990), it has been recognized only recently that the progressive hyperpolarization of thalamocortical neurons engenders fluctuations in the membrane potential in the frequency range of the sleep EEG. Thus, at moderate stages of hyperpolarization, oscillations are observed in the frequency range of sleep spindles, whereas the membrane potential of strongly hyperpolarized neurons fluctuates in the delta range (Steriade *et al.*, 1993a; Steriade, Nuñez, & Amzica, 1993b; Steriade *et al.*, 1994). There is evidence that the progression from waking to deep nonREM sleep is associated with an increased level of hyperpolarization (Hirsch, Fourment, & Marc, 1983). The progression of sleep stages from "spindle sleep" (stage 2) to slow-wave sleep (stages 3 and 4), and the partly inverse relationship between EEG power density in the spindle and slow-wave ranges (Aeschbach & Borbély, 1993; Dijk, Hayes, & Czeisler, 1993), have therefore a counterpart at the level of thalamocortical neurons. The recognition of these relationships opens the possibility to investigate sleep homeostasis at the cellular level.

REM Sleep Homeostasis

Not only nonREM sleep, but also REM sleep is regulated by homeostatic mechanisms. However, there are differences in the regulatory processes of the two sleep states. Early sleep deprivation experiments in humans revealed a higher priority for slow-wave sleep (Dement & Greenberg, 1966; Jones & Oswald, 1968; Webb & Friel, 1970). However, although the rebound of REM sleep after a prolonged sleep deprivation was delayed, it persisted longer than the slow-wave sleep rebound (Berger & Oswald, 1962; Williams *et al.*, 1964). A selective REM sleep deprivation augments REM sleep propensity which is manifested by the increasing number of interventions required to prevent REM sleep episodes (Dement, 1960; Kales, Hoedemaker, Jacobson, & Lichtenstein, 1964). However, the REM sleep rebound during recovery sleep does not occur consistently (Cartwright, Monroe, & Palmer, 1967). These results have led to the conclusion that REM sleep is not as finely regulated as slow-wave sleep. However, recent findings do not support this interpretation: A selective REM sleep deprivation during the first 5 hours of sleep induced an REM sleep rebound in the remaining 2.25 hour interval of the sleep episode (Beersma, Dijk,

Blok, & Everhardus, 1990). In another study, the induction of a substantial REM sleep deficit by curtailment of sleep duration during either two or four consecutive nights resulted in a REM sleep rebound in the recovery nights (Brunner, Dijk, Tobler, & Borbély, 1990; Brunner, Dijk, & Borbély, 1993). In these different experimental situations, the REM sleep rebound occurred either at a time when slow-wave propensity was low at the end of the sleep episode (Beersma *et al.*, 1990) or, in the partial sleep deprivation paradigm, when it was much less increased than REM sleep propensity (Brunner *et al.*, 1990, 1993). Finally, a sleep-dependent disinhibition of REM sleep in the course of the sleep episode has been demonstrated in a study using the forced desynchrony protocol (Dijk & Czeisler, 1995). Taken together, the results show that not only slow-wave sleep and slow-wave activity in nonREM sleep, but also REM sleep is homeostatically regulated. However, if slow-wave propensity is high, the manifestation of REM sleep homeostasis is suppressed.

Similar to humans, moderate total sleep deprivation in the rat (i.e., less than 12 hours) does not induce a significant increase of REM sleep, whereas a more severe sleep deficit (i.e., 12–24 hours) causes a massive REM sleep rebound (Borbély and Neuhaus, 1979; Franken, *et al.*, 1991a, 1993; Tobler and Borbély, 1986). The question is still open whether REM sleep-regulating processes are activated only by a substantial deficit and to what extent, as in humans, the high priority of the compensatory response of slow-wave activity prevents the manifestation of REM sleep homeostasis after a moderate sleep loss. In the rat (Borbély *et al.*, 1984; Franken *et al.*, 1991a) and in the rabbit (Tobler *et al.*, 1990a), 24-hour sleep deprivation induced an increase in power density in the theta band in REM sleep. This could be an indication of an intensity dimension also in REM sleep.

Circadian Aspects of Sleep Regulation

Conceptual Basis and Models

The main objectives of models of circadian rhythms have been (1) to simulate spontaneously occurring phenomena in free-running rhythms such as changing phase relations between the sleep–wake cycle and the temperature rhythm, or the splitting of the rest–activity rhythm into two components, and (2) to account for the effects of light, the major zeitgeber, on the phase, amplitude and period of circadian rhythms. The database underlying the models consists in general of records of rest and activity rather than of polygraphically recorded sleep and waking. Moreover, the models attempt to describe general features of the circadian system rather than focus specifically on the sleep–wake rhythm.

Various models of circadian rhythms postulated more than one circadian oscillator. Whereas the influence of these putative oscillators is little apparent under entrained conditions, it may become manifest during freerun. Interactions between different oscillators have been proposed to account for phenomena such as splitting in rodents as well as for the phase dependence of sleep duration and the biphasic sleep pattern in humans (Daan & Berde, 1978; Kronauer, 1984, 1987, 1990; Kronauer, Czeisler, Pilato, Moore-Ede, & Weitzmann, 1982; Wever, 1984).

The assumption of multiple self-sustaining oscillators is not mandatory for accounting for various features of free-running rhythms. As pointed out by Eastman (1984), the pattern of internal desynchronization between the sleep–wake cycle and the temperature rhythm can be simulated by periodic or occasional phase shifts of the former rhythm. Daan, Beersma and associates (Beersma *et al.*, 1987; Daan & Beersma, 1984; Daan *et al.*, 1984) showed that the interaction of a relaxation oscillator (Process S) with a single circadian oscillator (Process C) can account for various features of free-running rhythms such as the pattern of internal desynchronization, "phase-trapping," and circabidian cycles as well as polyphasic sleep. A single pacemaker was shown to be sufficient to generate different rates of reentrainment in different overt rhythms (Daan & Beersma, 1992). However, the combined influence of two virtual identical pacemaker components must be assumed to account for specific rest–activity patterns in animals such as splitting (Beersma & Daan, 1992). The possibility of the presence of multiple circadian components within the pacemaker gained support from a report that the suprachiasmatic nucleus contains a large population of autonomous, single-cell, circadian oscillators (Welsh, Logothetis, Meister, & Reppert, 1995). However, the usefulness of circadian models for exploring sleep regulation is limited if the homeostatic facet is not incorporated.

Experimental Approaches

Various experimental approaches have been taken to quantify the role of the circadian pacemaker in the regulation of sleep.

FREE-RUNNING SLEEP–WAKE CYCLES DURING TEMPORAL ISOLATION. In these experiments, human subjects are studied under free-running conditions in temporal isolation for several weeks or months. Subjects typically self-select their bedtimes and thus the light–dark cycle to which they are exposed. Under these conditions, they develop free-running sleep–wake and core body temperature cycles, both of which typically remain synchronized with a period which is close to 25 hours (Aschoff, 1965; Czeisler, 1978; Wever, 1979). Compared to the entrained state, (i.e. when subjects are synchronized to the 24-hour day), the phase angle between the core body temperature rhythm and the sleep–wake cycle differs under free-running conditions. During freerun, humans select to go to sleep close to the circadian temperature nadir, whereas in the entrained state, the core body temperature minimum is located 1–3 hours before the end of the habitual sleep episode (Czeisler, 1978; Czeisler *et al.*, 1992; Zulley, 1979; Zulley, Weaver, & Aschoff, 1981). Concurrent with the advance of the temperature rhythm relative to sleep onset, a redistribution of REM sleep across the sleep episode occurs (Czeisler, Zimmerman Ronda, Moore-Ede, & Weitzman, 1980a; Zulley *et al.*, 1981).

In approximately 30% of all approximately 1-month-long studies, the sleep–wake cycle becomes desynchronized from the circadian rhythm of core body temperature (i.e., the period of the sleep–wake cycle is no longer identical to the period of the core body temperature rhythm) (Aschoff, Gerecke, & Wever, 1967; Czeisler, 1978).

During periods of synchronized or desynchronized freerun, neither the sleep initiations nor the sleep terminations are distributed uniformly over the circadian cycle of core body temperature. Sleep initiations primarily occur on the falling limb of the endogenous component of the temperature rhythm, whereas the majority of sleep terminations are located on the rising limb of this rhythm (Czeisler, 1978; Strogatz, Knonaur, & Czeisler, 1986; Zulley, 1979; Zulley *et al.*, 1981). In addition to this circadian modulation of the propensity to initiate and terminate sleep, a dependence of the duration of sleep episodes on the circadian phase at which sleep is initiated has been demonstrated. Long sleep episodes are initiated close to the maximum of the endogenous rhythm of core body temperature and short episodes close to the minimum (Czeisler, Weitzman Moore-Ede, Zimmerman, & Kronauer, 1980b; Zulley & Wever, 1982; Zulley *et al.*, 1981). This variation in sleep duration (range ~7–18 hours) cannot be explained on the basis of the considerable variation in the duration of wakefulness preceding sleep (range ~18–35 hours) (Strogatz *et al.*, 1986).

In a limited number of cases showing spontaneous desynchronization, sleep was recorded polysomnographically, which allowed the analysis of the dependence of sleep structure on the circadian phase at which sleep occurred. These analyses revealed that REM sleep is strongly dependent on circadian phase with the maximum of REM sleep propensity occurring shortly after the minimum of the endogenous rhythm of body temperature (Czeisler *et al.*, 1980a; Zulley, 1979). Thus the endogenous circadian pacemaker which drives the rhythm of core body temperature is a major determinant of sleep propensity, sleep duration, and REM sleep propensity. An unequivocal circadian modulation of slow-wave sleep has not been demonstrated (Weitzman, Czeisler, Zimmermann, & Ronda, 1980). However, in these experiments, slow-wave activity in nonREM sleep was not analyzed.

SLEEP DISPLACEMENT STUDIES. In such experiments, the duration of wakefulness prior to sleep and the timing of sleep within the 24-hour period are manipulated simultaneously. These studies have demonstrated that sleep can be initiated at all circadian phases, provided that the duration of prior wakefulness is of sufficient length. When sleep is initiated on the initial part of the rising limb of the core body temperature rhythm by extending the duration of prior wakefulness, sleep duration is reduced compared to baseline (Åkerstedt & Gillberg, 1981; Berger, Walker, Scott, Magnusson, & Pollack, 1971). SWA and slow-wave sleep in the initial part of sleep and total sleep time remain responsive to variations in the duration of prior wakefulness even when sleep is initiated during the early morning hours (Åkerstedt & Gillberg, 1986a, b). SWA and slow-wave sleep decrease in the course of sleep episodes independent of circadian phase (Åkerstedt & Gillberg, 1981; Dijk, Brunner, & Borbély, 1991a). This decrease of SWA is also observed when sleep is extended for more than 12 hours (Dijk, Cajochen, Tobler, & Borbély, 1991b). In contrast, the time course of REM sleep is strongly dependent on circadian phase (Åkerstedt & Gillberg 1981; Berger *et al.*, 1971; Dijk *et al.*, 1991a).

The sleep displacement studies show that homeostatic mechanisms are of sufficient strength to allow initiation of sleep outside the phase of habitual sleep timing and that circadian processes continue to modulate some aspects of sleep such as sleep duration and REM sleep even when the homeostatic drive for sleep is elevated due to sleep deprivation.

FORCED DESYNCHRONY PROTOCOLS. Circadian and sleep-dependent components of sleep propensity can be studied by a scheduled uncoupling of the rest–activity cycle from the output of the circadian pacemaker in so-called forced–desynchrony protocols. In these protocols, the sleep–waking cycle is scheduled to be either shorter or longer than the period of the pacemaker while maintaining the usual 1:2 ratio between sleep and waking, thereby minimizing the confounding effects of sleep deprivation. In various experimental protocols, subjects were scheduled to short sleep-wake cycles (20, 90 180, 240 minutes) (Carskadon & Dement, 1975; Czeisler, 1978; Dantz, Edgar, & Dement, 1994; Lavie, 1986; Webb & Agnew, 1975; Weitzman *et al.*, 1974; Zulley, 1990) and consequently sleep was initiated at many different circadian phases. The results of these studies showed that sleep propensity as indexed by sleep latency and total sleep time is highest near the minimum of the core body temperature rhythm, which under entrained conditions occurs 1–3 hours before the end of the habitual sleep episode (Czeisler *et al.*, 1992). Sleep propensity is at a minimum prior to the habitual bedtime. Not only sleep propensity, but also REM sleep propensity peaks close to the nadir of the body temperature rhythm (Carskadon & Dement, 1980; Weitzman *et al.* 1974; Zulley, 1990).

Circadian and homeostatic aspects of sleep regulation were quantified in an extensive, approximately 1-month-long protocol in which subjects were scheduled to a long (28 hour) sleep–wake cycle of which one third (i.e. 9.33 hours) was reserved for sleep in darkness (Dijk & Czeisler, 1994, 1995). This sleep episode is of sufficient duration to quantify sleep-dependent processes as a function of circadian phase. Light intensity during the waking period was kept at a low level to minimize its effect on the circadian pacemaker, and endogenous circadian phase was assessed by continuous recording of core body temperature and by determining the circadian rhythms of cortisol and melatonin. Under such conditions, the endogenous circadian rhythms of core body temperature and melatonin freerun with a period of 24.1–24.2 hours (Czeisler *et al.*, 1995). The results confirmed that sleep propensity is highest at the minimum of the endogenous circadian rhythm of core body temperature. Sleep propensity gradually decreases on the rising limb of the core body temperature rhythm and reaches its nadir 16 hours after its minimum. Under entrained conditions this phase corresponds to 22 hours, i.e., a time point close to habitual bedtime. Analysis of sleep consolidation as a function of both circadian phase and time elapsed since start of the sleep episode revealed that sleep consolidation can be maintained for approximately 8 hours when sleep is initiated close to the minimum of the circadian rhythm of sleep propensity and ends shortly after the crest of the circadian rhythm of sleep propensity (Figure 4). In contrast, sleep episodes initiated close to the maximum of the circadian rhythm of sleep propensity are characterized by poor sleep consolidation in the second half of the sleep episode. From these data, it thus appears that the paradoxical phase relationship between the endogenous circadian rhythm of sleep propensity and the habitual timing of the sleep–wake cycle serves to consolidate sleep and wakefulness. In addition, the study showed that sleep spindle activity exhibits a marked circadian rhythm. This circadian rhythm, with its crest coinciding with the beginning of the habitual sleep episode, in interaction with the sleep-dependent increase in sleep spindle activity results in high levels of sleep spindle activity when sleep coincides with the timing of the habitual sleep episode. In contrast, SWA and its time course during sleep were shown to be primarily determined by homeostatic processes.

A protocol resembling the human forced-desynchrony protocol has been applied in rats by imposing a short rest–activity cycle by means of a rapid succession of light and dark periods (Borbély, 1975; Borbély, Huston, & Waser, 1975). The light-induced variations of sleep were superimposed upon the circadian rhythm. The circadian variations of sleep consisted in a progressive decline of total sleep time during the habitual light period and a large part of the habitual dark period, and a subsequent rapid increase. The sleep-modulating effect of light was largest during the habitual dark period. Moreover, a phase relation between the light–dark-induced changes of nonREM sleep and REM sleep itself exhibited a circadian modulation.

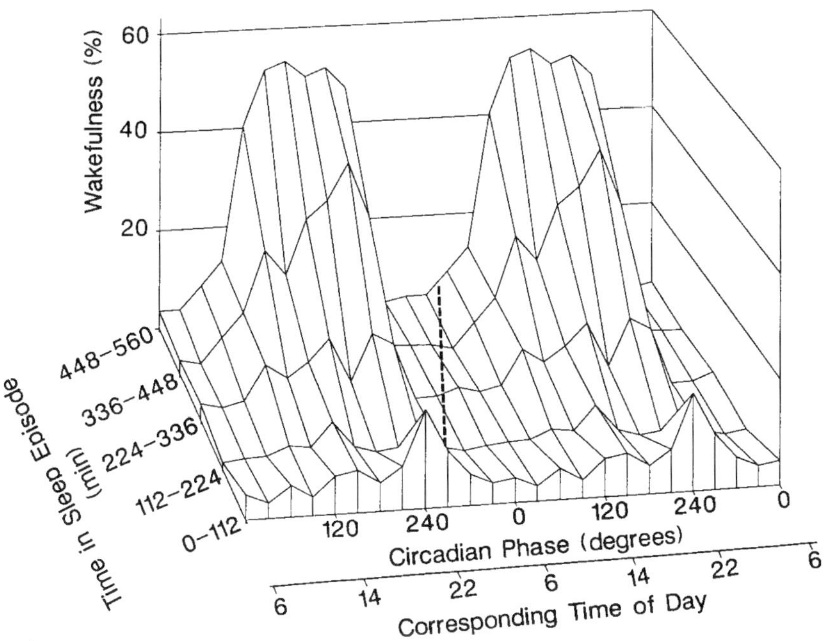

Figure 4. Three-dimensional plot of wakefulness in scheduled sleep episodes, circadian phase, and time elapsed since start of sleep episode. Data were assigned to 12 circadian bins (30 degs each) and five bins of time since start of the sleep episode (112 minutes each). Each point represents wakefulness expressed as percentage of recording time in a bin. Thus, the data point at (0–112 minutes, 0 degs) represents wakefulness in all first 112 minutes of sleep episodes which occur within the circadian phase interval of −15.0 to 15.0 degs. Note that the first 112-minute bin starts at lights-out and wakefulness prior to sleep onset is included in this bin. To facilitate interpretation of the data, clock times which under entrained conditions would correspond to the circadian phase of the core body temperature rhythm are indicated. The dashed line represents the trajectory of circadian phase and time passed since start of scheduled sleep time. This trajectory corresponds to the trajectory of circadian phase and time passed since start of sleep episode during a nocturnal sleep episode under entrained conditions. (From Dijk and Czeisler, 1994, with permission. Copyright 1994, Elsevier Science.)

Ultradian Aspects of Sleep Regulation

The nonREM-REM sleep cycle is a salient feature of mammalian sleep. In adult humans, the duration of the cycle is 90–100 minutes, whereas in the rat, it is approximately 8–14 min (Figure 5). Models have been proposed to account for the cyclic recurrence of REM sleep. The reciprocal interaction model posited on the basis of neurophysiologic data in animals that the nonREM–REM sleep cycle is generated by the mutual interaction of two distinct neuronal systems (McCarley & Hobson, 1975). The original hypothesis was later transposed to humans, and was further elaborated into the limit cycle reciprocal interaction model (McCarley & Massaquoi, 1986; Massaquoi & McCarley, 1990, 1992). The authors showed that by varying one of the initial values, a bimodal distribution of REM sleep latency could be obtained. Moreover, the model was applied to predict the ultradian phase response to a cholinergic agent and to forced awakenings. In the most recent version,

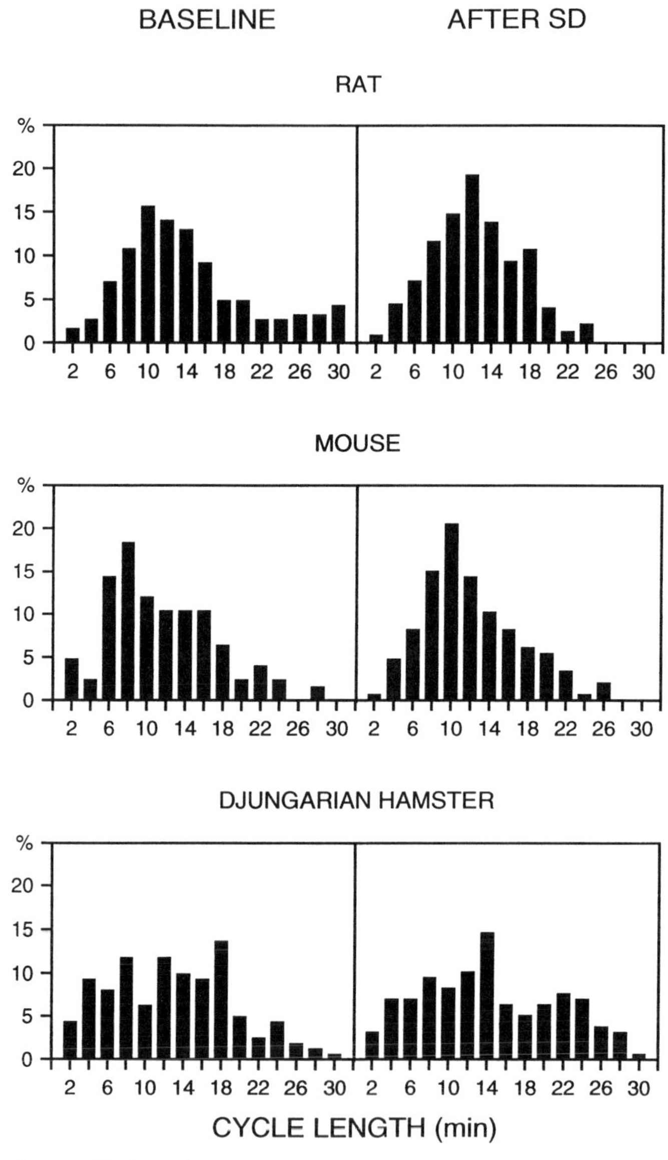

Figure 5. Ultradian nonREM–REM sleep cycle. Histogram of sleep cycle length in three rodents during baseline (BL) conditions (left) and after sleep deprivation (SD) (right). Rat (24-hour SD; BL and recovery 8 hours; $n = 10$), mouse (6-hour SD; BL and recovery 6 hours; $n = 8$); Djungarian hamster (4-hour SD; BL and recovery 8 hours; $n = 7$). Ordinate: 100% is the sum of cycles contained in the recording period. Sleep cycles are plotted for 2-minute bins with endpoints indicated on the abscissa. Note the similar distributions for baseline and SD. (Rat: from Trachsel *et al.*, 1991; others: T. Deboer and I. Tobler, unpublished.)

homeostatic processes similar to those proposed by Achermann *et al.* (1990) as well as an arousal process were incorporated (Massaquoi & McCarley, 1992).

Beersma, Daan, and Van den Hoofdakker (1984) used the concepts of the original reciprocal interaction model to account for the altered distribution of REM sleep latency in depression. They showed that sleep abnormalities in depressive patients can be accounted for by the decrease in the initial value of a single variable which may be taken to represent the extent of REM sleep inhibition.

Whereas the nonREM–REM sleep cycle was not part of the original version of the two-process model, it has been incorporated in more recent versions. This was achieved by introducing an oscillating variable that triggers REM sleep whenever slow-wave activity falls below a certain threshold (Achermann *et al.*, 1990). In simulations of empirical data, the actual occurrence of REM sleep was used as a trigger (Achermann & Borbély, 1990; Achermann *et al.*, 1993). There is an obvious convergence between this approach and the limit cycle reciprocal interaction model, since a combination of the key parts of the two models has been proposed (Achermann & Borbély, 1992; Massaquoi & McCarley, 1992).

Studies in the cat which addressed the question of whether changes of sleep propensity affect the nonREM–REM sleep cycle yielded inconsistent results (Lydic, McCarley, & Hobson, 1984; Ursin, Moses, Naitoh, & Johnson, 1983). In the rat, mouse, and Djungarian hamster, the distribution of the sleep cycle was analyzed under baseline conditions and after sleep deprivation (Figure 5). Although various parameters of the distribution were affected by enforced waking, major changes were not observed.

Integration and Perspectives

Interaction of a Homeostatic and Circadian Process

The observation that the human sleep–wake cycle can spontaneously desynchronize from the circadian rhythm of core body temperature (Aschoff *et al.*, 1967) as well as other findings led to the proposition that the human circadian system is constituted by multiple circadian oscillators (see Borbély & Achermann, 1992, for a review). However, as claims that the mammalian brain contains multiple circadian pacemakers were not confirmed, models of sleep regulation based on two or more oscillators lost their appeal. The notion has gained general acceptance that a single circadian oscillator interacts with a relaxation oscillator, and that this interaction can account for many aspects of sleep under a variety of different schedules (Daan *et al.*, 1984; Dijk & Czeisler, 1994, 1995; Eastman, 1984). Nevertheless, as stated in a previous section, multiple circadian components may be present within the pacemaker.

The two-process model postulates that homeostatic Process S, whose level rises during waking and declines during sleep, interacts with circadian Process C, which is independent of sleep and waking. Whereas the time course of Process S was derived from the changes of EEG slow-wave activity (see above), the characteristics of Process C were more difficult to specify. In the original qualitative version, a sine function was assumed. Subsequently, on the basis of sleep duration data, a skewed sinusoidal function (rapid rise, slow decline) was postulated to modulate the thresholds defining the onset and termination of sleep (Daan *et al.*, 1984). Based on the necessity to account for a constant sleep latency during most of the daytime waking period, a skewing of the sine wave in the opposite direction (slow rise, rapid decline) was envisaged (Borbély, Achermann, Trachsel, & Tobler, 1989). This assumption is supported by data obtained in a forced desynchrony protocol in which the circadian component of sleep latency showed a slow rise and rapid decline (Dijk & Czeisler, 1994). One of the problems is that the pattern of the rhythm may not reflect the output of the pacemaker, but rather the coupling of the pacemaker with secondary processes (Achermann & Borbély, 1994). In the model of Kronauer (1990), for example, Process C is represented by a van der Pol oscillator whose output approximates a sine wave and whose phase and amplitude are modified by light. However, to simulate empirical data, a further transformation (e.g., phase shift, change of the shape) of the pacemaker output must be postulated.

Type of Interaction

How does the homeostatic and circadian processes interact? According to the initial versions of the two-process model, interactions between the two processes occur at the intersections of Process S with the lower threshold (Borbély, 1982b) or with the upper and lower thresholds (Daan *et al.*, 1984) that are modulated by the circadian Process C. These interactions define the times of onset and termination of sleep. However, for physiologic and psychophysiologic variables, a continuous interaction of S and C appears to be plausible. In mathematical terms, the discontinuous and continuous interaction models are equivalent (Achermann & Borbély, 1994). Already in the first quantitative model (Daan *et al.*, 1984) as well as in later propositions by different groups, a continuous interaction of the two processes was assumed to underlie the variations in daytime vigilance (Achermann & Borbély, 1992; Åkerstedt & Folkard, 1995; Borbély *et al.*, 1989; Folkard & Åkerstedt, 1987).Changes of vigilance during a short photoperiod and a forced-desynchrony protocol were simulated on the basis of an additive interaction of these processes (Achermann & Borbély, 1994). The data from the latter protocol showed that nonlinear interactions must also be present. On the basis of sleep studies in squirrel monkeys which had sustained lesions of the suprachiasmatic nuclei (SCN), Edgar *et al.* (1993) postulated a continuous interaction between a circadian and a homeostatic process. These authors proposed that the SCN generates a circadian waking process which opposes a homeostatic sleep process. However, the results of a human forced-desynchrony study (Dijk & Czeisler, 1994, 1995) indicate that the circadian process may not only promote waking, but also favor sleep. Thus, depending on the phase relation, S and C may act either synergistically or antagonistically in promoting sleep or waking.

Independent Control of the Homeostatic and Circadian Process

Homeostatic and circadian facets of sleep regulation can be independently manipulated and therefore may be controlled by separate mechanisms. Thus, during a 72-hour sleep deprivation period, the subjective alertness ratings showed a prominent circadian rhythm (Åkerstedt & Fröberg, 1977). Conversely, in a study in which the phase of the circadian process (as indexed by body temperature and plasma melatonin) was shifted by bright light in the morning, the time course of slow-wave activity remained unaffected (Dijk, Beersma, & Daan, 1989). In subjects recorded under a forced-desynchrony protocol, both a circadian and a sleep-wake dependent, homeostatic factor determining the time course of alertness and performance (Dijk, Duffy, & Czeisler, 1992; Johnson *et al.*, 1992) and sleep variables (Dijk & Czeisler, 1995) were identified.

Two lines of evidence from animal experiments indicate that the homeostatic and circadian facets of sleep regulation are mediated by separate processes. First, a conflict can be created between the homeostatic and circadian tendencies in the rat by terminating a 24-hour sleep deprivation period at the onset or in the middle of the dark period, which is the circadian period of predominant waking (Borbély & Neuhaus, 1979; Mistlberger, Bergmann, Waldenar, & Rechtschaffen, 1983; Trachsel, Tobler, & Borbély, 1986). In one of these studies, slow-wave activity showed a rebound in two stages: an immediate increase followed by waking and a second, delayed increase at light onset which may have been facilitated by a "circadian gate" (Borbély & Neuhaus, 1979).

The second line of evidence stems from experiments in which the circadian facet of sleep regulation was disrupted by lesions of the suprachiasmatic nuclei (Mistlberger *et al.* 1983; Tobler, Borbély, & Groos, 1983b). Arrhythmic rats were subjected to 24-hour sleep deprivation to test their ability to compensate for sleep loss. Both slow-wave activity in nonREM sleep and the amount of REM sleep were increased. Similar results were published in the same year by another group (Mistlberger *et al.*, 1983) and were confirmed in a more recent study (Trachsel, Edgar, Seidel, Heller, & Dement, 1992). Conversely, there are data suggesting that sleep deprivation affects neither the phase nor the period of the free-running rest–activity rhythm of the rat (Beersma *et al.*, 1987; Borbély, 1982a). These results indicate that the homeostatic component of sleep regulation is morphologically and functionally distinct from the circadian component.

In intact animals, the variables serving as potential markers of the circadian pacemaker are usually "contaminated" by behavioral variations. Cortical temperature in the rat was recently analyzed in an attempt to separate the circadian from the sleep/wake-dependent component (Franken, Tobler, & Borbély, 1992). Two approaches were used: For one, measurements were made during a 24-hour period of enforced wakefulness in which the sleep-dependent variations of temperature could be eliminated. Second, a correlation analysis was performed to estimate the influence of sleep and waking on cortical temperature. The residual variance exhibited a 24-hour periodicity with an amplitude of 0.15° C, which was similar to the rhythm recorded during enforced wakefulness. The maximum of the temperature rhythm was situated in the first half of the dark period, and its amplitude did not change after transition from LD 12:12 to continuous dark.

Finally, the robustness of the homeostatic sleep regulatory process was impressively evident in a rat study in which baseline sleep and recovery sleep after sleep deprivation were analyzed under long and short photoperiods (Franken, Tobler, & Borbély, 1995). Although the distribution of sleep and waking were markedly altered by the lighting schedules, the level of sleep and its response to prolonged waking remained unaffected. Also, in the chipmunk, a diurnal rodent, maintained in either a long or short photoperiod, the 24-hour value of sleep did not differ (Dijk & Daan, 1989). Similar findings were obtained in the Djungarian hamster (Tobler & Deboer, 1994). These results in animals can be related to results in humans showing that the photoperiod has a marked influence on the pattern of body temperature, whereas the 24-hour level of temperature remains constant (Wehr *et al.*, 1993).

References

Achermann, P. (1988). *Schlafregulation des Menschen: Modelle und Computersimulationen.* Ph.D. thesis, ETH Zürich, Zürich.

Achermann, P., & Borbély, A. A. (1990). Simulation of human sleep: ultradian dynamics of EEG slow-wave activity. *Journal of Biological Rhythms, 5,* 141–157.

Achermann, P., & Borbély, A. A. (1992). Combining different models of sleep regulation. *Journal of Sleep Research, 1,* 144–147.

Achermann, P., & Borbély, A. A. (1994). Simulation of daytime vigilance by additive interaction of a homeostatic and a circadian process. *Biological Cybernetics, 71,* 115–121.

Achermann, P., Beersma, D. G. M., & Borbély, A. A. (1990) The two-process model: Ultradian dynamics of sleep. In J. A. Horne (Ed.), *Sleep '90* (pp. 296–300). Bochum, Germany: Pontenagel Press.

Achermann, P., Dijk, D.-J., Brunner, D. P., & Borbély, A. A. (1993). A model of human sleep homeostasis based on EEG slow-wave activity: Quantitative comparison of data and simulations. *Brain Research Bulletin, 31,* 97–113.

Aeschbach, D., & Borbély, A. A. (1993). All-night dynamics of the human sleep EEG. *Journal of Sleep Research, 2,* 70–81.

Åkerstedt, T., & Folkard, S. (1990). A model of human sleepiness. In J. A. Horne (Ed.), *Sleep '90* (pp. 310–313). Bochum, Germany: Pontenagel Press.

Åkerstedt, T., & Folkard, S. (1995). Validation of the S and C components of the three-process model of alertness regulation. *Sleep, 18,* 1–6.

Åkerstedt, T., & Fröberg, J. E. (1977). Physiological circadian rhythms in women during 72 h of sleep deprivation. *Waking and Sleeping, 1,* 387–394.

Åkerstedt, T., & Gillberg, M. (1981). The circadian variation of experimentally displaced sleep. *Sleep, 4,* 159-169.

Åkerstedt, T., & Gillberg, M. (1986a). Sleep duration and the power spectral density of the EEG. *Electroencephalography and Clinical Neurophysiology, 64,* 119–222.

Åkerstedt, T., & Gillberg, M. (1986b). A dose–response study of sleep loss and spontaneous sleep termination. *Psychophysiology, 64,* 119–222.

Alföldi, P., Tobler, I., & Borbély, A. A. (1990). Sleep regulation in rats during early development. *American Journal of Physiology, 258,* 634–644.

Aschoff, J. (1965). Circadian rhythms in man. *Science, 148,* 1427–1432.

Aschoff, J., Gerecke, U., & Wever, R. (1967). Desynchronization of human circadian rhythms. *Japanese Journal of Physiology, 17,* 450–457.

Beersma, D. G. M., & Daan, S. (1992). Generation of activity–rest patterns by dual circadian pacemaker systems: A model. *Journal of Sleep Research, 1,* 84–87.

Beersma, D. G. M., Daan, S., & Van den Hoofdakker, R. H. (1984). Distribution of REM latencies and other sleep phenomena in depression as explained by a single ultradian rhythm disturbance. *Sleep, 7,* 126–136.

Beersma, D. G. M., Daan, S., & Dijk, D.-J. (1987). Sleep intensity and timing—A model for their circadian control. *Lectures on Mathematics in the Life Sciences, 19,* 39–62.

Beersma, D. G. M., Dijk, D.-J., Blok, C. G. H., & Everhardus, I. (1990). REM sleep deprivation during five hours leads to an immediate REM sleep rebound and to suppression of non-REM sleep intensity. *Electroencephalography and Clinical Neurophysiology, 76,* 114–122.

Berger, R. J., & Oswald, I. (1962). Effects of sleep deprivation on behavior, subsequent sleep, and dreaming. *Journal of Mental Science, 108,* 457–465.

Berger, R. J., Walker, J. M., Scott, T. D., Magnusson, L. J., & Pollack, S. L. (1971). Diurnal and nocturnal sleep stage patterns following sleep deprivation. *Psychonomic Science, 23,* 273–275.

Blake, H., & Gerard, R. W. (1937). Brain potentials during sleep. *American Journal of Physiology, 119,* 692–703.

Borbély, A. A. (1975). Circadian rhythm of vigilance in rat: Modulation by short light–dark cycles. *Neuroscience Letters, 1,* 67-71.

Borbély, A. A. (1980). Sleep: Circadian rhythm versus recovery process. In M. Koukkou, D. Lehmann & J. Angst (Eds.), *Functional states of the brain: Their determinants* (pp. 151–161). Amsterdam: Elsevier.

Borbély, A. A. (1982a). Sleep regulation: Circadian rhythm and homeostasis. In D. Ganten & D. Pfaff (Eds.), *Current topics in neuroendocrinology. Sleep: clinical and experimental aspects* (Vol. 1, pp. 83–103). Berlin: Springer-Verlag.

Borbély, A. A. (1982b). A two-process model of sleep. *Human Neurobiology, 1,* 195-204.

Borbély, A. A. (1994). Sleep homeostasis and models of sleep regulation. In M. H. Kryger, T. Roth, & W. C. Dement (Eds.), *Principles and practice of sleep medicine,* (2nd ed., pp. 309–320). Philadelphia: Saunders.

Borbély, A. A., & Achermann, P. (1992). Concepts and models of sleep regulation: An overview. *Journal of Sleep Research, 1,* 63–79.

Borbély, A. A., & Neuhaus, H. U. (1979). Sleep-deprivation: Effect on sleep and EEG in the rat. *Journal of Comparative Physiology, 133,* 71–87.

Borbély, A. A., Huston, J. P., & Waser, P. G. (1975). Control of sleep states in the rat by short light–dark cycles. *Brain Research, 95,* 89–101.

Borbély, A. A., Baumann, F., Brandeis, D., Strauch, I., & Lehmann, D. (1981). Sleep deprivation; effect on sleep stages and EEG power density in man. *Electroencephalography and Clinical Neurophysiology, 51,* 483–493.

Borbély, A. A., Tobler, I., & Hanagasioglu, M. (1984). Effect of sleep deprivation on sleep and EEG power spectra in the rat. *Behavioral Brain Research, 14,* 171–182.

Borbély, A. A., Achermann, P., Trachsel, L., & Tobler, I. (1989). Sleep initiation and sleep intensity: Interaction of homeostatic and circadian mechanisms. *Journal of Biological Rhythms, 4,* 149–160.

Brunner, D. P., Dijk, D.-J., Tobler, I., & Borbély, A. A. (1990). Effect of partial sleep deprivation on sleep stages and EEG power spectra: Evidence for nonREM and REM sleep homeostasis. *Electroencephalography and Clinical Neurophysiology, 75,* 492–499.

Brunner, D. P., Dijk, D.-J., & Borbély, A. A. (1993). Repeated partial sleep deprivation progressively changes the EEG during sleep and wakefulness. *Sleep, 16,* 100–113.

Bunnell, D. E., & Horvath, S. M. (1985). Effects of body heating during sleep interruption. *Sleep,* 8, 274–282.

Cannon, W. B. (1939). *The wisdom of the body.* New York: Norton.

Carskadon, M. A., & Dement, W. C. (1975). Sleep studies on a 90–minute day. *Electroencephalography and Clinical Neurophysiology, 39,* 145–155.

Carskadon, M. A., & Dement, W. C. (1980). Distribution of REM sleep on a 90 min sleep–wake schedule. *Sleep, 2,* 309–317.

Cartwright, R. D., Monroe, L. J., & Palmer, C. (1967). Individual differences in response to REM deprivation. *Archives of General Psychiatry, 16,* 297–303.

Church, M. W., March, J. D., Hibi, S., Cavness, C., & Feinberg, I. (1975). Changes in frequency and amplitude of delta activity during sleep. *Electroencephalography and Clinical Neurophysiology, 39,* 1–7.

Czeisler, C. A. (1978). *Human circadian physiology: Internal organization of temperature, sleep–wake and neuroendocrine rhythms monitored in an environment free of time cues.* Ph.D. dissertation, Stanford University, Stanford, California.

Czeisler, C. A., Zimmerman, J. C. Ronda, J. M., Moore-Ede, M. C., & Weitzman, E. D. (1980a). Timing of REM sleep is coupled to the circadian rhythm of body temperature in man. *Sleep, 2,* 329–346.

Czeisler, C. A., Weitzman, E. D., Moore-Ede, M. C., Zimmerman, J. C., & Kronauer, R. S. (1980b). Human sleep: Its duration and organization depend on its circadian phase. *Science, 210,* 1264–1267.

Czeisler, C. A., Dumont, M., Duffy, J. F., Steinberg, J. D., Richardson, G. S., Brown, E. N., Sánchez, R., Ríos, C. D., & Ronda, J. M. (1992). Association of sleep–wake habits in older people with changes in output of circadian pacemaker. *Lancet, 340,* 933–936.

Czeisler, C. A., Duffy, J. F., Shanahan, T. L., Brown, E. N., Mitchell, J. F., Dijk, D.-J., Rimmer, D. W., Ronda, J. M., Allan, J. S., Emens J. S., & Kronauer, R. E. (1995). Reassessment of the intrinsic period of the human circadian pacemaker in young and older subjects. *Sleep Research, 24A,* 505.

Daan, S., & Beersma, D. G. M. (1984). Circadian gating of human sleep–wake cycles. In M. C. Moore-Ede & C. A. Czeisler (Eds.), *Mathematical models of the circadian sleep–wake cycle* (pp. 129–158). New York: Raven Press.

Daan, S., & Beersma, D. G. M. (1992). A single pacemaker can produce different rates of reentrainment in different overt rhythms. *Journal of Sleep Research, 1,* 80–83.

Daan, S., & Berde, C. (1978). Two coupled oscillators: Simulations of the circadian pacemaker in mammalian activity rhythms. *Journal of Theoretical Biology, 70,* 297–313.

Daan, S., Beersma, D. G. M., & Borbély, A. A. (1984). The timing of human sleep: Recovery process gated by a circadian pacemaker. *American Journal of Physiology, 246,* R161–R178.

Daan, S., Beersma, D. G. M., Dijk, D.-J., Åkerstedt, T. A., & Gillberg, M. (1988). Kinetics of an hourglass component involved in the regulation of human sleep and wakefulness. In W. Th. Hekkens, M. G. A. Kerkhof, and W. J. Rietveld (Eds.), *Trends in chronobiology* (pp. 183–193). Oxford: Pergamon Press.

Dantz, B., Edgar, D. M., & Dement, W. C. (1994). Circadian rhythms in narcolepsy: Studies on a 90 minute day. *Electroencephalography and Clinical Neurophysiology, 90,* 24–35.

Deboer, T., Franken, P., & Tobler, I. (1994). Sleep and cortical temperature in the Djungarian hamster under baseline conditions and after sleep deprivation. *Journal of Comparative Physiology, 174,* 145–155.

Dement, W. C. (1960). The effects of dream deprivation. *Science, 131,* 1705–1707.

Dement, W., & Greenberg, S. (1966). Changes in total amount of stage four sleep as a function of partial sleep deprivation. *Electroencephalography and Clinical Neurophysiology, 20,* 523–526.

Dement, W., & Kleitman, N. (1957). Cyclic variations in EEG during sleep and their relation to body movements, body motility, and dreaming. *Electroencephalography and Clinical Neurophysiology, 9,* 673–690.

Dijk, D.-J. (1995). EEG slow-waves and sleep spindles: Windows on the sleeping brain. *Behavioral Brain Research, 69,* 109–116.

Dijk, D.-J., & Czeisler, C. A. (1993). Body temperature is elevated during the rebound of slow wave sleep following 40-h of sleep deprivation on a constant routine. *Journal of Sleep Research, 2,* 117–120.

Dijk, D.-J., & Czeisler, C. A. (1994). Paradoxical timing of the circadian rhythm of sleep propensity serves to consolidate sleep and wakefulness in humans. *Neuroscience Letters, 166,* 63–68.

Dijk, D.-J., & Czeisler, C. A. (1995). Contribution of the circadian pacemaker and the sleep homeostat to sleep propensity, sleep structure and electroencephalographic slow waves and sleep spindle activity in humans. *Journal of Neuroscience, 15,* 3526–3538.

Dijk, D.-J., & Daan, S. (1989). Sleep EEG spectral analysis in a diurnal rodent: *Eutamias sibiricus. Journal of Comparative Physiology A, 165,* 205–215.

Dijk, D.-J, Beersma, D. G. M., & Daan, S. (1987). EEG power density during nap sleep: Reflection of an hourglass measuring the duration of prior wakefulness. *Journal of Biological Rhythms, 2,* 207–219.

Dijk, D. J., Beersma, D. G. M., Daan, S. (1989). Bright morning light advances the human circadian system without affecting NREM sleep homeostasis. *American Journal of Physiology, 256,* R106–R111.

Dijk, D.-J., Brunner, D. P., & Borbély, A. A. (1990a). Time course of EEG power density during long sleep in humans. *American Journal of Physiology, 258,* R650–R661.

Dijk, D.-J., Brunner, D. P., Beersma, D. G. M., & Borbély, A. A. (1990b) Slow wave sleep and electroencephalogram power density as a function of prior waking and circadian phase. *Sleep, 13,* 430–440.

Dijk, D.-J., Brunner, D. P., & Borbély, A. A. (1991a). EEG power density during recovery sleep in the morning. *Electroencephalography and Clinical Neurophysiology, 78,* 203–214.

Dijk, D.-J., Cajochen, C., Tobler, I., & Borbély, A. A. (1991b). Sleep extension in humans: Sleep stages, EEG power spectra, and body temperature. *Sleep, 14,* 294–306.

Dijk, D.-J., Duffy, J. F., & Czeisler, C. A. (1992). Circadian and sleep–wake dependent aspects of subjective alertness and cognitive performance. *Journal of Sleep Research, 1,* 112–117.

Dijk, D.-J., Hayes, B., & Czeisler, C. A. (1993). Dynamics of electroencephalographic sleep spindles and slow wave activity in men: Effect of sleep deprivation. *Brain Research, 626,* 190–199.

Eastman, C. (1984). Are separate temperature and activity oscillators necessary to explain the phenomena of human circadian rhythms? In M. C. Moore-Ede & C. A. Czeisler (Eds.), *Mathematical models of the circadian sleep–wake cycle* (pp. 81–103). New York: Raven Press.

Edgar, D. M., Dement, W. C., & Fuller, C. A. (1993). Effect of SCN lesions on sleep in squirrel monkeys: Evidence for opponent processes in sleep-wake regulation. *Journal of Neuroscience, 13,* 1065–1079.

Feinberg, I., March, J. D., Floyd, T. C., Jimison, R., Bossom-Demitrack, L., & Katz, P. H. (1985). Homeostatic changes during post-nap sleep maintain baseline levels of delta EEG. *Electroencephalography and Clinical Neurophysiology, 61,* 134–137.

Feinberg I., Maloney, T., & March, J. D. (1992). Precise conservation of nonREM period 1 (nonREMP1) delta across naps and nocturnal sleep: Implications for REM latency and nonREM/REM alternation. *Sleep, 15,* 400–403.

Folkard, S., & Åkerstedt, T. (1987). Towards a model for the prediction of alertness and/or fatigue on different sleep/wake schedules. In A. Oginski, J. Polorski, & J. Rutenfranz (Eds.), *Contemporary advances in shiftwork research: Theoretical and practical aspects in the late eighties* (pp. 231–240). Krakow, Poland: Medical Academy.

Folkard, S., & Åkerstedt, T. (1989). Towards the prediction of alertness on abnormal sleep/wake schedules. In A. Coblentz (Ed.), *Vigilance and performance in automatized systems* (pp. 287–296). Dordrecht, The Netherlands: Kluwer.

Folkard, S., & Åkerstedt, T. (1992). A three-process model of the regulation of alertness–sleepiness. In R. J. Broughton & R. D. Ogilvie (Eds.), *Sleep, arousal, and performance* (pp. 11–26). Boston: Birkhäuser.

Franken, P., Dijk, D.-J., Tobler, I., & Borbély, A. A. (1991a). Sleep deprivation in the rat: Effects on electroencephalogram power spectra, vigilance states, and cortical temperature. *American Journal of Physiology, 261,* R198–R208.

Franken, P., Tobler, I., & Borbély, A. A. (1991b). Sleep homeostasis in the rat: Simulations of the time course of EEG slow-wave activity. *Neuroscience Letters, 130,* 141–144.

Franken, P., Tobler, I., & Borbély, A. A. (1992). Sleep and waking have a major effect on the 24-h rhythm of cortical temperature in the rat. *Journal of Biological Rhythms, 7,* 341–352.

Franken, P., Tobler, I., & Borbély, A. A. (1993). Effects of 12-h sleep deprivation and of 12-h cold exposure on sleep regulation and cortical temperature in the rat. *Physiology and Behavior, 54,* 885–894.

Franken, P., Tobler, I., & Borbély, A. A. (1995). Varying the photoperiod in the rat: Profound effect on the 24-h sleep pattern but no effect on sleep homeostasis. *American Journal of Physiology, 269,* R691–R701.

Friedman, L., Bergmann, B. M., & Rechtschaffen, A. (1979). Effects of sleep deprivation on sleepiness, sleep intensity, and subsequent sleep in the rat. *Sleep, 1,* 369–391.

Geering, B. A., Achermann, P., Eggimann, F., & Borbély, A. A. (1993). Period–amplitude and power spectral analysis: A comparison based on all-night sleep EEG recordings. *Journal of Sleep Research, 2,* 121–129.

Hanagasioglu, M., & Borbély, A. A. (1982). Effect of voluntary motor activity on sleep in the rat. *Behavioral Brain Research, 4,* 359–368.

Hirsch, J. C., Fourment, A., & Marc, M. E. (1983). Sleep-related variations of membrane potential in the lateral geniculate body relay neurons of the cat. *Brain Research, 259,* 308–312.

Horne, J. A., & Shackell, B. S. (1987). Slow wave sleep elevations after body heating: Proximity to sleep and effects of aspirin. *Sleep, 10,* 383–392.

Horne, J. A., & Staff, L. H. E. (1983). Exercise and sleep: Body heating effects. *Sleeps, 6,* 36–46.

Johnson, M. P., Duffy, J. F., Dijk, D.-J., Ronda, J. M., Dyal, C. M., & Czeisler, C. A. (1992). Short-term memory, alertness and performance: A reappraisal of their relationship to body temperature. *Journal of Sleep Research, 1,* 24–29.

Jones, H. S., & Oswald, I. (1968). Two cases of healthy insomnia. *Electroencephalography and Clinical Neurophysiology, 24,* 378–380.

Kales, A., Hoedemaker, F. S., Jacobson, A., & Lichtenstein, E. L. (1964). Dream deprivation: An experimental reappraisal. *Nature, 204,* 1337–1338.

Karacan, I., Williams, R. L., Finley, W. W., & Hursch, C. J. (1970). Effects of naps on nocturnal sleep: Influence on the need for stage 1-REM and stage 4 sleep. *Biological Psychiatry, 2,* 391–399.

Kattler, H., Dijk, D.-J., & Borbély, A. A. (1994). Effect of unilateral somatosensory stimulation prior to sleep on the sleep EEG in humans. *Journal of Sleep Research, 3,* 159–164.

Knowles, J. B., MacLean, A. W., Salem, L., Vetere, C., & Coutler, M. (1986). Slow-wave sleep in daytime and nocturnal sleep: An estimation of the time course of 'Process S.' *Journal of Biological Rhythms, 1,* 303–308.

Knowles, J. B., Maclean, A. W., Brunet, D., & Coulter, M. (1990). Nap-induced changes in the time course of Process S. Effects on nocturnal slow wave activity. In J. A. Horne (Ed.), *Sleep '90* (pp. 68–70). Bochum; Germany: Pontenagel Press.

Kronauer, R. E. (1984). Modeling principles for human circadian rhythms. In M. C. Moore-Ede & C. A. Czeisler (Eds.), *Mathematical models of the circadian sleep–wake cycle* (pp. 105–128). New York: Raven Press.

Kronauer, R. E. (1987). Temporal subdivision of the circadian cycle. *Lectures on Mathematics in the Life Sciences, 19,* 63–120.

Kronauer, R. E. (1990). A quantitative model for the effects of light on the amplitude and phase of the deep circadian pacemaker, based on human data. In J. A. Horne (Ed.), *Sleep '90* (pp. 306–309). Bochum, Germany: Pontenagel Press.

Kronauer, R. E., Czeisler, C. A., Pilato, S. F., Moore-Ede, M. C., & Weitzmann, E. D. (1982). Mathematical model of the human circadian system with two interacting oscillators. *American Journal of Physiology, 242,* R3–R17.

Krueger, J. M., & Obál, F. (1993). A neuronal group theory of sleep function. *Journal of Sleep Research, 2,* 63–69.

Lancel, M., van Riezen, H., & Glatt, A. (1991). Effects of circadian phase and duration of sleep deprivation on sleep and EEG power spectra in the cat. *Brain Research, 548,* 206–214.

Lavie, P. (1986). Ultrashort sleep–waking schedule. III. "Gates" and "forbidden zones" for sleep. *Electroencephalography and Clinical Neurophysiology, 63,* 414–425.

Lydic, R., McCarley, R. W., & Hobson, J. A. (1984). Forced activity alters sleep cycle periodicity and dorsal raphe discharge rhythm. *American Journal Physiology, 247,* R135–R145.

Maron, L., Rechtschaffen, A., & Wolpert, E. A. (1964). Sleep cycle during napping. *Archives of General Psychiatry, 11,* 503–507.

Massaquoi, S., & McCarley, R. W. (1990). Resetting the REM sleep oscillator. In J. A. Horne (Ed.), *Sleep '90* (pp. 301–305). Bochum, Germany: Pontenagel Press.

Massaquoi, S., & McCarley, R. W. (1992). Extension of the limit cycle reciprocal interaction model of REM cycle control: An integrated sleep control model. *Journal of Sleep Research, 1,* 138–143

McCarley, R. W., & Hobson, J. A. (1975). Neuronal excitability modulation over the sleep cycle: A structural and mathematical model. *Science, 189,* 58–60.

McCarley, R. W., & Massaquoi, S. (1986). A limit cycle mathematical model of the REM sleep oscillator system. *American Journal of Physiology, 251,* R1011–R1029.

McGinty, D., Szymusiak, R., & Thomson, D. (1994). Preoptic/anterior hypothalamic warming increases EEG delta frequency within non-rapid eye movement sleep. *Brain Research, 667,* 273–277.

Mistlberger, R. E., Bergmann, B. M., Waldenar, W., & Rechtschaffen, A. (1983). Recovery sleep following sleep deprivation in intact and suprachiasmatic nuclei lesioned rats. *Sleep, 6,* 217–233.

Mistlberger, R., Bergmann, B., & Rechtschaffen, A. (1987). Period–amplitude analysis of rat electroencephalogram: Effects of sleep deprivation and exercise. *Sleep, 10,* 508–522.

Morairty, S. R., Szymusiak, R., Thomson, D., & McGinty, D. J. (1993). Selective increases in non-rapid eye movement sleep following whole body heating in rats. *Brain Research, 617,* 10–16.

Neckelmann, D., & Ursin, R. (1993). Sleep stages and EEG power spectrum in relation to acoustical stimulus arousal threshold in the rat. *Sleep, 16,* 467–477.

Obál, Jr., F., Alföldi, P., & Rubicsek, G. (1995). Promotion of sleep by heat in young rats. *Pflügers Archiv, 430,* 729–738.

Oleksenko, A. I., Mukhametov, L. M., Polyakova, I. G., Supin, A. Y., & Kovalzon, V. M. (1992). Unihemispheric sleep deprivation in bottlenose dolphins. *Journal of Sleep Research, 1,* 40–44.

Opp, M. R., & Krueger, J. M. (1994). Interleukin-1 is involved in responses to sleep deprivation in the rabbit. *Brain Research, 639,* 57–65.

Pappenheimer, J. R., Koski, G., Fencl, V., Karnovsky, M. L., & Krueger, J. (1975). Extraction of sleep-promoting factor S from cerebrospinal fluid and from brains of sleep-deprived animals. *Journal of Neurophysiology, 38,* 1299–1311.

Steriade, M., Jones, E. G., & Llinás, R. R. (1990). *Thalamic oscillations and signaling.* New York: Wiley.

Steriade, M., McCormick, D. A., & Sejnowski, T. J. (1993a). Thalamocortical oscillations in the sleeping and aroused brain. *Science, 262,* 679–685.

Steriade, M., Nuñez, A., & Amzica, F. (1993b). A novel slow (<1 Hz) oscillation of neocortical neurons *in vivo*: Depolarizing and hyperpolarizing components. *Journal of Neuroscience, 13,* 3252–3265.

Steriade, M., Contreras, D., & Amzica, F. (1994). Synchronized sleep oscillations and their paroxysmal developments. *Trends in Neurosciences, 17,* 199–208.

Strogatz, S. H., Kronauer, R. E., & Czeisler, C. A. (1986). Circadian regulation dominates homeostatic control of sleep length and prior wake length in humans. *Sleep, 9,* 353–364.

Tobler, I. (1985). Deprivation of sleep and rest in vertebrates and invertebrates. In S. Inoué & A. A. Borbély (Eds.), *Endogenous sleep substances and sleep regulation* (pp. 57–66). Utrecht: VNU Science Press.

Tobler, I., & Borbély, A. A. (1986). Sleep EEG in the rat as a function of prior waking. *Electroencephalography and Clinical Neurophysiology, 64,* 74–76.

Tobler, I., & Borbély, A. A. (1990). The effect of 3-h and 6-h sleep deprivation on sleep and EEG spectra of the rat. *Behavioral Brain Research, 36,* 73–78.

Tobler, I., & Deboer, T. (1994). Sleep, sleep regulation and cortical temperature in a photoperiodic rodent. *Journal of Sleep Research, 3(Supplement 1),* 254.

Tobler, I., & Franken, P. (1993). Sleep homeostasis in the guinea pig: Similar response to sleep deprivation in the light and dark period. *Neuroscience Letters, 164,* 105–108.

Tobler, I., & Jaggi, K. (1987). Sleep and EEG spectra in the Syrian hamster (*Mesocricetus auratus*) under baseline conditions and following sleep deprivation. *Journal of Comparative Physiology A, 161,* 449-459.

Tobler, I., & Scherschlicht, R. (1990). Sleep and EEG slow-wave activity in the domestic cat: Effect of sleep deprivation. *Behavioral Brain Research, 37,* 109–118.

Tobler, I., Murison, R., Ursin, R., Ursin, H., & Borbély, A. A. (1983a). The effect of sleep deprivation and recovery sleep on plasma corticosterone in the rat. *Neuroscience Letters, 35,* 297–300.

Tobler, I., Borbély, A. A., & Groos, G. (1983b). The effect of sleep deprivation on sleep in rats with suprachiasmatic lesions. *Neuroscience Letters, 42,* 49–54.

Tobler, I., Franken, P., & Scherschlicht, R. (1990a). Sleep and EEG spectra in the rabbit under baseline conditions and following sleep deprivation. *Physiology and Behavior, 48,* 121–129.

Tobler, I., Dijk, D.-J., & Borbély, A. A. (1990b). Comparative aspects of sleep regulation in three species. In J. A. Horne (Ed.), *Sleep '90* (pp. 349–351). Bochum, Germany: Pontenagel Press.

Tobler, I., Franken, P., Trachsel, L., & Borbély, A. A. (1992). Models of sleep regulation in mammals. *Journal of Sleep Research, 1,* 125–127.

Tobler, I., Franken, P., & Jaggi, K. (1993). Vigilance states, EEG spectra and cortical temperature in the guinea pig. *American Journal of Physiology, 264,* R1125–R1132.

Tobler, I., Franken, P., Gao, B., Jaggi, K., & Borbély, A. A. (1994). Sleep deprivation in the rat at different ambient temperatures: Effect on sleep, EEG spectra and brain temperature. *Archives Italienne de Biologie, 132,* 39–52.

Trachsel, L., Tobler, I., & Borbély, A. A. (1986). Sleep regulation in rats: effects of sleep deprivation, light, and circadian phase. *American Journal of Physiology, 251,* R1037–R1044.

Trachsel, L., Tobler, I., Achermann, P., & Borbély, A. A. (1991). Sleep continuity and the REM–nonREM cycle in the rat under baseline conditions and after sleep deprivation. *Physiology and Behavior, 49,* 575–580.

Trachsel, L., Edgar, D. M., Seidel, W. F., Heller, W. C., & Dement, W. C. (1992). Sleep homeostasis in suprachiasmatic nuclei-lesioned rats: Effects of sleep deprivation and triazolam administration. *Brain Research, 589,* 253–261

Ursin, R., Moses, J., Naitoh, P., & Johnson, L. C. (1983). REM–NREM cycle in cat may be sleep-dependent. *Sleep, 6,* 1–9.

Webb, W. B., & Agnew, H. W. (1971). Stage 4 sleep: Influence on time course variables. *Science, 174,* 1354–1356.

Webb, W. B., & Agnew, H. W. (1975). Sleep efficiency for sleep–wake cycles of varied length. *Psychophysiology, 12,* 637–641.

Webb, W. B., & Friel, J. (1970). Characteristics of "natural" long and short sleepers: A preliminary report. *Psychological Reports, 27,* 63–66.

Wehr, T. A., Moul, D. E., Barbato, G., Giesen, H. A., Seidel, J. A., Barker, C., & Bender, C. (1993). Conservation of photoperiod-responsive mechanisms in humans. *American Journal of Physiology, 265,* R846–R857.

Weitzman, E. D., Nogeire, C., Perlow, M., Fukushima, D., Sassin, J., McGregor, P., Gallagher, T. F., & Hellman, L. (1974). Effects of prolonged 3-hour sleep-wake cycle on sleep stages, plasma cortisol, growth hormone and body temperature in man. *Journal of Clinical Endocrinology and Metabolism, 38,* 1018–1030.

Weitzman, E. D., Czeisler, C. A., Zimmermann, J. C., & Ronda, J. M. (1980). Timing of REM and stages 3 + 4 sleep during temporal isolation in man. *Sleep, 2,* 391–407.

Welsh, D. K., Logothetis, D. E., Meister, M., & Reppert, S. M. (1995). Individual neurons dissociated from rat suprachiasmatic nucleus express independently phased circadian firing rhythms. *Neuron, 14,* 697–706.

Werth, E., Dijk, D.-J., Achermann, P., & Borbély, A. A. (1996). Dynamics of the sleep EEG after n early evening nap: Experimental data and simulations. *American Journal of Physiology–Regulatory Integrative and Comparative Physiology, 40*(3), R501–R510.

Wever, R. A. (1979). *The circadian system of man. Results of experiments under temporal isolation.* New York: Springer-Verlag.

Wever, R. A. (1984). Toward a mathematical model of circadian rhythmicity. In M. C. Moore-Ede & C. A. Czeisler (Eds.), *Mathematical models of the circadian sleep-wake cycle* (pp. 17–79). New York: Raven Press.

Williams, H. L., Hammack, J. T., Daly, R. L., Dement, W. C., & Lubin, A. (1964). Responses to auditory stimulation, sleep loss and the EEG stages of sleep. *Electroencephalography and Clinical Neurophysiology, 16,* 269–279.

Zulley, J. (1979). *Der Einfluss von Zeitgebern auf den Schlaf des Menschen.* Frankfurt: Fischer.

Zulley, J. (1990). Day and night sleep: The bedrest condition. In J. Horne (Ed.), *Sleep '90* (pp. 319–323). Bochum, Germany: Pontenagel Press.

Zulley, J., & Wever, R. A. (1982). Interaction between the sleep–wake cycle and the rhythm of rectal temperature. In J. Aschoff, S. Daan, & G. Groos (Eds.), *Vertebrate circadian systems: Structure and physiology* (pp. 253–261). Berlin: Springer.

Zulley, J., Wever, R., & Aschoff, J. (1981). The dependence of onset and duration of sleep on the circadian rhythm of rectal temperature. *Pflügers Archiv, 391,* 314–318.

18

Mammalian Photoperiodism

Michael R. Gorman, Bruce D. Goldman, and Irving Zucker

Introduction

Many organisms have evolved physiologic and behavioral adaptations that are presumed to increase reproductive fitness in highly seasonal environments. This review will focus on mammals, the group in which perhaps the most progress has been made in understanding mechanisms of seasonal adjustments at the neuroendocrine level. Seasonal modulation in a typical mammal may involve several traits, including reproductive capacity and related behaviors, increases and decreases in energy storage, and changes in pelage density. Food availability, precipitation, and ambient temperature vary seasonally in a more or less predictable fashion, and are potential environmental zeitgebers. Day length (DL) is, however, the most noise free and probably the most frequently used cue for phasing seasonal responses among mammals in mid and higher latitudes. This use of DL is termed photoperiodism and should be distinguished from the use of photic cues for the entrainment of circadian rhythms. Several review articles summarize recent progress in understanding mammalian photoperiodism (Bartness & Goldman, 1989; Goldman & Elliott, 1988; Goldman & Nelson, 1993; Karsch *et al.*, 1984; Nelson, Badura, & Goldman, 1990). A recapitulation of the extensive corpus of findings is beyond our present scope; instead, we selectively review a few extensively studied model systems. We emphasize ways in which the natural progression of DLs in nature provides information used by animals to achieve seasonally appropriate adjustments. Our emphasis is on species, e.g., hamsters, mice, and voles, in which seasonal transitions do not recur spontaneously in the absence of seasonal changes in DL. These Type I rhythms (Zucker,

---

Michael R. Gorman Department of Psychology, University of California, San Diego, La Jolla, California 92093. Bruce D. Goldman Department of Physiology and Neurobiology, University of Connecticut, Storrs, Connecticut 06269. Irving Zucker Departments of Psychology and Integrative Biology, University of California, Berkeley, California 94720.

Circadian Clocks, Volume 12 of *Handbook of Behavioral Neurobiology*, edited by Joseph S. Takahashi, Fred W. Turek, and R. Y. Moore, Kluwer Academic / Plenum Publishers, New York, 2001.

Lee, & Dark, 1991) are not fully endogenous, and their recurrence in mammals is contingent on seasonal variations in DL and associated changes in the pineal melatonin rhythm. Several species with fully endogenous circannual rhythms are considered in Chapter 19. It should be emphasized that despite the emphasis in this review on Type I rhythms, there is no evidence to suggest that the fundamental mechanisms of photoperiodism are different in Type I and Type II rhythms. Indeed, both types of rhythms appear to depend on a circadian mechanism to measure DL, and the pineal gland is an important part of the photoperiodic mechanism in both (see below).

Seasonal Patterns in Type I Mammals

The formalisms of Type I seasonal rhythms were well understood at the time of Hoffmann's review in this series (Hoffmann, 1981): When DLs decrease in late summer or early autumn, the reproductive organs of long-day breeding mammals undergo involution (Figure 1). In controlled laboratory conditions, the same reproductive response is evoked after a sudden shift from long to short day lengths, or as a consequence of blinding or exposure to continuous darkness. In a more general sense, short DLs trigger a shift to the physiologic state that is characteristic of autumn/winter for a given species (Goldman, 1983). Thus, for autumn breeders, reproduction is stimulated by short DLs. It is of particular interest that the winter state is not maintained indefinitely when animals are held under short days in the laboratory. After several months of short day exposure, the reproductive system returns to the state that is typical of spring/summer. The same phenomenon is observed in mammals that are blinded or exposed to continuous darkness (DD). In the case of long-day breeding mammals, the growth of the gonads that occurs after several months of short DLs has been termed "spontaneous" recrudescence to emphasize that this change is not triggered by a corresponding change in DL. Rather, it appears to depend on the action of an internal time-keeping mechanism—an interval timer—that is set in motion by exposure to short days. Nonreproductive traits exhibit a similar pattern of change; for example, a molt from summer to winter pelage can be induced by exposure to short DLs, and a molt back to summer pelage occurs several months later even in animals held under a fixed, short DL (Duncan & Goldman, 1984).

Critical Day Length

Early studies that quantified the response to decreasing or increasing DLs led to the concept that the changes from summer to winter state were triggered by the crossing of a critical DL. In male Syrian hamsters that had been kept in long days, a one-step decrease to a DL of 12L:12D (or to shorter DLs) resulted in testicular regression, whereas DLs of 12.5 hours (12.5L) or longer supported testis maintenance and spermatogenesis. Also, if male hamsters were first exposed to a very short DL so that their testes became involuted, testicular growth could be initiated by shifting the animals to DLs of 12.5L or longer, but not to DLs of 12L or less (J. A. Elliott, 1976). Similar results were obtained in female hamsters. These observations supported the contention that 12.5L was the critical DL for the maintenance of reproductive function in Syrian hamsters (J. A. Elliott, 1976). The critical DL varies between species and also intraspecifically with latitude of origin (Dark, Johnston, Healy, & Zucker, 1983; Heath & Lynch, 1982), strain, and perhaps domestication

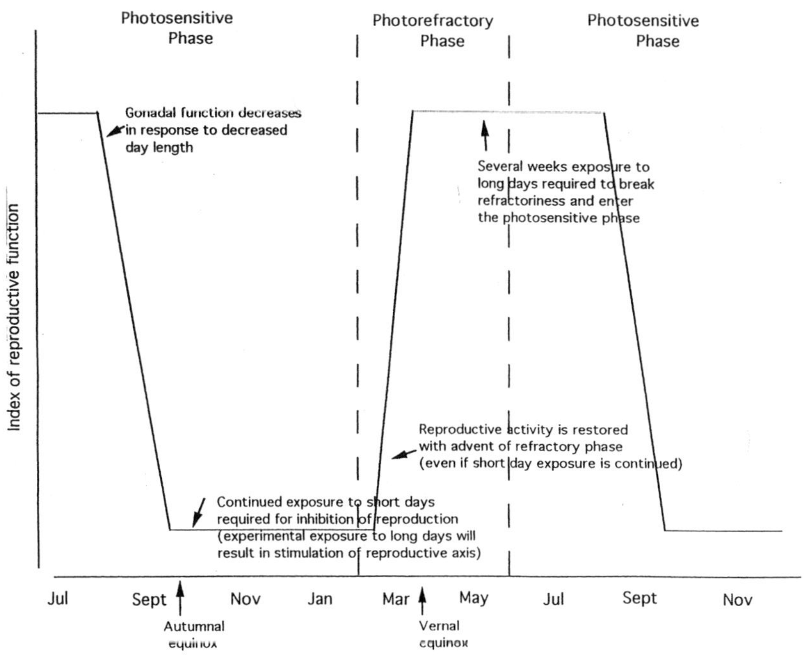

Figure 1. Photoperiodic regulation of annual gonadal changes in a typical Type I mammal, the Syrian hamster. Note that day length changes are an absolute requirement at two stages of the annual cycle: (1) decreased day length initiates gonadal regression (i.e., winter condition) in late summer/autumn and (2) increased day length is needed to restore photosensitivity after spontaneous gonadal recrudescence at the onset of the refractory phase. The transition from the winter to the spring state can be initiated by internal processes and does not require exposure to increasing day lengths. Some Type II (circannual) mammals can, under experimental conditions of controlled lighting, exhibit responses to decreased and increased day lengths that closely resemble those shown by Type I animals. However, in Type II species, (1) decreased day length is not an absolute requirement for the transition from summer to spring state and (2) exposure to long (summer) days is not required to restore responsiveness to short days. Thus, the two stages that must be triggered by day length changes in Type I species can be driven by internal mechanisms in Type II animals. This added element of internal control may account for the ability of Type II species to express circannual rhythms when held under constant conditions. In Type II mammals, photoperiod changes are not needed for sustaining repeated annual cycles, but are required for entraining the circannual rhythm to local annual time.

(Vitaterna & Turek, 1993). Few species have been examined for reproductive responses across a full range of DLs. In Turkish hamsters, *Mesocricetus brandti,* gonadal function is inhibited by DLs less than 15L or greater than 17L (Hong, Rollag, & Stetson, 1986), suggesting that dependence on a photoperiodic threshold is not a universal mammalian pattern; some species may sustain reproduction only within a narrow range of long DLs.

Several more recent observations have forced revisions of the earliest version of the critical DL concept. It is generally inaccurate to refer to a critical DL for a particular species, as values may be trait-specific. Thus, in male Siberian hamsters, transfer from 16L to 14L leads to testicular regression, but does not evoke a molt to winter pelage. Exposure to shorter DLs results in both gonadal involution and the development of the winter coat; the critical DLs for maintenance of summer pelage and reproductive activity evidently differ (Duncan, Goldman, DiPinto, & Stetson, 1985).

CHANGES IN DAY LENGTH

Modification of the critical DL concept was compelled by the recognition that the response to a particular DL depends on photoperiodic history. Male Siberian hamsters transferred from 16L to 14L underwent testicular involution, but males that had undergone testicular regression in 8L and were then exposed to 14L manifested testicular growth. For this species, 14L can either inhibit or stimulate the reproductive axis, depending on the animals' preceding photoperiod experience (Hoffmann, Illnerova, & Vanecek, 1986). A similar type of photoperiod history effect has been described for Syrian hamsters (Hastings *et al.*, 1989b). In Siberian hamsters, a large fraction of DLs experienced in nature are potentially ambiguous: DLs as long as 16L induced partial gonadal regression when they followed longer DLs, whereas 10L induced partial gonadal growth after shorter DLs (Rivkees, Hall, Weaver, & Reppert, 1988). This photoperiod history effect, moreover, also obtains under more naturalistic conditions where DLs change gradually, and thus is not an artifact of abrupt changes in DL or melatonin duration (see below) (Gorman, 1995; Gorman & Zucker, 1995a; Gorman & Zucker, 1997b).

PHOTOREFRACTORINESS

In one sense, the phenomenon of photorefractoriness represents another type of photoperiod history effect. All photoperiodic mammals show refractoriness to short days. After a period of 20–25 weeks of exposure to short days, animals spontaneously revert from the typical winter state to that characteristic of spring and summer and are then said to be refractory to short photoperiods. Photorefractoriness is thought to be triggered by the action of an endogenous neuroendocrine timing mechanism that remains to be identified (J. A. Elliott & Goldman, 1981).

In some respects, photorefractory animals retain responsiveness to DL signals. When photorefractory hamsters were held under short photoperiod for more than 1 year, they remained reproductively active; however, when such animals were exposed to long days for approximately 10 weeks and were then returned to short photoperiod, a second period of gonadal regression ensued (Reiter, 1969; Stetson, Watson-Whitmyre, & Matt, 1977). The ability of long-day exposure to break photorefractoriness (i.e., to allow a return to the photosensitive state) is a clear indication that the photorefractory animal still discriminates long from short photoperiods. It seems likely that a similar mechanism for measuring DL may be operative during the photosensitive and photorefractory phases of the annual cycle. Indeed, the pineal melatonin rhythm, which is a part of the mechanism for photoperiodic responses during the photosensitive phase, is also important in the process of breaking refractoriness in Syrian hamsters (Bittman & Zucker, 1981).

The terms photorefractoriness and photosensitivity do not apply only to reproductive responses. A similar pattern of changes has been described for virtually all types of photoperiodic traits in various species, including spring and fall molts, seasonal thermoregulatory adaptations, and seasonal changes in body weight and lipid metabolism.

Several species of rodents (hamsters, mice, voles) become photorefractory only in response to prolonged exposure to short photoperiod. These species never maintain a winter state indefinitely despite continued short-day treatment. They do maintain a summer state indefinitely during continuous exposure to long photoperiod. In contrast, a number of mammals exhibit both short-day and long-day photorefractoriness. Ferrets initiate reproductive activity during the spring as they become refractory to the reproductive inhibition associated with the short photoperiods of winter. In this sense, the winter/spring phase of the ferret's annual reproductive cycle is similar to that described above for hamsters. Ferrets maintain stimulated reproductive systems during the long days of summer, but, unlike hamsters, this state cannot be sustained indefinitely by continued exposure to artificially lengthened photoperiod. Rather, after several months in long days, the ferret 'spontaneously' enters a period of reproductive inhibition (Herbert & Klinowska, 1978). Sheep also exhibit the ability to become refractory after prolonged exposure to either long or short days, and this can account for the transition both into and out of the reproductive season under natural conditions (Robinson & Karsch, 1984; Robinson, Wayne, & Karsch, 1985). The property of exhibiting refractoriness to both types of photoperiod—implying the capacity for "spontaneous" changes in both the spring and the fall—may well be functionally coupled with the ability to express circannual rhythms. Interestingly, no mammals have been found to become refractory only to long days; mammals become refractory either to short days or to both short days and long days.

Yet another type of photoperiod history effect may be related to variation in photoresponsiveness within a population. Thus, Siberian hamsters that were exposed to a very long DL (18L) for 10 weeks were subsequently nonresponsive to short-day exposure; they failed to exhibit the typical short-day responses. Hamsters born early in the breeding season would be exposed to long days for several weeks prior to encountering the short days of late summer/early fall, whereas hamsters born later in the season would have experienced fewer long days; the animals born earlier may be more likely to remain reproductively active during the short days of fall/winter (Gorman & Zucker, 1997a).

Transmission of Day Length Information from Mother to Offspring

Perhaps the most fascinating illustration of the impact of photoperiodic history on response to DL is revealed in studies of maternal communication of photic information to offspring. In montane voles maintained from birth in an "intermediate" DL of 14L, the rates of body growth and reproductive maturation varied depending on whether the mother had been exposed to a longer versus a shorter photoperiod during gestation. Cross-fostering experiments revealed that information transmitted by mother voles to the fetuses influenced the photoperiodic response to the 14L postnatal day length (Horton, 1984, 1985). Similar relations obtain for Siberian hamsters (Stetson, Elliott, & Goldman, 1986).

All absolute DLs, with the exception of the longest and shortest days of the year, occur twice during the annual cycle. It is important that photoperiodic organisms distinguish between the two different times of year at which an 'intermediate' DL, such as 14L, occurs. The transmission of DL cues from mother to fetus, as described above, may facilitate this process. Newborn rodents that spend the first weeks of life in a burrow may not be directly exposed to photic cues until they become mobile and leave the nest. Cues received from their mother prior to birth could enable them more rapidly to assess the significance of a 14L photoperiod encountered around the time of weaning; they might be able to assess whether DLs have been increasing or decreasing in the interval between birth and weaning. The nature of the responses observed in montane voles and Siberian hamsters suggests that this may in fact be a reasonable portrayal of the function of the system for transmission of photoperiod cues from mother to fetus.

A similar system has been described for meadow voles (*Microtus pennsylvanicus*) as well, but here the mother continues to influence her offspring after birth; both prenatal and postnatal influences of the dam have been reported (Lee & Zucker, 1988). Not only the ambient photoperiod, but also the mother's state of photosensitivity/photorefractoriness affects her offspring. Dams that were exposed to short days for only a few weeks prior to and during gestation transmitted short day cues to their offspring, but dams that were in short days long enough to achieve a photorefractory state before mating transmitted signals to their pups that produced seasonal adaptations intermediate between those typical of short-day and long-day modes (Lee, 1993).

The neuroendocrine mechanism that mediates maternal–fetal communication has been investigated most extensively in Siberian hamsters, where the mother's pineal gland is essential for transmission of photoperiod information to the fetuses (J. A. Elliott & Goldman, 1989). Daily infusions of melatonin to pinealectomized dams during late pregnancy mimicked the effects of exposing pineal-intact dams to different photoperiods during gestation. Thus, long-duration melatonin infusions and short-day exposure during pregnancy exerted similar effects on fetuses, whereas short-duration infusions mimicked the actions of long-day exposure (Weaver & Reppert, 1986; Weaver, Keohan, & Reppert, 1987). This is consistent with the role of the pineal melatonin rhythm in adult mammals, as discussed below.

Pineal Gland and Mammalian Photoperiodism

Pineal Melatonin and Photoperiodic Responses

In the 1960s and 1970s, the pineal gland and its hormone, melatonin, were identified as major components of the photoperiodic mechanism in mammals. In Syrian hamsters, pinealectomy completely prevented the gonadal regression normally elicited by exposure to short days or under various forms of light deprivation (Czyba, Girod, & Durand, 1964; Hoffman & Reiter, 1965; Reiter, 1980). Such observations were later extended to a number of other species; in virtually all cases, pinealectomized animals did not exhibit the species-typical photoperiodic responses (Goldman, 1983).

McCord and Allen (1917) reported that a factor present in bovine pineal tissue acted upon the dermal melanophores of amphibians to cause a lightening of the skin color. Following up on this observation, Lerner and colleagues extracted and characterized the substance responsible for this skin lightening effect and named it melatonin (Lerner, Case, Lee, Takahashi, & Mori, 1958). Melatonin is synthesized from serotonin by the action of two pineal enzymes, N-acetyltransferase (NAT) and hydroxyindole-O-methyltransferase (HIOMT). Pineal melatonin biosynthesis occurs almost exclusively at night (Minneman & Wurtman, 1976). The timing of the pineal melatonin rhythm is determined by the circadian system (Moore & Klein, 1974; Ralph, Mull, Lynch, & Hedlund, 1971) and entrained by the ambient light–dark cycle.

Studies that employed timed daily injections of melatonin yielded convincing evidence that melatonin is the hormonal mediator of hamster photoperiodism.

Most significantly, injections of melatonin in pinealectomized Syrian hamsters induced gonadal regression similar to that observed when pineal-intact hamsters are exposed to short photoperiods (Tamarkin, Hollister, Lefebvre, & Goldman, 1977). The evidence accumulated through many such experiments implicated melatonin as a part of the photoperiodic mechanism, but failed to identify which features of the melatonin signal were functionally important. The circadian phase of the nocturnal melatonin elevation, its duration, and its amplitude all have been proposed as critical aspects of the signal. Timed infusions of melatonin, first administered to pinealectomized Siberian hamsters and sheep, permitted precise experimental control over phase, duration, and amplitude of blood concentrations of melatonin. These studies clearly pointed to the duration of each daily melatonin elevation as the principal determinant of photoperiodic responses (Figure 2). In both sheep and hamsters, long-duration melatonin elevations simulated effects of short-photoperiods, and short duration melatonin elevations induced long-day type responses (Bittman & Karsch, 1984; D. S. Carter & Goldman, 1983a, b). Later studies extended these findings to Syrian hamsters (Maywood, Buttery, Vance, Herbert, & Hastings, 1990). These observations are consistent with the seasonal changes in melatonin secretory patterns; the duration of nocturnal melatonin secretion increases in a graded fashion as DL decreases (Goldman, 1983). The comparison between hamsters and sheep is instructive because these species are long-day and short-day breeders, respectively. Correspondingly, the reproductive axis of hamsters is stimulated by short-duration melatonin signals and inhibited by long-duration signals, whereas in sheep, long-duration melatonin signals stimulate reproduction and short-duration signals are inhibitory.

The timed melatonin infusion procedure permits reasonably close simulation of actual patterns of melatonin secretion. Our understanding of the action of melatonin also has been furthered by employing two separate infusions of melatonin/day. Testis regression was induced in Syrian hamsters by twice-daily 2.5-hour melatonin infusions separated by an interval of 3 hours (Pitrosky *et al.*, 1995). This

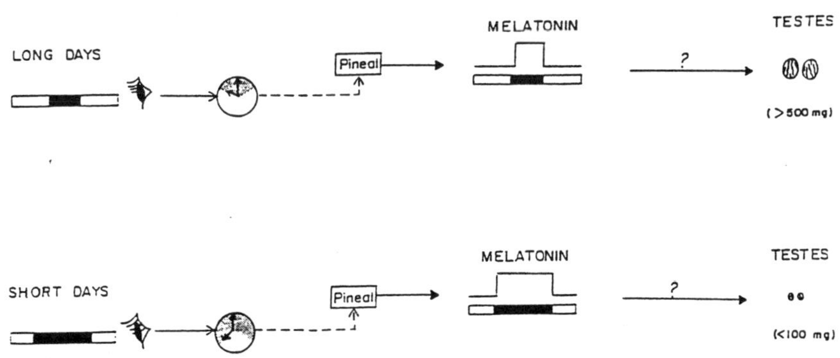

Figure 2. Above: The "duration hypothesis" of melatonin action as applied to a long-day breeding mammal. Light cues received by the retina and transmitted to the circadian system are used to determine day length. This information is transmitted to the pineal gland, where the neural signals are transduced to an endocrine message of day length in the form of the circadian rhythm of pineal melatonin secretion. In long days, the nightly elevation of melatonin secretion is of relatively short duration, and associated with stimulation of the reproductive system in long-day breeders. Below: In short days, the duration of the nightly melatonin elevation is increased, and leads to inhibition of the reproductive system. The question marks indicate that the target sites for these photoperiodic actions of melatonin remain unknown.

suggests that the "skeleton" of a long-duration melatonin elevation may be adequate to evoke a short-day response. In Siberian hamsters, two daily infusions of melatonin did not have additive effects when they were separated by an interval of 2 or 3 hours; a single daily infusion of 10 hour duration completely inhibited testis growth, but two daily 5-hour melatonin infusions separated by an interval of 2 hours without infusion did not inhibit testicular growth. A gap of 1 hour between the two 5-hour infusions resulted in a partial additive effect, as indicated by incomplete inhibition of gonadal growth (Goldman, Darrow, & Yogev, 1984). In Siberian hamsters, if not in Syrian hamsters, the long-duration melatonin signal must be relatively continuous (i.e., not interrupted by a period greater than 1 hour) to evoke typical short-day responses. Significant species variations may exist in the degree to which a long-duration melatonin signal may be interrupted by a melatonin-free interval without impairing its effectiveness as a short-day signal. The observation that it may be possible in Syrian hamsters to evoke a short-day effect with an interrupted melatonin infusion does not contradict the hypothesis that the duration of the endogenous melatonin elevation normally codes for DL. Rather, these results may reflect the mechanism by which melatonin acts at the level of the target tissues; though melatonin is normally secreted in one continuous episode each night, the hormone may not need to be present throughout the entire duration of the signal in order to effect its actions. Two reviews summarize the results of timed melatonin infusion studies (Bartness, Powers, Hastings, Bittman, & Goldman, 1993; Goldman, 1991).

Circadian Mechanism for Photoperiodic Time Measurement

It is interesting to consider the importance of the melatonin signal in relation to earlier models that focused on the precise role of light in photoperiodic time measurement (PTM). In all mammals studied, PTM depends on an interaction of light with a circadian mechanism: The *timing* of light exposure, as opposed to its total *duration*, is paramount (J. A. Elliott, 1976). By use of skeleton photoperiods (twice-daily light pulses simulating dawn and dusk) and night-break experiments (interruption of a long night with a brief light pulse), long-day responses were elicited with very short light exposures, provided that light fell during certain parts of the subjective night (the so-called photoinductive phase). Further evidence of this type was obtained from experiments using resonance and T-cycle paradigms in which circadian entrainment patterns were such that light, at least intermittently, fell during the animals' subjective night (J. A. Elliott, 1976; Nelson, Bamat, & Zucker, 1982; Sullivan & Lynch, 1986). In external coincidence models, light is posited both to entrain a rhythm of photosensitivity and to induce long day responses when it coincides with a so-called photosensitive phase. In the internal coincidence model, light entrains two separate circadian oscillators and the phase relation between these oscillators determines whether long- or short-day responses occur.

The interaction of light with circadian mechanisms to measure DL is best understood in terms of the joint actions of light to (1) entrain a circadian rhythm of melatonin production and (2) suppress melatonin secretion acutely. *T*-Cycle, resonance, and night-interruption experiments that result in gonadal stimulation are often characterized by entrainment patterns that generate short-duration melatonin signals. Alternatively, these treatments can produce entrainment patterns congruent with long-duration melatonin signals, but those signals are interrupted by light, temporarily suppressing melatonin secretion (Goldman & Elliott, 1988). In rats, the pineal secretory rhythm may be regulated by two separate circadian oscillators, one of which controls the onset and the other the termination of each nightly episode of melatonin secretion. The phase relations between these two oscillators may determine the duration of nocturnal melatonin secretion and, ultimately, the nature of the photoperiodic response. In this model, both oscillators can be phase shifted by light pulses, and each oscillator can influence the phase of the other (Illnerova & Vanacek, 1989). This interesting hypothesis is most congruent with an internal coincidence model of PTM.

Controversy also exists as to the role of the circadian system in determining the response to a melatonin signal. Early observations that melatonin injections to intact long-day hamsters induced gonadal regression only at specific times of day pointed to a circadian rhythm of sensitivity to melatonin. The duration hypothesis accounted, at least in part, for these findings by positing that effective injections extended the endogenous nocturnal melatonin signal. Because melatonin signals which contain breaks above a certain threshold (see above) are not summated, injections in late morning or early afternoon did not add to the endogenous melatonin signal. A role of the circadian system in response to melatonin signals was suggested by the observation that daily injections over several weeks induced gonadal regression in pinealectomized male hamsters only when the injections occurred 2 hours after dark onset in long-DL hamsters (Stetson & Watson-Whitmyre, 1986). Others, however, have not replicated this result (Karp, Hastings, & Powers, 1991). Gonadal regression has also been reported in pinealectomized hamsters after twice-daily melatonin injections separated by 8–9 hours in a particular phase relation to the light/dark cycle (Watson-Whitmyre & Stetson, 1983).

The many timed melatonin infusion experiments in sheep and in Siberian and Syrian hamsters have failed to reveal evidence that the time of day of melatonin infusion is important; nevertheless, a few observations suggest the circadian system may play some role in the response to this hormone. In Syrian hamsters, long-duration melatonin infusions induced gonadal regression when presented at frequencies of 24 or 25 hours, but not at a frequency of 16 or 28 hours (Maywood *et al.*, 1992). A circadian mechanism may underlie this frequency dependence. Alternatively, more frequent melatonin infusions may compromise responsiveness by imposing too short a melatonin-free interval, whereas less frequent infusions may be ineffective because they provide excessively long melatonin-free intervals. This interpretation seems inconsistent with findings in *tau*-mutant hamsters, in which the period of the free-running circadian rhythm is approximately 20% shorter than in the wild type. Here the critical DL and minimum melatonin duration that induce gonadal regression are shortened by approximately 20% compared to wild type (Stirland, Grosse, Loudon, Hastings, & Maywood, 1995; Stirland, Mohammad, & Loudon, 1996b), and the range of inhibitory melatonin signals is shifted to higher frequencies (Stirland, Hastings, Loudon, & Maywood, 1996a). Furthermore, though ablation of the suprachiasmatic nucleus (SCN) failed to prevent melatonin-induced gonadal regression in Syrian hamsters (see below), it does abolish this response when the circadian phase of the infusions was varied over successive days (Grosse & Hastings, 1996). In pinealectomized Siberian hamsters, SCN lesions did prevent the action of exogenous melatonin to inhibit reproductive activity (Bartness, Goldman, & Bittman, 1990).

Photoperiod History Effects and the Pineal Melatonin Rhythm

Can the duration hypothesis of melatonin action be reconciled with photoperiod history effects? It is possible that photoperiod history modifies the response to a melatonin signal of given duration. Alternatively, photoperiod history may affect the melatonin-rhythm generating system so that different melatonin signals are produced under a given DL as a function of photoperiod history. Some evidence supports each of these hypotheses.

Because refractoriness only develops after several months of exposure to short DL, it may be considered a type of photoperiod history effect. Reports of changes in the melatonin rhythm in photorefractory Syrian hamsters (Lerchl & Nieschlag, 1992) and sheep (Almeida & Lincoln, 1984) have not been confirmed in two other studies in sheep in which the melatonin rhythm did not change with the onset of photorefractoriness (Malpaux, Moenter, Wayne, Woodfill, & Karsch, 1988; Malpaux, Robinson, & Karsch, 1987). Clearly, responsiveness to melatonin changes with photorefractoriness; for both Syrian hamsters and sheep, the typical short-day responses to long-duration melatonin signals is absent in animals that have become refractory to short days. Similarly, animals that are treated chronically with short-day type melatonin signals eventually become "refractory" to these signals (Bittman, 1978; Karsch *et al.*, 1986). In summary, photorefractoriness in Syrian hamsters and sheep appears to be primarily associated with the loss of typical short-day type responses to long-duration melatonin signals and does not require a change in the pattern of melatonin secretion.

Melatonin profiles do not appear to vary as a function of photoperiod history in adult mammals. The duration of elevated melatonin signals was similar in hamsters and sheep undergoing gonadal stimulation or inhibition as a function of their prior DL (Hastings *et al.*, 1989b; Hoffmann *et al.*, 1986; Niklowitz, Lerchl, & Nieschlag, 1994; Robinson & Karsch, 1987). These melatonin profiles, however, were measured 7–10 weeks after transfer to a common DL, and the possibility remains that melatonin duration varied between groups at some earlier point, or that other aspects of the melatonin signal, e.g. amplitude, might be important in these intermediate photoperiods (Niklowitz *et al.*, 1994).

Manipulation of melatonin patterns with injections or infusions further suggests that prior conditions may modify the response to a melatonin signal of a given duration. Adult male Syrian hamsters respond differently to a given melatonin signal as a function of previous melatonin/DL history; animals injected with melatonin 1 hour before dark onset in 16L, to lengthen melatonin duration, were reproductively stimulated when transferred to 12L and injections were discontinued (Karp, Dixon, & Powers, 1990). In contrast, saline-injected controls underwent gonadal regression upon transfer to the same 12L, a condition presumed to generate comparable melatonin patterns in the two groups. Similarly, in adult Siberian hamsters, short-DL males exposed to melatonin infusions gradually decreasing in duration from 10 to 7.5 hours over 12 weeks underwent gonadal stimulation, whereas an identical infusion schedule delivered to long-DL hamsters resulted in complete gonadal regression (Gorman & Zucker, 1997b).

Changes in generation of the melatonin signal, however, may underlie the maternally communicated photoperiod history effect observed in juvenile hamsters. Maternal DL during gestation did not alter responsiveness of offspring to melatonin signals of given durations, but did affect the duration of the melatonin signal generated by the pups (Shaw & Goldman, 1995a, c). At 18 days of age, male pups exhibited a longer duration pineal melatonin elevation in 14L if their mothers had been under 16L during gestation as compared to 10L. This may explain why juvenile testis development is slower in the pups from 16L than from 10L dams. Similar differences in pineal melatonin patterns related to the photoperiod of gestation were not observed in 18-day-old female pups (Shaw & Goldman, 1995a).

Three approaches—identification of melatonin uptake sites, localized melatonin infusions or implants, and ablation studies—have been used to define potential sites of melatonin action vis-à-vis photoperiodism. Several melatonin-uptake sites have been reported in mammals, and these vary among species. Most interest has centered on the pars tuberalis, the medial basal hypothalamus, and the SCN as possible sites of action.

Use of a radioiodinated analog of melatonin, I-MEL, with high specific activity permits *in vitro* autoradiographic mapping of melatonin-uptake sites. Among mammalian species, the variation in I-MEL-binding sites ranges from no binding in the central nervous system of ferrets to widespread binding in the diencephalon and telencephalon of sheep (Bittman, 1993; Morgan, Barrett, Howell, & Helliwell, 1994). In several myomorph rodents, dense neural I-MEL binding is found in the SCN, the paraventricular (PVT) and reunions nuclei of the thalamus, and the area postrema. Binding in the SCN was also seen in rabbits, little brown bats, and humans. All species examined showed high levels of I-MEL binding in the pars tuberalis (PT), the only binding site common to all mammals (Bittman, 1993; Morgan *et al.*, 1994).

The anterior medial hypothalamus may be an important site of melatonin action in several species. Unilateral melatonin implants in the supra- and retrochiasmatic areas of the hypothalamus, but not in the preoptic or posterior hypothalamic regions, induced uterine regression in female white-footed mice (Glass & Lynch, 1981, 1982). In males, removable implants in the preoptic/anterior hypothalamic region induced gonadal regression when presented for 10 hours daily, but not when presented for only 5 hours daily or when two 5-hour exposures daily were separated by an interval of 3 hours (Dowell & Lynch, 1987). In Syrian hamsters, bilateral implants of melatonin in the anterior or medial basal hypothalamus prevented short-day induced gonadal regression; the effects of these continuous-release brain implants was thus similar to the effects of subcutaneous continuous-release melatonin implants in this species (Hastings, Walker, Roberts, & Herbert, 1988). In juvenile Siberian hamsters, daily 10 hour infusions of melatonin (75 pg/day) unilaterally in the SCN, PVT, or reunions nucleus—all of which show uptake of labeled melatonin in this species (Duncan, Takahashi, & Dubocovich, 1989)—inhibited testis growth (Badura & Goldman, 1992). At a lower dose (20 pg/day), only microinfusions into the SCN were effective.

Ablation studies have also contributed to current information on sites of melatonin action. In several species, the SCN are not essential target sites for the reproductive inhibitory actions of melatonin; ablation of the SCN in male Syrian hamsters disrupts melatonin secretion (Lehman, Bittman, & Newman, 1984), but does not prevent gonadal regression in response to three daily melatonin injections (Bittman, Crandell, & Lehman, 1989; Bittman, Goldman, & Zucker, 1979) nor to melatonin infusion (Maywood *et al.*, 1990). Integrity of the SCN also is not required for the action of melatonin to delay implantation in the spotted skunk (Berria, DeSantis, & Mead, 1988), or for inhibition of prolactin (PRL) and progesterone secretion in mink (Bonnefond, Martinet, & Monnerie, 1990). In contrast, ablation of the SCN in Siberian hamsters blocked effects of exogenous melatonin infusions on reproductive parameters and body mass (Bartness *et al.*, 1990; Song & Bartness, 1996). Responsiveness to long-duration melatonin infusions was not compromised by destruction of the anterior hypothalamic region in Siberian hamsters (Song & Bartness, 1996), although these ablations prevented gonadal responsiveness to short DLs or melatonin infusions in Syrian hamsters (Bonnefond, *et al.*, 1989; Hastings,

Maywood, Ebling, Williams, & Titchener, 1991). Lesions at the juncture of the ventromedial and dorsomedial hypothalamus prevented gonadal regression in Syrian hamsters exposed to short DLs or given long-duration melatonin infusions, although photoperiodic regulation of PRL secretion was unaffected (Maywood & Hastings, 1995). This result suggests that pathways for melatonin action in the regulation of gonadotropins and PRL are separable in this species.

Studies in sheep have been particularly informative, and this is the only species in which an attempt has been made directly to assess the role of the PT. Melatonin implants in the mediobasal hypothalamus (MBH), but not other areas of the brain induced short-day changes in PRL and gonadotropin secretion in ewes and rams (Lincoln, 1992; Lincoln & Maeda, 1992; Malpaux, Daveau, Maurice, Gayrard, & Thiery, 1993; Malpaux, Skinner, & Maurice, 1995). Similar results from implants in the adjacent PT suggest that melatonin may be acting at either site (Lincoln, 1994), although one study detected no effect of PT implants on luteinizing hormone (LH) or PRL secretion in ewes (Malpaux, Daveau, Maurice, Locatelli, & Thiéry, 1994). A PT site of melatonin action on PRL secretion was suggested by experiments on pituitary-disconnected rams, where the anterior pituitary was presumably isolated from neural control, but not from the PT; hypothalamic control of PRL secretion was eliminated, but photoperiodic control of PRL was maintained (Lincoln & Clarke, 1994). The nature of the preparation precludes assessment of effects of photoperiod on gonadotrophin secretion. As with the results cited above for Syrian hamsters, these results also suggest separable sites of melatonin action for influencing the secretion of gonadotrophins and PRL, respectively.

Each of the methods described above has serious limitations for identifying critical sites of melatonin action. I-MEL-binding studies may fail to detect critical sites if melatonin substrates contain few cells, are spatially dispersed, or express lower affinity melatonin receptors (Bittman, 1993; Krause & Dubocovich, 1990). Localized application of melatonin may be compromised by diffusion to other brain regions, especially from implants to midline nuclei from which melatonin may diffuse to the ventricular system. The potentially nonphysiologic concentrations of hormone at the implant site also may complicate interpretation of the results. Negative findings are problematic where unilateral implants are employed; bilaterally paired structures may be receiving discrepant photoperiodic signals, and the results may thus differ from those obtained with bilateral implants (Hastings *et al.*, 1988). Elimination of photoperiodic responses after brain damage may reflect elimination of a site of melatonin action, but could just as plausibly derive from disruption of a part of a neural effector pathway that is essential for melatonin action, yet does not itself bind melatonin. Brain lesions might also interrupt pathways for the regulation of the pituitary gland so as to interfere with photoperiodic responses by causing chronically elevated or depressed secretion of relevant hormones. Given these limitations, confidence in the functional implications of a particular melatonin-binding site may only be justified if convergent analysis with several experimental methods yields compatible outcomes. This goal has not been reached for any putative melatonin target site. For example, melatonin implants in the MBH of Syrian hamsters prevented short-day decreases in PRL, but destruction of this region did not compromise the PRL response to melatonin; anterior hypothalamic lesions prevented delayed implantation in skunks, yet melatonin uptake was not detected in this region.

Despite the ubiquitous importance of melatonin for mediating photoperiodic effects in mammals, interspecific differences in the sites of melatonin action might be related to species differences in the proximate roles of melatonin on seasonal responses. In the golden-mantled ground squirrel and in the ferret, for example, pineal melatonin is not necessary to effect transitions in and out of reproductive condition, although it plays a large role in synchronizing such changes (Herbert, Stacey, & Thorpe, 1978; Zucker, 1985). The latter species, notably, exhibits no concentrated brain I-MEL binding; I-MEL binding is seen only in the PT in ferrets. The pineal of Syrian hamsters is necessary for gonadal inhibition in short DLs and for the development of refractoriness in short days and breaking refractoriness in long days. The pineal gland of Turkish hamsters, in contrast, is necessary for gonadal maintenance in long DLs, but not for timing of spontaneous recrudescence in pinealectomized animals (D. S. Carter & Goldman, 1982). These examples illustrate just a few species differences in proximate roles of the pineal gland in mediating seasonal responses. These various roles of melatonin may or may not be mediated by separate substrates. The lack of coherence in the sites of action of melatonin may be reduced after more comprehensive assessment of effects of lesions or implants and by clarification of the different roles of melatonin between species and among traits within a species.

Photoperiodic Regulation of Multiple Traits

The heavy emphasis on reproduction and relative neglect of other photoperiodic traits leaves open the question of whether the several photoperiodic responses in mammals are mediated by a common mechanism. This issue can be approached by asking two specific questions: (1) Is the pineal melatonin rhythm an essential component for PTM in all cases; i.e., are all mammalian photoperiodic responses mediated by the pineal gland? (2) Are the formal properties of the melatonin dependence (e.g., critical DL, development of refractoriness, influence of photoperiod history) identical for gonadal and nonreproductive traits? Some evidence bearing on the second question has been reviewed above. Because several traits are influenced by gonadal hormone concentrations which are under pineal-dependent photoperiodic control, we consider the extent to which various traits are photoperiodically regulated independently of seasonal changes in gonadal hormone secretion.

Body Weight

Changes in body weight and fat content are common tactics for coping with winter energy shortages. In some species, fat stores are accumulated in summer and depleted to supplement reduced food intake in winter. In other species, fat stores and body weight decline prior to winter, and this strategy is thought to reduce overall energy expenditure by decreasing the need for foraging to maintain the larger body mass characteristic of the summer condition (Dark & Zucker, 1985).

Although body weights are influenced by gonadal hormones (Wade, 1986), seasonal changes in gonadal hormone secretion account for only a part of the photoperiodic modulation of this trait in several species (Bartness & Wade, 1984; Dark & Zucker, 1984; Loudon, 1994). Gonadectomized voles and hamsters exhibit decreases in body weight upon transfer to short DLs, and in voles, these decrements are followed by a spontaneous increase in body weight 15–25 weeks later (Dark & Zucker, 1985; Vitale, Darrow, Duncan, Shustak, & Goldman, 1985). Seasonally regulated changes in PRL secretion may contribute to changes in body weight in Siberian hamsters (Niklowitz & Hoffmann, 1988), although another study failed to reveal a direct relationship between PRL and body weight in this species (Bartness, Wade, & Goldman, 1987).

Few studies have addressed the question of critical DL for body weight. The seasonal decline in body weight occurs gradually over 12–14 weeks in Siberian hamsters maintained in a natural photoperiod. After weight loss was accelerated by food restriction, the return to ad libitum feeding caused hamsters to increase body weights to match those of control animals. These results suggest that a variable body-weight set point that changes gradually with ambient DL or with time of exposure to short DLs is defended in this species (Dark & Zucker, 1985; Steinlechner, Heldmaier, & Becker, 1983). A graded response to DL is suggested by the finding of greater weight gain in female Syrian hamsters upon transfer to 8L than 10L, even though both short DLs are equally effective in terminating estrous cycles (Wade, Bartness, & Alexander, 1986).

In some rodents, a component of the photoperiod-driven change in body weight may be pineal independent. The weight increase observed in Syrian hamsters transferred from long to short DLs was not abolished by pinealectomy (Bartness & Wade, 1985; Hoffman, Davidson, & Steinberg, 1982). In contrast, pinealectomy completely blocked photoperiod-dependent changes in body weight in Siberian hamsters (Vitale *et al.*, 1985).

Pelage

Whereas the optimal season to mate varies between species as a function of gestation length and environmental factors, the increased energetic challenges in winter faced by temperate-zone mammals can be ameliorated by development of a thicker winter pelage. Short- and long-day breeders both undergo winter molts, suggesting that the endocrine mechanisms mediating reproductive activity and seasonal molt are separable, at least interspecifically. The molt to winter pelage depends on a decrease in circulating PRL concentrations (reviewed in Goldman & Nelson, 1993) and is largely independent of gonadal hormones. The role of photoperiod-mediated changes in PRL secretion in regulation of seasonal changes in pelage has been most thoroughly studied in Siberian hamsters and in mink (Duncan & Goldman, 1984, 1985; Martinet, Allain, & Weiner, 1984), and in these species the pelage rhythm is pineal dependent. In male meadow voles, pinealectomy eliminated short-day induced changes in guard hair length, but not in underhair growth (Smale, Dark, & Zucker, 1988a). Among prairie voles, however, pinealectomy eliminated short-day responses for both types of fur (Smale, Nelson, & Zucker, 1988b).

Prolactin

Most photoperiodic mammals exhibit large seasonal fluctuations in circulating PRL, with higher concentrations in summer (long DL) than winter (short DL). In male Syrian hamsters, the short-day decline in circulating PRL augments the effects of reduced LH and follicle-stimulating hormone (FSH) in causing testicular regression (Bartke, Croft, & Dalterio, 1975). This reproduction-related action of PRL in hamsters does not, obviously, pertain to winter-breeding mammals in which circulating PRL concentrations also decline in short days. The consistent relation between DL and PRL secretion across species may reflect the important role of this hormone as a mediator of seasonal pelage changes (above). For the most part, photoperiod-driven changes in PRL are pineal dependent, but a pineal-independent component has been implicated in some species. Pinealectomy failed to eliminate decreases in PRL after blinding in female, but not male Syrian hamsters (Blask, Leadem, Orstead, & Larsen, 1986). Similarly, knife cuts ventral to the paraventricular nucleus (PVN) of the hypothalamus, which are believed to interrupt pineal melatonin secretion, do not prevent short-day induced decreases in PRL (Badura & Goldman, 1994), nor does ablation of the PVN prevent short-day decreases in PRL in Siberian hamsters (Bittman, Bartness, & Goldman, 1991).

The critical DLs for PRL and gonadotrophin secretion in male Syrian hamsters are similar (Hastings *et al.*, 1989b). Secretion of these hormones, however, is not similarly dependent on photoperiod history: Transfer from 16L to 12L reduced PRL concentrations to typical SD levels, whereas transfer from 8L to 12L modestly increased titers in only one of three experiments (Hastings *et al.*, 1989b). Gonadotropin concentrations in 12L were a function of photoperiod history. Moreover, PRL concentrations in male Syrian hamsters are affected by photoperiod by 5 weeks of age, whereas regulation of FSH concentrations by DL appears only several weeks later (Donham, Palacio, & Stetson, 1994). Gonadal recrudescence and increases in FSH and PRL also follow different time courses under permissive DLs: In 12L, gonadal recrudescence and FSH concentrations increased after 20 weeks, but PRL remained depressed for at least 30 weeks. Dissociated time courses for changes in PRL and FSH secretion were not apparent in shorter DLs (Donham *et al.*, 1994). Different critical DLs for inhibition of PRL and FSH may exist, moreover, for juvenile Siberian hamsters gestated in long DLs. Blood PRL concentrations were decreased by postnatal DLs of 12L, whereas suppression of FSH was reported in 13L (Shaw & Goldman, 1995b).

Reproductive Behavior

Fewer male Syrian hamsters achieve intromission and ejaculation in short compared to long DLs (Morin & Zucker, 1978; Power *et al.*, 1989). Because reproductive behaviors largely depend on the presence of gonadal hormones, much of the short-day effect may be due to the reduced titers of sex steroid hormones. There is also evidence for a steroid-independent modulation of reproductive behaviors by DL: In the absence of gonadal hormones, copulatory deficits developed more rapidly among SD than LD hamsters, and restoration of copulatory behaviors is less complete in the former animals after administration of testosterone (Miernicki, Karp, & Powers, 1990); normal mating behavior is maintained by lower testosterone concentrations in castrated hamsters housed in long than in short DLs (Campbell, Finkelstein, & Turek, 1978; Powers *et al.*, 1989). Short DLs also decrease the proportion of ovariectomized hamsters that display lordosis and aggressive behaviors during steroid hormone treatment (Badura, Sisk, & Nunez, 1987a; Badura, Yant, & Nunez, 1987b; Bittman, Hegarty, Layden, & Jonassen, 1990; A. S. Elliott & Nunez, 1992; Karp & Powers, 1993).

Steroid-independent and steroid-dependent photoperiodic traits respond differently to DL: Transfer of intact male hamsters from 8L to 12L stimulated gonadal growth and elevated testosterone concentrations but failed to restore copulatory behavior (Powers *et al.*, 1989). Moreover, the gonadal response to 12L varied as a function of photoperiod history, but 12L was uniformly inhibitory as assessed in terms of reproductive behavior (Hastings *et al.*, 1989b; Powers *et al.*, 1989). Finally, the time course for the development of photorefractoriness appears to differ for gonadal regression and behavioral responsiveness to steroid hormones (Honrado, Bird, & Fleming, 1991a; Powers *et al.*, 1989). In female hamsters, for example, lordosis, approach, and aggressive behaviors all returned to initial values 20 weeks after exposure to short DLs, whereas uterine weights and percentage of animals displaying estrous cycles reached initial values only after 27 weeks (Honrado, Paclik, & Fleming, 1991b).

There is disagreement whether reduced sensitivity to gonadal hormones in short DLs requires participation of the pineal gland. In female Syrian hamsters, short DLs reduced behavioral sensitivity to EB after knife cuts ventral to the PVN, which disrupt pineal melatonin secretion (Badura *et al.*, 1987a); pinealectomy, however, did not prevent a short-day reduction in behavioral responsiveness to estradiol, nor did daily melatonin injections induce short-day loss of responsiveness (Badura & Nunez, 1989). In another strain of Syrian hamsters, pinealectomy did eliminate the effect of short DL on this measure (Karp & Powers, 1993). In male hamsters, pinealectomy eliminates the short-day induced changes in behavioral responsiveness to gonadal hormones (Miernicki *et al.*, 1990).

In summary, the pineal gland comprises an essential part of the mechanism for photoperiodic responses in mammals, except in a few instances in some species where particular photoperiodic responses occur independently of the pineal. The formal properties of the pineal dependence of several traits, however, vary. Critical DLs may differ, as may the influence of photoperiodic history and the time course of photorefractoriness.

Sex Differences in Seasonality

Because the role of melatonin as a photoperiodic hormone has been most intensively studied in relation to seasonal breeding, there naturally has been some interest in exploring possible sex differences in physiologic actions of this hormone. The basic mechanism for involvement of melatonin in photoperiodic time measurement appears to be similar between sexes. Asymmetries exist between males and females with respect to the timing of their investments in the reproductive process. For most mammals, the major investment on the part of the male precedes that of the female. Several weeks are required to complete the process of spermatogenesis as compared to only a few days for the maturation of ovarian follicles and ova. Where both sexes are starting from a condition of gonadal regression, males will require more time than females to achieve breeding status. Further, in many species, males may invest considerable effort in establishing territories before mating occurs. In contrast, the major reproductive effort on the part of the female usually begins only after mating, and the male's investment frequently ends at this time (Bronson, 1989).

Consider the case of the prairie vole, *Microtus ochrogaster*. In this species, males exhibit testicular regression in response to short days, but the females show little reproductive response to photoperiod. Rather, the females are induced into estrus by pheromonal stimulation from a reproductively active male; the odor of a castrated or a short-day male will not provide this stimulation (Nelson, 1985). Thus, female voles are seasonal, but only because they tend to follow the photoperiod-cued seasonal rhythm of the males. In this example, the ability of females to respond very rapidly to a social stimulus may have freed them from the need to rely on photoperiod as a predictive cue (C. S. Carter, Getz, Gavish, McDermott, & Arnold, 1980; Nelson, 1985).

Castrated male Turkish hamsters exhibited an extended hibernation season as compared to intact males (Hall, Bartke, & Goldman, 1982). In contrast, ovariectomy failed to alter the duration of the hibernation season in female Turkish hamsters (Hall & Goldman, 1982). This may reflect basic sex differences in the length of the hibernation season and the need for males to become active and establish territories before the emergence of females from hibernacula (French, 1988).

Seasonality in an Ecophysiologic Context

Gradual and Abrupt Photoperiod Transitions

In Siberian hamsters, 8 weeks after a transfer from 16L to 8L the duration of elevated nocturnal melatonin secretion remains compressed and the gonads remain large if the transition is accomplished solely by advancing the time of dark onset; the melatonin signal expands more rapidly and the gonads undergo regression sooner if the same DL transition is effected by delaying light onset (Gorman, Freeman, & Zucker, 1997; Hoffmann & Illnerova, 1986). Thus, the nature of the transition from long to short days can affect the rate of expression of short-day traits. DLs in nature change continuously and gradually, never more than 5–6 minutes per day, with the magnitude of the daily change varying with latitude and time of year. The vast majority of studies of mammalian photoperiodism eliminates this pattern of daily photoperiodic change and instead employs experimental paradigms involving static photoperiods and abrupt shifts (e.g., 2–8 hours) to a longer or shorter DL accomplished on a single day. These simplified photoperiodic regimens, although useful as analytical probes, lack much of the predictive information inherent in the natural photocycle. It is of interest to assess whether the photoperiodic relations defined under artificial conditions obtain in more naturalistic contexts and to determine whether additional types of information are extracted by animals exposed to simulated natural photoperiods.

Gradual and abrupt changes in DL elicit different patterns of seasonal change in several contexts. Abrupt decreases in DL result in greater between-hamster synchronization of testicular regression than do gradual transfers (Heideman & Bronson, 1993) and also appear to extend the range of DLs over which photoperiod history affects gonadal function (Gorman & Zucker, 1995a; Rivkees *et al.*, 1988). Rates of gonadal regression and body weight loss, moreover, differ in hamsters exposed to a pattern of DLs decreasing gradually versus abruptly (Gorman *et al.*, 1997). Shorter, but increasing DLs can induce greater somatic and gonadal growth than fixed, long DLs, further demonstrating that the pattern of change in natural DLs carries information which animals may use in addition to absolute DL (Gorman, 1995).

In male Siberian hamsters with a short DL photoperiod history, a pattern of melatonin infusions gradually decreasing in duration from 10 to 7.5 hours stimulates gonadal growth, but the identical pattern of melatonin infusions induces gonadal regression in hamsters with a long DL photoperiod history (Gorman & Zucker, 1997b). The photoperiod history effect may result, in part, from the abrupt change in melatonin duration experienced by the latter group. The pattern of long, but decreasing melatonin durations is experienced in nature only by animals with short DL photoperiod histories, and thus would elicit gonadal growth. That animals with a long-day history respond in an opposite fashion suggests that unnatural DL and melatonin transitions can obscure the ecologically relevant interpretation of the

DL/melatonin signal. Abrupt changes in DL and melatonin duration, which are the norm in photoperiodism research, are powerful stimuli that override subtler signals intrinsic to the natural pattern of change in DL and melatonin duration.

Phasing of Seasonal Traits

If various seasonal traits have different critical DLs, photoperiod history dependences, and graded responses to DL, then abrupt transfers between DLs will alter the relative timing of changes in these traits. For example, Siberian hamsters in a 60° latitude photoperiod initiate gonadal regression when DLs decrease gradually to approximately 13.6L in late summer (Gorman & Zucker, 1995a), most likely as a consequence of decreased FSH concentrations (Milette, Schwartz, & Turek, 1988). If control of PRL secretion depends on a second, shorter critical DL that is independent of photoperiodic history, then initiation of decreased PRL secretion will not occur until several weeks later, when ambient DL reaches the critical value. PRL-dependent adaptations that result in the winter molt might thus be initiated many weeks after onset of gonadal regression. In contrast, abrupt transfer from long to short DLs will induce rapid and nearly simultaneous decreases in concentrations of FSH and PRL, and the latency between FSH- and PRL-dependent changes will be shorter after abrupt than gradual decreases in DL.

Plasticity in the timing of gonadal recrudescence relative to gonadal regression also supports this formulation. The interval between onset of testicular regression and gonadal recrudescence is 4 weeks shorter in Siberian hamsters housed in a 60° latitude simulated natural photoperiod (SNP) than those maintained in a 40° SNP (Gorman & Zucker, 1995a). In a third condition in which 2 hours of DL were subtracted from every day of a 40° SNP (SNP 40° − 2 hours), the interval between regression and recrudescence was shorter by an additional 4 weeks than in the natural 40° SNP.

Two models may explain these findings, both of which suggest differential responses to various types of short DLs. First, gonadal regression and the triggering of an interval timer of fixed duration may occur at different decreasing DLs in late summer or fall. The interval between regression and recrudescence would be shorter, other things being equal, when DLs decreased more rapidly as in SNP 60° compared to SNP 40°. When regression begins at comparatively shorter DLs, as was the case in SNP 40° − 2 hours, the interval between regression and the DL necessary to trigger the interval timer may be shorter still than that observed in SNP 40°. Other evidence supports the idea that DLs capable of inducing gonadal regression (e.g., 14L) do not trigger the interval timer underlying recrudescence in the same manner as do other shorter DLs (Duncan *et al.*, 1985; Gorman & Zucker, 1995b). Alternatively, the duration of the interval timer may not be fixed, but instead may depend upon the duration of the short DL. The latter possibility is suggested by the finding that spontaneous gonadal development occurs 4 weeks earlier in hamsters gestated and raised in 10L than those housed throughout in 12L (Gorman & Zucker, 1995a).

Evolution of Photoperiodism in Mammals

Construction of a plausible model of the evolution of mammalian photoperiodism is informed by two lines of evidence: the neuroendocrine basis for PTM in various vertebrates and the genetic basis for photoperiodism. These are considered below.

Remarkably similar involvement of the pineal gland and its hormone, melatonin, has been observed for the photoperiodic responses of a variety of mammals. Well-studied species include representatives of the rodents, mustelids, ungulates, and two species of marsupials (Goldman, 1983). In all cases, pinealectomy results in complete or nearly complete loss of the animal's abilities to display the species-typical seasonal physiologic adjustments to changes in photoperiod; appropriate treatment with melatonin evokes these seasonal responses. The photoperiod-related change in the duration of the nocturnal elevation of circulating melatonin also has been reported in diversely related mammals. The causal relation between long-duration melatonin elevation and short-day-type (or winter) responses has been critically tested in only three species—sheep, Siberian hamsters, and Syrian hamsters—but there is less extensive evidence that this relation obtains in other species, including one marsupial. The nature of the pineal melatonin signal and its mode of action may be similar in all photoperiodic mammals.

The commonality of neuroendocrine substrates outlined above supports the concept of a single ancestral origin for the mammalian photoperiodic mechanism. This is reinforced by the observation that the pineal glands of birds and lizards do not appear to be a similarly essential component of their photoperiodic systems (Hastings, Vance, & Maywood, 1989a; Underwood & Goldman, 1987). Thus, the pineal route is apparently not the only means that has evolved for vertebrate photoperiodism, and its near universal presence among photoperiodic mammals, but not in their vertebrate progenitors, suggests that this particular version of the neuroendocrine substrate for photoperiodism became fixed during early mammalian history, or perhaps in a line of reptilian ancestry that led to the appearance of mammals.

VARIABILITY IN PHOTOPERIODIC RESPONSIVENESS WITHIN SPECIES

GENETIC BASIS. Variability in photoperiodic responsiveness has been observed between different breeding populations in a single species. For example, white-footed mice captured in Connecticut exhibit gonadal regression and development of winter pelage when exposed to short photoperiod, but these responses are not apparent in mice captured in Georgia (Carlson, Zimmermann, & Lynch, 1989; Lynch, Heath, & Johnston, 1981). Selection experiments demonstrate that variation in photoresponsiveness has a heritable basis in deer mice and Siberian hamsters (Desjardins, Bronson, & Blank, 1986; Kliman & Lynch, 1992). This type of variation in photoperiodic responsiveness is of considerable interest because when members of a species invade new habitats, they are likely to be subjected to altered climatic conditions that may favor changes in seasonal strategies. Heritable variations in photoresponsiveness within breeding populations could serve as the basis for evolution of adaptations to new environmental conditions.

Different photoperiodic phenotypes have also been observed within breeding populations in the field in deer mice and prairie voles. In both species, a small proportion of the population in the field engages in winter breeding; this is probably related to the observation that some wild-caught individuals fail to undergo reproductive inhibition when exposed to short days in the laboratory (Johnston & Zucker, 1980; Kerbeshian, Bronson, & Bellis, 1994). The maintenance of such alternative phenotypes suggests that resistance to winter inhibition of reproduction may sometimes result in increased genetic fitness. In small rodents with relatively short life expectancies in the field (3–4 months is not atypical), most adults may not survive the winter. In relatively mild winters, individuals that "gamble" on reproduction may meet with some success and thereby contribute disproportionately to the gene pool of the following spring's population as compared to their more conservative conspecifics (Nelson, 1987).

PHYSIOLOGIC BASIS. The neuroendocrine basis for variations in photoperiodic responsiveness within a species has been investigated in a few instances. In wild caught white-footed mice and deer mice, individuals that failed to exhibit gonadal regression in short days also failed to undergo gonadal involution when treated with melatonin, using regimens that did evoke regression in photoresponsive mice (Blank & Freeman, 1991; Heath & Lynch, 1982). For these species, individual variations in photoperiodic responsiveness appear to be based on differences in the melatonin response system rather than differences in patterns of melatonin secretion; this is also suggested by the observation that short-day-nonresponsive white-footed mice from Georgia exhibited pineal melatonin rhythms that were similar to those of their photoresponsive conspecifics from Connecticut (Lynch, Sullivan, Heath, & Tamarkin, 1982).

Variations in short-day responsiveness have also been observed among individuals in laboratory-bred populations of Siberian hamsters. In this case, the neuroendocrine basis for the variation differs from that described for the two species of mice described above. In Siberian hamsters, circadian timing differs between individuals that do and do not respond to DL. Thus, the nonresponders and responders almost invariably have free-running periods in constant darkness that are longer and shorter than 24 hours, respectively. Furthermore, the phase angle of entrainment of the circadian activity rhythm to short photoperiod is delayed by several hours in nonresponders as compared to that of responders (Gorman & Zucker, 1997a; Puchalski & Lynch, 1986, 1988). In Siberian hamsters, the failure of a subset of the population to respond to short days reflects this difference in phase angle of entrainment, which prevents the expansion of the nocturnal melatonin elevation that is required for induction of short-day responses (Freeman & Goldman, 1997a, b; Puchalski & Lynch, 1988). The genetic basis for the interindividual variations in photoperiodic responsiveness in Siberian hamsters was demonstrated in artificial selection studies that produced breeding lines with much higher proportions of nonresponders than recorded in the original breeding stock (Freeman & Goldman, 1997a; Kliman & Lynch, 1992; Goldman, Dhandapani, & Goldman 2000).

The short-day-nonresponsive phenotype in laboratory-bred Siberian hamsters can be reversed by allowing the animals access to running wheels at an early age. In one breeding line achieved through selection for the nonresponsive trait, about 80% of the male offspring fail to exhibit testis regression or molt to winter pelage when exposed to short days for several weeks, beginning at 19–20 days of age. However, when such animals are provided free access to running wheels during short-day exposure, they exhibit the typical short-photoperiod responses. The daily act of wheel-running appears to cause a phase advance of 4–7 hours in the time of activity onset, probably reflecting a general shift in the phase relation between the circadian system and the light/dark cycle (Freeman & Goldman, 1997b).

The different neuroendocrine bases for variations in photoperiodic responsiveness in white-footed mice and deer mice on the one hand and Siberian hamsters on the other could reflect fundamental species differences; alternatively, the fact that the mouse populations, unlike the Siberian hamsters, were not subjected to artificial selection to accentuate the differences in responsiveness and had not been maintained in the laboratory for several generations may be most relevant. Laboratory selection studies provide powerful tools for revealing the reservoir of genetic variation upon which natural selection may act; we cannot, however, assume that the genetic basis for the results of a particular episode of artificial selection will be identical to those favored by natural selection.

The probability that an individual rodent will respond to short days may also vary with age, as older individuals are less likely to exhibit gonadal regression in short photoperiods (Christian, 1980; Donham, Horton, Rollag, & Stetson, 1989). This phenomenon has been observed in Siberian hamsters (Freeman & Goldman, 1997a), but in this species, it appears that it is a period of exposure to very long DLs, rather than increased age itself, that renders individuals unresponsive to short days (Gorman & Zucker, 1997a; Goldman, Dhandapani, & Goldman, 2000). Thus, hamsters born early in the breeding season will experience the maximal photoperiods of summer and may thereby remain reproductively active despite exposure to short days later in the year. Hamsters born later during the breeding season will not experience such long DLs and will consistently exhibit inhibition of reproduction as photoperiod decreases in late summer/fall.

Summary

Many temperate zone mammals rely on photoperiodic cues to regulate seasonal cycles in behavior and physiology; it is questionable whether photoperiodism is used by tropical species to phase seasonal rhythms (Bronson & Heideman, 1994). Other cues—food and water availability, ambient temperature, availability of specific plant compounds—also can effectively synchronize seasonal rhythms, but day length is probably the most common zeitgeber; by virtue of its unique reliability as an indicator of time of year, day length is an excellent predictor of future environmental conditions. The neuroendocrine basis of photoperiodism is perhaps best understood in mammals. Only for this vertebrate class has a specific hormone, melatonin, been identified as the neuroendocrine transducer of day length on a host of behavioral and physiologic functions. The identification of several neuroendocrine components of the mammalian photoperiodic mechanism provides a basis for future cellular and molecular analysis.

Most experimental analyses have probed mammalian photoperiodic systems with abrupt and large changes in day length, typically 4- to 6-hour shifts achieved in a single day. In nature, animals appear to track much smaller and gradual daily changes in day length and are capable of extracting predictive information not accessed in the classical experimental paradigms. Studies that utilize simulated natural photoperiods have explored this issue and should prove useful in specifying the significance of photoperiodic history on photoperiodic responses.

Many of the basic principles of mammalian photoperiodism, if not the concrete physiologic details, may apply to nonmammalian vertebrates. Notably, the use of the circadian system to measure day length is widespread among diverse groups of animals and plants. The ubiquity of circadian organization of photoperiodism may reflect evolutionary continuity between the photoperiodic mechanisms of the several vertebrate classes and perhaps also among other phyla.

Almeida, O. F. X., & Lincoln, G. A. (1984). Reproductive refractoriness in rams and accompanying changes in the patterns of melatonin and prolactin secretion. *Biology of Reproduction, 30,* 143–158.

Badura, L. L., & Goldman, B. D. (1992). Central sites mediating reproductive responses to melatonin in juvenile male Siberian hamsters. *Brain Research, 598,* 98–106.

Badura, L. L., & Goldman, B. D. (1994). Prolactin secretion in female Siberian hamsters following hypothalamic deafferentation: Role of photoperiod and dopamine. *Neuroendocrinology, 59,* 49–56.

Badura, L. L., & Nunez, A. A. (1989). Photoperiodic modulation of sexual and aggressive behavior in female golden hamsters (*Mesocricetus auratus*): Role of the pineal gland. *Hormones and Behavior, 23,* 27–42.

Badura, L. L., Sisk, C. L., & Nunez, A. A. (1987a). Neural pathways involved in the photoperiodic control of reproductive physiology and behavior in female hamsters (*Mesocricetus auratus*). *Neuroendocrinology, 46,* 339–344.

Badura, L. L., Yant, W. R., & Nunez, A. A. (1987b). Photoperiodic modulation of steroid-induced lordosis in golden hamsters. *Physiology and Behavior, 40,* 551–554.

Bartke, A., Croft, B. T., & Dalterio, S. (1975). Prolactin restores plasma testosterone levels and stimulates testicular growth in hamsters exposed to short day-length. *Endocrinology, 97,* 1601–1604.

Bartness, T. J., & Goldman, B. D. (1989). Mammalian pineal melatonin: A clock for all seasons. *Experientia, 45,* 939–945.

Bartness, T. J., & Wade, G. N. (1984). Photoperiodic control of body weight and energy metabolism in Syrian hamsters (*Mesocricetus auratus*): Role of pineal gland, melatonin, gonads and diet. *Endocrinology, 114,* 492–498.

Bartness, T. J., & Wade, G. N. (1985). Photoperiodic control of seasonal body weight cycles in hamsters. *Neuroscience and Biobehavioral Reviews, 9,* 599–612.

Bartness, T. J., Wade, G. N., & Goldman, B. D. (1987). Are the short-photoperiod-induced decreases in serum prolactin responsible for the seasonal changes in energy balance in Syrian and Siberian hamsters? *Journal of Experimental Zoology, 244,* 437–454.

Bartness, T. J., Goldman, B. D., & Bittman, E. L. (1990). SCN lesions block the reception of melatonin daylength signals in Siberian hamsters. *American Journal of Physiology, 260,* R102–R112.

Bartness, T. J., Powers, J. B., Hastings, M. H., Bittman, E. L., & Goldman, B. D. (1993). The timed infusion paradigm for melatonin delivery: What has it taught us about the melatonin signal, its reception, and the photoperiodic control of seasonal responses? *Journal of Pineal Research, 15,* 161–190.

Berria, M., DeSantis, M., & Mead, R. A. (1988). Effects of suprachiasmatic nuclear ablation and melatonin on delayed implantation in the spotted skunk. *Neuroendocrinology, 48,* 371–375.

Bittman, E. L. (1978). Hamster refractoriness: The role of insensitivity of pineal target tissues. *Science, 202,* 648–650.

Bittman, E. L. (1993). The sites and consequences of melatonin binding in mammals. *American Zoologist, 33,* 200–211.

Bittman, E. L., & Karsch, F. J. (1984). Nightly duration of pineal melatonin secretion determines the reproductive response to inhibitory day length in the ewe. *Biology of Reproduction, 30,* 585–593.

Bittman, E. L., & Zucker, I. (1981). Photoperiodic termination of hamster refractoriness: Participation of the pineal gland. *Biology of Reproduction, 24,* 568–572.

Bittman, E. L., Goldman, B. D., & Zucker, I. (1979). Testicular responses to melatonin are altered by lesions of the suprachiasmatic nuclei in golden hamsters. *Biology of Reproduction, 21,* 647–656.

Bittman, E. L., Crandell, R. G., & Lehman, M. N. (1989). Influences of the paraventricular and suprachiasmatic nuclei and olfactory bulbs on melatonin responses in the golden hamster. *Biology of Reproduction, 40,* 118–126.

Bittman, E. L., Hegarty, C. M., Layden, M. Q., & Jonassen, J. A. (1990). Influences of photoperiod on sexual behaviour, neuroendocrine steroid receptors and adenohypophysial hormone secretion and gene expression in female golden hamsters. *Journal of Molecular Endocrinology, 5,* 15–20.

Bittman, E. L., Bartness, T. J., & Goldman, B. D. (1991). Suprachiasmatic and paraventricular control of photoperiodism in Siberian hamsters. *American Journal of Physiology, 260,* R90–R101.

Blank, J. L., & Freeman, D. A. (1991). Differential reproductive response to short photoperiod in deer mice: Role of melatonin. *Journal of Comparative Physiology A, 169,* 501–506.

Blask, D. E., Leadem, C. A., Orstead, K. M., & Larsen, B. R. (1986). Prolactin cell activity in female and male Syrian hamsters: An apparent sexually dimorphic response to light deprivation and pinealectomy. *Neuroendocrinology, 42,* 15–20.

Bonnefond, C., Walker, A. P., Stutz, J. A., Maywood, E., Juss, T. S., Herbert, J., & Hastings, M. H. (1989).

The hypothalamus and photoperiodic control of FSH secretion by melatonin in the male Syrian hamster. *Journal of Endocrinology, 122,* 247–254.
Bonnefond, C., Martinet, L., & Monnerie, R. (1990). Effects of timed melatonin infusions and lesions of the suprachiasmatic nuclei on prolactin and progesterone secretions in pregnant or pseudopregnant mink (*Mustela vison*). *Journal of Neuroendocrinology, 2,* 583–591.
Bronson, F. H. (1989). *Mammalian reproductive biology.* Chicago: University of Chicago Press.
Bronson, F. H., & Heideman, P. D. (1994). Seasonal regulation of reproduction in mammals. In E. Knobil & J. D. Neill (Eds.), *Physiology of reproduction* (pp. 541–583). New York: Raven Press.
Campbell, C. S., Finkelstein, J. S., & Turek, F. W. (1978). The interaction of photoperiod and testosterone on the development of copulatory behavior in castrated male hamsters. *Physiology and Behavior, 21,* 409–415.
Carlson, L. L., Zimmermann, A., & Lynch, G. R. (1989). Geographic differences for delay of sexual maturation in *Peromyscus leucopus*: Effects of photoperiod, pinealectomy, and melatonin. *Biology of Reproduction, 41,* 1004–1013.
Carter, C. S., Getz, L. L., Gavish, L., McDermott, J. L., & Arnold, P. (1980). Male-related pheromones and the activation of female reproduction in the prairie vole (*Microtus ochrogaster*). *Biology of Reproduction, 23,* 1038–1045.
Carter, D. S., and Goldman, B. D. (1982). Pineal is required for testicular maintenance in the Turkish hamster (*Mesocricetus brandti*). *Endocrinology, 111,* 863–871.
Carter, D. S., & Goldman, B. D. (1983a). Antigonadal effects of timed melatonin infusion in pinealectomized male Djungarian hamsters (*Phodopus sungorus sungorus*): Duration is the critical parameter. *Endocrinology, 113,* 1261–1267.
Carter, D. S., & Goldman, B. D. (1983b). Progonadal role of the pineal in the Djungarian hamster (*Phodopus sungorus sungorus*): Mediation by melatonin. *Endocrinology, 113,* 1268–1273.
Christian, J. J. (1980). Regulation of annual rhythms of reproduction in temperate small rodents. In A. Steinberger & E. Steinberger (Eds.), *Testicular development, structure, and function* (pp. 367–380). New York: Raven Press.
Czyba, J. C., Girod, C., & Durand, N. (1964). Sur l'antagonisme épiphysohypophysiore et les variations saisonniers de la spermatogenese chez le Hamster doré (*Mesocricetus auratus*). *Comptes Rendus des Séances de la Societé de Biologie et de ses Filiales, 158,* 742–745.
Dark, J., & Zucker, I. (1984). Gonadal and photoperiodic control of seasonal body weight changes in male voles. *American Journal of Physiology, 247,* R84–R88.
Dark, J., & Zucker, I. (1985). Seasonal cycles in energy balance: Regulation by light. *Annals of the New York Academy of Sciences, 453,* 170–181.
Dark, J., Johnston, P. G., Healy, M., & Zucker, I. (1983). Latitude of origin influences photoperiodic control of reproduction of deer mice (*Peromyscus maniculatus*). *Biology of Reproduction, 28,* 213–220.
Desjardins, C., Bronson, F. H., & Blank, J. L. (1986). Genetic selection for reproductive photoresponsiveness in deer mice. *Nature, 322,* 172–173.
Donham, R. S., Horton, T. H., Rollag, M. D., & Stetson, M. H. (1989). Age, photoperiodic responses, and pineal function in meadow voles, *Microtus pennsylvanicus. Journal of Pineal Research, 7,* 243–252.
Donham, R. S., Palacio, E., & Stetson, M. H. (1994). Dissociation of the reproductive and prolactin photoperiodic responses in male golden hamsters. *Biology of Reproduction, 51,* 366–372.
Dowell, S. F., & Lynch, G. R. (1987). Duration of the melatonin pulse in the hypothalamus controls testicular function in pinealectomized mice (*Peromyscus leucopus*). *Biology of Reproduction, 36,* 1095–1101.
Duncan, M. J., & Goldman, B. D. (1984). Hormonal regulation of the pelage color cycle in the Djungarian hamster, *Phodopus sungorus.* II. Role of prolactin. *Journal of Experimental Zoology, 230,* 97–103.
Duncan, M. J., & Goldman, B. D. (1985). Physiological doses of prolactin stimulate pelage pigmentation in Djungarian hamster. *American Journal of Physiology, 248,* R664–R667.
Duncan, M. J., Goldman, B. D., DiPinto, M. N., & Stetson, M. H. (1985). Testicular function and pelage color have different critical daylengths in the Djungarian hamster, *Phodopus sungorus sungorus. Endocrinology, 116,* 424–430.
Duncan, M. J., Takahashi, J. S., & Dubocovich, M. L. (1989). Characteristics and autoradiographic localization of 2-[125I]-iodomelatonin binding sites in Djungarian hamster brain. *Endocrinology, 125,* 1011–1018.
Elliott, A. S., & Nunez, A. A. (1992). Photoperiod modulates the effects of steroids on sociosexual behaviors of hamsters. *Physiology and Behavior, 51,* 1189–1193.
Elliott, J. A. (1976). Circadian rhythms and photoperiodic time measurement in mammals. *Federation Proceedings, 35,* 2339–2346.
Elliott, J. A., & Goldman, B. D. (1981). Seasonal reproduction: Photoperiodism and biological clocks. In N. T. Adler (Ed.), *Neuroendocrinology of reproduction* (pp. 377–423). New York: Plenum Press.

Elliott, J. A., & Goldman, B. D. (1989). Reception of photoperiodic information by fetal Siberian hamsters: Role of the mother's pineal gland. *Journal of Experimental Zoology, 252,* 237–244.

Freeman, D. A., & Goldman, B. D. (1997a). Evidence that the circadian system mediates photoperiodic nonresponsiveness in Siberian hamsters. *Journal of Biological Rhythms, 12,* 100–109.

Freeman, D. A., & Goldman, B. D. (1997b). Photoperiodic nonresponsive Siberian hamsters: Effects of age on the probability of nonresponsiveness. *Journal of Biological Rhythms, 12,* 110–121.

French, A. R. (1988). The patterns of mammalian hibernation. *American Scientist, 76,* 569–575.

Glass, J. D., & Lynch, G. R. (1981). Melatonin: Identification of sites of antigonadal action in mouse brain. *Science, 214,* 821–823.

Glass, J. D., & Lynch, G. R. (1982). Diurnal rhythm of response to chronic intrahypothalamic melatonin injections in the white-footed mouse, *Peromyscus leucopus. Neuroendocrinology, 35,* 117–122.

Goldman, B. D. (1983). The physiology of melatonin in mammals. In R. J. Reiter (Ed.), *Pineal research reviews* (Vol. 1, pp. 145–182). New York: Liss.

Goldman, B. D. (1991). Parameters of the circadian rhythm of pineal melatonin secretion affecting reproductive responses in Siberian hamsters. *Steroids, 56,* 218–225.

Goldman, B. D., & Elliott, J. A. (1988). Photoperiodism and seasonality in hamsters: role of the pineal gland. In M. H. Stetson (Ed.), *Processing of environmental information in vertebrates* (pp. 203–218). New York: Springer-Verlag.

Goldman, B. D., & Nelson, R. J. (1993). Melatonin and seasonality in mammals. In H.-S. Yu & R. J. Reiter (Eds.), *Melatonin: Biosynthesis, physiological effects, and clinical applications* (pp. 225–252). Baco Raton, FL CRC Press.

Goldman, B. D., Darrow, J. M., & Yogev, L. (1984). Effects of timed melatonin infusions on reproductive development in the Djungarian hamster (*Phodopus sungorus*). *Endocrinology, 114,* 2074–2083.

Goldman, S. L., Dhandapani, K., & Goldman, B. D. (2000). Genetic and environmental influences on short-day responsiveness in Siberian hamsters. *Journal of Biological Rhythms, 15,* 417–428.

Gorman, M. R. (1995). Seasonal adaptations of Siberian hamsters: I. Accelerated gonadal and somatic development in increasing versus static long day lengths. *Biology of Reproduction, 53,* 110–115.

Gorman, M. R., & Zucker, I. (1995a). Seasonal adaptations of Siberian hamsters. II. Pattern of change in day length controls annual testicular and body weight rhythms. *Biology of Reproduction, 53,* 116–125.

Gorman, M. R., & Zucker, I. (1995b). Testicular regression and recrudescence without subsequent photorefractoriness in Siberian hamsters. *American Journal of Physiology, 269,* R800–R806.

Gorman, M. R., & Zucker, I. (1997a). Environmental induction of photononresponsiveness in the Siberian hamster, *Phodopus sungorus. American Journal of Physiology, 272,* R887–R895.

Gorman, M. R., & Zucker, I. (1997b). Pattern of change in melatonin duration determines testicular responses in Siberian hamsters, *Phodopus sungorus. Biology of Reproduction, 56,* 668–673.

Gorman, M. R., Freeman, D. A., & Zucker, I. (1997). Photoperiodism in hamsters: Abrupt versus gradual changes in day length differentially entrain morning and evening circadian oscillators. *Journal of Biological Rhythms, 12,* 122–135.

Grosse, J., & Hastings, M. H. (1996). A role for the circadian clock of the suprachiasmatic nuclei in the interpretation of serial melatonin signals in the Syrian hamster. *Journal of Biological Rhythms, 11,* 317–324.

Hall, V. D., & Goldman, B. D. (1982). Hibernation in the female Turkish hamster (*Mesocricetus brandti*): An investigation of the role of the ovaries and of photoperiod. *Biology of Reproduction, 27,* 811–815.

Hall, V. D., Bartke, A., & Goldman, B. D. (1982). Role of the testes in regulating duration of hibernation in the Turkish hamster (*Mesocricetus auratus*). *Biology of Reproduction, 27,* 802–810.

Hastings, M. H., Walker, A. P., Roberts, A. C., & Herbert, J. (1988). Intra-hypothalamic melatonin blocks photoperiodic responsiveness in the male Syrian hamster. *Neuroscience, 24,* 987–991.

Hastings, M. H., Vance, G., & Maywood, E. (1989a). Some reflections on the phylogeny and function of the pineal. *Experientia, 45,* 903–909.

Hastings, M. H., Walker, A. P., Powers, J. B., Hutchison, J., Steel, E. A., & Herbert, J. (1989b). Differential effects of photoperiodic history on the response of gonadotrophins and prolactin to intermediate daylengths in the male Syrian hamster. *Journal of Biological Rhythms, 4,* 335–350.

Hastings, M. H., Maywood, E. S., Ebling, F. J. P., Williams, L. M., & Titchener, L. (1991). Sites and mechanism of action of melatonin in the photoperiodic control of reproduction. In J. Arendt & P. Pevet (Eds.), *Advances in pineal research* (pp. 147–157). London: Libbey.

Heath, H. W., & Lynch, G. R. (1982). Intraspecific differences for melatonin-induced reproductive regression and the seasonal molt in *Peromyscus leucopus. General and Comparative Endocrinology, 48,* 289–295.

Heideman, P. D., & Bronson, F. H. (1993). Sensitivity of Syrian hamsters (*Mesocricetus auratus*) to amplitudes and rates of photoperiodic change typical of the tropics. *Journal of Biological Rhythms, 8,* 325–337.

Herbert, J., & Klinowska, M. (1978). Day length and the annual reproductive cycle in the ferret (*Mustelo furo*): The role of the pineal body. In I. Assenmacher & D. S. Farner (Eds.), *Environmental endocrinology* (pp. 87–93). Berlin: Springer-Verlag.

Herbert, J., Stacey, P. M., & Thorpe, D. H. (1978). Recurrent breeding seasons in pinealectomized or optic-nerve-sectioned ferrets. *Journal of Endocrinology, 78,* 389–397.

Hoffman, R. A., & Reiter, R. J. (1965). Pineal gland: Influence on gonads of male hamsters. *Science, 148,* 1609–1611.

Hoffman, R. A., Davidson, K., & Steinberg, K. (1982). Influence of photoperiod and temperature on weight gain, food consumption, fat pads and thyroxine in male golden hamsters. *Growth, 46,* 150–162.

Hoffmann, K. (1981). Photoperiodism in vertebrates. In J. Aschoff (Ed.), *Handbook of behavioral neurobiology. Biological rhythms* (pp. 449–473). New York: Plenum Press.

Hoffmann, K., & Illnerova, H. (1986). Photoperiodic effects in the Djungarian hamster. Rate of testicular regression and extension of pineal melatonin pattern depend on the way of change from long to short photoperiods. *Neuroendocrinology, 43,* 317–321.

Hoffmann, K., Illnerova, H., & Vanecek, J. (1986). Change in duration of the nighttime melatonin peak may be a signal driving photoperiodic response in the Djungarian hamster (*Phodopus sungorus*). *Neuroscience Letters, 67,* 68–72.

Hong, S. M., Rollag, M. D., & Stetson, M. H. (1986). Maintenance of testicular function in Turkish hamsters: Interaction of photoperiod and the pineal gland. *Biology of Reproduction, 34,* 527–531.

Honrado, G. I., Bird, M., & Fleming, A. S. (1991a). The effects of short day exposure on seasonal and circadian reproductive rhythms in male golden hamsters. *Physiology and Behavior, 49,* 277–287.

Honrado, G. I., Paclik, L., & Fleming, A. S. (1991b). The effects of short day exposure on the seasonal and reproductive rhythms of female golden hamsters. *Physiology and Behavior, 50,* 357–363.

Horton, T. H. (1984). Growth and maturation in *Microtus montanus*: Effects of photoperiods before and after weaning. *Canadian Journal of Zoology, 62,* 1741–1746.

Horton, T. H. (1985). Cross-fostering of voles demonstrates *in utero* effect of photoperiod. *Biology of Reproduction, 33,* 934–939.

Illnerova, H., & Vanacek, J. (1989). Complex control of the circadian rhythm in pineal melatonin production. In B. Mess, C. Ruzsas, L. Tima, & P. Pevet (Eds.), *The pineal gland: Current state of pineal research* (pp. 137–153). Amsterdam: Elsevier, and Budapest: Akademiai Kiado.

Johnston, P. G., & Zucker, I. (1980). Photoperiodic regulation of the testes of adult white-footed mice (*Peromyscus leucopus*). *Biology of Reproduction, 23,* 859–866.

Karp, J. D., & Powers, J. B. (1993). Photoperiodic and pineal influences on estrogen-stimulated behaviors in female Syrian hamsters. *Physiology and Behavior, 54,* 19–28.

Karp, J. D., Dixon, M. E., & Powers, J. B. (1990). Photoperiod history, melatonin, and reproductive responses of male Syrian hamsters. *Journal of Pineal Research, 8,* 137–152.

Karp, J. D., Hastings, M. H., & Powers, J. B. (1991). Melatonin and the coding of day length in male Syrian hamsters. *Journal of Pineal Research, 10,* 210–217.

Karsch, F. J., Bittman, E. L., Foster, D. L., Goodman, R. L., Legan, S. J., & Robinson, J. E. (1984). Neuroendocrine basis of seasonal reproduction. *Recent Progress in Hormone Research, 40,* 185–225.

Karsch, F. J., Bittman, E. L., Robinson, J. E., Yellon, S. M., Wayne, N. L., Olster, D. H., & Kaynard, A. H. (1986). Melatonin and photorefractoriness: Loss of response to the melatonin signal leads to seasonal reproductive transitions in the ewe. *Biology of Reproduction, 34,* 265–274.

Kerbeshian, M. C., Bronson, F. H., & Bellis, E. D. (1994). Variation in reproductive photoresponsiveness in a wild population of meadow voles. *Biology of Reproduction, 50,* 745–750.

Kliman, R. M., & Lynch, G. R. (1992). Evidence for genetic variation in the occurrence of the photoresponse of the Djungarian hamster, *Phodopus sungorus. Journal of Biological Rhythms, 7,* 161–175.

Krause, D. N., & Dubocovich, M. L. (1990). Regulatory sites in the melatonin system of mammals. *Trends in Neuroscience, 13,* 161–170.

Lee, T. M. (1993). Development of meadow voles is influenced postnatally by maternal photoperiodic history. *American Journal of Physiology, 265,* R749–R755.

Lee, T. M., & Zucker, I. (1988). Vole infant development is influenced perinatally by maternal photoperiodic history. *American Journal of Physiology, 255,* R831–R838.

Lehman, M. N., Bittman, E. L., & Newman, S. W. (1984). Role of the hypothalamic paraventricular nucleus in neuroendocrine responses to daylength in the golden hamster. *Brain Research, 308,* 25–32.

Lerchl, A., & Nieschlag, E. (1992). Interruption of nocturnal pineal melatonin synthesis in spontaneous recrudescent Djungarian hamsters (*Phodopus sungorus*). *Journal of Pineal Research, 13,* 36–41.

Lerner, A. B., Case, J. D., Lee, T. H., Takahashi, Y., & Mori, W. (1958). Isolation of melatonin, the pineal factor that lightens melanocytes. *Journal of the American Chemical Society, 80,* 2587–2594.

Lincoln, G. A. (1992). Administration of melatonin into the mediobasal hypothalamus as a continuous or intermittent signal affects the secretion of follicle stimulating hormone and prolactin in the ram. *Journal of Pineal Research, 12,* 135–144.

Lincoln, G. A. (1994). Effects of placing micro-implants of melatonin in the pars tuberalis, pars distalis and the lateral septum of the forebrain on the secretion of FSH and prolactin, and testicular size in rams. *Journal of Endocrinology, 142,* 267–276.

Lincoln, G. A., & Clarke, I. J. (1994). Photoperiodically-induced cycles in the secretion of prolactin in hypothalamo-pituitary disconnected rams: Evidence for translation of the melatonin signal in the pituitary gland. *Journal of Neuroendocrinology, 6,* 251–260.

Lincoln, G. A., & Maeda, K. -I. (1992). Reproductive effects of placing micro-implants of melatonin in the mediobasal hypothalamus and preoptic area in rams. *Journal of Endocrinology, 132,* 201–215.

Loudon, A. S. I. (1994). Photoperiod and the regulation of annual and circannual cycles of food intake. *Proceedings of the Nutrition Society, 53,* 495–507.

Lynch, G. R., Heath, H. W., & Johnston, C. M. (1981). Effect of geographical origin on the photoperiodic control of reproduction in the white-footed mouse, *Peromyscus leucopus. Biology of Reproduction, 25,* 475–480.

Lynch, G. R., Sullivan, J. K., Heath, H. W., & Tamarkin, L. (1982). Daily melatonin rhythms in photoperiod sensitive and insensitive while-footed mice (*Peromyscus leucopus*). In R. J. Reiter (Ed.), *The pineal and its hormones* (pp. 67–73). New York: Liss.

Malpaux, B., Robinson, J. E., & Karsch, F. J. (1987). Reproductive refractoriness of the ewe to inductive photoperiod is not caused by inappropriate secretion of melatonin. *Biology of Reproduction, 36,* 1333–1341.

Malpaux, B., Moenter, S. M., Wayne, N. L., Woodfill, C. J. I., & Karsch, F. J. (1988). Reproductive refractoriness of the ewe to inhibitory photoperiod is not caused by alteration of the circadian secretion of melatonin. *Neuroendocrinology, 48,* 264–270.

Malpaux, B., Daveau, A., Maurice, F., Gayrard, V., & Thiery, J.-C. (1993). Short-day effects of melatonin on luteinizing hormone secretion in the ewe: Evidence for central sites of action in the mediobasal hypothalamus. *Biology of Reproduction, 48,* 752–760.

Malpaux, B., Daveau, A., Maurice, F., Locatelli, A., & Thiéry, J. C. (1994). Evidence that melatonin binding sites in the pars tuberalis do not mediate the photoperiodic actions of melatonin on LH and prolactin secretion in ewes. *Journal of Reproduction and Fertility, 101,* 625–632.

Malpaux, B., Skinner, D. C., & Maurice, F. (1995). The ovine pars tuberalis does not appear to be targeted by melatonin to modulate luteinizing hormone secretion, but may be important for prolactin release. *Journal of Neuroendocrinology, 7,* 199–206.

Martinet, L., Allain, D., & Weiner, C. (1984). Role of prolactin in the photoperiodic control of moulting in the mink (*Mustela vison*). *Journal of Endocrinology, 103,* 9–15.

Maywood, E. S., & Hastings, M. H. (1995). Lesions of the iodomelatonin-binding sites of the mediobasal hypothalamus spare the lactotropic, but block the gonadotropic response of male Syrian hamsters to short photoperiod and to melatonin. *Endocrinology, 136,* 144–153.

Maywood, E. S., Buttery, R. C., Vance, G. H. S., Herbert, J., & Hastings, M. H. (1990). Gonadal responses of the male Syrian hamster to programmed infusions of melatonin are sensitive to signal duration and frequency but not to signal phase nor to lesions of the suprachiasmatic nuclei. *Biology of Reproduction, 43,* 174–182.

Maywood, E. S., Grosse, J., Lindsay, J. O., Karp, J. D., Powers, J. B., Ebling, F. J. P., Herbert, J., & Hastings, M. H. (1992). The effect of signal frequency on the gonadal response of male Syrian hamsters to programmed melatonin infusions. *Journal of Neuroendocrinology, 4,* 37–43.

McCord, C. P., & Allen, F. P. (1917). Evidences associating pineal gland function with alterations in pigmentation. *Journal of Experimental Zoology, 23,* 207.

Miernicki, M., Karp, J. D., & Powers, J. B. (1990). Pinealectomy prevents short photoperiod inhibition of male hamster sexual behavior. *Physiology and Behavior, 47,* 293–299.

Milette, J. J., Schwartz, N. B., & Turek, F. W. (1988). Importance of follicle-stimulating hormone in the initiation of testicular growth in photostimulated Djungarian hamsters. *Endocrinology, 122,* 1060–1066.

Minneman, K. P., & Wurtman, R. J. (1976). The pharmacology of the pineal gland. *Annual Review of Pharmacology and Toxicology, 16,* 33–51.

Moore, R. Y., & Klein, D. C. (1974). Visual pathways and the central neural control of a circadian rhythm in pineal serotonin N-acetyltransferase activity. *Brain Research, 71,* 17–33.

Morgan, P. J., Barrett, P., Howell, H. E., & Helliwell, R. (1994). Melatonin receptors: Localization, molecular pharmacology and physiological significance. *Neurochemistry International, 24,* 101–146.

Morin, L. P., & Zucker, I. (1978). Photoperiodic regulation of copulatory behaviour in the male hamster. *Journal of Endocrinology, 77,* 244–258.

Nelson, R. J. (1985). Photoperiod influences reproduction in the prairie vole (*Microtus ochrogaster*). *Biology of Reproduction, 33,* 596–602.

Nelson, R. J. (1987). Photoperiod-nonresponsive morphs: A possible variable in microtine population-density fluctuations. *American Naturalist, 130,* 350–369.

Nelson, R. J., Bamat, M. K., & Zucker, I. (1982). Photoperiodic regulation of testis function in rats: Mediation by a circadian mechanism. *Biology of Reproduction, 26,* 329–335.

Nelson, R. J., Badura, L. L., & Goldman, B. D. (1990). Mechanisms of seasonal cycles of behavior. *Annual Review of Psychology, 41,* 81–108.

Niklowitz, P., & Hoffmann, K. (1988). Pineal and pituitary involvement in the photoperiodic regulation of body weight, coat color and testicular size of the Djungarian hamster, *Phodopus sungorus. Biology of Reproduction, 39,* 489–498.

Niklowitz, P., Lerchl, A., & Nieschlag, E. (1994). Photoperiodic responses in Djungarian hamsters (*Phodopus sungorus*): Importance of light history for pineal and serum melatonin profiles. *Biology of Reproduction, 51,* 714–724.

Pitrosky, B., Kirsch, R., Vivien-Roels, B., Georg-Bentz, I., Canguilhem, B., & Pevet, P. (1995). The photoperiodic response in Syrian hamster depends upon a melatonin-driven circadian rhythm of sensitivity to melatonin. *Journal of Neuroendocrinology, 7,* 889–895.

Powers, J. B., Steel, E. A., Hutchison, J. B., Hastings, M. H., Herbert, J., & Walker, A. P. (1989). Photoperiodic influences on sexual behavior in male Syrian hamsters. *Journal of Biological Rhythms, 4*(1), 61–78.

Puchalski, W., & Lynch, G. R. (1986). Evidence for differences in the circadian organization of hamsters exposed to short day photoperiod. *Journal of Comparative Physiology A, 159,* 7–11.

Puchalski, W., & Lynch, G. R. (1988). Characterization of circadian function in Djungarian hamsters insensitive to short day photoperiod. *Journal of Comparative Physiology A, 162,* 309–316.

Ralph, C. L., Mull, D., Lynch, H. J., & Hedlund, L. (1971). A melatonin rhythm persists in rat pineals in darkness. *Endocrinology, 89,* 1361–1366.

Reiter, R. J. (1969). Pineal function in long term blinded male and female golden hamsters. *General and Comparative Endocrinology, 12,* 460–468.

Reiter, R. J. (1980). The pineal and its hormones in the control of reproduction. *Endocrine Reviews, 1,* 109–131.

Rivkees, S. A., Hall, D. A., Weaver, D. R., & Reppert, S. M. (1988). Djungarian hamsters exhibit reproductive responses to changes in daylength at extreme photoperiods. *Endocrinology, 122,* 2634–2638.

Robinson, J. E., & Karsch, F. J. (1984). Refractoriness to inductive day lengths terminates the breeding season of the suffolk ewe. *Biology of Reproduction, 31,* 656–663.

Robinson, J. E., & Karsch, F. J. (1987). Photoperiodic history and a changing melatonin pattern can determine the neuroendocrine response of the ewe to daylength. *Journal of Reproduction and Fertility, 80,* 159–165.

Robinson, J. E., Wayne, N. L., & Karsch, F. J. (1985). Refractoriness to inhibitory day lengths initiates the breeding season of the suffolk ewe. *Biology of Reproduction, 32,* 1024–1030.

Shaw, D., & Goldman, B. D. (1995a). Gender differences in influence of prenatal photoperiods on postnatal pineal melatonin rhythms and serum prolactin and follicle-stimulating hormone in the Siberian hamster (*Phodopus sungorus*). *Endocrinology, 136,* 4237–4246.

Shaw, D., & Goldman, B. D. (1995b). Influence of prenatal and postnatal photoperiods on postnatal testis development in the Siberian hamster (*Phodopus sungorus*). *Biology of Reproduction, 52,* 833–838.

Shaw, D., & Goldman, B. D. (1995c). Influence of prenatal photoperiods on postnatal reproductive responses to daily infusions of melatonin in the Siberian hamster (*Phodopus sungorus*). *Endocrinology, 136,* 4231–4236.

Smale, L., Dark, J., & Zucker, I. (1988a). Pineal and photoperiod influences on fat deposition, pelage, and testicular activity in male meadow voles. *Journal of Biological Rhythms, 3,* 349–355.

Smale, L., Nelson, R. J., & Zucker, I. (1988b). Daylength influences pelage and plasma prolactin concentrations but not reproduction in the prairie vole, *Microtus ochrogaster. Journal of Reproduction and Fertility, 83,* 99–106.

Song, C. K., & Bartness, T. J. (1996). The effects of anterior hypothalamic lesions on short-day responses in Siberian hamsters given timed melatonin infusions. *Journal of Biological Rhythms, 11*(1), 14–26.

Steinlechner, S., Heldmaier, G., & Becker, H. (1983). The seasonal cycle of body weight in the Djungarian hamster: Photoperiodic control and the influence of starvation and melatonin. *Oecologia, 60,* 401–405.

Stetson, M. H., & Watson-Whitmyre, M. (1986). Effects of exogenous and endogenous melatonin on gonadal function in hamsters. *Journal of Neural Transmission Supplement, 21,* 55–80.

Stetson, M. H., Watson-Whitmyre, M., & Matt, K. S. (1977). Termination of photorefractoriness in golden hamsters: Photoperiodic requirements. *Journal of Experimental Zoology, 202,* 81–88.

Stetson, M. H., Elliott, J. A., & Goldman, B. D. (1986). Maternal transfer of photoperiodic information influences the photoperiodic response of prepubertal Djungarian hamsters (*Phodopus sungorus*). *Biology of Reproduction, 34,* 664–669.

Stirland, J. A., Grosse, J., Loudon, A. S. I., Hastings, M. H., & Maywood, E. S. (1995). Gonadal responses of the male *tau* mutant Syrian hamster to short-day-like programmed infusions of melatonin. *Biology of Reproduction, 53,* 361–367.

Stirland, J. A., Hastings, M. H., Loudon, A. S. I., & Maywood, E. S. (1996a). The *tau* mutation in the Syrian hamster alters the photoperiodic responsiveness of the gonadal axis to melatonin signal frequency. *Endocrinology, 137,* 2183–2186.

Stirland, J. A., Mohammad, Y. N., & Loudon, A. S. I. (1996b). A mutation of the circadian timing system (*tau* gene) in the seasonally breeding Syrian hamster alters the reproductve response to photoperiod change. *Proceedings of the Royal Society of London. B. Biological Sciences, 263,* 345–350.

Sullivan, J. K., & Lynch, G. R. (1986). Photoperiod time measurement for activity torpor, molt and reproduction in mice. *Physiology and Behavior, 36,* 167–174.

Tamarkin, L., Hollister, C. W., Lefebvre, N. G., & Goldman, B. D. (1977). Melatonin induction of gonadal quiesence in pinealectomized Syrian hamsters. *Science, 198,* 953–955.

Underwood, H., & Goldman, B. D. (1987). Vertebrate circadian and photoperiodic systems: Role of the pineal gland and melatonin. *Journal of Biological Rhythms, 2,* 279–315.

Vitale, P. M., Darrow, J. M., Duncan, M. J., Shustak, C. A., & Goldman, B. D. (1985). Effects of photoperiod, pinealectomy and castration on body weight and daily torpor in Djungarian hamsters (*Phodopus sungorus*). *Journal of Endocrinology, 106,* 367–375.

Vitaterna, M. H., & Turek, F. W. (1993). Photoperiodic responses differ among inbred strains of golden hamsters (*Mesocricetus auratus*). *Biology of Reproduction, 49,* 496–501.

Wade, G. N. (1986). Sex steroids and energy balance: Sites and mechanisms of action. *Annals of the New York Academy of Sciences, 474,* 389–399.

Wade, G. N., Bartness, T. J., & Alexander, J. R. (1986). Photoperiod and body weight in female Syrian hamsters: Skeleton photoperiods, response magnitude, and development of refractoriness. *Physiology and Behavior, 37,* 863–868.

Watson-Whitmyre, M., & Stetson, M. H. (1983). Simulation of peak pineal melatonin release restores sensitivity to evening melatonin injections in pinealectomized hamsters. *Endocrinology, 112,* 763–765.

Weaver, D. R., & Reppert, S. M. (1986). Maternal melatonin communicates daylength to the fetus in Djungarian hamsters. *Endocrinology, 119,* 2861–2863.

Weaver, D. R., Keohan, J. T., & Reppert, S. M. (1987). Definition of a prenatal sensitive period for maternal-fetal communication of daylength. *American Journal of Physiology, 253,* E701–E704.

Zucker, I. (1985). Pineal gland influences period of circannual rhythms of ground squirrels. *American Journal of Physiology, 249,* R111–R115.

Zucker, I., Lee, T. M., & Dark, J. (1991). Suprachiasmatic nucleus and annual rhythms of mammals. In D. C. Klein, R. Y. Moore, & S. M. Reppert (Eds.), *Suprachiasmatic nucleus: The mind's clock* (pp. 246–259). New York: Oxford University Press.

19

Circannual Rhythms
Mammals

IRVING ZUCKER

Orientation in time, which ranks among the most important of organismal adaptations, is apparent in the ubiquitous distribution of seasonal rhythms: mammals anticipate and take advantage of predictable changes in the external environment imposed by the 12-month geophysical cycle. Striking seasonal changes in temperature, humidity, and day length, as well as prominent seasonal variation in biotic factors such as food availability, presence of predators, and access to mates, place a premium on accurate timing. The array of mammalian functions that varies seasonally includes, among others, reproduction, food intake, sleep, body and brain mass, thermoregulation, affiliative and agonistic social behaviors, pelage characteristics, secretion of hormones, and neurotransmitter concentrations in brain (Bartness & Wade, 1985; Buijs *et al.*, 1986; Dark, Dark, & Zucker, 1987; Ferkin & Zucker, 1991; Goldman & Nelson, 1993; Heldmaier & Steinlechner, 1981; Sadlier, 1969; Turek & Van Cauter, 1994; Walker, Haskell, Berger, & Heller, 1980). Seasonal rhythms are evident at all levels of biological organization.

Three types of seasonal rhythms have been distinguished (Zucker, Lee, & Dark, 1991). Type I rhythms contain an endogenous component, but do not persist beyond a single cycle in the absence of environmental input; they are common among short-lived temperate and boreal mammals. Their endocrine and neural bases have been elaborated in considerable detail and reviewed from several perspectives (Bartness, Powers, Hastings, Bittman, & Goldman, 1993; Bronson & Heideman, 1994; Nelson, Badura, & Goldman, 1990; Turek & Van Cauter, 1994; also see Chapter 18).

Type II rhythms are fully endogenous, persist for two or more cycles with a period that differs from 12 months in individuals maintained in constant conditions, and are not contingent on periodic environmental input. These "true" circannual

---

IRVING ZUCKER Departments of Psychology and Integrative Biology, University of California, Berkeley, California 94720.

Circadian Clocks, Volume 12 of *Handbook of Behavioral Neurobiology*, edited by Joseph S. Takahashi, Fred W. Turek, and R. Y. Moore, Kluwer Academic / Plenum Publishers, New York, 2001.

rhythms (CARs), are displayed by long-lived species from several orders, including rodents, carnivores, ungulates, bats, and primates. The general properties of CARs, which were well reviewed by Gwinner (1986), are not recapitulated here. Descriptive studies of Type II rhythms also are neglected, as are Type III rhythms; the latter, exemplified by seasonal allergic reactions in humans, are devoid of major endogenous components and are manifestations of environmental periodicities (e.g., production of pollens). The focus here is on elaboration of physiologic substrates underlying generation and entrainment of Type II rhythms.

Common and distinctive features of Type I and II rodent reproductive rhythms are summarized in Figure 1. In both cases, spermatogenesis and ovulation are suppressed during the short day lengths (DLs) of winter. The duration of reproductive quiescence is controlled by a species-specific interval timer; 15–25 weeks after the gonads have undergone involution, and while DLs are still short, gonadal recrudescence is initiated, culminating several weeks later in full recovery of gametogenesis. In Type I rhythms, this "spontaneous" recrudescence is accompanied by refractoriness to short DLs and indefinite maintenance of gametogenesis; exposure to long DLs for 3–4 months is required to break refractoriness to short days. Only then are animals able to undergo a second gonadal regression during exposure to decreasing day lengths. In rodents with Type I rhythms, reproduction is maintained indefinitely in long DLs; of the many species tested, none becomes refractory to long days. By contrast, rodents with Type II rhythms (e.g., squirrels and marmots) that have undergone spontaneous gonadal recrudescence do not require exposure

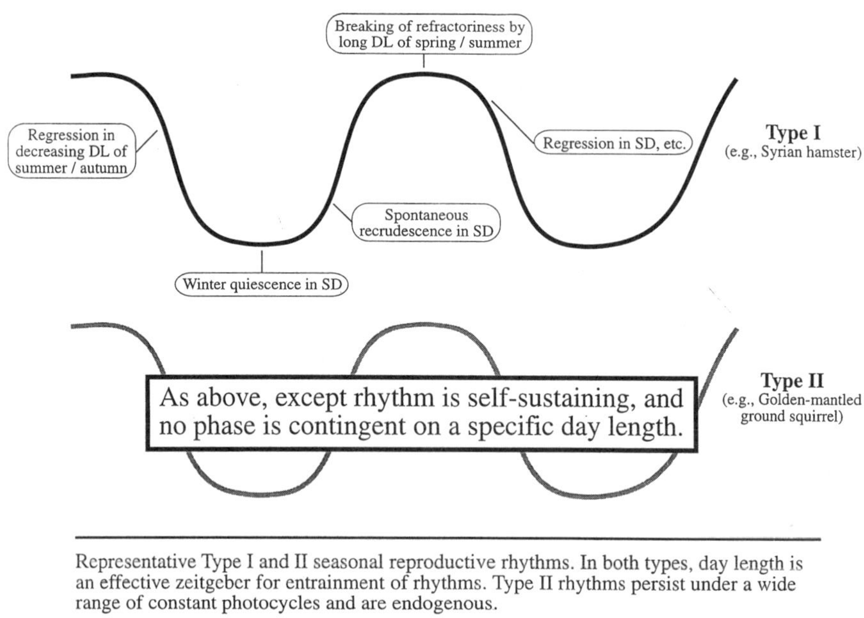

Figure 1. Schematic representation of annual reproductive rhythm of Type I and Type II rodent reproductive cycles. Day length (DL) is essential to induce gonadal regression in the autumn and to break refractoriness in the spring in Type I rodents. In Type II rodents, the cycle is self-sustained and there are no particular day length requirements to induce regression or break refractoriness (if any). SD, Short days.

to any particular day length to permit subsequent gonadal involution. A fixed number of weeks after gonadal recrudescence has occurred, a second regression ensues, regardless of the ambient day length. The end result is a self-sustained rhythm in which the reproductive apparatus waxes and wanes on an annual basis. Refractoriness to DL either is not a feature of Type II rhythms or, if it occurs, is broken spontaneously. No DLs prevent seasonal involution of the reproductive apparatus in rodents with Type II rhythms. Where adequate tests have been performed, the differences described above for reproduction apply to nonreproductive Type I and II seasonal rhythms (Dark & Zucker, 1985).

Evolution of Mammalian Cirannual Rhythms (CARs)

Timing of reproduction to coincide with food availability may be the primary target of natural selection in the evolution of mammalian seasonal rhythms (Bronson & Heideman, 1994; Farner, 1985). Energy requirements of lactating dams are high (Bronson, 1989) and rapid, sustained tissue growth and low energy reserves render newly weaned small rodents especially vulnerable to food shortages. Consequently, females that deliver young coincident with plentiful food and minimally challenging abiotic conditions (e.g., ambient temperatures near or within the range of thermoneutrality) presumably are more fit than those that reproduce at other times. In temperate and boreal regions, spring and early summer are the seasons during which "the living is easy" and disproportionately more babies are born (Bronson, 1989).

Nonreproductive processes that concurrently benefit reproduction (e.g., establishment and defense of territories, or increased food consumption) may be linked to a common clock mechanism that also phases reproduction. Conversely, behaviors incompatible with reproduction (e.g., hibernation; Wimsatt, 1969) may have evolved with separate temporal controllers or linkage to a common timer that maintains them in antiphase to reproduction.

Circannual rhythms evidently arose independently several times during the evolution of mammals. In most cases, the period of the free-running rhythm is 1 or 2 months shorter than 1 year (Gwinner, 1986). In nature, both Type I and II rhythms are entrained to the annual geophysical cycle, typically the variation in day length (Gwinner, 1986). The nearly universal dependence on external synchronization by day length prompts a question: Why did different forms of seasonality evolve? The earliest mammals most likely were small and nocturnal (Crompton & Jenkins, 1979) and, if seasonal, possessed Type I rhythms (Farner, 1985). The life history characteristics and ecological factors that favored evolution of Type II rhythms remain unknown, except that increased longevity is an obvious prerequisite. For animals exposed to a potentially confusing array of photic cues (e.g., long-distance migrants in the course of their annual migrations; Gwinner, 1996), the potential advantage of circannual organization as a buffer against environmental "noise" seems clear. Similarly, hibernating mammals in lightless hibernacula for half of each year can benefit from an annual clock that provides temporal information during dormancy. This does not, however, account for the evolution of robust CARs in ungulates that have continuous access to environmental zeitgebers. The Type II organization, once it arose, may have been retained in species with continuous access to zeitgebers, perhaps because it was no more expensive than the more fundamental Type I mechanism. Lack of information on the distribution of Type I and Type II rhythms in recent mammals constrains analysis of factors that favor evolution of one or the other of these temporal programs.

Innateness of Type II Rhythms

This question can be addressed by tracking rhythms in animals that have never been exposed to seasonal cues, be they of environmental or maternal origin. Golden-mantled ground squirrels maintained from birth in one of several fixed light–dark cycles and temperatures display CARs in body mass and reproduction (Pengelley & Asmundson, 1974; Zucker, 1988); these findings do not provide evidence relevant to rhythm innateness because in every instance animals were gestated by dams that mated in the field and were exposed to seasonal changes in day length and temperature. Because female mammals communicate day length information to their fetuses (Lee & Zucker, 1988; Stetson, Ray, Creyaufmiller & Horton, 1989; Weaver & Reppert, 1986), maternal signals may "jump-start" CARs or influence their entrainment.

The type of experiment that needs to be implemented is exemplified by studies in the circadian domain. During maintenance in constant light, hamster dams that have sustained ablation of the suprachiasmatic nucleus (SCN) show no evidence of coherent circadian organization, but their offspring manifest normal circadian rhythms postnatally (Davis & Gorski, 1988). Innateness is inferred because neither maternal circadian nor postnatal photic information is necessary for the genesis of circadian rhythms. Comparable analyses have not been performed on circannual rhythms.

Prenatal day length does influence postnatal phasing and amplitude of CARs in red deer, *Cervus elaphus* (Adam, Kyle & Young, 1994; Adam, Kyle, Young, & Hirst, 1995). Stags kept during part of gestation in 18L versus 6L photocycles, and postnatally in identical 12L photoperiods, differed markedly. Puberty occurred at an earlier age and lower body mass, and antlers hardened earlier in 18L than 6L animals. The prenatal day length exerts a long-lasting influence on patterns of secretion of pituitary and gonadal hormones nominally under circannual control.

Entrainment of CARs

Under natural conditions, the period of CARs is approximately 12 months. Entrainment is demonstrated if one or more of the following conditions is satisfied: (1) concordance of the period of the free-running rhythm with that of the zeitgeber, (2) the zeitgeber phase shifts the CAR within several cycles, or (3) the zeitgeber phase shifts the CAR differentially at several phases of the annual cycle (Gwinner, 1986).

Not only sinusoidal, but also rectangular symmetric and asymmetric sawtooth changes in DL effectively entrain CARs in primates and ungulates, as reviewed by Gwinner (1986), who concludes that "among the thoroughly studied species there are only very few for which photoperiod appears to be ineffective as a circannual zeitgeber." The notable exception among mammals has been the most common model circannual species, the golden-mantled ground squirrel, *Spermophilus lateralis.* The statement that "light did not affect the free-running period and did not appear to act as a zeitgeber" for squirrel CARs (Pengelley, Asmundson, Barnes, & Aloia,

1976) was based on studies that employed fixed photoperiods devoid of much of the predictive information contained in natural photocycles (e.g., gradual change in DL). Such manipulations do not definitively test the efficacy of light as a zeitgeber. Photocycles that more nearly simulate natural annual variations in day length entrain the CARs of hibernation, body mass, and estrus of golden-mantled ground squirrels with a period that does not differ from 12 months (Lee & Zucker, 1991; Manley, 1982). Several squirrels even entrained their body mass cycles to a compressed 6-month sinusoidal photocycle (Lee & Zucker, 1991). Squirrel CARs, in common with those of other mammals, are responsive to day length. Photic entrainment also has been demonstrated in a second sciurid rodent, the woodchuck, *Marmota monax.* Animals were maintained beginning at the summer solstice in austral or boreal photoperiods. The austral photoperiod altered growth rate in males and females and gonadal activity in males within 3–6 months; after 18 months, rhythms of austral animals were phase shifted by 5 months relative to those of boreal counterparts (Concannon, Roberts, Baldwin, Erb & Tennant, 1993) (Figure 2). The utility of light as a circannual zeitgeber is not called into question by the observation that 3–4 years might be required to effect complete reentrainment of woodchuck CARs (Davis & Finnie, 1975). Squirrels and woodchucks do not face the extreme or abrupt entrainment challenges imposed in boreal-to-austral translocation experiments. The much smaller day length changes these species experience during aboveground sojourns each year between April and September are gradual and thus pose less of a challenge for entrainment (Gorman & Zucker, 1995; 1998). The question of interest, whether and when DL during each annual cycle corrects the 1.5-month

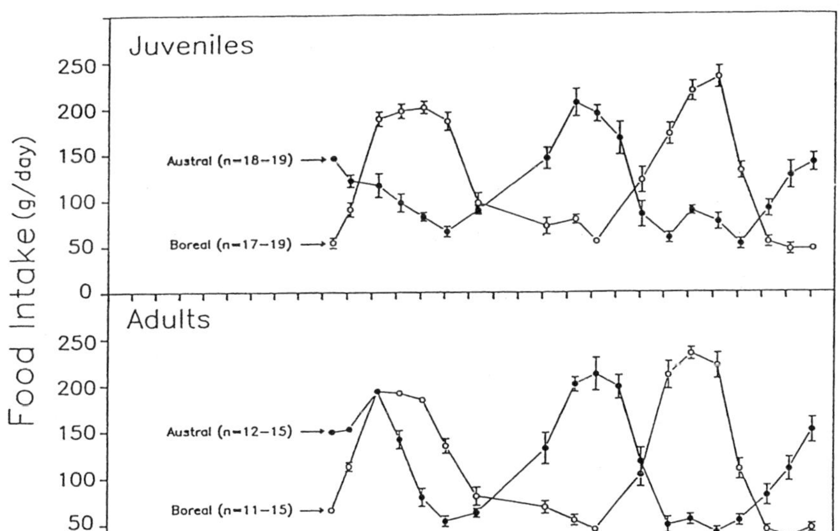

Figure 2. Mean ± SEM daily food intake in woodchucks exposed to boreal (open symbols) and austral (closed symbols) photoperiods. The first two observations on adults involved use of only 300 g of available food per cage per day, all of which was consumed by most pairs of austral adults. (From Concannon *et al.*, 1993, with permission.)

difference between the period of CARs and the 12-month geophysical cycle, is addressed in the next section.

LIGHT

The exegesis of circannual rhythms appropriates insights gained from analyses of circadian rhythms. Differential responsiveness to light at several phases of the circadian cycle (as documented in phase response curves) accounts for the effectiveness of light as a circadian zeitgeber. By analogy, one might expect circannual oscillators to be phase advanced or delayed, respectively, by light at different phases of the annual cycle. Because of the very long time constants involved and other intrinsic difficulties, specification of circannual phase response curves to light has not been attempted. The minimal photic cues adequate for entrainment of CARs also remain to be specified for any mammal. In golden-mantled ground squirrels, the period of the free-running CAR is 10.5 months and entrainment requires phase delays of approximately 45 days each year. Squirrels born in June and housed in a simulated natural photoperiod for 42°N were exposed to a 3-month block of short days (8 hours of light/day) either in autumn or spring. Because the ambient photocycle entrains CARs (Lee & Zucker, 1991), interspersing of short days engages the same circannual phase in each squirrel. Beginning at 1 year of age, after squirrels had been exposed to short days in autumn or spring, they were transferred to an LD 14:10 photocycle that permits CARs to freerun (Zucker & Boshes, 1982). The interposition of short days for 3 months in the summer advanced the body mass rhythm by 36 and 24 days relative to control values in the first and second years after treatment, respectively (S. M. Hiebert, E. M. Thomas, T. M. Lee & I. Zucker, in preparation). Short-day treatments in the spring did not affect CARs. We infer that long DLs during the summer phase-delay CARs and facilitate entrainment to a period of 12 months.

Woodfill, Wayne, Moenter, and Karsch (1994) approached the same issue by providing pinealectomized ewes with seasonally restricted melatonin (Mel) infusions. Because the effects of day length on the ewe neuroendocrine axis are transduced by the pattern of Mel secretion (Bittman & Karsch, 1984), a 3-month Mel infusion regimen can be designed that physiologically approximates a 3-month exposure to spring DLs. Melatonin was infused in pinx ewes in a sequence that mimicked endogenous Mel secretion for a specific quadrant of the annual DL cycle; e.g., the spring group received progressively shorter nightly Mel infusions and no Mel was provided for the ensuing 9 months each year.

Spring Mel treatments synchronized the annual rhythm of luteinizing hormone (LH) secretion with a period of 12 months, but onset and end of elevated LH titers were advanced relative to those of pineal-intact controls in a natural photocycle. Melatonin infusions from the summer solstice to the autumnal equinox also effectively entrained the LH rhythm, with period and phase similar to those of intact animals. Autumn Mel infusions were less effective, but still provided greater synchronization than was evident in the noninfused control sheep; the latter failed to entrain to the natural photocycle. Synchronization did not occur in ewes infused with a winter pattern of Mel. Exposure to Mel for as few as 3 months each year effectively entrains the annual reproductive rhythm; maximal zeitgeber effectiveness occurs during the spring and summer. The minimal number of spring days effective in entraining circannual rhythms is not known, nor have "skeleton" paradigms been used to probe the circannual system (Pittendrigh & Daan, 1976). Absolute duration of DL and direction of DL change, both of which are transduced by the pattern of nocturnal Mel secretion, as well as the stage of the circannual cycle during which photic signals are experienced, influence entrainment of sheep CARs (Woodfill, Robinson, Malpaux & Karsch, 1991; Woodfill *et al.*, 1994).

Photic Entrainment Pathways. The neural pathways by which light entrains CARs have not been specified for any species. The retinohypothalamic tract, a direct projection from retinal ganglion cells to the SCN that mediates photic entrainment of circadian rhythms (Morin, 1994), also is sufficient to mediate light entrainment of Type I rhythms of hamsters (Rusak & Morin, 1976; Smale & Morin, 1990). The retinohypothalamic tract thus is a plausible candidate pathway for entrainment of Type II rhythms. SCN ablations that interrupted the retinohypothalamic tract of squirrels compromised entrainment to a simulated natural photoperiod (Lee & Zucker, 1991). In comparison to intact controls, a significantly lower percentage of squirrels that sustained damage to the retinohypothalamic tract had circannual periods of 12 months after exposure for 3 years to the simulated natural photoperiod; interindividual variability in circannual markers also was greatly increased after SCN ablation. Because SCN ablation also eliminates or disrupts pineal melatonin secretion, failure to entrain may be a consequence of abnormal or absent melatonin secretion. Whether the retinohypothalamic projection to the SCN is sufficient for entrainment of CARs in the absence of other visual projections remains to be established.

Mediation by the Pineal Gland

In both squirrels and sheep, entrainment is abolished after pinealectomy, and CARs freerun in DL regimens that entrain rhythms of intact animals. Thus, the period of the body mass CAR did not differ from 365 days in intact animals, although it was substantially shorter than 1 year among pinealectomized squirrels maintained in simulated natural DLs (Figure 3) (Hiebert, *et al.*, 2000). Photic entrainment of reproductive rhythms of ewes is similarly dependent on the pineal gland (Woodfill *et al.*, 1994). Changes in duration of nocturnal Mel secretion over the course of several months presumably provide the signals that entrain circannual oscillators. Neurons with high concentrations of Mel receptors are likely components of the entrainment pathway for Type II rhythms. The diversity in distribution of Mel receptors in species with Type II rhythms (Bittman, 1993) does not point to any one structure as critical for entrainment. Thus, Mel binding is sparse in the SCN, but prominent in septal and preoptic areas and in the hippocampus of ewes (Bittman & Weaver, 1990) and nearly completely absent from the brains of ferrets (Weaver & Reppert, 1990). In golden-mantled ground squirrels, specific Mel binding is evident in several limbic system structures and in the SCN (Bittman, Thomas, & Zucker, 1994). In these species with Type II rhythms, as in every rodent with Type I seasonal rhythms, Mel binding is most dense in the pars tuberalis of the pituitary (Bittman, 1993).

Pineal Mel is not necessary for entrainment of *circadian* rhythms; after a phase shift in the light–dark cycle, pinealectomized and intact squirrels reentrain circadian locomotor rhythms at the same rate (Martinet & Zucker, 1985). Melatonin provides a humoral link between retinorecipient SCN neurons that monitor light signals and oscillators that generate CARs. Whereas *circadian* oscillators receive direct visual afferance, and a humoral link is not necessary for entrainment, *circ-*

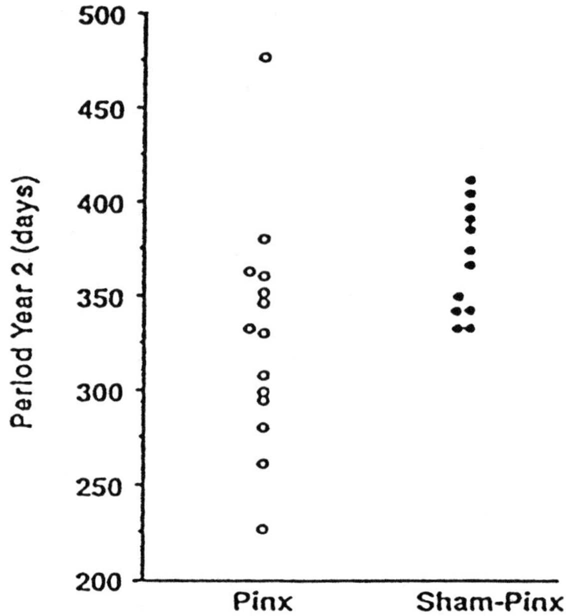

Figure 3. Period of body mass rhythm for individual pinealectomized (pinx) and sham-pinx squirrels in the second year in the simulated natural photoperiod. Note increased variance for the pinx group. The period did not differ significantly from 365 days for sham-pinx animals and was significantly shorter than 1 year for the pinx animals. Data based on results in Hiebert *et al.*, (2000).

annual clocks are not necessarily retinorecipient and humoral factors are implicated in the entrainment process.

Temperature

Phase delays of 4.5 months in peak body mass were reported in golden-mantled ground squirrels kept for 9 months at −3°C (Mrosovsky, 1980). A true phaseshift was inferred from the persistence of differences in timing 1 year after treatment. Because of potential masking effects of long-duration, low ambient temperatures (T_a) on the body mass cycle, the most compelling evidence for a true phase shift was the delay of about 5 months in onset of reproductive condition in males 1 year after they were returned to 21°C. Less consistent results were obtained with temperature challenges typically existing in hibernacula. A 20-week cold treatment (6°C) in the spring delayed the subsequent peak body mass and molt termination; a similar intervention in the fall resulted in earlier peak body masses and acceleration of testicular descent. In virtually all instances, however, there was no evidence in subsequent years that the rhythms had been phase advanced; Mrosovsky (1990; p. 687) concluded that "the question of thermal influences on circannual periodicity of squirrels remains unresolved." Simulated natural temperature cycles have not been demonstrated to entrain squirrel CARs, so their effectiveness as circannual zeitgebers remains questionable. Changes in body temperature (T_b) during hibernation lengthen the period of the free-running CAR in *Spermophilus lateralis* in a consistent manner (reviewed by

Mrosovsky, 1986) and reduce the phase delays that zeitgebers such as day length need to induce to effect entrainment to a period of 12 months.

Circannual Phase Response Curves to Hormones

Neurotransmitter and drug effectiveness varies with the stage of the circadian cycle (Rusak & Bina, 1990; Turek, 1987); in extreme cases, opposite effects are observed during subjective day and night, respectively (Margules, Lewis, Dragovich & Margules, 1972). Seasonal fluctuations in responsiveness to drugs also have been reported (Lacoste & Wirz-Justice, 1989), but endogenous circannual processes have not been implicated. Results described in the following section point to the existence of circannual phase response curves to hormones.

Estradiol

Blood concentrations of gonadal hormones are highly correlated with the circannual body mass cycle in golden-mantled squirrels. Peak hormone concentrations occur shortly after the termination of hibernation in the spring, when body mass is at its nadir, and trough values are recorded in the late summer, coincident with body mass peaks (e.g., Licht, Zucker, Hubbard & Boshes, 1982). The waveform of the body mass cycle is changed in ovariectomized squirrels treated continuously with estradiol (Zucker & Boshes, 1982). The possibility that estradiol might produce phase-specific advances and delays was confirmed in ovariectomized females monitored for at least 1 year before, during, and after treatment. When estradiol administration was restricted to the weight-gain phase, the period of the CAR in body mass was lengthened by 36 days. Similar treatment restricted to the weight-loss phase shortened the circannual period by 66 days. These phase delays and advances were preserved during the year after discontinuation of hormone treatment (Figure 4) (Lee & Zucker, 1992; Hiebert, Lee, Licht, & Zucker, 1998).

Melatonin

Three months of Mel treatment beginning in late summer (autumn melatonin) produced durable phase shifts of the CARs of juvenile female squirrels. During the year of treatment, onset of estrus was advanced by 25 days; this difference was amplified during the first 2 years after treatment (Figure 5) (S. M. Hiebert, E. M. Thomas, E. M. Lee & I. Zucker, unpublished observations). Similar effects were observed for the CAR in body mass. Identical Mel treatment in the spring did not affect the body mass rhythm and onset of estrus was delayed during the first cycle; however, this phase delay waned in the 2 years after treatment was terminated. Spring Mel treatment may mask rather than phase shift the reproductive rhythm. Circannual oscillators evidently are differentially responsive to melatonin at discrete phases of the annual cycle. Short Mel signals in late spring and summer ostensibly produce effects opposite to those of long signals, i.e., phase delays that entrain squirrel CARs to a period of 12 months.

These observations raise the possibility that drug treatments of several months' duration owe part of their effectiveness to phase shifting of circannual rhythms. An endogenous component in the seasonal rhythm of brain neurotransmitter content

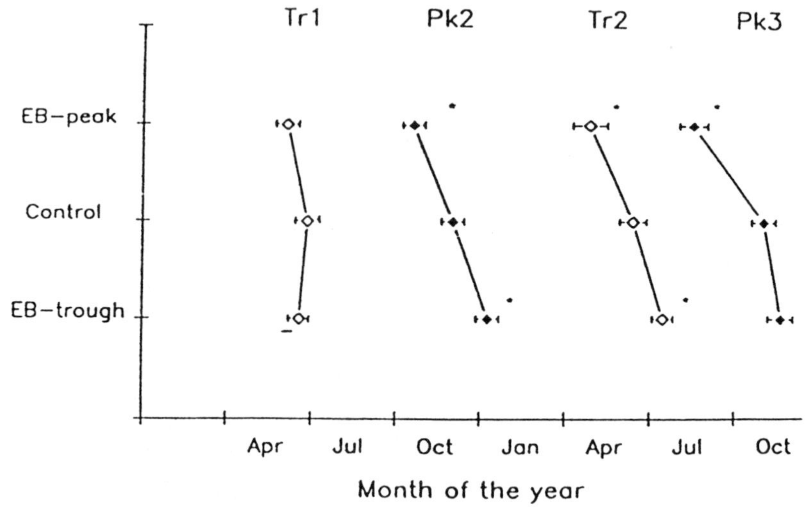

Figure 4. Mean annual dates of midpeak or midtrough body mass clusters for first (Tr1) and second mass troughs (Tr2) and second (Pk2) and third mass peaks (Pk3). Silastic implants that contained estradiol benzoate (EB) or were empty (control) were implanted at the trough or peak phase of the body mass cycle. Bars indicate SEs. Asterisk indicates significant difference from control value. (From Lee & Zucker, 1992, with permission.)

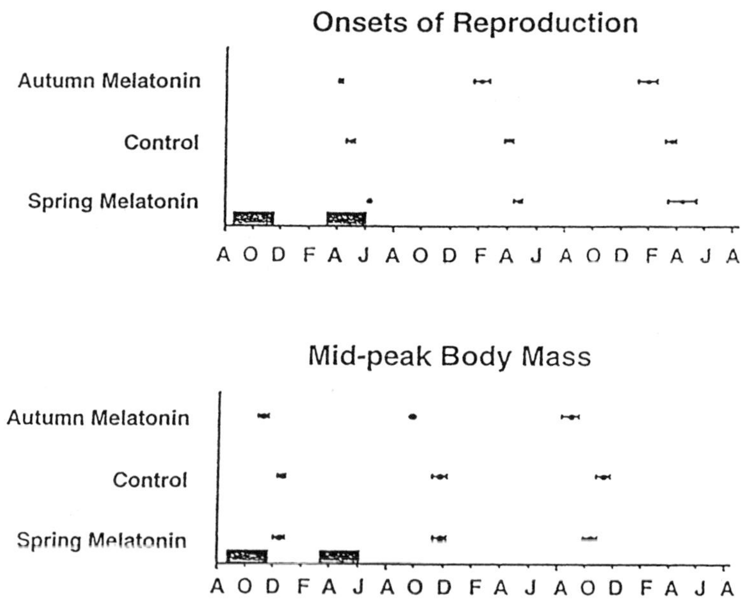

Figure 5. Onsets of estrus (top panel) and mid-peaks in body mass (lower panel) of squirrels treated with melatonin (Mel) capsules from August through November (Autumn Melatonin) or March through June (Spring Melatonin) as indicated with bars on the abscissa. Control animals were implanted with empty capsules. Values show are means ± SEM. Note the permanent phase advance in CARs of body mass and estrus induced by autumn melatonin treatment. (S. M. Hiebert, E. M. Thomas, E. M. Lee, & I. Zucker, unpublished observations.)

(Carlsson, Svennerholm, & Winblad, 1980) could account for a substantial portion of the seasonal variance in responsiveness to drugs.

Role of Hormones in Generation of Type II Rhythms

Ablation of the several endocrine organs influences expression of circannual rhythms and can prevent entrainment to the illumination cycle, but in no instance have hormones been implicated in the generation of CARs. Thus, removal of the ovaries, testes, and pineal glands of golden-mantled ground squirrels is compatible with normal CAR generation (Hiebert *et al.*, 2000; Lee & Zucker, 1995; Zucker, 1985; Zucker & Boshes, 1982). Tests of CAR generation after hypophysectomy, adrenalectomy, or thyroidectomy have not been reported in this species. The substantial shortening in the period of CARs after pinealectomy in ground squirrels (Zucker, 1985) reflects elimination of the entraining influence of light (and melatonin; Hiebert *et al.*, 2000), rather than a direct hormonal action on circannual clocks (see above).

Reports that thyroidectomy prevents the development of photorefractoriness and extends the breeding season in ewes and rams with Type II rhythms are restricted to the reproductive axis (Karsch, Dahl, Hachigian & Thrun, 1995; Parkinson & Follett, 1994); other circannual rhythms (e.g., in blood prolactin concentrations) are not affected. Similar extension of the reproductive season has been reported in ewes after extensive damage to the anterior hypothalamic region (Hileman, Kuehl & Jackson, 1994); both phenomena may indicate a specialized thyroid–brain system that affects photic control of reproduction.

Generation of Circannual Rhythms: Neural Pacemakers

Systematic attempts to localize circannual functions in the mammalian central nervous system date from circa 1980 and have failed to identify a circumscribed population of neurons that generates signals that ultimately find overt expression in circannual rhythms (reviewed in Zucker, 1988; Zucker *et al.*, 1991). In theory, disparate neural structures and pathways may mediate the several circannual rhythms in a particular species (e.g., body mass, sleep, reproduction, hibernation, molt, food intake, and metabolism in golden-mantled ground squirrels). Alternatively, redundant pacemakers and neural networks each may be capable of generating all or one of these CARs. In the simplest organization, a single localized pacemaker might be both necessary and sufficient for generation of all circannual rhythms. The success of ablation experiments depends on the verisimilitude of these scenarios to actual neural organization. In our present state of ignorance, ablation methodology, primitive though it is, may constitute the best avenue for identification of circannual pacemakers.

Role of the Suprachiasmatic Nucleus (SCN)

The conjecture that circannual rhythms are derived by "counting" circadian cycles (Gwinner, 1973) was evaluated by monitoring CARs of squirrels in which the SCN had been ablated (SCNx). Circadian wheel-running and body temperature rhythms were eliminated or severely disrupted in SCNx squirrels maintained in

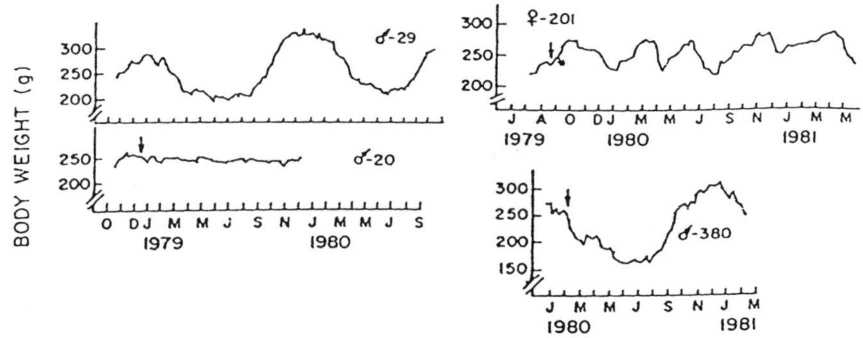

Figure 6. Circannual body mass cycles of an unoperated squirrel (no. 29) and of three squirrels with complete ablation of the SCN. The body mass rhythm was eliminated in no. 20, normal in no. 380, and replaced with a 3- to 5-month cycle in no. 201. The arrows designate the times at which the animals sustained SCN lesions. (Modified from Zucker *et al.*, 1983).

constant light at 23°C (Zucker, Boshes, & Dark, 1983) or in a fixed LD 14:10 photocycle (Zucker, Ruby, & Dark, 1993). Most SCNx squirrels generated normal CARs, but a minority did not manifest body mass rhythms or generated rhythms with periods of 3–5 months (Figure 6). Variable effects of SCN damage were not attributable to differences in lesion size or location, nor to the stage of the circannual cycle at which the neural insult was sustained (Dark, Pickard, & Zucker, 1985). Generation of CARs was not contingent on normal circadian organization. Several squirrels with severely disrupted circadian rhythms manifested CARs in reproduction and body mass (Zucker *et al.*, 1983). Independence of circannual from circadian rhythms also was congruent with the lack of influence of entrainment to circadian T-cycles of 23 versus 25 hours on the period of CARs of squirrels (Carmichael & Zucker, 1986). These observations point to a separation in the neural substrates that generate circadian and circannual rhythms. Some of the brain lesions that eliminated circadian rhythms spared circannual rhythms and vice versa (Zucker *et al.*, 1983). The SCN may contribute to coupling among several remotely located circannual oscillators, but its neurons and glia do not constitute a hierarchical "master" circannual pacemaker.

The aforementioned experiments assessed CARs in squirrels kept at 23°C. The SCN, unlike most other squirrel neuropil, preserves relatively high metabolic rates during deep hibernation, and undergoes phase-specific changes in 2-deoxy-D-glucose uptake during the hibernation cycle (Kilduff, Miller, Radeke, Sharp, & Heller, 1990). Ground squirrels normally are exposed to low T_a and spend much of the fall and winter months in deep hibernation at T_b only a few degrees above freezing (Lyman, Willis, Malan, & Wang, 1982). We reinvestigated the role of the SCN in circannual organization of squirrels afforded the opportunity to enter hibernation during continuous maintenance for several years at 6°C. Intact squirrels tested under these conditions express circannual body mass cycles with a hibernation season that typically lasts 5–7 months. Animals do not remain torpid continuously, but arouse at intervals of 2–14 days, remain euthermic ($T_b \approx 37$°C) for less than 1 day, and then reenter hibernation (Twente & Twente, 1967). A circannual clock with a period of 10.5 months regulates the onset and termination of the hibernation season each year (Heller & Poulson, 1970; Pengelley & Asmundson, 1974). SCN

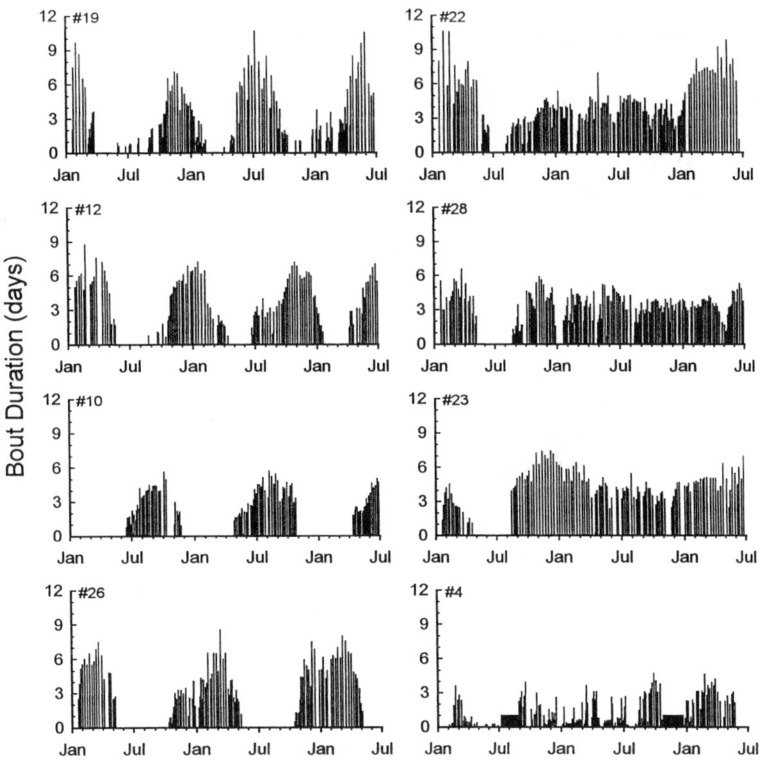

Figure 7. Hibernation patterns for squirrels that sustained ablation of the SCN (SCNx; four panels on the right and top two panels on the left) and for control animals without damage to the SCN (#10, #26). The ordinate on the left side of each plot indicates the duration of each hibernation bout. Missing data for #10 in the first hibernation season reflect transmitter failure, battery replacement, and postsurgical recovery time. The right four panels illustrate continuous hibernation beginning in August of the first year of testing and continuing until the experiment was terminated; during this time, none of these SCNx squirrels was euthermic for more than 10 consecutive days. The prolonged euthermic intervals that occurred around July of the first year of testing were associated with lower than normal body mass values; hibernation resumed when body mass returned to baseline values. Two black rectangles in the hibernation record of animal #4 indicate occasions when hibernation was assessed by visual inspection; on all occasions this animal was in the curled-up posture typical of hibernating squirrels and cold to the touch. (From Ruby *et al.*, 1996, with permission. Copyright 1996, National Academy of Sciences, U.S.A.)

ablation is associated with marked changes in temporal organization of hibernation (Figure 7) (Dark, Kilduff, Heller, Licht, & Zucker, 1990; Ruby, Dark, Heller, & Zucker, 1996). Half the animals manifested esentially continuous hibernation, except for the usual periodic arousals from torpor. For almost 2 years individual torpor bouts were never separated by more than 10 days and circannual body mass rhythms were eliminated or severely disrupted. SCNx squirrels that did not hibernate continuously nevertheless terminated hibernation 1 month later than control animals, although the circannual period for this subset of SCNx animals was similar to that of control squirrels. SCN ablation altered temporal organization of virtually every aspect of hibernation. In intact squirrels, the SCN may tonically inhibit hibernation throughout the late spring and summer months, when reproduction and accumulation of fat stores essential for overwinter survival require squirrels to be active above ground. Alternatively, the SCN may only generate or respond to signals that terminate torpor in the spring. The SCN evidently is more crucial for circannual timing in squirrels that undergo hibernation than in those prevented from undergoing torpor. It is unclear whether disruption of the circannual rhythm of body mass after SCN ablation represents a general disturbance of circannual timing or masking effects of changes in hibernation behavior.

Structures Other Than the SCN

Ablation of the lateral hypothalamic area (Mrosovsky & Hallonquist, 1986) and the hypothalamic paraventricular nucleus (Dark & Zucker, 1986) had little or no effect on the generation of circannual rhythms of reproduction or body mass of golden-mantled ground squirrels. Destruction of the hypothalamic ventromedial nucleus, although it profoundly increased the amplitude of the body mass rhythm, did not perturb rhythm generation (Mrosovsky, 1975). Similar negative results were obtained after destruction of the olfactory bulbs in squirrels (Ruby, Zucker, Licht, & Dark, 1993), although bulbectomy was associated with a 2-month increase in the period of circannual body mass and testosterone rhythms. Subdiaphragmatic vagotomy, despite its influence on several aspects of energy balance, had little or no effect on the CAR in body mass (Powley & Fox, 1986). To this litany may be added the lateral septal area and the anterior hypothalamic area, destruction of which similarly spares CARs (J. Dark & I. Zucker, unpublished observations).

Circannual Modulation of Circadian Rhythms

Seasonal variation in circadian rhythms has been described for birds and mammals maintained in natural photocycles for 1 or more years (Daan & Aschoff, 1975; Gwinner, 1986). Amounts of locomotor activity and phase angles of entrainment to the light–dark cycle vary over the course of a year; some animals revert from mainly nocturnal activity in summer to primarily diurnal activity in winter (Erkinaro, 1971; Rowesmitt *et al.*, 1982). Such changes in behavior have been viewed as responses to seasonal changes in day length, temperature, and food availability, but in some species they occur in the absence of changes in the external environment. For squirrels maintained in constant light at 21–23°C, the period of the free-running circadian locomotor rhythm changes from less than to greater than 24 hours at intervals of approximately 10.5 months (Mrosovsky, Boshes, Hallonquist & Lang, 1976; Zucker *et al.*, 1983). Among squirrels kept in a constant LD 14:10 photocycle, locomotor activity remains entrained by the illumination cycle, but begins earlier and ends later in the day during the spring and summer than in autumn and winter (Freeman & Zucker, 2000; Lee, Carmichael, & Zucker, 1986; Lee, Holmes & Zucker, 1990). The interval between successive recurrence of summer or winter patterns of entrainment is approximately 10.5 months. A circannual clock modulates circadian rhythms both under free-running (Zucker *et al.*, 1983) and entrained conditions (Lee *et al.*, 1986) in this species. Similar results have been described in fish and birds (Müller, 1978; Rutledge, 1974). Increases in the duration of the daily active phase and marked phase advances in daily activity onset occur each year coincident with onset of reproductive activity in male and female squirrels. The association of seasonal changes in circadian rhythms with the reproductive cycle has long been thought to reflect "the action of reproductive hormones on the circadian system, rather than vice versa" and "there are only very few data that suggest that seasonal variations in overt circadian functions may occur independently of concurrent changes in reproductive condition" (Gwinner, 1986, pp. 80–81). It is notable, therefore, that long-term gonadectomized as well as intact squirrels of both sexes, manifested similar seasonal changes in the circadian system (Figure 8) (Lee & Zucker, 1995). Seasonal variations in locomotor activity are generated by a circannual clock independent of and possibly not influenced by, concurrent changes in secretion of gonadal hormones. By inference, the neural substrates that generate circannual oscillations receive input from a steroid-independent seasonal mechanism that gates expression of circadian rhythms.

An alternative hypothesis posits that circannual changes in body temperature cause the circannual variation in circadian rhythms (Lee *et al.*, 1990). During the spring and summer, squirrels are homeothermic and defend a T_b of 37°C; during the autumn and winter heterothermic phase, T_b declines to 1–2°C above T_a. Squirrels kept at 23°C maintain a T_b of 37°C for half the year and decrease T_b intermittently to 24°C during the remaining 6 months (Lee *et al.*, 1990). A constant T_a is therefore compatible with marked seasonal fluctuations in T_b. The decrease in T_b that occurs intermittently during the heterothermic phase of the annual cycle is associated with lengthening in the period of circadian oscillations and changed patterns of entrainment to the illumination cycle (Lee *et al.*, 1990). When T_a was increased and decreased by 10°C during the squirrels' heterothermic phase, the period of the activity rhythm became shorter and longer, respectively. Identical manipulations of T_a were without effect on circadian rhythms during the squirrels' homeothermic phase. T_a appears to influence circadian rhythms only when it effects changes in T_b (Lee *et al.*, 1990). More recent work questions this interpretation; squirrels in which seasonal variations in T_b were eliminated nevertheless continued to manifest circannual rhythms in circadian organization. Changes in T_b may be sufficient but are not necessary to promote expression of circannual changes in circadian patterns of entrainment (Freeman & Zucker, 2000). Because homeothermic birds show similar seasonal changes in circadian organization (Rutledge, 1974), body temperature is at most one of several factors that can induce changes in circadian timing.

Prospect

Pessimists can point to the ever-increasing gap in understanding of circannual versus circadian rhythms. The long time constants of circannual rhythms limit progress and raise practical concerns. Between 1989 and 1995, the Medline Index of the National Library of Medicine listed 144 circannual and 8427 circadian rhythm publications. This discrepancy in research activity does not reflect negatively on the intrinsic biological interest, medical significance, or aesthetics of circannual rhythms.

Optimists can celebrate the substantial recent progress in the specification of external factors that synchronize and phase shift CARs. Identification of central nervous system sites at which melatonin transduces effects of light on circannual rhythms is on the horizon and may, coincidentally, help localize circannual clocks. The latter outcome would provide a major boost and might provoke exponential growth in research on circannual rhythms.

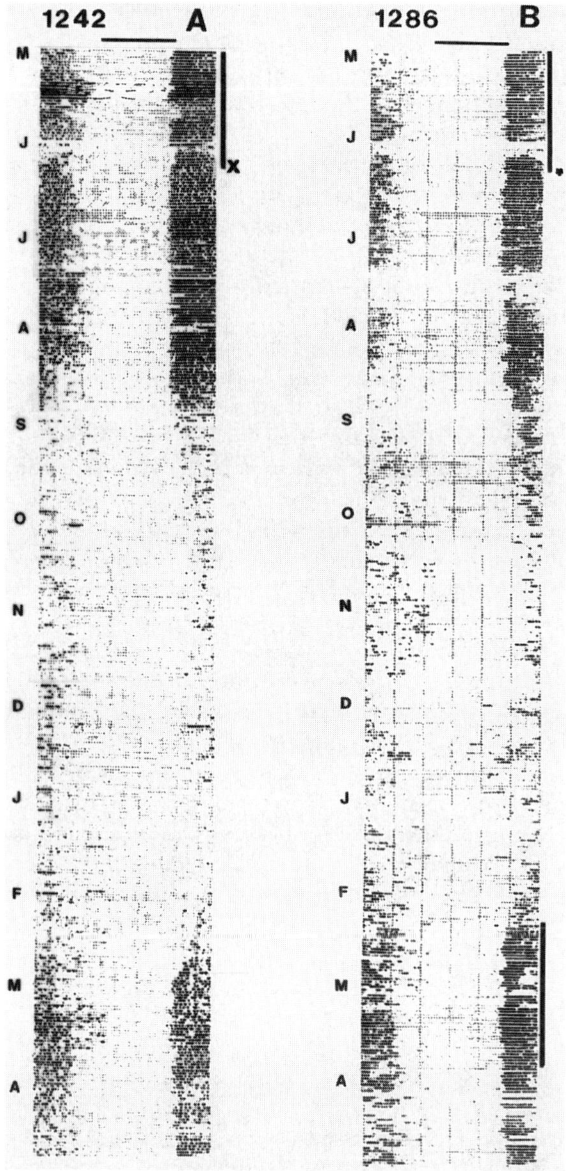

Figure 8. Daily wheel-running activity of representative female squirrels between 1 May 1983 and April 1984. Activity for each day is plotted beneath that from the preceding day in the conventional manner. Females were ovariectomized (A, at ×) or sham-ovariectomized (B, at *) in June 1983. The daily dark phase is indicated by the horizontal bar at the top of each figure; intervals of reproductive activity are marked by vertical bars in the right margins of each actogram. Calendar months are noted in the left margins. (From Lee & Zucker, 1995, with permission. Copyright 1995, Sage Publications.)

Acknowledgments

I am grateful to John Dark, Bruce Goldman, and Michael Gorman for comments on the manuscript and Molly Lojo and Brian Prendergast for assistance with manuscript preparation. Supported by Grant HD-14595 from the NICHD.

References

Adam, C. L., Kyle, C. E., & Young, P. (1994). Influence of prenatal photoperiod on postnatal reproductive development in male red deer (*Cervus elaphus*). *Journal of Reproduction and Fertility, 100,* 607–611.

Adam, C. L., Kyle, C. E., Young, P., & Hirst, D. J. (1995). Influence of prenatal photoperiod on postnatal plasma concentrations of progesterone and prolactin in female red deer (*Cervus elaphus*) reared in constant equatorial photoperiod. *Journal of Pineal Research, 18,* 77–83.

Bartness, T. J., & Wade, G. N. (1985). Photoperiodic control of seasonal body weight cycles in hamsters. *Neuroscience and Biobehavioral Reviews, 9,* 599–612.

Bartness, T. J., Powers, J. B., Hastings, M. H., Bittman, E. L., & Goldman, B. D. (1993). The timed infusion paradigm for melatonin delivery: What has it taught us about the melatonin signal, its reception, and the photoperiodic control of seasonal responses? *Journal of Pineal Research, 15,* 161–190.

Bittman, E. L. (1993). The sites and consequences of melatonin binding in mammals. *American Zoologist, 33,* 200–211.

Bittman, E. L., & Karsch, F. J. (1984). Nightly duration of pineal melatonin secretion determines the reproductive response to inhibitory day length in the ewe. *Biology of Reproduction, 30,* 585–593.

Bittman, E. L., & Weaver, D. R. (1990). The distribution of melatonin binding sites in neuroendocrine tissues of the ewe. *Biology of Reproduction, 43,* 986–993.

Bittman, E. L., Thomas, E. M., & Zucker, I. (1994). Melatonin binding sites in sciurid and hystricomorph rodents: studies on ground squirrels and guinea pigs. *Brain Research, 648,* 73–79.

Bronson, F. H. (1989). *Mammalian reproductive biology.* Chicago: University of Chicago Press.

Bronson, F. H., & Heideman, P. D. (1994). Seasonal regulation of reproduction in mammals. In E. Knobil & J. D. Neill (Eds.), *The physiology of reproduction* (pp. 541–583). New York: Raven Press.

Buijs, R., Pevet, P., Masson-Pevet, M., Pool, C., De Vries, G., Canguilhem, B., & Vivien-Roels, B. (1986). Seasonal variation in vasopressin innervation in the brain of the European hamster (*Cricetus cricetus*). *Brain Research, 371,* 193–196.

Carlsson, A., Svennerholm, L., & Winblad, L. (1980). Seasonal and circadian monoamine variations in human brains examined post mortem. *Acta Psychiatrica Scandinavica, 61,* 75–85.

Carmichael, M. S., & Zucker, I. (1986). Circannual rhythms of ground squirrels: A test of the frequency demultiplication hypothesis. *Journal of Biological Rhythms, 1,* 277–284.

Concannon, P., Roberts, P., Baldwin, B., Erb, H., & Tennant, B. (1993). Alteration of growth, advancement of puberty, and seasonal-appropriate circannual breeding during 28 months of photoperiod reversal in woodchucks. *Biology of Reproduction, 48,* 1057–1070.

Crompton, A. W. & Jenkins, Jr., F. A. (1979). Origin of mammals. In J. A. Lillegraven, Z. Kielen-Jaworowska, & W. A. Clemens (Eds.), *Mesozoic mammals: The first two thirds of mammalian history* (pp. 59–73). Berkeley: University of California Press.

Daan, S., & Aschoff, J. (1975). Circadian rhythms of locomotor activity in captive birds and mammals: Their variations with season and latitude. *Oecologia, 18,* 269–316.

Dark, J., & Zucker, I. (1985). Seasonal cycles in energy balance: Regulation by light. *Annals of the New York Academy of Sciences, 453,* 170–181.

Dark, J., & Zucker, I. (1986). Circannual rhythms of ground squirrels: Role of the hypothalamic paraventricular nucleus. *Journal of Biological Rhythms, 1,* 17–23.

Dark, J., Pickard, G. E., & Zucker, I. (1985). Persistence of circannual rhythms in ground squirrels with lesions of the suprachiasmatic nuclei. *Brain Research, 332,* 201–207.

Dark, J., Dark, K. A., & Zucker, I. (1987). Long day lengths increase brain weight and DNA content in the meadow vole, *Microtus pennsylvanicus. Brain Research, 409,* 302–307.

Dark, J., Kilduff, T. S., Heller, H. C., Licht, P., & Zucker, I. (1990). Suprachiasmatic nuclei influence hibernation rhythms of golden-mantled ground squirrels. *Brain Research, 509,* 111–118.

Davis, D. E., & Finnie, E. P. (1975). Entrainment of circannual rhythms in weight of woodchucks. *Journal of Mammalogy, 56,* 199–203.

Davis, F. C., & Gorski, R. A. (1988). Development of hamster circadian rhythms: Role of the maternal suprachiasmatic nucleus. *Journal of Comparative Physiology, 162,* 601–610.

Erkinaro, E. (1971). Seasonal changes in the phase position of circadian activity rhythms in some voles, and their endogenous component. *Aquilo Serie Zoolologica, 13,* 87–91.

Farner, D. S. (1985). Annual rhythms. *Annual Review of Physiology, 47,* 65–82.

Freeman, D. A. & Zucker, I. (2000). Temperature-independence of circannual variations in circadian rhythms of golden-mantled ground squirrels. *Journal of Biological Rhythms, 15,* 336–343.

Ferkin, M. H., & Zucker, I. (1991). Seasonal control of odour preferences of meadow voles, *Microtus pennsylvanicus,* by photoperiod and ovarian hormones. *Journal of Reproduction and Fertility, 92,* 433–441.

Goldman, B. D., & Nelson, R. J. (1993). Melatonin and seasonality in mammals. In H. S. Yu & R. J. Reiter (Eds.), *Melatonin* (pp. 226–252). Boca Raton, FL: CRC Press.

Gorman, M. R., & Zucker, I. (1995). Seasonal adaptations of Siberian hamsters. II. Pattern of change in day length controls annual testicular and body weight rhythms. *Biology of Reproduction, 53,* 116–125.

Gorman, M. R., & Zucker, I. (1998). Mammalian seasonal rhythms: New perspectives gained from the use of simulated natural photoperiods. In Y. Touitou (Ed.), *Biological clocks. mechanisms and applications.* (pp. 195–204). Amsterdam: Elsevier.

Gwinner, E. (1973). Circannual rhythms in birds: Their interaction with circadian rhythms and environmental photoperiods. *Journal of Reproduction and Fertility, Supplement 19,* 51–65.

Gwinner, E. (1986). *Circannual rhythms.* Berlin: Springer-Verlag.

Gwinner, E. (1996). Circadian and circannual programmes in avian migration. *Journal of Experimental Biology, 199,* 39–48.

Heldmaier, G., & Steinlechner, S. (1981). Seasonal control of energy requirement for thermoregulation in the Djungarian hamster, living in natural photoperiod. *Journal of Comparative Physiology, 142,* 429–437.

Heller, H. C., & Poulson, T. L. (1970). Circannian rhythms. II. Endogenous and exogenous factors controlling reproduction and hibernation in chipmunks (*Eutamias*) and ground squirrels (*Spermophilus*). *Comparative Biochemistry and Physiology, 33,* 357–383.

Hiebert, S. M., Lee, T. M., Licht, P., & Zucker, I. (1998). Estradiol phase shifts circannual body mass rhythms of male ground squirrels. *American Journal of Physiology, 274,* R754–R759.

Hiebert, S. M., Thomas, E. M., Lee, T. M., Pelz, K. M., Yellon, S. M., & Zucker, I. (2000). Photic entrainment of circannual rhythms in golden-mantled ground squirrels: Role of the pineal gland. *Journal of Biological Rhythms, 15,* 126–134.

Hileman, S. M., Kuehl, D. E., & Jackson, G. L. (1994). Effect of anterior hypothalamic area lesions on photoperiod-induced shifts in reproductive activity of the ewe. *Endocrinology, 135,* 1816–1823.

Karsch, F. J., Dahl, G. E., Hachigian, T. M., & Thrun, L. A. (1995). Involvement of thyroid hormones in seasonal reproduction. *Journal of Reproduction and Fertility Supplement, 49,* 409–422.

Kilduff, T. S., Miller, J. D., Radeke, C. M., Sharp, F. R., & Heller, H. C. (1990). 14C-2-Deoxyglucose uptake in the ground squirrel brain during entrance to and arousal from hibernation. *Journal of Neuroscience, 10,* 2463–2475.

Lacoste, V., & Wirz-Justice, A. (1989). Seasonal variation in normal subjects: An update of variables current in depression research. In N. Rosenthal & M. Blehar (Eds.), *Seasonal affective disorders and phototherapy* (pp. 167–229). New York: Guilford Press.

Lee, T. M., & Zucker, I. (1988). Vole infant development is influenced perinatally by maternal photoperiodic history. *American Journal of Physiology, 255,* R831–R838.

Lee, T. M., & Zucker, I. (1991). Suprachiasmatic nucleus and photic entrainment of circannual rhythms in ground squirrels. *Journal of Biological Rhythms, 6,* 315–330.

Lee, T. M., & Zucker, I. (1992). Estradiol phase-shifts circannual rhythms of golden mantled ground squirrels. *American Journal of Physiology, 262,* R1096–R1099.

Lee, T. M., & Zucker, I. (1995). Seasonal variations in circadian rhythms persist in gonadectomized golden-mantled ground squirrels. *Journal of Biological Rhythms, 10,* 188–195.

Lee, T. M., Carmichael, M. S., & Zucker, I. (1986). Circannual variations in circadian rhythms of ground squirrels. *American Journal of Physiology, 250,* R831–R836.

Lee, T. M., Holmes, W. G., & Zucker, I. (1990). Temperature-dependence of circadian rhythms in golden-mantled ground squirrels. *Journal of Biological Rhythms, 5,* 25–34.

Licht, P., Zucker, I., Hubbard, G., & Boshes, M. (1982). Circannual rhythms of plasma testosterone and luteinizing hormone levels in golden-mantled ground squirrels (*Spermophilus lateralis*). *Biology of Reproduction, 27,* 411–418.

Lyman, C. P., Willis, J. S., Malan, A., & Wang, L. C. H. (1982). *Hibernation and torpor in mammals and birds.* New York: Academic Press.

Manley, K. L. S. (1982). *The effect of photoperiod on torpor and weight cycles in the golden-mantled ground squirrel,* Spermophilus lateralis. M.S. thesis, University of California, Davis, California.

Margules, D. L., Lewis, M. J., Dragovich, J. A., & Margules, A. S. (1972). Hypothalamic norepinephrine: Circadian rhythms and the control of feeding behavior. *Science, 178,* 640–642.

Martinet, L., & Zucker, I. (1985). Role of the pineal gland in circadian organization of diurnal ground squirrels. *Physiology and Behavior, 34,* 799–803.

Morin, L. P. (1994). The circadian visual system. *Brain Research Reviews, 67,* 102–127.

Mrosovsky, N. (1975). The amplitude and period of circannual cycles of body weight in golden-mantled ground squirrels with medial hypothalamic lesions. *Brain Research, 99,* 97–116.

Mrosovsky, N. (1980). Circannual cycles in golden-mantled ground squirrels: Phase shift produced by low temperatures. *Journal of Comparative Physiology, 136,* 349–353.

Mrosovsky, N. (1986). Thermal effects on the periodicity, phasing, and persistence of circannual cycles. In H. C. Heller, X. J. Musacchia, & L. C. H. Wang (Eds.), *Living in the cold: Physiological and biochemical adaptations* (pp. 403–410). New York: Elsevier.

Mrosovsky, N. (1990). Circannual cycles in golden-mantled ground squirrels: Fall and spring cold pulses. *Journal of Comparative Physiology, 167,* 683–689.

Mrosovsky, N., & Hallonquist, J. (1986). Lateral hypothalamic lesions in ground squirrels: Rapid recovery and residual deficits. *Brain Research, 399,* 282–295.

Mrosovsky, N., Boshes, M., Hallonquist, J. D., & Lang, K. (1976). Circannual cycle of circadian cycles in a golden-mantled ground squirrel. *Naturwissenschaften, 63,* 298–299.

Müller, K. (1978). The flexibility of the circadian system of fish at different latitudes. In J. Thorpe (Ed.), *Rhythmic activity of fish* (pp. 91–104). London: Academic Press.

Nelson, R. J., Badura, L. L., & Goldman, B. D. (1990). Mechanisms of seasonal cycles of behavior. *Annual Review of Psychology, 41,* 81–108.

Parkinson, T. J., & Follett, B. K. (1994). Effect of thyroidectomy upon seasonality in rams. *Journal of Reproduction and Fertility, 101,* 51–58.

Pengelley, E. T., & Asmundson, S. J. (1974). Circannual rhythmicity in hibernating mammals. In E. T. Pengelley (Ed.), *Circannual clocks* (pp. 95–160). New York: Academic Press.

Pengelley, E. T., Asmundson, S. J., Barnes, B., & Aloia, R. C. (1976). Relationship of light intensity and photoperiod to circannual rhythmicity in the hibernating ground squirrel, *Citellus lateralis. Comparative Biochemistry and Physiology, 53,* 273–277.

Pittendrigh, C. S., & Daan, S. (1976). A functional analysis of circadian pacemakers in nocturnal rodents. IV. Entrainment: Pacemaker as clock. *Journal of Comparative Physiology, 106,* 291–331.

Powley, T. L., & Fox, E. A. (1986). Vagotomy does not alter circannual body weight cycles in the hibernator *Citellus lateralis. Brain Research, 364,* 159–161.

Rowsemitt, C. N., Petterborg, L. J., Claypool, L. E., Hoppensteadt, F. C., Negus, N. C., & Berger, P. J. (1982). Photoperiodic induction of diurnal locomotor activity in *Microtus montanus,* the montane vole. *Canadian Journal of Zoology, 60,* 2798–2803.

Ruby, N. F., Zucker, I., Licht, P., & Dark, J. (1993). Olfactory bulb removal lengthens the period of circannual rhythms and disrupts hibernation in golden-mantled ground squirrels. *Brain Research, 608,* 1–6.

Ruby, N. F., Dark, J., Heller, H. C., & Zucker, I. (1996). Ablation of the suprachiasmatic nucleus alters timing of hibernation in ground squirrels. *Proceedings of the National Academy of Sciences of the USA, 93,* 9864–9868.

Rusak, B., & Bina, K. G. (1990). Neurotransmitters in the mammalian circadian system. *Annual Review of Neuroscience, 13,* 387–401.

Rusak, B., & Morin, L. P. (1976). Testicular responses to photoperiod are blocked by lesions of the suprachiasmatic nuclei in golden hamsters. *Biology of Reproduction, 15,* 366–374.

Rutledge, J. T. (1974). Circannual rhythms of reproduction in male European starlings (*Sturnus vulgaris*). In E. T. Pengelley (Ed.), *Circannual clocks: Annual biological rhythms* (pp. 297–345). New York: Academic Press.

Sadlier, R. M. F. S. (1969). *The ecology of reproduction in wild and domestic mammals.* London: Methuen.

Smale, L., & Morin, L. P. (1990). Photoperiodic responsiveness of hamsters with lesions of the lateral geniculate nucleus is related to hippocampal damage. *Brain Research Bulletin, 24,* 185–190.

Stetson, M. H., Ray, S. L., Creyaufmiller, N., & Horton, T. H. (1989). Maternal transfer of photoperiodic information in Siberian hamsters. II. The nature of the maternal signal, time of signal transfer, and the effect of the maternal signal on peripubertal reproductive development in the absence of photoperiodic input. *Biology of Reproduction, 40,* 458–465.

Turek, F. W. (1987). Pharmacological probes of the mammalian circadian clock: Use of the phase response curve approach. *Trends in Pharmacological Sciences, 8,* 212–217.

Turek, F. W., & Van Cauter, E. (1994). Rhythms in reproduction. In E. Knobil & J. D. Neill (Eds.), *The physiology of reproduction* (pp. 487–540). New York: Raven Press.

Twente, J. W., & Twente, J. A. (1967). Seasonal variation in the hibernating behaviour of *Citellus lateralis.* In K. C. Fisher, A. R. Dawe, C. P. Lyman, E. Schonbaum, & F. E. South, Jr. (Eds.), *Mammalian hibernation, III* (pp. 47–63). Edinburgh: Oliver & Boyd.

Walker, J. M., Haskell, E. H., Berger, R. J., & Heller, H. C. (1980). Hibernation and circannual rhythms of sleep. *Physiological Zoology, 53,* 8–11.

Weaver, D. R., & Reppert, S. M. (1986). Maternal melatonin communicates daylength to the fetus in Djungarian hamsters. *Endocrinology, 119,* 2861–2863.

Weaver, D. R., & Reppert, S. M. (1990). Melatonin receptors are present in the ferret pars tuberalis and pars distalis, but not in brain. *Endocrinology, 127,* 2607–2609.

Wimsatt, W. A. (1969). Some interrelations of reproduction and hibernation in mammals. *Symposia of the Society for Experimental Biology, 23,* 511–549.

Woodfill, C. J. I., Robinson, J. E., Malpaux, B., & Karsch, F. J. (1991). Synchronization of the circannual reproductive rhythm of the ewe by discrete photoperiodic signals. *Biology of Reproduction, 45,* 110–121.

Woodfill, C. J. I., Wayne, N. L., Moenter, S. M., & Karsch, F. J. (1994). Photoperiodic synchronization of a circannual reproductive rhythm in sheep: Identification of season-specific time cues. *Biology of Reproduction, 50,* 965–976.

Zucker, I. (1985). Pineal gland influences period of circannual rhythms of ground squirrels. *American Journal of Physiology, 249,* R111–R115.

Zucker, I. (1988). Neuroendocrine substrates of circannual rhythms. In D. J. Kupfer, T. H. Monk, & J. D. Barches (Eds.), *Biological rhythms and mental disorders* (pp. 219–251). New York: Guilford Press.

Zucker, I., & Boshes, M. (1982). Circannual body weight rhythms of ground squirrels: Role of gonadal hormones. *American Journal of Physiology, 243,* R546–R551.

Zucker, I., Boshes, M., & Dark, J. (1983). Suprachiasmatic nuclei influence circannual and circadian rhythms of ground squirrels. *American Journal of Physiology, 244,* R472–R480.

Zucker, I., Lee, T. M., & Dark, J. (1991). The suprachiasmatic nucleus and annual rhythms of mammals. In D.C. Klein, R.Y. Moore, & S.M. Reppert (Eds.), *The suprachiasmatic nucleus: The mind's clock* (pp. 246–259). New York: Oxford University Press.

Zucker, I., Ruby, N. F., & Dark, J. (1993). The suprachiasmatic nucleus mediates rhythms of hibernation and daily torpor in rodents. In C. Carey, G. L. Florant, B. A. Wunder, & B. Horwitz (Eds.), *Life in the cold* (pp. 277–289). Boulder, CO: Westview Press.

Part VI

Human Circadian Rhythms

20

Human Circadian Physiology and Sleep–Wake Regulation

CHARLES A. CZEISLER AND DERK-JAN DIJK

PHENOMENOLOGY AND SCOPE OF CIRCADIAN RHYTHMICITY IN HUMANS

MANIFEST RHYTHMICITY CHARACTERIZES NEARLY ALL PHYSIOLOGIC AND BEHAVIORAL VARIABLES

Periodic daily cycles have been observed in a wide range of physiologic and behavioral variables in humans. These include cyclic variations in sleep propensity, neurobehavioral performance, motor performance, as well as thermoregulatory, endocrine, cardiac, gastrointestinal, pulmonary, renal, and neurologic functions, including sensory sensitivity. The rhythms of all of these variables exhibit a specific temporal relationship with each other and with the environmental light–dark cycle. The mechanisms underlying these rhythms, their temporal relationships, and their functional significance have been the subject of numerous investigations (Conroy & Mills, 1970).

Human body temperature was one of the first such periodic functions identified, having been recognized more than two centuries ago (Aschoff, 1955; Benedict, 1904; Hunter, 1778). Typically, body temperature peaks in the early evening hours and reaches a nadir in the latter half of the nightly sleep episode. The adrenal hormone, cortisol, on the other hand, is generally at or near its lowest level shortly before an individual's habitual bedtime, and rises to its highest level of the 24-hour day just after an individual's habitual wake time (Czeisler *et al.*, 1976; Hellman *et al.*,

CHARLES A. CZEISLER Circadian, Neuroendocrine and Sleep Disorders Section, Division of Endocrinology, Department of Medicine, Brigham and Women's Hospital, Harvard Medical School, Boston, Massachusetts 02115. DERK-JAN DIJK Institute of Pharmacology, University of Zürich, CH-8057 Zürich, Switzerland and Center for Chronobiology, University of Surrey, Guilford GU27XH United Kingdom.

Circadian Clocks, Volume 12 of *Handbook of Behavioral Neurobiology*, edited by Joseph S. Takahashi, Fred W. Turek, and R. Y. Moore, Kluwer Academic / Plenum Publishers, New York, 2001.

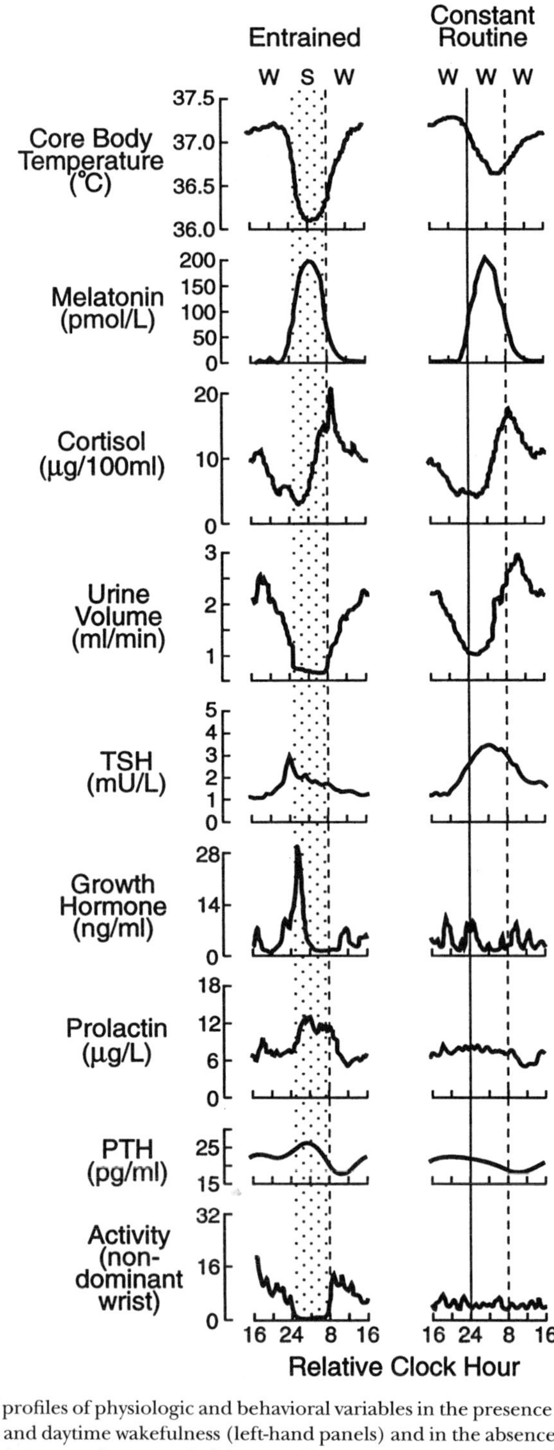

Figure 1. Temporal profiles of physiologic and behavioral variables in the presence of a sleep–wake cycle, i.e., nocturnal sleep and daytime wakefulness (left-hand panels) and in the absence of a sleep–wake cycle (right-hand patterns), i.e., continuous wakefulness under constant routine conditions. The timing of the sleep episode is indicated by the stippled area (left-hand panels) and the timing of habitual bedtime and

1970; Pincus, 1943; Reynold, Quigley, Kennard, & Thorn, 1951; Weitzman, Schaumburg, & Fishbein, 1966; Weitzman, Zimmerman, Czeisler, & Ronda, 1983). Secretion of the pineal hormone, melatonin, begins to increase in the late evening, reaches a peak in the middle of the night, and declines after wake time (Pelham, Vaughan, Sandock, & Vaughan, 1973; Ralph, Mull, Lynch, & Hedlund, 1971; Weitzman *et al.*, 1978). Characteristic circadian profiles of body temperature, melatonin and cortisol secretion, and their phase relationships persist even in the absence of sleep (Åkerstedt, Palmblad, de la Torre, Marana, & Gillberg, 1980; Czeisler *et al.*, 1990b) (Figure 1).

Heart rate and blood pressure fluctuate about relatively high levels during active waking hours and drop abruptly at bedtime upon the assumption of the supine posture, remaining low throughout the sleep episode and rising abruptly upon awakening and resumption of the upright posture. Subjective alertness and many aspects of human performance (e.g., vigilance, memory, speed of decision making) also display circadian variation, with daytime peaks and nocturnal nadirs. These circadian profiles in waking neurobehavioral variables persist during extended periods of wakefulness, although they are overlaid on a pattern of linear decline across days due to cumulative elevation of the homeostatic drive for sleep (Åkerstedt *et al.*, 1979; Åkerstedt & Fröberg, 1977; Åkerstedt, Fröberg, Friberg, & Wetterberg, 1979; Dijk, Duffy, & Czeisler, 1992; Dinges, Orne, Whitehouse, & Orne, 1987; Johnson *et al.*, 1992).

Finally, the alternation of sleep and waking within each day can be viewed as a cyclic phenomenon. However, the timing and temporal relationship between the sleep–wake cycle and the geophysical light–dark cycle vary with photoperiod, latitude, availability of artificial sources of light, work schedule, as well as social demands, age, and cultural traditions. The typical subject in most circadian studies conducted to date, i.e., healthy, young, urban adult males living in industrialized societies, is day-active, with one or two major bouts of activity per 24 hours and a major sleep episode at night.

Although all of phenomena described above can be considered rhythmic phenomena, the mechanisms underlying these rhythmic variations may be quite diverse. A key issue in research on human circadian rhythms has been the extent to which these daily variations reflect (1) responses evoked by periodic environmental/behavioral stimuli, (2) responses evoked by an appetitive or homeostatic process, or (3) oscillations driven by an endogenous circadian pacemaker (Aschoff, 1960; Miners & Waterhouse, 1981, 1989; Wever, 1973, 1985).

Rhythmicity Persists in the Absence of Periodic External Cycles

Humans, like nearly all other species, continue to behave in a periodic manner when studied in the absence of regular time cues from the external environment. Nearly 60 years ago, Kleitman found that an approximately 24-hour rhythm of body temperature could persist in a human subject living deep within a cave, isolated from

waketime during the weeks prior to the constant routine are indicated by the solid and stippled lines (right-hand panels). All data are from normal young healthy men, 18–30 years old, studied under similar conditions. Not all variables were monitored in the same subjects. TSH, thyroid-stimulating hormone; PTH, parathyroid hormone. Data reproduced with permission from the following sources: Melatonin from Shanahan (1995); TSH from Allen and Czeisler (1994); prolactin from Waldstreicher *et al.* (1996); PTH from El-Hajj *et al.* (1987). (From Czeisler and Khalsa, 2000, with permission.)

periodic changes in the external environment and living on an imposed non-24-hour routine of sleep and wakefulness (Kleitman & Kleitman, 1953). Twenty-five years later, Aschoff and colleagues, as well as other scientists in Germany, France, England, and the United States, studied the timing of rest and activity (which in healthy young people usually coincides closely with polysomnographically defined sleep and wakefulness) in human subjects living in underground bunkers, subterranean caverns, and soundproof laboratories, also isolated from the usual 24-hour cycles of light, temperature, and social interaction (Aschoff & Weaver, 1962; Findley, 1966; Mills, 1964; Siffre, 1964; Webb & Agnew, 1972). Subjects in those later experiments were free to choose when they wished to turn out the lights and go to sleep and when they wished to begin their active, waking "day" (so-called "free-running conditions"). Under these conditions, humans maintain in general a monophasic rest–activity cycle and the ratio of rest to activity remains approximately 1:2, even in the absence of an imposed sleep–wake schedule. The main finding from those studies was that the rest–activity cycle, which remained synchronized with the core body temperature rhythm and with the self-selected light–dark cycle, exhibited a free-running period that was desynchronized from the 24-hour geophysical day (*external desynchronization*). Under such conditions, the rest–activity cycle oscillated with a period longer than 24 hours, but also displayed considerable interindividual variability. These data indicated that human circadian rhythms were not solely generated by geophysical cycles (e.g., ambient illumination, electromagnetic field strength, ambient temperature, etc.), which all have a period of 24 hours. Thus, this finding demonstrated the endogenous origin of the rest–activity cycle, together with rhythms of body temperature, renal, and endocrine function in the adult human.

Under free-running conditions, the internal phase relationships among many variables, such as core body temperature, plasma cortisol secretion, urine production, and alertness, remain stable; however, the self-selected behavioral rest–activity/sleep–wake cycle changes its phase relationship with this ensemble of physiologic rhythms from that which is observed under entrained, 24-hour conditions (Wever, 1973). Thus, subjects typically choose to go to sleep at or after the nadirs of the *endogenous* components of their body temperature and subjective alertness rhythms under free-running conditions, 6–8 hours later in the endogenous temperature and alertness cycles than under entrained conditions. Similarly, the self-selected bedtime of free-running subjects generally occurs after levels of the hormone cortisol have begun to rise, rather than at the cortisol nadir, as occurs under entrained conditions (Weitzman, Czeisler, & Moore-Ede, 1979).

Despite this change in internal phase relationships, most physiologic functions continue to oscillate in synchrony with each other and with a near 24-hour period under such free-running conditions. However, human subjects do not always choose to sleep and wake in synchrony with the output of these biological rhythms. Sometimes the timing of the sleep–wake cycle becomes labile, with high cycle-to-cycle variability in the duration of sleep and wake episodes. In such cases, the average period of the sleep–wake cycle typically deviates considerably from 24 hours averaging > 30 hours in most such cases and < 16 hours in some cases) (see Figure 2). Components of a number of rhythms [such as core body temperature, cortisol, urine volume, urinary potassium, and rapid eye movement (REM) sleep propensity] nonetheless continue to maintain a fixed phase relationship to each other and to oscillate with a common near-24-hour period. This phenomenon was dubbed "*spontaneous internal desynchronization*" by Aschoff and his colleagues in Germany (Aschoff, Gerecke, & Wever, 1967). Because of the difference in their periods of oscillation, there

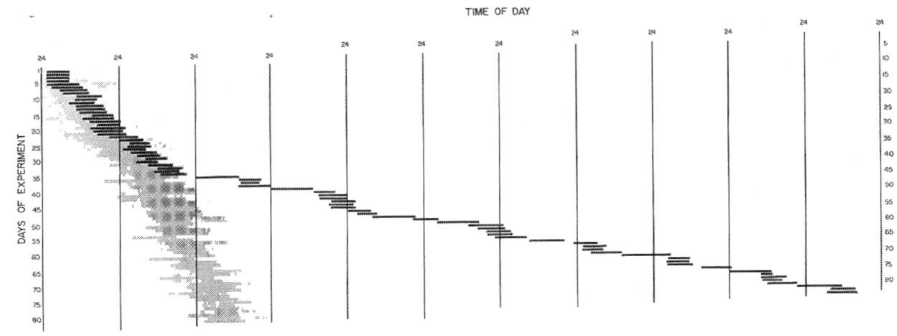

Figure 2. Synchronized and desynchronized freerun. Times of sleep (black bars) and times during which body temperature was below the mean (36.68°C) are plotted for subject PR01, who lived without knowledge of clock time in the temporal isolation facility of the Laboratory of Human Chronophysiology at Montefiore Hospital and Medical Center. Bed rests on days 1–5 were scheduled to regular times; the remainder were self-selected. During the first segment (35 days) of the study, the sleep–wake cycle was synchronized with the core body temperature rhythm. Thereafter, the sleep–wake cycle desynchronized from the core body temperature rhythm. Note that in the desynchronized section, long sleep episodes and long sleep–wake cycles (usually one) are followed by a number of short sleep episodes and short sleep-wake cycles. During these clusters of short sleep–wake cycles the period of the sleep–wake cycle is close to the body temperature rhythm. (From Czeisler *et al.*, 1980c, with permission. Copyright 1980, American Sleep Disorders Association, Rochester, MN.)

is no longer a fixed phase relationship between the sleep–wake cycle and the rhythms coupled to the core body temperature cycle in such cases. Yet, the timing of the behavioral sleep–wake cycle, with its associated changes in activity level, posture, light exposure, and meal timing, appears to evoke corresponding, highly reproducible changes in a number of physiologic rhythms. For example, during spontaneous internal desynchronization, body temperature is lower during sleep than waking, irrespective of the phase relationship of the sleep–wake cycle to the near-24-hour (circadian) component of the temperature rhythm. Similarly, cortisol secretion is inhibited, growth hormone secretion is augmented, and slow-wave sleep peaks shortly after the onset of sleep, even when sleep onset occurs at an atypical phase relationship to the circadian component of the cortisol and temperature rhythms (Weitzman, Czeisler, Zimmerman, & Ronda, 1981).

Furthermore, the sleep–wake cycle and the evoked components of rhythms associated with it continue to interact with endogenous circadian phase, such that the spontaneous duration of sleep varies systematically with circadian phase, and the temperature declines evoked by sleep appears to vary with circadian phase (Czeisler *et al.*, 1980a; Zulley & Wever, 1982). In fact, the internal structure of sleep itself, as well as subjective alertness and cognitive performance, vary with the circadian phase of that underlaying, near-24-hour oscillation even during desynchrony. These complex interactions, initially recognized during spontaneous desynchrony, have led to the development of a number of mathematical models attempting to characterize the relationship of the behavioral sleep–wake cycle to the underlying physiologic oscillations (Borbély, 1982a, 1994; Daan, Beersma, & Borbély, 1984; Kronauer, Czeisler, Pilato, Moore-Ede, & Weitzman, 1982; Moore-Ede & Czeisler, 1984; Strogatz, 1986; Wever, 1965). These complex interactions have also inspired experiments aimed at quantifying the extent to which rhythmic variations persist in the absence of sleep and associated changes of posture, light exposure, activity, and meal timing.

Temporal Distribution of Behavior Induces Apparent Rhythmicity in Other Variables

Studies in which the rest–activity cycle spontaneously desynchronized from endogenous near-24-hour rhythms allowed quantitative characterization of the component of daily rhythmicity that was evoked by the rest–activity cycle and associated behaviors. However, because of the interaction between the rest–activity cycle and the physiologic rhythms coupled to the near-24-hour component of the body temperature cycle, these behaviors are distributed nonuniformly across endogenous circadian phases during *spontaneous* internal desynchrony (Strogatz, Kronauer, & Czeisler, 1987). Therefore, using the techniques pioneered by Kleitman and Kleitman (1953), a number of groups have attempted to quantify more accurately the component of these rhythms evoked by sleep–wake behaviors through use of experimental protocols involving continuous sleep deprivation (Åkerstedt *et al.*, 1979; Johnson *et al.*, 1992; Patrick & Gilbert, 1896) or by uniformly scheduling the timing of rest and activity across all circadian phases using *forced-desynchrony* protocols (Carskadon & Dement, 1975; Czeisler, Allan, & Kronauer, 1990a; Dijk & Czeisler, 1991; Hiddinga, Beersma, & van den Hoofdakker, 1997; Lavie & Scherson, 1981; Minors & Waterhouse, 1981; Weitzman *et al.*, 1974). In the latter experiments, the rest–activity cycle length is scheduled outside the range of entrainment of the endogenous circadian pacemaker. Artificial "day" lengths ranging from as short as 20 minutes (7 minutes of rest followed by 13 minutes awake per cycle) to as long as 30 hours (10 hours in bed followed by 20 hours awake per cycle) have been imposed for durations of as short as 1 day to as long as 1 month.

While endogenous near-24-hour rhythms persist under conditions of forced desynchrony, these forced-desynchrony studies have also demonstrated that periodic changes in behavior (e.g., sleep and waking, eating, etc.) exert powerful and consistent effects on a number of manifest rhythms, effects that are often of the same magnitude as the endogenous component of such rhythms. For example, just as in spontaneous desynchrony, body temperature is higher during wakefulness than it is during sleep (Beersma & Hiddinga, 1996; Dijk & Czeisler, 1985; Dijk, Shanahan, & Czeisler, 1997b). This evoked component is, on average, equivalent in magnitude to the endogenous component of the temperature cycle. Cortisol secretion is another example of this dual influence. Cortisol secretion is inhibited just after bedtime and augmented at the transition between rest in darkness and waking activity in room light, although the magnitude of this evoked component is much smaller than the endogenous circadian component of the cortisol rhythm (Weitzman *et al.*, 1983).

Furthermore, the evoked components of alertness and cognitive performance decline throughout the waking day and are restored by sleep. Since subjective alertness and cognitive performance exhibit a progressive linear decline superimposed on rhythmic near-24-hour variations during sleep deprivation, the evoked component of these variables is associated with the *duration* of prior wakefulness rather than wakefulness itself. Similarly, the restoration of subjective alertness and cognitive performance is not instantaneous, but depends on the duration of sleep. These and other observations have led to the notion that sleep serves a restorative function and that a process (called "sleep homeostasis" or the "sleep homeostat") maintains a running tally of the duration of sleep and wakefulness. The magnitude of these rhythms in alertness and cognitive performance, evoked by the durations of wakefulness and sleep, is comparable to the amplitude attributable to the circadian pacemaker even during the first 16 hours of the waking day (Dijk *et al.*, 1992; Johnson *et al.*, 1992).

All these data also indicate that the changes in manifest rhythms associated with the rest–activity cycle are observed independent of the period of the rest–activity cycle. This suggests that these effects are indeed a *direct* consequence of the behavioral/physiologic changes associated with the periodic rest–activity cycle (the timing of which can be readily manipulated experimentally) rather than a consequence of an unidentified oscillatory process driving the rest–activity cycle.

Rhythmicity Persists in the Absence of Periodic Changes in Behavior

As described in the previous section, during internal desynchrony, there are components of a number of endogenous circadian rhythms that maintain a persistent, near-24-hour oscillation even in the absence of synchrony with the behavioral rest–activity cycle. This suggests that at least a component of each of these physiologic rhythms is not simply a passive response evoked by periodic changes in the sleep–wake cycle and associated behaviors, but rather is generated by an endogenous circadian oscillatory process. However, characterization of the near-24-hour oscillatory component of such rhythms during spontaneous internal desynchrony requires lengthy segments of data collected during desynchrony and mathematical procedures for extraction of the endogenous circadian component from observed data. This is because during long-term studies in which behavioral activity is desynchronized from the endogenous circadian process, the evoked effects of activity, sleep, meal timing, light exposure, showers, etc. (also called masking effects) are distributed across a variety of circadian phases. The length of study necessary to extract the average waveform of the circadian component of these rhythms depends on a number of factors, including the relative strengths of the endogenous circadian and the evoked components for a given variable. In practice, at least 3- to 4-week experiments are required to estimate the relative contribution of both components with precision. Even during internal desynchrony, these evoked effects obscure the contribution of the oscillatory component to any *single* day of recording. Therefore, a technique called a *constant routine* was proposed by John Mills and his colleagues and developed to characterize circadian rhythmicity of physiologic variables, within a day or two, in the absence of periodic changes in behavior (Mills, Minors, & Waterhouse, 1978).

Using a modification of this technique (Czeisler *et al.*, 1985), it has been demonstrated that even when subjects remain continuously awake at constant bed rest in constant indoor room light for extended (40–70 hour) periods with meals and activity distributed uniformly throughout the 24-hour day, near-24-hour rhythms persist in a number of different physiologic and cognitive variables (see Figure 1). In fact, robust oscillations are observed in levels of body temperature, cortisol, melatonin, thyroid-stimulating hormone (TSH), subjective alertness, cognitive performance, short-term memory, urine output, urinary electrolyte excretion, and sleep propensity under such constant behavioral and environmental conditions. Furthermore, the phase relationship among the endogenous circadian rhythms observed during the constant routine are consistent with the fixed phase relationships among the endogenous near-24-hour components of those rhythms observed during internal desynchrony.

Taken together, these results suggest that an endogenous pacemaker with an intrinsic near-24-hour period drives a component of each of these rhythms and that the outputs of that pacemaker organize the phase relationships among those endogenous near-24-hour rhythms.

Manifest Rhythmicity Is Comprised of an Endogenous Component Combined with Environmentally and Behaviorally (Including Homeostatically Determined) Evoked Components

Comparison of the waveform of the endogenous component of rhythms unmasked during the constant routine protocol with the manifest rhythm observed under conditions of ordinary entrainment to the 24-hour day indicates that during entrainment to the 24-hour day, the phase relationship between the endogenous circadian component and the components of each rhythm evoked by environmental and behavioral stimuli interact to produce manifest rhythmicity.

Most often, this interaction serves to enhance the magnitude of the daily changes. For example, the entrained phase of the endogenous component of the circadian temperature cycle is such that the body temperature rhythm reaches its nadir during the time of the habitual sleep episode even if subjects remain awake throughout the night. However, in the absence of sleep and darkness at night, the magnitude of the observed drop in temperature is approximately *half* of that observed when subjects are asleep at night due to the absence of the thermolytic effect of rest, the supine posture, and sleep on body temperature. Bright light exposure at night—ordinarily limited to the waking day—elevates body temperature (Cajochen, Dijk, & Borbély, 1992; Myers & Badia, 1993). Thus, under ordinary entrained conditions, the endogenous and evoked components of the body temperature rhythm appear to interact to produce a manifest temperature rhythm that is much larger in magnitude than either the endogenous or evoked components. The detailed explanation of the causes of these effects is unknown and likely to be complex. Thus, the effect of light could be indirect, by increasing alertness and by decreasing the release of the thermolytic hormone, melatonin (Cagnacci, Elliott, & Yen, 1992; Dawson, Gibbon, & Singh, 1996), or more direct through one or more inputs to the central nervous system. The act of lying down, which normally precedes sleep, lowers body temperature (Kleitman & Doktorsky, 1933), and sleep itself profoundly affects physiology, endocrinology, etc. To distinguish the roles of these possible factors would entail a series of experiments in which only one factor at a time was changed.

A mutual amplification of the manifest rhythm by the endogenous and evoked components is also true for the rhythm of plasma prolactin (Waldstreicher *et al.*, 1996) concentrations and the rhythm of urine production. The situation of mutual influence from endogenous circadian and evoked components does not always invoke a synergy (potentiation) and is more complex for some physiologic variables. For example, the amplitude of the circadian component of the rhythm in plasma TSH levels is much *larger* than the magnitude of the manifest rhythm of TSH observed under ordinary conditions of entrainment of the rest–activity cycle to the 24-hour day. This is because the phase of the endogenous circadian TSH rhythm is such that TSH levels peak near the midpoint of the habitual sleep episode when subjects remain awake throughout the night. The evoked effect of nightly sleep episode results in an abrupt decline in TSH levels, coincident with the rising phase of the endogenous component of the circadian TSH rhythm (Allan & Czeisler, 1994). The combined effect of these opposing influences mutes the manifest 24-hour TSH rhythm under conditions of ordinary entrainment to the 24-hour day.

Finally, there are other variables, such as growth hormone secretion, heart rate, and blood pressure, for which the evoked component associated with the rest–activity cycle is much larger than the endogenous component. In general, complex interactions among an endogenous circadian rhythm, responses evoked by periodic behaviors associated with the rest–activity cycle, and responses to external environmental stimuli such as the light–dark cycle determine the nature of manifest rhythmicity observed during entrainment to the 24-hour day. For some of these variables, it has been shown that the relative contributions of endogenous and evoked components vary between subjects and that the effects of the evoked components vary over the circadian cycle (Dijk *et al.*, 1997b; Wever, 1985).

Although many physiologic variables have been shown to exhibit circadian variation, in only few cases has the mechanism by which the circadian pacemaker controls the variation in these variables been elucidated. In addition, very little is known about the physiologic mechanisms underlying the evoked effects of the sleep–wake cycle and associated behavioral changes on the variables discussed.

Sleep–Wake Regulation: A Single Circadian Pacemaker or Multiple Oscillatory Processes?

The cornerstone of the hypothesis that multiple oscillatory processes must underlie the human circadian timing system was derived from the observation that after an abrupt phase shift of the rest–activity and light–dark cycle, different physiologic variables appear to reentrain to the shifted schedule at different rates (Moore-Ede, Kass, & Herd, 1977). Furthermore, it was reported that different rhythms lose synchronization at different day lengths during a fractional desynchrony protocol in which the period of the imposed rest–activity/light–dark cycle changes gradually (Folkard, Wever, & Wildgruber, 1983). These findings, however, do not demonstrate the existence of multiple self-sustaining oscillatory processes, since such apparent dissociation of manifest rhythms could occur (1) if a single circadian pacemaker were synchronously driving the endogenous components of all of these variables, *and* (2) the evoked effects of the abruptly shifted or gradually shifting rest–activity cycle were to vary in magnitude across variables, as is known to occur (see previous section) (Daan & Beersma, 1992). Indeed, phase shift studies in which constant routines were used to evaluate the endogenous component of a number of observed rhythms before and after an abrupt phase shift indicate that, on average, the endogenous components of these rhythms maintain a fixed phase relationship to each other before and after a light-induced phase shift (see below). Thus, phase shift studies in which the evoked component was controlled during the evaluation suggest that a single light-sensitive pacemaker drives the endogenous, circadian component of those rhythms that persist during a constant routine (Czeisler *et al.*, 1990b; Shanahan & Czeisler, 1991).

Yet, the amplitude of cortisol and body temperature have been shown to dissociate during constant routines after protocols designed to suppress endogenous circadian amplitude (Jewett, Kronauer, & Czeisler, 1991). Furthermore, the remarkable phenomenon of spontaneous internal desynchronization between the behavioral rest–activity/sleep–wake cycle and the group of autonomous physiologic and cognitive rhythms marked by the endogenous component of the body temperature cycle can only be explained by models that incorporate two oscillatory processes that interact to generate the rest–activity and temperature cycles observed in such cases

(Aschoff *et al.*, 1967). The fact that mathematical description of this phenomenon requires two oscillatory processes has led some scientists to search for an additional circadian pacemaker (other than the suprachiasmatic nucleus, SCN) as the site of a second circadian pacemaker in humans. However, the persistent oscillation of rhythmic components with a near-24-hour period in humans whose self-selected rest–activity cycle has an average period that deviates substantially from 24 hours does not necessarily imply the existence of two similar, but anatomically discrete circadian pacemakers.

First of all, the properties of these two oscillatory processes are quite different. The endogenous components of the rhythms of core body temperature, plasma cortisol and melatonin, urine volume, urinary potassium, sleep propensity, REM sleep propensity, subjective alertness, and cognitive performance consistently oscillate with a very stable, near 24-hour period under a wide variety of experimental conditions. Furthermore, behavioral manipulations, such as sleep deprivation, do not arrest the near-24-hour oscillations even in sleep propensity itself. In contrast, the period of the behavioral rest–activity (i.e., sleep–wake) cycle is highly labile, with considerable cycle-to-cycle variability, can be directly manipulated by scheduling the timing of the rest–activity cycle, and is highly dependent on the availability of behavioral options. In fact, there is no evidence of the persistence of rhythmicity at the period of the rest–activity cycle when the rest–activity cycle is itself held constant, such as during extended sleep deprivation.

Second, the observation of two oscillatory processes during spontaneous internal desynchrony does not necessarily indicate that there are two biological "clocks" or circadian pacemakers timing those processes (Daan *et al.*, 1984). For example, systematic underestimation of the passage of time by human subjects living in the absence of external time cues, which has been noted to occur and which depends on the behavioral options available to the subjects, could lead subjects to tend to extend their waking day and delay their subsequent sleep. As a consequence, the spontaneous awakening would occur at a later phase of the endogenous circadian temperature and sleep propensity rhythm (Aschoff, 1985). This is especially true since under conditions of temporal isolation, waking up is not facilitated by external morning light or an alarm clock. Since manifest sleep propensity and subjective alertness are codetermined by (1) a homeostatic process that increases the drive for sleep as a function of the duration of prior waking and (2) the phase of the endogenous circadian pacemaker (see section below), waking up at a later phase of the endogenous circadian component of the sleep propensity rhythm would reduce manifest sleep propensity at the crest of the near-24-hour component of the sleep propensity rhythm. Such a change in internal phase relations would thereby reduce the magnitude of the manifest sleep propensity rhythm. The strength of the residual magnitude of the endogenous near-24-hour sleep propensity rhythm versus the level of confidence in the subjective estimation of the time of day would then determine whether or not the self-selected behavioral rest–activity cycle would remain internally synchronized to the endogenous, near-24-hour component of sleep propensity rhythm during long-term temporal isolation studies. If the residual magnitude of this endogenous near-24-hour sleep propensity rhythm were of insufficient strength to capture or entrain the behavioral rest–activity cycle in some individuals, then the rest–activity cycle could lose synchronization with the endogenous, near-24-hour component of the body temperature and sleep propensity rhythms (Zulley & Campbell, 1985). Under such conditions, the rest–activity cycle would be expected to exhibit relative internal coordination to an internal synchronizer that is of insufficient strength to entrain the behavioral rhythm, which matches the observed data. Thus, the observation of two oscillatory processes during spontaneous internal desynchrony does not require the existence of two anatomically discrete circadian pacemakers timing those processes. However, recent data on the persistence of a circadian rhythm of melatonin in the isolated hamster retina *in vitro* (Tosini & Menaker, 1996) and earlier observations of a persistent circadian rhythm of disk shedding in the retina of the rat (Terman & Terman, 1985), which persists in SCN-lesioned rats, demonstrated the existence of at least one circadian oscillator outside the SCN in mammals.

Entrainment of the Human Circadian Pacemaker

Endogenous Circadian Rhythms Are Synchronized by the Environmental Light–Dark Cycle in Humans

In nearly all eukaryotic organisms studied, the periodic light–dark cycle is the principal environmental stimulus that synchronizes (or entrains) the non-24-hour intrinsic period of endogenous circadian rhythms to a period of exactly 24 hours. Twenty-five years ago it was reported that the environmental light–dark cycle alone was not a strong enough zeitgeber to entrain human circadian rhythms to the 24-hour day and that social interaction, rather than environmental light, was the principal synchronizer of the human circadian pacemaker to the 24-hour day (Aschoff *et al.*, 1971; Aschoff, Hoffmann, Pohl, & Wever, 1975; Wever, 1970). There were even reports that weak electromagnetic fields had more of an impact on human circadian rhythms than light (Wever, 1970). These and other results led to the general conclusion that the light–dark cycle was relatively unimportant for the entrainment of human circadian rhythms, and that periodic social cues were necessary for our entrainment to the period of the earth's rotation.

It was concluded that circadian rhythm synchronization was primarily dependent on higher cortical functioning in humans, in contrast to other mammals, in which it has since been shown that synchronization occurs by photic input to a light-sensitive pacemaker in the hypothalamus. Unfortunately, there were technical flaws in both of the key experiments that served as the basis for the conclusion that human circadian rhythms could not be entrained by light. Free access to multiple sources of auxiliary lighting during scheduled episodes of "darkness" made the first set of human experiments that claimed a preeminent role for social contacts in entrainment more similar to animals studied on a self-selected illumination routine. Under such conditions, it has been shown that both birds and mammals will freerun, undermining the conclusion that humans were different from other species in their response to an imposed light–dark cycle. The second set of experiments was confounded by the effects of scheduled activity, which, as described above, exerts powerful masking effects on all of the observed rhythms. Furthermore, the 4-day experiment was much too short to evaluate entrainment of the human circadian pacemaker, and therefore did not warrant the conclusion that human circadian rhythms could be entrained by social cues in constant darkness. Finally, the effect of social contact per se was not evaluated as distinct from the scheduling of the rest–activity/sleep–wake cycle alone, as could be accomplished with a mechanical alarm clock devoid of social interaction. Thus, the conclusion that periodic social cues were the primary synchronizers of human circadian rhythms, making humans different from nearly all other species studied in the mechanism by which their circadian rhythms were synchronized to the 24-hour day, was not justified on the basis of these two sets of experiments.

Later experiments indicated that human circadian temperature and activity rhythms could be entrained by a strict light–dark cycle comparable to that used in entrainment studies of plants and animals. In fact, when studied under comparable conditions, it was found that periodic exposure to indoor room light alternating with darkness was sufficient to entrain the human circadian system, just as in other mammals (Czeisler, Richardson, Moore-Ede, & Weitzman, 1978). However, the early studies of photic entrainment in human subjects, like those in other mammals, could not distinguish whether synchronization occurred (1) directly through and action of light on the endogenous circadian pacemaker or (2) indirectly by an influence of lighting conditions on the subjects' self-selected scheduling of their behavioral rest–activity/sleep–wake cycle.

More recent evaluation of the light sensitivity of the human circadian pacemaker using both the constant routine protocol and the classic free-running protocol have unequivocally demonstrated that endogenous circadian components of the rhythms of core body temperature, plasma cortisol, TSH and melatonin, subjective alertness and cognitive performance, urine production, and sleep propensity can be reset by scheduled exposure to light, independent of changes in the timing of the sleep–wake schedule or in the timing of social interaction (Arendt & Broadway, 1986; Carrier & Dumont, 1995; Czeisler *et al.*, 1986; Czeisler *et al.*, 1989, 1990b; Dijk, Visscher, Bloem, Beersma, & Daan, 1987; Dijk, Beersma, Daan, & Lewy, 1989; Drennan, Kripke, & Gillan, 1989; Duffy, Kronauer, & Czeisler, 1996; Shanahan & Czeisler, 1991) (Figure 3). Furthermore, the internal phase relationships among the endogenous components of these variables remains stable even after 4- to 12-hour light-induced phase shifts, suggesting that a light-sensitive circadian pacemaker, probably located in the suprachiasmatic nucleus of the hypothalamus, underlies this ensemble of rhythmic components.

Figure 3. Differential effects of light exposure and behavioral events (sleep–wake cycle and associated cycles of activity, social interactions, etc.) on phase and amplitude of the core-body temperature rhythm in humans in a 16-day protocol while living in dim light (10–15 lux). In groups 1 and 2, the sleep–wake cycle was shifted 12.2 hours later than the subjects' habitual time. This shift of the sleep–wake cycle was accompanied by either exposure to 5 hours of darkness centered 1.5 hours before the temperature minimum (group 1, closed symbols) or exposure to 5 hours of bright light (group 2, open squares). In groups 3 and 4, the sleep–wake cycle was shifted 8.4 hours earlier than the subjects' habitual time. This shift of the sleep–wake cycle was accompanied by either exposure to 5 hours of darkness centered 1.5 hours after the temperature minimum (group 3, closed symbols) or exposure to 5 hours of bright light (group 4, open circles). Endogenous circadian phase and amplitude were assessed on an initial constant routine (CR1), after 3 consecutive days of bright light/darkness exposure, or after another 3 consecutive days of this treatment regimen. In this polar representation of phase and amplitude on thr first, second, and third constant routines, circadian phase is represented in degrees, with the minimum of core body temperature (ECP_{min}) on CR1 normalized to 0 deg. Amplitude is indicated on the radius, with temperature amplitude on CR1 normalized to a value of 1.00. Data are means (SEM). Time proceeds in a clockwise direction and phase advance shifts are plotted counterclockwise. Dashed lines indicate the change in relative phase and amplitude in subsequent constant routines. Large phase shifts were observed in the bright-light-treated groups (open symbols). In those subjects in whom light exposure was centered before the temperature minimum (open circles), a phase advance was observed, whereas in the subjects in

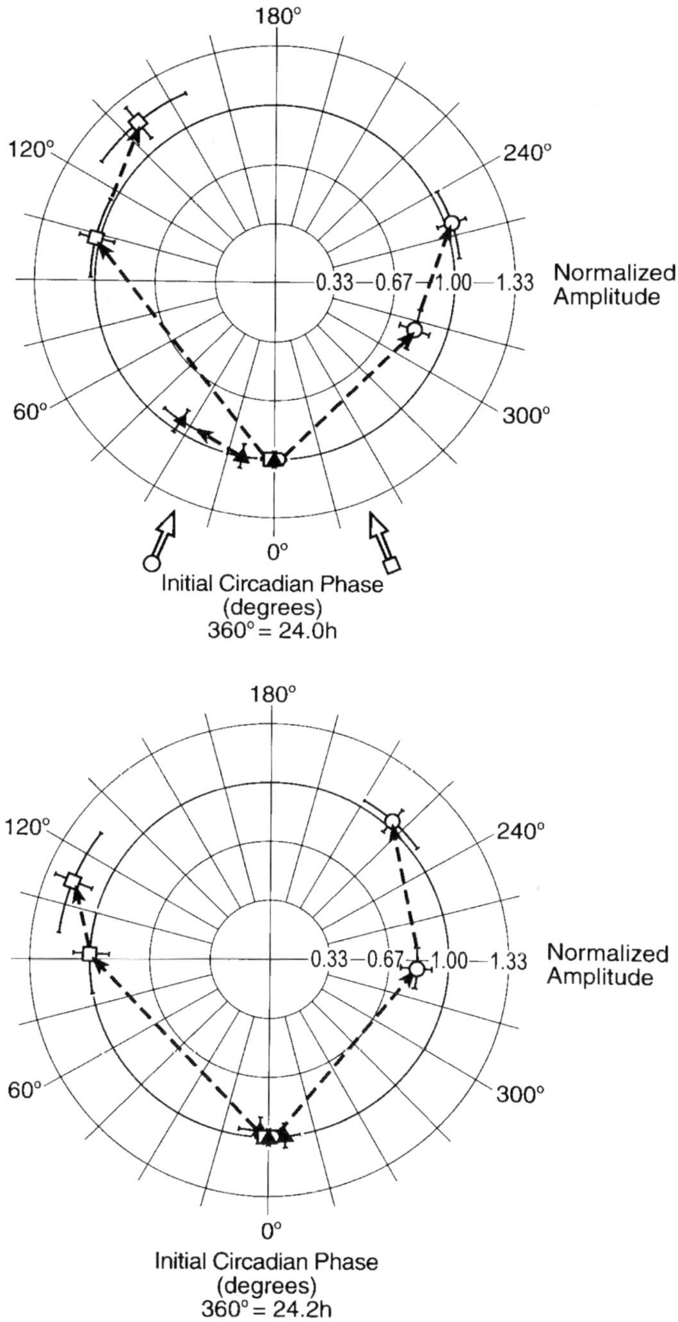

whom light was centered after the temperature minimum (open squares) large phase delays were observed. When 360 deg was set to 24 hours, i.e., the period of the core body temperature cycle = 24.0 h (panel A), the subjects exposed to darkness (closed triangles) shifted to a slightly later phase. When the drift of phase due to the slightly longer than 24 hours intrinsic period was taken into account (i.e., 360 deg = 24.2 hours) (panel B), the shift associated with inversion of the sleep–wake cycle was minimal. (From Duffy *et al.*, 1996, with permission.)

Phase Response Curve to Light in Humans: Type 0 Resetting to Three-Cycle stimulus, Type 1 Resetting to One-cycle Stimulus, and Amplitude Reduction after a Two-Cycle Stimulus

In the late 1950s, Hastings and Sweeney first demonstrated in the blue-green alga *Gonyaulax polyedra* that single light exposures could reset circadian phase and that the magnitude and direction of these phase shifts were dependent on the intensity and the duration of the light exposure (described by a dose–response curve, DRC) as well as on the circadian phase of the light exposure (described by a phase response curve, PRC) (Hastings & Sweeney, 1958).

Circadian PRCs to light have since been described in nearly every eukaryotic organism studied (Johnson, 1997). The resetting response to light described by the PRC has been observed to share the following properties across a wide array of species, from unicellular algae to primates:

1. Light stimuli early in the subjective night induce phase delay shifts.
2. Light stimuli late in the subjective night induce phase advance shifts.
3. Light stimuli during the subjective day induce minimal phase shifts.

The main difference between light-induced PRCs observed across species is the magnitude of the resetting response. There are two types of PRCs: a low-amplitude (Winfree Type 1) PRC with maximal phase shifts of only a few hours, and a high-amplitude (Winfree Type 0) PRC, with phase shifts as great as 12 hours (Winfree, 1980). In Type 0 resetting, large phase shifts are achieved through an antecedent suppression of the amplitude of oscillation of the circadian pacemaker, whereas in Type 1 resetting, there is little effect of the stimulus on amplitude.

In humans, both the magnitude and the direction of the resetting response to three cycles of exposure to 5-hour episodes of bright light depend on the initial circadian phase at which the light exposure occurs. A phase response curve was constructed to a cyclic bright light stimulus consisting of three cycles of exposure to 5 hours of bright light (~7000–12000 lux) per cycle, 11 waking hours of exposure to room light (~150 lux) per cycle, and 8 hours of exposure to darkness (<0.02 lux) during scheduled sleep episodes per cycle (Czeisler *et al.*, 1989). The largest phase shifts in response to that complex stimulus occurred when the bright light episodes were centered near the initial temperature nadir, while smaller shifts were observed when the bright light episodes were centered 180 deg away from the initial temperature nadir. Phase advances occurred when the bright light episodes were centered after the endogenous temperature nadir, whereas phase delays occurred when light pulses were administered before the temperature nadir (see Figure 4).

While the phase dependence of the resetting response to light is consistent with data from other species, there are some unexpected features of light-induced phase resetting in humans. First of all, the so-called "break point" when resetting responses to a light stimulus switch from delays to advances occurs closer to dawn than to midnight in humans living on a 24-hour day. This is substantially later than had been predicted from data in nocturnal mammals, and may be related to our exposure to artificial light in the evening hours (see below). This timing of the so-called "break point" has implications for the use of light exposure to shift circadian phase. Thus, for the average young man, light exposure from midnight to 4 a.m. will actually induce a phase delay shift (required for westward travel) rather than the phase advance shift (required for eastward travel) which had been anticipated from rodent data. Second, the large magnitude of the phase shifts induced by the three-

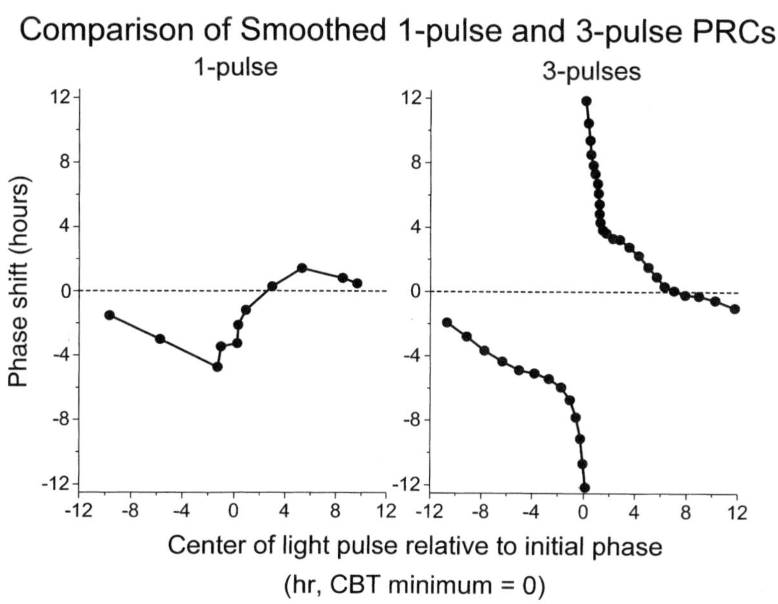

Figure 4. Induction of a Type 1 PRC by a one-cycle stimulus (left) and a Type 0 PRC by a three-cycle stimulus (right). CBT, Core body temperature. The nadir of the fitted endogenous temperature cycle is assigned a phase of 0 deg. The magnitude and direction of the phase shift achieved are plotted against the initial phase of the stimuli. Left: Smoothed phase response curve for a one-cycle bright-light stimulus (intensity 7000–12,000 lux; duration 5.2 hours). Right: Smoothed PRC to a three-cycle bright-light stimulus (for each cycle, the intensity was 7000–12,000 lux and the duration was 5 hours). (Modified from Jewett *et al.*, 1994.)

cycle stimulus together with the shape of the PRC indicate that a three-cycle bright light stimulus can induce a strong (Winfree Type 0) resetting response in humans, as had been anticipated by Winfree (1987). Third, the human circadian pacemaker appears to be sensitive throughout the day and the so-called dead zone is virtually absent (Jewett *et al.*, 1997).

Subsequent studies from a number of laboratories have indicated that exposure to a weaker stimulus (e.g., one cycle of bright light exposure) induces weak, Type 1 resetting of the human circadian pacemaker (Honma, Honma, & Wada, 1987; Jewett, Kronauer, & Czeisler, 1994; Minors, Waterhouse, & Wirz-Justice, 1991). Oscillator theory predicts that if a strong stimulus can induce Type 0 resetting and a weak stimulus induces Type 1 resetting, then there exists a stimulus intermediate in strength which, if given at the exact circadian phase between that which induces phase advances and that which induces phase delays, will bring the oscillating system to a theoretical point of singularity where the amplitude is markedly reduced. In fact, two cycles of exposure to bright light have been shown to induce substantial reduction of the amplitude of endogenous circadian components of the temperature, cortisol, and melatonin rhythms (Jewett *et al.*, 1991; Shanahan, 1995; Shanahan, Zeitzer, & Czeisler, 1997), as would be anticipated given the data on Type 0 resetting. These data are all well fit by Kronauer's model of the effect of light on the human circadian pacemaker (Kronauer, 1990) and refinements of this model (Jewett & Kronauer, 1997).

The results described in the previous paragraph challenge the long-held belief that only lower organisms, such as unicellular algae, plants, and insects, are sufficiently sensitive to light to be capable of strong (Winfree's Type 0) phase resetting. However, that belief stems from Pittendrigh's classic studies of the circadian rhythm of pupal emergence in *Drosophila pseudoobscura*, which is so exquisitely sensitive to the phase-resetting effects of light that less than 1 minute of exposure to dim blue light can induce strong Type 0 resetting. However, adult flies of that same species require 12 hours of exposure to bright light (~7000 lux) to achieve comparable Type 0 resetting of the circadian pacemaker driving their activity rhythm. In fact, the human circadian pacemaker is more sensitive to the phase-shifting effects of light than that of some insects, such as the cockroach, *Leucophaea maderae*, which requires exposure to 80,000 lux for 12 hours to achieve strong resetting.

In 1987, Kronauer and colleagues proposed a mathematical model of the effect of light on the human circadian pacemaker, which has since been refined (Kronauer, 1987, 1990; Kronauer, Forger, & Jewett, 1999). That model postulated that light, even ordinary indoor room light, exerted a direct action on both the phase and amplitude of the human circadian pacemaker and that there is a nonlinear relationship between light intensity and its biological effectiveness as a circadian synchronizer. Experimental data gathered since that time have been consistent with the predictions derived from Kronauer's model and are inconsistent with the view that exposure to bright light (>500 lux) is required to phase shift or entrain the circadian pacemaker in humans. Under laboratory conditions, the human circadian system can be shifted by a simple cycle of ordinary room light/wakefulness and darkness/sleep. However, in the outdoor world, exposure to ordinary room light at night (as occurs in some shift workers) is often insufficient to overcome the strength of the prevailing synchronizer (i.e., daytime sunlight exposure) needed to phase shift the human circadian pacemaker. Although it was previously thought that the conflicting timing of social zeitgebers presented such phase shifts, it is more likely that shift workers often do not exhibit a phase shift of their circadian timing system because they are exposed to complex and conflicting light–dark cycles (room light at night, bright light during commute). In fact, dose–response curve data indicate that exposure to ordinary indoor room light can exert a *direct* physiologic effect on the human circadian pacemaker that is sufficient to achieve the small daily phase shifts necessary to entrain the human circadian pacemaker to the 24-hour day (Boivin, Duffy, Kronauer, & Czeisler, 1996; Zeitzer, Kronauer, & Czeisler, 1997) (see Figure 5). It has not been established which photoreceptor mediates the effect of light on the human circadian pacemaker. Dose–response curves in which differential effects of light intensities about the levels at which the scotopic system saturates and studies in which the effects of bright red light (to which the scotopic system is insensitive) were investigated suggest that the photopic system is involved (Zeitzer *et al.*, 1997) (see Figure 5). The data exclude neither the involvement of the scotopic system nor the involvement of novel photoreceptors.

HIGH LIGHT SENSITIVITY OF THE HUMAN CIRCADIAN PACEMAKER: IMPLICATIONS FOR THE ASSESSMENT OF INTRINSIC PERIOD AND EFFECTIVENESS OF NONPHOTIC ZEITGEBERS

This high light sensitivity implies that during the 16-hour waking day, sighted subjects are exposed continuously to light of sufficient intensity to influence the

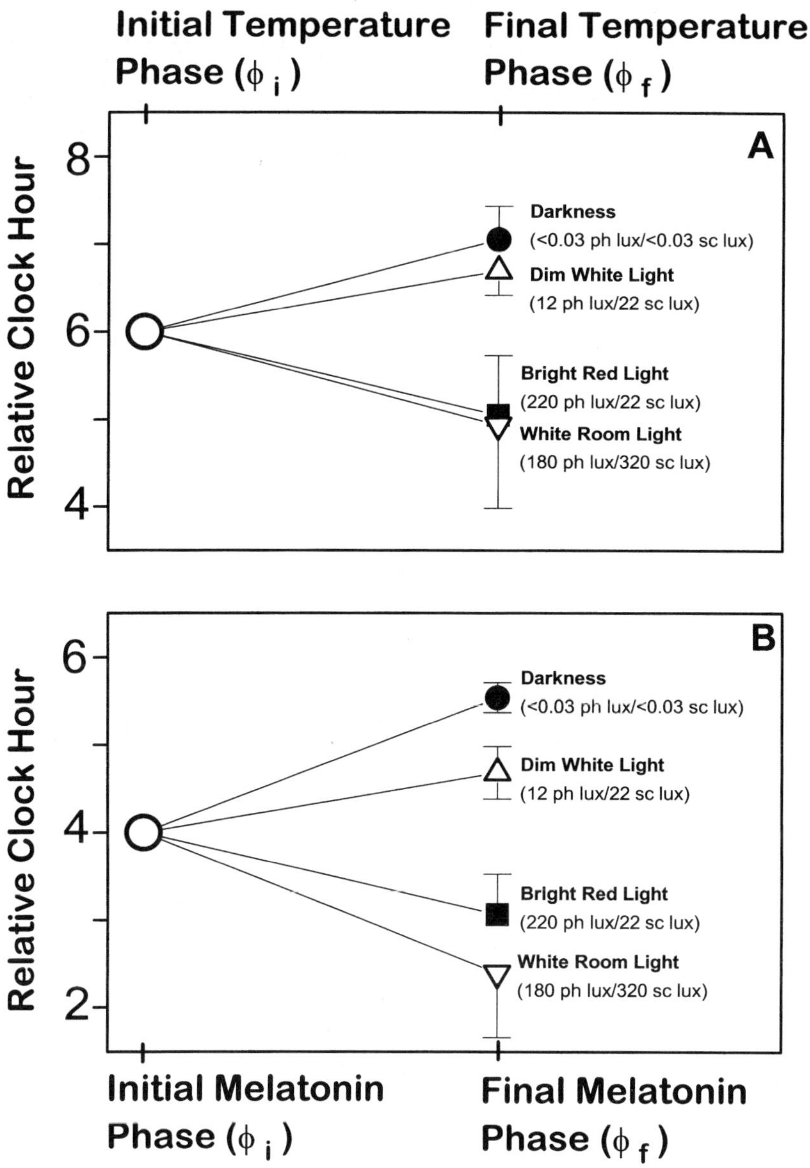

Figure 5. Photopic transduction implicated in human circadian entrainment of the phase of the endogenous components of the core body temperature (A) and plasma melatonin (B) rhythms. Phase changes under darkness conditions (<0.03 photopic lux; <0.03 scotopic lux), very dim white light conditions (12 photopic lux; 22 scotopic lux), white room light conditions (180 photopic lux; 320 scotopic lux), and bright red light conditions (220 photopic lux; 22 scotopic lux) are plotted. Note that bright red light-induced phase shifts in the core body temperature rhythm and the melatonin rhythm are similar to each other and similar to those induced by white room light. (From Zeiter *et al.*, 1997, with permission. Copyright 1997, Elsevier Science.)

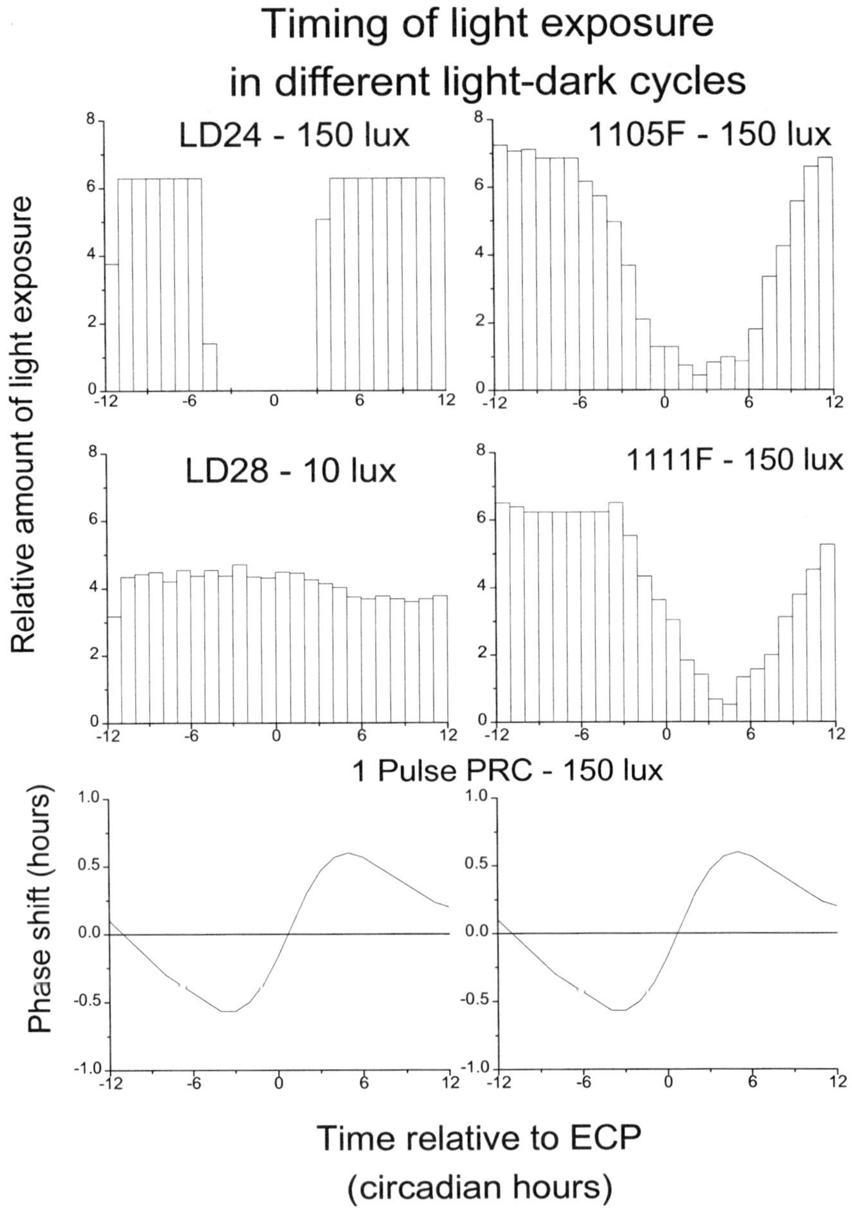

Figure 6. Schematic representation of distribution of light exposure over the circadian cycle during entrainment (LD24; 150 lux), freerun, i.e., self-selected sleep–wake cycle and associated light–dark cycle (subject 1105F and subject 1111F), and a 28-hour forced desynchrony protocol (LD28; 10 lux), i.e., when the sleep–wake cycle and associated light–dark cycle were scheduled to a 28-hour cycle. The histograms are normalized so that their sum equals 100%. To illustrate the potential effect of these light distributions on the manifest circadian periodicity, a one-pulse PRC to light is plotted twice in the two bottom panels. Note that during the freerun, light primarily falls on the phase delay of the PRC, whereas in the forced-desynchrony protocol, light is distributed near uniformly across the PRC. (From Klerman *et al.*, 1996, with permission. Copyright, 1996, American Physiological Society.)

endogenous circadian pacemaker. Under entrained conditions, both the phase advance and phase delay portion of the PRC are exposed to light (see Figure 6). However, during classic free-running studies, in which subjects can self-select their sleep–wake cycle and associated light–dark cycle, the phase angle between the sleep–wake cycle and body temperature cycle changes such that the phase delay portion of the PRC to light is exposed to light, whereas the phase advance portion coincides with the sleep episode and is thus shielded from light. Theoretically, such a self-selection of light–dark cycles should thus result in an observed period that is longer than the "true" intrinsic period (Klerman, Dijk, Kronauer, & Czeisler, 1996). This prediction has been confirmed. Thus, whereas in the classic free-running studies the observed period of the body temperature rhythm was close to 25 hours (Wever, 1979), the period observed in experiments in which the confounding effect of nonuniform light exposure was controlled for was much closer to 24 hours (Beersma & Hiddinga, 1998; Czeisler *et al.*, 1995a; Middleton, Arendt, & Stone, 1996).

In blind subjects who are free-running in society, the reported average period is also significantly shorter than 25 hours (Czeisler *et al.*, 1995b; Sack, Lewy, Blood, Keith, & Nakagawa, 1992). The fact that the human circadian pacemaker is sensitive to the light intensities experienced under normal conditions has many practical applications for the treatment of shift workers, transmeridian travelers, and patients with other circadian rhythm sleep disorders. Furthermore, given that light acts directly on the endogenous circadian pacemaker in humans, the apparent role of the rest–activity cycle as an intermediary in conveying synchronizing input from the external environment to the endogenous circadian pacemaker largely results from the fact that humans are generally exposed to light only when they are awake and are exposed to darkness whenever they are asleep. Urban dwellers receive minimal exposure to bright light during the day, largely the result of shielding from natural light within modern residential and commercial buildings (Savides, Messin, Senger, & Kripke, 1986). Ironically, due to the widespread availability of artificial electric light, humans are now exposed to a longer photoperiod and much higher levels of light during the night. It thus appears that an entire class of circadian rhythm sleep disorders may be largely the result of these fundamental behavior-induced changes in the periodic environment to which we, as a species, are ordinarily exposed.

The high sensitivity to light also implies that assessment of the putative synchronizing effects of nonphotic zeitgebers requires careful control of light exposure. Actually, light exposure associated with exposure to other putative synchronizers such as social cues or activity–rest cycles is likely to have been responsible for the observed synchronizing effects of these nonphotic zeitgebers. For example, any observed effect of a shift of the sleep–wake cycle on endogenous circadian rhythms in a protocol in which subjects were exposed to dim light during the wake episode can be attributed to light rather than phase shifting effects of rest or activity. The most compelling evidence for the existence of nonphotic zeitgebers has been derived from the study of blind subjects.

Blind Subjects: Freerun and Entrainment?

Given the high light sensitivity of the human circadian pacemaker to light, it is not surprising that many blind subjects without conscious light perception are not synchronized to the 24-hour day while living in society (Miles, Raynal, & Wilson, 1977; Sack *et al.*, 1992). In some of these blind subjects, it has been demonstrated that the rhythms of melatonin, cortisol, sleep propensity, urine production, etc., are not entrained to the 24-hour day despite the fact that these subjects are maintaining a 24-hour rest–activity cycle (Klein *et al.*, 1993). These data are consistent with the data described above demonstrating that light is the principal zeitgeber in humans.

More puzzling are reports of those blind subjects without conscious light perception who apparently are entrained to the 24-hour day. These data may be explained by assuming that despite the absence of conscious light perception in these subjects, light still reaches the light-sensitive circadian pacemaker. That this is a realistic possibility has been derived from experiments in which it was demonstrated that in some of these blind subjects, bright light can elicit suppression of plasma melatonin, a response thought to be mediated by the suprachiasmatic nuclei of the hypothalamus, which are the loci of a light-sensitive circadian pacemaker (Czeisler *et al.*, 1995b). However, the apparent entrainment of some blind subjects without eyes cannot be explained by residual light input to the light-sensitive pacemaker, and constitutes the strongest evidence that nonphotic zeitgebers may be effective circadian synchronizers in humans. A recent study in a completely blind subject demonstrated that the endogenous circadian rhythms of core body temperature and plasma melatonin can be affected significantly by a sleep–wake schedule with a period slightly shorter than 24 hours (see Figure 7) (Klerman *et al.*, 1998). Previously, some evidence that scheduled activity could affect human circadian rhythms had been published (Van Reeth *et al.*, 1994). Taking together, these data demonstrate that in humans, a light-sensitive circadian pacemaker similar to that in other mammals drives rhythmicity in many variables. Unidentified nonphotic stimuli, however, do exert a significant, albeit weak, influence on these circadian rhythms.

Functional Organization of the Human Circadian System: Sleep and Subjective Alertness

Phenomenology of Sleep in Humans

The alternation between sleep and wakefulness and associated changes in consciousness is one of the most fascinating aspects of the circadian organization of human physiology, probably because this is one of the few circadian rhythms of which we are directly aware. The timing and structure of sleep during normal entrainment to the 24-hour day has been described in great detail in the young adult. Noninvasive polysomnographic recordings of the electroencephalogram, eye movements, and muscle tone have demonstrated that sleep is not a unitary state. After an initial episode of quiet wakefulness during which the eyes are closed and the electroencephalogram is dominated by alpha (8–12 Hz) oscillations, slow, rolling eye movements and a slowing of the EEG define the stage 1 transition to sleep onset. The duration of the interval from lights-out to sleep onset varies widely, but on average is between 10 and 20 minutes. After sleep onset, the EEG exhibits further characteristic changes. First, spindle-shaped 12- to 15-Hz oscillations (i.e., sleep spindles) and sharply negative EEG deflections, so-called K-complexes, characterize stage 2 sleep. Hereafter, high-amplitude slow waves (also called delta waves; 0.2–4.5 Hz) become more abundant and define slow-wave sleep (SWS; or stages 3+4). During the sleep stages 2, 3, and 4, rapid eye movements are absent and therefore these sleep stages are also called nonREM sleep. After approximately 1 hour of nonREM sleep, the EEG suddenly desynchronizes, i.e., slow waves, sleep spindles,

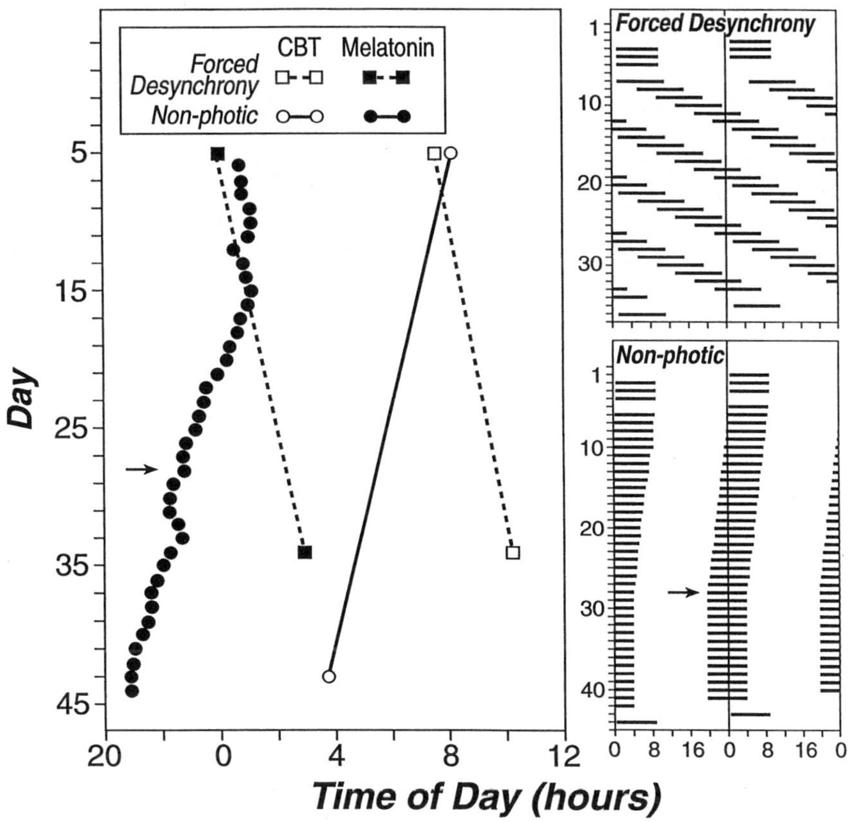

Figure 7. Evidence for nonphotic entrainment of the melatonin and core body temperature (CBT) rhythms in a blind subject. The periods of the melatonin and temperature rhythms were assessed during a protocol in which the sleep–wake cycle and associated activities were scheduled to a 28-hour cycle (referred to as forced desynchrony) and to a 23.8-hour cycle (referred to as nonphotic). During the 28-hour forced desynchrony protocol the period of melatonin (closed squares) and core body temperature (open squares) both were 24.1 hours, i.e., these rhythms desynchronized from the sleep–wake cycle. During the nonphotic protocol, the rhythms of melatonin (closed circles) and body temperature (open circles) advanced and remained synchronized with the sleep–wake cycle. Arrows indicate when the schedule changed from a 23.8-hour day to a 24.0-hour day. (From Klerman *et al.*, 1998, with permission. Copyright 1998, American Physiology Society.)

and K-complexes disappear, while muscle tone remains low or is even absent. Because during this sleep state, rapid eye movement is present and the EEG is almost as activated as the EEG in wakefulness, this sleep state is called REM sleep or paradoxical sleep. In the adult human, sleep is characterized by a cyclic alternation between nonREM sleep and REM sleep with a period of approximately 80–110 minutes. In the course of the four to five nonREM–REM cycles that occur during the nocturnal sleep episode, the two sleep stages and the EEG exhibit further characteristic changes; the nonREM sleep episodes become progressively shorter, whereas REM sleep episode duration increases. Within REM sleep, the density of phasic events such as the rapid eye movements and muscle twitches become more abundant over consecutive REM episodes. Slow EEG components—either quantified by visual scoring or computerized EEG analysis—decline exponentially over consecutive nonREM sleep episodes. In contrast, sleep spindle activity increases over consecutive nonREM sleep episodes (Carskadon & Dement, 1996; Dijk, 1995).

Numerous other physiologic processes, such as heart rate and heart rate variability, growth hormone, prolactin, and renin–angiotensin, exhibit characteristic changes either in association with the nonREM–REM sleep cycle or the progression of sleep (Brandenberger, Follenius, Simon, & Libert, 1988; Van Cauter, 1990).

In the young, healthy adult, nocturnal sleep is highly consolidated and wakefulness typically comprises less than 10% of the total time in bed. The final awakening, i.e., the beginning of the waking day, occurs after on average 7–9 hours of sleep, which under entrained conditions, will occur approximately 1–2 hours after the minimum of the endogenous circadian temperature rhythm.

During the nocturnal sleep episode, both time elapsed since start of the sleep episode and circadian phase change simultaneously. Therefore, the changes that occur across the nocturnal sleep episode, as described for SWS, REM sleep, and sleep propensity, may either be due to the progression of the sleep process or to the change in endogenous circadian phase. Much research has been devoted to assess the extent to which the characteristics of sleep are determined by the progression of the sleep process, i.e., time elapsed since sleep onset, and by endogenous circadian phase.

Circadian Pacemaker Is a Codeterminant of Subjective Alertness, Cognitive Performance, Sleep Propensity, and Sleep Structure

The role of the circadian pacemaker in the regulation of sleep propensity and sleep structure has been elucidated in various protocols. When the human circadian system is entrained to the 24-hour light–dark cycle, subjective alertness and cognitive performance remain at high and fairly stable levels across the entire 16-hour waking episode. Sleep latencies, assessed at multiple times across the waking day, exhibit either a fairly stable level or a bimodal pattern. Typically, sleep latencies are longest in the first hours of the waking day. In the afternoon, i.e., after 6–8 hours of wakefulness, sleep latencies are somewhat shorter, especially when the subjects carry a sleep debt (Carskadon & Dement, 1981, 1992; Dijk, Beersma, & Daan, 1987a; Lack & Lushington, 1997; Richardson, Carskadon, Orav, & Dement, 1982). Despite the greater homeostatic sleep pressure, sleep latencies during the evening hours are longer than in the afternoon. The latter observation demonstrates that sleep latencies are not solely determined by the time elapsed since the onset of the wake episode and indicates that endogenous circadian timing mechanisms contribute to sleep propensity.

In the young adult, the major rest episode is initiated between 22:00 and 01:00 hours, i.e., approximately 5–6 hours before the timing of the endogenous circadian minimum of the core body temperature rhythm (and approximately 1–2 hours after the evening rise in plasma melatonin levels) (Duffy, Dijk, Klerman, & Czeisler, 1998). Under free-running conditions in which the sleep–wake cycle remains synchronized to the core body temperature rhythm, sleep is typically initiated close to the minimum of the endogenous component of the temperature cycle. Concurrent with this 6-hour delay of the beginning of the sleep episode relative to the temperature rhythm, subjective alertness exhibits a decline beginning several hours before the self-selected sleep time (Czeisler, 1978). In addition, the typical increase of REM sleep across the sleep episode is no longer observed, and high amounts of REM sleep are already present in the first part of sleep, coinciding with the minimum of the core body temperature rhythm (Czeisler, Zimmerman, Ronda, Moore-Ede, & Weitzman, 1980c; Zulley, Wever, & Aschoff, 1981). The progressive increase in the density of rapid eye movements across consecutive REM sleep episodes, is, however, not affected by the change in phase relationship of the sleep–wake cycle and the endogenous circadian component of the body temperature rhythm (Zimmerman, Czeisler, Laxminarayan, Knauer, & Weitzman, 1980). Likewise, the typical declining trend of SWS across the sleep episode is not affected by the change in the phase relationship between the sleep–wake cycle and the circadian core body temperature rhythm (Weitzman, Czeisler, Zimmerman, & Ronda, 1980).

These data suggest that the circadian pacemaker plays an important role in the regulation of some aspects of sleep, like sleep propensity and REM sleep, whereas other aspects of sleep, like SWS and the density of rapid eye movements, are primarily determined by sleep–wake history of sleep or the sleep homeostat.

Sleep–Wake Cycle and Cycles in Alertness and Cognitive Performance Are Only Loosely Coupled to the Endogenous Circadian Pacemaker

Studies in which sleep was displaced experimentally and thus occurred at various circadian phases illustrate that sleep can be initiated at all phases of the endogenous circadian cycle, provided that the wake episode preceding sleep has been of sufficient length (Åkerstedt & Gillberg, 1981; Dijk, Brunner, & Borbély, 1991; Webb, Agnew, & Williams, 1971). In addition, a large body of data illustrate that under entrained conditions, when sleep is scheduled to coincide with daytime hours (during some shiftwork schedules), both sleep efficiency and sleep duration are reduced (Åkerstedt, Torsvall, & Gillberg, 1982). Fractional desynchronization studies in which the period of the imposed sleep–wake cycle gradually increased or decreased until it desynchronized from the endogenous circadian cycle of body temperature illustrate that a component of cycles in subjective alertness and cognitive performance remain coupled to the sleep–wake cycle (Folkard & Åkerstedt, 1987). Studies under free-running conditions in which the sleep–wake cycle spontaneously desynchronized from the endogenous circadian core body temperature rhythm show that whereas the self-selected sleep–wake cycle is only coupled loosely to endogenous circadian rhythms, the endogenous circadian timing system continues to modulate both sleep propensity and sleep structure (Czeisler, Weitzman, Moore-Ede, Zimmerman, & Knauer, 1980b; Zulley *et al.*, 1981). Thus during internal desynchronization, sleep initiations and sleep terminations are neither locked in a fixed phase relationship to the endogenous temperature rhythm nor are they distributed randomly with respect to the circadian temperature rhythm. Sleep initiations instead occur across a wide variety of temperature phases, with bedtimes most often occurring on the falling limb of the endogenous component of the temperature rhythm. The majority of sleep terminations, on the other hand, occur on the rising limb of this rhythm (Strogatz, Kronauer, & Czeisler, 1987).

In addition to this circadian modulation of the propensity to initiate and terminate sleep, a strong dependence of the duration of sleep episodes on the circadian phase at which sleep is initiated has been demonstrated. When the duration of the waking episode preceding sleep exceeds 16 hours and sleep is initiated close to the maximum of the endogenous rhythm of core body temperature (i.e., near the nadir of the endogenous sleep propensity rhythm as indexed by the probability to initiate sleep), sleep episodes are very long. In contrast, sleep episodes initiated around the minimum of the body temperature (i.e., near the maximum of the circadian sleep propensity rhythm) are shorter. This variation in sleep duration (range 7–18 hours) cannot be explained by the considerable variation in the duration of wakefulness preceding sleep (range 18–35 hours), nor can a single circadian pacemaker driving the rhythm of sleep propensity account for this observation (Strogatz, 1986).

The limited number of cases of spontaneous desynchronization in which sleep was recorded polysomnographically have confirmed the results from both the sleep displacement studies and free-running studies in which the sleep–wake cycle remained synchronized with the endogenous circadian rhythms. REM sleep propensity peaks at or shortly after the minimum of the core body temperature rhythm and SWS declines in the course of sleep, apparently independent of the endogenous circadian phase of the core body temperature rhythm.

Although both sleep displacement and spontaneous desynchrony studies have contributed greatly to our understanding of the circadian regulation of sleep, the interpretation of the data is limited by the considerable variation in the duration of wakefulness preceding sleep episode. The data demonstrating that this variation in the history of wakefulness contributes to variations in sleep propensity and sleep structure, and therefore can be a confounding factor in the assessment of the role of the circadian pacemaker in the regulation of sleep, are summarized in the next section of this chapter.

Homeostatic Regulation (i.e., the Duration of Wakefulness and Sleep) Has Powerful Effects on Sleep Propensity, Sleep Structure, and Subjective Alertness

The data described in the previous section support an important role for the circadian pacemaker in the regulation of sleep timing and sleep structure. However, at least one other regulatory mechanism exerts a major influence on sleep propensity and sleep structure. Sleep deprivation studies have unequivocally demonstrated that after extended periods of wakefulness of multiple days with reduced sleep, subjective alertness and cognitive performance deteriorate and sleep propensity, as indexed by sleep latency, increases (Brunner, Dijk, Tobler, & Borbély, 1990; Carskadon & Dement, 1979, 1981; Dinges *et al.*, 1997; Monk, Buysse, Reynolds, Jarrett, & Kupfer, 1992). Furthermore, stages 3+4 of nonREM sleep and computer-detected slow-wave activity increase as a function of the duration of prior wakefulness (Dijk *et al.*, 1987a; Dijk, Brunner, Beersma, & Borbély, 1990; Webb & Agnew, 1971). Conversely, both sleep extension and daytime napping lead to higher levels of subjective alertness and reduce sleep propensity and SWS during the sleep episode following the extended sleep or daytime nap (Werth, Dijk, Achermann, & Borbély, 1996). Actually, dose–response relationships have been obtained for wake and sleep durations within a physiologic range and sleep propensity, sleep duration, and EEG slow-wave activity during nonREM sleep (Dijk, 1995). Furthermore, selective deprivation of either SWS or REM sleep results in compensatory rebounds of these sleep stages (Agnew, Webb, & Williams, 1964; Beersma, Dijk, Blok, & Everhardus, 1990; Dement, 1960; Dijk, Beersma, Daan, Bloem, & van den Hoofdakker, 1987b). These data demonstrate that sleep structure is actively regulated and that during wakefulness, sleep propensity builds up and is relieved during sleep. This aspect of sleep regulation, i.e., the memory for the history of sleep and waking, is referred to as sleep homeostasis (Borbély, 1982b). This mechanism, which is somewhat similar to an hourglass mechanism or relaxation oscillator, is thought to keep "sleep debt" within certain limits and implies that sleep serves to recover from the "wear and tear" of wakefulness. Sleep deprivation studies in SCN-lesioned animals demonstrated that the process of sleep homeostasis is not dependent on an intact SCN, since sleep deprivation still results in an increase of sleep propensity and an enhancement of slow-wave sleep in the absence of the SCN (Tobler, Borbély, & Groos, 1983; Trachsel, Edgar, Seidel, Heller, & Dement, 1992).

The presence of homeostatic mechanisms in the regulation of sleep raises the question of how in the intact organism, this aspect of sleep regulation is integrated with the circadian regulation of sleep–wake cycles. Various conceptual and mathematical models (Borbély & Achermann, 1992) and elaborate experimental protocols have been designed to address this question and they will be discussed in the following sections.

Interaction between Circadian and Homeostatic Mechanisms and Their Phase Relationships Serve to Consolidate Sleep and Wakefulness

In adult humans, and in some other primates, such as the squirrel monkey, sleep and wakefulness are highly consolidated during conditions of normal entrainment. This implies that despite the increase of sleep propensity associated with sustained wakefulness, sleep initiation is postponed and alertness and cognitive performance are maintained for a full 16 hours. The role of the endogenous circadian pacemaker in this consolidation of wakefulness and high levels of subjective alertness has been investigated by forced desynchronization of the sleep–wake cycle and endogenous circadian rhythms (Dijk & Czeisler, 1995; Hume & Mills, 1977). In these experiments, the rest–activity cycle and associated dim-light–dark cycle is scheduled to a period (e.g., 20 or 28 hours) that is outside the range of entrainment of the endogenous circadian pacemaker. When the experiment is of sufficient length, sleep will be scheduled to occur at virtually all circadian phases, as indexed by the phase of the melatonin rhythm or core body temperature rhythm, while at the same time, variations in the duration of wakefulness preceding the scheduled sleep episodes and variations in the duration of sleep preceding scheduled waking episodes are minimized. These protocols allow for the study of the relative contribution of circadian and homeostatic mechanisms and their interaction because for any variable of interest, e.g., alertness or REM sleep, a large number of combinations of endogenous circadian phase and prior wake or sleep duration is realized (see Figure 8). These experiments have generated some surprising and, at first glance, paradoxical results. The phase relationship between the endogenous circadian pacemaker and the sleep–wake cycle during entrainment is finely tuned and achieves consolidation of sleep and wakefulness by synergistic and nonadditive interactions of homeostatic and circadian mechanisms (Dijk & Czeisler, 1994, 1995). A first perplexing results concerns sleep initiation. After approximately 18 hours of wakefulness, sleep can be readily initiated at all circadian phases. However, a pronounced circadian variation of sleep latencies is observed. Shortest sleep latencies are present at the minimum of the endogenous core body temperature rhythm, which under entrained conditions corresponds to approximately 6 a.m. (i.e., close to habitual wake time). When sleep is initiated later in the cycle (i.e., during those phases that under entrained conditions would correspond to the habitual waking day), sleep latencies gradually become longer until a maximum is reached approximately 8 hours before the endogenous circadian temperature minimum. This

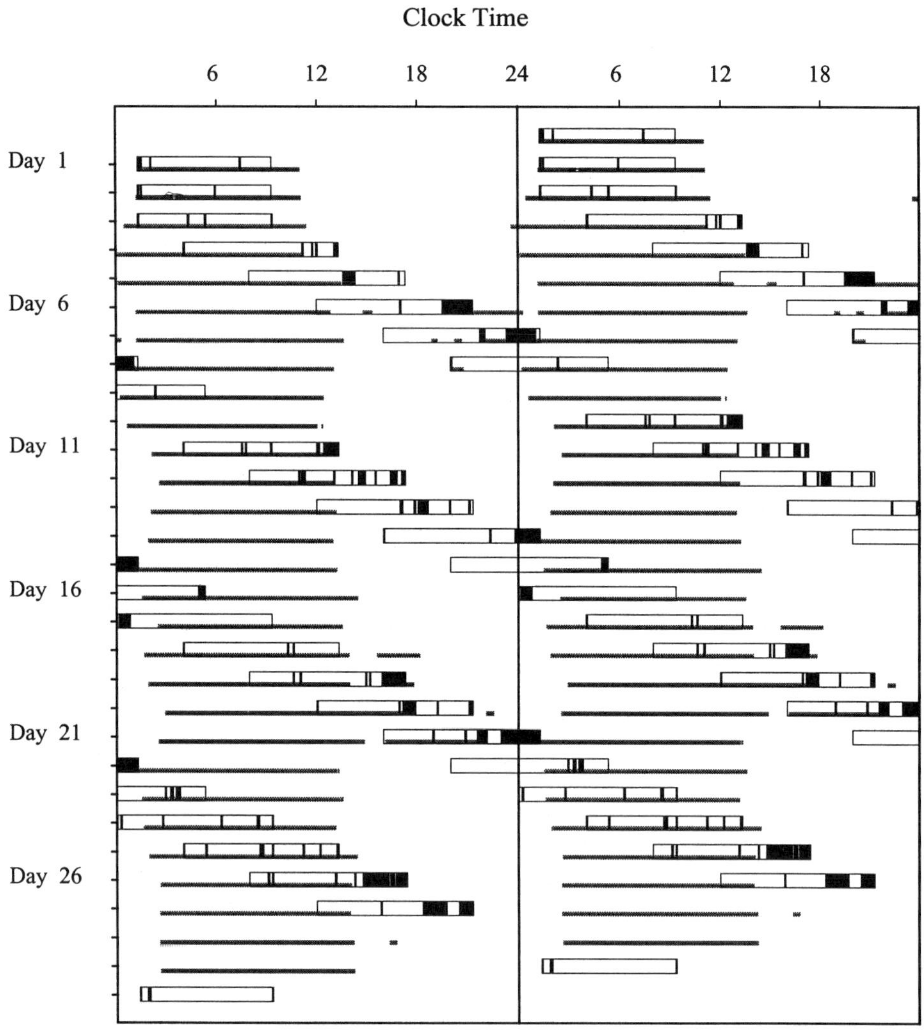

Figure 8. Forced desynchrony of the sleep–wake cycle and plasma melatonin rhythm and associated variations in sleep consolidation in a young healthy male subject living in time isolation. After 3 baseline days the subject (1209) was scheduled to a 28-hour rest–activity cycle. Scheduled sleep episodes during which sleep was succesfully recorded are indicated by open boxes. Polysomnographically recorded wakefulness within the scheduled sleep episodes is indicated by black vertical bars within the open boxes. Times during which the plasma melatonin concentration was above 10 pg/ml are indicated by black lines. Note that the period of the plasma melatonin rhythm is slightly longer than 24 hours and that sleep is disrupted when scheduled to occur outside the phase of mleatonin secretion. This disruption is most pronounced at the end of scheduled sleep episodes, i.e. when homeostatic sleep pressure has dissipated. (Previously unpublished figure based on data collected by T. L. Shanahan, J. F. Duffy and C. A. Czeisler, personal communication, that were presented in part in Shanahan, 1995, and reanalyzed by H. G. Wei and D.-J. Dijk.)

circadian phase corresponds to the beginning of the habitual sleep episode under entrained conditions. Thus, from a circadian point of view, it is most difficult to fall asleep just before the circadian phase at which we habitually go to sleep.

After 18 hours of wakefulness, sleep consolidation in the initial part of sleep episodes is high, almost independent of endogenous circadian phase. The impact of the endogenous circadian pacemaker on sleep consolidation becomes apparent in the second part of sleep episodes, when the homeostatic drive for sleep has subsided. When the second half of the scheduled sleep episodes coincides with the trough of the temperature cycle and the phase of melatonin secretion, sleep remains highly consolidated (Dijk, Shanahan, Duffy, Ronda, & Czeisler, 1997c). In contrast, when the second half of the scheduled sleep episodes coincides with the latter half of the rising limb of the temperature cycle, sleep is severely disrupted. In fact, maximum wake propensity as indexed by sleep disruption is located just prior to the circadian phase of habitual sleep initiation (see Figure 9).

Sleep consolidation thus results from an interaction of the homeostatic mechanisms (i.e., time elapsed since start of sleep episode) and endogenous circadian phase. Plots of sleep consolidation as function of circadian phase and time elapsed since start of sleep episode allow for a quantitative analysis of this interaction. It reveals, not surprisingly, that sleep consolidation decreases with the progression of sleep for most circadian phases at which sleep is initiated. However, this decrease of sleep consolidation is not observed when sleep is initiated at approximately 225–285 deg of the body temperature cycle, which under entrained conditions, corresponds to 23–01 hours (see Figure 10). This analysis suggests sleep consolidation can only be achieved by a fine-tuned interaction of the circadian and sleep-dependent contributions to sleep propensity. This suggests a functional significance of the seemingly paradoxical entrained phase relationship between sleep and the endogenous circadian variation of sleep propensity (Dijk & Czeisler, 1994).

According to this concept, sleep remains consolidated at night because the circadian drive for sleep becomes progressively stronger, with a maximum close to habitual wake time (Dijk & Czeisler, 1994, 1995). Conversely, wakefulness is thought to remain consolidated during the day because the circadian drive for wakefulness become progressively stronger as the day progresses, thereby opposing the progressive increase in the homeostatic drive for sleep associated with sustained wakefulness (Borbély, Achermann, Trachsel, & Tobler, 1989; Dijk & Czeisler, 1994, 1995; Edgar, Dement, & Fuller, 1993). The opening of the sleep gate is then heralded by the onset of melatonin secretion (Lavie, 1997).

Further evidence in support of this view has been derived from data on subjective alertness and cognitive performance during forced and fractional desynchronization protocols (Dijk *et al.*, 1992; Folkard & Åkerstadt, 1989; Johnson *et al.*, 1992; Monk, Moline, Fookson, & Peetz, 1989; Monke, Buysse, Reynolds, & Kupfer, 1996). The circadian rhythm of alertness as assessed in these protocols reaches a minimum close to the minimum of the core body temperature rhythm. When the beginning of the waking day coincides with the initial part of the rising limb of the core body temperature rhythm, high levels of alertness and cognitive performance can be maintained for the full 18-hour day. However, when the initial part of the waking day coincides with the maximum of the core body temperature rhythm, alertness will rapidly deteriorate, and very low levels will be reached when the end of the 18-hour waking day coincides with the minimum of the core body temperature rhythm. However, when alertness is assessed at the beginning of a waking day coinciding with the minimum of the core body temperature rhythm, alertness is at moderate to high

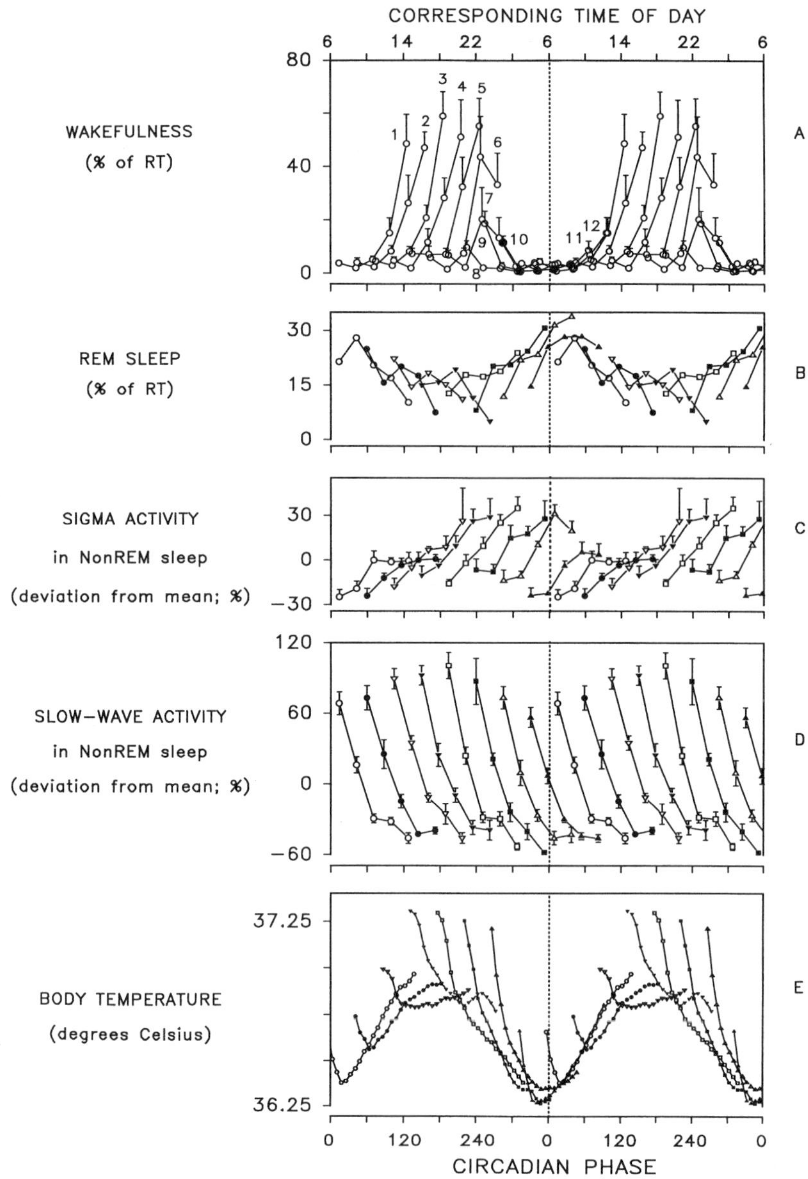

Figure 9. Phase relationships between the circadian rhythm of plasma melatonin, sleep consolidation, nonREM sleep, REM sleep, and EEG activity during a 28-hour forced desynchrony-protocol. Data are plotted against circadian phase of the plasma melatonin rhythm (0 deg corresponds to the fitted maximum, bottom *x* axis. To facilitate comparison with the situation in which the circadian system is entrained to the 24-hour day, the top *x* axis indicates the average clock time of the circadian melatonin rhythm during the first day of the forced desynchrony protocol, i.e., immediately upon release from entrainment. Plasma melatonin data were expressed as *z* scores to correct for interindividual differences in mean values. Wakefulness is expressed as percentage of recording time (RT) within each scheduled sleep episode. NonREM sleep and REM sleep are expressed as percentage of total sleep time (TST). EEG low-frequency sleep spindle activity in nonREM sleep and alpha activity in REM sleep are expressed as percentage deviation from the mean. Data are double plotted, i.e., all data plotted left from the dashed vertical line are repeated to the right of this vertical line. (From Dijk *et al.*, 1988c, with permission.)

levels. These data demonstrate that alertness and cognitive performance as well as mood (Bolvin *et al.*, 1997) are codetermined by endogenous circadian phase and time elapsed since start of the sleep episode. During the habitual waking day, high levels of alertness can be maintained despite sustained wakefulness because the circadian system counteracts the deterioration of alertness associated with wakefulness. When subjects have to be awake during the biological night after having been awake for 12 hours or more (a situation often experienced by shift workers during their first night on the night shift), the homeostatic deterioration of alertness and the circadian drive for sleep interact to have a significant adverse effect on alertness.

According to this concept, homeostatic mechanisms are primarily acting downstream from the circadian pacemaker in the regulation of subjective alertness, sleep propensity, and sleep timing. However, since during sleep, light input to the pacemaker is necessarily reduced, homeostatic mechanisms feed back onto the light-sensitive pacemaker and thereby do play a role in entrainment to external light–dark cycles (Beersma, Daan, & Dijk, 1987).

Other aspects of sleep, such as REM sleep and EEG sleep spindle activity, are also affected by both circadian phase and time elapsed since start of the sleep episode. In contrast, EEG slow-wave activity seems to be primarily affected by the sleep–wake history. This is in accordance with the hypothesis that this electrophysiologic signal is primarily affected by sleep homeostatic processes (Achermann, Dijk, Brunner, & Borbély, 1993).

Neuroanatomic and Physiologic Aspects of the Interaction of Circadian Rhythmicity and Homeostasis

The pathways and mechanisms, e.g., humoral or neural, by which the circadian pacemaker modulates sleep propensity are to a large extent unknown. Likewise, the mechanisms of homeostatic control have not been identified unequivocally. Projections from the SCN and rhythmic variables that may play a role in the regulation of sleep propensity include the body temperature rhythm, the melatonin rhythm, direct projections from the SCN to the ventral lateral preoptic area, and indirect projections from the SCN to areas such as the dorsal raphe nuclei. Putative homeostatic factors include adenosine and prostaglandin D2. These neuroanatomic and physiologic aspects of the interaction between the circadian and homeostatic process are currently under intense investigation, but will not be discussed here.

Effects of Age

Many factors contribute to the variability in circadian rhythmicity in humans. Some of these factors include changes in photoperiod, gender, and age. Of these factors, the variability related to age has probably been investigated most extensively.

One of the most common complaints of older people is frequent sleep interruption and early morning awakening (Foley *et al.*, 1995; Mant & Eyland, 1988). Polygraphic sleep recordings indicate that the amplitude of slow waves during sleep is decreased with advancing age (Feinberg, Koresko, & Heller, 1967; Miles & Dement, 1980; Reynolds *et al.*, 1985). The loss of sleep consolidation has been hypothesized to be related to a decreasing buildup for the homeostatic drive for sleep during a less active daytime (which could explain the fall in SWS), or a decreasing rate of dissipation of the homeostatic drive during sleep, causing individuals to have more difficulty in maintaining alertness and activity during the daytime. Thirty percent of

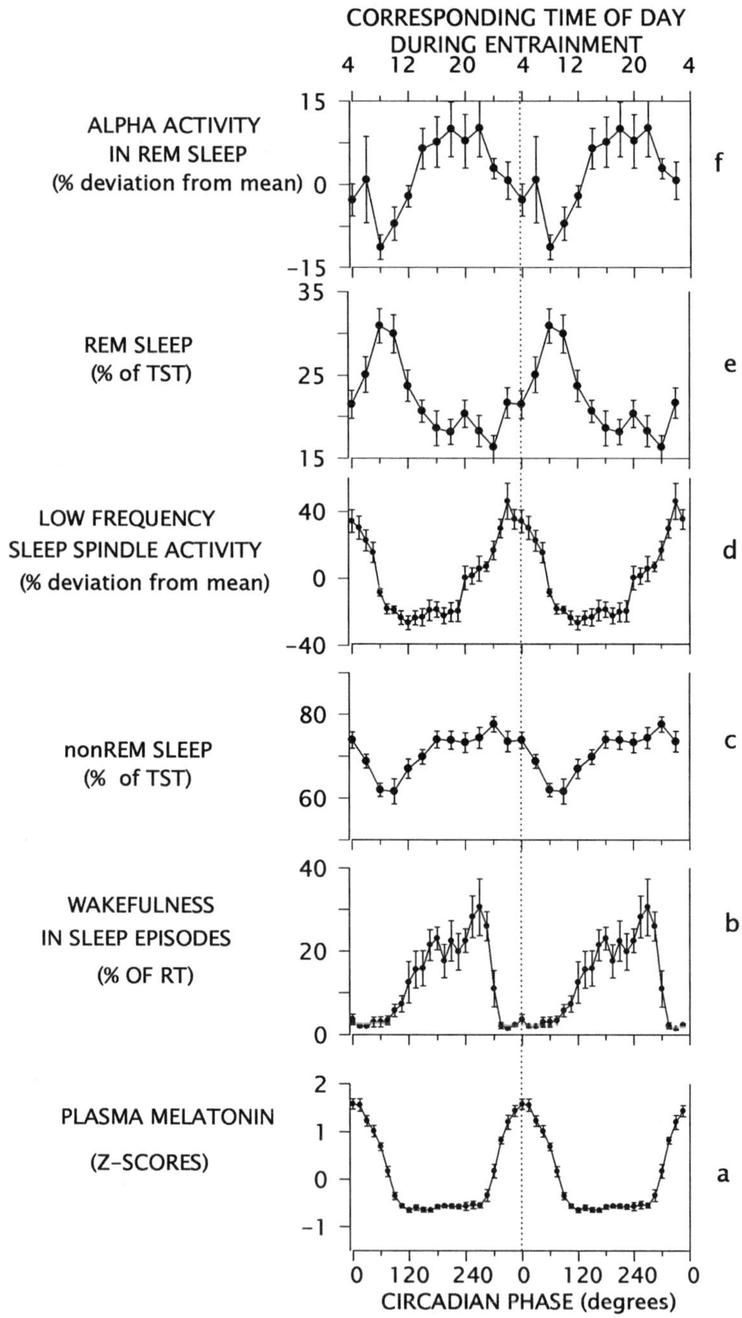

Figure 10. Interaction between circadian and sleep-dependent regulation of sleep and core body temperature. The time courses of wakefulness (A), REM sleep (B), sleep spindle (sigma) activity (C), slow-wave activity (D), and core body temperature (E) during sleep episodes are plotted for sleep episodes initiated at different circadian phases of the core body temperature cycle in a forced-desynchrony protocol. (0 deg corresponds to the fitted minimum of the core body temperature rhythm, bottom *x* axis. To facilitate comparison with the situation in which the circadian system is entrained to the 24-hour day, the top *x* axis indicates the average clock time of the circadian core body temperature rhythm during the

older people report waking up before 5 a.m. at least once per week, with 15% doing so nearly every morning. This earlier time of awakening is associated with an earlier entrained phase of the endogenous component of the body temperature rhythm (Czeisler *et al.*, 1992; Duffy *et al.*, 1998; Myers & Badia, 1995). However, it is not known whether this circadian phase advance is a result or a cause of the early morning awakening and associated earlier exposure to light reported among older people.

Classical entrainment theory states that an advanced phase during entrainment to an external light–dark cycle may be related to a shortening of the intrinsic period of the endogenous circadian pacemaker or an increased sensitivity to the phase-setting effects of light. This was consistent with prior reports that the intrinsic period of the circadian pacemaker shortened with age (Weitzman, Moline, Czeisler, & Zimmerman, 1982). However, recent data indicate that, when light exposure and activity level are controlled in a forced-desynchrony protocol, the intrinsic period of the circadian pacemaker observed in healthy older people does not differ from that observed in healthy young adults (Czeisler *et al.*, 1995a). Therefore, the cause of the earlier entrained phase of the body temperature rhythm is not known. Other factors contributing to the disrupted sleep and early morning awakening observed in older people could include age-related changes in light sensitivity, homeostatic drive for sleep, endogenous circadian amplitude, and/or sleep-generating mechanisms (Duffy *et al.*, 1998; Edgar, 1994; Myers & Badia, 1995). Furthermore the age-related increase in sleep pathologies, including sleep apnea and periodic limb movements in sleep, may contribute to the disrupted sleep reported by older people. Forced-desynchrony studies have demonstrated that profound age-related changes in sleep can be observed at virtually all circadian phases even in healthy older people with no significant sleep pathologies (Dijk, Duffy, Riel, & Czeisler, 1997a; Duffy *et al.*, 1998), suggesting that aging is associated with neurobiological changes that are not secondary to other ailments (Duffy, Dijk, Klerman, & Czeisler, 1997).

In summary, age-related changes in sleep–wake organization are profound, and understanding the neurobiological correlates of these changes may shed more light on the circadian and homeostatic regulation of sleep structure and sleep consolidation.

first day of the forced desynchrony protocol, i.e., immediately upon release from entrainment. (A) Time course of wakefulness in scheduled sleep episodes. Sleep-dependent resolution is 112 min, i.e., fifths of sleep episode. Sleep episodes were binned in 30-deg bins according to the circadian phase at lights-out. Curve 1: sleep episodes initiated between −15 and 15 degrees; curve 2: 15–45 deg; etc. Note that during sleep episodes represented in curve 10 (filled symbols), wakefulness remains low throughout the sleep episode. These sleep episodes were initiated between 225 and 285 deg, which corresponds to 23–01 hours under entrained conditions, i.e., habitual bedtimes. In other sleep episodes, sleep becomes disrupted toward the end. (B) Time course of REM sleep (expressed as percentage of recording time, RT) in scheduled sleep episodes. Sleep-dependent resolution is 112 min, i.e., fifths of sleep episode. Sleep episodes were binned in 45-deg bins, according to the circadian phase at lights-out. First curve (open circles): sleep episodes initiated between −22.5 and 22.5 deg; second curve (closed circles): sleep episodes initiated between 22.5 and 67.5 deg; etc. (C) Time course of EEG sigma activity (12.75–15.0 Hz) during nonREM sleep in scheduled sleep episodes. The sleep-dependent and circadian resolutions are identical to those in panel B. (D) Time course of EEG slow wave activity (0.75–4.5 Hz) during nonREM sleep in scheduled sleep episodes. The sleep-dependent and circadian resolutions are identical to those in panel B. (E) Time course of core body temperature. Sleep-dependent resolution is 30 min. Note that the first value of each temperature curve represents the average core body temperature during the last 30 minutes before lights-out. Circadian resolution is identical to that in panel B. (From Dijk & Czeisler, 1995, with permission.)

The profound impact of the circadian timing system on human behavior and physiology has only recently been widely recognized. Comparative studies have demonstrated that circadian rhythms are a fundamental characteristic of all living systems and there is no reason to assume that the behavior and physiology of our species is not subject to circadian control mechanisms similar to those described in other species. More detailed studies of the species-invariant and species-dependent aspects of circadian physiology and their biological basis is needed to deepen our understanding of human circadian physiology. Circadian physiology provides a unique model for the study of the complex interactions among the environment, the biological substrate, and behavior. Within this context, the challenges posed by our geophysical, social, cultural, and biological environment can be confronted and rational/empirical therapies for dysregulations of our circadian system interacting with these environments may be designed. Light treatment (Campbell *et al.*, 1995) and melatonin administration (Hughes, Sack, & Lewy, 1998) in older people are two examples of the potential of interventions derived from the study of human circadian physiology and sleep–wake regulation.

Acknowledgments

Drs. T. Åkerstedt, D. F. Dinges, and J. M. Waterhouse provided helpful comments on earlier versions of the manuscript. The research conducted by the Laboratory for Circadian and Sleep Disorders Medicine has been supported by grants from NIA, NIH, HLBI, NASA, and NSBRI and a GCRC grant to the Brigham and Women's Hospital. The authors are grateful to Nicole Bruno and Nina Arnfeld for their help in the preparation of the manuscript.

REFERENCES

Achermann, P., Dijk, D.-J., Brunner, D. P., & Borbély, A. A. (1993). A model of human sleep homeostasis based on EEG slow-wave activity: Quantitative comparison of data and simulations. *Brain Research Bulletin, 31,* 97–113.

Agnew, H. W., Webb, W. B., & Williams, R. L. (1964). The effects of stage four sleep deprivation. *Electroencephalography and Clinical Neurophysiology, 17,* 68–70.

Akerstedt, T. A., Fröberg, J. E., Friberg, Y., & Wetterberg, L. (1979). Melatonin excretion, body temperature, and subjective arousal during 64 hours of sleep deprivation. *Psychoneuroendocrinology, 4,* 219–225.

Allan, J. S., & Czeisler, C. A. (1994). Persistence of the circadian thyrotropin rhythm under constant conditions and after light-induced shifts of circadian phase. *Journal of Clinical Endocrinology and Metabolism, 79,* 508–512.

Arendt, J., & Broadway, J. (1986). Phase response of human melatonin rhythms to bright light in Antarctica. *Journal of Physiology, 377,* 68P–68P.

Aschoff, J. (1955). Der tagesgang der korpertemperatur beim menschen. *Klinische Wochenschrift, 33,* 545–551.

Aschoff, J. (1960). Exogenous and endogenous components in circadian rhythms. *Cold Spring Harbor Symposia on Quantitative Biology, 25,* 11–28.

Aschoff, J. (1985). On the perception of time during prolonged temporal isolation. *Human Neurobiology, 4,* 41–52.

Aschoff, J., & Wever, R. (1962). Spontanperiodik des Menschen bei Ausschluss aller Zeitgeber. *Naturwissenschaften, 49,* 337–342.

Aschoff, J., Gerecke, U., & Wever, R. (1967). Desynchronization of human circadian rhythm. *Japanese Journal of Physiology, 17,* 450–457.

Aschoff, J., Fatranská, M., Giedke, H., Doerr, P., Stamm, D., & Wisser, H. (1971). Human circadian rhythms in continuous darkness: Entrainment by social cues. *Science, 171,* 213–215.
Aschoff, J., Hoffmann, K., Pohl, H., & Wever, R. (1975). Re-entrainment of circadian rhythms after phase-shifts of the zeitgeber. *Chronobiologia, 2,* 23–78.
Åkerstedt, T., & Fröberg, J. E. (1977). Psychophysiological circadian rhythms in women during 72 h of sleep deprivation. *Waking and Sleeping, 1,* 387–394.
Åkerstedt, T., & Gillberg, M. (1981). The circadian variation of experimentally displaced sleep. *Sleep, 4,* 159–169.
Åkerstedt, T. A., Fröberg, J. E., Friberg, Y., & Wetterberg, L. (1979). Melatonin excretion, body temperature and subjective arousal during 64 hours of sleep deprivation. *Psychoneuroendocrinology, 4,* 219–225.
Åkerstedt, T. A., Palmblad, J., de la Torre, B., Marana, R., & Gillberg, M. (1980). Adrenocortical and gonadal steroids during sleep deprivation. *Sleep, 3,* 23–30.
Åkerstedt, T., Torsvall, L., & Gillberg M. (1982). Sleepiness and shift work: Field studies. *Sleep, 5,* S95–S106.
Beersma, D. G. M., & Hiddinga, A. E. (1996). The impact of motor activity on the sleep–wake and circadian pacemaker components of body temperature in humans. *Journal of Sleep Research, 5,* 13.
Beersma, D. G. M., & Hiddinga, A. E. (1998). No impact of physical activity on the period of the circadian pacemaker in humans. *Chronobiology International, 15,* 49–57.
Beersma, D. G. M., Daan, S., & Dijk, D.-J. (1987). Sleep intensity and timing: A model for their circadian control. *Lecture on Mathematics in the Life Sciences, 19,* 39–62.
Beersma, D. G. M., Dijk, D.-J., Blok, C. G. H., & Everhardus, I. (1990). REM sleep deprivation during 5 hours leads to an immediate REM sleep rebound and to suppression of non-REM sleep intensity. *Electroencephalography and Clinical Neurophysiology, 76,* 114–122.
Benedict, F. G. (1904). Studies in body-temperature I. Influence of the inversion of the daily routine; the temperature of night-workers. *American Journal of Physiology, 11,* 145–169.
Boivin, D. B., Duffy, J. F., Kronauer, R. E., & Czeisler, C. A. (1996). Dose–response relationships for resetting of human circadian clock by light. *Nature, 379,* 540–542.
Boivin, D. B., Czeisler, C. A., Dijk, D.-J., Duffy, J. F., Folkard, S., Minors, D. S., Totterdell, P., & Waterhouse, J. M. (1997). Complex interaction of the sleep–wake cycle and circadian phase modulates mood in healthy subjects. *Archives of General Psychiatry, 54,* 145–152.
Borbély, A. A. (1982a). A two process model of sleep regulation. *Human Neurobiology, 1,* 195–204.
Borbély, A. A. (1982b). Sleep regulation: Circadian rhythm and homeostasis. In D. Ganten & D. Pfaff (Eds.), *Sleep: Clinical and experimental aspects* (pp. 83–103). Berlin: Springer-Verlag.
Borbély, A. A. (1994). Sleep homeostasis and models of sleep regulation. In M. H. Kryger, T. Roth, & W. C. Dement (Eds.), *Principles and practice in sleep medicine* (pp. 309–320). Philadelphia: Saunders.
Borbély, A. A. & Achermann, P. (1992). Concepts and models of sleep regulation: An overview. *Journal of Sleep Research, 1,* 63–79.
Borbély, A. A., Achermann, P., Trachsel, L., & Tobler, I. (1989). Sleep initiation and initial sleep intensity: Interaction of homeostatic and circadian mechanisms. *Journal of Biological Rhythms, 4,* 149–160.
Brandenberger, G., Follenius, M., Simon, C., Ehrhart, J., & Libert, J. P. (1988). Nocturnal oscillations in plasma renin activity and REM–NREM sleep cycles in humans: A common regulatory mechanism? *Sleep, 11,* 242–250.
Brunner, D. P., Dijk, D.-J., Tobler, I., & Borbély, A. A. (1990). Effect of partial sleep deprivation on sleep stages and EEG power spectra: Evidence for non-REM and REM sleep homeostasis. *Electroencephalography and Clinical Neurophysiology, 75,* 492–499.
Cagnacci, A., Elliott, J. A., & Yen, S. S. C. (1992). Melatonin: A major regulator of the circadian rhythm of core temperature in humans. *Journal of Clinical Endocrinology and Metabolism, 75,* 447–452.
Cajochen, C., Dijk, D.-J., & Borbély, A. A. (1992). Dynamics of EEG slow-wave activity and core body temperature in human sleep after exposure to bright light. *Sleep, 15,* 337–343.
Campbell, S. S., Terman, M., Lewy, A. J., Dijk, D.-J., Eastman, C. I., & Boulos, Z. (1995). Light treatment for sleep disorders: Consensus report. V. Age-related disturbances. *Journal of Biological Rhythms, 10,* 151–154.
Carrier, J., & Dumont, M. (1995). Sleep propensity and sleep architecture after bright light exposure at three different times of day. *Journal of Sleep Research, 4,* 202–211.
Carskadon, M. A., & Dement, W. C. (1975). Sleep studies on a 90-minute day. *Electroencephalography and Clinical Neurophysiology, 39,* 145–155.
Carskadon, M. A., & Dement, W. C. (1979). Effects of total sleep loss on sleep tendency. *Perceptual and Motor Skills, 48,* 495–506.
Carskadon, M. A., & Dement, W. C. (1981). Cumulative effects of sleep restriction on daytime sleepiness. *Psychophysiology, 18,* 107–113.
Carskadon, M. A., & Dement, W. C. (1992). Multiple sleep latency tests during the constant routine. *Sleep, 15,* 396–399.

Carskadon, M. A., & Dement, W. C. (1996). Normal human sleep: An overview. In M. H. Kryger, T. Roth, & W. C. Dement (Eds.), *Principles and practice of sleep medicine*, (pp. 16–25). Philadelphia: Saunders.

Conroy, R. T. W. L., & Mills, J. N. (1970). *Human circadian rhythms.* London: Churchill.

Czeisler, C. A. (1978). *Human circadian physiology: Internal organization of temperature, sleep–wake, and neuroendocrine rhythms monitored in an environment free of time cues.* Thesis, Stanford University, Stanford, California.

Czeisler, C. A. & Khalsa, S. B. S. (2000). The human circadian timing system and sleep-wake regulation. In M. H. Kryger, T. Roth, & W. C. Dement (Eds.), *Principles and Practice of Sleep Medicine* (3rd ed., pp. 353–374). Philadelphia: W. B. Saunders Company.

Czeisler, C. A., Moore-Ede, M. C., Regestein, Q. R., Kisch, E. S., Fang, V. S., & Ehrlich, E. N. (1976). Episodic 24-hour cortisol secretory patterns in patients awaiting elective cardiac surgery. *Journal of Clinical Endocrinology and Metabolism, 42,* 273–283.

Czeisler, C. A., Richardson, G. S., Moore-Ede, M. C., & Weitzman, E. D. (1978). Entrainment of human circadian rhythms by light-dark cycles: A reassessment. In *6th Annual Meeting, American Society for Photobiology*, (p. 73).

Czeisler, C. A., Weitzman, E. D., Moore-Ede, M. C., Kronauer, R. E., Zimmerman, J. C., & Campbell, C. (1980a). Human sleep: Its duration and structure depend on the interaction of two separate circadian oscillators. *Sleep Research, 9,* 270.

Czeisler, C. A., Weitzman, E. D., Moore-Ede, M. C., Zimmerman, J. C., & Knauer, R. S. (1980b). Human sleep: Its duration and organization depend on its circadian phase. *Science, 210,* 1264–1267.

Czeisler, C. A., Zimmerman, J. C., Ronda, J. M., Moore-Ede, M. C., & Weitzman, E. D. (1980c). Timing of REM sleep is coupled to the circadian rhythm of body temperature in man. *Sleep, 2,* 329–346.

Czeisler, C. A., Brown, E. N., Ronda, J. M., Kronauer, R. E., Richardson, G. S., & Freitag, W. O. (1985). A clinical method to assess the endogenous circadian phase (ECP) of the deep circadian oscillator in man. *Sleep Research, 14,* 295–295.

Czeisler, C. A., Allan, J. S., Strogatz, S. H., Ronda, J. M., Sánchez, R., Ríos, C. D., Freitag, W. O., Richardson, G. S., & Kronauer, R. E. (1986). Bright light can phase delay the human circadian pacemaker independent of the timing of the sleep-wake cycle. *American Physiological Society Abstract Issue, 29,* 139.

Czeisler, C. A., Kronauer, R. E., Allan, J. S., Duffy, J. F., Jewett, M. E., Brown, E. N., & Ronda, J. M. (1989). Bright light induction of strong (Type 0) resetting of the human circadian pacemaker. *Science 244,* 1328–1333.

Czeisler, C. A., Allan, J. S., & Kronauer, R. E. (1990a). A method for assaying the effects of therapeutic agents on the period of the endogenous circadian pacemaker in man. In J. Montplaisir & R. Godbout (Eds.), *Sleep and biological rhythms: Basic mechanisms and applications to psychiatry* (pp. 87–98). New York: Oxford University Press.

Czeisler, C. A., Johnson, M. P., Duffy, J. F., Brown, E. N., Ronda, J. M., & Kronauer, R. E. (1990b). Exposure to bright light and darkness to treat physiologic maladaptation to night work. *New England Journal of Medicine, 322,* 1253–1259.

Czeisler, C. A., Dumont, M., Duffy, J. F., Steinberg, J. D., Richardson, G. S., Brown, E. N., Sánchez, R., Ríos, C. D., & Ronda, J. M. (1992). Association of sleep–wake habits in older people with changes in output of circadian pacemaker. *Lancet, 340,* 933–936.

Czeisler, C. A., Duffy, J. F., Shanahan, T. L., Brown, E. N., Mitchell, J. F., Dijk, D.-J., Rimmer, D. W., Ronda, J. M., Allan, J. S., Emens, J. S., & Kronauer, R. E. (1995a). Reassessment of the intrinsic period (*t*) of the human circadian pacemaker in young and older subjects. *Sleep Research, 24A,* 505.

Czeisler, C. A., Shanahan, T. L., Klerman, E. B., Martens, H., Brotman, D. J., Emens, J. S., Klein, T., & Rizzo, III, J. F. (1995b). Suppression of melatonin secretion in some blind patients by exposure to bright light. *New England Journal of Medicine, 332,* 6–11.

Daan, S., & Beersma, D. G. M. (1992). A single pacemaker can produce different rates of reentrainment in different overt rhythms. *Journal of Sleep Research, 1,* 80–83.

Daan, S., Beersma, D. G. M., & Borbély, A. A. (1984). Timing of human sleep: Recovery process gated by a circadian pacemaker. *American Journal of Physiology, 246,* R161–R178.

Dawson, D. S., Gibbon, S., & Singh, P. (1996). The hypothermic effect of melatonin on core body temperature: Is more better? *Journal of Pineal Research, 20,* 192–197.

Dement, W. C. (1960). The effect of dream deprivation. *Science, 131,* 1705–1707.

Dijk, D.-J. (1995). EEG slow waves and sleep spindles: Windows on the sleeping brain. *Behavioral Brain Research, 69,* 109–116.

Dijk, D.-J., & Czeisler, C. A. (1991). A quantitative assessment of the circadian influence on sleep propensity and sleep structure by forced desynchrony of the sleep–wake cycle and core–body-temperature rhythm. *Sleep Research, 20A,* 531.

Dijk, D.-J., & Czeisler, C. A. (1994). Paradoxical timing of the circadian rhythm of sleep propensity serves to consolidate sleep and wakefulness in humans. *Neuroscience Letters, 166*, 63–68.

Dijk, D.-J., Czeisler, C. A. (1995). Contribution of the circadian pacemaker and the sleep homeostat to sleep propensity, sleep structure, electroencephalographic slow waves and sleep spindle activity in humans. *Journal of Neuroscience, 15*, 3526–3538.

Dijk, D.-J. & Lavie, P. (1997). Review: The rhythms of human sleep propensity and core body temperature. *Sleep Research, 26,* 763 & 773.

Dijk, D.-J., Beersma, D. G. M., & Daan, S. (1987a). EEG power density during nap sleep: Reflection of an hourglass measuring the duration of prior wakefulness. *Journal of Biological Rhythms, 2,* 207–219.

Dijk, D.-J., Beersma, D. G. M., Daan, S., Bloem, G. M., & van den Hoofdakker, R. H. (1987b). Quantitative analysis of the effects of slow wave sleep deprivation during the first 3 h of sleep on subsequent EEG power density. *European Archive of Psychiatry and Neurological Sciences, 236,* 323–328.

Dijk, D.-J., Visscher, C. A., Bloem, G. M., Beersma, D. G. M., & Daan, S. (1987). Reduction of human sleep duration after bright light exposure in the morning. *Neuroscience Letters, 73,* 181–186.

Dijk, D.-J., Beersma, D. G. M., Daan, S., & Lewy, A. J. (1989). Bright morning light advances the human circadian system without affecting REM sleep homeostasis. *American Journal of Physiology, 256,* R106–R111.

Dijk, D.-J., Brunner, D. P., Beersma, D. G. M., & Borbély, A. A. (1990). Electroencephalogram power density and slow wave sleep as a function of prior waking and circadian phase. *Sleep, 13,* 430–440.

Dijk, D.-J., Brunner, D. P., & Borbély, A. A. (1991). EEG power density during recovery sleep in the morning. *Electroencephalography and Clinical Neurophysiology, 78,* 203–214.

Dijk, D.-J., Duffy, J. F., & Czeisler, C. A. (1992). Circadian and sleep/wake dependent aspects of subjective alertness and cognitive performance. *Journal of Sleep Research, 1,* 112–117.

Dijk, D.-J., Duffy, J. F., Riel, E., & Czeisler, C. A. (1997a). Altered interaction of circadian and homeostatic aspects of sleep propensity results in awakening at an earlier circadian phase in older people. *Sleep Research, 26,* 710.

Dijk, D.-J., Shanahan, T. L., & Czeisler, C. A. (1997b). Individual differences and non-additive effect of rest–activity cycle and circadian melatonin phase on core body temperature. *Sleep Research, 26,* 709.

Dijk, D.-J., Shanahan, T. L., Duffy, J. F., Ronda, J. M., & Czeisler, C. A. (1997c). Variation of electroencephalographic activity during nonREM and REM sleep with phase circadian melatonin rhythm in humans. *Journal of Physiology, 505,* 851–858.

Dinges, D. F., Orne, M. T., Whitehouse, W. G., & Orne, E. C. (1987). Temporal placement of a nap for alertness: Contributions of circadian phase and prior wakefulness. *Sleep, 10,* 313–329.

Dinges, D. F., Pack, F., Williams, K., Gillen, K. A., Powell, J. W., Ott, G. E., Aptowicz, C., & Pack, A. I. (1997). Cumulative sleepiness, mood disturbance, and psychomotor vigilance performance decrements during a week of sleep restricted to 4–5 hours per night. *Sleep, 20,* 267–277.

Drennan, M., Kripke, D. F., & Gillin, J. C. (1989). Bright light can delay human temperature rhythm independent of sleep. *American Journal of Physiology, 257,* R136–R141.

Duffy, J. F., Kronauer, R. E., & Czeisler, C. A. (1996). Phase-shifting human circadian rhythms: Influence of sleep timing, social contact and light exposure. *Journal of Physiology, 495,* 289–297.

Duffy, J. F., Dijk, D.-J., Klerman, E. B., & Czeisler, C. A. (1997). Altered phase relationship between body temperature cycle and habitual awakening in older subjects. *Sleep Research, 26,* 711.

Duffy, J. F., Dijk, D.-J., Klerman, E. B., & Czeisler, C. A. (1998). Later endogenous circadian temperature nadir relative to an earlier waketime in older people. *American Journal of Physiology, 275,* R1478–1487.

Edgar, D. M. (1994). Sleep–wake circadian rhythms and aging: Potential etiologies and relevance to age-related changes in integrated physiological systems. *Neurobiology of Aging, 15,* 499–501.

Edgar, D. M., Dement, W. C., & Fuller, C. A. (1993). Effect of SCN lesions on sleep in squirrel monkeys: Evidence for opponent processes in sleep–wake regulation. *Journal of Neuroscience, 13,* 1065–1079.

El Hajj, G., Klerman, E. B., Brown, E. N., Choe, Y., Brown, E. M., & Czeisler, C. A. (1997). Parathyroid hormone circadian rhythm is truly endogenous. *Journal of Clinical Endocrinology and Metabolism, 82,* 281–286.

Feinberg, I., Koresko, R. L., & Heller, N. (1967). EEG sleep patterns as a function of normal and pathological aging in man. *Journal of Psychiatric Research, 5,* 107–144.

Findley, J. D. (1966). Programmed environments for the experimental analysis of human behavior. In W. K. Hong (Ed.), *Operant behavior: Areas of research and application* (pp. 827–848). New York: Appleton-Century-Crofts.

Foley, D. J., Monjan, A. A., Brown, S. L., Simonsick, E. M., Wallace, R. B., & Blazer, D. G. (1995). Sleep complaints among elderly persons: An epidemiologic study of three communities. *Sleep, 18,* 425–432.

Folkard, S. & Åkerstedt, T. (1987). Towards a model for the prediction of alertness and/or fatigue on different sleep/wake schedules. In A. Oginski, J. Pokorski, & J. Rutenfranz (Eds.), *Contemporary advances in shiftwork research* (pp. 231–240). Krakow, Poland: Medical Academy.

Folkard, S., & Åkerstedt, T. (1989). Towards the prediction of alertness on abnormal sleep/wake schedules. In A. Coblentz (Ed.), *Vigilance and performance in automated systems,* (pp. 287–296). Dordrecht: Kluwer.

Folkard, S., Wever, R. A., & Wildgruber, C. M. (1983). Multi-oscillatory control of circadian rhythms in human performance. *Nature, 305,* 223–226.

Hastings, J. W., & Sweeney, B. M. (1958). A persistent diurnal rhythm of luminescence in *Gonyaulax polyedra. Biological Bulletin, 115,* 440–458.

Hellman, L., Nakada, F., Curti, J., Weitzman, E. D., Kream, J., Roffwarg, H., Ellman, S., & Fukushima, D. K. G. (1970). Cortisol is secreted episodically by normal man. *Journal of Clinical Endocrinology and Metabolism, 30,* 411–422.

Hiddinga, A. E., Beersma, D. G., & van den Hoofdakker, R. H. (1997). Endogenous and exogenous components in the circadian variation of core body temperature in humans. *Journal of Sleep Research, 6,* 156–163.

Honma, K., Honma, S., & Wada, T. (1987). Phase-dependent shift of free-running human circadian rhythms in response to a single bright light pulse. *Experientia, 43,* 1205–1207.

Hughes, R. J., Sack, R. L., & Lewy, A. J. (1998). The role of melatonin and circadian phase in age-related sleep-maintenance insomnia: Assessment in a clinical trial of melatonin replacement. *Sleep, 21,* 52–68.

Hume, K., & Mills, J. (1977). Rhythms of REM and slow-wave sleep in subjects living on abnormal time schedules. *Waking and Sleeping, 1,* 291–296.

Hunter, J. (1778). Of the heat of animals and vegetables. *Philosophical Transactions of the Royal Society of London, 68,* 7–49.

Jewett, M. E., & Kronauer, R. E. (1998). Refinement of a limit cycle oscillator model of the effects of light on the human circadian pacemaker. *Journal of Theoretical Biology, 192,* 455–465.

Jewett, M. E., Kronauer, R. E., & Czeisler, C. A. (1991). Light-induced suppression of endogneous circadian amplitude in humans. *Nature, 350,* 59–62.

Jewett, M. E., Kronauer, R. E., & Czeisler, C. A. (1994). Phase/amplitude resetting of the human circadian pacemaker via bright light: A further analysis. *Journal of Biological Rhythms, 9,* 295–314.

Jewett, M. E., Rimmer, D. W., Duffy, J. F., Klerman, E. B., Kronauer, R. E., & Czeisler, C. A. (1997). The human circadian pacemaker is sensitive to light throughout subjective day without evidence of transients. *American Journal of Physiology, 273,* R1800–R1809.

Johnson, C. H. (1997). *An atlas of phase repsonse curves for circadian and circatidal rhythms.* Nashille, TN: Department of Biology, Vanderbilt University.

Johnson, M. P., Duffy, J. F., Dijk, D.-J., Ronda, J. M., Dyal, C. M., & Czeisler, C. A. (1992). Short-term memory, alertness and performance: A reappraisal of their relationship to body temperature. *Journal of Sleep Research, 1,* 24–29.

Klein, T., Martens, H., Dijk, D.-J., Kronauer, R. E., Seely, E. W., & Czeisler, C. A. (1993). Chronic non-24-hour circadian rhythm sleep disorder in a blind man with a regular 24-hour sleep–wake schedule. *Sleep, 16,* 333–343.

Kleitman, N., & Doktorsky, A. (1933). Studies on the physiology of sleep. VII. The effect of the position of the body and of sleep on rectal temperature in man. *American Journal of Physiology, 104,* 340–343.

Kleitman, N., & Kleitman, E. (1953). Effect of non-twenty-four-hour routines of living on oral temperature and heart rate. *Journal of Applied Physiology, 6,* 283–291.

Klerman, E. B., Dijk, D.-J., Kronauer, R. E., & Czeisler, C. A. (1996). Simulations of light effects on the human circadian pacemaker: Implications for assessment of intrinsic period. *American Journal of Physiology, 270,* R271–282.

Klerman, E. B., Rimmer, D. W., Dijk, D.-J., Kronauer, R. E., Rizzo, III, J. F., & Czeisler, C. A. (1998). Nonphotic entrainment of the human circadian pacemaker. *American Journal of Physiology, 43,* R991–R996.

Kronauer, R. E. (1987). A model for the effect of light on the human "deep" circadian pacemaker. *Sleep Research, 16,* 621.

Kronauer, R. E. (1990). A quantitative model for the effects of light on the amplitude and phase of the deep circadian pacemaker, based on human data. In J. Horne (Ed.), *Sleep '90, Proceedings of the Tenth European Congress on Sleep Research* (pp. 306–309). Dusseldorf: Pontenagel Press.

Kronauer, R. E., Czeisler, C. A., Pilato, S. F., Moore-Ede, M. C., & Weitzman, E. D. (1982). Mathematical model of the human circadian system with two interacting oscillators. *American Journal of Physiology, 242,* R3–R17.

Kronauer, R. E., Forger, D., & Jewett, M. E. (1999). Quantifying human circadian pacemaker response to brief, extended, and repeated light stimuli over the photopic range. *Journal of Biological Rhythms, 14,* 500–515.

Lack, L. C. & Lushington, K. (1996). The rhythms of human sleep propensity and core body temperature. *Journal of Sleep Research, 5,* 1–11.

Lavie, P. (1997). Melatonin: Role in gating nocturnal rise in sleep propensity. *Journal of Biological Rhythms, 12,* 657–665.

Lavie, P., & Scherson, A. (1981). Ultrashort sleep–waking schedule. I. Evidence of ultradian rhythmicity in 'sleep ability'. *Electroencephalography and Clinical Neurophysiology, 52,* 163–174.

Mant, A., & Eyland, E. A. (1988). Sleep patterns and problems in elderly general practice attenders: An Australian survey. *Community Health Studies, 12,* 192–199.

Middleton, B., Arendt, J., & Stone, B. M. (1996). Human circadian rhythms in constant dim light (8 lux) with knowledge of clock time. *Journal of Sleep Research, 5,* 69–76.

Miles, L. E., & Dement, W. C. (1980). Sleep and aging. *Sleep, 3,* 119–220.

Miles, L. E. M., Raynal, D. M., & Wilson, M. A. (1977). Blind man living in normal society has circadian rhythms of 24.9 hours. *Science, 198,* 421–423.

Mills, J. N. (1964). Circadian rhythms during and after three months in solitude underground. *Journal of Physiology, 174,* 217–231.

Mills, J. N., Minors, D. S., & Waterhouse, J. M. (1978). Adaptation to abrupt time shifts of the oscillator[s] controlling human circadian rhythms. *Journal of Physiology, 285,* 455–470.

Minors, D. S., & Waterhouse, J. M. (1981). Endogenous and exogenous components of circadian rhythms when living on a 21-hour day. *International Journal of Chronobiology, 8,* 31–48.

Minors, D. S., & Waterhouse, J. M. (1989). Masking in humans: The problem and some attempts to solve it. *Chronobiology International, 6,* 29–53.

Minors, D. S., Waterhouse, J. M., & Wirz-Justice, A. (1991). A human phase-response curve to light. *Neuroscience Letters, 133,* 36–40.

Monk, T. H., Moline, M. L., Fookson, J. E., & Peetz, S. M. (1989). Circadian determinants of subjective alertness. *Journal of Biological Rhythms, 4,* 393–404.

Monk, T. H., Buysse, D. J., Reynolds III, C. F., Jarrett, D. B., & Kupfer, D. J. (1992). Rhythmic vs. homeostatic influences on mood, activation, and performance in young and old men. *Journal of Gerontology, 47,* P221–P227.

Monk, T. H., Buysse, D. J., Reynolds III, C. F., & Kupfer, D. J. (1996). Circadian determinants of the postlunch dip in performance. *Chronobiology International, 13,* 123–133.

Moore-Ede, M. C., & Czeisler, C. A. (1984). Mathematical models of the circadian sleep–wake cycle. *Biological Abstracts, 27,* 3085–3085.

Moore-Ede, M. C., Kass, D. A., & Herd, J. A. (1977). Transient circadian internal desynchronization after light–dark phase shift in monkeys. *American Journal of Physiology, 232,* R31–R37.

Myers, B. L., & Badia, P. (1993). Immediate effects of different light intensities on body temperature and alertness. *Physiology and Behavior, 54,* 199–202.

Myers, B. L., & Badia, P. (1995). Changes in circadian rhythms and sleep quality with aging: Mechanisms and interventions. *Neuroscience and Biobehavioral Reviews, 19,* 553–571.

Patrick, G. T. W., & Gilbert, J. A. (1896). Studies from the Psychological Laboratory of the University of Iowa. *Psychology Review, III,* 468–483.

Pelham, R. W., Vaughan, G. M., Sandock, K. L., & Vaughan, M. K. (1973). Twenty-four-hour cycle of a melatonin-like substance in the plasma of human males. *Journal of Clinical Endocrinology and Metabolism, 37,* 341–344.

Pincus, G. (1943). A diurnal rhythm in the excretion of urinary ketosteroids by young men. *Journal of Clinical Endocrinology and Metaoblism, 3,* 195–199.

Ralph, C. L., Mull, D., Lynch, H. J., & Hedlund, L. (1971). A melatonin rhythm persists in rat pineals in darkness. *Endocrinology, 89,* 1361–1366.

Reynolds, A. E., Quigley, T. B., Kennard, H. E., & Thorn, G. W. (1951). Reaction of the adrenal cortex to physical and emotional stress in college oarsmen. *New England Journal of Medicine, 244,* 754–757.

Reynolds III, C. F., Kupfer, D. J., Taska, L. S., Hoch, C. C., Sewitch, D. E., & Spiker, D. G. (1985). Sleep of healthy seniors: A revisit. *Sleep, 8,* 20–29.

Richardson, G. S., Carskadon, M. A., Orav, E. J., & Dement, W. C. (1982). Circadian variation of sleep tendency in elderly and young adult subjects. *Sleep, 5,* S82–S94.

Sack, R. L., Lewy, A. J., Blood, M. L., Keith, L. D., & Nakagawa, H. (1992). Circadian rhythm abnormalities in totally blind people: Incidence and clinical significance. *Journal of Clinical Endocrinology and Metabolism, 75,* 127–134.

Savides, T. J., Messin, S., Senger, C., & Kripke, D. F. (1986). Natural light exposure of young adults. *Physiology and Behavior, 38,* 571–574.

Shanahan, T. L. (1995). *Circadian physiology and the plasma melatonin rhythm in humans.* Thesis, Harvard Medical School, Cambridge, Massachusetts.

Shanahan, T. L., & Czeisler, C. A. (1991). Light expousre induces equivalent phase shifts of the endogenous circadian rhythms of circulating plasma melatonin and core body temperature in men. *Journal of Clinical Endocrinology and Metabolism, 73,* 227–235.

Shanahan, T. L., Zeitzer, J. M., & Czeisler, C. A. (1997). Resetting the melatonin rhythm with light in humans. *Journal of Biological Rhythms, 12,* 556–567.

Siffre, M. (1964). *Beyond time.* New York: McGraw-Hill.

Strogatz, S. H. (1986). *The mathematical structure of the human sleep wake cycle.* New York: Springer-Verlag.

Strogatz, S. H., Kronauer, R. E., & Czeisler, C. A. (1987). Circadian pacemaker interferes with sleep onset at specific times each day: Role in insomnia. *American Journal of Physiology, 253,* R172–R178.

Terman, M., & Terman. J. (1985). A circadian pacemaker for visual sensitivity. *Annals of the New York Academy of Sciences, 453,* 147–161.

Tobler, I., Borbély, A. A., & Groos, G. (1983). The effect of sleep deprivation on sleep in rats with suprachiasmatic lesions. *Neuroscience Letters, 42,* 49–54.

Tosini, G., & Menaker, M. (1996). Circadian rhythms in cultured mammalian retina. *Science, 272,* 419–421.

Trachsel, L., Edgar, D. M., Seidel, W. F., Heller, H. C., & Dement, W. C. (1992). Sleep homeostasis in suprachiasmatic nuclei-lesioned rats: Effects of sleep deprivation and triazolam administration. *Brain Research, 589,* 253–261.

Van Cauter, E. (1990). Diurnal and ultradian rhythms in human endocrine function: A minireview. *Hormone Research, 34,* 45–53.

Van Reeth, O., Sturis, J., Byrne, M. M., Blackman, J. D., L'Hermite-Balériaux, M., Leproult, R., Oliner, C., Refetoff, S., Turek, F. W., & Van Cauter, E. (1994). Nocturnal exercise phase delays circadian rhythms of melatonin and thyrotropin secretion in normal men. *American Journal of Physiology, 266,* E964–E974.

Waldstreicher, J., Duffy, J. F., Brown, E. N., Rogacz, S., Allan, J. S., & Czeisler, C. A. (1996). Gender differences in the temporal organization of prolactin (PRL) secretion: Evidence for a sleep-independent circadian rhythm of circulating PRL levels—A clinical research center study. *Journal of Clinical Endocrinology and Metabolism, 81,* 1483–1487.

Webb, W. B., & Agnew, H. W. (1971). Stage 4 sleep: Influence of time course variables. *Science, 174,* 1354–1356.

Webb, W. B., & Agnew, Jr., H. W. (1972). Sleep and waking in an environment free of cues to time. *Psychophysiology, 9,* 133–133.

Webb, W. B., Agnew, H. W., & Williams, R. L. (1971). Effect on sleep of a sleep period time displacement. *Aerospace Medicine, 42,* 152–155.

Weitzman, E. D., Nogeire, C., Perlow, M., Fukushima, D., Sassin, J., McGregor, P., Gallagher, T. F., & Hellman, L. (1974). Effects of a prolonged 3-hour sleep–wake cycle on sleep stages, plasma cortisol, growth hormone, and body temperature in man. *Journal of Clinical Endocrinology and Metabolism, 38,* 1018–1030.

Weitzman, E. D., Schaumburg, H., & Fishbein, W. (1966). Plasma 17-hydrocorticosteroid levels during sleep in man. *Journal of Clinical Endocrinology and Metabolism, 26,* 121–127.

Weitzman, E. D., Weinberg, U., D'Eletto, R., Lynch, H., Wurtman, R. J., Czeisler, C. A., & Erlich, S. (1978). Studies of the 24 hour rhythm of melatonin in man. *Journal of Transmitters, 13,* 325–337.

Weitzman, E. D., Czeisler, C. A., & Moore-Ede, M. C. (1979). Sleep–wake, neuroendocrine and body temperature circadian rhythms under entrained and non-entrained (free-running) conditions in man. In M. Suda, O. Hayaishi, & H. Nakagawa (Eds.), *Biological rhythms and their central mechanism* (pp. 199–227). Amsterdam: Elsevier/North-Holland.

Weitzman, E. D., Czeisler, C. A., Zimmerman, J. C., & Ronda, J. M. (1980). Timing of REM and stages 3 and 4 sleep during temporal isolation in man. *Sleep, 2,* 391–407.

Weitzman, E. D., Czeisler, C. A., Zimmerman, J. C., & Ronda, J. M. (1981). The sleep–wake pattern of cortisol and growth hormone secretion during non-entrained (free-running) conditions in man. In E. Van Cauter & G. Copinschi (Eds.), *Human pituitary hormones: Developments in endocrinology* (pp. 29–41).

Weitzman, E. D., Moline, M. L., Czeisler, C. A., & Zimmerman, J. C. (1982). Chronobiology of aging: Temperature, sleep–wake rhythms and entrainment. *Neurobiology of Aging, 3,* 299–309.

Weitzman, E. D., Zimmerman, J. C., Czeisler, C. A., & Ronda, J. (1983). Cortisol secretion is inhibited during sleep in normal man. *Journal of Clinical Endocrinology and Metabolism, 56,* 352–358.

Werth, E., Dijk, D.-J., Achermann, P., & Borbély, A. A. (1996). Dynamics of the sleep EEG after an early evening nap: Experimental data and simulations. *American Journal of Physiology, 271,* R501–R510.

Wever, R. (1965). A mathematical model for circadian rhythms. In J. Aschoff (Ed.), *Circadian clocks* (pp. 47–63). Amsterdam: North-Holland.

Wever, R. (1970). Zur Zeitgeber-Stärke eines Licht–Dunkel-Wechsels für die circadiane Periodik des Menschen. *European Journal of Physiology, 321,* 133–142.

Wever, R. (1973). Internal phase-angle differences in human circadian rhythms: Causes of changes and problems of determinations. *International Journal of Chronobiology, 1,* 371–390.

Wever, R. A. (1979). *The circadian system of man: Results of experiments under temporal isolation.* New York: Springer-Verlag.

Wever, R. A. (1985). Internal interactions within the human circadian system: The masking effect. *Experientia, 41,* 332–342.

Winfree, A. T. (1980). *The geometry of biological time.* New York: Springer-Verlag.

Winfree, A. T. (1987). *The timing of biological clocks.* New York: Scientific American.

Zeitzer, J. M., Kronauer, R. E., & Czeisler, C. A. (1997). Photopic transduction implicated in human circadian entrainment. *Neuroscience Letters, 232,* 135–138.

Zimmerman, J. C., Czeisler, C. A., Laxminarayan, S., Knauer, R. S., & Weitzman, E. D. (1980). REM density is dissociated from REM sleep timing during free-running sleep episodes. *Sleep, 2,* 409–415.

Zulley, J., & Campbell, S. S. (1985). Napping behavior during "spontaneous internal desynchronization": Sleep remains in synchrony with body temperature. *Human Neurobiology, 4,* 123–126.

Zulley, J., & Wever, R. A. (1982). Interaction between the sleep–wake cycle and the rhythm of rectal temperature. In J. Aschoff, S. Dann, & G. Gross (Eds.), *Vertebrate circadian systems* (pp. 253–261). Berlin: Springer-Verlag.

Zulley, J., Wever, R., & Aschoff, J. (1981). The dependence of onset and duration of sleep on the circadian rhythm of rectal temperature. *Pflügers Archive, 391,* 314–318.

21

Rhythms of Human Performance

J. M. WATERHOUSE, D. S. MINORS, T. ÅKERSTEDT,
T. REILLY, AND G. ATKINSON

ACCIDENTS, ERRORS, OR POOR PERFORMANCE?

Humans are diurnal by nature but modern industrialized societies must function throughout the 24 hours of the day. Not only are emergency services required to be able to act at any time, but communications, commerce and industrial processes also never cease their operations. In the military sphere in particular, round-the-clock capability is required. Whereas these processes have been accompanied by an increasing replacement of humans by technology, if only because humans are too slow, expensive, and unreliable, our species still is required to tend such systems and act in relation to the information they give or require. The role of humans has tended to change, therefore, from the that of providing brute strength and even intellectual input to one of vigilance and the minding of machines.

Systems do go wrong however, and errors occur attributable both to equipment and to the people who control it. In recent reviews of accidents (Åkerstedt, 1995a,b; Andlauer *et al.*, 1982; Costa, 1997; Dinges, 1995; Folkard, 1990; Horne and Reyner, 1995; Novak, Smolensky, Fairchild, & Reves, 1990; Smith, Folkard and Poole, 1994; Wojtczak-Jaroszowa & Jarosz, 1987), a common concern has been to establish how much is due to equipment failure and how much to human error. This concern is to protect not only the individuals involved, but also the public at large, who, directly or indirectly, can be affected by such disasters, as in the cases of the *Exxon Valdez*, the Bhopal chemical refinery, and the nuclear plants at Three Mile Island and Chernobyl.

Studies of road traffic accidents and of accidents in the workplace, in both military and civilian contexts, have stressed that human error can arise for a multi-

---

J. M. WATERHOUSE, T. REILLY, AND G. ATKINSON Research Institute for Sport & Exercise Sciences, Liverpool John Moores University, Liverpool, L3 2ET United Kingdom. D. S. MINORS School of Biological Sciences, University of Manchester, Manchester, M13 9PT United Kingdom. T. ÅKERSTEDT Laboratory of Clinical Stress Research, Karolinksa Institute, Stockholm, Sweden.

Circadian Clocks, Volume 12 of *Handbook of Behavioral Neurobiology*, edited by Joseph S. Takahashi, Fred W. Turek, and R. Y. Moore, Kluwer Academic / Plenum Publishers, New York, 2001.

tude of possible reasons, with several often interacting with one another in any particular case (Folkard, Condon, & Herbert, 1984; Gerbert & Kemmler, 1986; Hamelin, 1987; Langan-Fox & Empson, 1985; Lauridsen & Tonnesen, 1990; Pokorny, Blom, & Van Leeuwen, 1987; Ribak *et al.*, 1983; Torsvall, Åkerstedt, Gillander, & Knutsson, 1989). Some of the causes commonly cited include the following:

1. Environmental conditions, such as lighting, temperature, noise, and weather (Hamelin, 1987; Langan-Fox & Empson, 1985; Wojtczak-Jaroszowa & Jarosz, 1987).
2. Time-of-day (circadian) factors (Hamelin, 1987; Lauridsen & Tonnesen, 1990; Pokorny *et al.*, 1987).
3. Time awake, particularly time on duty, (Gerbert & Kemmler, 1986; Hamelin, 1987).
4. Work load and other aspects of stress (Gerbert & Kemmler, 1986; Langan-Fox & Empson, 1985; Ribak *et al.*, 1983).
5. Sleep loss, particularly due to daytime sleep and resulting cumulative loss (Folkard *et al.*, 1984; Torsvall *et al.*, 1989).
6. Inability to respond to the environment, or even sleeping in the workplace (Folkard *et al.*, 1984; Torsvall *et al.*, 1989).
7. Working immediately after waking up from sleep, particularly when sleeping near to workplace (Ribak *et al.*, 1983; Pokorny *et al.*, 1987).
8. Sleep quality, i.e., loss of slow-wave and rapid eye movement sleep (Åkerstedt, Torsvall, & Gillberg, 1987).

Many studies have concentrated on accidents during night work if only because, intuitively, this is when the body is better prepared for sleep than work, and conditions for work might be worst and supervision least, and so it might be predicted that the accident rate would be highest. Even so, the advent of rapid transmeridian flight has meant that most people experience jet lag at first hand, at least occasionally, and the circumstance of working at the wrong "body time" will occur after such flights. Air crew in particular will be subject to a combination of time-zone transitions and shiftwork, any of those errors could have catastrophic consequences. Therefore, considerable interest has been shown in investigating this group during a series of transglobal flights, particularly with regard to the effects of loss of sleep and motivation.

It is often reported that individuals differ in their susceptibilities to these factors (Folkard, 1990; Hamelin, 1987). Thus, many night workers can shrug off feelings of fatigue and only a fraction of the work force contributes to the increased incidence of ulcers or cardiovascular disorders in night workers (Waterhouse, Folkard, & Minors, 1992); thus, it has been suggested that some individuals develop adequate coping mechanisms or even possess some factors that protect them from the effects of work at abnormal times.

Accidents are rare and this, coupled with the fact that they cannot be predicted, makes them very difficult to study. Also, accidents must be investigated retrospectively, with all the interpretative difficulties that this entails. As a result, attention has been directed at the measurement of errors in the workplace, these being more common (Hildebrandt, Rohmert, & Rutenfranz, 1975). The conclusions that have been drawn from these studies are very similar to those drawn from studies on accidents, namely, that the same types of human failure are involved in a complex, interactive way.

Catastrophic errors in the workplace are comparatively rare and their complex causal nexus, sometimes with components that are difficult to quantify, does not make them easily amenable to scientific study and interpretation. In practice, therefore, it is more expedient and common to attempt to measure mental performance, physical performance, and subjective assessments of mood. This line of reasoning—that accidents are due to errors, and that errors can be predicted from mood and different types of performance—will receive further comment later.

Having considered the factors that affect human performance, we need to build them into a general model that enables us to predict combinations of circumstances that would be particularly inauspicious. Moreover, we also need to give advice to individuals on how to minimize and deal with the low points that will occur in any work schedule.

Performance During the Course of the Nychthemeron

Measuring Mental Performance

Measuring performance in field studies is beset by problems (Craig, Wilkinson, & Colquhoun, 1981b; Folkard, 1990; Folkard & Monk, 1985; Kirkcaldy, 1984). First, many tasks are difficult to quantify, such as those involving decision-making, where a quantitative assessment can seldom be made, or when there is no end product to examine, as in teaching, or keeping watch on board a ship. Often the conditions where the task is being carried out vary very considerably (driving, outdoor work) or provide distractions that differ from one work site to another (e.g., lighting, noise), making comparisons between studies very difficult. This difficulty is increased by a lack of standardization on the task itself, there being different demands of the work force according to whether the work is self-paced or externally paced (the latter applies to tasks involving inspection of articles on a conveyor belt, for example). Finally, the physiologic state of workers—how much sleep loss they have suffered or the length of their duty period—can rarely be controlled.

Some of these studies, often requiring the painstaking collection of huge amounts of data, have been summarized by Hildebrandt *et al.* (1975). In general, errors are higher at night than during the daytime, but a dip in performance (i.e., increase in errors) is found in the early afternoon. The results also show that performance is worse in the second half of the shift compared with the first, and that external factors also are important. As an illustration of this last point, ignoring a warning light was least frequent in the evening in a group of locomotive drivers and this was attributed to its ease of visibility in the darkness at this time (Hildebrandt *et al.*, 1975). Presumably, during the night hours, the ease of visibility of the warning light was offset by operator decrement and, during the daytime, the drivers were more alert and so poorer visibility was unimportant.

Many of the difficulties presented by field studies can be controlled or quantified (and hence corrected for statistically) in a laboratory environment; subjects' conditions of work and, to some extent, their lifestyle can be controlled more closely, as can the detailed nature of the tasks they perform.

The earlier tasks that were used in laboratory experiments tended to be pencil-and-paper tests. More recently, tests have tended to become more automated, recorded on portable tape records or personal computers, and often set automatically by these devices. This has the added advantage that the tests can be scored automatically. Many tasks have been used (summarized in Craig *et al.*, 1981b; Folkard,

1990; Folkard & Monk, 1985; Kirkcaldy, 1984; Minors & Waterhouse, 1981), but they can be classified into the following main types (see also Adan, 1993).

SUBJECTIVE ASSESSMENTS. These measure mood and feelings, often by requiring the subject to mark a line somewhere between its ends which indicate "not at all" or "as much as possible" to a question like "How alert do you feel?"

SIMPLE OBJECTIVE ASSESSMENTS. These tend to focus upon some aspect of sensory, cognitive, or manipulative ability, short-term memory, or some combination of these. Visual search tasks, such as those which require a single letter to be found from blocks of random letters, are examples of tests dominated by the speed of sensory input and processing. Manipulative skill can be assessed by having a subject threading beads onto a string, for example, or insert ball bearings into holes of the appropriate size. Logical reasoning can be assessed by syllogisms or by Baddeley's logical reasoning test in which the pair of letters A, B is presented in either order and the subject is required to assess the truth of a statement such as "B is not followed by A."

Short-term memory has been assessed by means of Folkard's search and memory test (Folkard, 1990; Folkard & Monk, 1985). A block of random letters is given and each line has to be scanned separately for a "target" of between two and six letters. As the length of the target increases, so does the extent to which short-term memory is required to accomplish the task.

Reaction time has been measured in several ways. *Simple reaction time* is the delay between receiving a signal (visual or auditory) and responding to it, generally by pressing a button. The task can be made more complex by requiring, in *choice reaction time*, the subject to respond to a choice of stimuli and responses, for example, by extinguishing the correct light from a choice of four when it becomes illuminated. Either of these tests of reaction time can be modified by adding a variable amount of warning for the subject before the stimulus is given.

MORE COMPLEX ASSESSMENTS. There are several ways in which complexity can be added to tests such as those described above. First, the task can be made more difficult, involving more thought or a combination of more skills. Examples of the former possibility are mental addition and plotting Cartesian coordinates; examples of the latter are tracking a moving object, copying meaningless symbols, and substituting digits for symbols. The use of video games has been suggested, with separate scores being kept of the various skills demanded by the game (Kennedy, Bittner, Harbeson, & Jones, 1982). Another possibility is to mix together a battery of tasks (Gillooly, Smolensky, Albright, Hsi, & Thorne, 1990), the components of which can remain simple. This raises the concept of "work load," which can be increased by putting time constraints on individuals, particularly if they are required to act upon or choose priorities when several tasks are present simultaneously. An excess of work load can give rise to stress, and tests to produce this (for example, the Stroop test) have been designed. The ultimate example of such a test is a flight simulator. Whereas this, will be more "real" than the other tasks described, it is expensive, and its sheer complexity will pose problems if it is required to separate out the different factors involved in performance.

One realistic aspect that can be duplicated to a greater extent in the laboratory is vigilance (Campbell & Wilkinson, 1981; Craig *et al.*, 1981b), which can be assessed by inserting abnormal stimuli into repetitive stimuli that form a "normal" background. The stimuli can be auditory or visual and realism can be added by including additional tasks that tend to distract the attention. Vigilance tasks can last for extended periods of time, a few hours or more, and are key compnents of the task of minding machines.

Circadian Rhythms in Mental Performance

Figure 1 shows the rhythms of oral temperature, mood, and some mental performance tasks measured diurnally in healthy subjects living a conventional sleep–activity schedule. Clearly, there are rhythmic changes, with lowest values in the early morning and late evening for most variables (Folkard, 1990). When these measures are extended through the night, not unexpectedly, the values are lower (Monk *et al.*, 1997); that is, there are negative and unfavorable effects upon mood and several types of mental task associated with work at night.

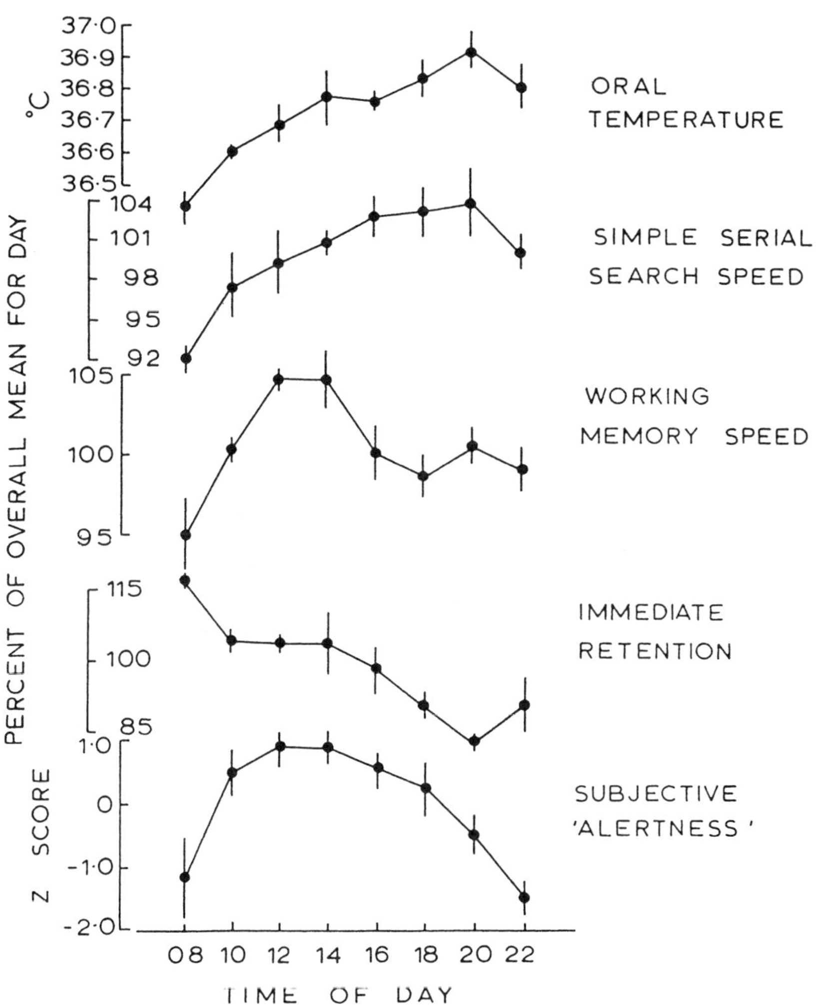

Figure 1. The relationship between time of day and oral temperature, mental performance and mood in normal subjects. (From Folkard, 1990, with permission. Copyright 1990, The Royal Society.)

A closer inspection of Figure 1 shows that the tests differ in detail when their time courses are considered (Folkard, 1990). The differences can be seen as variations in the rate at which a decrement occurs as a result of time elapsed while on duty or awake. As a general rule, those tasks which are simplest and involve the input of much data rather than much mental processing of it—visual search, for example—seem to decrement least. As the amount of central processing increases, so does the speed of deterioration. This deterioration is often referred to as "fatigue" and, as already described, also has been found in field studies (Hamelin, 1987; Pokorny *et al.*, 1987; Wojtczak-Jaroszowa & Jarosz, 1987). Indeed, a common conclusion from field and laboratory studies is that, whatever other factors might exist, mood and mental performance are determined very considerably by the interaction between circadian rhythms and time on task and, moreover, that circumstances where both factors are unfavorable should be avoided. Therefore, as will become clear later, these two factors dominate much of attempts to model mood and performance that has been carried out.

Figure 1 also indicates that the diurnal changes in short-term memory are different from those in the other variables: short-term memory decreases throughout the daytime, with a maximum in the early morning (Folkard, 1990; Folkard & Monk, 1985). In tasks that require different amounts of short-term memory, there will be different interactions between these two opposing effects. As short-term memory increases in importance for a particular task, so the trough of the rhythm shifts from the late daytime toward the nighttime, with intermediate use of short-term memory producing a bi-peaked performance curve (Folkard & Monk, 1985).

Ultradian Components of Mental Performance

In addition to circadian rhythms of performance, ultradian rhythms also exist. The best known is the temporary decline in performance in the early afternoon—the "post-lunch dip." Controversy exists as to whether the dip derives from having eaten lunch or whether other circadian rhythms, e.g., those in epinephrine level or body temperature, are involved also. Views are summarized by Craig, Baer, and Diekmann (1981a) Monk, Buysse, Reynolds, and Kupfer (1996), and Smith and Miles (1986). There seems to be a decrement for some people and some performance tasks after having eating lunch, but the dip is still present to some extent if no lunch is eaten. That is, it also is an effect of the time of day. This effect is not mirrored by body temperature. The reader is referred to the original papers for more details on this issue.

Another ultradian rhythm in mental performance has been described (Klein & Armitage, 1979) that involves the division of function between the left and right hemispheres, the former dealing more with numbers and words, the latter with patterns of sound or images. The two hemispheres have been suggested to show rhythmic changes in their processing power, the rhythms being in antiphase and with a period of about 90 minutes, although later work has not always confirmed this (Ortega & Cabrera, 1990).

Work Load and Motivation

Like other measures of performance, auditory and visual vigilance show circadian changes and deteriorate with the amount of time spent on the task (Craig, 1985). It is also clear that vigilance is much affected by several other factors, among which are motivation, work load, sleep loss, and personality type (Davies, 1985).

Since all mental performance tests seem to be affected by sleep loss and personality type, these factors will be considered later; since vigilance is particularly affected by motivation and work load, these two factors will be considered now.

Mental performance tests need some practice, during which sessions, the individual develops a strategy for dealing with the task. It has to be assumed that this strategy is adhered to during experiments designed to measure circadian and time-on-duty effects. However, this assumption need not be true, particularly when work loads are insufficiently high or excessive (Craig, 1985). In the former case, individuals might become bored; in the latter, they might be forced to sacrifice accuracy for speed.

Evaluating work load has been attempted in several ways. First, the severity of task can be calculated, for example, the number of aircraft dealt with per hour by an air-traffic controller (Costa, 1993) or assessments of military pilots by independent observers (Gerbert & Kemmler, 1986; Ribak *et al.*, 1983). Second, the assessment can be made by the work force itself (Braby, Harris, & Muir, 1993). Third, physiologic changes associated with mental effort can be measured (the difference between "alert" and "stressed" seems rather arbitrary here). Possible variables are heart rate and its variability (Braby *et al.*, 1993; Honzikova, Penaz, & Fiser, 1988), blood pressure (Honzikova *et al.*, 1988), body temperature (Pokorny, Blom, & Opmeer, 1988), and plasma and urinary catecholamines (Dimsdale & Ziegler, 1991; Vivoli, Bergomi, Rovesti, Carrozzi, & Vezzori, 1993). Correlations between these variables and between these and the other estimates of work load can be high and positive (Braby *et al.*, 1993) but this close matching is not invariably the case (Dimsdale & Zeigler, 1991).

The lack of a high correlation between a stimulus and the individual's responses to it can be a useful indicator. Thus, as the work load increases, so, too, should the subjective and objective responses of the individual. If the response is too high (and "stressed" seems a useful concept here), or if it is too low—through overconfidence, lack of attention, or boredom—then accidents and errors become more likely. A highly motivated individual appears to balance effort and work load (Gaillard, 1993).

Other Factors Affecting Mental Performance

Several other factors influence mental performance. The effect of type of task, particularly how much short-term memory it entails, has already been mentioned, but the same task can be associated with a different frequency of errors according to the circumstances under which it is performed. For example, the number of errors made by air-traffic controllers increased with increasing work load imposed externally, but there was no increase in error frequency if the increased workload were self-imposed (Langan-Fox & Empson, 1985). The possibility that this difference reflected motivation was not investigated. Such a finding, if widely applicable, would have considerable implications for e.g., nurses keeping vigil or for those whose job is paced by a conveyer belt.

In studies attempting to assess human performance, the effects of the environment are generally controlled or corrected for, of course. Nevertheless, the role of the environment should not be underestimated. Effects of lighting and noise have already been mentioned, but the further example (Vivoli *et al.*, 1993) of rises in urinary catecholamines and blood pressure in truck drivers in fog and traffic jams—levels high enough to be considered as indicators of stress—serves to emphasize this point.

The parallelism between simple repetitive tasks (for example, visual search) and body temperature has been mentioned already (see Figure 1). Is this parallelism evidence of a causal link? Certainly it would be conceivable that a raised temperature would promote the biochemical reactions occurring in the brain, but does the evidence support the causal hypothesis? Many researchers believe that it does, but Wilkinson (1982) argued that there is not a causal link, basing his case on the failure to find a significant positive correlation between body temperature and mental performance scores taken at the same time, particularly during night work. However, if brain temperature were reflected in performance only after a delay (for biochemical reactions to occur, for instance), then such a positive correlation need not exist, and, as noted below, the interference caused by fatigue might be greater during night work. Moreover, experiments involving raising or lowering body temperature to produce predictable changes of mental performance are complex to interpret due to possible effects of stress (Minors & Waterhouse, 1981). Thus, both producing a low-grade fever in individuals and cooling them by immersion in cold water are likely to change motivation! A similar conclusion that there might be a causal relationship, but that it has not been proved, can be drawn when the time courses of plasma catecholamines level and mental performance tests are compared.

Considering body temperature, catecholamine and mental performance together, the concept arises of "general arousal" being a link between them all. This can be coordinated quite readily with another widely held concept, namely the "inverted-U" relationship between arousal and mental performance (Yerkes–Dodson law). In the absence of other factors such as fatigue, therefore, performance will increase until late in the afternoon, when body temperature and catecholamine levels peak in parallel to raised arousal at this time. As already described (and see Gaillard, 1993), estimating the arousal that should be produced by a stimulus, thereby drawing attention to an abnormally high or low response for a particular task, might be a better predictor of situations likely to produce human error.

Tasks with a high short-term memory component show inverse circadian rhythms (Figure 1). Does this indicate that this task is controlled by a different oscillator? Such a conclusion cannot be drawn on the basis of data obtained from entrained subjects since the difference in phase might reflect different time courses in the output pathways. By contrast, on non-24-hour "days," a forced internal desynchronization has been reported (Folkard, Hume, Minors, Waterhouse, & Watson, 1985). Normally this would be regarded as a strong argument for the presence of more than one oscillator, but mental performance data were collected only during waking hours, and the effect of this upon estimates of phase and period can be difficult to determine precisely. It is exactly this problem that has led to a reappraisal of the relationship. This was achieved by testing subjects during 36 hours of continuous wakefulness on a constant routine, and during forced desynchronization on non-24-hours "days" (Boivin *et al.*, 1997; Johnson *et al.*, 1992). In the first protocol, changes in temperature and performance are influenced less than normal by interference from sleep and activity; in the second, the circadian and imposed rhythms take up all possible phase relationships to each other and so the two rhythms can be separated. These studies found a parallelism between temperature and short-term memory. In short, therefore, the case for short-term memory involving a separate internal oscillator is not as strong as once believed.

An alternative approach to the problem of establishing links, but this time between different mental performance tests, has used sets of data obtained under the same circumstances—ideally simultaneously—but for different mental tasks (Folkard, Marks, & Froberg, 1986). Such a factor or cluster analysis draws attention to those variables that appear to behave similarly, but it cannot determine if this is due to causal links or to some other factor (fatigue, for example) common to them. Perhaps it is worth remembering that the demonstration of an apparent ultradian rhythm shifting between left and right hemisphere dominance indicates that mental performance tests, at least on these criteria, could be divided into two main groups. That is, it might be possible to group mental performance tests according to more than one criterion, but how far this establishes causality is unknown.

Physical Performance

In many occupations, there is a strong physical component involving short-term activity that is often mainly anaerobic (gripping objects or lifting them) or longer term activity that has a largely aerobic component (for example, mining and construction industries). There do not seem to have been systematic studies of such work forces in the field, though studies upon athletes and nonathletes performing exercise are far more common. They have been reviewed several times (Atkinson & Reilly, 1996; Reilly, Atkinson, & Waterhouse, 1997b; Shephard, 1984; Winget, deRoshia, & Holley, 1985).

Studies upon the effects of time of day on the responses to exercise have normally concentrated upon maximal or near-maximal performance; their application to working conditions is not direct, therefore, since the physical work load is generally well below maximal values but is continued intermittently for extended periods of time. Even so, the results are likely to be some guide.

There is agreement that athletic performance capability is poorest at night and improves throughout the day to peak in the late afternoon or early evening. While this is conducive to best results at times when most competitions are held, it suggests that normal hours of work are taking place before this time of peak ability. (However, it is not clear to what extent there might be an earlier falloff in physical performance due to fatigue when the physical performance is submaximal and intermittent.) With anaerobic power (leg muscle or arm muscle all-out performance, for example), there is a tendency for a circadian rhythm to parallel that of rectal temperature. (Reilly & Down, 1992; Reinberg *et al.*, 1988). This is true also of the rhythm in maximum isometric strength of both arm and leg muscles.

Another common, but not invariable, finding is that perceived exertion, whether working maximally or submaximally, is greatest at night and least in the late afternoon (Coldwells, Atkinson, & Reilly, 1994; Faria & Drummond, 1982; Taylor, Gibson, Edwards, & Reilly, 1994). The perception of effort increases after successive nights of sleep loss, at least when effort is elicited by weight-lifting exercises (Reilly & Piercy, 1994). Furthermore, when the work rate is self-selected by subjects, the power output follows closely the normal circadian curve in core temperature (Atkinson, Coldwells, Reilly, & Waterhouse, 1993). It has also been claimed (O'Connor & Davis, 1992) that physical exercise improves mood and decreases anxiety at all times of the day.

One result which applies more directly to the work place is the finding that the length of the spinal column is greatest just after waking (Tyrrell, Reilly, & Troup, 1985). This is attributed to swelling of intervertebral discs during sleep, when they do not support the weight of the body. The discs normally shorten within 1 hour or so of getting up due to compressive loading of the spine. The spine further shortens during the course of a working day unless it is periodically unloaded during rest or recovery periods. The overall loss of height during the working day amounts to about 1% of total body length.

Joint flexibility (range of movement) shows marked circadian rhythmicity across a wide variety of human movements, including lumbar flexion and extension, glenohumeral lateral rotation and whole-body forward flexion (Reilly, Atkinson, & Waterhouse, 1997b). Amplitudes of these rhythms can vary by up to 20% from the 24-hour mean. Values drop to their minimum levels in the early morning and can influence physical performance in activities such as swimming (Baxter & Reilly, 1983).

In one study (Reilly & Smith, 1986), performance at a tracking task (a task where the subject is required to move a joystick to keep a point on a screen within some target area) was measured in subjects cycling at loads varying from 0% to 85% of maximal. The results showed an "inverted-U" curve, with best performance at the task being obtained with a cycling load eliciting about 40% of the subjects' maximal oxygen uptake, and cycling loads greater or less than this producing poorer results. (It is interesting to note that physical tasks often require about 40% maximal exertion to be used.) If this finding is a general one, then there is the possibility that incorporating some degree of physical activity into a task might improve performance at that task. There is also a possible link with the concept that too high or low mental work loads also produce submaximal performance.

Sustained Duty Periods

Performance Rhythms

So far, we have considered mental and physical performance almost exclusively in subjects who slept for about 8 hour during each 24-hour period. What happens to performance if work periods are prolonged with little or no sleep, as might occur in emergency and military contexts?

One group of studies has been performed upon medical staff. One of the earliest (Poulton, Hunt, Carpenter, & Edwards, 1978) showed that performance in a logical reasoning test fell if the subject had worked for more than 18 hour or had lost more than 3 hours of sleep. In another test, a series of hematology reports was devised for assessment, one fourth of which had grossly abnormal results. Sleep deficits led to a more variable rate of working rather than to an increase in errors when assessing these reports, and this lack of errors was attributed to the process of compensation. This process also is invoked with night workers, as discussed below, and implies that some extra effort has been required to achieve such compensation. A similar conclusion, that the effects of a limited degree of sleep loss could be compensated for, was reached in another study which used several mood and mental tests (Orton & Gruzelier, 1989). In a third study, requiring the assessment of electrocardiograms, efficiency again fell, with doctors taking longer to diagnose faults, rather than there being a significant increase in wrong or missed diagnoses (Friedman, Bigger, & Kornfeld, 1971). Night shift also resulted in slower task performance (Smith-Coggins, Rosekind, Hurd, & Buccino, 1994), but both day and night shifts showed increasing problems with deciding correctly the order of priorities in a complex situation as time on the shift increased.

In military circumstances, the amounts of sleep deprivation can be far higher.

In one study (Haslam, 1984), three platoons of soldiers were on maneuvers for 9 days; their sleep patterns varied, including no sleep for 3 days followed by 4 hours per day for the next 6 days, or 0, 1.5, or 3 hours of sleep per 24 hours throughout the 9-day operation. Observers measured military performance, mood ratings, and performance at cognitive tests. Those soldiers having no sleep became ineffective by the third day; any form of cooperative action or one that required sustained attention was not possible. By contrast, physical and mental tasks that were simple and well learned were not affected. Those soldiers with 1.5 hours of sleep per day became ineffective after 6 days. The other groups of soldiers continued to be effective throughout the 9 days, but performance was sustained least well in the middle of the night.

Laboratory-based tests have found similar, if not identical, results (see, for example, Rogers *et al.*, 1995). That is, circadian rhythms continue, but with falling peaks of mood and performance, and especially deep troughs of mood and performance in the middle of the night. These effects are more marked if the task is complex or demands vigilance, and rather less marked if the tasks are more cognitively demanding such as visual search. An overall decreased speed of performance and an increase in erratic performance are also common findings (compare this result with that from medical personnel, as noted above).

Summaries of field and laboratory results generally conclude that sleep loss leads to decrement in mental performance subject to the following generalizations (Angus, Pigeon, & Heslegrave, 1992; Bohnen & Gaillard, 1994; Koslowsky & Babkoff, 1992; Krueger, 1989).

1. Sleep loss of as little as 2 hours can produce decrements in performance.
2. It is often the speed of performance rather than its accuracy that is affected.
3. The type of task is important, with complex and vigilance-requiring tasks being affected most.
4. Nocturnal performance can show lapses—extreme deterioration to the point of inactivity—associated with microsleeps.
5. The decline of mood (fatigue, etc.) exacerbates the situation by causing a loss of motivation.
6. Compensation by the individual can occur, but the cost of this, particularly in the longer term, is unclear.

As has been suggested already, laboratory-based experiments do not necessarily show such a decline in performance as those in the field, a point that has been stressed by Angus *et al.* (1992). They and others have drawn attention to the fact that in field conditions, there is often no opportunity, or the circumstances are inappropriate, for the chance to recover, and so the cumulative work load is higher. Accordingly, this group devised a protocol involving continuous testing so that the repeated batteries of tests appeared to the subjects to be a long task rather than a series of short ones. In accord with their hypothesis, performance in this protocol declined much more markedly, to 50% of baseline values, after 2 days without sleep or respite.

It seems that the effects of sleep loss and the decrement in performance due to time spent on a task interact. Thus, toward the end of a 16-hour work period, the accident rate begins to rise steeply (Rosa, 1995), and deterioration in complex tasks can occur after as little as 0.5 hour if sleep loss has occurred also (Bohnen & Gaillard, 1994).

Physical performance, by contrast, seems to be less affected by sleep loss. Thus, in a football marathon for charity, activity showed a trough during the night but recovered the next day (Reilly, Atkinson, & Waterhouse, 1997b). However, this analysis concentrated on the physical component, to the exclusion of skill and tactics. In practice, these are important also and in this as in other sports (especially, for example, basketball and rifle shooting), factors in addition to muscle strength are crucial. It would be predicted that these components would deteriorate with sleep loss, though field studies have not been carried out.

The Role of Naps

The above studies stress that with some tasks, the decrement can be considerable. The most obvious way to reduce this decrement would seem to be by taking short sleep or naps. This has been stressed by Haslam (1984), and applies equally in the laboratory environment.

It is not unusual for adults, particularly elderly ones, to nap during the daytime, but it seems that all subjects benefit from naps during prolonged work hours. Thus, a 2-hour nap after a vigil of 45 or 53 hours resulted in a considerable degree of recuperation (Naitoh, 1981), and a 4-hour nap substantially reversed the decrement during the daytime following a night without sleep. Dividing a recovery sleep of 8 hours into two 4-hour naps taken 12 hours apart was even more effective (Nicholson *et al.*, 1985). Naps much shorter than this can be of some use; a 1-hour nap taken at 0200 during a night without sleep reduced the overnight decrement in auditory vigilance, though not in more complex visual vigilance and logical thought tests (Rogers, Spencer, Stone, & Nicholson, 1989). Since there are many types of mental performance, and their rates of decrement vary with increasing duration of performance (see Figure 1), it is not surprising that not all decrements appear to be reversed or reduced equally easily.

There have been attempts to see if a routine of more frequent, shorter naps is as effective as less frequent, longer sleeps (Naitoh, 1992; Stampi, Broughton, Mullington, Rivers, & Campos, 1990) in maintaining mental performance. Put differently, is our 8 hour/16 hour sleep–active day the ideal, or would two equal 12-hour "days" (4 hour/8 hour sleep–active day) or even shorter days (involving multiple sleeps each of 2 hours, 1 hour, or even 20 minutes) be equally effective? In one sense, one would predict that the shorter the day, the less would the decrement in mental performance, this reflecting the earlier point that performance, particularly in very repetitive or complex tasks or in those requiring vigilance, falls off with time spent on the task. There is, however, another effect of sleep and naps that tends to offset the advantage of many short sleeps; this is sleep inertia, (Campbell, 1992; Dinges, 1992; Naitoh, Kelly, & Babkoff, 1993; Stampi *et al.*, 1990), which appears to consist of two components. The more immediate one is the phenomenon whereby, immediately after waking from a nap, particularly one during the night, performance falls for some minutes after being awakened. The effect seems to be more marked if the subject awakens from slow-wave sleep and/or in a quiet rather than noisy environment (Stampi *et al.*, 1990). It tends to be much less important after longer sleeps or spontaneous awakening since these are rarely from slow-wave sleep. The effect is generally found to wear off about 10 minutes after waking up but there are clear implications here for workers sleeping in the workplace. The second effect of sleep inertia is that improved mental performance (due to the loss of fatigue) is achieved only after a longer period, about 1 hour, after waking up. This is an important component in some models of mental performance, as discussed later. Thus, days that are too short show poorer performance because sleep inertia effects offset the decrease of fatigue due to having slept. Nevertheless, a routine involving multiple naps might be useful in some contexts, particularly night work; the timing of such naps then becomes an important factor, and will also be discussed later.

Naps can be described as "recuperative" (the sense used here so far), but also as "appetitive or "prophylactic." Recuperative naps are taken to relieve fatigue due to lost sleep or activity, or both; appetitive naps are those taken due to habit, even if the subject does not feel sleepy; prophylactic naps are those taken before a period of work in order to start as refreshed as possible (Dinges, 1992). There is no evidence as to whether they differ in their effects (described above). There is a further role that short sleeps can play; they can act as an "anchor sleep" (Minors & Waterhouse, 1982; Stampi *et al.*, 1990). This role applies far more to air crews and so will be considered later.

Some studies have considered the value and ease of taking naps at different times of the day and any differences in the sleep-stage architecture that this produces. This will be considered after a brief account of these factors in sleeps of a more conventional length.

Factors Influencing Sleep

Assessing Sleep. Since sleep loss appears to be an important factor in determining mental performance, assessing the amount and quality of sleep is highly desirable. The simplest means is to use a sleep diary, but this suffers from the obvious disadvantages that it is subjective and filled in retrospectively. Even so, it is often the only means available in field conditions. Alternatively, a Stanford Sleepiness Scale can be used (Carskadon & Dement, 1975). At the other extreme is polygraphic recording of sleep and its stages (the EEG), and this has become less impracticable recently with the development of ambulatory devices for recording the EEG (see, for example, Torsvall *et al.*, 1989). Somewhere between these extremes is the use of movement sensors that are worn on the nondominant wrist and record movement (Brown, Smolensky, Alonzo, & Redman, 1990; Middelkoop, Van Dam, Smilde Vanden Doel, & Van Dijk, 1997). These have been used to estimate sleep times and, particularly if the times of retiring and rising are known independently, the amount of waking and movement during a sleep. It can be seen that, as with the measurement of mental performance, a compromise often has to be struck between what is possible in practice and desirable in theory.

Sleep Quantity and Quality. This topic has been dealt with fully in several review articles (Borbély & Ackermann, 1992; Dijk & Czeisler, 1995; Strogatz, Kronauer, & Czeisler, 1986). The salient features as they apply to individuals on normal and abnormal sleep/wake schedules are summarized below.

1. The ease of getting to sleep and staying asleep shows a circadian variation that is linked to the rhythm of body temperature (Zulley, Wever, & Aschoff, 1981). Sleep onset is easiest to initiate at or just before the temperature trough and most difficult at or just before the peak, when waking is most likely to occur spontaneously.

2. The propensity to fall asleep has been investigated in finer detail by the protocol known as the Multiple Sleep Latency Test (MSLT). This requires subjects to live ultrashort "days," of 20 minutes or 4 hours, for example, and to remain in

constant bed-rest conditions, but to divide the time between sleep and awake (Lavie, 1986; Zulley, 1990), for example, 7 minutes/13 minutes in the first protocol and 1.5 hours/2.5 hours in the second. The total amount of sleep achieved as a proportion of that available can then be assessed over the 24 hours. These protocols confirm that nighttime (when the body temperature is low) is most propitious for sleep, but there is a smaller peak in the earlier afternoon (Borbély, Ackermann, Trachsel, & Tobler, 1989). (Compare this with the post-lunch dip and other ultradian rhythms of mental performance noted earlier.)

3. The amount of slow-wave sleep has been shown on many occasions to be a function of prior wake time rather than circadian phase (Daan, Beersma, & Borbély, 1984). More recent investigations have tended to use delta wave activity, assessed from the sleep EEG, with the same results. This has led to sleep models that propose the accumulation of a factor during waking and its dissipation, in association with delta wave activity, during sleep.

4. The distribution of REM (rapid eye movement) sleep shows a circadian rhythm, the propensity being inversely related to body temperature (Borbély Ackermann, 1992). As a consequence of this and feature 3 above, during a normal night's sleep, the early REM–nonREM cycles tend to be richer in slow-wave sleep and the later ones in REM sleep.

Naps appear to obey very similar rules with regard to ease of initiating sleep and its composition. However, brief naps will be too short to allow slow-wave sleep to be reached and so contain mainly stage 2 sleep (Campbell, 1992). Therefore, taking naps, while an very effective way of reducing performance decrement, might be an inefficient way of doing so if they are too brief. Accepting that sleep loss might occur, it is also appropriate to note that Wilkinson (1972) found that deterioration in performance of a vigilance test did not occur in a group of subjects until their regular sleep duration at night had been reduced from 8 hours to 5.5 hours. This can be interpreted to indicate that we normally sleep longer than necessary and this would fit in with the concept of dividing sleep into obligatory and facultative portions, the optional part coming second in a normal night (Horne, 1985). However, mental performance data are notoriously noisy, and decrements have to be substantial before they become statistically significant.

Irregular and Changed Sleep–Wake Schedules

Can these general rules about sleep and naps, and the earlier concepts relating to other factors affecting mental and physical performance, be applied to those occasions when individuals work and sleep at abnormal hours (that is, not nocturnal sleep and diurnal rest and activity), when their sleep activity cycles change frequently, or when a change is imposed upon them by time-zone transitions?

At the outset it must be realized that such studies are uncommon and/or suffer from lack of frequent and regular data sampling; the complexity of organization involved in performing such field studies should not be underestimated. As before, therefore, it is often necessary to supplement these studies with laboratory-based simulations. The difficulty of assessing mental performance accurately in the field has meant that most assessments are of mood or use sleep diaries, though the situation is beginning to change as portable means of measuring mental performance and wrist movement and the use of sleep EEG become more common.

Work hours on nuclear submarines are unusual in that a 6-hours-on/12-hour-off system is sometimes adopted. This acknowledges that an 8-hour turn of duty can lead to performance decrement near the end, particularly in demanding tasks associated with high degrees of responsibility (Beare, Bondi, Biersner, & Naitoh, 1981). Unfortunately, such a schedule is incompatible with entrainment of the body clock, and the quantity of sleep as well as self-assessed mood both decline. The decrement in performance at complex tasks that these results imply has not been reported in the scientific literature.

A similar schedule (insofar as it involves short duty periods and a cycle length that differs from 24-hours) is followed by watch-keepers in some merchant ships. There are generally two duty periods, and two opportunities to sleep, each "day." Results are similar to those found in laboratory simulations. Thus, sleep is divided into two unequal lengths, the longer tending to coincide with nighttime. Mental performance at a variety of tasks and alertness both paralleled the temperature rhythm to some extent, and so took on a period different from 24 hours. However, the parallelism was not exact, with mood and performance peaking at different times, and values at the beginning of a shift being lower than those at the same time of day, but at the end of a shift. (This indicates that the effects of sleep inertia temporarily outweighed the recuperative effect of sleep.) Alertness tended to be lower than under control conditions, possibly due to cumulative sleep loss. Whereas the shorter duty hours should be advantageous, it was concluded (Colquhoun, Watson, & Candon, 1987) that the disrupted sleep pattern and sleep inertia offset this as far as overall performance was concerned.

TIME-ZONE TRANSITIONS

Accounts of the effects of time-zone transitions on sleep, mood, and mental and physical performance stress that physiologic maladjustment and jet lag, the negative feelings associated with this, are almost universal, and that they are more marked and last longer after eastward flights (for reviews, see Graeber, 1982; Reilly, Atkinson, & Waterhouse, 1997c; Waterhouse, Reilly, & Atkinson, 1997; Winget, deRoshia, Markley, & Holley, 1984). Mental and physical performance both deteriorate because the trough of the temperature rhythm coincides with the new local day, but mental performance also suffers because of poorer sleep (again associated with the unadjusted temperature rhythm). The double negative of performance after sleep loss and at the trough of the temperature rhythm can cause substantial decrements. Different rhythms of mood and performance adjust to the new time zone at different rates and often adjust to an eastward time-zone transition of more than nine time zones by a delay rather than an advance. It seems likely that these differences reflect different influences of sleep loss and time awake upon the rhythms rather than the presence of more than one endogenous oscillator.

These difficulties are not uniformly seen in all travelers. Thus, they are worse in middle age than in youth (Moline *et al.*, 1992) and in some personality types (neurotic introverts) than others (neurotic extroverts) (Colquhoun, 1984). Both issues apply to shift workers also and will be considered further in the next subsections.

Even though the few studies on physical and exercise performance that have been carried out suggest that any effects are small (O'Connor & Morgan, 1990), two other factors should be borne in mind. First, very small changes in maximum performance can be the difference between losing a contest and creating a new world record; second, negative effects upon mood and perceived exertion (that is, a loss of "feeling good" and an increased sense of effort) can detract from training and undermine the process of building up psychologically for an event (Reilly, 1990; Reilly & Mellor, 1988; Shephard, 1984; Winget *et al.*, 1985).

Night Work

The possibility that a night worker might sleep on duty at night is as real as it is natural, and produces the fear that it might contribute to major accidents. There is evidence that this does occur. First, EEG recordings indicate that drowsing can occur, even overt sleep episodes (Åkerstedt *et al.*, 1987; Kecklund & Åkerstedt, 1993; Torsvall *et al.*, 1989; Wedderburn, 1987). Second, a rather unusual phenomenon—night shift paralysis—can occur, by which an individual, though apparently awake, nevertheless become immobile and is unable to respond to the environment. It lasts from a few second to minutes and has been described in nurses (Folkard *et al.*, 1984) and air-traffic controllers (Folkard & Condon, 1987). It appears to affect 3–6% of the working population, and is considered to be a form of sleep paralysis. These phenomena do not seem to have been observed in laboratory studies, though extremely long delays in tests of reaction time or missed signals in vigilance tests are both found and might share some affinity. (This failure to identify unambiguously an exact equivalence serves as reminder that field and laboratory studies are not identical.) This paralysis, coupled with the evidence of microsleeps, is a clear indication that humans are diurnal rather than nocturnal creatures. Less dramatic, but almost universal, is the finding that fatigue is higher and motivation lower on the night shift. This reflects not only the troughs of plasma epinephrine and body temperature rhythms at this time, but also the fact that daytime sleep is more difficult, due to both external (noise, light) and internal (body clock) factors. As a result of this combination, the common finding that productivity falls and performance declines during night shifts is hardly surprising (Monk & Folkard, 1985a). Since a morning shift requires a very early time of waking up, this, too, can produce sleep loss and has been associated with an increased frequency of accidents. Sleep, performance, and alertness, deteriorated in another study in which 8-hour shifts were changed to a shorter number of 12-hour shifts, presumably due to the increased possibility of "fatigue" during longer shifts (Rosa & Bonnet, 1993). All these results would be predicted from the factors discussed above, but others have stressed that changing shift patterns in accord with the wishes of the work force can lead to greater tolerance and job satisfaction in spite of these factors (Barton, Smith, Totterdell, Spelter, & Folkard, 1993; Williamson, Gower, & Clarke, 1994).

Night workers attempt to overcome these problems, particularly if they work at night frequently or permanently. One method is to increase the frequency of naps–to "top up" sleep–during the daytime, at a time or times that are most acceptable to the individual's lifestyle. These naps, particularly if taken before the night shift, improve alertness and mental performance during the night shift (Harma, Knauth, & Ilmarinen, 1989). Another method which has alleriated problems arising from fatigue during night work has been to schedule a 2-hour nap during the work period (Matsumota & Harada, 1994; Muzet, Nicholas, Tassi, Dewasmes, & Bonneau, 1995). A third possibility would be to adjust circadian rhythms to night work. This is generally accepted as being a slow process (Waterhouse *et al.*, 1992) and so is not viable unless individuals are on long periods of night work and/or means to promote adjustment are implemented. However, reports of a rapid, even immediate, adjustment of some variables (heart rate, blood pressure) to night work exist (see, for example, Sundberg, Kohvakka, & Gordin, 1988). This result is not really incompatible with the conventional view; it is likely that the body clock adjusts slowly, but that rhythms which have large exogenous components (heart rate and blood pressure, for example) will follow the individual's lifestyle with regard to overall timing. This rapid apparent adjustment will be beneficial to nocturnal performance, as has been commented on with air-traffic controllers (Lille & Cheliont, 1982). It has also been reported (Folkard, Knauth, Monk, & Rutenfranz, 1976) that tasks involving a high short-term memory load adjust more rapidly to night work.

In shift work, as with individuals who are sleep-deprived but on a conventional schedule, therefore, many factors influence mental performance. Some idea of the complexity can be inferred from a study in which adjustment during the course of a week of night work was investigated (Vidacek, Kaliterna, Radosevic-Vidacek, & Folkard, 1986). A group of 53 workers making capacitors (a repetitive task requiring manual dexterity and perceptual–motor coordination) was studied with respect to their cumulative sleep loss and productivity. Initially, productivity was low, but this improved to day-shift values by the third night, only to decrease on nights four and five. The interpretation placed on these results, which would accord with the concepts expressed above, was that the initial improvement was due to some adjustment to night work, but that this was later overridden by the effects of cumulative sleep loss. Indeed, this detrimental effect of accumulating sleep loss is often adduced as a reason for preferring the rapid rotation of shifts.

Physical performance on the night shift is not likely to deteriorate sufficiently for it to cause any problems (because maximum muscle power is rarely, if ever, used). Nevertheless, the negative effects of mood and the increase in perceived exertion for a given task are likely to add to the difficulties of the night shift.

Pilots and Air Crew

Finally, we consider those whose occupation requires them to undergo continual time-zone transitions coupled with irregular hours of work—air crew on long-haul flights.

Laboratory-based simulations of the irregular schedules worked by these groups have found a decreased efficiency of sleep, desynchronization between circadian rhythms, and decline of performance. This decline depended on the type of mental performance test and was attributed to the combination of working near the temperature trough, working extended hours, and cumulative sleep loss. However, the relationship between such results and those in the field has been questioned, particularly when the nature of the tasks required of pilots (a large amount of vigilance) and the relationship between poor performance in a laboratory test and the possibility of error or accident are considered (Craig, Wilkinson, & Colquhoun, 1981b; Folkard, 1990; Folkard & Monk, 1985; Hamelin, 1987; Kirkcaldy, 1984; Nicholson, Stone, Borland, & Spencer, 1984; Pokorny *et al.*, 1987; Wojtczak-Jaroszowa & Jarosz, 1987).

In his review of the reports from the confidential reporting system for airborne incidents, Green (1985) stressed the potential importance of sleep loss and sleep disruption, as do all reviews of the area (see for example, Graeber, Lauber, Connell, & Gander, 1986). Sleep during layovers has been assessed from air crews' logs, but also by the MSLT and by wrist monitors. Nevertheless, the preferred method is the EEG. All studies find that fatigue increases, sleep is poorer and fractionated, and sleep efficiency is decreased (for example, Graeber *et al.*, 1986; Samel & Wegmann, 1988; Sasaki, Kurosaki, Spinweber, Graeber, & Takahashi, 1993; Spencer, Stone, Rogers, & Nicholson, 1991); in addition, the cumulative loss of sleep is more marked with age (Gander, Nguyen, Rosekind, & Connell, 1993). As with passengers, there is also the common finding that problems are worse after an eastward flight than after one to a west. Two factors are thought to contribute to this. First, an eastward flight presents the opportunity of a sleep that is too early, and air crew are undecided whether or not to take this; after the westward flight, by contrast, a delayed sleep is invariably taken (this is shown very clearly in one large study, Graeber *et al.*, 1986). Second, adjustment to an eastward shift requires an advance of the body clock, which is more difficult, possibly because the natural period of the body clock is slightly in excess of 24 hours.

When sleep-stage architecture is considered, the normal associations between slow-wave sleep and prior wake time, and between REM sleep and the temperature minimum, tend to hold, though fractionated sleep tends to decrease the total amount of these (at least temporarily) in favor of stages 1 and 2 (Sasaki *et al.*, 1993).

In spite of these difficulties with sleep, it is clear that air crew make every attempt to minimize the cumulative loss. This is achieved to varying extents by the almost uniform practice of taking naps. Such naps appear to be taken just after a duty period ("recuperative"), just before ("prophylactic"), or at nighttime or home time ("anchor sleep," or "appetitive") (Rosekind, Gander, & Dinges, 1991). "Anchor sleep" (Minors & Waterhouse, 1982; Stampi *et al.*, 1990) is the regular timing of a portion of sleep, generally at a time coincident with night on home time. It seems to stabilize circadian rhythms, though whether it is the regular sleep itself or other regularities of lifestyle that it will produce (particularly the light/dark cycle) is not known.

Scheduled in-flight sleeps have been introduced for different parts of the crew at different times (Rosekind *et al.*, 1992). This improved performance on tests of reaction time and attention, and there were fewer attention lapses (the tests were performed in-flight during a period of low work load) than with air crew who took no naps. But there was also the problem of "sleep inertia." It remains to be seen how naps will be incorporated into such schemes if they become more widespread. Which aspect of sleep loss (recuperative, anchor sleep, etc.) will they address, and how will sleep inertia be dealt with?

Performance testing is rare in these studies for reasons of practicality, though one study has just been described (above). Other studies made during air crew layovers (Beh & McLaughlin, 1991) or in a simulator (Foushee, Lauber, Baetge & Acomb, 1986; French, Bisson, Neville, Mitcha, & Storm, 1994) tend to support predictions about decrements following sleep loss and/or measured at the trough of the circadian rhythm of body temperature. In addition, the warning of Green (1985) given above indicates clearly enough the problems as seen by the air crew members themselves. The subjective element was important also in a simulator study (Foushee *et al.*, 1986) and has been mentioned elsewhere (Rosekind *et al.*, 1992); namely, a tired crew might be irritable and cooperate inefficiently, whereas a motivated and friendly crew (unless it becomes too familiar and careless) will be the converse.

A study by Samel, Wegmann, and Vejvoda (1995) found objective evidence (EEG measurements) of "critical decrements" in air crew during long-haul flights, particularly ones to the east.

Differences between Individuals in Adjusting to Irregular Sleep–Activity Schedules

Many studies on night work stress that individuals respond differently to its demands. In part, this might indicate the different degrees to which coping mechanisms are employed successfully (see below), but it might also reflect the view that individuals differ intrinsically in their ability to adjust to night work (Costa, Lievore, Casaletti, Gaffuri, & Folkard, 1989; Harma, 1995; Monk & Folkard, 1985b). Some examples of these differences are as follows.

Rigidity versus Flexibility of Sleeping Habits. The advantage of being a "napper" rather than a "nonnapper" can be inferred from what has already been described; more generally, it is disadvantageous if an individual cannot take the opportunity to sleep whenever it is presented.

Ability to Throw off Feelings of Fatigue. Inability in this is disadvantageous to night workers. There is a dearth of reports from workers describing how they combat such feelings of fatigue. In one study of lorry drivers, methods included taking a brief break or nap, opening windows, scratching or slapping oneself, or eating sweets or slices of lemon (Wedderburn, 1987).

Morning versus Evening Types. These are defined with reference to the Horne and Ostberg questionnaire, which deals with individuals' preferred habits with respect to times of sleep, times when they feel most or least able to perform mentally or physically demanding tasks, and times when they prefer to be relaxing. Though there is evidence that the interpretation of an individual's score requires a knowledge of his or her cultural surroundings, it has been a consistent finding that morning types are less suited to night work, but more suited to the morning shift, than are evening types. Thus, evening types show a greater sleep propensity in the morning (as assessed by the MSLT) and a rise in sleep propensity later in the evening (Lavie & Segal, 1989), a later phasing of temperature, cardiovascular, and performance circadian rhythms (Kerkhof, 1985; Vidacek, Kaliterna, Radosevic-Vidacek, & Folkard, 1988), and a smaller incidence of health disorders when on night shifts (Hildebrandt & Strattman, 1979).

Age, Coping Mechanisms, and Commitment. Aging is associated with a decreasing tolerance to shift work, with decreasing flexibility in sleep times and decreasing physical fitness no doubt being contributory factors (Harma, 1995; Harma, Hakola, & Laitinen, 1992; Harma, Hakola, Åkerstedt, & Laitinen, 1994; Reilly, Waterhouse, & Atkinson, 1997d). Against this, however, must be set the value of increased experience, particularly if coping strategies are considered. These strategies include an acceptance of a nonconventional social and family life (Adams, Folkard, & Young, 1986), the adoption of suitable methods to gain enough sleep (Tepas, 1982), and methods for overcoming tiredness (Wedderburn, 1987).

These factors have also been described as a "commitment" to night work (Adams *et al.*, 1986).

In summary, a worker who initially possessed the appropriate attributes and/or developed the necessary coping mechanisms would not only be predicted to be the most tolerant of night work, but also to be safest and most productive, even though these last two points do not seem to have been investigated. However, such compensatory adjustments might extract a cost from the individual and lead to a decline in health (Waterhouse *et al.*, 1992).

A few words of caution are appropriate here. In the absence of long-term follow-up studies, it is unclear whether an individual who is tolerant of night work or time-zone transitions is such because of an innate ability to adjust or because the lack of such an ability has been compensated by the development of appropriate changes in lifestyle. What is clear is the converse, namely, that those who need to develop such mechanisms but cannot, or do not wish to, are intolerant of night work and, if tested, will show inflexible sleeping habits, etc. Another implication of this is that tolerant night workers and time-zone travelers are survivors of a self-selection process and that this makes them unrepresentative of the population as a whole. The statistical problems produced by this are complex and point to the need to study individuals longitudinally.

Predicting Low Alertness, High Work Load, and Poor Performance

As already discussed, there is considerable agreement between field and laboratory studies with regard to the factors that influence performance in individuals on conventional or altered sleep–wake schedules. How accurately, therefore, can we predict circumstances which would be likely to lead to "double negatives" or, worse, to a poor and even dangerous performance?

There have been several attempts to construct mathematical models to predict performance or alertness or both, all based on a circadian component plus, usually, a homeostatic one and, in one case, also a sleep inertia component. Since the model based on alertness is somewhat more comprehensive, including also predictions of sleep length, sleep latency and risk levels, we will choose that model for presentation here.

The work on this "three-process model of alertness regulation" was inspired by the "two-process model of sleep regulation," which had shown that sleep length and slow-wave activity could be described by a combination of homeostatic (prior time awake or amount of prior sleep) and circadian influences (Borbély, 1982). Using subjective alertness data from a number of experiments of altered sleep/wake patterns, it was found that alertness also was predicted from the combination of a circadian and a homeostatic component, plus a component for sleep inertia (see below for details). The output of the model has been validated against subjective and objective ratings of performance and EEG and electrooculogram (EOG) measures of sleepiness in the laboratory and in the field, and has shown considerable accuracy (Folkard & Åkerstedt, 1987, 1991). It also includes estimates of the risk of sleep and thus when total performance failure may ensue. The model has been modified to predict also sleep length and sleep latency in laboratory and field studies (Åkerstedt, 1995a,b; Åkerstedt & Folkard, 1995).

Three parameters were obtained from the modeling work (for details, see Figure 2). Process C represents sleepiness due to circadian influences and has a sinusoidal form with an afternoon peak. Process S is an exponential function of time-since-awakening, is high on awakening, falls rapidly initially, and gradually approaches a lower asymptote. At sleep onset, process S is reversed and called S′ and recovery occurs in an exponential fashion that initially increases very rapidly, but subsequently levels off toward an upper asymptote. Total recovery is usually accom-

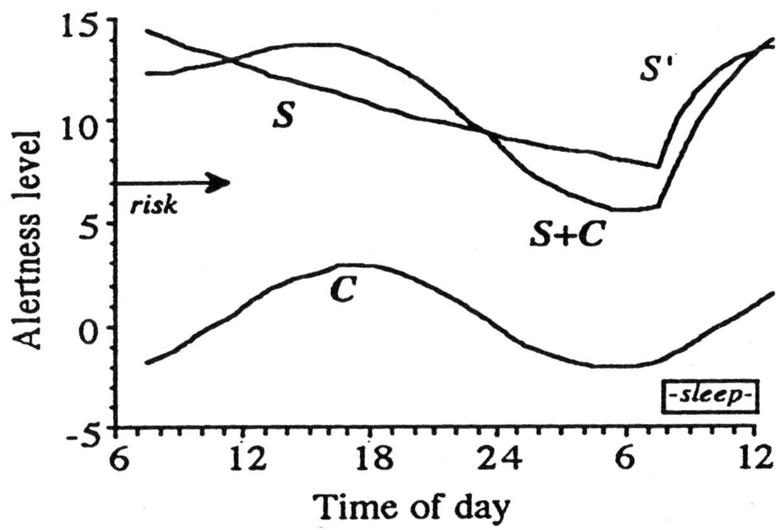

Figure 2. Parameters of the three-process model of alertness regulation. S, homeostatic component during waking; S′, homeostatic component during sleep; C, circadian component; S + C, alertness prediction (excluding sleep inertia), with values at or below 7 indicating severe risk.

plished in 8 hours. The final component is the exponential wakeup process or sleep inertia, which shows a steep initial rise during the first minutes after awakening and then levels out toward an asymptote. The input to the model is the times of rising and going to bed.

The predicted alertness is expressed as the arithmetic sum of S and C (Figure 2). The scale of the model ranges normally from 1 to 16; in practice, a score of 3 corresponds to extreme sleepiness, 14 to high alertness, and 7 to a sleepiness threshold (Folkard & Åkerstedt, 1991). The last refers to increased risk of sleep-onset changes in the EEG and EOG, as well as accompanying performance failure. Figure 3 shows predicted alertness when wakefulness is extended by 8 hours (to 24 hours), as is frequently the case, for example, with a first night shift. This prediction assumes that awakening occurs at 0700 in the morning after an 8–hour sleep, and thereafter no sleep occurs (due to the night shift) until 0700 the following day. The combined effect of S+C (long time awake and the circadian downswing) yields a fall of alertness during the night, with a trough in the early morning, to the risk level. After sleep is started, the steep recovery of factor S′, together with the circadian upswing, causes a rapid increase in (latent) alertness during sleep.

Several other authors have developed more or less elaborate models of the regulation of alertness and/or performance. Thus, Monk (1987), working with subjective alertness and forced desynchronization, found one circadian influence related to body temperature and another related to time-since-sleep. It appears, however, that the latter could be interpreted as a combination of time-since-sleep and sleep inertia effects (factors S and W above). Minors and Waterhouse (1987) presented a model similar to the three-process model, based on time of day, time-since-sleep, and subjective quality of prior sleep or nap. The homeostatic effects were linear, however, and the parameters have not been further validated. Folkard and Totterdell (1993) applied an approach similar to that of the three-process model

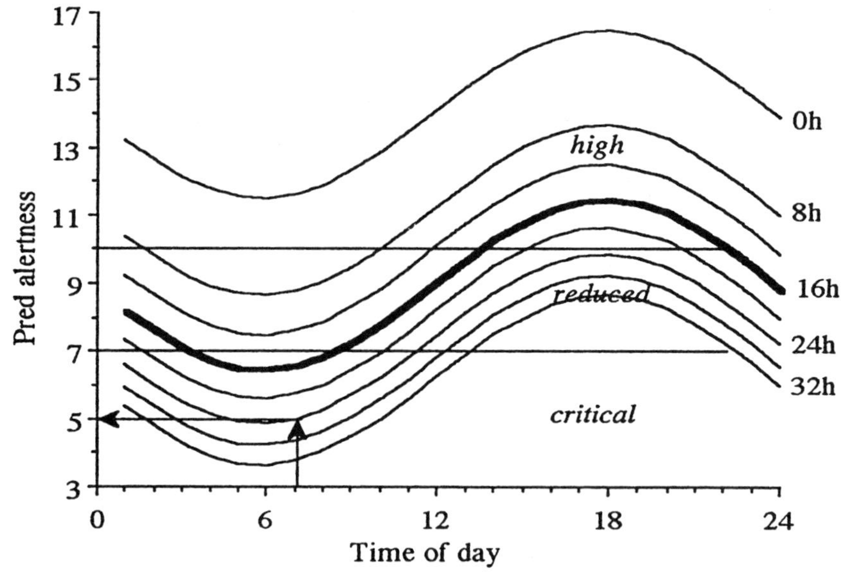

Figure 3. Alertness nomogram. Each curve represents the circadian variation of alertness after a certain number of hours of prior waking. The arrows describe the use of the nomogram. Labels "critical" to "high" refer to verbal interpretation of predicted levels.

of alertness to performance data. In a somewhat different approach, Spencer (1987) found that performance on a "digit symbol substitution test" could be explained mainly by a combination of sinusoidal and cubic components. The latter appears to be a combination of factors S and W of the alertness model (S' was not estimated). It has not been validated further, however.

The value of such models is their predictive power. The predictions of the two-factor model (S + C, Figure 3) have been compared with the results of the field studies. Even though prediction and field observations have generally been in accord (Åkerstedt, 1998), the main conclusion has been that field data are sparse and/or not of the highest quality. Recently, we have attempted a systematic survey of the effects of phase adjustment of the circadian oscillator to night work and of taking two sleep periods during the daytime rather than a single sleep period (Waterhouse, J. M., Åkerstedt T., and Fischer, F. unpublished). The main predictions have been, first, to stress the value of phase adjustment in reducing the duration of alertness below the sleepiness threshold and, second, to stress the value of taking at least one sleep period as close as possible before the start of the night shift. In practice, such a nap would end more than 1 hour before the shift start, and so effects of sleep inertia would be finished, and the sleep would be "prophylactic" in nature.

Summary: Overcoming the Difficulties

By way of a summary of the factors affecting performance, we offer an outline of some of the means of reducing this decrement. They have often been covered or suggested in other sections of this chapter, and comment and citations will be added only if the issue has not already been covered here. Reviews of methods used in military aviation (Ferrer, Bisson, & French, 1995; Rosekind *et al.*, 1995), civil aviation

(Samel *et al*, 1995), and space flight (Santy, Faulk, & Davis, 1994; Stampi, 1994) contexts have recently been published.

The main approaches to reducing performance decrement, which necessarily stress night work, are as follows. Some of the points are mutually antagonistic, but this stresses the complexity of the problem and the need for discussion and compromise.

1. Improve conditions of work (lighting, noise).
2. Alternate between tasks, particularly if vigilance is required and the task is repetitive.
3. Warn some individuals (e.g., rigid sleepers, morning types) against night work [but see Harrison, Bright, & Horne (1996), for evidence that individuals can be motivated to fall asleep more quickly].
4. Make the morning shift start later and the night shift end earlier (Keckland & Åkerstedt, 1995).
5. Reduce hours of work, particularly overnight, unless the work load is light.
6. Offer advice on how to overcome feelings of fatigue during a period of work.
7. Reduce sleep loss by the use of short-acting hypnotics. Initiating sleep is less of a problem than ensuring that, at its termination, whether spontaneously or not, performance and alertness are unimpaired. This would be particularly important with unpredictable sleep/activity schedules as in military contexts, for example. Drugs of the benzodiazepine group have been tried with considerable success (see for example, Kanno, Watanabe, & Kazamatsuri, 1992; Reilly, Atkinson, & Budgett, 1997a; Stone & Turner, 1997; Walsh *et al.*, 1991). Promoting alertness during the activity span is a natural adjunct, but research here is still in its infancy and a suitable drug, one that does not have undesirable side effects and is nonaddictive, has yet to be found (Åkerstedt & Ficca, 1997). Caffeine is widely used, its effectiveness having been established in conjunction with naps (Bonnet & Arand, 1994), hypnotics (Walsh, Muehlbach, & Schweitzer, 1995), or bright light (Wright, Badia, Myers, & Plenzler, 1997).
8. Reduce sleep loss by the use of naps during hours off duty. The timing of these must depend upon the individual's convenience, but advice as outlined at the end of the preceding section should be valuable.
9. Reduce performance decrement and sleep loss by scheduling a nap time during a work period (Smith & Wilson, 1990).
10. Where possible, produce a shift system that is least antagonistic to the properties of the circadian system. In practice, this will either mean a rapid rotation of shifts (to which there can be minimal cumulative sleep loss and adjustment) or a slow rotation in the delaying direction (morning shift/afternoon shift/night shift) with rest days after night work (to which adjustment can be made) (Czeisler, Moore-Ede, & Coleman, 1982).
11. Promote adjustment of circadian rhythms to a slow rotation of shifts, or stabilize circadian rhythms while on irregular or rapidly rotating shift schedules.

Whichever option is chosen, this requires the appropriate timing of zeitgebers, as has been described in another chapter. So far, the major method has been that of scheduling exposure to bright light and dark (see for example, Boulos *et al.*, 1995; Czeisler & Dijk, 1995; Czeisler *et al.*, 1990; Dawson, Encel, & Lushington, 1995; Eastman *et al.*, 1995a; Englund, Loving, & Kripke, 1990). As alternatives, timed ingestion of melatonin (see, for example, Arendt *et al.*, 1987; Comperatore, Lieberman, Kirby, Adams, & Crowley, 1996; Folkard, Arendt, & Clark 1993; Petrie, Dawson, Thompson, & Brook, 1993) or regular times of activity (Eastman, Hoese, Youngstedt, & Liu, 1995b) have been used. Results suggest that melatonin (which might be acting as a hypnotic) gives better results for alertness than for improvements at measures of mental or physical performance. By contrast, light appears to adjust the circadian oscillator and so shift all rhythms and improve mood and performance. Even so, it is important also to realize that part of the protocol advocates dim light, and this can clash with safety considerations in the work place (see suggestion 1, above).

Short sleeps (together with the necessary concomitant of activity and even meal times) can also be used in some circumstances, when they can be considered to be acting as "anchor sleeps"; this is particularly the case on rest days in the middle of a spell of night shifts, when loss of adjustment to night work is undesirable.

In high-risk circumstances, when public safety is involved or when a combination of negative factors can occur, it is advisable to have some backup system lest the worker fails or falls asleep. Devices to assess when an individual falls asleep, or at least is failing to respond to the environment, have been tested (Gardner-Medwin, 1985). Before they receive general acceptance, the devices will need to fulfil two important criteria. First, there must be no occasions when a nap, night-shift paralysis or lapse of awareness is missed, and yet there must be very few "false positives" (when the device raises the alarm unnecessarily) for the device to retain credibility. This places considerable demands on the technology. Second, the device must be easy to operate, convenient to wear, and unobtrusive. Devices making use of the EEG, galvanic skin resistance, or eye movement are all under test, but none is yet regarded as sufficiently reliable.

Postscript: Human Error or System Error?

The assumptions that have been made in relating laboratory results, field productivity, subjective mood assessment and performance at mental and physical tasks to errors and even accidents in the workplace have been described above, but are often stressed in the literature and deserve reiteration now. In addition, it is important to remember that this causal sequence is forced upon us, not only because accidents are so rare and unpredictable, but also because measurements in the field can be impracticable and erratic.

It should also be remembered that, in efficient systems, human error should trigger backup systems and alarm systems, rather than precipitate an accident (Dorner, 1990).

Even though we must accept that submaximal mental performance is commonplace, the fact that gross errors and accidents are so much rarer indicates that such backup systems operate, even if they amount to little more than the individual being able to perceive and react to what is happening. Clearly, the problem becomes vastly more serious when marked sleepiness, or even sleep itself, intervenes. Equally clearly, detailed models to predict the likelihood of sleep onset and, particularly, means to indicate the moment when it occurs, or the individual otherwise becomes unaware of the immediate environment, are urgently needed.

References

Adams, J., Folkard, S., & Young, M. (1986). Coping strategies used by nurses on night duty. *Ergonomics, 29,* 185–196.

Adan, A. (1993). Circadian variations in psychological measures: A new classification. *Chronobiologia, 20,* 145–162.

Åkerstedt, T. (1995a). Work hours, sleepiness and accidents. Introduction and Summary. *Journal of Sleep Research, 4*(Supplement 2), 1–3.

Åkerstedt, T. (1995b). Work hours, sleepiness and the underlying mechanisms. *Journal of Sleep Research, 4*(Supplement 2), 15–22.

Åkerstedt, T. (1998). Is there an optimum sleep/wake pattern in shiftwork? *Scandinavian Journal of Work and Environmental Health, 24*(S3), 18–27.

Åkerstedt, T., & Ficca, G. (1997). Alertness-enhancing drugs as a countermeasure to fatigue in irregular work hours. *Chronobiology International, 14,* 145–158.

Åkerstedt, T., & Folkard, S. (1995). Validation of the S and C components of the three-process model of alertness regulation. *Sleep, 18,* 1–6.

Åkerstedt, T., Torsvall, L., & Gillberg, M. (1987). Sleepiness in shiftwork. A review with emphasis on continuous monitoring of EEG and ECG. *Chronobiology International, 4,* 129–140.

Andlauer, P., Rutenfranz, J., Kogi, K., Thierry, H., Vieux, N., & Duverneuil, G. (1982). Organization of night shifts in industries where public safety is at stake. *International Archives of Occupational and Environmental Health, 49,* 353–355.

Angus, R. G., Pigeon, R. A., & Heslegrave, R. J. (1992). Sustained operations studies: From the field to the laboratory. In C. Stampi (Ed.), *Why we nap* (pp. 217–241). Boston: Birkhauser.

Arendt, J., Aldhous, M., English, J., Marks, V., Arendt, J. H., Marks, M., & Folkard, S. (1987). Some effects of jet-lag and their alleviation by melatonin. *Ergonomics, 30,* 1379–1393.

Atkinson, G., & Reilly, T. (1996). Circadian variation in sports performance. *Sports Medicine, 21,* 292–312.

Atkinson, T., Coldwells, A., Reilly, T., & Waterhouse, J. (1993). Circadian rhythmicity in self-chosen work-rate. In C. Gutenbrunner, G. Hildebrandt, & R. Moog (Ed.), *Chronobiology and chronomedicine: Basic research and applications* (pp. 478–484). Frankfurt: Lang-Verlag.

Barton, J., Smith, L., Totterdell, P., Spelten, E., & Folkard, S. (1993). Does individual choice determine shift system acceptability? *Ergonomics, 36,* 93–9.

Baxter, C., & Reilly, T. (1983). Influence of time of day on all-out swimming. *British Journal of Sports Medicine, 17,* 122–127.

Beare, A. N., Bondi, K. R., Biersner, R. J., & Naitoh, P. (1981). Work and rest on nuclear submarines. *Ergonomics, 24,* 593–610.

Beh, H. C., & McLaughlin, P. J. (1991). Mental performance of air crew following layovers on transzonal flights. *Ergonomics, 34,* 123–135.

Bohnen, H. G. M., & Gaillard, A. W. K. (1994). The effects of sleep loss in a combined tracking and time estimation task. *Ergonomics, 37,* 1021–30.

Boivin, D., Czeisler, C., Dijk, D. J., Duffy, C., Folkard, S., Minors, D., Totterdell, P., & Waterhouse, J. (1997). Complex interaction of the sleep–wake cycle and circadian phase modulates mood in healthy subjects. *Archives of General Psychiatry, 54,* 145–152.

Bonnet, M. H., & Arand, D. L. (1994). The use of prophylactic naps and caffeine to maintain performance during a continuous operation. *Ergonomics, 37,* 1009–1020.

Borbély, A. A. (1982). Sleep regulation: Circadian rhythm and homeostasis. In: D. Ganten & D. Pfaff (Eds.), *Sleep. clinical and experimental aspects* (pp. 83–104). Berlin: Springer-Verlag.

Borbély, A. A., & Ackermann, P. (1992). Concepts and models of sleep regulation: An overview. *Journal of Sleep Research, 1,* 63–79.

Borbély, A. A., Ackermann, P., Trachsel, L., & Tobler, I. (1989). Sleep initiation and initial sleep intensity: Interaction of homeostatic and circadian mechanisms. *Journal of Biological Rhythms, 4,* 149–160.

Boulos, Z., Campbell, S. S., Lewy, A. J., Terman, M., Dijk, D. J., & Eastman, C. I. (1995). Light treatment for sleep disorders—Consensus report 7; Jet-lag. *Journal of Biological Rhythms, 10,* 167–176.

Braby, C. D., Harris, D., & Muir, H. C. (1993). A psychophysiological approach to the assessment of work underload. *Ergonomics, 36,* 1035–1042.

Brown, A. C., Smolensky, M. H., Alonzo, G. E., & Redman, D. P. (1990). Actigraphy: A means of assessing circadian patterns in human activity. *Chronobiology International, 7,* 125–133.

Campbell, K. B., & Wilkinson, R. T. (1981). Sleep in the natural environment: Physiological and psychological recording and analysing techniques. In L. C. Johnson, D. I. Tepas, W. P. Colquhoun, & M. J. Colligan (Eds.), *Biological rhythms, sleep and shift work* (pp. 581–606). New York: Spectrum.

Campbell, S. S. (1992). The timing and structure of spontaneous naps. In C. Stampi (Ed.), *Why we nap* (pp. 71–81). Boston: Birkhauser.

Carskadon, M. A., & Dement, W. C. (1975). Sleep studies on a 90-min day. *Electroencephalography and Clinical Neurophysiology, 39,* 145–155.

Coldwells, A., Atkinson, G. and Reilly, T. (1994). Scores of variation in back and leg dynamometry. *Ergonomics, 37,* 79–86.

Colquhoun, W. P. (1984). Effects of personality on body temperature and mental efficiency following transmeridian flight. *Aviation, Space and Environmental Medicine, 55,* 493–496.

Colquhoun, W. P., Watson, K. J., & Gordon, D. S. (1987). A shipboard study of a four-crew rotating watchkeeping system. *Ergonomics, 30,* 1341–1352.

Comperatore, C., Lieberman, H., Kirby, A., Adams, B., & Crowley, J (1996). Melatonin efficacy in aviation missions requiring rapid deployment and night operations. *Aviation, Space and Environmental Medicine, 67,* 520–524.

Costa, G. (1993). Evaluation of workload in air-traffic controllers. *Ergonomics, 36,* 1111–1120.

Costa, G. (1997). The problem: Shiftwork. *Chronobiology International, 14,* 89–98.

Costa, G., Lievore, F., Casaletti, G., Gaffuri, E., & Folkard, S. (1989). Circadian characteristics influencing interindividual differences in tolerance and adjustment to shiftwork. *Ergonomics, 32,* 373–385.

Craig, A. (1985). Vigilance: Theories and laboratory studies. In S. Folkard & T. H. Monk (Eds.), *Hours of work. temporal factors in work scheduling* (pp. 107–121). Chichester, UK: Wiley.

Craig, A., Baer, K., & Diekmann, A. (1981a). The effects of lunch on sensory–perceptual functioning in man. *International Archives of Occupational and Environmental Health, 49,* 105–114.

Craig, A., Wilkinson, R. T & Colquhoun, W. P. (1981b). Diurnal variation in vigilance efficiency. *Ergonomics, 24,* 641–651.

Czeisler, C. A., & Dijk, D. J. (1995). Use of bright light to treat maladaption to night shift work and circadian rhythm sleep disorders. *Journal of Sleep Research, 4*(Supplement 2), 70–73.

Czeisler, C. A., Johnson, M. P., Duffy, J. F., Brown, E. N., Ronda, J. M., & Kronauer, R. E. (1990). Exposure to bright light and darkness to treat physiologic maladaptation to night work. *New England Journal of Medicine, 322,* 1253–1259.

Czeisler, C. A., Moore-Ede, M. C., & Coleman, R. M. (1982). Rotating shiftwork schedules that disrupt sleep are improved by applying circadian principles. *Science, 21,* 460–463.

Daan, S., Beersma, D. G. M., & Borbély, A. A. (1984). Timing of human sleep: Recovery process gated by a circadian pacemaker. *American Journal of Physiology, 246,* R161–R178.

Davies, D. R. (1985). Individual and group differences in sustained attention. In S. Folkard & T. H. Monk (Eds.), *Hours of work. Temporal factors in work scheduling* (pp. 123–132). Chichester, UK: Wiley.

Dawson, D., Encel, N., & Lushington, K. (1995). Improving adaptation to simulated night shift: Time exposure to bright light versus daytime melatonin administration. *Sleep, 18,* 11–21.

Dijk, D. J., & Czeisler, C. A. (1995). Contribution of the circadian pacemaker and the sleep homeostat to sleep propensity, sleep structure, electroencephalographic slow waves, and sleep spindle activity in humans. *Journal of Neuroscience, 15,* 3525–3538.

Dimsdale, J. E., & Ziegler, M. G. (1991). What do plasma and urinary measures of catecholamines tell us about human response to stressors? *Circulation, 83*(Supplement 11), 36–42.

Dinges, D. F. (1992). Adult napping and its effect on ability to function. In C. Stampi (Ed.), *Why we nap* (pp. 118–134). Boston: Birkhauser.

Dinges, D. F. (1995). An overview of sleepiness and accidents. *Journal of Sleep Research, 4*(Supplement 2), 4–14.

Dorner, D. (1990). The logical failure. *Philosophical Transactions of the Royal Society of London B, 327,* 463–473.

Eastman, C. I., Boulos, Z., Terman, M., Campbell, S. S., Dijk, D. J., & Lewy, A. J. (1995a). Light treatment for sleep disorders—Consensus report 6; shift work. *Journal of Biological Rhythms, 10,* 157–164.

Eastman, C. I., Hoese, E. K., Youngstedt, S. D., & Liu, L. W. (1995b). Phase-shifting human circadian rhythms with exercise during night shift. *Physiology and Behaviour, 58,* 1287–1291.

Englund, C. E., Loving, R., & Kripke, D. P. (1990). Bright light amelioration of shift work. *Annual Review of Chronopharmacology, 7,* 33–36.

Faria, I. E., & Drummond, B. J. (1982). Circadian changes in resting heart rate and body temperature, maximal oxygen consumption and perceived exertion. *Ergonomics, 25,* 381–386.

Ferrer, C. F., Bisson, R. U., & French, J. (1995). Circadian rhythm desynchronosis in military deployments: A review of current strategies. *Aviation Space and Environmental Medicine, 66,* 571–8.

Folkard, S. (1990). Circadian performance rhythms: Some practical and theoretical implications. *Philosophical Transactions of the Royal Society of London B, 327,* 543–553.

Folkard, S., & Åkerstedt, T. (1987). Towards a model for the prediction of alertness and/or fatigue on different sleep wake schedules. In A.Oginski, & J. Pokorski, J. Rutenfranz (Eds.), *Contemporary advances in shiftwork research* (pp. 231–240) Krakow, Poland: Medical Academy.

Folkard, S., & Åkerstedt, T. (1991). A three process model of the regulation of alertness and sleepiness. In R. Ogilvie & R. Broughton (Eds.), *Sleep, arousal and performance: Problems and promises* (pp. 11–16). Boston: Birkhauser.

Folkard, S., & Condon, R. (1987). Night shift paralysis in air-traffic control officers. *Ergonomics, 30,* 1353–1363.

Folkard, S., & Monk, T. H. (1985). Circadian performance rhythms. In S. Folkard & T. H. Monk (Eds.), *Hours of work. Temporal factors in work scheduling* (pp. 37–52). Chichester, UK: Wiley.

Folkard, S., & Totterdell, P. (1993). Dissecting circadian performance rhythms: Implications for shiftwork. *Ergonomics, 36*, 283–288.

Folkard, S., Knauth, P., Monk, T. J., & Rutenfranz, J. (1976). The effect of memory load on the circadian variation in performance efficiency under a rapidly rotating shift system. *Ergonomics, 19*, 479–488.

Folkard, S., Condon, R., & Herbert, M. (1984). Night shift paralysis. *Experientia, 40*, 510–512.

Folkard, S., Hume, K. I., Minors, D. S., Waterhouse, J. M., & Watson, P. L. (1985). Independence of the circadian rhythm in alertness from the sleep/wake cycle. *Nature, 313*, 678–679.

Folkard, S., Marks, M., & Froberg, J. C. (1986). Towards a causal nexus of human psychophysiological variables based on their circadian rhythmicity. *Revija za psihologiju, 16*, 1–9.

Folkard, S., Arendt, J., & Clark, M. (1993). Can melatonin improve shift workers' tolerance of the night shift? Some preliminary findings. *Chronobiology International, 10*, 315–320.

Foushee, H. C., Lauber, J. K., Baetge, M. M., & Acomb, D. B. (1986). Operational significance of exposure to short-haul air transport operations. *NASA Technical Memorandum* 88322.

French, J., Bisson, R. V., Neville, K. J., Mitcha, J., & Storm, W. F. (1994). Crew fatigue during simulated, long duration B-1B bomber missions. *Aviatian, Space and Environmental Medicine, 65*(5 Supplement), A1–A6.

Friedman, R. C., Bigger, J. T., & Kornfeld, D. S. (1971). The intern and sleep loss. *New England Journal of Medicine, 285*, 201–203.

Gaillard, A. W. L. (1993). Comparing the concepts of mental load and stress. *Ergonomics, 36*, 991–1005.

Gander, P. H., Nguyen, D., Rosekind, M. R., & Connell, L. J. (1993). Age, circadian rhythms and sleep loss in flight crews. *Aviation, Space and Environmental Medicine, 64*, 189–195.

Gardner-Medwin, A. R. (1985). Device to prevent vehicle drivers falling asleep. *Journal of Physiology, 371*, 22P.

Gerbert, R., & Kemmler, R. (1986). The causes of causes: Determinants and background variables of human factor incidents and accidents. *Ergonomics, 29*, 1439–1453.

Gillooly, P. B., Smolensky, M. H., Albright, D. L., Hsi, B., & Thorne, D. R. (1990). Circadian variation in human performance evaluated by the Walter Reed performance assessment battery. *Chronobiology International, 7*, 143–153.

Graeber, R. C. (1982). Alterations in performance following rapid transmeridian flight. In F. M. Brown & R. C. Graeber (Eds.), *Rhythmic aspects of behaviour* (pp. 173–212). Hillsdale, NJ: Erlbaum.

Graeber, R. C., Lauber, J. K., Connell, L. J., & Gander, P. H. (1986). International aircrew sleep and wakefulness after multiple time zone flights: A co-operative study. *Aviation, Space and Environmental Medicine, 57*(Supplement), B3–B9.

Green, R. G. (1985). Stress and accidents. *Aviation, Space and Environmental Medicine, 56*, 638–641.

Hamelin, P. (1987). Lorry drivers' time habits in work and their involvement in traffic accidents. *Ergonomics, 30*, 1323–1333.

Harma, M. (1995). Sleepiness and shiftwork: Individual differences. *Journal of Sleep Research, 4*(Supplement 2), 57–61,

Harma, M., Knauth, P., & Ilmarinen, J. (1989). Daytime napping and its effects on alertness and short-term memory performance in shiftworkers. *International Archives of Occupational and Environmental Health, 61*, 341–345.

Harma, M. I., Hakola, T., & Laitinen, J. H. (1992). Relation of age to circadian adjustment to night work. *Scandinavian Journal of Work and Environmental Health, 18*(Supplement 2), 116–118.

Harma, M., Hakola, T., Åkerstedt, T., & Laitinen, J. (1994). Age and adjustment to nightwork. *Ocupational and Environmental Medicine, 51* 568–573.

Harrison, Y., Bright, V., & Horne, J. A. (1996). Can normal subjects be motivated to fall asleep faster? *Physiology and Behaviour, 60*, 681–684.

Haslam, D. R. (1984). The military performance of soldiers in sustained operations. *Aviation, Space and Environmental Medicine, 55*, 216–221.

Hildebrandt, G., & Strattman, I. (1979). Circadian system response to night work in relation to the individual circadian phase position. *International Archives of Occupational and Environmental Health, 3*, 73–83.

Hildebrandt, G., Rohmert, W., & Rutenfranz, J. (1975). The influence of fatigue and rest period on the circadian variation of error frequency in shift workers (engine drivers). In P. Colquhoun, S. Folkard, P. Knauth, & J Rutenfranz (Eds.), *Experimental studies on shiftwork* (pp. 177–187). Opladen, Germany: Westdeutschen Verlag.

Honzikova, N., Penaz, J., & Fiser, B. (1988). Power spectra of blood pressure and heart rate fluctuations during mental load. *Journal of Interdisciplinary Cycle Research, 19*, 75–79.

Horne, J. (1985). Sleep loss: Underlying mechanisms and tiredness. In S. Folkard & T. H. Monk (Eds.), *Hours of work. Temporal factors in work scheduling* (pp. 53–65). Chichester, UK: Wiley.

Horne, J. A., & Reyner, L. A. (1995). Driver sleepiness. *Journal of Sleep Research, 4*(Supplement 2), 23–29.

Johnson, M. P., Duffy, J. F., Dijk, D. J., Ronda, J. M., Dyal, C. M., & Czeisler, C. A. (1992). Short-term memory, alertness and performance: A reappraisal of their relationship to body temperature. *Journal of Sleep Research, 1,* 24–29.

Kanno, O., Watanabe, H., & Kazamatsuri, H. (1992). Effect of zopiclone, flunitrazepam, triazolam and levomepromazine on the transient change in sleep–wake schedule: Polygraphic study and the evaluation of sleep and daytime condition. *Progress in Neuro-psychopharmacology and Biological Psychiatry, 17,* 229–239.

Kecklund, G., & Åkerstedt, T. (1993). Sleepiness in long distance truck driving: An ambulatory EEG study of night driving. *Ergonomics, 36,* 1007–1017.

Kecklund, G., & Åkerstedt, T. (1995). Effects of timing of shifts on sleepiness and sleep duration. *Journal of Sleep Research, 4*(Supplement 2), 47–50.

Kennedy, R. S., Bittner, A. C., Harbeson, M., & Jones, B. M. (1982). Television computer games: A new look in performance testing. *Aviation, Space and Environmental Medicine, 53,* 49–53.

Kerkhof, G. A. (1985). Inter-individual differences in the human circadian system: A review. *Biological Psychology, 20,* 83–112.

Kirkcaldy, B. D. (1984). Performance and circadian rhythms. *European Journal of Applied Physiology, 52,* 375–379.

Klein, T., & Armitage, R. (1979). Rhythms in human performance: 1h-hour oscillations in cognitive style. *Science, 204,* 1326–1327.

Koslowsky, M., & Babkoff, H. (1992). Meta-analysis of the relationship between total sleep deprivation and performance. *Chronobiology International, 9,* 132–136.

Krueger, G. P. (1989). Sustained work, fatigue, sleep loss and performance: A review of the issues. *Work and Stress, 3,* 129–141.

Langan-Fox, C. P., & Empson, J. A. C. (1985). Actions not as planned in military air-traffic. *Ergonomics, 28,* 1509–1521.

Lauridsen, O., & Tonnesen, T. (1990). Injuries related to the aspects of shift working. A comparison of different offshore shift arrangements. *Journal of Occupational Accidents, 12,* 167–176.

Lavie, P. (1986). Ultrashort sleep–waking schedule. III. "Gates" and "forbidden zones." *Electroencephalography and Clinical Neurophysiology, 63,* 414–425.

Lavie, P., & Segal, S. (1989). Twenty-four-hour structure of sleepiness in morning and evening persons investigated by ultrashort sleep–wake cycle. *Sleep, 12,* 522–528.

Lille, F., & Cheliont, F. (1982). Variation in diurnal and nocturnal waking state in air-traffic controllers. *European Journal of Applied Physiology, 49,* 319–328.

Matsumoto, K., & Harada, M. (1994). The effect of night-time naps on recovery from fatigue following night work. *Ergonomics, 37,* 899–907.

Middelkoop, H. A. M., Van Dam, E. M., Smilde Vanden Doel, D. A., & Van Dijk, G. (1997). 45–hour continuous quintuple-site actimetry: Relations between trunk and limb movements and effects of circadian sleep–wake rhythmicity. *Psychophysiology, 34,* 199–203.

Minors, D. S., & Waterhouse, J. M. (1981). *Circadian rhythms and the human.* Bristol, UK: Wright PSG.

Minors, D. S., & Waterhouse, J. M. (1982). Anchor sleep as a synchronizer of rhythms on abnormal routines. *International Journal of Chronobiology, 7,* 165–188.

Minors, D. S., & Waterhouse, J. M. (1987). The role of naps in alleviating sleepiness during an irregular sleep–wake schedule. *Ergonomics, 30,* 1261–1273.

Moline, M. L., Pollack, C. P., Monk, T. H., Lester, L. S., Wagner, D. R., Zendell, S. M., Graeber, R. C., Salter, C. A., & Hirsch, E. (1992). Age-related differences in recovery from simulated jet-lag. *Sleep, 15,* 28–40.

Monk, M. (1987). Subjective ratings of sleepiness—The underlying circadian mechanisms. *Sleep, 10,* 343–353.

Monk, T. H., & Folkard, S. (1985a). Shift work and performance. In S. Folkard & T. H. Monk (Eds.), *Hours of work. Temporal factors in work scheduling* (pp. 239–252). Chichester, UK: Wiley.

Monk, T. H., & Folkard, S. (1985b). Individual differences in shift work adjustment. In S. Folkard & T. H. Monk (Eds.), *Hours of work. Temporal factors in work scheduling,* (pp. 227–237). Chichester, UK: Wiley

Monk, T. H., Buysse, D. J., Reynolds, C. F., Berga, S. L., Jarrett, D. B., Begley, A. E., & Kupfer, D. J. (1997). Circadian rhythms in human performance and mood under constant conditions. *Journal of Sleep Research, 6,* 9–18.

Monk, T. H., Buysse, D. J., Reynolds H. C. F., & Kupfer, D. J. (1996). Circadian determinants of the post lunch dip in performance. *Chronobiology International, 13,* 123–133.

Muzet, A., Nicolas, A., Tassi, P., Dewasmes, G., & Bonneau, A. (1995). Implementation of napping in industry and the problem of sleep inertia. *Journal of Sleep Research, 4*(Supplement 2), 67–69.

Naitoh, P. (1981). Circadian cycles and the restorative power of naps. In L. C. Johnson, D. I. Tepas, W. P. Colquhoun, & M. I. Colligan (Eds.), *Biological rhythms, sleep and shift work* (pp. 553–580). New York: Spectrum.

Naitoh, P. (1992). Minimal sleep to maintain performance: The search for sleep quantum in sustained operations. In C. Stampi (Ed.), *Why we nap,* (pp. 199–216). Boston: Birkhauser.

Naitoh, P., Kelly, T., & Babkoff, H. (1993). Sleep inertia: Best time not to wake up? *Chronobiology International, 10,* 109–118.

Nicholson, A. N., Stone, B. M., Borland, R. G., & Spencer, M. B. (1984). Adaptation to irregularity of rest and activity. *Aviation, Space and Environmental Medicine, 55,* 102–112

Nicholson, A. N., Pascoe, P. A., Roehrs, T., Roth, T., Spencer, M. B., Stone, B. M., & Zorick, F. (1985). Sustained performance with short evening and morning sleeps. *Aviation, Space and Environmental Medicine, 56,* 105–114.

Novak, R. D., Smolensky, M. H., Fairchild, E. J., & Reves, R. R. (1990). Shiftwork and industrial injuries at a chemical plant in southeast Texas. *Chronobiology International, 7,* 155–164.

O'Connor, P. J., & Davis, J. C. (1992). Psychobiologic responses to exercise at different times of day. *Medicine and Science in Sports and Exercise, 24,* 714–719.

O'Connor, P. J., & Morgan, W. P. (1990). Athletic performance following rapid traversal of multiple time zones. *Sports Medicine, 10,* 20–30.

Ortega, S. M., & Cabrera, M. C. (1990). Ultradian rhythms in the EEG and task performance. *Chronobiologia, 17,* 183–194.

Orton, D. I., & Gruzelier, J. H. (1989). Adverse changes in mood and cognitive performance of house officers after night duty. *British Medical Journal, 298,* 21–23.

Petrie, K., Dawson, A. G., Thompson, L., & Brook, R. (1993). A doubt-blind trial of melatonin as a treatment for jet-lag in international cabin crew. *Biological Psychiatry, 33,* 526–530.

Pokorny, M. L. I., Blom, D. H. J., & Opmeer, C. H. J. M. (1988). Effects of work and circadian rhythm on bus drivers' oral temperature. *Chronobiology International, 5,* 425–432.

Pokorny, M. L. I., Blom, D. H. J., & Van Leeuwen, P. (1987). Shifts, duration of work and accident risk of bus drivers. *Ergonomics, 30,* 61–88.

Poulton, E. C., Hunt, G. M., Carpenter, A., & Edwards, R. S. (1978). The performance of junior hospital doctors following reduced sleep and long hours of work. *Ergonomics, 21,* 279–295.

Reilly, T. (1990). Human circadian rhythms and exercise. *Critical Reviews of Biomedical Engineering, 18,* 165–180.

Reilly, T., & Down, A. (1992). Investigation of circadian rhythms in anaerobic power and capacity of the legs. *Journal of Sports Medicine and Physical Fitness, 32,* 343–347.

Reilly, T., & Mellor, S. (1988). Jet-lag in student Rugby League players following a near maximal time-zone shift. In T. Reilly, A. Lees, K. Davids, & W.J. Murphy (Eds.), *Science and football* (pp. 249–256). London: E. and F. N. Spon.

Reilly, T., & Piercy, M. (1994). The effect of partial sleep deprivation on weight-lifting performance. *Ergonomics, 37,* 107–116.

Reilly, T., & Smith, D. (1986). Effect of work intensity on performance in a psychomotor task during exercise. *Ergonomics, 29,* 601–606.

Reilly, T., Atkinson, G., & Budgett, R. (1997a). Effects of temazepam on physiological and performance variables following a westerly flight across five time zones. *Journal of Sports Sciences, 15,* 62.

Reilly, T., Atkinson, G., & Waterhouse, J. (1997b). *Biological rhythms and exercise.* Oxford: Oxford University Press.

Reilly, T., Atkinson, G., & Waterhouse, J. (1997c). Travel fatigue and jet-lag. *Journal of Sports Science, 15,* 265–369.

Reilly, T., Waterhouse, J., & Atkinson, G. (1997d). Aging, rhythms of physical performance and adjustment to changes in the sleep–activity cycle. *Occupational and Environmental Medicine, 54,* 812–816.

Reinberg, A., Motohashi, Y., Bourdeleau, P., Andlauer, P., Levi, F., & Bicakova-Rocher, A. (1988). Alteration of period and amplitude of circadian rhythms in shift workers, with special reference to temperature, right and left hand grip strength. *European Journal of Applied Physiology, 57,* 15–25.

Ribak, J., Ashkenazi, I. E., Klepfish, A., Augar, D., Tall, J., Kallner, B., & Noyman Y. (1983). Diurnal rhythmicity and air force flight accidents due to pilot error. *Aviation, Space and Environmental Medicine, 54,* 1096–1099.

Rogers, A. S., Spencer, M. B., Stone, B. M., & Nicholson, A. N. (1989). The influence of a 1h nap on performance overnight. *Ergonomics, 32,* 1193–1205.

Rogers, C. D., Paterson, D. H., Cunningham, D. A., Noble, E. G., Pettigrew, F. P., Myles, W. S., & Taylor, A. W. (1995). Sleep deprivation: Effects on work capacity self-paced walking, contractile properties and perceived exertion. *Sleep, 18,* 30–38.

Rosa, R. R. (1995). Extended workshifts and excessive fatigue. *Journal of Sleep Research, 4*(Supplement 2), 51–56.

Rosa, R. R., & Bonnet, M. H. (1993). Performance and alertness on 8h and 12h rotating shifts at a natural gas utility. *Ergonomics, 36*, 1177–1193.

Rosekind, M. P., Gander, P. H., & Dinges, D. F. (1991). Alertness management in flight operations: strategic napping. *SAE Technical Paper Series*, No. 912138.

Rosekind, M. R., Graeber, R. C., Dinges, D. F., Connell, L. J., Pountree, M. S., Spinweber, C. L., & Gillen, K. A. (1992). Crew factors in flight operations: IX Effects of planned cockpit rest on crew performance and alertness in long-haul operations. *NASA Technical Memorandum*, No. 103884.

Rosekind, M. R., Smith, R. M., Miller, D. L., Co, E. L., Gregory, K. B., Webbon, L. L., Gander, P. H., & Lebacqz, J. V. (1995). *Journal of Sleep Research, 4*(Supplement 2), 62–66.

Samel, A., & Wegmann, H. M. (1988). Sleep and circadian rhythms of an airline pilot operating on the polar route: a case study. *Aviation, Space and Environmental Medicine, 59*, 443–447.

Samel, A., Wegmann, H. M., & Vejvoda, M. (1995). Jet-lag and sleepiness in aircrew. *Journal of Sleep Research, 4*(Supplement 2), 30–36.

Santy, P. A., Faulk, D. M., & Davis, J. R. (1994). Strategies for the preflight circadian shifting of space shuttle crews. *Journal of Clinical Pharmacology, 34*, 535–542.

Sasaki, M., Kurosaki, Y. S., Spinweber, C. L., Graeber, R. C., & Takahashi, T. (1993). Flight crew sleep during multiple layover polar flights. *Aviation, Space and Environmental Medicine, 64*, 641–667.

Shephard, R. J. (1984). Sleep, biorhythms and human performance. *Sports Medicine, 1*, 11–37.

Smith, A. P., & Miles, C. (1986). The effects of lunch on cognitive vigilance tasks. *Ergonomics, 29*, 1251–1261.

Smith, A. P., & Wilson, M. (1990). The effects of naps during night duty on the performance and mood of female nurses working in an intensive care unit. In G. Costa, G. Cesana, K. Kagi, & A. Wedderburn (Eds.), *Shiftwork: Health, sleep and performance* (pp. 147–152). Frankfurt: Peter Lang.

Smith, L., Folkard, S., & Poole, C. J. M. (1994). Increased injuries on night shift. *Lancet, 344*, 1137–1139.

Smith-Coggins, R., Rosekind, M. R., Hurd, S., & Buccino, K. R. (1994). Relationship of day versus night sleep to physician performance and mood. *Annals of Emergency Medicine, 24*, 928–934.

Spencer, M. B. (1987). The influence of irregularity of rest and activity on performance: A model based on time since sleep and time of day. *Ergonomics, 30*, 1275–1276.

Spencer, M. B., Stone, B. M., Rogers, A. S., & Nicholson, A. N. (1991). Circadian rhythmicity and sleep of aircrew during polar schedules. *Aviation, Space and Environmental Medicine, 62*, 3–13.

Stampi, C. (1994). Sleep and circadian rhythms in space. *Journal of Clinical Pharmacology, 34*, 518–534.

Stampi, C., Broughton, R., Mullington, J. Rivers, M., & Campos, J. (1990). Ultrashort sleep strategies during sustained operations: the recuperative value of multiple 80-, 50- and 20-min naps. In G. Costa, G. Cesana, K. Kogi, & A. Wedderburn (Eds.), *Shiftwork: Health, sleep and performance* (pp. 623–628). Frankfurt: Peter Lang.

Stone, B., & Turner, C. (1997). Promoting sleep in shiftworkers and intercontinental travellers. *Chronobiology International, 14*, 133–143.

Strogatz, S. H., Kronauer, R. E., & Czeisler, C. A. (1986). Circadian regulation dominates homeostatic control of sleep length and prior wake length in humans. *Sleep, 9*, 353–364.

Sundberg, S., Kohvakka, A., & Gordin, A. (1988). Rapid reversal of circadian blood pressure rhythm in shift workers. *Journal of Hypertension, 6*, 393–396.

Taylor, D., Gibson, H., Edwards, R., & Reilly, T. (1994). Correction of isometric leg strength tests for time of day. *European Journal of Musculoskeletal Research, 3*, 25–27.

Tepas, D. I. (1982). Shift worker sleep strategies. *Journal of Human Ergology, 11*(Supplement), 325–336

Torsvall, L., Åkerstedt, T., Gillander, K., & Knutsson, A. (1989). Sleep on the night shift: 24-hour EEG monitoring of spontaneous sleep/wake behaviour. *Psychophysiology, 26*, 352–358.

Tyrrell, A. R., Reilly, R., & Troup, J. O. G. (1985). Circadian variation in stature and the effects of spinal loading. *Spine, 10*, 161–164.

Vidacek, S., Kaliterna, L., Radosevic-Vidacek, B., & Folkard, S. (1986). Productivity on a weekly rotating shift system: Circadian adjustment and sleep deprivation effects. *Ergonomics, 29*, 1583–1590.

Vidacek, S., Kaliterna, L., Radosevic-Vidacek, B., & Folkard, S. (1988). Personality differences in the phase of circadian rhythms: A comparison of morningness and extraversion. *Ergonomics, 31*, 873–888.

Vivoli, G., Bergomi, M., Rovosti, S., Carrozzi, G., & Vezzori, A. (1993). Biochemical and haemodynamic indicators of stress in truck drivers. *Ergonomics, 36*, 1089–1097.

Walsh, J. K., Muehlbach, M. J., & Schweitzer, P. K. (1995). Hypnotics and caffeine as countermeasures for shiftwork-related sleepiness and sleep disturbance. *Journal of Sleep Research, 4*(Supplement), 80–83.

Walsh, J. K., Schweitzer, P. K., Anch, A. M., Muehlbach, M. J., Jenkins, N. A., & Dickins, Q. S. (1991). Sleepiness alertness on a simulated night shift following sleep at home with triazolam. *Sleep, 14*, 140–146.

Waterhouse, J. M., Folkard, S., & Minors, D. S. (1991). *Shiftwork, healthy and safety. An overview of the scientific literature 1978–1990.* London: HMSO.

Waterhouse, J. M., Reilly, T., & Atkinson, G. (1997). Jet-lag. *Lancet, 350*(9091), 1611–1616.

Wedderburn, A. A. I. (1987). Sleeping on the job: The use of anecdotes for recording rare but serious events. *Ergonomics, 30,* 1229–1233.

Wilkinson, R. T. (1972). Sleep deprivation—Eight questions. In W. P. Colquhoun (Ed.), *Aspects of human efficiency* (pp. 25–30). London: English Universities Press.

Wilkinson, R. T. (1982). The relationship between body temperature and performance across circadian phase shifts. In F. M. Brown & R. C. Graeber (Eds.), *Rhythmic Aspects of Behaviour* (pp. 213–240). London: Erlbaum.

Williamson, A. M., Gower, C. G. I., & Clarke, B. C. (1994). Changing the hours of shiftwork: A comparison of 8- and 12- hour shift rosters in a group of computer operators. *Ergonomics, 37,* 287–298.

Winget, C. M., deRoshia, C. W., & Holley, D. C. (1985). Circadian rhythms and athletic performance. Medicine and Science in Sports and Exercise, 17, 498–516.

Winget, C. M., deRoshia, C. W., Markley, C. L., & Holley, D. C. (1984). A review of human physiological and performance changes associated with desynchronosis of biological rhythms. *Aviation, Space and Environmental Medicine, 55,* 1085–1096.

Wojtczak-Jaroszowa, J., & Jarosz, D. (1987). Chronohygenic and chronosocial aspects of industrial accidents. In J. E. Pauly & L. E. Scheving (Eds.), *Advances in Chronobiology* (pp. 415–426). New York: Liss.

Wright, K. P., Badia, P., Myers, B. L., Plenzler, S. C. (1997). Combination of bright light and caffeine as a countermeasure for impaired alertness and performance during extended sleep deprivation. *Journal of Sleep Research, 6,* 26–35.

Zulley, J. (1990). Day and night sleep; the bedrest condition. In J. Horne (Ed.), *Sleep, 90* (pp. 319–323). Bochum, Germany: Pontenagel Press.

Zulley, J., Wever, R., & Aschoff, J. (1981). The dependence of onset and duration of sleep on the circadian rhythm of rectal temperature. *Pflügers Archives, 391,* 314–318.

22

Circadian Rhythms, Aging and Dementia

TIMOTHY H. MONK AND PHYLLIS C. ZEE

INTRODUCTION

It is really quite surprising how many circadian cycles people get through in their lives. By the time a person reaches the age of 80 years, she or he will have generated a total of more than 29,000 circadian cycles. Recently, while one of the authors was speaking with an elderly volunteer subject about to take part in a sleep deprivation study, the older gentlemen remarked that the last time he had stayed awake all night was to get a good view of a July 14th parade in Paris, at the end of the Great War, 67 years, or 24,000 circadian cycles, prior to the conversation! He was concerned that he would be unable to repeat the task, but actually did quite well (in fact, better than the young adult control subjects; see below).

In view of the large number of cycles involved, it would not be surprising if the human circadian system would start to "run down," doing its job less well than it did in young adulthood. Both anecdotal observation and empirical study suggest that circadian dysfunction and sleep disruption may indeed be an unwelcome byproduct of advancing age (Miles and Dement, 1980; Prinz, Obrist & Wang, 1975; Vitiello *et al.*, 1986; Weitzman, Moline, Czeisler, & Zimmerman, 1982). Not only do older people sleep at night less well than they did when they were young, but they also appear to have more problems staying awake during the day. This seems to be particularly problematic in the institutionalized demented elderly, who have been shown to exhibit sleep patterns in the "twilight zone" with no 1 hour of the 24 hours associ-

---

TIMOTHY H. MONK Sleep and Chronobiology Center, Western Psychiatric Institute and Clinic, University of Pittsburgh School of Medicine, Pittsburgh, Pennyslvania 15213. PHYLLIS C. ZEE Department of Neurology, Northwestern University Medical Faculty Foundation, Inc., Chicago, Illinois 60611.

Circadian Clocks, Volume 12 of *Handbook of Behavioral Neurobiology*, edited by Joseph S. Takahashi, Fred W. Turek, and R. Y. Moore, Kluwer Academic / Plenum Publishers, New York, 2001.

ated with 60 minutes of sleep or 60 minutes of wakefulness (Jacobs, Ancoli-Israel, Parker, & Kripke, 1989).

Because of these problems with sleep, the elderly (particularly when institutionalized) consume a disproportionate number of sleeping pills, often at an inappropriate dose level and regimen. A NIH Consensus Development Conference (1991) has highlighted these problems and has concluded that nonpharmacologic approaches must be developed to ameliorate the sleep disorders of the elderly. For society as a whole, such problems will inevitably escalate. In the year 1989, there were 6.7 million Americans, for example, in the 80 years and older age group (16.2% of the elderly population). Currently, elderly Americans are consuming more than $36.4 billion in annual health care costs, averaging 16.3% of total income for the 75 years and older age group (U.S. Bureau of the Census, 1992).

Sleep-related (or more accurately night-related) problems are often cited as reasons for institutionalizing elderly people (Pollak, Perlick, Linsner, Wenston, & Hsieh, 1990; Rabins, Mace, & Lucas, 1982; Sanford, 1975). As well as being very distressing to the individual concerned, the process of institutionalization is highly expensive, often financially crippling the entire family and/or requiring significant infusions of tax dollars. Any intervention to reduce the percentage of seniors who are institutionalized would be enormously cost-effective for society. Moreover, not only *nocturnal* sleep problems are troubling to the elderly. Daytime sleep, by its very presence, can also be socially crippling and embarrassing to the older person, as well as dangerous when linked to the driving of automobiles.

There are a number of different ways by which the circadian system might "go wrong" with advanced age. First, the endogenous circadian pacemaker (ECP) may generate a less robust circadian signal. Second, that signal may be adequately generated, but fail to "get through" to be expressed as rhythms of sleepiness and wakefulness. Third, the ECP's entrainment mechanisms may fail to work properly, allowing the circadian pacemaker to become overly labile or inappropriately phased. Fourth, behavioral isolation may lead to a solitary and/or sedentary lifestyle with impoverished social and physical (daylight/dark) zeitgebers impinging on the individual. This may render the circadian system unable properly to entrain the ECP to a healthy nycthemeral routine. This chapter will draw upon studies involving both human and animal subjects in order to develop an understanding of how sleep and circadian rhythms might change with advanced age.

Animal Studies

Introduction

A number of age-related changes in physiologic and behavioral circadian rhythms have been observed in several animal species (Brock, 1991; Ingram, London, & Reynolds, 1982). There is evidence to indicate that age-related changes in the circadian system can be due to alterations in the generation, entrainment, and expression of circadian rhythms.

Amplitude

The most-studied and consistent change in aging is a reduction in the amplitude of circadian rhythms. For example, the amplitude of the circadian rhythms in temperature (Sacher and Duffy, 1978), corticosterone (Nicolau and Milcu, 1979), testosterone (Tenover, Matsumoto, Clifton, & Bremner, 1988), melatonin (Reiter *et al.*, 1981), and locomotor activity (Martin, Fuchs, Bender, & Harting, 1986; Peng, Jiang, & Hsu, 1980; Sacher and Duffy, 1978) is decreased in old rodents. In some old rats, the circadian rhythm of drinking and locomotor activity have been reported to become arrhythmic (Peng, Yaung, & Yang, 1984). A reduction of the amplitude of the circadian fluctuation of sleep and wakefulness has also been reported (Rosenberg, Zepelin, & Rechtschaffen, 1979; Van Gool & Mirmiran, 1986). Old mice spent more time asleep during the usual active period and were awake more during the usual sleep period (Welsh, Richardson, & Dement, 1986) and old rats showed a decrease in the amplitude of the circadian variation of slow-wave sleep (Rosenberg *et al.*, 1979). Dampening of the amplitude of circadian rhythms may result from an alteration of the clock itself, input pathways to the clock or output systems regulated by the circadian clock.

Phase and Entrainment

Age-related changes in the entrainment of circadian rhythms have been reported. A commonly observed alteration with advancing age is an advanced phase angle of entrainment of the locomotor activity rhythm during exposure to a light/dark (LD) cycle. Old hamsters demonstrated an earlier onset of activity than young animals during entrainment to an LD 14:10 cycle (Zee, Rosenberg, & Turek, 1992). Aging also affects the rate of resynchronization of the locomotor activity rhythm (Zee *et al.*, 1992) and the sleep/wake cycle (Rosenberg *et al.*, 1979) in rodents. Overall, the rate of entrainment following a delay shift of the LD cycle was slower in old animals when compared to young animals.

These age-related changes in the phase angle of entrainment and rate of synchronization of circadian rhythms suggest that the response of the circadian system to synchronizing agents (zeitgebers) is also altered. An important approach to studying how the process of aging affects the circadian system has been to study its response to particular zeitgebers in free-running conditions at various circadian times. Such studies have shown that old hamsters are less sensitive to the phase shifting effects of submaximal light pulses (Zhang *et al.*, 1996) as well as to a variety of nonphotic agents (Penev, Zee, Wallen, & Turek, 1995; van Reeth, Zhang, Zee, & Turek, 1992; van Reeth, Zee, & Turek 1993). The phase-shifting effect of pharmacologic agents such as triazolam (van Reeth *et al.*, 1992), as well as nonpharmacologic zeitgebers such as dark pulses (van Reeth *et al.*, 1993) is markedly decreased or lost with advanced age in hamsters. For example, the phase shifting effects of injections of a serotonin agonist (8-OH-DPAT) on the locomotor activity rhythm was greatly attenuated in old hamsters, indicating that aging of the circadian system may be associated with impaired serotonergic control (Penev *et al.*, 1995).

Period

Age-related alterations in the amplitude and entrainment of circadian rhythms do not necessarily imply a change in the circadian clock itself, but could be due to a decrease in sensory input or alterations in the processes that lie between the clock and its output rhythms. More convincing evidence that the clock itself may be altered in senescence comes from the finding that the free-running period of the circadian rhythms of locomotor activity (Davis & Menaker, 1980; Morin, 1988; Pittendrigh & Daan, 1974; Rosenberg, Zee, & Turek, 1991) and sleep–wakefulness (Van Gool, Witting, & Mirmiran, 1987) shortens with age in hamsters and rats. A shortening of the free-running period is consistent with the findings of an advanced phase angle of entrainment of the locomotor activity rhythm in old hamsters. However, in mice, aging is associated with a *lengthening* of the free-running period (Possidente, McEldwoney, & Pabon, 1995).

Lability

A characteristic feature of the locomotor activity rhythm in rodents is the consolidation of the active period during the night and the rest period during the day. However, in old animals, the activity rhythm becomes fragmented, such that there is more activity during the normal rest period and more inactivity during the usual active time of the rest/activity cycle (Brock, 1991; Davis & Menaker, 1980; Peng & Kang, 1984; Turek *et al.*, 1995). This age-related fragmentation of the activity–rest cycle is in good agreement with the reported deterioration of the sleep–wake cycle in old rodents (Naylor *et al.*, 1998; Van Gool *et al.*, 1987; Welsh *et al.*, 1986). Old mice, rats and hamsters spend more time asleep during the normal period of activity and are more awake during the normal period of rest. In addition to fragmentation of circadian rhythms, old animals exhibit large day to day variability in both the onset and amplitude of the activity/rest cycle (Brock, 1991; Richardson, Moore-Ede, Czeisler, & Dement, 1985; Van Gool & Mirmiran, 1986).

Further support for an increase in the lability of the circadian system with age comes from the observation that some old hamsters show unusually large (>300 minutes) phase shifts of the locomotor activity rhythm in response to light pulses administered near the transition region from phase delays to advances, indicating that this phase position may be unstable in old animals (Rosenberg *et al.*, 1991).

Evidence for Problems of Circadian Signal Generation

Introduction

Whether age-related changes in the expression of overt circadian rhythmicity is a consequence of alterations in the ECP, the suprachiasmatic nuclei (SCN), its afferent or efferent pathways, or other nonspecific processes in the senescent organism remains an area of active investigation. Changes in sensory input, motor output, and physiologic processes that lie between the circadian pacemaker and the overt rhythm may be responsible for some of the disturbances in circadian rhythms during old age. However, the consistency of a reduction in the amplitude of many endocrine, physiologic, and behavioral variables may indicate changes in the clock itself (Richardson *et al.*, 1985; Van Gool & Mirmiran, 1986). Indeed, recent findings have shown a decrease in the amplitude and aberrant circadian pattern of neuronal firing at the level of the SCN in rats (Satinoff *et al.*, 1993).

Anatomical Evidence for Age-Related Changes

Further support for age-related changes in both the structure and function of the circadian clock is indicated by the finding of cell degeneration and selective loss of vasopressin- and vasoactive intestinal polypeptide-immunoreactive neurons in the SCN of old rats (Van Someren, Mirmiran, & Swaab, 1993). However, other investigators have found no significant age-related changes in the SCN of rats up to 3 years of age (Turek *et al.*, 1995). Limited evidence also suggests the presence of neuronal loss in the aging SCN of humans. Neuronal loss in the SCN appears to be even more prominent in patients with dementia of the Alzheimer's type (Swaab, Fliers, & Partiman, 1985; Van Someren *et al.*, 1993), although negative findings have been reported recently by Robert Y. Moore and colleagues (personal communication).

Evidence for Age-Related Functional Changes

At a functional level, aging alters the circadian rhythms of glucose utilization and alpha adrenergic receptor levels in the SCN (Weiland and Wise, 1990; Wise, Cohen, Weiland, & London, 1988). Recently, it has been shown that, with aging, the thresholds for both the phase shifting effects of light on the circadian rhythms of locomotor activity and its induction of immediate early gene expression in the SCN are higher than in young animals (Sutin, Dement, Heller, & Kilduff, 1993; Zhang *et al.*, 1996; Sutin *et al.*, 1993), indicating that the aged circadian pacemaker is less sensitive to photic signal. Finally, age-related changes in the free-running period of circadian rhythms, an intrinsic property of the SCN, have been reported in several species of rodents and in a few human studies (Brock, 1991; Morin, 1988; Swaab *et al.*, 1985; Richardson *et al.*, 1985; Rosenberg *et al.*, 1991; Van Gool & Mirmiram, 1986; Witting, Mirmiran, Bos, & Swaab, 1994).

In both animals and humans, alterations in the expression of circadian rhythms with aging most likely reflect changes in functional integrity of the ECP itself coupled with physiologic, hormonal, behavioral and environmental changes that accompany old age. Decreased signal transmission and responsiveness of the circadian system to both internal and external synchronizing agents may result in age-related disruption of temporal organization. However, the primary defects underlying the effects of aging on the circadian system remain to be elucidated.

Evidence for Problems of Rhythm Expression

As we mentioned previously, sleep and its timing create significant problems for the elderly. Although many elderly patients complain to their physicians of an inability to sleep, in reality, the problem is often one of sleepiness intruding into the daytime (and, particularly, early evening) hours when it is *not* wanted, with sleepiness being absent in the early hours of the morning, after midnight, when it *is* earnestly desired. Thus, the impression arises that the *expression* of the circadian system in feelings of alertness and sleepiness (and in the physiologic rhythms underlying them) is being disrupted in the aged. Pursuing the analogy of the clock versus the hands of the clock given by Moore-Ede, Sulzman, and Fuller (1982), one might argue that the clock (SCN) might be working fairly well in advanced age, but that the circadian system simply fails to "get through" into rhythm expression as the subject advances in age. In this section, we shall discuss age-related declines in the ability of the circadian system to be expressed in either physiological variables, psychologic variables, or sleep (including both its timing and its architecture).

Physiologic Rhythms

Many different rhythms have been used as physiologic markers for the human circadian system and as indications of how that system might change with advancing age. Early studies made use of measures derived from urinary electrolytes. Lobban and Tredre (1967) showed age-related abnormalities in the circadian rhythms of urinary potassium, and concluded that the mechanisms controlling renal diurnal rhythms undergo some deterioration in old age.

As we mentioned earlier, the "benchmark" circadian rhythm for human subjects is that in body temperature measured either orally or (more usually) rectally. A number of different authors have studied the circadian temperature rhythms of young and old, comparing estimates of phase and amplitude. During the early 1980s there were two landmark studies. At Stanford University, Richardson, Carskadon, Orav, and Dement (1982) compared oral temperatures from 10 older subjects (60–83 years) with those from 8 young controls, finding in the former shallower nocturnal troughs both in temperature and in subjective sleepiness (Figure 1). The New York investigation by Weitzman *et al.* (1982) was unique in studying older and younger subjects under both entrained and free-running conditions. Six older (53–70 years) males were compared with 6 young male controls. Weitzman *et al.* not only showed reduced (rectal) temperature rhythm amplitudes under both entrained and free-running conditions, with evidence for increased phase lability in the elderly, but also showed a reduction in free-running τ with increased age, confirming extrapolations made from animal studies (discussed earlier in this chapter).

Other studies of age and gender effects on temperature rhythms were carried out by Vitiello, Prinz, and colleagues and by Campbell and colleagues. Vitiello *et al.* (1986) showed amplitude reductions in 8 older men compared to 10 young men, and Moe, Prinz, Vitiello, Marks, and Larsen (1991) showed older women to have higher amplitude (and earlier peaking) circadian temperature rhythms than older men, confirming a finding of Campbell, Gillin, Kripke, Erikson, and Clopton (1989).

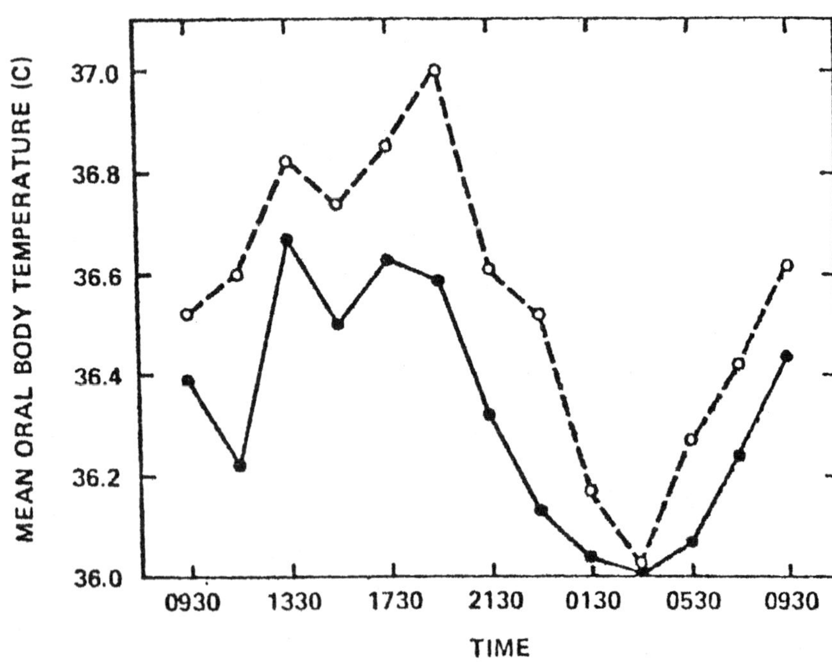

Figure 1. Oral temperature curves from 10 older men (solid line) and 8 young men (broken line) plotted as a function of time. (After Richardson *et al.*, 1982.)

Using a "constant routine" protocol, Czeisler *et al.* (1992) showed phase advances and reduced amplitudes in the elderly, and confirmed the amplitude reduction to be less apparent in older women than in older men.

While the "conventional wisdom" with regard to the aging human circadian temperature rhythm has been that amplitude declines, the evidence has until recently been based upon but a handful of subjects. The two most often cited publications, for example, are those of Weitzman and colleagues and Vitiello and colleagues, who each had only six and eight elders in their sample, respectively, with neither study including women. The Pittsburgh group (Monk, Buysse, Reynolds, Kupfer, & Houck, 1995b) attempted to rectify this with a study of 48 healthy seniors in the "old old" age range (20 men, 28 women; age 77 years and older). Rectal temperatures were recorded on nychthemeral routines, in both laboratory and field settings, and also using an "unmasking protocol" involving 36 hours of constant wakeful bed rest. Interestingly, despite the advanced age of this sample, no general amplitude diminution with advancing age was observed. Indeed, the temperature rhythms of women in their 80s looked strikingly similar to those of young men in their 20s (see Figure 2). Only when the amplitude of older men was compared with that of younger men did *any* statistically reliable age effect in amplitude arise. Thus, we suggest that caution be exercised in asserting that age necessarily results in circadian temperature rhythm amplitude reduction, particularly for women (who comprise the majority in this age group).

Psychological Measures

While there may be doubts about amplitude reductions in temperature rhythms, there appears to be little doubt that circadian *alertness* rhythms show lower amplitudes with advanced age (Monk, Buysse, Reynolds, Kupfer, & Houck, 1996). The first experiment in this study used 5-day baseline (preshift) segments from a simulated "jet-lag" study to compare the phase and amplitude of circadian alertness rhythms from 13 younger (37–52 years) men, 15 older (79–91 years) women and 10 older (71–86 years) men. Subjective alertness was assessed as "global vigor" using a visual analogue scale technique (Monk, 1989) which was given approximately seven times per day during wakefulness. Subjects were unaware of clock time and isolated from physical time cues, but were required to live their normal habitual nycthemeral routine. Each subject's time series was fitted with five cycles of a 24-hour sinusoid and the resulting estimates of mesor (mean level), amplitude, and acrophase then subjected to analysis of variance. This revealed older males to have significantly lower amplitudes ($p < .01$) than the other two groups, with a significant trend towards an earlier phase ($p < .001$). The effects are illustrated by regenerating sinusoids with the relevant mean group estimates of mesor, amplitude and acrophase (Figure 3), and appeared despite relatively robust circadian temperature rhythms in the older groups.

Possible mechanisms for age-related amplitude reductions in subjective alertness rhythms can be gleaned from the second experiment of the Monk *et al.* (1996) study which used an "unmasking" protocol, involving 36 hours of constant wakeful bed rest. Under such conditions, one can assume that any rhythmicity observed must be attributable to endogenous process (i.e. the circadian pacemaker) since *external* conditions have been homogenized. Seven older women (79–88 years) and 11 older men (79–86 years) were compared with 17 young controls (9 men, 8 women; 19–28 years). Again, despite robust rectal temperature rhythms, when the linear decline

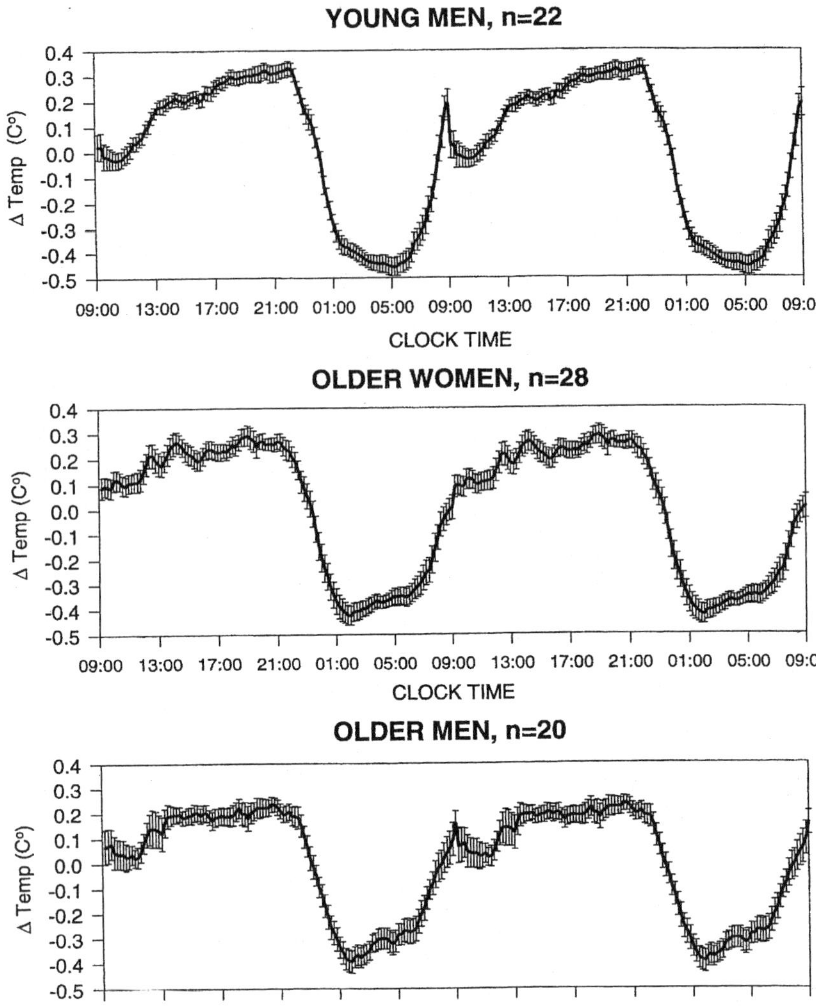

Figure 2. Double-plotted 24-hour rectal temperature rhythms from 22 young men (upper panel), 28 older women (middle panel), and 20 older men (bottom panel). Plotted is mean ± one standard error, expressing each datum as deviation from that subject's 24-hour mean. (After Monk *et al.*, 1995b.)

over the 36 hours of the vigil was factored out (Figure 4), there was a dramatic reduction in rhythmic amplitude in the older subjects' global vigor rhythms, particularly for the older men. Thus, it would appear that although subjective alertness is normally governed by both rhythmic (circadian pacemaker) and homeostatic (time since waking) processes (Folkard & Akerstedt, 1992; Monk, Moline, Fookson, & Peetz, 1989), in advancing age (particularly for men), the homeostatic forces may predominate. This implies that perhaps there is some dysfunction in the efferent pathways running from the circadian pacemaker. Similar patterns of findings occurred when objective performance was considered, rather than subjective alertness (Monk, Buysse, Reynolds, Jarrett, & Kupfer, 1992b). Thus, it would appear that the

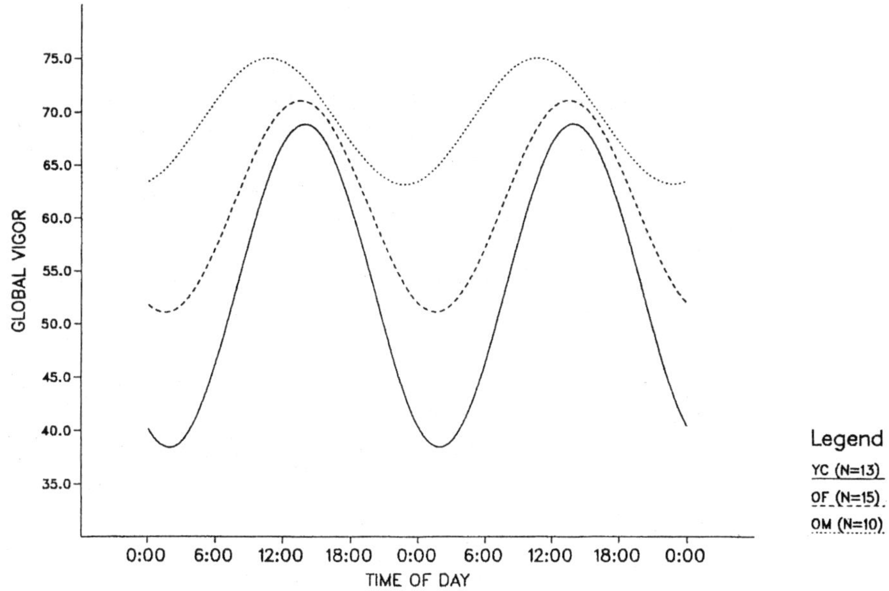

Figure 3. Double plotted 24-hour sinusoids generated with phase, amplitude and mean level parameters as estimated from 5-day individual baseline alertness (global vigor) time series (approximately seven times per day) in three groups: (—) young controls ($N = 13$), (– –) older women ($N = 15$, and (- - -) older men ($N = 10$). (After Monk *et al.*, 1996.)

subjective alertness rhythms of the elderly (especially elderly men) are less robust than those of the young. This may result from a diminished influence of the ECP as a determinant of subjective alertness, even in those elderly people for whom body temperature rhythms are still quite robust. It also may explain why so many elderly people suffer from misplaced sleepiness occurring at times of day when it is unwanted and absent at night when it is needed.

Sleep

Sleep pattern problems are always the most salient for the person involved, and the presumption of the previous sections has been that the rhythm expression problems for the elderly described above might underlie some of their distress in terms of daytime hypersomnia and nighttime insomnia. Such patterns are particularly problematic for the demented. In a classic study of the sleep patterns of nursing home residents, Jacobs *et al.* (1989) dramatically illustrated (using wrist actigraphy) that there was no 1 hour of the day or night when everyone was either awake for the full 60 minutes or asleep for the full 60 minutes. Thus, there appeared to be a distinct flattening of the sleep–wake cycle, similar to the tendency toward multiphasic sleep observed with advancing age in lower animals (Brock, 1985).

Turning to cognitively intact older people, the evidence is less dramatic, but there is still some evidence for daytime napping to increase with advancing age (Buysse *et al.*, 1992; Tune, 1969). Studies of healthy older adults in their eighth or even ninth decade under constant routine conditions or following a 6-hour phase

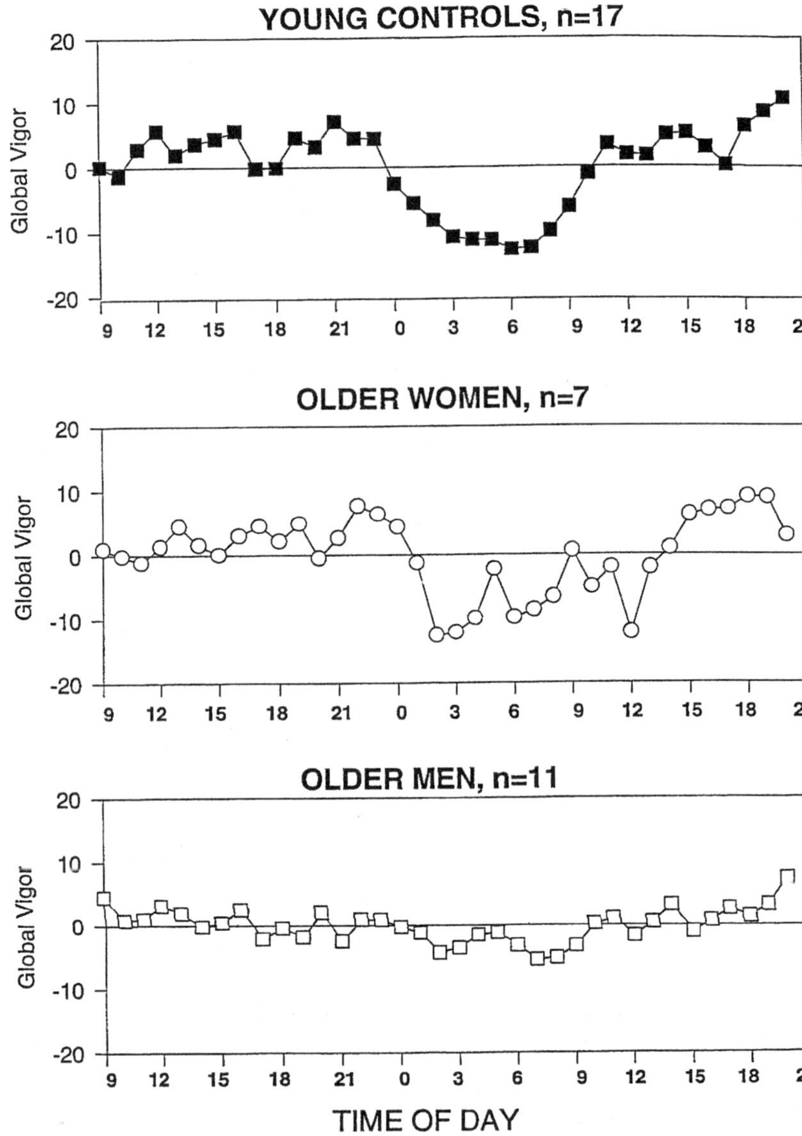

Figure 4. Mean hourly ratings of alertness (global vigor) plotted as a function of time over 35 hours of constant wakeful bed rest, expressing each datum as a deviation from that subject's linear decrease in alertness over the vigil. (After Monk *et al.*, 1996.)

advance of bedtime indicated that age-related changes in sleep and sleepiness may result primarily from alterations in the homeostatic, rather than circadian control mechanisms (Monk *et al.*, 1992b; Monk, Buysse, Reynolds, & Kupfer, 1993).

Phase Shift Tolerance in Elderly People

Introduction

An important point to note when considering the phase shift tolerance of the elderly is that use of the circadian entrainment mechanisms is *not* simply an isolated occurrence precipitated when seniors visit friends or family in another time zone or happen to pull a night shift. On the contrary, entrainment mechanisms are in use every day ensuring that the circadian system remains properly entrained to the 24-hour routines of earth and society. Arguably, some of the non-24-hour behaviors observed in the institutionalized elderly (e.g., the polyphasic sleep patterns observed by Jacobs *et al.*, 1989) may be precipitated by insufficient zeitgebers (not enough daylight and/or social contacts) which then fail properly to entrain the circadian system (see later section). Thus, before considering the challenges posed to the entrainment mechanisms by shift work, jet-lag, and laboratory phase shift simulations, we shall first consider the role of entrainment mechanisms in everyday nychthemeral routines.

Everyday Normal Entrainment

Although the human biological clock is endogenous and self-sustaining, the "natural" period at which it will run in the absence of all time cues and temporal constraints is not 24 hours. For healthy young men who are forbidden to nap, this "free-running" period (denoted τ) is usually considered to be centered upon 25.0 hours with a standard deviation of about 0.5 hours around that mean (Wever, 1979). For young women, and for those who are allowed to nap, τ may be rather shorter, and there is a debate within the field as to what the true value of τ for human beings might be. From forced-desynchrony studies, for example, estimates of free-running τ have often been closer to 24.2 hours than to 25.0 hours (Dijk, Duffy, & Czeisler, 1992). Whatever the precise value of the figure, one can conclude, though, that it is usually longer than 24.0 hours (Campbell, Dawson, & Zulley, 1993b; Wever 1979). In lower animals, τ has been shown to shorten with advancing age (see above), and it is thus natural to ask whether the same is true for humans. Due to a paucity of data, the evidence is mixed, Wever (1979) concluded that there was no τ shortening with age, but others (Monk, 1991; Weitzman *et al.*, 1982) have concluded otherwise. Monk and Moline (1989) found a statistically reliable correlation of -0.69 between temperature rhythm τ and age in a mixed sample of 12 free-running subjects. Evidence from Czeisler's group which previously supported a τ reduction with advancing age is now (with a larger *n*) inconclusive (C. A. Czeisler, personal communication, 1995).

Were τ to indeed shorten with age, then this shortening could be advanced as a possible explanation for why older people tend to retire to bed and wake up earlier than the young (Miles and Dement, 1980). Diary studies have confirmed that not only bed times and wake times move to an earlier position, but also many of the everyday events in between. A recent study compared older (71–91 years) men and women with young controls using the Social Rhythm Metric (SRM), which is discussed later in this chapter. This study revealed that all of the following events were performed earlier by the elderly, usually by about 1 hour: out of bed, morning beverage, breakfast, lunch, dinner, exercise, evening snack, TV viewing, return home, go to bed (Monk, Reynolds, Machen, & Kupfer, 1992a). Not surprisingly, when scales measuring morningness–eveningness are administered to the elderly, their scores usually come out in the "morning-type" end of the spectrum, in contrast to younger controls (Czeisler *et al.*, 1992; Monk *et al.*, 1991). Thus it would appear that it is natural for everyday life of the elderly to be phased earlier than the young, possibly due to age-related changes in the entrainment mechanism and/or τ.

Shift Work

One of the most frequent, and most chronic, major challenges to the entrainment mechanisms of the circadian system is posed by abnormal work hours, often referred to as "shift work." Particularly troubling are shift work systems that require duty at night or which rotate between different duty times on a weekly basis. Typically, 20–25% of a nation's work force are engaged in shift work of one form or another, and large numbers of people are therefore affected (Kogi, 1985; Tasto & Colligan, 1978). Many of these persons are no longer young adults, and with changing labor markets and job tenure uncertainty, many late-middle-age people are finding themselves having to work abnormal hours at a time of life when they did not expect to. Moreover, as the baby-boom generation reaches their half century in age, this age segment of the work force will undoubtedly increase in size. In addition, many hitherto retired people are also returning to the workplace, raising the number of elderly persons engaged in shift work. The question of how well the late-middle-aged and elderly can cope with shift work is thus not a trivial one.

Since increasing age is usually associated with increasing shift work experience and with fewer child-rearing responsibilities, one might expect that shift work coping would become easier in one's later years. Unfortunately, the reverse is true, and many people who coped reasonably well with abnormal work schedules in their young adult years begin to have serious problems in coping during their late 40s or early 50s (Foret, Bensimon, Benoit, & Vieux, 1981). These problems concern insomnia during day sleeps and health problems, particularly in the gastrointestinal and cardiovascular areas (Scott & LaDou, 1990).

It is natural to assume that these difficulties arise because the aging circadian system is no longer as agile in being able to phase shift to a temporal orientation appropriate to the work schedule, i.e., has become *less* phase labile. However, this may be an oversimplification since Reinberg, Vieux, and Andlauer (1981) found elderly intolerant shift workers to be *more* phase labile than their tolerant counterparts. In a later study Reinberg, Andlauer, Bourdeleau, and Levi (1984) found quite deviant circadian periods being exhibited by these French rotating-shift refinery workers. The issue of phase lability is further discussed in a later section on laboratory studies.

More likely explanations for age-related intolerance to shift work are also related to changes in the circadian system, but not to its phase lability. As mentioned earlier, advancing age is associated with an increasing "morning-type" disposition, and morningness as a trait has been implicated as a negative factor for shift work tolerance (Hildebrandt & Stratmann, 1979). Moreover, sleep also becomes more fragile, or easily disrupted, in the later years of life, a fragility that may be compounded with that inherent in trying to sleep at the "wrong" circadian phase.

As many aging politicians and business executives have learnt to their cost, jet lag can create more of a burden to the middle-aged and elderly than to the young adult. Even airline pilots are subject to this effect, as noted by Preston and Bateman (1970) in their pioneering study. This was confirmed by Graeber, Lauber, Connell, and Gander's (1986) multinational study involving several hundred cockpit personnel engaged in a variety of transmeridianal flight schedules. Both Dement, Seidel, Cohen, Bliwise, and Carskadon (1986) and Nicholson, Pascoe, Spencer, Stone, and Green (1986) noted that older subjects obtained less overall sleep, and that the sleep they did obtain was of poorer quality.

Perhaps because of the extremely high cost of jet-lag studies, there have been none that have made direct comparisons of young and old groups of subjects flown to a different time zone. Indeed most jet-lag field studies (e.g., Klein, Wegmann, & Hunt, 1972) have been concerned with young people (usually young men). Perhaps as a consequence, field studies of jet-lag have underestimated the size of the jet lag effect. This was advanced by Graeber, Sing, and Cuthbert (1981) as an explanation of why rather little jet-lag effect was seen in a study of the transatlantic deployment of soldiers.

LABORATORY PHASE SHIFT STUDIES

In two laboratory studies involving a 6-hour phase advance in routine, age comparisons were made between middle-aged (37–52 years) men and young (18–25 years) men (Moline *et al.*, 1992) and between (the same) middle-aged men and 15 older women (79–91 years) and 10 older men (71–86 years) (Monk *et al.*, 1993). Both sets of studies involved body temperature recording as the prime circadian rhythm and polygraphically recorded sleep as the other major dependent variable. The Moline *et al.* studies indicated that although there were *baseline* differences between middle-aged and young men in phase (young earlier) and amplitude (young larger), there were no age differences in either the *magnitude* of the phase shift effect or in the *rate* of temperature rhythm adjustment in either phase or amplitude. In contrast, the sleep variables showed greater disruption in the middle-aged men than in the young men, with more wakefulness during the sleep episode and more light (Stage 1) sleep consequent upon the phase shift. Thus, although the circadian pacemakers of the middle-aged responded similarly to those of the young, their sleep appeared to be much more vulnerable to the disruptive effects of the phase shift.

A similar dissociation between circadian rhythm and sleep effects were observed when older men and women were compared to middle-aged men (Monk *et al.*, 1993). Older women, in particular, showed temperature rhythm adjustment patterns in both phase and amplitude that were almost identical to those of men about half their age, but significantly greater disruptions in sleep that were slow to recover from the phase shift. Older men showed temperature rhythm phase adjustment patterns that were similar to the middle-aged men, but a less pronounced postshift recovery in temperature rhythm amplitude and, again, profound and long lasting disruption in sleep. This study was also interesting in showing that the amount of rapid eye movement (REM) sleep (a stage of sleep tied to the phase of the circadian pacemaker; Moore-Ede *et al.*, 1982) showed the expected disruption and monotonic postshift recovery in the middle-aged, but not in the elderly. Thus, the link between REM sleep and the circadian pacemaker appeared to weaken with advanced age.

Combining the sleep findings from the two sets of studies (Figure 5), we can see that both baseline and postshift sleep are affected by age, but that the magnitude of the disruption appears to increase as one advances in age beyond young adulthood. An interesting case study has also been reported (Monk, Buysse, Reynolds, & Kupfer, 1995a) of an 85-year-old healthy woman who experienced both 6-hour phase advance and phase delay conditions in separate runs of the 15-day protocol outlined above. Interestingly, circadian rhythm amplitude and sleep efficiency were found to be less disrupted in the delay than in the advance condition, in contrast to circadian rhythm phase, which appeared to show more rapid adjustment in the advance condition than in the delay condition. This may relate to changes in the phase response curve (PRC) with advancing age.

Zeitgebers and Their Enhancement

Introduction

As has been pointed out elsewhere in this chapter, aberrant circadian rhythms in the elderly may be the result of poor *alignment* of the circadian system. Poor alignment may either result from dysfunction of the circadian entrainment system per se (i.e., a comparatively flat PRC) or from insufficient time cue information for

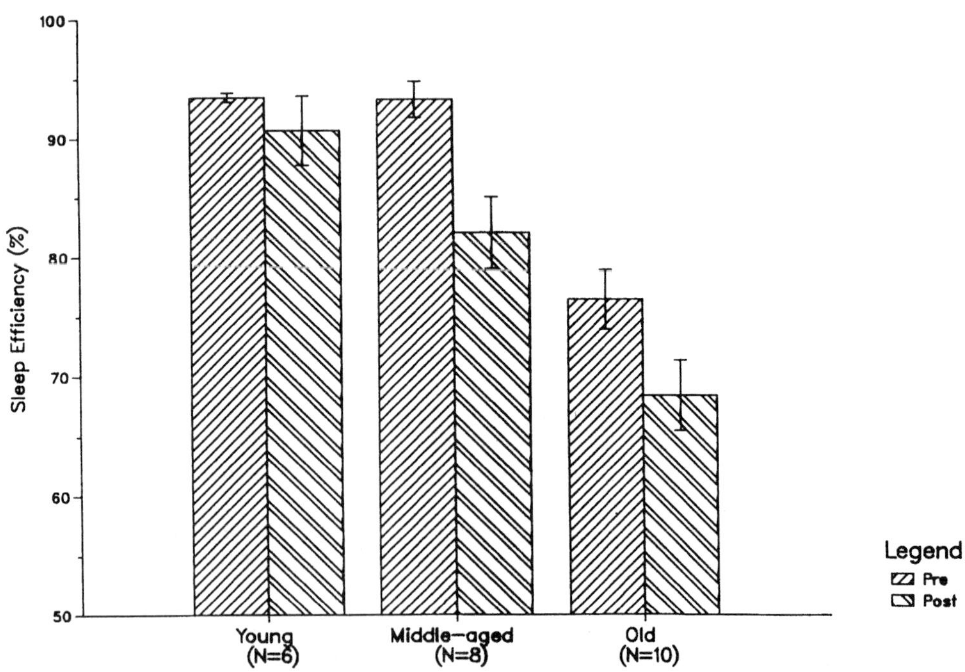

Figure 5. Mean sleep efficiency from three nights before and three nights after a 6-hour phase advance in routine from young, middle-aged, and older men. The young men's data are from Moline *et al.* (1992) and the middle-aged and older men's data are from Monk *et al.* (1988, 1993).

that system to work on (i.e., insufficient zeitgebers actually impinging upon the individual). It is to this second possibility that the present section is addressed.

Daily Social Rhythms in the Elderly

Without doubt, the daily temporal structure in the life of a 30-year-old is very different from that of an 80-year-old. For the 30-year-old, there are a myriad of temporal demands stemming from the need to attend to one's job, care for one's children and interact with spouse, friends, family, and coworkers. In contrast, the 80-year-old is often widowed, retired, and freed from parental responsibilities. Any temporal constraints that do occur are usually self-imposed, resulting from volunteer work, club membership, and appointments with friends. For the demented, the patterning of social-based temporal structure may be almost entirely absent, with little cognitive awareness of what time it is and what can be expected to happen next. It is therefore reasonable to hypothesize that the pattern of social-based zeitgebers impinging upon the individual might vary greatly between the three groups.

An instrument, the Social Rhythm Metric (SRM), has been developed (Monk, Flaherty, Frank, Hoskinson, & Kupfer, 1990) which allows investigators to quantify the level of regularity of the daily rhythms making up a person's life. The SRM is a diary-like instrument which requires the respondent to record "when" and "with whom" each of 17 events (including bedtimes, wake times, meals, social/work activities, and recreation) are performed, the diary being completed each evening for a week. An algorithm is then used to derive (among other measures) an SRM score which varies on a continuum between 0 (most irregular) and 7 (most regular).

When we administered the SRM to 45 healthy elders (71–91 years), comparing them to 21 young adults (19–28 years), we expected to find that the elderly had lower (i.e., less regular) scores. On the contrary, however, we found the older people to have significantly *more* regular social rhythms than the young (Monk *et al.*, 1992a). This was particularly surprising in view of the inferiority we measured in the elderly subjects' sleep. It is possible, however, that a less healthy elderly sample would have showed less regularity. The high SRM scores we observed in this healthy sample may either have represented a positive, adaptive response to changes in the circadian system with advanced age or a trait-like predisposition toward regularity that had served these people well over the years.

More recently we published a cross-sectional study of SRM scores over the entire adult life span (20–89 years) in a sample of 239 healthy individuals (Monk *et al.*, 1997). As is clear from Figure 6, the age-related increase we had observed previously can be seen as part of a more general increasing trend (toward greater lifestyle regularity) over the entire adult years. This trend undoubtedly has both biological and psychosocial antecedents. The apparent increase in lifestyle regularity over the adult years does indicate, however, that the interplay between the circadian system and the life of the individual is not static, but changes as the individual progresses through life.

Habitual Daylight Exposure in the Elderly and Demented

While the *social* zeitgebers impinging upon the elderly may still be fairly robust, there is increasing evidence from light measurement studies to suggest that an elderly person's exposure to daylight levels of illumination might be rather minimal. The reason why the word "daylight" is important here is that, in contrast to

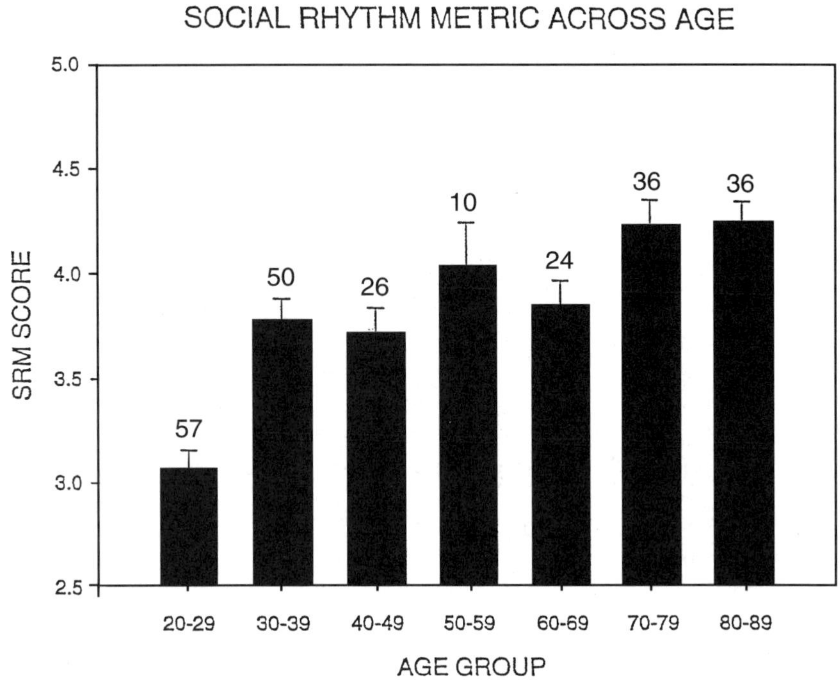

Figure 6. Mean Social Rhythm Metric (SRM) score (higher number represents greater regularity) plotted as a function of age group, the number within the bar represents sample size. (After Monk *et al.*, 1997.)

lower animals, human beings tend to respond in a qualitatively different way to daylight levels of illumination (>2000 lux) as compared to the much lower levels typically found indoors. Thus, daylight illumination levels permit a much broader range of entrainment of the circadian system (Wever, Poiasek, & Wildgruber, 1983) and may allow a stronger ("Type 0") phase resetting of the circadian system than allowed for by conventional artificial illumination (Boivin, Duffy, Kronauer, & Czeisler, 1996; Czeisler & Dijk, 1995; Czeisler *et al.*, 1986). This may result from the fact that such high (>2500 lux) levels are needed to suppress the nocturnal production of the pineal hormone melatonin (Lewy, Wehr, Goodwin, Newsome, & Markey, 1980). The question of how much daylight illumination is experienced by the elderly and the demented is thus of more than simply academic interest.

Studies using a device which recorded light levels automatically on an around-the-clock basis indicate that even in younger adults, remarkably little time (typically around 1.5 hours per day) is normally spent experiencing daylight (>2000 lux) illumination levels (Savides, Messin, Senger, & Kripke, 1986). Campbell, Kripke, Gillin, and Hrubovcak (1988) studied 10 healthy elderly (55–81 years) and 13 demented elderly (62–84 years) subjects whose ambient light levels (at the wrist) were recorded from 2 to 7 consecutive days. Campbell *et al.* showed that the healthy elderly spent significantly more time in daylight (>2000 lux) illumination than the demented (0.99 vs. 0.49 hours), but that there was a far larger effect of gender (males 1.23 hours, females 0.31 hours). It should be noted that despite being located in Southern California, neither group experienced much daylight, although it was no less, on average, than that reported in younger adults.

Enhancement of Light Exposure

The Campbell *et al.* data are intriguing in that they suggest a possible therapeutic regimen involving increases in daylight exposure either through natural effects (being outdoors more), or using banks of bright artificial light ("light boxes"). However, caution should be exercised when prescribing such therapies, since for women, at least, Genhart, Kelly, Coursey, Datiles, and Rosenthal (1993) showed that, in those *not* suffering from seasonal affective disorder, light therapy led to a significant worsening in mood and affect. This finding may explain why Campbell *et al.*'s (1988) female subjects showed so little daylight exposure—they may have found the light aversive in regard to their mood.

Phototherapy using evening sessions with a light box have proved effective however in two situations. In the institutionalized demented, Satlin, Volicer, Ross, Herz, and Campbell (1992) showed amelioration in some (but not all) of the symptoms of "sundowning" (nocturnal restlessness and agitation) from a week of evening (1900–2100) light therapy (1500–2000 lux). This confirms early positive findings from Ancoli-Israel's group. Also, both Campbell, Dawson, and Anderson (1993a) and Lack and Wright (1993) showed that early-morning insomnia in the elderly can be improved by fairly intensive evening phototherapy. Both studies showed a delay in circadian body temperature rhythms consequent upon the therapy, suggesting a chronobiological mechanism for the effect.

In conclusion, it seems to be the case that many old people get little daylight exposure (particularly women) and that augmentation of light exposure might thus prove of chronobiological benefit. However, it should be recognized that many older people do not like bright light, and that negative mood symptoms may result from the therapy itself.

Conclusions

While age-related changes in a variety of overt circadian rhythms have been clearly established, the mechanisms underlying these changes remain to be elucidated. Certainly, the changes should *not* be considered to represent only a simple degeneration of the SCN. Observations in both humans and animals indicate that the changes in circadian rhythmicity during aging may occur at several different levels within the circadian system, such as at the pacemaker itself, at the input pathways to the pacemaker, and/or at the output pathways for the actual expression of the circadian rhythm in question. All of these processes may combine to affect the sleep and daytime functioning of the aging individual. In addition, age-related behavioral and environmental changes, such as decreased social and professional constraints, decreased exposure to light, and decreased physical activity, may weaken synchronizing signals received by the circadian clock. The decreased exposure to synchronizing agents such as bright light and social/physical activity is particularly pronounced in the housebound elderly, in residents of nursing homes, and in patients with dementia. Because disruptions in temporal organization have adverse effects on the health and well-being of older individuals and because the elderly are often overprescribed with hypnotic medications, it is important to determine how the response of the circadian system to internal and external zeitgebers might change with age and to design new clinical approaches for treating sleep and circadian rhythm disorders in the elderly.

Acknowledgments

This work was supported in part by AG 13396, AG 15136, AG 09297, and AG 11412.

References

Boivin, D. B., Duffy, J. F., Kronauer, R. E., & Czeisler, C. A. (1996) Dose–response relationships for resetting of human circadian clock by light. *Nature, 379,* 540–542.

Brock, M. A. (1985). Biological clocks and aging. *Review of Biological Research in Aging, 2,* 445–462.

Brock, M. A. (1991). Chronobiology and aging. *Journal of the American Geriatrics Society, 39,* 74–91.

Buysse, D. J., Browman, B. A., Monk, T. H., Reynolds, C. F., Fascizka, A. L., & Kupfer, D. J. (1992). Napping and 24-hour sleep/wake patterns in healthy elderly and young adults. *Journal of the American Geriatrics Society, 40,* 779–786.

Campbell, S. S., Kripke, D. F., Gillin, J. C., & Hrubovcak, J. C. (1988). Exposure to light in healthy elderly subjects and Alzheimer's patients. *Physiology and Behavior, 42,* 141–144.

Campbell, S. S., Gillin, J. C., Kripke, D..F., Erikson, P., & Clopton, P. (1989). Gender differences in the circadian temperature rhythms of healthy elderly subjects: Relationships to sleep quality. *Sleep, 12,* 529–536.

Campbell, S. S., Dawson, D., & Anderson, M. W. (1993b). Alleviation of sleep maintenance insomnia with timed exposure to bright light. *Journal of the American Geriatrics Society, 41,* 829–836.

Campbell, S. S., Dawson, D., & Zulley, J. (1993b). When the human circadian system is caught napping: Evidence for endogenous rhythms close to twenty-four hours. *Sleep, 16,* 638–640.

Czeisler, C. A., & Dijk, D. J. (1995). Use of bright light to treat maladaptation to shift work and circadian rhythm sleep disorders. *Journal of Sleep Research, 4,* 70–73.

Czeisler, C. A., Allan, J. S., Strogatz, S..H., Ronda, J. M., Sanchez, R., Rios, C. D., Freitag, W. D., Richardson, G. S., & Kronauer, R. E. (1986). Bright light resets the human circadian pacemaker independent of the timing of the sleep–wake cycle. *Science, 233,* 667–671.

Czeisler, C. A., Dumont, M., Duffy, J. F., Steinberg, J. D., Richardson, G. S., Brown, E. N., Sanchez, R., Rios, C. D., & Ronda, J. M. (1992). Association of sleep–wake habits in older people with changes in output of circadian pacemaker. *Lancet, 340,* 933–936.

Davis, F. C., & Menaker, M. (1980). Hamsters through time's window: temporal structure of hamster locomotor rhythmicity. *American Journal of Physiology, 239,* R149–R155.

Dement, W. C., Seidel, W. F., Cohen, S. A., Bliwise, N. G., & Carskadon, M. A. (1986). Sleep and wakefulness in aircrew before and after transoceanic flights. *Aviation Space & Environmental Medicine, 57,* B14–B28.

Dijk, D. J., Duffy, J. F., & Czeisler, C. A. (1992). Circadian and sleep/wake dependent aspects of subjective alertness and cognitive performance. *Journal of Sleep Research, 1,* 112–117.

Folkard, S., & Akerstedt, T. (1992). A three-process model of the regulation of alertness-sleepiness. In R. J. Broughton & R. D. Ogilvie (Eds.) *Sleep, arousal, and performance: A tribute to Bob Wilkinson* (pp. 11–26). Boston: Birkhauser.

Foret, J., Bensimon, G., Benoit, O., & Vieux, N. (1981). Quality of sleep as a function of age and shift work. In A. Reinberg, N. Vieux, & P. Andlauer (Eds.) *Night and shift work: Biological and social aspects* (pp. 149–160). Oxford: Pergamon Press.

Genhart, M. J., Kelly, K. A., Coursey, R. D., Datiles, M., & Rosenthal, N. E. (1993). Effects of bright light on mood in normal elderly women. *Psychiatry Research, 47,* 87–97.

Graeber, R. C., Sing, H. C., & Cuthbert, B. N. (1981). The impact of transmeridian flight on deploying soldiers. In L. C. Johnson, D. I. Tepas, W. P. Colquhoun & M. J. Colligan (Eds.) *Variations in work–sleep schedules: Effects on health and performance* (Vol. 7, pp. 513–537). New York: Spectrum.

Graeber, R. C., Lauber, J. K., Connell, J. L., & Gander, P. H. (1986). International aircrew sleep and wakefulness after multiple time zone flights: A cooperative study. *Sleep Research, 15,* 273.

Hildebrandt, G., & Stratmann, I. (1979). Circadian system response to night work in relation to the individual circadian phase position. *International Archives of Occupational and Environmental Health, 43,* 73–83.

Ingram, D. K., London, E. D., & Reynolds, M. A. (1982). Circadian rhythmicity and sleep: Effects of aging in laboratory animals. *Neurobiology of Aging, 3,* 287–297.

Jacobs, D., Ancoli-Israel, S., Parker, L., & Kripke, D. F. (1989). Twenty-four-hour sleep–wake patterns in a nursing home population. *Psychology and Aging, 4,* 352–356.

Klein, K. E., Wegmann, H. M., & Hunt, B. I. (1972). Desynchronization of body temperature and performance circadian rhythms as a results of out-going and homegoing transmeridian flights. *Aerospace Medicine, 43,* 119–132.

Kogi, K. (1985). Introduction to the problems of shift work. In S. Folkard & T. H. Monk (Eds.) *Hours of work: Temporal factors in work scheduling* (pp. 165–184). New York: Wiley.

Lack, L., & Wright, H. (1993). The effect of evening bright light in delaying the circadian rhythms and lengthening the sleep of early morning awakening insomniacs. *Sleep, 16,* 436–443.

Lewy, A. J., Wehr, T. A., Goodwin, F. K., Newsome, D. A., & Markey, S. P. (1980). Light suppresses melatonin secretion in humans. *Science, 210,* 1267–1269.

Lobban, M. C., & Tredre, B. E. (1967). Diurnal rhythms of renal excretion and of body temperature in aged subjects. *Journal of Physiology, 188,* 48P–49P.

Martin, J. R., Fuchs, A., Bender, R., & Harting, J. (1986). Altered light/dark activity difference with aging in two rat strains. *Journal of Gerontology, 41,* 2–7.

Miles, L., & Dement, W. C. (1980). Sleep and aging. *Sleep, 3,* 119–220.

Moe, K. E., Prinz, P. N., Vitiello, M. V., Marks, A. L., & Larsen, L. H. (1991). Healthy elderly women and men have different entrained circadian temperature rhythms. *Journal of the American Geriatrics Society, 39,* 383–387.

Moline, M. L., Pollak, C. P., Monk, T. H., Lester, L. S., Wagner, D. R., Zendell, S. M., Graeber, R. C., Salter, C. A., & Hirsch, E. (1992). Age-related differences in recovery from simulated jet lag. *Sleep, 15,* 28–40.

Monk, T. H. (1989). A visual analogue scale technique to measure Global Vigor and Affect (GVA). *Psychiatry Research, 27,* 89–99.

Monk, T. H. (1991). Sleep and circadian rhythms. *Experimental Gerontology, 26,* 233–243.

Monk, T. H., & Moline, M. L. (1989). The timing of bedtime and waketime decisions in free-running subjects. *Psychophysiology, 26,* 304–310.

Monk, T. H., Moline, M. L., & Graeber, R. C. (1988). Inducing jet-lag in the laboratory. Patterns of adjustment to an acute shift in routine. *Aviation, Space, and Environmental Medicine, 59,* 703–710.

Monk, T. H., Moline, M. L., Fookson, J. E., & Peetz, S. M. (1989). Circadian determinants of subjective alertness. *Journal of Biological Rhythms, 4,* 393–404.

Monk, T. H., Flaherty, J. F., Frank, E., Hoskinson, K., & Kupfer, D. J. (1990). The social rhythm metric: An instrument to quantify the daily rhythms of life. *Journal of Nervous and Mental Disease, 178,* 120–126.

Monk, T. H., Reynolds, C. F., Buysse, D. J., Hoch, C. C., Jarrett, D. B., Jennings, J. R., & Kupfer, D. J. (1991). Circadian characteristics of healthy 80 year olds and their relationship to objectively recorded sleep. *Journal of Gerontology: Medical Sciences, 46,* M171–M175.

Monk, T. H., Reynolds, C. F., Machen, M. A., & Kupfer, D. J. (1992a). Daily social rhythms in the elderly and their relation to objectively recorded sleep. *Sleep, 15,* 322–329.

Monk, T. H., Buysse, D. J., Reynolds, C. F., Jarrett, D., & Kupfer, D. J. (1992b). Rhythmic vs homeostatic influences on mood, activation, and performance in young and old men. *Journal of Gerontology: Psychological Sciences, 47,* 221–227.

Monk, T. H., Buysse, D. J., Reynolds, C. F., & Kupfer, D. J. (1993). Inducing jet lag in older people: Adjusting to a 6-hour phase advance in routine. *Experimental Gerontology, 28,* 119–133.

Monk, T. H., Buysse, D. J., Reynolds, C. F., & Kupfer, D. J. (1995a). Inducing jet lag in an older person: Directional asymmetry. *Experimental Gerontology, 30,* 137–145.

Monk, T. H., Buysse, D. J., Reynolds, C. F., Kupfer, D. J., & Houck, P. R. (1995b). Circadian temperature rhythms of older people. *Experimental Gerontology, 30,* 455–474.

Monk, T. H., Buysse, D. J., Reynolds, C. F., Kupfer, D. J., & Houck, P. R. (1996). Subjective alertness rhythms in elderly people. *Journal of Biological Rhythms, 11,* 268–276.

Monk, T. H., Reynolds, C. F., Kupfer, D. J., Hoch, C. C., Carrier, J., & Houck, P. R. (1997). Differences over the life span in daily life-style regularity. *Chronobiology International, 14,* 295–306.

Moore-Ede, M. C., Sulzman, F. M., & Fuller, C. A. (1982). *The clocks that time us.* Cambridge, MA: Harvard University Press.

Morin, L. P. (1988). Age-related changes in hamster circadian period, entrainment and rhythm splitting. *Journal of Biological Rhythms, 3,* 237–248.

National Institutes of Health Consensus Development Conference Statement: The treatment of sleep disorders of older people. (1991). *Sleep, 14,* 169–177.

Naylor, E., Buxton, O. M., Bergman, B., Easton, A., Zee, P. C., & Turek, F. W. (1998). Effects of aging on sleepin the golden hamster. *Sleep, 21,* 687–693.

Nicholson, A. N,. Pascoe, P. A., Spencer, M. B., Stone, B. M., & Green, R. L. (1986). Nocturnal sleep and daytime alertness of aircrew after transmeridian flights. *Aviation, Space, and Environmental Medicine, 57,* B43–B52.

Nicolau, G. Y., & Milcu, S. M. (1979). Circadian rhythms of corticosterone and nucleic acids in the rat adrenal in relation to age. *Endocrinologie, 17,* 245–250.

Penev, P. D., Zee, P. C., Wallen, E. P., & Turek, F. W. (1995). Aging alters the phase-resetting properties of a serotonin agonist on hamster circadian rhythmicity. *American Journal of Physiology, 268,* R293–R298.

Peng, M. T., & Kang, M. (1984). Circadian rhythms and patterns of running-wheel activity, feeding and drinking behaviors of old male rats. *Physiology and Behavior, 33,* 615–620.

Peng, M. T., Jiang, M. J., & Hsu, H. K. (1980). Changes in running-wheel activity, eating and drinking and their day/night distributions throughout the life span of the rat. *Journal of Gerontology, 35,* 339–347.

Peng, M. T., Yaung, C. L., & Yang, W.Z. (1984). Circadian organization of running-wheel activity, food intake and drinking of old male rats. *Proceedings of the National Science Council, Republic of China. Part B, Life Sciences, 8,* 308–318.

Pittendrigh, C., & Daan, S. (1974). Circadian oscillations in rodents: A systematic increase of their frequency with age. *Science, 186,* 548–550.

Pollak, C. P., Perlick, D., Linsner, J. P., Wenston, J., & Hsieh, F. (1990). Sleep problems in the community elderly as predictors of death and nursing home placement. *Journal of Community Health, 15,* 123–135.

Possidente, B., McEldowney, S., & Pabon, A. (1995). Aging lengthens circadian period for wheel-running activity in C57BL mice. *Physiology and Behavior, 57,* 575–579.

Preston, F. S., & Batemen, S. (1970). Effect of time zone changes on the sleep pattern of BOAC 707 crews on world-wide schedules. *Aerospace Medicine, 41,* 1409–1415.

Prinz, P. N., Obrist, W. D., & Wang, H. S. (1975). Sleep patterns in healthy elderly subjects: Individual differences as related to other neurobiological variables. *Sleep Research, 4,* 132

Rabins, R. V., Mace, N. L., & Lucas, M. J. (1982). The impact of dementia on the family. *Journal of the American Medical Association, 248,* 333–335.

Reinberg, A., Andlauer, P., Bourdeleau, P., & Levi, F. (1984). Rythms circadian de la force des mains droites et gauches: Desynchronisation chez certain travaillers postes. [Right and left hand grip circadian rhythms: desynchronization in some shift workers]. *Comptes Rendus de l'Academie des Sciences, 299,* 633–666.

Reinberg, A., Vieux, N., & Andlauer, P. (Eds.). (1981). *Night and shift: Biological and social aspects.* Oxford: Pergamon Press.

Reiter, R. J., Craft, C. M., Johnson, J. E., Jr., King, T. S., Richardson, B. A., Vaughan, G. M., & Vaughan, M. K. (1981). Age-associated reduction in nocturnal pineal melatonin levels in female rats. *Endocrinology, 109,* 1295–1297.

Richardson, G. S., Carskadon, M. A., Orav, E. J., & Dement, W.C. (1982). Circadian variation of sleep tendency in elderly and young adult subjects. *Sleep, 5,* S82–S94.

Richardson, G. S., Moore-Ede, M. C., Czeisler, C. A., & Dement, W. C. (1985). Circadian rhythms of sleep and wakefulness in mice: analysis using long-term automated recording of sleep. *American Journal of Physiology, 248,* R320–R330.

Rosenberg, R. S., Zepelin, H., & Rechtschaffen, A. (1979). Sleep in young and old rats. *Journal of Gerontology, 34,* 525–532.

Rosenberg, R. S., Zee, P. C., & Turek, F. W. (1991). Phase response curves to light in young and old hamsters. *American Journal of Physiology, 261,* R491–R495.

Sacher, G. A., & Duffy, P. H. (1978). Age changes in rhythms of energy metabolism, activity, and body temperature in *Mus* and *Peromyscus. Advances in Experimental Medicine and Biology, 108,* 105–124.

Sanford, J. R. A (1975). Tolerance of debility in elderly dependents by supporters at home: Significance for hospital practice. *British Medical Journal, 3,* 471–473.

Satinoff, E., Li, H., Tcheng, T. K., Liu, C., McArthur, A. J., Medanic, M., Gillette, M. U. (1993). Do the suprachiasmatic nuclei oscillate in old rats as they do in young ones? *American Journal of Physiology, 265,* R1216–R1222.

Satlin,.A., Volicer, L., Ross, V., Herz, L., & Campbell, S. S. (1992). Bright light treatment of behavioral and sleep disturbance in patients with Alzheimer's disease. *American Journal of Psychiatry, 149,* 1028–1032.

Savides, T. J., Messin, S., Senger, C., & Kripke, D. F. (1985). Natural light exposure of young adults. *Sleep Research, 14,* 310.

Scott, A. J., & LaDou, J. (1990). Shiftwork: Effects on sleep and health with recommendations for medical surveillance and screening. *Occupational Medicine, 5,* 273–299.

Sutin, E. L., Dement, W. C., Heller, H. C., & Kilduff, T. S. (1993). Light-induced gene expression in the suprachiasmatic nucleus of young and aging rats. *Neurobiology of Aging, 14*, 441–446.

Swaab, D. F., Fliers, E., & Partiman, T. S. (1985). The suprachiasmatic nucleus of the human brain in relation to sex, age and senile dementia. *Brain Research, 342*, 37–44.

Tasto, D. L., & Colligan, M. J. (1978). *Health consequences of shift work (Project URI1–4426).* Menlo Park, CA: Stanford Research Institute.

Tenover, J. S., Matsumoto, A. M., Clifton, D. K., & Bremner, W. J. (1988). Age-related alterations in the circadian rhythms of pulsatile luteinizing hormone and testosterone secretion in healthy men. *Journal of Gerontology, 43*, M163–M169.

Tune, G. S . (1969). The influence of age and temperament on the adult human sleep-wakefulness pattern. *British Journal of Psychology, 60*, 431–441.

Turek, F. W., Pinto, L. H., Vitaterna, M. H., Penev, P. D., Zee, P. C., & Takahashi, J. S. (1995). Pharmacological and genetic approaches for the study of circadian rhythms in mammals. *Frontiers in Neuroendocrinology, 16*, 191–223.

U.S. Bureau of the Census. (1992). 1990 census of population and housing: Alphabetic index of industries and occupations. Washington, D.C.: U.S. Bureau of the Census.

Van Gool, W. A., & Mirmiran, M. (1986). Effects of aging and housing in an enriched environment on sleep–wake patterns in rats. *Sleep, 9*, 335–347.

Van Gool, W. A., Witting, W., & Mirmiran, M. (1987). Age-related changes in circadian sleep–wakefulness rhythms in male rats isolated from time cues. *Brain Research, 413*, 384–387.

van Reeth, O., Zhang, Y., Zee, P. C., & Turek, F. W. (1992). Aging alters feedback effects of the activity-rest cycle on the circadian clock. *American Journal of Physiology, 263*, R981–R986.

van Reeth, O., Zhang, Y., Reddy, A., Zee, P., & Turek, F. W. (1993). Aging alters the entraining effects of an activity-inducing stimulus on the circadian clock. *Brain Research, 607*, 286–292.

van Someren, E. J., Mirmiran, M., & Swaab, D. F. (1993). Non-pharmacological treatment of sleep and wake disturbances in aging and Alzheimer's disease: Chronobiological perspectives. *Behavioral Brain Research, 57*, 235–253.

Vitiello, M. V., Smallwood, R. G., Avery, D. H., Pascualy, R. A., Martin, D. C., & Prinz, P. A. (1986). Circadian temperature rhythms in young adult and aged men. *Neurobiology of Aging, 7*, 97–100.

Weiland, N. G., & Wise, P. M. (1990). Aging progressively decreases the densities and alters the diurnal rhythms of alpha 1-adrenergic receptors in selected hypothalamic regions. *Endocrinology, 126*, 2392–2397.

Weitzman, E. D., Moline, M. L., Czeisler, C. A., & Zimmerman, J. C. (1982). Chronobiology of aging: Temperature, sleep–wake rhythms and entrainment. *Neurobiology of Aging, 3*, 299–309.

Welsh, D. K., Richardson, G. S., & Dement, W. C. (1986). Effect of age on the circadian pattern of sleep and wakefulness in the mouse. *Journal of Gerontology, 41*, 579–586.

Wever, R. A. (1979). *The circadian system of man: results of experiments under temporal isolation.* New York: Springer Verlag.

Wever, R. A., Poiasek, J., & Wildgruber, C. M. (1983). Bright light affects human circadian rhythms. *Pfulgers Archiv European Journal of Physiology, 396*, 85–87.

Wise, P. M., Cohen, I. R., Weiland, N. G., & London, E. D. (1988). Aging alters the circadian rhythm of glucose utilization in the suprachiasmatic nucleus. *Proceedings of the National Academy of Sciences of the USA, 85*, 5305–5309.

Witting, W., Mirmiran, M., Bos, N. P., & Swaab, D. F. (1994). The effect of old age on the free-running period of circadian rhythms in rat. *Chronobiology International, 11*, 103–112.

Zee, P. C., Rosenberg, R. S., & Turek, F. W. (1992). Effects of aging on entrainment and rate of resynchronization of circadian locomotor activity. *American Journal of Physiology, 263*, R1099–R1103.

Zhang, Y., Kornhauser, J. M., Zee, P. C., Mayo, K. E., Takahashi, J. S., & Turek, F. W. (1996). Effects of aging on light-induced phase-shifting of circadian behavioral rhythms, Fos expression, and CREB-phosphorylation in the hamster suprchiasmatic nucleus. *Neuroscience, 70*, 951–961.

23

Circadian Rhythms in Affective Disorders

ELLEN LEIBENLUFT AND ELLEN FRANK

INTRODUCTION

The affective or mood disorders are among the most common psychiatric conditions. They cause considerable morbidity and mortality, and therefore have received a great deal of research attention. For reasons which will be described below, researchers with an interest in circadian rhythms have been among those working to understand the pathophysiology and etiology of mood disorders. In this chapter, we will review their work, along with the conceptual and methodological issues that have complicated it.

Following a brief, general description of the affective syndromes, we will describe the clinical features that have led researchers to suspect that circadian rhythm dysregulation may play a role in their pathogenesis. Next, we will discuss the methodological and conceptual challenges confronting workers in this field. Finally, we will describe the various hypotheses that have been suggested concerning a possible role for circadian rhythm dysregulation in the pathogenesis of mood disorders and the data that support or refute these hypotheses.

CLINICAL DESCRIPTION OF THE AFFECTIVE SYNDROMES

Although we distinguish affective disorders from other psychiatric illnesses on the basis of the mood changes observed, other behavioral changes, such as alterations in sleep, appetite, and activity level, are also characteristic of these syndromes.

ELLEN LEIBENLUFT Pediatrics and Developmental Neuropsychology Branch, National Institute of Mental Health, Bethesda, Maryland 20892-1255. ELLEN FRANK Western Psychiatric Institute and Clinic, University of Pittsburgh, Pittsburgh, Pennsylvania 15213.

Circadian Clocks, Volume 12 of *Handbook of Behavioral Neurobiology*, edited by Joseph S. Takahashi, Fred W. Turek, and R. Y. Moore, Kluwer Academic / Plenum Publishers, New York, 2001.

In addition, each of the mood disorders has a characteristic (and frequently cyclic) course. For the purposes of this chapter, it is worth considering both the symptoms of these illnesses and their usual course since both have relevance for the relationship between circadian timing and affective disorders. The following disorders have been discussed in the literature on circadian rhythms and mood disorders and will therefore be described here: major depressive disorder, bipolar disorder (characterized by one or more major depressive episodes and one or more episodes of mania or hypomania), rapid cycling bipolar disorder, seasonal affective disorder, and premenstrual dysphoric disorder.

The diagnostic criteria for all psychiatric illnesses are specified in the *Diagnostic and Statistical Manual* of the American Psychiatric Association, 4th edition (DSM-IV; American Psychiatric Association, 1994). The DSM-IV criteria for a *major depressive episode* are summarized in Table 1. From the perspective of the circadian rhythm researcher, it is important to note that a major depressive episode can be characterized by either increased activity (psychomotor agitation and insomnia, often accompanied by loss of appetite and weight) or by hypoactivity (psychomotor retardation and hypersomnia, often accompanied by carbohydrate craving and weight gain). The former constellation of symptoms is frequently called a "typical" depression, whereas the latter is called "atypical."

Although *major depressive disorder* can be diagnosed following a single episode of major depression, in most instances, the disorder takes a recurrent form. The course of the illness is variable; episodes may return after only months of remission or, in other cases, after many years of wellness. Major depressive disorder was formerly called unipolar depression to distinguish it from bipolar illness, described below. It is one of the most common forms of affective illness, with a lifetime prevalence of 21% in women and 17% overall (Kessler *et al.*, 1994).

The diagnostic criteria for *manic* and *hypomanic episodes* are summarized in Table 2. In addition to elevated or irritable mood, these episodes are characterized by a general increase in activity level, with racing thoughts, rapid speech, increased psychomotor activity, and decreased sleep. *Bipolar illness* (formerly called manic-

TABLE 1. DSM-IV DIAGNOSTIC CRITERIA FOR MAJOR DEPRESSIVE EPISODE

A. 2-week period during whic at least five of the following symptoms have been present nearly every day and represent a change from previous functioning; at least one of the symptoms is either depressed mood or loss of interest or pleasure:
   1. Depressed mood most of the day
   2. Markedly diminished loss of interest or pleasure in all, or almost all, activities most of the day
   3. Significant weight loss or gain; or decrease or increase in appetite
   4. Insomnia or hypersomnia
   5. Psychomotor agitation or retardation (observable by others)
   6. Fatigue or loss of energy
   7. Feelings of worthlessness or excessive or inappropriate guilt
   8. Diminished ability to think or concentrate, or indecisiveness
   9. Recurrent thoughts of death or recurrent suicidal ideation, or a suicide attempt or specific plan for committing suicide

B. The symptoms cause clinically significant distress or impairment in social, occupational, or other important areas of functioning

C. The symptoms are not due to the direct physiologic effects of a substance or a general medical condition

TABLE 2. DSM-IV DIAGNOSTIC CRITERIA FOR MANIC EPISODE AND HYPOMANIC EPISODE

A. A distinct period of abnormally or persistently elevated, expansive, or irritable mood lasting at least 1 week (or any duration if hospitalization is required)

B. During the period of mood disturbance, three (or more) of the following symptoms have persisted and have been present to a significant degree:
   1. Inflated self-esteem or grandiosity
   2. Decreased need for sleep (e.g., feels rested after only 3 hours of sleep)
   3. Talkativeness
   4. Flight of ideas or a sense that thoughts are racing
   5. Distractibility
   6. Increase in goal-directed activity or psychomotor agitation
   7. Excessive involvement in pleasurable activities that have a high potential for painful consequences, e.g., unrestrained buying sprees, sexual indiscretions, foolish business investments, etc.

C. In a manic episode, the mood disturbance is sufficiently severe to cause marked impairment in occupational functioning or in usual activities or relationships with others, or to necessitate hospitalization, or is associated with the presence or psychosis

D. In contrast to a manic episode, a hypomanic episode requires a minimum of only 4 days of symptoms; in addition, a hypomanic episode is associated with an unequivocal change in function observable by others, but it is not associated with marked impairment, hospitalization, or psychosis

E. The symptoms are not due to the direct physiologic effects of a substance or a general medical condition

depressive disorder) is characterized by recurrent episodes of major depression and mania or hypomania. Since bipolar depressive episodes generally involve "atypical" symptoms (i.e., hypersomnia and psychomotor retardation), the alternation of these depressive episodes with hypomanic or manic episodes makes it possible to argue that bipolar illness is as much a dysregulation of activity level as it is of mood.

As in major depressive disorder, the course of bipolar disorder is variable; episodes may be closely spaced or separated by years of remission. One form of bipolar illness which has received much attention in the literature on circadian rhythms and mood disorder is *rapid cycling bipolar disorder* (RCBD). According to the DSM-IV, a bipolar patient is considered to be rapid cycling when he or she has four or more episodes of mania, hypomania, or major depression in 1 year (American Psychiatric Association, 1994). Most rapid cycling patients, however, show a much more rapid alternation of mood states; indeed, in what is now called ultrarapid cycling, more than one mood state may be observed within a single day (Kramlinger & Post, 1996).

In addition to these illnesses, other mood disorders are of particular interest to circadian rhythm researchers because of their characteristic cyclic pattern. *Seasonal affective disorder* (SAD) is a recurrent mood disorder with a regular onset at a specific time of year (Rosenthal *et al.*, 1984). The most common form of SAD consists of a major depressive episode in the fall or winter, coupled with a remission or hypomania in the spring or summer. However, a pattern of summer depression has also been observed (Wehr, Sack, & Rosenthal, 1987b). The symptom picture of patients with SAD is usually that of an atypical depression.

Finally, *premenstrual dysphoric disorder* (PMDD) is a mood disorder characterized by depressive symptoms during the luteal phase, or second half, of the menstrual cycle, and remission within a few days of the start of menses (American Psychiatric Association, 1994). According to DSM-IV, the clinical features of PMDD are quite similar to those of major depression, with the obvious exception of a shorter duration. In addition, the sleep, appetite, and weight changes in PMDD are rarely as marked as in major depression. However, physical symptoms such as breast tenderness, bloating, and headaches may occur.

Why Think about Circadian Rhythms in Affective Illness?

Several characteristic clinical features have caused researchers to hypothesize that abnormalities in circadian rhythm regulation may be involved in the pathogenesis of mood disorders. First, most mood disorders are recurrent, and some are clearly periodic. A minority of bipolar patients, such as those with 48-hour cycles (Wehr, Goodwin, Wirz-Justice, Breitmaier, & Craig, 1982), have episodes that recur with a regular periodicity. Both SAD and PMDD are regularly cyclic, and the efficacy of bright light therapy in SAD indicates that the relationship between SAD patients' mood cycles and the seasonal cycle may be causal, rather than merely correlational. In the case of PMDD, the mood disorder appears to be entrained to the menstrual cycle rather than caused by hormonal fluctuations, since the disruption of patients' luteal phase with the antiprogestin RU-486 does not prevent the development of mood symptoms (Schmidt *et al.*, 1991).

Researchers have also studied circadian rhythms in patients with mood disorders because these patients typically experience marked changes in their sleep–wake cycle and in their activity level as they cycle between hypomania or mania and depression, or between depression and euthymia (normal mood). Patients with bipolar disorder often sleep 12 hours or more when depressed, and then experience almost total insomnia when manic. In addition, the sleep deprivation literature clearly demonstrates that changes in the sleep–wake cycle are not merely symptoms of mood disorders, but also have pathogenic significance. That is, numerous studies have shown that approximately 50% of depressed patients will become euthymic if kept awake for all or part of the night (Wehr, 1990a). In bipolar patients, sleep deprivation can precipitate a switch into mania (Wehr, 1990a). Patients whose depression remits after a night of sleep deprivation frequently relapse after a night of recovery sleep, leading to the formulation that sleep deprivation is antidepressant whereas sleep is depressogenic (Wehr, 1990a). In addition, evidence that sleep deprivation in the second half of the night may be a more effective antidepressant than sleep deprivation in the first half of the night indicates that a circadian mechanism may be involved (Sack, Rosenthal, Duncan, Mendelson, & Wehr, 1988).

The third clinical feature of mood disorders that has caused researchers to focus on circadian rhythms is the symptom of diurnal variation. Diurnal variation is defined as a systematic change in mood over the course of the day, with typical diurnal variation being characterized by a gradual lifting of depressed mood as the day goes on (Haug & Wirz-Justice, 1993). It is quite common for patients with severe depressive episodes to report nearly catatonic incapacitation early in the day, with noticeable improvement in mood and energy as the day wears on. In a sense, this phenomenon can be seen as an extension of the sleep deprivation literature, since once again, increased duration of wakefulness is associated with an antidepressant response. Indeed, the presence of diurnal variation has in some, but not all, studies been associated with a positive response to sleep deprivation (Reinink, Bouhuys, Wirz-Justice, & van den Hoofdakker, 1990). The concept of diurnal variation can also be extended to bipolar illness, since it has been shown that, in RCBD, there is a systematic relationship between time of day and the direction of mood switches. That is, rapid cycling patients tend to experience not just a lifting in mood as the day wears on, but an actual switch up, out of depression into hypomania or euthymia, and then (consistent with a depressogenic effect of sleep) switch back into depression overnight (Feldman-Naim, Turner, & Leibenluft, 1996).

Conceptual and Methodological Issues

Several conceptual issues complicate the design and interpretation of circadian rhythm studies in patients with affective illnesses. Prime among these is the fact that the symptoms of mood disorders are likely to feed back on the circadian system and affect the phase, amplitude, and stability of circadian rhythms. Thus, when patients and controls are found to differ on a circadian measure, it is unclear whether the intergroup difference is the cause or the result of the mood disorder. For example, compared to a euthymic control, a depressed, hypersomnic patient is less active and has less exposure to light (as a result of the "gating" effect of closing one's eyes to sleep). The investigator who finds phase or amplitude differences between patient and control in one or more circadian rhythms is then unable to discern whether these differences preceded, and somehow contributed to, the onset of symptoms; whether they are an irrelevant epiphenomenon; or whether they are secondary to the patient's symptoms, but nonetheless contribute to the development, or perpetuation, of the affective episode (Wehr, 1990b).

In order to determine whether circadian rhythm differences between patients and controls are the cause (rather than the result) of the patients' mood disorder, some investigators have standardized the sleep–wake schedules of both patients and controls for 1 week before conducting a research procedure such as the constant routine (e.g., Dahl *et al.*, 1993; Sack, Lewy, & White, 1990). However, this strategy has disadvantages that might limit the study's generalizability. Since it has been proposed that prescribing regular sleep–wake cycles to patients can be a therapeutic intervention (Frank, Shear, Essex, Leibenluft, & Monk, 1995), the study protocol might alter the clinical phenomenon being investigated. In addition, the design would introduce a selection bias, since only a subsample of patients would be willing and able to comply with such a protocol. Nonetheless, this strategy does eliminate the possibility that observed physiologic differences are simply epiphenomena resulting from the influence of the patient's disordered sleep–wake cycle on circadian rhythms.

Another technique for untangling cause from effect involves the use of experimental, rather than descriptive or correlative, designs. One can manipulate a patient's circadian rhythms with, for example, exogenous melatonin or exposure to bright light, and then measure the effect of the intervention both on the patient's rhythms and on his or her symptoms. This technique has been used in studies of patients with SAD (Sack *et al.*, 1990) and, to a lesser extent, PMDD (Parry *et al.*, 1994). These studies are, in essence, both clinical efficacy trials (testing the utility of light or melatonin as a treatment for the disorder) as well as tests of the hypothesis that circadian dysfunction plays an etiologic role in the disorder. Such approaches have not been applied to patients with bipolar or major depressive disorder, although we currently are conducting trials of melatonin and phototherapy in RCBD (Leibenluft *et al.*, 1995), with circadian measures after the treatment and after a control intervention.

In addition to the increased use of experimental designs, other methodological techniques could enhance our understanding of the possible role of circadian rhythm dysfunction in the pathogenesis of affective disorders. One such method involves complementing cross-sectional studies with longitudinal approaches so that the sequential stages in the development of the illness can be observed. For example, investigators have studied a few bipolar patients' circadian rhythms continuously as they cycle in and out of depression and mania (Wehr & Goodwin, 1983). Figure 1 shows sleep, mood, and phase of onset of nocturnal melatonin secretion in a patient with RCBD as she progresses through a depressive episode. In this subject, the onset of nocturnal melatonin secretion is relatively phase delayed early in the episode and then gradually advances as the episode wears on, reaching its most advanced position just before the switch into hypomania. Thus, the phase of melatonin onset varies, and it appears to bear a systematic relationship both to mood state and to the stage of the depressive episode. It is important to note that the results of a cross-sectional study of circadian phase done early in this depressive episode would differ from one done late in the episode. Dew *et al.* (1996) presented similar data concerning EEG sleep profiles in elderly depressives, demonstrating that patients whose sleep was studied early in an episode showed more severe abnormalities than those

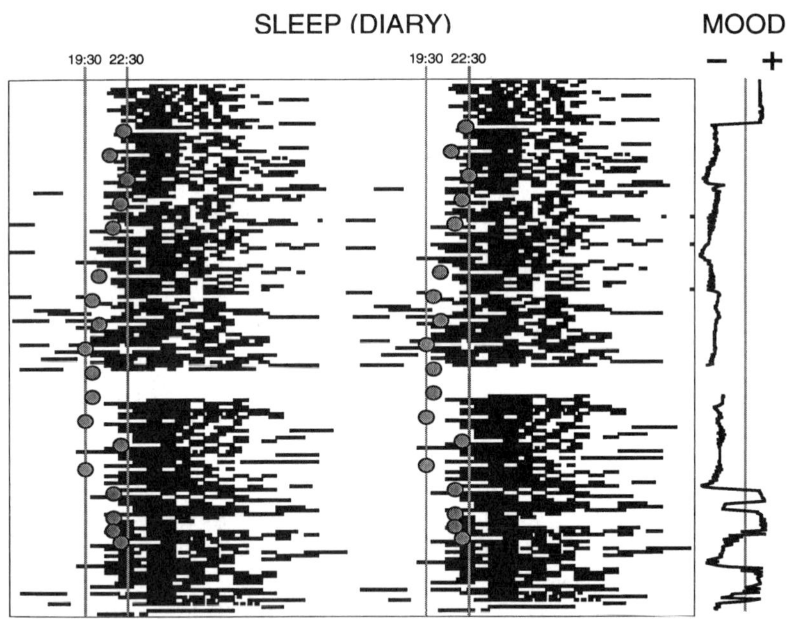

Figure 1. Data were obtained from a 40-year-old woman with rapid cycling bipolar disorder being treated with fluoxetine and lithium carbonate. Left: Self-reported sleep is double-plotted, with consecutive 24-hour (14:00–14:00) periods displayed in raster format. Superimposed red circles indicate the time of melatonin onset, as measured in blood samples drawn in dim light (<1 lux) every 30 minutes from 5:00 p.m. until midnight. Melatonin on set was defined as the time of appearance of sustained detectable (>5 pg/ml) melatonin levels; the assay was performed by Stockgrand, Ltd., Surrey, U.K. Right: Self-reported mood is displayed. Each morning upon awakening, the patient rated her mood on a 100-mm line, with 100 being "most manic I've ever felt" and 0 being "most depressed I've ever felt." See text for discussion of the data.

whose sleep was studied later. Unfortunately, the cost and inconvenience of longitudinal studies make it unlikely that large numbers of patients will ever be studied in this way. Alternatively, investigators conducting cross-sectional studies could be more attentive to longitudinal course and, for example, standardize the stage of illness (e.g., early vs. late in an episode) at which patients are studied.

Standardizing the patient's stage of illness is one means of increasing the homogeneity of a study sample. Heterogeneous patient samples may account, at least in part, for some of the conflicting results of circadian studies in patients with mood disorders. While bipolar and unipolar depressed patients appear to differ on a number of physiologic parameters (Goodwin & Jamison, 1990), they are frequently combined into a single group in circadian rhythm studies. Similarly, inpatients and outpatients are frequently combined, although inpatients living on a hospital routine are probably exposed to more consistent zeitgebers. In addition, it seems possible (if not likely) that the circadian rhythms of anergic, hypersomnic depressives might well differ from those of agitated, insomnic depressives since the sleep–activity cycles of the two groups are so different. Therefore, investigators should consider either including only one subtype of patient in a study or analyzing the data from subgroups separately. Patients with relatively frequent cycles, such as those with SAD, PMDD, or RCBD, may be particularly good subjects for circadian studies because intraindividual comparisons can be made (i.e., comparing the euthymic with the depressed or hypomanic state), further decreasing variability.

Finally, it is important to note that many studies of circadian rhythms in patients with mood disorders are difficult to interpret because the investigators failed to use appropriate unmasking techniques. For example, no studies exist in which the circadian rhythms of patients with the most common affective illness, major depressive disorder, have been studied under constant routine conditions before and after treatment. Even when the constant routine is used in depressed patients, the interpretation of the data may not be straightforward. Sleep deprivation, a necessary component of the procedure, may induce remission (Wu & Bunney, 1990), so that a patient who started the study depressed may end it euthymic or hypomanic.

Circadian Models of Affective Illness

Investigators have suggested a number of theories concerning the possible role of circadian dysfunction in the pathophysiology of mood disorders. We have organized our discussion of these according to the circadian parameter that is hypothesized to be the primary site of pathology. However, since the parameters are interrelated, there will obviously be overlap between hypotheses, and more than one hypothesis could be invoked to explain a study's result. For example, it has been suggested that instability in the phase of one or more circadian rhythms may contribute to the pathophysiology of bipolar illness and/or major depressive disorder (Halberg, 1968). However, since low-amplitude rhythms are particularly susceptible to phase shifts (Aschoff, 1983), the observation of unstable phase in mood-disordered patients leaves open the question of whether the primary abnormality is the unstable phase itself or, rather, a low amplitude. With this caveat in mind, we will discuss the following hypotheses for the possible role of circadian abnormalities in the pathogenesis of affective disorders: phase abnormalities (including phase advance, phase delay, unstable phase, and an altered alpha:rho ratio), decreased amplitude, period abnormalities, and the two-process model of sleep.

PHASE ADVANCE. Of all circadian models for affective illness, the phase-advance hypothesis has probably received the most attention. While investigators have studied the phase of several circadian rhythms in depressed patients, including temperature, thyrotropin secretion, motor activity, melatonin, and urinary 3-methoxy-4-hydroxyphenylglycol (MPHG) excretion, the greatest support for the phase-advance hypothesis comes from studies of cortisol secretion and of the timing of rapid eye movement (REM) sleep.

Georgi (1947) suggested that early morning awakening, as well as other sleep-wake abnormalities commonly seen in depressed patients, could be caused by a phase disturbance in one or more circadian rhythms. Beginning in the mid-1960s, investigators noted that depressed patients, compared to controls, had a higher proportion of REM sleep in the early part of the night (Gresham, Agnew, & Williams, 1965). Kupfer and Foster (1972) published data suggesting that decreased REM latency was a specific biological marker for depression. Drawing on these observations, Papousek (1975) proposed that the sleep abnormalities of depression result from a phase advance of the REM sleep propensity rhythm relative to the onset of sleep. Wehr, Wirz-Justice, Goodwin, Duncan, and Gillin (1979) hypothesized that, in depressed patients, the rhythm of the "strong" oscillator driving REM sleep, temperature, and cortisol was abnormally advanced relative to the rhythm of the "weak" oscillator controlling sleep onset. According to this model, sleep is depressogenic when it occurs during a sleep-sensitive circadian interval early in the morning, and sleep deprivation is antidepressant because it removes sleep from this sleep-sensitive interval. Wehr *et al.* (1979) tested the phase-advance hypothesis experimentally by advancing the sleep of a bipolar woman 6 hours, an intervention which restored a normal phase angle between sleep onset and REM onset and induced a clinical remission.

The observation that REM latency is decreased in depressed patients compared to normals has received a great deal of support in the literature, although shortened REM latency is also found in other psychiatric patients. In a recent meta-analysis of 177 investigations of EEG sleep including a total of 7151 patients, the authors noted that REM latency was generally decreased in affective patients relative to controls and to other psychiatric patients, especially in the age group of 20–50 years (Benca, Obermeyer, Thisted, & Gillin, 1992). While the analysis also showed decreased REM latency in patients with borderline personality disorder, eating disorders, and narcolepsy (compared to controls), relatively few patients in these diagnostic categories have been studied, and the comorbidity of these illnesses with depression is high.

Several lines of evidence indicate that the phase-advanced REM sleep of depressed patients may be more than an epiphenomenon. First, recent data indicate that shortened REM latency may be a trait in depressed patients, persisting even after they become euthymic (Kupfer, Frank, Grochocincki, Gregor, & McEachran, 1988; Rush *et al.*, 1986). Second, there is evidence that the asymptomatic first-degree relatives of depressed patients may also exhibit decreased REM latency compared to controls (Giles, Roffwarg, & Rush, 1987; Giles, Biggs, Rush, & Roffwarg, 1988; Giles *et al.*, 1989). Third, experimental interventions in which the timing of sleep is shifted and patients' depressive symptoms remit support the pathogenic importance of phase-advanced REM sleep. The initial case of experimentally phase-advanced sleep (Wehr *et al.*, 1979), described above, has been followed by other studies documenting the antidepressant effects of this intervention (Sack, Nurnberger, Rosenthal, &

Wehr, 1985). Recently, investigators have prolonged the antidepressant effects of sleep deprivation by coupling it with a phase advance of the sleep–wake cycle (Vollmann & Berger, 1993).

It is important to note that decreased REM latency is only one of several abnormalities that have been consistently documented in the sleep of depressed patients. Patients complain of frequent awakenings and of either increased or decreased sleep duration, complaints which have been confirmed by polysomnographic studies (Benca *et al.*, 1992). In addition, studies show that the REM sleep of depressed patients is abnormal not only in its timing, but also in its amount. That is, depressed patients, compared with controls, tend to have an increased ratio of REM sleep to total sleep time, increased REM density and activity, and a decreased ratio of slow-wave sleep (SWS) to total sleep time (Benca *et al.*, 1992). Antidepressant medications appear to suppress REM sleep and increase REM latency (Wehr & Wirz-Justice, 1982). As discussed below, Borbély and Wirz-Justice (1982) used the two-process model of sleep regulation to attempt to explain these observed abnormalities in the REM and slow-wave sleep of depressed patients.

Researchers testing the phase-advance hypothesis have also measured cortisol secretion since the "strong" circadian oscillator driving REM sleep was also thought to control the timing of cortisol secretion. Souetre, Wehr, and Rosenthal (unpublished) reviewed data from 25 studies that documented the cortisol rhythm in a total of 435 patients with major depressive disorder or bipolar depression. Fourteen of these studies found that the mean nadir of cortisol secretion was shifted earlier in depressed patients compared to controls, with the shift ranging in magnitude from 10 minutes to 6 hours. However, in only 3 of these studies were the differences statistically significant, a fact which could be due to small sample sizes and/or high interindividual variability. Interestingly, only 2 of the 25 studies found that the nadir or cortisol secretion was phase delayed in depressed patients compared to controls. An unpublished meta-analysis of cortisol data appears to confirm the finding of an early nadir in the cortisol secretion of patients compared to controls (Wirz-Justice, 1995). Like shortened REM latency, the timing of the cortisol nadir appears to be genetically influenced (Linkowski *et al.*, 1993), so the relationship of both these traits to the heritability of mood disorders bears further exploration.

Since none of the cortisol studies employed the constant routine procedure, it is possible that observed differences between patients and controls are due to the masking effects of sleep. Indeed, whereas the phase-advance hypothesis was originally articulated by Wehr and Wirz-Justice (1980) as an internal coincidence model, these cortisol studies do not typically measure the phase angle between cortisol secretion and other rhythms (such as the time of sleep onset). Instead, investigators have simply compared the absolute time of the cortisol nadir in patients to that in normals. In addition, data from patients with major depressive disorder have often been combined with those from bipolar depressed patients. Since one study which analyzed data from the two groups separately found that the nadir of cortisol secretion was more likely to be phase advanced in major depressives than in the bipolar patients (Linkowski *et al.*, 1985), the fact that they are frequently pooled may further contribute to the lack of statistically significant findings.

In the more limited number of studies that have examined the phase of temperature, thyroptropin, or melatonin in patients with affective illnesses, unmasking techniques have not generally been employed. For example, the studies of melatonin and temperature in nonseasonal depressives were designed so that patients were placed in dim light only at bedtime and were allowed to sleep. Several of the temperature studies did find that the nocturnal minimum was phase advanced in patients compared with controls, although again the differences were generally not statistically significant.

In sum, the phase-advance hypothesis has received some support from studies that show that, in depressed patients compared with controls, REM latency may be decreased and the timing of the nocturnal cortisol nadir may be advanced. However, negative studies certainly exist, so it is unclear how consistent or pronounced the phase advance is and whether it has pathogenic significance. In addition, if these two rhythms are indeed advanced in patients with mood disorders, it is unclear whether the abnormality is limited to these rhythms or is part of a more widespread pattern, since Wehr *et al.* (1993) demonstrated that the phases of various rhythms (as well as the nadir and peak of, for example, the cortisol rhythm) can be dissociated in control subjects.

Finally, if rhythms in depressed patients are pathologically advanced, it is possible that this finding is secondary to other abnormalities, such as an unusually short tau or increased sensitivity to phase-advancing zeitgebers. With respect to the latter, it is possible that mood-disordered patients have a higher amplitude phase response curve to light than do normals. Some indirect support for this hypothesis can be derived from the finding by Lewy, Wehr, Goodwin, Newsome, and Rosenthal (1981) that bipolar patients may have increased sensitivity to melatonin suppression by light, although other investigators have been unable to replicate this (Lam *et al.*, 1990).

PHASE DELAY. The hypothesis that abnormally delayed rhythms may cause affective illness has been suggested in the case of SAD, with the rhythm in question being that of nocturnal melatonin secretion. Specifically, Lewy, Sack, and Singer (1984) hypothesized that the onset of nocturnal melatonin secretion was phase delayed in SAD patients compared to normals. Since this abnormality could be reversed by morning light, the phase-delay hypothesis would also explain the effectiveness of morning phototherapy in treating SAD. In contrast to the studies of phase markers in nonseasonal depressives, both Sack *et al.*'s (1990) original finding and the attempts to replicate it have used melatonin (rather than REM sleep or cortisol) as a phase marker and have employed appropriate unmasking techniques. Most, but not all, studies have found a phase-delayed nocturnal melatonin onset (and, in some instances, a phase-delayed nocturnal temperature rhythm) in SAD patients when they are depressed compared to when they are euthymic and, in some instances, compared to controls (Avery *et al.*, 1990; Thalen, Kjellman, Morkrid, & Wetterberg, 1995). However, it is now well established that bright light exposure at times other than morning also has therapeutic effects in SAD (Wirz-Justice *et al.*, 1993), a finding which is inconsistent with the phase-delay hypothesis. In addition, while morning light does phase advance rhythms in patients with SAD, there is no correlation between the presence or amount of phase advance and the therapeutic effect (Checkley, 1989).

A seasonal pattern is not sustained in all patients who receive the diagnosis of SAD (Schwartz, Brown, Wehr, & Rosenthal, 1996); it is possible that the subgroup of patients who remain consistently seasonal have a phase-delayed melatonin rhythm. In addition, it may be worthwhile exploring whether patients with nonseasonal depression who share the atypical features of SAD patients (e.g., hypersomnia, carbohydrate craving) have one or more phase-delayed rhythms and respond to phototherapy.

UNSTABLE PHASE. Unstable phase is a central feature of one of the first hypotheses concerning the possible role of abnormal circadian rhythms in the pathogenesis of mood disorders. Halberg (1968) suggested that some, but not all, circadian rhythms in patients with RCBD are unentrained to the 24-hour day–night cycle and freerun. According to this hypothesis, the free-running rhythm goes in and out of phase with other, entrained circadian rhythms. This creates a beat phenomenon such that mood switches occur when the rhythms are in phase or 180 deg out of phase with each other. Kripke, Mellaney, Atkinson, and Wolf (1978) presented data to support this hypothesis when they reported that the temperature rhythm of five of seven patients with rapid cycling bipolar disorder appeared to be free-running, with an abnormally short tau. However, other investigators failed to detect free-running rhythms in patients with mood disorders (Mrosovsky, Reebs, Honrado, & Salmon, 1989; Wehr *et al.*, 1985).

An alternative to the free-running hypothesis states that patients with mood disorders are weakly entrained to external zeitgebers, so that their circadian rhythms exhibit an unstable phase. According to this model, the majority of patients would not have free-running rhythms, although the most severely affected might. Two studies reported that the phase of the naturalistic temperature rhythm was less stable in patients than in controls (Daimon, Yamada, Tsujimoto, & Takahashi, 1992; Tsujimoto, Yamada, Shimoda, Hanada, & Takahashi, 1990); these studies also found decreased amplitude in the patients' temperature rhythm. In addition, preliminary data from Leibenluft, Schwartz, Turner, Rosenthal, and Wehr (1993) indicate that the unmasked phase of nocturnal melatonin onset may be advanced approximately 90 minutes in patients with RCBD in the hypomanic state compared to its timing in the depressed state. However, as noted above, such descriptive studies cannot determine whether the observed phase shifts are epiphenomena of, for example, the patients' shifting sleep schedules, or whether these shifting schedules are caused by circadian dysfunction.

There are several possible causes for unstable circadian rhythms in patients with mood disorders. We will discuss below the data suggesting that the circadian rhythms of depressed patients have abnormally low amplitude, a characteristic that could lead to phase instability. Alternatively, patients with mood disorders may have unstable rhythms because they are exposed to fewer (or less powerful) zeitgebers than unaffected individuals and/or because they are less sensitive to those zeitgebers. Ehlers, Frank, and Kupfer (1988) suggested that psychosocial stress, such as marital separation or the loss of a job, may precipitate depression through the loss of social zeitgebers, i.e., people or routines that act as entraining agents. Alternatively, psychosocial stress could precipitate depression through the introduction of "zeitstorers," i.e., events or people (such as a newborn baby) that actively disrupt existing routines (Ehlers *et al.*, 1988). Similarly, Healy and Waterhouse (1995) hypothesized that depression is triggered by altered social routines that cause the patient's rhythms to become desynchronized with those of the environment. In this model, depression resembles pathologic syndromes caused by shift work or jet lag. Of course, depressed patients probably differ from control subjects in both the social and the biological zeitgebers to which they are exposed. For example, compared to controls, depressed individuals may have decreased exposure to both light and activity as well as unstable social routines (Kupfer *et al.*, 1974; Wolff, Putnam, & Post, 1985). Indeed, social and biological zeitgebers covary in humans since bed and awakening times dictate light exposure and activity level, but are also at least partly dependent on social routines.

Altered Alpha:Rho Ratio. According to the two-oscillator model of circadian regulation, separate oscillators track dusk and dawn and control, respectively, the onset and offset of the sleep–rest period. This regulatory mechanism thus allows the animal's alpha:rho ratio to wax and wane seasonally (Illnerova & Vanecek, 1982). Recent studies demonstrate that humans retain this photoperiodically responsive mechanism, raising the theoretical possibility that the pathophysiology of affective illness involves an alteration in the alpha:rho ratio of one or more circadian rhythms (Wehr *et al.*, 1993).

In three types of depression, major depressive disorder, PMDD, and SAD, there are descriptive data to support this hypothesis. In the case of premenstrual and seasonal depression, the rhythm in question is nocturnal melatonin secretion, and the alpha:rho ratio is increased in patients compared to controls. Parry *et al.* (1990) reported that, compared to controls, women with PMDD have a phase-advanced offset of melatonin secretion, leading to a shorter secretion duration. Wehr, Giesen, Moul, Turner, and Schwartz (1995) have preliminary, unpublished data indicating that in normal women, but not men, the duration of nocturnal melatonin secretion increases in winter. However, women with SAD were less likely than unaffected women to demonstrate this winter expansion. Therefore, in the studies of patients with SAD or PMDD, the mood-disordered patients (who are largely female in the first instance and exclusively so in the second) have a decreased duration of nocturnal melatonin secretion compared to controls.

In contrast, the data in patients with major depressive disorder involve cortisol secretion, and the finding suggests a decrease in the alpha:rho ratio in depressed patients. Specifically, a meta-analysis of the cortisol data shows that the nadir, but not the peak, of cortisol secretion is phase advanced in depressed patients compared with controls (Wirz-Justice, 1995), so that the patients have an increased duration of the nocturnal period of rising cortisol secretion. These findings indicate the importance of studying rhythms in depressed patients for at least an entire 24-hour period.

Decreased Amplitude

A number of studies have found that the amplitude of one or more circadian rhythms is diminished in depressed patients compared with normals. In particular, studies of temperature, melatonin, and thyrotropin consistently report decreased amplitude in depressed patients. For example, in the literature reviewed by Souetre *et al.* (unpublished), 13 of 17 temperature studies, 12 of 13 melatonin studies, and all 9 thyrotropin studies showed a blunted amplitude of the nocturnal rhythm in patients compared with normals. As mentioned above, two more recent studies of temperature in depressed patients (Daimon *et al.*, 1992; Tsujimoto *et al.*, 1990), as well as one of melatonin in patients with premenstrual dysphoric disorder (Parry *et al.*, 1990), also show reduced amplitude in the depressed subjects. The consistency of this literature has led several authors to suggest that decreased amplitude, rather than shifted or unstable phase, is the most consistent chronobiological abnormality in affective illness (Czeisler, Kronauer, Mooney, Anderson, & Allan, 1987; Souetre *et al.*, 1989). However, depressed patients have a more blunted rest–activity cycle than controls (Kupfer *et al.*, 1974; Wehr, Muscettola, & Goodwin, 1980); this rest–activity pattern or the well-documented sleep abnormalities in depression could decrease the amplitude of patients' nocturnal rhythms.

It should be noted that recent improvements in assay sensitivity and specificity for both thyrotropin and melatonin may invalidate findings reported in older studies. In the case of melatonin, for example, two studies used highly specific assays as well as individually matched control subjects (Rubin, Heist, McGeoy, Hanada, & Lesser, 1992; Thompson, Franey, Arendt, & Checkley, 1988). In both of these studies, there was no significant difference in amplitude between patients and controls. Indeed, in the second study, which was not conducted under constant routine conditions, there was a trend toward increased secretion in the patients, particularly the premenopausal women (Rubin *et al.*, 1992).

Antidepressant medications appear to increase the amplitude of nocturnal rhythms in depressed patients (Goetze & Tolle, 1987; Linkowski *et al.*, 1987), but the myriad physiologic effects of these medications make it difficult to determine whether the increased amplitude is a cause or a correlate of their antidepressant properties. Given the known effects of bright light on melatonin (Lewy, Wehr, Goodwin, Newsome, & Rosenthal, 1980) and the close relationship between nocturnal temperature and melatonin (Shanahan & Czeisler, 1991), the fact that phototherapy increases the amplitude of the nocturnal temperature rhythm in SAD patients is perhaps more direct evidence that decreased amplitude has pathogenic significance (Rosenthal *et al.*, 1990). However, interventions that increase amplitude could be effective not because of their amplitude-enhancing effects per se, but rather because increasing the rhythm's amplitude makes it less susceptible to phase shifts. For example, Leibenluft *et al.* (1995) published preliminary data suggesting that midday (but not morning) light has a mood-stabilizing effect in patients with RCBD, and hypothesized that this therapeutic effect occurs because the midday light enhances the amplitude of the nocturnal melatonin rhythm and makes it more resistant to the phase shifts that appear to accompany mood switches.

Period Abnormalities

In patients with RCBD, there is an intuitive appeal to the suggestion that the illness is caused by rapidly cycling circadian rhythms, i.e., rhythms with an abnormally short tau. Since mutations in nonhuman species cause abnormally long or short circadian periods (Vitaterna *et al.*, 1994), there is further appeal to the suggestion that bipolar disorder, an illness with well-documented heritability (Goodwin & Jamison, 1990), is caused by a tau mutation in humans. In order to test this hypothesis, one would have to place a number of patients in temporal isolation. Such a study would be extremely costly and difficult to carry out and the interpretation of its results would be complicated by the fact that temporal isolation might itself act as a clinical intervention. Medicating patients during such a procedure would probably confound the results, whereas withholding medication could be problematic for both clinical and ethical reasons.

To our knowledge, a total of six patients with affective illnesses (two unipolar depressives, two depressed bipolar patients, and two patients with RCBD) have been placed in temporal isolation (Dirlich *et al.*, 1981; Wehr *et al.*, 1985). In two of these patients (neither of them rapid cycling; Wehr *et al.*, 1985), the sleep–wake cycle exhibited an abnormally short tau. The effects of the intervention on the patients' clinical status was variable. Therefore, to the extent that data are available, tau does not appear to be abnormal in patients with RCBD.

Depression and the Two-Process Model of Sleep

According to Borbély's (1982) model of sleep regulation, the timing of sleep is the result of an interaction between two processes: one circadian (process C) and one homeostatic (process S). Process C is the circadian rhythm controlling REM sleep, whereas process S is reflected in the level of slow-wave activity. Borbély and Wirz-Justice (1982) extended the two-process model to affective illness by hypothesizing that depression results from a deficiency of process S. Such a deficiency could explain the decreased sleep duration, increased sleep latency, decreased sleep efficiency, and decreased delta sleep observed in depressed patients. In addition, because processes S and C are reciprocal, the deficiency in process S causes an accentuation of process C, resulting in decreased REM latency and increased total REM time. The model can be used to explain the antidepressant effects of sleep deprivation since sleep deprivation would increase process S and temporarily reverse the underlying abnormality.

Investigators have attempted to operationalize and test Borbély and Wirz-Justice's model. Achermann & Borbély (1990) used the model to simulate changes in SWS induced by the antidepressant medication clomipramine and by sleep deprivation in two patients. Ehlers, Havstad, and Kupfer (1996) found that, compared to nonresponders, responders to clomipramine had a redistribution of SWS to the earlier part of the night. However, contrary to what the model would predict, total or average slow-wave activity did not distinguish responders from nonresponders (Ehlers *et al.*, 1996). These same investigators suggested a modification of the two-process model in which the deficient process S is associated not only with SWS, but also with growth hormone-releasing hormone (GHRH) or with the ratio of GHRH to corticotropin-releasing hormone (CRH). The latter association would then explain the finding of GH hyposecretion after sleep onset in depressed patients (Ehlers & Kupfer, 1987). Similarly, the amplified process C is associated not only with REM sleep, but also with CRH, thereby accounting for hypercortisolemia in depression. This model recently received some support in a study showing a correlation between the GH/cortisol ratio and slow-wave activity in normal volunteers (Seifritz *et al.*, 1996).

Discussion and Conclusions

The literature does not clearly support any one of the proposed hypotheses concerning the role of abnormal circadian rhythms in the pathogenesis of mood disorders. While this lack of consistent findings may indicate that there are no causal connections between circadian rhythm dysfunction and mood disorders, it is also possible that the inconsistent findings are due to methodological shortcomings in the studies.

Our review suggests that there may be merit to the argument, first advanced by Lewy *et al.* (1984), that circadian rhythm patterns might be used to distinguish one type of mood disorder from another. According to Lewy *et al.*'s (1984) model, phase-advanced rhythms characterize nonseasonal major depressive disorder (which usually involves "typical" symptomatology), while phase-delayed rhythms characterize SAD (with its "atypical" symptomatology). We would add to the model the suggestion that phase instability may characterize RCBD (as well, perhaps, as other forms of bipolar disorder) and that the circadian rhythms of patients with nonseasonal major depressive disorder may be low amplitude (as well as relatively phase advanced). In addition, it is possible that, within one clinical syndrome, different rhythms might exhibit different abnormalities; thus, for example, one might observe a low-amplitude temperature rhythm coupled with a phase-advanced cortisol nadir in patients with nonseasonal major depression.

The notion that circadian rhythm abnormalities distinguish one subtype of depression from another has face validity, since the subtypes differ predominately in their cyclic characteristics. That is, depressive subtypes differ in the frequency of mood cycles (e.g., seasonal, monthly) and/or in the precise perturbations of the sleep–wake cycle that occur when patients become symptomatic (i.e., whether sleep duration is increased, decreased, or, in the case of bipolar disorder, variable). However, subtyping the depressive disorders in this way and then observing that they differ in their circadian characteristics once again leaves open the possibility that differences in circadian rhythms are epiphenoma.

In the final analysis, observations of the sleep–wake cycle constitute the most compelling data to support the contention that circadian rhythms play a role in the pathogenesis of affective disorders. That is, each of the disorders is characterized by specific sleep symptoms, and the sleep deprivation literature convincingly demonstrates that these changes in the sleep–wake cycle are not merely symptoms, but have the potential to affect the course of the illness. However, whether the sleep–wake cycle's pathogenic potential stems from circadian mechanisms or from other mechanisms involved in sleep regulation remains an open question.

Most investigators have made the implicit or explicit assumption that, if circadian rhythm dysfunction is causally involved in affective disorders, it plays a primary role. For example, investigators have debated whether mood disorders are caused by an abnormality in the clock or instead by a mismatch between the timing of the patient's rhythms and those of his or her environment (Healy & Waterhouse, 1990; Wehr, 1990b). An alternative formulation is that mood disorders arise from cause(s) that are entirely distinct from circadian mechanisms and from the sleep–wake cycle, but that circadian rhythms and/or the sleep–wake cycle nonetheless play a role in the development of the clinical syndrome. For example, Wehr, Sack, and Rosenthal (1987a) suggested that mania involves a positive feedback process in which sleep deprivation resulting from the manic patient's sleeplessness increases the severity of the illness. Thus, some or all manic episodes may arise from a cause separate from the sleep–wake cycle, but changes in the latter influence the clinical syndrome, contributing to its characteristically explosive course. Similarly, the unstable sleep–wake cycle experienced by patients with RCBD could induce instability in the phase of melatonin or other circadian rhythms; this instability could then further destabilize the sleep–wake cycle and lead to a magnification of the patient's symptoms.

An example of this principle is our experimental treatment of a patient with RCBD, shown in Figure 2. This patient, who had a longstanding pattern of severe mood cycles accompanied by dramatic shifts in the timing and duration of his sleep, was placed first in an LD 10:14 and then in an LD 12:12 condition. This intervention stabilized the timing and duration of his sleep, and also greatly dampened the severity of his mood cycles. Even if the patient's abnormal sleep–wake cycle was the result, rather than the cause, of his mood cycles, the stabilization of his sleep–wake cycle appears to have had a beneficial effect on his mood. It is possible that a particularly important element of this therapeutic regimen was the maintenance of the shortened photoperiod even when the patient was hypomanic, an intervention which probably served to increase the amount of sleep, or at least rest, that occurred during this time. Preventing sleep deprivation may thus have prevented the development of a more florid hypomanic episode. Furthermore, if, as some have argued (Kukopulos *et al.*, 1980), the occurrence of mania is causally related to subsequent depression, then truncating the hypomanic episode could conceivably have dampened the next depression and therefore had more long-range effects on the course of the illness.

Thus, it is possible that disturbances in circadian rhythms or the sleep–wake

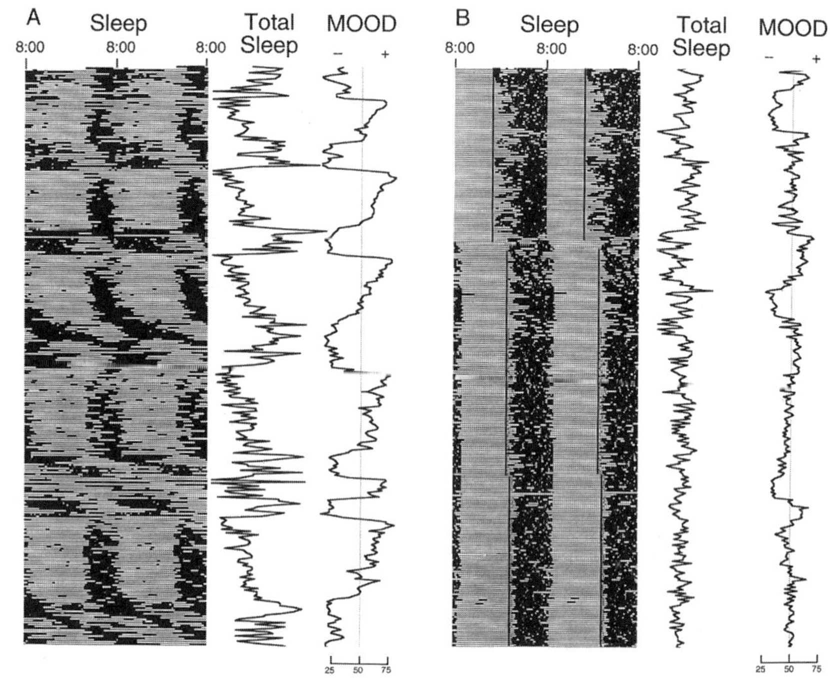

Figure 2. Data were obtained from a 50-year-old man with rapid cycling bipolar disorder being treated with sodium valproate and sertraline throughout. (A) Data obtained when the patient was living on a naturalistic schedule. Left: Self-reported sleep is double-plotted, with consecutive 24-hour (8:00a.m.–8:00a.m) periods displayed in raster format. Middle: Total sleep time for each 24-hour period is graphed. Right: Self-reported mood is displayed. Each morning upon awakening, the patient rated his mood on a 100-mm line, with 100 being "most manic I've ever felt" and 0 being "most depressed I've ever felt." (B) Data obtained when the patient was adhering to a prescribed LD schedule, which varied from LD 10:14 to LD 13:11. The dark period is indicated by gray lines on the raster plot of sleep. Total sleep time and mood are displayed as in panel A. See text for discussion on the data. (Wehr, T. A., Turner, E. H., Shirrada, J., Clark, C. H., Barker, C., & Liebenluft, E. (1988) Treatment of a rapidly cycling bipolar patient by using extended bedrest and darkness to promote sleep. *Biological Psychiatry, 43,* 822–828.

cycle are secondary to other mechanisms creating mood disorders, but are still more than epiphenomena. That is, such circadian disturbances may be either necessary or sufficient to cause incipient symptoms to develop into a full-blown syndrome, or they may convert a moderate syndrome into a severe, disabling one. As demonstrated by the case above, this formulation leaves open the possibility that circadian mechanisms, even if they are not primarily responsible for the development of mood disorders, might be manipulated to therapeutic advantage. Compared to the medications now currently used to treat mood disorders, both bright light and exogenous melatonin act rapidly on their target system and both have a rapid offset of action (Rosenthal *et al.*, 1984). Furthermore, since the time of administration determines their clinical effects, these circadian interventions are more flexible and can be used in a more targeted way than classical antidepressants (Leibenluft *et al.*, 1995). Thus, it may well be worthwhile to continue to explore the role of circadian mechanisms in the pathogenesis of mood disorders, even if it is unlikely that either such mechanisms or disturbances in the sleep–wake cycle are the primary causes of illness.

Achermann, P., & Borbély, A. A. (1990). Simulation of human sleep: Ultradian dynamics of electroencephalographic slow-wave activity. *Journal of Biological Rhythms, 5,* 141–157.

American Psychiatric Association (1994). *Diagnostic and statistical manual of mental disorders* (4th ed.). Washington, DC: Author.

Aschoff, J. (1983). Disorders of the circadian system as discussed in psychiatric research. In T. Wehr & F. Goodwin (Eds.), *Circadian rhythms in psychiatry* (pp. 33–39). Pacific Grove, CA: Boxwood Press.

Avery, D., Dahl, K., Savage, M., Brenglemann, G., Larson, L., Bitiello, M., & Prinz, P. (1990). The temperature rhythm is phase-delayed in winter depression (Abstract). *Biological Psychiatry, 27,* 99A.

Benca, R. M., Obermeyer, W. H., Thisted, R. A., & Gillin, J. C. (1992). Sleep and psychiatric disorders: A meta-analysis. *Archives of General Psychiatry, 49,* 651–668.

Borbély, A. A. (1982). A two process model for sleep regulation. *Human Neurobiology, 1,* 200–205.

Borbély, A. A., & Wirz-Justice, A. (1982). Sleep, sleep deprivation and depression: A hypothesis derived from a model of sleep regulation. *Human Neurobiology, 1,* 205–210.

Checkley, S. (1989). The relationship between biological rhythms and the affective disorders. In J. Arendt, D. S. Minors, & J. Waterhouse (Eds.), *Biological rhythms in clinical practice.* London, Butterworth.

Czeisler, C. A., Kronauer, R. E., Mooney, J. J., Anderson, J. L., & Allan, J. S. (1987). Biologic rhythm disorders, depression and phototherapy. *Psychiatric Clincs of North America, 10,* 687–709.

Dahl, K., Avery, D. H., Lewy, A. J., Savage, M. V., Brenglemann, G. L., Larsen, L. H., Vitiello, M. V., & Prinz, P. N. (1993). Dim light melatonin onset and circadian temperature during a constant routine in hypersomnic winter depression. *Acta Psychiatrica Scandinavica, 88,* 60–66.

Daimon, K., Yamada, N., Tsujimoto, T., & Takahashi, S. (1992). Circadian rhythm abnormalities of deep body temperature in depressive disorders. *Journal of Affective Disorders, 26,* 191–198.

Dew, M. A., Reynolds, C. F., Buysse, D. J., Houck, P. R., Hoch, C. C., Monk, T. H., & Kupfer, D. J. (1996). Electroencephalographic sleep profiles during depression. *Archives of General Psychiatry, 53,* 148–156.

Dirlich, G., Kammerloher, A., Schulz, H., Lund, R., Doerr, P., & von Zerssen, D. (1981). Temporal coordination of rest–activity cycle, body temperature, urinary free cortisol and mood in a patient with 48-hour unipolar–depressive cycles in clinical and time-cue-free environments. *Biological Psychiatry, 16,* 163–179.

Ehlers, C. L., & Kupfer, D. J. (1987). Hypothalamic peptide modulation of EEG sleep in depression: A further application of the S-process hypothesis. *Biological Psychiatry, 22,* 513–517.

Ehlers, C. L., Frank, E., & Kupfer, D. J. (1988). Social zeitgebers and biological rhythms: A unified approach to understanding the etiology of depression. *Archives of General Psychiatry, 45,* 948–952.

Ehlers, C. L., Havstad, J. W., & Kupfer, D. J. (1996). Estimation of the time course of slow-wave sleep over the night in depressed patients: Effects of clomipramine and clinical response. *Biological Psychiatry, 39,* 171–181.

Feldman-Naim, S., Turner, E. H., & Leibenluft, E. (1996). *Circadian variation in the direction of mood switches in patients with rapid-cycling bipolar disorder.* Poster presented at the 149th Annual Meeting of the American Psychiatric Association, New York.

Frank, E., Shear, M. K., Essex, M. J., Leibenluft, E., & Monk, T. H. (1995). *Routines, rhythms, and affective states: The relationship of lifestyle regularity to mood and anxiety disorders.* Paper presented at the 34th Annual Meeting of the American College of Neuropsychopharmacology, San Juan, Puerto Rico.

Georgi, F. (1947). Psychophysische korrelationen: III. Psychiatrische probleme im licte der rhythmusforschung. *Schweizerische Medizinische Wochenschrift, 49,* 1276–1280.

Giles, D. E., Roffwarg, H. P., & Rush, A. J. (1987). REM latency concordance in depressed family members. *Biological Psychiatry, 22,* 910–914.

Giles, D. E., Biggs, M. M., Rush, A. J., & Roffwarg, H. P. (1988). Risk factors in families of unipolar depression: I. Psychiatric illness and reduced REM latency. *Journal of Affective Disorders, 14,* 51–59.

Giles, D. E., Kupfer, D. J., Roffwarg, H. P., Rush, A. J., Biggs, M. M., & Etzel, B. A. (1989). Polysomnographic parameters in first-degree relatives of unipolar probands. *Psychiatry Research, 27,* 127–136.

Goetze, U., & Tolle, R. (1987). Circadian rhythms of free urinary cortisol, temperature, and heart rate in endogenous depressives and under antidepressant therapy. *Neuropsychobiology, 18,* 175–184.

Goodwin, F. K., & Jamison, K. R. (1990). *Manic-depressive illness.* Oxford: Oxford University Press.

Gresham, S. C., Agnew, H. W., & Williams, R. L. (1965). The sleep of depressed patients: An EEG and eye movement study. *Archives of General Psychiatry, 13,* 503–507.

Halberg, F. (1968). Physiologic considerations underlying rhythmometry with special reference to emotional illness. In J. de Ajuriaguerra (Ed.), *Cycles biologiques et psychiatrie* (pp. 73–126). Paris: Masson.

Haug, H.-J., & Wirz-Justice, A. (1993). Diurnal variation of mood in depression: Important or irrelevant? *Biological Psychiatry, 34,* 201–203.

Healy, D., & Waterhouse, J. M. (1990). The circadian system and affective disorders: Clocks or rhythms? *Chronobiology International, 7,* 5–10.

Healy, D., & Waterhouse, J. M. (1995). The circadian system and the therapeutics of the affective disorders. *Pharmacology and Therapeutics, 65,* 241–263.

Illnerova, H., & Vanecek, J. (1982). Two-oscillator structure of the pacemaker controlling the circadian rhythm of N-acetyltransferase in the rat pineal gland. *Journal of Comparative Physiology, 145,* 539–548.

Kessler, R. C., McGonagle, K. A., Zhao, S., Nelson, C. B., Hughes, M., Eshleman, S., Wittchen, H.-U., & Kendler, K. S. (1994). Lifetime and 12-month prevalence of DSM-III-R psychiatric disorders in the United States. *Archives of General Psychiatry, 51,* 8–19.

Kramlinger, K. G., & Post, R. M. (1996). Ultra-rapid and ultradian cycling in bipolar affective illness. *British Journal of Psychiatry, 168,* 314–323.

Kripke, D. F., Mellaney, D. J., Atkinson, M., & Wolf, S. R. (1978). Circadian rhythm disorders in manic-depressives. *Biological Psychiatry, 13,* 335–351.

Kukopulos, A., Reginaldi, D., Laddomada, G., Floris, G., Serra, G., & Tondo, L. (1980). Course of the manic-depressive cycle and changes caused by treatments. *Pharmacopsychiatry, 13,* 156–157.

Kupfer, D. J., & Foster, G. (1972). Interval between onset of sleep and rapid-eye-movement sleep as an indicator of depression. *Lancet, 2,* 684–686.

Kupfer, D. J., Weiss, B. L., Foster, F. G., Detre, T. P., Delgado, J., & McPartland, R. (1974). Psychomotor activity in affective states. *Archives of General Psychiatry, 132,* 928–931.

Kupfer, D. J., Frank, E., Grochocinski, V. J., Gregor, M., & McEachran, A. B. (1988). EEG sleep profiles in recurrent depression: A longitudinal investigation. *Archives of General Psychiatry, 45,* 678–681.

Lam, R. W., Berkowitz, A. L., Berga, S. L., Clark, C. M., Kripke, D. F., & Gillin, J. C. (1990). Melatonin suppression in bipolar and unipolar mood disorders. *Psychiatry Research, 33,* 129–134.

Leibenluft, E., Schwartz, P. J., Turner, E. H., Rosenthal, N. E., & Wehr, T. A. (1993). *State-dependent changes in pineal and thyroid axis function in patients with rapid-cycling bipolar disorder.* Poster presented at the 32nd Annual Meeting of the American College of Neuropsychopharmacology, Honolulu, Hawaii.

Leibenluft, E., Turner, E. H., Feldman-Naim, S., Schwartz, P. J., Wehr, T. A., & Rosenthal, N. E. (1995). Light therapy in patients with rapid cycling bipolar disorder: Preliminary results. *Psychopharmacology Bulletin, 31,* 705–710.

Lewy, A. J., Wehr, T. A., Goodwin, F. K., Newsome, D. A., & Rosenthal, N. E. (1980). Light suppresses melatonin secretion in humans. *Science, 210,* 1267–1269.

Lewy, A. J., Wehr, T. A., Goodwin, F. K., Newsome, D. A., & Rosenthal, N. E. (1981). Manic-depressive patients may be supersensitive to light. *Lancet, 1,* 383–384.

Lewy, A. J., Sack, D. A., & Singer, C. M. (1984). Assessment and treatment of chronobiologic disorders using plasma melatonin levels and bright light exposure: The clock-gate model and the phase response curve. *Psychopharmacology Bulletin, 20,* 561–565.

Linkowski, P., Mendlewicz, J., Leclercq, R., Brasseur, M., Hubain, P., Golstein, J., Copinschi, G., & Van Cauter, E. (1985). The 24-hour profile of adrenocorticotropin and cortisol in major depressive illness. *Journal of Clinical Endocrinology and Metabolism, 61,* 429–438.

Linkowski, P., Mendlewicz, J., Kerhofs, M., Leclercq, R., Golstein, J., Brasseur, M., & Copinschi, G. (1987). 24-hour profile of adrenocorticotropin, cortisol, and growth hormone in major depressive illness: Effect of antidepressant treatment. *Journal of Clinical Endocrinology and Metabolism, 65,* 141–152.

Linkowski, P., Van Onderbergen, A., Kerkhofs, M., Bosson, D., Mendlewicz, J., & Van Cauter, E. (1993). A twin study of the 24-hour cortisol profile: Evidence for genetic control of the human circadian clock. *American Journal of Physiology, 264,* E173–E181.

Mrosovsky, N., Reebs, S. G., Honrado, G. I., & Salmon, P. A. (1989). Behavioral entrainment of circadian rhythms. *Experientia, 45,* 696–702.

Papousek, M. (1975). Chronobiologische aspekte der zyklothymie. *Fortschritte der Neurologie Psychiatrie, 43,* 381–440.

Parry, B. L., Berga, S. L., Kripke, D. F., Klauber, M. R., Laughlin, G. A., Yen, S. S. C., & Gillin, J. C. (1990). Altered waveform of plasma nocturnal melatonin secretion in premenstrual depression. *Archives of General Psychiatry, 47,* 1139–1146.

Parry, B. L., Hauger, R., Lin, E., LeVeau, B., Mostofi, N., Clopton, P. L., & Gillin, J. C. (1994). Neuroendocrine effects of light therapy in late luteal phase dysphoric disorder. *Biological Psychiatry, 36,* 356–364.

Reinink, E., Bouhuys, N., Wirz-Justice, A., & van den Hoofdakker, R. (1990). Prediction of the antidepressant response to total sleep deprivation by diurnal variation of mood. *Psychiatry Research, 32,* 113–124.

Rosenthal, N. E., Sack, D. A., Gillin, J. C., Lewy, A. J., Goodwin, F. K., Davenport, Y., Mueller, P. S., Newsome, D. A., & Wehr, T. A. (1984). Seasonal affective disorder: A description of the syndrome and preliminary findings with light therapy. *Archives of General Psychiatry, 41,* 72–80.

Rosenthal, N. E., Levendosky, A. A., Skwerer, R. G., Joseph-Vanderpool, J. R., Kelly, K. A., Hardin, T., Kasper, S., DellaBella, P., & Wehr, T. A. (1990). Effects of light treatment on core body temperature in seasonal affective disorder. *Biological Psychiatry, 27,* 39–50.

Rubin, R. T., Heist, E. K., McGeoy, S. S., Hanada, K., & Lesser, I. M. (1992). Neuroendocrine aspects of primary endogenous depression. *Archives of General Psychiatry, 49,* 558–567.

Rush, A. J., Erman, M. K., Giles, D. E., Schlesser, M. A., Carpenter, G., Vasavada, N., & Roffwarg, H. P. (1986). Polysomnographic findings in recently drug-free and clinically remitted depressed patients. *Archives of General Psychiatry, 43,* 878–884.

Sack, D. A., Nurnberger, J., Rosenthal, N. E., & Wehr, T. A. (1985). Potentiation of antidepressant medications by phase-advance of the sleep–wake cycle. *American Journal of Psychiatry, 142,* 606–608.

Sack, D. A., Rosenthal, N. E., Duncan, W. C., Mendelson, W. B., & Wehr, T. A. (1988). Early versus late partial sleep deprivation therapy of depression. *Acta Psychiatrica Scandinavica, 77,* 219–224.

Sack, R. L., Lewy, A. J., & White, D. M. (1990). Morning versus evening light treatment for winter depression: Evidence that the therapeutic effects of light are mediated by circadian phase shifting. *Archives of General Psychiatry, 47,* 343–351.

Schmidt, P. J., Nieman, L. K., Grover, G. N., Muller, K. L., Merriam, G. R., & Rubinow, D. R. (1991). Lack of effect of induced menses on symptoms in women with premenstrual syndrome. *New England Journal of Medicine, 324,* 1174–1179.

Schwartz, P. J., Brown, C., Wehr, T. A., & Rosenthal, N. E. (1996). Winter-seasonal affective disorder: A follow-up study of the first 59 patients of the National Institute of Mental Health seasonal studies program. *American Journal of Psychiatry, 153,* 1028–1036.

Seifritz, E., Muller, M. J., Trachsel, L., Lauer, C. J., Hemmeter, U., Hatzinger, M., Moore, P., & Holsboer-Trachsler, E. (1996). Revisiting the Ehlers and Kupfer hypothesis: The growth hormone cortisol secretion ratio during sleep is correlated with electroencephalographic slow wave activity in normal volunteers. *Biological Psychiatry, 39,* 139–142.

Shanahan, T. L., & Czeisler, C. A. (1991). Light exposure induces equivalent phase shifts of the endogenous circadian rhythms of circulating plasma melatonin and core body temperature in men. *Journal of Clinical Endocrinology and Metabolism, 73,* 227–235.

Souetre, E., Salvati, E., Belugou, J. L., Pringuey, D., Candito, M., Krebs, B., Ardisson, J. L., & Darcourt, G. (1989). Circadian rhythms in depression and recovery: Evidence for blunted amplitude as the main chronobiological abnormality. *Psychiatry Research, 28,* 262–278.

Souetre, E., Wehr, T. A., & Rosenthal, N. E. (Unpublished). *Biological clocks, depression and environment.*

Thalen, B. E., Kjellman, F., Morkrid, L., & Wetterberg, L. (1995). Melatonin in light treatment of patients with seasonal and nonseasonal depression. *Acta Psychiatrica Scandinavica, 92,* 274–284.

Thompson, C., Franey, C., Arendt, J., & Checkley, S. A. (1988). A comparison of melatonin secretion in depressed patients and normal subjects. *British Journal of Psychiatry, 152,* 260–265.

Tsujimoto, T., Yamada, N., Shimoda, K., Hanada, K., & Takahashi, S. (1990). Circadian rhythms in depression. Part II: Circadian rhythms in inpatients with various mental disorders. *Journal of Affective Disorders, 18,* 199–210.

Vitaterna, M. H., King, D. P., Chang, A.-M., Kornhauser, J. M., Lowrey, P. L., McDonald, J. D., Dove, W. F., Pinto, L. H., Turek, F. W., & Takahashi, J. S. (1994). Mutagenesis and mapping of mouse gene, *Clock,* essential for circadian behavior. *Science, 264,* 719–725.

Vollmann, J., & Berger, M. (1993). Sleep deprivation with consecutive sleep-phase advance therapy in patients with major depression: A pilot study. *Biological Psychiatry, 33,* 54–57.

Wehr, T. A. (1990a). Effects of sleep and wakefulness in depression and mania. In J. Montplaisir & R. Godbout (Eds.), *Sleep and biological rhythms* (pp. 42–86). Oxford: Oxford University Press.

Wehr, T. A. (1990b). Reply to Healy D and Waterhouse JM: The circadian system and affective disorders: Clocks or rhythms? *Chronobiology International, 7,* 11–14.

Wehr, T. A., & Goodwin, F. K. (1983). Biological rhythms in manic-depressive illness. In T. A. Wehr & F. K. Goodwin (Eds.), *Circadian rhythms in psychiatry* (pp. 129–184). Pacific Grove, CA: Boxwood Press.

Wehr, T. A., & Wirz-Justice, A. (1980). Internal coincidence model for sleep deprivation and depression. *Sleep,* 26–33.

Wehr, T. A., & Wirz-Justice, A. (1982). Circadian rhythm mechanisms in affective illness and in antidepressant drug action. *Pharmacopsychiatry, 15,* 31–39.

Wehr, T. A., Wirz-Justice, A., Goodwin, F. K., Duncan, W. C., & Gillin, J. C. (1979). Phase advance of the circadian sleep–wake cycle as an antidepressant. *Science, 206,* 710–713.

Wehr, T. A., Muscettola, G., & Goodwin, F. K. (1980). Urinary 3-methoxy-4-hydroxy-phenylglycol circadian rhythms: Early timing (phase-advance) in manic depressives compared with normal subjects. *Archives of General Psychiatry, 37,* 257–263.

Wehr, T. A., Goodwin, F. K., Wirz-Justice, A., Breitmaier, J., & Craig, C. (1982). 48 hour sleep–wake cycle in manic-depressive illness: Naturalistic observations and sleep deprivation experiments. *Archives of General Psychiatry, 39,* 559–565.

Wehr, T. A., Sack, D. A., Duncan, W. C., Mendelson, W. B., Rosenthal, N. E., Gillin, J. C., & Goodwin, F. K. (1985). Sleep and circadian rhythms in affective patients isolated from external time cues. *Psychiatry Research, 15,* 327–339.

Wehr, T. A., Sack, D. A., & Rosenthal, N. E. (1987a). Sleep reduction as a final common pathway in the genesis of mania. *American Journal of Psychiatry, 144,* 201–204.

Wehr, T. A., Sack, D. A., & Rosenthal, N. E. (1987b). Seasonal affective disorder with summer depression and winter hypomania. *American Journal of Psychiatry, 144,* 1403–1411.

Wehr, T. A., Moul, D. E., Barbato, G., Giesen, H. A., Seidel, J. A., Barker, C., & Bender, C. (1993). Conservation of photoperiod-responsive mechanisms in humans. *American Journal of Physiology, 265,* R846–R857.

Wehr, T. A., Giesen, H. A., Moul, D. E., Turner, E. H., & Schwartz, P. J. (1995). Suppression of human responses to seasonal changes in day-length by modern artificial lighting. *American Journal of Physiology, 269,* R173–R178.

Wirz-Justice, A. (1995). Biological rhythms in mood disorders. In F. E. Bloom & D. J. Kupfer (Eds.), *Psychpharmacology: The fourth generation of progress* (pp. 999–1016). New York: Raven Press.

Wirz-Justice, A., Graw, P., Krauchi, K., Gisin, B., Jochum, A., Arendt, J., Fisch, H. U., & Buddeberg, C. (1993). Light therapy in seasonal affective disorder is independent of time of day or circadian phase. *Archives of General Psychiatry, 50,* 929–937.

Wolff, E. A., Putnam, F. W., & Post, R. M. (1985). Motor activity in affective illness: The relationship of amplitude and temporal distribution to changes in the affective state. *Archives of General Psychiatry, 42,* 288–294.

Wu, J., & Bunney, W. (1990). The biological basis of an antidepressant response to sleep deprivation and relapse: Review and hypothesis. *American Journal of Psychiatry, 147,* 14–21.

24

Sleep Disorders

DANIEL J. BUYSSE, JULIE CARRIER, MARY AMANDA DEW, MARTICA HALL, TIMOTHY H. MONK, PETER NOWELL, CHARLES F. REYNOLDS III, AND DAVID J. KUPFER

INTRODUCTION

One of the major functions of the circadian timing system is to generate robust rhythms of wakefulness and sleep, whether the dominant pattern is one of short sleep bouts during a portion of the 24-hour day (as in many nocturnal rodents) or one of sustained and consolidated sleep (as in humans). Dysfunction of the circadian system, or a mismatch between demands of the circadian system and volitional behavior, can lead to sleep disruption. Indeed, models of human sleep regulation and laboratory studies predict that sleep attempts at unfavorable circadian times will result in difficulty falling asleep or maintaining sleep (Dijk & Czeisler, 1995). The sleep/wake problems of jet lag and shift work verify this prediction. Numerous other factors besides circadian regulation can also produce sleep problems, including "primary" disturbances of sleep mechanisms, psychiatric and medical problems, and sleep-related breathing disorders. Even among these common "noncircadian" sleep disorders, disturbances in biological rhythms may play a role in symptom patterns and pathophysiology. For instance, the sleep onset difficulties of some patients with chronic insomnia may be related to a phase delay of circadian rhythms, and the daytime sleepiness of patients with narcolepsy may indicate reduced amplitude of circadian rhythms.

Almost all individuals experience sleep complaints at some point during their lives. A growing number of community studies document high rates of moderate to severe sleep disturbances across all sociodemographic sectors of the population.

DANIEL J. BUYSSE, JULIE CARRIER, MARY AMANDA DEW, MARTICA HALL, TIMOTHY H. MONK, PETER NOWELL, CHARLES F. REYNOLDS III, AND DAVID J. KUPFER Sleep and Chronobiology Center, Western Psychiatric Institute and Clinic, University of Pittsburgh School of Medicine, Pittsburgh, Pennsylvania 15213.

Circadian Clocks, Volume 12 of *Handbook of Behavioral Neurobiology*, edited by Joseph S. Takahashi, Fred W. Turek, and R. Y. Moore, Kluwer Academic / Plenum Publishers, New York, 2001.

Moreover, the public health burden associated with sleep disturbances is significant (National Commission on Sleep Disorders Research, 1993): sleep disturbances have been repeatedly linked to mortality (Shapiro & Dement, 1993; Wingard & Berkman, 1983), psychological and physical morbidity and associated treatment expenses (Bixler, Kales, Soldatos, Kales & Healey, 1979; Hyyppä & Kronholm, 1989; Jacquinet-Salord, Lang, Fourlaud, Nicoulet, & Bingham, 1993; Kuppermann *et al.*, 1995; Weyerer & Dilling, 1991), work and functional status (Johnson & Spinweber, 1983; Kuppermann *et al.*, 1995), and quality of life (Kuppermann *et al.*, 1995; Shapiro & Dement, 1993; Urponen, Vuori, Hasan, & Partinen, 1988).

In this chapter, we will briefly outline the features of normal human sleep, the major categories of sleep problems and disorders, and their epidemiology and impact. We will then describe sleep disorders which directly involve circadian dysfunction, such as jet lag and shift work, as well as their treatment. Next, we will focus on those sleep disorders which, although they do not involve such obvious circadian dysfunction, may involve more subtle disturbances in circadian rhythms. For each of these disorders, we will provide a brief description of clinical features, circadian aspects, and treatment.

Normal Human Sleep

Normal adult human sleep occurs as a single consolidated bout of nocturnal sleep lasting 6–9 hours. In many cultures, nocturnal sleep is supplemented by a shorter "siesta" sleep which occurs about 11–12 hours after the nocturnal temperature nadir. Evidence from studies of extended nocturnal darkness suggest that our conception of highly consolidated nocturnal sleep as "normal" may be at odds with our evolutionary heritage (Wehr *et al.*, 1993), but personal and societal demands make this conception compelling nonetheless. Human sleep is categorized into rapid eye movement (REM) sleep and non-rapid eye movement (NREM) sleep, which in turn is subdivided into NREM stages 1–4. REM sleep is characterized by relatively fast, low-amplitude EEG activity, voluntary muscle atonia, and "phasic" activity including rapid eye movements, variations in heart rate and blood pressure, and isolated muscle twitches. The NREM stages correspond to progressively "deeper" NREM sleep, characterized by increasing delta EEG activity and increasing arousal thresholds. NREM and REM sleep follow a characteristic temporal organization: sleep onset is through NREM; NREM and REM sleep cycle approximately every 90 minutes during the night; most stage 3/4 sleep occurs in the first third of the night and REM episodes become progressively longer, with greater amounts of phasic activity (e.g., eye movements) later in the sleep episode (see Carskadon & Dement, 1994, for further description). Figure 1 displays a "sleep histogram" of a healthy adult subject, demonstrating the progression of sleep sages across the night. Monitoring and staging human sleep requires a minimum of one channel of EEG (usually derived from the vertex), one channel of electromyogram (typically submentalis), and measurement of eye movements. Clinical sleep studies also include monitoring of the electrocardiogram, airflow and respiratory effort, oximetry, and peripheral electromyogram (typically from the anterior tibialis).

The endogenous circadian pacemaker plays an important role in the regulation of sleep propensity and of sleep architecture in humans. Sleep propensity increases during the falling limb of the circadian temperature curve and is maximal near its trough; it decreases on the rising limb of the circadian curve and is minimal near its

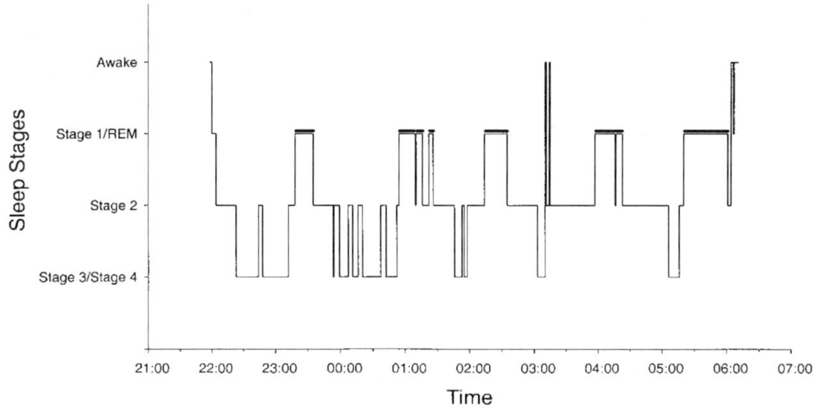

Figure 1. A "Sleep histogram" for a healthy young adult's nocturnal sleep. By convention, "deeper" NREM stages are indicated closer to the abscissa, and REM sleep is indicated by dark blocks. Note that most deep NREM sleep occurs early in the night, and that NREM and REM sleep alternate with a period of approximately 90 minutes. A small number of brief awakenings are normal events during nocturnal sleep.

peak (Czeisler, Weitzman, Moore-Ede, Zimmerman, & Knauer, 1980a; Dijk & Czeisler, 1994; Lavie, 1986; Zulley, Wever, & Aschoff, 1981). REM sleep propensity shows a strong circadian influence, being maximal shortly after the temperature minimum and minimal near the temperature acrophase (Carskadon & Dement, 1980; Czeisler, Zimmerman, Ronda, Moore-Ede, & Weitzman, 1980b; Dijk & Czeisler, 1995; Karacan, Finley, Williams, & Hursch, 1970; Zulley, 1980). Slow-wave sleep (SWS) propensity is less affected by circadian rhythmicity, but rather increases as a function of prior wakefulness, and is thus controlled primarily by a homeostatic mechanism (Karacan *et al.*, 1970; Webb and Agnew, 1977; Zulley, 1992).

Classification of Sleep Symptoms and Sleep Disorders

The major types of sleep/wake *symptoms* can be grouped into four categories: difficulty falling or staying asleep (insomnia), excessive sleepiness (hypersomnia), difficulty sleeping and waking at the desired times (sleep–wake schedule or circadian rhythm sleep problems), and abnormal behavior during sleep (parasomnias). These categories make intuitive and heuristic sense, and have been used widely in epidemiologic studies. However, each broad type of sleep complaint may arise from a number of distinct causes. In some cases, a single etiologic agent may produce different types of sleep complaints in different individuals. For instance, periodic limb movements during sleep can produce symptoms of either insomnia or hypersomnia.

Sleep disorders represent an attempt to categorize sleep complaints according to more specific clinical features or etiologies, functional impairments, and exclusion criteria. Although the precise pathophysiologic basis remains unidentified for many sleep disorders, classification according to *disorders* rather than *complaints* leads to more homogeneous groups for both clinical and research purposes. Broadly speaking, sleep disorders may be grouped into those which arise from primary dysfunction of sleep/wake mechanisms, producing insomnia or hypersomnia symptoms (dyssomnias), those which involve abnormal behavior or physiologic events during sleep (parasomnias), and those which are symptomatic of broader psychiatric, medical, or substance–induced disorders.

Several different classification systems have been used for sleep disorders. The three classifications with the most widespread current use are the *International Classification of Sleep Disorders* (ICSD; Diagnostic and Classification Steering Committee [DCSC], 1990), the *Diagnostic and Statistical Manual of Mental Disorders,* 4th ed. (DSM-IV; American Psychiatric Association, 1994), and the *International Statistical Classification of Diseases and Related Health Problems-10th Revision* (ICD-10; World Health Organization, 1992). Table 1 outlines these classifications. Each system includes a number of different diagnoses which are associated with complaints of insomnia or hypersomnia, and each identifies several types of parasomnias. Although each system also includes a section for circadian rhythm sleep disorders, the ICSD specifies the greatest number of specific types. In ICD-10, circadian disorders are included among both the "nonorganic" and the "organic" sleep disorders.

Epidemiology of Sleep Symptoms and Disorders

Most epidemiologic studies have focusesd on the *symptoms* of insomnia, hypersomnia, and snoring, and the *disorder* of obstructive sleep apnea. Some data are also available regarding the prevalence of narcolepsy and circadian rhythm sleep disorders. Prevalence estimates for these sleep disturbances have been based primarily on self-report surveys (with the exception of sleep apnea, as noted below).

Prevalence of Sleep Symptoms

The point prevalence of insomnia (i.e., the proportion of respondents reporting a current, ongoing problem in this area) ranges from 10% to 35% in adult populations (Ancoli-Israel, 1989; Ganguli, Reynolds, & Gilby, 1996; Henderson *et al.*, 1995; Partinen, 1994), but some reports have documented rates of up to 54% of men and 61% of women (Hyyppä & Kronholm, 1989; Urponen *et al.*, 1988). Insomnia complaints have a higher prevalence in women and in older persons. There is also strong evidence that it is highly comorbid with psychiatric status, particularly depression and anxiety-related disorders, and with preexisting physical health conditions (Ford & Kamerow, 1989; Weyerer & Dilling, 1991). Insomnia is strongly associated with the likelihood of taking hypnotics and with more frequent use of health services (Ford & Kamerow, 1989; Henderson *et al.*, 1995; Rosekind, 1992; Weyerer & Dilling, 1991).

Hypersomnia symptoms also appear to be relatively common, present in 5–20% of the adult population (Bixler *et al.*, 1979; Ganguli *et al.*, 1996; Wetter & Young, 1994; see Partinen, 1994, for review). Hypersomnia is somewhat more common in older persons (Ancoli-Israel, 1989), but its association with age appears weaker than the age–insomnia relationship. Hypersomnia also bears a weaker relationship to gender than does insomnia, but in the opposite direction; hypersomnia is more prevalent in men. This effect may result from its strong association with apnea, which is more common in men (Partinen, 1994).

Snoring is reported as ongoing problem by 5–25% of the adult population

(Partinen, 1994). Men are 1.5–2 times as likely as women to suffer from this problem. Snoring appears to increase with age up until about age 60 years, but is more strongly related to body mass and obesity than to age (Ancoli-Israel, 1989; Partinen, 1994).

Prevalence of Sleep Disorders

There are very few studies on the prevalence of circadian rhythm sleep disorders. Schrader, Bovim, and Sand (1993) performed a cross-sectional nationwide epidemiological study in Norway using a random sample of 10,000 individuals of ages 18–67 years. Of the 129 possible delayed sleep phase syndrome cases identified from the screening questionnaires, the diagnosis was confirmed in 17 cases, or a prevalence of 0.17%. The mean age of onset was 15.4 years, confirming that adolescence seems to be the most common period for the onset of the disorder (see below). Using criteria from the ICSD, no cases of advanced sleep phase syndrome were detected, but the exclusion of individuals above 67 years old might have diminished the probability of detecting such cases. To our knowledge, there have been no true epidemiologic studies on the hypernycthemeral syndrome. However, Sack, Lewy, Blood, Keith, and Nakagawa (1992) reported free-running circadian rhythms in about half of the totally blind subjects included in their study.

In contrast to prevalence studies of other sleep disturbances, those focused on sleep apnea typically combine self-report surveys with follow-up clinical assessments on a subsample of the surveyed group; prevalence of apnea for the entire sample is then estimated. Apnea prevalence estimates under this approach have been fairly well established, at least for conventional indices (Stradling, 1995). For example, depending on the severity criteria used to define apnea, estimated prevalence rates range from 0.3% to 5% (Bresnitz, Goldberg, & Kosinski, 1994; Partinen, 1994; Stradling, 1995). The prevalence of obstructive sleep apnea syndrome, i.e., sleep apnea events accompanied by symptoms, has been estimated at 2% of adult women and 4% of adult men (Young *et al.*, 1993).

Epidemiologic studies of narcolepsy have been conducted in several countries, leading to lifetime prevalence estimates of 0.01–0.1% in most adult populations (Dement, Carskadon, & Ley, 1973; Roth, 1980). The prevalence of narcolepsy varies with race and ethnicity. For instance, narcolepsy is rare in Israeli Jews (0.0002%) (Lavie & Peled, 1987) and more common among Japanese populations (0.6%) (Honda, Asaka, Tanimura, & Furusho, 1983). Although differences in case definition lead to some of this discrepancy, differences in prevalence may also be related to genetic factors. Narcolepsy is strongly linked to histocompatibility antigens HLA-DR2, HLA-DQw1 (also called Dqw15), and Dqw6 (Billiard & Seignalet, 1985; Juji, Matsuki, Naohara, Satake, & Honda, 1985) and the prevalence of these antigens varies in different ethnic and racial groups.

Circadian Rhythm Sleep Disorders: Clinical Features

As noted above, one major function of the endogenous circadian pacemaker is to prepare the individual, from both physiologic and psychological perspectives, for a restful night of sleep. It is therefore not surprising that dysfunction of the circadian system from either external (e.g., night work, jet lag) or internal (e.g., aging) causes is likely to result in sleep disturbance. Although these disorders may produce

TABLE 1. SLEEP DISORDER CLASSIFICATIONS

International Classification of Sleep Disorders (*ICSD*)	*Diagnostic and Statistical Manual of Mental Disorders, 4th ed.* (*DSM-IV*)	*International Statistical Classification of Diseases, 10th revision* (*ICD-10*)
Dyssomnias	Primary sleep disorders	Nonorganic sleep disorders
A. Intrinsic sleep disorders (selected examples)	A. Dyssomnias	1. Nonorganic insomnia
1. Psychologic insomnia	1. Primary insomnia	2. Nonorganic hypersomnia
2. Narcolepsy	2. Primary hypersomnia	3. Nonorganic disorder of the sleep–wake schedule
3. Obstructive sleep apnea syndrome	3. Narcolepsy	
4. Central sleep apnea syndrome	4. Breathing-related sleep disorder	4. Sleepwalking (somnambulism)
5. Periodic limb movement disorder	5. Circadian rhythm sleep disorder	5. Sleep terrors (night terrors)
6. Restless legs syndrome	a. Delayed sleep phase type	6. Nightmares
B. Extrinsic sleep disorders (selected examples)	b. Jet lag type	7. Other nonorganic sleep disorder
1. Environmental sleep disorder	c. Shift work type	8. Nonorganic sleep disorder, unspecified
2. Adjustment sleep disorder	d. Unspecified type	Sleep disorders (organic)
3. Insufficient sleep syndrome	6. Dyssomnia not otherwise specified	1. Disorders of initiating and maintaining sleep (insomnias)
4. Hypnotic-dependent sleep disorder	B. Parasomnias	
5. Alcohol-dependent sleep disorder	1. Nightmare disorder	2. Disorders of excessive somnolence (hypersomnias)
C. Circadian rhythm sleep disorders	2. Sleep terror disorder	
1. Time-zone change (jet lag) syndrome	3. Sleepwalking disorder	3. Disorders of the sleep–wake schedule
2. Shift work sleep disorder	4. Parasomnia not otherwise specified	a. Delayed sleep phase syndrome
3. Irregular sleep–wake pattern	Sleep disorders related to another mental disorder	b. Irregular sleep–wake pattern
4. Delayed sleep phase syndrome		4. Sleep apnea
5. Advanced sleep phase syndrome	A. Insomnia related to another mental disorder	a. Central
6. Non-24-hour sleep–wake disorder		b. Obstructive
Parasomnias	B. Hypersomnia related to another mental disorder	5. Narcolepsy and cataplexy
A. Arousal disorders		6. Other sleep disorders
1. Confusional arousals		Kleine–Levin syndrome
2. Sleepwalking		7. Sleep disorder, unspecified
3. Sleep terrors		

- B. Sleep–wake transition disorders (selected examples)
  - 1. Rhythmic movement disorder
  - 2. Sleep starts
- C. Parasomnias usually associated with REM sleep (selected examples)
  - 1. Nightmares
  - 2. Sleep paralysis
  - 3. REM sleep behavior disorder
- D. Other parasomnias (selected examples)
  - 1. Sleep bruxism
  - 2. Sleep enuresis
  - 3. Infant sleep apnea

Sleep disorders associated with medical/psychiatric disorders

- A. Associated with mental disorders (selected examples)
  - 1. Mood disorders
  - 2. Anxiety disorders
- B. Associated with neurological disorders (selected examples)
  - 1. Dementia
  - 2. Sleep-related epilepsy
- C. Associated with other medical disorders (selected examples)
  - 1. Nocturnal cardiac ischemia
  - 2. Sleep-related asthma
  - 3. Fibrositis (fibromyalgia) syndrome

Proposed sleep disorders (selected examples)

- 1. Short sleeper
- 2. Subwakefulness syndrome

Other sleep disorders

- A. Sleep disorder due to a general medical condition
  - 1. Insomnia type
  - 2. Hypersomnia type
  - 3. Parasomnia type
  - 4. Mixed type
- B. Substance-induced sleep disorder
  - 1. Insomnia type
  - 2. Hypersomnia type
  - 3. Parasomnia type
  - 4. Mixed type

symptoms of insomnia or hypersomnia, they are distinguished from other sleep disorders by the additional presence of prominent abnormalities in sleep *timing*.

Delayed Sleep Phase Syndrome

Delayed sleep phase syndrome (DSPS) is, as its name implies, primarily a disorder of sleep timing. Classically, the patient is able to get a satisfactory 7–8 hours of sleep, with normal sleep architecture, but is only able to obtain that sleep at an aberrantly late time relative to societal norms (e.g., from 0400 to noon). This usually results in conflicts with societal demands for presence at school or work, since the individual is unable to advance his/her bedtime appropriately (having sleep onset latencies of several hours whatever he/she tries). The disorder is particularly prevalent in young adolescents (Thorpy, Korman, Spielman, & Glovinsky, 1988), and it may be difficult for the clinician to differentiate between true DSPS and conduct or anxiety disorders which include school avoidance as a symptom (Regestein & Monk, 1995).

In DSPS, sleep duration and architecture are assumed to be normal, and sleep is assumed to occur at a normal phase angle with regard to other circadian rhythms (Terman *et al.*, 1995). Empirical studies demonstrate that polysomnographic sleep architecture of DSPS patients is normal except for delayed sleep onset (Alvarez, Dahlitz, Vignau, & Parkes, 1992). There are very few studies on the circadian characteristics of patients with DSPS, and they yield conflicting results. Alvarez *et al.* (1992) described normal profiles (i.e., occurring at clock times typical for healthy subjects) of plasma melatonin and urinary melatonin metabolites in a group of patients with DSPS studied under entrained conditions. However, Oren, Turner, and Wehr (1995) reported delayed timing of the plasma melatonin peak and the rectal temperature minimum in a patient studied under constant conditions. Elliot Weitzman conceptualized DSPS as resulting from a phase response curve (PRC) with no advance portion (Weitzman *et al.*, 1981). The hypothesis of an absent phase advance portion of the PRC in DSPS patients has yet to be tested. Even with a normal PRC, DSPS patients might fail to entrain normally if their extended morning sleep prevents them from being exposed to light and social zeitgebers, which normally effect a corrective phase advance (Terman *et al.*, 1995).

Advanced Sleep Phase Syndrome

The converse of DSPS, advanced sleep phase syndrome (ASPS), is posited to be a disorder in which pathologically early bedtimes and waketimes (e.g., 1930–0330) result from an inability to phase *delay* to a more socially appropriate bedtime. Individuals with ASPS are likely to complain of insomnia with early morning awakening, but not difficulty falling asleep. Their early sleep schedules may interfere with evening social obligations, but do not usually interfere with daytime work hours. ASPS is more common among the elderly. It is important to note that, despite symmetry in having DSPS and ASPS at either end of the adult lifespan, the evidence for an absent or dysfunctional advance portion of the PRC underlying ASPS is by no means compelling. As discussed in more detail in Chapter 22, many older people do indeed experience early morning awakening and choose earlier bedtimes, which renders them more "morning-type" in circadian disposition than younger adults (Czeisler *et al.*, 1992; Monk, Reynolds, Machen, & Kupfer, 1992). However, unlike DSPS, the sleep of the ASPS patient is clearly disrupted, and as Roehrs and Roth

(1994) discuss, it is questionable whether ASPS is truly a circadian pathology as opposed to an age-related sleep phenomenon.

Hypernycthemeral Syndrome (Non-24-Hour Sleep–Wake Syndrome)

Because the human circadian system does not naturally freerun to exactly 24.0 hours, but usually to a slower period, most individuals effect a small phase advance every day in order to maintain entrainment to a 24-hour period. Individuals with hypernycthemeral syndrome show a failure of entrainment, in which there is a mismatch between the 24-hour geophysical and societal period and the non-24.0-hour tau of the individual. The two rhythmic processes then beat into and out of phase with each other, leading to clusters of nights with sleep disruption alternating with clusters of comparatively good nights of sleep. In essence, these individuals are free-running in the real world, despite the presence of normal zeitgebers. Hypernycthemeral syndrome is seen most commonly among blind individuals with no photic input to the suprachiasmatic nucleus (SCN), but may also afflict sighted individuals who, for either behavioral or constitutional reasons, are unable to entrain their circadian system to a 24.0-hour cycle.

Hypernycthemeral syndrome in blind individuals has been enormously important in aiding our understanding of human circadian entrainment processes. Quite early in the history of the discipline, Lobban and Tredre (1967) and Hollwich and Dieckhues (1974) showed that blind individuals' circadian rhythms can be profoundly disrupted. Through the 1970s, when the role of photic zeitgebers were downplayed in human circadian rhythm research, this result was rather neglected. However, following the increased emphasis on photic zeitgebers that has occurred since 1980, a number of investigators have studied the circadian rhythms of blind individuals and shown a very high incidence of hypernycthemeral syndrome.

Sack, Lewy, and colleagues have shown a lack of entrainment in about half of their blind subjects, using dim melatonin onset (DLMO) as a circadian marker. For a number of blind individuals, free-running physiologic rhythms are associated with cyclic occurrence of severe subjective and objective sleep disturbances (Klein *et al.*, 1993; Sack *et al.*, 1992). On the other hand, some totally blind persons are able to maintain a roughly 24-hour cycle even when their other rhythms are free-running (Sack *et al.*, 1992). Interestingly, Nakagawa, Sack, and Lewy (1992) have shown that even in the absence of sleep disturbance complaints, the sleep propensity rhythm (as measured during an ultrashort sleep–wake cycle) can freerun with the melatonin, cortisol, and temperature rhythms in a totally blind subject. In a particularly elegant study, Czeisler *et al.* (1995) showed that the small proportion of profoundly blind subjects who *were* successfully entrained to 24.0 hours did, in fact, retain the ability to suppress melatonin by bright lights. Despite their absence of any *conscious* light/dark discrimination, these patients appeared to have an intact retinohypothalamic tract (RHT) linkage between the retina and the circadian timekeeping system (CTS). Thus, the CTS remained "sighted" and continued to be entrained to the 24.0-hour daylight/darkness cycle.

Irregular Sleep–Wake Pattern

Individuals with DSPS, ASPS, and hypernycthemeral syndrome have unusual, but *predictable*, sleep timing. By contrast, individuals with irregular sleep–wake pattern show no discernible regularity to their sleep periods. Rather, several brief bouts of sleep occur at different times of day, with no clear clustering across days. The irregularity and unpredictability of sleep may lead to complaints of either insomnia or hypersomnia.

Irregular sleep–wake pattern may be seen in individuals with dementia or other serious neurologic disorders, but it may also result from social and occupational demands. Increasingly, modern society, either through choice or the needs of employment (see below), has adopted a "napping" approach to sleep, akin to the "grazing" approach to food which has often replaced the standard three meals per day. This is a very dangerous trend because of the sleep deprivation and sleep fragmentation which can result. This sleep disruption can, in turn, affect daytime sleepiness in a serious manner (Mitler, Gujavarty, & Browman, 1982). Thus, individuals may slip into a vicious cycle where increasing sleep disruption leads to an increasingly irregular sleep schedule which further disrupts their sleep.

Jet Lag

"Jet lag" includes a constellation of sleep disturbances, impaired alertness and/or performance, and somatic symptoms which occur after travel across time zones. These symptoms arise from a mismatch between the timing of physical and social zeitgebers in the local environment and the individual's endogenous circadian rhythms. Because the circadian system cannot, under normal conditions, adjust instantaneously to an abrupt shift in routine, the circadian system will be inappropriately aligned for a number of days following arrival as it adjusts phase from the old routine to the new one (Aschoff, Hoffman, Pohl, & Wever, 1975; Graeber, 1989; Winget, DeRoshia, Markley, & Holley, 1984). During this adjustment process, sleep may be impaired. Nocturnal awakenings may occur, related to hunger, the need to void, or a simple lack of sleepiness. During the following day, this sleep disruption may lead to fatigue and performance decrements, but such effects will also appear because the individual is being required to be awake during the "down phase" of his or her circadian cycles. Malaise and loss of concentration may stem from the circadian desynchrony characteristic of the reentrainment process per se (Klein, Wegmann, & Hunt, 1972).

Recovery from jet lag may not be monotonic. Particularly following eastbound flights, there may be a strong "second night" effect whereby the second night is worse than the first, and the recovery function takes a "zigzag" form (Monk, Moline, & Graeber, 1988; Sasaki, Endo, Nakagawa, Kitahara, & Mori, 1985).

An increasing segment of the population (often late-middle-aged males) have occupations which require them to spend up to 2 weeks per month in time zones radically different from "home time." The term "chronic jet-lag syndrome" (CJLS) has been coined for this disorder (Monk, 1994). Because of its chronic, rather than acute, nature, CJLS is much more similar to rotating shift work (see below) in its effects, which may include gastrointestinal disturbance, heart disease, and marital disharmony in addition to the expected disruptions of sleep and daytime functioning.

Shift Work

One fifth of the work force is required to work outside the normal "daylight" 7 a.m. to 7 p.m. time frame and can thus be fairly described as shift workers. Approximately 60% of shift workers complain of sleep disturbances (Rutenfranz, Colquhon,

Knauth, & Ghata, 1977). The worst sleep disruption occurs in those required to work at night, either permanently or as part of a rotating shift schedule. Adaptation to shift work generally declines with age. Sleep disruption springs both from circadian rhythm disruptions and from the competing demands of society and the worker's family who run on a day-oriented schedule.

Moore-Ede (1986) coined the term shiftwork maladaptation syndrome (SMS) to describe the disorder of failing to cope with abnormal work hours. In addition to sleep disruption and excessive sleepiness at work and when traveling home from work (Richardson, Miner, & Czeisler, 1990), SMS has been shown to be associated with gastrointestinal symptoms (Scott & LaDou, 1990), heart disease (Knutsson, Akerstedt, Orth-Gomer, & Jonsson, 1986), and an increased divorce rate (White & Keith, 1990). As more individuals are exposed to abnormal work hours and the population ages, there will be an increasing need to address this disorder from educational, research, and legislative perspectives (Office of Technology Assessment, 1991). See Chapter 21 for more discussion of shift work issues.

Treatment of Circadian Rhythm Sleep Disorders

Despite their different clinical manifestations, the circadian rhythm sleep disorders share common pathophysiologic mechanisms, i.e., difficulty generating or entraining circadian rhythms to match environmental time cues. Because of this underlying similarity, treatments for the various circadian rhythm sleep disorders also share many common features. Broadly speaking, treating circadian rhythm sleep disorders follows three basic principles: (1) maximizing exposure to appropriately timed bright light, (2) implementing behavioral strategies to protect sleep, enforce wakefulness, and regularize the sleep–wake schedule, and (3) judicious use of pharmacologic treatments. Treatment approaches for circadian rhythm sleep disorders are summarized in Table 2.

Bright Light

As in other mammals, the effect of bright light on the phase of human circadian rhythms depends on the phase of the circadian cycle at which the bright light exposure is applied. This characteristic is described by a phase response curve with the following properties: (1) phase delays are obtained when bright light exposure is centered early in the subjective night of the subject (before the minimum of the temperature circadian rhythm), (2) phase advances are obtained when bright light exposure is centered late in the subjective night of the subject (after the minimum the temperature of the circadian rhythm), and (3) no phase shift is obtained when bright light exposure is centered during the subjective day of the subject (Czeisler *et al.*, 1989; Minors, Waterhouse, & Wirz-Justice, 1991).

Treatment strategies for sleep and circadian rhythm disorders have been developed from these basic observations. In 1992, a task force was commissioned by the American Sleep Disorders Association (ASDA) and the Society for Light Treatment and Biological Rhythms (SLTBR) to review the experimental and clinical evidence for the efficacy of light treatment of sleep disorders. The task force report has been published (Boulos *et al.*, 1995; Campbell *et al.*, 1995a–c; Dijk *et al.*, 1995; Eastman *et al.*, 1995; Terman *et al.*, 1995) and will be summarized here.

Table 2. Summary of Treatments for Circadian Rhythm Sleep Disorders

Disorder	Light treatment	Behavioral treatment	Pharmacologic Treatment
Delayed sleep phase syndrome	1. Bright a.m. light 2. Dark exposure in p.m.[a]	1. Chronotherapy (progressive 3-hour delay in sleep) 2. Sleep hygiene (regular sleep times, minimize alcohol)	1. Benzodiazepine hypnotic to aid sleep onset 2. Cyanobalamin (vitamin B_{12}) 3. Evening melatonin
Advanced sleep phase syndrome	Bright p.m. light	1. Advance chronotherapy (progressive 3-hour advance in sleep times)[a] 2. Sleep hygiene (regular sleep times)	None described
Hypernycthemeral (non-24-hour sleep–wake) syndrome	1. Regularly scheduled bright a.m. light?[a] 2. Ensure light exposure during day, dark exposure at night	1. Rigorously scheduled and enforced bedtimes 2. Increase daytime activity	1. Melatonin successful in some blind individuals 2. Benzodiazepine for periodic insomniaa
Irregular sleep–wake pattern	1. Bright a.m. light[a] 2. Ensure light exposure during day, dark exposure at night	1. Rigorously scheduled and enforced bedtimes 2. Increase daytime activity	1. Benzodiazepine hypnotic at night[a]
Jet lag sleep disorder	1. Bright light to shift circadian rhythms prior to departure 2. Exposure to daylight, darkness in destination time zone	1. Rapidly assume sleep–wake schedule of destination time zone 2. Avoid excessive alcohol, caffeine	1. Benzodiazepine hypnotic to avoid sleep deprivation, assist sleep in destination time zone 2 Melatonin pre- and postflight at destination bedtime
Shift work sleep disorder	1. Bright light during scheduled shift 2. Avoid light on trip home from work 3. Provide darkness in sleep environment	1. Avoid social, physical disruptions to sleep 2. Sleep hygiene (regular sleep hours, comfortable sleep environment, minimize alcohol and caffeine) 3. Nap before shifts	1. Benzodiazepine for disrupted daytime sleep 2. Melatonin[a] 3. Caffeine during night-shift work hours

[a]Treatment with limited empirical evidence. See text for details.

BRIGHT LIGHT EFFECTS ON SLEEP AND ACTIVATION. Several studies have examined the effects of bright light on sleep in healthy young subjects. In these studies, the phase of the circadian rhythms is changed while the timing of the sleep episode is kept constant. Morning bright light exposure *increases* early-night sleep propensity (as shown by a decreased sleep latency and a decreased amount of wakefulness at the beginning of the night), whereas evening bright light exposure *decreases* early-night sleep propensity (Campbell & Dawson, 1992; Carrier & Dumont, 1995; Drennan, Kripke, & Gillin, 1989). Some studies have also reported that morning bright light exposure reduces late-night sleep propensity as reflected by a faster accumulation of wakefulness at the end of the night and an earlier sleep termination (Campbell & Dawson, 1992; Dijk, Visscher, Bloem, Beersma, & Daan, 1987; Dijk, Beersma, Daan, & Lewy, 1989), although this latter result was not replicated in a later study (Carrier & Dumont, 1995). The moderate phase angle disturbance induced by bright light exposure in normal subjects does not affect sleep architecture, including slow-wave sleep.

In addition to these effects on sleep, some studies have suggested that bright light produces a *direct alerting or activating effect* (for review, see Campbell *et al.*, 1995b). For example, bright light exposure during the night increases the alertness of subjects undergoing sleep deprivation (Badia, Myers, Boecker, Culpepper, & Harsh, 1991; Campbell & Dawson, 1990; French, Hannon, & Brainard, 1990; Myers & Badia, 1993). Another study reported that a single bright light exposure just before bedtime increases sleep latency (Cajochen, Dijk, & Borbely, 1992; Dijk, Cajochen, & Borbely, 1991). Guilleminault *et al.* (1995) reported that bright light may help to improve sleep in patients with psychophysiologic insomnia, presumably through a direct alerting effect during the daytime. It has not been determined whether this alerting/activating property is an "intrinsic" effect of bright light or whether it is mediated by another system or mechanism (e.g., phase shift of circadian rhythms, melatonin suppression).

BRIGHT LIGHT TREATMENT FOR DELAYED AND ADVANCED SLEEP PHASE SYNDROMES. Lewy, Sack, and Singer (1985) first suggested that timed bright light exposure may be effective in treating DSPS and ASPS. Patients with DSPS are generally exposed to bright light in the early morning (e.g., 0600–0900 hours). This timing of bright light exposure has been shown to advance the phase of the temperature circadian rhythms of patients with DSPS, to adjust their sleep–wake cycle to an earlier time, and to increase their subjective sleep quality (Czeisler *et al.*, 1988; Rosenthal *et al.*, 1990). There are very few studies on the therapeutic value of evening bright light exposure for "pure" ASPS. Lack and Wright (1993) studied the effects of bright light for two consecutive evenings on sleep and circadian rhythms in a group of patients suffering from early morning insomnia. Following evening bright light treatment, circadian rhythms of temperature and melatonin were phase delayed by 1–4 hours and mean final wakeup time was delayed by more than 1 hour. Campbell, Dawson, and Anderson (1993) also demonstrated that evening bright light exposure is effective in the treatment of sleep maintenance insomnia in the elderly. In this study, exposure to 2 hours of evening bright light produced a phase delay of the circadian rhythm and increases in sleep efficiency, stage 2, REM, and slow-wave sleep. The use of evening bright lights has had some success in delaying temperature rhythms and diminishing wakefulness late in the sleep period (Campbell, Kripke, Gillin, & Hrubovcak, 1988; Czeisler *et al.*, 1986).

Bright Light Treatment for Jet Lag and Shift Work. Appropriately timed exposure to bright light may accelerate circadian adaptation and alleviate symptoms in jet lag and shift work (see Boulos *et al.*, 1995 and Eastman *et al.*, 1995 for reviews). In the real world, jet lag and shift work confront the individual with different types of zeitgebers: congruent to the desired sleep–wake schedule in the case of jet lag, and conflicting in the case of shift work. In laboratory simulation experiments, the distinction between jet lag and shift work is often arbitrary, since subjects are not submitted to strongly congruent *or* conflicting time cues. Some laboratory jet-lag or shift-work simulations have reported faster adaptation of the circadian rhythms with bright light exposure compared to dim light following a phase advance (Honma *et al.*, 1995) or a phase delay of the sleep–wake cycle (Dawson & Campbell, 1991; Wever, 1985). There have been few bright-light field studies of shift work, but all have shown that scheduled exposure to bright light during the night and exposure to darkness during the day improve the adaptation of circadian rhythms to night shift work (Czeisler *et al.*, 1990; Eastman, 1992; Eastman, Stewart, Mahoney, Liu, & Fogg, 1994). A study assessing the relative contribution of bright light exposure during the night and of wearing dark goggles (0.35% transmission) when returning home in the morning showed that both bright light and darkness exposure were important for producing the adaptation of the temperature circadian rhythm to the new schedule (Eastman, 1992).

Behavioral Treatments

Although sleep initiation is not under volitional control, wakefulness can (within certain limits) be enforced. As a result, voluntary behavioral control can be exercised over the timing of wakefulness and *opportunities* for sleep. This principle underlies many of the behavioral techniques available for treating circadian rhythm sleep disorders.

Sleep Diaries. A record of sleep–wake schedules and factors which affect sleep is an important part of the assessment and treatment of circadian rhythm sleep disorders. Prospective daily diaries convey much more accurate data regarding sleep–wake hours than simple retrospective reports, which run the risk of biasing toward "average" days or "worst-case" days. A number of types of sleep–wake diaries are available. The "visual" type of sleep diary uses a graphic format on which the individual marks his/her time of going to bed, time of arising, wakefulness during the night, and naps. This type of diary is very useful for quickly examining sleep patterns over a number of successive days, and can clearly demonstrate phase shifts or treatment effects (Figure 2). An alternative is a text-based sleep diary, which asks for times of going to bed, awakenings during the night, and final wakeup time (Monk *et al.*, 1994a) (Figure 3). This type of diary has the advantage of being able to evaluate additional information, such as subjective ratings of sleep quality, medications, and exercise.

Another diary-like instrument, the Social Rhythm Metric (SRM), has been developed to provide a numerical estimate of the irregularity or regularity of a person's lifestyle (Monk, Flaherty, Frank, Hoskinson, & Kupfer, 1990; Monk, Petrie, Hayes, & Kupfer, 1994b). The SRM has proven useful in quantifying lifestyle disruptions resulting from depression (Szuba, Yager, Guze, Allen, & Baxter, 1992), anxiety disorders (Shear *et al.*, 1996), bereavement (Prigerson *et al.*, 1994), and the birth of a

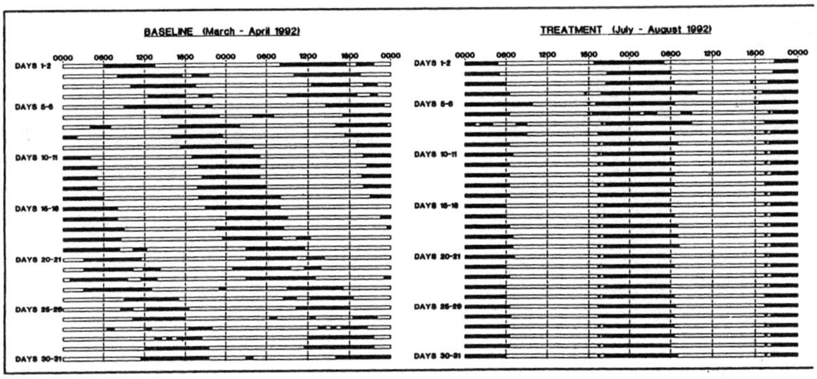

Figure 2. Sleep–wake patterns in a 5-year-old blind child with hypernycthemeral syndrome (non-24-hour sleep–wake cycle) before and after treatment with melatonin. Sleep in indicated in black bars, wake in white. Data are double-plotted. Treatment with melatonin 0.5 mg was initiated on day 4 of the Treatment block. The patient's 25.2-hour-sleep–wake cycle at baseline was entrained to a 24-hour cycle with melatonin. (From Lapierre and Dumont, 1995, with permission. Copyright 1995, Elsevier Science.)

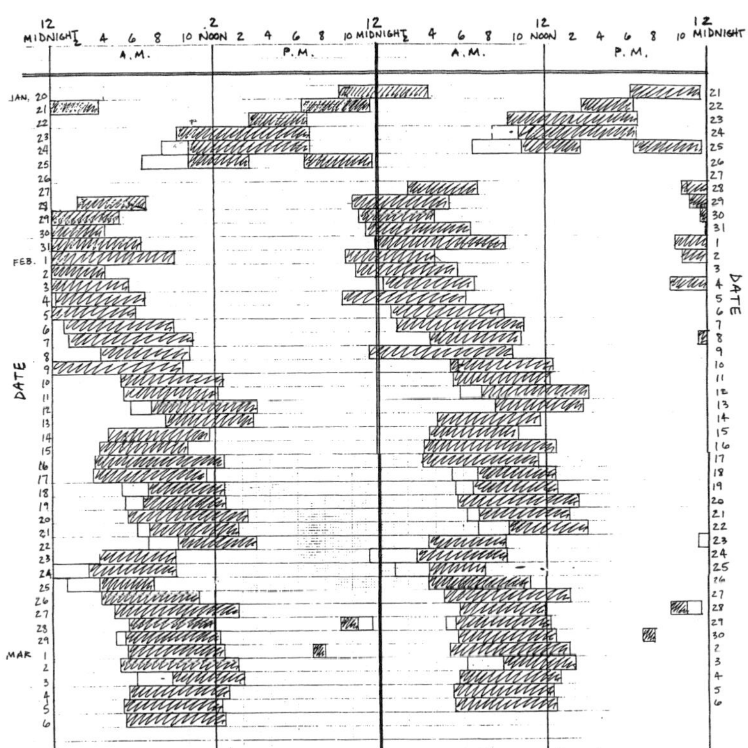

Figure 3. A typical "graphic" sleep–wake diary, with hatched areas denoting sleep and open areas wakefulness. This graph shows the sleep–wake pattern of a 40-year-old man with obstructive sleep apnea and an irregular sleep–wake schedule. Following several days of progressive phase advances in sleep, the patient appeared to show a hypernycthemeral pattern. However, his sleep eventually stabilized at a delayed phase.

baby (Monk *et al.*, 1996). It may also prove to be a useful tool in the diagnosis and treatment of circadian rhythm sleep disorders.

CHRONOTHERAPY. As discussed below, Elliot Weitzman conceptualized DSPS as resulting from a PRC with no advance portion (Weitzman *et al.*, 1981). He likened it to living on a "one-way street" where phase delays were allowed, but phase advances were not. This suggested the therapy ("chronotherapy") which Weitzman and Czeisler pioneered, namely that of "driving around the block," i.e., scheduling repeated phase delays in sleep timing (typically 3 hours per day) until desirable bedtimes and waketimes had been obtained (Czeisler *et al.*, 1981). After chronotherapy, the patient has to maintain a strict regimen (even on weekends) to ensure that no phase "slippage" occurs. In contrast to phase delay chronotherapy, "reverse chronotherapy" (consecutive 3-hour phase *advances*) has *not* been reported as a successful treatment for ASPS.

Chronotherapy may exert a beneficial effect through several mechanisms: (1) gradual entrainment of the circadian pacemaker indirectly through a hypothesized "weak" (sleep–wake) oscillator, (2) accumulated sleep deprivation, which promotes sleep onset at desired times, and (3) changes in the timing of exposure to light and social zeitgebers, which favor reentrainment at more "normal" times. In clinical practice, chronotherapy is often combined with pharmacotherapy or bright light treatment in order to ensure rigid entrainment to the desired sleep–wake schedule.

OTHER BEHAVIORAL TECHNIQUES. Behavioral treatment of circadian rhythm sleep disorders relies upon the individual adopting a specified wakeup time and "protected" time for sleep. Enforcing a wakeup time can be very difficult in practice, particularly in DSPS or irregular sleep–wake pattern, where the individual may not be highly motivated to change his/her schedule. Multiple alarm clocks and light exposure, exercise, and social contact upon awakening may all be helpful. Conversely, ensuring an environment conducive to sleep is also extremely important, particularly since the required sleep times may be unusual. For instance, shiftworkers and individuals undergoing chronotherapy need to sleep during daylight hours on at least some days. Blocking out ambient light with heavy draperies, ensuring a cool bedroom, enlisting other family or household members in noise suppression, and disconnecting telephones (or using answering machines) may all be necessary to protect the sleep environment.

Ensuring a comfortable sleep environment is one of the techniques of good "sleep hygiene" (Hauri, 1977; see below), which were originally developed in conjunction with insomnia treatment, but which are also useful behavioral adjuncts in treating circadian rhythm sleep disorders. Additional sleep hygiene measures, such as avoiding alcohol and excessive use of caffeine, may also help to optimize sleep. However, empirical evidence for the effectiveness of sleep hygiene education in circadian rhythm sleep disorders is sparse. Holbrook, White, and Hutt (1994) attempted to use sleep hygiene education to help law enforcement officers reverse impairments attributed to rotating shift work. While knowledge of sleep hygiene principles improved, follow-up assessments found little or no change in sleep hygiene practices.

Specific sets of behavioral techniques have been devised to help individuals cope with jet lag (Graeber, 1989; Monk, 1987) and shift work (Monk, 1994). These are described in greater detail in Chapter 21.

Drugs from many classes, as well as hormones and dietary agents, have been used to treat individuals with circadian rhythm sleep disorders. However, the effects of most of these agents on circadian rhythms, or their effects on the sleep disturbances associated with circadian rhythm sleep disorders, have not been systematically evaluated. We will discuss data available for three classes of pharmacologic agents: melatonin, benzodiazepines, and stimulants.

MELATONIN. Lewy, Ahmed, Jackson, and Sack (1992) showed that physiologic doses of orally administered melatonin shift the melatonin circadian rhythm according to a phase response curve. Specifically, afternoon or early evening administration of 0.5 mg of melatonin advances the onset of melatonin secretion, whereas morning administration delays melatonin rhythm. Melatonin may therefore be used as an entraining agent in circadian rhythm disorders. However, other studies have shown that melatonin may also produce a direct sedating effect (Dawson & Encel, 1993; Dollins, Zhdanova, Wurtman, Lynch, & Deng, 1994). Accordingly, in many studies, especially those using pharmacologic doses, it is not clear if the therapeutic effect is achieved through a hypnotic effect, a circadian effect, or a combination of the two.

Several studies have shown that late afternoon or evening administration of melatonin can entrain the irregular sleep–wake pattern of blind people (Figure 4) (Arendt, Aldhous, & Wright, 1988; Folkard, Arendt, Aldhous, & Kennett, 1990; Lapierre & Dumont, 1995; Palm, Blennow, & Wetterberg, 1991; Tomoda, Miike, Vezono, & Kawasaki, 1994). Melatonin administration can also advance the sleep–wake cycle of patients suffering from DSPS. Dahlitz *et al.* (1991) first demonstrated that daily administration of 5 mg of melatonin at 2200 hours for 1 month advanced sleep onset times and wake times of patients with delayed sleep syndrome. Oldalni *et al.* (1994) used ambulatory polysomnography to evaluate the efficacy of a physiologic dose of oral melatonin (0.5 mg/day) administered in the late afternoon in advancing the sleep–wake cycle of six DSPS patients. Following 1 month of melatonin treatment, mean sleep onset time was advanced by 115 minutes and mean final wake time by 106 minutes. Several studies have also shown melatonin to alleviate some of the detrimental effects of jet lag experienced by passengers (Arendt, Aldhous, & Marks, 1986; Arendt *et al.*, 1987; Petrie, Dawson, Thompson, & Brook, 1993), flight crews (Petrie, Conaglen, Thompson, & Chamberlain, 1989) or subjects submitted to a laboratory phase shift of the sleep–wake cycle (Samel *et al.*, 1991).

Preliminary findings on the efficacy of melatonin for treating symptoms related to shift work have suggested that melatonin might produce beneficial effects on sleep and alertness, but its effects on performance have yet to be evaluated systematically (Folkard, Arendt, & Clark, 1993). In a study comparing the adaptation to night shifts in groups of subjects receiving either bright light treatment, exogenous melatonin, or placebo, Dawson, Encel, and Lushington (1995) showed that bright light was superior to melatonin or placebo for phase shifting the circadian system and improving sleep and performance.

BENZODIAZEPINES. Benzodiazepines may affect sleep and circadian rhythms in three basic ways: (1) through direct hypnotic effects, (2) through direct effects on the suprachiasmatic nucleus, and (3) through changes in circadian rhythms such as

SLEEP DIARY **BEDTIME** KEEP BY BED

Please fill out this part of the diary last thing at night.

day .... date ....

Today, when did you have: (if none, write 'none')
breakfast ....
lunch ....
dinner ....

How many of the following did you have in each time period? *(if none, leave blank)*

	before or with breakfast	after breakfast before/with lunch	after lunch before/with dinner	after dinner
caffeinated drinks	....	....	....	....
alcoholic drinks	....	....	....	....
cigarettes	....	....	....	....
cigars/pipes/plugs (of chewing tobacco)	....	....	....	....

Which drugs and medications did you take today? *(prescribed & over the counter)*

name	time	dose
....	....	....
....	....	....
....	....	....
....	....	....

What exercise did you take today? (if none, check here) ☐

start .... end .... type ....
start .... end .... type ....

How many daytime naps did you take today? (if none, write 0) ....
give times for each:

start .... end .... start .... end ....

SLEEP DIARY **WAKETIME** KEEP BY BED

Please fill out this part of the diary first thing in the morning.

day .... date ....

went to bed last night at ....
lights out at ....
minutes until fell asleep ....
finally woke at ....

Awakened by *(check one)*:
alarm clock/radio ☐
someone whom I asked to wake me ☐
noises ☐
just woke ☐

After falling asleep, woke up this many times during the night (circle)
0 1 2 3 4 5 or more

total number of minutes awake ....

- woke to use bathroom (circle # times)
  0 1 2 3 4 5 or more
- awakened by noises/child/bedpartner (circle # times)
  0 1 2 3 4 5 or more
- awakened due to discomfort or physical complaint (circle # times)
  0 1 2 3 4 5 or more
- just woke (circle # times)
  0 1 2 3 4 5 or more

Ratings (place a mark somewhere along the line):

Sleep Quality:
very bad ______________________ very good

Mood on Final Wakening:
very tense ______________________ very calm

Alertness on Final Wakening:
very sleepy ______________________ very alert

Figure 4. Page from the Pittsburgh Sleep Diary (Monk *et al.*, 1994a). The subject/patient completes the left side of the page upon retiring to bed and the right side of the page upon awakening.

core body temperature, motor activity, or hormone levels. There is little controversy regarding the direct hypnotic effect of benzodiazepines (Gaillard, 1994): these drugs reduce sleep latency, reduce wakefulness during sleep, and increase sleep duration. As a result, benzodiazepines may be helpful for aiding sleep onset and maintaining the desired sleep–wake schedule in any of the circadian rhythm sleep disorders.

Direct effects of benzodiazepines on SCN function have been suspected because of the high concentration of γ-amino butyric acid (GABA) receptors in the SCN and in brain regions which receive efferents from the SCN (Moore & Speh, 1993). As these observations predict, high doses of triazolam produce phase shifts of the circadian wheel-running rhythm in golden hamsters, apparently mediated by enhanced motor activity (Turek & Losee-Olson, 1986). Although similar phase-shifting effects have not been observed in humans under entrained conditions, no laboratory studies using unmasking or phase shift paradigms have been reported.

Contradictory findings have been reported for the effect of benzodiazepines on various entrained circadian rhythms. For example, alprazolam 2 mg given to six healthy volunteers at 9:00 p.m. significantly suppressed melatonin and cortisol throughout the night as compared to baseline measures in one study, although the effect was not consistently observed in all subjects (McIntyre, Norman, Burrows, & Armstrong, 1993). On the other hand, studies with triazolam 0.5 mg, temazepam 20 mg, and zolpidem 10 mg (an imidazopyridine which acts at the benzodiazepine recognition site) did not show significant effects on melatonin or cortisol rhythms (N. H. P. Allen, Smythe, Martin, Plaskett, & Osola, 1994; Copinschi *et al.*, 1990; Copinschi, Akseki, Morenoreyes, Leproult, & L'Hermite-Baleriaux, 1995). Temazepam also did not affect the sensitivity of melatonin to suppression by bright light in one study (N. H. P. Allen *et al.*, 1994). Thus, evidence for direct benzodiazepine effects on SCN function or on the expression of specific circadian rhythms has not been clearly demonstrated in humans. The contradiction between human and animal studies in this regard may be related to differences in dose, routes of administration, and direct effects on sleep/wakefulness.

Short-acting benzodiazepines may improve sleep quality after an acute phase shift of the sleep–wake cycle (Kanno, Watanabe, & Kazamatsuri, 1993; Nicholson, Roth, & Stone, 1985; Seidel, Roth, Roehrs, Zorick, & Dement, 1984) and increase alertness during the following waking period (Bonnet *et al.*, 1988). By analogy, benzodiazepines may improve the sleep and alertness of night shift workers. Although triazolam did improve daytime sleep in one study of simulated night shift work (Walsh, Muehlbach, & Schweitzer, 1984), it did not improve alertness during the nighttime working period. The role of benzodiazepines in treating insomnia related to actual shift work has been controversial, not only because of questions of effectiveness, but also because shift work is a chronic situation and benzodiazepines are approved only for short-term administration. A conservative recommendation is that shift workers use hypnotics for not more than one or two sleep periods immediately after a sleep schedule change (Roehrs, Zorick, & Roth, 1994).

STIMULANTS Stimulants such as dextro-amphetamine, methylphenidate, and pemoline are indicated to manage excessive daytime sleepiness in disorders such as narcolepsy. They are not routinely recommended for treating sleepiness in circadian rhythm sleep disorders. Although the alerting effects of these drugs have been well-studied (see Mitler, Aldrich, Koob, & Zarcone, 1994, for review), their circadian effects have not been studied as thoroughly. In one study, dextro-amphetamine 20 mg increased body temperature in 41 healthy subjects compared to body temperature in a placebo control group (Pigeau, Naitoh, Buguet, & McCann, 1995). However, this is likely to represent a masking effect of the drug, rather than a true chronobiotic effect on phase or amplitude.

While amphetamines and methylphenidate are widely used in sleep medicine, caffeine is also a central nervous system stimulant with potential effects on circadian rhythms. There has been no experimental evidence in humans examining the direct effects of caffeine on endogenous circadian rhythms. However, caffeine has been demonstrated to raise metabolic rate and boy temperature, probably as a masking effect (Bonnet & Arand, 1992; Zahorska-Markiewicz, 1980).

A number of studies have examined the direct alerting effect of caffeine to enhance vigilance during sustained operations (Smith, Kendrick, & Maben, 1992). These effects may help to manage normal circadian variations in sleepiness and subjective alertness, as well as the impaired alertness characteristic of circadian rhythm sleep disorders such as shift work and jet lag (Dekker, Paley, Popkin, & Tepas, 1993; Miller, Lombardo, & Fowler, 1995; Smith, Brockman, Flynn, Maben, & Thomas, 1993; Walsh *et al.*, 1990).

The effects of caffeine have been examined in a number of simulated night shift experiments. Most of these demonstrate that caffeine increases alertness and improves psychomotor performance during usual night-shift hours when taken between 2230 and 0120 hours (reviewed in Walsh, Muehlbach, & Schweitzer, 1995). At dosages of approximately 250–400 mg, the beneficial effects of caffeine persist until at least 0530 hours. For most subjects, caffeine taken at the start of the night shift does not interfere significantly with daytime sleep beginning at 0900 hours. There is also some evidence that single doses of caffeine at the beginning of a night shift may be more alerting than divided doses throughout the shift. Combining caffeine with a

4-hour afternoon nap before a simulated night shift may further improve performance (Bonnet & Arand, 1994).

The effects of caffeine on alertness and performance do not consistently mirror one another. For example, Muehlbach and Walsh (1995) found that caffeine 2 mg/kg decreased physiologic sleep tendency on a simulated night shift compared with placebo. However, the caffeine and placebo groups performed at equivalent levels on a simulated assembly-line task. It is also important to note that, despite the positive results of laboratory research examining caffeine as a shift-work countermeasure, controlled trials in actual shift-working populations have not been conducted. Finally, the possibility of sleep disruption due to excessive caffeine intake in actual shift workers is a very real one, which could have the paradoxical effect of *impairing* alertness due sleep loss. In short, specific guidelines for the use of caffeine during night shift work are not available.

Other Primary Sleep Disorders

Narcolepsy

Clinical Features. Narcolepsy, as the name implies, is characterized by sudden sleep attacks. Three other features of the narcolepsy tetrad include cataplexy (the sudden bilateral loss of muscle tone, typically following a strong emotional stimulus), sleep-related hallucinations, and sleep paralysis. Other common symptoms include disrupted nocturnal sleep and "automatic" waking behavior, which results from profound sleepiness. Daytime and nocturnal sleep in narcolepsy is characterized by the appearance of REM sleep at, or soon after, sleep onset. The precise pathophysiology of narcolepsy is unknown, but it appears to involve altered brainstem cholinergic and monoaminergic receptor concentrations/activity, leading to abnormal REM sleep phenomena (Aldrich, 1990; Guilleminault, 1994).

Circadian Aspects. The presence of frequent daytime sleep episodes and nocturnal awakenings suggests a reduced amplitude of the circadian sleep–wake rhythm, and raises the possibility of broader circadian dysfunction in narcolepsy. Studies have investigated a variety of circadian measures in narcoleptic patients under entrained conditions, in temporal isolation, and on altered sleep–wake schedules. The total amount of sleep per 24 hours is neither increased nor decreased in patients with narcolepsy (Broughton *et al.*, 1988), and the response to sleep deprivation indicates intact homeostatic sleep drive (Tafti, Rondouin, Besset, & Billiard, 1992). However, ambulatory monitoring studies and a study using a 90-minute sleep–wake schedule demonstrated a significant reduction in the amplitude of sleep propensity and motor activity rhythms (Broughton *et al.*, 1988; Dantz, Edgar, & Dement, 1994; Middlekoop *et al.*, 1995). Despite decreased circadian variation in sleep latencies, patients with narcolepsy and healthy controls demonstrated identical rhythms of *subjective* sleepiness in the 90-minute study (Dantz *et al.*, 1994), suggesting a different "calibration" of subjective sleepiness in patients with narcolepsy.

Studies of other rhythms in narcolepsy have not shown consistent abnormalities. Although Mosko and colleagues reported reduced amplitude of core body temperature in patients with narcolepsy compared to controls under entrained conditions, the higher nocturnal temperature nadir in patients may have been due to greater wakefulness during sleep (Mosko, Holowach, & Sassin, 1983). Subsequent studies found no difference in any temperature measure of narcoleptic patients compared to controls under entrained, temporal isolation (Pollak & Wagner, 1994) and 90-minute sleep–wake schedule (Dantz *et al.*, 1994) conditions. Similarly, normal rhythms have been found for cortisol (Schulz, Kiss, & Hasse, 1990; Takahashi *et al.*, 1976; Weitzman, 1976), melatonin (Ahmed, Sack, Rich, & Lewey, 1991; Bourgine, Claustrat, Besset, & Billiard, 1986), and cognitive and motor performance (Pollak, Wagner, Moline, & Monk, 1992).

In summary, patients with narcolepsy clearly demonstrate reduced amplitude of circadian rhythms in sleep propensity and motor activity, with no consistent evidence of circadian dysfunction in temperature, endocrine, or performance rhythms. These findings suggest that the circadian clock per se functions normally in narcolepsy, but that the effector mechanisms which mediate circadian rhythms of alertness are impaired (Dantz *et al.*, 1994; Broughton & Mullington, 1994).

TREATMENT. Stimulant medications are the mainstay of treatment for the excessive sleepiness in narcolepsy (Aldrich, 1990; Mitler *et al.*, 1994). Methylphenidate, dextro-amphetamine, methamphetamine, and pemoline are the most commonly used medications, although modafinil is also widely used in Europe. The subjective improvement noted by most patients is reflected in more modest improvements in sleep latency tests. Tricyclic antidepressants and selective serotonin reuptake inhibitors are useful for treating the cataplexy, sleep-related hallucinations, and sleep paralysis which often accompany narcolepsy.

PRIMARY INSOMNIA

CLINICAL FEATURES. Primary insomnia refers to a complaint of difficulty falling asleep or staying asleep, or nonrestorative sleep, which lasts for at least 1 month and impairs daytime function. Other related terms for chronic insomnia, such as psychophysiologic insomnia, include more specific concepts of conditioned arousal or somatized tension/anxiety as clinical features and etiologic features (Buysse & Reynolds, 1992; DCSC, 1990). Primary insomnia excludes cases of chronic insomnia associated with medical conditions, psychiatric disorders, or medication/substance use.

CIRCADIAN ASPECTS. Primary insomnia, by definition, is not diagnosed if it occurs exclusively during the course of a circadian rhythm sleep disorder. However, the extent to which disturbances in circadian rhythms predispose, precipitate, or perpetuate primary insomnia is unclear. For instance, studies of chronic insomnia that select patients based on sleep-onset difficulties or on early-morning awakening may bias selection of subjects toward those with phase-delayed or phase-advanced rhythms, respectively. In practical terms, therefore, the distinction between primary insomnia and circadian rhythms sleep disorders is not always straightforward. Features which suggest the diagnosis of a circadian rhythm sleep disorder, as opposed to primary insomnia, include (1) the presence of a relatively undisturbed sleep episode, once sleep is initiated, (2) a phase delay or advance in *both* bedtime and rise time, (3) a sleep–wake schedule which changes in a consistent direction across days (e.g., bedtime and rise times become progressively later), and (4) the presence of an external factor which forces sleep to unfavorable circadian phases (e.g., shift work, travel across time zones).

Temporal isolation and unmasking studies have not been conducted in patients with well-defined primary insomnia. A constant-routine study of individuals with chronic sleep-onset insomnia (which is probably similar to primary insomnia) and healthy controls found evidence for delayed temperature acrophase and nadir in the former group (Morris, Lack, & Dawson, 1990). MacFarlane *et al.* (1984) reported that sleep–onset insomnia was associated with phase-delayed temperature rhythms, and early-morning awakening insomnia with phase-advanced rhythms. Under entrained conditions, core body temperature declined less at night in 10 patients with primary insomnia than in age- and sex-matched controls, but phase was not altered (Mendelson, Garnett, & Linnoila, 1984). In another study under entrained conditions, hourly nocturnal plasma melatonin concentrations in 10 patients with primary insomnia were compared to age-matched controls. Melatonin levels in the patient group tended to rise earlier in the evening, and peak levels were significantly lower compared to controls. The patients with the most severely reduced melatonin levels were also the patients with the most chronic sleep disturbance (Hajak *et al.*, 1995).

Irregular rest–activity cycles such as irregular bedtimes may reflect an underlying circadian disturbance, and such poor sleep hygiene has been demonstrated in patients with primary insomnia (Lacks & Rotert, 1986). Pollak, Perlick, & Linsner (1992) reported increased motor activity at night and decreased activity during the day in elderly insomniacs versus control subjects. Whether this flattening of motor activity reflects underlying circadian dysfunction, an epiphenomenon of disrupted sleep, or voluntary changes in activity level cannot be determined.

In summary, patients with chronic insomnia have not shown consistent evidence of circadian dysfunction in terms of core body temperature or endocrine rhythms. The lack of well-controlled data in this population, however, prevents any definitive statements regarding their circadian function.

TREATMENT. Because the exact pathophysiology of primary insomnia is unknown, behavioral and pharmacologic treatments are directed at symptoms and presumed etiologies (see Buysse, Morin, & Reynolds, 1995 for review). A variety of behavioral treatments have been developed for chronic insomnia (including primary insomnia). These include the following.

Sleep hygiene education is concerned with modifying health practices (such as diet, exercise, and substance use) and environmental factors (such as light, noise, and temperature) that may be either detrimental or beneficial to sleep (Hauri, 1977). Although sleep hygiene education by itself does not produce robust improvements in chronic insomnia (Morin, Culbert, & Schwartz, 1994), such education is usually an implicit part of all behavioral treatments.

Stimulus control therapy consists of a set of instructional procedures designed to curtail sleep-incompatible behaviors. This method targets both conditioning and circadian factors. These procedures are (1) go to bed only when sleepy, (2) use the bed and bedroom only for sleep and sex (i.e., no reading, television watching, eating, or working during the day or at night), (3) get out of bed and go into another room whenever you are unable to sleep for 15–20 minutes, and return only when sleepy again, (4) arise in the morning at the same time regardless of the amount of sleep during the previous night, and (5) do not nap during the day (Bootzin, 1972).

Sleep restriction therapy consists in initially curtailing the amount of time spent in bed to match the actual amount of sleep the individual obtains, as determined from a 2-week sleep diary. Once a patient's sleep becomes more efficient, then the allowable time in bed is increased gradually to allow for the greatest total amount of sleep time while preserving newly achieved capacity for more efficient sleep (Spielman, Saskin, & Thorpy, 1987). This method relies on the homeostatic regulation of sleep to facilitate sleep initiation and maintenance after an increased duration of wakefulness.

Relaxation therapies are designed to alleviate somatic or cognitive arousal and consist of techniques such as progressive muscular relaxation, autogenic training, and biofeedback (Hauri, 1981). *Cognitive–behavioral treatments* target the maladaptive cognitions that perpetuate insomnia (Morin, Colecchi, Stone, Sood, & Brink, 1995). These behavioral interventions have been used as single interventions or, more typically, as combined or multimodal approaches.

Although these behavioral treatments for chronic insomnia are not conceptualized as circadian interventions, they share the common feature of stabilizing daily sleep–wake rhythms. Sleep hygiene, stimulus control, and sleep restriction, in particular, stress the importance of setting regular bed and rising times. In addition to providing social and behavioral zeitgebers, stable wake times presumably regularize exposure to daylight, which serves further to entrain the circadian timing system.

Benzodiazepine hypnotics are the most common pharmacologic treatment for chronic insomnia (including primary insomnia). These medications should be used in the lowest effective dose and not on a chronic basis due to concerns about side effects such as memory impairment, falls, and tolerance (loss of effectiveness with repeated dosing). However, a substantial minority of patients will continue to require adjuvant hypnotic medication on a continuous or episodic schedule. Alternative medication strategies have included sedating antidepressants, melatonin, tryptophan, and antihistamines. Again, the chronicity of primary insomnia suggests that medication strategies be used in conjunction with educational and behavioral approaches (Dingemanse, 1995; Espie, 1993; Hartmann, 1995; Maczaj, 1993).

Obstructive Sleep Apnea Syndrome

Clinical Features. Obstructive sleep apnea syndrome (OSAS) and its related syndromes, sleep hypopnea and upper airway resistance, share two cardinal features: Daytime sleepiness, and repeated awakenings from sleep due to partial or complete upper airway closures (Strollo & Rogers, 1996). Episodes of airway obstruction and arousal may occur hundreds of times per night, leading to decreased oxyhemoglobin saturation, severely disrupted sleep, and, thereby, excessive sleepiness. Loud snoring usually accompanies apnea syndromes, and daytime fatigue, cognitive deficits, depression, morning headaches, and nocturia are common. Other clinical features associated with apnea syndromes include male sex, obesity, age >40 years, and hypertension. The pathophysiology of sleep apnea involves physical narrowing of the upper airway, which is influenced by neuromuscular tone and sleep stage; apnea events are most common during REM sleep, presumably because of atonia during REM sleep (Sullivan & Issa, 1985).

Circadian Aspects. Circadian variation appears to play little role in the core clinical features of sleep apnea. For instance, the frequency of apneas during daytime sleep or during the first 4 hours of the night is highly correlated with the frequency of apneas across the entire night, and absolute numbers of apnea events are also very similar at these different times (Roberts & Hooper, 1985; Sériès, Cormier, & La Forge, 1991).

Circadian patterns of various measures have been assessed in patients with

OSAS, but most of these show no clear abnormalities. For instance, circadian patterns of melatonin, cortisol, interleukins, and γ-interferon showed no difference between OSAS patients and healthy controls either before or after continuous positive airway pressure (CPAP) treatment in one recent study (Entzian, Linnemann, Schlaak, & Zabel, 1996). *Reduced amplitude* has been demonstrated for other physiologic measures, most notably blood pressure. Approximately 30–60% of patients fail to demonstrate the usual nocturnal decrease in blood pressure (Lapiński *et al.*, 1993; Noda, Okada, Hayashi, Yasuma, & Yokota, 1993). Elevations in nocturnal blood pressure correlate significantly with hypoxemia, body mass index, age, and severity of apnea (Hoffstein & Mateika, 1992; Lavie, Yoffe, Berger, & Peled, 1993). Other measures which show reduced circadian variation in OSAS patients include urinary catecholamines (Lapiński *et al.*, 1993), urine volume and sodium excretion (Rodenstein, D'Odemont, Pieters, & Aubert-Tulkens, 1992; Warley and Stradling, 1988), and tumor necrosis factor-α (Entzian *et al.*, 1996). In most of these cases, reduced amplitudes in OSAS patients result from a failure to demonstrate the normal sleep-related decline in the measure. By contrast, the mesor and amplitude of serum erythropoietin rhythms are *increased,* and the acrophase advanced, in apnea patients (Cahan *et al.*, 1992). Erythropoetin maintains red blood cell mass and oxygen-carrying capacity, and increases in response to hypoxemia.

Altered circadian rhythms in OSAS patients are more likely to result from factors such as increased awakenings during sleep and hypoxemia (with compensatory physicologic responses) than from primary dysfunction of the circadian pacemaker. The methodology necessary to identify a primary dysfunction (i.e., constant-routine studies) would be difficult to conduct in apnea patients, given their degree of daytime sleepiness and nocturnal sleep disruption. Nevertheless, the reduced amplitude of circadian rhythms in apnea patients may be an important measure of pretreatment symptoms and therapeutic response (e.g., nocturia, hypertension). Furthermore, any individuals enrolled in studies of circadian function should be carefully screened to exclude apnea, since apnea can affect the amplitude of various circadian rhythms.

TREATMENT The initial treatment for OSAS consists of continuous positive airway pressure (CPAP) administered via nasal mask. This creates a "pressure splint" which prevents airway collapse. Patients may not dramatic improvements in nocturnal sleep quality and daytime sleepiness with CPAP. Other treatments for OSAS include the use of oral appliances to maintain upper airway patency during sleep and various surgical approaches (including uvulopalatopharyngoplasty and maxillomandibular advancement) designed to increase the size of the airway. Finally, behavioral measures, such as weight loss and training to ensure sleep in the lateral recumbent position, may provide some symptomatic relief, but are generally insufficient treatment in themselves (Sanders, 1994; Strollo & Rogers, 1996).

RESTLESS LEGS SYNDROME AND PERIODIC LIMB MOVEMENTS

CLINICAL FEATURES. Restless legs syndrome (RLS) is characterized by four clinical features: (1) the desire to move the extremities, often associated with uncomfortable sensations, (2) motor restlessness, (3) worsening of symptoms at rest and temporary relief with movement, and (4) worsening of the symptoms in the evening or night (Walters & The International Restless Legs Syndrome Study Group, 1995). Approximately 80% of patients with RLS also have periodic limb movements (PLMs) during sleep, which can lead to frequent brief arousals (Montplaisir, Lapierre, Warnes, & Pelletier, 1992). RLS can cause sleep-onset insomnia, and PLMs (with their associated arousals) can cause either insomnia or daytime sleepiness. The etiology of these related disorders is unknown, although alteration in dopaminergic receptor function is suggested by decreased central D2 dopamine receptor occupancy and the beneficial response to dopaminergic drugs (Montplaisir, Lorrain, & Godbout, 1991; Staedt *et al.*, 1995). Altered opioid receptor function (Hening *et al.*, 1986) and peripheral axonal neuropathy (Iannaccone *et al.*, 1995) have also been implicated.

Circadian Aspects. The reliable worsening of RLS symptoms in the evening and night hours is not related to sleep, but follows a circadian pattern with a nocturnal peak. Treatment with dopaminergic drugs sometimes alters this pattern: Nocturnal symptoms are reduced, but morning and daytime symptoms increase in as many as 30–50% of patients (R. P. Allen, Kaplan, Buchholz, & Walters, 1990; Becker, Jamieson, & Brown, 1993; Montplaisir, Goodbout, Poirier, & Bedard, 1986). The actual distribution of PLMs during sleep shows two distinct patterns in different individuals, one with a peak near sleep onset and progressively declining levels during sleep, and the other with a maximum number of PLMs in the middle of the sleep period (Culpepper, Badia, & Shaffer, 1992). No studies have been conducted to determine whether the worsening of RLS shows a circadian pattern independent of any masking effects inherent in a normal nycthemeral routine.

Treatment. Treatment for RLS and PLMs is almost exclusively pharmacologic. Dopaminergic drugs such as levodopa/carbidopa combinations, bromocriptine, or amantadine are considered the treatment of choice, and can diminish the subjective sensations of RLS, the number of periodic limb movements, and arousals associated with PLMs (Montplaisir *et al.*, 1992). Benzodiazepines such as clonazepam and temazepam also effectively treat PLMs, and may be particularly helpful in decreasing arousals associated with PLMs. Other medications, such as opiate drugs and gabapentin, have also been shown to be effective for some patients (O'Keefe, 1996). Finally, drugs such as selective serotonin reuptake inhibitors may aggravate RLS and PLMs (Dorsey, Lukas, & Cunningham, 1996), and may need to be avoided or discontinued.

Parasomnias

Sleep disorders characterized by abnormal behavior, rather than by insomnia or hypersomnia, may arise from either NREM sleep or REM sleep. The most common NREM parasomnias are sleep terrors and sleepwalking. In each condition, the individual has a partial arousal from deep NREM sleep (usually stage 3 or 4), which is accompanied by difficulty awakening fully and partial or complete amnesia for the event (DCSC, 1990; Keefauver and Guilleminault, 1994). The partial arousal may lead to either a panicked scream and heightened autonomic activity (sleep terror), walking or other motor behaviors (sleepwalking), or some combination of the two. One of the hallmarks of NREM parasomnias is their timing within the first third of the night. The predictability of these episodes is not related to circadian factors, but rather to the preponderance of slow-wave sleep at that time of night; episodes can occur later in sleep or during daytime sleep if a sufficiently long bout of NREM sleep is achieved. Treatment for NREM parasomnias includes conservative steps, such as ensuring a safe environment and locking doors and windows; preventing sleep deprivation, which may worsen episodes; hypnosis and psychotherapy; and judicious use of benzodiazepine or antidepressant drugs (Broughton, 1994; Thorpy, 1990).

The most common REM sleep parasomnias are nightmares and REM behavior disorder (RBD). Nightmares, or dream anxiety attacks, are frightening dreams accompanied by autonomic arousal. RBD involves the physical enactment of dreaming mentation, and frequently leads to injury of the affected individual or his/her bedpartner. The hallmark of RBD is partial or complete release from the usual atonia of REM sleep, which may be identified through repeated bursts of muscle activity during REM sleep on the polysomnogram (Mahowald and Schenck, 1994). Although REM sleep parasomnias may also show characteristic temporal patterning across the night, this is related to the circadian variation of REM sleep. No studies have formally reported on the temporal distribution of REM parasomnias, although clinical evidence suggests that these episodes are more common in later REM periods, which are longer and contain more phasic activity. Treatment for severe nightmares often includes psychotherapy and treatment with antidepressants for comorbid depression or anxiety disorders. RBD is often successfully treated with benzodiazepines (e.g., clonazepam); some reports have also suggested efficacy in some patients of dopaminergic drugs, tricyclic antidepressants, or carbamazepine (Mahowald & Schenck, 1994).

Medical and Psychiatric Sleep Disorders

Medical and psychiatric illnesses are a frequent cause of sleep complaints, including insomnia and hypersomnia. The source of sleep disruption may be nonspecific (e.g., pain and anxiety) or more specific (e.g., nocturnal exacerbation of asthma, or early-morning awakening associated with depression). In addition, specific symptoms and events associated with medical or psychiatric disorders can occur during or immediately following sleep. For instance, angina pectoris and panic attacks can both begin during specific sleep stages. The extent to which these symptoms and events are related to endogenous *circadian rhythms,* as opposed to *sleep,* is difficult to determine. For example, circadian oscillations within the neuroendocrine or immune systems may increase susceptibility to clinical events during sleep such as myocardial infarction or asthma attacks. Conversely, events related to sleep such as irregular heart rate and breathing during REM sleep may trigger cardiovascular events, producing a circadian variation in such events. There are essentially no data pertaining to pathologic physical symptoms or events during constant-routine or temporal isolation studies, which would be necessary to tease apart the independent influences of circadian and sleep factors.

In this section we will briefly summarize the best-characterized relationships between medical symptoms/events, circadian rhythms, and sleep. We will also summarize the circadian–sleep–mood relationship observed in major depressive disorder (MDD) and one of its subtypes, seasonal effective disorder (SAD). Although sleep disturbances are common in most psychiatric disorders, there is little evidence linking circadian dysregulation to the sleep disturbances seen in psychiatric disorders other than depression. Diurnal variations in symptom patterns or in physiologic measures (e.g., hormone secretion patterns) in psychiatric patients may suggest circadian dysregulation, but since these are not related to patients' sleep disturbances per se, we will not review them here.

There is considerable epidemiologic evidence for a circadian rhythm in myocardial infarction. The majority of first-time infarcts occur between 6 a.m. and noon (e.g., Muller, Stone, & Turi, 1985; Ridker, Manson, & Buring, 1990) and this time frame has also been associated with a peak in angina, silent ischemia, strokes, and sudden cardiac death (see George, 1994, for review). Of the significant cardiovascular events with circadian variation in time of occurrence, only atypical angina peaks during the night (as reviewed by George, 1994), and angina can become a source of sleep disturbance. Myocardial ischemia may be precipitated by events related to sleep or by circadian patterns of hormone release. For example, REM sleep may be associated with an increased risk of myocardial ischemia (Murao, Marumi, & Katayama, 1972). Conversely, neuroendocrine rhythms, such as the rise in circulating cortisol levels in the early morning hours, may also contribute to ischemic events (Thurton & Degan, 1974). As noted by George (1994), the research conducted to date does not allow conclusions to be drawn about the independent and relative contribution of sleep-related phenomena and circadian rhythms on the incidence and consequences of cardiovascular ischemia.

NOCTURNAL ASTHMA

Nocturnal bronchoconstriction and wheezing are reported by the majority of individuals with asthma (e.g., Chhabra, 1995; Connolly, 1979). Frequency and severity of nocturnal symptoms are associated with severity of asthma and are unrelated to age and gender (e.g., Chhabra, 1995). In addition to causing discomfort and sleep disruption, nocturnal bronchoconstriction may be related to the increased risk of mortality between midnight and 8 a.m. observed among individuals with asthma (see Douglas, 1994, for review). Circadian factors may influence nocturnal asthma via two pathways. First the normal drop in circadian-mediated flow rates is greatly exaggerated in individuals with asthma (Hetzel & Clark, 1979). Second, circadian oscillations in immune function, which appears to differ between individuals with and without asthma, contribute to airway inflammation in the early morning hours (e.g., Jarjour & Busse, 1995). It therefore appears that circadian factors are important mediators of nocturnal bronchoconstriction and wheezing in individuals with asthma.

DEPRESSION

Major depression has been characterized by alterations in rhythms of mood, sleep, hormones, and core body temperature (see Wirz-Justice, 1995, for review). This observation has led some researchers to hypothesize that depression may be driven in part by a dysregulation of the circadian system. Wehr and Wirz-Justice (1981) were among the first to propose that the circadian system of individuals with depression was phase advanced, leading to a desynchronization of sleep and other endogenous rhythms. From this perspective, the sleep disturbances characteristic of depression, especially short REM latency and early-morning awakening, were seen as a direct result of a phase advance in other circadian rhythms, such as core body temperature and cortisol. Tests of the phase advance hypothesis have revealed inconsistent results. Although some studies have found a phase advance in depression (e.g., Gwirtsman *et al.*, 1989), others have failed to find such evidence (e.g.,

Tsujimoto, Yamada, Shimoda, Hanada, & Takahashi, 1990). With regard to sleep itself, two studies have reported that REM sleep is advanced relative to sleep onset, but not relative to clock time in depressed patients (Buysse, Jarrett, Miewald, Kupfer, & Greenhouse, 1990; Reynolds, Newton, Shaw, Coble, & Kupfer, 1982). These results argue against a phase advance of REM sleep (and, by implication, other endogenous rhythms) in depression, although these studies did not directly assess core body temperature or other rhythms. Computer simulations and phase-shift studies in healthy control subjects have also been conducted to determine whether delaying sleep produces "depressive" sleep patterns. Computer simulations demonstrate that normal subjects' REM latency is shortened and first REM period duration is lengthened to values similar to those seen in depression only after a phase shift of 5–6 hours (MacLean, Knowles, & Vetere, 1986). Phase-shift experiments in healthy young subjects showed that REM latency and first REM period duration were *not* appreciably altered by phase shifts of 3 hours in sleep onset time (Knowles, MacLean, & Cairns, 1982). These results also suggest that simple phase angle disturbances are unlikely to underlie the sleep disturbances seen in depression.

Seasonal Affective Disorder

Depressed mood and complaints of hypersomnia are hallmark symptoms of seasonal affective disorder (SAD). Symptoms of SAD are clustered within winter months and at geographic latitudes where exposure to light is at its lowest (Rosen *et al.*, 1990). The SAD symptom profile suggests that circadian dysregulation may play a role in the pathophysiology of this disorder, as light is an important circadian rhythm signal. Various hypotheses suggest that SAD results from a phase delay (Lewy, Sack, Miller, & Hoban, 1987), circadian amplitude changes in response to diminishing light exposure during winter months (Czeisler *et al.*, 1989), or abnormal photoperiodic responses to diminishing light (Rosenthal *et al.*, 1984) (for reviews, see Blehar & Lewy, 1990; Carrier & Dumont, 1995). Evidence that exposure to bright light is effective in ameliorating the symptoms of SAD also supports the hypothesis that this is a chronobiological disorder (Terman *et al.*, 1989).

The importance of sleep as a mediator of the relationship between circadian rhythms and mood in SAD has not been demonstrated. First, pretreatment EEG sleep patterns in individuals with SAD do not resemble those typical of depression (e.g., Anderson *et al.*, 1994; Rosenthal *et al.*, 1989; Thase, 1989). Second, these studies have shown that sleep patterns are also, for the most part, unchanged after bright light therapy despite a reduction in symptoms of SAD (Patonen, Appelberg, & Patinen, 1993). Finally, a study of the effects of light therapy and total sleep deprivation on EEG sleep under a constant-routine condition reported that sleep responses in women with SAD were almost identical to those of healthy control subjects (Brunner *et al.*, 1996).

A summary of treatments for noncircadian sleep disorders appears in Table 3.

Conclusions

Sleep complaints and sleep disorders are widely prevalent and they cause substantial morbidity. Knowledge regarding the regulation of circadian rhythms has helped to identify a set of sleep disorders which arise directly from circadian dysfunction or from societal disregard for circadian function. Other "primary" sleep dis-

TABLE 3. SUMMARY OF TREATMENTS FOR NONCIRCADIAN SLEEP DISORDERS

Disorder	Behavioral treatments	Pharmacologic treatments	Other treatments
Narcolepsy	1. Sleep hygiene 2. Scheduled daytime naps	1. Stimulants for sleepiness 2. Antidepressants for cataplexy, sleep hallucinations, sleep paralysis	—
Primary insomnia	1. Sleep hygiene 2. Stimulus control 3. Sleep Restriction 4. Relaxation 5. Cognitive–behavior therapy	1. Intermittent benzodiazepines for short periods of time 2. Antidepressants[a] 3. Melatonin[a]	Bright a.m. light[a]
Obstructive sleep apnea syndrome	1. Weight loss 2. Sleep hygiene	Low-dose antidepressants if needed for insomnia	1. Continuous positive airway pressure (CPAP) 2. Upper airway surgery
Restless legs syndrome/ periodic limb movements	—	1. Dopaminergic drugs 2. Benzodiazepines 3. Opioids	Treat underlying iron deficiency, renal failure if present
NREM parasomnias	1. Sleep hygiene 2. Avoid sleep deprivation 3. Hypnosis 4. Psychotherapy	1. Benzodiazepine hypnotic 2. Antidepressant	—
REM parasomnias	Psychotherapy, hypnosis for nightmares	1. Benzodiazepine hypnotic (clonazepam for REM behavior disorder) 2. Antidepressant	Evaluate and treat underlying neurological conditions in REM behavior disorder
Sleep disorder due to medical/ neurological disorders	Sleep hygiene	1. Medication specific for underlying condition 2. Low-dose short-acting benzodiazepine	—
Sleep disorder due to psychiatric disorder	1. Sleep hygiene 2. Psychotherapy 3. Cognitive–behavior therapy	1. Medication specific for underlying condition 2. Benzodiazepine, sedating antidepressant as adjuncts	

[a]Treatment with limited empirical evidence. See text for details.

orders, as well as those secondary to medical or psychiatric disorders, may also include abnormal expression or control of circadian rhythms. However, potential circadian aspects of these disorders are not well understood due to the small number of studies which have directly examined these aspects. Understanding the regulation of human circadian rhythms has also led to the development of somatic, pharmacologic, and behavioral treatments for sleep disorders which exert their effects, at least in part, through the circadian timing system. Among the most pressing issues for sleep disorders medicine are a need to better understand the pathophysiology of these disorders and a need to develop more effective long-term treatments. Continued awareness of the contribution of circadian factors to sleep regulation will assist in both of these endeavors.

Acknowledgments

This work was supported in part by MH30915, MH24652, MH52247, MH37869, MH16804, MH01235, MH48891, AG 15138, and AG 00972.

References

Ahmed, S., Sack, R. L., Rich, G. B., & Lewey, A. J. (1991). Twenty-four hour secretion of melatonin is normal in narcoleptics. *Sleep Research, 20,* 194.

Aldrich, M. S. (1990). Narcolepsy. *New England Journal of Medicine, 323,* 389–394.

Allen, N. H. P., Smythe, P. J., Martin, N., Plaskett, E., & Osola, K. (1994). The effect of temazepam on melatonin secretion and light sensitivity. *Journal of Psychopharmacology, 8,* 94–97.

Allen, R. P., Kaplan, P. W., Buchholz, D. W., & Walters, J. K. (1990). Double-blind, placebo controlled comparison of propoxyphene and carbidopa/levodopa for treatment of periodic limb movements in sleep. *Sleep Research, 20,* 199.

Alvarez, B., Dahlitz, M. J., Vignau, J., & Parkes, J. D. (1992). The delayed sleep phase syndrome: Clinical and investigative findings in 14 subjects. *Journal of Neurology, Neurosurgery and Psychiatry, 55,* 665–670.

American Psychiatric Association. (1994). *Diagnostic and statistical manual of mental disorders* (4th ed.). Washington, DC: Author.

Ancoli-Israel, S. (1989). Epidemiology of sleep disorders. *Clinics in Geriatric Medicine, 5,* 347–362.

Anderson, J. L., Rosen, L. N., Mendelson, W. B., Jacobsen, F. M., Skwerer, R. G., Joseph-Vanderpool, J. R., Duncan, C. C., Wehr, T. A., & Rosenthal, N. E. (1994). Sleep in fall/winter seasonal affective disorder: Effects of light and changing seasons. *Journal of Psychosomatic Research, 38,* 323–337.

Arendt, J., Aldhous, M., & Marks, V. (1986). Alleviation of jet lag by melatonin: Preliminary results of controlled double blind trial. *British Medical Journal, 292,* 1170.

Arendt, J. H., Aldhous, M., English, J., Marks, V., Marks, M., & Folkard, S. (1987). Some effects of jet-lag and their alleviation by melatonin. *Ergonomics, 30,* 1379–1393.

Arendt, J., Aldhous, M., & Wright, J. (1988). Synchronisation of a disturbed sleep–wake cycle in a blind man by melatonin treatment. *Lancet, 1,* 772–773.

Aschoff, J., Hoffman, K., Pohl, H., & Weaver, R. A. (1975). Re-entrainment of circadian rhythms after phase-shifts of the zeitgeber. *Chronobiologia, 2,* 23–78.

Badia, P., Myers, B., Boecker, M., Culpepper, J., & Harsh, J. R. (1991). Bright light effects on body temperature, alertness, EEG and behavior. *Physiology and Behavior, 50,* 583–588.

Becker, P. M., Jamieson, A. O., & Brown, W. D. (1993). Dopaminergic agents in restless legs syndrome and periodic limb movements of sleep: Response and complications of extended treatment in 49 cases. *Sleep, 16,* 713–716.

Billiard, M., & Seignalet, J. (1985). Extraordinary association between HLA-DR2 and narcolepsy. *Lancet, 1,* 226–227.

Bixler, E. O., Kales, A., Soldatos, C. R., Kales, J. D., & Healey, S. (1979). Prevalence of sleep disorders in the Los Angeles metropolitan area. *American Journal of Psychiatry, 136,* 1257–1262.

Blehar, M. C., & Lewy, A. J. (1990). Seasonal mood disorders: Consensus and controversy. *Psychopharmacology Bulletin, 26,* 465–493.

Bonnet, M. H., & Arand, D. L. (1992). Caffeine use as a model of acute and chronic insomnia. *Sleep, 15,* 526–536.

Bonnet, M. H., & Arand, D. L. (1994). Impact of naps and caffeine on extended nocturnal performance. *Physiology and Behavior, 56,* 103–109.

Bonnet, M. H., Dexter, J. R., Gillin, J. C., James, S. P., Kripke, D., Mendelson, W., & Mitler, M. (1988). The use of triazolam in phase-advanced sleep. *Neuropsychopharmacology, 1,* 225–234.

Bootzin, R. R. (1972). A stimulus control treatment for insomnia. *Proceedings of the American Psychological Association,* 395–396.

Borkovec, T. D., & Fowles, D. C. (1973). Controlled investigation of the effects of progressive and hypnotic relaxation on insomnia. *Journal of Abnormal Psychology, 82,* 153–158.

Boulos, Z., Campbell, S. S., Lewy, A. J., Terman, M., Dijk, D.-J., & Eastman, C. I. (1995). Light treatment for sleep disorders; Consensus report. VII. Jet lag. *Journal of Biological Rhythms, 10,* 167–176.

Bourgine, N., Claustrat, B., Besset, A., & Billiard, M. (1986). Melatonin plasma concentration during daytime and nighttime in narcoleptic subjects and controls. *Sleep Research, 15,* 46.

Bresnitz, E. A., Goldberg, R., & Kosinski, R. M. (1994). Epidemiology of obstructive sleep apnea. *Epidemiologic Reviews, 16,* 210–227.

Broughton, R. J. (1994). Parasomnias. In S. Chokroverty (Ed.), *Sleep disorders medicine: Basic science, technical considerations, and clinical aspects* (pp. 381–399). Boston: Butterworth-Heinemann.

Broughton, R., & Mullington, J. (1994). Chronobiological aspects of narcolepsy. *Sleep, 17,* S35–S44.

Broughton, R., Dunham, W., Newman, J., Lutley, K., Duschesne, P., & Rivers, M. (1988). Ambulatory 24 hour sleep–wake monitoring in narcolepsy–cataplexy compared to matched controls. *Electroencephalography and Clinical Neurophysiology, 70,* 473–481.

Brunner, D. P., Krauchi, K., Dijk, D., Leonhardt, G., Haugh, H., & Wirz-Justice, A. (1996). The sleep EEG in seasonal affective disorder and in control women: Effects of midday light treatment and sleep deprivation. *Biological Psychiatry, 40,* 485–496.

Buysse, D. J., & Reynolds, C. F. (1990). Insomnia. In M. J. Thorpy (Ed.), *Handbook of sleep disorders* (pp. 375–433). New York: Marcel Dekker.

Buysse, D. J., Jarrett, D. B., Miewald, J. M., Kupfer, D. J., & Greenhouse, J. B. (1990). Minute-by-minute analysis of REM sleep timing in major depression. *Biological Psychiatry, 28,* 911–925.

Buysse, D. J., Morin, C. M., & Reynolds, C. F. (1995). Sleep disorders. In G. O. Gabbard (Ed.), *Treatments of psychiatric disorders* (pp. 2397–2453). Washington, DC: American Psychiatric Press.

Cahan, C., Decker, M. J., Arnold, J. L., Washington, L. H., Veldhuis, J. D., Goldwasser, E., & Strohl, K. P. (1992). Diurnal variations in serum erythropoietin levels in healthy subjects and sleep apnea patients. *Journal of Applied Physiology, 72,* 2112–2117.

Cajochen, C., Dijk, D.-J., & Borbely, A. A. (1992). Dynamics of EEG slow-wave activity and core body temperature in human sleep after exposure to bright light. *Sleep, 15,* 337–343.

Campbell, S. S., & Dawson, D. (1990). Enhancement of nighttime alertness and performance with bright ambient light. *Physiology and Behavior, 48,* 317–320.

Campbell, S. S., & Dawson, D. (1992). Aging young sleep: A test of the phase advance hypothesis of sleep disturbance in the elderly. *Sleep Research, 1,* 205–210.

Campbell, S. S., Kripke, D. F., Gillin, J. C., & Hrubovcak, J. C. (1988). Exposure to light in healthy elderly subjects and Alzheimer's patients. *Physiology and Behavior, 42,* 141–144.

Campbell, S. S., Dawson, D., & Anderson, M. W. (1993). Alleviation of sleep maintenance insomnia with time exposure to bright light. *Journal of the American Geriatrics Society, 41,* 829–836.

Campbell, S. S., Eastman, C. I., Terman, M., Lewy, A. J., Boulos, Z., & Dijk, D.-J. (1995a). Light treatment for sleep disorders: Consensus report. I. Chronology of seminal studies in humans. *Journal of Biological Rhythms, 10,* 105–109.

Campbell, S. S., Dijk, D.-J., Boulos, Z., Eastman, C. I., Lewy, A. J., & Terman, M. (1995b). Light treatment for sleep disorders: Consensus report. III. Alerting and activation effects. *Journal of Biological Rhythms, 10,* 129–132.

Campbell, S. S., Terman, M., Lewy, A. J., Dijk, D.-J., Eastman, C. I., & Boulos, Z. (1995c). Light treatment for sleep disorders: Consensus report. V. Age related disturbances. *Journal of Biological Rhythms, 10,* 151–154.

Carrier, J., & Dumont, M. (1995). Sleep propensity and sleep architecture after bright light exposure at three different times of day. *Journal of Sleep Research, 4,* 202–211.

Carskadon, M. A., & Dement, W. C. (1980). Distribution of REM sleep on a 90 minute sleep–wake schedule. *Sleep, 2,* 309–317.

Carskadon, M. A., & Dement, W. C. (1994). Normal human sleep: An overview. In M. H. Kryger, T. Roth, & W. C. Dement (Eds.), *Principles and practice of sleep medicine* (pp. 16–25). Philadelphia: Saunders.

Chhabra, S. K. (1995). An epidemiological investigation into nocturnal asthma. *Journal of Asthma, 32,* 147–150.

Connolly, C. K. (1979). Diurnal rhythms in airway obstruction. *British Journal of Diseases of the Chest, 73,* 357–366.

Copinschi, G., Van Onderbergen, A., L'Hermite-Baleriaux, M., Szyper, M., Caufriez, A., L'Hermite, M., Robyn, C., Turek, F. W., & Van Cauter, E. (1990). Effects of the short-actiing benzodiazepine triazolam, taken at bedtime, on circadian and sleep-related hormonal profiles in normal men. *Sleep, 13,* 232–244.

Copinschi, G., Akseki, E., Morenoreyes, R., Leproult, R., & L'Hermite-Baleriaux, M. (1995). Effects of bedtime administration of zolpidem on circadian and sleep. *Sleep, 18,* 417–424.

Culpepper, W. J., Badia, P., & Shaffer, J. I. (1992). Time-of-night patterns in PLMS activity. *Sleep, 15,* 306–311.

Czeisler, C. A., Weitzman, E. D., Moore-Ede, M. C., Zimmerman, J. C., & Knauer, R. S. (1980a). Human sleep: Its duration and organization depend on its circadian phase. *Science, 210,* 1264–1267.

Czeisler, C. A., Zimmerman, J. C., Ronda, J. M., Moore-Ede, M. C., & Weitzman, E. D. (1980b). Timing of REM sleep is coupled to the circadian rhythm of body temperature in man. *Sleep, 2,* 329–346.

Czeisler, C. A., Richardson, G. S., Coleman, R. M., Zimmerman, J. C., Moore-Ede, M. C., Dement, W. C., & Weitzman, E. D. (1981). Chronotherapy: Resetting the circadian clocks of patients with delayed sleep phase insomnia. *Sleep, 4,* 1–21.

Czeisler, C. A., Allan, J. S., Strogatz, S. H., Ronda, J. M., Sanchez, R., Rios, C. D., Freitag, W. O., Richardson, G. S., & Kronauer, R. E. (1986). Bright light resets the human circadian pacemaker independent of the timing of the sleep–wake cycle. *Science, 233,* 667–670.

Czeisler, C. A., Kronauer, R. E., Johnson, M. P., Allan, J. S., Johnson, T. S., & Dumont, M. (1988). Action of light on the human circadian pacemaker: Treatment of patients with circadian disorders. In J. Horne (Ed.), *Sleep, 88* (pp. 42–47). Stuttgart: Gustav Fischer Verlag.

Czeisler, C. A., Kronauer, R. E., Allan, J. S., Duffy, J. F., Jewett, M. E., Brown, E. N., & Ronda, J. M. (1989). Bright light induction of strong (type 0) resetting of the human circadian pacemaker. *Science, 244,* 1328–1333.

Czeisler, C. A., Johnson, M. P., Duffy, J. F., Brown, E. N., Ronda, J. M., & Kronauer, R. E. (1990). Exposure to bright light and darkness to treat physiologic maladaptation to night work. *New England Journal of Medicine, 322,* 1253–1259.

Czeisler, C. A., Dumont, M., Duffy, J. F., Steinberg, J. D., Richardson, G. S., Brown, E. N., Sanchez, R., Rios, C. D., & Ronda, J. M. (1992). Association of sleep–wake habits in older people with changes in output of circadian pacemaker. *Lancet, 340,* 933–936.

Czeisler, C. A., Shanahan, T. L., Klerman, E. B., Martens, H., Brotman, D. J., Emens, J. S., Klein, T., & Rizzo, J. F. (1995). Suppression of melatonin secretion in some blind patients by exposure to bright light. *New England Journal of Medicine, 332,* 6–11.

Dahlitz, M., Alvarez, B., Vignau, J., English, J., Arendt, J., & Parkes, J. D. (1991). Delayed sleep phase syndrome response to melatonin. *Lancet, 337,* 1121–1124.

Dantz, B., Edgar, D. M., & Dement, W. C. (1994). Circadian rhythms in narcolepsy: Studies on a 90 minute day. *Electroencephalography and Clinical Neurophysiology, 90,* 24–35.

Dawson, D., & Campbell, S. S. (1991). Timed exposure to bright light improves sleep and alertness during simulated night shifts. *Sleep, 14,* 511–516.

Dawson, D., & Encel, N. (1993). Melatonin and sleep in humans. *Journal of Pineal Research, 15,* 1–12.

Dawson, D., Encel, N., & Lushington, K. (1995). Improving adaptation to simulated night shift: Timed exposure to bright light versus daytime melatonin administration. *Sleep, 18,* 11–21.

Dekker, D. K., Paley, M. J., Popkin, S. M., & Tepas, D. I. (1993). Locomotive engineers and their spouses: Coffee consumption, mood, and sleep reports. *Ergonomics, 36,* 233–238.

Dement, W. C., Carskadon, M., & Ley, R. (1973). The prevalence of narcolepsy. *Sleep Research, 2,* 147.

Diagnostic Classification Steering Committee, Thorpy, M. J. (Chairman). (1990). *International Classification of Sleep Disorders: Diagnostic and coding manual,* Rochester, MN: American Sleep Disorders Association.

Dijk, D.-J., & Czeisler, C. A. (1994). Paradoxical timing of the circadian rhythm of sleep propensity serves to consolidate sleep and wakefulness in humans. *Neuroscience Letters, 166,* 63–68.

Dijk, D.-J., & Czeisler, C. A. (1995). Contribution of the circadian pacemaker and the sleep homeostat to sleep propensity, sleep structure, electroencephalographic slow waves, and sleep spindle activity in humans. *Journal of Neuroscience, 15,* 3526–3538.

Dijk, D.-J., Visscher, C. A., Bloem, G. M., Beersma, D. G. M., & Daan, S. (1987). Reduction of human sleep duration after bright light exposure in the morning. *Neuroscience Letters, 73,* 181–186.

Dijk, D.-J., Beersma, D. G. M., Daan, S., & Lewy, A. J. (1989). Bright morning light advances in human circadian system without affecting NREM sleep homeostasis. *American Journal of Physiology, 256,* R106–R111.

Dijk, D.-J., Cajochen, C., & Borbely, A. A. (1991). Effect of a single 3-hour exposure to bright light on core body temperature and sleep in humans. *Neuroscience Letters, 121,* 59–62.

Dijk, D.-J., Boulos, Z., Eastman, C. I., Lewy, A. J., Campbell, S. S., & Terman, M. (1995). Light treatment for sleep disorders: Consensus report II. Basic properties of circadian physiology and sleep regulation. *Journal of Biological Rhythms, 10,* 113–125.

Dingemanse, J. (1995). Pharmacotherapy of insomnia: Practice and prospects. *Pharmacy World and Science, 17,* 67–75.

Dollins, A. B., Zhdanova, I. V., Wurtman, R. J., Lynch, H. J., & Deng, M. H. (1994). Effect of inducing nocturnal serum melatonin concentrations in daytime on sleep, mood, body temperature and performance. *Proceedings of the National Academy of Sciences of the USA, 91,* 1824–1828.

Dorsey, C. M., Lukas, S. E., & Cunningham, S. L. (1996). Fluoxetine-inducted sleep disturbance in depressed patients. *Neuropsychopharmacology, 14,* 437–442.

Douglas, N. J. (1994). Asthma. In M. H. Kryger, T. Roth, & W. C. Dement (Eds.), *Principles and practice of sleep medicine* (pp. 748–757). Philadelphia: Saunders.

Drennan, M., Kripke, D. F., & Gillin, J. C. (1989). Bright light can delay human temperature rhythm independent of sleep. *American Journal of Physiology, 257*(1 Part 2), R136–R141.

Eastman, C. I. (1992). High-intensity light for circadian adaptation to a 12-h shift of the sleep schedule. *American Journal of Physiology, 263,* R428–R436.

Eastman, C. I., Stewart, K. T., Mahoney, M. P., Liu, L., & Fogg, L. F. (1994). Dark goggles and bright light improve circadian rhythm adaptation to night-shift work. *Sleep, 17,* 535–543.

Eastman, C. I., Boulos, Z., Terman, M., Campbell, S. S., Dijk, D.-J., & Lewy, A. J. (1995). Light treatment for sleep disorders: Consensus report. VI. Shift work. *Journal of Biological Rhythms, 10,* 157–164.

Entzian, P., Linnemann, K., Schlaak, M., & Zabel, P. (1996). Obstructive sleep apnea syndrome and circadian rhythms of hormones and cytokines. *American Journal of Respiratory and Critical Care Medicine, 153,* 1080–1086.

Espie, C. A. (1993). ABC of sleep disorders. Practical management of insomnia: Behavioural and cognitive techniques. *British Medical Journal, 306,* 509–511.

Folkard, S., Arendt, J., Aldhous, M., & Kennett, H. (1990). Melatonin stabilises sleep onset time in a blind man without entrainment of cortisol or temperature rhythms. *Neuroscience Letters, 113,* 193–198.

Folkard, S., Arendt, J., & Clark, M. (1993). Can melatonin improve shift workers' tolerance of the night shift? Some preliminary findings. *Chronobiology International, 10,* 315–320.

Ford, D. E., & Kamerow, D. B. (1989). Epidemiologic study of sleep disturbances and psychiatric disorders. *Journal of the American Medical Association, 262,* 1479–1484.

French, J., Hannon, P., & Brainard, G. C. (1990). Effects of bright light illuminance on body temperature and human performance. *Annual Review of Chronopharmacology, 7,* 37–40.

Gaillard, J. M. (1994). Benzodiazepines and GABA-ergic transmission. In M. H. Kryger, T. Roth, & W. C. Dement (Eds.), *Principles and practice of sleep medicine* (pp. 349–354). Philadelphia: Saunders.

Ganguli, M., Reynolds, C. F., & Gilby, J. E. (1996). Prevalence and persistence of sleep complaints in a rural elderly community sample: The MoVIES project. *Journal of the American Geriatrics Society, 44,* 778–784.

George, C. F. P. (1994). Cardiovascular disease and sleep. In M. H. Kryger, T. Roth, & W. C. Dement (Eds.), *Principles and practice of sleep medicine* (pp. 835–846). Philadelphia: Saunders.

Graeber, R. C. (1989). Jet lag and sleep disruption. In M. H. Kryger, T. Roth, & W. C. Dement (Eds.), *Principles and practice of sleep medicine* (pp. 324–331). Philadelphia: Saunders.

Guilleminault, C. (1994). Narcolepsy syndrome. In M. H. Kryger, T. Roth, & W. C. Dement (Eds.), *Principles and practice of sleep medicine* (pp. 549–561). Philadelphia: Saunders.

Guilleminault, C., Clerk, A., Black, J., Labanowski, M., Pelayo, R., & Claman, D. (1995). Nondrug treatment trials in psychophysiologic insomnia. *Archives of Internal Medicine, 155,* 838–844.

Gwirtsman, H. E., Halaris, A. E., Wold, A. W., DeMet, E., Piletz, J. E., & Marler, M. (1989). Apparent phase advance in diurnal MHPG rhythm in depression. *American Journal of Psychiatry, 146,* 1427–1433.

Hajak, G., Rodenbeck, A., Staedt, J., Bandelow, B., Huether, G., & Ruther, E. (1995). Nocturnal plasma melatonin levels in patients suffering from chronic primary insomnia. *Journal of Pineal Research, 19,* 116–122.

Hartmann, P. M. (1995). Drug treatment of insomnia: Indications and newer agents. *American Family Physician, 51,* 191–194.

Hauri, P. (1977). *The sleep disorders. Current concepts.* Kalamazoo, MI: Scope Publications, Upjohn.

Hauri, P. (1981). Treating psychophysiologic insomnia with biofeedback. *Archives of General Psychiatry, 38,* 752–758.

Henderson, S., Jorm, A. F., Scott, L. R., Mackinnon, A. J., Christensen, H., & Korten, A. E. (1995).

Insomnia in the elderly: Its prevalence and correlates in the general population. *Medical Journal of Australia, 162,* 22–24.

Hening, W. A., Walters, A., Kavey, N., Gidro-Frank, S., Cote, L., & Fahn, S. (1986). Dyskinesias while awake and periodic movements in sleep in restless legs syndrome: Treatment with opioids. *Neurology, 36,* 1363–1366.

Hetzel, M. R., & Clark, T. J. H. (1979). Comparison of normal and asthmatic circadian rhythms in peak expiratory flow rate. *Thorax, 35,* 732–738.

Hoffstein, V., & Mateika, J. (1992). Evening-to-morning blood pressure variations in snoring patients with and without obstructive sleep apnea. *Chest, 101,* 379–384.

Holbrook, M. I., White, M. H., & Hutt, M. J. (1994). Increasing awareness of sleep hygiene in rotating shift workers. *Perceptual and Motor Skills, 79,* 520–522.

Hollwich, F., & Dieckhues, B. (1974). Changes in the circadian rhythm of blind people. In L. E. Scheuing, F. Halberg, & J. E. Pauly (Eds.), *Chronobiology,* (pp. 285–292). Tokyo: Igaku Shoin.

Honda, Y., Asaka, A., Tanimura, M., & Furusho, T. (1983). A genetic study of narcolepsy and excessive daytime sleepiness in 308 families with a narcolepsy or hypersomnia proband. In C. Guilleminault & E. Lugaresi (Eds.), *Sleep/wake disorders: Natural history, epidemiology and long-term evolution* (pp. 187–199). New York: Raven Press.

Honma, K. I., Honma, S., Nakamura, K., Sasaki, M., Endo, T., & Takahashi, T. (1995). Differential effects of bright light and social cues on reentrainment of human circadian rhythms. *American Journal of Physiology, 268,* R528–R535.

Hyyppä, M. T., & Kronholm, E. (1989). Quality of sleep and chronic illnesses. *Journal of Clinical Epidemiology, 42,* 633–638.

Iannaccone, S., Zucconi, M., Marchettini, P., Ferini-Strambi, L., Nemni, R., Quattrini, A., Palazzi, S., Lacerenza, M., Formaglio, F., & Smirne, S. (1995). Evidence of peripheral axonal neuropathy in primary restless legs syndrome. *Movement Disorders, 10,* 2–9.

Jacquinet-Salord, M. C., Lang, T., Fourlaud, C., Nicoulet, I., & Bingham, A. (1993). Sleeping tablet consumption, self-reported quality of sleep, and working conditions. *Journal of Epidemilogy and Community Health, 47,* 64–68.

Jarjour, N. N., & Busse, W. W. (1995). Cytokines in bronchoalveolar lavage fluid of patients with nocturnal asthma. *American Journal of Respiratory and Critical Care Medicine, 152,* 1474–1477.

Johnson, L. C., & Spinweber, C. L. (1983). Good and poor sleepers differ in Navy performance. *Military Medicine, 148,* 727–731.

Juji, T., Matsuki, K., Naohara, T., Satake, M., & Honda, Y. (1985). Extraordinary association between HLA-DR2 and narcolepsy. *Lancet, 1,* 227.

Kanno, O., Watanabe, H., & Kazamatsuri, H. (1993). Effects of zopiclone, flunitrazepam, triazolam, and levomepromazine on the transient change in sleep–wake schedule: Polygraphic study, and the evaluation of sleep and daytime condition. *Progress in Neuro-Psychopharmacology and Biological Psychiatry, 17,* 229–239.

Karacan, I., Finley, W. W., Williams, R. L., & Hursch, C. J. (1970). Changes in stage 1-REM and stage 4 sleep during naps. *Biological Psychiatry, 2,* 261–265.

Keefauver, S. P., & Guilleminault, C. (1994). Sleep terrors and sleepwalking. In M. H. Kryger, T. Roth, & W. C. Dement (Eds.), *Principles and practice of sleep medicine* (pp. 567–573). Philadelphia: Saunders.

Klein, K. E., Wegmann, H. M., & Hunt, B. I. (1972). Desynchronization of body temperature and performance circadian rhythms as a result of out-going and home-going transmeridian flights. *Aerospace Medicine, 43,* 119–132.

Klein, T., Martens, H., Dijk, D.-J., Kronauer, R. E., Seely, E. W., & Czeisler, C. A. (1993). Circadian sleep regulation in the absence of light perception: Chronic non-24-hour circadian rhythm sleep disorder in a blind man with a regular 24-hour circadian schedule. *Sleep, 16,* 333–343.

Knowles, J. B., MacLean, A. W., & Cairns, J. (1982). REM sleep abnormalities in depression: A test of phase-advanced hypothesis. *Biological Psychiatry, 17,* 605–609.

Knutsson, A., Akerstedt, T., Orth-Gomer, K., & Jonsson, B. G. (1986). Increased risk of ischaemic heart disease in shift workers. *Lancet, 2,* 89–92.

Kuppermann, M., Lubeck, D. P., Mazonson, P. D., Patrick, D. L., Stewart, A. L., Buesching, D. P., & Fifer, S. K. (1995). Sleep problems and their correlates in a working population. *Journal of General Internal Medicine, 10,* 25–32.

Lack, L., & Wright, H. (1993). The effect of evening bright light in delaying the circadian rhythms and lengthening the sleep of early morning awakening insomniacs. *Sleep, 16,* 436–443.

Lacks, P., & Rotert, M. (1986). Knowledge and practice of sleep hygiene techniques in insomniacs and good sleepers. *Behaviour Research and Therapy, 24,* 365–368.

Lapierre, O., & Dumont, M. (1995). Melatonin treatment of a non-24-hour sleep–wake cycle in a blind retarded child. *Biological Psychiatry, 38,* 119–122.

Lapínski, M., Przybylowski, T., Lewandowski, J., Januszewicz, A., Wocial, B., Kowalski, J., Droszcz, W., & Januszewicz, W. (1993). Diurnal blood pressure rhythm and urinary catecholamine excretion in obstructive sleep apnoea and essential hypertension. *Journal of Hypertension, 11,* S292–S293.

Lavie, P. (1986). Ultrashort sleep–waking schedule. III. Gates and "forbidden zones" for sleep. *Electroencephalography and Clinical Neurophysiology, 63,* 414–425.

Lavie, P., & Peled, R. (1987). Narcolepsy is a rare disease in Israel. *Sleep, 10,* 608–609.

Lavie, P., Yoffe, N., Berger, I., & Peled, R. (1993). The relationship between the severity of sleep apnea syndrome and 24-h blood pressure values in patients with obstructive sleep apnea. *Chest, 103,* 717–721.

Lewy, A. J., Sack, R. L., & Singer, C. M. (1985). Treating phase typed chronobiologic sleep and mood disorders using appropriately timed bright artificial light. *Psychopharmacology Bulletin, 21,* 368–372.

Lewy, A. J., Sack, R., Miller, L., & Hoban, T. (1987). Antidepressant and circadian phase-shifting effects of light. *Science, 235,* 352–354.

Lewy, A. J., Ahmed, S., Jackson, J. M., & Sack, R. L. (1992). Melatonin shifts human circadian rhythms according to a phase-response curve. *Chronobiology International, 9,* 380–392.

Lobban, M. C., & Tredre, B. E. (1967). Diurnal rhythms of renal excretion and of body temperature in aged subjects. *Journal of Physiology, 188,* 48P–49P.

MacFarlane, J., Cleghorn, J. M., Brown, G. M., Kaplan, R., Brown, P., & Mittorn, J. (1984). Circadian rhythms in chronic insomnia. *Sleep Research, 13,* 223.

MacLean, A. W., Knowles, J. B., & Vetere, C. (1986). REM sleep and depression: Further use of computer simulation to test the phase-advance hypothesis. *Psychiatry Research, 19,* 25–36.

Maczaj, M. (1993). Pharmacological treatment of insomnia. *Drugs, 45,* 44–55.

Mahowald, M. W., & Schenck, C. H. (1994). REM sleep behavior disorder. In M. H. Kryger, T. Roth, & W. C. Dement (Eds.), *Principles and practice of sleep medicine* (pp. 574–588). Philadelphia: Saunders.

McIntyre, I. M., Norman, T. R., Burrows, G. D., & Armstrong, S. M. (1993). Alterations to plasma melatonin and cortisol after evening alprazolam administration in humans. *Chronobiology International, 10,* 205–213.

Mendelson, W. B., Garnett, D., & Linnoila, M. (1984). Do insomniacs have impaired daytime functioning? *Biological Psychiatry, 19,* 1261–1264.

Middelkoop, H. A. M., Lammers, G. J., Van Hilten, B. J., Ruwhof, C., Pijl, H., & Kamphuisen, H. A. C. (1995). Circadian distribution of motor activity and immobility in narcolepsy: Assessment with continuous motor activity monitoring. *Psychophysiology, 32,* 286–291.

Miller, L. S., Lombardo, T. W., & Fowler, S. C. (1995). Caffeine and time of day effects on a forced discrimination task in humans. *Physiology and Behavior, 57,* 1117–1125.

Minors, D. S., Waterhouse, J. M., & Wirz-Justice, A. (1991). A human phase-response curve to light. *Neuroscience Letters, 133,* 36–40.

Mitler, M. M., Gujavarty, K. S., & Browman, C. P. (1982). Maintenance of wakefulness test: A polysomnographic technique for evaluating efficacy in patients with excessive somnolence. *Electroencephalography and Clinical Neurophysiology, 53,* 658–661.

Mitler, M. M., Aldrich, M. S., Koob, G. F., & Zarcone, V. P. (1994). Narcolepsy and its treatment with stimulants. ASDA standards of practice. *Sleep, 17,* 352–371.

Monk, T. H. (1987). Coping with the stress of jet-lag. *Work and Stress, 1,* 162–166.

Monk, T. H. (1994). Disorders relating to shift work and jet-lag. In J. M. Oldham (Ed.), *Annual Review of Psychiatry* (pp. 729–756). Washington, DC: American Psychiatric Press.

Monk, T. H., Moline, M. L., & Graeber, R. C. (1988). Inducing jet lag in the laboratory: Patterns of adjustment to an acute shift in routine. *Aviation, Space and Environmental Medicine, 59,* 703–710.

Monk, T. H., Flaherty, J. F., Frank, E., Hoskinson, K., & Kupfer, D. J. (1990). The social rhythm metric (SRM): An instrument to quantify the daily rhythms of life. *Journal of Nervous and Mental Disease, 178,* 120–126.

Monk, T. H., Reynolds, C. F., Machen, M. A., & Kupfer, D. J. (1992). Daily social rhythms in the elderly and their relation to objectively recorded sleep. *Sleep, 15,* 322–329.

Monk, T. H., Reynolds, C. F., Kupfer, D. J., Coble, P. A., Hayes, A. J., Machen, M. A., Petrie, S. R., & Ritenour, A. M. (1994a). The Pittsburgh Sleep Diary (PghSD). *Journal of Sleep Research, 3,* 111–120.

Monk, T. H., Petrie, S. R., Hayes, A. J., & Kupfer, D. J. (1994b). Regularity of daily life in relation to personality, age, gender, sleep quality and circadian rhythms. *Journal of Sleep Research, 3,* 196–205.

Monk, T. H., Essex, M. J., Snider, N. A., Klein, M. H., Lowe, K. K., & Kupfer, D. J. (1996). The impact of the birth of a baby on the time structure and social mixture of a couple's daily life and its consequences for well-being. *Journal of Applied Social Psychology, 26,* 1237–1258.

Montplaisir, J., Godbout, R., Poirier, G., & Bedard, M. A. (1986). Restless legs syndrome and periodic movements in sleep: Physiopathology and treatment with L-Dopa. *Clinical Neuropharmacology, 9,* 456–463.

Montplaisir, J., Lorrain, D., & Godbout, R. (1991). Restless legs syndrome and periodic leg movements in sleep: The primary role of dopaminergic mechanism. *European Neurology, 31,* 41–43.

Montplaisir, J., Lapierre, O., Warnes, H., & Pelletier, G. (1992). The treatment of the restless leg syndrome with or without periodic leg movements in sleep. *Sleep, 15,* 391–395.

Moore, R. Y., & Speh, J. C. (1993). GABA is the principal neurotransmitter of the circadian system. *Neuroscience Letters, 150,* 112–116.

Moore-Ede, M. C. (1986). Jet lag, shift work and maladaption. *News in Physiological Sciences, 1,* 156–160.

Morin, C. M., Culbert, J. P., & Schwartz, S. M. (1994). Nonpharmacological interventions for insomnia: A meta-analysis of treatment efficacy. *American Journal of Psychiatry, 151,* 1172–1180.

Morin, C. M., Colecchi, C., Stone, J., Sood, R. K., & Brink, D. (1999). Behavioral and pharmacological therapies for late-life insomnia: A randomized controlled trial. *Journal of the American Medical Association, 281,* 991–999.

Morris, M., Lack, L., & Dawson, D. (1990). Sleep-onset insomniacs have delayed temperature rhythms. *Sleep, 13,* 1–14.

Mosko, S. S., Holowach, J. B., & Sassin, J. F. (1983). The 24-hour rhythm of core temperature in narcolepsy. *Sleep, 6,* 137–146.

Muehlbach, M. J., & Walsh, J. K. (1995). The effects of caffeine on simulated night-shift work and subsequent daytime sleep. *Sleep, 18,* 22–29.

Muller, J. E., Stone, P. H., & Turi, Z. G. (1985). Circadian variation in the frequency of onset of acute myocardial infarction. *New England Journal of Medicine, 313,* 1315–1322.

Murao, S., Marumi, K., & Katayama, S. (1972). All night polysomnographic studies of natural angina. *Japanese Heart Journal, 13,* 295–305.

Myers, B. L., & Badia, P. (1993). Immediate effects of different light intensities on body temperature an alertness. *Physiology and Behavior, 54,* 199–202.

Nakagawa, H., Sack, R. L., & Lewy, A. L. (1992). Sleep propensity free-runs with the temperature, melatonin and cortisol rhythms in a totally blind person. *Sleep, 15,* 330–336.

National Commission on Sleep Disorders Research. (1993). *Wake up America: A national sleep alert.* Washington, DC: U.S. Government Printing Office.

Nicholson, A. M., Roth, T., & Stone, B. M. (1985). Hypnotics and aircrew. *Aviation, Space and Environmental Medicine, 56,* 299.

Noda, A., Okada, T., Hayashi, H., Yasuma, F., & Yokota, M. (1993). 24-Hour ambulatory blood pressure variability in obstructive sleep apnea syndrome. *Chest, 103,* 1343–1347.

Office of Technology Assessment, U.S. Congress (1991). *Biological rhythms: Implications for the worker.* Washington, DC: U.S. Government Printing Office.

O'Keeffe, S. T. (1996). Restless legs syndrome. A review. *Archives of Internal Medicine, 156,* 243–248.

Oldani, A., Ferini-Strambi, L., Zucconi, M., Stankov, B., Fraschini, F., & Smirne, S. (1994). Melatonin and delayed sleep phase syndrome: Ambulatory polygraphic evaluation. *Neuroreport, 6,* 132–134.

Oren, D. A., Turner, E. H., & Wehr, T. A. (1995). Abnormal circadian rhythms of plasma melatonin and body temperature rhythms in the delayed sleep phase syndrome. *Journal of Neurology, Neurosurgery and Psychiatry, 58,* 379.

Palm, L., Blennow, G., & Wetterberg, L. (1991). Correction of non-24-hour sleep/wake cycle by melatonin in a blind retarded boy. *Annals of Neurology, 29,* 336–339.

Partinen, M. (1994). Epidemiology of sleep disorders. In M. H. Kryger, T. Roth, & W. C. Dement (Eds.), *Principles and practice of sleep medicine* (pp. 437–452). Philadelphia: Saunders.

Patonen, T., Appelberg, B., & Patinen, M. (1993). Effects of light treatment on sleep structure in seasonal affective disorder. *European Archives of Psychiatry and Clinical Neuroscience, 242,* 310–313.

Petrie, K., Conaglen, J. V., Thompson, L., & Chamberlain, K. (1989). Effect of melatonin on jet lag after long haul flights. *British Medical Journal, 298,* 705–707.

Petrie, K., Dawson, A. G., Thompson, L., & Brook, R. (1993). A double-blind trial of melatonin as a treatment for jet lag in international cabin crew. *Biological Psychiatry, 33,* 526–530.

Pigeau, R., Naitoh, P., Buguet, A., & McCann, C. (1995). Modafinil, d-amphetamine and placebo during 64 hours of sustained mental work: Effects on mood, fatigue, cognitive performance and body-temperature. *Journal of Sleep Research, 4,* 212–228.

Pollak, C. P., Perlick, D., & Linsner, J. P. (1992). Daily sleep reports and circadian rest–activity cycles of elderly community residents with insomnia. *Biological Psychiatry, 32,* 1019–1027.

Pollak, C. P., Wagner, D. R., Moline, M. L., & Monk, T. H. (1992). Cognitive and motor performance of narcoleptic and normal subjects living in temporal isolation. *Sleep, 15,* 202–211.

Pollak, C. P., & Wagner, D. R. (1994). Core body temperature in narcoleptic and normal subjects living in temporal isolation. *Pharmacology, Biochemistry and Behavior, 47,* 65–71.

Prigerson, H. G., Reynolds, C. F., Frank, E., Kupfer, D. J., George, C. J., & Houck, P. R. (1994). Stressful life events, social rhythms, and depressive symptoms among the elderly: An examination of hypothesized causal linkages. *Psychiatry Research, 51,* 33–49.

Regestein, Q. R., & Monk, T. H. (1995). Clinical aspects of delayed sleep phase syndrome: A review. *American Journal of Psychiatry, 152,* 602–608.

Reynolds, C. F., Newton, T. F., Shaw, D. H., Coble, P. A., & Kupfer, D. J. (1982). Electroencephalographic sleep findings in depressed outpatients. *Psychiatry Research, 6,* 65–75.

Richardson, G. S., Miner, J. D., & Czeisler, C. A. (1990). Impaired driving performance in shiftworkers: The role of the circadian system in a multifactorial model. *Alcohol, Drugs and Driving, 5 & 6*(4 & 1), 265–273.

Ridker, P. M., Manson, J. E., & Buring, J. E. (1990). Circadian variation of acute myocardial infarction and the effect of low dose aspirin in a randomized trial of physicians. *Circulation, 82,* 897–902.

Roberts, C. J., & Hooper, R. G. (1985). Prediction of polysomnography results by abbreviated testing. *Chest, 88*(Supplement), 43.

Rodenstein, D. O., D'Odemont, J. P., Pieters, T., & Aubert-Tulkens, G. (1992). Diurnal and nocturnal diuresis and natriuresis in obstructive sleep apnea. *American Review of Respiratory Disease, 145,* 1367–1371.

Roehrs, T., & Roth, T. (1994). Chronic insomnias associated with circadian rhythm disorders. In M. H. Kryger, T. Roth, & W. C. Dement (Eds.), *Principles and practice of sleep medicine* (pp. 477–481). Philadelphia: Saunders.

Roehrs, T., Zorick, F., & Roth, T. (1994). Transient and short-term insomnia. In M. H. Kryger, T. Roth, & W. C. Dement (Eds.), *Principles and practice of sleep medicine* (pp. 486–493). Philadelphia: Saunders.

Rosekind, M. R. (1992). The epidemiology and occurrence of insomnia. *Journal of Clinical Psychiatry, 53*(Supplement), 4–6.

Rosen, L. N., Targum, S. D., Terman, M., Bryant, M. J., Hoffman, H., Kasper, S. F., Hamovit, J. R., Docherty, J. P., Weech, B., & Rosenthal, N. E. (1990). Prevalence of seasonal affective disorder at four latitudes. *Psychiatry Research, 31,* 131–144.

Rosenthal, N. E., Sack, D. A., Gillin, J. C., Lewy, A. J., Goodwin, F. K., Davenport, Y., Mueller, P. S., Newsome, D. A., & Wehr, T. A. (1984). Seasonal affective disorder: A description of the syndrome and preliminary findings with light therapy. *Archives of General Psychiatry, 41,* 72–80.

Rosenthal, N. E., Skwerer, R. G., Levendosky, B. A., Joseph-Vanderpool, J. R., Jacobsen, F. M., Duncan, C. C., Gaist, P. A., & Wehr, T. A. (1989). Sleep architecture in seasonal affective disorder: The effects of light therapy and changing seasons. *Sleep Research, 18,* 440.

Rosenthal, N. E., Joseph-Vanderpool, J. R., Levendosky, A. A., Johnston, S. H., Allen, R., Kelley, K. A., Souetre, E., Schultz, P. M., & Starz, K. E. (1990). Phase-shifting effects of bright morning light as treatment for delayed sleep phase syndrome. *Sleep, 13,* 354–361.

Roth, B. (1980). *Narcolepsy and hypersomnia.* Basel: Springer-Verlag.

Rutenfranz, J., Colquhoun, W. P., Knauth, P., & Ghata, J. N. (1977). Biomedical and psychosocial aspects of shift work: A review. *Scandinavian Journal of Work, Environment and Health, 3,* 165–182.

Sack, R. L., Lewy, A. J., Blood, M. L., Keith, L. D., & Nakagawa, H. (1992). Circadian rhythm abnormalities in totally blind people: Incidence and clinical significance. *Journal of Clinical Endocrinology and Metabolism, 75,* 127–134.

Samel, A., Wegmann, H. M., Vejvoda, M., Maab, H., Gundel, A., & Schutz, M. (1991). Influence of melatonin treatment on human circadian rhythmicity before and after a simulated 9-hr time shift. *Journal of Biological Rhythms, 6,* 235–248.

Sanders, M. H. (1994). Medical therapy for sleep apnea. In M. H. Kryger, T. Roth, & W. C. Dement (Eds.), *Principles and practice of sleep medicine* (pp. 678–693). Philadelphia: Saunders.

Sasaki, M., Endo, S., Nakagawa, S., Kitahara, T., & Mori, A. (1985). A chronobiological study on the relation between time zone changes and sleep. *Jikeikai Medical Journal, 32,* 83–100.

Schrader, H., Bovim, G., & Sand, T. (1993). The prevalence of delayed and advanced sleep phase syndrome. *Journal of Sleep Research, 2,* 51–55.

Schulz, H., Kiss, E., & Hasse, S. (1990). The 24-h hormonal secretion pattern of narcoleptic patients and control subjects. In J. Horne (Ed.), *Sleep, 90* (pp. 375–378). Bochum, Germany: Pontenagel Press.

Scott, A. J., & LaDou, J. (1990). Shiftwork: Effects on sleep and health with recommendations for medical surveillance and screening. *Occupational Medicine, 5,* 273–299.

Seidel, W. F., Roth, T., Roehrs, T., Zorick, F., & Dement, W. C. (1984). Treatment of a 12-hour shift of sleep schedule with benzodiazepines. *Science, 224,* 1262–1264.

Sériès, F., Cormier, Y., & La Forge, J. (1991). Validity of diurnal sleep recording in the diagnosis of sleep apnea syndrome. *American Review of Respiratory Disease, 143,* 947–949.

Shapiro, C. M., & Dement, W. C. (1993). Impact and epidemiology of sleep disorders. *British Medical Journal, 1893,* 19–22.

Shear, M. K., Randall, J., Monk, T. H., Ritenour, A., Tu, X., Frank, E., Reynolds, C. F., & Kupfer, D. J. (1994). Social rhythm in anxiety disorder patients. *Anxiety, 1,* 90–95.

Smith, A. P. Kendrick, A. M., & Maben, A. L. (1992). Effects of breakfast and caffeine on performance and mood in the late morning and after lunch. *Neuropsychobiology, 26,* 198–204.

Smith, A. P., Brockman, P., Flynn, R., Maben, A., & Thomas, M. (1993). Investigation of the effects of coffee on alertness and performance during the day and night. *Neuropsychobiology, 27,* 217–223.

Spielman, A. J., Saskin, P., & Thorpy, M. J. (1987). Treatment of chronic insomnia by restriction of time in bed. *Sleep, 10,* 45–56.

Stradling, J. R. (1995). Obstructive sleep apnea: Definitions, epidemiology, and natural history. *Thorax, 50,* 683–689.

Staedt, J., Stoppe, G., Kogler, A., Riemann, H., Hajak, G., Munz, D. L., Emrich, D., & Ruther, E. (1995). Nocturnal myoclonus syndrome (periodic movements in sleep) related to central dopamine D2-receptor alteration. *European Archives of Psychiatry and Clinical Neuroscience, 245,* 8–10.

Strollo, P. J., & Rogers, R. M. (1996). Obstructive sleep apnea. *New England Journal of Medicine, 334,* 99–104.

Sullivan, C. E., & Issa, F. G. (1985). Obstructive sleep apnea. *Clinics in Chest Medicine, 6,* 633–650.

Szuba, M. P., Yager, A., Guze, B. H., Allen, E. M., & Baxter, J. R. (1992). Disruption of social circadian rhythms in major depression: A preliminary report. *Psychiatry Research, 42,* 221–230.

Tafti, M., Rondouin, G., Besset, A., & Billiard, M. (1992). Sleep deprivation in narcoleptic subjects: Effect on sleep stages and EEG power density. *Electroencephalography and Clinical Neurophysiology, 83,* 339–349.

Takahashi, Y., Takahashi, K., Huguchi, T., Niimi, Y., Miyasita, A., & Ishii, Y. (1976). Pituitary hormone secretions and narcolepsy. In C. Guilleminault, W. C. Dement, & P. Passouant (Eds.), *Narcolepsy* (pp. 77–96). New York: Spectrum.

Terman, M., Terman, J. S., Quitkin, F. M., McGrath, P. J., Stewart, J. W., & Rafferty, B. (1989). Light therapy for seasonal affective disorder: A review of efficacy. *Neuropsychopharmacology, 2,* 1–22.

Terman, M., Lewy, A. J., Dijk, D.-J., Boulos, Z., Eastman, C. I., & Campbell, S. S. (1995). Light treatment for sleep disorders: Consensus report. IV. Sleep phase and duration disturbances. *Journal of Biological Rhythms, 10,* 135–147.

Thase, M. E. (1989). Comparison between seasonal affective disorder and other forms of recurrent depression. In N. E. Rosenthal & M. C. Blehar (Eds.), *Seasonal affective disorders and phototherapy* (pp. 64–78). New York: Guilford Press.

Thorpy, M. J. (1990). Disorders of arousal. In M. J. Thorpy (Ed.), *Handbook of sleep disorders* (pp. 531–549). New York: Marcel Dekker.

Thorpy, M. J., Korman, E., Spielman, A. J., & Glovinsky, P. B. (1988). Delayed sleep phase syndrome in adolescents. *Journal of Adolescent Health Care, 9,* 22–27.

Thurton, M. B., & Deegan, T. (1974). Circadian variations of plasma catecholamines, cortisol, and immunoreactive insulin concentrates in supine subjects. *Clinica Chimica Acta, 55,* 389.

Tomoda, A., Miike, T., Uezono, K., & Kawasaki, T. (1994). A school refusal case with biological rhythm disturbance and melatonin therapy. *Brain and Development, 16,* 71–76.

Tsujimoto, T., Yamada, N., Shimoda, K., Hanada, K., & Takahashi, S. (1990). Circadian rhythms in depression. Part II: Circadian rhythms in inpatients with various mental disorders. *Journal of Affective Disorders, 18,* 199–210.

Turek, F. W., & Losee-Olson, S. H. (1986). A benzodiazepine used in the treatment of insomnia phase-shifts the mammalian circadian clock. *Nature, 321,* 167–168.

Urponen, H., Vuori, I., Hasan, J., & Partinen, M. (1988). Self-evaluations of factors promoting and disturbing sleep: An epidemiological survey in Finland. *Social Science and Medicine, 26,* 443–450.

Walsh, J. K., Muehlbach, M. J., & Schweitzer, P. K. (1984). Acute administration of triazolam for the daytime sleep of rotating shift workers. *Sleep, 7,* 223–229.

Walsh, J. K., Muehlbach, M. J., Humm, T. M., Dickins, Q. S., Sugerman, J. L., & Schweitzer, P. K. (1990). Effect of caffeine on physiological sleep tendency and ability to sustain wakefulness at night. *Psychopharmacology, 101,* 271–273.

Walsh, J. K., Muehlbach, M. J., & Schweitzer, P. K. (1995). Hypnotics and caffeine as countermeasures for shiftwork-related sleepiness and sleep disturbance. *Journal of Sleep Research, 4,* 80–83.

Walters, A. S., & The International Restless Legs Syndrome Study Group. (1995). Toward a better definition of the restless legs syndrome. *Movement Disorders, 10,* 634–642.

Warley, A. R. H., & Stradling, J. R. (1988). Abnormal diurnal variation in salt and water excretion in patients with obstructive sleep apnea. *Clinical Science, 74,* 183–185.

Webb, W. B., & Agnew, H. W. (1977). Analysis of the sleep stages in sleep–wakefulness regimens of varied length. *Psychophysiology, 14,* 445–450.

Wehr, T. A., & Wirz-Justice, A. (1981). Internal coincidence model for sleep deprivation and depression. In *Sleep 1980, 5th European Congress on Sleep Research* (pp. 26–33). Basel: Karger.

Wehr, T. A., Moul, D. E., Barbato, G., Giesen, H. A., Seidel, J. A., & Barker, C. (1993). Conservation of photoperiod-responsive mechanisms in humans. *American Journal of Physiology, 265*(4 Part 2), R846–R857.

Weitzman, E. (1976). Twenty-four hour neuroendocrine secretory patterns: Observations on patients with narcolepsy. In C. Guilleminault, W. C. Dement, & P. Passouant (Eds.), *Narcolepsy* (pp. 521–542). New York: Spectrum.

Weitzman, E. D., Czeisler, C. A., Coleman, R. M., Spielman, A. J., Zimmerman, J. C., & Dement, W. (1981). Delayed sleep phase syndrome. *Archives of General Psychiatry, 38,* 737–746.

Wetter, D. W., & Young, T. B. (1994). The relation between cigarette smoking and sleep disturbance. *Preventive Medicine, 23,* 328–334.

Wever, R. A. (1985). Use of light to treat jet lag: Differential effects of normal and bright artificial light on human circadian rhythms. *Annals of the New York Academy of Sciences, 453,* 282–304.

Weyerer, S., & Dilling, H. (1991). Prevalence and treatment of insomnia in the community: Results from the Upper Bavarian Field Study. *Sleep, 14,* 392–398.

White, L., & Keith, B. (1990). The effect of shift work on the quality and stability of marital relations. *Journal of Marriage and Family, 52,* 453–462.

Wingard, D. K., & Berkman, L. F. (1983). Mortality risk associated with sleeping patterns among adults. *Sleep, 6,* 102–107.

Winget, C. M., DeRoshia, C. W., Markley, C. L., & Holley, D. C. (1984). A review of human physiological and performance changes associated with desynchronosis of biological rhythms. *Aviation, Space and Environmental Medicine, 55,* 1085–1096.

Wirz-Justice, A. (1995). Biological rhythms in mood disorders. In F. E. Bloom & D. J. Kupfer (Eds.), *Psychopharmacology: The third generation of progress* (pp. 999–1017). New York: Raven Press.

World Health Organization. (1992). *International statistical classification of diseases and related health problems-10th revision.* Geneva: World Health Organization.

Young, T., Palta, M., Dempsey, J., Skatrud, J., Weber, S., & Badr, S. (1993). The occurrence of sleep-disordered breathing among middle-aged adults. *New England Journal of Medicine, 328,* 1203–1205.

Zahorska-Markiewicz, B. (1980). The thermic effect of caffeinated and decaffeinated coffee ingested with breakfast. *Acta Physiologica Polonica, 31,* 17–20.

Zulley, J. (1980). Distribution of REM sleep in entrained 24-hour and free-running sleep–wake cycles. *Sleep,* 2, 377–389.

Zulley, J. (1992). Circadian aspects of sleep structure. *Journal of Interdisciplinary Cycle Research, 23,* 167–172.

Zulley, J., Wever, R., & Aschoff, J. (1981). The dependence of onset and duration of sleep on the circadian rhythm of rectal temperature. *Pflugers Archiv European Journal of Physiology, 391,* 314–318.

25

Circadian Modulation of Endocrine Secretion

EVE VAN CAUTER AND ORFEU M. BUXTON

GENERAL CHARACTERISTICS OF 24-HOUR HORMONAL RHYTHMS

A high degree of temporal organization exists in the endocrine system and is reflected by distinct 24-hour temporal patterns of secretory activity for each endocrine axis. One hormone may be actively secreted in a complex, pulsatile pattern while another may be in a quiescent period. Much of the temporal variability and organization of endocrine activity results from modulatory effects of circadian rhythmicity, a near-24-hour endogenous oscillation originating in the suprachiasmatic nucleus of the hypothalamus and from the effects of sleep/wake homeostasis controlled by diverse and diffuse neural centers, as depicted in Figure 1. These rhythmic patterns of hormonal secretion are thought to provide internal temporal organization through the coordination of physiologic processes in the time domain, and to also provide an adaptive synchronization of the organism relative to predictable variations of the external environment.

The peripheral concentrations of the majority of hormones vary systematically across the 24-hour span, and well-documented studies have demonstrated that these daily variations do not occur solely in response to the light–dark cycle or sleep–wake transitions. Figure 2 illustrates, as representative examples, the 24-hour profiles of cortisol, thyrotropin (TSH), growth hormone (GH), and prolactin (PRL). Each of these hormones exhibits periodicity of secretion over the 24-hour cycle and detailed studies have been needed to tease apart the relative contributions to these patterns of the circadian clock, the sleep/wake cycle, and exogenous stimuli. Understanding of the clinical implications of this temporal organization has become increasingly

---

EVE VAN CAUTER AND ORFEU M. BUXTON Department of Medicine, University of Chicago, Chicago, Illinois 60637.

Circadian Clocks, Volume 12 of *Handbook of Behavioral Neurobiology*, edited by Joseph S. Takahashi, Fred W. Turek, and R. Y. Moore, Kluwer Academic / Plenum Publishers, New York, 2001.

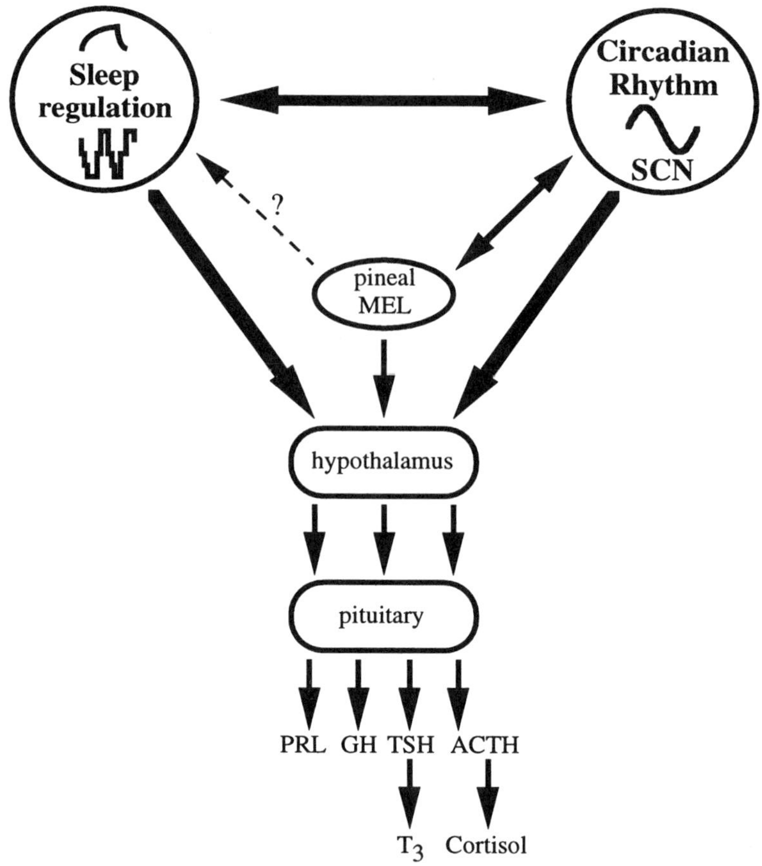

Figure 1. Schematic diagram of the interrelationships between circadian rhythmicity, sleep regulatory processes, and the endocrine secretions of the pineal gland and the pituitary gland. PRL, Prolactin; GH, growth hormone; TSH, thyrotropin; ACTH, adrenocorticotropin.

important as the advent of the industrial and postindustrial eras has brought new artificial light–dark environments and abnormal sleep patterns to increasing numbers of people.

The next section of this chapter describes methodologies that have been used to differentiate the modulatory effects of the circadian system per se from the effects of sleeping, waking, and transitions between these states and the methods most commonly used to quantify hormonal profiles. Later sections review the characteristics of the 24-hour patterns of hormonal secretion starting with the pineal and following with the hypothalamic/pituitary corticotropic, thyrotropic, somatotropic, and lactotropic axes. The order of presentation, from most affected by circadian rhythmicity to most affected by sleep, reflects our current understanding of the degree of modulation by the circadian system relative to the influence of sleep/wake regulatory systems. The scope of this chapter does not permit the review of the modulatory effects of circadian rhythmicity on other endocrine systems. Circadian rhythms in the gonadotropic axis, osmoregulation, and glucose regulation are

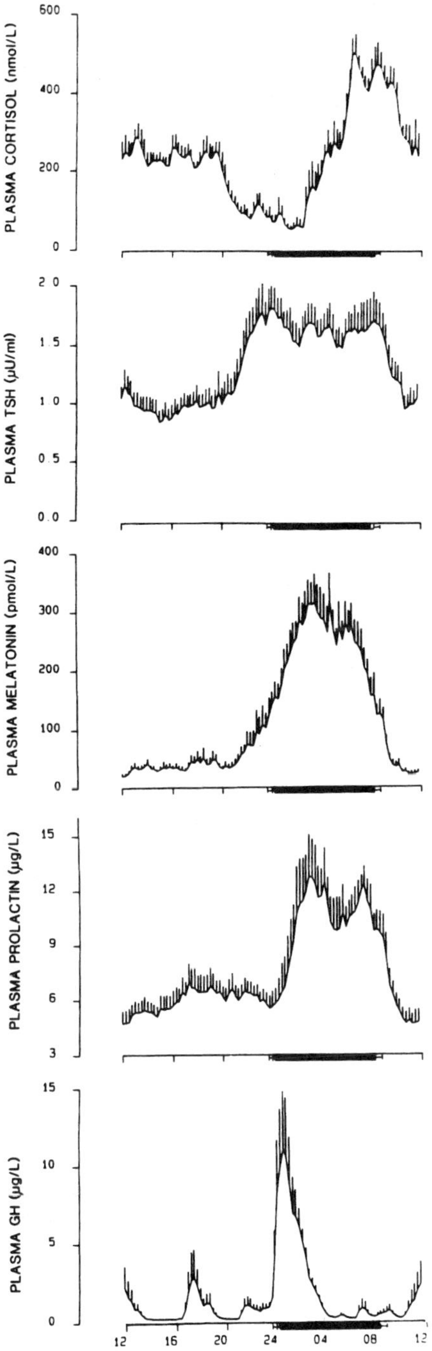

Figure 2. Mean 24-hour profiles (+SEM) of cortisol, thyrotropin (TSH), melatonin, prolactin, and growth hormone (GH) in eight normal young males sampled at 15-minute intervals. Black bars represent the mean sleep period. (Data previously published; van Coevorden *et al.*, 1991.)

reviewed elsewhere (Brandenberger, Charloux, Gronfier, & Otzenberger, 1998; Turek & Van Cauter, 1993; Van Cauter, Polonsky & Scheen, 1997b).

Methodologies for Studying 24-Hour Hormonal Patterns

Experimental Protocols

The demonstration of 24-hour rhythmicity in endocrine parameters is usually based on the observation of consistently reproducible characteristics in a set of 24-hour profiles of peripheral hormonal levels from a group of individuals studied over a period of time which rarely exceeds 72 hours due to limitations of the maximum safely extractable blood volume. Subject groups typically are as homogeneous as possible in terms in age, gender, sleep–wake cycles, and meal schedules. Prior to the beginning of the experiment, the volunteers need to adhere to a standardized schedule of meals and bedtimes for several days in order to maximize interindividual synchronization. Night workers, shift workers, subjects with irregular sleep–wake cycles, or subjects having made a transmeridian flight less than 2 months before the experiment must be excluded.

Blood sampling is usually done at regular intervals through an intravenous catheter in the forearm. The catheter should be inserted at least 2 hours before the collection of the first sample in order to avoid possible artifactual effects related to the venipuncture stress. If the quality of sleep is expected to play an important role in determining the pattern of the hormonal profile, extending the tubing through a hole in a wall to an adjoining room during sleep periods minimizes disruptions of sleep due to the sampling procedure, and two or three prestudy habituation nights allow the subjects to adapt to experimental procedures and sleep alterations to subside. Despite these precautions, the duration of sleep and the total duration of rapid eye movement (REM) stages may be significantly reduced by the presence of the catheter (Kerkhofs, Linkowski, & Mendlewicz, 1989).

To quantify circadian components in hormonal profiles, it is necessary to sample at intervals not exceeding 1 hour since less frequent sampling intervals may obscure the pulsatile pattern of hormonal release and bias the quantification of the circadian rhythm. Sampling frequencies of 15–20 minutes for at least 24 hours allow for simultaneous quantification of both circadian variation and ultradian pulsatile variations recurring at 1- to 4-hour intervals. The use of frequent sampling is essential for hormones that are highly pulsatile, such as cortisol (Van Cauter, 1987), but is less crucial for those that are secreted in a tonic fashion, such as melatonin.

Factors other than the endogenous circadian signal, such as sleep–wake transitions, meals, physical or mental activity, and postural changes may all be reflected in increases or decreases of hormonal secretion rates. To eliminate these "masking" effects, experimental protocols referred to as "constant routines" are used to reliably derive estimates of the circadian amplitude and phase of peripheral hormones and of other physiologic variables, such as body temperature (Czeisler, Chiasera & Duffy, 1991). Constant-routine conditions involve the maintenance of continuous wakefulness, a constant recumbent posture, constant levels of illumination, and caloric intake via an intravenous glucose infusion or hourly ingestion of identical caloric loads, which is generally maintained for 30–40 hours. Constant-routine conditions thus require sleep deprivation, a condition that may itself alter the levels of some hormones, such as TSH or cortisol, as described below. The 24-hour profile of melatonin secretion under dim light conditions is not affected by sleep or meals and has been shown to provide estimations of circadian phase and amplitude highly correlated with those derived from the temperature rhythm or from the cortisol rhythm observed under constant-routine conditions. To dissociate the effects of circadian rhythmicity from the effects of the sleep period, it is also possible to take advantage of the fact that the circadian system and the overt rhythms driven by it do not adapt rapidly to abrupt shifts in the light–dark and sleep–wake cycles. This relative inertia of the circadian system allows for the delineation of the effects of sleep per se, reflected in hormonal events that remain associated with the shifted sleep period, whereas hormonal events controlled by the circadian system will remain close to the synchronization observed prior to the shift. This is illustrated in Figure 3, which shows the effects of an abrupt 12-hour shift of sleep– wake and dark–light cycles on the profiles of cortisol, TSH, prolactin, and GH.

Time- and Frequency-Domain Analyses

Numerous procedures to quantify circadian variations have been proposed. The Cosinor and related tests (Halberg, Tong & Johson, 1967) are based on the assumption that the hormonal profile must be adequately described by a single sinusoid. This assumption is rarely met for biological rhythms, especially for human rhythms, and therefore the Cosinor test generally provides unreliable estimations of human circadian rhythmicity. Other procedures for the detection and quantification of circadian variations have been based on periodogram calculations or on nonlinear regression procedures (Cleveland, 1979; Van Cauter, 1979). As shown in Figure 4, these methods provide an adequate description of asymmetric waveshapes. The times of occurrence of the maximum and the minimum of the best-fit curve are often referred as the acrophase and the nadir, respectively. The amplitude of the rhythm may be estimated as 50% of the difference between the maximum and minimum of the best-fit curve. With the periodogram procedure, confidence intervals for the amplitude, acrophase, and nadir may be calculated.

The analysis of pulsatile variations may be considered at two levels. Significant variations in plasma concentrations may be defined and characterized based on estimations on the size of measurement error, primarily consisting of assay error. If sampling is sufficiently frequent, it is also possible mathematically to derive secretory rates from the plasma concentrations by "deconvolution." This procedure often reveals more pulses of secretion than a direct analysis of peripheral concentrations and more accurately defines the temporal limits of single pulses. However, deconvolution also amplifies measurement error.

To quantify ultradian pulsatile oscillations of peripheral concentrations or secretory rates, either time-domain analysis, in which the data are plotted versus time and pulses are identified, or frequency-domain analysis, in which amplitude is plotted versus frequency or period, may be used. These two approaches differ fundamentally in the mathematical and statistical treatment of the data and in the questions they address and should therefore be considered as complementary. A number of computer algorithms for identification of hormonal pulse identification are available, e.g. ULTRA (Van Cauter, 1988), PULSAR (Merriam & Wachter, 1982), CLUSTER (Veldhuis & Johnson, 1986), and DETECT (Oerter, Guardabasso, & Rodbard, 1986) and their properties have been reviewed in detail elsewhere (Evans *et al.*, 1992; Royston, 1989; Urban *et al.*, 1988). No universal, ready-to-use procedure exists which is applicable to all types of data. The investigator must choose the

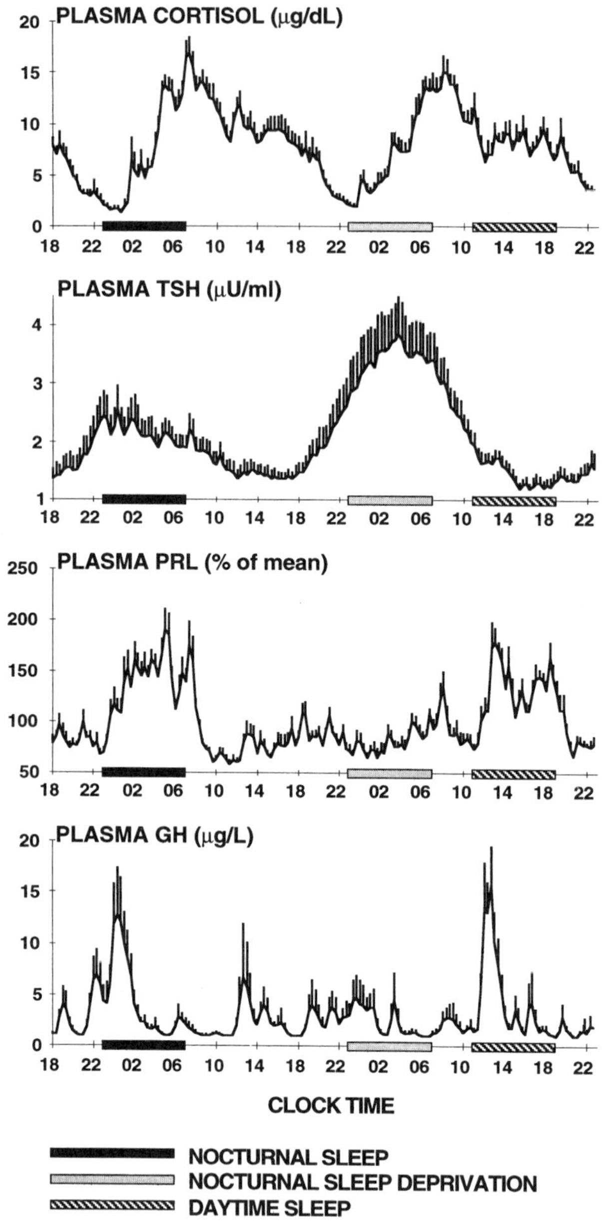

Figure 3. Mean plasma cortisol, thyrotropin (TSH), growth hormone (GH), and prolactin (PRL) profiles (+SEM) during a 12-hour shift of the light–dark and sleep–wake cycles from eight young adult males sampled at 20-minute intervals. The black bar represents the normal sleep period in darkness (2300–0700), the shaded bar the period of nocturnal sleep deprivation, and the striped bar the the shifted sleep period in darkness (1100–0700). (Cortisol and GH data previously published; Van Cauter *et al.*, 1991. TSH data previously published; Van Cauter & Aschoff, 1989. Unpublished prolactin data kindly provided by M. L'Hermite-Balériaux.)

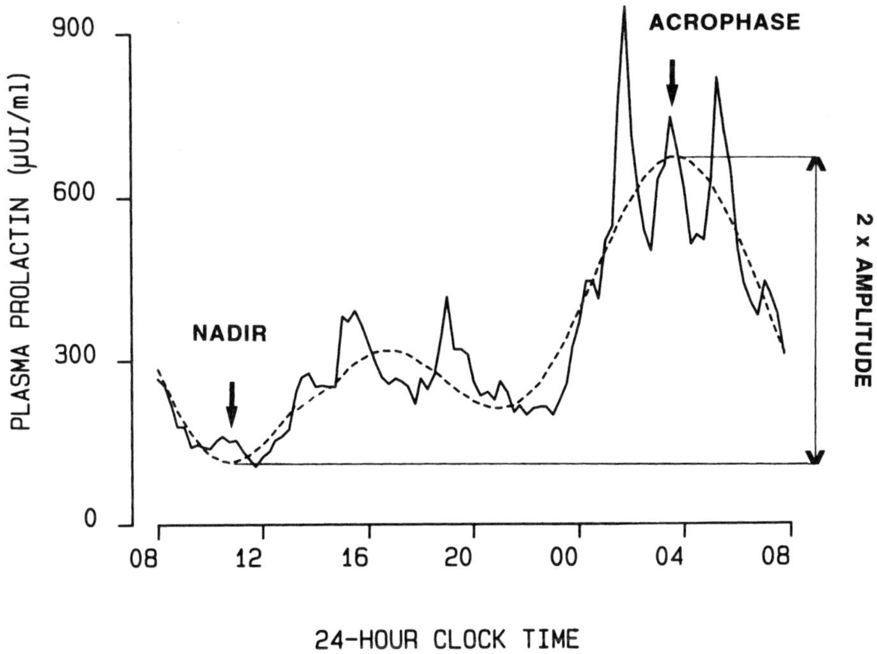

Figure 4. Individual 24-hour profile of plasma prolactin from a normal young male sampled at 15-minute intervals. The dashed line represents the best-fit curve obtained with a regression procedure which is used to define the nadir, acrophase, and amplitude, as indicated by arrows. (Data previously published; Van Cauter *et al.*, 1981.)

procedure whose underlying assumptions best match the data in ha2d. Comparisons of the performances of several procedures are often quite informative. In contrast to time-domain analysis, frequency-domain analysis may reveal the periodic, quasi-periodic, or random nature of underlying mechanisms (Matthews, 1988; Van Cauter, 1987). Descriptions of the principles and applications of spectral analysis are outside the scope of this chapter, but are described in studies which employed the technique for analyses of endocrine oscillation (McLeod & Craigon, 1985; O'Meara, Sturis, Van Cauter, & Polonsky, 1993; Sturis *et al.*, 1992).

Circadian Control of Pineal Melatonin

Profiles of Plasma and Salivary Melatonin as Circadian Markers

The 24-hour profile of plasma melatonin transduces light–dark information into a humoral signal and appears to be a robust marker of circadian rhythmicity that persists in constant conditions and can be shifted by zeitgebers of the circadian clock (Rosenthal, 1991). Figure 5 shows the temporal profile of plasma melatonin levels in a subject studied during a 36-hour period of continuous wakefulness under dim light conditions (<300 lux). The nocturnal rise starts in the early evening, between 2000 and 2200, which is before the normal sleep period, and the maximum occurs around the middle of the sleep period. The dim light melatonin onset (DLMO) has proven useful as a circadian marker and does not require sampling for

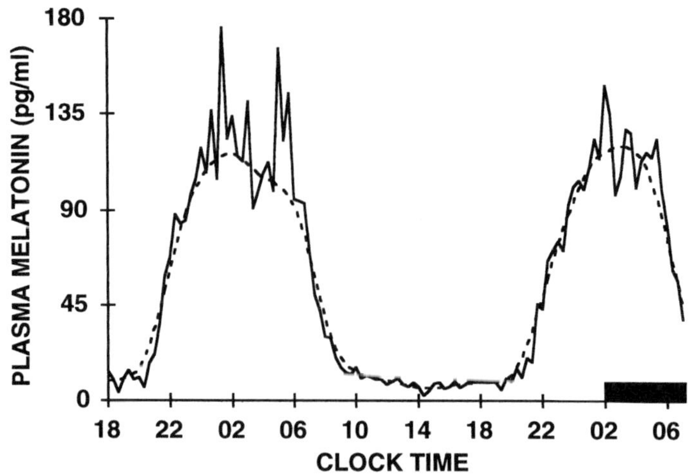

Figure 5. Individual plasma melatonin profile from a normal young adult male during a 34 hour constant routine. The dashed line represents the best-fit curve calculated by a regression procedure, and the dark bar the recovery sleep period in darkness. (Data previously published; Van Cauter *et al.*, 1994.)

a 24-hour period (Lewy & Sack, 1989). Salivary sampling of melatonin has become a convenient method for assessing the DLMO (English, Middleton, Arendt, & Wirz-Justice, 1993) since it obviates the need for an intravenous catheter, provides an additional noninvasive marker of circadian phase, and allows for field studies. In addition to timing of the rise of melatonin levels before sleep onset, the fact that the melatonin rhythm and the rhythm of body temperature undergo equivalent phase shifts following a regimen of scheduled exposures to bright light and darkness (Shanahan & Czeisler, 1991) or acute phase shifts to exercise (Buxton *et al.*, 1997; Van Reeth *et al.*, 1994) also supports the concept that both rhythms are controlled by the central circadian pacemaker. Partial phase response curves to single pulses of bright light and single pulses of exercise have been determined using the DLMO and are depicted in Figure 6 (Van Cauter *et al.*, 1994; Van Reeth *et al.*, 1994). The extent to which the amplitude of the melatonin rhythm, with its characteristically high inter-individual variability, reflects the amplitude of the circadian clock is a matter of speculation. The melatonin rhythm appears to be largely insensitive to stimuli other than light, such as sleep and food intake. Exposure to light of sufficient intensity (>200 lux) results in an acute "dose-dependent" inhibition of nocturnal melatonin secretion, with a rapid return to high nighttime levels when darkness resumes (Laakso, Hätönen, Stenberg, Alila, & Smith, 1993; Lewy, Wehr, Goodwin, Newsome, & Markey, 1989; Lewy *et al.*, 1995). Postural changes may also acutely and rapidly influence melatonin levels to a small extent, so postural changes should therefore be avoided during sampling for estimation of the DLMO (Deacon & Arendt, 1994a).

The role of melatonin in the human has received intense attention. Many studies have examined the effects on sleep and circadian rhythmicity. Pineal melatonin may play a permissive role for sleep (Dollins, Zhdanova, Wurtman, Lynch, & Deng, 1994; Tzischinsky & Lavie, 1994), possibly through a suppressive effect on body temperature (Cagnacci, Elliott, & Yen, 1992; Deacon & Arendt, 1994b) Indeed, nocturnal melatonin secretion has been estimated to contribute approximately 40%

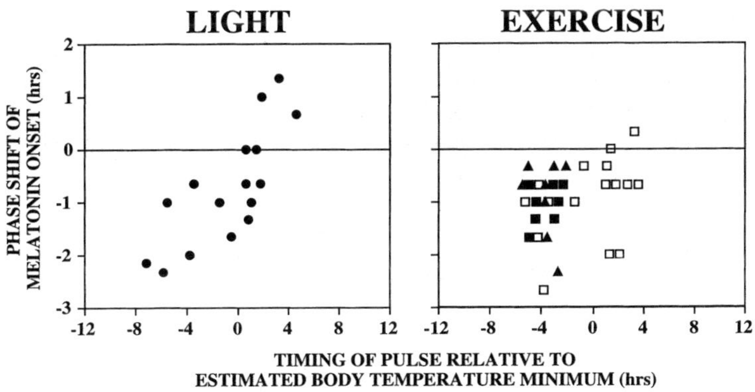

Figure 6. Partial phase response curves to single pulses of light (left) and exercise (right) in humans. The phase shift of the melatonin onset from the evening before to the evening after stimulus presentation is plotted versus the timing of the estimated body temperature minimum. Light pulses (closed circles) consisted of a 5000-lux stimulus over 3 hours on a background of dim light (<300 lux). Exercise pulses consisted of either low intensity (40–60% VO_2max), 3-hour arm-and-leg cycling activity [n = 16, open squares (Van Reeth *et al.*, 1994); n = 8, filled squares (Buxton *et al.*, 1997)] or high-intensity (75% VO_2max), 1-hour stepper exercise (n = 8, filled triangles (Buxton *et al.*, 1997)]. Data at the same point are slightly offset for clarity.

of the amplitude of the circadian regulation of body temperature. (Cagnacci *et al.*, 1992).

Use of Melatonin in Jet Lag and Shift Work

There is good evidence to indicate that, while the rhythm of melatonin secretion is driven by the circadian pacemaker, this rhythm also feeds back on the clock. (Armstrong, 1989, 1991; Cassone, 1990). Melatonin 1a receptors have been identified by *in situ* hybridization in the human suprachiasmatic nucleus (Weaver, Rivkees, Carlson, & Reppert, 1991). Exogenous administration of melatonin has also been shown to resynchronize certain overt rhythms, and in particular, the sleep–wake cycle, in a variety of conditions, including jet lag in normal subjects and free-running rhythms in blind subjects. The existence of a phase response curve of the DLMO to repeated (four consecutive days) oral administration of a low dose of melatonin (0.5 mg) has been demonstrated in human subjects (Lewy, Ahmed, Jackson, & Sack, 1992). In this paradigm, melatonin administration causes phase advances when administered in the afternoon and evening and phase delays when given in the morning. Nocturnal administration does not result in any phase shift. These characteristics are similar to those of the phase response curve to dark pulses given to animals free-running in constant light, which constitutes a mirror image of the shape of the phase response curve to light. These human data support a role for melatonin as an internal zeitgeber.

Another condition for which melatonin may be a useful treatment is for the facilitation of adaptation to night work. This approach would theoretically combine the phase-shifting effects of the drug with the facilitation of daytime sleep. However, the few controlled trials give a mixed result (Dawson, Encel, & Lushington, 1995; Folkard, Arendt, & Clark, 1993). As noted by Arendt (Arendt & Deacon, 1997; Dawson *et al.*, 1995; Folkard *et al.*, 1993), optimization of timing of administration, dosage, and formulation needs to be further evaluated. However, the potential of melatonin and melatonin analogs for the treatment of sleep disorders and circadian maladaptation in night work should not be underestimated. The need for new strategies to cope with shift work is especially imperative considering that the chronic use of benzodiazepines and other hypontics in shift work, a common treatment choice, has deleterious rather than beneficial effects.

Circadian Control and Sleep Modulation of the Corticoptroic Axis

24-Hour Profiles of Corticotropic Activity

Activity of the corticotropic axis may be monitored via plasma levels of the pituitary hormone, adrenocorticotropin (ACTH), and of the adrenal hormone directly controlled by ACTH stimulation, cortisol. The 24-hour profiles of plasma ACTH and cortisol shown in Figure 7 show an early-morning maximum, declining levels throughout the daytime, a quiescent period centered around midnight, and a rapid elevation which usually starts 2–3 hours after sleep onset. The 24-hour rhythm of adrenal secretion is primarily dependent on the circadian pattern of ACTH release, which is amplified by a daily variation in adrenal responsiveness to ACTH. The rhythm in ACTH release results, in turn, from periodic changes in level of stimulation by corticotropin-releasing hormone (CRH). A circadian variation parallel to that of cortisol has been demonstrated for the plasma levels of the adrenal steroids (Lejeune-Lenain, Van Cauter, Desir, Beyloos, & Franckson, 1987). Figure 8 shows the mean 24-hour profiles of cortisol, 11-hydroxyandrostenedione (11- OAD), dehydroepiandrosterone (DHEA), and androstenedione (AD) from 10 normal young men. The amplitude of the circadian variation in 11-OAD levels, a steroid

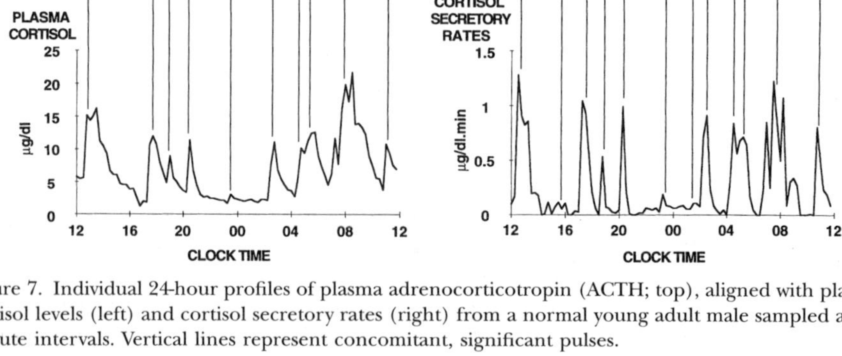

Figure 7. Individual 24-hour profiles of plasma adrenocorticotropin (ACTH; top), aligned with plasma cortisol levels (left) and cortisol secretory rates (right) from a normal young adult male sampled at 15-minute intervals. Vertical lines represent concomitant, significant pulses.

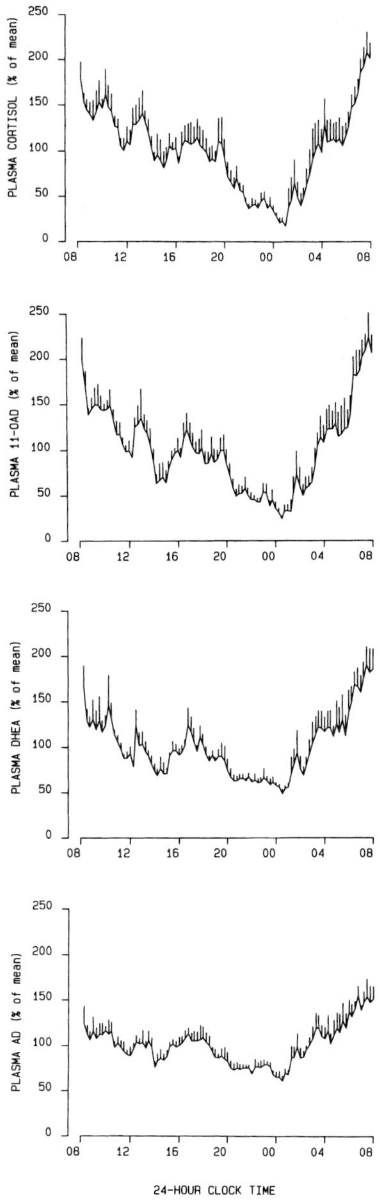

Figure 8. Mean 24-hour profiles of plasma cortisol, 11-hydroxyandrostenedione (11-OAD), dehydroepiandrosterone (DHEA), and androstenedione (AD) from 10 normal young males sampled at 15-minute intervals. Data are expressed as the percentage of the individual's 24-hour mean. Group data were then expressed as mean + SEM. (Data from Lejeune-Lenain *et al.*, 1987.)

derived from adrenal AD, is essentially similar to that of cortisol, whereas the amplitude of the variation in DHEA and AD levels, two steroids of partially gonadal origin, is much lower (Lejeune-Lenain *et al.*, 1987). Pulses of plasma concentration of adrenal steroids occur in remarkable synchrony with bursts of cortisol secretion, indicating the pulsatile ACTH stimulation is reflected in the temporal organization of all adrenal secretions. The circadian profile of cortisol secretion is produced by modulation of the height of successive secretory pulses (Van Cauter, 1990; Van Cauter, van Coevorden, & Blackman, 1990; Veldhuis, Iranmanesh, Lizarralde & Johnson, 1989a; Veldhuis, Iranmanesh, Johnson, & Lizarralde, 1989b). When a smooth curve is fitted to the 24-hour profile of plasma concentrations to estimate the parameters of the circadian rhythm, the amplitude of the cortisol rhythm averages 75% of the 24-hour mean. Estimations of the amplitude of the circadian rhythm of ACTH are more variable due to the highly pulsatile nature of the pattern. In normal conditions, the acrophase of the pituitary–adrenal periodicity occurs between 0600 and 1000. The quiescent period typically extends from 2000 to 0200 (Van Cauter, Leproult & Kupfer, 1996). The end of the quiescent period of cortisol secretion in the early morning hours, when observed under constant-routine conditions, is a robust and widely used marker of circadian phase.

Interactive Effects of Circadian Rhythmicity and Sleep

The cortisol profiles shown in Figure 3 illustrate the remarkable persistence of the waveshape of the cortisol profile when sleep is manipulated. The study period extended over 53 hours and included 8 hours of nocturnal sleep (i.e., from 2300 until 0700), followed by 28 hours of sleep deprivation in recumbent positions, and 8 hours of daytime sleep (i.e., from 1100 until 1900). This experimental design thus involved a 12-hour delay shift of the sleep period (Van Cauter *et al.*, 1991). The overall waveshape of the temporal profile of cortisol levels was not markedly affected by the absence of sleep or the presence of sleep at an abnormal time of day. Studies using similar experimental designs have indicated that the 24-hour periodicity of corticotropic activity is primarily controlled by circadian rhythmicity. Transient evoked responses may affect the waveshape of the cortisol rhythm. In particular, brief elevations of plasma levels are associated with meals, prticularly at lunch time (Follenius, Brandenberger, Hietter, Simeoni, & Reinhardt, 1982; Van Cauter, Shapiro, Tillil, & Polonsky, 1992b) and stressful events, such as venipuncture (Van Cauter & Honinckx, 1985) or exercise (Luger *et al.*, 1987), may cause a 1- to 2-hour secretory episode.

In addition, a modest modulation by sleep/wake transitions has been clearly demonstrated. Indeed, a number of studies have indicated that sleep onset is reliably associated with a short-term inhibition of cortisol secretion (Born, Muth, & Fehm, 1988; Van Cauter *et al.*, 1991; Weitzman, Zimmerman, Czeisler, & Ronda, 1983), although this effect may not be detectable when sleep is initiated at the time of the daily maximum of corticotropic activity, i.e., in the morning (Weibel, Follenius, Spiegel, Ehrhart, & Brandenberger, 1995b). Under normal conditions, since cortisol secretion is already quiescent in the late evening, this inhibitory effect of slow-wave sleep stages (Follenius, Brandenberger, Badasept, Libert, & Ehrhart, 1992) prolongs the quiescent period of cortisol secretion. During sleep deprivation, the nadir of cortisol secretion is elevated and occurs earlier than under normal nocturnal sleep conditions. Conversely, awakening at the end of the sleep period is consistently followed by a pulse of cortisol secretion (Spath-Schwalbe, Gofferje, Kern, Born, &

Fehm, 1991; Van Cauter *et al.*, 1990) During continued wakefulness, the immediate effects of sleep onset and sleep offset on corticotropic activity are absent and thus the nadir of cortisol levels is higher and the morning acrophase is lower, resulting in a reduction of cortisol rhythm amplitude of 15% compared to normal conditions. Several studies have shown that awakenings interrupting the sleep period consistently trigger pulses of cortisol secretion (Follenius *et al.*, 1992; Spath-Schwalbe *et al.*, 1991; Van Cauter *et al.*, 1990) Thus, sleep fragmentation, as occurs in aging (Bliwise, 1994), is likely to be associated with increased nocturnal corticotropic activity.

Health Implications of Cortisol Quiescent Period Reduction in Aging

In adults of both sexes, the evening nadir of cortisol concentrations increases with aging (Van Cauter *et al.*, 1996). As shown in Figure 9, the duration of the quiescent period is reduced because of a phase delay in its onset and a phase advance in the offset. i.e., in the onset of the subsequent circadian cortisol rise. These alterations develop slowly and progressively in adulthood concomitant with decrements in parameters of sleep quality which are thought to be partially dependent on the circadian system (e.g., sleep fragmentation, REM sleep) (Copinschi & Van Cauter, 1995; Van Cauter *et al.*, 1996). This is in contrast with the effects of aging on the somatotropic axis, which follow the chronology of aging of characteristics of sleep–wake homeostasis, with age-related alterations already fully experienced by midlife (Copinschi & Van Cauter, 1995). In recent studies, we have observed that the age-related delay in the onset of the quiescent period of cortisol may be mimicked in normal young adults by one night of total or partial sleep deprivation (Leproult, Buxton, & Van Cauter, 1997). We hypothesize that the normal course of aging may result in accumulated sleep loss resulting from an inability to obtain adequate quality and amount of sleep, which may be partially responsible for the elevation in evening cortisol levels. Thus, alterations in hypothalamo–pituitary–adrenal activity in aging could partially reflect disturbances of both sleep and circadian rhythmicity.

Studies in rodents have indicated that a slight degree of hyperactivity of the HPA axis at the usual trough of corticosterone levels is capable of exerting deleterious metabolic effects, including increased adiposity and insulin resistance (Dallman *et al.*, 1993). Consistent with these findings, data obtained in human subjects indicate that a short-term elevation of cortisol in the evening results in a prolonged state of insulin resistance, whereas the same degree of elevation in the morning has insignificant effects on glucose and insulin levels (Plat, Féry, L'Hermite-Balériaux, Mockel & Van Cauter, 1996). Thus, age-related alterations of evening cortisol secretion could play a role in the development of reduced glucose tolerance in the elderly.

Circadian and Sleep Control of the Thyrotropic Axis

24-Hour Profiles of Thyrotropin and Thyroid Hormones

Examples of profiles of plasma TSH are shown in Figures 2 and 3 (second panels from the top). Daytime TSH levels are low and relatively stable and are followed by a rapid elevation starting in the early evening (1900–2100) culiminating in a nocturnal maximum occurring around the beginning of the sleep period (Brabant *et al.*, 1990; Veldhuis, Iranmanesch, Johnson, & Lizarralde, 1990). The later

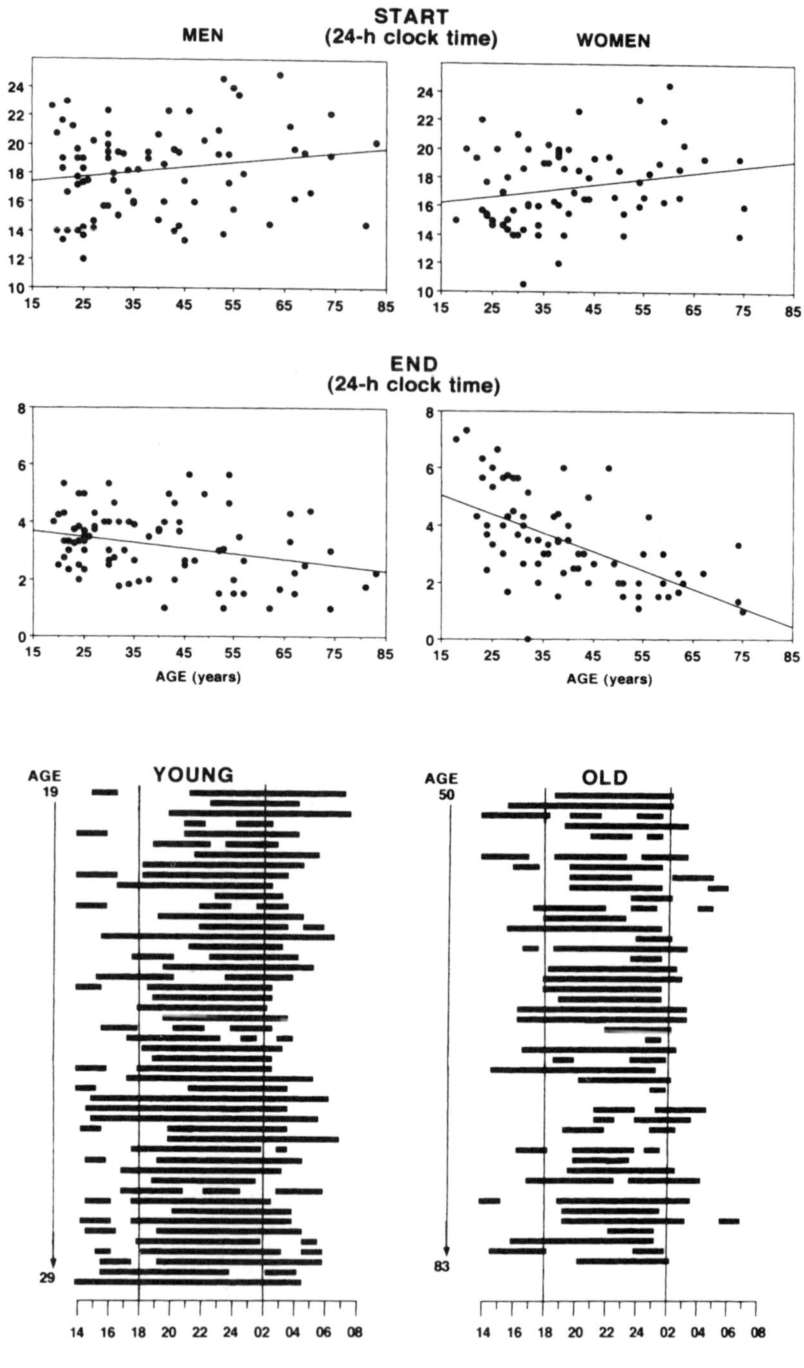

Figure 9. Onset (top) and offset (middle) of the evening cortisol quiescent period (QP) in men (top two left panels) and women (top two right panels) versus the age of the subject. Timing and duration of the cortisol QP within a single subject are shown for each horizontal line in young (left, bottom) and old (right, bottom). The vertical lines at 1800 and 0200 are provided for a visual reference. (Data previously published; Van Cauter *et al.*, 1996.)

part of sleep is associated with a progressive decline in TSH levels, and daytime values resume shortly after morning awakening. The 24-hour temporal pattern of TSH secretion seems to reflect both tonic and pulsatile release, with both the frequency and the amplitude of the pulses increasing during the nighttime (Brabant *et al.*, 1990; Greenspan, Klibanski, Schoenfeld, & Ridgway, 1986; Veldhuis *et al.*, 1990; Veldhuis, Johnson, Lizarralde, & Iranmanesh, 1992). Pulses of TSH secretion persist under both somatostatin and dopamine treatment, suggesting that the control of pulsatility is largerly TRH dependent (Brabant, Prank, Hoang-Vu, Hesch, & von zür Mülhen, 1991). The 24-hour profile of plasma TSH has been used as a marker of the human circadian clock. Indeed, the timing of the nocturnal TSH elevation appears to accurately reflect circadian timing and shifts with the rhythms of body temperature and melatonin (Allan & Czeisler, 1994; Van Cauter *et al.*, 1994).

Circadian and/or sleep-related variations in thyroid hormones have been difficult to demonstrate, probably because these hormones are largely bound to serum proteins and thus their peripheral concentrations are primarily modulated by diurnal variations in hemodilution caused by postural changes (Brabant *et al.*, 1987; Greenspan *et al.*, 1986). However, under conditions of sleep deprivation, the increased amplitude of the TSH rhythm may result in a detectable increase in the amplitude of the rhythm of plasma triiodotyronine (T3) levels, which parallels the nocturnal rise of TSH (Van Cauter & Turek, 1995; Figure 10). If sleep deprivation is prolonged for a second night, the nocturnal rise of TSH is markedly diminished compared to that occurring during the first night (Allan & Czeisler, 1994; Van Cauter *et al.*, 1994). It is likely that the elevated thyroid hormone levels observed following the first night of sleep deprivation persist during the daytime period because of the prolonged half-life of these hormones and may limit the subsequent TSH rise at the beginning of the next night-time period.

Interactive Effects of Circadian Rhythmicity and Sleep on Thyrotropin and Triiodotyronine

Sleep has an important impact on the thyrotropic axis. An inhibitory effect of sleep on TSH secretion may be revealed during sleep deprivation, when nocturnal TSH secretion is increased by as much as 200% over the levels observed during nocturnal sleep (Brabant *et al.*, 1990; Parker, Rossman, Pekary, & Hershman, 1987). However, when sleep occurs during daytime hours, TSH secretion is not suppressed significantly below normal daytime levels (Hirschfeld *et al.*, 1996). Thus, the inhibitory effect of sleep on TSH secretion appears to be operative when the nighttime elevation has taken place, indicating once again the interaction of effects of circadian time and effects of sleep. Elevations of nocturnal T3 and T4 levels have also been observed in a 64-hour sleep deprivation paradigm (Gary *et al.*, 1996). In an examination of the relationship of TSH and specific sleep events, a pulse-by-pulse analysis of TSH profiles and sleep stages has revealed a consistent association between descending slopes of TSH concentrations and stages of slow-wave (SW; stages I, II, and IV) sleep (Goichot, Brandenberger, Sainio, Wittersheim, & Follenius, 1992). When assessed with spectral analysis of the sleep EEG, this temporal relationship was quantified as a negative cross-correlation between TSH fluctuations and relative spectral power in the delta range (Gronfier *et al.*, 1995).

Alterations during Simulated Shift Work and Jet Lag

The fact that the inhibitory effects of sleep on TSH secretion are time dependent may cause, under certain circumstances, elevations of plasma TSH levels which

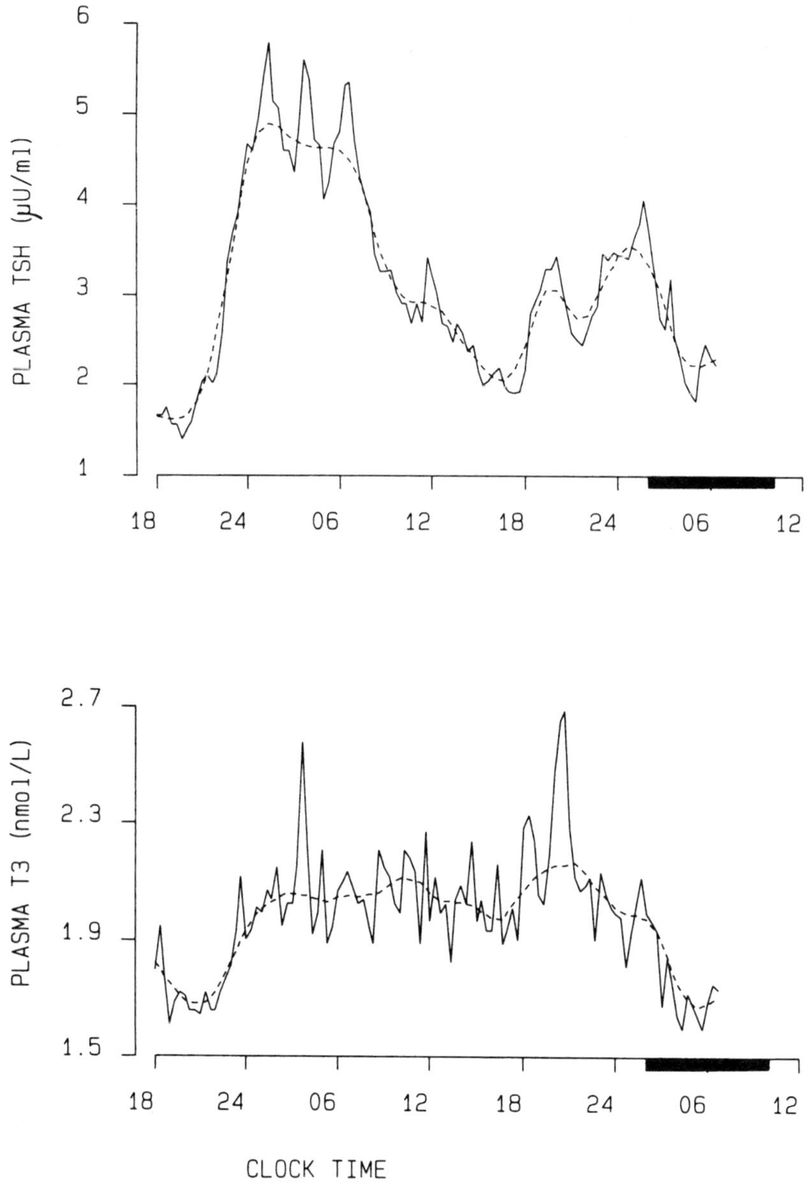

Figure 10. Individual profiles of plasma thyrotropin (TSH) and triiodotyronine (T3) from a single normal young male sampled at 20-minute intervals over a 38-hour period beginning with a 32-hour period of wakefulness. The black bar represents the recovery sleep period in darkness. The dashed line represents the best-fit curve obtained with a regression procedure. (From Van Cauter & Aschoff, 1989, with permission.)

reflect the misalignment of sleep and circadian timing. An example is shown in the lower panel of Figure 11, which shows the mean TSH profiles observed in a group of normal young men in the course of adaptation to simulated "jet lag" (Hirschfeld *et al.*, 1996). Following a 24-hour baseline period, the sleep–wake and dark–light cycles were abruptly advanced by 8 hours (i.e., from 2300–0700 to 1500–2300). In the course of adaptation to this 8-hour advance shift, TSH levels increased progressively

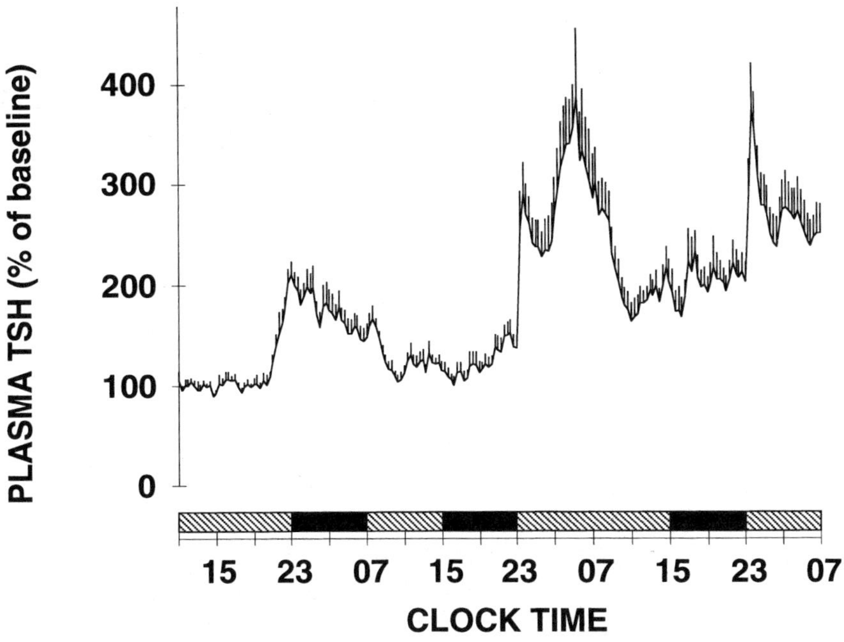

Figure 11. Mean profiles (+SEM) of four young females and four young males submitted to an 8-hour advance of the sleep–wake and light–dark cycles on a background of a constant routine. Dark bars (2300–0700, 1500–2300, 1500–2300) represent the sleep periods in darkness. Thyrotropin (TSH) levels are normalized by individual and expressed as a percentage of the mean daytime baseline levels. Note the recurring circadian nocturnal rise of TSH secretion. (Data from Hirschfeld *et al.*, 1996.)

because daytime sleep failed to inhibit TSH, and nighttime wakefulness was associated with large circadian-dependent TSH elevations. As a result, mean TSH levels following awakening from the second shifted sleep period were more than two fold higher than during the same time interval following normal nocturnal sleep. This overall elevation of TSH levels was paralleled by a small increase in T3, but not free thyroxine concentrations (Hirschfeld *et al.*, 1996).

For permanent night workers, the TSH rhythm in the eight subjects studied was markedly different from the control group of regular day workers (Weibel *et al.*, 1995a). The timings of the TSH onset, acrophase, and nadir maintained a fixed phase relationship with the body temperature rhythm and were significantly shifted relative to the circadian markers in the day workers, reflecting a degree of adaptation to the work schedule by the circadain system in these individuals. An alteration of the waveshape of the TSH rhythm across the sleep period was, however, evident in the duration of the declining TSH levels, which began 5 hours earlier in the night workers and therefore resulted in a reduction in the amount of TSH secreted per day. The clinical significance of such a chronic reduction is not clear.

Sleep Control and Circadian Modulation of the Lactotropic Axis

Role of Sleep in the Control of Prolactin Secretion

Under normal conditions, PRL levels are minimal at midday, increase slightly during the afternoon, and then exhibit a nocturnal elevation starting shortly after sleep onset and culminating around midsleep. In adults, both males and females, the nocturnal maximum corresponds to an average increase of more than 200% above the minimum level (Spiegel *et al.*, 1994; Van Cauter *et al.*, 1981; Waldstreicher *et al.*, 1996). Morning awakening is also consistently associated with a brief PRL pulse (Spiegel *et al.*, 1994; Waldstreicher *et al.*, 1996). Pulses occur during the daytime as well as during the night (Van Cauter *et al.*, 1981; Veldhuis & Johnson, 1988). Deconvolution analysis indicates that the peripheral profile of prolactin levels reflects both tonic and intermittent release (Veldhuis *et al.*, 1992).

Meal ingestion, especially at lunch time, and high-intensity exercise may result in short-term elevations of prolactin secretion (Galbo, 1995; Quigley, Ropert, & Yen, 1981). In older adults, the nocturnal prolactin elevation is reduced by approximately 50% (van Coevorden *et al.*, 1991).

Circadian Modulation of Prolactin Secretion

A number of experiments involving advances or delays of sleep times have shown that the sleep-related rise of PRL may still be present, although with a reduced amplitude, when sleep does not occur at the normal nocturnal time and that maximal stimulation is observed only when sleep and circadian effects are superimposed (Desir *et al.*, 1982; Spiegel *et al.*, 1994; Waldstreicher *et al.*, 1996). The sleep-independent circadian component of PRL secretion (Desir *et al.*, 1982; Radomski *et al.*, 1995) is seen in the progressive increase across the late afternoon and the hours preceding the usual bedtime, and effect which is much more pronounced in women than in men (Waldstreicher *et al.*, 1996). Decreased dopaminergic inhibition of PRL during sleep (Spiegel *et al.*, 1994) and increased sensitivity of the lactotrophs to Gonadotropin-Releasing Hormone (GnRH) stimulation (Rossmanith, Boscher, Ulrich, & Benz, 1993) have been proposed as mechanisms underlying the nocturnal PRL elevation under normal sleep conditions. In addition to this circadian increase in PRL in the later part of the day, two studies have indicated that noctural PRL levels may increase in the absence of sleep (Okatani & Sagara, 1993; Wehr *et al.*, 1993). One study showed that PRL increases when the subject is awake in a dark environment, but not in the presence of light, and therefore suggested that the nocturnal elevation of PRL is partially dependent on melatonin (Okatani & Sagara, 1993). Oral administration of low doses of melatonin during the daytime indeed results in an elevation of plasma PRL levels. The other study indicated that nocturnal increases of PRL in the absence of sleep are dependent on a state of "quiet rest" and do not occur if the subjects expect to be disturbed (Wehr *et al.*, 1993). An understanding of the etiology of the circadian component of PRL secretion requires further elucidation.

Adaptation in Jet Lag and Shift Work

Figure 12 depicts the changes in prolactin levels observed in subjects submitted to an abrupt 8-hour delay shift of the light–dark and sleep–wake cycles, simulating westward jet lag (Van Cauter & Turek, 1995). By the fifth day after the shift, the 24-hour profile of PRL appears totally synchronized to the new schedule. During intermediate stages of adaptation, the profile becomes asymmetric, with concentrations peaking at the beginning of the night period rather than being maintained at a high level across the entire night period. Similar observations have been obtained in subjects undergoing actual jet lag following transmeridian travel (Desir *et al.*, 1982).

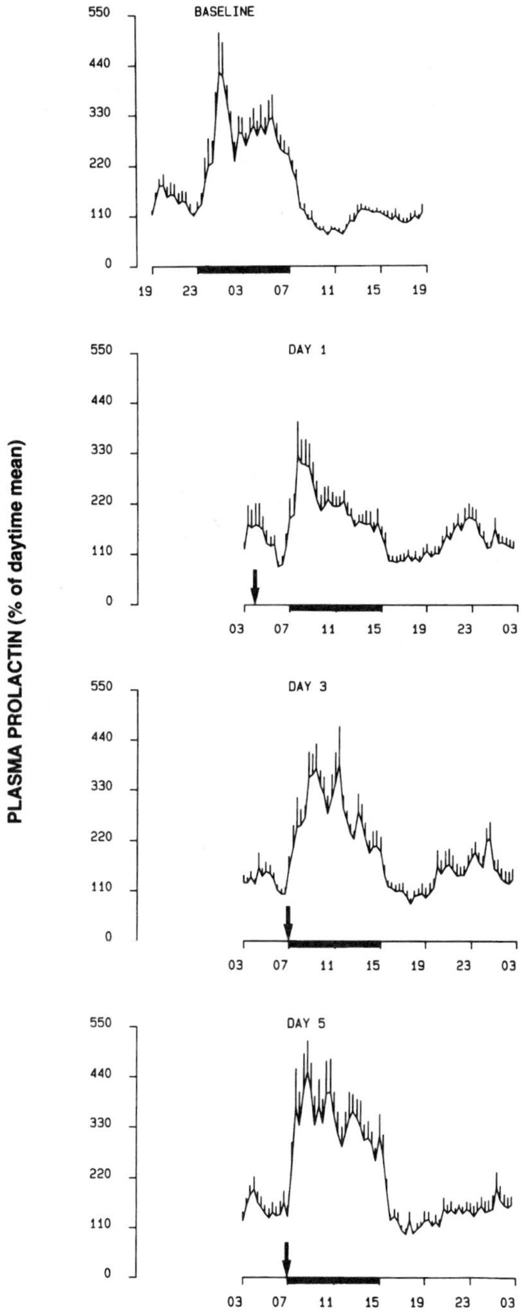

Figure 12. Mean 24-hour profiles (+SEM) of plasma prolactin from six normal young men sampled at 20-minute intervals during baseline, and 1, 3, and 5 days after an abrupt 8-hour delay of the light–dark and sleep–wake cycles simulating the jet lag of westward travel. Black bars represent the time in bed and dark. Light levels after the shift were 1500–2000 lux. (Unpublished data from E. Van Cauter, M. L'Hermite-Balériaux, D. Bosson, A. Van Onderbergen, and G. Copinschi.)

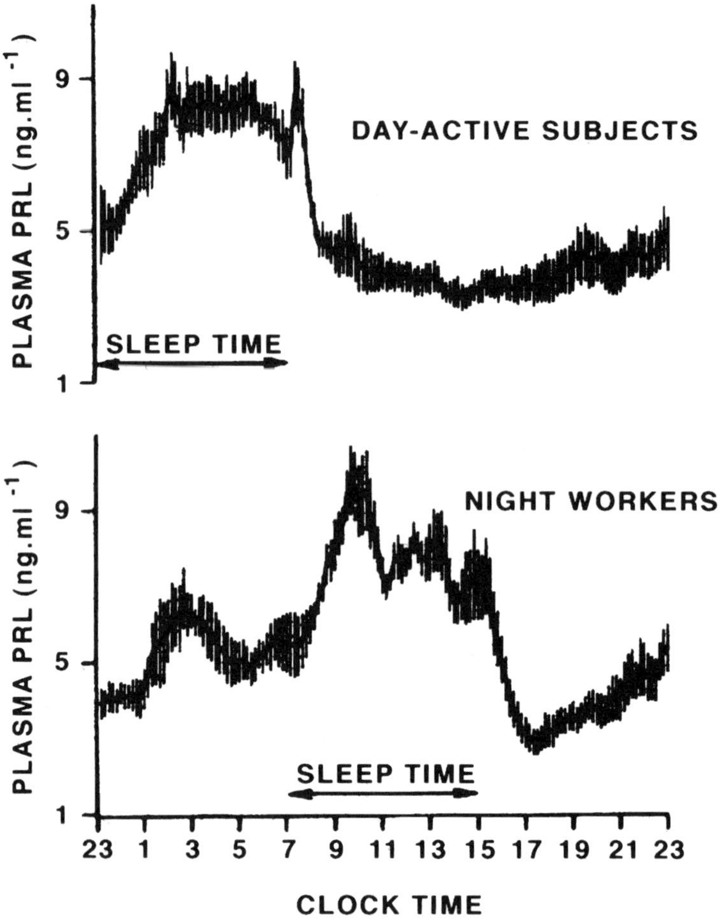

Figure 13. Mean profiles (±SEM) of plasma prolactin (PRL) from two groups of normal young males sampled at 10-minute intervals. Left: dayworkers ($n=8$); right: habitual night workers ($n=10$). (Data from Spiegel *et al.*, 1996.)

Figure 13 compares the diurnal rhythms in the chronic situation of permanent night workers as compared to permanent day workers (Spiegel, Weibel, Gronfier, Brandenberger, & Follenius, 1996) and illustrates the chronic state of internal desynchrony that characterizes night shift workers. Sleep-related prolactin secretion adapts to the night schedule, whereas components reflecting circadian timing show little adaptation (Spiegel *et al.*, 1996).

Sleep Control and Circadian Modulation of the Somatotropic Axis

Role of Sleep in the Control of Growth Hormone Secretion

In normal adult subjects, the 24-hour profile of plasma GH consists of low levels interrupted by large secretory bursts. The most reproducible pulse occurs shortly after sleep onset (Sassin *et al.*, 1969; Takahashi, Kipnis, & Daughaday, 1968). This relationship between sleep onset and GH secretion appears to be most consistent in the human, perhaps because human sleep is consolidated in a single 7- to 9-hour period, whereas sleep is fragmented into multiple bouts in most other mammals. In men, the sleep-onset GH pulse is generally the largest and, after the third decade of age, often the only pulse observed over the 24-hour span. In women, daytime GH pulses are more frequent and the sleep-associated pulse, while still present in most cases, does not account for the majority of the 24-hour GH release. These gender differences in GH secretion are depicted in Figure 14.

Sleep onset will elicit a pulse in GH secretion whether sleep is advanced, delayed, or interrupted and reinitiated. A pulse of GH secretion following sleep onset has been observed in subjects submitted to a variety of manipulations of the sleep–wake cycle, including delays by 3, 5, 8, 12, and 16 hours, daytime recovery sleep following 28 hours of continuous wakefulness, nocturnal recovery sleep following 40 hours of continuous wakefulness, a 7-hour advance due to transmeridian travel, and a 7-hour advance following 33 hours of sleep deprivation in the laboratory (reviewed in Van Cauter & Copinschi, 1999). A recent study of night workers indicated that the main GH secretory episode occurs during the first half of the daytime sleep period (Weibel, Spiegel, Gronfier, Follenius, & Brandenberger, 1997b).

The sleep-onset GH pulse is correlated temporally with the first period of SW sleep and a relationship between the amount of SW sleep and the amount of concomitant GH secretion has been demonstrated in young adults (Van Cauter *et al.*, 1992a). Studies using pharmacologic stimulation of SW sleep have further demonstrated that this relationship reflects common mechanisms rather than a fortuitous temporal association (Gronfier *et al.*, 1996; Van Cauter *et al.*, 1997a). Administration of ritanserin, a selective serotonin ($5HT_2$)-receptor antagonist, results in parallel and highly correlated small-amplitude increases of delta wave activity and nocturnal GH release (Gronfier *et al.*, 1996). Reliable stimulation of SW sleep in normal subjects has also been obtained with oral administration of low doses of gamma-hydroxybutyrate (GHB), a metabolite of gamma-aminobutyric acid which is normally present in the mammalian brain and is an investigational drug for the treatment of narcolepsy (Van Cauter et al., 1997a). In Figure 15, the amount of SW activity sleep and GH levels after GHB or placebo treatment at 2300 is depicted.

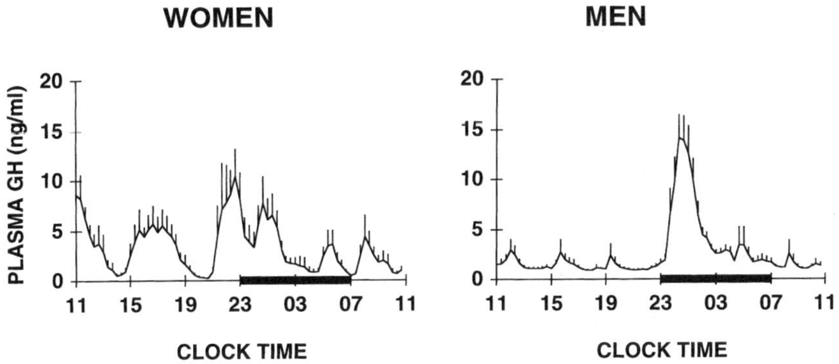

Figure 14. Mean profiles (+SEM) of plasma growth hormone (GH) in (left) normal young adult women ($n = 8$) and (right) men ($n = 6$). The dark bar represents the normal sleep period in darkness (2300–0700).

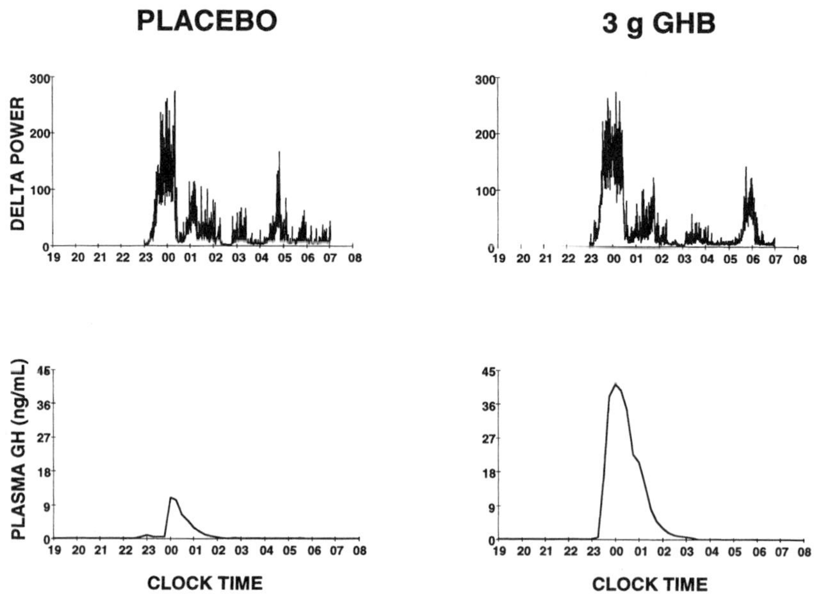

Figure 15. Individual profiles of delta power (top) and plasma growth hormone (GH) from a normal young adult male sampled at 15-minute intervals after ingestion of either placebo (right panels) or a 3-dose of gamma-hydroxybutyrate (GHB) just prior to sleep onset. Note the increase in delta power and the concomitant increase in GH secretion with GHB. (Data previously published; Van Cauter *et al.*, 1997a.)

There was a high correlation between increases in the amount of stage IV after GHB administration, and corresponding increases in GH secretion during the same period. The relationship between SW activity and GH secretion is not one-to-one, however. Indeed, nocturnal GH secretion can occur in the absence of SW sleep and approximately one third of the SW periods are not associated with detectable GH secretion. Because GH secretion is also under inhibitory control by somatostastin, variability of somatostatinergic tone may cause a dissociation between SW sleep and nocturnal GH release. The short-term negative feedback inhibition exerted by GH on its own secretion may also explain observations of the occasional lack of a GH pulse during the first SW period when a secretory pulse occurred prior to sleep onset.

Circadian Modulation of Growth Hormone Secretory Propensity

While sleep is clearly a major determinant of the 24-hour profile of GH secretion in humans, there is also evidence for the existence of circadian modulation of GH secretion. First, daytime naps are more consistently associated with GH release when they occur in the afternoon, when the propensity for SW sleep is increased, than in the morning, when REM sleep predominates (Karacan, Rosenbloom, Londono, Williams, & Salis, 1974; Othmer, Mendelson, Levine, Malarkey, & Daughaday, 1974). Second, it has been shown that in normal subjects submitted to a 3-hour sleep–wake cycle (60 minutes of sleep every 3 hours) for a prolonged period of time, the elevation of GH concentration following sleep onset is largest when the

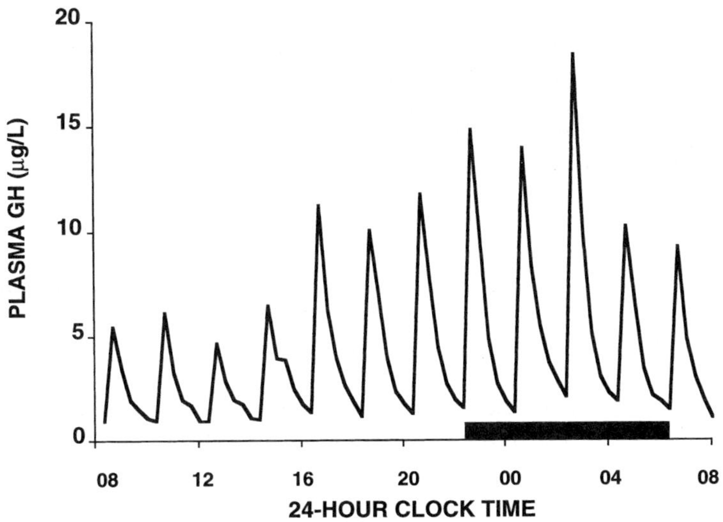

Figure 16. Mean profiles of plasma growth hormone (GH) from seven normal middle-aged male subjects sampled at 20-minute intervals. Growth hormone releasing hormone (GHRH) was given as an iv bolus (0.33 μg/kg) at 2 hour intervals. Daytime naps were not allowed. The bar represents the sleep period in darkness from 2300–0700. (Data from Jaffe *et al.*, 1995.)

sleep episode begins in the late evening, i.e., around the usual bedtime (Aschoff, 1979). The late evening and early part of the night appear to represent a period of increased propensity for GH secretion, as studies involving abrupt delays of the sleep period in very young adults have shown increases in GH pulsatility at this time of day, suggesting the existence of a circadian modulation of GH secretion (Mullington, Hermann, Holsboer, & Pollmacher, 1996; Seifritz *et al.*, 1995; Van Cauter *et al.*, 1992a; Weibel *et al.*, 1997b). The most convincing evidence for a circadian influence has been obtained in a recent study using repeated injections of growth-hormone releasing hormone at 3-hour intervals. The mean profiles of plasma GH observed under this treatment are shown in Figure 16. A diurnal variation in peak GH response is clearly demonstrated and was interpreted as reflecting a diurnal rhythm in somatostatinergic tone (Jaffe, Turgeon, DeMott Friberg, Watkins, & Barkan, 1995). Elevated GH responses were apparent in the early evening (before sleep onset), indicating that this diurnal variation partially reflects circadian rhythmicity.

Impact of Aging and Age-Related Sleep Alterations

Aging is associated with dramatically diminished GH and insulin-like growth factor I (IGF-I) secretion and reduced sleep quality (Bliwise, 1993; Ho *et al.*, 1987; Landin-Wilhelmsen *et al.*, 1994; Prinz, 1995; Prinz, Vitiello, Raskind, & Thorpy, 1990; van Coevorden *et al.*, 1991). In healthy elderly men over 65 years of age, the total GH secreted over a 24-hour period is generally less than one third of the daily GH secreted by men under 30 years of age. (Finkelstein, Roffwarg, Boyar, Kream, & Hellman, 1972; Frank *et al.*, 1995; Ho *et al.*, 1987; Iranmanesh, Lizarralde, & Veldhuis, 1991; van Coevorden *et al.*, 1991; Vermeulen, 1987). Interestingly, the amount of SW sleep in older adults is reduced in the same proportion. This decline in overall GH secretion is likely related to a decrease in the amplitude, rather than the frequency of GH pulses. In a retrospective analysis involving nearly 100 simultaneous recordings of sleep and 24-hour GH secretion in adult men 18–82 years of age, we concluded that these dramatic aging effects to diminish SW sleep and GH secretion begin an exponential pattern of reduction early in adulthood that is essentially complete by the beginning of the fifth decade (Copinschi & Van Cauter, 1995). Sleep-related GH pulses in aging are detectable, but reduced in magnitude (Frank *et al.*, 1995; Ho *et al.*, 1987; van Coevorden *et al.*, 1991; Vermeulen, 1987). From the results of our retrospective analysis, we observed that the proportion of daily GH output occurring during the first few hours of sleep does not decrease with age, but remains stable or even slightly increases. A significant correlation between levels of IGF-I and delta power has been reported in older adults (Prinz *et al.*, 1995). The parallelism between decreased amount and quality of deep sleep and diminished somatotropic activity suggests that some of the peripheral effects of the hyposomatotropism in the elderly, such as the reduction in lean body mass, may partially reflect a central alteration in sleep control.

Adaptation in Jet Lag and Shift Work

Studies of the 24-hour GH profile in jet lag and shift work have confirmed that the major nocturnal GH pulse rapidly adapts to the shifted sleep time (Golstein *et al.*, 1983; Weibel, Follenius, Spiegel, Gronfier, & Brandenberger, 1997a). An example is shown in the lowest panel of Figure 3. Misalignments between endogenous circadian rhythmicity and scheduled sleep periods such as occur in jet lag and shift work may have deleterious consequences on GH secretion insofar as the duration and quality of abnormally timed sleep is generally diminished and therefore SW sleep-associated GH secretion may also be diminished. It remains to be determined whether long-term sleep disruptions may affect levels and body composition in a clinically significant manner.

Conclusions

The temporal variation of hormonal secretion may be affected primarily by circadian rhythmicity, such as is the case for melatonin, or by sleep, such as is the case for GH. Indeed, hormonal secretion may be affected in a variety of ways by both circadian rhythmicity and sleep–wake homeostasis. The remarkable degree of temporal organization in the endocrine system provides a useful model for assessing the impact of altered temporal patterns of activity in humans. The understanding of the role of the endocrine system in mediating the physiopathologic implications of alterations in light exposure and sleep–wake habits associated with modern living conditions in industrial societies is a challenging and fertile area of research in chronobiology.

Acknowledgments

This work was supported by grants from the Air Force Office of Scientific Research (94-1-0203 and F49620-94-0203), supplemented by a U.S. Department of Defense Augmentation Award for Science and Engineering Research Training grant (F49620-96-1-0252), and by grants from the National Institutes of Health (NIDDK

DK-41814 and NIA AG-11412). We thank Rachel Leproult for assistance with computer graphics and data analysis.

References

Allan, J., & Czeisler, C. A. (1994). Persistence of the circadian thyrotropin rhythm under constant conditions and after light-induced shifts of circadian phase. *Journal of Clinical Endocrinology and Metabolism, 79,* 508–512.

Arendt, J., & Deacon, S. (1997). Treatment of circadian rhythm disorders—Melatonin. *Chronobiology International, 14,* 185–204.

Armstrong, S. M. (1989). Melatonin and circadian control in mammals. *Experientia, 45,* 932–938.

Armstrong, S. M. (1991). Melatonin: A chronobiotic with anti-aging properties? *Medical Hypotheses, 34,* 300–309.

Aschoff, J. (1979). Circadian rhythms: General features and endocrinological aspects. In D. T. Krieger (Ed.), *Endocrine rhythms* (pp. 1–61). New York: Raven Press.

Bliwise, D. L. (1993). Sleep in normal aging and dementia. *Sleep, 16,* 40–81.

Bliwise, D. L. (1994). Normal aging. In M. H. Kryger, T. Roth, & W. C. Dement (Eds.), *Principles and practice of sleep medicine* (pp. 26–39). Philadephia: Saunders.

Born, J., Muth, S., & Fehm, H. L. (1988). The significance of sleep onset and slow wave sleep for nocturnal release of growth hormone (GH) and cortisol. *Psychoneuroendocrinology, 13,* 233–243.

Brabant, G., Brabant, A., Ranft, U., Ocran, K., Köhrle, J., Hesch, R. D., & von zür Mühlen, A. (1987). Circadian and pulsatile thyrotropin secretion in euthyroid man under influence of thyroid hormone and glucocorticoid administration. *Journal of Clinical Endocrinology and Metabolism, 65,* 83–88.

Brabant, G., Prank, K., Ranft, U., Schuermeyer, T., Wagner, T. O. F., Hauser, H., Kummer, B., Feistner, H., Hesch, R. D., & von zür Mühlen, A. (1990). Physiological regulation of circadian and pulsatile thyrotropin secretion in normal man and women. *Journal of Clinical Endocrinology and Metabolism, 70,* 403–409.

Brabant, G., Prank, K., Hoang-Vu, C., Hesch, R. D., & von zür Mülhen, A. (1991). Hypothalamic regulation of pulsatile thyrotropin secretion. *Journal of Clinical Endocrinology and Metabolism, 72,* 145–150.

Brandenberger, G., Charloux, A., Gronfier, C., & Otzenberger, H. (1998). Ultradian rhythms in hydromineral hormones. *Hormone Research, 49* (3–4), 131–135.

Buxton, O. M. Frank, S. A., L'Hermite-Balériaux, M., Leproult, R., Turek, F. W., & Van Cauter, E. (1997). Roles of intensity and duration of nocturnal exercise in causing phase-shifts of human circadian rhythms. *American Journal of Physiology (Endocrinology and Metabolism), 273,* E536–E542.

Cagnacci, A., Elliott, J. A., & Yen, S. S. C. (1992). Melatonin: A major regulator of the circadian rhythm of core temperature in humans. *Journal of Clinical Endocrinology and Metabolism, 75,* 447–452.

Cassone, V. M. (1990). Effects of melatonin on vertebrate circadian systems. *TINS, 13,* 457–464.

Cleveland, W. S. (1979). Robust locally weighted regression and smoothing scatterplots. *Journal of the American Statistical Association, 74,* 829–836.

Copinschi, G., & Van Cauter, E. (1995). Effects of ageing on modulation of hormonal secretions by sleep and circadian rhythmicity. *Hormone Research, 43,* 20– 24.

Czeisler, C. A., Chiasera, A. J., & Duffy, J. F. (1991). Research on sleep, circadian rhythms and aging: Applications to manned spaceflight. *Experimental Gerontology, 26,* 217–232.

Dallman, M. F., Strack, A. L., Akana, S. F., Bradbury, M. J., Hanson, E. S., Scribner, K. A., & Smith, M. (1993). Feast and famine: Critical role of glucocorticoids with insulin in daily energy flow. *Frontiers in Neuroendocrinology, 14,* 303–347.

Dawson, D., Encel, N., & Lushington, K. (1995). Improving adaptation to simulated night shift: Timed exposure to bright light versus daytime melatonin administration. *Sleep, 18,* 11–21.

Deacon, S., & Arendt, J. (1994a). Posture influences melatonin concentrations in plasma and saliva in humans. *Neuroscience Letters, 167,* 191–194.

Deacon, S. J. & Arendt, J. (1994b). Phase-shifts in melatonin 6-sulphatoxymelatonin and alertness rhythms after treatment with moderately bright light at night. *Clinical Endocrinology, 40,* 413–420.

Desir, D., Van Cauter, E. L'Hermite, M., Refetoff, S., Jadot, C., Caufriez, A., Copinschi, G., & Robyn, C. (1982). Effects of "jet lag" on hormonal patterns. III. Demonstration of an intrinsic circadian rhythmicity in plasma prolactin. *Journal of Clinical Endocrinology and Metabolism, 55,* 849– 857.

Dollins, A. B. Zhdanova, I. V., Wurtman, R. J., Lynch, H. J., & Deng, M. H. (1994). Effects of inducing nocturnal serum melatonin concentrations in daytime on sleep, mood, body temperature, and performance. *Proceedings of the National Academy of Sciences of the USA, 91,* 1824–1828.

English, J., Middleton, B. A. Arendt, J., & Wirz-Justice, A. (1993). Rapid direct measurement of melatonin in saliva using an iodinated tracer and solid phase second antibody. *Annals of Clinical Biochemistry, 30,* 415–416.

Evans, W. S., Sollenberger, M. J., Booth, R. A. J., Rogol, A. D., Urban, R. J., Carlsen, E. C., Johnson, M. L., & Veldhuis, J. D. (1992). Contemporary aspects of discrete peak-detection algorithms. II. The paradigm of the luteinizing hormone signal in women. *Endocrine Reviews, 13,* 81–104.

Finkelstein, J. W., Roffwarg, H. P., Boyar, R. M., Kream, J., & Hellman, L. (1972). Age-related change in the twenty-four-hour spontaneous secretion of growth hormone. *Journal of Clinical Endocrinology and Metabolism, 35,* 665–670.

Folkard, S., Arendt, J., & Clark, M. (1993). Can melatonin improve shift workers' tolerance of the night shift? Some preliminary findings. *Chronobiology International, 10,* 315–320.

Follenius, M., Brandenberger, G., Hietter, B., Simeoni, M., & Reinhardt, B. (1982). Diurnal cortisol peaks and their relationship to meals. *Journal of Clinical Endocrinology and Metabolism, 55,* 757–761.

Follenius, M., Brandenberger, G., Badasept, J., Libert, J., & Ehrhart, J. (1992). Nocturnal cortisol release in relation to sleep structure. *Sleep, 15,* 21–27.

Frank, S., Roland D. C., Sturis, J., Byrne, M. M. Spire, J. P., Refetoff, S., Polonsky, K. S., & Van Cauter, E. (1995). Effects of aging on glucose regulation during wakefulness and sleep. *American Journal of Physiology, 269,* E1006–E1016.

Galbo, H. (1995). Integrated endocrine responses and exercise. In L. J. DeGroot (Ed.), *Endocrinology* (Vol. 3, pp. 2692–2701). Philadelphia: Saunders.

Gary, K. A., Winokur, A., Douglas, S. D., Kapoor, S., Zaugg, L., & Dinges, D. F. (1996). Total sleep deprivation and the thyroid axis: Effects of sleep and waking activity. *Aviation, Space and Environmental Medicine, 67,* 513–519.

Goichot, B., Brandenberger, G., Sainio, J., Wittersheim, G., & Follenius, M. (1992). Nocturnal plasma thyrotropin variations are related to slow-wave sleep. *Journal of Sleep Research, 1,* 186–190.

Golstein, J., Van Cauter, E., Desir, D., Noel, P., Spire, J.-P., Refetoff, S., & Copinschi, G. (1983). Effects of "jet lag" on hormonal patterns. IV. Time shifts increase growth hormone release. *Journal of Clinical Endocrinology and Metabolism, 56,* 433–440.

Greenspan, S. L., Klibanski, A., Schoenfeld, D., & Ridgway, E. C. (1986). Pulsatile secretion of thyrotropin in man. *Journal of Clinical Endocrinology and Metabolism, 63,* 661–668.

Gronfier, C., Luthringer, R., Follenius, M., Schaltenbrand, N., Macher, J. P., Muzet, A., & Brandenberger, G. (1995). Temporal link between plasma thyrotropin levels and electroencephalographic activity in man. *Neuroscience Letters, 200,* 97–100.

Gronfier, C., Luthringer, R., Follenius, M., Schaltenbrand, N., Macher, J. P., Muzet, A., & Brandenberger, G. (1996). A quantitative evaluation of the relationships between growth hormone secretion and delta wave electroencephalographic activity during normal sleep and after enrichment in delta waves. *Sleep, 19,* 817–824.

Halberg, F., Tong, Y. L., & Johnson, E. A. (1967). Circadian system phase: An aspect of temporal morphology; procedures and illustrative examples. In *Cellular aspects of biorhythms* (pp. 20–48). Berlin: Springer-Verlag.

Hirschfeld, U., Moreno-Reyes, R., Akseki, E., L'Hermite-Balériaux, M., Leproult, R., Copinschi, G., & Van Cauter, E. (1996). Progressive elevation of plasma thyrotropin during adaptation to simulated jet lag: Effects of treatment with bright light or zolpidem. *Journal of Clinical Endocrinology and Metabolism, 81,* 3270–3277.

Ho, K. Y., Evans, W. S., Blizzard, R. M., Veldhuis, J. D., Merriam, G. R., Samojlik, E., Furlanetto, R., Rogol, A. D., Kaiser, D. L., & Thorner, M. O. (1987). Effects of sex and age on the 24-hour profile of growth hormone secretion in man: Importance of endogenous estradiol concentrations. *Journal of Clinical Endocrinology and Metabolism, 64,* 51–58.

Iranmanesh, A., Lizarralde, G., & Veldhuis, J. D., (1991). Age and relative adiposity are specific negative determinants of the frequency and amplitude of growth hormone (GH) secretory bursts and the half-life of endogenous GH in healthy men. *Journal of Clinical Endocrinology and Metabolism, 73,* 1081–1088.

Jaffe, C., Turgeon, D., DeMott Friberg, R., Watkins, P., & Barkan, A. (1995). Nocturnal augmentation of growth hormone (GH) secretion is preserved during repetitive bolus administration of GH-releasing hormone: Potential involvement of endogenous somatostatin—A clinical research center study. *Journal of Clinical Endocrinology and Metabolism, 80,* 3321–3326.

Karacan, I., Rosenbloom, A. L., Londono, J. H., Williams, R. L., & Salis, P. J. (1974). Growth hormone levels during morning and afternoon naps. *Behavioral Neuropsychiatry, 6,* 67–70.

Kerkhofs, M., Linkowski, P., & Mendlewicz, J. (1989). Effects of intravenous catheter on sleep in healthy men and in depressed patients. *Sleep, 12,* 113–119.

Laakso, M. L., Hätönen, T., Stenberg, D., Alila, A., & Smith, S. (1993). One-hour exposure to moderate illuminance (500 lux) shifts the human melatonin rhythm. *Journal of Pineal Research, 15*, 21–26.

Landin-Wilhelmsen, K., Wilhelmsen, L., Lappas, G., Rosen, T., Lindstedt, G., Lundberg, P., & Bengtsson, B. (1994). Serum insulin-like growth factor I in a random population sample of men and women: Relation to age, sex, smoking habits, coffee consumption and physical activity, blood pressure and concentrations of plasma lipids, fibrinogen, parathyroid hormone and osteocalcin. *Clinical Endocrinology, 41*, 351–357.

Lejeune-Lenain, C., Van Cauter, E., Desir, D., Beyloos, M., & Franckson, J. R. M. (1987). Control of circadian and episodic variations of adrenal androgens secretion in man. *Journal of Endocrinological Investigation, 10*, 267–276.

Leproult, R., Buxton, O. M., & Van Cauter, E. (1997). Nocturnal sleep deprivation results in an elevation of cortisol levels the next evening. *Sleep, 20*, 865–870.

Lewy, A. J., & Sack, R. L. (1989). The dim light melatonin onset as a marker for circadian phase position. *Chronobiology International, 6*, 93–102.

Lewy, A. J., Wehr, T. A., Goodwin, F. K., Newsome, D. A., & Markey, S. P. (1980). Light suppresses melatonin secretion in humans. *Science, 210*, 1267–1269.

Lewy, A. J., Ahmed, S., Jackson, J. M. L., & Sack, R. L. (1992). Melatonin shifts human circadian rhythms according to a phase-response curve. *Chronobiology International, 9*, 380–392.

Lewy, A. J., Sack, R. L., Blood, M. L. Bauer, V. K., Cutler, N. L., & Thomas, K. T. (1995). Melatonin marks circadian phase position and resets the endogenous circadian pacemaker in humans. In *Circadian clocks and their adjustment* (pp. 303– 321). Chichester, UK: Wiley.

Luger, A., Deuster, P. A., Kyle, S. B., Galucci, W. T., Montgomery, L. C., Gold, P. W., Loriaux, D. L. & Chrousos, G. P. (1987). Acute hypothalamic–pituitary–adrenal responses to the stress of treadmill exercise. *New England Journal of Medicine, 316*, 1309–1315.

Matthews, D. R. (1988). Time series analysis in endocrinology. *Acta Paediatrica Scandinavica, Supplement, 347*, 55–62.

McLeod, B. J., & Craigon, J. (1985). Time series analysis of plasma LH and FSH concentrations as a method of assessing episodic secretion. *Journal of Reproduction and Fertility, 74*, 575–587.

Merriam, G. R., & Wachter, K. W. (1982). Algorithms for the study of episodic hormone secretion. *American Journal of Physiology, 243*, E310–E318.

Mullington, J., Hermann, D., Holsboer, F., & Pollmacher, T. (1996). Age-dependent suppression of nocturnal growth hormone levels during sleep deprivation. *Neuroendocrinology, 64*, 233–241.

O'Meara, N. M., Sturis, J., Van Cauter, E., & Polonsky, K. S. (1993). Lack of control by glucose of ultradian insulin secretory oscillations in impaired glucose tolerance and in non-insulin-dependent diabetes mellitus. *Journal of Clinical Investigation, 92*, 262–271.

Oerter, K. E., Guardabasso, V., & Rodbard, D. (1986). Detection and characterization of peaks and estimation of instantaneous secretory rate for episodic pulsatile hormone secretion. *Computers and Biomedical Research, 19*, 170–191.

Okatani, Y., & Sagara, Y. (1993). Role of melatonin in nocturnal prolactin secretion in women with normoprolactinemia and mild hyperprolactinemia. *American Journal of Obstetrics and Gynecology, 168*, 854–861.

Othmer, E. Mendelson, W. B., Levine, W. R., Malarkey, W. B., & Daughaday, W. H. (1974). Sleep-related growth hormone secretion and morning naps. *Steroids and Lipids Research, 5*, 380–386.

Parker, D. C., Rossman, L. G., Pekary, A. E., & Hershman, J. M. (1987). Effect of 64-hour sleep deprivation on the circadian waveform of thyrotropin (TSH): Further evidence of sleep-related inhibition of TSH release. *Journal of Clinical Endocrinology and Metabolism, 64*, 157–161.

Plat, L., Féry, F., L'Hermite-Balériaux, M., Mockel, J., & Van Cauter, E. (1996). Metabolic effects of short-term physiological elevations of plasma cortisol are more pronounced in the evening than in the morning. Paper presented at the Annual Meeting of the American Diabetes Association, San Francisco, California.

Prinz, P. N. (1995). Sleep and sleep disorders in older adults. *Journal of Clinical Neurophysiology, 12*, 139–146.

Prinz, P. N., Vitiello, M. V., Raskind, M. A., & Thorpy, M. J. (1990). Geriatrics: Sleep disorders and aging. *New England Journal of Medicine, 323*, 520–526.

Prinz, P., Moe, K., Dulberg, E., Larsen, L., Vitiello, M., Toivola, B., & Merriam, G. (1995). Higher plasma IGF-1 levels are associated with increased delta sleep in healthy older men. *Journal of Gerontology, 50A*, M222–M226.

Quigley, M. E., Ropert, J. F., & Yen, S. S. C. (1981). Acute prolactin release triggered by feeding. *Journal of Clinical Endocrinology and Metabolism, 52*, 1043–1045.

Radomski, M. W., Buguet, A., Montmayeur, A., Bogui, P., Bourdon, L., Doua, F., Lonsdorfer, A., Tapie, P., & Dumas, M. (1995). Twenty-four-hour plasma cortisol and prolactin in human African trypanosomiasis patients and healthy African controls. *American Journal of Tropical Medicine and Hygiene, 52,* 281–286.

Rosenthal, N. E. (1991). Plasma melatonin as a measure of the human clock. *Journal of Clinical Endocrinology and Metabolism, 73,* 225–226.

Rossmanith, W. G., Boscher, S., Ulrich, U., & Benz, R. (1993). Chronobiology of prolactin secretion in women: Diurnal and sleep-related variations in the pituitary lactotroph sensitivity. *Neuroendocrinology, 58,* 263–271.

Royston, J. P. (1989). The statistical analysis of pulsatile hormone secretion data. *Clinical Endocrinology, 30,* 201–210.

Sassin, J. F., Parker, D. C., Mace, J. W., Gotlin, R. W., Johnson, L. C., & Rossman, L. G. (1969). Human growth hormone release: Relation to slow-wave sleep and sleep–waking cycles. *Science, 165,* 513–515.

Seifritz, E., Hemmeter, U., Trachsel, L., Lauer, C., Hatzinger, M., Emrich, H., Holsboer, F., & Holsboer-Trachsler, E. (1995). Effects of flumazenil on recovery sleep and hormonal secretion after sleep deprivation in male controls. *Psychopharmacology* (Berlin), *120,* 449–456.

Shanahan, T. L., & Czeisler, C. A. (1991). Light exposure induces equivalent phase shifts of the endogenous circadian rhythms of circulating plasma melatonin and core body temperature in men. *Journal of Clinical Endocrinology and Metabolism, 73,* 227–235.

Spath-Schwalbe, E., Gofferje, M., Kern, W., Born, J., & Fehm, H. L. (1991). Sleep disruption alters nocturnal ACTH and cortisol secretory patterns. *Biological Psychiatry, 29,* 575–584.

Spiegel, K., Follenius, M., Simon, C., Saini, J., Ehrhart, J., & Brandenberger, G. (1994). Prolactin secretion and sleep. *Sleep, 17,* 20–27.

Spiegel, K., Weibel, L., Gronfier, C., Brandenberger, G., & Follenius, M. (1996). Twenty-four hour prolactin profiles in night workers. *Chronobiology International, 13,* 283–293.

Sturis, J., Polonsky, K. S., Shapiro, E. T., Blackman, J. D., O'Meara, N. M., & Van Cauter, E. (1992). Abnormalities in the ultradian oscillations of insulin secretion and glucose levels in type 2 (non-insulin-dependent) diabetic patients. *Diabetologia, 35,* 681–689.

Takahashi, Y., Kipnis, D. M., & Daughaday, W. H. (1968). Growth hormone secretion during sleep. *Journal of Clinical Investigation, 47,* 2079–2090.

Turek, F. W., & Van Cauter, E. (1993). Rhythms in reproduction. In E. Knobil, J. J. Neill, *et al.*, (Eds.), *The physiology of reproduction,* (pp. 1789–1830). New York: Raven Press.

Tzischinsky, O., & Lavie, P. (1994). Melatonin possesses time-dependent hypnotic effects. *Sleep, 17,* 638–645.

Urban, R. J., Evans, W. S., Rogol, A. D., Kaiser, D. L., Johnson, M. L., & Veldhuis, J. D. (1988). Contemporary aspects of discrete peak-detection algorithms. I. The paradigm of the luteinizing hormone pulse signal in men. *Endocrine Reviews, 9,* 3–37.

Van Cauter, E. (1979). Method for characterization of 24-h temporal variation blood constituents. *American Journal of Physiology, 237,* E255–E264.

Van Cauter, E. (1987). Time series analysis in pulse detection techniques. In T. O. F. Wagner & M. Filicori (Eds.), *Episodic hormone secretion: From basic science to clinical application* (pp. 51–60). Hameln, Germany: TM-Verlag.

Van Cauter, E. (1988). Estimating false-positive and false-negative errors in analyses of hormonal pulsatility. *American Journal of Physiology, 254,* E786–E794.

Van Cauter, E. (1990). Computer-assisted analysis of endocrine rhythms. In D. Rodbard & G. Forti (Eds.), *Computers in endocrinology* (pp. 59–70). New York: Raven Press.

Van Cauter, E., & Aschoff, J. (1989). Endocrine and other biological rhythms. In L. J. DeGroot (Ed.), *Endocrinology* (Vol. 3, pp. 2658–2705). Philadelphia: Saunders.

Van Cauter, E., & Copinschi, G. (1999). Interactions between growth hormone secretion and sleep. In R. G. Smith & M. O. Thorner (Eds.), *Human growth hormone: Research and clinical practice* (pp. 261–284). Totowa, NJ: Humana Press.

Van Cauter, E., & Honinckx, E. (1985). Pulsatility of pituitary hormones. *Experimental Brain Research Supplement, 12,* 41–60.

Van Cauter, E., & Turek, F. W. (1995). Endocrine and other biological rhythms. In L. J. DeGroot (Ed.), *Endocrinology* (pp. 2487–2548). Philadelphia: Saunders.

Van Cauter, E., L'Hermite, M., Copinschi, G., Refetoff, S., Desir, D., & Robyn, C. (1981). Quantitative analysis of spontaneous variations of plasma prolactin in normal man. *American Journal of Physiology, 241,* E355–E363.

Van Cauter, E., van Coevorden, A., & Blackman, J. D. (1990). Modulation of neuroendocrine release by sleep and circadian rhythmicity. In S. Yen & W. Vale (Eds.), *Advances in neuroendocrine regulation of reproduction,* (pp. 113–122). Norwell, MA: Serono Symposia, USA.

Van Cauter, E., Blackman, J. D., Roland, D., Spire, J. P., Refetoff, S., & Polonsky, K. S. (1991). Modulation of glucose regulation and insulin secretion by circadian rhythmicity and sleep. *Journal of Clinical Investigation, 88,* 934–942.

Van Cauter, E., Kerkhofs, M., Caufriez, A., Van Onderbergen, A., Thorner, M. O., & Copinschi, G. (1992a). A quantitative estimation of GH secretion in normal man: Reproducibility and relation to sleep and time of day. *Journal of Clinical Endocrinology and Metabolism, 74,* 1441.

Van Cauter, E., Shapiro, E. T., Tillil, H., & Polonsky, K. S. (1992b). Circadian modulation of glucose and insulin responses to meals: Relationship to cortisol rhythm. *American Journal of Physiology, 262,* E467–E475.

Van Cauter, E., Sturis, J., Byrne, M. M., Blackman, J. D., Leproult, R., Ofek, G., L'Hermite-Balériaux, M., Refetoff, S., Turek, F. W., & Van Reeth, O. (1994). Demonstration of rapid light-induced advances and delays of the human circadian clock using hormonal phase markers. *American Journal of Physiology, 266,* E953–E963.

Van Cauter, E., Leproult, R., & Kupfer, D. J. (1996). Effects of gender and age on the levels and circadian rhythmicity of plasma cortisol. *Journal of Clinical Endocrinology and Metabolism, 81,* 2468–2473.

Van Cauter, E., Plat, L., Scharf, M., Leproult, R., Cespedes, S., L'Hermite-Balériaux, M., & Copinschi, G. (1997a). Simultaneous stimulation of slow-wave sleep and growth hormone secretion by gamma-hydroxybutyrate in normal young men. *Journal of Clinical Investigation, 100,* 745–753.

Van Cauter, E., Polonsky, K. S., & Scheen, A. J. (1997b). Roles of circadian rhythmicity and sleep in human glucose regulation. *Endocrine Reviews, 18,* 716–738.

van Coevorden, A., Mockel, J., Laurent, E., Kerkhofs, M., L'Hermite-Balériaux, M., Decoster, C., Nève, P., & Van Cauter, E. (1991). Neuroendocrine rhythms and sleep in aging. *American Journal of Physiology, 260,* E651– E661.

Van Reeth, O., Sturis, J., Byrne, M. M., Blackman, J. D., L'Hermite-Balériaux, M., Leproult, R., Oliner, C., Refetoff, S., Turek, F. W., & Van Cauter, E. (1994). Nocturnal exercise phase delays circadian rhythms of melatonin and thyrotropin in normal men. *American Journal of Physiology, 266,* E964–E974.

Veldhuis, J. D., & Johnson, M. L. (1986). Cluster analysis: A simple, versatile, and robust algorithm for endocrine pulse detection. *American Journal of Physiology, 250,* E486–E493.

Veldhuis, J. D., & Johnson, M. L. (1988). Operating characteristics of the hypothalamo–pituitary–gonadal axis in men: Circadian, ultradian, and pulsatile release of prolactin and its temporal coupling with luteinizing hormone. *Journal of Clinical Endocrinology and Metabolism, 67,* 116–123.

Veldhuis, J., Iranmanesh, A., Lizarralde, G., & Johnson, M. (1989a). Amplitude modulation of a burstlike mode of cortisol secretion subserves the circadian glucocorticoid rhythm. *American Journal of Physiology, 257,* E6– E14.

Veldhuis, J. D., Iranmanesh, A., Johnson, M. L., & Lizarralde, G. (1989b). Amplitude, but not frequency, modulation of adrenocorticotropin secretory bursts gives rise to the nyctohemeral rhythm of the corticotropic axis in man. *Journal of Clinical Endocrinology and Metabolism, 71,* 452–463.

Veldhuis, J. D., Iranmanesh, A., Johnson, M. L., & Lizarralde, G. (1990). Twenty-four-hour rhythms in plasma concentrations of adenohypophyseal hormones are generated by distinct amplitude and/or frequency modulation of underlying pituitary secretory bursts. *Journal of Clinical Endocrinology and Metabolism, 71,* 1616–1623.

Veldhuis, J. D., Johnson, M. L., Lizarralde, G., & Iranmanesh, A. (1992). Rhythmic and nonrhythmic modes of anterior pituitary gland secretion. *Chronobiology International, 9,* 371–379.

Vermeulen, A. (1987). Nyctohemeral growth hormone profiles in young and aged men: Correlation with somatomedin-C levels. *Journal of Clinical Endocrinology and Metabolism, 64,* 884–888.

Waldstreicher, J., Duffy, J. F., Brown, E. N., Rogacz, S., Allan, J. S., & Czeisler, C. A. (1996). Gender differences in the temporal organization of prolactin (PRL) secretion: Evidence for a sleep-independent circadian rhythm of circulating PRL levels—A Clinical Research Center study. *Journal of Clinical Endocrinology and Metabolism, 81,* 1483–1487.

Weaver, D. R., Rivkees, S. A., Carlson, L. L., & Reppert, S. M. (1991). Localization of melatonin receptors in mammalian brain. In D. C. Klein, R. Y. Moore, & S. M. Reppert (Eds.), *Suprachiasmatic nucleus: The mind's clock* (pp. 289–308). New York: Oxford University Press.

Wehr, T., Moul, D., Barbato, G., Giesen, H., Seidel, J., Barker, C., & Bender, C. (1993). Conservation of photoperiod-responsive mechanisms in humans. *American Journal of Physiology, 265,* R846–R857.

Weibel, L., Brandenberger, G., Goichot, B., Spiegel, K., Ehrhart, J., & Follenius, M. (1995a). The circadian thyrotropin rhythm is delayed in regular night workers. *Neuroscience Letters, 187,* 83–86.

Weibel, L., Follenius, M., Spiegel, K., Ehrhart, J., & Brandenberger, G. (1995b). Comparative effect of night and daytime sleep on the 24-hour cortisol secretory profile. *Sleep, 18,* 549–556.

Weibel, L., Follenius, M., Spiegel, K., Gronfier, C., & Brandenberger, G. (1997a). Growth hormone secretion in night workers. *Chronobiology International, 14,* 49–60.

Weibel, L., Spiegel, K., Gronfier, C., Follenius, M., & Brandenberger, G. (1997b). Twenty-four-hour melatonin and core body temperature rhythms: Their adaptation in night workers. *American Journal of Physiology, 272,* R948–R954.

Weitzman, E. D., Zimmerman, J. C., Czeisler, C. A., & Ronda, J. M. (1983). Cortisol secretion is inhibited during sleep in normal man. *Journal of Clinical Endocrinology and Metabolism, 56,* 352–358.

26

Seasonal Photoperiodic Responses of the Human Circadian System

THOMAS A. WEHR

The two-oscillator structure we propose for the pacemaker offers a basis for accommodation to seasonal change in two quite different respects. On the one hand it meets the challenge of seasonal change in the pattern of the external day...Its two components, responding differentially to sunrise and sunset, would make the pacemaker an adequate clock for every differently patterned day in the year. On the other hand that same complexity is a potential sensor for the time of year. It is a clock for all seasons.

Collin S. Pittendrigh and Serge Daan, 1976c

In most considerations of melatonin rhythms in nonhuman mammals, it is usual to consider the duration of the melatonin peak or the phasing of the rhythm as being important...while less attention is given to the magnitude of the nocturnal melatonin increase. Conversely, in the human the primary concern of many investigators...is the height of the melatonin peak with less attention being paid to the duration of the peak or the phasing of the rhythm. Why the emphasis varies according to species is not clear.

Russel J. Reiter, 1987

THE "DAY" AND "NIGHT" WITHIN

Among the many external changes to which organisms must adapt is the special class of changes that predictably recur in association with daily and seasonal cycles in the environment. In the course of their evolution, most forms of life have taken advantage of the predictability of these changes by developing temporal programs that trigger adaptive responses to them in anticipation of their periodic recurrence.

THOMAS A. WEHR Clinical Psychobiology Branch, Intramural Research Program, National Institute of Mental Health, Bethesda, Maryland 20892.

Circadian Clocks, Volume 12 of *Handbook of Behavioral Neurobiology*, edited by Joseph S. Takahashi, Fred W. Turek, and R. Y. Moore, Kluwer Academic / Plenum Publishers, New York, 2001.

With regard to daily cycles in the environment, the world of sunlight and the world of darkness present challenges and opportunities that are sufficiently different from one another that most organisms have specialized in the active engagement of the environment in only one of these two worlds. During the portion of the day to which they are less well adapted, they withdraw into a secure refuge, presumably to maximize efficiency of energy expenditure and to minimize exposure to threats to survival. Thus, as day alternates with night, most organisms alternate between two physiologic and behavioral states, one that facilitates active engagement of their field environment and one that facilitates quiescence and immobility in their home environment (Rusak, 1989). Since these states and the processes that are responsible for their cyclic recurrence are endogenous and self-sustaining, they can be conceptualized as a "biological day" and a "biological night" within the organism that mirror and anticipate the day and night outside.

This evolutionary and ecological perspective may help to explain why circadian rhythms frequently exhibit waveforms that are characterized by distinct diurnal and nocturnal periods with relatively discrete transitions between them (corresponding

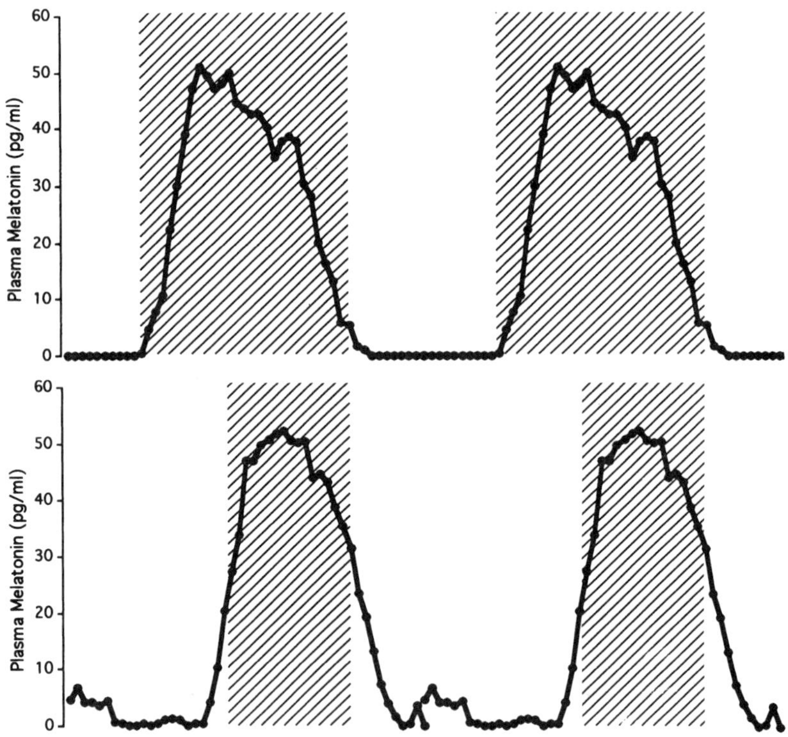

Figure 1. The duration of nocturnal melatonin secretion measured in continuous darkness in a "constant routine" protocol is longer after exposure to long nights than after exposure to short nights ($N = 15$ individuals). Hatched areas indicate timing of dark periods to which individuals had been exposed during the nights preceding the 24-hour period of continuous darkness in the constant routine. Dark periods were 14 hours (6 p.m.–8 a.m.) and 8 hours (midnight–8 a.m.) long, respectively. Twenty-four-hour profiles are shown in duplicate to facilitate visual inspection of the waveform. (Reprinted from Wehr, 1996.)

to a biological day, a biological dusk, a biological night, and a biological dawn). In humans, for example, diurnal periods of waking activity, high body temperature, absence of melatonin secretion, low prolactin secretion, and falling levels of cortisol alternate with nocturnal periods of sleep, low body temperature, active melatonin secretion, high prolactin secretion, and rising levels of cortisol (Wehr, 1996) (see Figures 1–3). The transition from diurnal to nocturnal periods, when sleep begins, body temperature decreases, melatonin secretion switches on, prolactin secretion doubles, and cortisol levels begin to rise, is relatively abrupt and can be conceptualized as a biological dusk, while the transition from nocturnal to diurnal periods, when wakefulness begins, body temperature rises, melatonin secretion switches off, prolactin secretion halves, and cortisol levels begin to fall, can be conceptualized as a biological dawn.

The temporal programs that give rise to the biological day and night originate in chemical processes that take place within pacemaker cells of the suprachiasmatic nucleus (SCN) of the hypothalamus, and (to judge from emerging findings in molecular neurobiology) probably arise from negative feedback cycles of gene

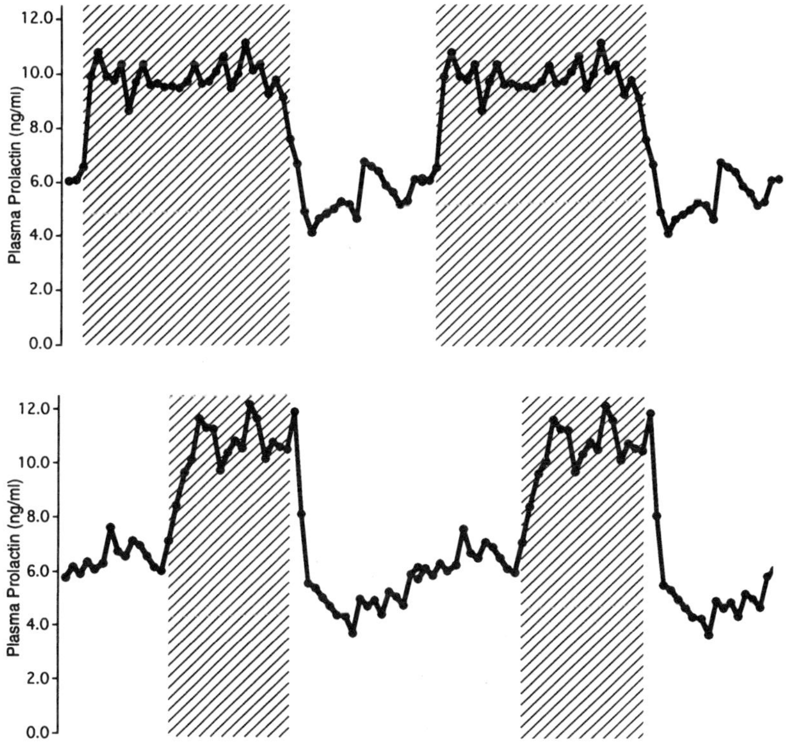

Figure 2. The duration of the nocturnal period of increased prolactin secretion is longer during long nights than during short nights ($N = 12$ individuals). Hatched areas indicate timing of dark periods to which individuals were exposed during the 24-hour period when measurements were made. Dark periods were 14 hours (6 p.m.–8 a.m.) and 8 hours (midnight–8 a.m.) long, respectively. Twenty-four-hour profiles are shown in duplicate to facilitate visual inspection of the waveform. (Reprinted from Wehr 1996.)

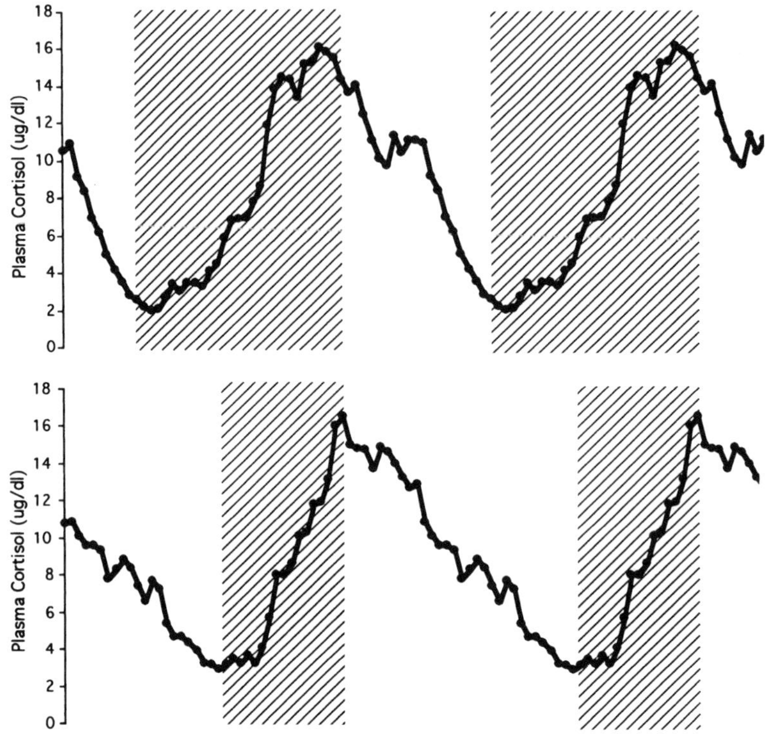

Figure 3. The duration of the nocturnal period of rising plasma cortisol levels measured in a "constant routine" protocol is longer after exposure to long nights than after exposure to short nights ($N = 12$ individuals). Hatched areas indicate timing of dark periods to which individuals had been exposed during the nights preceding the 24-hour period of continuous darkness in the constant routine. Dark periods were 14 hours (6 p.m.–8 a.m.) and 8 hours (midnight–8 a.m.) long, respectively. Twenty-four-hour profiles are shown in duplicate to facilitate visual inspection of the waveform. (Reprinted from Wehr 1996.)

expression. The timing of these processes responds to light stimuli that reach the SCN via the retinohypothalamic tract and serve as time cues, so that the timing of the biological day and night becomes appropriately synchronized with the timing of the external day–night cycle. A period that corresponds to the biological night can be detected at the molecular level within the cells of the SCN: during the biological night, but not during the biological day, expression of the immediate-early gene c-*fos* can be elicited in these cells by brief exposures to light (Rusak, Robertson, Wisden, & Hunt, 1990).

The "Night Within" Adjusts to Seasonal Changes in the Length of the Night Outside

Because the earth's axis of rotation is tilted, the duration of night is not constant, but varies during the course of the year. In many animals, the temporal programs in the SCN respond to information about changes in night length (conveyed by the retinohypothalamic tract) by making proportional adjustments in the duration of the biological night so that it becomes longer in winter and shorter in summer. Thus, for example, the nocturnal period during which melatonin is secreted, or when nocturnal animals are active, becomes longer after animals are exposed to long nights and shorter after they are exposed to short nights (Illnerova & Vanecek, 1982; Pittendrigh & Daan, 1976c). It was the capacity of the circadian pacemaker to make these adjustments that led Pittendrigh and Daan (1976c), in their classic paper on a pacemaker physiology, to call it a "clock for all seasons."

The capacity of the SCN to adjust the duration of the biological night in response to changes in duration of the solar night can be detected at the molecular level within its cells. The nocturnal period corresponding to the biological night, during which c-*fos* expression can be elicited in these cells by brief light pulses, becomes longer after exposure to long nights and shorter after exposure to short nights (Sumova, Travnickova, Peters, Schwartz, & Illnerova, 1995; Vuillez, Jacob, Teclemariam-Mesbah, & Pévet, 1996).

Experiments show that the human circadian pacemaker, in the course of its evolution, has retained a capacity to detect changes in the length of the night and to make corresponding adjustments in the duration of the biological night, as occurs in other animals (Wehr, 1991, 1996; Wehr *et al.*, 1993). Thus, when humans are transferred from short (8-hour) "nights" to long (14-hour) "nights," the duration of nocturnal periods of sleep, low body temperature, active melatonin secretion, high prolactin secretion, and rising levels of cortisol become longer (Figures 1–4).

Some of these responses to changes in duration of the scotoperiod (e.g., those in melatonin, cortisol, and body temperature) persist for at least one circadian cycle in constant conditions in which subjects remain continuously awake and at rest in constant dim (< 1 lux) light (Wehr *et al.*, 1993). This finding demonstrates that the duration of the biological night in humans, as in other animals, is governed by a process that (1) is endogenous and self-sustaining, (2) is capable of adjusting the duration of the biological night in response to changes in duration of the scotoperiod, and (3) possesses the property of inertia, or memory, so that its adjustments to the scotoperiod to which it was most recently exposed persist for at least one cycle in constant conditions.

Cagnacci, Soldani, and Yen (1997) published further evidence that processes that control the duration of the human biological night are capable of modifying its duration in response to certain stimuli. When they simultaneously administered light and melatonin to experimental subjects in the morning, the intrinsic time of offset of melatonin secretion was unaffected, but the intrinsic time of onset was advanced (Figure 5). Therefore, the intrinsic duration of secretion was increased by this intervention. Morning light alone advanced the intrinsic timing of both onset and offset of secretion. The addition of exogenous melatonin to the morning light treatments enhanced the light-induced phase advance of onset of secretion and prevented the light-induced phase advance of offset of secretion. This result shows that melatonin secretion (at least in the morning) feeds back on the processes that regulate its intrinsic duration, and that it exerts opposite effects on the intrinsic timing of its onset and offset. Presumably, melatonin acts directly on the SCN to achieve these effects, because the SCN controls the intrinsic timing of onset and offset of secretion and it is richly supplied with melatonin receptors (Morgan, Barrett, Howell, & Kelliwell, 1994).

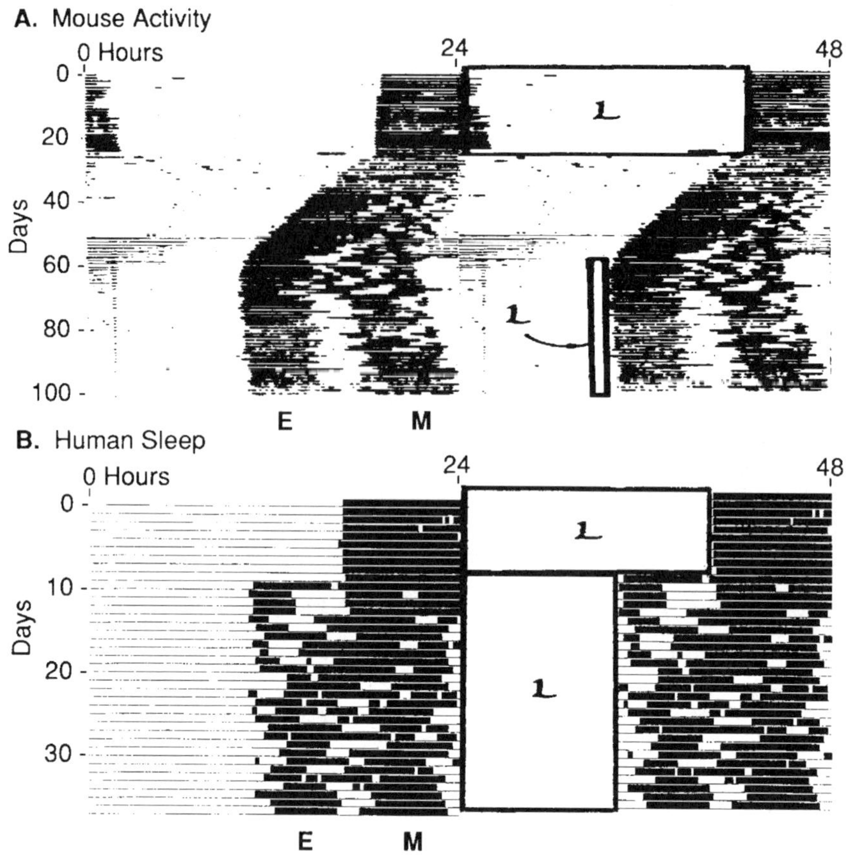

Figure 4. When schedules of daily light exposures (L) were shifted from long days to short days, the nocturnal sleep phase of a human subject (panel B) expanded and separated into an evening component (E) and a morning component (M). Shortening the photoperiod caused similar changes to occur in the activity phase of nocturnal rodents, as the record of mouse wheel-running activity (panel A) illustrates. (Mouse [Peromyscus leukopus] data were adapted from Pittendrigh and Daan, 1976a; human data are from Wehr, 1991, from which the figure is reprinted.)

Responses to the Natural Scotoperiod in an Urban Environment

While it is clear that humans have conserved a capacity to detect and respond to seasonal changes in duration of the natural scotoperiod, it is also clear that they have so altered their environment with technical innovations that produce artificial light at night and artificial darkness in the daytime that they are seldom exposed to a purely natural scotoperiod. Therefore, it is necessary to ask whether, and to what extent, the human SCN detects and responds to seasonal changes in duration of the natural scotoperiod in the modern urban environment.

To address this question, we measured the duration of the biological night, as indexed by the duration of nocturnal melatonin secretion during a 24-hour period of constant dim (< 1 lux) light, in winter and summer in individuals who were going about their usual activities in their usual environments in the Washington, D.C.,

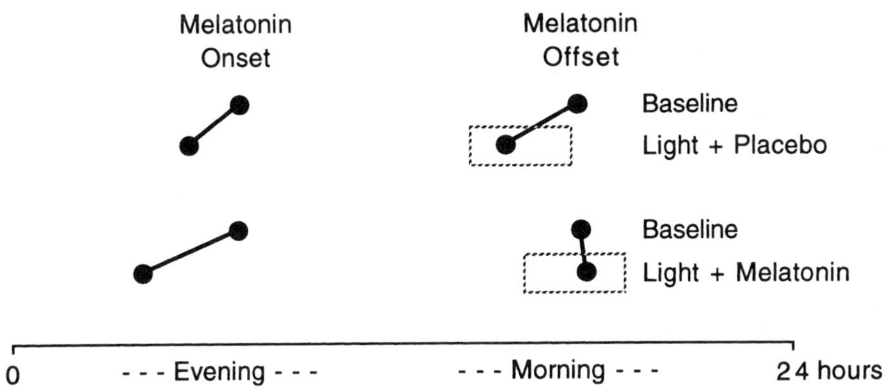

Figure 5. When bright light and melatonin are administered together in the morning, the intrinsic duration of nocturnal human melatonin secretion rapidly increases. Morning exposure to bright light alone advances the intrinsic time of both onset and offset of nocturnal melatonin secretion on the following night (top). Simultaneous administration of melatonin during morning exposure to bright light enhances the advance of onset and prevents the advance of offset on the following night, causing the duration of secretion to increase (bottom). (Figure is based on data of Cagnacci *et al.*, 1997.) To facilitate comparison of the effects of the interventions, the timing of events in the two experimental groups has been adjusted so that onset and offset of secretion at baseline prior to the interventions (upper portion of each dataset) occur at the same points along the time axis.

metropolitan area (39° N) (Wehr, Giesen, Moul, Turner, & Schwartz, 1995a). Based on results of our investigations of humans' responses to changes in duration of artificial scotoperiods (Wehr *et al.*, 1993), we assumed that the duration of melatonin secretion in constant dim light (hereafter referred to as its "intrinsic duration") would be an indirect measure of the duration of the biological night that is programmed by cells in the SCN and would be governed by, and would reflect, these cells' perceptions of, and responses to, the effective duration of the scotoperiod to which an individual was most recently exposed in their microenvironment.

We found that, on average, healthy individuals fail to respond to seasonal changes in duration of the natural scotoperiod. In these individuals, the mean intrinsic duration of nocturnal melatonin secretion does not change across these seasons (Table 1). Since healthy men have been shown to respond to experimental

TABLE 1. ONSET, OFFSET, AND DURATION OF NOCTURNAL MELATONIN SECRETION IN CONSTANT DIM LIGHT IN WINTER AND SUMMER[a]

	Onset time	Offset time	Duration (hours)	Winter–summer difference (hours)
Healthy men ($N = 22$)				
Summer	20:26 ± 1.03	5:45 ± 1:35	9:19 ± 1:11	
Winter	20:40 ± 1:07	5:56 ± 1:27	9:16 ± 1:16	−0:03 ± 1:17[b]
Healthy women ($N = 33$)				
Summer	20:49 ± 1:05	5:23 ± 1:16	8:33 ± 1:11	
Winter	20:35 ± 1:14	5:23 ± 1:54	8:49 ± 1:46	0:15 ± 1:29[b]

[a]Data from Wehr *et al.*, 2001
[b]Not significantly different from zero.

manipulations of artificial scotoperiods (Wehr *et al.*, 1993), it would seem that their failure to respond to seasonal changes in the natural scotoperiod must be attributed to external factors, such as exposure to artificial light and artificial darkness, that mask the natural scotoperiod in their urban microenvironments.

A gender difference emerged from these studies. In individuals in whom the intrinsic duration of nocturnal melatonin increased after exposure to long nights, the maximum level of melatonin secretion increased in women, but not in men (Figures 6–8).

Interestingly, seasonal and experimentally induced changes in the intrinsic duration of nocturnal melatonin secretion in humans are almost entirely a function of changes in the intrinsic time of morning offset of secretion and hardly at all a function of changes in the intrinsic time of evening onset of secretion (Figure 9). Hence, responsiveness of processes that control biological dawn to the phase-advancing effects of morning light seems to account for interindividual variation in responsiveness to changes in duration of the scotoperiod.

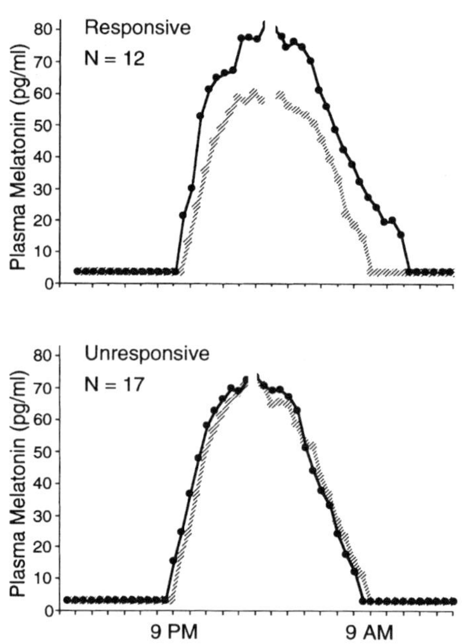

Figure 6. Winter (solid black line) and summer (hatched gray line) profiles of the intrinsic pattern of melatonin secretion in constant dim (<1 lux) light in women living in a modern urban environment. To preserve the waveform of the melatonin rhythm, average profiles were calculated by referencing the individuals' data to time of onset of secretion (left half of each profile) and disappearance of melatonin (right half of each profile). The times of onset of secretion and disappearance of melatonin in the profiles correspond to the average times of onset and offset in each group in each season. The upper portion of the figure shows profiles of melatonin secretion in women who responded to seasonal changes in duration of the natural scotoperiod by lengthening the intrinsic duration of melatonin secretion by $\geq$2.0 hours; the lower portion shows profiles in women who did not respond (winter–summer difference = 0.0 ± 0.5 hours). In women in whom the duration of secretion increased in winter, the maximum level of secretion also increased (top). As can be seen, seasonal changes in the intrinsic duration of melatonin secretion were largely due to changes in the intrinsic timing of disappearance of melatonin. (Data in the figure combine data from healthy women and women with seasonal affectiv)e disorder; Wehr 1997.)

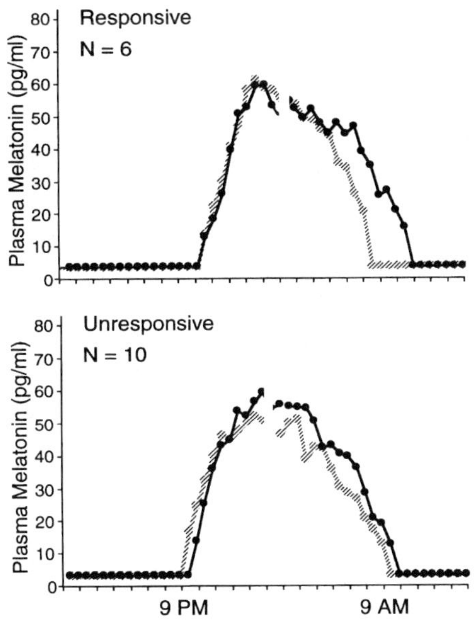

Figure 7. Winter (solid black line) and summer (hatched gray line) profiles of the intrinsic pattern of melatonin secretion in constant dim (<1 lux) light in men living in a modern urban environment. Methods for calculating average profiles and other indications are the same as in Figure 6. (Data in the figure combine data from healthy men and men with seasonal affective disorder; Wehr 1997.)

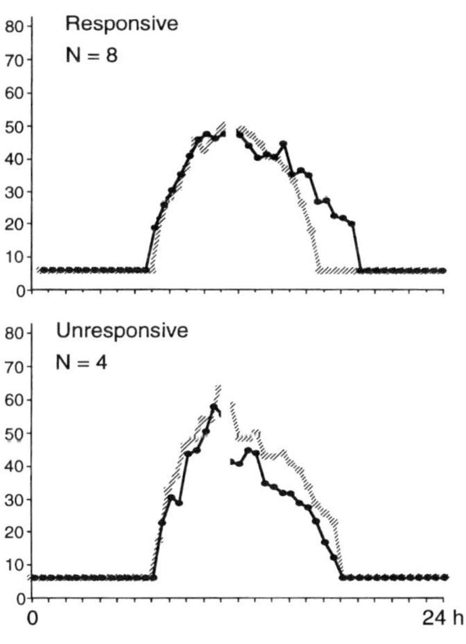

Figure 8. Long- "night" (solid black line) and short- "night" (hatched gray line) profiles of the intrinsic pattern of melatonin secretion in constant dim (<1 lux) light in men who participated in an experiment in which they were exposed to artificial scotoperiods of different duration for several weeks (see Figure 1) Methods for calculating average profiles and other indications are the same as in Figure 6. (Data are from Wehr *et al.*, 1993.)

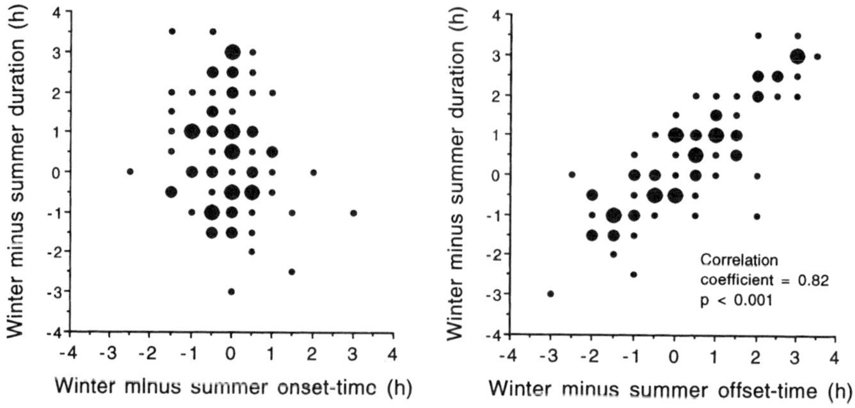

Figure 9. Relationship of seasonal changes in the intrinsic duration of nocturnal melatonin secretion to seasonal changes in the intrinsic timing of evening onset of secretion (left) and morning offset of secretion (right). Changes in duration are largely related to changes in timing of offset of secretion (see also Figures 6–8). (Data are from Wehr 1997a.)

A similar and possibly related finding has been reported in other animals. When animals are free-running in constant darkness, brief exposure to a light pulse transiently compresses the duration of the biological night, but only if it is applied near the expected time of dawn (Elliott & Tamarkin, 1994; Illnerova & Vanecek, 1987). Light pulses administered at that time induce phase advances in the biological dawn (as indexed by offset of locomotor activity or melatonin secretion) that are four times greater than phase advances in the timing of biological dusk (as indexed by onset of locomotor activity or melatonin secretion) and thus compress the duration of the biological night (Figure 10).

The critical role of individuals' responses to morning light in determining their responses to seasonal changes in the natural scotoperiod may be accentuated in modern environments in which artificial light is used to extend daytime activities into the evening hours. This practice causes the melatonin rhythm to assume a later phase position than it would assume in a purely natural scotoperiod (Wehr, 1997) (Figure 11). Consequently, the circadian phase during which onset of secretion would be affected by the last rays of the sun in the natural world is shifted into the dim and relatively uniform artificial light to which humans are exposed after dark in the modern world. In contrast, the circadian phase during which offset of secretion would be affected by the first rays of the sun at dawn in the natural world is shifted into the full light of day in the modern world. Perhaps those individuals who fail to exhibit responses to seasonal changes in the natural scotoperiod in an urban environment are those in whom this situation leads to a perpetual phase advance of offset of secretion and therefore a perpetual compression of the biological night.

The nature of the human response to change in scotoperiod duration is most clearly illustrated when melatonin profiles in individuals who unequivocally respond are compared with those in individuals who unequivocally do not respond (Figures 6–8). Based on an analysis of the frequency distribution of winter–summer differences in the intrinsic duration of nocturnal melatonin secretion in nearly 100 individuals, we defined an unequivocal response as a winter–summer difference of + 2.0 hours or more and an unequivocal nonresponse as a winter–summer differ-

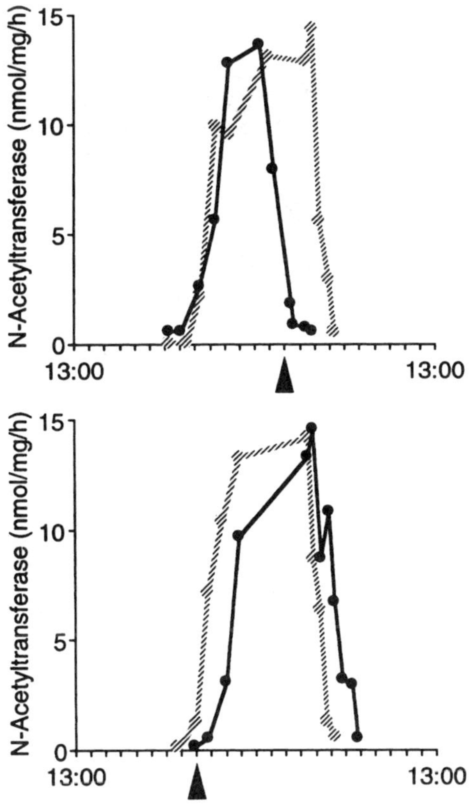

Figure 10. In rats in continuous darkness, a 1-minute light stimulus administered late in the biological night (arrow in upper panel) selectively phase advances the offset of the nocturnal period of increased pineal N-acetyltransferase (NAT) activity during the next circadian cycle (solid line in upper panel), thereby decreasing the duration of this period and of nocturnal melatonin secretion. In contrast, a light stimulus administered early in the biological night (arrow in lower panel) phase delays both the onset and the offset of the nocturnal period of increased pineal N-acetyltransferase activity during the next circadian cycle (solid line in lower panel), thereby leaving the duration of this period and of nocturnal melatonin secretion unchanged. Data from control animals that were not exposed to the light stimuli are also shown in each panel (hatched lines). Times of light administration and times of NAT activity measurements during the next circadian cycle are shown relative to clock times of these events. (Adapted from Illnerova and Vanecek, 1987.)

ence of 0.0 ± 0.5 hours (Wehr, 1997). This analysis suggests that nonseasonal variation due to week-to-week changes and to measurement error is a considerable source of noise in this type of data.

Using these criteria, we found that approximately one in three healthy women living in modern urban environments unequivocally respond to seasonal changes in the length of the solar night (Figure 6). In contrast, only one in eight healthy men unequivocally respond (Figure 7). Since one in two men unequivocally respond to experimental manipulations of artificial scotoperiods (Figure 8), external factors, such as exposures to artificial light and darkness, must be largely responsible for the failure of urban men to respond to seasonal changes in the natural scotoperiod.

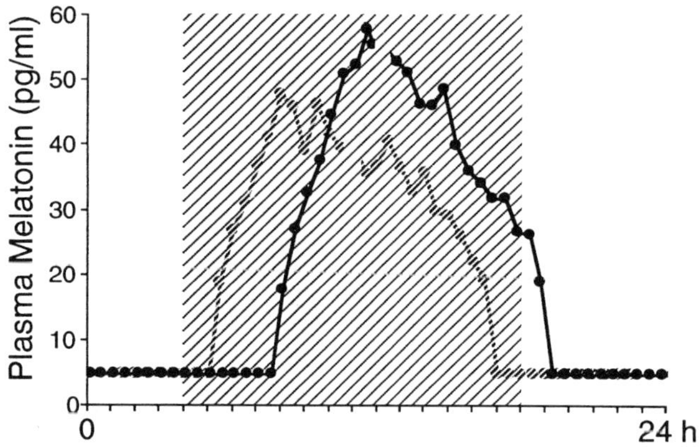

Figure 11. Effect of modern urban lighting conditions on the intrinsic pattern of human melatonin secretion in winter. The graph shows mean levels of plasma melatonin in constant dim light in six individuals who were studied twice, once after they were exposed to defined 14-hour dark periods (hatched area) in experimental conditions (hatched line) and once after they were exposed to natural 14-hour winter nights (same hatched area) in their usual living conditions (solid line). The individuals' use of artificial light in the evening, after dark, appears to have shifted the melatonin rhythm several hours later than would be expected relative to the solar night. Methods for calculating average profiles and other indications are the same as in Figure 6. (Data are from Wehr *et al.* 1993, 1995a.)

Models and Mechanisms: A Complex Circadian Pacemaker

Extensive research on scotoperiod responses of the circadian system in animals suggests that it is very likely that the temporal programs that generate the human biological night and govern its duration derive from chemical processes that take place within cells of the SCN, and that the impact of scotoperiod on these processes is mediated by the retinohypothalamic tract. If this is so, how does the SCN program the duration of the biological night, and how does it modify this program in response to changes in the length of the night outside?

A considerable body of research in *Drosophila* and *Neurospora* and recent research in mice suggests that this question ultimately will be answered at the molecular level, and will involve cycles of gene expression and the modification of these cycles by light-triggered molecular events. As already mentioned, a manifestation of the temporal program for the biological night, and of its proportional adjustments to change in scotoperiod, can be detected within the SCN at the molecular level in the expression of c-*fos* that is triggered by brief exposure to light. This expression can be triggered only during the biological night, and the duration of the period during which it can be triggered is governed by the duration of the scotoperiod to which the animal was most recently exposed. Although it is unclear whether c-*fos* expression plays a role in the scotoperiod responses of the SCN, it is clearly connected with a scotoperiod-responsive process in that structure.

The temporal program for the biological night and for its proportional adjustments to change in scotoperiod can also be detected in the electrical activity of SCN cells *in vitro*. Mrugala, Ziomanczuk, Jagota, and Schwartz (1998) found that spontaneous neuronal firing rates from single-unit extracellular recordings of SCN cells in hypothalamic slices exhibit robust peaks during the daytime, and that the duration of these peaks was longer when animals previously had been exposed to long days than when they had been exposed to short days.

In the absence of concrete knowledge of the molecular substrates of pacemaker function, Pittendrigh and Daan (1976c) developed a model of the vertebrate circadian system that was based on formal analyses of the behavior of the pacemaker as revealed in long-term recordings of rodent motor activity in a variety of experimental conditions. In this classic model, scotoperiod responses of the circadian pacemaker are affected by changes in the mutual phase relationships between two separate circadian oscillators, one of which (the evening oscillator, "E") is entrained to dusk and controls the time of onset of the biological night, the other of which (the morning oscillator, "M") is entrained to dawn and controls the time of offset of the biological night. In Illnerova and Vanecek's (1982) extension of this model, E controls the intrinsic time of onset of nocturnal melatonin secretion and M controls the intrinsic time of offset of nocturnal melatonin secretion. According to the model, as the interval between dusk and dawn increases, the phase angle between the oscillations of E and M increases. Consequently, the interval between the onset and offset of nocturnal melatonin secretion increases.

The Pittendrigh–Daan model was partly inspired by the observation that the behavior of many animals exhibits distinct evening and morning peaks that might represent separate expressions of two oscillators (Aschoff, 1996; Tobler, 1989). This bimodal pattern of circadian oscillations has been detected at the cellular level within the SCN *in vitro.* In hypothalamic slices taken from animals that previously had been exposed to long days, Mrugala *et al.* (1998) found that firing rates in single-unit records of SCN cells often exhibited dual morning and evening peaks. In slices taken from animals that previously had been exposed to short days, they found that firing rates exhibited a single daytime peak. Consistent with these observations, we found that most humans exhibit dual evening and morning peaks in the distribution of their sleep when they sleep in long artificial scotoperiods (Wehr, 1992; Wehr *et al.*, 1993) (Figures 4 and 12) and that some exhibit dual evening and morning peaks in nocturnal melatonin secretion (Wehr *et al.*, 1995b), as also occurs in sheep (Arendt, 1979) and trout (Zachmann, Knijff, Ali, & Anctil, 1992) (Figure 13). These findings suggest that the Pittendrigh–Daan two-oscillator model could be applied to the human circadian system, too.

The strongest evidence supporting the dual-oscillator model in rodents is "splitting" of the activity–rest cycle that sometimes occurs when it freeruns in constant light, a phenomenon in which the nocturnal activity period separates into separate evening (e) and morning (m) bouts which oscillate independently of one another and then recouple in a metastable phase relationship that is shifted 180° from the usual phase relationship (Pittendrigh & Daan, 1976c) (Figure 14). When the dual-oscillator system shifts from its usual coupling mode to its metastable one, the intrinsic period of its free-running oscillations becomes shorter. We observed a phenomenon that resembles splitting in all of these respects in humans whose circadian systems freeran while they were isolated from external time cues (Wehr, 1992) (Figure 14).

If a dual-oscillator model provides a valid description of the human circadian system it will surely prove to be an oversimplification. Subsequent to the development of this model, Pittendrigh (1988) called attention to "increasing evidence that the circadian system of an animal comprises a multiplicity of oscillators—even several that are independently coupled to the light cycle." He went on to suggest that

> a complex circadian program could be constructed on the basis of relatively few (perhaps even one) light-coupled pacemakers that themselves function as drivers—or zeitgebers, if you will—for a populations of slave oscillations in peripheral organs. Such a pacemaker–

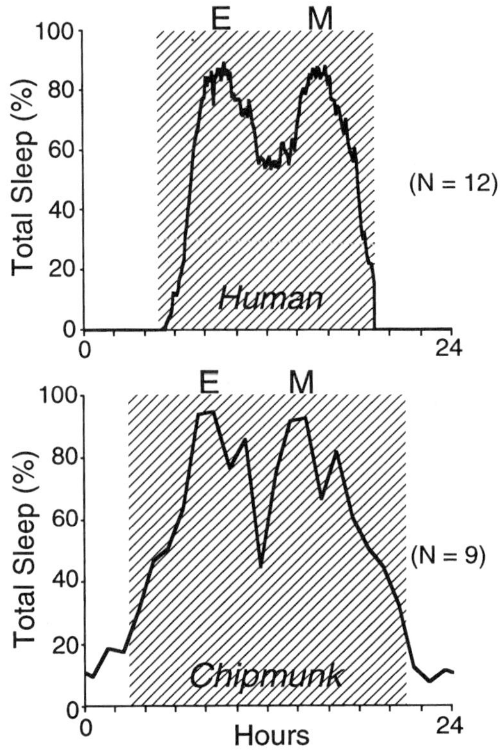

Figure 12. In long nights, Siberian chipmunks (bottom) and humans (top) exhibit homologous, symmetrically bimodal profiles of electroencephalographically monitored sleep. (Human data are from Wehr *et al.* 1993; chipmunk [Eutamias sibiricus] data are adapted from Dijk and Daan, 1989.)

> slave architecture to the program has the attractive feature of permitting selection to adjust the daily timing of one function (controlled by one slave oscillation) without interfering with others...The importance of these conceptualizations of a circadian program—of the 'day within'—lies in their multioscillator foundation and the associated assumption that the constituent oscillators (either as pacemakers or slaves) differ in their intrinsic periods (frequencies) and coupling, directly or indirectly, to the light cycle. It follows that the temporal fine structure of any such daily program will change seasonally as photoperiod (and hence the coupling strength of zeitgeber to pacemaker) changes.

Some of the ways that the human circadian system responds to experimental manipulations of the scotoperiod appear to be consistent with this more complex model. For example, when subjects are transferred from short (8-hour) "nights" to long (14-hour) "nights," the intrinsic timing of the evening decline in core body temperature systematically advances about 1 hour relative to the intrinsic timing of the evening onset of melatonin secretion and the evening rise of thyrotropin secretion, as if changes in the different variables were controlled by separate evening oscillators or slave oscillators (Wehr *et al.*, 1993) (Figure 15). Furthermore, none of the scotoperiod-induced changes in any of the variables is correlated with scotoperiod-induced changes in any of the other variables.

Whether there are separate neuroanatomic substrates for evening and morning oscillators is unknown. Possibly consistent with this idea, there is evidence that

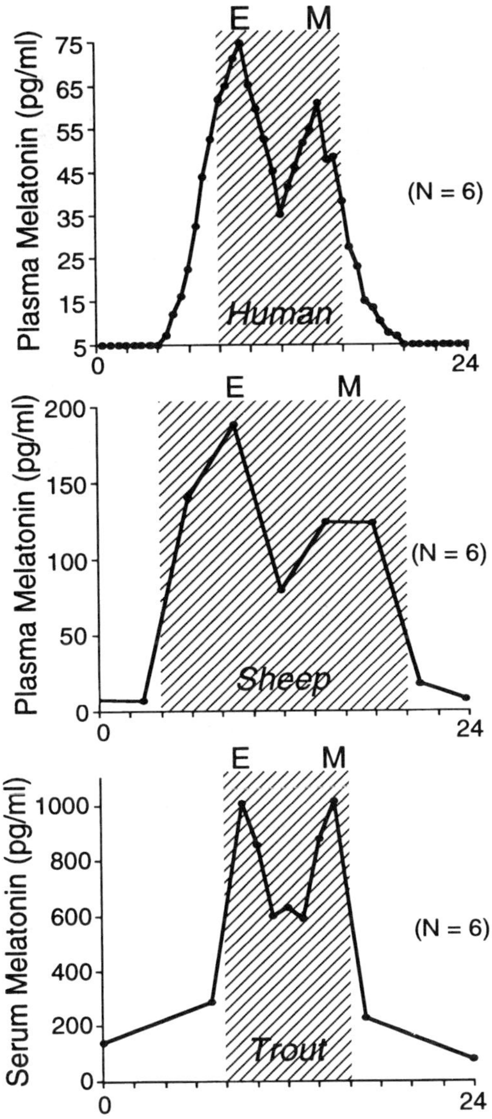

Figure 13. The pattern of nocturnal melatonin secretion is bimodal, with evening and morning peaks, in trout (data adapted from Zachman *et al.*, 1992), sheep (data adapted from Arendt, 1979), and in some humans (data from Wehr *et al.*, 1995b).

different populations of SCN cells express c-*fos* in response to evening and morning light exposures (Rea, 1992) and that phase responses to evening and morning light may be mediated by different input pathways (Harrington & Rusak, 1986). There is also evidence that serotonergic input to the SCN may selectively affect functions, such as onset of nocturnal activity, that are hypothesized to be controlled by an evening-type oscillator (Meyer-Bernstein, Blanchard, & Morin, 1997).

There is some evidence that there are separate molecular substrates for evening and morning oscillators. It appears likely that circadian rhythms are generated by

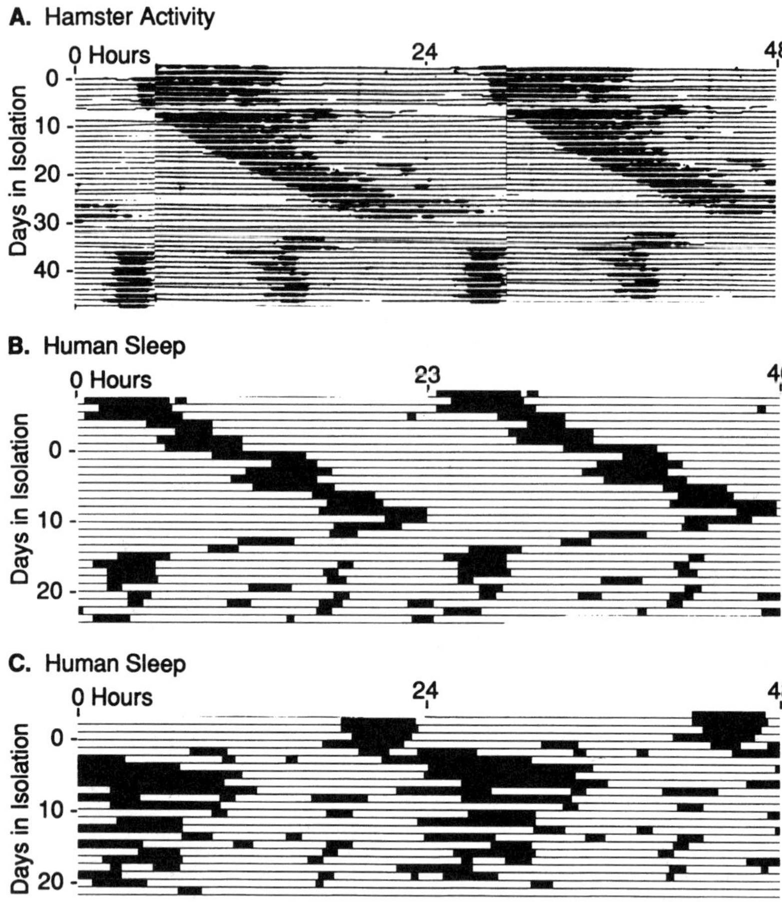

Figure 14. "Splitting" of free-running circadian rhythms of hamster running-wheel activity (top) and electroencephalographically monitored human sleep (middle and bottom) into two components that become stably recoupled approximately 180° apart. A 23-hour plotting interval was chosen for one of the human sleep records (middle) to facilitate comparison with the hamster activity record (top). Prior to splitting, the rhythms exhibit a single component with an intrinsic period greater than 24 hours. After splitting, the rhythms exhibit two components nearly 180° apart with an intrinsic period that is less than 24 hours. (Hamster activity data were adapted from Pickard and Turek, 1983; human sleep data were adapted from Wehr *et al.*, 1985; figure is reprinted from Wehr, 1992.)

negative feedback loops between one or more "clock" genes and the products of their expression (Dunlap, 1998). In *Drosophila*, for example, cyclic expression of the *per* gene seems to play a role in the generation of circadian rhythms. Certain mutations in *per* are associated with disturbances in the timing of circadian rhythms, such as advances or delays in their phase position when they are entrained to a light–dark cycle. Like many other animals, *Drosophila* exhibits dual morning and evening bouts of locomotor activity. Of relevance to the present discussion, mutations in *per* selectively affect the timing of the evening bout without altering the timing of the morning bout (Hamblen-Coyle, Wheeler, Rutila, Rosbash, & Hall, 1992). This observation seems to indicate that cyclic expression of *per* is part of a molecular mecha-

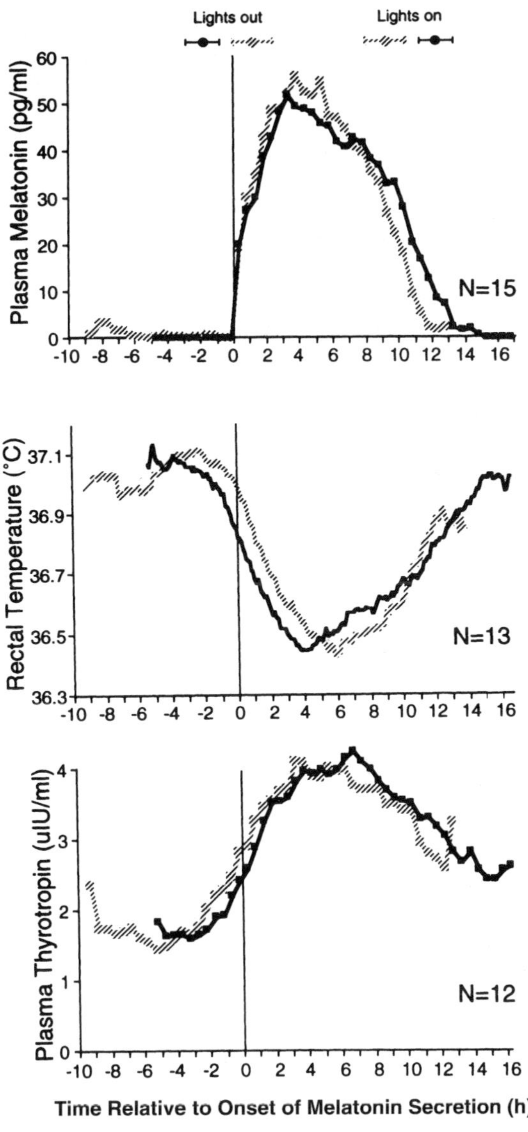

Figure 15. The nightly decline in rectal temperature occurs earlier, relative to the onset of melatonin secretion and the rise of thyrotropin secretion, after exposure to long (14-hour) nights than after exposure to short (8-hour) nights. Mean profiles of 30-minute plasma levels of melatonin (top), 6-minute levels of rectal temperature (middle), and 30-minute plasma levels of thyrotropin (bottom) were measured while individuals remained continuously awake in constant dim light after they had been chronically exposed to short (8-hour) nights (hatched lines) and long (14-hour) nights (solid lines). For each photoperiod schedule, data for all three variables are referenced to the individuals' times of onset of nocturnal melatonin secretion. Mean (± SD) times of onsets ("lights-out") and offsets ("lights-on") of dark periods to which the individuals had been exposed are shown at the top of the figure (the light–dark cycle was suspended during the "constant routine" measurement period). (Reprinted from Wehr *et al.*, 1993.)

nism that generates the rhythm of an evening-type oscillator in *Drosophila.* Recently, homologues of *per* were identified and cloned in humans. The findings in *Drosophila* raise the possibility that negative feedback cycles in the expression of *per* could be part of a molecular mechanism that selectively generates either evening- or morning-type circadian oscillations in humans, such as those that appear to govern the intrinsic timing of the onset or offset, respectively, of nocturnal melatonin secretion (Figure 16).

The Clock Is a Calendar, and Melatonin Is Its Messenger

In the discussion so far, we have focused on cyclic temporal programs that periodically alter the organism's physiology and behavior in such a way that it is able to anticipate and prepare for environmental conditions that recur on a daily basis. Many organisms have developed analogous temporal programs that periodically alter their physiology and behavior in such a way that they are able to anticipate and prepare for environmental conditions that recur on a seasonal basis. They use such programs, for example, to confine breeding activity to a time of year when subsequent conditions will be most favorable for reproductive success, or to withdraw from the environment when conditions are unfavorable for their survival, by migrating or hibernating.

Extensive research in animals has shown that these seasonal temporal programs are built on the circadian temporal programs that so far have been the focus of our discussion (Pittendrigh, 1988). These animals take advantage of the fact that the circadian system can detect and respond to changes in night length by using it as a sensor to detect change in season and to trigger seasonal changes in their behavior.

Melatonin, which is secreted exclusively during the biological night, plays a special role as a chemical mediator of these responses (Bartness & Goldman, 1989; Karsch, Woodfill, Malpaux, Robinson, & Wayne, 1991). Cells that regulate breeding and other seasonal behaviors possess receptors with which they can detect the nightly presence of melatonin and interval-timing mechanisms with which they can measure the duration of its presence. When the duration increases, indicating that nights are becoming longer, these cells initiate fall–winter responses. When the duration decreases, indicating that nights are becoming shorter, they initiate spring–summer responses. In this way, temporal programs in the SCN that regulate circadian rhythms serve not only as a clock, but also as a calendar.

As we have seen, the human retinohypothalamic–pineal axis can similarly detect seasonal changes in night length and make proportional adjustments in the intrinsic duration of nocturnal melatonin secretion, producing the melatonin message that other animals use to trigger seasonal changes in their behavior (Wehr, 1991, 1997). The downstream effects of this message and the systems that respond to it, however, have not been identified in humans.

Downstream Responses to the Seasonal Melatonin Message in Humans

The downstream response to the melatonin message that has been most extensively studied in animals is seasonal breeding (Bartness & Goldman, 1989; Karsch *et al.*, 1991). Scotoperiod-induced changes in the duration of nocturnal melatonin

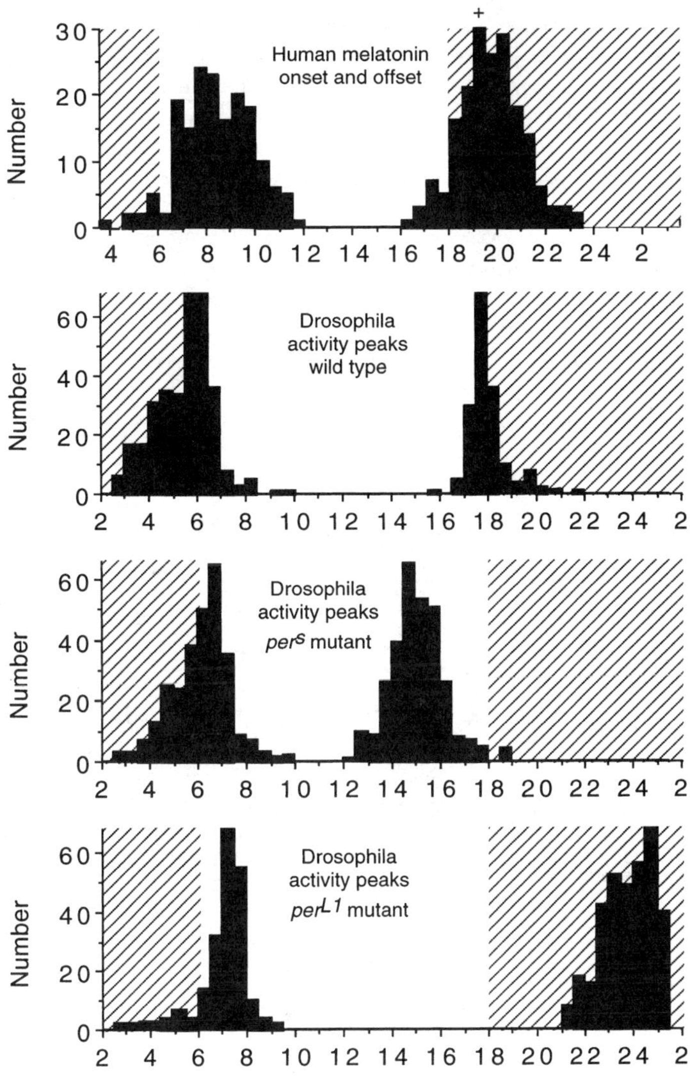

Figure 16. Frequency distribution histograms of the intrinsic times of onset and offset of melatonin secretion in humans (top) and of the times of morning and evening peaks in locomotor activity in wild-type and *per*^{S} and *per*^{L1} mutant *Drosophila* (bottom three panels, respectively). Mutations in *per* selectively affect the timing of the evening activity peak, perhaps indicating that cyclic expression of *per* plays a role in the generation of the rhythm of an evening (E)-type circadian oscillator that is entrained to dusk. The recent identification and cloning *per* homologues in humans raises the possibility that cyclic expression of one or more *per* genes is part of a molecular mechanism that selectively generates the rhythm of either an evening (E)-type circadian oscillator that is entrained to dusk and controls the onset of nocturnal human melatonin secretion or a morning (M)-type circadian oscillator that is entrained to dawn and controls the offset of nocturnal human meltonin secretion (Qiu & Hardin, 1996). (Human data are previously unpublished data of T. A. Wehr; *Drosophila* data are from Hamblen-Coyle *et al.*, 1992.)

secretion trigger breeding-related changes in hormones and behavior in many different types of animals. The exact temporal relationship between breeding and changes in duration of nocturnal melatonin varies according to species and seems to be determined by the nature of seasonal changes in the animal's environment and by the duration of the animal's gestational period. Thus, lengthening of the duration of melatonin secretion in winter triggers breeding in sheep, while shortening of the duration in spring triggers breeding in rodents. Because the gestational period of sheep is long and that of rodents is short, the young of both types of animals are born in the spring, when conditions are most favorable for their survival. The opposite effects of melatonin on reproductive function in sheep and rodents highlights the fact that melatonin is generically neither progonadal nor antigonadal, but serves to convey information about time of year that different animals use in different ways, according to their particular requirements.

It is unclear whether human reproductive function responds to scotoperiod-induced changes in the duration of nocturnal melatonin secretion as occurs in other animals. Humans do exhibit seasonal variation in rates of conception (Roenneberg & Aschoff, 1990a,b) (Figure 17) and the administration of melatonin can modify

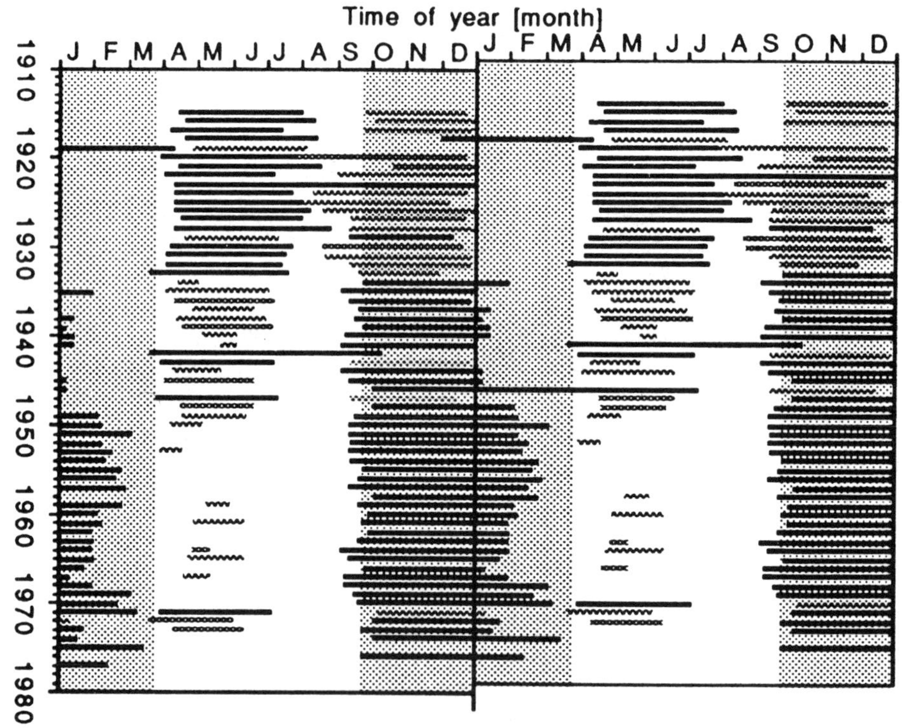

Figure 17. Trends in annual patterns of variation in rates of conception in humans living in the United States during the twentieth century. Horizontal bars show extent of periods each year when conception rates were above the annual mean. Dark bars indicate major peaks and hatched bars indicate minor peaks. The pattern is bimodal, with a shift in predominance from spring/summer peaks to fall/winter peaks over the course of the century. (From Roenneberg & Aschoff, 1990a, with permission. Copyright 1990, Sage Publications.)

human reproductive neuroendocrine function (reviewed in Cagnacci & Volpe, 1996). Furthermore, alterations in human reproductive function are sometimes associated with alterations in melatonin secretion (reviewed in Cagnacci & Volpe, 1996). Nevertheless, it has not been demonstrated that scotoperiod-induced changes in the duration of melatonin secretion can modify human reproductive function.

Although there are many local variations and exceptions, several generalizations can be made about seasonal variations in rates of human conceptions (Roenneberg & Aschoff, 1990a,b) (Figure 17). The conception rate increases at two opposite times of year, in the spring and the fall (Figure 17). Traditionally, the spring increase was greater than the fall increase, but this pattern has reversed in postindustrial societies, and the magnitude of seasonal variation has decreased.

Human spermatozoa possess melatonin receptors (van Vuuren, Pitout, van Aswegen, & Theron, 1992), and melatonin reduces sperm motility (Irez *et al.*, 1992). During the follicular phase of the menstrual cycle, the amplitude of the luteinizing hormone (LH) secretory pulse increases spontaneously at night, when melatonin is secreted, and can be augmented in the daytime, when melatonin is absent, if melatonin is administered (reviewed in Cagnacci & Volpe, 1996). Furthermore, in the Arctic region during the dark season, the duration of melatonin secretion and the level of LH secretion increase, and ovulation is defective (Stokkan & Reiter, 1994; Kauppila, Kivela, Pakarinen, & Vakkuri, 1987).

All of these findings suggest that human reproductive function might respond to changes in duration of nocturnal melatonin secretion that are induced by seasonal changes in duration of the natural scotoperiod. Since the retinohypothalamic–pineal axis responds to seasonal changes in duration of the natural scotoperiod in only a minority of humans who live in the modern urban environment, it might prove useful to identify these responsive individuals and confine future studies of this question to this subgroup.

In animals, scotoperiod-induced changes in the duration of nocturnal melatonin secretion can induce changes not only in reproductive function, but also in activity, aggression, sleep, food intake, weight, metabolism, and other behaviors and functions. Thus, at one time of year, an animal may become more active and aggressive, eat more, sleep less, become interested in sex, and interact more with its physical and social environment. At another time of year, it may become less active and aggressive, eat less, sleep more, lose interest in sex, and withdraw from its physical and social environment. It may be relevant that similar seasonal changes in behavior occur as pathology in humans who are afflicted with seasonal affective disorder (SAD).

Seasonal affective disorder is characterized by recurrent depressions that usually begin in November and end in March (Wehr & Rosenthal, 1989). Symptoms of winter depression commonly include lethargy, social withdrawal, loss of interest in sex, oversleeping, overeating, carbohydrate craving, and weight gain. Sometimes, the picture changes to an opposite one in the spring, with increased energy, social engagement, increased libido, decreased need for sleep, decreased appetite, and weight loss. The ubiquitous role of melatonin as a chemical mediator of seasonal change in animals and the similarities between patients' symptoms and behaviors that melatonin triggers in animals led clinical investigators to hypothesize that scotoperiod-induced changes in the duration of melatonin secretion comprise the mechanism that drives the seasonal cycles of this disorder (Rosenthal *et al.*, 1984).

This hypothesis has been tested in a number of naturalistic studies and experiments. For the hypothesis to be tenable, however, its assumption that the duration of nocturnal melatonin secretion is longer in winter than in summer in patients with SAD would have to be shown to be true. Our recent research shows that this assumption is true (T. A. Wehr, unpublished data).

Thus, the intrinsic duration of melatonin secretion increases in winter in patients with SAD, but not in healthy individuals. These results are consistent with (but do not prove) the classic melatonin hypothesis of SAD: patients with SAD generate a biological signal of change of scotoperiod that is absent in healthy individuals and that is identical to one that animals use to detect change of season and induce seasonal changes in their behavior.

The classic melatonin-duration hypothesis of the pathogenesis of SAD has been tested in a number of experiments, with inconsistent results. Supporting the hypothesis are findings that exposure to light improves winter depression (Rosenthal *et al.*, 1984), and that exposure of the eyes, and not the skin, to light is necessary for this improvement to occur (Wehr, Skwerer, Jacobsen, Sack, & Rosenthal, 1987). Also consistent with the hypothesis is the finding that morning treatment with the beta-blocker drug propranolol, which suppresses the terminal portion of nocturnal melatonin secretion and thereby shortens its duration, is associated with improvement of winter depression (Schlager, 1994), and the finding that administration of melatonin in patients who have improved during light treatments induces some of the most characteristic symptoms of winter depression (Rosenthal *et al.*, 1986). Observations that suppression of melatonin secretion by light is not necessary for improvement to occur during light treatment, however, are inconsistent with the hypothesis (Checkly, Franey, Winton, Corn, & Arendt, 1989; Wehr *et al.*, 1986).

In future studies of the melatonin-duration hypothesis of SAD, it will be necessary to determine why patients respond to seasonal changes in the natural scotoperiod differently than healthy individuals. In this effort, responses of the M oscillator to morning light are of particular interest, because, as was mentioned previously, changes in the intrinsic duration of nocturnal melatonin secretion are mostly a function of changes in the timing of the morning offset of secretion that are induced by the phase-advancing effect of morning light.

Other Effects of Photoperiod on Human Biology

In the course of our experiments, we found other, unexpected effects of changes in the duration of artificial scotoperiods on human neuroendocrine, sleep, and the thermoregulatory physiology (Wehr, 1996; Wehr *et al.*, 1993). The sleep-related peak of growth hormone secretion that occurs about 1 hour after the onset of sleep was twice as high in short (14-hour) "nights" as it was in long (14-hour) "nights" (Figure 18). The level of the daytime plateau of core body temperature was 0.3°C lower in long (16- hour) "days" than in short (10-hour) "days." The amount of sleep and rest was longer in long nights than in short nights. Furthermore, sleep was fragmented and exhibited a bimodal pattern of distribution in long nights, but was consolidated and exhibited a unimodal pattern of distribution in short nights (Figures 4, 12, and 19). Although the consolidated pattern of sleep that humans exhibit in short nights has been considered to be unique when compared with patterns of sleep in other animals, the polyphasic, bimodal pattern of sleep that humans exhibit in long nights is quite similar to patterns of sleep or activity that other animals exhibit in long nights (Wehr, 1992).

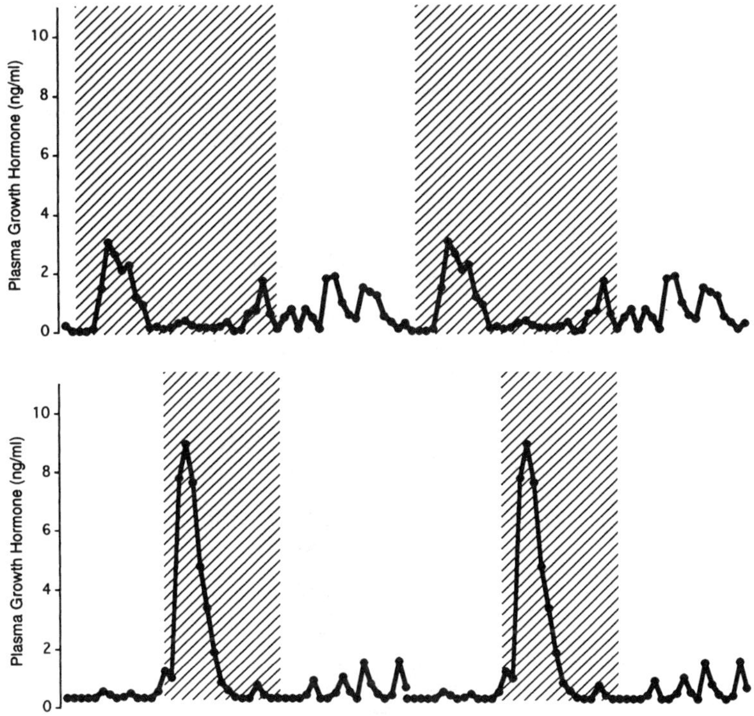

Figure 18. The peak of sleep-related growth hormone secretion is twice as high during short nights as during long nights (N = 12 individuals). Hatched areas indicate timing of dark periods to which individuals were exposed during the 24-hour period when measurements were made. Dark periods were 14 hours (6 p.m.–8 a.m.) and 8 hours (midnight–8 a.m.) long, respectively. Twenty-four-hour profiles are shown in duplicate to facilitate visual inspection of the waveform. (Reprinted from Wehr 1996.)

Impact of Artificial Light and Artificial Darkness on Photoperiod-Responsive Elements of the Human Circadian System

Modern humans use artificial light primarily to extend the light phase, and their daily period of active wakefulness, into the first hours of the night. Consequently, the biological night has become delayed relative to its usual phase relationship to the natural scotoperiod (Figure 11), and, in many individuals, it has become permanently compressed in a short, summer mode year-round (Wehr *et al.*, 1995a). Several inferences about the consequences of this change can be drawn from the results of experiments in which humans have been exposed to artificial scotoperiods of different durations (Wehr *et al.*, 1993): The nocturnal period during which modern humans rest and sleep and are exposed to melatonin, high levels of prolactin, low levels of core body temperature, and rising levels of cortisol is probably shorter at most times of year than it would be in a purely natural scotoperiod, and it no longer varies in duration as the seasons change (Figures 1–4 and 15). In addition, the nocturnal peak of sleep-related growth hormone secretion is probably higher (Figure 18) and the diurnal plateau of core body temperature is probably lower at most times of year than they would be in a purely natural scotoperiod. Finally,

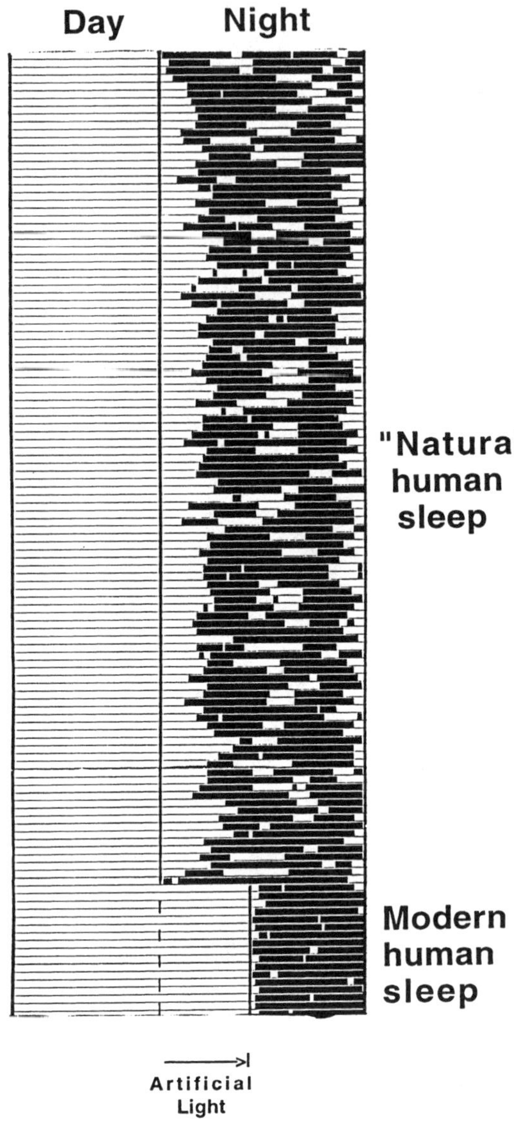

Figure 19. Extending the illuminated portion of the day into the evening hours reduces total sleep time and virtually abolishes rest. (Data are from Wehr *et al.*, 1993.)

humans no longer sleep in bouts of several hours' duration that are interspersed with periods of quiet wakefulness and that exhibit a bimodal pattern of distribution as they may have once done at most times of year in a natural scotoperiod (Figures 4, 12, and 19).

The long-term consequences of these changes and their full impact on human behavior and biology are not well understood. Results of animal experiments suggest that chronic compression of the biological night might, for example, interfere with entrainment of the sleep–wake cycle to the 24-hour day in some individuals. When animals are transferred from long nights to short nights, the light-responsive portion of the phase response curve (PRC) for light becomes shorter in duration and its amplitude becomes reduced (Pittendrigh, Elliot, & Takamura, 1984) (Figure 20). In humans who live in artificially lengthened days, a reduction in magnitude of phase-advance responses to morning light could lead to extreme phase delays of the sleep–wake cycle in some individuals (delayed sleep phase syndrome), or even free-

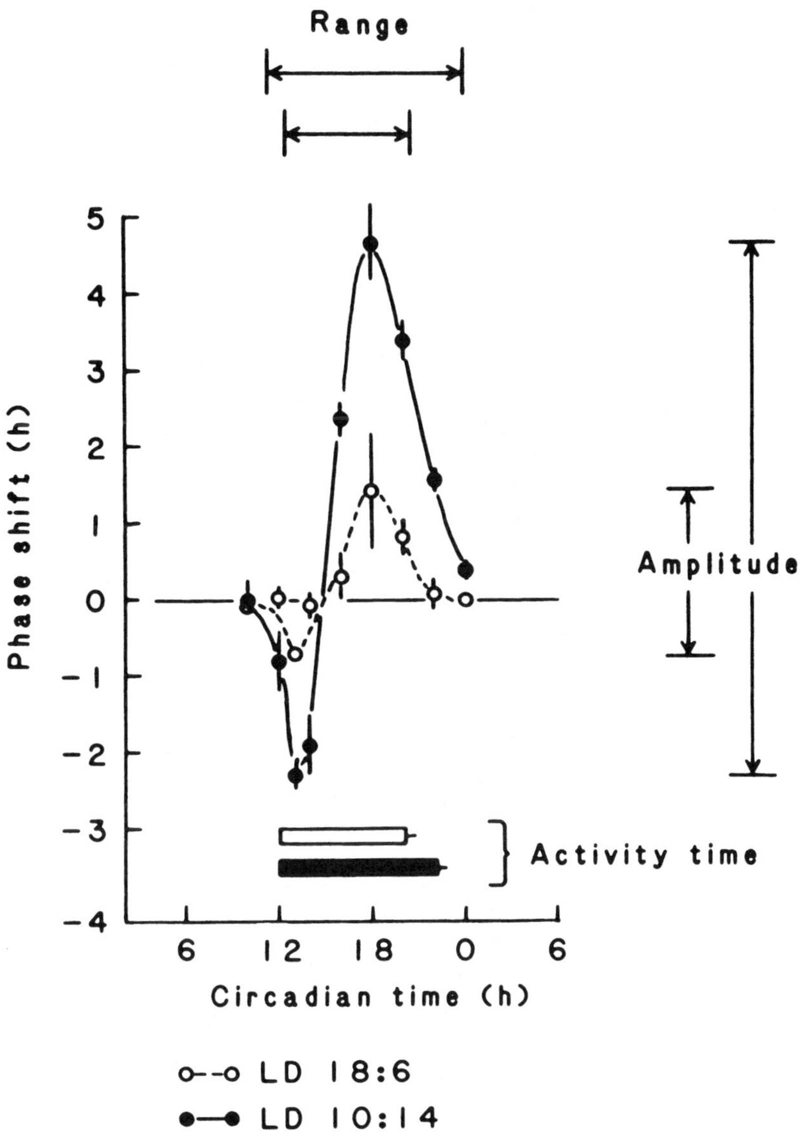

Figure 20. Photoperiod dependence of the hamster's phase response curve for light. Each point plots the mean phase shift for hamsters pulsed with light at the indicated circadian times on the 10th cycle in constant darkness (DD) after release from entrainment to either a long photoperiod (open symbols) or a short photoperiod (closed symbols). Following entrainment to the short photoperiod (LD 10:14), phase shift responses ($\geqslant$ 0.5 hour) occur over a broader range of the circadian cycle and the phase response curve has a larger amplitude than it does following entrainment to the long photoperiod. (From Pittendrigh *et al.*, 1984, with permission. Copyright 1984, The Novartis Foundation.)

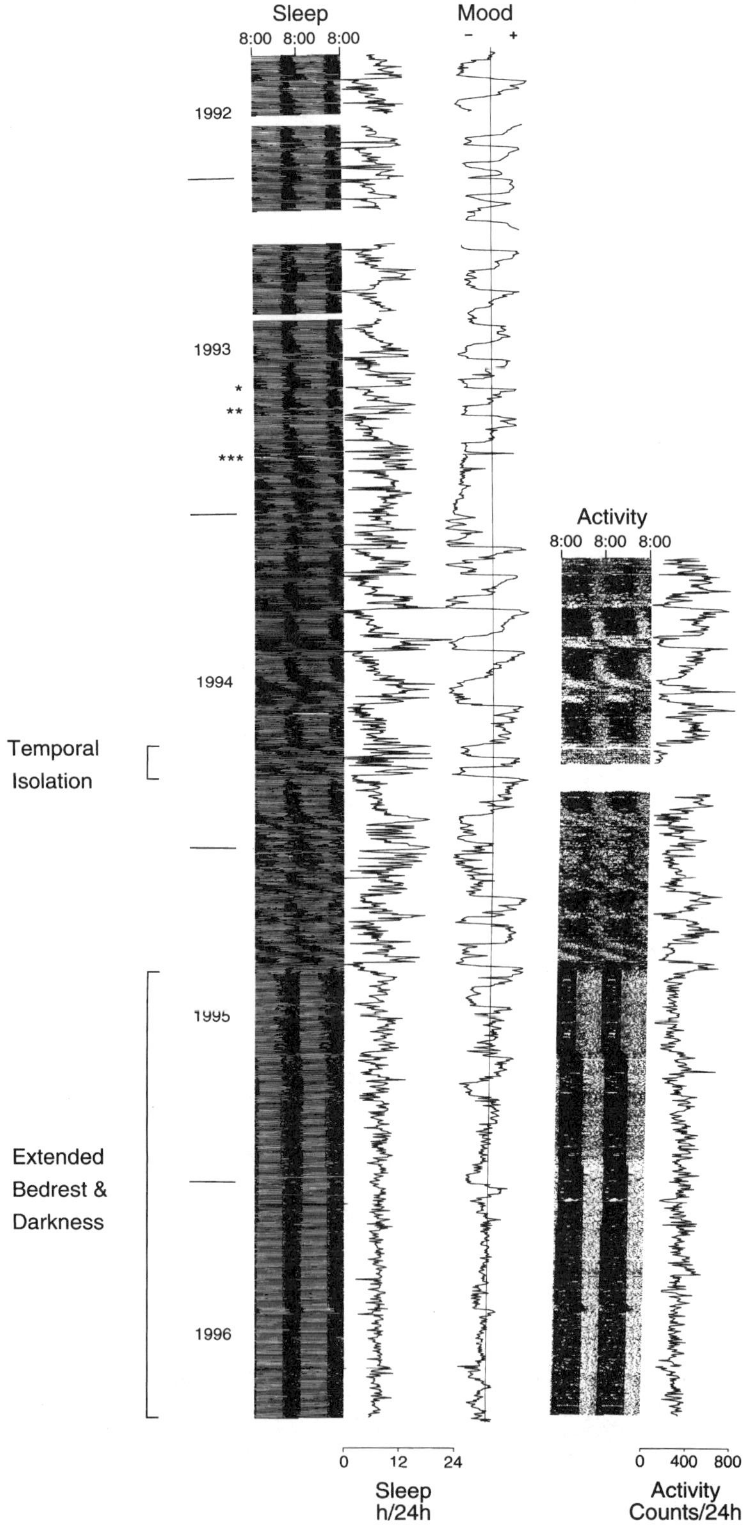
Sleep
8:00 8:00 8:00
Mood
- +
1992
1993
*
**
***
1994
Temporal Isolation
1995
Extended Bedrest & Darkness
1996
Activity
8:00 8:00 8:00
0 12 24
Sleep h/24h
0 400 800
Activity Counts/24h

running of the sleep–wake cycle (hypernyctohemeral syndrome). In animals, compressing the duration of the nightly dark period can also result in unstable entrainment of the sleep–wake cycle with multiweek cycles characterized by shifts back and forth and dramatic "phase jumps" in the timing of sleep (Pittendrigh and Daan, 1976b). It may be relevant that similar disturbances of the sleep–wake cycle occur in some forms of manic-depressive (bipolar) illness (Figure 20). These disturbances probably play an important role in the pathogenesis of the illness because experimental phase shifts of the sleep–wake cycle and experimental deprivation of sleep are capable of triggering dramatic changes in the clinical state of such patients (reviewed in Wehr, Turner, Clark, & Barker, & Leibenluft, 1998).

Since decreasing the duration of the scotoperiod reduces the amplitude of the PRC for light in animals (Figure 20), it is possible that using artificial light to compress the nightly dark period in modern times has contributed to the emergence of the types of sleep–wake cycle disturbances that occur in delayed sleep-phase syndrome, hypernyctohemeral syndrome, and/or bipolar illness. Consistent with this hypothesis, we have shown in pilot studies that lengthening the duration of the nightly dark period can be used successfuly to treat these disorders (Oren, Giesen, Wehr, 1997; Wehr *et al.*, 1998b) (Figure 21).

Prospects for Future Research on Human Photoperiodism

Two questions seem especially important for future research on photoperiod responses of the retinohypothalamic-pituitary-axis in humans. First, why do some individuals living in a modern urban environment respond to seasonal changes in duration of the natural scotoperiod whereas other do not? The fact that a small subgroup of subjects fails to respond even to experimental manipulations of the duration of carefully defined artificial scotoperiods suggests that some individuals may be absolutely refractory (Figure 8). However, other individuals who show no seasonal response might be capable of responding if they were exposed to appropriate lighting conditions. The fact that scotoperiod-induced changes in the intrinsic duration of melatonin secretion are largely a function of changes in the intrinsic timing of offset of secretion (Figure 9) suggests that factors influencing exposure to, or processing of, morning light may be important determinants of whether an individual responds or fails to respond to seasonal changes in duration of the natural scotoperiod. In some unresponsive individuals, for example, the M oscillator, or its functional equivalent, might be blind.

Second, what are the downstream effects of the seasonal melatonin message on human biology and behavior, especially on human reproductive function? In future

←

Figure 21. Recordings of sleep, mood, and wrist motor activity in a patient whose rapidly cycling form of manic-depressive (bipolar) illness improved after he adhered to a regimen of long, regularly scheduled nightly periods of darkness, rest, and sleep from June 1995 onward. During the regimen, the duration of the nightly dark period was gradually reduced from 14 hours (LD 10:14) to 10 hours (LD 14:10). Data for sleep (15-minute sampling frequency) and activity (12-minute sampling frequency) are shown in a raster format in which 24-hour (8 a.m. to 8 a.m.) segments of data are plotted successively beneath one another and double-plotted to the right as a visual aid. Total daily sleep, total daily activity counts, and daily mood self-ratings as recorded on a 100-mm line are shown beside the raster plots. A vertical line indicates the center (50 mm) of the mood rating scale. The alternating dark and light areas in the first half of the activity raster plot correspond to periods of increased and decreased activity that were associated with periods of hypomania and depression, respectively. (For details, see Wehr *et al.* (1998), from which figure is reprinted.

research, this question could be addressed more effectively if investigators focused on the minority of individuals in whom the retinohypothalamic–pituitary axis exhibits robust responses to seasonal changes in the natural scotoperiod in spite of their being exposed to ambient artificial light in the urban environment (Figures 6 and 7).

References

Arendt, J. (1979). Radioimmunoassayable melatonin: Circulating patterns in man and sheep. In J. Ariens Kappers & P. Pévet (Eds.), *Progress in brain research* (Vol. 52, pp. 249–258). Amsterdam: Elsevier.

Aschoff, J. (1966). Circadian activity pattern with two peaks. *Ecology, 47*, 657–662.

Bartness, T. J., & Goldman, B. D. (1989). Mammalian pineal melatonin: A clock for all seasons. *Experientia, 45*, 939–945.

Cagnacci, A., & Volpe, A. (1996). Influence of melatonin and photoperiod on animal and human reproduction. *Journal of Endocrinological Investigation, 19*, 382–411.

Cagnacci, A., Soldani, R., & Yen, S. S. C. (1997). Contemporaneous melatonin administration modifies the circadian response to nocturnal bright light stimuli. *American Journal of Physiology, 272*, R482–R486.

Checkley, S. A., Franey, C., Winton, F., Corn, T., & Arendt, J. (1989). A neuroendocrine study of the mechanism of action of phototherapy in seasonal affective disorder. In C. Thompson & T. Silverstone (Eds.), *Seasonal affective disorder* (pp. 223–232). London: Clinical Neuroscience Publications.

Dijk, D.-J., & Daan, S. (1989). Sleep EEG spectral analysis in a diurnal rodent, *Eutamias sibiricus*. *Journal of Comparative Physiology A, 165*, 205–215.

Dunlap, J. (1998). Circadian rhythms: An end in the beginning. *Science, 280*, 1548–1549.

Elliott, J. A., & Tamarkin, L. (1994). Complex circadian regulation of pineal melatonin and wheel-running in Syrian hamsters. *Journal of Comparative Physiology A, 174*, 469–484.

Hamblen-Coyle, M. J., Wheeler, D. A., Rutila, J. E., Rosbash, M., & Hall, J. C. (1992). Behavior of period-altered circadian rhythm mutants of *Drosophila* in light:dark cycles (Diptera: Drosophilidae). *Journal of Insect Behavior, 5*, 417–445.

Harrington, M. E., & Rusak, B. (1986). Lesions of the thalamic intergeniculate leaflet alter hamster circadian rhythms. *Journal of Biological Rhythms, 1*, 309–325.

Illnerova, H., & Vanecek, J. (1982). Two-oscillator structure of the pacemaker controlling the circadian rhythm of N-acetyltransferase in the rat pineal gland. *Journal of Comparative Physiology, 145*, 539–548.

Illnerova, H., & Vanecek, J. (1987). Dynamics of discrete entrainment of the circadian rhythm in the rat pineal N- acetyltransferase activity during transient cycles. *Journal of Biological Rhythms, 2*, 95–108.

Irez, T. O., Senol, H., Alagoz, M., Basmaciogullari, C., Turan, F., Kuru, D., & Ertungealp, E. (1992). Effects of indoleamines on sperm motility *in vitro*. *Human Reproduction, 7*, 987–990.

Karsch, F. J., Woodfill, C. J. I., Malpaux, B., Robinson, J. E., & Wayne, N. L. (1991). Melatonin and mammalian photoperiodism: Synchronization of annual reproductive cycles. In D. C. Klein, R. Y. Moore, & S. M. Reppert (Eds.), *Suprachiasmatic nucleus: The mind's clock* (pp. 217–232). New York: Oxford University Press.

Kauppila, A., Kivela, A., Pakarinen, A., & Vakkuri, O. (1987). Inverse seasonal relationship between melatonin and ovarian activity in humans in a region with a strong seasonal contrast in luminosity. *Journal of Clinical Endocrinology and Metabolism, 65*, 823–828.

Meyer-Bernstein, E. L., Blanchard, J. H., & Morin, L. P. (1997). The serotonergic projection from the median raphe nucleus to the suprachiasmatic nucleus modulates activity phase onset, but not other circadian rhythm parameters. *Brain Research, 755*, 112–120.

Morgan, P. J., Barrett, P., Howell, H. E., & Kelliwell, R. (1994). Melatonin receptors: Localization, molecular pharmacology and physiological significance. *Neurochemistry International, 242*, 101–146.

Mrugala, M. M., Ziomanczuk, P., Jagota, A., & Schwartz, W. J. (2000). Rhythmic multiunit neural activity in slices of hamster suprachiasmatic nucleus reflect prior photoperiod. *American Journal of Physiology, 278*, R987–R994.

Oren, D. A., Giesen, H. A., & Wehr, T. A. (1997). Restoration of detectable melatonin secretion after entrainment to a 24-hour schedule in a "free-running" man. *Psychoneuroendocrinology, 22*, 39–52.

Pickard, G. E., & Turek, F. W. (1983). The suprachiasmatic nuclei: Two circadian clocks? *Brain Research, 268*, 201–210.

Pittendrigh, C. S. (1988). The photoperiodic phenomena: Seasonal modulation of the 'day within.' *Journal of Biological Rhythms, 3,* 173–188.
Pittendrigh, C. S., & Daan, S. (1976a). A functional analysis of circadian pacemakers in nocturnal rodents. I. The stability and lability of spontaneous frequency. *Journal of Comparative Physiology A, 106,* 223–252.
Pittendrigh, C. S., & Daan, S. (1976b). A functional analysis of circadian pacemakers in nocturnal rodents. IV. Entrainment. *Journal of Comparative Physiology A, 106,* 291–331.
Pittendrigh, C. S., & Daan, S. (1976c). A functional analysis of circadian pacemakers in nocturnal rodents. V. Pacemaker structure: A clock for all seasons. *Journal of Comparative Physiology A, 106,* 333–355.
Pittendrigh, C. S., Elliott, J., & Takamura, T. (1984). The circadian component in photoperiodic induction. In *Photoperiodic regulation of insect and molluscan hormones* (pp. 26–47). London: Pitman.
Qiu, J., & Hardin, P. E. (1996). *per* mRNA cycling is locked to lights-off under photoperiodic conditions that support circadian feedback function. *Molecular and Cellular Biology, 16,* 4182–4188.
Rea, M. A. (1992). Different populations of cells in the suprachiasmatic nuclei express c-*fos* in association with light-induced phase delays and advances of the free-running activity rhythm in hamsters. *Brain Research, 579,* 107–112.
Reiter, R. J. (1987). The melatonin message: Duration versus conicidence hypotheses. *Life Science, 40,* 2119–2131.
Roenneberg, T., & Aschoff, J. (1990a). Annual rhythm of human reproduction: I. Biology, sociology, or both? *Journal of Biological Rhythms, 5,* 217–239.
Roenneberg, T., & Aschoff, J. (1990b). Annual rhythm of human reproduction: II. Environmental correlations. *Journal of Biological Rhythms, 5,* 217–239.
Rosenthal, N. E., Sack, D. A., Gillin, J. C., Lewy, A. J., Goodwin, F. K., Davenport, Y., Mueller, P. S., Newsome, D. A., & Wehr, T. A. (1984). Seasonal affective disorder: A description of the syndrome and preliminary findings with light therapy. *Archives of General Psychiatry, 41,* 72– 80.
Rosenthal, N. E., Sack, D. A., Jacobsen, F. M., James, S. P., Parry, B. L., Arendt, J., Tamarkin, L., & Wehr, T. A. (1986). Melatonin in seasonal affective disorder. *Journal of Neural Transmission, 21,* 257–267.
Rusak, B. (1989). The mammalian circadian system: Models and physiology. *Journal of Biological Rhythms, 4,* 121–134.
Rusak, B., Robertson, H. A., Wisden, W., & Hunt, S. P. (1990). Light pulses that shift rhythms induce gene expression in the suprachiasmatic nucleus. *Science, 248,* 1237–1240.
Schlager, D. S. (1994). Early-morning administration of short- acting beta-blockers for treatment of winter depression. *American Journal of Psychiatry, 151,* 1383–1385.
Stokkan, K. -A., & Reiter, R. J. (1994). Melatonin levels in Arctic urban residents. *Journal of Pineal Research, 16,* 33–36.
Sumova, A., Travnickova, Z., Peters, R., Schwartz, W. J., & Illnerova, H. (1995). The rat suprachiasmatic nucleus is a clock for all seasons. *Proceedings of the National Academy of Sciences of the USA, 92,* 7754–7758.
Tobler, I. (1989). Napping and polyphasic sleep in mammals. In D. F. Dinges & R. J. Broughton (Eds.), *Sleep and alertness: Chronobiological, behavioral, and medical aspects of napping* (pp. 9–30). New York: Raven Press.
van Vuuren, R. J., Pitout, M. J., van Aswegen, C. H., & Theron, J. J. (1992). Putative melatonin receptor in human spermatozoa. *Clinical Biochemistry, 25,* 125–127.
Vuillez, P., Jacob, N., Teclemariam-Mesbah, R., & Pévet, P. (1996). In Syrian and European hamsters, the duration of sensitive phase to light of the suprachiasmatic nuclei depends on the photoperiod. *Neuroscience Letters, 208,* 37–40.
Wehr, T. A. (1991). The durations of human melatonin secretion and sleep respond to changes in daylength (photoperiod). *Journal of Clinical Endocrinology and Metabolism, 73,* 1276–1280.
Wehr, T. A. (1992). In short photoperiods, human sleep is biphasic. *Journal of Sleep Research, 1,* 103–107.
Wehr, T. A. (1996). A 'clock for all seasons' in the human brain. In R. M. Buijs, A. Kalsbeek, J. H. Romihn, C. M. A. Pennartz, & M. Mirmiran (Eds.), *Progress in brain research* (Vol. 111, pp. 319–340). Amsterdam: Elsevier.
Wehr, T. A. (1997). Melatonin and seasonal rhythms. *Journal of Biological Rhythms, 12,* 517–526.
Wehr, T. A., & Rosenthal, N. E. (1989). Seasonality and affective illness. *American Journal of Psychiatry, 146,* 829–839.
Wehr, T. A., Sack, D. A., Duncan, W. C. Jr., Rosenthal, N. E., Mendelson, W. B., Gillin, J. C., & Goodwin, T. K. (1985). Sleep and circadian rhythms in affective patients isolated from external time cues. *Psychiatry Research, 15,* 327–339.
Wehr, T. A., Sack, D. A., Jacobsen, F. M., Tamarkin, L., Arendt, J., & Rosenthal, N. E. (1986). Phototherapy of seasonal affective disorder: Time of day and suppression of melatonin are not critical for antidepressant effects. *Archives of General Psychiatry, 43,* 870–875.

Wehr, T. A., Skwerer, R., Jacobsen, F. M., Sack, D. A., & Rosenthal, N. E. (1987). Eye- versus skin-phototherapy of seasonal affective disorder. *American Journal of Psychiatry, 144,* 753–757.

Wehr, T. A., Moul, D. E., Giesen, H. A., Seidel, J. A., Barker, C., & Bender, C. (1993). Conservation of photoperiod-responsive mechanisms in humans. *American Journal of Physiology, 265,* R846–R857.

Wehr, T. A., Giesen, H. A., Moul, D. E., Turner, E. H., & Schwartz, P. J. (1995a). Suppression of human responses to seasonal changes in day-length by modern artificial lighting. *American Journal of Physiology, 269,* R173– R178.

Wehr, T. A., Schwartz, P. J., Turner, E. H., Feldman-Naim, S., Drake, C., & Rosenthal, N. E. (1995b). Bimodal patterns of human melatonin secretion consistent with a two-oscillator model of regulation. *Neuroscience Letters, 194,* 105–108.

Wehr, T. A., Turner, E. H., Clark, C. H., Barker, C., & Leibenluft, E. (1998). Treatment of a rapidly cycling bipolar patient by using extended bedrest and darkness to stabilize sleep. *Biological Psychiatry, 43,* 822–828.

Wehr, T. A., Duncan, W. C. Jr., Sher, L., Aeschbach, D., Schwartz, P. J., Turner, E. H., Postolache, T. T., & Rosenthal, N. E. A circadian signal of change of season in seasonal affective disorder. Submitted for publication.

Zachmann, A., Knijff, S. C. M., Ali, M. A., & Anctil, M. (1992). Effects of photoperiod and different intensities of light exposure on melatonin levels in the blood, pineal organ, and retina of the brook trout (*Salvelinus fontinalis* Mitchill). *Canadian Journal of Zoology, 70,* 25–29.

Index